A Century of Plant Virology in India

Bikash Mandal • Govind Pratap Rao
Virendra Kumar Baranwal
Rakesh Kumar Jain
Editors

A Century of Plant Virology in India

Springer

Editors
Bikash Mandal
Advanced Center for Plant Virology
Indian Agricultural Research Institute
New Delhi, India

Govind Pratap Rao
Advanced Center for Plant Virology
Indian Agricultural Research Institute
New Delhi, India

Virendra Kumar Baranwal
Advanced Center for Plant Virology
Indian Agricultural Research Institute
New Delhi, India

Rakesh Kumar Jain
Advanced Centre for Plant Virology
Indian Agricultural Research Institute
New Delhi, India

ISBN 978-981-10-5671-0 ISBN 978-981-10-5672-7 (eBook)
https://doi.org/10.1007/978-981-10-5672-7

Library of Congress Control Number: 2017954384

Printed on acid-free paper

This Springer imprint is published by Springer Nature
The registered company is Springer Nature Singapore Pte Ltd.
The registered company address is: 152 Beach Road, #21-01/04 Gateway East, Singapore 189721, Singapore

Foreword

It is a real pleasure to contribute a short foreword to this important book, which is birthday present to Professor Anupam Varma from his grateful colleagues.

Anupam and I were Ph.D. students of the late Sir Frederick (Fred) Bawden in the 1950s and 1960s at Rothamsted Experimental Station in the U.K. At that time Fred was the Head of the Plant Pathology Department, and one of the world's leading virologists; he was the author of the only plant virus textbook available, and, together with Brill Pirie, had purified tobacco mosaic virus virions, and shown that they were composed of ribonucleic acid and protein. Fred suggested that Anupam and I work together to become plant virologists, it was a great experience, especially as Anupam is so much better organised than me, and we have been firm friends and colleagues ever since.

In the mid 1960s Anupam returned to India to a career at IARI, and, as a leader in that fine institution, he will have contributed to a great or lesser extent, directly or collegially, to most of the contents of this book.

Books like this are of great value as they provide a 'line in the sands of time'. It will be interesting to compare its contents with those of, say, Fred Bawden's 'Plant Viruses and Virus Diseases' (3rd or 4th Editions), and to see the enormous strides made by science over the past half century, but younger colleagues will also be pleased to find that many of the questions posed by Fred Bawden still remain to be answered.

Adrian Gibbs

ANU Emeritus Faculty
'Molony Room', 24 Balmain Crescent Building 1 C,
Canberra, ACT, 0200 Australia
June 2017

Preface

The virus-associated plant diseases have a history of more than 100 years in India. Viruses have undoubtedly infected plants and caused diseases for centuries before they are described and proven to be the causal agents. But important progress related to the identification of plant viruses only began after the 1970s in India. Plant viruses have emerged as the most serious constraints in the production of several crops in India during the last four decades. Virus diseases constitute a major limiting factor to the quality and productivity of cereals, horticulture crops, and many other economically important crops all over the country. Annual yield losses caused by virus diseases may vary, but under the favourable conditions, virus disease may lead to disastrous consequences to farming and industry community. The scientific literature concerning occurrence, characterization, diagnosis, detection and management of plant viruses is growing at a fast pace.

India has made significant advancement in the last century on diagnostic, biological and molecular properties, epidemiology, host-pathogen-insect interactions as well as management of plant viruses. To date, no authentic compilation is available to know the progress of plant virus disease research in India. Hence, we planned to compile the major findings on plant viruses and diseases occurring in India in the form of a book entitled *A Century of Plant Virology in India*. This volume contains 31 chapters contributed by the experienced and recognized experts working on the different aspects of plant virology in India. The information on various topics is at advanced as well as comprehensive levels. The book has been divided into four important sections. Section I comprises comprehensive information on the plant viruses, and descriptions have been provided on genera-wise distribution, occurrence and properties of different viruses. The major and minor virus genera covered in this section are allexi-, ampelo-, babu-, badna-, begomo-, carla-, carmo-, clostero-, cucumo-, emara-, ilar-, luteo-, maclura-, mandari-, mastre-, peclu-, polero-, poty-, sobemo-, tospo-, tobamo-, and tungroviruses. Besides, a chapter on update information of viroids is also included. Section II covers an update information of insect vectors such as aphids, whitefly and thrips occurring in India and their virus-vector relationships. Section III discusses the advancement on the diagnosis of viruses based on serological and nucleic acid-based technologies. Section IV is focused on the management of plant viruses, which covered conventional, biological and transgenic approaches.

We most sincerely acknowledge the contribution of the authors for their efforts in synthesizing the most updated reviews. We also like to thank the support and input of the publisher, Springer (India) Pvt. Ltd., New Delhi, for their determined effort to publish this book. We strongly hope that this book will be useful to everyone interested in plant virology, plant pathology, plant biology and molecular biology and serve as an exhaustive and up-to-date reference on various aspects of plant viruses studied during the past more than a century in India.

New Delhi, India

Bikash Mandal
Govind Pratap Rao
Virendra Kumar Baranwal
Rakesh Kumar Jain

Contents

About the Editors

Dr. Bikash Mandal is a principal scientist in the Advanced Centre for Plant Virology, Division of Plant Pathology, Indian Agricultural Research Institute (IARI), New Delhi. His major research areas are molecular characterization of plant viruses, diagnosis and application of plant virus genome as useful genetic resource. Dr. Mandal has a total of 90 peer-reviewed research papers in the national and international journals and 2 patents. He is a fellow of the Indian Virological Society, Indian Phytopathological Society and National Academy of Biological Sciences, and a member of the National Academy of Sciences, Allahabad, India. He is a member in the Study Group on *Nanoviridae* of the International Committee on Taxonomy of Viruses. Dr. Mandal is the editor-in-chief of *VirusDisease*, an international journal on virology, which is published through Springer. Dr. Mandal teaches plant virology and guides master and PhD students at IARI, New Delhi. His major contribution is in developing technologies for the diagnosis of plant viruses that are important in Indian agriculture. During last 23 years, Dr. Mandal characterized as many as 24 different plant viruses belonging to 10 genera and generated genome sequence resources, which were utilized to design molecular tools to generate recombinant antigens and antibodies. Dr. Mandal has developed several prototypes of plant virus diagnostic kits. Currently, Dr. Mandal's lab has initiated work on utilising plant virus genome as useful genetic resources and developed plant virus-based vector for the expression of foreign protein in plant.

Dr. Govind Pratap Rao is working as a principal scientist (plant pathology) at the Indian Agricultural Research Institute, New Delhi. Dr. Rao has 29 years of research experience on plant pathology especially on plant virology and phytoplasmas. He did significant contributions in the characterization of viruses infecting cucurbits, sugarcane, maize and sorghum. He also developed polyclonal antibodies against *Sugarcane mosaic virus*, *Sugarcane streak mosaic virus*, *Sugarcane yellow leaf virus*, *Maize dwarf mosaic virus* and *Rice tungro bacilliform* and *Rice tungro spherical viruses*. He has published over 130 research publications and authored and edited nearly 17 books to his credit. He has also guided 3 MSc and 11 PhD students on different aspects of plant pathology. He has been working in different capacities as scientific officer, Sr. scientific officer (plant pathology), head of the Division of Plant Pathology and officer in charge at research stations of UP Council of Sugarcane Research centres at Seorahi, Gorakhpur and Shahjahanpur from 1987 to 2010. He has been awarded several prestigious awards to his credit. The most important ones are the National Biotechnology Associateship Award (1991–1992), DBT, Govt. of India; Young Scientist Award (1994–1995) from DST, Govt. of India; Overseas BOYSCAST Award (1996) from DST, Govt. of India; President Award, Society for General Microbiology, UK, in 1998; Best UP Agriculture Scientist Award (UPCAR), Govt. of Uttar Pradesh, in 2002; Vigyan Ratna Award by CST, Govt. of Uttar Pradesh, for the years 2003–2004; Jin Xiu Qiu Award in 2006 by Guangxi Province, Nanning, China; Global Award of Excellence (2008), IS, Al-Arish, Egypt; and Dr. Ram Badan Singh Vishisht Krishi Vaigyanik Puraskar (2014) by UPCAR, Lucknow, India. Dr. Rao is editor-in-chief of *Sugar Tech*, an international journal of sugar crops and related industries, and *Phytopathogenic Mollicutes*, an international journal of phloem-limited microorganisms. Dr. Rao is also secretary general of Indian Virological Society, New Delhi, and member of several prestigious scientific societies and organizations like APS, USA; ASM, USA; ISSCT, Mauritius; IPWG, Italy; SSRP, New Delhi; and IPS, New Delhi. Besides, Dr. Rao has visited 27 countries as visiting scientist, for invited talks, research training, panel discussion and attending workshop and conferences, and as postdoc fellow. At present Dr. Rao is working on the characterization, epidemiology and management of viruses infecting cereal crops, millets and maize and phytoplasmas infecting important agriculture crops in India.

Dr. Virendra Kumar Baranwal is a professor of plant pathology at the Division of Plant Pathology, Indian Agricultural Research Institute (IARI), New Delhi. He teaches postgraduate courses in plant pathology and has guided three MSc and seven PhD students at IARI. He has led several externally funded projects on genomics and diagnostics of plant viruses and has published more than 90 research papers in peer-reviewed journals. He is a fellow of the National Academy of Agricultural Sciences and National Academy of Sciences, India. His research group has made significant contribution on the characterization, diagnostics and evolutionary relationship of badnaviruses infecting banana, citrus and sugarcane and determined the immunodominant region in the N-terminal of coat protein that led to the development of coat protein construct and synthetic peptide for the production of polyclonal antibodies against a badnavirus infecting banana. His group developed a partial dimer construct of banana badnavirus which caused infection in banana when co-inoculated with viral suppressors. His group identified and characterized for the first time in India ampeloviruses associated with grapevine and developed diagnostic reagents. He demonstrated occurrence of new isolates of whitefly-transmitted carlavirus in legume crops using next-generation sequencing. He led a group which demonstrated multiplex RT-PCR for simultaneous detection of multiple virus infection caused by carla-, poty- and allexiviruses in allium crops and for mixed infection of different viruses and greening bacterium in citrus. His latest contribution includes development of RPA-based PCR detection of banana bunchy top virus using crude extract and without the use of thermal cycler. Development of novel protocol of virus elimination in garlic using solar therapy coupled with apical meristem developed by his team will help improve garlic yield. Development of microarray chip having probe sets for more than 1,100 viruses at genus and species level and 40 viroids is another contribution made by his group.

Dr. Rakesh Kumar Jain is presently dean and joint director (education) at IARI, New Delhi. He did his postdoctorate at CSIRO, Melbourne, Australia; University of Florida, Gainesville; and University of Georgia, Tifton, USA. As a plant virologist, his research canvass includes emerging plant viruses such as tobacco streak virus affecting sunflower and groundnut, papaya ringspot virus affecting papaya, groundnut bud necrosis virus affecting tomato and watermelon, bud necrosis virus affecting cucurbits with reference to their distribution profile and genetic diversity, development of diagnostics and virus-resistant transgenic plants. He has guided several MSc and PhD students. Besides, he has hosted scientists from Australia, Bangladesh, Spain, the UK and the USA in his laboratory. He is a fellow of the National Academy of Agricultural Sciences, Indian Phytopathological Society and Indian Virological Society.

Introduction: A Century of Plant Virology in India

1

Bikash Mandal, Govind Pratap Rao, Virendra Kumar Baranwal, and Rakesh Kumar Jain

Abstract

Plant viruses are important constraints in Indian Agriculture. There are as many as 168 plant virus species documented in India. The viruses belonging to the genera, *Babuvirus, Badnavirus, Begomovirus, Closterovirus, Cucumovirus, Emaravirus, Ilarvirus, Luteovirus, Macluravirus, Polerovirus, Potyvirus and Tospovirus*, are economically important. The insects, aphid, thrips and whitefly are the important vectors in India. Virus diseases are more problematic in vegetable pulse and fiber crops. The investigation of plant viruses began in India a few years after the discovery of virus. Plant Virology in India has a long and remarkable history. In this book, we bring out the research findings on plant viruses that were carried out in India during the past more than 100 years. The book contains 31 chapters of which 20 are dealt with the characterization of the viruses belonging to 22 genera, one chapter is on viroids, three chapters are on virus vectors, two on diagnosis and four on management of the viruses.

Keywords

History • Plant virology • India

1.1 Introduction

Viruses are molecular pathogens and infect cellular organisms. They are a unique class of pathogens that are difficult to control. Since the discovery of virus in tobacco mosaic disease at the end of the Nineteenth century in The Netherlands

B. Mandal (✉) • G.P. Rao • V.K. Baranwal • R.K. Jain
Advanced Centre for Plant Virology, Division of Plant Pathology, Indian Agricultural Research Institute, New Delhi 110012, India
e-mail: leafcurl@rediffmail.com

B. Mandal et al. (eds.), *A Century of Plant Virology in India*,
https://doi.org/10.1007/978-981-10-5672-7_1

(Beijerinck 1898), the subject of Plant Virology has considerably evolved during more than the last 100 years. Beside the academic interest, the control of viruses to save the agricultural produces has been a major objective of studies of plant viruses. During 1960–1970, attempts to identify inhibitors of plant viruses as achieved in case of the other pathogens like fungi, were so far largely unsuccessful. Resistant cultivars developed through classical breeding were successfully deployed to manage other pathogens. Although, breeding for resistance against viruses too is considered as the best way to manage them, there are limited sustained successes due to lack of availability of useful sources of resistance and large diversity of rapidly evolving viruses. In the 1980s, the pathogen derived resistance similar to cross protection was demonstrated using genetic engineering approach (Powell et al. 1986). The engineering resistance involving genetic modification of crop plant using parts of virus genome, which is and popularly known as transgenic resistance, has been proved to be successful against numerous viruses and plant species. However, socio-political issues largely discouraged adoption of the transgenic technology in many countries including India. In the present millennium, the study of plant viruses reached the new depth, where the understanding of virus genomics and functional genomics opened up new opportunities to develop better strategies to strengthen the plant's ability to defend against virus infection. However, it is now increasingly understood that the virus disease develops in a plant system following a highly complex network of interactions of plant and viral proteins. In this process, further complexities are added with the interactions with vectors that spread the virus from one plant to another. The tripartite interactions among virus, host plant and vector differ based on the kind of each interacting partners. Gene silencing and identification of interacting protein partners in plant or vector and application of genome editing are emerging areas in the plant virology for achieving resistance in plant. Understanding of the role of micro RNA in virus infection and its modulation of expression has been shown as another emerging approach of prevention of plant virus infection (Pérez-Quintero et al. 2010). It is expected that the 'plant virus medicine' will soon be a reality for the preventive and prophylactic measures against plant virus infection through the topical application of gene silencing therapeutics through nanomaterials (Mitter et al. 2017).

Plant viruses are one of the most important classes of pathogens in Indian Agriculture. The majority of the agricultural areas in India are under the tropical and sub-tropical climate that favours prevalence of viruses and their vectors. The population pressure in India is increasingly influencing intensive cultivation of high-yielding cultivars throughout the year. This provides opportunities to the virus and vectors to establish in an agro-ecosystem challenging the harvest of the full potential of the crop yield. The plant virus diseases in India have evolved as more complex problems simultaneously with the changes in both agriculture system as well as climate. With reference to the development of Plant Virology at global level, the studies of plant viruses in India too have a long and remarkable history.

The investigation of virus diseases in India began a few years after the discovery of virus. The research in Plant Pathology started in India with the establishment of Indian Agricultural Research Institute (IARI) during 1905 in Pusa Bihar. The early historical account of Plant Pathology in India has been documented (Raychaudhuri et al. 1972). The mosaic or katte of small cardamom (*Elettaria cardamomum*) was perhaps the first virus disease recorded during 1900 in southern India (Mollison 1900), which was later identified as a virus disease based on the transmission by an aphid vector (Uppal et al. 1945).

1.2 The Developmental Phages

The first systematic investigation of virus diseases began on sugarcane mosaic during 1922 at IARI, Pusa, Bihar (Dastur 1923). Later, a pioneering work on tobacco leaf curl and its transmission studies through whitefly (*Bemisia tabaci*) was published from IARI (Pal and Tandon 1937; Pruthi and Samuel 1937). During the next two decades, several virus diseases were recorded in cereals, pulses, plantation crops and vegetables. The historical milestones of plant virus research in India have been documented (Raychaudhuri et al. 1972; Sastry and Sai-Gopal 2010).

The research laboratory specifically to conduct plant virus studies was first established in Pune in 1938 by the then Bombay Government in India. Later during 1956, the laboratory was transferred to IARI. In 1950s, IARI was the major research institute to conduct research on plant viruses. During this period, two more research stations on plant virology were created in Shimla and Kalimpong. The Advanced Center for Plant Virology (Fig. 1.1) came into existence in 1988 at IARI, New Delhi, which played an important role in the modern era of Plant Virology in India.

The subject of Plant Virology in India evolved through broadly four distinct eras, (i) The empirical era (1900–1940), when the viral diseases were documented based on the preliminary studies on symptoms and transmission by sap and vector; (ii) the biological era (1940–1970), when studies were conducted mainly on the biological properties such as host range, source of resistance, virus-vector relationships, virus inhibition and disease dissemination; (iii) the serological era (1960–1990), when the emphasis of the work was on virus diagnosis. Among the several methodologies, serology dominated as the most convincing technique for the identification of

Fig. 1.1 Advanced Centre for Plant Virology at Indian Agricultural Research Institute, New Delhi, the major seat for plant virus research in India

viruses and (iv) the molecular era (1990 onward), when the studies of virus began at genomic level that included isolation of viral nucleic acids, cloning, amplification and sequencing of the gene and genome of viruses. Subsequently, in the recent time, studies were conducted to understand the infectivity of the cloned DNA, transgenic resistance, gene function and host-pathogen interactions at cellular level. The generation of viral genome sequence resources and development of infectious clones of DNA and RNA plant viruses opened up the opportunity to exploit the plant viruses for the useful purposes.

1.3 Design and Objectives of the Book

In the ninth report of the International Committee on Taxonomy of Viruses, 1016 virus species and 309 tentative virus species were documented globally. The Indian Plant Virus database has been developed in 2015, which documented 168 plant virus species occurring in India (http://220.227.138.213/virusdb/). Over the past more than 100 years an enormous amount of information was generated in the large body of literature. The objective of this book is to bring this wealth of information in one consolidated platform so as to understand how the subject of Plant Virology evolved in India and how to position the present and the next generation of scientists to deal with the problems of plant viruses in Indian agriculture. The book is designed with the four parts covering characterisation, virus-vectors, diagnosis and management.

Part I: Virus Characterization This is the major part of the book that deals with the properties of the viruses. There are 22 articles that describe the virus genera wise accomplishment of research work. The genera of plant viruses included in this part are *Allexivirus, Ampelovirus, Babuvirus, Badnavirus, Begomovirus, Carlavirus, Carmovirus, Closterovirus, Cucumovirus, Emaravirus, Ilarvirus, Luteovirus, Macluravirus, Mandarivirus, Mastrevirus, Pecluvirus, Polerovirus, Potyvirus, Sobemovirus, Tobamovirus, Tospovirus* and *Tungrovirus*. Among all these genera of viruses, the viruses of the genera *Begomovirus* and *Tospovirus* are highly aggressive viral pathogens in many important crops and have the history of recurrent epidemic episodes, and as a result they received maximum attention to research investigation in India. Viroids were discovered in 1971 and the work in India commenced in 1980s. In India, viroids have been identified in citrus, tomato, apple, ornamentals, rubber and grapes. One chapter of the research finding on viroids occurring in India has been included in this part.

Part II: Virus-Vectors In the early stage of Plant Virology (1915–1940), several insect vectors were discovered to transmit plant viruses with extraordinary specificity. The vectoring property became an important criterion to differentiate the virus disease from those caused by fungi or bacteria. In India, the first systematic study on the vector transmission was conducted with tobacco leaf curl virus and whitefly. The major virus vectors in India are aphids, whitefly and thrips. This part provides the up-to-date work conducted in India on these important virus-vectors.

Part III: Virus Diagnosis Diagnosis of plant viruses gained momentum in 1970 onward when electron microscopy, serology and subsequently nucleic acid based techniques were used for the diagnosis of plant viruses. Research on diagnosis significantly contributed to identification and classification of viruses. Of all the techniques, enzyme-linked immunosorbent assay and polymerase chain reaction were extensively used in diagnosis of viruses. Two chapters one each on serology and nucleic acid based diagnosis approaches were included in this part.

Part IV: Virus Management The final aim of understanding plant viruses is to develop strategies to prevent crop yield losses. There is no valid estimate to figure out the losses caused plant viruses in India. However, some viruses are known either to cause crop failure in a season or some causes gradual degradation of the potential yield. Over all, it is perceived that plant viruses are responsible for a significant crop yield losses in India and thus management solutions of viruses are pressing demand of the crop growers and practitioners. This part brings together the different areas of research e.g., conventional approaches, antiviral defence, quarantine and transgenics, that were investigated for the management of virus diseases in India.

References

Beijerinck MW (1898) Concerning a contagium vivum fluidum as a cause of the spot-disease of tobacco leaves. Verh Akad Wet Amsterdam II(6):3–21

Dastur J (1923) The mosaic disease of sugarcane in India. Agric J India 18:505–509

Mitter N, Worrall EA, Robinson KE, Li P, Jain RG, Taochy C, Fletcher SJ, Carroll BJ, Lu GQ, Xu ZP (2017) Clay nanosheets for topical delivery of RNAi for sustained protection against plant viruses. Nat Plants 3:16207

Mollison JW (1900) Cultivation of the betel palm, cardamom and pepper in the Kanara district of Bombay presidency with notes on the manures used in spice gardens of that tract. Department of Land Record and Agriculture, Bombay. Bulletin No 20, 24 p.

Pal BP, Tandon RK (1937) Types of leaf-curl in Northern India. Indian J Agric Sci 7:363–393

Pérez-Quintero AL, Neme R, Zapata A, López C (2010) Plant microRNAs and their role in defense against viruses: a bioinformatics approach. BMC Plant Biol 10:138

Powell AP, Nelson RS, De B, Hoffmann N, Roger SG, Fraley RT, Beachy RN (1986) Delay of disease development in transgenic plants that express the tobacco mosaic virus coat protein gene. Science 232:738–743

Pruthi HS, Samuel CK (1937) Entomological investigations on the leaf curl diseases of tobacco in North Bihar. I. Transmission experiments with some suspected insect vectors. II An alternate host of the virus and the insect transmitter. Indian J Agric Sci 7:659–670

Raychaudhuri SP, Verma JP, Nariani TK, Sen B (1972) The history of plant pathology in India. Annu Rev Phytopathol 10:21–36

Sastry KS, Sai-Gopal DVR 2010 Glimpses of plant virus research in India. In: Compendium of plant viruses, phytoplasma and viroid diseases research in India (1903–2008). Tirupati: Violoy Publication, pp 1–8

Uppal BN, Varma PM, Capoor SP (1945) Mosaic disease of cardamom. Curr Sci 14:208–209

Part I

Virus Characterisation

2 Ampeloviruses Associated with Grapevine Leafroll Disease: A New Group of Viruses in India

Sandeep Kumar, Richa Rai, and Virendra Kumar Baranwal

Abstract

Ampeloviruses (family *Closteroviridae*) are filamentous monopartite, single-stranded, positive-sense RNA genome. They are transmitted by mealybugs in semi-persistent manner and vegetative propagating material remains the major route of spread. Ampeloviruses are recent addition to the plant viruses in India. *Grapevine leafroll-associated virus* 3 (GLRaV-3) was first ampelovirus to be recorded from India in the year 2012. Of the nine distinct species of the genus *Ampelovirus*, only three, *Grapevine leafroll-associated virus* 1 (GLRaV-1), GLRaV-3, GLRaV-4 infecting grapevine have been reported from India. The isolates of GLRaV-3 and GLRaV-4 are diverse, a few being the recombinant ones. This chapter describes the grapevine leafroll disease caused by different ampeloviruses, their geographical distribution, characterization, diversity, management strategies and also discusses about the future course of works to be taken.

Keywords

Ampeloviruses • Grapevine • Diversity • Diagnostics • India

S. Kumar (✉)
AICRP on Medicinal & Aromatic Plants and Betelvine, Orissa University of Agriculture & Technology, Bhubaneswar 751003, India

Advanced Centre for Plant Virology, Division of Plant Pathology, ICAR-Indian Agricultural Research Institute, New Delhi 110012, India
e-mail: sandeepkumariari@gmail.com

R. Rai • V.K. Baranwal
Advanced Centre for Plant Virology, Division of Plant Pathology, ICAR-Indian Agricultural Research Institute, New Delhi 110012, India

B. Mandal et al. (eds.), *A Century of Plant Virology in India*,
https://doi.org/10.1007/978-981-10-5672-7_2

2.1 Introduction

The term *Ampelovirus* is derived from an ancient Greek word *ampelos* meaning grapevine, the host for the type species. It includes the virus species with flexuous filamentous particles of size 1400–2000 nm long, monopartite, single-stranded, positive-sense RNA genome of 13.0–18.5 kb size, transmitted by pseudococcid mealybugs and soft scale insects. *Ampelovirus* is one of the four genera of the virus family *Closteroviridae*, others three being *Closterovirus*, *Crinivirus* and *Velarivirus* (Fig. 2.1). Additionally, the family consists of five unassigned viruses. Despite being named after grapevine the genus *Ampelovirus* also includes non-grapevine infecting viruses. Majority of the ampleoviruses are recorded from woody plants such as grapevine, plum, fig and pineapple. The virus species list of the genus *Ampelovirus* recognized by International Committee on Taxonomy of Viruses

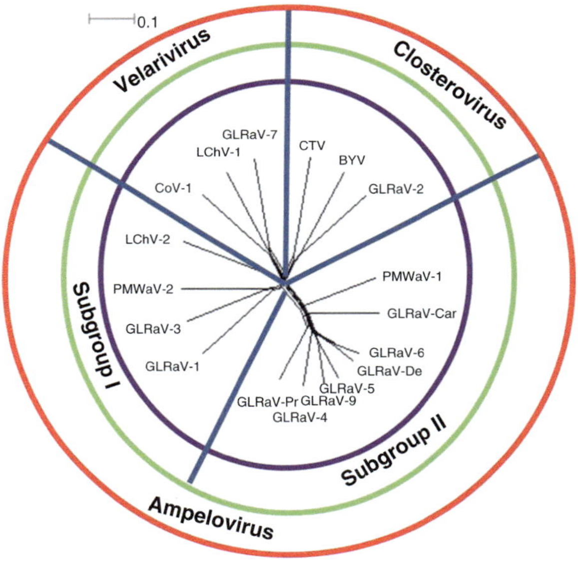

Fig. 2.1 Neighbour network reconstruction of the complete HSP70h genes of grapevine leafroll disease associated viruses. Nucleotide sequences were taken from GenBank and the network was constructed using SplitsTreev4 (Huson and Bryant 2006). Sequences used for constructing the network are: *GLRaV-1* Grapevine leafroll-associated virus 1, AF195822, *GLRaV-2* Grapevine leafroll-associated virus 2, AF039204, *GLRaV-3* Grapevine leafroll-associated virus 3, NC_004667, *GLRaV-4* Grapevine leafroll-associated virus 4, FJ467503, *GLRaV-5* Grapevine leafroll-associated virus 4 strain 5, NC_016081, *GLRaV-6* Grapevine leafroll-associated virus 4 strain 6, FJ467504, *GLRaV-9* Grapevine leafroll-associated virus 4 strain 9, AY297819, *GLRaV-De* Grapevine leafroll-associated virus 4 strain De, AM494935, *GLRaV-Car* Grapevine leafroll-associated virus 4 strain Car, FJ907331, *GLRaV-Pr* Grapevine leafroll-associated virus 4 strain Pr, AM182328, *GLRaV-7* Grapevine leafroll-associated virus 7, HE588185, *PMWaV-1* Pineapple mealybug wilt-associated virus 1, *PMWaV-2* Pineapple mealybug wilt-associated virus 2, *LChV-1* Little cherry virus 1, NC_001836, *LChV-2* Little cherry virus 2, AF416335, *CoV-1* Cordyline virus 1, HM588723, *CTV* Citrus tristeza virus, NC_001661, *BYV* Beet yellows virus

(ICTV) consists of nine species *Blackberry vein banding-associated virus* (BVBaV), *Grapevine leafroll-associated virus* 1 (GLRaV-1), *Grapevine leafroll-associated virus* 3 (GLRaV-3), *Grapevine leafroll-associated virus* 4 (GLRaV-4), *Little cherry virus* 2 (LChV-2), *Pineapple mealybug wilt-associated virus* 1 (PMWaV-1), *Pineapple mealybug wilt-associated virus* 2 (PMWaV-2), *Pineapple mealybug wilt-associated virus* 3 (PMWaV-3) and *Plum bark necrosis stem pitting-associated virus* (PBNSPaV) (www.ictvonline.org/virusTaxonomy.asp).

Replication of ampeloviruses takes place in cytoplasm in association with membranous vesicles. The membranous vesicles may be derived either from endoplasmic reticulum or from peripheral vesiculation and disruption of mitochondria (GLRaV-1, GLRaV-3). The gene expression strategy happens to be ribosomal shifting for ORF1a and ORF1b. Other ORFs produces their respective proteins by translation of a set of nested 3′ co-terminal subgenomic RNAs (King et al. 2012).

2.2 Subgroups of *Ampelovirus*

Viruses belonging to the genus *Ampelovirus* show wide and distinct variations in genome size and organization. Accordingly they are grouped in two subgroups (Fig. 2.1). The subgroup I includes viruses with large (in excess of 17,000 nt) and complex (9–12 ORFs) genome *viz.* GLRaV-3, GLRaV-1, PMWaV-2, LChV-2 and BVBaV (Martelli et al. 2012; King et al. 2012; Naidu et al. 2015). GLRaV-3, the type species of the genus, has the largest genome in the genus comprising 12 ORFs (13 genes). The difference in genome size between isolates depends on the length of 5′ NTR (Naidu et al. 2015; Jarugula et al. 2010; Maree et al. 2008). Contrastingly, 3′ NTR of all isolates of GLRaV-3 is comparatively shorter in length having a consistent length of 277 nt and remain more conserved. The subgroup II comprises of smaller (approximately 13,000–14,000 nts) and simpler (6 ORFs, 7 genes) genome viral species *viz.* GLRaV-4, PMWaV-1, PMWaV-3 and PBNSPaV. One of the salient features of this subgroup is that they lack CPm. PMWaV-1 of the subgroup has a genome length of 13,071. Its seven ORFs (including ORF 1a and ORF 1b) express the replication related proteins, a 6 kDa hydrophobic protein, the HSP70h, the ~60 kDa protein, the CP and a 24 kDa protein, respectively (Fig. 2.2) (Martelli et al. 2012; King et al. 2012).

2.3 Symptoms and Transmission

Ampeloviruses cause a range of symptoms such as rolling, yellowing and reddening of the leaves (grapevine), stem pitting (plum), wilting and produce no symptom in pineapple. In natural condition these viruses are transmitted by mealy bugs (family *Pseudococcidae*) and scale insects (family *Coccidae*) in a semipersistent manner. The vector species and its range vary from virus to virus. Pineapple infecting ampeloviruses are transmitted by two species of the genus *Dysmicoccus* while LChV-2 is vectored by *Phenacoccus aceris*. None of the ampeloviruses is reported

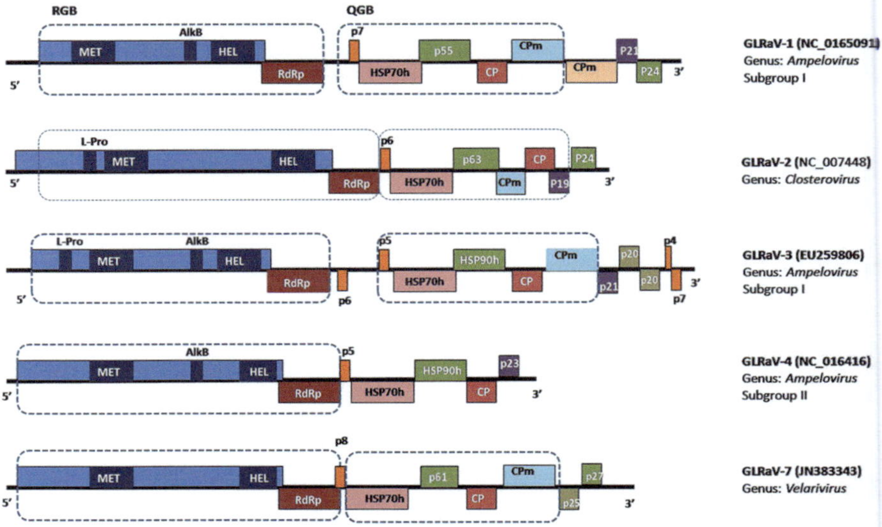

Fig. 2.2 Schematic representation of the genome organizations of grapevine leafroll disease associated viruses. *GLRaV-1* Grapevine leafroll-associated virus 1, NC_0165091, *GLRaV-2* Grapevine leafroll-associated virus 2, NC_007448, *GLRaV-3* Grapevine leafroll-associated virus 3, EU259806, *GLRaV-4* Grapevine leafroll-associated virus 4, NC_016416, *GLRaV-7* Grapevine leafroll-associated virus 7, JN383343. Corresponding genera, subgroups and accession numbers are indicated to the right side of the genome maps. The open reading frames (ORFs) are shown as *boxes* with designated protein domains such as *L-Pro* papain-like leader protease, *AlkB* AlkB domain, *MET* methyltransferase, *HEL* RNA helicase and *POL* RNA dependent RNA polymerase domains of the replicase. Conserved ORFs form the replication gene block (*RGB*) and quintuple gene block (*QGB*) and they are denoted by *dotted line boxes*. Abbreviations indicating ORFs are: *CP* coat protein, *CPm* minor coat protein, *RdRp* RNA-dependent RNA polymerase. The other ORFs are designated with approximate molecular weight and a common "p" designator. Figures drawn are not to the scale

to be transmitted through seed or by mechanical means (King et al. 2012). These viruses can be carried over in the vegetative cuttings used for propagation of their respective host plants and thus vegetative propagating materials become the primary source of virus spread over long distance (Kumar 2013; King et al. 2012).

2.4 Ampeloviruses in India

The occurrence of ampeloviruses in India is reported recently. Before 2012, there was no authentic information on virus or virus like diseases of grapevine in India. A news report appeared in a daily *The Indian Express* (4th November, 2007) indicated the presence of grapevine leafroll disease (GLD) in the vineyards of Maharashtra, which accounts for 94% of country's wine production. It further mentioned how this disease has started a debate and blame game among the various stakeholders of viticulture and related industries (Jadhav and Sonawane 2007). This disease had started creating havoc among famers and wine and raisin industries. Few farmers had already removed their vineyards because of GLD. Subsequently in the year 2012, Indian Agricultural Research Institute (IARI), New Delhi in collaboration with National Research Centre for Grapes (NRCG), Pune found the association of GLRaV-1 and GLRaV-3 in the vineyards of Nashik and Pune regions of Maharashtra (Kumar et al. 2012a, b). Till date, out of nine ICTV recognized ampeloviruses, only three viral species have been reported from India, all associated with grapevine leafroll disease. In this chapter a comprehensive account of work done on ampeloviruses in India *vis-a-vis* their global stand has been discussed and a way forward for the work has also been outlined.

2.5 Disease and Virus Description

2.5.1 Grapevine Leafroll Disease (GLD)

Globally, the first descriptions of grapevine leafroll date back to the mid nineteenth century. It got several synonyms in different languages such as White Emperor disease (English), Rollkranheit and Blattrollkranheit (German), Rugeau and Enrolument (French), Rossore and Accartocciamento fogliare (Italian), enrollamiento de la hoja and enrollado (Spanish), Enrolamento de la folha (Portuguese) (Martelli and Boudon-Padieu 2006). Scheu (1935) demonstrated the graft transmission of leafroll from diseased to healthy vines and hypothesized the viral origin of the disease. However, Harold Olmo, a viticulturist of University of California, Davis and his colleagues in 1943 reported that the concerned problem was perpetuated by vegetative propagation and proposed that a virus was involved with the disease (Olmo and Rizzi 1943). Further, scientists demonstrated that the disease was also transmissible via grafts, which in turn provided strong evidence that a virus is the causal organism (Alley and Golino 2000; Harmon and Snyder 1946). In India, though said to be present since 2002, the first authentic report of the disease appeared in 2012 (Kumar 2013; Kumar et al. 2012a, b; Jadhav and Sonawane 2007).

2.5.2 Symptoms

GLD is said to be a complex disease with asymptomatic and symptomatic phases. It is unique in its symptomatology as the exhibition of symptoms begins on mature leaves which is in contradiction to many virus diseases where the exhibition of symptoms take place on newly developing parts (Naidu et al. 2015). Expression of symptoms is highly variable from cultivar to cultivar and from season to season. Exhibition of red and reddish-purple discolourations in the interveinal areas of mature leaves at the basal part of the shoots in late spring or summer, depending on the climate and geographic location, is one of the early sign in dark-berried cultivars. In Indian condition the typical symptoms have been observed from November–December to February. Symptoms are more expressive in dark-fruited/red-fruited cultivars than in light-fruited/white-fruited cultivars. As the season advances, in dark-fruited cultivars the red to reddish-purple colour in interveinal lamina become prominent, leaf blades become thick, brittle and the margins of the infected leaves roll downward (Fig. 2.3). In severe cases, the whole leaf surface becomes deep purple (Martelli and Boudon-Padieu 2006; Rayapati et al. 2008). The symptoms are similar in light-fruited cultivars, but the leaves become chlorotic to yellowish, instead of reddish to reddish-purple (Fig. 2.3). Some white-fruited cultivars show no visual sign of infection (i.e. latent infection). In advanced stages of infection, the margins of the leaves of both kinds of cultivars roll downward, expressing the symptom that gives the

Fig. 2.3 Symptoms of grapevine leafroll disease (*GLD*) observed during the survey conducted for the study. (**a**) Vines of cultivar Cabernet Souvignon in a vineyard of Nashik; (**b**) Vine of a cultivar Pinot Noir at experimental farm of ICAR-National Research Centre for Grapes (ICAR-NRCG), Pune; (**c**, **d**) The close-up views of the leaves of two different vines of cultivar Pinot Noir (from ICAR-NRCG, Pune) found to be positive for GLRaV-1 and GLRaV-3, respectively; (**e**) Close up view of leaves of a vine of cultivar Shiraj from Nashik found to be positive for both GLRaV-1 and GLRaV-3; (**f**) Close up view of leaves of a vine of light-fruited cultivar Thompson Seedless (from ICAR-NRCG, Pune) found to be positive for GLRaV-3

disease its common name, i.e. "leafroll" (Rayapati et al. 2008; Martelli and Boudon-Padieu 2006). Most grape rootstocks, particularly American hybrids, do not show symptoms of leafroll even though they may carry the virus (Kovacs et al. 2001; Pietersen 2004). GLRaV-4 and related viruses elicit milder symptomatology compared to GLRaV-1 and GLRaV-3 (Martelli et al. 2012). Some strains of GLRaV-2 and -7 cause asymptomatic infection. Association of different GLRaVs and their strains with the disease further amplifies the complexity in symptomatology. Additionally, mixed infections among GLRaVs and with other viruses and viroids could be one of the factors in many intrigues of the disease (Naidu et al. 2015). Synthesis of two classes of anthocynins namely, *cyanidin-3-glucoside* and *malvidin-3-glucoside* has been reported to contribute in the expression of reddish-purple colour of virus-infected leaves of dark-fruit grapevine (Gutha et al. 2010).

2.5.3 Impacts of the Disease

The disease reduces yields, delays fruit ripening, reduces soluble solids, delays crop maturity, reduces berry anthocyanin & berry weight, and increases titratable acidity in fruit juice ultimately resulting in reduced wine quality (Atallah et al. 2012; Rayapati et al. 2008; Charles et al. 2006; Mannini et al. 1998). Degeneration of the phloem vessels and loss of photosynthetic potential of the leaves of infected vines are the major reason for decrease in quantity and quality (Freeborough and Burger 2008). As reviewed by Kumar (2013), GLRaV-3 reduces photosynthesis by 25–65 % depending upon the cultivar and environment. Bertamini et al. (2004) carried out a well designed research work showing the impact of disease on photosynthetic aspects of the host. In this study the virus-infected leaves showed reduced level of total chlorophyll (Chl), carotenoids (Car), soluble proteins and RuBP activity. An increase of Chl/Car ratio and a reduction of Chl a/Chl b ratio (ratio between chlorophyll a and chlorophyll b) were observed which could be due to the relatively faster decrease of Chl than Car. Photosynthetic study conducted in isolated thylakoids showed that because of leafroll infection there was marked inhibition of whole chain and photosystem (PS) II activity but only minimal inhibition of PS I activity was observed. It was inferred that the marked loss of PS II activity in infected leaves could be due to the loss of 47, 43, 33, 28–25, 23 and 17 kDa polypeptides as demonstrated by decrease in the amount of these polypeptides in SDS-PAGE analysis. The inhibition of donor side of PS II was also confirmed by immunological studies showing the significantly diminished content of 33 kDa protein of the water-splitting complex in infected leaves (Bertamini et al. 2004). Based on sensory descriptive analysis of 2010 wines it was suggested that GLD significantly affects the colour, aroma and astringency of wines. The study further suggested the influence of host × environment interactions on overall impact of the disease, causing maximum impact during cooler seasons (Alabi et al. 2016).

Globally, GLD is considered as the most economically destructive disease amongst the virus and virus like diseases of grapevines. Yield reductions due to GLD may vary, but reductions of around 50 % (or ≥60 % if the disease is severe) are commonly reported on a worldwide basis (Rayapati et al. 2008). As per several reports, reduction in quantity produced of grapevines may be in the tune of 30–68 % (Atallah et al. 2012). Practically, even a small decrease in annual yields due to GLD has a cumulative impact on the long-term viability and profitability of a vineyard (Rayapati et al. 2008). The estimated economic impact of GLD ranges from approximately $25,000 to $40,000 per hectare in the absence of any control measure (Atallah et al. 2012).

2.5.4 Causal Agents: A Chronological Perspective

Despite confirmation of the nature of the disease as of viral origin by California based scientist Harmon and Snyder (1946), the causal agent remained unknown until the late 1970s. Namba et al. (1979) found closterovirus like particles in Japanese vines with leafroll symptoms, and reported the association of ampelovirus with the disease. A few years afterwards, two serologically different viruses from Switzerland were partially characterized and referred as "type I" and "type II" (Gugerli et al. 1984). Later, a number of new putative closteroviruses identified from vines with leafroll symptoms in Europe and USA. After 1995, Roman numerals were replaced by Arabic numerals to differentiate the different viruses (Martelli and Boudon-Padieu 2006). Till 2008, ten different viruses with filamentous particles, called grapevine leafroll-associated viruses (GLRaVs) were found associated with grapevine leafroll disease (GLD) and they were differentiated from one another by a number in increasing order as GLRaV-1 to -10 in the order of their discovery and were reported to be serologically distinct from each other (Martelli et al. 2002; Karthikeyan et al. 2008; Martelli 2009). By 2011, the number of GLRaVs had gone up to 12 but by the end of 2011, the number had been reduced to 11 due to withdrawing of GLRaV-8 from the ninth ICTV report because it proved to be the part of grapevine genome rather than being of viral origin (Martelli et al. 2012). The 11 filamentous viruses belonging to family *Closteroviridae* have been found associated with the leafroll disease of grapevines are GLRaV-1, GLRaV-2, GLRaV-3, GLRaV-4, GLRaV-5, GLRaV-6, GLRaV-7, GLRaV-9, GLRaV-Pr (sequence originally deposited in GenBank under the name of GLRaV-10), GLRaV-De (sequence originally deposited in GenBank under the name of GLRaV-11) and GLRaV-Car (Martelli et al. 2012). Very recently, a novel ampelovirus has been detected in grapevines showing typical symptoms of GLD from Japan and it has been tentatively named as GLRaV-13 (Ito and Nakaune 2016). It showed closest but significantly distant relationship to GLRaV-1 in the subgroup I cluster of the genus *Ampelovirus*. But the name of GLRaV-13 might be controversial as its pathogenicity remains unclear; therefore, further study is needed in this regard (Ito and Nakaune 2016).

In ninth report of ICTV, out of eleven viruses associated with GLD, one (GLRaV-2) has been approved as the member of the genus *Closterovirus*, three (GLRaV-1, -3, and -5) have been placed in the genus *Ampelovirus* and six (GLRaV-4, -6, -9, GLRaV-Pr, GLRaV-De and GLRaV-Car) have been putatively assigned to the genus *Ampelovirus*, whereas one GLD causing virus (GLRaV-7) could not be assigned to any genus of the family *Closteroviridae* (King et al. 2012). As per the studies of various researchers, ratification vote on taxonomic proposal of ICTV-2013 abolished the species GLRaV-5 and floated a new species GLRaV-4 which was earlier putatively assigned to the genus *Ampelovirus* (Adam et al. 2013). In the ratification vote on taxonomic proposal of ICTV-2014, a new genus *Velarivirus* was created and GLRaV-7, which earlier remained unassigned to any genus of the family *Closteroviridae*, has been given the status of type species of the genus *Velarivirus* (Adam et al. 2014). Recent studies based on genome size, structure and shared biological, epidemiological and serological characteristics suggested to consider GLRaV-5, GLRaV-6, GLRaV-9, GLRaV-Pr, GLRaV-De and GLRaV-Car as the strains of GLRaV-4 and thus they are written as GLRaV-4 strain 5, GLRaV-4 strain 6, GLRaV-4 strain Pr, GLRaV-4 strain De and GLRaV-4 strain Car, respectively. Together these viruses are known as GLRaV-4 like viruses i.e. GLRaV-4 LV (Naidu et al. 2015; Martelli et al. 2012). It can be noted that all grapevine infecting ampeloviruses can cause grapevine leafroll disease (GLD) whereas all GLD causing viruses cannot be ampeloviruses, such as GLRaV-2 (genus *Closterovirus*) and GLRaV-7 (genus *Velarivirus*). Therefore, the recent taxonomy, as available on ICTV website, has grouped GLD causing viruses into five species namely, GLRaV-1, GLRaV-2, GLRaV-3, GLRaV-4 (and its strains) and GLRaV-7 (www.ictvonline.org/virusTaxonomy.asp). Further studies on tentative GLRaV-13 may lead to minor alteration in the taxonomy of GLD causing viruses.

2.5.5 Genome Organization of GLRaVs

The genome size of GLRaVs range from 13, 626 nt in GLRaV-4 strain Car to 18, 671 nt in GLRaV-3 encoding six ORFs to 12 ORFs, respectively (Naidu et al. 2015). A major portion of 5′ end of genome of GLRaVs encoding a characteristic core of replication-associated genes is referred as replication gene block (RGB) (Fig. 2.2). The RGB constitutes ORF 1a and 1b encoding replication-associated proteins containing important domains such as methyltransferase (MET), RNA helicase (HEL) and RNA-dependent RNA polymerase (RdRp). Except GLRaV-7 and GLRaV-2 (i.e. ampeloviruses associated with GLD), ORF 1a of GLRaVs uniquely harbours an AlkB domain, which is a characteristic feature of many RNA viruses infecting woody plants. This domain has role in reversal of alkylation damage through RNA demethylation. ORFs located downstream to RGB are responsible for encoding structural and accessory proteins of GLRaVs. Unlike RGB, downstream ORFs are more variable and do not possess the same level of organizational conservation. In this portion of genome of GLRaVs (except GLRaV-4 LV), there occurs a block of five ORFs known as quintuple gene block (QGB), a hallmark of the family *Closteroviridae* (Fig. 2.2). In QGB the first ORF is a small transmembrane protein having role in cell-to-cell movement, second is homologous to cellular heat shock protein 70 (HSP70h), third in the QGB is ~60 kDa protein, sometimes denoted as HSP90h (as in GLRaV-3 and GLRaV-4 LV). Both second and third genes of QGB cooperate in cell-to-cell movement and virion head assembly. CP and coat protein minor (CPm) are the last two genes of QGB. CP gene encodes for coat protein and gives the characteristic elongated morphology to the virion. CPm is responsible for the formation of main component of the virion head in other closteroviruses (Naidu et al. 2014, 2015; Martelli et al. 2012).

There is a marked difference between the arrangement of CP and CPm genes in QGB between GLRaV-2 and GLRaVs-1, -3, and -7. Like other members of the genus *Closterovirus*, in GLRaV-2 CPm gene is followed by CP gene whereas in GLRaVs-1, -3, and -7 i.e. CP gene is followed by CPm gene. Interestingly, two divergent copies of CPm is found in GLRaV-1 whereas CPm is conspicuous by its absence in GLRaV-4 LV (Fig. 2.2). GLRaV-3 is unique by the presence of an additional ORF encoding 6 k-Da (ORF 2) protein and a GC-rich intergenic region between ORF 2 and ORF 3 which is unlike the other members of the family *Closteroviridae*. Presence of ORF 11 (p4) and ORF 12 (p7) further add to the uniqueness of GLRaV-3 as they are absent in other closteroviruses (Naidu et al. 2015; Martelli et al. 2012). ORFs proximal to 3′ end of GLRaVs are more versatile and their functions are yet to be known. Still, based on analogies it has been suggested that these ORFs could be responsible for suppression of the host RNA silencing and long distance transport of the virus (Naidu et al. 2015). Replication of ampeloviruses in general has been briefly discussed in the beginning of this chapter however; lack of universally conserved QGB in GLRaVs not only suggests the likely differences in replication but also indicates the possibility of different host-virus interactions between individual GLRaVs. Additionally, lack of a CPm in GLRaV-4 LV and its duplication in GLRaV-1 suggests the probable dissimilarities in head segmentation patterns among GLRaVs. As far as 5′ UTR is concerned GLRaVs stand unique because of remarkable diversity in its sequence and predicted secondary structure (Naidu et al. 2015).

2.5.6 Transmission and Host-Range

GLD, once thought to be only graft transmissible, was found to be spreading within vineyards and mealybugs were first shown to be responsible for transmitting associated viruses in 1990 (Tsai et al. 2010; Engelbrecht and Kasdorf 1990). Since then, some mealybug (family *Pseudococcidae*) and soft-scale (family *Coccidae*) species have been shown to transmit different GLRaVs (Tsai et al. 2010). Transmission of GLRaVs seems to occur in a semi-persistent modality (Tsai et al. 2008). So far, vectors of GLRaV-1, -3, -4, -5, -6, -9 and GLRaV-Pr have been identified (Martelli and Boudon-Padieu 2006; Martelli et al. 2012). GLRaV-1, -3, and -4 and its strains are transmitted by several species of mealybugs of the genera *Heliococcus* (GLRaV-1, and -3), *Phenacoccus* (GLRaV-1, and -3), *Pseudococcus* (GLRaV-1, and -3) and *Planococcus* (GLRaV-3, -4 and its strains) and scale insects of the genera *Pulvinaria* (GLRaV-1, and -3), *Neopulvinaria* (GLRaV-1, and -3), *Parthenolecanium* (GLRaV-1, and -3), *Coccus* (only GLRaV-3), *Saissetia* (only GLRaV-3), *Parasaissetia* (only GLRaV-3), *Ceroplastes* (GLRaV-3, −4 and its strains) (Naidu et al. 2014; Kumar 2013; King et al. 2012). There is very limited knowledge of transmission biology of these viruses as far as scale insects are concerned and based on mealybugs transmission, lack of virus-vector specificity has been suggested. Further, till date no insect vector has been identified for GLRaV-2 and −7. Vegetative cuttings of grapevine are transient and can carry their virus payload along with them and because of this fact viruses associated with GLD are sometimes called as "suitcase" or "samsonite" viruses (Rayapati et al. 2008). Mechanical transmission of ampeloviruses is not reported but GLRaV-2 has been experimentally shown to be mechanically transmitted from grapevine tissues to *Nocotiana benthamiana* (Naidu et al. 2014). Use of infected plant materials, while establishing new vineyards or during replacing vines in an established vineyard is the principal means of spread of GLD. The associated viruses do not have any natural hosts other than *Vitis* species. However, very recently there has been a report of natural infection of GLRaV-1 to pomegranate trees in Turkey. Thus, pomegranate (*Punica granatum* L.) could be an alternate host for GLRaV-1 (Caglayan et al. 2016). Further studies in this regard may give an in-depth understanding of the host range of GLD associated viruses.

2.5.7 Geographical Distribution

GLD is new to India and found in all grape-growing regions of the world, including Europe, South and North America, Middle East, Africa and Oceania (Sharma et al. 2011). Because of its wide presence it has been said that wherever grapevines are grown, occurrence of grapevine leafroll disease can be seen (Goheen 1988). In India the disease was first reported from the vineyards of Nashik and Pune regions of Maharashtra. Kumar (2013) suggested the presence of disease in the vineyards of Nashik and Pune regions which eventually fall in hot-tropical agro-climate but the study could not find GLD in Koppal district of Karnataka (mild-tropical agro-climate) and in Jammu and Kashmir (temperate agro-climate). However, in the same year another group of researchers proved the presence of GLD in another part of temperate region of India i.e. in Himachal Pradesh (Kumar et al. 2013). In a recent study disease has also been found in Manipur, a North-Eastern state of India. The associated virus in Manipur has been detected to be as GLRaV-4 (Fig. 2.4). GLRaV-3 and GLRaV-1 are the two most common viruses associated with the leafroll disease of grapevine not only at Indian condition but also at global level (Kumar 2013; Fuchs et al. 2009).

2.5.8 Virus Characterization, Recombination and Selection Pressure Analyses

Nucleotide data of NCBI suggest the availability of 44 full genome sequences of GLRaVs and their isolates. But, till date the complete genome sequencing of any ampelovirus has not been done in India. Partial characterization of GLRaV-1, GLRaV-3 and GLRaV-4 from India has been attempted. GLRaV-1 and -3 have been

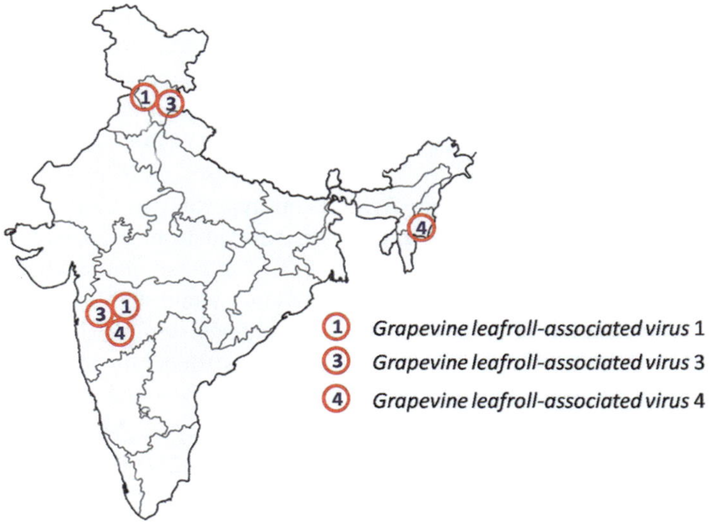

Fig. 2.4 Distribution of Grapevine leafroll-associated viruses (GLRaVs) in different states of India

characterized following one-step RT-PCR while GLRaV-4 has been characterized using two-step RT-PCR. The p24 gene of two isolates of GLRaV-1 was characterized to be of 630 base pairs (bp) and based on p24 gene phylogeny, the global isolates of GLRaV-1 segregated into three distinct groups. Two Indian isolates of GLRaV-1 clustered in group 1 with Claretvine and RRG isolates from USA (Kumar et al. 2012b). However, based on CP and HSP70h (heat shock protein 70 homologue) genes a recent study showed that global isolates of GLRaV-1 clustered into eight and seven groups, respectively (Fan et al. 2015). Partial HSP70h and entire p19.7 genes of 546 bp and 540 bp, respectively were characterized from the eleven isolates of GLRaV-3. The global isolates of GLRaV-3 segregated into eight clusters irrespective of their geographic origins (Naidu et al. 2015; Maree et al. 2015). Most of the Indian isolates clustered in group 2 of the global isolates but isolates Revella-4/12, Revella-4/14, KS-B-7 and Nashik showed discordant grouping behaviour based on different gene based phylogenies. Globally, this was the first such report of incongruent grouping patterns of isolates of GLRaV-3 based on different genes (Kumar 2013). On the basis of CP, HSP70h and p23 phylogenies, GLRaV-4 isolates from India grouped in group 1 with LR106 isolate of USA. In p23 phylogeny two isolates were closely related to LR106 isolate while other two isolates were distantly related to the same isolate.

Turturo et al. (2005) was the first to indicate the phenomenon of recombination in GLRaV-3 population. Later, Farooq et al. (2013) confirmed the recombination events in GLRaV-3 and proved that CP gene acts as one of the recombination hotspots in GLRaV-3 genome. However, based on p19.7 gene recombinant analysis, the Nashik isolate of GLRaV-3 from India was noted to be a recombinant isolate, having parental sequences of 6–18 isolate from USA and Manjri- A2–38/36 isolate from India. It was also hypothesized that the recombination events could be the reason behind phylogenic incongruence and evolutionary process (Kumar 2013). The normalized values for the ratio of nonsynonymous substitutions per nonsynonymous site (*dN*) to synonymous substitutions per synonymous site (*dS*) indicated that HSP70h and p19.7, despite being under strong purifying selection pressures to preserve the amino acid sequences encoded by them and thereby retaining the biological functions, showed the contrasting patterns of evolution with their differential selection pressures. HSP70h gene (69.06 %) was under more purifying or negative selection pressure than p19.7 gene (49.16 %) and thus HSP70h gene of GLRaV-3 was subjected to stronger functional constraints which is nothing but the amount of intolerance towards nucleotide substitution. The relative higher value of normalized *dN-dS* for p19.7 indicates the comparatively flexible nature of the gene to accommodate the non-synonymous changes (Kumar 2013).

F-Pachore vani, GRP-G, GDR-I and GRP-G isolates of GLRaV-4 from India were observed to be the recombinant ones. Further, GDR-I, GRP-G and TS-N isolates from India contributed their genomic region either as major parents or minor parents in the evolution of some GLRaV-4 isolates from other countries. In case of GLRaV-4, 46 % of the codons in CP, 58.8 % of the codons in HSP70h genomic regions and 23.4 % of codons in the p23 genomic region were under purifying selection pressure. The HSP70h gene of GLRaV-4 isolates exhibited 1.5–2.7 times lower dN/dS values compared to the CP and p23 genes, indicating a stronger negative or purifying selection pressure acting upon HSP70h compared to CP and p23 genes.

2.5.9 Management

Because of the graft transmissibility nature of GLD, the best way of its management lies in the fact of employing the first line of defence i.e. to use the virus free propagating materials at the time of vineyard establishment or replacement of diseased vines. Screening for virus free vines at nursery stage is an essential step for producing the GLD free propagating materials. Robust diagnostics make the screening process easier. Globally different kinds of diagnostics have been developed and used for producing the disease free planting materials. Amongst them serology with ELISA has been remained the method of choice and thus it has been used widely. In India, polyclonal antisera using expressed fusion coat protein have been used to develop the sensitive diagnostics against GLRaV-3 and GLRaV-4. Such diagnostics can be used by certified nurseries for production of clonally selected and sanitized propagation material which is very effective and the only preventive method available for leafroll management (Martelli and Boudon-Padieu 2006; Rayapati et al. 2008). In recent years micrografting of shoot apices onto hypocotyls from Vialla seeds has been proved effective against seven grapevine viruses including GLRaV-1, -2, and -3 (Spilmont et al. 2012). Virus elimination from grapevine selections using tissue culture could be used for certification purpose (Sim et al. 2012).

The various tissue culture techniques either alone or in combination with others have been used to eliminate several viruses from different plants. Meristem tip culture has been used to eliminate GLRaV-1 along with GFLV (Fayek et al. 2009; Youssef et al. 2009). Somatic embryogenesis has also been used to eliminate several phloem limited grapevine viruses including GLRaV-1 and GLRaV-3 (Gambino et al. 2006). Efforts are also being made to develop resistance against GLD using transgenic approach but till date no transgenic has been released for cultivation purpose (Ling et al. 2008; Gouveia and Nolasco 2012; Kumar 2013).

Roguing i.e. selective removal of infected vines is the least costly method to manage the GLD. Level of infection, timing of removal in relation to age of the vineyard, and the cost-benefit ratio of replanting are the factors which must be taken into account while selectively removing the infected vines. But, in general "roguing and replanting" the individual vines is more effective in the formative years of vineyards i.e. much before the establishment of infection at large scale (Rayapati et al. 2008). Sensitive diagnostic assay based annual rouging would always be better (Naidu et al. 2014). It has been suggested that rouging can give an additional benefit of $17,000–$22,000/ha to the growers (Atallah et al. 2012). Further, Fuller et al. (2013) has suggested that the economical benefits from using certified virus-free planting materials is more than $50 million per year for the North Coast region of California. Vector management is another important strategy to manage the leafroll diseases of grapevines especially when vineyards are susceptible to sustained immigration of mealybugs (Charles et al. 2006). Managing grape mealybug is most effective when the insects are in their crawling stage. Chloronicotinyl insecticides such as imidacloprid can be used as along with irrigation water. Chemigation with thiamethoxam and dinotefuran has shown their effectiveness in deficit irrigation situations. Foliar sprays of chloropyriphos can also be used for dormant applications (Rayapati et al. 2008). Using a combination of systemic and contact insecticides would be better strategy for vector management (Tsai et al. 2008). Wallingord et al. (2015) tested the efficacy of horticultural oil and two classes of insecticides namely, acetamiprid and spirotetramat on grape mealy bug (*Pseudococcus maritimus*), primary vector for GLRaVs in North America and they found that the tested materials slowed the spread of vector with varied efficacy. Following the hygienic practices by the workers and use of sanitized equipments would also reduce the spread of mealybugs and scale insects which in turn will check the spread of the disease (Pietersen et al. 2013; Naidu et al. 2014).

2.6 Concluding Remarks

Ampeloviruses are group of viruses named after grapevine but the group also includes non-grapevine infecting viruses. Out of nine ampeloviruses reported worldwide, only three grapevine infecting viruses i.e. GLRaV-1, GLRaV-3 and GLRaV-4 have been recently reported from India. Grapevine leafroll disease is an important and complex disease of grapevine. Further investigation is needed to look for other associated viruses in India. It is also needed to explore the other grapevine growing areas of the country for the associated viruses. Partial characterization of the viruses discovered from Indian vineyards has been carried out but complete sequence of any ampelovirus from India has not been done so far. Thus, there is a need to go for complete sequencing of these viruses so that we can have a broader understanding of viruses and the disease in Indian scenario. The scope of diversity study can be widened to include more number of isolates which in turn will lead towards a better understanding of genetic diversity, population structure and evolution of these viruses. The elucidation of biological and epidemiological implications of knowledge generated from such diversity studies will help in improving the sanitary status of grapevine planting materials. It will finally provide the avenues for development of robust strategies for mitigating the negative impacts of the disease.

In India the study of GLD is of recent origin but globally the disease has been discovered in mid-nineteenth century and mid-twentieth century in Europe and United States, respectively. Despite the fact of having a long history of its discovery at global level our knowledge on various aspects of the diseases and the associated viruses is quite limited (Naidu et al. 2014). A multidisciplinary system biology approach using modern tools of molecular biology, -omics, cell biology and other related disciplines along with the available genome sequence of the grapevine can shed more light on the disease, associated viruses and unparalleled complexity of the disease. Further investigations should be focussed to decipher the unknown functional genomics, host-pathogen interactome, gap between genomics and phenomics of the disease and transmission specificity of GLRaVs with their specific vectors (Naidu et al. 2014, 2015). Viral suppressors of RNA silencing (VSR) of GLRaV-3 (ORF 10) from India has been studied by Kumar (2013) but there is a need to widen the study as the detailed research into VSRs of GLRaVs will lead towards deciphering the mechanisms of silencing suppression (Naidu et al. 2015).

Further research is needed to decide the situations under which chemical control of vectors either alone or in combination with other measures such as rouging can be recommended to manage GLD (Wallingford et al. 2015). Additionally, research is also needed to have a deeper understanding of ecology and epidemiology of GLD. The discovery of pomegranate being as a natural alternate host of GLRaV-1 in Turkey (Caglayan et al. 2016) has added another dimension of complexity in the disease. Further investigation is required to see the implications of alternate host in the ecology and epidemiology of the GLD. In coming years a due vigilance is anticipated from the growers of the regions where both pomegranate and grapevine are cultivated in neighbourhood of each other. Proper hygienic condition and sanitary measures would also be required from the nurseries while producing the planting materials for pomegranate trees and grapevines both. In Indian condition there is a dire need to make efforts so that the knowledge generated from research can be translated for practical purpose which requires a powerful and enduring togetherness between research and extension personnel. The diagnostics develop in laboratories must help in producing the certified virus-free planting materials by recognized nurseries. Quarantine is an important aspect for disease like GLD as it has been suggested that the disease has been introduced to India through imported planting materials (Kumar 2013). Sensitive diagnostics can help in quarantine certification of imported planting materials and thus will check the further introduction other associated viruses and their strains. Therefore, in India the researchers should also strive to keep on developing more sensitive diagnostics against GLRaVs. As suggested by Naidu et al. (2014), use of certified virus-free planting materials in combination of roguing and sanitation on regular basis along with environmentally safe vector management strategies would lead towards sustainable management of GLD.

References

Adam MJ, King AMQ, Carstens EB (2013) Ratification vote on taxonomic proposals to the International Committee on Taxonomy of Viruses (2013). Arch Virol 158:2023–2030

Adam MJ, King AMQ, Lefkowitz EJ, Carstens EB (2014) Ratification vote on taxonomic proposals to the International Committee on Taxonomy of Viruses (2014). Arch Virol 158:2023–2030

Alabi AO, Federico Casassa L, Gutha LR, Larsen RC, Henick-Kling T, Harbertson JF, Naidu RA (2016) Impacts of grapevine leafroll disease on fruit yield and grape and wine chemistry in a wine grape (*Vitis vinifera* L.) cultivar. PLoS One. doi:10.1371/journal.pone.0149666

Alley L, Golino D (2000) The origins of the grape programat Foundation Plant Materials Service. In: Proceedings of 50th annual ASEV meeting, Seattle, WA. Am J Enol Vitic 51:222–230

Anonymous. www.ictvonline.org/virusTaxonomy.asp. As accessed on 25 Jan 2017

Atallah SS, Gómez MI, Fuchs MF, Martinson TE (2012) Economic impact of grapevine leafroll disease on *Vitis vinifera* cv. Cabernet franc in Finger Lakes Vineyards of New York. Am J Enol Vitic 63:73–79

Bertamini M, Muthuchelian K, Nedunchezhian N (2004) Effect of grapevine leafroll on the photosynthesis of field grown grapevine plants (*Vitis vinifera* L. cv. Lagrein). J Phytopathol 152:145–152

Caglayan K, Elci E, Gazel M (2016) Detection and partial characterization of *grapevine leafroll-associated virus* 1 in pomegranate trees in Turkey. Eur J Plant Pathol 145:199–202

Charles JG, Cohen D, Walker JTS, Forgie SA, Bell VA, Breen KC (2006) A review of the ecology of Grapevine leafroll-associated virus type 3 (GLRAV-3). N Z Plant Protect-se 59:330–337

Engelbrecht DJ, Kasdorf GGF (1990) Transmission of grapevine leafroll disease and associated closteroviruses by the vine mealybug *Planococcus ficus*. Phytophylactica 22:341–346

Fan X, Hong N, Dong Y, Ma Y, Zhang ZP, Ren F, Hu G, Zhou J, Wang G (2015) Genetic diversity and recombination analysis of *Grapevine leafroll-associated virus* 1 from China. Arch Virol 160:1669–1678

Farooq AB, Ma YX, Wang Z, Zhuo N, Wenxing X, Wang GP, Hong N (2013) Genetic diversity analyses reveal novel recombination events in Grapevine leafroll-associated virus 3 in China. Virus Res 171:15–21

Fayek MA, Jomaa AH, Shalaby AA, Al-Dhaher MMA (2009) Meristem tip culture for in vitro eradication of Grapevine leafroll-associated virus 1 (GLRaV-1) and grapevine fan leaf virus (GFLV) from infected flame seedless grapevine plantlets. Ini Inv. 4: a. URL: http://revistaselectronicas.ujaen.es/index.php/ininv/article/view/303/290

Freeborough MJ, Burger JT (2008) Leafroll: economic implications. Wynboer –a technical guide for wine producers. http://www.wynboer.co.za/recentarticles/200812-leafroll.php3

Fuchs M, Martinson TE, Loeb GM, Hoch HC (2009) Survey for the three major leafroll disease-associated viruses in Finger Lakes Vineyards in New York. Plant Dis 93:395–401

Fuller KB, Alston JM, Golino DA (2013) The benefits from certified virus-free nursery stock: a case study of grapevine leafroll-3 in the North Coast region of California. Robert Mondavi Institute-Center for Wine Economics Working Paper number 1306, UC-Davis, p 35

Gambino G, Bondaz J, Gribaudo I (2006) Detection and elimination of viruses in callus, somatic embryos and regenerated plantlets of grapevine. Eur J Plant Pathol 114:397–404

Goheen AC (1988) Leafroll. In: Pearson RC, Goheen AC (eds) Compendium of grape diseases. St Paul, APS, p 52

Gouveia P, Nolasco G (2012) The p19.7 RNA silencing suppressor from *Grapevine leafroll-associated virus 3* shows different levels of activity across phylogenetic groups. Virus Genes 45:333–339

Gugerli P, Brugger JJ, Bovey R (1984) L'enroulement de la vigne: mise en évidence de particules virales et développement d'une méthode immunoenzymatique pour le diagnostic rapide. Rev Suisse Vitic Arboric 16:299–304

Gutha LR, Casassa LF, Harbertson JF, Naidu RA (2010) Modulation of flavonoid biosynthetic pathway genes and anthocyanins due to virus infection in grapevine (*Vitis vinifera* L.) leaves. BMC Plant Biol 10:187

Harmon FN, Snyder E (1946) Investigations on the occurrence, transmission, spread, and effect of "white" fruit color in the Emperor grape. Proc Am Soc Hort Sci 47:190–194

Huson DH, Bryant D (2006) Application of phylogenetic networks in evolutionary studies. Mol Biol Evol 23:254–267

Ito T, Nakaune R (2016) Molecular characterization of a novel putative ampelovirus tentatively named grapevine leafroll-associated virus 13. Arch Virol 161:2555–2559

Jadhav R, Sonawane R (2007) Virus strikes Maharashtra vineyards, imports stopped. The Indian Express. URL: http://www.indianexpress.com/news/virus-strikes-maharashtra-vineyards-imports-stopped/236138/0

Jarugula S, Gowda S, Dawson WO, Naidu RA (2010) 3′-coterminal subgenomic RNAs and putative cis-acting elements of *Grapevine leafroll-associated virus 3* reveals 'unique' features of gene expression strategy in the genus *Ampelovirus*. Virol J 7:180

Karthikeyan G, Alabi OJ, Mekuria T, Martin RR, Naidu RA (2008) Occurrence of two distinct molecular variants of *Grapevine leafroll-associated virus* 1 in the Pacific Northwest vineyards. In: Proceedings of the 2nd annual national viticulture research conference, Davis, California, pp 39–40

King AMQ, Adams MJ, Carstens EB, Lefkowitz EJ (2012) Virus taxonomy: classification and nomenclature of viruses. Ninth report of international committee on taxonomy of viruses. Elsevier Academic Press, San Diego

Kovacs LG, Hanami H, Fortenberry M, Kaps ML (2001) Latent infection by leafroll agent GLRaV-3 is linked to lower fruit quality in French-American hybrid grapevines Vidal blanc and St. Vincent. Am J Enol Vitic 52:254–259

Kumar S (2013) Studies on virus(es) associated with grapevine leafroll disease in India. PhD thesis, Indian Agricultural Research Institute, New Delhi

Kumar S, Baranwal VK, Singh P, Jain RK, Sawant SD, Singh SK (2012) Characterization of a *Grapevine leafroll-associated virus 3* from India showing incongruence in its phylogeny. Virus Genes 45:195–200

Kumar S, Sawant SD, Sawant IS, Prabha K, Jain RK, Baranwal VK (2012) First report of *Grapevine leafroll-associated virus 1* infecting grapevines in India. Plant Dis 96:1828

Kumar S, Singh L, Ferretti L, Barba M, Zaidi AA, Hallan V (2013) Evidence of *Grapevine leafroll associated virus-1–3*, *Grapevine fleck virus* and *Grapevine virus B* occurring in Himachal Pradesh. India Indian J Virol 24:66–69

Ling KS, Zhu HY, Gonsalves D (2008) Resistance to *Grapevine leafroll associated virus-2* is conferred by post-transcriptional gene silencing in transgenic *Nicotiana benthamiana*. Transgenic Res 17:733–740

Mannini F, Gerb V, Credi R (1998) Heat-treated virus-infected grapevine clones: agronomical and enological modifications. Acta Hortic 473:155–163

Maree HJ, Freeborough MJ, Burger JT (2008) Complete nucleotide sequence of a South African isolate of *Grapevine leafroll-associated virus* 3 reveals a 5′UTR of 737 nucleotides. Arch Virol 153:755–757

Maree HJ, Pirie MD, Bester R, Oosthuizen K, Burger JT (2015) Phylogenomic analysis reveals deep divergence and recombination in an economically important grapevine virus. PLoS One 10:e0126819

Martelli GP (2009) Grapevine virology highlights 2006–2009. In: Extended abstracts of 16th meeting of ICVG, Dijon, France, pp 15–23

Martelli GP, Abou Ghanem-Sabanadzovic N, Agranovsky AA, Al Rwahnih M, Dolja VV, Dovas CI, Fuchs M, Gugerli P, Hu JS, Jelkmann W, Katis NI, Maliogka VI, Melzer MJ, Menzel W, Minafra A, Rott ME, Rowhani A, Sabanadzovic S, Saldarelli P (2012) Taxonomic revision of the family Closteroviridae with special reference to the grapevine leafroll-associated members of the genus Ampleovirus and the putative species unassigned to the family. J Plant Pathol 94:7–19

Martelli GP, Agranovsky AA, Bar-Joseph M, Boscia D, Candresse T, Coutts RHA, Dolja VV, Falk BW, Gonsalves D, Jelkmann W, Karasev AV, Minafra A, Namba S, Vetten HJ, Wisler CG, Yoshikawa N (2002) The family *Closteroviridae* revised. Arch Virol 147:2039–2044

Martelli GP, Boudon-Padieu E (2006) Directory of infectious diseases of grapevines viruses and virus–like diseases of the grapevine: Bibliographic report 1998–2004. Options Méditérr B55:11–201

Naidu RA, Maree HJ, Burger JT (2015) Grapevine leafroll disease and associated viruses: a unique pathosystem. Annu Rev Phytopathol 53:613–634

Naidu RA, Rowhani A, Fuchs M, Golino D, Martelli GP (2014) Grapevine leafroll: a complex viral disease affecting a high-value fruit crop. Plant Dis 98:1172–1185

Namba S, Yamashita S, Doi Y, Yora K, Terai Y, Yano R (1979) Grapevine leafroll virus, a possible member of closteroviruses. Ann Phytopathol Soc Jpn 45:497–502

Olmo HP, Rizzi AD (1943) Selection for fruit color in the Emperor grape. Proc Am Soc Hort Sci 42:395–400

Pietersen G (2004) Spread of Grapevine leafroll disease in South Africa – a difficult, but not insurmountable problem. Wynboer –a technical guide for wine producers. http://www.wynboer.co.za/recentarticles/articles.php3

Pietersen G, Spreeth N, Oosthuizen T, Van Rensburg A, Van Rensburg M, Lottering D, Rossouw N, Tooth D (2013) Control of grapevine leafroll disease spread at a commercial wine estate in South Africa: a case study. Am J Enol Vitic 64:296–305

Rayapati AN, O'Neil S, Walsh D (2008) Grapevine leafroll disease. Washington State University Extension Bulletin. EB2027E. http://cru.cahe.wsu.edu/CEPublications/eb2027e/eb2027e.pdf

Scheu G (1935) Die Rollkranheit des Rebstocks. Der Deutsche Wenbau 14:222–223, 345–346, 356–358

Sharma AM, Wang JB, Duffy S, Zhang SM, Wong MK, Rashed A, Cooper ML, Daane KM, Almeida RPP (2011) Occurrence of grapevine leafroll-associated virus complex in Napa valley. PLoS One 6:e26227

Sim ST, Al Rwahnih M, Rowhani A, Golino DA (2012) Virus elimination from grape selections using tissue culture at Foundation Plant Services, University of California, Davis. In: Extended abstracts of 17th meeting of ICVG, Davis, CA, USA, pp 262–263

Spilmont A-S, Ruiz A, Grenan S (2012) Efficiency of micrografting of shoot apices as a sanitation method against seven grapevine viruses (ArMV, GFLV, GLRaV-1, -2, -3, GFkV, GVA). In: Extended abstracts of 17th meeting of ICVG, Davis, CA, USA, pp 270–271

Tsai CW, Chau J, Fernandez L, Bosco D, Daane KM, Almeida RPP (2008) Transmission of *Grapevine leafroll-associated virus 3* by the vine mealybug (*Planococcus ficus*). Phytopathology 98:1093–1098

Tsai CW, Rowhani A, Golino DA, Daane KM, Almeida RPP (2010) Mealybug transmission of grapevine leafroll viruses: an analysis of virus-vector specificity. Phytopathology 100:830–834

Turturo C, Saldarelli P, Yafeng D, Digiaro M, Minafra A, Savino V, Martelli GP (2005) Genetic variability and population structure of *Grapevine leafroll-associated virus 3* isolates. J Gen Virol 86:217–224

Wallingford AK, Fuchs MF, Martinson T, Hesler S, Loeb GM (2015) Slowing the spread of grapevine leafroll-associated viruses in commercial vineyards with insecticide control of the vector, *Pseudococcus maritimus* (Hemiptera: Pseudococcidae). J Insect Sci 15:112. doi:10.1093/jisesa/iev094

Youssef SA, Al-Dhaher MMA, Shalaby AA (2009) Elimination of *Grapevine fanleaf virus* (GFLV) and *Grapevine leafroll-associated virus 1* (GLRaV-1) from infected grapevine plants using meristem tip culture. Int J Virol 5:89–99

Biology and Molecular Biology of Babuviruses Occurring in India

3

R. Selvarajan, Bikash Mandal, V. Balasubramanian, Amrita Banerjee, S. Vijayanandraj, and Amalendu Ghosh

Abstract

The genus, *Babuvirus* of the family *Nanoviridae* contains three virus species, *Abaca bunchy top virus* (ABTV), *Banana bunchy top virus* (BBTV) and *Cardamom bushy dwarf virus* (CBDV). In India, only two babuviruses, BBTV and CBDV are known to affect banana and large cardamom, respectively. BBTV, which causes bunchy top disease in banana, is a nationally important virus as it is widely prevalent in all the banana growing states including North-East region. Whereas, CBDV, which causes foorkey disease of large cardamom, is of regional importance as it is restricted only in the North-Eastern sub-Himalayan mountains. Early infection of these babuviruses cause 100% yield loss. Bunchy top of banana and foorkey disease of large cardamom are known in India for a long time and considerable information has been generated. This chapter summarises the biological and molecular properties of both the babuviruses occurring in India.

R. Selvarajan (✉) • V. Balasubramanian
Molecular Virology Laboratory, ICAR-NRC for Banana,
Tiruchirapalli 620102, Tamil Nadu, India
e-mail: selvarajanr@gmail.com

B. Mandal (✉) • S. Vijayanandraj • A. Ghosh
Advanced Center for Plant Virology, Division of Plant Pathology, IARI,
New Delhi 110012, India
e-mail: leafcurl@rediffmail.com

A. Banerjee
Division of Crop protection, ICAR-National Rice Research Institute,
Hazaribag, Jharkhand 825301, India

B. Mandal et al. (eds.), *A Century of Plant Virology in India*,
https://doi.org/10.1007/978-981-10-5672-7_3

Keywords
Babuviruses in India • Banana bunchy top virus • Cardamom bushy dwarf virus • Banana • Large cardamom

3.1 Introduction

Multi-component ssDNA containing small isometric plant viruses were initially classified under the genus *Nanovirus*. The legume infecting nanoviruses being substantially different from banana bunchy top virus (BBTV), the nanoviruses were reclassified in the eighth report of International Committee on Taxonomy of Viruses (ICTV) where a family, *Nanoviridae* containing two genera, *Nanovirus* and *Babuvirus* were created (Vetten et al. 2005). The genus *Nanovirus,* included all the legume infecting nanoviruses and the genus *Babuvirus* included only, BBTV. Later, two more babuviruses were discovered, Abaca bunchy top virus (ABTV) from Philippines (Sharman et al. 2008) and Cardamom bushy dwarf virus (CBDV) from India (Mandal et al. 2013).

Foorkey disease of large cardamom was the first babuvirus associated disease known in India, which was recorded in 1935 in Darjeeling hills. Bunchy top disease of banana, although was known in Oceania region in early twentieth century, it was recorded in India during 1940s. In India, only two babuviruses so far has been identified, BBTV causing banana bunchy top disease and CBDV causing foorkey disease of large cardamom. Banana is one of the most important fruit crop cultivated throughout India. BBTV has been characterised from at least from three geographic regions of India (Vishnoi et al. 2009; Banerjee et al. 2014; Selvarajan et al. 2010). The wide spread occurrence and serious yield limiting ability of BBTV made it to be one of the top most important phytopathogens in India. The other babuvirus, CBDV that caused foorkey disease of large cardamom is prevalent only in the North-East sub-Himalayan mountains. Large cardamom is an important spice plantation crop grown under the special organic environmental conditions prevailing in the North-East sub-Himalayan mountains. CBDV infection results into complete loss of large cardamom clump. As CBDV is prevalent in a particular agro-ecological area of the country, it is of regional significance. Babuviruses were known to naturally infect the plant species of the family Musaceae. The current research work on CBDV provided the evidence of a new babuvirus species that infects plant species of the family Zingiberaceae. CBDV is the only member of the family *Nanoviridae* that was discovered in India. A few review papers are available for the nanoviruses and babuviruses (Mandal 2010; Selvarajan 2015). As both the babuviruses, BBTV and CBDV are economically significant viral pathogens in India, a large body of literature has been generated. This chapter presents the work on the biological and molecular characterization of these babuviruses occurring in India.

3.2 Banana Bunchy Top Virus

3.2.1 Occurrence and Significance

Banana bunchy top disease (BBTD) caused by BBTV was first recorded in Fiji in 1889. It is believed that BBTD spread to Australia and Sri Lanka in 1913 through infected suckers from Fiji and later during 1940's it was presumed to be introduced to Kerala state of India (Jones 2000). Upon gaining entry in the southern tip of India, the virus has spread to all the banana growing states of the country. BBTD remains a major problem in Kerala, Andhra Pradesh, Tamil Nadu, Orissa, Maharashtra, Madhya Pradesh, Gujarat, Bihar, Karnataka, West Bengal, Assam and Uttar Pradesh (Singh 2003). It has also been recorded in wild and cultivated bananas in Nagaland, Meghalaya, Arunachal Pradesh, Mizoram, Sikkim and Tripura.

Of all the viruses known to infect banana in India, BBTV is the most serious and destructive virus. In lower Pulney hills of Tamil Nadu, a very famous elite dessert banana cultivar, Virupakshi (Pome group, AAB) having unique flavor and distinct aroma, has been near extinct due to BBTV since 1970's and the area under this banana has been reduced from 18,000 ha to 2000 ha (Kesavamoorthy 1980). A survey conducted during May 2009 in Lower Pulney hills (Kodaikanal) recorded 15.26–83.88% incidence in Hill banana across the plantations (Selvarajan et al. 2011). Metha et al. (1964) reported a loss of about ₹ 40 million annually reported in Kerala. Emergence of BBTV in tissue culture plantations during 2007–2011 in Jalgaon, Maharashtra and Kodur, Andhra Pradesh caused an annual loss of production worth of US$50 million (Selvarajan and Balasubramanian 2014). An outbreak of BBTD in 2011 in Theni district of Tamil Nadu recorded an infection in 0.3 million plants of both tissue culture and conventional sucker grown plants. As BBTD causes significant yield reduction and reduces the productivity in India and other parts of the world, it is considered as one of the most destructive diseases of banana in the world.

3.3 Biology of BBTV

3.3.1 Disease Symptom and Dissemination

BBTV infected plants express discontinuous dark green flecks and streaks of variable length on leaf sheath, midrib, leaf veins and petioles of infected plants (Fig. 3.1). Leaves that are produced after the infection are progressively shorter both in width and length with limited elongation of petioles and remain abnormally erect (Sharma 1988). Infected leaves are narrow and brittle in texture, display marginal yellowing or chlorosis and leaves bunches at the top, hence the name is "bunchy top" disease (Fig. 3.1). BBTV occurs in the phloem tissues of banana and incites symptoms such as leaf chlorosis, vein clearing, dwarfing and leaf atrophy (Wu and Su 1990; Su et al. 2007). Mostly, BBTV infected plants fail to produce bunch, however, in late infections the plant may produce bunch but the fingers are malformed and not fit

Fig. 3.1 Disease symptom of BBTV. (**a**) Field view of hill banana cultivation in lower pulney hills. (*Insert*) Severe symptom of BBTD; (**b**) Vein flecking on leaf lamina; (**c**) *Dark green dots* and streaks on petiole; (**d**) Chlorosis; (**e**) *Greenish* leafy tips in male bud of BBTV infected Cavendish banana plant; (**f**) Aphids vector – *Pentalonia nigronervosa*

for sale. Occasionally, bracts of male flower bud turn to leafy green structures and exhibit dark green dots and streaks (Thomas et al. 1994). Any daughter suckers emerging from infected plant exhibit severe symptoms. Late infection of BBTV in cultivar Grand Naine banana leads to throw bunch with extremely long or very short peduncle. Sometimes affected Grand Naine banana fingers appear like a non Cavendish type.

In BBTV infection, the virus resides in the plant without exhibiting any visible symptom which is termed as latency. BBTV has been found to express visual symptoms only 23–25 days after inoculation but the virus could be detected early from young roots or cortex tissue even before the symptom expression (Hafner et al. 1995). The shortest time for the diagnosis of BBTV using polymerase chain reaction (PCR) is reported as 15 days after infection (Hooks et al. 2008). Sometimes, the infected plants (PCR positive) do not show any symptoms even up to 560 days under pot culture experiment (R. Selvarajan, unpublished). Samraj et al. (1970) have reported that a minimum time of 5 days and maximum 10–15 days is required for the down ward movement of the virus after inoculation with the aphid and this might change depending upon the vigour of the plant. A short, but significant, latent period of 20–28 h is required for vector transmission (Anhalt and Almeida 2008) and it has been found that the optimum temperature range for acquisition of virus by the vector was 25–27°C.

The only confirmed hosts of BBTV are species within the genus *Musa* (*M. balbisiana*) (Espino et al. 1993; Magee 1927), *M. ornate* (Thomas and Dietzgen 1991), *M. acuminate ssp zebrina*, *M. velutina*, *M. coccinia* (Thomas and Iskra-Caruana 2000), *M. sinensis*, *M. paradisica* and their hybrids and Fei' bananas and *Ensete ventricosum* (Selvarajan and Balasubramanian 2013). This disease has been observed on a wild species, *Musa itenerans* that occurs as a feral in hills of Arunachal Pradesh (Selvarajan et al. 2010). *Colocasia esculenta* has been reported to be a host for BBTV from Pune (Ram and Summanwar 1984) but it was disproved later (Geering and Thomas 1997; Hu et al. 1996).

BBTV is primarily spread through the use of infected vegetative propagules, including the suckers, corms, (Magee 1948) and tissue-cultured plants (Drew et al. 1989). Drew et al. (1989) have demonstrated that BBTD is transmitted through micro propagated plantlets established from infected plants. When these tissue culture plantlets were established in the greenhouse, only 73% of plant developed characteristic symptoms and the remaining 27% of plants appeared healthy without typical symptoms of the BBTD. BBTV is secondarily, naturally transmitted by the banana black aphid vector, *Pentalonia nigronervosa* (Hemiptera, Aphididae) (Fig. 3.1) in a persistent circulative manner (Anhalt and Almeida 2008; Selvarajan et al. 2006). Recently, another closely related species, *P. caladii*, has been shown to transmit BBTV under experimental conditions with a lower level of efficiency than *P. nigronervosa* (Watanabe et al. 2013).

The aphids are usually found clustered around the unfurled heart-leaf and the sheathing leaf base of petioles which are ideal locations for feeding and protection. They are also found on the base of the pseudostem and on very young suckers. Menon and Christudas (1967) reported the life history and population dynamics of *Pentalonia nigronervosa* in Kerala, India and stated that climatic conditions existing during the summer months and rainy weather are unfavourable for the banana aphid in Kerala.

3.4 Genomic Properties of BBTV in India

BBTV genome consists of six circular single stranded DNA individually packed in six particles (Table 3.1). The genome of each component is about 1.1 kbp in size and each encodes for a single open reading frame (ORF) for different functional proteins from the virion sense strand such as DNA-R (replication initiation protein), -S (coat protein), -M (movement protein), -C (cell cycle link protein), -N (nuclear shuttle protein), and -U3 (a protein of unknown function) (Vetten et al. 2012).

In India, complete genome of three BBTV isolates, one each from Lucknow (UP), Bhagalpur (Bihar) and Lower Pulney Hills (TN) have been sequenced and characterised (Fig. 3.2) and found that they belong to Pacific-Indian Oceans (PIO) group (Islam et al. 2010; Selvarajan et al. 2010; Vishnoi et al. 2009). Complete genome sequences of these Indian isolates showed high degree of similarity with the corresponding sequences of BBTV isolates originating from Fiji, Egypt,

Table 3.1 The major genome components of cardamom bushy dwarf virus (CBDV) and Indian isolates of banana bunchy top virus (BBTV)

DNA component	Protein[a]	Potential function	CBDV (Nucleotide size)	BBTV(Nucleotide size)
DNA-R	M-Rep	*R*eplication initiator protein for all DNAs	1091–1110	861–1111
DNA-S	CP	*S*tructural protein, virion formation	1085–1087	1074–1076
DNA-C	Clink	*C*ell-cycle regulation	1025–1028	1014–1018
DNA-M	MP	Cell-to-cell *m*ovement	1067–1093	1040–1048
DNA-N	NSP	Presumed *N*SP	1073–1116	1080–1096
DNA-U1	U1	*U*nknown	1076–1080	–
DNA-U2	U2	*U*nknown	1077–1079	–
DNA-U3	U3	*U*nknown	1066–1109	1035–1064
DNA-U4	U4	*U*nknown	1106	–

[a]*M-Rep* Master replication initiator protein, *CP* coat protein, *Clink* cell-cycle link protein, *MP* movement protein, *NSP* nuclear shuttle protein. U1 to U4 are of unknown function
– indicates that protein has not been described

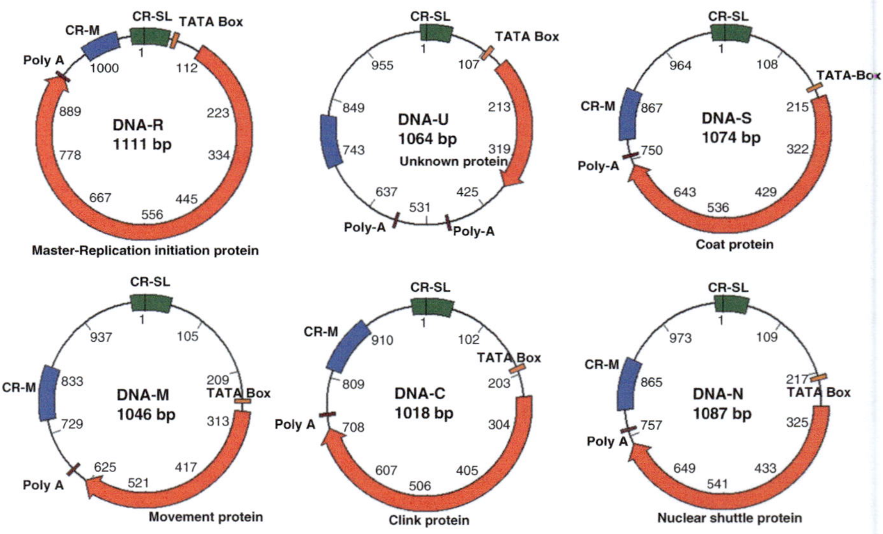

Fig. 3.2 Organization of the genomic components of the BBTV-Hill banana isolate. The DNA component- R, -U, -S, -M, -C and -N are illustrated diagrammatically. The positions and orientations of genes are indicated with *shaded arrows*. Also the positions of the stem loop common region (CR-SL), consensus TATA box and polyadenylation signal sequences are shown

Pakistan, and Australia. An analysis of the coat protein sequences of 16 Indian isolates with distinct geographical origins revealed that they belong to the PIO group, except the isolates from Shevroy and Kodaikanal hills of Tamil Nadu (Selvarajan et al. 2010). Recently a novel BBTV isolate from Umiam (Meghalaya) which has a deletion of about 20 nucleotides in the DNA-R component has been characterized (Banerjee et al. 2014).

3.5 Sequence Diversity

Earlier, BBTV was grouped into the South Pacific group comprising isolates from Australia, Burundi, Egypt, Fiji, India, Tonga and Western Samoa and the Asian group, comprising isolates of Vietnam, Philippines and Taiwan based on sequence analysis of BBTV DNA-R, -S and -N (Karan et al. 1994, 1997; Wanitchakorn et al. 2000). Karan et al. (1994) determined the maximum variability of South Pacific isolates to be 3.8% with a mean of 1.9%, whereas the value for Asian isolates was found to be 4.2% with a mean of 3% for DNA-R components. The two groups of BBTV differ by an average of 90.6% (DNA-R), 11.86% (DNA-S) and 14.5% (DNA-N) over the entire nucleotide sequence with an average difference of 32% (DNA-R), 38.6% (DNA-S) and 27% (DNA-N) within CR-M (Karan et al. 1994, 1997; Wanitchakorn et al. 2000). A more recent analysis, based on a much larger sample set, determined the mean variation to be approximately 1.6% and 2.9% for South Pacific and Asian isolates respectively (Hu et al. 2007). The genetic diversity of BBTV isolates among countries is very low, *viz.*, India (except isolates from north eastern region) (Vishnoi et al. 2009; Islam et al. 2010; Selvarajan et al. 2010), Pakistan (Amin et al. 2008), Africa (Niyongere et al. 2013, 2015), and Oceania (Stainton et al. 2012). However, in India, relatively a greater diversity for BBTV was observed in the North-eastern region (Banerjee et al. 2014). Recently, based on the phylogenetic relationships among the DNA-R component sequences, various BBTV isolates were grouped into two different lineages: (i) the Pacific-Indian Oceans (PIO) group (formerly South Pacific group) comprising isolates in Africa, Australia, Hawaii, south Asia, Myanmar, and Tonga; and (ii) the South-East Asian (SEA) group (formerly Asian group) comprising isolates from China, Indonesia, Japan, the Philippines, Taiwan, and Vietnam (Fig. 3.3) (Stainton et al. 2012; Yu et al. 2012; Banerjee et al. 2014; Kumar et al. 2015). Motif-based analysis revealed that several unusual recombination events occurred and those events have contributed to the evolution of BBTV genome components (Wang et al. 2013).

3.6 BBTD in North-East (NE) India

The North-eastern region of India, comprising eight states *viz.*, Assam, Arunachal Pradesh, Meghalaya, Manipur, Mizoram, Nagaland, Tripura and Sikkim, represents a distinct agro-climatic zone of the country. The NE India is a bio-diversity hotspot and possesses diverse germplasm of banana of both wild and cultivated. However, little was known about BBTV occurring in the NE India, except the preliminary information on BBTV coat protein (Selvarajan et al. 2010). Recently, a new isolate of the virus (BBTV-Umiam) was identified and characterized from banana growing in mid-hills of Meghalaya in NE India (Fig. 3.3) (Banerjee et al. 2014). The overall genome organization of BBTV-Umaim was mostly identical with previously reported isolates from India except having some distinct features *viz.*, deletion of 20 nucleotides in the intergenic region of DNA R, absence of predicted ORF in DNA U3 and probability for a small ORF in DNA U3. The BBTV-Umiam is supposed to

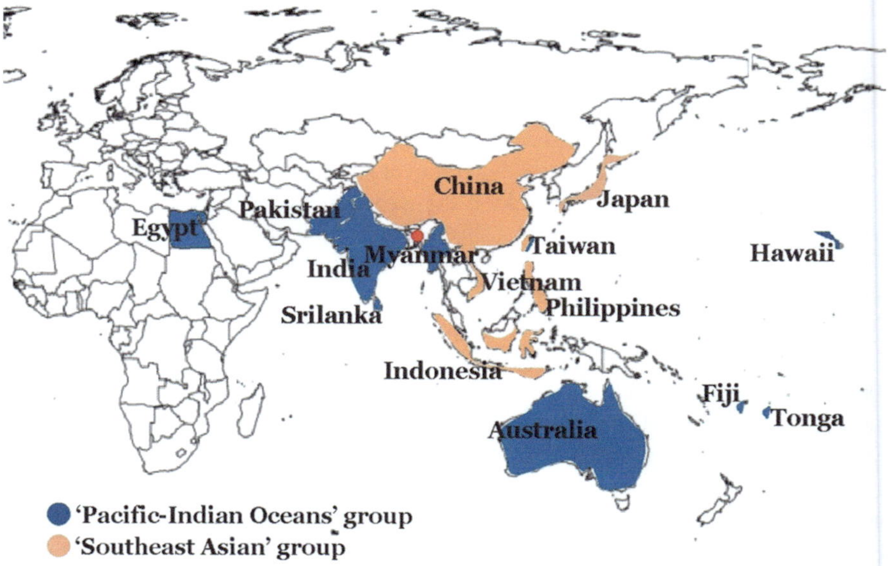

Fig. 3.3 World wide distribution of *Banana bunchy top virus* (BBTV) isolates grouped into "Pacific-Indian Oceans" (PIO) and "South-east Asian" (SEA). The site of collection of the new BBTV isolate (BBTV-Umiam) from Meghalaya, India is indicated by a *red circle*

be a PIO group member due to its geographical origin (Umiam, Meghalaya, India). The earlier reported Indian BBTV isolates always grouped within PIO cluster instead of SEA cluster indicating the possible introduction of BBTV into southern India from Australia through Fiji and Srilanka by infected planting materials (Wardlaw 1972). Similarly, in phylogenetic grouping the BBTV isolate from Meghalaya (BBTV-Umiam) grouped within PIO cluster sharing 95.5% nucleotide sequence identity (Fig. 3.4). However, all the BBTV isolates from plains of India clustered together suggesting their separate and independent evolution, but BBTV-Umiam was found to be the most distinct member among the PIO isolates identified so far (Fig. 3.4). Although, overall sequence analysis of genomic components, as well as, ORFs clearly indicated strong similarity of BBTV-Umiam with PIO group, but BBTV-Umiam shared relatively less nucleotide identity with PIO group for each genomic component (85.0–95.4%) and corresponding ORFs (93.8–97.5%) than that of earlier PIO isolates (91.5–99.6% and 96.0–99.3%, respectively) (Banerjee et al. 2014). Recombination analysis revealed two intra-component (in DNA U3 and DNA N) and five inter-component recombination events [DNA U3 (1), DNA S (2), DNA M (1) and DNA N (1)] around CR-M and CR-SL region of BBTV-Umiam, but none of them was unique (Banerjee et al. 2014). The BBTV DNA U3 being the most recombined one (Hyder et al. 2011; Stainton et al. 2012) showed similar trend in BBTV-Umiam, while an event involving the transfer of a DNA N fragment from PIO group virus resembling those found in India, Pakistan and Australia into the progenitor of a group of viruses from Tonga was reported earlier (Stainton et al.

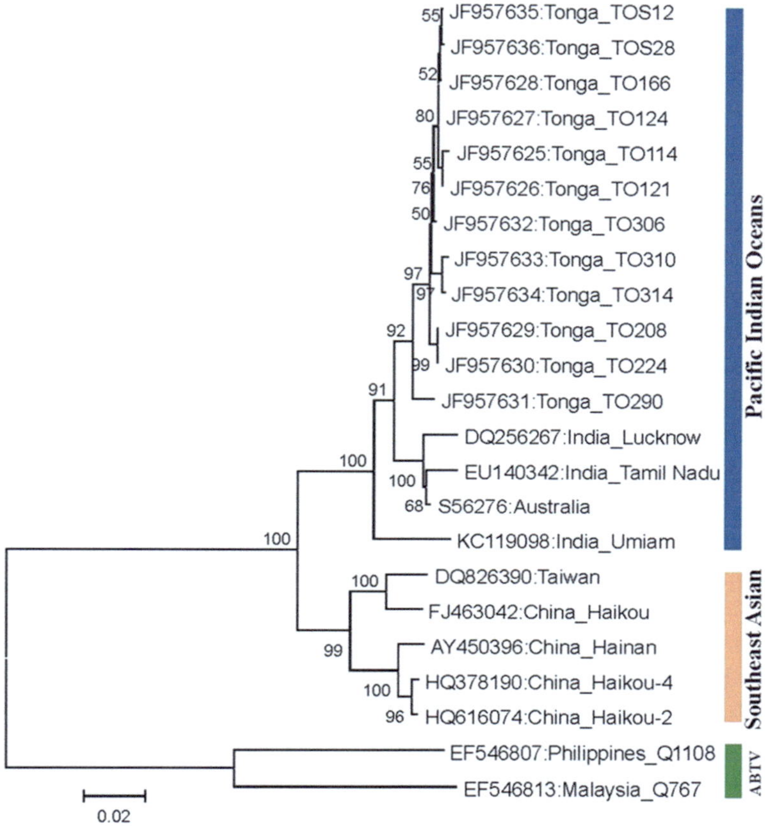

Fig. 3.4 Phylogenetic relationships based on nucleotide sequences of DNA-R of BBTV-Umiam with previously reported PIO and SEA isolates of BBTV. The evolutionary history was inferred using the Neighbor-Joining method. The percentage of replicate trees in which the associated taxa clustered together in the bootstrap test (1000 replicates) are shown next to the branches (shown only when > 50%). The tree is drawn to scale, with branch lengths in the same units as those of the evolutionary distances used to infer the phylogenetic tree. The evolutionary distances were computed using the 'Maximum Composite Likelihood' method and are in the units of the number of base substitutions per site. Each sequence is labelled with the GenBank accession number followed by origin and isolate name

2012). Earlier workers indicated CR-M region as a major recombination hotspot and CR-SL region as a minor hotspot (Stainton et al. 2012). Though in case of BBTV-Umiam, maximum inter-component recombination events were identified around CR-SL (Banerjee et al. 2014). Moreover, recombination had no role in the deletion of 20 nucleotides in the intergenic region of BBTV-Umiam DNA R. Generally, in plant pararetroviruses (*Caulimovirus*), this kind of natural deletion is thought to be the outcome of recombination (Howarth et al. 1981). Thus, the genetic distinctness of BBTV-Umiam was not the outcome of genetic cross-over. On the other hand, all the PIO DNA R components were reported to have descended from a common recombinant ancestor deriving a proportion of its Rep encoding sequence from an unknown babuvirus (Stainton et al. 2012). Banerjee et al. (2014) provided strong evidence for considering BBTV-Umiam (KC119098) and BBTV-HaiKou-4 (HQ378190) as the hypothetical major and minor parental sequences, respectively for intra-component recombination of DNA R of all Tonga isolates of BBTV (JF957625-JF957636) (Fig. 3.5).

Further studies has confirmed occurrence of BBTV in Assam, Arunachal Pradesh, Nagaland, Manipur, Mizoram, Sikkim and Tripura including samples from commercial orchards, road side banana mats and even in tissue culture raised plant materials (Banerjee et al. 2015). Altogether, ten BBTV isolates distributed throughout the surveyed areas were characterized based on DNA R segment. The full DNA R sequences of each isolate except the isolate from Mizoram shared >97.0% similarity with BBTV isolates reported from plains of India. However, these isolates showed relatively less similarity (~95.0%) with BBTV-Umiam. Interestingly, the Mizoram isolate shared only 91.0–92.0% similarity with both PIO and SEA group members. While, during phylogenetic analysis the Mizoram isolate including other isolates from NE India clustered within PIO group, but the clustering pattern indicated the distinctiveness of Mizoram isolate as of previously reported BBTV-Umiam from Meghalaya (Banerjee et al. 2015). Recent survey in seven districts of Tripura *viz.*, North Tripura, Dhalai, Khowai, West Tripura, Shipahijala, Gomati, and South Tripura showed prevalence of BBTV in all parts of Tripura. Phylogenetic analysis based on complete nucleotide sequence of BBTV DNA R confirmed Tripura isolates of BBTV as PIO group members. However, Tripura isolates formed two different clades within the PIO group. Tripura isolates showed on an average 10.19% variation with SEA group and 2.79% variation with PIO groups. Moreover, Tripura isolates were more identical with the isolates reported from plains of India rather than the distinct isolate reported from Meghalaya (BBTV-Umiam). Thus, the source of planting material from plains of India, as well as, from neighboring countries could have contributed to the geographical expansion of PIO isolates of BBTV within NE India. However, the existence of distinct PIO isolates in naturally growing banana mats of Meghalaya and Mizoram further strengthened the possibility of differential evolution of BBTV in this isolated region. The natural occurrence of hybrids of *Musa balbasiana* and *M. acuminate* in this geographically isolated region could be the contributing factor in accumulating genetic distinctiveness in BBTV-Umiam which needs further characterization.

3.7 Large Cardamom Bushy Dwarf Virus

Large cardamom (*Amomum subulatum*), a perennial herb belongs to the family Zingiberaceae, order Scitaminae, is an important spice crop of India. Large cardamom is cultivated in the eastern sub-Himalayan mountains at altitudes ranging between 800–3000 m above mean sea level under of the forest cover conditions. Sikkim and Darjeeling hills of West Bengal have the major areas (30,000 ha) under large cardamom cultivation. Usually, the large cardamom plantations are

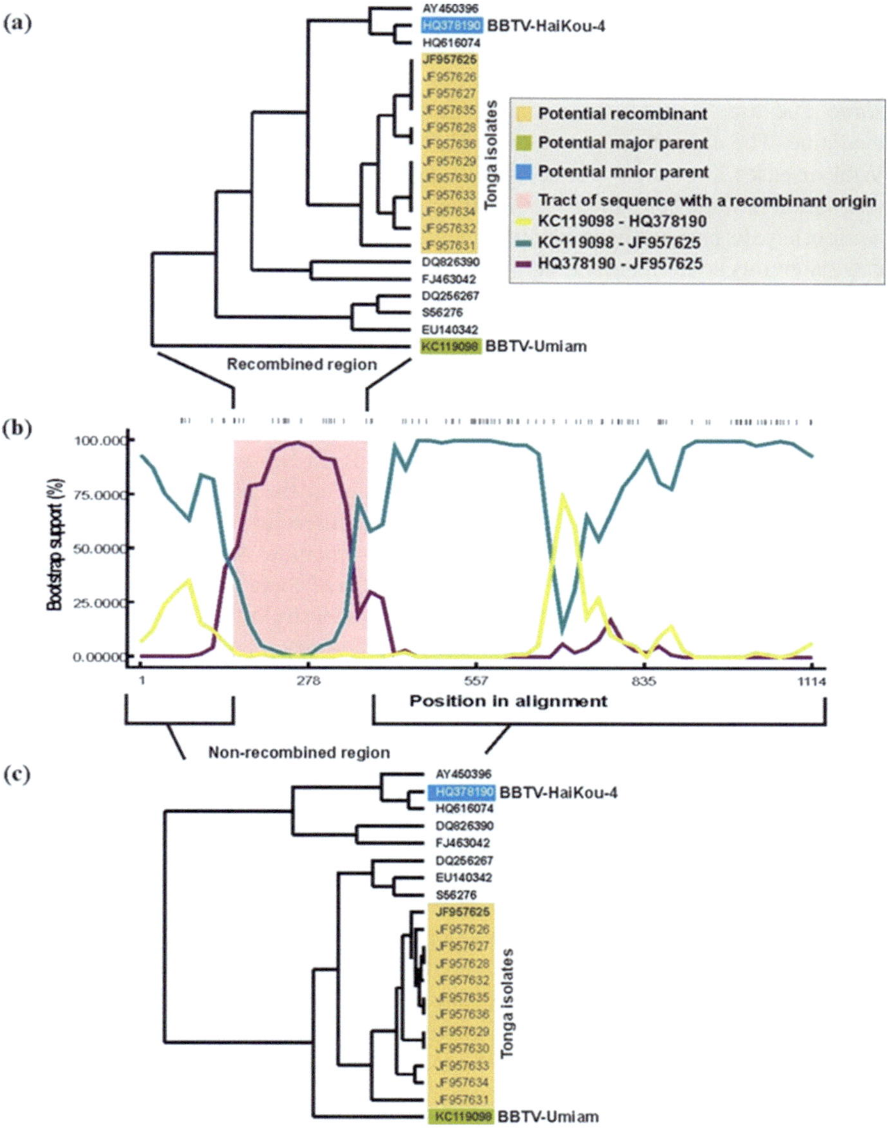

Fig. 3.5 Intra-component recombination analysis of BBTV DNA-R of Tonga isolates. (**a**) Phylogenetic tree of nucleotide sequence of the recombined region showing clustering of recombinant isolates JF957625-JF957636 (Tonga isolates) with HQ378190 (BBTV-HaiKou-4); (**b**) Bootstrap support plot showing a graphical overview of the recombination event in JF957625 (the representative Tonga isolate) at position 150–373 nucleotide. The recombinant isolate shows similarity to KC119098 (BBTV-Umiam, major parent) throughout the entire genome but shares very high identity with HQ378190 (BBTV-HaiKou-4, minor parent) only in the recombined region; (**c**) Phylogenetic tree of the nucleotide sequence of the non-recombined region showing clustering of Tonga isolates with KC119098 (BBTV-Umiam).

established with seedlings or suckers (tiller with roots separated from rhizome) or tissue culture raised plants and the plantations are maintained for many years. The plant produces large red or black coloured aromatic capsules at the base near rhizome. The dried capsules containing seeds are used for flavoring food and preparing medicine. The capsule size, colour and aroma are distinct from small cardamom. Viral diseases known as 'chirke' and 'foorkey' are the major limiting factors in cultivation of large cardamom (Varma and Capoor 1964). The virus associated with the foorkey disease has been identified as cardamom bushy dwarf virus (CBDV), a new babuvirus so far recorded only in India.

3.7.1 The Foorkey Disease

Foorkey is one of the earliest known viral diseases in India, which was first documented in 1936 from Darjeeling hills (Mitra 1936). In Nepalese language, foorkey means bushy dwarf. The typical symptoms appear in the newly developed tillers from the infected rhizome. Initially, the affected tillers grow to the height up to 6–12 in. with slight bunchy top appearance and sometimes leaves are not unfurled (Fig. 3.6). Subsequently, numerous stunted plantlets of about 2–3 in. with small pale green leave proliferate from the rhizome giving a bushy dwarf appearance at the base. The CBDV infected clumps survive for a few years but the clumps become sterile and unproductive. The recognition of initial disease symptoms of foorkey under field conditions is difficult, however, when the symptoms are fully expressed, it is easy to spot in the plantation. Most of the commonly grown cultivars such as Golsey, Ramsey, Sawaney and Varlangey are susceptible. The disease incidence varies from place to place and higher incidence is generally observed in lower altitudes ranging from 300 to 1380 m above MSL. The high incidence of foorkey up to 39.3% was observed in Kooldhara and Khaptali Gaon in Darjeeling hills (Mandal et al. 2013).

The virus associated can be readily detected by PCR or dot-blot hybridization using the primers or probe from the replication associated protein gene (*Rep*) of DNA-R. PCR is most effective in detecting the CBDV. The primer pair AV5F: tggcgcgatatgtggtatgc and AV6R: tcagcaagaaaccaactttattc derived from the *Rep* gene amplify a ~ 0.84 kb DNA fragment. The PCR successfully detected the virus in the different plant parts of large cardamom *viz.*, leaf, leaf sheath, meristem, stem, root and mother rhizome. Another primer pair from stem and loop (S&L) region of the Rep genome component of the virus (AV32F: ggggcttattattacccccagcg and AV33R: agcgcttacgtggcgcactaact) amplified 1.1 kb fragment. The Rep or S&L primers are useful tools for the monitoring of the occurrence of the virus by PCR.

3.7.2 Resolving the Etiology of Foorkey

The existence of foorkey disease was known in large cardamom for a long time, however, the virus associated with the disease was not known. Foorkey was considered as a viral disease based on the fact that the disease could be transmitted through

Fig. 3.6 Symptoms of foorkey disease of large cardamom cv. Varlangey. (**a**) A healthy clump bearing flower; (**b**) Large cardamom spike and capsules from healthy and diseased plants; (**c**) A clump showing initial symptoms of shorter lateral pseudostems containing pale *yellow* leaves; (**d**) The emerging leaves from infected plants are unfurled and twisted; (**e**) Excessive stunted shoots developing in a clump; Proliferation of stunted shoots giving a bushy appearance of a clump

an aphid vector. The virus identity was elusive because no virus particle could be readily observed in transmission electron microscopy. However, a few isometric virus particles of 17–20 nm were observed in the partially purified preparation as well as in some field samples collected from Kalimpong during 2003–2004. During this time, BBTV was considered as a member of the genus *Nanovirus*, containing the similar virion morphology to that observed in the foorkey samples. Therefore, the observation of the virion morphology was an important clue that directed to further examination of foorkey samples by ELISA and PCR with the antiserum and primer of BBTV, respectively. The ELISA showed weak serological relationship. The sequence (859 nucleotides) generated from the PCR product showed 82%

sequence identity with the DNA-R of BBTV and 47.6–48.5% identity with the other nanoviruses. The previous studies showed that the virus associated with the foorkey disease was different from BBTV by the fact that it did not infect banana and was vectored by a different aphid species (Basu and Ganguly 1968; Varma and Capoor 1964). The serological and partial genome sequence information further provided evidence of existence of a new nanovirus, which was named as CBDV (Mandal et al. 2004).

3.7.3 Mode of Spread

3.7.3.1 Through Planting Materials

CBDV is not known to spread through seed or contact of the infected plant materials. In the experimental conditions, the virus is also not transmitted through sap inoculation. The most important mode of spread of the virus is through vegetative propagative materials derived through the infected clump. The suckers, which are commonly used for raising new plantation, are the initial source of CBDV as it is difficult to judge the initial infection by the visual observation of clump. Tissue culture planting materials also can potentially circulate the virus if the plantlets are derived from the infected clump.

3.7.3.2 Through the Aphid Vector

The banana aphid, *Pentalonia nigronervosa* was initially reported as vector of CBDV (Varma and Capoor 1964). Later, Basu and Ganguly (1968) published a brief note that the aphid, *Mycromyzus kalimpongensis* transmitted CBDV but not *P. nigronervosa*. As both the studies reported contradictory findings, it was necessary to confirm the findings. Further, there was no studies on the natural occurrence of the aphid species on large cardamom as well as the role of aphid vector in natural dissemination of CBDV. The study was initiated during 2012–2015 in the Regional Station of IARI, Kalimpong to understand the temporal occurrence of aphid species on large cardamom in Darjeeling and Sikkim hills (Ghosh et al. 2016a). This study for the first time documented the natural occurrence and seasonal dynamics of three aphid species, *M. kalimpongensis*, *P. nigronervosa* and *Aulacorthum solani* on large cardamom. The abundance and colonization habits of these aphids on large cardamom are different and it was demonstrated that only *M. kalimpongensis*, which persists in the plantation throughout the year and colonises mainly in the roots and other underground parts of large cardamom, could transmit CBDV. This study (Ghosh et al. 2016a) further confirmed *M. kalimpongensis* but not *P. nigronervosa* is the vector of CBDV. In 2012, while surveying a large cardamom plantation in the Darjeeling hill, an interesting observation was encountered that the foorkey affected clumps contained the higher number of *M. kalimpongensis* in the underground plant parts compared to that in the healthy clumps. Further surveys confirmed the consistence association of *M. kalimpongensis* with the foorkey affected plants irrespective of seasons, altitudes and large cardamom cultivars. This specific behavior of *M. kalimpongensis* was studied in the laboratory and contained field experiments, which revealed that the infection of large cardamom by CBDV influences the aphid to migrate the infected plants (Ghosh et al. 2016b). In the contained field experiment, it was observed that the aphids colonizing on the infected plants had reduced nymphal period and increased longevity and fecundity compared to those grown on the healthy plants. Therefore, CBDV infected plants facilitates emergence of more number of *M. kalimpongensis*. When, foorkey affected plants gradually dries, the viruliferous aphids migrate to the nearby clumps and eventually spread CBDV. This study suggests a general pattern of dissemination of CBDV by *M. kalimpongensis* in the plantation, where CBDV infected plants attract and stimulate emergence of more viruliferous aphids. The most of the members of the family *Nanoviridae* are vectored by aphids, but, the alteration of behavior of aphid vector by the virus infected plants that favors its dissemination of the virus was first demonstrated through the interaction of CBDV, large cardamom and *M. kalimpongensis* (Ghosh et al. 2016b).

CBDV appears to be naturally transmitted through aphid from large cardamom to large cardamom as there is no alternate host of CBDV is known so far. CBDV could not be transmitted from large cardamom to *Musa sapientum*, *Gladiolus* sp, *Canna indica*, *Zingiber officinale*, *Triticum aestivum*, *Sorghum vulgare* and *Zea mays* (Varma and Capoor 1964). The only experimental host of CBDV known is small cardamom (*Elettaria cardamom*), which although is not grown in the area where large cardamom is cultivated.

3.7.4 Molecular Properties of CBDV

3.7.4.1 Cloning of the Genome Components

CBDV was considered as a new member of the genus *Babuvirus* based on the distinct sequence of Rep gene of DNA-R genome component. In order to establish it as a distinct virus species, Mandal et al. (2013) cloned and analysed the complete set of genome components that established CBDV as the third distinct species of the genus *Babuvirus*, where the other two species were BBTV and ABTV. DNA-R was the first genome component that was cloned by designing a pair of abutting primers based on the sequence of the Rep gene of CBDV. Babuviruses are known to contain six DNA components eg., DNA-R, -S, -M, -N, -C and U3. Therefore, CBDV was also expected to contain similar set of genomic components. The each genome components of the members of the family *Nanoviridae* have common features containing a coding region of a single protein and a non-coding region containing a stem and loop (SL) structure and major common region (CR-M), TATA box and poly-A signal. The SL structure contains the nonanucleotides is highly conserved among the genome components. A pair of abutting primer (AV32F&AV33R) designed from the SL region of the DNA-R, was expected to amplify the other genome components of CBDV. However, this approach resulted in cloning only DNA-S, DNA-M and an unknown component (DNA-U1). To obtain the DNA-N and -C components, partial genome fragment was amplified based on the primers designed based on the respective components of BBTV and ABTV and the specific abutting primers from this partial sequence resulted in amplification and cloning of the full-length DNA. Further the rolling circle amplification (RCA) followed by restriction digestion resulted in several putative clones with unit-length inserts. Sequencing of these RCA clones led to identify the satellite Rep component (Sat-Rep), DNA-U2 and DNA-U3.

3.7.4.2 Major Genome Components

The sequence analysis of all the clones obtained through PCR and RCA revealed existence of six DNA components (DNA-R, -S, -M, -N, -C and U3), which in the analogy of the BBTV and ABTV, were considered as the integral components of CBDV (Fig. 3.7, Table 3.1). In addition, three more DNA components (U1, U2 and Sat-Rep) that were associated with the foorkey affected samples were also identified. Association of these novel nine DNA components with the foorkey disease established CBDV as a new virus species of the genus *Babuvirus* family *Nanoviridae* (Mandal et al. 2013).

The CBDV DNA components were composed of 1079–1134 nucleotides. The sequence comparison between the DNA components showed the CR-SL was 49–70

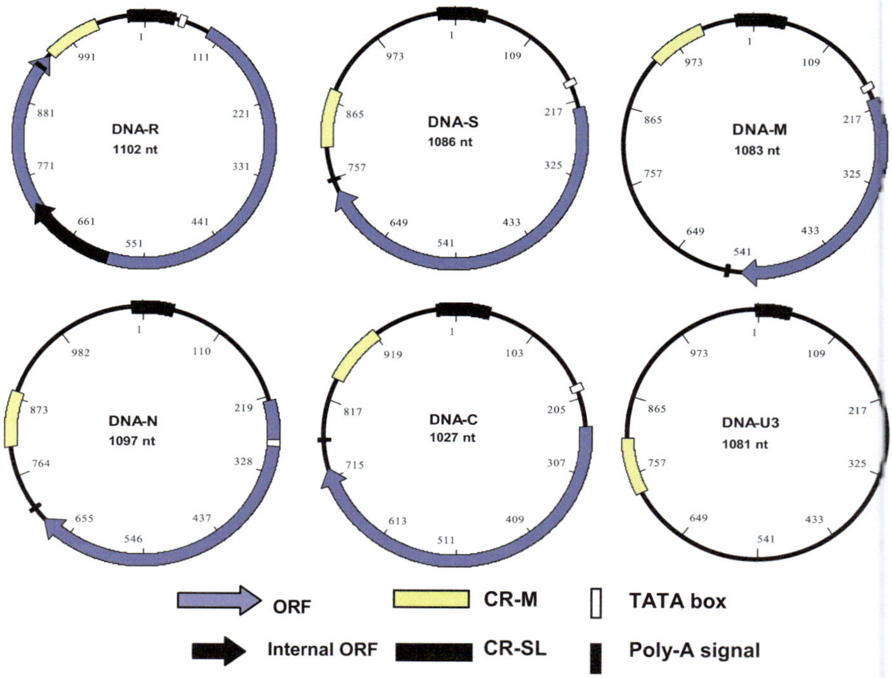

Fig. 3.7 Genome organization of Cardamom bushy dwarf virus. DNA-R encoding replication initiation protein, DNA-S encoding coat protein, DNA-M encoding movement protein, DNA-N encoding nuclear shuttle protein, DNA-C encoding cell cycle link protein. *ORF* open reading frame, *SL* stem and loop structure, *CR-SL* stem-loop common region, *CR-M* major common region. Nucleotide count was shown from the beginning of 5′ region of SL (Adopted from Mandal et al. 2013)

nucleotides, which contained the nonanucleotide, TATTATTAC in the loop region. The sequence in the loop of SL region was slightly different among the DNA components. DNA-R, -S and -M were different from DNA-N, -C, -U3 and -U2 by AC in place of CT. The CR-SL of CBDV genome components was highly different sharing only 46.3–97.1% sequence identity with BBTV and ABTV. The iteron sequences, R (GTCCC), F1 (GGGAC) and F2 (GGAAC) adjacent to the stem region were present in all the components of CBDV. The CR-M was 73–80 nucleotides long sharing 55.9–86.0% similarities among the components and contained two highly conserved regions, the 17 nt GC rich identical sequence (AAGGGCCGAAGGCCCGT) and the near identical 17 nt sequence (CGC/AAA/CTTAT/A/CGACCTGTC). The CR-M of CBDV is also highly divergent from BBTV and ABTV with only 18.8–56.0% sequence identity.

The DNA-R contained 1102 nucleotides that encoded a major protein of 33.6 kDa, potentially associated with replication (Rep). In addition, a small ORF encoding a 5.1 kDa protein of unknown function was present within the Rep ORF. The Rep protein of CBDV shared the closest sequence identity of 85.3–87.7% with the master-Rep (m-Rep) of BBTV and ABTV, whereas 53.8–56.9% with the other species of the member of the family *Nanoviridae*. The DNA-S contains 570–571 nucleotides intergenic region and 513 nucleotides region that encodes a 19.5 kDa capsid protein. CBDV shares a closer sequence identity (80.5%) with ABTV than BBTV (75.8%) in the amino acids sequence of capsid protein. The DNA-M component encodes 13.49 kDa movement protein, which is highly different from that of BBTV and CBDV (59.8–66.6%). The DNA-N genome encodes 18.2 kDa nuclear shuttle protein (NSP). Interestingly, a TATA box is located 57 nt after the initiation codon of the NSP ORF. The NSP of CBDV is distantly related to that of both ABTV and BBTV with 68.3–69.0% amino acid sequence identity.

The DNA-C component encodes a 18.5 kDa cell cycle linking protein (Clink) that contains retinoblastoma like-binding motif LFCDE (LXCXE). The DNA-U3 contained CR-SL and CR-M structures similar to that in the other major components but do not contain any significant ORF. The U3 of CBDV shared low sequence similarities (43.8–45.1%) with BBTV and ABTV.

3.7.4.3 Satellite Component

A Sat-Rep DNA component was isolated through RCA from the foorkey affected sample. Sat-Rep is composed of 1134 nucleotides, which is 6 nucleotides shorter than DNA-R. The DNA contains a major ORF of 855 nucleotides encoding 33kDa protein. The dNTP binding motif, GNEGKS, which is present in the Sat-Rep protein, is different from that in DNA-R of CBDV. The nucleotide sequence length of loop and stem and the nonanucleotides of Sat-Rep are different from that of DNA-R of CBDV. The nucleotide sequence of Sat-Rep is significantly different (42.1–43.5%) from DNA-R of babuvirus including CBDV.

3.7.4.4 The Other Unknown Components

In addition to U3 genome component, two other unknown components, U1 (1080 nucleotides) and U2 (1078 nucleotides) were isolated from foorkey affected plant. None of these DNA contains any major ORF. These unknown components, however, contained CR-SL and CR-M as found in the major genome components of CBDV. The SL structure of these components is almost identical to the DNA-R, -S and -M components. The BLAST search showed no sequence similarity in the region other than CR-SL and CR-M with the members of *Nanoviridae* in the database.

3.7.4.5 Phylogenetic Relationships and Diversity in CBDV

The mean genetic distance of CBDV of six major genome components is 38.0–40.2% from the other two babuviruses, BBTV and ABTV. Phylogenetically, CBDV shows distinct divergence from the other two babuviruses. There is a variation in phylogenetic relationships with reference to the DNA components of CBDV. DNA-R, -N and -U3 showed closer phylogenetic relatedness with BBTV, DNA-M and -S with ABTV, whereas DNA-C is highly different from both the babuviruses. The Sat-Rep component of CBDV shares a closer phylogenetic relations with S1 satellite DNA of BBTV. The unknown components, U1 and U2 of CBDV are distinctly related from U3 of ABTV, BBTV and CBDV.

CBDV having multipartite genomes may lead to reassortment of whereby entire genome components of different strains. Savory and Ramakrishnan (2014) have extensively analyzed the reassortment among 163 CBDV isolates collected in North-East India. They have showed the evidence of recombination, which might have played a role in the evolutionary dynamics of populations. By sequencing six discrete genome components for each isolate, they demonstrated that over 40% of the isolates displayed evidence of at least one reassortment event during their evolutionary histories. They also observed that DNA-M and DNA-N components are the most predisposed to reassortment. The comparisons of the common regions of different genome components revealed signatures of concerted evolution mediated by frequent inter-component homologous recombinations.

3.8 Concluding Remarks

BBTV is internationally and nationally a serious threat to banana and plantains and has been characterized at molecular level in India. CBDV is restricted to large cardamom growing regions of Sikkim and West Bengal. The genomic properties of BBTV and CBDV have been studied well and so far a total of 1325 accession numbers (324 for BBTV and 1001 for CBDV) have been contributed to the GenBank database from India (Fig. 3.8). However, the infectivity of the cloned DNA components is required to confirm that these genome components are necessary to cause bunchy top and foorkey disease, respectively in these crop species. So far, infectivity of the cloned genome components has been demonstrated only for Faba bean necrotic yellows virus (FBNYV), a, nanovirus, but not for any babuviruses (Mandal 2010). The genome of babuviruses (BBTV, ABTV and CBDV) is composed of six DNA components, DNA-R, -S, -M, -N, -C and U3 (Mandal et al. 2004; Vetten et al. 2012). The DNA-U3 is considered as an integral part of the babuvirus genome, but CBDV U3 did not contain any major ORF as in case of the other two babuviruses.

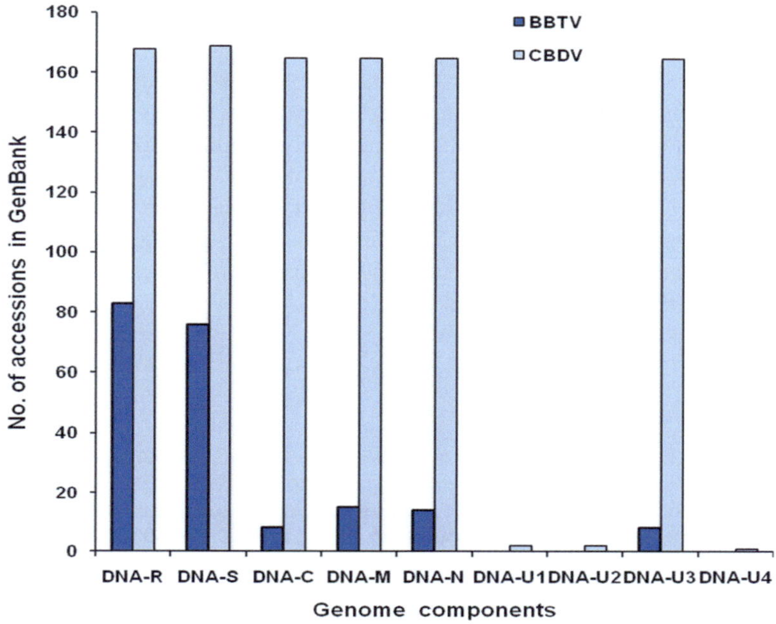

Fig. 3.8 The genome sequence numbers for the different genome components of cardamom bushy dwarf virus (CBDV) and Indian isolates of banana bunchy top virus (BBTV) reported from India in the GenBank database

However, some of the BBTV isolates were shown to contain a small ORF in U3 genome encoding 9–10 kDa protein (Beetham et al. 1999; Vishnoi et al. 2009). Two additional unknown components (U1 and U2) were found in CBDV affected large cardamom, which did not contain any major protein coding sequence. The virus associated with Cardamom bushy dwarf disease (CBDD) is biologically distinct from BBTV that its vector is *M. kalimpongensis* (Basu and Ganguly 1968) whereas banana black aphid, *Pentalonia nigronervosa* transmits BBTV and banana is a non-host for CBDV (Varma and Capoor 1964). Serologically, CBDV is different from BBTV (Mandal et al. 2004). Nine novel genomic components of CBDV are associated with the CBDD of large cardamom whereas only eight components including two satellite DNA's are associated with BBTV. The overall genome sequence identity and phylogeny showed evolutionary divergence of CBDV from the existing members within the genus *Babuvirus*. The genus, *Babuvirus* was created with the sole species, BBTV (Vetten et al. 2005). In 2008, another babuvirus species, ABTV infecting abaca was reported from Philippines (Sharman et al. 2008) and so far it has not been known in India. The study of Mandal et al. (2013) showed the evidence of the third member under genus *Babuvirus* and thrown light on the further diversity in babuvirus. Further, eradication of these viruses in India need integrated approach combining surveillance, monitoring, destruction of infected material and replanting with virus free plants and controlling vector using systemic insecticides would bring down the disease incidence in the area where plantations are perennial for both of these two crops. In case of banana, the implementation of National certification system for tissue culture raised plants (NCS-TCP) by DBT is a major step in containing the incidences of BBTV in tissue culture raised plants. At present, 270 million tissue culture banana plants are supplied by the 95 tissue culture recognized units across India. Among 270 million plants, only 35 million plants are certified during 2015–16. In 2009–11, a severe outbreak of BBTV in Jalgaon, Maharastra and Kodur, AP should be taken as lesion by the TC banana producers and all the TCPUs must enter into the NCS-TCP system and must get virus free certification before dispatching the plants to the farmers. The impact of the certification system in managing BBTV in banana needs to be studied.

References

Amin I, Qazi J, Mansoor S, Ilyas M, Briddon RW (2008) Molecular characterisation of banana bunchy top virus (BBTV) from Pakistan. Virus Genes 36:191–198

Anhalt MD, Almeida RPP (2008) Effect of temperature, vector life stage, and plant access period on transmission of banana bunchy top virus to banana. Phytopathology 98:743–748

Banerjee A, Roy S, Beherea GT, Roy SS, Dutta SK, Ngachana SV (2014) Identification and characterization of a distinct banana bunchy top virus isolate of Pacific-Indian Oceans group from North-East India. Virus Res 183:41–49

Banerjee A, Singh R, Roy S, Behere GT, Roy SS, Dutta SK, Ranebennur H, Ngachan SV (2015) Genetic diversity of Banana bunchy top virus from Northeast India showed existence of distinct PIO isolates in naturally growing banana mats. In: Proceedings of IPS zonal symposium on Holistic Plant Health Management in Organic Agriculture.10–11 February. ICAR Research Complex for NEH Region, Umiam, Meghalaya, p 54

Basu AN, Ganguly B (1968) A note on the transmission of foorkey disease of large cardamom by the aphid, *Micromyzus kalimpongensis* Basu. Ind Phytopathol 21:127

Beetham PR, Harding RM, Dale JL (1999) Banana bunchy top virus DNA-2 to -6 are monocistronic. Arch Virol 144:89–105

Drew RA, Moisander JA, Smith MK (1989) The transmission of banana bunchy top virus in micropropagated bananas. Plant Cell Tissue Org Cult 16:187–193

Espino RC, Magnaye LV, Johns AP, Juanillo C (1993) Evaluation of Philippine banana cultivars for resistance to bunchy-top and fusarium wilt. In: Valmayor RV, Hwang SC, Ploetz R, Lee SC, Roa NV (eds) Proceedings: International symposium on recent developments in banana cultivation technology, 14–18 December 1992, Chiuju, Pingtung, Taiwan, Philippines: Taiwan Banana Research Institute. INIBAP/ASPNET, Los Banos, pp 89–102

Geering ADW, Thomas JE (1997) Search for alternative hosts of banana bunchy top virus in Australia. Aus Plant Pathol 26:250–254

Ghosh A, Das A, Lepcha R, Mandal B (2016a) Identification, distribution and temporal occurrence of aphids infesting large cardamom and their efficiency in transmitting large cardamom viruses in northeastern sub-Himalayan region. Aus Plant Pathol. doi:10.1007/s13313-016-0437-0

Ghosh A, Das A, Vijayanandra S, Mandal B (2016b) Cardamom bushy dwarf virus infection in large cardamom alters plant selection preference, life stages and fecundity of aphid vector, *Micromyzus kalimpongensis* (Aphididae: Hemiptera). Environ Entomol 45:178–184

Hafner GJ, Harding RM, Dale JL (1995) Movement and transmission of banana bunchy top virus DNA component one in bananas. J Gen Virol 76:2279–2285

Hooks CRR, Wright MG, Kabasawa DS, Manandhar R, Almeida RPP (2008) Effect of banana bunchy top virus infection on morphology and growth characteristics of banana. Ann Appl Biol 153:1–9

Howarth AJ, Gardner RC, Messing J, Shepherd R (1981) Nucleotide sequence of naturally occurring deletion mutants of cauliflower mosaic virus. Virology 112:678–685

Hu JS, Wang M, Sether D, Xie W, Leonhardt KW (1996) Use of polymerase chain reaction (PCR) to study transmission of *banana bunchy top virus* by the banana aphid (*Pentalonia nigronervosa*). Ann Appl Biol 128:55–64

Hu JM, Fu HC, Lin CH, Su HJ, Yeh HH (2007) Reassortment and Concerted Evolution in *Banana Bunchy Top Virus* Genomes. J Virol 81(4):1746–1761

Hyder MZ, Shah SH, Hameed S, Naqvi SMS (2011) Evidence of recombination in the banana bunchy top virus genome. Infect Genet Evol 11:1293–1300

Islam MN, Naqvi AR, Jan AT, Mohd Q, Haq R (2010) Genetic diversity and possible evidence of recombination among Banana Bunchy Top Virus (BBTV) Isolates. Int Res J Microbiol 1(1):001–012

Jones DR (2000) Diseases of Banana. Abaca and Enset, CAB International, Wallingford

Karan M, Harding RM, Dale JL (1994) Evidence for two groups of banana bunchy top virus isolates. J Gen Virol 75:3541–3546

Karan M, Harding RM, Dale JL (1997) Association of banana bunchy top virus DNA components 2 to 6 with banana bunchy top disease. Mol Plant Pathol:1–16

Kesavamoorthy RC (1980) Radical changes in ecosystem in the Pulney hills. In: Muthukrishnan CR, Abdul Chaser, J B. M. Md. Proc. 13th national seminar on banana production technology, TNAU, Coimbatore. pp 23–28

Kumar PL, Selvarajan R, Iskra-Caruana ML, Chabannes M, Hanna R (2015) Biology, Etiology, and Control of Virus Diseases of Banana and Plantain. Adv Virus Res 91:229–269

Magee CJP (1927) Investigations of the bunchy top disease of bananas. Bulletin of the Council for Scientific and Industrial Research (Australia). 30: 64

Magee CJP (1948) Transmission of banana bunchy top to banana varieties. J Aus Inst Agri Sci 14:18–24

Mandal B, Mandal S, Pun KB, Varma A (2004) First report of the association of a Nanovirus with ‘Foorkey’ disease of large cardamom in India. Plant Dis 88:428

Mandal B (2010) Advances in small isometric multicomponent ssDNA viruses infecting plants. Ind J Virol 21:18–30

Mandal B, Shilpi S, Roy-Barman A, Mandal S, Varma A (2013) Nine novel DNA components associated with the foorkey disease of large cardamom: Evidence of a distinct babuvirus species in *Nanoviridae*. Virus Res 178:297–305

Menon MR, Christudas SP (1967) Studies on the population of aphid Pentalonia nigronervosa Coq. On banana plants in Kerala. Agri Res J Kerala 5:84–86

Metha PR, Joshi NC, Rao MH, Renjhen PL (1964) Bunchy top serious disease of banana in India. Sci Cult *30*:259–263

Mitra M (1936) Report of the imperial mycologist. Scientific Report of Agricultural Research Institute, Pusa. 1933–1934

Niyongere C, Lepoint P, LosengeT BG, Ateka EM (2015) Towards understanding the diversity of banana bunchy top virus in the Great Lakes region of Africa. Afr J Agri Res 10(7):702–709

Niyongere C, Losenge T, Ateka EM, Ntukamazina N, Ndayiragije P, Simbare A et al (2013) Understanding banana bunchy top disease epidemiology in Burundi for an enhanced and integrated management approach. Plant Pathol 62:562–570

Ram RD, Summanwar AS (1984) *Colocasia esculenta* (L) Schott. A reservoir of bunchy top disease of banana. Curr Sci 53:145–146

Samraj J, Menon MR, Christudas SP (1970) The movement of banana bunchy top virus in plant. Agri Res J Kerala 8:106–108

Savory FR, Ramakrishnan U (2014) Asymmetric patterns of reassortment and concerted evolution in Cardamom bushy dwarf virus. Infect Genet Evol 24:15–24

Selvarajan R, Balasubramanian V (2013) Natural occurrence of banana bunchy top virus in Ensete superbum in India. Ind J Virol 24:97–98

Selvarajan R, Balasubramanian V (2014) Host interaction host–virus interactions in banana-infecting viruses. In: Gaur RK, Hohn T, Sharma P (eds) Plant virus–host interaction molecular approaches and viral evolution. Elsevier Academic Press, Waltham, pp 57–78

Selvarajan R, Balasubramanian V, Sathiamoorthy S (2006) Vector transmission of banana bract mosaic and banana streak viruses in India. In: Abstracts of XVI annual convention and international symposium on management of vector-borne viruses, ICRISAT, 7–10 February 2006, p 110

Selvarajan R, Mary Sheeba M, Balasubramanian V, Rajmohan R, Lakshmi Dhevi N, Sasireka T (2010) Molecular characterization of geographically different banana bunchy top virus isolates in India. Ind J Virol 21:110–116

Selvarajan R, Balasubramanian V, Sheeba MM, Raj Mohan R, Mustaffa MM (2011) Virus-indexing technology for production of quality banana planting material: a boon to the tissue-culture industry and banana growers in India. In: Van den Bergh I et al. (eds) Acta Horticulturae: 897. Proceedings of international ISHS-ProMusa symposium on global perspectives on Asian challenges, pp 463–469

Selvarajan R (2015) Viral diseases of banana. CAB Rev 10. No. 050

Sharma SR (1988) Banana bunchy top virus. Int J Trop Plant Dis 6:19–41

Sharman M, Thomas JE, Skabo S, Holton TA (2008) Abaca' bunchy top virus, a new member of the genus Babuvirus (family Nanoviridae). Arch Virol 153:135–147

Singh SJ (2003) Viral diseases of Banana, 1st edn. Kalyani Publishers, Ludhiana, pp 10–40

Stainton D, Kraberger S, Walters M, Wiltshire EJ, Rosario K, Halafihi M et al (2012) Evidence of inter-component recombination, intra-component recombination and reassortment in banana bunchy top virus. J Gen Virol 93:1103–1119

Su HJ, Hwang AS, Lee SY, Chao CP (2007) Conservation, disease indexing and utilization of pathogen free citrus and banana genetic resources in Taiwan. International training workshop on the conservation and utilization of Tropical/ Subtropical plant genetic resources, In, pp 1–24

Thomas JE, Dietzgen RG (1991) Purification, characterization and serological detection of virus-like particles associated with banana bunchy top virus in Australia. J Gen Virol 72:217–224

Thomas JE, Iskra-Caruana ML (2000) Bunchy top. In: Jones DR (ed) Diseases of banana, abacá and enset. CABI Publishing, Wallingford, pp 241–253

Thomas JE, Iskra-Caruana ML, Jones DR (1994) Banana bunchy top disease, *Musa* disease fact sheet No.4, INIBAP, Montpellier, p 2

Varma PM, Capoor SP (1964) 'Foorkey' disease of large cardamom. Ind J Agri Sci 34:56–62

Vetten HJ, Chu PWG, Dale JL, Harding R, Hu J, Katul L, Kojima M, Randles JW, Sano Y, Thomas JE (2005) Family *Nanoviridae*. In: Virus taxonomy: eighth report of the International Committee on Taxonomy of Viruses. Elsevier, London, pp 343–352

Vetten HJ, Dale JL, Grigoras I, Gronenborn B, Harding R, Randles JW., et al (2012) Family Nanoviridae. In: Virus taxonomy: ninth report of the international committee on taxonomy of viruses, pp 395–404

Vishnoi R, Raj SK, Prasad V (2009) Molecular characterization of an Indian isolate of banana bunchy top virus based on six genomic DNA components. Virus Genes 38:334–344

Wang HI, Chang CH, Lin PH, Fu HC, Tang C, Yeh HH (2013) Application of Motif-Based Tools on Evolutionary Analysis of Multipartite Single-Stranded DNA Viruses. PLoS One 8(8):e71565. doi:10.1371/journal.pone.0071565

Wanitchakorn R, Harding RM, Dale JL (2000) Sequence variability in the coat protein gene of two groups of banana bunchy top isolates. Arch Virol 145:593–602

Wardlaw CN (1972) Banana Diseases: Including plantains and abaca. William Clowes and Sons Ltd., London, pp 68–115

Watanabe S, Greenwell AM, Bressan A (2013) Localization, concentration, and transmission efficiency of banana bunchy top virus in four asexual lineages of Pentalonia aphids. Viruses 5:758–775

Wu RY, Su HJ (1990) Purification and characterization of banana bunchy top virus. J Phytopathol 128:153–160

Yu NT, Zhang YL, Feng TC, Wang JH, Kulye M, Yang WJ et al (2012) Cloning and sequence analysis of two banana bunchy top virus genomes in Hainan. Virus Genes 44:488–494

Emergence and Diversity of Badnaviruses in India

4

Virendra Kumar Baranwal and Susheel Kumar Sharma

Abstract

Badnaviruses (family *Caulimoviridae*) have bacilliform virion morphology containing circular double stranded DNA genome. They are transmitted by mealybugs in a semi-persistent manner. Badnaviruses are emerging as serious viral pathogens affecting many crops in India. Banana streak virus (BSV), sugarcane bacilliform virus (SCBV), citrus yellow mosaic virus (CYMV), piper yellow mottle virus (PYMoV) and bougainvillea infecting badnavirus are the five badnaviruses known to occur in India. Badnaviruses reported from India are genetically diverse. This article highlights the research work carried out on badnaviruses affecting different crops in India.

Keywords

Badnaviruses • Diversity • Diagnostics • India • Banana streak viruses • Sugarcane bacilliform viruses • Citrus yellow mosaic virus • Piper yellow mottle virus • Bougainvillea badnaviruses

4.1 Introduction

The first record of a badnavirus in India was made by Ahlawat et al. (1996a). Subsequently it was recorded in other crops. The term 'badna' has been derived from the bacilliform DNA virus. *Commelina yellow mottle virus* (ComYMV) is the

V.K. Baranwal (✉)
Advanced Centre for Plant Virology, Division of Plant Pathology, ICAR-Indian Agricultural Research Institute, New Delhi 110012, India
e-mail: vbaranwal2001@yahoo.com

S.K. Sharma
ICAR Research Complex for NEH Region, Manipur Centre, Lamphelpat, Imphal 795004, India

B. Mandal et al. (eds.), *A Century of Plant Virology in India*,
https://doi.org/10.1007/978-981-10-5672-7_4

type species of the genus *Badnavirus*. Badnaviruses are known as pararetroviruses, which replicate *via* transcription/reverse transcription like mammalian retroviruses. The major difference between retroviruses and pararetroviruses is that pararetroviruses do not integrate into the host genome during replication; however, integration to the host genomic DNA is an essential step in the replication of retroviruses (Pfeiffer and Hohn 1983). The replication cycle of pararetroviruses is thus exclusively episomal. For synthesis of minus-strand DNA, host cytosolic tRNA^{met} act as primer for reverse transcription by viral RT (Boeke and Corces, 1989). The opposite strand is primed from a purine-rich region and polymerized by viral encoded reverse transcriptase (RT) and RNase H (Hull et al. 1986; Verver et al. 1987). The site-specific discontinuities found in badnavirus genome are formed at the priming sites for both minus and plus-strand DNA synthesis and are made by the oncoming strand displacing the existing strand for a short distance, where ligation does not occur to form a closed circle (King et al. 2012). The pararetroviruses like that of retroviruses encodes a POL gene comprising of protease, reverse transcriptase and RNase H coding functions (Toh et al. 1983), however they lack the integrase gene.

Typically badnavirus genome contains three open reading frames (ORFs). The convention is to start the numbering of the viral genomes of reverse transcribing dsDNA pararetroviruses at the 5′ end of the minus-strand tRNA^{met} primer-binding site. The ORF1 of some badnaviruses species starts with non-AUG start codon, whereas in other *Badnavirus* species the AUG start codon of ORF1 as well the AUG start codon of ORF2 are in weak context. These suboptimal contexts of the ORF1 and ORF2 start codons allows the leaky scanning of the ribosomes, which can then reach to the ORF2 and ORF3 respectively and translate them (Geering et al. 2005b; Lheureux et al. 2007). Protein encoded by ORF1 is P1, function of which is not known. P2 protein encoded by ORF2 is virion-associated protein (VAP). ORF3 encodes for a polyprotein which is cleaved post-translationally to yield proteins involved in cell-to-cell movement (movement protein: MP), coat protein (CP), aspartic protease, reverse transcriptase (RT) and ribonuclease H (RNaseH) functions (King et al. 2012).

Badnaviruses are highly diverse with regard to the biological properties as they infect either dicotyledonous or monocotyledonous plants (Borah et al. 2013). Geographically, badnaviruses are distributed across the different tropical countries in Asia, Africa and America continents and in the Pacific Islands (Borah et al. 2013). Most of the viral species in the genus *Badnavirus* are primarily associated with the diseases of tropical and subtropical plant species with some viral species associated with the temperate and sub-Antarctic host species. They are highly heterogeneous both at genetic and serological level (Lockhart and Olszewski 1993; Geering et al. 2000; Harper et al. 2004) as they share very low nucleotide sequence identity even between the species of same genus. There is a pronounced antigenic variability within different species of the genus *Badnavirus* (Lockhart and Olszewski 1993). Interveinal chlorotic mottling and streaking are the most frequent symptoms associated with the members of genus *Badnavirus* (King et al. 2012; Borah et al. 2013). Badnaviruses are generally restricted to the vascular tissues of their host plants (Dahal et al. 1998a; Dahal et al. 2000). Their virions occur in the cytoplasm without forming any inclusion bodies (Borah et al. 2013).

Another very peculiar characteristics of plant pararetroviruses is their ability to occur as both host genome integrated form (endogenous pararetroviruses: EPRV) and exogenous episomal forms (Chabannes and Iskra-Caruana 2013). Some *Badnavirus* species are discovered to be present both as endogenous (viral DNA integrated to the host nuclear genome) as well as exogenous episomal forms (Geering et al. 2010). Three badnaviruses (banana streak OL virus: BSOLV, banana streak GF virus: BSGFV and banana streak IM virus: BSIMV) have been reported to be existing in endogenous as well as episomal forms and are known as activable endogenous badnaviruses. In addition, other banana streak viruses, Dioscorea infecting badnaviruses (dioscorea bacilliform AL virus: DBALV and dioscorea bacilliform SN virus: DBSNV), fig badnavirus 1 (FBV-1), kalanchoe top-spotting virus (KTSV), taro bacilliform virus (TaBV) and dracaena mottle virus (DrMV) have also been reported to be integrated in the respective host genome, however conclusive evidence for their integration has not been reported so far (Bhat et al. 2016; Geering et al. 2014). The viral DNA has become integrated into the host genome by illegitimate recombination process through the non-homologous end-joining (Gayral et al. 2008; Iskra-Caruana et al. 2010; Iskra-Caruana et al. 2014a, b). The replication competent EPRV sequences are found in the genomes of *Musa balbisiana* (Chabannes and Iskra-Caruana 2013; Iskra-Caruana et al. 2010). In addition, the EPRV sequences are also found in many plant species which are replication-defective, thus are not able to activate to cause the episomal virus infection (Chabannes et al. 2013; Chabannes and Iskra-Caruana 2013).

Transmission of badnaviruses occurs through different means. Badnaviruses in general are not mechanically transmissible, probably because of their specific ability to infect vascular tissues. Vegetative propagation of the infected planting materials also contributes to the spread of viruses over large geographical areas and introduction to new areas where viruses are not known to occur. The horizontal spread of badnaviruses refers to the transmission by mealybugs (family Pseudococcidae) under field conditions. Mealybugs transmit badnaviruses in semi-persistent manner. Mealybug mediated transmission is not considered to be an efficient mean of spread under field conditions due to the slow movement of mealybugs (Daniells et al. 2001). The other means of badnaviruses spread particularly in case of banana infecting badnaviruses (banana streak viruses: BSV) is through activation of homologous integrated BSV sequences present in the *Musa balbisiana* (B) genome (known as vertical spread) (Harper et al. 1999; Ndowora et al. 1999; Chabannes et al. 2013).

As per the International Committee on Taxonomy of Viruses (ICTV) the genus *Badnavirus* comprises of 37 species and at least nine tentative species (Adams et al. 2015; King et al. 2012) of all these badnaviruses only six virus species and three new tentative species are recorded from India. Badnaviruses measuring 30 nm in width and 60–900 nm in length, with modal particle length of 130 nm (King et al. 2012) are emerging as important viral pathogens of diverse crops in tropics. In this chapter, the research work conducted on the badnaviruses known to infect various crops in India, is presented.

4.2 Disease and Virus Description

4.2.1 Banana Streak Disease

Symptoms of streak disease of banana generally appear as chlorotic and necrotic streaks on leaf lamina running parallel to the veins. Severe infections may result in the death of growing point, pseudostem cracking, internal necrosis of pseudostem and abnormal bunch development (Dahal et al. 2000; Sharma et al. 2014b) (Fig. 4.1a–d). The symptom expression is intermittent and there are periods of symptomless growth between before disease first appears and in reemergence of disease, similarly sometimes the newly developed leaves may remain symptomless for some times. Development and expression of symptoms depends on a number of factors like badnavirus isolate, host genotype or cultivar, level of crop management and prevailing environmental conditions. Depending on the severity of infection, the estimated yield loss due to the BSV infection ranges from 7 to 90% (Dahal et al. 2000; Daniells et al. 2001). The level of yield loss due to virus infection depends on banana genotypes and cultural management.

In India, although the streak disease of banana was present since 1970s (Dekmann and Putter 1996; Jones and Lockhart 1993; Wardlaw 1972), the authentic reports on association of badnaviruses with this disease came much later. The banana cultivar 'Mysore' (*Musa* AAB group) is a natural banana hybrid originated in India. All the

Fig. 4.1 Disease symptoms of badnaviruses prevalent in India. Typical streak symptoms on banana varieties due to BSMYV infection, (**a**) Chini champa in Meghalaya; (**b**) Chini champa; (**c**) Malbhog; (**d**) Chini champa in Kahikuchi, Assam; prominent yellowish freckling to chlorotic freckling with striate mosaic pattern and yellow chlorotic streaks on sugracane due to SCBV infection, (**e**) variety CoBIN 4172 in Assam; (**f**) variety BO 150 in Bihar; (**g**) variety 01235 in Uttar Pradesh; yellow mosaic symptoms of CYMV on (**h**) sweet orange; (**i**) sweet orange; yellow mottling and leaf deformation due to PYMoV on (**j**) black pepper; severe yellow mosaic and chorotic vein-banding due to badnavirus infection on (**k**) bougainvillea in Tirupati; (**l**) bougainvillea in Delhi

plants of these cultivars were initially observed to uniformly exhibit the chlorotic and necrotic streaks symptoms on entire leaf lamina, and it was initially thought to be a genetic disorder in this cultivar as there were no signs of the transmission of the symptoms to nearby grown plants of other cultivars (Wardlaw 1972). Later in Australia association of an isolate of banana streak virus was reported with the chlorotic and necrotic streak symptoms of Mysore cultivar of banana (Lockhart and Olszewski 1993; Thomas et al. 1994). The full genome sequences of BSV-Mys was later characterized from the banana cultivar Mysore and now known as *Banana streak MY virus* (BSMYV) species (Geering et al. 2005b).

From India first authentic evidence on the association of a badnavirus with streak disease came in the year 2004 from the *Musa* AAB hybrids (Cherian et al. 2004). Further there were some fragmentary records available on the outbreaks of streak disease from different parts particularly in South and North-East, however there is no authentic report available on epidemics (Singh et al. 2011; Vision 2030 2011, NRCB). With the discovery of association of *Badnavirus* species complex with streak disease of banana and plantains worldwide, it is thought that a genetically and serologically heterogeneous badnaviruses complex are causing this disease (Geering et al. 2000, 2005a; Iskra-Caruana et al. 2014b; Lockhart and Olszewski 1993). The term 'banana streak virus' therefore represent a *Badnavirus* species complex which is all associated with streak disease of banana. Since not all the BSV species have yet been explored, the term 'cryptic species' has been used by Thomas et al. (2013). At present ICTV recognizes nine species to be associated with banana streak disease i.e. *Banana streak MY virus* (BSMYV), *Banana streak GF virus* (BSGFV), *Banana streak OL virus* (BSOLV), *Banana streak VN virus* (BSVNV), *Banana streak IM virus* (BSIMV), *Banana streak UA virus* (BSUAV), *Banana streak UI virus* (BSUIV), *Banana streak UL virus* (BSULV) and *Banana streak UM virus* (BSUMV) (King et al. 2012; Adams et al. 2015). Banana streak disease has been known to occur on the natural banana hybrids in India (Cherian et al. 2004; Selvarajan et al. 2008).

4.2.1.1 Geographical Distribution

The first authentic report on association of badnaviruses with streak disease of banana in India came in 2004 (Cherian et al. 2004). Based on electron microscopy and sequencing of RT/RNase H genomic region, they reported the association of BSOLV (that time referred as BSV-Onnne) with the symptomatic samples of banana cultivars *Musa* sp. BRS-1 (AAB group) and Mysore Poovan (AAB group) grown in Kerala. These isolates (BSV-K1 and BSV-K3) shared 97–99% sequence identity with BSOLV species from Nigeria for RT/RNase H sequences (Cherian et al. 2004). BSMYV was later detected in Poovan banana from South India (Selvarajan et al. 2008). Electron microscopy, long PCR and rolling circle amplification (RCA) confirmed association of short BSOLV variants with leaf streak disease of two banana cultivars Poovan (AAB) and Safed velchi (AB) grown in Kerala and Pune, Maharashtra (Baranwal et al. 2014). An extensive survey was done during 2010–2013 for streak disease in different banana fields of North East India (Assam), North India (Himachal Pradesh), East India (Bihar), West

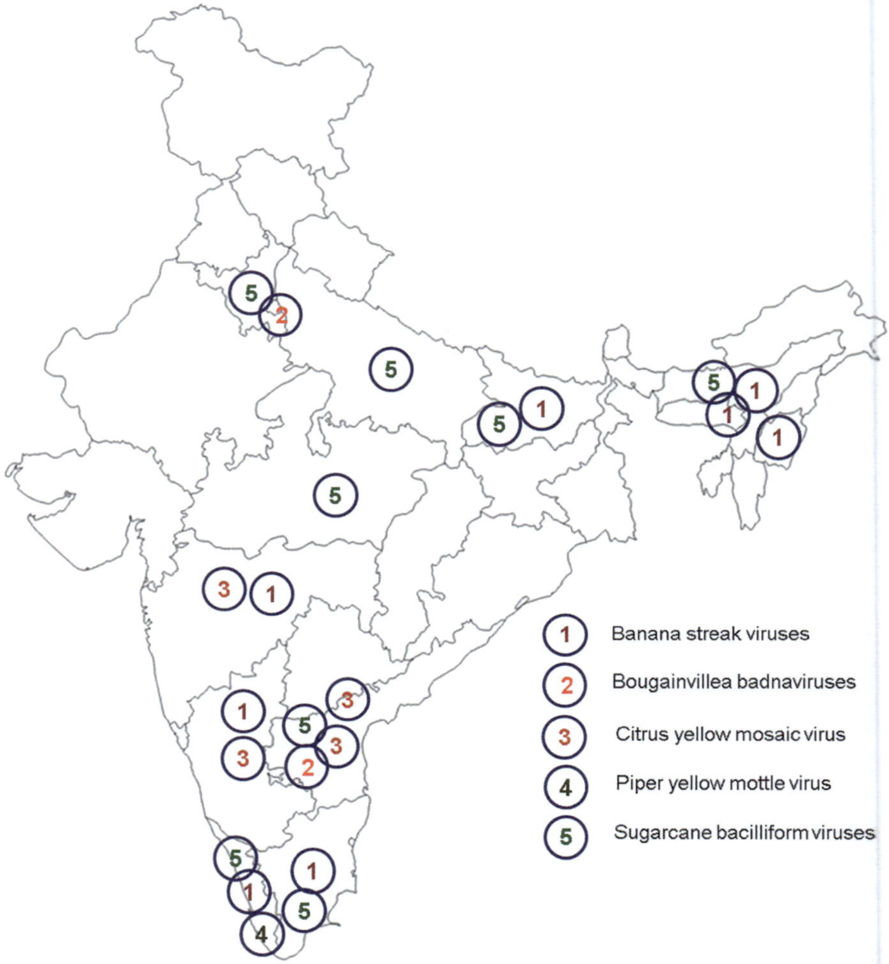

Fig. 4.2 Distribution of badnaviruses in the different states of India

India (Maharashtra) and South India (Karnataka, Tamil Nadu and Kerala) (Sharma et al. 2014b) (Fig. 4.2). Out of 158 samples indexed, 46% were tested positive for BSV infection based on the combined results of antigen-coated plate-enzyme linked immunosorbent assay (ACP-ELISA), duplex immunocapture-PCR (D-IC-PCR), rolling circle amplification (RCA) and sequence analysis of amplified fragments. A high incidence of BSMYV was found in the tested samples (with 41% samples positive). This indicated high prevalence of streak disease in India. Recently we have detected BSMYV from the natural banana mats grown in Meghalaya and Manipur (unpublished results). Based on the field symptoms, a high incidence (40–50%) of streak disease in South India compared to North-East (25–30%) was recorded (Sharma et al. 2015). In addition different tissue culture banana batches obtained from different commercial units were also indexed for BSMYV. Out of the 140 tissue culture raised banana plant batches tested, 16 were detected positive for BSMYV infection (11.5%).

4.2.1.2 Transmission and Spread

In India mealybug species *Ferrisia virgata* was reported to transmit BSMYV under experimental conditions (Selvarajan et al. 2011). Under screen house, mealybugs acquire BSV once after feeding on infected plant and optimum acquisition access period was reached after third day. After that the mealybugs remained infective (capable of transmitting the BSV) up to 5 days from the day of transfer from the virus source, indicating the semi-persistent type of transmission (Kubiriba et al. 2001). Badnaviruses in general and BSV in particular are not mechanically transmissible, probably because of their specific ability to infect vascular tissues. In India virus mainly spread through infected planting materials (Sharma et al. 2014b).

4.2.1.3 Virus Characterization

As discussed previously BSV in recent literature refers to a generic name for different badnaviruses which share up to 50 and 38% nucleotide diversity for core and complete RT/RNase H region, respectively however they all cause symptomatically similar disease. Out of the nine ICTV recognized *Badnavirus* species (BSMYV, BSGFV, BSOLV, BSVNV, BSIMV, BSUAV, BSUIV, BSULV and BSUMV) (King et al. 2012; Adams et al. 2015) and at least five tentative species; banana streak CA virus (BSCAV), banana streak YN virus (BSYNV, previously named banana streak Acuminata Yunnan virus), banana streak UJ virus (BSUJV), banana streak UK virus (BSUKV) and banana streak PE virus (BSPEV) (Harper et al. 2005); only two species i.e. BSMYV and BSOLV has been reported from India (Baranwal et al. 2014; Sharma et al. 2014b, 2015). Based on rolling circle based and end-to-end PCR amplification, full genome sequences of two shorter than full length episomal BSOLV variants (BSOLV-IN1 and BSOLV-IN2 isolated from Poovan and Safed velchi banana respectively) were reported (Baranwal et al. 2014). Both the isolates had a genome size of 6950 bp which was shorter than typical BSOLV genome of 7389 bp reported from Nigeria. Open reading frames (ORFs) 1 and 2 of these shorter than genome length isolates were nearly identical (>99%) to that of BSOLV but contained deletions in ORF3 and intergenic region (IGR). All known conserved domains in short ORF3 and promoter elements in IGR were present which indicated that these shorter BSOLV variants were replicationally competent viruses. An identical shorter BSOLV sequence was submitted in NCBI GenBank database (BSV-TRY: DQ859899). In addition full genome of three BSMYV isolates; BSMYV-IN1, BSMYV-IN2 and BSMYV-IN3 obtained from triploid banana hybrids Chini champa (AAB), Malbhog (AAB) and Monthan (ABB) respectively from North-East and South India were sequenced using sequence-independent improved RCA (Sharma et al. 2015). Genome size of BSMYV-IN1 and BSMYV-IN3 was 7650 bp whereas that of BSMYV-IN2 was 7641 bp (Fig. 4.3). Phylogenetically Indian BSV isolates forms two clusters, one comprising of BSOLV-IN1 and BSOLV-IN2 and other BSMYV-IN1, BSMYV-IN2 and BSMYV-IN3 (Sharma et al. 2015) (Fig. 4.4).

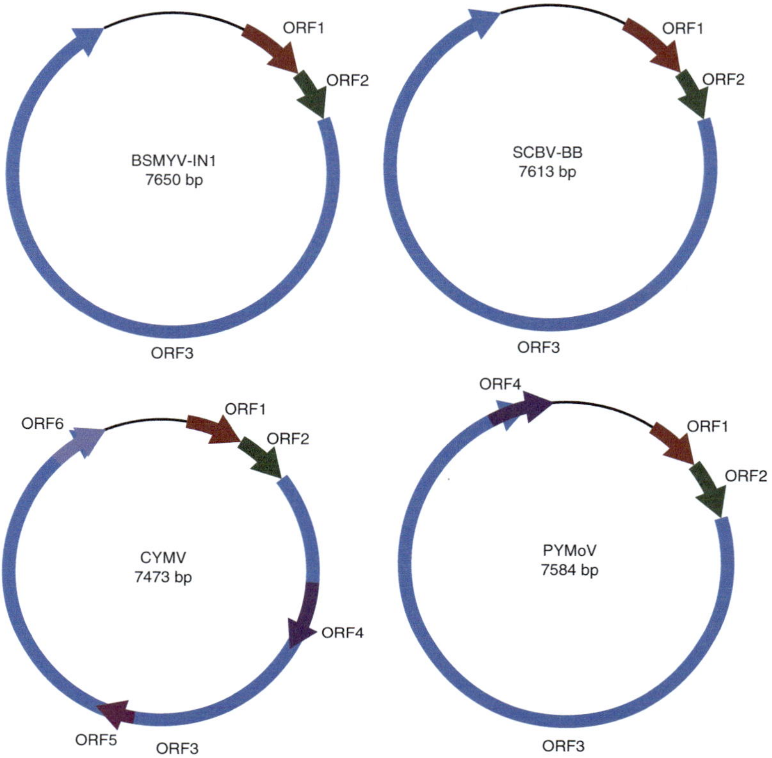

Fig. 4.3 Genome maps and position of different open reading frames (ORFs) in badnaviruses known to occur in India. (**a**) Banana streak MY virus (BSMYV-IN1: KF724854), (**b**) Sugarcane bacilliform virus-BB (SCBV-BB: JN377535), (**c**) Citrus yellow mosaic virus (CYMV-AL: EU489744), (**d**) Piper yellow mottle virus (PYMoV: KJ873041). The open reading frames (ORFs) are mentioned with directional *arrows* and intergenic region as *solid line*

Fig. 4.4 (continued) banana streak OL virus (BSOLV), banana streak MY virus (BSMYV), banana streak MY virus-IN1 (BSMYV-IN1), banana streak MY virus-IN2 (BSMYV-IN2), banana streak MY virus-IN3 (BSMYV-IN3), banana streak OL virus-IN1 (BSOLV-IN1), banana streak OL virus-IN2 (BSOLV-IN2), banana streak GF virus (BSGFV), banana streak UA virus (BSUAV), banana streak UI virus (BSUIV), banana streak UL virus (BSULV), banana streak UM virus (BSUMV), banana streak CA virus (BSCAV), banana streak IM virus (BSIMV), banana YN streak virus (BSYNV, previously named banana streak acuminata Yunan virus), banana streak VN virus (BSVNV), sugarcane bacilliform IM virus (SCBIMV), sugarcane bacilliform MO virus (SCBMOV), sugarcane bacilliform GD virus (SCBGDV), sugarcane bacilliform GA virus (SCBGAV), sugarcane bacilliform virus isolate BO91 (SCBV-BO91), sugarcane bacilliform virus isolate Iscam (SCBV-Iscam), sugarcane bacilliform virus isolate BB (SCBV-BB), sugarcane bacilliform virus isolate BT (SCBV-BT), sugarcane bacilliform virus isolate BRU (SCBV-BRU), dioscorea bacilliform SN virus (DBSNV), citrus yellow mosaic virus (CYMV), commelina yellow mottle virus (ComYMV), cacao swollen shoot virus (CSSV), taro bacilliform virus (TaBV), kalanchoe top-spotting virus (KTSV), gooseberry vein banding associated virus (GVBAV), pineapple bacilliform CO virus (PBCOV), grapevine vein clearing virus (GVCV), bougainvillea chlorotic vein banding virus (BCVBV) and rice tungro bacilliform virus (RTBV)

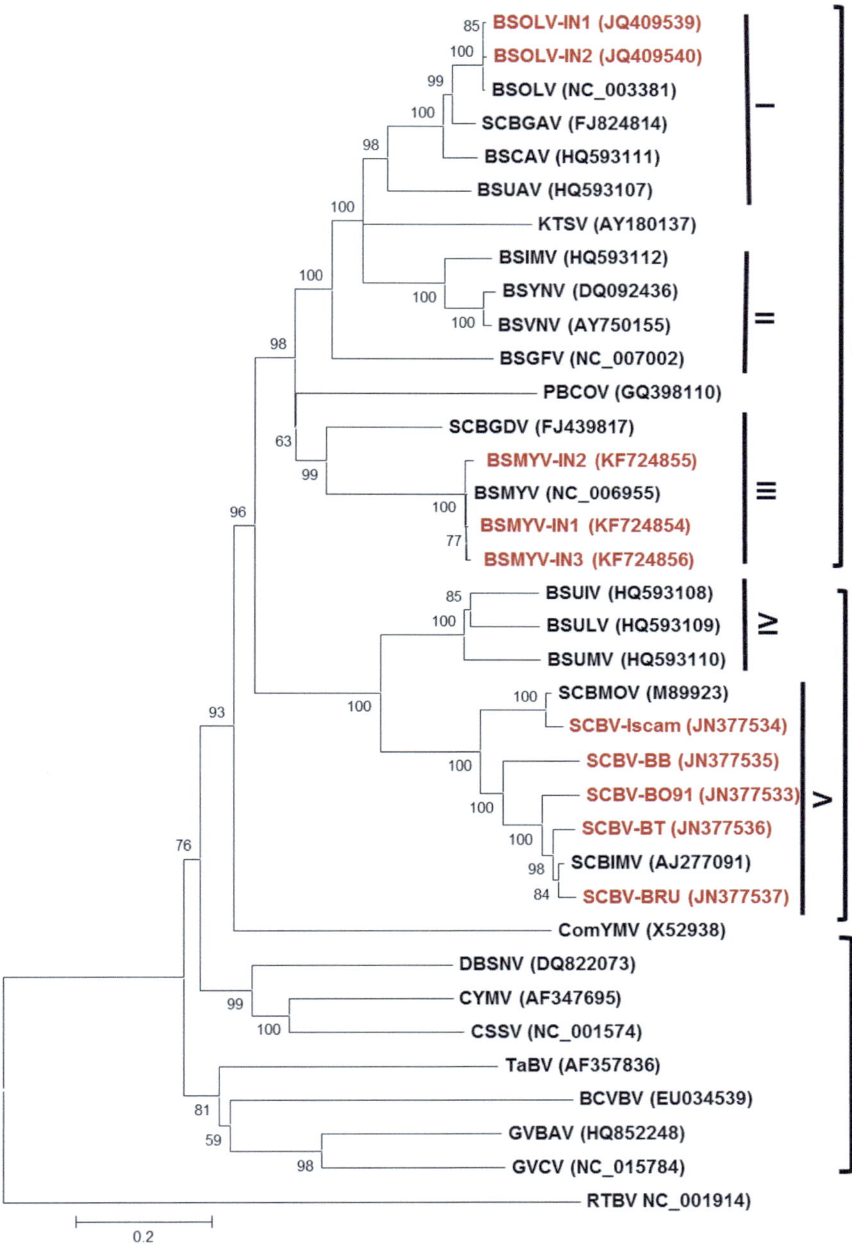

Fig. 4.4 Maximum likelihood (ML) phylogentic tree illustrating complete ORF3 polyprotein based relationship of Indian BSV and SCBV isolates (all marked with red colour) with other badnaviruses. Bootstrap support values from a ML search with 1000 replicates are shown at the nodes of the branches. A species under the sister genus *Tungrovirs*: *Rice tungro bacilliform virus* (RTBV) was used as the out group. The branches showing a bootstrap values less than 60% were collapsed. The scale bar indicates the number of substitutions per site. Acronyms are as follows:

Management of banana streak disease largely depends on production, supply and planting of disease free propagating materials. Electron microscopy (EM) has been used to observe the bacilliform virus particles associated with the streak disease of banana (Cherian et al. 2004; Baranwal et al. 2014; Sharma et al. 2014a). As the virus concentration in field samples is usually very low; and therefore, EM may give false negative results. Immunosorbent electron microscopy (ISEM) was developed for BSMYV detection (Sharma et al. 2014b). Serology and nucleic acid based detection procedures are available for BSV, however profound serological and genetic heterogeneity poses a serious challenge. Integrated badnavirus sequence counterparts having significant sequence homology to the episomal BSVs are reported in banana genome (Harper et al. 1999; Geering et al. 2005a), which makes PCR based diagnostics unreliable as both integrated and episomal sequences are equally amplified. Thus immunocapture-PCR (IC-PCR) combined with ISEM are currently considered standard methods of BSV indexing worldwide. For both these methods, availability and continuous supply of quality virus specific antisera is a prerequisite. Badnaviruses in general and BSV in particular occur at very low concentration in their respective host, moreover there is no laboratory propagation hosts available for their large scale multiplication. The only alternative remained is to use in vitro expressed viral structural proteins for antibodies production. The precise location of coat protein (CP) sequences was not known in any of the badnaviruses genome, which limited the use of expressed coat protein as antigen for polyclonal antibodies production. For the first time putative CP coding sequence was identified in silico in BSMYV polyprotein (which is first report on identification of putative CP in any of badnaviruses) (Sharma et al. 2014a). The identified putative CP p37 (37 kDa protein) was in vitro expressed in *E. coli* and purified in large quantity (Sharma et al. 2014a). Purified in vitro expressed CP was used for production of high titre polyclonal antiserum in rabbit (Sharma et al. 2014a). Developed antiserum was successfully used for specific detection of BSMYV in field samples at an effective dilution of 1:1000 or 1:2000 in antigen-coated plate-enzyme linked immunosorbent assay (ACP-ELISA). Antiserum was also used in ISEM and Western blotting. Using the developed antiserum a duplex-immunocapture-PCR (D-IC-PCR) was standardized for routine indexing of field and tissue cultured plantlets (Sharma et al. 2014b). When different detection methods were compared, direct-PCR and direct binding-PCR (DB-PCR) were not found to yield confirmatory and accurate results for the presence of episomal BSV infection. However, D-IC-PCR employing *Musa* genome specific internal primers and BSMYV specific primers with an immunocapture time of 3 h was found to be a reliable robust method for detection of episomal BSMYV. Tissue cultured banana plantlets has gained a popularity in India and at present taround more than 300 million tissue culture raised banana plants are planted every year in different parts of India. To get optimum production from tissue cultured banana plants, they are necessarily required to be free from viruses and are to be tested as per the standards of National Certification System for Tissue Culture Raised Plants (NCS-TCP), Department of Biotechnology, Government of India. Due to the interference of *Musa* genome integrated endogenous BSV sequences (eBSV) and low virus titre, indexing for BSV has to be followed very strictly. Thus a system of indexing was proposed to be carried out at least at two stages (i) indexing of mother plant or during initial culture establishment and (ii) during primary or ideally at secondary hardening stage. This will reduce risk of spread of BSV from infected tissue culture planting materials under field conditions (Sharma et al. 2014b). Improved random primed-RCA was also developed for high amplification of badnaviruses genome (Sharma et al. 2015). DNA isolated from symptomatic banana leaves collected during lower temperature regimes (winter or rainy season) containing high concentration of replicative form of viral DNA (sealed minichromosomal DNA) when used in RCA followed by another addition of phi29 polymerase at the mid of incubation time, leads to higher *Badnavirus* genome specific amplification (Sharma et al. 2015). Further, sequence independent random primed-RCA coupled with restriction fragment length polymorphism (RCA-RFLP) could also be used for reliable detection of episomal banana streak viruses.

Earlier, Selvarajan et al. (2011) developed a multiplex-reverse transcription-PCR (mRT-PCR) for the simultaneous detection of BSMYV and banana bunchy top virus (BBTV) in same sample. The first strand cDNA was synthesized using random primers followed by the mRT-PCR using primers specific to the BSMYV (targeting RT/RNase H region), BBTV (partial master rep) and Musa genome specific internal sequence tagged microsatellite site (STMS) primers. Developed mRT-PCR was successfully used for indexing of field and tissue culture plants.

4.2.1.4 Subpopulation Variation, Genetic Diversity and Sequence Analysis

Phylogenetically badnaviruses infecting banana and sugarcane which are the closely related groups of badnaviruses infecting monocots, forms two major clusters (Sharma et al. 2015) (Fig. 4.4). Indian BSMYV isolates along with the BSMYV from Australia were in cluster 1. The other members of cluster 1 were badnaviruses infecting banana and sugarcane infecting badnaviruses from Guadeloupe. Phylogenetically BSMYV isolates are more closely related to sugarcane bacilliform GD virus (SCBGDV) compared to other badnaviruses. Shorter BSOLV variants (BSOLV-IN1 and BSOLV-IN2) were also in cluster 1. However, Indian BSV isolates further formed two subclusters, with subcluster I comprising of BSOLV-IN1 and BSOLV-IN2, while subcluster III comprising of BSMYV-IN1, BSMYV-IN2 and BSMYV-IN3 (Fig. 4.4).

RCA coupled with RFLP (RCA/RFLP) successfully demonstrated subpopulation level variation in BSMYV isolates infecting different natural banana hybrids. Out of the ten restriction enzymes tested, eight successfully yielded informative restriction patterns for the five BSMYV isolates (Sharma et al. 2015). Thus genetic variability exists within in species also. Estimated rate of evolution in case of BSMYV subpopulation was highest for ORF2, ORF3 and IGR and least for ORF1, whereas in BSOLV subpopulation the highest evolution rate was observed for IGR and lowest for ORF2 (Sharma et al. 2015). Tajima's D statistical showed negative D values for all three ORFs and IGR for episomal BSMYV subpopulation set, thus presenting a continuum of genetic variability at intraspecies or subpopulation level. Although at overall sequence level low genetic polymorphism was observed for Indian BSMYV subpopulation, the RCA/RFLP could reliably detect sequence variants.

4.2.2 Sugarcane Freckling Disease

The other related groups of badnaviruses which also infect a monocotyledonous crop are sugarcane bacilliform viruses (SCBV) which causes the freckling disease of sugarcane, are reported to infect a number of species of sugarcane (*Saccharum officinarum*, *S. barberi*, *S. sinense*, *S. robustum* and *Saccharum* interspecific hybrids). Like that of banana infecting badnaviruses, bacilliform viruses infecting sugarcane are collectively known as sugarcane bacilliform viruses (SCBV). Worldwide, two recognized species; *Sugarcane bacilliform IM virus* (SCBIMV), *Sugarcane bacilliform MO virus* (SCBMOV) and more than six tentative species are associated with leaf freckle disease of sugarcane (Muller et al. 2011; King et al. 2012; Karuppaiah et al. 2013). In India, during the year 1992, SCBV infection was suspected in *Saccharum officinarum* genotypes (Black Tanna, Listada and D1135) (Karuppaiah et al. 2013). However, first confirmatory report based on EM and ELISA came later (Viswanathan et al. 1996). Since then although the disease was observed on many germplasm/breeding lines, very few detailed studies had been taken. SCBV generally induce symptoms of chlorotic specks which eventually turns to stripes with mild mottling (Fig. 4.1e–g), however very specific symptoms are not always induced and expression of symptoms is largely determined by sugarcane cultivar and prevailing environmental conditions (Rao et al. 2014). Symptoms ranging from yellowish to chlorotic freckling, chlorosis and interveinal chlorosis have been recorded on different commercial sugarcane cultivars grown in India (Viswanathan et al. 1999; Rao et al. 2002; Rao et al. 2014). Various experiments on characterization have shown that virus concentration is generally very low in the sugarcane tissues and even goes below the detection limits during hot summers (Rao et al. unpublished data). Many a times SCBV infection remains symptomless and there are reports of >90% incidence on noble cane (*S. officinarum*) germplasm under field conditions (Viswanathan et al. 1999).

4.2.2.1 Geographical Distribution

SCBV infection was also found in *S. barberi*, *S. robustrum*, *S. spontaneum*, *S. sinense*, *Saccharum* hybrids and other commonly cultivated cultivars (Viswanathan et al. 1996; Singh et al. 2003; Rao et al. 2014). In survey conducted during 2006–2009, all the 13 symptomatic and nonsymptomatic samples of sugarcane varieties grown in Uttar Pradesh, Andhra Pradesh, Tamil Nadu, Madhya Pradesh, Kerala, Bihar and Haryana states were detected positive for SCBV in PCR (amplifying 580 bp RT/RNase H region of genome) (Singh et al. 2009). Mixed infection of SCBV along with other viruses was reported in the samples from all these states (Singh et al. 2009). In an extensive survey for SCBV infecting different sugarcane cultivars grown in different states of India (Assam, Bihar, Tamil Nadu, Haryana, Kerala, Andhra Pradesh, Madhya Pradesh and Uttar Pradesh) all the 28 varieties were tested positive in EM and PCR (Rao et al. 2014) (Fig. 4.2).

4.2.2.2 Transmission and Spread

Sugarcane infecting badnaviruses are transmitted by mealybugs. In India virus has spread in major sugarcane growing pockets through the infected planting materials (Viswanathan and Rao 2011).

4.2.2.3 Virus Characterization

RT/RNase H nucleotide sequence analysis of eight representative isolates originating from five Indian states exhibited sequence diversity up to 27% (Rao et al. 2014). Five isolates (from Assam, Bihar and Uttar Pradesh) shared maximum sequence identity of 86–88% with SCBIMV. Whereas other three isolates (from Kerala, Tamil Nadu and Uttar Pradesh) shared maximum identity of 75–79% with SBMOV for RT/RNase H region, which is below the ICTV species demarcation criteria of 80% identity (King et al. 2012). This indicated the existence of high genetic diversity in SCBV population under field conditions in India. Phylogenetically these eight isolates segregated into three subclusters, out of which two subclades were novel. Karuppaiah et al. (2013) reported full genome sequences of genetically diverse SCBV isolates. A full genome of SCBV-B091 (infecting commercial cultivar BO91, Coimbatore) and other four isolates, SCBV-BB, SCBV-BT, SCBV-BRU and SCBV-Iscam originated from *S. officinarum* germplasm Boetatoe Bilatoe, Black Tannna, Black Reunion and Iscambine respectively maintained at sugarcane germplasm collection, Kannur, Kerala. Genome length of these SCBV isolates ranged from 7553 to 7884 bp (Karuppaiah et al. 2013) (Fig. 4.3). These Indian SCBV isolates shared identities of 69–85% for full genome sequences. Phylogenetically these isolates formed four new groups. Based on ICTV species demarcation criteria, SCBV-BT and SCBV-Iscam fell in SCBIMV and SCBMOV genotypes respectively, however SCBV-BB, SCBV-BO and SCBV-BRU represent novel *Badnavirus* species (Karuppaiah et al. 2013). Based on the sequence analysis for different genomic regions, isolate SCBV-BB was most distinct. SCBV-BRU has the largest genome size (7884 bp) among all sugarcane infecting badnaviruses and has probably evolved recently. Since earlier reported primers for SCBV were not able to detect these newly reported isolates from India, a new set of primers were designed which could detect all known SCBV species/variants prevalent in Indian subcontinent (Karuppaiah et al. 2013).

Detection of SCBV was initially done based on electron microscopy and ELISA (Viswanathan et al. 1996; Singh et al. 2003). PCR based detection was then standardized for screening of sugarcane varieties and germplasm (Singh et al. 2009; Rao et al. 2014). Generally primers to RT/RNase H region of genome or core RT/RNase H region were used in PCR based detection (Singh et al. 2009; Karuppaiah et al. 2013 Rao et al. 2014). Since SCBV is not known to occur as integrated form in the host genome, hence PCR is considered as standard indexing procedure. Recently Sharma et al. (2015) reported that RCA concentrated DNA can be used as a template in PCR for sensitive detection of SCBV in field samples.

4.2.2.4 Recombination and Evolutionary Analysis

Recombination is expected to be a frequent phenomenon in badnaviruses genome. RT/RNase H region of badnavirus genome is conserved and recombination cold spot. A diverse field isolate from Uttar Pradesh (SCBV-UP, CoSe92423) was identified to be a recombinant for RT/RNase H region where two other Indian SCBV isolates had contributed sequences as donors (Rao et al. 2014). This was the first report worldwide on evolution of genetically diverse variant of SCBV through recombination. Studies have indicated the relatedness of badnaviruses infecting banana and sugarcane and it is proposed that badnaviruses has made a host shift among these two plant species. Sharma et al. (2015) analyzed the sequence sets of BSV and SCBV to identify intra-BSV, intra-SCBV and inter BSV-SCBV recombination events. A total of 26 intra-BSV and 22 intra-SCBV recombination events were identified with almost all genomes of these badnaviruses sampled till date to have contained recombination traces. Thirty-two unique inter BSV-SCBV recombination events were detected which has emerged due to the sequence fragment exchange among these two groups of badnaviruses. Recombination hot spots are concentrated in IGR (3′ end after ORF3 and 5′ end after +1 start site) followed by ORF1, whereas, the ORF3 genomic region were recombination cold-spot in badnavirus genome (Sharma et al. 2015). Distribution of recombination traces among BSV and SCBV genomes and sequence exchange among these two groups of badnaviruses further supported their relatedness and possible host shift.

4.2.3 Citrus Yellow Mosaic Disease

Mosaic disease of citrus was first described in southern part of India by Dakshinamurti and Reddy (1975). The disease and associated virus was characterized much letter in detail (Ahlawat et al. 1984; Ahlawat et al. 1996a, b). Disease was reported to be widely prevalent on citrus species like sweet orange (*Citrus sinensis*) and pummelo (*Citrus grandis*) (Ahlawat et al. 1984; Reddy and Murti 1985). In Andhra Pradesh area, an incidence of 10–70% was recorded on Satgudi sweet orange (Ahlawat et al. 1996a). Virus infection led to the reduction in fruit yield to an extent of 77% in 10 years old citrus trees and fruits obtained from infected trees had 10% less juice and ascorbic acid (Reddy and Murti 1985). Virus infection leads to characteristic symptoms of yellow mosaic in leaves, yellow flecking along the veins and vein-banding (Ahlawat et al. 1996a) (Fig. 4.1h, i). Disease was graft transmitted and association of bacilliform virus was reported (Ahlawat et al. 1996a, b). Out of the 10 *Citrus decumana* plants sap inoculated with virus, only one plant was infected and observed to have bacilliform particles under EM 3 months post-inoculation. Virus was graft and dodder transmitted to *C. limonia*, *C. volkameriana*, *C. jambhiri*, *C. sinensis* cvs. Satgudi, Mosambi, Chini, *C. reticulata*, *C. limettoides*, *C. grandis*, *C. paradisi* cv. Duncan, *C. medica*, *C. aurantium*, *C. mitis* etc. but not transmitted to *C. aurantifolia* (Ahlawat et al. 1996a). Using badnavirus-specific primers, genomic fragment of associated virus was first PCR amplified and based on its serological relationship, the virus was first proposed as a member of the genus

Badnavirus by Ahlawat et al. (1996a). Subsequently the disease was reported from acid lime in Andhra Pradesh (Sai Gopal et al. 2000).

Leaves of Rangpur lime root stocks in Tirupati showing yellow mosaic symptoms showed bacilliform virus particles under EM (Baranwal et al. 2005a). Sequences of a fragment of intergenic region and ORF3 region of viral genome from CYMV infected Rangpur lime showed high identity to that of Indian CYMV isolate infecting sweet orange. CYMV incidence of 7–20% on sweet orange in different districts of Andhra Pradesh was recoded (Gopi et al. 2010).

In different citrus growing belts of Andhra Pradesh, CYMV incidence of 14–17% on sweet orange and 2–11% on acid lime was reported (Gaddam et al. 2012a).

4.2.3.1 Geographical Distribution

CYMV is widely prevalent in sweet orange orchards in South India. High disease incidence has been reported in different districts of Andhra Pradesh, Karnataka (Reddy and Murti 1985) (Fig. 4.2). Disease has also been reported from Maharashtra (Nagpur) (Ghosh et al. 2014).

4.2.3.2 Transmission and Spread

CYMV was transmitted by citrus mealy bug (*Planococcus citri*) and mechanical inoculation (Pant and Ahlawat 1997) under experimental conditions (Reddy 1997). CYMV is graft transmissible and mealybug (*Planococcus citri*) acts as its natural vector (Reddy and Ahlawat 1997).

4.2.3.3 Virus Characterization

First full genome sequence of a CYMV isolate from naturally infected sweet orange in Andhra Pradesh which was maintained on Madame Vinous sweet orange by graft inoculation at USA was reported by Huang and Hartung (2001). Genome length was 7559 bp and encodes for six putative ORFs, unlike the three ORFs in other badnaviruses (Fig. 4.3). They also constructed 1.4-genome length partial dimer of this CYMV isolate which was proved infectious on sweet orange by Agrobacterium-mediated inoculation. Five months-post inoculation inoculated sweet orange plants showed yellow mosaic symptoms, observed to have bacilliform particles under EM and detected CYMV positive in PCR (Huang and Hartung 2001). Subsequently a full genome sequence of CYMV from Rangpur lime (DQ875213: 7522 bp) was reported. Borah et al. (2009) reported the full genome sequences of two CYMV isolates from naturally infected acid lime (AL isolate) and pummelo (PM isolate) which were 7473 and 7487 nucleotides long respectively. Although isolated from different citrus species, these CYMV isolates showed high sequence similarity with each other for all the genomic regions. Four full genome sequences of CYMV isolates infecting sweet orange (two isolates), acid lime (one isolate) and Rangpur lime (one isolate) collected from Andhra Pradesh were reported (Gupta et al. 2009), genomes of which varied from 7497 to 7558 bp and containing deletions/additions in ORF3 and IGR of viral genome. Later, Anthony Johnson et al. (2012) reported full genome sequences of two more isolates infecting rough lemon (*Citrus jambhiri*) and nursery plant of sweet orange in different areas of Andhra Pradesh. These isolates contained insertions and deletions at many place in the genome. Phylogenetic analysis based on complete genomes showed two clusters of CYMV isolates reported from India. CYMV isolates infecting sweet orange were in one group whereas those infecting other citrus species like acid lime, pumello and rough lemon were in another group (Anthony Johnson et al. 2012). Recently two more CYMV isolates collected from field-infected mandarin and mosambi sweet orange plants from Nagpur were characterized based on the sequences of ORF3 fragment and IGR (Ghosh et al. 2014). These isolates were graft-transmitted to healthy seedlings of Nagpur mandarin, mosambi sweet orange, Rangpur lime, and acid lime. Phylogenetic analysis of all reported CYMV isolates indicated the existence of two strains in India, one of those which infect acid lime and Rangpur lime and other of those which infects other citrus species (Ghosh et al. 2014).

Initially polyclonal antisera against purified virus preparation of CYMV were developed by Pant and Ahlawat (1997). A dot-blot hybridization based detection procedure for CYMV was first developed by Reddy and Ahlawat (1997). CYMV showed serological relatedness to SCBV (Ahlawat et al. 1996a; Pant and Ahlawat 1997). However, CYMV is weakly immunogenic (Baranwal et al. 2003) and therefore PCR based detection has been widely employed for its indexing. Presence of high levels of polyphenolics and tannins in the leaves of different citrus species leads to oxidation of nucleic acids on the release of sap from cells and thus its degradation. This nucleic acid therefore does not lead to confirmatory results in PCR. Baranwal et al. (2003) developed an improved sodium sulphite based DNA extraction procedure which could lead to improved yield, quality and stability of isolated DNA as compared to normal DNA extraction procedure and kit. Even at lower concentration, CYMV was amplified from the DNA isolated through sodium sulphite based procedure.

A multiplex-PCR based detection was developed to amplify both CYMV (fragment of RT/RNase H region of genome) and citrus greening bacterium *Candidatus* liberibacter asiaticus; CLa (Baranwal et al. 2005b). DNA template from midrib of citrus leaf samples which are infected with CYMV and greening bacterium could be used as common target for the DNA isolation to be used for simultaneous detection of both CLa and CYMV (Baranwal et al. 2005b). PCR was employed for detection of CYMV infection in Vidarbha region of Maharashtra (Ghosh et al. 2007). Ghosh et al. (2008) developed a sensitive duplex-PCR technique for the simultaneous detection of CYMV (targeting fragment of ORF3) and citrus tristeza virus (CTV) (targeting fragment of ORF8) in infected samples. The duplex-PCR could detect the CYMV and CTV nucleic acid up to 10^{-5} and 10^{-6} dilutions. Four methods of DNA isolation were further compared by Borah et al. (2008) to determine their suitability for PCR and dot-blot hybridization based detection of CYMV in acid lime and pummelo. Dot-blot hybridization was not influenced by the method of DNA extraction and irrespective of the method followed; similar results were obtained for dot-blot hybridization. However, PCR based detection was much influenced by the method of DNA isolation and sarkosyl method was reported to be the most suitable method when DNA is to be used for PCR detection of CYMV (Borah et al. 2008). Thus method of DNA isolation largely influences the reliability of PCR based detection of CYMV. CYMV specific DNA was amplified from leaf, twig bark, fruit-rind and fruit juice from the infected sweet orange plant, however no amplification was observed in fruit rag (Gopi et al. 2010). Polyclonal antisera raised against rangpur strain of CYMV showed positive reaction with CYMV infecting rangpur lime, acid lime and sweet orange in agarose double diffusion test and dot blot ELISA (Gaddam et al. 2012b).

Recently for more sensitive, robust and specific detection of CYMV in infected samples, loop-mediated isothermal amplification (LAMP) and SYBR green real-time PCR (SGRTPCR) based detection techniques were developed (Anthony Johnson et al. 2014). In these methods primers targeting putative coat protein domain in ORF3 were used. Both the protocols were successfully used for detection of CYMV in rough lemon (*Citrus jambhiri*), Nagpur Mandarin (*Citrus reticulata*), Pumello (*C. grandis*), sweet orange (*C. sinensis*) and acid lime (*C. aurantifolia*) samples. PCR and LAMP were found equally sensitive in detection of CYMV, capable of detecting in 10 ng of total DNA. However, SGRTPCR assay was ten-fold more sensitive and were able to detect virus in 1 ng of total DNA (Anthony Johnson et al. 2014). Using the sensitive SGRTPCR assay, CYMV titres were compared in different citrus species. The highest virus titre was detected in rough lemon and lowest in Nagpur mandarin with 7.98×10^{10} and 5.55×10^{7} of viral load in dsDNA copies/10 ng of total DNA respectively.

4.2.4 Pepper Yellow Mottle Disease

Piper yellow mottle virus (PYMoV) a species under the genus *Badnavirus* which causes yellow mottle disease of pepper has been reported first time from Kerala by Bhat et al. (2003) from black pepper (*Piper nigrum*). During 2002, black pepper plantations in Kozhikode and Wyanad districts of Kerala were reported to have 100% incidence (Bhat et al. 2003).

The symptoms of disease are characterized by the chlorotic mottling, vein clearing, leaf distortion, reduced plant vigor, short internodes, stunted vines and poor fruit setting (Bhat et al. 2003; Hareesh and Bhat 2008). Some cultivars exhibit vein banding, vein thickening and green island-like symptoms (Fig. 4.1j). Virus was graft transmitted from diseased to healthy black pepper and showed serological relatedness to banana streak viruses and sugarcane bacilliform viruses in DAC-ELISA (Bhat et al. 2003). Bacilliform virus particles were found associated with the disease under EM (Bhat et al. 2003).

4.2.4.1 Geographical Distribution

Most of the reports on occurrence of PYMoV are from Kerala. However, possibly virus occurs in other parts also where black pepper, Indian long pepper and other susceptible host are being cultivated (Fig. 4.2). Infection of PYMoV from betel vine (*Piper betle*) and Indian long pepper (*P. longum*) was first reported from Kerala (Siju et al. 2008). Sequences of ORF3 fragments of betel vine and long pepper isolates shared nucleotide identity of 89% with that of black pepper isolate. In some cultivars PYMoV infection does not exhibits typical symptoms. Symptomless infection of PYMoV was reported on black pepper cultivars (Panniyur 1, Panniyur 5 and Panchami) and on a wild species of piper (*Piper colubrinum*) (Bhat et al. 2012).

4.2.4.2 Transmission and Spread

The transmission of PYMoV is primarily through the vegetative means (stem cuttings) and secondarily through different mealybug species (Hareesh and Bhat 2008). Mealybug (*Ferrisia virgata*) acts as vector of PYMoV (Bhat et al. 2003). PYMoV was also shown to be seed transmitted in black pepper (Hareesh and Bhat 2010). Virus was demonstrated to be transmitted from these symptomless cultivars to other healthy cultivars by mealybug and graft transmission.

4.2.4.3 Virus Characterization

Partial sequences of ORF1 and ORF3 of three PYMoV isolates from black pepper were reported (Hareesh and Bhat 2008). These isolates shared a high sequence identity (95%) among themselves. The first complete genome sequence of PYMoV from the black pepper plant (cv. Karimunda from Calicut, Kerala, India) has been recently identified using a combination of deep sequencing and Sanger sequencing method by Hany et al. (2013). The genome of PYMoV is 7.6 kb, putatively encoding four ORFs (Fig. 4.3). Phylogenetically PYMoV was closely related to cacao swollen shoot virus (CSSV), citrus yellow mosaic virus (CYMV) and dioscorea bacilliform viruses (DBV) (Hany et al. 2013).

The genome sequences of three other isolates infecting black pepper, betelvine and Indian long pepper which shared sequence similarity of 89–99% with the other PYMoV isolate were also reported (Deeshma and Bhat 2015). Phylogenetic analysis based on ORF3 polyprotein indicated that PYMoV isolates segregated to a separate cluster which was distinct from other known badnaviruses and fig badnavirus 1 (FBV-1) was the nearest relative to this cluster (Deeshma and Bhat 2015).

Hareesh and Bhat (2008) reported a protocol for DNA isolation from black pepper and PCR detection of PYMoV. RT/RNase H based degenerate primers were used for PCR detection of PYMoV infecting betel vine and Indian long pepper (Hareesh and Bhat 2008). Bhat and Siju (2007) reported a modified method of total nucleic acid (DNA and RNA) extraction from black pepper. A multiplex-RT-PCR detection assay was developed for simultaneous detection of PYMoV (primers for partial ORF1) and cucumber mosaic virus (primers for CP) in nursery and field samples of black pepper (Bhat and Siju 2007). DAS-ELISA was used for large scale indexing of black pepper plants for PYMoV (Bhadramurthy et al. 2005). During winter months (October to February) high concentration of virus was detected. Within a plant, young leaves contain high concentration of PYMoV whereas root region contain the least.

A PCR detection assay was developed for indexing of black pepper planting material for PYMoV (Bhat et al. 2009). Primers for ORF3 genomic fragment were used. This PCR assay was used for indexing of 845 plants of black pepper (14 cultivars) and 82% of which were tested positive for PYMoV. At the time of indexing most of the plants were not exhibiting any external symptoms however, after 1–3 months some positive plants were found to exhibit symptoms of PYMoV, this indicated the suitability of developed PCR assay in routine indexing of black pepper plants even when they are symptomless carrier.

Real-time PCR was developed for sensitive detection of PYMoV in black pepper plants (Bhat and Siljo 2014). Primers for real-time assay were designed based on the conserved ORF3 genomic fragment (600 bp) of different PYMoV isolates. This detection assay was 1000 times more sensitive than the conventional PCR. Real-time PCR could be a method of choice for indexing of mother plants.

Recently more sensitive loop-mediated isothermal amplification (LAMP) assay was developed for the detection of PYMoV (Bhat et al. 2013). In developed LAMP assay, positive reaction could be detected by turbidity, green fluorescence, formation of pellet in the reaction tube and by gel electrophoresis. This assay was 100 times more sensitive compared to conventional PCR.

4.3 Badnaviruses Infecting Bougainvillea

There is a single report on characterization of badnaviruses associated with *Bougainvillea spectabilis* from India. Severe yellow mosaic symptoms were observed on *B. spectabilis* in Tirupati during 2008. Subsequently *B. spectabilis* plants at ICAR-Indian Agricultural Research Institute, New Delhi campus were observed to exhibit chlorotic vein-banding symptoms during 2009 (Figs. 4.1k, l and 4.2). In leaf dip EM, bacilliform virus particles measuring 120–150 × 20 nm were observed (Baranwal et al. 2010). RT/RNase H genomic region of both the isolates was PCR amplified and sequenced. These two isolates shared 76.2% nucleotide and 88.6% amino acid identity with each other and 72.3–73.5% and 83.4–84.9% nucleotide and amino acid identities with the *Bougainvillea chlorotic vein banding virus* (BCVBV) (species name as recognized in the ratification vote on taxonomic proposals to ninth ICTV report: Adams and Carstens 2012), which was earlier described as bougainvillea spectabilis chlorotic vein banding virus isolates from Taiwan. Since these isolates showed less than 80% nucleotide sequence identities with characterized badnaviruses, which is below the ICTV species demarcation criteria for badnaviruses, hence these represent new tentative species (Baranwal et al. 2010). Phylogenetically these isolates formed a separate cluster which was near to the BCVBV isolates from Taiwan.

4.4 Concluding Remarks

Very less targeted efforts on the management of badnaviruses infecting different crops in India has been undertaken. Mealybugs although are vectors of badnaviruses, however their role in field spread is not very significant. All the five badnaviruses reported from India are known to infect the crops which are by and large propagated through vegetative means. In addition, PYMoV was reported to be seed transmitted in black pepper. Hence the management tactics for these badnaviruses is through the use of virus free planting materials. Through NCS-TCP certification system, the spread of BSV in case of tissue cultured plants has largely been managed under Indian conditions. However, these viruses still spread through transportation of infected rhizomes/corms and other propagating materials if not properly indexed. SCBV since does not exhibits typical symptoms in many sugarcane clones, and has now become widely prevalent in all the sugarcane growing belts of India. Similar is the situation with PYMoV and badnaviruses infecting bougainvillea. CYMV has emerged as one of the major constraint to citriculture industry in Southern part of India. With the efficient and robust diagnostic tools available for BSV, CYMV and PYMoV indexing of mother plants, vegetatively propagated planting materials are to be strictly followed which can help in managing the diseases caused by badnaviruses.

Of the five badnaviruses known to occur in India, BSVs, SCBVs, CYMV and PYMoV are widely prevalent and becoming increasingly important. The diversity and extent of losses in crop production makes them emerging problem under Indian conditions. Streak disease of banana with which more than 12 genetically heterogenous badnaviruses are reported to be associated worldwide, need to be further investigated with large number of samples originating from different banana species/cultivars in different banana growing pockets including North-East India (where *Musa balbisiana* is considered to have originated) so as to have comprehensive profile of all BSV species occurring in India. Only two species, BSMYV and BSOLV are encountered in the samples tested till date, however possibility of existence of other BSV species cannot be ruled out. Once the extent of genetic diversity of all BSV species is known, potential introduction of other BSV species which are not present in India needs to be monitored. For sugarcane bacilliform viruses maximum number of virus species and genetic heterogeneity occurs in India. Since many of these species are found associated with the germplasm collection, possibility of their spread to fields through crop improvement programmes and other means need to be checked. Moreover there is a need to develop simple, robust detection procedures for routine indexing of these viruses in sugarcane germplasm, newly developed cultivars and field samples. In case of CYMV efforts are to be strengthened to check its spread to the areas where it has not entered yet. Extent of genetic diversity of PYMoV need to studied and more number of full genome sequences originating from different species/cultivars need to be completed. Bougainvillea vines in different parts of India are found to exhibit the symptoms of badnaviruses infection, hence extensive surveys need to be carried out to characterize the associated viruses and their genetic and biological diversity.

There are many other crops like pineapple, grapevine, taro, yam etc. which are grown in India and badnaviruses are reported on these crops in different parts of world. Sampling of these crops for possible infection of badnaviruses are to be undertaken and if they are free from infection, strict quarantine is to be followed to prevent any possibility of their introduction to India.

Badnaviruses are not mechanically transmitted. Non-availability of infectious clones in case of most of badnaviruses makes it difficult to understand the possible role of ORFs, pathogenicity and host-virus interactions. Hence infectious clones of badnaviruses occurring in India need to be made and are to be used to study host-virus interactions and possible resistance/tolerance mechanisms. Role of different proteins encoded by badnavirus genomes, exact location of functional proteins in ORF3 polyprotein, mechanism of proteolytic cleavage of polyprotein and function of extra ORFs (particularly in case of CYMV and PYMoV) are to be a part of active research in future.

Badnaviruses are genetically highly heterogeneous. Possible mechanisms of such a high variability need to be elucidated. Based on the pathogenicity analysis and biological consequences of such a high genetic variability, we have to come to a stable taxonomy of these viruses. Keeping in view the genetic heterogeneity and existence of integrated counterparts in case of some badnaviruses (particularly in case of BSVs), robust and simplified diagnostics need to be developed for routine indexing of mother plants and planting materials before supply to the farmers. This will help in devising durable molecular management strategy for badnaviruses present in India.

Acknowledgements Authors are grateful to all badnavirus workers in India for sharing the information. We are grateful to Dr. G.P. Rao, ICAR-IARI and Dr. A.I. Bhat, ICAR-IISR for sharing the diseased photographs of SCBV and PYMoV, respectively.

References

Adam MJ, Carstens EB (2012) Ratification vote on taxonomic proposals to the international committee on taxonomy of viruses. Arch Virol 157:1411–1422

Adams MJ, Lefkowitz EJ, King AMQ, Bamford DH, Breitbart M, Davison AJ, Ghabrial SA, Gorbalenya AE, Knowles NJ, Krell P, Lavigne R, Prangishvili D, Sanfacon H, Siddell SG, Simmonds P, Carstens EB (2015) Ratification vote on taxonomic proposals to the international committee on taxonomy of viruses (2015). Arch Virol 160:1837–1850

Ahlawat YS, Chenulu VV, Vishwanath SM, Pandey PK (1984) Studies on a mosaic disease of citrus in India. Curr Sci 54:873–874

Ahlawat YS, Pant RP, Lockhart BEL, Srivastava M, Chakraborty NK, Varma A (1996a) Association of *Badnavirus* with citrus mosaic disease in India. Plant Dis 80:590–592

Ahlawat YS, Varma A, Pant RP, Shukla A, Lockhart BEL (1996b) Partial characterization of a *Badnavirus* associated with citrus yellow mosaic disease in India, 13th Conference of the International Organization of Citrus Virologists. University of California Press, California, pp 208–217

Anthony Johnson AM, Borah BK, Sai Gopal DVR, Dasgupta I (2012) Analysis of full-length sequences of two Citrus yellow mosaic badnavirus isolates infecting *Citrus jambhiri* (Rough Lemon) and *Citrus sinensis* L. Osbeck (Sweet Orange) from a nursery in India. Virus Genes. doi:10.1007/s11262-012-0808-8

Anthony Johnson AM, Dasgupta I, Sai Gopal DVR (2014) Development of loop-mediated isothermal amplification and SYBR green real-time PCR methods for the detection of citrus yellow mosaic badnavirus in citrus species. J Virol Methods 203:9–14

Baranwal VK, Majumdar S, Ahlawat YS, Singh RP (2003) Sodium sulphite yields improved DNA of higher stability for PCR detection of citrus yellow mosaic virus from citrus leaves. J Virol Methods 112:153–156

Baranwal VK, Singh J, Ahlawat YS, Gopal K, Charaya MU (2005a) *Citrus yellow mosaic virus* is associated with mosaic disease in Rangpur lime rootstock of citrus. Curr Sci 89:1596–1599

Baranwal VK, Majumder S, Ahlawat YS, Singh RP (2005b) A novel approach for simultaneous detection of *Citrus yellow mosaic virus* and citrus greening bacterium by multiplex polymerase chain reaction. Indian J Biotechnol 4:528–533

Baranwal VK, Arya M, Singh J (2010) First report of two distinct badnaviruses associated with *Bougainvillea spectabilis* in India. J Gen Plant Pathol 76:236–239

Baranwal VK, Sharma SK, Khurana D, Verma R (2014) Sequence analysis of shorter than genome length episomal *Banana streak OL virus* like sequences isolated from banana in India. Virus Genes 48:120–127

Bhadramurthy V, Retheesh ST, Bhat AI, Madhubala R, Hareesh PS, Pant RP (2005) Development of ELISA-based technique for the detection of a putative *Badnavirus* infecting black pepper (*Piper nigrum* L.) Indian Phytopath 58:314–318

Bhat AI, Siju S (2007) Development of a single tube multiplex RT-PCR for the simultaneous detection of *Cucumber mosaic virus* and *Piper yellow mottle virus* associated with stunt disease of black pepper. Curr Sci 93:973–976

Bhat AI, Siljo A (2014) Detection of viruses infecting black pepper by SYBR green-based real time PCR assay. J Plant Pathol 96:105–109

Bhat AI, Devasahayam S, Sharma YR, Pant RP (2003) Association of a *Badnavirus* in black pepper (*Piper nigrum* L.) transmitted by mealybug (*Ferrisia virgata*) in India. Curr Sci 84:1547–1550

Bhat AI, Siljo A, Jiby MV, Thankamani CK, Mathew PA (2009) Polymerase chain reaction (PCR) based indexing of black pepper (*Piper nigrum* L.) plants against *Piper yellow mottle virus*. J Spices Aromat Crop 18:28–32

Bhat AI, Siljo A, Devasahayam S (2012) Occurrence of symptomless source of *Piper yellow mottle virus* in black pepper (*Piper nigrum* L.) varieties and a wild piper species. Arch Phytopathol Plant Prot 45:1000–1009

Bhat AI, Siljo A, Deeshma KP (2013) Rapid detection of *Piper yellow mottle virus* and *Cucumber mosaic virus* infecting black pepper (*Piper nigrum*) by loop-mediated isothermal amplification (LAMP). J Virol Methods 193:190–196

Bhat AI, Hohn T, Selvarajan R (2016) Badnaviruses: the current global scenario. Virus 8:177. doi:10.3390/v8060177

Boeke JD, Corces VG (1989) Transcription and reverse transcription of retrotransposons. Annu Rev Microbiol 43:403–434

Borah BK, Johnson AMA, Sai Gopal DVR, Dasgupta I (2008) A comparison of four DNA extraction methods for the detection of citrus yellow mosaic badnavirus from two species of citrus using PCR and dot-blot hybridization. J Virol Methods 151:321–324

Borah BK, Johnson AMA, Sai Gopal DVR, Dasgupta I (2009) Cloning and computational analysis of complete genome sequences of citrus yellow mosaic badnavirus (CMBV) from acid lime and pummelo. Virus Genes 39:137–140

Borah BK, Sharma S, Kant R, Johnson AMA, Sai Gopal DVR, Dasgupta I (2013) Bacilliform DNA-containing plant viruses in the tropics: commonalities within a genetically diverse group. Mol Plant Pathol 14:759–771

Chabannes M, Iskra-Caruana M-L (2013) Endogenous pararetroviruses-a reservoir of virus infection in plants. Curr Opin Virol 3:615–620

Chabannes M, Baurens FC, Duroy P-O, Bocs S, Vernerey M-S, Rodier-Goud M, Barbe V, Gayral P, Iskra-Caruana M-L (2013) Three infectious viral species lying in wait in the banana genome. J Virol 87:8624–8637

Cherian AK, Baranwal VK, Malathi VG, Pant RP, Ahlawat YS (2004) Banana streak virus from India and its detection by polymerase chain reaction. Indian J Biotechnol 3:409–413

Dahal G, Hugues J'A, Thottapilly G, Lockhart BEL (1998) Effect of temperature on symptom expression and reliability of banana streak badnavirus detection from naturally-infected plantain and banana (*Musa* spp.) Plant Dis 82:16–21

Dahal G, Ortiz R, Tenkouano A, Hughes JA, Thottappilly G, Vuylsteke D, Lockhart BEL (2000) Relationship between natural occurrence of banana streak badnavirus and symptom expression, relative concentration of viral antigen, and yield characteristics of some micropropagated *Musa* spp. Plant Pathol 49:68–79

Dakshinamurti V, Reddy GS (1975) Mosaic: a transmissible disorder of sweet oranges. Indian Phytopath 28:398–399

Daniells JW, Geering ADW, Bryde NJ, Thomas JE (2001) The effect of banana streak virus on the growth and yield of dessert bananas in tropical Australia. Ann Appl Biol 139:51–60

Deeshma KP, Bhat AI (2015) Complete genome sequencing of *Piper yellow mottle virus* infecting black pepper, betelvine, and Indian long pepper. Virus Genes 50:172–175

Dekmann M, Putter CAJ (1996) FAO/IPGRI technical guidelines for the safe movement of germplasm. Food and Agriculture Organization of the United nations/International Plant Genetic Resources Institute, Rome

Gaddam SA, Kotakadi VS, Reddy MN, Saigopal DVR (2012a) Survey and indexing of citrus yellow mosaic virus infecting citrus species in Rayalaseema region of Andhra Pradesh. Arch Appl Sci Res 4:1821–1824

Gaddam SA, Kotakadi VS, Reddy MN, Saigopal DVR (2012b) Antigenic relationships of *Citrus yellow mosaic virus* by immunological methods. Asian J Plant Sci Res 2:566–569

Gayral P, Noa C, Lescot M, Lheureux F, Lockhart BEL, Matsumoto T, Piffanelli P, Iskra-Caruana M-L (2008) A single banana streak virus integration event in the banana genome as the origin of infectious endogenous pararetrovirus. J Virol 82:6697–6710

Geering ADW, McMichael LA, Dietzgen RG, Thomas JE (2000) Genetic diversity among banana streak virus isolates from Australia. Phytopathology 90:921–927

Geering ADW, Olszewski NE, Harper G, Lockhart BEL, Hull R, Thomas JE (2005a) Banana contains a diverse array of endogenous badnaviruses. J Gen Virol 86:511–520

Geering ADW, Pooggin MM, Olszewski NE, Lockhart BEL, Thomas JE (2005b) Characterisation of *Banana streak Mysore virus* and evidence that its DNA is integrated in the B genome of cultivated *Musa*. Arch Virol 150:787–796

Geering ADW, Scharaschkin T, Teycheney P-Y (2010) The classification and nomenclature of endogenous viruses of the family *Caulimoviridae*. Arch Virol 155:123–131

Geering ADW, Maumus F, Copetti D, Choisne N, Zwickl DJ, Zytnicki M, McTaggart AR, Scalabrin S, Vezzulli S, Wing et al (2014) Endogenous florendoviruses are major components of plant genomes and hallmarks of virus evolution. Nat Commun 5:5269. doi:10.1038/ncomms6269

Ghosh DK, Aglave B, Bhanare K, Baranwal VK (2007) PCR based detection of citrus yellow mosaic disease from Vidarbha region of Maharashtra. Indian Phytopathol 60:520–526

Ghosh DK, Aglave B, Baranwal VK (2008) Simultaneous detection of one RNA and one DNA virus from naturally infected citrus plants using duplex PCR technique. Curr Sci 94:1314–1318

Ghosh DK, Bhose S, Mukherjee K, Aglave B, Warghane AJ, Motghare M, Baranwal VK, Dhar AK (2014) Molecular characterization of citrus yellow mosaic Badnavirus (CMBV) isolates revealed the presence of two distinct strains infecting citrus in India. Phytoparasitica 42:681–689

Gopi V, Sankar TG, Kumar AG, Gopal K (2010) Diagnosis of *Citrus yellow mosaic virus* by PCR and ELISA in sweet orange (*Citrus sinensis* Osbeck). J Pl Dis Sci 5:158–162

Gupta KN, Baranwal VK, Prasanna BK, Singh J, Haq QMR, Gopal K (2009) Genome sequencing, comparison and phylogenetic analysis of citrus yellow mosaic virus isolates originating from different citrus species in India. J Virol 5:143–153

Hany U, Adams IP, Glover R, Bhat AI, Boonham N (2013) The complete genome sequence of *Piper yellow mottle virus* (PYMoV). Arch Virol 159:385–388

Hareesh PS, Bhat AI (2008) Detection and partial nucleotide sequence analysis of *Piper yellow mottle virus* infecting black pepper (*Piper nigrum* L.) in India. Indian J Virol 19:160–167

Hareesh PS, Bhat AL (2010) Seed transmission of *Piper yellow mottle virus* in black pepper (*Piper nigrum* L). J Plant Crop 38:62–65

Harper G, Osuji JO, Heslop-Harrison P, Hull R, Harper G, Lockhart BEL (1999) Integration of banana streak badnavirus into the *Musa* genome: molecular evidence. Virology 255:207–213

Harper G, Hart D, Moult S, Hull R (2004) Banana streak virus is very diverse in Uganda. Virus Res 100:51–56

Harper G, Hart D, Moult S, Hull R, Geering A, Thomas J (2005) The diversity of banana streak virus isolates in Uganda. Arch Virol 150:2407–2420

Huang Q, Hartung JS (2001) Cloning and sequence analysis of an infectious clone of *Citrus yellow mosaic virus* that can infect sweet orange *via Agrobacterium*-mediated inoculation. J Gen Virol 82:2549–2558

Hull R, Sadler J, Longstaff M (1986) Sequence of *Carnation etched ring virus* DNA: comparison with *Cauliflower mosaic virus* and *retroviruses*. EMBO J 5:3083–3090

Iskra-Caruana M-L, Baurens FC, Gayral P, Chabannes M (2010) A four-partner plant-virus interaction: enemies can also come from within. Mol Plant-Microbe Interact 23:1394–1402

Iskra-Caruana M-L, Duroy P-O, Chabannes M, Muller E (2014a) The common evolutionary history of badnaviruses and banana. Infect Genet Evol 21:83–89

Iskra-Caruana M-L, Chabannes M, Duroy P-O, Muller E (2014b) A possible scenario for the evolution of banana streak virus in banana. Virus Res 186:155–162

Jones DR, Lockhart BEL (1993) Banana streak disease. Musa Disease fact Sheet No. 1. INIBAP, Montpellier

Karuppaiah R, Viswanathan R, Ganesh KV (2013) Genetic diversity of sugarcane bacilliform virus isolates infecting *Saccharum* spp. in India. Virus Genes 46:505–516

King AMQ, Adams MJ, Carstens EB, Lefkowitz EJ (2012) Ninth report of international committee on taxonomy of viruses. In: Virus taxonomy: classification and nomenclature of viruses. Elsevier Academic Press, San Diego

Kubiriba J, Legg JP, Tushemereirwe W, Adipala E (2001) Vector transmission of *Banana streak virus* in the screenhouse in Uganda. Ann Appl Biol 139:37–43

Lheureux F, Laboureau N, Muller E, Lockhart BEL, Iskra-Caruana ML (2007) Molecular characterization of banana streak acuminata Vietnam virus isolated from *Musa acuminata* siamea (banana cultivar). Arch Virol 152:1409–1416

Lockhart BEL, Olszewski NE (1993) Serological and genomic heterogeneity of banana streak badnavirus: implications for virus detection in *Musa* germplasm. In: Ganry J (ed) Breeding banana and plantain for resistance to diseases and pests. CIRAD/INIBAP, Montpellier, pp 105–113

Muller E, Dupuy V, Blondin L, Bauffe F, Daugrois J, Nathalie L, Iskra-Caruana M-L (2011) High molecular variability of sugarcane bacilliform viruses in Guadeloupe implying the existence of atleast three new species. Virus Res 160:414–419

Ndowora T, Dahal G, LaFleur D, Harper G, Hull R, Olzsewski N, Lockhart BEL (1999) Evidence that *Badnavirus* infection in *Musa* can originate from integrated pararetroviral sequences. Virol 255:214–220

Pant RP, Ahlawat YS (1997) Partial characterization of citrus mosaic virus. Indian Phytopathol 50:557–564

Pfeiffer P, Hohn T (1983) Involvement of reverse transcription in the replication of *Cauliflower mosaic virus*: a detailed model and test of some aspects. Cell 33:781–789

Rao GP, Singh M, Rishi N, Bhargava KS (2002) Century status of sugarcane virus diseases research in India. In: Singh SB, Rao GP, Easwaramoorthy S (eds) Sugarcane Crop Management . Sci Tech Publishing, Houston, pp 223–254, ISBN 1-930813163

Rao GP, Sharma SK, Singh D, Arya M, Singh P, Baranwal VK (2014) Genetically diverse variants of sugarcane bacilliform virus infecting sugarcane in India and evidence of a novel recombinant Badnavirus variant. J Phytopathol 162:779–787

Reddy BVB (1997) Characterization of citrus mosaic virus and to develop methods for its quick detection. Ph.D. Thesis. Indian Agricultural Research Institute. New Delhi, India

Reddy BVB, Ahlawat YS (1997) Detection of the Citrus mosaic badnavirus by dot-blot hybridization. In: National Symposium on Citriculture, Nagpur, India. Abstract 82

Reddy GS, Murti VD (1985) Citrus diseases and their control. Indian Council of Agricultural Research Publication, New Delhi

Sai Gopal DVR, Sreenivasulu M, Raghavendra Rao G, Venkata Prasanna TM, Subbaiah KV (2000) Characterization and identification of a Citrus mosaic *Badnavirus* infecting acid lime (*Citrus aurantifolia*) in Andhra Pradesh. HI-TECH Citrus Management. In: Proceedings of International Symposium on Citriculture, Nagpur, 23–27 Nov 845–852

Selvarajan R, Balasubramanian V, Kavitha K, Kavitha KS, Sathiamoorthy S, Ahlawat YS (2008) Detection of *Banana bunchy top virus* and *Banana streak Mysore virus* by PCR: impact of storing virus infected banana samples. Indian J Virol. 19:155–159

Selvarajan R, Mary SM, Balasubramanian V (2011) Simultaneous detection of episomal *Banana streak Mysore virus* and *Banana bunchy top virus* using multiplex RT-PCR. Curr Sci 100:31–34
Sharma SK, Vignesh Kumar P, Baranwal VK (2014a) Immunodiagnosis of episomal *Banana streak MY virus* using polyclonal antibodies to an expressed putative coat protein. J Virol Methods 207:86–94
Sharma SK, Vignesh Kumar P, Poswal R, Rai R, Geetanjali AS, Prabha K, Jain RK, Baranwal VK (2014b) Occurrence and distribution of banana streak disease and standardization of a reliable detection procedure for routine indexing of banana streak viruses in India. Sci Hortic 179:277–283
Sharma SK, Vignesh Kumar P, Geetanjali AS, Pun KB, Baranwal VK (2015) Subpopulation level variation of banana streak viruses in India and common evolution of banana and sugarcane badnaviruses. Virus Genes 50:450–465
Siju S, Bhat AI, Hareesh PS (2008) Identification and characterization of a *Badnavirus* infecting betel vine and Indian long pepper. J Plant Biochem Biotechnol 17:73–76
Singh M, Gaur RK, Rao GP (2003) Distribution of sugarcane bacilliform badnavirus in India. J Mycol Pl Pathol 33:406–410
Singh D, Tewari AK, Rao GP, Karuppaiah R, Viswanathan R, Arya M, Baranwal VK (2009) RT-PCR/PCR analysis detected mixed infection of DNA and RNA viruses infecting sugarcane crops in different states of India. Sugar Tech 11:373–380
Singh HP, Uma S, Selvarajan R, Karihaloo JL (2011) Micropropagation for production of quality banana planting material in Asia-Pacific. Asia-Pacific Consortium on Agricultural Biotechnology. APCoAB, New Delhi
Thomas JE, McMichael LA, Dietzgen RG, Searle C, Matalevea S, Osasa A (1994) Banana streak virus in Australia, Western Samoa and Tonga. In: 4th ISSCT (International Society of Sugarcane Technologists) Pathology Workshop
Thomas JE, Vo J, Geering ADW (2013) Banana virus diseases in Australia, current strategies. In: XX Reunião Internacional da Associação para a Cooperação em Pesquisa e Desenvolvimento Integral das Musáceas (Bananas e Plátanos) 9 a 13 de setembro de-Fortaleza, CE. 2013
Toh H, Hayashida H, Miyata T (1983) Sequence homology between retroviral reverse transcriptase and putative polymerases of *Hepatitis B virus* and *Cauliflower mosaic virus*. Nature 305:827–829
Verver J, Schuns P, Hibi T, Goldbach R (1987) Characterization of the genome of *Soybean chlorotic mottle virus*. J Gen Virol 68:159–167
Vision 2030 (2011) National Research Centre for banana. Tiruchirapalli, Tamil Nadu, India
Viswanathan R, Rao GP (2011) Disease scenario and management of major sugarcane diseases in India. Sugar Tech 13:336–353
Viswanathan R, Alexander KC, Garg ID (1996) Detection of sugarcane bacilliform virus in sugarcane germplasm. Acta Virol 40:5–8
Viswanathan R, Balamuralikrishnan M, Premachandran MN, Tripathi BK (1999) Sugarcane Bacilliform virus: symptoms, detection and distribution in the world germplasm collection at Cannanore. Proc Internl Soc Sugar Cane Technol 23:347–354
Wardlaw CN (1972) Banana diseases: including plantains and abaca. William Clowes and Sons Ltd., London, pp 68–115

Begomoviruses and Their Satellites Occurring in India: Distribution, Diversity and Pathogenesis

5

V.G. Malathi, P. Renukadevi, S. Chakraborty, K.K. Biswas, Anirban Roy, P.N. Sivalingam, V. Venkataravanappa, and Bikash Mandal

Abstract

The begomoviruses (genus *Begomovirus*, family *Geminiviridae*) constitute the largest group of plant viruses causing devastating crop diseases in India. About 16% geminiviruses recorded worldwide occur in India. Begomovirus associated disease was recorded in India as early as 1924. During 1980s, begomovirus disease and the vector whitefly, *Bemisia tabaci* emerged as serious problems in many vegetable and pulse crops. The molecular characterisation studies on begomoviruses in India began during late 1980s and during last decade, a large number of begomoviruses were discovered causing diseases such as mosaic, yellow mosaic, yellow vein mosaic and leaf curl in numerous crops and wild plants. Currently, 322 begomovirus species have been officially recognised all over the world, of which about 82 begomovirus species are known to occur in India. The Indian subcontinent represents one of the important centres for begomoviruses origin and diversity. In this chapter, research work on identification and characterization of begomoviruses in diverse crops in India is presented. Geographic distribution, phylogenetic relationship, infectivities of cloned genome components, recombination, replication events, suppression of RNAi silencing, viral gene funtions as pathogenicity determinants and association of satellites and

V.G. Malathi (✉) • P. Renukadevi
Tamil Nadu Agricultural University, Coimbatore 641003, India
e-mail: vgmalathi@rediffmail.com

S. Chakraborty
School of Life Sciences, Jawaharlal Nehru University, New Delhi 110067, India

K.K. Biswas • A. Roy • B. Mandal
Indian Agricultural Research Institute, New Delhi 110012, India

P.N. Sivalingam • V. Venkataravanappa
National Institute of Biotic Stress Management, Raipur 493225, India

B. Mandal et al. (eds.), *A Century of Plant Virology in India*,
https://doi.org/10.1007/978-981-10-5672-7_5

their role in disease development with reference to the Indian begomoviruses are discussed.

Kewords
Begomovirus • Betasatellite • Alphasatellite • Whitefly • Diversity • Begomovirus in India

5.1 Introduction

Begomoviruses (genus *Begomovirus,* family *Geminiviridae*) are highly prolific plant viruses causing significant economic losses in the agricultural produces in the tropical and subtropical environments of the world. The name of the genus *Begomovirus* is derived from the first two letters of the each word of the the type species, *Bean golden mosaic virus* (BGMV) causing golden mosaic disease in bean in Central America. The family name, *Geminiviridae* is derived from the zodiac sign 'Gemini', as the virion (22 × 38 nm) has characteristic twinned isometric morphology. The virion consists of two incomplete icosahedra (T = 1) containing a total of 110 coat protein subunits organized as 22 pentameric capsomers, encapsidating a circular single stranded DNA (ssDNA) genome of 2.5–2.9 kb (Harrison et al. 1977; Stanley 1985). Based on the genome organization, insect vector and host range, geminiviruses (family *Geminiviridae*) are classified into nine genera, *Becurtovirus, Begomovirus, Topocuvirus, Turncurtovirus, Capulavirus, Curtovirus, Grablovirus, Mastrevirus* and *Eragrovirus* (Zerbini et al. 2017). Begomoviruses are transmitted by only one vector species, whitefly (*Bemisia tabaci* Genn.). About 68.1% of geminiviruses belong to the genus *Begomovirus.* Presently, 322 virus species have been officially recgnised under the genus *Begomovirus*, which is the maximum number of members so far known in any genera of plant viruses. The virus species, *citrus chlorotic dwarf associated virus* and *mulberry mosaic dwarf associated virus* are included in the unassigned genus as their replication associated protein and genome are distinct from the other nine genera.

The diseases caused by begomoviruses are, yellow mosaic, veinal yellowing, leaf distortion, curling and stunting. Infection in early seedling stage, leads to poor fruit set and infertile seeds, resulting in severe yield loss. Begomoviruses affect a large number of dicotyledonous crops such as cassava, sweet potato, cotton, grain legumes and vegetables in the tropical and subtropical countries. Some of the diseases like cotton leaf curl and cassava mosaic are known for a century and cause huge economic loss (Varma and Malathi 2003).

All the eight genera of the family *Geminiviridae* have monopartite genome, whereas begomoviruses contain either monopartite or bipartite genome (Fig. 5.1). Almost all the begomoviruses of the New World are bipartite and have DNA A and DNA B component encapsidated separately in a geminate particle. Both DNA A

Fig. 5.1 Genome organisation of genera of the family *Geminiviridae* (Adapted from Zerbini et al. 2017)

and DNA B are essential for infectivity. The basic features of genome organization are the presence of coat protein gene (ORF V1/AV1-coat protein) in the viral sense strand (right half) and the replication associated protein gene (C1/AC1/Rep) in the complementary strand (left half). Both the ORFs diverge from a non-coding intergenic region, consisting origin of replication of viral sense strand and promoters of CP and Rep gene. The genera differ in number of genes encoded in the complementary strand. The complementary strands of mastre and becurtoviruses have only Rep which is translated from ORFs C1:C2 by transcript splicing in the complementary strand. However, three ORFs, transcriptional activator protein gene which also functions as silencing suppressor gene (C2/AC2/TrAP/*ss*), replication enhancer gene (C3/AC3/REn) and symptom determinant gene (C4/AC4/*sd*) are present in the complementary strand in curto, topocu, turncurto and begomoviruses. The genus *Eragrovirus* encodes a gene for homologue of C2 and the genera *Capulovirus* and *Grablovirus* encode for another gene whose function is not yet resolved. In DNA B, the gene encoding nuclear shuttle protein is present in the viral sense strand (BV1, NSP) and the gene encoding cell to cell movement protein is in the complementary sense strand (BC1, MP). Both the ORFs are separated by an intergenic region having origin of replication in viral sense strand. The DNA A component is dependent on DNA B for intracellular and intercellular movement. DNA B is dependent on DNA A for replication and encapsidation. In order to facilitate recognition of DNA B by the Rep encoded by DNA A, within the intergenic region is present a segment

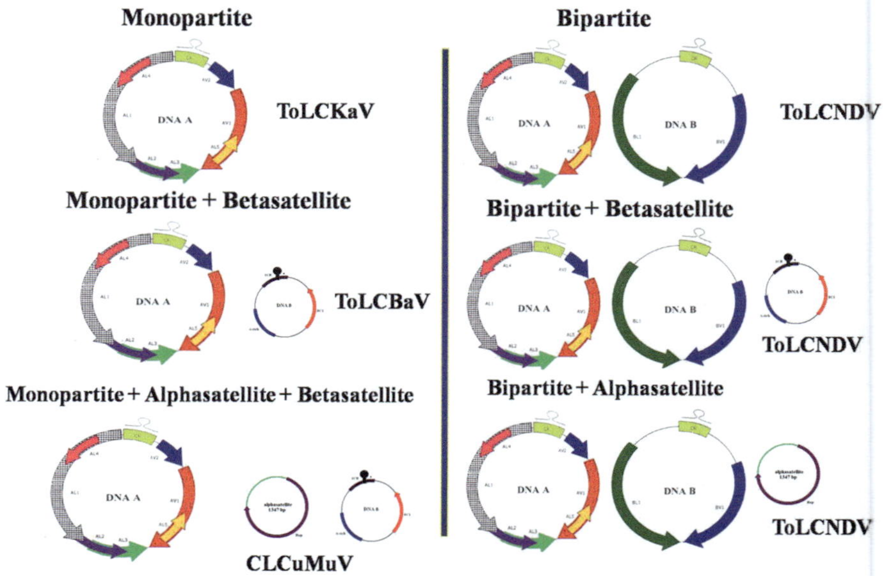

Fig. 5.2 Genomic components and genome organisation of the Old World begomoviruses

of 110–200 nt sequences which are highly conserved between the DNA A and DNA B component. This region labeled as common region (CR) is near identical between DNA A and DNA B component. The intergenic region consists of repetitive elements called iterons upstream of the highly conserved stem-loop structure. The iterons, represent the binding sites of Rep. The invariant nonanucleotide sequence TAA TAT TAC in the loop is conserved in all geminiviruses and the nicking between seventh and eighth nucleotide by Rep is proved to initiate replication.

Begomoviruses in the Old World have both monopartite and bipartite genome organization (Fig. 5.2). The monopartite begomoviruses in general and few of the bipartite begomoviruses of the Old World are associated with additional circular ssDNA components refered as satellites (1.3 kb). There are three types of satellites, alphasatellite, betasatellite and deltasatellite. The alphasatellites encode only one Rep gene having similarities with Rep protein of nanovirus. The betasatellite, shares the origin of replication sequence with the helper begomoviruses, it's replication is facilitated by the Rep protein encoded by DNA A of helper begomoviruses. There is one ORF (beta C1) encoded in the complementary sense DNA of betasatellite, which is the pathogenicity determinant and functions as silencing suppressor. All betasatellites have an extremely conserved region referred to as satellite conserved region (SCR), upstream of origin of replication which is essential for replication. Among monopartite begomoviruses, though DNA A alone can infect plant and systemically move, inoculation along with betasatellites lead to severe symptom production like enation, leaf malformation twisting and stunting. The alphasatellites replicate autonomously by it's own Rep. They are dependent on helper begomovirus for their spread. A new set of non-coding sub viral molecules (633–750) designated as deltasatellites (Lozano et al. 2016) have been identified recently with begomoviruses infecting sweet potato (sweepoviruses). They are structurally similar to subgenomic betasatellite associated with tomato leaf curl virus (ToLCV) from Australia; they have the conserved stem and loop structure with nononucleotide sequence TAATATAC and SCR similar to betasatellites. The contribution of alphasatellites and deltasatellites to viral pathogenicity is yet to be understood.

The process of infection is initiated with the introduction of the virus in protophloem cell by the deep probing of vector whitefly. From the ssDNA/CP complex, ssDNA is released into the nucleus, where ssDNA becomes double stranded DNA. The replication is facilitated by the host DNA polymerase. Detection of primers with 5′ ribonucleotides complementary to SIR in mastrevirus are suggestive of initiation at the short intergenic region site; In a bipartitie begomovirus, African cassava mosaic virus (ACMV) ribonucleotides priming has been shown. The dsDNA is transcribed by the host RNA polymerase II and the earliest gene transcribed is the C1/AC1 or replication initiation/ associated protein (Rep). The replication is by a combination of rolling circle replication and recombination dependant replication. The Rep protein initiates replication by niking at the nononucleotide sequence TAATATT↓AC. The newly synthesized + strand is copied into dsDNA again by host DNA polymerase which may enter the replication cycle. Alternatively, the ssDNA may get encapsidated by the coat protein. The movement of viral DNA from the infection foci is mediated by the movement protein V2 in monopartite viruses or by BV1/BC1 in bipartite viruses. The viral DNA is transported out of the nucleus into the periphery of the cell from where they are docked on to plasmodesmata and transported into adjacent cells, finally the viral DNA (either ss or ds) enter into phloem parenchyma and companion cell. It is hypothesized that geminiviruses move as ss/ds DNA/movement protein complex, spread to young unfurling leaves from where they are acquired by the vector.

5.1.1 Begomovirus species in India

It is not an exaggeration to state that Indian subcontinent represents the key centre for the origin, diversity and spread of begomoviruses. Approximately 16% of world geminiviruses occur in India which tops the list of numbers of begomoviruses in different countries. Pakistan and China have nearly 13% and 8% of the geminiviruses recorded. There are approximately 1500 gene sequence entries in database which is the highest record worldwide. At present, the NCBI database has 810 full-length genome sequences of Indian geminiviruses. They belong to two genera, *Mastrevirus* (15 sequences) and *Begomovirus* (795 sequences). There are approximately 82 species of begomoviruses (Table 5.1), 2 species of mastreviurses, about 545 betasatellites and 146 alphasatellites.

Applying 91% identity as the threshold value for the demarcation of species in the DNA A nucleotide component using SDT analysis with MUSCLE option, the virus members of the genus *Begomovirus* have been reanalyzed and 322 species have been recognized by ICTV (Brown et al. 2015). Of these, 80 species are known

Table 5.1 Indian begomoviruses and their hosts

S.No	DNA-A Virus	Acronyms	Hosts
1.	Abutilon mosaic virus	AbMV	Abutilon
2.	Ageratum enation virus	AEV	Poppy, *Cleome gynandra,* Amaranthus, *Ageratum Crassocephalum*, Soybean, Carrot, Ornamental, Ageratum, Pointed gourd, Calendula, Tagetes, Tomato, Zinnia, Fenugreek, Papaya
3.	Ageratum leaf curl virus	ALCuV	Calotropis
4.	Allamanda leaf mottle distortion virus	All LMDV	Allamanda
5.	Alternanthera yellow vein virus	AlYVV	Rumex, Picrorhiza
6.	Andrographis yellow vein leaf curl	AYVLCV	Andrographis
7.	Bhendi yellow vein Bhubaneswar virus	BYVBhV	Bhendi
8.	Bhendi yellow vein mosaic virus	BYVMV	Bhendi
9.	Bhendi yellow vein Haryana virus	BYVHarV	Bhendi
10.	Chayote enation yellow mosaic virus	ChEYMV	Chayote
11.	Chilli leaf curl virus	ChiLCV	Chilli, Kenaf, Amaranthus, Tomato, *Solanum nigrum*, Petunia, Mentha, Papaya, Eggplant, *Phaseolus aureus*
12.	Chilli leaf curl Ahmedabad virus	ChiLCAV	Chilli
13.	Chilli leaf curl India virus	ChiLCINV	Chilli, Tomato
14.	Chilli leaf curl Kanpur virus	ChiLCKaV	Chilli
15.	Chilli Leaf curl Vellanad virus	ChiLCVV	Chilli
16.	Clerodendron yellow mosaic virus	ClYMV	Bougainvillea, Clerodendron
17.	Coccinia mosaic Tamil Nadu virus	CocMTNV	Ivy gourd
18.	Corchorus golden mosaic virus	CoGMV	*Corchorus capsularis*, Boehmeria.
19.	Corchorus yellow vein mosaic virus	CoYV	*Corchorus oletorius*
20.	Cotton leaf curl Alabad virus	CLCuAlV	Bhendi
21.	Cotton leaf curl Bangalore virus	CLCuBaV	Cotton, Bhendi, *Hibiscus cannabinus*
22.	Cotton leaf curl Barasat virus	CLCuBrV	*Malachra capitata*
23.	Cotton leaf curl Kokhran virus	CLCuKoV	*Cyamopsis tetragonoloba*, cotton
24.	Cotton leaf curl Multan virus- Rajasthan	CLCuMuV-Ra	Cotton, *Hibiscus rosa-sinensis, Hibiscus cannabinus*
25.	Croton yellow vein mosaic virus	CroYVMV	Croton bonplandianum
26.	Dolichos yellow mosaic virus	DoYMV	Dolichos
27.	Eclipta yellow vein virus	EYVV	Hibiscus
28.	French bean leaf curl virus	FbLCV	French bean
29.	Hemidesmus yellow mosaic virus	HemYMV	*Hemidesmus indicus*

(continued)

Table 5.1 (continued)

S.No	DNA-A Virus	Acronyms	Hosts
30.	Hollyhock leaf curl virus	HoLCV	*Andrographis paniculata*
31.	Hollyhock yellow vein mosaic virus	HoYVMV	Hollyhock, Bhendi
32.	Horsegram yellow mosaic virus	HgYMV	French bean, Lima bean, Horse gram
33.	Indian cassava mosaic virus	ICMV	*Jatropha curcus*, Cassava
34.	Jatropha leaf crumple India virus	JLCrIV	*Jatropha curcus*
35.	Jatropha leaf crumple virus	JLCrV	*Jatropha curcus*
36.	Jatropha leaf curl virus	JLCuV	*Jatropha integerrima, Jatropha multifida, Jatropha podagrica, Jatropha gossipifolia, Jatropha curcus, Ludwigia parviflora*
37.	Jatropha yellow mosaic India virus	JYMV	*Jatropha gossipifolia*
38.	Jatropha mosaic India virus Katarniaghat	JMINV	*Jatropha curcus, Jatropha gossipifolia*
39.	Malvastrum yellow vein virus	MaYVV	*Hibiscus cannabinus*
40.	Mesta yellow vein mosaic Bahraich virus	MeYVMBaV	*Hibiscus cannabinus*
41.	Mesta yellow vein mosaic virus	MeYVMV	*Hibiscus cannabinus, Hibiscus sabdariffa*, Bhendi
42.	Mirabilis leaf curl India virus	MiLCV	*Mirabilis jalaba*
43.	Mungbean yellow mosaic India virus	MYMV	Mungbean, Dolichos, Cowpea, Blackgram, Soybean, French bean.
44.	Mungbean yellow mosaic virus	MYMIV	Soybean, Blackgram, Mungbean, Moth bean,
45.	Okra enation leaf curl virus	OELCuV	Bhendi, Wild Bhendi, *Hibiscus cannabinus*
46.	Papaya leaf curl virus	PaLCuV	*Croton bonplandianum,* Acalypha, *Jatropha gossypifolia*, Cluster bean, Chilli, *Crambe abyssinica,* Turnip, Radish, Papaya, Sunnhemp, *N.glutinosa*, Aster, *Amaranthus cruentus*, Soybean, *Brassica rapa.*
47.	Papaya leaf crumple virus	PaLCrV	*Solanum nigrum, Androdrographis paniculata, Cathranthus roseus*, Cowpea, Soybean, Papaya
48.	Pedilanthus leaf curl virus	PeLCV	Crape Jasmine, *Cestrum nocturnum*
49.	Pepper leaf curl Bangladesh virus	PepLCBV	*Solanum capsicastrum*, Chilli,
50.	Pepper leaf curl Lahore virus	PepLCLaV	Tomato
51.	Radish leaf curl virus	RaLCuV	Radish, Tobacco, Chilli, Bhendi
52.	Rhynchosia yellow mosaic India virus	RhYMIV	Rhynchosia
53.	Rhynchosia yellow mosaic virus	RhYMV	French bean

(continued)

Table 5.1 (continued)

S.No	DNA-A Virus	Acronyms	Hosts
54.	Rose leaf curl virus	RoLCuV	Rose
55.	Senna leaf curl virus	SeLCuV	Senna
56.	Sida leaf curl virus	SiLCuV	Sida
57.	Spinach yellow vein Sikar virus	SpiYVV	Spinach
58.	Squash leaf curl China virus	SLCCNV	Pumpkin, Ash gourd
59.	Sri Lankan cassava mosaic virus	SLCMV	Cassava
60.	Sunn hemp leaf distortion virus	SHLDV	Sunn hemp
61.	Sweet potato leaf curl virus	SPLCV	Sweet potato
62.	Synedrella leaf curl virus	SyLCuV	Synedrella
63.	Synedrella yellow vein clearing virus	SyYVV	Synedrella
64.	Tobacco curly shoot virus	TbCSV	Tomato, French bean, Wild sunflower
65.	Tobacco leaf curl Pusa virus	TbLCPuV	Tobacco
66.	Tomato enation leaf curl virus	ToELCV	Tomato
67.	Tomato leaf curl virus	ToLCV	Tomato, Mentha, Parthenium, Ocimum, sunflower, French bean, Papaya, Cherry tomato
68.	Tomato leaf curl Bangalore virus	ToLCBaV	Tomato, cotton
69.	Tomato leaf curl Bangladesh virus	ToLCBV	Gaillardia
70.	Tomato leaf curl Gujarat Virus	ToLCGuV	Tomato
71.	Tomato leaf curl Karnataka virus	ToLCKaV	Tomato
72.	Tomato leaf curl Joydebpur virus	ToLCJV	Chilli
73.	Tomato leaf curl Kerala virus	ToLCKeV	Tomato
74.	Tomato leaf curl New Delhi virus	ToLCNDV	Potato, Tomato, Bhendi, Chilli, Bittergourd, Eggplant, Ash gourd
			Cucumis, Ridge gourd, Chayote, Pumpkin, Sponge gourd, Papaya, Cotton
75.	Tomato leaf curl Palampur virus	ToLCPalV	Tomato, Rumex, Melon, Egg plant, Bitter cucumber, Pumpkin
76.	Tomato leaf curl Patna virus	ToLCPatV	Tomato, cotton, *Cassia tora*, Mentha, Tobacco
77.	Tomato leaf curl Pune virus	ToLCPuV	Tomato
78.	Tomato leaf curl Rajasthan virus	ToLCRaV	Tomato
79.	Tomato severe leaf curl virus	ToSLCV	Tomato
80.	Velvet bean severe mosaic virus	VBSMV	Velvet bean
81.	Vernonia yellow vein virus	VeYVV	*Vernonia cinerea*
82.	Vinca leaf curl virus	ViLCuV	*Vinca rosea*

Table 5.2 Proposed new virus species yet to be approved by ICTV

S.No	Name of the virus	Host	Isolation place/ state	Year
1.	Allamanda leaf mottle distortion virus isolate Al-K1, KC202818,	Allamanda	Kalyani/WB	2012
2.	Chayote enation yellow mosaic virus Adalur isolate-embryo AD1, KX.259336, KX.259339	Chayote	Adalur, Tamil Nadu	2015
3.	Chilli leaf curl Ahmedabad virus, KM880103	Chilli	Ahmedabad/Gu	2014
4.	Coccinia mosaic Tamil Nadu virus isolate TN TDV Coc 1, KM244719	Ivy gourd	Tindivanam/TN	2013
5.	Cotton leaf curl Barasat virus, LC080677	*Malachra capitata*	Barasat/WB	2014
6.	Eclipta yellow vein virus isolate WOK44, KT390456	Hibiscus	Mirzapur/UP	2014
7.	Hollyhock yellow vein mosaic virus isolate Alcea rosea:Lucknow, JQ911766 KT390462	Hollyhock	Lucknow/UP	2011
8.	Jatropha leaf crumple India virus, Jodhpur] isolate SKJ3, KM189819, KM189818, KM023146	*Jatropha curcus*	Jodhpur/RA	2011
9.	Mirabilis leaf curl India virus, LK054801	*Mirabilis jalaba*	Kangra/HP	2013
10.	Rhynchosia yellow vein mosaic virus clone pBdGn05, KP752090	French bean	Gandhinagar/GU	2014
11.	Senna leaf curl virus isolate Mohali, KU852742	Senna	Mohali/PU	2013
12.	Synedrella leaf curl virus, isolate Synd-1, KJ939345	Synedrella	Portblair/ Andaman	2009
13.	Synedrella yellow vein clearing virus, KX363443	Synedrella	Portblair/ Andaman	2013
14.	Tomato enation leaf curl virus isolate TC14, KP195260,	Tomato	Narasipura/KA	2008
15.	Tomato severe leaf curl virus isolate TC101, KP195267	Tomato	Kalakada/AP	2007

to occur in India (Table 5.1). The data entered in the NCBI database for the Indian begomoviruses after March 2015 to November 2016 were analyzed by the authors following the guidelines set by the study group on geminiviruses and 16 more new species were identified, which are listed in Table 5.2. The details on distribution of viruses in different states, the host range of some selected viruses and number of hosts Indian begomoviruses infect are given in Tables 5.3, 5.4, and 5.5.

Excellent reviews are available (Stanley 1985; Harrisson and Robinson 1999; Hanley-Bowdoin et al. 1999, 2013; Guiterrez 2000; Rojas et al. 2005; Briddon et al. 2003; Fondong 2013), which may be referred for understanding the geminivirus genome and gene functions. Begomoviruses being the most important viral pathogen in Indian agriculture (Varma and Malathi 2003), a great deal of attention has

Table 5.3 Number of begomoviruses recorded in the different states of India

S.No.	States	No. of viruses
1	Andhra Pradesh	8
2	Assam	1
3	Bihar	8
4	Chhattisgarh	1
5	Delhi	14
6	Gujarat	10
7	Haryana	10
8	Himachal Pradesh	10
9	Jharkhand	3
10	Karnataka	12
11	Kerala	9
12	Madhya Pradesh	6
13	Maharashtra	9
14	Meghalaya	1
15	Odisha	5
16	Punjab	10
17	Rajasthan	19
18	Tamil Nadu	16
19	Uttar Pradesh	33
20	Uttarakhand	1
21	West Bengal	11

Table 5.4 Natural infection of begomoviruses in major crop species in India

S.No.	Host (No. of viruses)	Viruses
1	Bhendi (11)	BYVMV, BYVBhV, ToLCDNV, OELCuV, OLCuV, CLCuAlV, CLCuBaV, RaLCuV, HoYVMV, MeYVMV, BVYMV-Har
2	Cassava (2)	ICMV, SLCMV
3	Chilli (11)	ChiLCAV, ChiLCV, ChiLCINV, ChiLCKV, ChiLCVV, PepLCBV, ToLCNDV, PaLCuV, RaLCuV, ToLCV, ToLCJV
4	Corchorus (2)	CoGMV, CoYMV
5	Cotton (6)	CLCuBaV, CLCuKoV, CLCuMuV, ToLCBaV, ToLCPTV, ToLCNDV
6	Cucurbits (6)	ChEYMV, CoMTNV, SLCCNV, ToLCNDV, ToLCPalV, AEV
7	Jatropha (7)	JLCrIV, JLCV, JMINV, JYMV, JLCrIV, ICMV, PaLCuV
8	Legumes (17)	AEV, MYMV, MYMIV, DoYMV, CLCuKoV, PaLCuV, FbLCV, VBSMV, ToLCNDV, ChiLCV, ToLCV, ToLCPtV, HgYMV, RhYMV, RhYMIV, TbSCV, PaLCrV
9	Mestha/Kenef (6)	ChiLCV, MeYVMV, MeYVMBaV, MaYVMV, OeLCuV, CLCuVaV
10	Papaya (5)	PaLCuV, PaLCrV, ChiLCV, ToLCNDV, ToLCV
11	Tobacco (3)	TbCSV, TbLCPuV, ToLCPtV
12	Tomato (18)	ChiLCV, ChiLCIV, PaLCuV, ToLCBaV, ToLCV, ToLCGuV, ToLCKaV, ToLCKeV, TbCSV, ToLCNDV, ToLCPalV, ToLCPtV, ToLCPuV, ToLCRaV, ToLCGaV, RaLCuV, AEV, PepLCULaV.
13	Radish (2)	RaLCuV, PaLCuV

Table 5.5 Host-range of the major begomoviruses in India

S.No.	Name of the virus	Hosts infected
1	AEV	Cleome, Amaranthus, Poppy, Ageratum, Carrot, Soybean, Zinnia, Tomato, Crassocephalum, Pointed gourd, Calendula, Tagetes, Fenugreek, Papaya
2	ChiLCV	Chilli, Kenaf, Amaranthus, *Solanum nigrum, Petunia,* Mentha, Egg plant, *Phaseolus aureus*, Papaya, Tomato
3	CLCuBaV	Cotton, Bhendi, Hibiscus
4	CLCuMuV	Hibiscus, Cotton, Hibiscus- rosa sinensis, *H.cannabinus*
5	MYMV	Blackgram, Mungbean, Soybean, Moth bean
6	MYMIV	Blackgram, Cowpea, Mungbean, French bean, Dolichos, Soybean
7	PaLCuV	Clusterbean, Papaya, Acalypha, Radish, Soybean, Tomato, Chilli, Amaranthus, Brassica, Aster, Jatropha, Crambe species, Turnip, Sunnhemp, Aster, Brassica
8	ToLCBaV	Cotton, Tomato
9	ToLCV	Tomato, Mentha, Chilli, Parthenium, Ocimum, Sunflower, French bean, papaya, cherry tomato
10	ToLCNDV	Potato, Luffa, Lagenaria, Pumpkin, Ash gourd, Cucumber, Chilli, Bhendi, Papaya, Cotton, Papaver, Egg plant, Bitter gourd, Chayote
11	ToLCPalV	Tomato, Pumpkin

been given to begomovirus research in India. As a result, a wealth of information has been generated. In the present chapter, efforts have been made to review the research work conducted on begomoviruses occurring in India. Details on full-length genome sequences of begomoviruses, discovery of new begomoviruses and recent changes in the nomenclature of the Indian begomoviruses are included. The investigations carried out to elucidate replication, suppression of RNAi and viral gene functions are briefly discussed. In this chapter, instead of crop based grouping, description of the work on the Indian begomoviruses are presented based on the four categories of disease symptoms: mosaic, yellow mosaic, yellow vein mosaic and leaf curl. The focus in the presentation is more on the significant biological and molecular features of the begomoviruses rather than the diseases.

5.2 Historical Developement

5.2.1 Global Scenario

The earliest recorded plant virus disease happens to be the one caused by whitefly transmitted geminivirus. The yellow vein virus symptoms in *Eupartorium chinense* was referred in a poem written by empress Koken in Manyoshu, a Japanese anthology prepared in AD 752. The virus described in the poem has been now identified as tobacco leaf curl virus belonging to the genus *Begomovirus* (Saunders et al. 2003). Almost thousand years later, the economically important geminivirus diseases were described in the nineteenth century such as African cassava mosaic

(Warburg 1894) and maize streak in Africa (Fuller 1901; Storey 1936), sugar beet curly top in western USA (Carsner and Stahl 1924), cotton leaf curl in Angola and Sudan (Tarr 1951) and bean golden yellow mosaic in South America (Costa 1976). In all these diseases, the role of whitefly or leafhopper as vector of the causal agent was established beyond doubt. However, the nature of the etiological agent remained elusive until 1980s. The studies on purification of the virus (Bock et al. 1974; Goodman et al. 1977; Larsen and Duffus 1984) revealed the consistent association of geminate particles with these diseases. The DNA genome was identified and on the basis of buoyant density gradient centrifugation, Francki and Hatta 1980 revealed the presence of one copy of circular DNA within the geminate particle. The viruses having ssDNA genome with the geminate morphology was proposed as a new group of plant viruses and named as geminivirus group (Harrison et al. 1977). On the basis of restriction mapping and infectivity dilution curve, Goodman 1977 predicted the genome of BGMV to be bipartite (Haber et al. 1981). However, other leafhopper transmitted viruses were found to be monopartite (Mullineaux et al. 1984). The bipartite nature of genome of the begomovirus was confirmed by complete nucleotide sequencing in the case of African cassava mosaic virus (ACMV) (Stanley and Gay 1983), tomato golden mosaic virus (TGMV) (Bisaro et al. 1982) and BGMV (Goodman 1977). Characterization of the Old World begomoviruses soon revealed the monopartitie nature of some begomoviruses such as tomato yellow leaf curl virus- (TYLCV) Thailand (Rochester et al. 1994) and TYLCV-Israel (Navot et al. 1991). The most innovative method of delivery of geminiviral genomic component into the host referred as agroinoculation was developed by Grimsley et al. (1986) for maize streak virus (MSV) in maize. This was picked up immediately and agroinoculation became the favoured technique of delivery of viral inoculum to seek answers to many questions on replication, viral gene function and suppression of RNAi defense and host resistance.

The absence of typical symptoms, when DNA A like components are inoculated in primary hosts for some begomoviruses such as Ageratum yellow vein mosaic virus (AYVMV) and cotton leaf curl viruses led to the discovery of alphasatellite and betasatellites associated with begomoviruses (Briddon et al. 2003; Saunders and Stanley 1999, 2000).

The difficulty in cloning the complete genomic component of unknown geminiviruses, especially those which are present in low concentration in plant tissues was overcome by the rolling circle amplification protocol (Haible et al. 2006). This method for enhancing the concentration of circular DNA molecule using Φ29 DNA polymerase made isolation and cloning of the geminiviral genomic components easy and thus the amplification technique revolutionized the geminivirus research.

5.2.2 Indian Scenario

The chronological development of begomovirus research can be catagorised into five distinct phases spanning nearly 100 years (Fig. 5.3). Yellow vein mosaic disease of bhendi occurring in Bombay, Maharashtra is the first whitefly transmitted

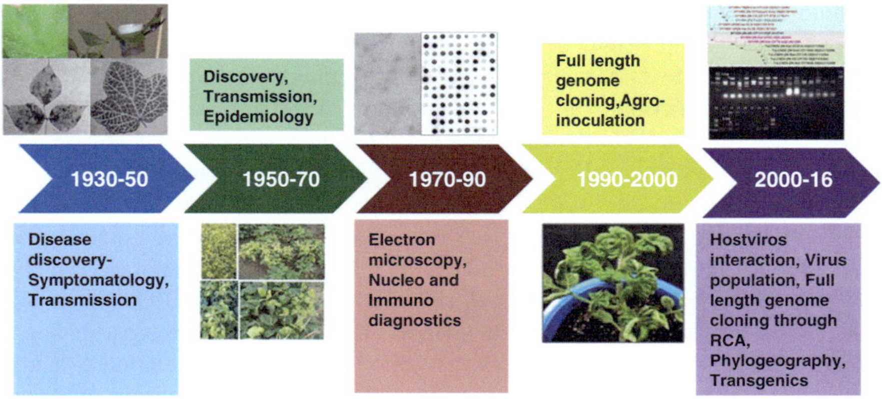

Fig. 5.3 Milestones in begomovirus research in India

geminivirus disease recorded in India (Kulkarni 1924). Subsequently, several diseases were described e.g., leaf curl disease of *Zinnia elegans* (Mathur 1932, 1933), tobacco (Pal and Tandon 1937; Pruthi and Samuel 1937) and tomato (Vasudeva and Samraj 1948) and yellow vein mosaic of pumpkin and yellow mosaic of limabean and dolichos (Capoor and Varma 1948a, b). While recording these diseases, the scientists also described the transmission characteristics of whitefly and host range of the virus. The period from 1950 to 1970, the investigations were focused on the vector, *Bemisia tabaci* and its management. To effectively manage the disease, the sources of resistance were identified such as *Carica cauliflora* for papaya leaf curl virus and *Abelmoschus manihot* var. *pungens* for yellow vein mosaic of bhendi. In this period, the most devastating chilli and papaya leaf curl disease (Mishra et al. 1963) and yellow mosaic disease (YMD) of grain legumes were recorded (Nariani 1960). It is interesting to note that, although yellow mosaic of limabean and dolichos was recorded as early as 1948, YMD of mungbean and blackgram was observed only in 1960s. In the year 1971, Seth et al. (1971) observed chlorotic stripes in self sown bajra plants. The virus could be easily transmitted by *Cicadullina mebila* and so they identified it as maize streak virus similar to one in Africa. The spread of yellow mosaic disease to different leguminous hosts was identified in northern and southern India. Muniyappa and Veeresh (1984) described whitefly transmitted viruses in several plant species, which indicated the potential of these viruses (Nene et al. 1971; Nene 1972) to emerge as serious pathogen.

The association of geminate particle with the disease was first demonstrated for Indian cassava mosaic virus (ICMV) (Malathi and Srinivasan 1983; Malathi et al. 1985). During the decade between 1980 and 1990, the major research area was diagnosis based on electron microscopy, serology and nucleic acid hybridization. Purification of horsegram yellow mosaic virus (HgYMV) (1987) and tomato leaf curl Bangalore virus (ToLCBaV), preparation of polyclonal antibody and its interaction with other begomoviruses were demonstrated (Muniyappa et al. 1991a, b). Immunosorbent electron microscopy was utilised to demonstrate the association of begomoviruses in legumes, vegetables and other crops (Varma et al. 1989, 1993). As the picture on geminiviruses became clear globally, using the polyclonal antibody and DNA A probe to ACMV, geminivirus etiology in several plant species in India was confirmed (Harrison et al. 1991). Using monoclonal antibodies to ICMV and ACMV, the differential epitope profile of viruses were generated for ICMV, bhendi yellow vein mosaic virus (BYVMV), dolichos yellow mosaic virus (DoYMV) and mungbean yellow mosaic virus (MYMV) (Swanson et a1. 1992).

The molecular characterization of begomoviruses by then gained momentum in India and the begomoviruses associated with the yellow mosaic and tomato leaf curl were cloned and sequenced (Varma et al. 1991; Ramachandran et al. 1996; Srivastava et al. 1995). The infectivity of cloned components of the Old World bipartite begomovirus MYMIV was shown for the first time through agroinoculation (Mandal et al. 1997). More or less at this time (1989–1993), the cotton leaf curl epidemic emerged in western India and the involvement of a distinct begomovirus was established (Varma et al. 1993). The period is also significant with the discovery of betasatellites in BYVMV infection in southern India (Jose and Usha 2003) and with cotton leaf curl virus in north western India (Radhakrishnan et al. 2004b). Betasatellites were further found to be ubiquitously associated with all the monopartitie and some bipartitie Old World begomoviruses. As more begomovirus disease problems emerged, there is a greater emphasis in begomovirus research in different Indian institutions; while on one side, molecular mechanism of viral pathogenicity was investigated, in parallel more begomoviruses from diverse plant species including crops and weeds were discovered and characterised.

5.3 Begomoviruses Causing Mosaic Diseases

The begomoviruses which are named after the mosaic symptom (Fig. 5.3) are Indian cassava mosaic virus (ICMV), Sri Lankan cassava mosaic virus (SLCMV), Jatropha mosaic India virus and Coccinia mosaic Tamil Nadu virus. Of these, ICMV and SLCMV are economically more important.

5.4 ICMV and SLCMV

5.4.1 Discovery and Distribution

The begomoviruses causing the severe cassava mosaic disease (CMD) in India were reported as ICMV (Malathi and Sreenivasan 1983; Hong et al. 1993; Dutt et al. 2005) and in Sri Lanka as SLCMV (Saunders et al. 2002b). Although, SLCMV was originally identified in Sri Lanka, the major characterization was carried out from the samples collected from Kerala and Tami Nadu, in India.

The CMD is a typical example of how crop introduced from another continent got infected by the virus occurring indigenously. Cassava, (*Manihot esculenta* Crantz) of the family *Euphorbiaceae* is a perennial shrub with tuberous roots and its origin lies within the boundaries of Amazon centring around Brazil. Though, it was domesticated 10,000 years ago, it remained confined to Latin America until sixteenth century. The Portuguese rulers introduced the crop into Kerala state in the eighteenth and nineteenth century mainly to overcome the famine that existed between 1870 and 1920. The introduced crop was popularized by the erstwhile rulers of Travancore state. The resilient nature of the crop to survive in drought and in acid soil led to its widespread adoption. The CMD constitutes the earliest recorded disease as 'Krauselkranheit' by Warburg (1894) in Africa. As early as 1935, Storey established the viral etiology the disease. The CMD was noticed even in 1940s and Abraham (1956) mentioned this as a challenging threat to the cultivation of the crop. However the first published record of the disease happened only in 1966 by Alagianagalingam and Ramakrishnan (1966). The disease was mainly restricted to important cassava growing regions of India, *viz.*, Kerala and Tamil Nadu. Manivasagam et al. (2006), surveyed and recorded more than 90% disease with disease severity ranging from 2.35 to 4.0. The crop was introduced in other states like Andhra Pradesh, North Eastern Indian States following which, incidence of the disease was noticed in these states too. The inadvertent distribution of virus borne setts has contributed to widespread of the disease where the crop is introduced. While both ICMV and SLCMV are present in Tamil Nadu and Kerala, their distribution in other states is not clear. There has been no report of these two viruses from other countries in Indian Subcontinent. Recently, in 2015, the cassava plants in Ratanakiri, Raunmonn, Cambodia were found infected by SLCMV (Wang et al. 2015).

5.4.2 Economic Loss

The yield loss due to the disease is dependent on the varieties infected. Yield loss ranges from 17 to 88%; in the highly susceptible cultivar Kalikalan, hairy roots like tubers were observed when the symptoms are severe. Malathi et al. (1985), reported upto 45% reduction in highly susceptible cultivars.

5.4.3 Symptomatology

The characteristics symptoms of the disease are discoloured pale green chlorotic area alternating with darker green tissue resulting in mosaic pattern; the leaves may be distorted and appear like leaf curl; in some cases even a shoe string like appearance is observed (Fig. 5.4e). The plant growth may be stunted and the cuttings from infected plants show deteoration.

Fig. 5.4 Mosaic and yellow mosaic diseases in different crops caused by begomoviruses. (**a**–**d**) Mungbean yellow mosaic India virus causing yellow mosaic disease in grain legumes; (**e**) Indian cassava mosaic virus causing mosaic disease in cassava (**f**) tomato leaf curl New Delhi virus causing chlorotic blistering in cucumber; (**g**) corchorus golden mosaic virus associated with mosaic disease of jute

5.4.4 Transmission and Host Range

ICMV is transmitted from cassava to cassava by *B. tabaci*, with acqucition access period (AAP) of 24 h and IAP of 24 h on cassava seedlings (seedlings raised from true seeds) (Mathew and Muniappa 1993). The efficiency of transmission was about 11%. They also showed transmission of ICMV from cassava to *N. benthamiana* to be about 8% and from *N. benthamiana* to cassava about 20%, the overall transmission efficiency was 22%. The low efficiency of whitefly transmission was understood when Lisha et al. 2003 revealed that the population of whiteflies occurring on cassava is very distinct and only these populations successfully transmitted ICMV from cassava to cassava. They showed that the cassava reared whiteflies do not breed on sweet potato and neither the sweet potato reared whiteflies on cassava. However, eggplant and tobacco were common hosts for both biotypes. The cassava adopted haplotype of *B. tabaci* is referred to as Indian cassava biotype. Antony et al. (2006), further investigated vector transmission and demonstrated that after 8 h AAP/IAP, the presence of ICMV could be detected through PCR in the salivary glands, stylet and digestive tract, using specific primers. They obtained 71.43% transmission of ICMV from cassava to cassava. Typical chlorotic mosaic and distortion of leaves were observed 9 days post inoculation (DPI).

Duraisamy et al. (2013) performed transmission experiments using meristem derived virus-free plants of *cv*H226, using cassava strain whitefly by giving 48 h of AAP and IAP. They found that starvation for 3 h before AAP, 48 h AAP resulted in 80.5% of transmission. Typical symptoms were produced 25th DPI. They confirmed the presence of ICMV and SLCMV in the vector through PCR using specific primers for replicase gene. From above results, it appears that whatever might have been bottlenecks in achieving transmission earlier, had been resolved by using cassava strain specific whitefly haplotype.

Both ICMV and SLCMV are highly sap transmissible. ICMV was transmitted to four species of *Nicotiana*, *Petunia hybrid* and *Nicandra physoides*. In all these hosts, leaf curling, crinkling and chlorotic lesion, leaf deformations were produced 6–10 days after inoculation. Severe stunting and reduction in leaf lamina were also observed (Mathew and Muniyappa 1993). However ICMV was not sap transmitted from cassava to cassava, or from any other hosts to cassava.

Jose et al. (2008), performed sap inoculation of SLCMV on 75 plant species, and found that the virus was sap transmitted easily to 39 species belonging to *Solanaceae*. Chlorotic spots, leaf curling and vein clearing were some of the symptoms. SLCMV differed from ICMV in being infectious on *N. longiflora*.Some of the hosts like, *N. amplexicaulis*, *N. benavidesii*, *N. nudicaulis* which are easily infected by ICMV, are not hosts for SLCMV. In all these host species presence of the virus was confirmed by SLCMV specific primers.

Though the experimental host range of ICMV and SLCMV is wide, natural occurrence of these two viruses in other hosts is met with rarely. On the basis of PCR results, association of ICMV has been reported from bittergourd (Rajinimala and Rabindran 2007), mulberry (Sherry 2016) and Jatropha (Aswathnaryana et al. 2007; Gao et al. 2016).

5.4.5 Serology

The symptoms of CMD in India, were not distinguishable from the disease symptoms in cassava in Africa. The electron microscopic evidence for the association of geminivirus with the CMD in India was given by Malathi and Srinivasan (1983). Malathi et al. (1985) showed the positive serological reaction in gel diffusion test with the PAb raised against "T" strain of ACMV. Using the panel of monoclonal antibodies to ICMV and ACMV in TAS – ELISA tests, Harrison et al. (1986) and Harrisson and Robinson (1999) differentiated the Cassava mosaic virus isolates from Africa, India and Sri Lanka into three groups as A, B, C respectively. Mathews and Muniyappa (1992) purified ICMV from *N. benthamiana* and gave clear evidence of the presence of geminate particles by electron microscopy. They produced antibody by purifying coat protein which was used in detecting the ICMV in ISEM tests in cassava, ceara rubber and many hosts. The polyclonal antibody also reacted with other begomoviruses associated with weeds.

5.4.6 Genome Comparison

The complete genome of ICMV and SLCMV were sequenced and their entity as separate species was established (Saunders et al. 2002a, b). The occurrence of SLCMV in India was first noted by Dutt et al. (2005). Using abutting primers, they cloned full-length genomic components from infected cassava samples form Kerala. The genome of ICMV and SLCMV consists of 2.7–2.8 kb circular ssDNA encapsidated within geminate particles. They are bipartite and genome organization is similar to the Old World bipartite begomoviruses with two virion sense (AV1, AV2) ORFs and four (AC1, AC2, AC3, AC4) in complementary sense. In one of the isolates of ICMV, ICMV-KerI, there is 41 bp triple repeats insertion in CR which has extended the CR region upto 267 nucleotides in DNA A, 318 in DNA B. There is also an insertion of unrelated sequence of 21 bp in ICMV-Mah isolate. The iteron sequences of ICMV were identified as GGTACTCA, whereas that of SLCMV was TTGGAGACA similar to iteron sequences of ACMV.

In the GenBank database, full length DNA A sequences is available for 9 isolates of ICMV and 15 isolates of SLCMV. Of nine sequences of ICMV available, four are derived from Jatropha samples. On the contrary, all SLCMV sequences are derived only from cassava. There is nearly 80–84% nucleotide identity in DNA A component between SLCMV and ICMV (Dutt et al. 2005; Saunders et al. 2002b). Both ICMV and SLCMV exhibit 72–74% identitiy with ACMV and less than 69% identity in DNA A component with other begomoviruses. The interesting feature is SLCMV and ICMV share nearly 94% identity in DNA B component, however this identity is more pronounced in the coding region, outside CR, the identity is nearly 97%. Within the noncoding region, identity of SLCMV with ICMV is very less upstream of stem-loop region (57–59%) but more in downstream of stem-loop region (80%). In this case it is speculated that recombination event might have occurred where ICMV DNA B sequences up stream of stem loop might have been replaced with the same region from SLCMV DNA A. As a result of this recombination event SLCMV DNA A clones have transferred *cis* acting iteron motif required for replication.

Rothenstein et al. (2006) employed PCR-RFLP strategy to assess the variability among ICMV and SLCMV isolates. Full length DNA A types were restricted by *EcoRI*, *Hpa II* and *Sau3 A1* and their profiles were compared to identify ICMV and SLCMV isolates. On the basis of restriction profile they could decipher six genotypes, which on complete nucleotide sequencing revealed the presence of more number of SLCMV isolates than ICMV. When Patil et al. (2005) investigated the biodiversity using PCR-RFLP, and found that ICMV and SLCMV though present in mosaic affected areas, ICMV was restricted to certain regions, whereas SLCMV was widespread. On the basis of RFLP pattern, they also recorded high proportion (40%) of samples showing novel patterns different from ICMV and SLCMV. In their extensive survey, extent of mixed infection was low and the randomly distributed point mutation gave rise to novel RFLP patterns and not any recombination event.

5.4.7 Phylogenetic Relationship

Rothenstein et al. (2006) performed analysis taking recombination into consideration and showed that the tree topologies generated vary when partitioning of analysis is done between DNA A segments 2778–857, 857–1957 and 1958–2777 nucleotides. Different tree topologies obtained when they analysed the DNA A dissecting into three segments are indicative of recombinatorial events. Especially in the case of SLCMV (*CO*) and ICMV Mah isolates. When full length DNA A is considered without partitioning, all ICMV and SLCMV isolates get distinctly separated as two clades. Interestingly, when analysis is performed for DNA B components, the ICMV and SLCMV cluster together. Legg et al. (2015) analyzed the relationship between all cassava infecting geminiviruses and found that the ICMV and SLCMV representing South Asian groups of Cassava geminiviruses are well separated from other African CMGs.

5.4.8 Recombination

Nucleotides comparison between ACMV, ICMV and SLCMV clearly established that SLCMV show high nucleotide identity outside CR, the identity especially upstream of loop region is very less. This region is highly conserved among all SLCMV isolates. Saunders et al. (2002b) suggested that recombinations have occurred between SLCMV DNA A and ICMV DNA B. This phenomenon referred to as regulon grafting (transfer of intergenic region harboring replicational and transcriptional control elements) so as to enable SLCMV DNA A to capture ICMV DNA B and introduce *cis* acting iteron sequences to achieve replication. Rothenstein et al. (2005), Patil et al. (2005) also suggested several recombinational events occurring in DNA A component.

5.4.9 Establishment of Koch's Postulate

The cloned DNA components of ICMV (lDNA A and DNA B) and SLCMV when introduced either through mechanical inoculation or through *Agrobacterium tumefaciens* produced symptom readily in *Nicotiana benthamiana*, SLCMV produced severe stunting, leaf curl and chlorosis and inoculation with ICMV led to leaf curl symptoms. SLCMV was more virulent and severe than ICMV. Saunders et al. (2002b) demonstrated that ICMV clones were infectious on *N. clevlandi* and *N. glutinosa*. In the case of SLCMV, DNA A alone induced leaf roll and vein swelling symptom in *N. benthamiana.* The cloned components of SLCMV successfully induced symptom expression in cassava by biolistic inoculation. Leaf curl and mosaic symptoms were observed, 6 weeks after inoculation (Saunders et al. 2002b). Dutt et al. (2005) also obtained a mild mosaic symptom expression in meristem derived cassava plants cv Ebwanateraka 6 month post inoculation after biolistic delivery.

Symptom expression following inoculation of cloned ICMV component on cassava continued to be elusive. This bottleneck was overcome, when Rothenstein et al. (2005) infected cassava plants by delivering ICMV cloned components using a hand held particle gun. All the plants showed mild CMD symptom within 60 DPI. Mittal et al. (2008) proved the infectivity of SLCMV on cassava and the model host *Arabidopsis* by agroinoculation. They observed that post inoculation, DNA extracted from *Arabidopsis* was essentially multimers and showed slow electrophoretic mobility. In all these infectivity experiments, the presence of the viral DNA was demonstrated by analysis of replicative forms in Southern blots.

5.4.10 Exchange of Components

The cassava geminiviruses, ACMV, SLCMV and ICMV offer interesting experimentation system to study the complementation/ re-assortment of genomic components between begomoviruses. The pseudorecombination produced by re-assortment of components of ACMV and ICMV did not result in infection in *N. benthamiana* proving that these two are distinct virus species and exchange of component is not possible. Contrastingly pseudorecombinant produced by inoculation with ACMV DNA A and SLCMV DNA B, was infectious on *N. benthamiana*. This further suggests that iteron conservation seen between ACMV and SLCMV has promoted the *trans* replication of SLCMV DNA B by ACMV DNA A. Karthikeyan et al. (2016) demonstrated pseudorecombination between SLCMV and ICMV in *N. benthamiana* which resulted in symptom expression and the replication of ICMV DNA B by SLCMV DNA A.

Interestingly, SLCMV DNA A alone when inoculated on *N. benthamiana* produced symptom phenotype similar to the phenotype produced by the monopartite begomovirus. In addition Saunders et al. (2002b) showed that SLCMV DNA A replicated betasatellite, Ageratum yellow vein mosaic betasatellite and produced severe downward leaf curl symptom in *N. glutinosa* and yellow vein symptom in *Ageratum conyzoides*. The symptom produced in ageratum was like those produced by AYVMV and beet curly top virus (BCTV). Hence, Saunders et al. (2002b) suggested that SLCMV perhaps is a monopartite virus, which evolved to become bipartite by capturing ICMV DNA B.

5.4.11 Molecular Basis of Pathogenicity

The role of coat protein in symptom production was examined by Kelkar et al. (2016), taking SLCMV/*N. benthamiana* system. They generated CP null mutants, single, double triple and quadruple amino acid replacement mutations and tested the infectivity in comparison with wild type. The coat protein null mutants were not infectious, other mutation in combination or alone produced mild symptoms and reduced viral titre. The mutants like K129P/F152E led to the absence of leaf rolling, which was restored when the mutant K128W was inoculated jointly. The viral titre as determined by qPCR was more than 100-folds reduced in some mutants like T128L, S134H, F152E. These changes in amino acid residues were predicted to affect the secondary structure of coat protein which may affect its other functions like interaction with V2.

Interestingly Resmi et al., (2014, 2015) observed a strange phenomenon that SLCMV, Rep gene induced high rate of transportation of IS426 elements in *Agrobacterium tumefaciens* when SLCMV – Rep gene under the transcriptional control of 35S promoter in sense orientation was introduced into *Agrobacterium tumefaciens* strain EHA105; the binary plasmid size increased in 15-folds. The sequences proximal to 35S promoter were rearranged, this region comprised of 1.3 kb IS426 element of *Agrobacterium tumefaciens*. Transfer of Rep gene in anti-sense orientation or non-functional version of Rep did not cause the transposition. They suggest that SLCMV Rep may act similar to RepA protein and may act along with transposans to trigger the transposition of 1S426.

Another dimension of interaction between *Agrobacterium tumefaciens* and begomoviruses was studied by Resmi et al. (2015), by raising transgenic *N. benthamiana* incorporated with vir E2 gene of *A. tumefaciens.*The three transgenic plants showed attenuated symptom after agroinoculation with SLCMV viral DNA and suggest that Vir E2 may be a good candidate gene to develop resistance against geminivirus.

5.4.12 Virus Prevalence in the Field

Survey conducted in Kerala (Jose et al. 2011) and Tamil Nadu (Rajinimala et al. 2011) clearly pointed out that, the disease incidence is very high, nearly 100% in some locations; in these locations there was high whitefly population of 17 or 15/plant. Jose et al. (2011) clearly showed that the infected planting material contribute to more disease spread than whitefly mediated transmission. In both the states, incidence of SLCMV infection was always higher than the ICMV, mixed infection of ICMV and SLCMV together were seen upto 12–16% that too only in two districts of Thiruvananthapuram and Malappuram. Jose et al. (2011) detected up to 33% infection of ACMV in the district Path in amthitta which needs to be carefully looked into. In the context of mixed infection Karthikeyan et al. 2016 made some interesting observation. The cassava plants infected by SLCMV collected from Malappuram and Thiruvananthapuram were maintained in glass house at MKU, India and University of Bazel, Switzerland. The plants from Thiruvananthapuram did not show persistent infection of SLCMV, exhibited symptom recovery phenomenon. The Malappuram isolate was persistent. Interestingly the SLCMV infected Thiruvananthapuram plant, exhibited the emergence of symptom after 6 month period, but the re emergent virus was found to be ICMV. All the three isolate, persistent SLCMV, non- persistent SLCMV and re-emerged ICMV were infective on *N. benthamiana.* Interestingly pseudorecombination occurred between SLCMV DNA A which transreplicated ICMV DNA B.

5.4.13 Resistance to SLCMV and ICMV

During the last five decades, continuous attempts are being made to identify resistant lines (Nair et al. 1998; Abraham et al. 2006; George et al. 2012). The cultivar SreePadmanaba (TMS30001 line Mnga-1 line) has been evaluated for resistance in field conditions and has been released for cultivation in Tamil Nadu. The open pollinated seedlings of Mnga-1 were evaluated from which 242 resistant lines have been identified. Two of three lines CMR-1 and CMR 129 are expected to perform well (Unnikrishnan et al. 2011). About 56 clones derived from West African land races having CMD-2 gene were crossed with inbred lines at CTCRI, India which are being evaluated (Sheela et al. 2012). Several interspecific hybrids involving species *Manihot glaziovii*, *M. caerulescens*, *M. peruviana* were generated of which, hybrids of *M. caerulescens* showed higher level of resistance. The hybrids are used as resistance donors to introgress the genes into elite Indian cultivars; One such derivative CMC-1 (Sheela et al. 2012) shows resistance. From above results it is not clear whether the resistance in Indian cassava cultivars to ICMV and SLCMV is governed by the polygenic recessive gene (CMD-1) or a major dominant gene (CMD-2).

5.5 Begomoviruses Causing Yellow Mosaic Disease

Begomoviruses causing yellow mosaic diseases are known to infect plant species of the families Leguminosae, Verbenaceae and Malvaceaeare/Tiliaceae in India. Mungbean yellow mosaic virus (MYMV) and mungbean yellow mosaic India virus (MYMIV) affect many grain legumes like cowpea, mungbean, blackgram and soybean (Fig. 5.4a–d). *Clerodendron inerme*, a common hedge plant grown in India is affected by a yellow mosaic disease, where Clerodendron yellow mosaic virus, a new begomovirus was identified (Sivalingam et al. 2011). The yellow mosaic disease of Jute (*Corchorus capsularis*) was described from West Bengal in 1978. Some of the elite cultivars of Jute like JRC 7447 and JRC212 showed nearly 50% of disease incidence (Ghosh et al. 2008). Corchorus golden mosaic virus, a New World begomovirus associated with yellow mosaic of Jute was characterized by Ghosh et al. 2008, 2012. Interestingly, the nonanucleotide sequence at the origin of replication of this virus was CATTATTAC instead of TAATATTAC, It is significant that a New World bipartite begomovirus has been described in a crop which has been cultivated since seventeenth century in India. MYMIV and MYMV are the two most important begomovirus causing causing widespread yield losses in grain legumes in India.

5.6 MYMV and MYMIV

The two virus species, MYMV (Morinaga et al. 1990; Ramachandran et al. 1996) and MYMIV (Varma et al. 1991) causing yellow mosaic disease in grain legumes constitute the most interesting group of viruses that are evolutionarily most diversed from all the so far known begomoviruses in India. Considering that most of the pulse crops have the centre of origin in the Indian subcontinent, yellow msaic viruses may be considered as the virus of Indian origin. The begomoviruses causing yellow mosaic disease in legumes in Asian and South East Asian countries are designated as legumoviruses (LYMVs) due to their uniqueness. Currently, LYMVs consist of seven members, MYMV, MYMIV, dolichos yellow mosaic virus (DoYMV), horsegram yellow mosaic virus (HgYMV), Rhynchosia yellow mosaic virus (RhYMV), Rhynchosia yellow mosaic India virus (RhYMIV) and velvet bean severe mosaic virus (VbSMV). Of these LYMVs, MYMV and MYMIV are important as they have wide host range and cause considerable economic loss.

5.6.1 Discovery and Distribution

The earliest record of yellow mosaic disease is in 1948–1950, much later than the record of yellow vein mosaic disease of bhendi by Kulkarni (1924). Capoor and Varma (1948a, b, 1950) observed yellow mosaic symptoms in 1940 at Poona, in Dolichos and Lima bean (*Phaseolus lunatus*). They traced the yellow mosaic disease upto Gujarat, Khandesi and Deccan regions. The virus causing the disease was transmitted by whitefly and they rightly identified the viruses which occurred in dolichos and Lima bean as two distinct viruses which were not transmissible to other leguminous hosts. About 5 years later in 1955, Nariani observed yellow mosaic symptoms in mungbean at an experimental farm at IARI New Delhi and identified the virus distinct from dolichos and designated it as a mungbean yellow mosaic virus, as it was not transmitted either to dolichos or Lima bean (Nariani 1960). Subsequently yellow mosaic disease symptoms were observed in many pulse crops such as blackgram, cowpea, cluster bean, French bean, groundnut, horsegram, hyacinth bean, moth bean, mungbean, Lima bean, pigeonpea and soybean (Table 5.6). The disease was described mainly as yellow mosaic, occasionally as yellow flecks. Due to overlapping host range it was presumed that the disease in all these leguminous hosts is caused by mungbean yellow mosaic virus (Nariani 1960). Meanwhile the disease was also reported from Pakistan in cowpea and mungbean (Ahmed and Harwood 1973) in mungbean from Bangladesh (Jalaluddin and Shaikh 1981), Sri Lanka (Joseph et al. 1998) and Thailand (Shimizu et al. 1987). MYMIV was recently found associated with tomato leaf curl betasatellite in kidney bean in Oman (Shahid et al. 2017).

5.6.2 Economic Loss

Reduction in yield in grain legumes is highly dependent on the time of infection. The decrease in yield is marked when infection strikes at early stage. Infection of blackgram at four, five, six, seven and eighth week resulted in yield reduction upto 85%, 60%, 44%, 28% and less than 10% respectively (Nene 1972, 1973; Vohra and Beniwal 1979; Dhingra and Chenulu 1985; Suteri and Srivastava 1979). Singh et al.

Table 5.6 Crops and areas affected by mungbean yellow mosaic India virus and mungbean yellow mosaic virus in India

S.No.	Crop	Disease	Area	Reference
1.	Blackgram (*Vigna mungo*)	Yellow mosaic (yellow mottle and necrotic mottle)	Northern India	Nene (1973), and Singh et al. (1979)
			South India (Coimbatore)	Murugesan et al. (1977)
2.	Cowpea (*Vigna unguiculata*)	Yellow Mosaic	Northern India	Nene (1972)
		Yellow fleck	Northern India	Sharma and Varma (1976)
		Golden mosaic	India	Varma and Reddy (1984) and Srivastava and Varma (1988)
3.	Clusterbean (*Cyamopsis tetragonoloba*)	Yellow mosaic	Southern India	Rao et al. (1982)
4.	French bean (*Phaseolus vulgaris*)	Yellow mosaic	Uttar Pradesh (Pantnagar)	Singh (1979)
			Western India (Pune)	Unpublished result
5.	Groundnut (*Arachis hypogea*)	Yellow mosaic	Southern India	Sudhakar Rao et al. (1979)
6.	Horsegram (*Macrotyloma uniflourum*)	Yellow mosaic	Southern India	Muniyappa et al. (1975) and Muniyappa and Reddy (1976)
7.	Hyacinth bean (*Lablab purpureus*)	Yellow mosaic	Several part of India	Capoor and Varma (1948a, b), Muniyappa et al. (1975)
8.	Moth bean (*Vigna aconitifolia*)	Yellow mosaic	Rajasthan	Satyavir (1980)
9.	Mungbean (*Vigna radiata*)	Yellow mosaic	In all parts of India	Mishra et al. (1978), Nariani (1960), Varma et al. (1992), Bansal et al. (1984), and Nene (1973)
10.	Lima bean (*Phaseolus lunatus*)	Yellow mosaic	Western India	Capoor and Varma (1948a, b)
11.	Pigeonpea (*Cajanus cajan*)	Yellow mosaic	Northern India	Williams et al. (1968)
12.	Soybean (*Glycine max*)	Yellow mosaic	Madhya Pradesh	Keshwal et al. (1988) and Suteri (1974)

(1979) observed 19–21% reduction in blackgram and mungbean cultivation. In the blackgram cultivar T-49, yellow mosaic virus infection affects seed quality too as protein profile and nutrient content get altered. Varma et al. (1992) predicted that the yield loss due to YMD could be as high as $300 million in an epidemic year taking blackgram, mungbean and soybean together.

5.6.3 Symptomatology

The typical symptoms caused by infection of the yellow mosaic viruses as the name implies are the characteristic bright yellow/or golden mosaic. To start with, infection appears as small yellow specks on the veinlet of the young unfurling leaves which enlarge to form mosaic patterns with irregular green patches alternating with each other. The yellow area increases, coalesces and produces complete yellowing of leaves. Nene (1973) also recorded necrotic mottle symptoms in resistant cultivars of blackgram.

The affected plants produce fever flowers and pods, they turn yellow in colour and size of pods and seeds are heavily reduced. In French bean, when infected by MYMV and MYMIV downward leaf curl and stunting symptoms are produced; in the field, natural infection by HgYMV, bright yellow mosaic symptoms are expressed in French bean.

5.6.4 Transmission and Host Range

Both MYMV and MYMIV are transmitted by Whitefly, *Bemisia tabaci* Genn. in a persistent circulative manner. While yellow mosaic virus isolates of India are not sap transmissible, the mungbean isolate from Thailand is mechanically transmissible (Honda et al. 1983). *B. tabaci* is able to acquire and inoculate the virus in minimum acquisition and inoculation access periods (AAP and IAP) of 10–15 min each and optimum AAP and IAP are between 4–6 and −4 h, respectively. After an AAP of 30 min the virus requires a latent period of more than 3 h in the vector for transmission to occur (Nair and Nene 1973). However Chenulu et al. 1979 reported that latent periods have no effect. After AAP and IAP of 24 h each, single whitefly could infect 25% of plants and for 100% transmission, four to ten whiteflies per plant are required (Nair and Nene 1973). Female whiteflies are better transmitters and also retain the virus for a longer period (10 days) than the male whiteflies (3 days) (Rathi and Nene 1974). Neither female nor male adults can retain infectivity throughout the lifespan. Nymphs can acquire the virus from the infected leaves but not the first instar (Rathi and Nene 1974; Murugesan et al. 1977). The virus does not pass through the eggs of *B. tabaci*. The HgYMV was transmitted to several leguminous hosts and required only 30 min of IAP and 10 min of AAP. Incubation period in the vector was 6 h and infectivity was retained for 12 days (Muniyappa and Reddy 1976).

The most characteristic feature of YMVs is its very limited narrow host range. The two virus species MYMV and MYMIV are recorded only in legumes and earlier reports of its probable occurrence in *Brachiaria ramosa, Eclipta alba, Cosmos bipinnatus* and *Xanthium strumarium* (Rathi and Nene 1974) have been disproved after molecular characterization of viruses infecting those hosts has been accomplished. Within the leguminous hosts too, the four virus species exhibit difference in infectivity. MYMV, MYMIV and HgYMV infect majority of the crops such as blackgram, mungbean, soybean, horsegram, moth bean, French bean and pigeonpea. These two species MYMV and MYMIV do not infect cowpea and dolichos;

however through agroinoculation MYMIV isolate was shown to infect cowpea (Malathi et al. 2005) and a MYMIV isolate has been characterized from dolichos (Singh et al. 2006) which are suggestive that MYMIV can infect these two hosts too. There are differences in symptoms caused by HgYMV in French bean. While MYMV and MYMIV cause leaf curl like symptoms, HgYMV produces yellow mosaic. The virus species DoYMV has a very restricted host range that it infects only dolichos from which it has not been transmitted to any other host.

5.6.5 Seed Borne Nature of MYMV

The yellow discoloration of pods and seeds of infected plants and symptom emergence in the very first trifoliate leaf of the plants in the field were suggestive that MYMV may be seed borne, which was investigated by Satya et al. (2015). The distribution of the virus in various parts of the seeds of blackgram (*Vigna mungo* L. Hepper) plants naturally infected in the field was determined by polymerase chain reaction (PCR). Southern blot analysis and nucleotide sequencing of the PCR amplicons from the seed parts from groups of ten seeds revealed the presence of MYMV in the seed coat, cotyledon, and embryonic axes.The presence of virion particles was confirmed through double antibody sandwich enzyme-linked immune sorbent assay (DAS-ELISA) and immunosorbent electron microscopy (ISEM) even in a single whole seed. In confocal microscopy, positive fluorescent signals were obtained using coat protein gene-specific primers in the embryonic axes. However, in the grow-out tests performed with the same batch of seeds, therewas no symptom development in the seedlings though the virus (both DNA A and B components) was detected in 32% of tested seedlings.

5.6.6 Serology

Presence of geminate particles, reaction to polyclonal antibodies to ACMV and ICMV, positive hybridization with DNA A probe to ICMV gave clear indication that viruses causing the YMD are begomoviruses. Muniyappa et al. (1987) purified HgYMV and gave clear evidence for association of geminate particles and produced polyclonal antibody to HgYMV. Epitope profile generated on the basis of reaction to monoclonals to ACMV and ICMV differentiated the yellow mosaic viruses into two groups, one group comprising YMV in dolichos and another group including viruses infecting all other legumes (Harrison et al. 1991; Swanson et al. 1992). Genomic components of a mungbean isolate of YMV from Thailand were cloned by Morinaga et al. 1990. Varma et al. (1991) cloned the genomic component of blackgram isolate of MYMIV at IARI, in parallel a blackgram isolate of MYMV was cloned at Madurai by Ramachandran et al. (1996). Subsequently YMV isolates from mungbean, moth bean, pigeonpea, cowpea, soybean, dolichos and horse gram have been cloned and sequenced (Malathi 2007).

5.6.7 Genome Comparison

The genome of LYMVs comprises of two components DNA A and DNA B and their organization is similar to Old World bipartite begomovirus. The common region spans from 110 to 180 nucleotide length and has the Rep binding iteron sequences, promoters of Rep and CP gene and the characteristic stem-loop region having nono-nucleotide sequence. The comparison of complete nucleotide sequence of DNA A component of yellow mosaic viruses with other begomoviruses, led to clear differentiation of four species, MYMV, MYMIV, DoYMV, HgYMV on the basis of 91% species demarcation. Three other yellow mosaic viruses described from India are RhYMIV (Jyothsna et al. 2011) and VBSMV (Zaim et al. 2011), rhynchosia yellow mosaic virus (unpublished GenBank accession no. KP752090). The most distinct virus is DoYMV which shares only 61% identity with all other YMVs. The identity between the two species MYMV and MYMIV is 81%. RhYMIV and VBSMV exhibit 71–72% identity with MYMV and MYMIV. The relationship is more or less similar with respective isolates of MYMV and MYMIV from Thailand, Pakistan, Bangladesh, Nepal and Indonesia. In the DNA B component, the cognate DNA B of Thailand isolate MYMV-[TH-Mg1] and one blackgram isolate of Vamban MYMV-[KA 27] identity with DNA B of MYMIV-Bg3, MYMIV-Cp, MYMIV-Mg and MYMI-Sb is only 67%.

5.6.8 Multiple DNA B Components

The most unusual feature of MYMV is association of one DNA A with multiple DNA B. One blackgram isolate of MYMV from south India, MYMV-[IN:Vig] is associated with two distinct types of DNA B component. One type of DNA B MYMV-[KA 27] which shows 97% sequence identity with DNA B of Thailand isolate; and other set of DNA B's, KA22, KA28, KA34 which show only 71–72% identity with Thai isolate, but exhibited nearly 90–92% identity with DNA B of MYMIV (Karthikeyan et al. 2004; Balaji et al. 2004). Two types of DNA B components are also associated with a soybean isolate of MYMV [IN: Mad:Sb], one being closely related (96%) to DNA B of HgYMV.

Critical comparison of nucleotide sequence of the DNA B components of MYMV along with one DNA B variant cloned from Gujarat MYMIV-IN Anand 25 (John et al. 2008) revealed how the components have evolved. The four DNA B components, three associated with MYMV, one with MYMIV between themselves share 96% identity in the coding region ORF BVI and ORF BCI. However they differ in the non coding region. While DNA B of MYMV- KA22, KA28, KA34 exhibited similarity with CR of MYMV-[IN Vig], Gujarat isolate showed maximum identity with CR of MYMIV (John et al. 2008). These DNA B molecules referred to as DNA B variants may represent molecules generated by exchange of components between MYMV and MYMIV. Swapping of CR could have occurred from MYMV to MYMIV ("origin donation or regulon grafting" when both the viruses were present together in mixed infection. This is well borne out by the divergence observed between A and B components in the CR region in MYMV and MYMIV.

5.6.9 Association of Satellites

Analyses of LYMV infected samples by RCA let to unexpected identification of both beta and alphasatellites. Over the past few years, betasatellite has been found associated with MYMIV (Rouhibakhsh and Malathi 2005) and MYMV infected plants (Sathya et al. 2013). The symptoms in the presence of betasatellites are severe like crumpling, and severe leaf curl. In all these cases the betasatellite was identified as papaya leaf curl betasatellite. Sathya et al. (2013) found that samples of MYMV infected blackgram samples revealed the presence of alphasatellites, which was identified to belong to the vernonia yellow vein alphasatellite species. The importance of association of these satellites in LYMV pathogenicity is not understood, whether such tri/tetrapartite association is stable and contributes to viral pathogenicity needs to be looked into.

5.6.10 Divergence in CR

The most characteristic feature of MYMV/MYMIV is the divergence in CR between DNA A and DNA B component. Unlike the NW begomoviruses, wherein CR is near identical, 15–23% divergence was observed in whole CR and 23–29% in the origin of replication (Usharani et al. 2004a, b; John et al. 2008; Girish and Usha 2005). Compared to DNA B of Thailand isolate, there were deletion and mismatches in all the DNA B components. The most prominent deletion observed was deletion of 18 nucleotides from the nucleotide co-ordinate 2632–2649 compared to MYMV KA 27 and MYMV-NAM. While the rep binding iteron sequence is ATCGGTGT in MYMV/MYMIV with HgYMV it is GGTAT.

5.6.11 Phylogenetic Relationships

The phylogenetic analysis of YMV genome with other begomoviruses revealed that the viruses infecting legumes, (from Asian and South Asian countries) are distinct from and basal to all other begomoviruses; When other begomoviruses are clearly separated into Old World and New World viruses, the two group of viruses legumoviruses and sweepoviruses are distinct and do not group with either OW or NW viruses. Within legumoviruses, MYMV and MYMIV cluster together and DoYMV occupies a separate clade.

5.6.12 Recombination in LYMVs

The recombination events were considered to be low among LYMVs as mixed infection of LYMVs with other viruses within a single host species has not yet been met with; the multiple DNA B components though is suggestive of mixed infection of MYMV and MYMIV followed by component exchange, no recombination events has been predicted. However, Ramesh and Chauhan (unpublished, personal communication) analyzed all the LYMVs and predicted four different recombination events in DNA A, of which the event 3, spanning from nucleotide co-ordinate 557–1074 in DNA A was detected in 48 isolates of MYMIV. Recombination in DNA A was higher in MYMIV (48 isolates) than MYMV and DoYMV. In DNA B component, three events were detected in MYMV and one in soybean isolate of MYMIV.

5.6.13 Establishment of Koch's Postulates

The technique of agroinoculation wherein *Agrobacterium tumefaciens* cells having more than one copy of DNA A and DNA B component in a binary vector are used to deliver the viral genome is widely used to prove the infectivity. For the first time, infectivity of OW bipartitie virus was demonstrated through agroinoculation. In the case of MYMV/MYMIV, the sprouting seeds are inoculated with slurry of bacterial cells, from which the replicating genome of the virus is rescued, which replicates actively resulting in systemic spread and typical symptoms (Mandal et al. 1997).

Adapting this technique Koch's postulates were established for cloned component of MYMIV isolates of blackgram (Mandal et al. 1997), mungbean (Biswas and Varma 2001), cowpea (Malathi et al. 2005), pigeonpea (Chakraborty 1996), soybean (Usharani et al. 2005); MYMV- blackgram isolate (Karthikeyan et al. 2004; Balaji et al. 2004; soybean isolate (Girish and Usha 2005) and HgYMV (Barnabas et al. 2010). It is interesting to note here that in all the clones of genomic components used, there was considerable divergence in CR between A and B components, but still, DNA A could *trans*-replicate DNA B and systemic symptoms were produced.

This technique was also used to prove that the yellow mosaic disease in cowpea is caused by a variant of MYMIV (Malathi et al. 2005) in north India. Normally MYMIV isolates of blackgram, mungbean, are not transmitted to cowpea by whitefly and *vice versa*. But Malathi et al. (2005) showed that MYMIV-cowpea isolate, agroinoculated onto blackgram and mungbean, produced typical symptoms from which it could be transmitted back to cowpea. The adaptation of cowpea isolate when inoculated on to blackgram and greengram plants was maintained by viral progeny which could be easily transmitted to cowpea. However the blackgram isolate could not infect cowpea, it only produced atypical leaf curl like symptoms. Though, blackgram and cowpea isolates of MYMIV shares 96% of sequence identity (Surendranath et al. 2005) reassortment of components (pseudorecombination) did not result in symptom production, suggesting, hindrance in the complementation both in replication and systemic movement of the components.

Jacob et al. (2003) suggested that presence of both DNA A and DNA B components in one strain of agrobacterium improved the infectivity of clones. Inoculations were performed with MYMV DNA A with different DNA B components (Balaji et al. 2004) and DNA B specific responses were observed. Inoculation of MYMV with cognate DNA B, KA27, which is closely related to DNA B of Thailand isolate of MYMV, resulted in severe symptoms of stunting and high viral titre in mungbean. However, inoculation with KA22 DNA B (more closely related to MYMIV DNA B) showed intense mosaic and high viral titre in blackgram. Mahajan et al. (2011) further analysed the response of blackgram plants to inoculation with combinations of different DNA B. when DNA A was co-inoculated with KA27+KA22 DNA B, there was amelioration of severe stunting, reduction in appearance of yellow mosaic symptoms typical of KA22 in blackgram plants. These results were obtained even if KA22 were inoculated post inoculation of KA27. However, such yellow mosaic symptoms caused by KA22 could not be ameliorated by KA27 inoculation. By doing exchange of viral proteins, Mahajan et al. (2011), concluded that, the severe stunting phenotype obtained while inoculating KA27 DNA B is contributed by ORF BV1, the nuclear shuttle protein which seems to be the major symptoms determinant of MYMV DNA B. Kuruba et al. (2016) explained the differential symptom expression on the basis of ORF BV1 (NSP). They demonstrated by leaf tissue hybridization that in KA 27 infected mungbean plant and KA22 infected blackgram plants, mesophyll spread of the virus was observed. On the contrary, KA27 infected blackgram plants, virus did not spread to mesophyll cells. By exchanging NSP fragments of KA22 and KA27, they hypothesized that NSP determines the mesophyll spread.

5.6.14 Molecular Basis of Pathogenicity

5.6.14.1 Viral Replication and Reprogramming Cell Cycle Machinery

The viral DNA replication takes place either by rolling circle replication or by recombination dependent way. The geminiviruses encoded Rep initiates the replication by cleaving at the nonanucleotide site, but does not have polymerase activity. The host DNA polymerase and many host factors together referred to as host replisome take part in viral DNA replication. Three phases are recognized in viral DNA replication. They are initiation, elongation and termination.

Replication initiation protein or Replication associated protein (Rep) is a multifunctional protein, which is highly conserved in all geminiviruses. Rep executes ATP dependent isomerase I, ATPase and Helicase activities (Hanley-Bowdoin et al. 2013; Pant et al. 2001; Yadava et al. 2010). Mukherjee and his associates expressed the MYMIV protein and studied the events of replication (Singh et al. 2008; Yadava et al. 2010). Rep binds to the unique iteron sequence in a co-operative manner which results in conformational changes in DNA, like formation of cruciform structure. More than one molecule of Rep is required to perform both nicking and ligating activities, hence Rep oligomerizes and further hetero-oligomers are formed between Rep and REn.

Following the initiation by nicking activity of Rep resulting in free 3'OH end, elongation phase occurs. During the elongation phase, Rep, REn and many host fork proteins get assembled at the 5'3' direction. In order to synthesize new strand, the existing viral strand need to be dislodged by DNA unwinding. Yadava et al. (2010) showed that MYMIV possess characteristics motive of Helicase, with limited processivity, MYMIV Rep helicase translocates in 3'–5' direction and requires ssDNA of minimum six nucleotide length. It was also found that mutation in oligomerzation domain results in abolishment of helicase activity. At the end of elongation phase, the nascent concatenated DNA is subjected to Rep mediated nicking and ligation releasing the full length ss circular DNA. The newly synthesized ss DNA either re-enters the replication process resulting in a more number of copies of viral DNA or gets encapsidated. To facilitate active replication, Rep interacts with several DNA machinery protein; both Rep and REn interact with proliferating cell nuclear antigen (PCNA) which is a processivity factor of host-DNA polymerase and is involved with DNA replication and repair; with large unit of replication factor C complex and subunit of replication protein A. MYMIV Rep interaction with both PCNA and Rep C complex, down regulate nicking and ligating activity. The most interesting interaction was shown with RAD54, a protein involved in homologous recombination. MYMIV-Rep interaction with RAD54 was examined in vitro using yeast system and *in planta* in *Arabidiopsis thaliana*, which showed that contrasting to PCNA, interaction with RAD54 enhances ATPase and Helicase activities.

Geminiviruses invade cells which are fully differentiated, they have exited the S phase and do not have conducive environment for DNA synthesis. The characteristic features of geminivirus infection are the reentry of the cell into S phase. Transcriptomics analysis of geminivirus infection revealed upregulation of genes associated with late G1, S and early G2 phase and downregulation of those associated with early G1 and Late G2 phase. Reentry into S phase and sustaining the environment at increased DNA synthesis phase is brought about by Rep binding to the key molecule of the cell cycles, the retinoblastoma related protein (RBR). Rep binding to RBR ensures release of the transcription factor E2F, which activates genes encoding plant DNA polymerase and other accessory factor.

The Rep interacts with kinases, GRIMP and GRIKI, serine threonine kinase and with histones H3. The host protein with which MYMIV- Rep protein interacts was studied by phage display library and pull-down assay. The MYMIV replicon was cloned in YrP34 vector and the rolling circle replication was demonstrated in yeast (Yadava et al. 2010). This replicon was used to examine eukaryotic host factors needed for replication and more than 150 factors were isolated and about two dozen factors have been found to be required for MYMIV DNA replication in yeast system.

5.6.15 Interaction Between Rep and CP

MYMIV-CP was expressed by Malik et al. (2005) and its interaction with MYMIV Rep was demonstrated and they showed that MYMIV Rep binding with CP caused downregulation of nicking activity. The MYMIV-CP by blocking RCA initiation regulates ssDNA level. The pre coat protein/AV2 of MYMIV was also shown to affect the nicking activity of Rep which was further confirmed by *in planta* studies wherein inoculations of viral genome with mutations in AV2 ended up in reduction in super coiled DNA (Rouhibakh and Malathi 2011, 2012).

Guerra-Peraza et al. (2005), analyzed the MYMV coat protein and revealed that, it has two nuclear localizing signals (reresidue 3 KR and 41 KRRR) and interacts with nuclear import factor import in α. Thus CP may get imported into nucleus through importin alpha dependent pathway.

5.6.16 Bidirectional Transcription and its Regulation

Geminivirues genome is tightly packed and bidirectionally transcribed with overlapping genes on either side of the intergenic region. They utilize the host RNA polymerase II for transcription and the complementary sense gene (Rep) represents the early phase of transcription and the viral sense genes, coat protein and movement protein are expressed later. Typically multiple polysistronic RNA are produced with precised 5′ end and common 3′ end. The mapping of polyadenylated transcripts have been completed for MYMV and MYMIV.

The bidirectional transcription of geminiviruses is regulated by an interesting viral encoded protein referred to as AC2/ or C2 as Transcriptional activator protein (TrAP). It is a multifunctional 15 kDa Zinc binding protein which has a C- terminal acidic type of activation domain and N terminal nucleic acid binding domain. It binds to ssDNA in a sequence non specific manner and binding is facilitated through Zn finger motifs (Bisaro 2006) It *trans* activates the right ward promoters in both DNA A and DNA B. In the case of New World begomoviruses, Bisaro (2006) suggested that TrAP activates CP promoters in the mesophyll cells, and in phloem tissues, removes a suppressor element that inhibited TrAP independent activity. Trink et al. (2005) demonstrated TrAP mediated transactivation of MYMV AV1, BV1 as well as BC1 promoters by transciant expression studies in *Nicotiana plambaginifolia* protoplast. The MYMV-TrAP can also activate host gene transcription. Constitutive expression of TrAP under its own promoter does not produce lethal phenotype, whereas under 35S promoter toxic effects are observed.

Shivaprasad et al. (2005) mapped the viral transcripts from MYMV- infected blackgram plants using the circularized RNA as template which facilitate mapping of both 5′ start and 3′ poly adenylation site. They isolated two major rightward transcripts initiation (~1 kb), ORF AV1 could be translated from both the transcripts and ORF AV2 from longer transcripts.

Study on transcription units of leftward ORFs revealed interesting results. A single major transcripts start site was located at position 2649, which may represent dicistronic mRNA from which both ORF AC1 and AC4 may be translated. Additionally two more transcripts (0.6 kb and 0.65 kb with start sites, 23 and 26 nt upstream of ORF AC2 were mapped. It is inferred that both ORF AC2 and ORF AC3 may be translated from this transcript. Interestingly while mapping the transcript, Shivaprasad et al. (2005) found that RNA transcripts starting upstream of TATAA box in the intergeneic region. These start sites were located for both rightward and leftward transcription, suggestive of readthrough transcription on the circular DNA. Annealing between such sense and antisense transcripts may give rise to dsRNA intermediates triggering RNAi silencing. This may also explain the siRNA derived from the MYMV promoter region by Pooggin et al. (2003).

In the DNA B component a 0.9 kb BV1transcription start site was mapped to A 410, A414 which is preceded by a short ORF. They also revealed that BC1 transcript (1.1 kb) exists in spliced and unspliced forms, having the consensus splice donor sites (AG/GU) and acceptor site (CAG/G). 123 nt length intron is located between 2359 and the BC1 start codon at 2117. This region also contains three shot ORFs which may facilitate in translation.

In MYMIV, Usharani et al. (2006) analyzed the transcripts of MYMIV infected French bean using 5′ and 3′ RACE and identified an additional transcript of 0.7 kb length on the rightward side, with start site mapping to 290. They speculated that ORF AV1 alone may be translated from such a transcript.

5.6.17 Promoters of LYMV

The intergenic region of DNA A and DNA B of begomoviruses share a common region of about 160–200 bp which contains *cis* elements that regulate the leftward transcription of Rep gene and rightward transcription of the CP gene. The TATAA box and G box present in CR constitute the core promoter elements for the transcription of Rep gene. Rep also functions as negative feedback regulator of its own transcription by binding to repeat sequences between the TATAA box and transcription start site. Unlike the Rep gene promoter the rightward CP gene promoter requires activation by TrAP which has been recorded by several workers (Hanley-Bowdoin et al. 1999).

The bidirectional promoter of MYMV in the intergenic region was studied by fusion of CAT reporter gene with putative promoter sequence. Shivaprasad et al. (2005) showed that except BV1 start site fusion, all other constructs expressed activity driven similar to what is driven by 35S promoter. Expression of CAT fused to start codon of AV2, AV1, BV1 and BC1 was enhanced 15 to 300-fold in the presence of MYMV- AC2. However AC2 activation on leftward ORFs was less pronounced (two to fourfold).

Sunitha et al. (2012) delineated a new 357-bp mono directional TrAP/Ren promoter. Besides stress- regulated motifs, several root specific motifs were also identified in this promoter which were further tested in transgenic tobacco lines. The transgenic plants having TrAP gene driven by its own promoters had normal phenotype, while toxicity was observed when TrAP gene was expressed under CaMV 35S promoter.

Usharani et al. (2006) studied the promoters of MYMIV by delivery of putative promoter constructs through *Agrobacterium tumifaciens* on *N. benthamiana* leaf and sprout seeds of French bean. They identified a TrAP independent activity of AV promoter and differential regulation of AC promoter. AV promoter had TATA box and initiator elements. Many transcription factor binding sites were also detected.

5.6.18 RNA Silencing Pathway and YMV Suppressors

RNA silencing refers to the highly conserved adaptive immunity response of the plants through siRNA to protect themselves from the invasion of viruses and transposans. The priming event is the formation of dsRNA intermediate which is cleaved into 21–24 nt siRNA by DICER like protein (DCLs).

The siRNA generated along with RISC complex target mRNA resulting in either its degradation or in translational arrest which is called as post transcriptional gene silencing. On the contrary the siRNA interact with argonaute 4 (AGO4), which directs the methylation of the promoter region resulting in transcriptional gene silencing. Upon geminivirus infection, it is presumed that the read through transcription at 3′ end of the virion sense and complementary sense transcripts produce dsRNA; however until now analysis of siRNA profiles upon geminivirus infection reveal abundance of 24 nt-siRNA (unlike RNA viruses) targeted to intergenic promoter region.

Countering this innate defense of the host plant, viruses encode proteins that interfere with the silencing pathway at any step and these diverse proteins are designated as viral suppressor (VSR). Geminiviruses encode three to four VSRs like AC2/C2/TrAp, C4/AC4/sd, AV2/V2/MP in monopartite viruses, and Beta C1 protein encoded by betasatellites.

Experiment conducted on the AC2 encoded by MYMV and MYMIV revealed interesting results contrasting to New World viruses. In NW virus TGMV, silencing activity is independent of transcription activation. Bisaro (2006) suggested that AL2 (AC2) protein interacts with Adenosine Kinase (ADK) which is required to maintain methyl cycles and S-adenosyl methionine dependent methyl transferase activity. AC2 interferes with ADK directed methylation and epigenetic modification of viral genome. They also suggested that AL2 interacts with SNF Kinase thereby limiting cellular AMP levels and basal defence of the plants.

The mechanism of action of AC2 protein of MYMV (Trinks et al. 2005) is different from that of NW viruses. Trinks et al. (2005) showed that silencing activity is dependent on transcriptional activation. They hypothesized that AC2 suppressor of MYMV activates expression of host endogenous gene like WEL-1 (WEL-1 3–5′ exonuclease). They suggest that AC2 may activate expression of cellular protein that may function as endogenous negative regulator in the host plants.

The intricacies of MYMIV-AC2 suppression of RNA silencing was investigated. Rahman et al. (2012) employed the AC2 of MYMIV for enhancing transgene expression, such as topoisomerase-II in the transgenic tobacco lines. They further elucidated the silencing suppression mechanism of MYMIV AC2 and showed that unlike several other suppressors, the AC2 of MYMIV does not bind to siRNA or dsRNA, but its suppression activity is mediated through interaction with the key components of the RNAi pathway, viz., RDR6 and AGO1 (Kumar et al. 2015). Interaction by AC2 inhibited the activity of RDR6, an essential component for biogenesis of siRNA and tasiRNA and also inhibited the function of AGO1, the major slicing factor of RISC (Kumar et al. 2015). They employed the AC2 of MYMIV to demonstrate the *in planta* activity of a hammerhead ribozyme designed to target rep-mRNA of MYMIV as an antiviral agent (Mishra et al. 2014). Consequent upon identification of suppressors in begomoviruses, their application in achieving transgenic resistance has been looked into. Sunitha et al. (2013) and Shanmugapriya et al. (2015) demonstrated that tobacco plants transformed with MYMV TrAP gene accumulated siRNA and led to reduction in MYMV-DNA accumulation in tobacco.

5.6.19 Transcriptome Profiling

Upon geminivirus infection, plant remodels its cellular components involved in the diverse processes, such as transcription, hormone signaling, metabolic pathway and defense related processes. In order to get insights into the molecular events leading to compatible/incompatible interaction between the *Vigna mungo* and MYMIV, Kundu et al. (2013), performed proteome analysis. The analysis at 3, 7, 14 days post MYMV inoculation, revealed 109 differentially expressed proteins which were identified by mass spectrometry. Among all the proteins, photosynthesis related proteins were the most affected in susceptible genotypes. Photosystem II electron transport was influenced and many networks of defense related proteins expressions were altered.

Using substractive hybridization technique, transcripts altered in *V. mungo* under MYMV pathogenesis were studied by Kundu et al. (2015). Enhanced expression of genes involved in phenylpropanoid pathway, ubiquitin proteosomal pathways were seen which could contribute to resistance. Whereas in susceptible genotype repression of photosynthesis related genes affecting chlorophyll synthesis and functions resulting in yield penalty were observed.

Yadav et al. (2009) while studying differential response of soybean genotyes to MYMV discovered that a soybean variety resistant to MYMIV showed rapid degradation of viral RNA compared to a susceptible variety. Yadav et al. (2009) found that in resistant soybean lines the viral siRNA were generated complementary to noncoding intergenic region, whereas in susceptible variety siRNA generation was targeted to coding region. Most of the IR specifc siRNA generated were of 24 nt length and by bisulphite sequencing it was also discovered that the sequence of IR were highly methylated thereby suggesting that methylation of viral genome may be probable mode of resistance action.

Yadav and Chattopadhyay (2014) studied host gene response in soybean plant susceptible to MYMIV. A high-throughput microarray analysis of MYMIV infected soybean probed with 17,000 genes revealed the enhanced expression of various genes linked with systemic acquired resistance, programmed cell death and disease resistance response.

Kundu et al. (2017) investigated microRNA profile of blackgram plants inoculated with MYMIV. They found out that miRNA belonging to the family of miR156, miR159, miR160, miR162, miR398, miR1511, miR1514, miR2118 and novel vmu-miRn7 vmu-miRn, vmu-miRn13 and vmu-miRn14. These miRNA targetted transcripts like NB-LRR, NAC, MYB and several transcription factors. Thus miRNA are speculated to be involved in MYMIV induced stress repsonse.

5.6.20 Resistance to Yellow Mosaic Viruses

LYMVs of grain legumes continue to be a big challenge as the genetics of resistance is not understood clearly. There are contradictory reports on the genetics of resistance to YMV in blackgram. It is suggested to be controlled by a single recessive gene (Singh and Patel 1977; Singh et al. 1988; Thakur et al. 1997; Saleem et al. 1998; Reddy and Singh 1995; Sudha et al. 2013). There are also reports of single dominant gene (Sandhu et al. 1985), two recessive genes (Verma and Singh 1988; Pal et al. 1991; Ammavasai et al. 2004; Muraleedhar et al. 2015) and complementary recessive genes (Shukla et al. 1985). Absence of any clear cut lead in the genetic analysis was attributed to difference in strains of viruses, vector population, vector biotypes and environmental conditions. At present, with more clarity regarding viruses, and the facilities to separate the viruses and screening with the specific viruses, genetic analysis may give reliable information.

For example Karthikeyan et al. (2011) and Sudha et al. (2013) employed agroinoculation besides field level screening for resistance to MYMV. They inoculated MYMV DNA A separately with KA22 and KA 27 type of DNA B and found that only one genotype ML 818 showed resistance reaction. Screening of the blackgram accession in the field showed that four accessions were moderately resistant showing <20% incidence. They were challenged through agroinoculation with MYMV constructs and the resistant phenotype was further corroborated by less virus accumulation.

Efforts have been initiated to identify the resistance gene and develop marker near to the locus which can be employed in marker assisted selection breeding program. Kundu and Pal (2012) and Maiti et al. (2011) developed resistance linked molecular markers for resistance to MYMIV in blackgram using R gene analogues. Two MYMIV resistance markers, Yr4 and *CYR1* were identified, of these two, *CYR1* locus is completely linked to MYMIV resistance. The *CYR1* (R) gene which co segregates with resistance was isolated and characterized (Maiti et al. 2012). The transcript sequences and the 3D model revealed it to be a typical CC-NBS-LRR type of R gene protein. They also predicted the interaction between CYR1- LRR and MYMIV-CP and suggested that CYR1 protein may recognize the CP-effector of the virus and contribute to incompatible interaction.

Markers have been identified and validated in blackgram (Gupta et al. 2015). The whole genome resequencing of MYMV resistant soybean cultivar UPSM-534 and the susceptible cultivar JS335 was performed by Yadav et al. (2015) which should help in generating reliable and robust markers associated with resistance.

5.7 Begomoviruses Associated with Yellow Vein Mosaic Disease

Yellow vein mosaic disease (Fig. 5.5) was known in India since 1924 in bhendi (Kulkarni 1924) and pumpkin (Varma 1955) in Maharashtra. Pumpkin yellow vein mosaic virus (PYVMV) now known as squash leaf curl China virus was

Fig. 5.5 Yellow vein mosaic disease in different crops caused by begomoviruses. (**a**) bhendi yellow vein mosaic virus in okra; (**b**) mesta yellow vein mosaic virus causing yellow vein mosaic disease in mesta; (**c**) squash leaf curl China virus causing yellow vein mosaic disease in pumpkin

characterized as the first begomovirus infecting cucurbit in India (Muniyappa et al. 2003). Now, more than seventeen begomoviruses (Tables 5.4 and 5.5) are known to be associated with the yellow vein mosaic disease in crops and weeds. In many cases, especially in bhendi and other weed hosts, the yellow vein viruses may also cause enation and leaf curl symptoms. For example, both BYVMV and okra enation leaf curl virus (OELCuV) can cause YVM or leaf curl symptom depending on the genotype they infect. Likewise, in *Ageratum conyzoides* and *Croton bonplandianum*, both YVM and enation leaf curl symptoms are observed. An Old World begomovirus, Corchorus yellow vein virus has been identified in a weed plant, *Corchorus olitorius* showing yellow vein symptoms in Maharashtra (Malathi VG, unpublished result, GenBank KC196077, KC223600).

5.8 BYVMV and OELCuV

In India, begomoviruses associated with yellow vein mosaic and leaf curl diseases of bhendi are bhendi yellow vein mosaic virus (BYVMV, Jose and Usha 2003), bhendi yellow vein Bhubeneshwar virus (BYVBhV, Venkatarvanappa et al. 2013b), bhendi yellow vein Haryana virus (BYVMV-Har, Venkataravanappa et al. 2014b) and the bipartite tomato leaf curl new Delhi virus (ToLCNDV, Venkataravanappa 2008) and raddish leaf curl virus (RaLCuV, Kumar et al. 2012a, b). These are the only virus species approved by the ICTV. Other important yellow vein mosaic viruses are, mesta yellow vein mosaic virus (MeYVMV, Roy et al. 2009) and hollyhock yellow vein mosaic virus (HoYVMV).

5.8.1 Discovery and Distribution

Bhendi yellow vein mosaic disease constitutes the earliest record of whitefly transmitted virus disease in India. It was first reported by Kulkarni in Bombay district in 1924. The viral etiological nature and transmission by whitefly were subsequently studied by Uppal et al. (1940) who named the disease as bhendi yellow vein mosaic disease. Subsequently Capoor and Varma (1950) and Varma (1952, 1955) studied the whitefly transmission of the virus. Soon incidence of the disease in other states was observed (Chelliah et al. 1975; Chelliah and Murugesan 1976; Khan and Mukhopadyay 1985; Bhugapati and Goswami 1992). Due to the earliest record, it can be reasonably assumed that the virus originated in India, (Nath et al. 1992). At present, the disease is widespread, occurring in all agroclimatic zones of India, in the absence of resistance sources, it is the most important constraint in productivity. A survey conducted on the incidence of yellow vein mosaic disease estimated the percentage disease incidence to range from 23 to 67 in Karnataka, 45–56 in Andhra Pradesh, 23–75 in Tamil Nadu, 42–75 in Kerala, 23–85 in Maharashtra, 24–65 in Haryana, 35–57in Uttar Pradesh, 45 in Delhi, 67 in Chandigarh and 45–66 in Rajasthan (Venkataravanappa 2008). Bhendi infected by BYVMV has also been recorded in Pakistan, Thailand, Sri Lanka and China (Tsai et al. 2013).

Contrasting to yellow vein mosaic disease, enation leaf curl disease was noticed only in 1980s (Singh and Dutta 1986) in Karnataka. The enation leaf curl was further detected in all the bhendi growing tracts of India (Venkataravanappa et al. 2016) and this disease has emerged as serious threat to cultivation. In some cultivars, yellow vein disease symptoms and enation leaf curl symptom appear separately, in some cultivars, both the symptoms appear together. The name of the virus as yellow vein mosaic and enation leaf curl virus does not imply that they cause only yellow vein mosaic or enation leaf curl. Thus the symptom expression is dependent on host genotype.

5.8.2 Economic Loss

If plants are infected at an early stage (20 days post germination) the fruits are severly malformed and yield loss is nearly 94–100%. If the infection occurs at later stage, the loss is reduced. There is 49–84% loss when infection occur at 50–60 days post germination (Sastry and Singh 1974; Pun and Doraiswamy 1999; Nath and Saikia 1992) recorded 84 and 49% yield loss when infection occur 50 and 65 days post germination. In the case of enation leaf curl disease the yield loss varies from 30 to 100%.

5.8.3 Symptomatology

The yellow vein mosaic disease has characteristics symptom of yellow veins surrounding green tissues. To begin with yellowing or vein clearing of small veinlets appear which intensify and the entire leaf becomes completely yellow or greenish in colour (Fig. 5.4a). Plants infected at early stage have small leaves and are extremely stunted. The fruits of infected plant are yellow in colour, deformed and are not marketable (Sastry and Singh 1974). Venkataravanappa et al. (2012a, b) differentiated three types symptoms, on the basis of developmental stage at which symptoms appear, Type I in which, plants get infected very early and develop symptoms, older leaves turn brown or wither, Type-II infection occurs after flowering, wherein only young leaves show symptom, older leaves are symptom free, fruits produced are hard and yellow, Type III, where the plants appear normal, produce marketable yield but emerging axillary shoots produce symptomatic leaves.

The enation leaf curl disease has much more devastating phenotype. Initially, small pin head like enations appear on the leaves, followed by leaf curling. The undersurface of the leaves show prominent enations and leaves become leathery and brittle. There is severe twisting of stem, lateral branches and petioles. As the twisting becomes severe, the entire plant appears spreading on the ground. In some cases the veinal enations become brown and leaves have corky appearance. The severely affected plants do not produce any fruits, if formed the fruits are deformed and are not marketable.

5.8.4 Prevalence of Viruses in Field

There is a considerable variation in the incidence of the diseases in different states. The disease is severe in north India in *Kharif* season. The ideal location which are considered as hots spots for screening are Karnal, Tarai region of Uttarakhand, Nadia district of West Bengal and Varanasi area of UP. In central and south India, the disease is pronounced in summer season, the hot spot locations being Guntur in AP. Jalgaon in Maharashtra, Surat in Gujarat, and Coimbatore in Tamil Nadu.

Venkataravanappa et al. (2014b), summed up that of the four major cluster or virus species BYVMV, BYVMaV, OYVMV, and OELCuV are predominant. BYVMV is present in southern, central, western and eastern and north India, BYVMaV occurs only in some location in AP, TN and Maharahtra. OELCuV also seen distributed through out the country but distribution is less compared to BYVMV.

5.8.5 Transmission and Host Range

BYVMV is efficiently transmitted by whiteflies. The transmission studies have been accomplished by Varma (1952) who showed that 12–24 h of AAP and 30 min of IAP and fasting prior to acquisition enhanced the efficiency of transmission. Raychaudhuri and Nariani 1977, made an interesting observation that *B. tabaci* is capable of harbouring different viruses and exhibit differential transmission of individual virus, Pun and Doraiswamy (1999), described that the age of the bhendi seedling used in the tests is important; 100% transmission was obtained when inoculation was done on 1 week old seedlings, only 31.7% infection occurred with 7 week old seedling. Venkataravanappa et al. (2012a), demonstrated whitefly transmission with 24 of AAP and IAP for BYVMV. They achieved 100% transmission on the susceptible cultivar 1685. Venkataravanappa et al. (2013b), investigated the transmission characteristics of BYVBhV and observed that with 24 h of IAP and AAP, symptoms were observed in the cultivar 1685 within 15 DPI.

The vector transmission experiment conducted by Venkataravanappa et al. (2014a) with OELCuV revealed very significant results; it showed that a minimum of two whiteflies can bring10% infection. Transmission increased with more number of whiteflies. The minimum IAP and AAP required were 1 h and 30 min respectively, the maximum transmission rate was observed with 24 h. The adult female whiteflies were more efficient than male. They also observed that efficiency of transmission was highest when 7 days old seedling were inoculated and it decreased upto 50% when 25 days old plants were inoculated. The begomoviruses infecting bhendi are not sap or seed transmissible.

Capoor and Varma (1950) found that in addition to bhendi (*A. esculentus*) *A. moschatus*, *A. manihot* and *Althaea rosea* were susceptible to BYVMV. Handa and Gupta (1993) showed the yellow vein net symptoms were produced in *Croton bonplandianum* using BYVMV from bhendi as inoculum. For the OELCuV, Venkataravanappa and coworkers inoculated 50 plant species and found that only two species in the family Malvaceae, *A. esculentus* and *Althaea rosea* expressed symptoms; in the solanaceous hosts symptoms were observed in the *Datura stramonium, N. glutinosa, N. clevlandii, N. occidentalis, N. tabacum* and *N. benthamiana.*

5.8.6 Serology

The whitefly transmission of BYVMV though was well known, presence of geminate particles was first time shown only in 1991 using ACMV antibody by immune specific electron microscopy by Harrison et al. (1991).They also showed the positive reaction of BYVMV to polyclonal antibody to ACMV in DAS – ELISA and positive hybridization to ACMV DNA probe. Handa and Gupta (1993) found that BYVMV reacted specifically to two ACMV monoclonal antibodies SCR 17 and SCR 18. Harrison et al. (1997) further showed that the epitope profile of cotton leaf curl viruses from Pakistan generated by reaction to 31 MAb to ACMV, ICMV, Okra leaf curl geminviruses were not indistinguishable from YVMV affected okra samples from India.

5.8.7 Genome Comparison

The DNA A genome of BYVMV was characterized by using begomovirus specific primers and DNA B equivalent component could not be detected in the infected bhendi, however, a betasatellite associated with BYVMV was identified; subsequently, it was demonstrated that both DNA A and betasatellite were necessary to cause yellow vein mosaic disease in bhendi (Jose and Usha 2003). During last 5 years, several bhendi begomoviruses have been characterized.

The genome Organization is basically similar to Old World monopartitie begomoviruses with two virion (V1, V2) and five complementary sense ORFs. The iteron sequences were incomplete direct repeats of GGTGT. Venkataravanappa et al. (2012a, b, 2013a, b, c, 2014a, b, 2016) characterized begomoviruses infecting bhendi throughout India, generated full length genome sequence of about 190 isolates and identified seven viruses, On the basis of 89% as the species threshold value, six species were identified, they were bhendi yellow vein Bhuvaneswar virus (BYVBhV) (Venkataravanappa et al. 2013a, b, c), bhendi yellow vein Delhi virus (Venkataravanappa et al. 2012b) bhendi yellow vein mosaic Maharastra virus (BYVMaV), bhendi yellow vein mosaic Harayana virus (BYVMV- Har) and bhendi yellow vein mosaic Karnal virus (BYVKaV), bhendi yellow vein India virus and ToLCNDV. However, the ICTV study group on geminiviruses reanalyzed the sequences and in the Xth report nomenclature of above mentioned viruses have been changed. As in the case of cotton leaf curl viruses, some of the new viruses were essentially recombinants, the committee perhaps decided to name them on the basis of the identity with the major parent, BYVMV.

The identity analysis performed by the authors (Venkataravanappa et al. 2016) by SDT, following the guidelines of the study group on taxonomy of geminivirus clearly justify separation of bhendi begomovirus isolates as distinct species on the basis of 91% identity. As the identity between some isolates is very low (<85%) in the future reports of ICTV, ambiguity may be cleared and the species status may be restored. In the present chapter the distinct status of species will be maintained as proposed by the authors when comparing the genome. Among the bhendi infecting begomoviruses 79–88% identity is observed between BYVMV, BYVIV, BYVMaV, BYVDV, BYVKnV, BYVHV viruses in DNA A component. However, bhendi yellow vein Bhuvaneswar virus is very distinct and shares only 66.9–76.6% identity with BYVMV. It is closely related to croton yellow vein virus (CYVMV) and 75–79% to mesta yellow vein mosaic virus (MeYVMV). In the case of OELCuV, highest percent identity of 82% is recorded with BYVMaV followed by cotton leaf curl Banglore virus (CLCuBaV) (81%) and MeYVMV (80%). Maximum identity was observed in the Rep region. The intergenic region of the genome shared very less identity (51–65%) between the viruses.

There is only one DNA B component sequence available, one DNA B of BYVDV (HQ 542082) which exhibits nearly 81–85% identity in the complete nucleotide sequence, 94% and 97% identity at amino acid level with ORF BV1 and BCI with DNA B component of ToLCNDV, respectively.

5.8.8 Phylogenetic Relationship

Venkataravanappa et al. (2014a, b) performed a comprehensive analysis involving 130 sequences of bhendi begomoviruses and observed four major clusters, consisting of BYVMV, BYVMaV, OELCuV, ToLCNDV. The BYVDV, BYVHV, BYVKnV belonged to the major cluster of BYVMV but occupied a separate group. The OELCuV and ToLCNDV clusters were distinct and well separated from BYVMV and BYVMaV. In all the analysis BYVBhV occupied a separate branch.

Venkataravanappa et al. (2014a, b, 2016) performed a neighbour net analysis of sequences using the split tree programme. The analysis involving bhendi infecting and related begomoviruses revealed extensive network structure rather than predominantly bifurcating tree like structures. The extremely networked tree is indicative of recombination suggesting how different parts of genome have different origin due to recombinations. Neverthless the network analysis confirms the distinct nature of OELCuV and BVYBhV.

The population structure analysis of okra infecting begomoviruses (110 sequences of different begomoviruses) using model-based algorithm indicates existence of at least four genetically different populations (Prasanna et al. 2010; Venkataravanappa et al. 2014b). Most of the isolates that related to four previously characterized species namely BYVMaV, BYVMV, OELCuV and ToLCNDV segregated together and formed four genetically cohesive populations. Among these, two predominant populations observed included the group of viruses genetically similar to BYVMV (52 sequences) and BYVMaV (42). Further four major populations was analyzed separately to determine the optimum number of sub-populations that best represented within population structure using ARLEQUIN Ver. 3.0. The BYVMaV population contained five and BYVMV contained three minor sub-populations. There was evidence of potentially three and two minor sub-populations in OELCuV and ToLCNDV populations respectively. Majority of the member isolates assigned to these subpopulations showed more than 60% support. With regard to haplotype distribution, a total of 107 haplotypes were detected out of the 110 sequences analyzed. The highest number of haplotypes was observed in BYVMV-I with 44 haplotypes followed by BYVMaV-I with 27 haplotypes. All the sub-populations recorded high haplotype diversity (hd of >0.95) with an exception of BYVMaV-II. The number of polymorphic sites was highest in BYVMV-I with 988 polymorphic sites and nearly 181 average nucleotide differences. A high level of nucleotide diversity (0.26351) with the average number of nucleotide differences of 717 was noted in ToLCNDV-II sub-population (Venkataravanappa et al. 2014b).

5.8.9 Recombination

The recombinational events have led to emergence of important variants of the bhendi viruses, which may even be considered as distinct species is further proved by indepth analysis of recombinational events in the RDP programme.

Venkataravanappa et al. (2014b) revealed that the bipartite begomovirus BYVDV has sequences derived from ToLCNDV and BYVMV viruses which were confirmed in all six detection methods in RDP programme. They observed that the only DNA B component associated with bhendi begomovirus is derived through recombination between different DNA B components of ToLCNDV. In the case of BVYBhV, a very distinct bhendi virus, two recombinational events confirmed by six methods were detected in IAC3,AC5 region with CYVMV and CLCuMV-Rajasthan and RhYVMV as major and minor parents. Venkataravanappa et al. (2014a, b), suggested that BYVHV has entire coat protein sequences derived from one isolate of BYVMV (previously referred to as OYVMV) and the rest of the genome from ToLCNDV. The BYVKnV is a recombinant between BYVMaV and OELCuV the event detected was in the right half of the genome extending upto AC3. The analysis of OELCuV showed that most of its sequences originate from BYVMV, BYVBhV and MeYVMV. For all the OELCuV isolates, the sequences containing the origin of replications originate from BYVMV, including the iteron sequences. For only one isolate of OELCuV (GU1119996) a small fragment is derived from ToLCNDV. Rishishwar et al. (2015) recorded presences of BYVMV in infected samples in Kalyani (west Bengal) and Aurangabad (Maharastra) but MeYVMV in samples from Varanasi (Uttar Pradesh) and Jalgaon (Maharastra).Interestingly they found that the Jalgaon isolate of MeVYMV showed a recombination event between MeYVMV as major parent and Malvastrum yellow vein Yunnan virus (MaYVYV, AJ786711) as a minor parent.The event located was between nt co-ordinate 2002 and 2020. The presence of sequence of (MaYVYV) is not reported earlier in India suggesting the possibility of its occurence in India potentially near to the location from where MeYVMV Jalgaon isolates were collected.

Vinoth-Kumar et al. (2017a) gave evidence for inter specific and inter strain recombination events and suggested that cotton infecting and bhendi infecting viruses may share a common ancestor. Different levels of recombination among BYVMV isolates have also been described. Serfraz et al. (2015) analyzed OELCuV isolates from Pakistan and found that OELCuV is a recombinant between BYVMV (isolates from south India) (at Rep region from 1522 to 1955) and OELCuV from northwest India as a minor parent. They suggest that cotton leaf curl Multan virus (CLCuMuV) and MeYVMV recombined to produce the entire region of rep present in OELCuV. They suggest that, BYVMV, MeYVMV and CLCuMuV are parents and as we move from north to north western regions CLCuMuV may dominate.

Vinoth-Kumar et al. (2016) found that distribution of SSR and recombinational events are linked, hence Vinoth-Kumar did an extensive analyzes of SSR and found that there is a good correlation between SSR and recombination events among BYVMV isolates. They found that 62 out of 67 begomovirus isolates had SSR sequences. Two types of SSR motifs (CG) 3-X1-T6–7 and $(AGA)_4$ –X1-A_6 were found frequently among begomoviruses.

5.8.10 Establishment of Koch's Postulates

For bhendi viruses, koch's postulates for the cloned components have been established for only one isolate of BYVMV-Madurai (Genbank accession no AF241479) by Jose and Usha (2003). Partial tandem repeat constructs of DNA A and betasatellite BYVB were agroinoculated to 125 plants in three different experiments of which 16 plants developed typical yellow vein symptoms 25 DPI. When DNA alone is inoculated, it resulted in mild leaf curl symptom.The functional role of betasatellite and C2 and C4 genes encoded in DNA A were further analyzed using the infectivity of these clones on to *Nicotiana benthamiana*.

For all other isolates belonging to different virus species, the whitefly transmission has been accomplished for every individual isolate before they have been subjected to molecular characterization. Therefore, it can be presumed that Koch's postulates have been established for these isolates also.

5.8.11 Molecular Basis of Pathogenicity

Various protein-protein interactions and the role of viral genes C2, C4 encoded by DNA A component and betasatellite C1 protein encoded by betasatellite were studied using the BYVMV/BYVMB combination. Kumar et al. (2006), studied subcellular localization, by fusing GFP to CP and βC1 coding sequences. They showed that CP was localized inside nucleus whereas βC1 is distributed in periphery. The nuclear localization signal in CP was predicted to be between amino acid residues 1 and 24, in βC1 nuclear export signal was located between aminoacids 105 and 115. They also demonstrated interaction between CP and βC1 in yeast two hybrid system. Transformation of *N. benthamiana* with βC1 ORF under the control of 35S promoter resulted in abnormal distorsion, twisting of stem and leaves and stunting of plants. Gopal et al. (2007) further extended their studies and showed strong suppression of gene silencing activities for C4 and βC1 but only a weak activity for C2. They showed that abnormal phenotypes were produced in *N. benthamiana* when transformed with C4 and βC1. They also analysed V sense and C sense promoters by both transient and stable expression in *N. benthamiana* that demonstrated efficiency of C sense promoters. Chandran et al. (2012), further looked into nuclear trafficking of C2 proteins. They found that BYVM2 C2 nuclear localization signal (NLS) was located in N terminus of the protein covering 17–31 amino acid. This NLS was recognized by the transport receptors, karyopherin α and βC2. Interaction through NLS with karopherin ensures its nuclear localization. The C2 protein was found both in cytoplasm and nucleus, which suggests that there is a NLS independent nuclear import of this protein. Interestingly, C2 was also demonstrated to play a role in symptom determination and virus replications by Chandran et al. (2014). When *N. benthaminana* plants were inoculated with BYVMV. DNA A engineered to have two stop codons in C2 ORF, plants did not develop any symptoms. The viral DNA level was drastically reduced. They concluded that C2 protein of BYVMV may have key role in symptom production and viral DNA replication (Chandran et al. 2014).

5.8.12 Resistance to BYVMV and OELCuV

Of all the virus diseases affecting vegetables, yellow vein mosaic disease of bhendi is the most devastating one as there is 100% infection even in elite varieties and hybrids and the yield losses range from 50 to 94%. The search for resistance and attempts to incorporate resistance genes have been many but with no impact. Sanwal et al. 2016 indicated contradiction between different research workers, which might also arise due to difference in virus species/strains, used as inoculum.

Screening of 941 indigenous and exotic germplasm of lines of Okra for YVMV field resistance under natural epiphytotic condition revealed that none of the accessions were immune or highly resistant, 43 were moderately resistant .The accessions IC 218887, IC 69286 and EC-305619 were resistant (Abdul et al. 2004). Out of eight wild species of bhendi, species having resistance to YVMV are *A. manihot, A. angulosus, A. crinitus, A. vitifolius, A. tuberculatus, A. panduraeformis, A.*

pungens and A. tetraphyllus (Dhankar et al. 1996). *A. manihot* ssp. *manihot* is widely used in developing resistant lines. Efforts are also being taken to induce mutation by gamma irradiation.

Venkataravanappa et al. (2012a, b), screened okra genotypes under both artificial and natural condition and found that the genotypes Nun 1145, Nun 1144, Nun 1142 and Nun 1140 showed resistance, genotypes M10, Nun 1142, Nun 1140 showed moderately resistance phenotype. In some selected genotypes symptoms free plants were found to contain virus by nucleic acid spot hybridization and PCR kits. Similarly Venkataravanappa et al. (2016), identified the genotypes Tulasi and Trisha also to have considerable resistance to BYVMaV.

About 36 betasatellites have been isolated from diseased okra which segregated into four groups (1) okra leaf curl betasatellite (OLCuB) (2) bhendi yellow vein betasatellite (BYVB) (3) bhendi yellow vein India betasatellite (BYVIB) and (4) croton yellow vein mosaic betasatellite (CroYVMB). Identification of CroYVMB, a "non-malvaceous betasatellite" in malvaceous okra might be the result of component reassortment between pathogens when they infect the hosts together (Venkataravanappa et al. 2011a, b).

5.9 Mesta Yellow Vein Mosaic Virus (MeYVMV) and Mesta Yellow Vein Mosaic Bahraich Virus (MeYVMBV)

MeYVMV and MeYVMBV are the two viruses causing yellow vein mosaic disease of mesta fibre crops, *Hibiscus cannabinus* (Kenaf) and *H. sabdariffa*. The disease in eastern India is caused by MeYVMV along with cotton leaf curl betasatellite, in northern India, the virus associated with the disease is mesta yellow vein Bahraich virus with Ludwigia leaf distortion betasatillite.

5.9.1 Discovery and Distribution

The yellow vein mosaic disease was first reported in mesta during 2005 from Eastern part of India (Chatterjee et al. 2005); similar symptoms were recorded in Mesta in U.P from Northern part of India in 2007 (Ghosh et al. 2007; Das et al. 2008a, b). Roy et al. (2009) surveyed for the disease incidence and observed that highest disease incidence occurred in southern West Bengal followed by eastern Uttar Pradesh and northern West Bengal; the lowest disease incidence was noted in northeastern Andhra Pradesh. Incidence of the disease at various locations ranged from 46 to 93% in eastern part of India, and 34–78% in Northern part.

5.9.2 Symptomatology

The symptoms of the disease are typical yellowing of veins, entire lamina turn yellow and form a yellow network (Fig. 5.5b). Entire lamina looks yellow, general vigour and height of the plant are affected. The Mesta cultivar HC – 583 was the most susceptible one.

5.9.3 Transmission and Host Range

The virus MeYVMV was efficiently transmitted by whitefly upto 85% in the case of *H. sabdariffa,* 78% in case of *Hibiscus cannabinus* and they have a very narrow host range. The symptoms were produced 8–10 days post inoculation (Chatterjee et al. 2008; Das et al. 2008a, b).

5.9.4 Genome Comparison

The virus isolates from eastern and northern region of India belong to MeYVMV and shared 92% identity between them and showed 83% identity with isolates from northern India which belong to the species MeYVMBV. The isolates from northern India exhibited 98–99% identity between them. The betasatillite associated with MeYVMBV of northern India belonged to CLCuMB species, contrasting to eastern India which had Ludwigia leaf distortion betasatillite. In a phlyogenetic analysis the east and south Indian isolates comprising MeYVMV clustered separately from north Indian isolates comprising MeYVMBV. MeYVMV also produced leaf curl like symptoms in kenaf. Besides MeYVMV and MeYVMBV, also CLCuMuV and ToLCJV were found to be associated with leaf curl in Northern India. Papaya leaf curl virus infect another fiber crop *Crotalaria juncea* along with radish leaf curl betasatellite. Recently ageratum enation virus (AgEV) was also described from *H.cannabinus.*

5.10 Begomoviruses Associated with Leaf Curl Disease

In India, maximum numbers of begomoviruses are known to cause leaf curl disease in economically important crops (Fig. 5.6). The leaf curl disease includes various types of symptoms like leaf distortion, enation, twisting and stunting of plants.

5.11 Tomato leaf curl New Delhi virus (ToLCNDV) and Tomato Leaf Curl Bangalore Virus (ToLCBaV)

ToLCNDV and ToLCBaV constitute two divergent lineages of begomoviruses causing leaf curl disease in tomato in India. ToLCNDV is a bipartite begomovirus and was characterized from New Delhi and Lucknow (Padidam et al. 1995; Srivastava et al. 1995). ToLCBaV is a monopartite begomovirus, which was described from Bengaluru, southern India (Muniyappa et al. 2000; Kirthi et al. 2002).

Fig. 5.6 Leaf curl disease of different crops caused by begomoviruses. (**a**) tomato leaf curl New Delhi virus (ToLCNDV) in tomato; (**b**) cotton leaf curl Multan virus-Rajasthan strain in cotton; (**c**) chilli leaf curl virus in chilli; (**d**) papaya leaf curl virus in papaya; (**e**) tobacco leaf curl virus in tobacco; (**f**) ToLCNDV causing apical leaf curl disease of potato; (**g**) okra enation leaf curl virus in okra; (**h**), (**i**) croton yellow vein mosaic virus in rapeseed; (**j**) ToLCNDV in pumpkin; (**k**) kenaf leaf curl virus in kenaf

There are 19 begomoviruses recorded so far in tomato from India. In addition to ToLCNDV, there is one more bipartite begomovirus, tomato leaf curl Palampur virus (ToLCPalV) affecting tomato in northern and sub-Himalayan region. There are eight monopartite begomoviruses, tomato leaf curl Bangalore virus (ToLCBaV), tomato leaf curl Kerala virus (ToLCKeV), tomato leaf curl Patna virus (ToLCPtV), tomato leaf curl Gujarat virus (ToLCGuV), tomato leaf curl Karnataka virus (ToLCKaV), tomato leaf curl virus (ToLCV), tomato leaf curl Joydebpur virus (ToLCJV), pepper leaf curl Lahore virus (PeLCLaV). Two more begomoviruses, tomato leaf curl Pune virus (ToLCPuV) and tomato leaf curl Rajasthan virus (ToLCRaV) have been characterized from Maharashtra and Rajasthan, respectively, however, it is not clear whether they have monopartite or bipartite genome. Besides these viruses, chilli leaf curl virus (ChiLCV), chilli leaf curl India virus (ChiLCIV), AEV, PaLCuV, and tobacco curly shoot virus also infect tomato under natural field conditions. Recently, two begomoviruses were recorded, tomato enation leaf curl virus- KP195260, Venkataravanappa unpublished and tomato severe leaf curl virus (ToSLCV)-KP195267 (Venkataravanappa unpublished), however infectivity of the cloned DNA of these viruses in tomato have not yet been demonstrated.

5.11.1 Discovery and Distribution

Tomato leaf curl disease was first reported from northern India in 1948 (Vasudeva and Samraj 1948), from central India in 1950, subsequently the disease emerged as a problem in tomato growing region of southern India. Since then the disease has been recognized as threat to cultivation of the crop throughout the country. Table 5.7 shows the details on incidence, and year of record of the disease from which it is evident, that tomato leaf curl disease is the most devastating disease affecting productivity. In 1990, it struck as epidemic in Kolar region of Karnataka. The disease incidence has increased with introduction of high yielding hybrid varieties. On the basis of virus characterization it can be reasonably assumed that the predominant viruses in southern region are ToLCBaV and different isolates of ToLCV, eastern region ChiLCV. In Northern, Western and Central India ToLCNDV are predominant.

ToLCNDV is the only begomovirus which has no trans-boundary limitations in movement across different countries in Asia and Europe. In Asian continent, ToLCNDV has been recorded in Bangladesh, Iran, Sri Lanka, Malaysia, Taiwan, Thailand and Indonesia. In Europe, its occurrence was first noticed in 2013 in Southern Spain (Lopez et al. 2015). In Tunisia, ToLCNDV affect zucchini, cucumber and melon (Mnari-Hattab et al. 2015) and in several vegetable crops in Scily and Southern Italy (Panno et al. 2016). At present, it appears that ToLCNDV is well spread in all these Asian, North African and Southern European countries.

Table 5.7 Geographical distribution and incidence of leaf curl in tomato in India

S.No.	Reported from (state/place)	Cropping season	Disease incidence (%)	Reference
1	Andhra Pradesh	–	–	Reddy and Yaraguntaiah (1981)
2	Assam	Winter	11–21.4	Borah and Bordoloi (1998)
3	Bihar	–	35	Dubey et al. (1986)
4	Chattisgarh	–	70–80	Singh et al. (1999)
5	Delhi	Winter	83–90	Vasudeva and Samraj (1948)
		Summer	14	Tripathi and Varma (2003)
6	Gujarat	–	–	Shih et al. (2003)
				Chakraborty et al. (2003)
7	Haryana	Summer	95–100	Banerjee and Kalloo (1990)
8	Himachal Pradesh	–	–	Gupta et al. (2001)
9	Jammu	Summer	52.3	Sastry et al. (1978)
		Winter	16	
10	Karnataka	Summer	78–99	Sastry and Singh (1973)
		Summer	18.5–55	Sastry et al. (1978)
		Summer	52–100	Saikia and Muniyappa (1986)
		Summer	4.5–100	
		Winter	15–35	Sastry et al. (1978)
		Winter	6.4–52.2	Saikia and Muniyappa (1986)
11	Kerala	–	–	Reddy et al. (2005)
12	Madhya Pradesh	Summer	1.3–70.7	Singh et al. (1999)
		Winter	3.4–86.3	
13	Maharashtra	–	–	Pimpale and Summanwar (1986), Mote (1978)
14	Punjab	Summer Winter	–	Butter and Rataul (1981)
15	Rajasthan	Winter	–	Bhardwaj (1992)
16	Tamil Nadu	Summer	80	Jeyarajan et al. (1986, 1988)
17	Uttar Pradesh	Winter	35	Saklani and Mathai (1977)
	Pantnagar	Summer	69	Saklani and Mathai (1977)
	Lucknow	–	–	Verma et al. (1975)
	Varanasi	Summer	100	Srivastava et al. (1995)
				Kalloo (1996)
	Kanpur	Winter	25	Singh and Lal (1964)
18	West Bengal (Kalyani)	–	30	Mukhopadhyay et al. (1994)
		Winter	48–54	Verma et al. (1989)

5.11.2 Economic Loss

Depending on the time of infection, disease incidence and the severity, the yield loss ranges from 17.6 to 99.7% (Butter and Rataul 1981; Kalloo 1996). Sasthri and Singh (1973) reported 92.3% loss when infection occurs at 30 days after transplanting. The yield reductions were 94.9, 90.0, 78.0 and 10.8% when plants get infected at 2, 4, 6, 10 weeks after planting (Sastry and Singh 1973). The tomato plants get infected at all developmental stages and Saikia and Muniyappa (1989) reported less yield loss in the summer planted crops (6.4–52.2%) compared to winter planted crops (52.5–100%).

5.11.3 Symptomatology

Typical symptoms of tomato leaf curl disease are curling, puckering of leaves, veinal yellowing, stunting, excessive branching, pale yellowing to deep yellowing of leaves (Vasudeva and Samraj 1948). In addition, the extreme distortion of leaves and stunting of plants are also observed. In severely stunted plants, flowers may drop off. In some genotypes, green vein banding, twisting, green enation are also seen on the under surface of the leaf. Sometimes upward rolling of margin and islands of golden colors scattered amidst the normal green tissue are also observed (Singh and Lal 1964). The type of symptoms produced is dependent on the genotype cultivated and the developmental stage at which infection occurred. The begomoviruses affecting tomato in Asian continent are referred by two generic names, tomato leaf curl viruses and tomato yellow leaf curl viruses; this gives impression as if yellow leaf curl symptoms are not caused by tomato leaf curl viruses, which is a misconception. In India, depending on the genotypes cultivated yellow leaf curl symptoms are caused by both mono and bipartite viruses in the field. Ultra structural changes like hypertrophy of nucleus and accumulation of dark granules and aggregate of virus like particles in the cytoplasm of matured sieve element was observed by Saikia and Muniyappa (1989).

5.11.4 Transmission and Host Range

A wealth of information is available regarding vector transmission of tomato leaf curl viruses. Since molecular characterization has been accomplished only in 1995s, it can be presumed that the details on transmission for the viruses from southern India may represent the data for monopartite begomoviruses mainly ToLCBaV; details from central and northern India may represent details for ToLCNDV. As early as 1948, Vasudeva and Samraj (1948) demonstrated whitefly transmission of leaf curl virus to several hosts. They observed that in winter, symptoms appeared 25 days post inoculation, while in summer it required only 15 days.

For ToLCBaV, Buttler and Rahul (1978) achieved 100% transmission with ten whiteflies/plant at optimum temperature of (33–39 °C). Muniyappa et al. (2000) reported minimum AAP of 10 min and IAP of 20 min for ToLCBaV-(Ban 4). They also described that geographically different isolates behaved in different manner. They reported that in one whitefly per plant inoculation tests, the females were more effective (95%) than males (25%) in transmitting the virus after 24 h of AAP, ToLCBaV persisted upto 12 days and not the entire life of vector (Muniappa et al. 2000). Similar results was reported for ToLCGuV by Chakraborty et al. (2003);

these results are contrasting to observations made by Reddy and Yaraguntaiah (1981) who found that the vector could retain the virus throughout the lifespan after 6 h of latent period.

In recent years, mechanical/sap transmission of ToLCNDV, ToLCGuV, ToLCKaV have been reported (Chatchawantcanphanich and Maxwell 2002; Chakraborty et al. 2003; Usharani et al. 2004a, b; Sohrab et al. 2004). The high virus titre in the host plants and their presence in mesophyll cells may explain the sap transmission.

The begomoviruses causing tomato leaf curl disease have a wide host range, affecting various dicotyledonous plants belonging to different families. Host range of the viruses has been determined by graft/whitefly transmission, agroinoculation/biolistic delivery of viral genome into tomato plants, or by detecting the viruses in naturally infected plants using specific primers or probes to virus species.

The tomato begomoviruses are known to infect economically important cultivated crops like chilli, papaya, sunnhemp, tobacco, *Physalis* sp., sesamum and potato (Vasudeva and Samraj 1948; Nariani 1968; Reddy and Yaraguntaiah 1981; Sastry et al. 1978; Rataul and Butter 1977; Saikia and Muniyappa 1986) and ornamental plants such as *Althaea rosea, Petunia hybrida, Phlox drumondii, Tithonia* sp., *Zinnia elegans, Vernonia cineria* (Rataul and Butter 1977; Sastry et al. 1978; Reddy and Yaraguntaiah 1979; Gupta et al. 2001). Other plant species infected are *Nicotiana sylvestris, N. glutinosa, N. rustica, Datura stramonium, Solanum nigrum, S. seaforthianum, Ageratum conyzoides, Acanthospermum hispidum, Centratherum anthelminticum, Cassia uniflora, Sida rhobombifolia, Euphoria hirta, E. geniculata, Scoparia dulcis, Shizanthus* sp., *Galinosoga parviflora, Nicandra physaloides, Flavaria australiasia, Sida mysoriensis, Xanthium strumarium* (Vasudeva and Samraj 1948; Nariani 1968; Singh and Lal 1964; Verma et al. 1975; Reddy and Yaraguntaiah 1981; Mariyappan and Narayanasamy 1986; Reddy and Ravi 1991; Sastry et al. 1978; Rataul and Butter 1977; Saikia and Muniyappa 1986; Gupta et al. 2001). Non host plants for these viruses were also identified. They are *Amaranthus caudatus, Achyranthus aspera. Gomphrena globosa, Catharanthus roseus, Chenopodium amaranticolor, Rhaphanus sativus, Brassica oleraceae, Cucumis sativus, Cucurbita pepo, Luffa acutangula, Momordica charantia, Acalypha indica, Phaseolus vulgaris, Dolichos lablab, Althaea rosea, Abelmoschus esculentus, Gossypium hirsutum, Solanum melongena, Solanum tuberosum, Datura metal, Withania sominifera, Vigna mungo, V. radiata, V. unguiculata, Lagenaria vulgaris, Clitoria sp. and Melilotus alba* (Reddy and Yaraguntaiah 1981; Verma et al. 1975).

The natural infection of begomoviruses has been identified in several plant species. Chilli (Hussain et al. 2004), potato (Usharani et al. 2004a, b), chayote (Mandal et al. 2004), *Luffa cylindrica* (Sohrab et al. 2004) and bitter gourd (Tahir and Haider 2005) are hosts for ToLCNDV; pepper is host for ToLCGuV (Chakraborty et al. 2003) and guar was identified as a host for TLCBaV (Khan et al. 2003).

5.11.5 Serology

Muniyappa et al. (1991a, b) purified the ToLCBaV virions from infected tomato sample and showed the association of geminate particles by immunosorbent electron microscopy. Using panels of monoclonal antibody to ICMV and ACMV Varma (1989) and Harrison and Robinson (1999) analyzed the epitope profile of tomato begomoviruses in TAS-ELISA and speculated that they are different from other Indian begemoviruses.

5.11.6 Genome Comparison

Full length genome characterization of ToLCNDV was initially accomplished by Padidam et al. (1995) and Srivastava et al. (1995). Subsequently more than 100 isolates of ToLCNDV have been characterized. Characterization of ToLCBaV was completed by Muniyappa et al. (2000) and Kirthi et al. (2002) and it was found that virus isolates which were labeled as tomato leaf curl viruses Ban1, Ban3, Ban4, Ban5 and Kolar belong to the species ToLCBaV. The genome organization of all the 16 begomoviruses (length varying from 2739 to 2759) resembles organization of OW begomoviruses, having two virion and sense and four complementary sense ORFs. In addition Padidam et al. (1999) predicted one small ORF, V3 on the viral strand, the function of which is not yet clear. In DNA B, there are two ORFs one each in viral and complementary strand.

From the details it is evident that leaf curl isolates from southern India have a monopartite genome, which is associated with betasatellite, while both mono (DNA-A with betasatellite) and bipartite begomoviruses have been found in leaf curl affected tomato in northern India. The characteristics feature of ToLCNDV, in Indian subcontinent is, it is invariably associated with either alpha or betasatellite, though it is bipartite.

The genomic component of the bipartite viruses, ToLCNDV and ToLCPalV when compared, around 80% and 89.4% identity was observed between DNA A and DNA B components respectively, the interesting point is 77–81% identity of the ToLCNDV is in the CR region.The sequence identity with the monopartite begomoviruses ranged from 69 to 73%. Interestingly with tomato leaf curl Rajasthan virus it was 86%. The ToLCBaV which is highly prevalent in the southern India shows 72–80% identity with other monopartite viruses, 70/71% with bipartite ToLCNDV and ToLCPalV.

The study group on taxonomy of geminivriuses has recommended 91% and 94% identity in the complete nucleotides sequence of DNA A component as the threshold values to demarcate the species and strains respectively. Consequently some of the tomato viruses have been regrouped as variants of existing species. Following are the isolates of ToLCNDV which show less than 90% identity; ToLCNDV, –1N-ND papaya, 2005- 89% (DQ989325), ToLCNDV- Bangladesh-cucumber-06- 89% (EF450316), ToLCNDV-India- Haryana, 2003-87% (FJ561298), ToLCNDV- 88% (FN645905), ToLCNDV Pakistan- Lahore- 89%

(HG316125), ToLCNDV- 90% (JX460805), ToLCNDV-90% (KC960492), ToLCNDV- 90% (KF002409), Parthenium -90% (JQ897969), ToLCNDV-2-71%(JQ897969), ToLCNDV-3- 85%(KC465466), ToLCNDV-484%(KF551592), ToLCNDV- spain - 90% (KT175406). All these isolates show less than 91% identity but have been included under ToLCNDV, which need to be rectified. Especially, the isolates ToLCNDV-2, ToLCNDV-3, ToLCNDV-4 which show very less identity need to be raised to species level.

ICTV in its ninth report grouped ToLCNDV isolates into three different strains; majority of ToLCNDV isolates as one strain, and other three isolates ToLCNDV-2,3,4 as three different strains, ICTV in its tenth report differentiates only two strains among ToLCNDV, one strain comprising all the isolates of ToLCNDV and another comprising ToLCNDV isolates from Spain (Zaidi et al. 2016). It appears that more stringent analysis need to be performed to group ToLCNDV isolates.

On the basis of infectivity Chatterji et al. (1999), identified severe and mild strains among ToLCNDV. They shared 94% nucleotide identity, but differed in the rep binding sites. The rep proteins of severe strain, the amino acid residue Asn10 was shown to specifically recognise the third base pair of the iteron sequences GGTGTCGGAGTC. However strainal differentiation based on symptom expression may not be conclusive as symptom expression is host genotype dependent. Therefore, categorization of ToLCNDV isolates into mild and severe strain is not being pursued presently.

5.11.7 Phylogenic Relationship

ToLCNDV represents unique begomovirus complex that comprises diverse isolates infecting hosts belonging to crop species, weeds and ornamentals distributed in diverse agroclimatic conditions. Whether, the grouping occurs in the context of geographical locations or on the basis of host species were examined by several workers.

Jyothsna et al. (2013a, b), proposed that ToLCNDV isolates can be categorised into three clusters, one major cluster comprising virus isolates from solanaceous hosts with exception of one isolate from ashgourd and pumpkin, a second cluster including isolates from cucurbitaceous hosts originating from South Asia and Indian subcontinent and a third cluster comprising isolates from okra, tomato and chilli pepper. Phylogenetic analysis clearly revealed that the three bipartite begomoviruses ToLCNDV, ToLCPalV and SLCNNV have originated together, in this group ToLCRaV is also present.

Zaidi et al. (2016), performed phylogenetic analysis including ToLCNDV isolates from Europe and from Southeast Asian countries. These isolates are evolutionarily distinct from ToLCNDV variants from India and they opined that, the ToLCNDV isolates from Europe and North African country had monophyletic origin.

An analysis performed including all the isolates (the present study) clearly showed that there is no association between host species and clustering of ToLCNDV isolates. The ToLCNDV isolates which showed less than 90% identity branched off independently akin to other monopartite viruses. The other major lineage consists of monopartite begomoviruses of which ToLCBaV represent the basal group. On the basis of 94% identity ToLCBaV isolates have been categorised into four strains A, B, C, D. Among the monopartite begomoviruses, ToLCKeV is distinct.

5.11.8 Recombination

Indian tomato begomoviruses share host range and are, often present in mixed infection (Kanakala et al. 2013), possibly co-exist with one another in an host, satisfying all the conditions required for recombination events. Prasanna and Rai (2007) analyzed the events by RDP and concluded very low or almost absence of recombination events for ToLCNDV isolates. Prasanna and Rai (2007) suggested that of the six isolates of ToLCNDV analyzed only three isolates showed recombination and there is a non-random distribution of events, the highest frequency being mapped in the N terminal portion of Rep. It is quite unexpected that despite wide host range and mixed infection only few recombination events are detected in ToLCNDV; that too only in some isolates. ToLCPalV was the major parent in the predicted events. On the contrary ToLCBaV was identified as the complex recombinant, arising from five to six events. Viruses from south India contained sequences closely related to isolates from Taiwan. In our analysis it was found that ToLCNDV was one of major parental sequences contributing to emergence of recombinants like ToLCRaV, radish leaf curl virus, ToLCPalV, papaya leaf crumple virus, ToLCPtV.

5.11.9 Component Complementation

Exchange of genetic components referred as pseudorecombination is common among bipartite viruses which may occur in mixed infection, for example, Kanakala et al. (2013) demonstrated pseudoreombination between ToLCNDV A and ToLCPalV DNA B, and *vice versa*. Interestingly, they also demonstrated whitefly transmission of pseudorecombinant progeny produced by inoculation of ToLCNDV-DNA A with ToLCPalV DNA B suggesting that hetero encapsidation of DNA B component of ToLCPalV by ToLCNDV DNA A. In an inoculation combining four components of both the viruses, the DNA B component of ToLCNDV dominates. Pseudorecombination between ToLCGuV and ToLCNDV has been shown by Chakraborty et al. (2008) and Jyothsna et al. (2013a, b) observed that there was an asymmetrical synergism that inoculation with ToLCGuV DNA A with ToLCNDV DNA B component led to expression of severe symptoms.

ToLCNDV DNA B is an interesting molecule which shares high identity with DNA B of atleast three viruses, ToLCGuV (Chakraborty et al. 2003) BYVMV (Venkataravanappa et al. 2015) and PepLCLaV (Shafiq et al. 2010). In all these cases identity between the viruses in DNA A is only 65% but DNA B shows 85–89% identity with ToLCNDV- DNA B. Infectivity and transreplication of these DNA B by ToLCNDV DNA A resulted in severe symptom expression. It is possible that these monopartite begomoviruses occurred in mixed infection with ToLCNDV and captured DNA B.

It is relevant to note here that ToLCGuV was found associated with tomato yellow leaf curl Thailand betesatellite (TYLCTHB) and produced very severe symptoms in *N. benthamiana* and tomato. Jyothsna et al. (2013a, b) suggested that occurrence of ToLCGuV in association with betasatellite is more frequent than its association with DNA B.

5.11.10 Virus Prevalence in the Field

The distribution of ToLCNDV and monopartite tomato begomoviruses were earlier thought to be geographical location specific (Chowdareddy et al. 2005). However, in recent years in northern India, the mixed infection of ToLCNDV and ToLCKaV or ToLCNDV and ToLCPalV are more frequent. In the southern India, ToLCNDV is recorded in almost all the cucurbitaceous hosts, okra and weeds. However ToLCNDV is not yet recorded from tomato in southern India. The tomato genotypes cultivated in southern India do not possess genetic resistance to ToLCNDV, as they succumb to infection when they are planted in north India. It is best speculated that, ToLCNDV in cucurbitaceous hosts in south India meet with some obstacles in getting transmitted to tomato, which need to be looked into.

The entry of ToLCNDV into cucurbits itself is of recent occurrence. ToLCNDV infection in okra has been identified in 2004, in majority of cucurbitaceous hosts between 2004 and 2007, later in ashgourd 2011, papaver 2012, cucumber 2012; in eggplant which was considered to be free of any WTG infection in India was found to be infected in 2009 (Pratap et al. 2011). Earlier researchers Vasudeva and Samraj (1948) and Singh and Lal (1964) considered cucurbitaceous hosts as non-host plants to ToLCNDV. Whether capturing betasatellite facilitates its expanding host range is one aspect which needs to be researched.

Whether the excessive spread of the virus in cucurbitaceous hosts could result due to its presence in seeds, need to be looked into in the context of seed borne nature and seed transmission of MYMV in India (Sathya et al. 2013).

5.11.11 Establishment of Koch's Postulate

Molecular characterization of nearly 100 isolates of ToLCNDV has been completed. However, Koch's postulates have been established through agroinoculation only for four isolates (U150115, AY428769, HQ264185, HQ141673). As a typical bipartite virus, though DNA A alone is infectious, systemic spread and expression of symptoms occur only when both DNA A and DNA B are inoculated. Padidam et al. (1995) observed that, mutation in the coat protein region (amino acids 65 or 172) did not affect systemic movement and symptoms development but affected single stranded DNA accumulation. They also showed the mutation in AV2 region to affect generation of ds DNA.

Pratap et al. (2011) recorded for first time a ToLCNDV in eggplant causing severe yellowing diseases in Nagpur in central India and Koch's postulates were established by inoculating dimeric clones of DNA A and DNA B components.

Sivalingam and Varma (2012), found that, when ToLCNDV A alone is inoculated, limited accumulation of viral DNA occurred, which increased several folds in plants co-infected with DNA B or betasatellites. The increase in virus titre very much reflected in symptom severity and transmissibility by whitefly.

Since ToLCNDV was always associated with various types of betasatellite, the pathogenicity of ToLCNDV in association with different betasatellites was examined (Jyothsna et al. 2013a, b). Plants co-inoculated with betasatellites showed enhanced symptom severity in both *N. benthamiana* and tomato and increased in helper viral DNA A and DNA B levels.

For monopartite begomoviruses ToLCGuV (Chakraborty et al. 2003; Jyothsna et al. 2013a, b), ToLCKaV (Chatchawankanphanich and Maxwell 2002), ToLCJV (Tiwari et al. 2012), ToLCBaV (Tiwari et al. 2012), Koch's postulates have been established by inoculating DNA A with respective betasatellites. ToLCBaV was found to replicate well at 23 °C and produce symptom, than at 25 °C. Of all the viruses studied, ToLCJV was the most virulent, wherein DNA A alone could cause very severe symptoms in both *N. benthamiana* and tomato (Tiwari et al. 2012). The DNA A component of all these monopartite viruses *trans* replicated different betesatellite confirming the promiscuity between DNA A and betasatellite.

5.11.12 Molecular Basis of Pathogenicity

The functions of coat protein were examined by Kirthi et al. (2002) by expressing the coat protein of ToLCBaV. The purified recombinant CP bound preferentially to ssDNA in sequence nonspecific manner. They proved that the Zinc finger motif in CP (corresponding to 65–68 amino acid residue in ToLCBaV is involved in binding to Zinc and DNA.

5.11.13 RNAi and PTGS Suppressor

The viral suppressor AC2 was studied by Yadava et al. (2010) and ToLCNDV AC2 was demonstrated to inhibit the enzyme activities of RdR6 and AGO4 protein which are needed for initiation and effector activities of RNAi. Consequent upon identification of suppressor in begomoviruses, their application in achieving transgenic resistantce has been looked into. Thus Praveen et al. (2010) made four different RNAi construct with varied length targeting AC4 gene of ToLCNDV, longer ds RNA constructs were more efficient in silencing of target gene. Using sense, antisense, self-complementary inverted repeats, non spliced hairpins and small hairpin constructs targeting Rep and AC4 segments they demonstrated suppression of viral symptoms.

5.11.14 Host Virus Interactions

A study by Sahu et al. (2010) highlighted the gene expression changes during incompatible interaction between tolerant tomato plant and ToLCNDV. A suppression substractive hybridization library (SSH) was prepared for a naturally tolerant cultivar of tomato, namely H-88-78-1. This study revealed that tolerant tomato plants have enhanced level of transcript related to cell cycle and DNA/RNA processing, signaling molecules, transporters, transcription factors along with the proteins of unknown functions. An interesting observation was that apart from these genes, classes of host ubiquitin proteasome pathway genes were also highly expressed in tolerant cultivar in comparison to a susceptible cultivar. Similar attempts were made to identify the differentially expressed genes during ToLCNDV-tomato interaction, (Naquvi et al. 2011a). The genes related to innate immunity, metabolism and ethylene signaling were implicated in the systemic infection during ToLCNDV infection in tomato (Naqvi et al. 2011a, b).

Kushwaha et al. (2015) inoculated ToLCNDV on to several solanaceous hosts, *Capsicum annuum*, *N. benthamiana*, and *Solanum lycopersicum,* and observed that *S. lycopersicum* and *N.tabaccum* developed symptoms and led to viral DNA accumulation more than other hosts. They observed that there were differential expression levels of RNAi pathway genes (RDR6, AGO1, and SGS3), as well as host defense pathway. The NBS-LRR type of protein and lipid transfer protein were upregulated and they concluded that, the expression levels of host defense genes, determined the viral DNA accumulation and symptom development. Sahu et al. (2016), subsequently attempted to characterize 26S proteosomal subunit RPT4a (SIRPT4) gene under ToLCNDV pathogenesis in the tolerant cultivar H-88-78-1. They showed specific binding of SIRPT4 at the stem loop region of IR in both DNA A and DNA B. They suggested that this binding is secondary structure specific and binding at IR inhibited RNA pol II activity, thereby reducing bidirectional transcription. When they silenced SIRPPT4 gene, the tolerant phenotype of H 88-78-1 was converted to susceptible phenotype. Overexpression of SIRPT4 gene resulted in programmed cell death and hypersensitive reactions. They suggest that SIRPT4 interference of viral pathogenicity is more due to specific binding and not due to any proteolytic function.

Mandal et al. (2015) studied the transcript level of SlTRNI gene which is important for cell expansion and vein formation. Though SITRN1 has two start sites, under viral pathogenesis there is a preferential use of one start site only. They found that the promoter sequences of SITRN1 have multiple W boxes which mediate induction of SITRN1 under ToLCNDV infection. They postulate that during stress SA pathway gets activated which induces WRKY16, leading to transcription of SITRN1gene.

5.11.15 Micro RNA and Their in Role In ToLCNDV Pathogenecity

Geminivirus proteins are known to modify the host PTGS pathways, which in turn deregulates normal cellular activities leading to disease development/resistance (Sahu et al. 2014b). In this regards, efforts have been made to identify the ToLCNDV-responsive-miRNAs responsible for either disease development or in providing tolerance (Naqvi et al. 2010; Pradhan et al. 2015). A NGS platform was used to identify the ToLCNDV-responsive miRNAs, which resulted in detection of 53 novel miRNAs (Pradhan et al. 2015). These novel miRNAs were not only involved in targeting leaf architecture and plant development related host genes, but were also implicated in plant defense response. For example, novel miRNAs such as Tom 14, Tom 43 have been shown to target transcription factors such as AP2/ERF; and teosinte branched1/cycloidea/PCF (TCP) transcription factor, respectively, and might lead to leaf curl phenotype in plant. Apart from this, disease resistance gene such as CC-NBS-LRR type protein was also shown to be targeted by a novel miRNA Tom 17. Besides these, role of conserved miRNAs has also been examined by Naqvi et al. (2010) which revealed that the differential accumulation of miR159/319 and miR172 have the correlation with leaf curl symptoms development in tomato. More interestingly, the authors have also postulated role of miR168 and miR162 in the alteration of global miRNA flux by targeting DCL1 and AGO1.

Moreover, numbers of reports on computational prediction of virus genome-derived miRNAs are also available, but few of them are reported to be involved in the disease resistance (Naqvi et al. 2011a, b; Shweta and Khan 2014).

5.11.16 Alpha and Betasatellites Associated with ToLCNDV and ToLCBaV

Though ToLCNDV is a bipartite virus in northern India, it is found associated with CLCuMuB (Sivalingam and Varma 2012; Jyothsna et al. 2013a, b) and can *trans* replicate ChLCB (Akhter et al. 2014), CLCuMuB, TYLCTHB (Jyothsna et al. 2013a, b) PaLCuB, BYVB and LuLDB. In all the cases there is enhancement in symptom expression and viral replication. ToLCBaV is associated only with ToLCBaB, but can trans replicate other betasatellites too (Tiwari et al. 2012). ToLCNDV is also found associated with alphasatellites.

5.11.17 Resistance to ToLCNDV and ToLCBaV

Identification of the resistant sources and deployment of the specific gene to counter the challenges of the specific virus in a selected location is the only strategy that will give rise to sustainable disease management. In the case of tomato leaf curl disease, the problem is more complicated as we are not sure how many viruses are causing the disease in a specific geographic location. Besides, the begomoviruses constantly evolve and either by component capturing or by recombination, they emerge as super virulent strain that it is difficult to develop a resistant line. Until now, six genes (*Ty* genes) derived from different tomato wild species have been identified. From the perusal of the data it is evident that *Ty-2* genes while confer resistance phenotype against monopartite begomovirus, it was not effective against bipartite ToLCNDV, Prasanna et al. (2015) attempted to combine *Ty-2* and *Ty-3* genes through marker assisted selection and screened the hybrid lines for resistance to viruses by challenging through agroinoculation of specific, monopartite and bipartite viruses. They also performed field screening to know the phenotype of genotypes response to unidentified viruses other than ToLCNDV, ToLCBaV, ToLCJV, ToLCGuV. They found that the lines and hybrids with *Ty-2* were susceptible to ToLCNDV. The *Ty-3* gene showed dosage effect with partial resistance of plants to ToLCNDV in heterozygotes stage. They also observed that *Ty-2* gene in either homozygous or heterozygous stage enhanced level of resistance of plants with heterozygous *Ty-3*. By pyramiding *Ty-2* and *Ty-3* genes considerable resistance toToL-CNDV can be achieved. The resistance of some of the tomato genotype Vaibav, Nandhini, having *Ty-2* gene to ToLCBaV were lost, when these genotypes were individually agroinoculated with ToLCBaV (Tiwari et al. 2012) and the cognate betasatellite.In the presence of betasatellites, the *Ty-2* gene effect seems to be neutralized which need to be looked into.

5.12 Chilli Leaf Curl Virus

Chilli leaf curl virus (ChiLCV) is the most predominant begomovirus affecting chilli production in India. This virus is present throughout India infecting wide range of solanaceous and non-solanaceous host. The virus is a typical Old World monopartite begomovirus associated with different betasatellites. Until now, 11 distinct begomovirus species are reported to be associated with chilli leaf curl disease in India e.g., ChiLCV, ChiLCINV, chilli leaf curl Kanpur virus (ChiLCKaV), chilli leaf curl Vellanad virus (ChiLCVV), PaLCuV, pepper leaf curl Bangladesh virus (PepLCBV), RaLCuV, ToLCJV, ToLCV, and ToLCNDV. Recently, a new recombinant virus, chilli leaf curl Ahmedabad virus has been recorded by Bhatt et al. (2016). Of all the viruses infecting chilli in India, ChiLCV is the most prevalent begomovirus species.

5.12.1 Discovery and Distribution

In India, the occurrence of leaf curl disease in chilli was first noticed in 1930 (Husain 1932). Successful whitefly transmission of the disease established it as a viral disease (Mishra et al. 1963; Dhanraj and Seth 1968; Shukla and Ram 1977). The incidence and severity of the disease has increased since 2007 (Chattopadhay et al. 2008). Due to high vector population, the incidence and spread of the disease are more in summer season. In the recent years (2014, 2015), chilli leaf curl emerged in epidemic proportion in the states of Madhya Pradesh and Maharashtra. Although, the first report of the disease was about half- a century ago, the association of begomovirus with this disease has been confirmed only in 2007 based on partial sequencing of viral genome (Senanayake et al. 2007).

5.12.2 Economic Loss

Leaf curl is a major constraint in chilli production in India. If the chilli plants get infected within 3–4 weeks after transplantation the yield loss reaches upto 90% and the quality of the fruits is also affected. In Jodhpur (Rajasthan), a major chilli an epidemic of leaf curl disease in chilli was recorded in 2004, where14–100% incidence of the disease was recorded (Senanayake et al. 2012). In recent years, epidemic of chilli leaf curl emerged in central India. Chilli is a major crop in Khargone, Dhar and Badwani districts of Madhya Pradesh, where leaf curl disease was not a major problem till 2013, however during 2014–2015, leaf curl emerged in epidemic proportions in most places resulting in widespread crop failure. In Yavatmal District, Maharashtra, the disease caused a serious yield loss in chilli during 2015, where leaf curl appeared within 3–4 weeks post transplanting and affected entire fields in most of the places by October–November. Many of the farmers were forced to abandon chilli crop midway and many uprooted or sown wheat in the chilli field.

5.12.3 Symptomatology

The infected chilli plant exhibits symptoms such as curling of leaves, leaf rolling and reduced leaf size. Puckering and blistering of inter veinal regions associated with thickening and swelling of veins (Dhanraj and Seth 1968; Mishra et al. 1963) may also occur. The infected plants may get stunted, bear very less fruits resulting in heavy loss.

5.12.4 Transmission and Host Range

The virus is not sap transmissible and so far seed transmission has not been reported. It is efficiently transmitted by whitefly. Senanayake et al. (2007) found that 60% transmission was brought out by a single whitefly and eight whiteflies per plant resulted in 100% transmission and the plants developed symptoms within 7-10dpi; the minimum AAP and IAP were determined as 180 and 60 min. In the serial inoculation experiment whitefly survived upto 5 days post AAP and became aviruliferous 1 day prior to death.

Chilli leaf curl virus has a wide host range and infects solanaceous and non solanaceous hosts in combinations with various betasatellites. Among the solanaceous hosts the most commonly infected species are chilli, tomato and rarely eggplant.

The malvaceous host recorded are kenaf, and hibiscus. Some of the other hosts in which ChiLCuV was recorded are petunia, mentha, amaranthus and *Phaseolus aureus, Solanum nigrum.*

5.12.5 Genome Comparison

Geminate particles of 18 × 30 mm were detected by Senanayaka et al. (2007) in chilli samples fom Rajasthan. This combined with whitefly transmission and hybridization of field infected samples with DNA A gave sure indication of involvement of a begomovirus. Subsequently genomic components have been cloned which showed nearly 96% identity with ChiLCuV from Pakistan and Varanasi. The genome organization was found to be typically that of OW begomoviruses.

5.12.6 Recombination

Majority of the begomovirus isolates infecting chilli have been identified to be recombinants and the recombination breakpoints have been mainly identified to be located around AC1 and AV1 regions (Kumar et al. 2015). The recombination fragments in these begomoviruses were mainly contributed by other chilli-infecting begomoviruses which confirms that intra-species recombination predominates among these viruses (Kumar et al. 2015). However, it is known that recombination also facilitates the transfer of genetic elements even between distantly related species. In this context, it is important to note that the emergence of ChiLCV-Salem isolate has been facilitated by other non-solanaeceous crop-infecting begomoviruses. Among ChiLCD associated begomoviruses, a single species, ChiLCV, has been reported to be non-recombinant in nature. Apart from causing leaf curl disease in chilli, these begomoviruses has been found to infect various economically important plants such as amaranthus, bitter gourd, mentha, papaya and petunia (Saeed et al. 2014; George et al. 2014; Nehra and Gaur 2014; Raj et al. 2010a, b; Senanayake et al. 2012). Recombination might have facilitated the genetic potential to cross host barriers. In addition to recombination, these begomoviruses and betasatellites possess higher genetic variability and high rate of nucleotide substitution (Kumar et al. 2015). Furthermore, adaptive selection has been reported to be acting on the coding regions of these begomoviruses and betasatellites causing ChiLCD (Kumar et al. 2015).

5.12.7 Establishment of Koch's Postulate

Chattopadhyay et al. (2008), agroinoculated dimeric constructs of ChiLCV and tomato leaf curl betasatellite and demonstrated pathogenicity in chilli and in *N. benthamiana,* Kumar et al. (2011) proved the expression of symptoms for a Palampur isolate. Further Kumar et al. (2015) inoculated different begomoviruses with different betasatellites, and established how association of the betasatellite is indispensable for the leaf curl disease development in chilli.

5.12.8 Distribution of Betasatellites Associated with ChiLCD

Recently, a survey conducted on the identification of the begomoviruses associated with ChiLCD reported the association of betasatellites with all the samples collected from major chilli growing regions in the country. Based on the revised species demarcation threshold for betasatellites (Briddon et al. 2008), a total of six different betasatellite groups have been found to be associated with this disease in India. These are chilli leaf curl betasatellite (ChiLCB), CroYVMB, radish leaf curl betasatellite (RaLCB), tomato leaf curl Bangladesh betasatellite (ToLCBDB), tomato leaf curl Joydebpur betasatellite (ToLCJB) and tomato leaf curl Ranchi betasatellite (ToLCRnB). The most prevalent betasatellites associated with these begomoviruses is ToLCBDB followed by ToLCJB (Kumar et al. 2015). Similar to the helper viruses, mixed infection of betasatellites also have been detected from ChiLCD infected samples. All these groups of betasatellites have been experimentally demonstrated to be trans-replicated by these begomoviruses (Chattopadhyay et al. 2008; Kumar et al. 2011, 2015). Further, the association of these betasatellites is indispensable for leaf curl disease development in chilli (Kumar et al. 2011, 2015).

5.13 Cotton Leaf Curl Multan (CLCuMuV) and Cotton Leaf Curl Kokhran Virus (CLCuKoV)

The most devastating epidemic outbreak of begomovirus is exemplified by cotton leaf curl disease in north western India in the years 1993–1996. In north western India, two important begomoviruses causing the cotton leaf curl disease are cotton leaf curl Multan virus (CLCuMuV) (Radhakrishnan et al. 2004a; Chowdareddy et al. 2005; Rajagopalan et al. 2012) and cotton leaf curl Kokhran virus (CLCuKoV); Kirthi et al. 2004). CLCuMuV includes the recombinant strain described from Rajasthan which was earlier referred to as separate species cotton lef curl Rajasthan virus (CLCuRaV) (Radhakrishnan et al. 2004a; Kumar et al. 2010b). CLCuKoV comprises two recombinant strains Burewala strain named earlier as separate species as cotton leaf curl Burewala virus (CLCuBuV) (Kumar et al. 2010b; Zaffalon et al. 2012; Rajagopalan et al. 2012) and Shadadapur strain earlier designated as separate species cotton leaf curl Shadadpur virus. Besides these two species, cotton leaf curl Bangalore virus (CLCuBaV), tomato leaf curl Bangalore virus (ToLCuBaV), tomato leaf curl Patna virus (ToLCuPtV), ToLCNDV viruses have also been described from cotton. A new virus named as cotton leaf curl Barasat virus has been reported from *Malachra capitata* from West Bengal. It is not yet clear whether it will infect cotton.The cotton leaf curl Alabad virus which occurs in cotton in Pakistan, is recorded only in okra in India. The geminiviruses, PaLCuV, ACMV, OELCuV and

CpCDV reported in cotton in Pakistan (Saleem et al. 2016) have not yet been isolated from cotton in India.

5.13.1 Discovery and Distribution

In India CLCuD was first reported at Indian Agricultural Research Institute (IARI), New Delhi in few isolated plants of *G. barbadense* in 1989 (Anonymous 1990); and in *G. hirsutum* plants near SriGanganagar, Rajasthan in 1993; Punjab in 1994 and Haryana in 1996 (Ajmera 1994, 1996). The disease spread further, in 1997 there was a severe epidemic, as per rough estimate an area about 2.19 Lakh/ha was affected by the disease (Narula et al. 1999). Within a short span of 4–5 years the disease spread to entire 15 lakh ha of the most potential irrigated cotton belts in north India. Elite cultivars like F846, Pakistani NIAB-72F505, LH1134, PL104, RST9, Somnath and Ganaganagar Ageti became susceptible and disease incidence in north west India ranged from 1 to 97% (Singh et al. 1994). During the past three decades cotton leaf curl struck as epidemics at least twice. In India the epidemic outbreak was between 1997 and 1998, when the recombinant strain of CLCuMuV, which was earlier referred to as CLCuRaV was the culprit. The recent epidemic outbreak in 2011 is attributed to a recombinant strain of CLCuKoV, which was earlier referred to as CLCuBuV.

Leaf curl disease was noticed in cotton grown in homestead garden in Bengaluru, Karnataka (Nateshan et al. 1996). In recent years CLCuMuV has been recorded in Okra, Cotton and *Hibiscus* plants in China. CLCuMuV also has been recorded in Hibuscus plants in Phillipines. It is possible that CLCuMuV has moved to other countries through virus borne Hibiscus plants.

5.13.2 Economic Loss

In India extensive field experiments have been conducted to estimate the yield loss due to CLCuD. In Rajasthan, at Sriganganager a reduction of 50.3% in seed cotton yield, 50.3% in number of opened bolls, 12.3% in boll weight and 16.1% in height per plant was estimated in the case of popular variety *G. hirsutum* F846. In the case of variety RST9, the reduction in seed cotton yield was 32.9%, in number of opened bolls 22.9%, and in height 3.5% but no reduction in boll weight was observed (Ajmera 1996). In Punjab, a reduction of 10.5–2.2% in seed yield/plant in variety F846 and 39.0–79.7% in Pakistani Narma (NIAB-72) were recorded (Singh et al. 1994).

Ten promising germplasm lines were evaluated to study the effect of CLCuD on yield and quality parameters. There was a drastic reduction in seed cotton yield in diseased plants (8.1–79.2%) of all the tested lines. Reduction in boll weight ranged from 1.6 to 37.1% whereas boll number reduction varied from 54.3 to 73.2%. Recently, the effect of CLCuD on seed cotton yield and fibre character of popular *Bt* cotton hybrids (RCH 134, MRC 6304) of Punjab have been studied; 46.1–52.7%

reduction in number of bolls, 43.4–54.2% in boll weight, 7.5% in fibre length 7.5% in fibre strength, 3.8% in micronaire value and 2.5% in span length were observed in all the cotton cultivars (Singh et al. 2013). During the period from 2008 to 2010, all the cotton varieties which were earlier resistant or tolerant succumbed to the disease and an average yield loss up to 53.6% were recorded in north western India.

5.13.3 Symptomatology

The initiation of disease is characterized by vein thickening or swelling on young upper leaves. Dark green bead like thickening of small veins can be clearly seen from the lower side against sun light. These irregular thickenings gradually extend and coalesce to form continuous reticulation of small veins. Veins of the leaves become thickened which are more pronounced on the underside. Affected leaves become thick, leathery brittle and greener than healthy leaves. Similar symptoms are also observed on bracts. The disease is further characterized by upward and downward curling of the leaves which occurs because of the uneven growth of veinal tissue on the abaxial side of the leaves. A single infected plant generally does not show both upward and downward curling in field as well as in green house condition (Godara et al. 2012). The disease also causes enation of the veins which frequently develop into cup shaped, leaf like outgrowths on underside of the leaves. In severe cases, spiral twisting of petiole, peduncle in the youngest part of the stem is also common. In severe cases buds are either totally supressed or die shortly after formation, causing partial or complete sterility of the plant. In plants affected at an earlier age, reduction of internodal length leading to stunting and reduced flowering- fruiting are observed (Narula et al. 1999).

5.13.4 Transmission and Host Range

The cotton leaf curl causal agent is transmissible naturally only by its whitefly vector (*Bemisia tabaci Genn*), but not by seed or by mechanical inoculation with sap or by soil (Singh et al. 1994).

In experimental studies, Nateshan et al. (1996) reported that a minimum of 1 h acquisition access period (AAP) and a minimum of 5 min inoculation access period (IAP) were required for transmission and the transmission threshold was 8 h for CLCuBaV isolate. The whole process of acquiring the virus and infecting a healthy plant was accomplished in 6 h 30 min to 8 h, suggesting that the virus is circulative and requires a relatively short latent period in the vector (Nateshan et al. 1996). If the AAP and IAP are increased up to 24 h each, efficiency of virus transmission increases which is shown by the increased proportion of *G. barbadense* seedlings that were infected from 20 to 87% (Nateshan et al. 1996).

Nateshan et al. (1996) reported that CLCuBaV was transmitted for at least 9 days after *B tabaci* left the virus source plants. Nateshan et al. (1996) also studied the transmission efficiency of male and female *B tabaci*. They reported that female

B. tabaci were efficient in transmitting CLCuBaV infecting 19/25 (76%), of the test plants than male *B. tabaci* in which 12/25 (48%) of the test plants were infected.

CLCuMuV-Rajasthan was easily transmitted by whitefly to cotton cv LH900 producing symptoms of vein thickening, leaf curling and enation. The virus could be transmitted by whitefly given a minimum AAP of 10 min and IAP of 10 min. The efficiency of transmission, increased with increase in AAP or IAP. Best transmission was obtained when the whiteflies were given AAP and IAP of 48 h (Radhakrishnan et al. 2004b).

Nateshan et al. (1996) determined the host range of CLCuBaV by whiteflies (*B. tabaci*) fed on infected *G. barbadense* plants. CLCuBaV is transmitted to 24 species in 6 plant families, (i) *Asteraceae (Acanthospermum hispidum, Ageratum conyzoides, Cosmos bipinnata, Dahlia sp., Sonchus brachyotis, Synedrella nodiflora and Zinnia elegans),* (ii) *Euphorbiaceae (Croton bonplandianum and Phyllanthus niruri),* (iii) *Fabaceae (Phaseolus vulgaris),* (iv) *Malvaceae (Althaea rosea, Gossypium barbadense* and *G. hirsutum*), *Oxalidaceae* (*Oxalis latifolia*), (v) *Solanaceae* (*Capsicum annuum, Datura stramonium, Solanum lycopersicum, Nicandra physaloides, Nicotiana benthamiana, N. glutinosa, N. occidentalis, N. rustica, N. sylvestris and N. tabacum*). All infected species developed leaf curl, leaf cupping or leaf rolling and in some cases enation or vein thickening. CLCuBaV could infect all *G. barbadense* cultivars and a few *G. hirsutum* cultivars tested, but the virus could not be transmitted to *G. arboretum* or *G. herbaceum* (Nateshan et al. 1996).

Radhakrishnan et al. (2004a) reported that CLCuMuV-Rajasthan is readily transmitted by whitefly to plants belonging to families *Malvaceae, Solanaceae* and *Fabaceae*. The plant species, which developed typical leaf curl symptoms are *A. esculentus, Alcea rosea, G. barbadense, G. hirsutum, S.lycopersicum, N. benthamiana, N. tabacum, P. vulgaris* and *Physalis floridana.*

5.13.5 Genome Comparison

The begomovirus association with CLCD was established in 1989 by ELISA and ISEM using polyclonal antibody to ICMV (Varma et al. 1993) and nucleo based diagonostic study was initiated by Radhakrishnan et al. (2004a); further begomoviruses associated with the disease in southern and north western India were characterized (Kirthi et al. 2004; Chowdareddy et al. 2005; Kumar et al. 2010b; Rajagopalan et al. 2012).

The genome organization of cotton leaf curl begomoviruses is similar to other OW monopartite begomoviruses with MP and CP genes on viral strand, Rep, REn, TrAP and *sd* genes on the complementary strand. The characteristic variation in Burewala strain of CLCuKoV is discussed in detail below. In addition to four ORFs in the complementary strand, 21 isolates of CLCuKoV -Burewala strain have an additional ORF C5 spanning from nucleotide co-ordinate 283–80. The function of this predicted protein is not yet deciphered. A extended version of this protein is also predicted in one of the isolates of Rajasthan strain of CLCuMuV. Some begomoviruses have been show to have this additional ORFs like, MYMIV in which Raghavan et al. (2004) identified AC5 protein contributing to replication function in yeast.

The striking deviation observed among isolates of Burewala strain of CLCuKoV is the absence of intact functional C2 gene. The C2 or AC2 gene encodes a transcription activator protein of 134 aa compostion which activates the V1/CP gene promoter and is very much required for gene expression regulation in all begomoviruses. It also has silencing suppressor activity contributing to viral pathogenesis. Amrao et al. (2010a, b) when analyzed the resistance breaking strains of cotton leaf curl viruses, identified the virus isolates to be recombinant ones between CLCuMuV and CLCuKoV and named it as separate species CLCuBuV which lacked the functional C2 gene. Since CLCuMuV and CLCuRaV are recombinants, despite less than 91% identity in the complete nucleotide sequence, they are considered as distinct strains of CLCuKoV, CLCuBuV respectively. The Burewala strain in Pakistan, mutation in C2 region were of three types; type one (Amrao et al. 2010a, b) with one inframe stop codon (1S) resulting in prematured terminated product of C2 having only 35aa residues, the second type of mutation comprises of two inframe stop codon (2S), one at the same position as the first site of mutation codon, the second inframe stop codon in three amino acid downstream of the first; the third type of mutation, a frame shift at nucleotide co-ordinate 1535, due to loss of Guanine, and a Guanine to Thymidine mutation which resuted in stop codon.

Interestingly in India, isolates of Burewala strain characterized from 2004 to 2010 contain only 2S type of mutation and lacked functional C2 protein. Rajagopalan et al. (2012) analyzed C2 region of more than 258 isolates and found out that the isolates from Fazilka and Bathinda region contained a mixture of isolates, majority of them are of 2S – inframe stop codon type and some isolates had intact C2. However in Sri Ganaganagar and Hanumangarh all isolates of Burewala strains were with 2S mutation. No isolate was found to have intact C2. Contrastingly in Hissar and Dabavali regions, isolates had intact C2 gene. In summary, isolates with defective C2 gene was more prevalent than with intact C2 gene. The 14 isolates which lacked functional C2 gene had an addition ORF C5, function of which is not yet determined.

Godara et al. (2016) characterized cotton leaf curl viruses at Delhi. They found out that three isolates belong to CLCuMuV-Rajasthan and two isolates to CLCuKoV. Association of one betasatellite, CLCuMB, and three alphasatellites related to cotton leaf curl Burewala alphasatellite and Gossypium darwini symptomless alphasatellites.

Interestingly, eight DNA- B molecules have been isolated from different wild species of cotton in Pakistan (Sattar et al. 2013).They shared 88–98% identity between them and exhibited very low identity (65–71%) with DNA B of other begomoviruses like SLCMV, ICMV. However, from India, until now DNA B molecule has not yet been detected. The two major begomoviruses CLCuMuV and CLCuKoV between them share only 72–84% identity in the complete nucleotide sequence. Both the viruses exhibit 77–83% identity with CLCuBaV from southern India. CLCuAlV which occurs only in okra in India, exihibits 66–68% identity with

CLCuKoV, 73–75% with CLCuBaV, and 76–83% with CLCuMuV. With other begomoviruses PaLCuV, ICMV 62–71% identity was observed.

5.13.6 Phylogenetic Relationship

Rajagopalan et al. (2012) performed the phylogenetic analyses of cotton begomoviruses from India and showed that the species CLCuKoV, CLCuMuV, CLCuAlV and CLCuBaV stand well separated from each other. The isolates of Burewala strains could be categorised into three types; Those isolates which lack C2 due to two inframe stop codons (2S) from Bathinda, Abohar Fazilka and Sri Ganganagar clustered in same clade along with 2S isolates of Burewala strain from Pakistan. The isolates of Burewala strain having uninterrupted intact C2 gene (four isolates) clustered with Burewala isolates of CLCuKoV from Pakistan having intact C2 gene. However four Indian isolates of Burewala strain with intact C2 gene occupied a separate clade.

Interesting results emerged from analysis of 191 full length of genomes of cotton and related begomoviruses (BYVMV) by Saleem et al. (2016). They observed that depending on the recombinational events occurring between CLCuMuV and CLCuKov, two major groups could be recognized among cotton leaf curl viruses. Group A represents recombinant viruses mainly derived from CLCuMuV and group B consists of isolates derived from CLCuKoV. Depending on the recombinational events, clade I to clade VII was differentiated.

5.13.7 Recombiantion

The contribution of recombination to emergence of newer super virulent strain – a theoretical prediction became a reality in the case of cotton leaf curl viruses. Between the two major viruses CLCuKoV and CLCuMuV, recombination occurred at different sites in genome and gave raise to Rajasthan, Burewala and Shadadpur strain. Kumar et al. (2010b) showed that Rajasthan strain of CLCuMuV they characterized had recombination in AV1 region; between CLCuMuV and CLCuKoV they also showed recombination in the Rep region with CLCuMuV and MeYVMBhV; another isolate of CLCuKoV they cloned, had recombination event in AC1 region with chilli leaf curl Pakistan virus. Interestingly when Rajagopalan et al. (2012) performed recombination analysis for isolates of Burewala strain from 2004 to 2010, no recombination events were detected in the genome from other begomoviruses. Kumar et al. (2015) identified recombination in an isolate of CLCuMuV at the 673–1981 region and the burewala strain of CLCuKoV exhibited recombination in the region between 1535 and 1817 nucleotides.

Saleem et al. (2016) observed two recombination events involving BYVMV; one in CLCuMuV – Sri Ganganagar and another in CLCuAlV viruses in CP region. All isolates of Rajasthan strains are recombinants between CLCuMuV and CLCuKoV with varying lengths of CLCuKoV sequences; in only one isolate of Rajasthan strain (CLCuMuV-HM037920 contribution from CLCuKoV spans both virion and complementary strand, CLCuMuV shares only a short stretch in CP region.

Rajasthan strain is the only strain where CLCuKoV contributes to the complementary sense strand. Saleem et al. (2016) suggest that since it is prevalent in India, this strain might have originated in India. On the contrary, the Burewala strain is dominant both in India and Pakistan, probably this strain originated first in Pakistan and got introduced through viruliferous whiteflies to India. The successful establishment of Burewala strain in both the countries suggest that successful and fittest recombination is the one in which complementary strand gene is from CLCuMuV and virion sense gene from CLCuKoV.

5.13.8 Establishment of Koch's Postulate

Radhakrishnan et al. (2004a) made complete tandem repeat constracts of Rajasthan strain of CLCuMuV and agroinoculated cotton cv LH 900. The plants showed enation and veinal thickening 28 days post inoculation. Kumar et al. (2015) made partial dimeric construct of CLCuMuV, CLCuKoV and Burewala strain of CLCuKoV and inoculated tobacco plants along with CLCuMB and CLCuMA. Eighty-five percent of the inoculated tobacco plants showed severe symptoms when CLCuKoV were inoculated together with beta and alphasatillite. Inoculation with alphasatellite and helper virus did not produce severe symptoms. The Burewala isolate with intact C2 gene induced severe symptom in comparison to the C2 mutants. Whitefly transmission of the progeny virus to cotton was performed using agroinoculated tobacco plants as inoculum. The cotton plants showed curling of leaves with Burewala isolate and with one isolate of CLCuMuV. The cotton plants inoculated with Burewala isolate showed severe symptoms. Sequencing of the viruses from the whitefly inoculated cotton plants revealed that the C2 gene mutation is retained.

5.13.9 Resistance

The first outbreak cotton leaf curl disease epidemic in Sri Ganganagar in 1992 was found to be caused by Rajasthan strain of CLCuMuV (Radhakrishnan 2002). CLCuKoV were also identified in adjacent field of Haryana and Punjab. However until 2004, Rajasthan strains of CLCuMuV was predominant in north western region of India. By 2005–2006, Burewala strain emerged, by 2010 Burewala strain displaced CLCuMuV and only in isolated places Rajasthan strain was located. The emergence of Burewala strains became obvious as most of the resistant lines succumbed to the disease. The resistant varieties derived from the resistant source LRA5166, RST9, RS875, RS810, RS2013, S1861, LH2076, H117, H1126, LH14144, CSH98, CSH238 and CSHH243 were highly susceptible to Burewala strain (Monga et al. 2008). The susceptibility of cotton cultivars, coinciding with build-up of whitefly population in the last 2 years resulted in increase in the incidence of the disease.

In this context, understanding genetics of resistance to leaf curl disease and deploying durable resistance gene become important. Ali (1997) found that resistance to cotton leaf curl viruses is controlled by single dominant gene, which can be introgressed into elite cultivars. Some of the wild species of Gossypium like *G. thurberi, G. anomalum, G. raimondii, G. armourianum, G. tomentosum* posses resistance genes to pests like whitefly which may be exploited (Azhar et al. 2010).

5.13.10 Molecular Basis of Pathogenecity

Khan et al. (2015a) examined the bidirectional promoter present in the IR of Burewala strains of CLCuKoV by transient agroinfiltration and transgeneic assay. Rep promoter in transformed tobacco plants showed two to fourfold higher activity than CaMV 35S promoter. Higher fluorescent identity of GFP was demonstrated in both tobacco and cotton leaves agroinfiltrated with CLCuKoV Rep promoter compared with CaMV 35S promoter and CLCuKoV CP promoter. The Rep promoter consisted of many cis- acting regulatory elements and growth regulator responsive transcription factor binding sites. Shukla et al. (2013a, b), identified the promoter sequence of CLCuMuB and confirmed the suppressor activity of the beta C1 protein by reversal of GFP silencing assay in *N. benthaminana.*

The functional attributes of CP and V2 of CLCuKoV was analyzed by over expression of the CP and V2 protein in bacteria, Sf21 cells and in planta by transient assay. Priyadharsini et al. (2011) purified the V2 and CP protein and showed the interaction between two proteins in ELISA and by surface plasmon resonance confocal microscopy studies of SF21 cells revealed the localization of CP in the nucleus and V2 in the periphery, when NLS signal of CP is abolished, CP is distributed in cytoplasm. They made V2- GFP and CP-YFP construct, expressed in *N. benthamiana* leaves and confirmed the localization of V2 in periphery and CP in the nucleus. When both constructs are co-infiltrated CP was found in the nucleus and in cytoplasm along with V2. The interaction between V2 and CP thus, may facilitate the movement of viral genome.

The CLCuMuV and CLCuKoV are associated with one predominant betasatellite CLCuMuB. In Pakistan, the resistance breaking Burewala strain was found associated with a recombinant CLCuMuB. The recombinant CLCuMuB is different from original sequence of CLCuMuB in having approximately 100 nucleotide fragment of the SCR originating from a betasatellite associated with tomato leaf curl disease (Amrao et al. 2010a, b). Kumar et al. (2015) identified Burewala strain in north western India and the satellites CLCuMuB and CLCuMA. CLCuBaV was found associated with Kenaf leaf curl betasatilitte .

Radhakrishnan (2002) identified the association of CLCuMuB with Rajasthan strains of CLCuMuV in Sri Ganganagar, Hissar, Delhi samples. Rajagopalan et al. (2012) analysed the samples from Abohar, Bathinda, Fasilika region and found that there was only one betasatellite CLCuMuB associated with burewala strains of CLCuKoV and Rajasthan strain of CLCuMuV. They could not isolate any alphasatellies. About 69 CLCUD associated betasatellites have been characterized which all belong to only one betasatellite species CLCuMuB. Zaffalon et al. (2012) detected CLCuMuB in many host like cotton, papaya, *Tribulus terrestris*; about seven alphasatellite were detected by Zaffalon et al. (2012) in leaf curl affected cotton samples. They belong to three different clades; cotton leaf curl Dabwali alphasatellite; cotton leaf curl Lucknow alphasatellite; gossypium Davidsoni symptomless alphasatellite (GDarSLCV) have been found associated with the disease. Their contribution to viral pathogenesis is not yet resolved.

5.14 Papaya Leaf Curl Virus

Papaya leaf curl virus (PaLCuV) has been recorded in minimum fifteen hosts belonging to diverse families of Apocyanaceae, Caricaceae, Malvaceae, Euphorbiaceae and Asteraceae (Table 5.3). It is also relevant to note here that, the type of symptoms PaLCuV produce ranges from yellow vein mosaic symptom to leaf crumpling depending on the host it infects. Other than PaLCuV, papaya is infected by AEV, ChiLCV, papaya crumple virus and ToLCNDV.

5.14.1 Discovery and Distribution

The leaf curl disease of papaya was first reported in Tamil Nadu by Thomas and Krishnaswamy (1939). On the basis of symptoms produced they called it as papaya leaf crumple disease. Subsequently leaf curl disease was observed in North India (Nariani 1956), eastern India (Sen et al. 1946). At present the disease is widespread in Haryana, UP (Singh 2006), Maharashtra, Andhra Pradesh, Karnataka (Govindu 1964) and Tamil Nadu (Surekha et al. 1977; Pandey and Marathe 1986; Verma 1996; Raj et al. 2008; Krishnareddy et al. 2010).

5.14.2 Symptomatology

The infected papaya plants show downward curling of leaves, vein thickening and extremely twisted petioles. The leaves appear leathery, deformed and plants are stunted. The affected plants do not bear any fruit, if any fruits are produced, they are small and distorted. Summanwar and Ram (1993) and Singh-Pant et al. (2012) observed upward rolling of lamina, enation mosaic and upward curling too. The plants irrespective of their age from 6 months to 3 years exhibited symptoms.

5.14.3 Transmission and Host Range

The causal agent of the disease is transmitted by grafting and by whitefly and not mechanically through sap. Though Sen et al. (1946) reported sap transmission of virus it was not confirmed by the other workers. Chandra and Samuel (1999) reported that graft inoculated plants show symptoms 5–7 weeks post inoculation.

The vector transmission of the virus has not been proved convincingly until recent years. Srivastava et al. (1977) showed more than 80% mortality of the whitefly, when allowed to feed on healthy papaya plants. Whitefly vector was thought unable to feed on papaya continuously. However Raichaudari (1977) reported that the virus can infect tomato, tobacco, sunnhemp, petunia and *Zinnia*. Summanwar and Ram (1993) recorded additional hosts of the viruses as chilli, *Datura* and hollyhock. Dubey et al. (2015), reported whitefly transmission of the virus, wherein symptoms appeared 4 weeks post inoculation. They showed that a minimum AAP of 30 min and IAP of 10 min are required for the transmission. The best transmission efficiency was obtained with IAP and AAP of 48 h. Raj et al. (2008) described successful transmission of PaLCuV from infected papaya to healthy papaya.

Of 12 test species tested by Dubey et al. (2015), 9 species, *Carica papaya, Carica annum, S. lycopersicum, N.tabacum, Crotolaria juncea, Petunia hybrida, Ageratum conyzoides, Datura stramonium* and *Zinnia elegans* plants expressed symptoms. These symptomatic plants were found positive in ELISA tests performed with polyclonal antiserum to ToLCNDV. On the basis of PCR detection and molecular characterization, PaLCuV has been detected in several hosts such as cotton (Mansoor et al. 2003), tomato, chilli, aster, tobacco (Kumar et al. 2009, 2012a, b), amaranthus (Srivastava et al. 2014a, b, c, d), radish, calotropis, cluster bean, Acalypha, soybean, and croton. In all above hosts, PaLCuV has been cloned and sequenced but none has been shown to be transmitted to papaya through whitefly transmission.

5.14.4 Genome Comparison

The association of begomovirus with papaya leaf curl symptom was established by Saxena et al. (1998) by Southern hybridization tests using DNA A probe to TGMV and ToLCNDV. They amplified DNA A fragments using Roja's primer and charecterised the virus, which was designated as PaLCuV. Subsequently, Krishnareddy et al. (2010) characterised PaLCuV isolates from Andhra Pradesh, Karnataka, Tamil Nadu and Uttar Pradesh. Singh-Panth et al. (2012) further extended the studies to Haryana and Delhi and characterised more isolates.

Saxena et al. (1998) characterised the PaLCuV from Lucknow and until 2010, it was thought PaLCuV is the only virus infecting papaya. However, Krishnareddy et al. (2010) characterised the leaf curl etiological agents from four different states and came out with interesting results. The isolate from UP, exhibited 94.7% identity with Lucknow isolate and so was designated as variant of PaLCuV. He identified the viruses in New Delhi and Andra Pradesh samples to harbour ToLCNDV and CYVMV, respectively.

5.14.5 Phylogenic Relationship

Krishnareddy et al. (2010), in their phylogenetic analysis performed on a multiple alignment of DNA A sequences found that Papaya leaf curl Coimbatore virus clustered with BYVMV and CLCuMuV-Rajasthan viruses, CroYVMV, ToLCNDV and PaLCuV grouped along with the isolates of respective species.

Phylogenetic analysis performed by Sing-Panth et al. (2012), revealed the position of ChiLCuV clustering with one ChiLCuV isolate infecting papaya in Punjab and another one infecting tomato; The Papaya leaf crumple virus was positioned close to TOLCNDV and ToLCV isolates from Pakistan. Singh-Panth (2012), detected high degree of recombination in seven DNA A sequences out of ten characterized by them. Recombination events were described in PaLCrV in AV1, AC3, AC2 region. Some of the events were between ToLCNDV and ChiLCuV. The AC4 ORF was highly variable and exhibited highest numbers of non synonymous mutations. Sinha et al. (2016) identified recombination events in PaLCuV in AV2, AV1 region and from AC1 to IR region between PaLCuV and BYVBhV and between CrYVMV and ChilCuV.

5.14.6 Establishment of Koch's Postulate

The complete tandem repeat constructs of DNA A when inoculated onto *N. benthamiana* plants, a mosaic pattern and slight curling were observed after 24 days (Sinha et al. 2016). Infectivity of papaya by agro- or biolistic-delivery has not been established consistently which needs to be focused. Singh-Panth et al. (2012) identified four different betasatellite species, tomato leaf curl betasatellite, papaya leaf curl betasatellite and CroYVMB. It is interesting to note here that these betasatellites are found associated with ToLCNDV and chilli leaf curl virus.

5.15 Begomoviruses Infecting Weeds

Characterization of begomoviruses infecting important crop plants had received attention owing to the economic importance of diseases caused by the viruses. Though the begomovirus causing yellow vein mosiac disease infecting the weed plants were known and recorded much earlier, only in the recent years, the diversity of begomoviruses occurring in weeds was looked into. The begomoviruses belong to the group of viruses referred as WILPAD viruses by Harrison (1981), as these viruses are more adapted to wild plants and are persistently transmitted by vectors contrary to CULPAD viruses which are more adapted to cultivated plant species like tobamoviruses. The begomoviruses in weed may represent the unchallenged genetic entity which need to be studied in detail. Some of the viruses characterized are discussed below.

In any discussion on ecological aspects on the begomoviruses, often it is interpreted that the weeds serve as reservoir of the viruses and serve as inoculum source. However perusal of the data show that only few viruses from weeds infect cultivated crop plants and *vice versa*, such as ICMV having cassava and Jatropha; PaLCuV infecting papaya, soybean and radish also occurring on weeds like, Calotropis and Acalypha. Pramesh et al. (2013) proved the infectivity of CroYVMV isolate through biolistic delivery of DNA A, along with betasatellite. Typical yellow vein symptoms were produced 15 days post inoculation in croton; the constructs also induced severe leaf curl in tomato plants. Ageratum enation virus infecting *Ageratum* also causes disease in tomato, soyabean, carrot. In all these cases, viruses have been characterised from the crop plants and weeds but whether transmission between the weeds and crop plants occur in nature has not been proved categorically. Therefore, the role of weeds as inoculum source, contributing to the primary spread of the virus continues to be an enigmatic question.

5.15.1 Jatropha Viruses

Jatropha species occur in wild conditions and is also cultivated as bio-diesel resource plant and as ornamental plant. In both wild and cultivated crops, severe leaf curl, yellow mottling, enation and yellow vein mosaic symptoms are observed. Incidence of such infection are more than 40% through out India even endangering commercial cultivation of the crop for biodiesel purpose. Snehi et al. (2016) had observed that there are more than five viruses causing these symptoms. As per the recent recommendation of study group on geminiviruses of ICTV 2015, four virus species are identified to infect Jatropha, they are ICMV (Aswathanarayana et al. 2007), Jatropha leaf curl, Jatropha mosaic India virus, Jatropha yellow mosaic virus. Recently two more new viruses is Jatropha leaf crumple virus and Jatropha leaf crumple India virus have been identified by Snehi et al. (2016). In phylogenetic analysis performed. Jatropha leaf crumple India virus, Jatropha mosaic India virus are closely related to ICMV. Jatropha leaf curl virus is closer to ChiLCuV, however JMINV is very distinct and separate.

The congress weed which grows abundantly in all agricultural fields *Parthenium hysterophorus* was shown to be infected by ToLCKaV. Kumar et al. (2016) cloned ToLCKaV and tomato leaf curl batasatellite, tomato leaf curl alphasatellite component from parthenium plants showing severe leaf curl symptoms through agroinfilteration, they confirmed Koch's postulates. Leaf curl symptoms were expressed in *Parthenium hysterophorus* and tomato plants.

5.15.1.1 Ageratum Enation Virus (AEV)

AEV was first characterized from Nepal, Fauquet et al. (1990) recognized it as separate species distinct from other begomovirus isolates characterized from *Ageratum conizoides* from southeast Asia. Subsequently, AEV has been characterized from *Ageratum* sp. and several other hosts as indicated in the Table 5.3. The satellites associated with the virus has been identified as Ageratum yellow leaf curl betasatellite and Ageratum enation alphasatellite. Besides AEV, Ageratum leaf curl virus and PaLCuV have been found infecting *Ageratum*. Interestingly Ageratum yellow vein virus, which is predominant in southern Asia is not yet detected in India.

5.15.2 Discovery and Distribution

The yellow vein, enation and leafcurl disease affected weed *Ageratum conzoides* and ornamental Ageratum are distributed throughout India. Though there is not yet an official record of the disease, the disease symptoms are observed thoughout India and the virus has been characterized in the year 2008.

5.15.3 Symptomatology

The disease plants show typical yellow vein symptom, accompanied by leaf curl, enation, leaf marginal rolling and reduction in leaf lamina and stunting. In *Cleome gynandra*, upward curling, crinkling, swelling of the petioles were observed.

5.15.4 Transmission and Host Range

The virus is readily transmitted by whitefly with AAP of 24 h and IAP of 24 h. The inoculated plants developed symptoms 25–30 DPI. The virus has been successfully transmitted from *Ageratum conzoides* to *Amaranthus, N. benthamiana* (Srivastava et al. 2013). The host range of the virus has been essentially revealed by the characterization of full length genome of DNA A sequences from plants showing yellow vein/ enation/ leaf curl symptoms. Thus besides Ageratum and ornamental species of Ageratum, AEV has been detected in pointed gourd (Raj et al. 2011), from grain *Amaranthus* (Srivastava et al. 2013), *Cleome gynandra* (Raj et al. 2010a, b), carrot (Kumar et al. 2013), and Zinnia (Kumar et al. 2010a, 2011). Presence of AEV in other hosts for which genome information is available in the database are tomato, soybean, *Papaver somniferum*, fenugreek, *Crassocephalum crepidioides* and papaya.

5.15.5 Genome Comparison

The genome organization of AEV is typically like that of Old World begomoviruses with intergenic region upto 281 nt. All the AEV isolates from India have iteron sequences GTACT and IRD sequences identified as FQIY. Between the AEV isolates from India, there is 92–99% identity at DNA-A nucleotide level, 97% with AEV from Nepal, 95% with AEV from Pakistan. The begomoviruses with which

AEV shares closest relationship (88.9%) is Tobacco curly shoot virus (Raj et al. 2011; Srivastava et al. 2013; Kumar et al. 2013).

On the basis of 94% identity for demarkation of strains within species, Tahir et al. (2015) categorised the AEV isolates into two strains, one Indian strain and one Nepal strain which characteristically differed in IRD and iteron sequences. The Koch's postulates with cloned components have been established for *Amaranthus* isolate (Srivastava et al. 2013) and for *Ageratum* and carrot isolates (Kumar et al. 2013).

5.15.5.1 Kenaf Leaf Curl

A disease causing leaf curl symptoms on kenaf has been observed in different parts of India. In eastern India, the begomovirus complex is composed of MeYVMV and CLCuMB while in southern India, ToLCJV and its associated betasatellite constitute the complex. In northern India, a new species of begomovirus, kenaf leaf curl virus, and an isolate of CLCuMuV were found associated with the disease (Paul et al. 2008).

5.16 Begomoviruses Infecting Ornamentals, Medicinal and Aromatic Plants

Efforts have also been made to characterize the begomoviruses infecting ornamental plants like hollyhock, (Srivastava et al. 2014a, b, c, d), *Calendula* sp. (Khan et al. 2007), *Dimorphothea* sp (Raj et al. 2007), jasmine (Srivastava et al. 2014a, b, c, d), *Duranta* (Jaidi et al. 2015a, b) and rose (Sahu et al. 2014a) and senna. However, further infectivity studies and cross inoculation studies are required to know their importance in the context of epidemic outbreak of crop diseases. In medicinal and aromatic plants several begomovirus associated diseases have been identified in India (Saeed and Samad 2017). *Withania somnifera* (ashwagandha), *Catharanthus roseus*, *Salvia hispanica* (chia) are known to be affected with yellow mosaic; *Andrographis paniculata* (Kalmegh) with yellow vein mosaic disease and *Mentha* spp, *Ocimum* spp, *Papaver somniferum* (Opium) and *Rosa* spp with leaf curl disease (Table 5.8).

5.17 Betasatellites Associated with Indian Begomoviruses and Their Role in Disease Development

The study group on geminiviruses submitted proposal in the year 2016 to ICTV on nomenclature and taxonomy of satellite DNA associated with begomo and mastreviruses. Accordingly the beta and deltasatellites have been included in the family designated as *Tolecusatellitidae,* comprising two genera *Betasatellite* and *Deltasatellite*. The decision regarding alphasatellites to be included as one of the genera under the family *Tolecusatellitidae* is deffered as they share lot of features

Table 5.8 Begomovirus reported in medicinal and aromatic plants in India

Name of the crop affected	Name of the virus/species	Nature of genome	Satellite/s molecule	Symptoms	References
Withania somnifera	Jatropha mosaic India virus	Monopartite	–	Yellow mosaic	Baghel et al. (2010)
Momordica charantia	Tomato leaf curl New Delhi virus and Pepper leaf curl Bangladesh virus	Monopartite	–	Yellow mosaic	Tiwari et al. (2012)
Andrographis paniculata	Eclipta yellow vein virus and Catharanthus yellow mosaic virus	Monopartite	Betasatellite	Vein yellowing and clearing, leaf curling.	Khan and Samad (2014), Khan et al. (2015b)
Solanum nigrum	Solanum leaf curl Lakshmangarh virus	Monopartite	–	Leaf curling, yellowing and stunting	Prajapat et al. (2013)
Tagetus patula	Ageratum enation virus isolate	Monopartite	Betasatellite and Alphasatellite	Leaf curling, crinkling and stunting	Marwal et al. (2013)
Mentha spp.	Tomato leaf curl Pakistan virus, Tomato leaf curl Karnataka virus and Chilli leaf curl India virus	Monopartite	Betasatellite	Leaf yellowing, mosaic, and crinkling	Samad et al. (2008)
Papaver somniferum	Tomato leaf curl New Delhi virus	Monopartite	–	Leaf curling and stunting	Srivastava et al. (2016)
Rosa spp.	Rose leaf curl virus	Monopartite	Betasatellite	Dwarfing and leaf curling	Sahu et al. (2014a)
Ocimum spp.	Tomato leaf curl virus, Chili leaf curl virus and Tomato leaf curl Albatinah virus	Monopartite	Betasatellite	Leaf curling, crinkling and yellowing	Gaur (2012)
Mucuna prureins	Velvet bean severe mosaic virus	Bipartite	–	Mosaic and yellowing	Zaim et al. (2011)

with nanoviruses. The family derives its name from the first satellite identified with Tomato leaf curl virus (Dry et al. 1997) which is a deltasatellite. The betasatellites and deltasatellites are believed to have common ancestor and so are placed together in one family. So far almost 1110 sequences of of full length betasatellites are available and they are differentiated into different species on the basis of nucleotide identity, <91% identity is recognized as threshold value to demarcate different species for both beta and deltasatellite. The deltasatellites are different from defective betasatellites in having more deletions or insertions in the SCR regions. While defective betasatellites occur with full length betasatellites, deltasatellites are not present along with betasatellites. So far only one deltasatellite, CroYVMB has been recorded from India, from *croton bonplandianum* (Dwakar et al. unpublished).

In India, the association of betasatellite in the begomovirus disease complex was first reported in bhendi yellow vein mosaic disease. In the absence of cognate betasatellite, BYVMV is unable to produce typical disease symptom in the host plant (Jose and Usha 2003). Extensive studies from several laboratories from India have confirmed association of betasatellites with begomoviruses infected crop plants of diverse types such as, *Abelmoschus esculentus* (okra), *Capsicum annuum* (chilli), *Citrullus lantus* (watermelon), *Cucumis sativus* (cucumber), *Gossypium hirsutum* (cotton), *Lagenaria siceraria* (long melon), *Luffa cylindrica* (sponge gourd), *Mentha arvensis* (mint), *Momordica charantia* (bitter gourd), *Raphanus sativus* (radish), *Solanum lycopersicum* (tomato), *Solanum tuberosum* (potato), *Phaseolus vulgaris* (French bean), *Vigna mungo* (blackgram) and *Vigna radiata* (mung bean) etc. (Varma and Malathi 2003; Borah et al. 2010; Singh et al. 2012; Jyosthna et al. 2013b; Kamaal et al. 2013; Satya et al. 2013). Betasatellite association has also been found in other plants like *Althaea rosea* (Hollyhock), *Kakimaeris indica* (Aster), *Hibiscus cannabinus* (Mesta), *Amaranthus hybridus* (Amaranth) which may act as reservoirs of infection (Chatterjee et al. 2007; George et al. 2014; Srivastava et al. 2013, 2014a, b, c, d). Totally about 545 betasatellites have been described from India.

Betasatellites are predominantly associated with the majority of the monopartite begomoviruses with notable exception of bipartite begomoviruses such as MYMIV (Rouhibakhsh and Malathi 2005) and ToLCNDV (Sivalingam et al. 2010; Jyothsna et al. 2013a, b). A monopartite begomovirus, ToLCBaV could transreplicate cognate betasatellite tomato leaf curl Bangalore betasatellite (ToLCBB) as well as non-cognate betasatellites namely, cotton leaf curl Multan betasatellite (CLCuMB) and luffa leaf distortion betasatellite (LuLDB). However, enhanced accumulation of viral DNA was found only in cognate betasatellite combination, suggesting a specific interaction between DNA A and its associated betasatellite (Tiwari et al. 2012). ToLCNDV DNA A and DNA B along with CLCuMuB or LuLDB betasatellite inoculated plants show 16-fold higher accumulation of DNA B and 60% reduced accumulation of betasatellite. In the case of association of betasatellite with bipartite begomovirus, DNA B and betasatellite act antagonistically (Jyothsna et al. 2013a, b).

Betasatellites, in contrast to its known association with begomoviruses, the association of betasatellite with mastrevirus was reported for the first time from wheat samples in the field infected by WDIV (Kumar et al. 2014). This betasatellite was identified as Ageratum yellow leaf curl betasatellite (AYLCB) and presence of this betasatellite induces more severe symptom as well enhances the accumulation of WDIV DNA (Kumar et al. 2014).

Ability of βC1 protein to suppress RNA silencing has been demonstrated for betasatellites such as ToLCBB, CLCuMB, LuLDB. The level of GFP siRNA was reduced in plants transiently expressing TOLCBB-βC1and not in CLCuMB-βC1, LuLDB-βC1 suggesting that βC1 encoded by different betasatellite interfere at different step of RNA silencing pathway (Shukla et al. 2013a, b). The interaction of βC1 with coat protein plays a collaborative role in inter- and intra-cellular dynamics of BYVMD complex (Kumar et al. 2006). BYVMV β-satellite based VIGS vector construction become possible by replacing the βC1 ORF with multiple cloning sites. The endogenous genes Su, PDS, PCNA and AGO1 could be successfully silenced using BYVMV β-satellite based VIGS vector (Jeyabharathy et al. 2015).

Radish, another important vegetable crop in India suffers from radish leaf curl disease (RaLCD) which has been found to be associated with radish leaf curl betasatellite. In subsequent study, this betasatellite was also found to be capable of substituting cognate DNA B's of ToLCNDV (Singh et al. 2012). The first evidences of chloroplast targeting by DNA virus encoded protein has been shown with the βC1 protein encoded by radish leaf curl betasatellite in *Nicotiana benthamiana*. The impediments caused by betasatellite at the different stages of chloroplast function causes photosynthetic inhibition and develops vein clearing symptom. In addition, the βC1 protein downregulates the expression of the genes involved in chlorophyll biosynthesis and chloroplast development (Bhattacharyya et al. 2015).

In 2009, Kumari et al. reported of a new tomato infecting virus from Patna in Northern India namely tomato leaf curl Patna Virus (ToLCPaV) and associated novel tomato leaf curl Patna betasatellite (ToLCPaB) which cause severe disease in tomato plants in Indo-Gangetic plain. Identification of ToLCV and associated tomato leaf curl Ranchi betsatellite further expands the list of tomato infecting begomoviruses and associated betasatellites in India (Kumari et al. 2011). Chilli leaf curl virus, causative agent of chilli leaf curl disease (ChLCD), has been shown to be associated with a betasatellite which is important to induce disease symptom in chilli (Chattopadhyay et al. 2008).

With increasing volume of systemic study involving begomovirus mediated disease complex, the importance of betasatellites in pathogenesis appears to be greater than ever. In India *Ipomoea purpurea* was a hitherto unknown natural host for sweet potato leaf curl virus (SPLCV). However, association of SPLCV with two different betasatellites i.e. croton yellow vein mosaic betasatellite (CroYVMB) and papaya leaf curl betasatellite (PaLCB) enables SPLCV to infect *Ipomoea purpurea* (Swapna Geetanjali et al. 2013). Similarly, a betasatellite associated with yellow vein mosaic disease of croton could infect radish and expand host range (Singh et al. 2012). Incidences of weed-infecting betasatellites infecting crop plants pose a serious threat for agro-economy in India as well as in world.

5.18 Alphasatellites Associated with Indian Begomoviruses and Their Role in Disease Development

The begomovirus/betasatellite complexes are often associated with a second type of circular ssDNA satellite, initially referred to as DNA-1 (Mansoor et al. 1999; Saunders and Stanley 1999; Briddon et al. 2004; Vinoth-Kumar et al. 2017b) and presently designated as alphasatellites. Alphasatellites components are about half the size (1350 nt) of begomovirus component and they have a highly conserved structure (Mubin et al. 2009). The alphasatellite Rep exhibits high levels of sequence identity to the Reps encoded by components of nanoviruses (Saunders et al. 2000; Saunders and Stanley 1999), a rolling-circle replication initiator protein encoded by viruses in the genus *Nanovirus*, family *Nanoviridae* that also have a genome of circular ssDNA (Gronenborn 2004). Consequently, alphasatellites are capable of autonomous replication, but require a helper begomovirus for spread in plants and for whitefly vector transmission. In addition to Rep, alphasatellites also have an A-rich region, ~200 nt long, down stream of the Rep-encoding region. In contrast to betasatellites, alphasatellites possess in their stem loop the nonanucleotide sequence, TAGTATTAC also found in the stem loop of viruses in the family *Nanoviridae*. Alphasatellites can affect both begomovirus titer and symptom development in host plants. Initially it was thought that the satellite molecules were limited to the OW, but recently, alphasatellites have been found associated with NW begomoviruses (Paprotka et al. 2010; Romay et al. 2010), thus expanding the geographical distribution of satellite molecules associated with begomoviruses. About 146 alphasatellites (Table-) belonging to 30 species have been described from India.

These molecules are not important in disease process and are not required for either infectivity or disease induction in host plants (Briddon et al. 2004). However these molecules reduce the accumulation of betasatellite DNA in plants and show little effect on accumulation of DNA A. Therefore they could have little minor modulating influence on the disease.

5.19 Concluding Remarks

In India, begomoviruses are distributed in all the agroclimatic zones from temperate sub-Himalayan foot hills to typical hot humid tropical climate zone. When the distribution of the virus (Table 5.3) is viewed, the maximum numbers of begomoviruses have been characterized from Uttar Pradesh (33) followed by Rajasthan (19) and Tamil Nadu (16). From this observation it is evident that some of the regions like north-eastern region, Gujarat and Odisha are not yet studied extensively. It is interesting to note that, there has been a gradual increase in the number of viruses described from 1990 to 2006; there is a spurt in the number of viruses recorded from 2006 to 2016, which may be due to discovery of RCA technique. The host plants from which viruses have been isolated, span from crop plants in 1990–2000 to weeds in 2005–2010.

Although as many as 80 begomviruses species have been indentified in India, the pan-Indian distribution is recorded only for BYVMV, ChiLCV and ToLCNDV. The wide distribution of ToLCNDV in all the states in southern India is in contrast to earlier understanding that ToLCNDV, a bipartite begomovirus is present only in northern India. However, it is to be noted that even now ToLCNDV in southern India is recorded only in cucurbitaceous hosts. MYMV and MYMIV occur both in northern and southern India; again the distribution pattern understood earlier has been negated. The most unexpected finding is that the leguminous host plants are infected by as many as 17 begomoviruses (Tables 5.3, 5.4, 5.6). Some of the viruses have been recorded only in the last 5 years. Whether, the viruses have moved to leguminous hosts in recent years or they have existed earlier but recorded now are not clear. The symptoms in these plants are not yellow mosaic but mosaic and leaf distortion. It is interesting that many begomoviruses are identified in soybean, cluster bean, French bean and *Phaseolus aureus*. Surprisingly, blackgram and mungbean are so far known to be affected only by MYMV and MYMIV. The vulnerability of leguminous plants especially soybean and French bean as revealed by infection by a large number of viruses is important in the context of possibility of recombination and emergence of new begomoviruses under the condition of mixed infection. Whether, these viruses will acquire ability to jump to blackgram, mungbean, pigeonpea and mothbean will have to be watched.

Solanaceous crops, tomato and chilli are infected by related viruses, those viruses which share more than 75% nucleotide identity. They are also infected by distantly related begomoviruses, RaLCuV, PaLCuv and AEV, which share less than 70% identity. The begomovirus, which infects a large number of hosts are PaLCuV (16 plant species), ToLCNDV (15 plant species) and AEV (14 plant species) (Fig. 5.7). Tobacco leaf curl is one of the earliest recorded diseases in India. The virus, tobacco leaf curl Pusa virus from Pusa, Bihar, may represent an isolate of the oldest virus reported. It is disheartening to see that though the disease is widespread in tobacco in different states, only one isolate has been characterized. Interestingly, tobacco curly shoot virus has been recorded from tomato, wild sunflower and French bean and not from tobacco. Some viruses like TbCSV, ToLCJV, ALYVV, ALCuV, HoLCV, MaYMV, SiLCuV have been reported earlier in other countries like China, Thailand and Pakistan. Whether, these viruses would have existed here in India or they gained entry in recent times needs to be investigated.

Phylogenetic relationship among Indian begomoviruses was inferred by analyzing the representative sequences of 80 virus species including the new species referred in the Table 5.2. The complete nucleotide sequences of DNA A component was analysed in CLUSTAL W programme in MEGA 6. It is evident, that Indian begomoviruses have polyphyletic origin and evolutionary pathway. The clustering of viruses is not dependent on geographical location; to a very limited extent grouping is governed by host species they infect. The begomoviruses analyzed fall into 14 major clusters; viruses clustered in IX and cluster XIII are of west Asian/African and American origin. The Indian begomoviruses could be grouped into 12 clusters, the most distinct one which do not share any common origin belong to cluster I comprising SLCMV, ICMV, JMINV, JLCrIV, cluster II, JYMV, cluster X- sweet

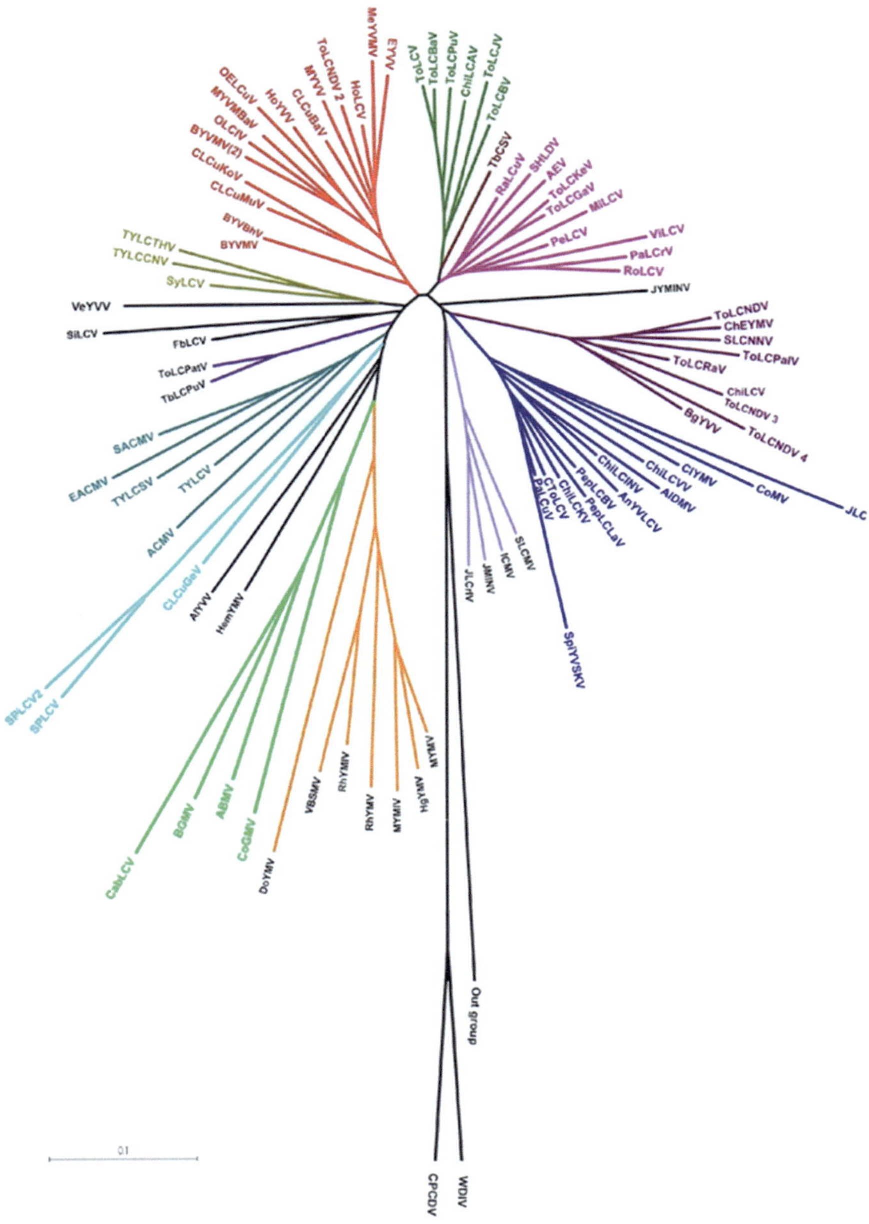

Fig. 5.7 Phylogenetic tree constructed on the basis of DNA A nucleotide sequence of Indian begomoviruses with other selected begomovirus sequences from NCBI database. The phylogenetic analyses were performed with MEGA 7.0 using neighbour joining method with 1000 bootstrap replications. Acronyms of viruses are as given in Table 5.1

potato leaf curl virus, cluster XIV- legume yellow mosaic virus. Of these, well separation of sweepoviruses and legomoviruses has been suggested by several workers (Qazi et al. 2007). The distinct nature of Jatropha viruses is an interesting finding, which further needs to be analyzed in detail as to which component of virus is divergent. Viruses in cluster II separate into two branches, one branch essentially having chilli viruses and second major branch having begomoviruses infecting Cucurbitaceae, ToLCNDV isolates and SLCCNV. It is relevant to note here that AEV the virus which infects cucurbits frequently is separated from cucurbitaceous cluster and belong to cluster IV. Of the four ToLCNDV isolates considered for analysis, ToLCNDV isolate infecting bhendi is grouped in cluster along with begomoviruses infecting Malvaceous and Teliaceous hosts (cotton, bhendi and mesta). Cluster IV contains array of leaf curl viruses in which, one branch is occupied by Solanaceous virus, other branch by AEV, RaLCuV, SHLDV and TbCSV. Positioning of some of the weed viruses are interesting. Synedrella leaf curl virus, which has been recorded from Andaman island, is grouped along with southeast Asian viruses (TYLCTHV, TYLCCNV). Vernonia yellowing virus is in the same cluster, but branches off independently. The weed infecting viruses seems to have independent origin and are well separated from crop infecting viruses.

International trade in agriculture, claimate change, resurgence of vector whitefly, deployment of new crop cultivars and intensive cultivations of crops throughout the seasons are important factors that aggravate begomoviral disease problems in the Indian subcontinent. The epidemic outbreak monitoring system is necessary to detect entry of an important begomovirus into new area or occurrence in new crop species, eg; whether ACMV occurs in cotton in India as in Pakistan, how extensive is ToLCNDV infection in cotton are some important questions which need to be answered. In this context, the cotton begomovirus movement from North-West India to South India needs to be monitored to avoid emergence of epidemic in a new production area.

Some of the begomoviruses like AEV occurring in weeds infect crop plants. Whether, the weed viruses contribute to crop disease development needs to be studied. The virus infecting ornamentals, especially *Hibiscus* sp., facilitate transboundary movement of begomovirus through cuttings, like cotton leaf curl viruses detected in *Hibiscus* in China. In the mixed infection situation, the transmission of individual virus by the vector, and its role either in eliminating the virus or making it more highly prevalent need to be studied. The genetics of resistance of begomoviruses is not addressed well, at present, there is ambiguity even in level of resistance as inferences have been drawn based on field infections. In most of the cases like bhendi there is no source of resistance available. Even, in some cases where resistance is observed, (e.g., *G. arboreum* to CLCuVs), the mechanism or the gene for gene interaction is not yet understood. The management involving exogenous application of dsRNA/siRNA will have to be strenghtened.

The classification of begomoviruses has become more challenging. The begomoviruses have been differentiated on the basis of 91% sequence identity in the DNA-A component. The current sequence based taxonomic method has resulted in a complex situation in begomovirus nomenclature and classification. The rapid changes in the constitution of begomovirus genome sequences are expected to generate more confusion and challenges in the taxonomy of begomovirus in future. An alternative robust system needs to be debated and evolved based on the basis of their infectivity on diffential hosts and their genetics of resistance in addition to the sequence identity.

Acknowledgement Authors thank the students, S. Rageshwari, S. Vinodkumar, C. Senthilraja and R. Priyanka, Department of Plant Pathology, TNAU, Coimbatore for their kind help in the preparation of this manscipt.

References

Abdul NM, Joseph JK, Karuppaiyan R (2004) Evaluation of okra germplasm for fruit yield, quality and field resistance to yellow vein mosaic virus. Indian J Plant Genet Resour 17:241–244

Abraham A (1956) Tapioca cultivation in India. Farm Bull. Indian Council of Agricultural Research, New Delhi. No. 17 – 20pp

Abraham K, Edison S, Unnikrishnan M, Sheela MN, Vimala B, Sreekumari MT et al (2006) Tuber crops varieties released byCTCR: Technical Bulletin No. 24. CTCRI, 64 p

Ahmed M, Harwood RF (1973) Studies on a whitefly transmitted yellow mosaic of urdbean (Phaseolus mungo). Plant Dis Rep 57:800–802

Ajmera, B. D. (1994). Occurrence of leaf curl virus on American Cotton (*G. hirsutum*) in North Rajasthan. Poster presentation, National seminar on Cotton Production. Challenges in 21st Century, April 18–20, Hisar, India

Ajmera, B. D. (1996) Satus and strategies of leaf curl of cotton in north India. Paper presented at National Seminar on Century of cotton in India. December 21st Surat, India

Akhter A, Akhtar S, Saeed M, Mansoor S (2014). Chilli leaf curl betasatellite enhances symptoms induced by tomato leaf curl New Delhi virus, a bipartite begomovirus. Int J Agric Biol 16:1225–1228

Alagianagalingam MN, Ramakrishnan K (1966) Cassava mosaic in India. S Indian Hortic 14:71–72

Ali M (1997) Breeding of cotton varieties for resistance of cotton leaf curl virus. Pakistan J Phytopathol 9(1):1–7

Ammavasai S, Phogat DS, Solanki IS (2004) Inheritance of resistance to mungbean yellow mosaic virus (MYMV) in greengram (*Vigna radiata* L. Wilczek). Indian. J Genet 64:146

Amrao L, Amin I, Shahid MS, Briddon RW, Mansoor S (2010a) Cotton leaf curl disease in resistant cotton is associated with a single begomovirus that lacks an intact transcriptional activator protein. Virology 66:333–355

Amrao L, Akhter S, Tahir MN, Amin I, Briddon RW, Mansoor S (2010b) Cotton leaf curl disease in Sindh province of Pakistan is associated with recombinant begomovirus components. Virus Res 153:161–165

Anonymous (1990) Cotton leaf curl virus is a whitefly transmitted geminivirus. Annual report, IARI

Antony B, Lisha VS, Palaniswami MS, Sugunan VS, Makeshkumar T, Henneberry TJ (2006) *Bemisia tabaci* and Indian cassava mosaic virus transmission. Int J Trop Insect Sci 26:176–182

Aswathanarayana DS, Rangaswamy KT, Shankarappa KS, Maruthi MN, Reddy LCN, Rekha AR, Murthy KKV (2007) Distinct begomoviruses closely related to cassava mosaic viruses cause Indian jatropha mosaic disease. Int J Virol 3:1–11

Azhar MT, Rehman MU, Aftab S, Zafar Y, Mansoor S (2010) Utilization of natural and genetically-engineered sources in *Gossypium hirsutum* for the development of tolerance against cotton leaf curl disease and fiber characteristics. Int J Agric Biol 12:744–748

Baghel G, Jahan T, Afreen B, Naqvi QA, Snehi SK, Raj SK (2010) Detection of a begomovirus associated with yellow mosaic disease of Ashwagandha (*Withania somnifera* L.) and its impact on biomass yield. Med Plants 2:219–223

Balaji V, Vanitharani R, Karthikeyan AS, Anbalagan S, Veluthambi K (2004) Infectivity analysis of two variable DNA B components of mungbean yellow mosaic virus-Vigna in *Vigna mungo* and *Vigna radiata*. J Biosci 29:297–308

Bansal RD, Khatri HL, Sharma OP, Singh IP (1984) Epidemiological studies on viral diseases of mung and moth in Punjab. J Res (PAU) 21:54–58

Barnabas AD, Radhakrishnan GK, Usha R (2010) Characterization of a begomovirus causing horsegram yellow mosaic disease in India. Eur J Plant Pathol 127:41–51

Banerjee MK, Kalloo G (1990) Nature of resistance to tomato leaf curl virus (TYLCV) in two species of *Lycopersicon*. Haryana Agric Univ J Res 20:225–228

Bhardwarj SC (1992) A note on the transmission of tomato leaf curl virus (TLCV) with Bemisia tabaci (Gennadius). Indian J Entomol 54:102–104

Bhatt BS, Chahwala FD, Rathod S, Singh AK (2016) Identification and molecular characterization of a new recombinant begomovirus and associated betasatellite DNA infecting *Capsicum annuum* in India. Arch Virol 161:1389–1394

Bhattacharyya D, Gnanasekaran P, Kumar RK, Kushwaha NK, Sharma VK, Yusuf MA, Chakraborty S (2015) A geminivirus betasatellite damages the structural and functional integrity of chloroplasts leading to symptom formation and inhibition of photosynthesis. J Exp Bot 66:5881–5895

Bhugabati KN, Goswami BK (1992) Incidence of yellow vein mosaic virus disease of okra in relation to whitefly (*Bemisia tabaci*) population under different sowing dates. Indian J Virol 8:37–39

Bisaro DM (2006) Silencing suppression by geminivirus proteins. Virology 344:158–168

Bisaro DM, Hamilton WDO, Coutts RHA, Buck KW (1982) Molecular cloning and characterization of the two DNA components of tomato golden mosaic virus. Nucleic Acids Res 10:4913–4922

Biswas K, Varma A (2001) Agroinoculation: a method of screening germplasm resistance to mungbean yellow mosaic geminivirus. Indian Phytopathol 54:240–245

Bock KR, Guthrie EJ, Woods RD (1974) Purification of maize streak virus and its relationship to viruses associated with streak diseases of sugarcane and *Panicum maximum*. Ann Appl Biol 77:289–296

Borah RK, Bordoloi DK (1998) Influence of planting time on the incidence of leaf curl virus disease and whitefly population on tomato. Indian J Virol 14:71–73

Borah BK, Cheema GS, Gill CK, Dasgupta I (2010) A geminivirus-satellite complex is associated with leaf deformity of mentha (Mint) plants in Punjab. Indian J Virol 21:103–109

Briddon RW, Bull SE, Amin I, Idris AM, Mansoor S, Bedford ID, Dhawan P, Rishi N, Siwatch SS, Abdel-Salam AM, Brown JK, Zafar Y, Markham PG (2003) Diversity of DNA β: a satellite molecule associated with some monopartite begomoviruses. Virology 312:106–121

Briddon RW, Bull SE, Amin I, Idris AM, Mansoor S, Bedford ID, Rishi N, Siwatch SS, Zafar Y, Abdel-Salam AM, Markham PG (2004) Diversity of DNA 1: a satellite-like molecule associated with monopartite begomovirus-DNA β complexes. Virology 324:462–474

Briddon RW, Brown JK, Moriones E, Stanley J, Zerbini FM, Zhou X, Fauquet CM (2008) Recommendations for the classification and nomenclature of the DNA-*β* satellites of begomoviruses. Arch Virol 153:763–781

Brown JK, Zerbini FM, Navas-Castillo J, Moriones E, Ramos-Sobrinho R, Silva JCF, Fiallo-Olivé E, Briddon RW, Hernández-Zepeda C, Idris A, Malathi VG, Martin DP, Rivera-Bustamante R, Ueda S, Varsani A (2015) Revision of begomovirus taxonomy based on pairwise sequence comparisons. Arch Virol 160:1593–1619

Butter NS, Rataul HS (1981) Nature and extent of losses in tomatoes due to tomato leaf curl virus (TLCV) transmitted by whitefly, *Bemisia tabaci* Gen. (Hemiptera, Aleyrodidae). Indian J Ecol 8:299–300

Capoor SP, Varma PM (1948a) Yellow mosaic of *Phaseolus lunatus*. Curr Sci 17:152–153

Capoor SP, Varma PM (1948b) A new virus disease of *Dolichos lablab*. Curr Sci 19:248–249

Capoor SP, Varma PM (1950) Yellow vein mosaic of *H. esculentus* L. *Indian*. J Agric Sci 20:217–230

Carsner E, Stahl CF (1924) Studies on curly-top disease of the sugar beet. J Agr Res 28:297–320

Chakraborty S (1996) Biological and molecular characterization of mungbean yellow mosaic geminivirus. Ph.D. thesis. Indian Agricultural Research Institute, New Delhi-110012, pp120

Chakraborty S, Pandey PK, Banerjee MK, Kalloo G, Fauguet CM (2003) Tomato leaf curl Gujarat virus, a new begomovirus species causing a severe leaf curl disease of tomato in Varanasi, India. Phytopathology 93:1485–1495

Chakraborty S, Vanitharani R, Chattopadhyay B, Fauquet CM (2008) Supervirulent pseudorecombination and asymmetric synergism between genomic components of two distinct species of Begomovirus associated with severe tomato leaf curl disease in India. J Gen Virol 89:818–828

Chandra J, Samuel DK (1999) Viral and phytoplasma diseases of papaya in India. In: Varma LR, Sharma RC (eds) Diseases of horticultural crops-fruits. Indus publishing Co., New Delhi, pp 493–496

Chandran SA, Levy Y, Mett A, Belausov E, Ramakrishnan U, Gafni Y (2012) Mapping of functional region conferring nuclear localization and karyopherin Â -binding activity of the C2 protein of bhendi yellow vein mosaic virus. J Gen Virol 93(Pt_6):1367–1374

Chandran SA, Jeyabharathy C, Usha R (2014) The C2 protein of bhendi yellow vein mosaic virus plays an important role in symptom determination and virus replication. Virus Genes 48(1):203–207

Chatchawankaphanich O, Maxwell DP (2002) Tomato leaf curl Karnataka virus from Bangalore, India, appears to be a recombinant Begomovirus. Phytopathology 12:637–645

Chatterjee A, Ghosh SK (2007) Association of a satellite DNA Î² molecule with mesta yellow vein mosaic disease. Virus Genes 35(3):835–844

Chatterjee A, Roy A, Padmalatha KV, Malathi VG, Ghosh SK (2005) Occurrence of a Begomovirus with yellow vein mosaic disease of mesta (*Hibiscus cannabinus* L. and *H. sabdariffa* L.) Aust Plant Pathol 34:609–610

Chatterjee A, Ghosh SK (2007) Association of a satellite DNA Î² molecule with mesta yellow vein mosaic disease. Virus Genes 35 (3):835–844

Chatterjee A, Roy A, Ghosh SK (2008) Acquisition, transmission and host range of a begomovirus associated with yellow vein mosaic disease of mesta (*Hibiscus cannabinus* and *H. sabdariffa*). Aust Plant Pathol 37:511–519

Chatterji A, Padidam M, Beachy RN, Fauquet CM (1999) Identification of replication specificity determinations in two strains of tomato leaf curl virus from New Delhi. J Virol 73:5481–5489

Chattopadhyay B, Singh AK, Yadav T, Fauquet CM, Sarin NB, Chakraborty S (2008) Infectivity of the cloned components of a begomovirus: DNA beta complex causing chilli leaf curl disease in India. Arch Virol 153(3):533–539

Chelliah S, Murugesan S (1976) Seasonal incidence of bhendi yellow vein mosaic. Madras Agric J 6:167–168

Chelliah S, Sellammal M, Murugesan S (1975) Influence of weather factors on the incidence of yellow vein mosaic disease of bhendi. Madras Agric J 62:412–419

Chenulu VV, Venkateswarlu V, Rangaraju R (1979) Studies on yellow mosaic disease of mungbean. Indian Phytopathol 32:230–235

Chowdareddy RV, Muniyappa V, Colvin J, Seal S (2005) A new begomovirus isolated from Gossypium barbadense in Southern India. Plant Pathol 54:570

Costa AS (1976) Whitefly-transmitted plant diseases. Annu Rev Phytopathol 16:429–447

Das S, Ghosh R, Paul S, Roy A, Ghosh SK (2008a) Complete nucleotide sequence of a monopartite begomovirus associated with yellow vein mosaic disease of mesta from north India. Arch Virol 153:1791–1796

Das S, Roy A, Ghosh R, Paul S, Acharyya S, Ghosh SK (2008b) Sequence variability and phylogenetic relationship of betasatellite isolates associated with yellow vein mosaic disease of mesta in India. Virus Genes 37:414–424

Dhankhar SK, Dhankhar BS, Saharan BS (1996) Screening of okra genotypes for resistance to yellow vein mosaic disease. Ann Biol 12:90–92

Dhanraj KS, Seth ML (1968) Enation in *Capsicum annuum* L (chilli) caused by a new strain of leaf curl virus. Indian J Hort 25:70–71

Dhingra KL, Chenulu VV (1985) Effect of yellow mosaic on yield and nodulation of soybean. Ind Phytopathol 38:248–251

Dry IB, Krake LR, Rigden JE, Rezaian MA (1997) A novel subviral agent associated with a geminivirus: the first report of a DNA satellite. Proc Natl Acad Sci USA 94:7088–7093

Dubey D, Pandey N, Tiwari AK, Upadhyaya PP (2015) Biological properties, transmission, serological characterization and varietal susceptibility of an isolate of Papaya leaf curl virus affecting papaya crops in Eastern Uttar Pradesh, India. Arch Phytopathol Plant Protect 48(7):611–621

Dubey GS, Yadav BP, Singh R (1986) Chemical control of leaf curl of tomato. In: National Seminar on Whitefly Transmited Plant Diseases. June 25–27. IARI, New Delhi, p 32

Duraisamy R, Natesan S, Muthurajan R, Gandhi K, Lakshmanan P, Karuppasamy N, Chokkappan M (2013) Molecular studies on the transmission of Indian cassava mosaic virus (ICMV) and Sri Lankan cassava mosaic virus (SLCMV) in cassava by *Bemisia tabaci* and cloning ICMV and SLCMV replicase gene from cassava. Mol Biotechnol 53:150–158

Dutt N, Briddon RW, Dasgupta I (2005) Identification of a second begomovirus, Sri Lankan cassava mosaic virus, causing cassava mosaic disease in India. Arch Virol 150:2101–2108

Fauquet CM, Briddon RW, Brown JK, Moriones E, Stanley J, Zerbini M, Zhou X (2008) Geminivirus strain demarcation and nomenclature. Arch Virol 153:783–821

Fondong VN (2013) Geminivirus protein structure and function. Mol Plant Pathol 14(6):635–649

Francki RB , Hatta T (1980) Chlorsis striate mosaic virus. Commonw. Mycol.lnst. Assoc. Appl. Bioi. Descriptions of Plant Viruses No. 221:4

Fuller C (1901) Mealie variegation. First report of the government entomologist. P. Davis & Sons, Government Printers, Pietermaritzburg, pp 1899–1900

Gao S, Qu J, Chua N, Ye J (2010) A new strain of Indian cassava mosaic virus causes a mosaic disease in the biodiesel crop *Jatropha curcas*. Arch Virol 155:607–612

George J, Suresh Kumar P, Unnikrishnan M (2012.; No. 51) Descriptions of recommended/released varieties under AICRP on tuber crops 1975–2011. Technical Bulletin. CTCRI, Trivandrum, p 122

George B, Kumar RV, Chakraborty S (2014) Molecular characterization of chilli leaf curl virus and satellite molecule associated with leaf curl disease of *Amaranthus* spp. Virus Genes 48:397–401

Ghosh R, Paul S, Roy A, Mir JI, Ghosh SK, Srivastava RK, Yadav US (2007) Occurrence of begomovirus associated with yellow vein mosaic disease of kenaf (*Hibiscus cannabinus*) in northern India. Online. Plant Health Prog. doi:10.1094/PHP-2007-0508-01-RS

Ghosh R, Paul S, Das S, Palit P, Acharyya S, Das A, Mir JI, Ghosh SK (2008) Molecular evidence for existence of a New World begomovirus associated with yellow mosaic disease of Corchorus capsularis in India. Aust Plant Dis Notes 3:59–62

Ghosh R, Paul S, Ghosh SK, Roy A (2009) An improved method of DNA isolation suitable for PCR-based detection of begomoviruses from jute and other mucilaginous plants. J Virol Methods 159:34–39

Ghosh R, Palit P, Das A, Ghosh SK, Roy A (2010) First report of an old world begomovirus infecting jute in India. J Plant Pathol 92:108

Ghosh R, Palit P, Paul S, Ghosh SK, Roy A (2012) Detection of Corchorus golden mosaic virus Associated with Yellow Mosaic Disease of Jute (Corchorus capsularis). Indian J Virol 23:70–74

Gaur RK (2012) First report of begomovirus infecting two ornamental plants: Ocimum Sanctum and Alternanthera variegata in India. J Clin Exp Pathol 2:46

Girish KR, Usha R (2005) Molecular characterization of two soybean-infecting begomoviruses from India and evidence for recombination among legume-infecting begomoviruses fromSouth-East Asia. Virus Res 108:167–176

Godara, S., Biswas, K.K., Singh, M. K., Singh, J., Monga, D., Praveen, S., Jain, R. K., Paul Khurana, S. M., Varma, A. (2012) Incidence, occurrence and characterization of begomoviruses associated with cotton leaf curl diseases in Northwest India. In: Int. Sym. on Food Security

Dilemma: Pl Health and Climate Change Issues, organized by Association for Advancement in Pl Protection, BCKV, Kalyani, West Bengal, Dec 7–9, pp-134–135 (Abs)
Godara S, Khurana SMP, Biswas KK (2016) Three variants of cotton leaf curl begomoviruses with theirsatellite molecules are associated with cotton leaf curl disease aggravation in New Delhi. J Plant Biochem Biotechnol. doi:10.1007/s13562-016-0370-x
Goodman RM (1977) Infectious DNA from a whitefly transmitted virus of *Phaseolus vulgaris*. Nature 266:54–55
Goodman RM, Bird J, Thongmeearkom P (1977) An unusual virus like particle associated with golden yellow mosaic of beans. Phytopathology 67:37–42
Gopal P, Kumar PP, Sinilal B, Jose J, Yadunandam AK, Usha R (2007) Differential roles of C4 and beta C1 in mediating suppression of post-transcriptional gene silencing: evidence for transactivation by the C2 of Bhendi yellow vein mosaic virus, a monopartite begomovirus. Virus Res 123:9–18
Govindu HC (1964) A review on virus diseases of crop plants. Information Pamphlet No. 2. Directorate of Agriculture, Bangalore. 13pp
Grimsley N, Hohn B, Hohn T, Walden R (1986) 'Agroinfection', an alternative route for viral infection of plants by using the Ti plasmid. Proc Natl Acad Sci USA 83:3282–3286
Gronenborn B (2004) Nanoviruses: genome organisation and protein function. Vet Microbiol 98:103–109
Guerra-Peraza O, Kirk D, Seltzer V, Veluthambi K, Schmit AC, Hohn T, Herzog E (2005) Coat proteins of rice tungro bacilliform virus and mungbean yellow mosaic virus contain multiple nuclear-localization signals and interact with importin alpha. J Gen Virol 86:1815–1826
Gupta D, Chowfla SC, Thakur PD, Sharma P, Gupta D, Sharma P (2001) Host range of tomato leaf curl virus from Himachal Pradesh. Ann Biol 17:199–201
Gupta SK, Souframanien J, Reddy KS (2015) Validation of molecular markers linked to yellow mosaic virus disease resistance in diverse genetic background of black gram [Vigna mungo (L.) Hepper]. Electron J Plant Breed 6:755–763
Gutierrez C (2000) DNA replication and cell cycle in plants: Learning from geminiviruses. EMBO J 19:792–799
Haber S, Ikegami M, Bajet NB, Goodman RM (1981) Evidence for a divided genome in bean golden mosaic virus, a geminivirus. Nature UK 289:324–326
Haible D, Kober S, Jeske H (2006) Rolling circle amplification revolutionizes diagnosis and genomics of geminiviruses. J Virol Methods 135:9–16
Handa A, Gupta MD (1993) Management of bhendi yellow vein mosaic virus disease. Indian Phytopathol 46(2):123–130
Hanley-Bowdoin L, Settlage SB, Orozco BM, Nagar S, Robertson D (1999) Geminviruses: models for plant DNA replication, transcription, and cell cycle regulation. Crit Rev Plant Sci 18:71–106
Hanley-Bowdoin L, Bejarano ER, Robertson D, Mansoor S (2013) Geminiviruses: masters at redirecting and reprogramming plant processes. Nat Rev Microbiol 11:777–788
Harrison BD (1981) Plant virus ecology: ingredients, interactions and environmental influences. Ann Appl Biol 99:195-209
Harrison BD, Lennon AM, Aiton MM (1986) Properties of four previously uncharacterized cassava viruses. Phytopathology 76:1075
Harrison BD, Liy YL, Khalid S, Hameed S, Otim-Nape GW, Robinson DJ (1997) Detection and relationships of cotton leaf curl virus and allied whitefly-transmitted geminiviruses occurring in Pakistan. Ann Appl Biol 130:61-75
Harrison BD, Robinson DJ (1999) Natural genomic and antigenic variation in whitefly-transmitted geminiviruses (begomoviruses). Ann Rev Phytopathol 37:369–398
Harrison BD, Barker H, Bock KR, Guthrie EJ, Meredith G, Atkinson M (1977) Plant viruses with circular single-stranded DNA. Nature (London) 270:760–762
Harrison BD, Muniyappa V, Swanson MM, Roberts IM, Robinson DJ (1991) Recognition and differentiation of seven whitefly transmitted geminiviruses from India and their relationships to African cassava mosaic and Thailand mung bean yellow mosaic viruses. Ann Appl Biol 118:299–308

Honda Y, Iwaki M, Saito Y, Thongmeearkom P, Kiratiya-Angul K, Deema N (1983) Mechanical transmission, purification, and some properties of whitefly borne mungbean yellow mosaic virus in Thailand. Plant Dis 67:801–804

Hong YG, Robinson DJ, Harrison BD (1993) Nucleotide sequence evidence for the occurrence of three distinct whitefly-transmitted geminiviruses in cassava. J Gen Virol 74:2437–2443

Husain M (1932) Leaf curl in cotton and other plants. Nature 129:792

Hussain M, Mansoor S, Iram S, Zafar Y, Briddon RW (2004) First report of Tomato leaf curl New Delhi virus affecting chilli pepper in Pakistan. Plant Pathol 53:794

Jacob SS, Vanitharani R, Karthikeyan AS, Chinchore Y, Thillaichidambaram P, Veluthambi K (2003) Mungbean yellow mosaic virus-Vi agroinfection by co delivery of DNA A and DNA B from one *Agrobacterium* strain. Plant Dis 87:247–251

Jaidi M, Kumar S, Srivastava A, Raj SK (2015a) First report of Clerondendron yellow mosaic virus on golden dewdrop (*Duranta erecta*) in India. New Dis Rep 32:12

Jaidi M, Srivastava A, Kumar S, Raj SK, Singh R (2015b) First report of natural occurrence of papaya leaf crumple virus on soyabean in India. New Dis Rep 32:15

Jalaluddin M, Shaikh MAQ (1981) Evaluation of Mungbean (*Vigna radiata* L. Wilczek) germplasm for resistance to yellow mosaic virus. SABRAO J 13:61–68

Jeyabharathy C, Shakila H, Usha R (2015) Development of a VIGS vector based on the betasatellite DNA associated with bhendi yellow vein mosaic virus. Virus Res 195:73–78

Jeyarajan R, Doraisamy S, Sivaprakasam K, Venkta Rao A, Ramakrishnan, L (1986) Incidence of whitefly transmitted viruses in Tamil Nadu. Madras. In: National Seminar on Whitefly Transmitted Plant Virus Diseases. June 25–27, IARI, New Delhi p 5

Jeyarajan R, Doraisamy S, Sivaprakasam K, Venkta Rao A, Ramakrishnan L (1988) Incidence of whitefly transmitted viruses in Tamil Nadu. Madras Agric J 75:212–213

John P, Sivalingam PN, Haq QMI, Kumar N, Mishra A, Briddon RW, Malathi VG (2008) Cowpea golden mosaic disease in Gujarat is caused by a mungbean yellow mosaic India virus isolate with a DNA B variant. Arch Virol 153:1359–1365

Jose A, Makeshkumar T, Edison S (2008) Host range of Sri Lankan cassava mosaic virus. J Root Crops 34:21–25

Jose A, Makeshkumar T, Edison S (2011) Survey of cassava mosaic disease in Kerala. J Root Crops 37:41–47

Jose J, Usha R (2003) Bhendi yellow vein mosaic disease in India is caused by association of a DNA Beta satellite with a begomovirus. Virology 305:310–317

Joseph K D S M, Parthipan S, Warnakulasooriya H U (1998) Status and strategies for development of mungbean in Sri Lanka. In : International consultation workshop on Mungbean: Proceedings of the Mungbean workshop, 7–11 September, 1997, New Delhi, India.. AVRDC, Shanhua, Taiwan, pp 198

Jyothsna P, Haq QMI, Singh P, Sumiya KV, Praveen S, Rawat R, Briddon RW, Malathi VG (2013a) Appl Microbiol Biotechnol 97:5457. doi:10.1007/s00253-012-4685-9

Jyothsna P, Haq QM, Singh P, Sumiya KV, Praveen S, Rawat R, Briddon RW, Malathi VG (2013b) Infection of tomato leaf curl New Delhi virus (ToLCNDV), a bipartite begomovirus with betasatellites, results in enhanced level of helper virus components and antagonistic interaction between DNA B and betasatellites. Appl Microbiol Biotechnol 97:5457–5471

Jyothsna P, Rawat R, Malathi VG (2011) Molecular characterization of a new begomovirus infecting a leguminous weed Rhynchosia minima in India. Virus Genes 42:407–414

Kalloo G (1996) Leaf curl virus of tomato and chilli in India. In: Proceeding of the phase I final workshop of the South Asian Vegetable Research Network. January 23–28. Kathmandu, Nepal, pp 229–234

Kamaal N, Akram M, Pratap A, Yadav P (2013) Characterization of a new begomovirus and a beta satellite associated with the leaf curl disease of French bean in northern India. Virus Genes 46:120–127

Kanakala S, Jyothsna P, Shukla R, Tiwari N, Veer BS, Swarnalatha P, Krishnareddy M, Malathi VG (2013) Asymmetric synergism and heteroencapsidation between two bipartite begomoviruses, tomato leaf curl New Delhi virus and tomato leaf curl Palampur virus. Virus Res 174:126–136

Karthikeyan AS, Vanitharani R, Balaji V, Veluthambi K (2004) Analysis of an isolate of mungbean yellow mosaic virus (MYMV) with a highly variable DNA B component. Arch Virol 149(8):1643–1652

Karthikeyan AK, Sudha M, Pandiyan M, Senthil N, Shobhana VG, Nagarajan P (2011) Screening of MYMV resistant mungbean (*Vigna radiata* (L) Wilczek) progenies through agroinoculation. Int J Plant Pathol 2:115–125

Karthikeyan C, Patil BL, Borah BK, Resmi TR, Turco S, Pooggin MM, Hohn T, Veluthambi K (2016) Emergence of a latent Indian cassava mosaic virus from cassava which recovered from infection by a non-persistent Sri Lankan cassava mosaic virus. Virus 8(10):264

Kelkar V, Kushawaha A, Dasgupta I (2016) Identification of amino acid residues of the coat protein of Sri Lankan cassava mosaic virus affecting symptom production and viral titer in *Nicotiana benthamiana*. Virus Res 217:38–46

Keshwal RL, Singh RP, Khare MN (1988) Effect of agronomical and ecological factors on incidence of soybean yellow mosaic. Indian J Virol 4:103–107

Khan MA, Mukhopadhyay S (1985) Studies on the seasonal spread of yellow vein mosaic virus disease of okra. Indian Phytopath 38:688–691

Khan A, Samad A (2014) A new isolate of along with a betasatellite associated with yellow vein leaf curl disease of in India. Plant Dis 98(5):698–698

Khan JA, Sohrab SS, Aminuddin (2003) Guar leaf curl disease from India is caused by Tomato leaf curl virus. New Dis Rep. http://www.bspp.org.uk/ndr/

Khan MS, Raj SK, Singh R (2006) First report of tomato leaf curl New Delhi virus infecting chilli in India. Plant Pathol 55:289

Khan AA, Khan MS, Raj SK, Naqvi QA (2007) Molecular identification of a begomovirus causing yellow vein disease on Calendula officinalis in India. EPPO Bull 37:420–426

Khan ZA, Abdin MZ, Khan JA (2015a) Functional characterization of a strong bi-directional constitutive plant promoter isolated from cotton leaf curl Burewalavirus. PLoS ONE 10(3):e0121656. doi: 10.1371/journal.pone.0121656

Khan A, Saeed ST, Samad A (2015b) New record of and a betasatellite associated with lethal leaf yellowing of Kalmegh (*Andrographis paniculata*) in northern India. Plant Dis 99:292–292

Kirthi N, Maiya SP, Murthy MRN, Savithri HS (2002) Evidence for recombination among the tomato leaf curl virus strains/species from Bangalore. India Arch Virol 147:255–272

Kirthi N, Priyadarshini CGP, Sharma P, Maiya SP, Hemalatha V, Sivaraman P, Dhawan P, Rishi N, Savithri HS (2004) Genetic variability of begomoviruses associated with cotton leaf curl disease originating from India. Arch Virol 149:2047–2057

Krishnareddy M, Venkataravanappa V, Madhuvanthi B, Jalali S (2010) Molecular characterization of begomoviruses associated with papaya leaf curl disease in India. Acta Hortic 851:465–47

Kulkarni GS (1924) Mosaic and other related diseases of crops in the bombay presidency. Poona Agric College Mag:6–12

Kumar PP, Usha R, Zrachya A, Levy Y, Spanov H, Gafni Y (2006) Protein-protein interactions and nuclear trafficking of coat protein and beta C1 protein associated with Bhendi yellow vein mosaic disease. Virus Res 122(1–2):127–136

Kumar J, Kumar A, Khan JA, Aminuddin A (2009) First report of papaya leaf curl virus naturally infecting tobacco in India. J Plant Pathol 91:97–112

Kumar Y, Bhardwaj P, Vipin H, Zaidi AA (2010a) Detection and characterization of Ageratum enation virus and a nanovirus-like satellite DNA1 from zinnia causing leaf curl symptoms in India. J Gen Plant Pathol 76(6):395–398

Kumar A, Kumar J, Khan JA (2010b) Sequence characterization of cotton leaf curl virus from Rajasthan: phylogenetic relationship with other members of geminiviruses and detection of recombination. Virus Genes 40:282–289

Kumar Y, Vipin H, Zaidi AA (2011) First report of Ageratum enation virus infecting *Crassocephalum crepidioides* (Benth.) S. Moore and Ageratum conyzoides L. in India. J Gen Plant Pathol 77(3):214–216

Kumar Y, Hallan V, Zaidi A (2011) Chilli leaf curl Palampur virus is a distinct begomovirus species associated with a betasatellite. Plant Pathol 60:1040–1047

Kumar J, Gunapati S, Singha SP, Gadreb R, Sharma NC, Tuli R (2013) Molecular characterization and pathogenicity of a carrot (*Daucus carota*) infecting begomovirus and associated betasatellite from India. Virus Res 178:478–485

Kumar J, Kumar J, Singh SP (2012a) Spatial and temporal diversity of begomoviral complexes in papayas with leaf curl disease. Arch Virol 157:1217–1232

Kumar J, Singh SP, Kumar J, Tuli R (2012b) A novel mastrevirus infecting wheat in India. Arch Virol 157(10):2031–2034

Kumar J, Kumar A, Singh SP, Roy JK, Lalit A, Parmar D, Sharma NC, Tuli R (2012) First report of infecting okra in India. New Dis Rep 25:9

Kumar J, Kumar J, Singh SP, Tuli R (2014) Association of satellites with a mastrevirus in natural infection: complexity of Wheat dwarf India virus disease. J Virol 88(12):7093–8104

Kumar R, Singh AK, Singh AK, Yadav T, Basu S, Kuswaha N, Chattopadhyay B, Chakraborty S (2015) Complexity of begomovirus and betasatellite populations associated with chilli leaf curl disease in India. J Gen Virol. doi:10.1099/jgv.0.000254

Kumar J, Kumar J, Singh S, Shukla V, Singh SP, Tuli R (2015) Prevalence of wheat dwarf India virus in wheat in India. Curr Sci 108:260–265

Kumar V, Mishra SK, Rahman J, Taneja J, Sundaresan G, Mishra NS, Mukherjee SK (2015) Mungbean yellow mosaic Indian virus encoded AC2 protein suppresses RNA silencing by inhibiting Arabidopsis RDR6 and AGO1 activities. Virology 486:158–172

Kumar S, Srivastava A, Jaidi M, Raj SK (2016) Molecular characterization of a begomovirus, alphasatellite and recombinant betasatellite associated with leaf curl disease of Parthenium hysterophorus. Plant Dis 100:2299–2305

Kumari P, Chattopadhyay B, Singh AK, Chakraborty S (2009) A new begomovirus species causing tomato leaf curl disease in Patna, India. Plant Dis 93(5):545

Kumari P, Singh AK, Chattopadhyay B, Chakraborty S (2011) A new begomovirus species and betasatellite causing severe tomato leaf curl disease in Ranchi. India New Dis Rep 23:11

Kundu A, Pal A (2012) Identification and characterization of elite inbred lines with MYMIV-resistance in *Vigna mungo*. Field Crop Res 135:116–125

Kundu A, Chakraborty D, Kundu A, Pal A (2013) Proteomics approach combined with biochemical attributesto elucidate compatible and incompatible plant-virus interactions between *Vigna mungo* and Mungbean Yellow Mosaic India Virus. BMC Proteome Sci 11:15

Kundu A, Patel A, Paul S, Pal A (2015) Transcript dynamics at early stages of molecular interactions of MYMIV with resistant and susceptible genotypes of the leguminous host, *Vigna mungo*. PLoS One 10(4):e0124687

Kundu A, Paul S, Deya A, Pal A (2017) High throughput sequencing reveals modulation of microRNAs in *Vigna mungo* upon mungbean yellow mosaic India virus inoculation highlighting stress regulation. Plant Sci 257:96–105

Kuruba BL, Buvani AP, Veluthambi K (2016) Yellow mosaic symptom caused by the nuclear shuttle protein gene of mungbean yellow mosaic virus is associated with single-stranded DNA accumulation and mesophyll spread of the virus. Acta Virol 60(3):224–233

Kushwaha N, Singh AK, Basu S, Chakraborty S (2015) Differential response of diverse solanaceous hosts to tomato leaf curl New Delhi virus infection indicates coordinated action of NBS-LRR and RNAi-mediated host defense. Arch Virol 160:1499–1509. doi:10.1007/s00705-015-2399-x

Larsen RC, Dufeus JE (1984) A simplified procedure for the purification of curly top virus and the isolation of its monomer and dimer particles. Phytopathology 73:114–118

Legg JP, Lava Kumar P, Makeshkumar T, Tripathi L, Ferguson M, Kanju E, Ntawuruhunga P, Cuellar W (2015) Cassava virus diseases: biology, epidemiology, and management. Adv Virus Res 91:85–142

Lisha VS, Antony B, Palaniswami MS, Henneberry TJ (2003) Bemisia tabaci (Homoptera: Aleyrodidae) biotypes in India. J Econ Entomol 96(2):322–327

López C, Ferriol M, Picó M (2015) Mechanical transmission of Tomato leaf curl New Delhi virus to cucurbit germplasm: Selection of tolerance sources in Cucumis melo. Euphytica 204:679–691

Lozano G, Trenado HP, Fiallo-Olive E, Chirinos D, Geraud-Pouey F, Briddon RW, Castilo JN (2016) Characterization of non-coding DNA satellites associated with sweepoviruses (genus *Begomovirus*, family *Geminiviridae*) – definition of a distinct class of begomovirus-associated satellites. Front Microbiol 7:1–13

Mahajan N, Parameswari C, Veluthambi K (2011) Severe stunting in blackgram caused by the Mungbean yellow mosaic virus (MYMV) KA27 DNA B component is ameliorated by co-infection or post-infection with the KA22 DNA B: MYMV nuclear shuttle protein is the symptom determinant. Virus Res 157:25–34

Maiti S, Basak J, Kundagrami S, Kundu A, Pal A (2011) Molecular marker-assisted genotyping of mungbean yellow mosaic India virus resistant germplasms of mungbean and urdbean. Mol Biotechnol 47:95–104

Maiti S, Paul S, Pal A (2012) Isolation, characterization, andstructure analysis of a non-TIR-NBS-LRR encoding candidategene from MYMIV-resistant *Vigna mungo*. Mol Biotechnol 52:217–233

Malathi VG (2007) Genetic identity of yellow mosaic viruses infecting legumes and their phylogenetic relationship. Indian Phytopathol 6:143–155

Malathi VG, Sreenivasan MA (1983) Association of gemini particles with cassava mosaic disease in India. J Root Crops 9:69–73

Malathi VG, Nair NG, Shantha P (1985) Cassava mosaic disease, Technical Bulletin Series, vol 5. Central Tuber Crops Research Institute, Trivandrum

Malathi VG, Surendranath B, Naghma A, Roy A (2005) Adaptation to new hosts shown by the cloned components of mungbean yellow mosaic India virus causing cowpea golden mosaic in northern India. Can J Plant Pathol 27:439–447

Malik PS, Kumar V, Bagewadi B, Mukherjee SK (2005) Interaction between coat protein and replication initiation protein of mungbean yellow mosaic India virus might lead to control of viral DNA replication. Virology 337:273–283

Mandal B, Varma A, Malathi VG (1997) Systemic infection of *Vigna mungo* usingthe cloned DNAs of the Blackgram isolate of Mungbean yellow mosaic geminivirus through agroinoculation and transmission of the progeny virus by whiteflies. J Phytopathol 145:505–510

Mandal B, Mandal S, Sohrab SS, Pun KB, Varma A (2004) A new yellow mosaic disease of Chayote in India. New Dis Rep. http://www.bspp.org.uk/ndr/

Mandal A, Sarkar D, Kundu S, Kundu P (2015) Mechanism of regulation of tomato TRN1 gene expression in late infection with tomato leaf curl New Delhi virus (ToLCNDV). Plant Sci 241:221–237

Manivasagam, S., Rabindran, R., Balasubramanian P., Natarajan, S. (2006) Studies on cassava mosaic disease with special reference to detection and virus variability. Abstracts of Papers of the 14th Triennial Symposium of International Society for Tropical Root Crops, Central Tuber Crops Research Institute, Sreekariyam, Thiruvananthapuram, Kerala, pp 153

Mansoor S, Khan SH, Bashir A, Saeed M, Zafar Y, Malik KA, Briddon RW, Stanley J, Markham PG (1999) Identification of a novel circular single-stranded DNA associated with cotton leaf curl disease in Pakistan. Virology 259:190–199

Mansoor S, Briddon RW, Bull SE, Bedford ID, Bashir A, Hussain M, Saeed M, Zafar Y, Malik KA, Fauquet C, Markham PG (2003) Cotton leaf curl disease is associated with multiple monopartite begomoviruses supported by single DNA β. Arch Virol 148:1969–1986

Mariappan V, Narayanasamy P (1986) Epidemiology of tomato leaf curl disease. In: National seminar on whitefly transmitted plant virus diseases. June 25–27. IARI, New Delhi, p 28

Marwal A, Sahu AK, Choudhary DK, Gaur RK (2013) Complete nucleotide sequence of a begomovirus associated with satellites molecules infecting a new host Tagetes patula in India. Virus Genes 47(1):194–198

Mathew AV, Muniyappa V (1992) Purification and characterization of Indian cassava mosaic virus. J Phytopathol 135(4):299–308

Mathew AV, Muniyappa V (1993) Host range of Indian cassava mosaic virus. Indian Phytopathol 46:16–23

Mathur RN (1932) Leaf-curl of cotton in garden zinnias in North India. Nature 129(3265):797, 2

Mathur RN (1933) Leaf-curl in Zinnia elegans at Dehra Dun. Indian J Agric Sci 111(I):89–96
Mishra MD, Raychaudhri SP, Jha A (1963) Virus causing leaf curl of chilli (*Capsicum annum* L.) Indian J Microbiol 3:73–76
Mishra RC, Tripathy D, Sahu RC (1978) YMv resistant mungbean for summer cultivation. Indian J Genet Plant Breed 38:103–105
Mishra SK, Ushasri C, Deb JK, Mukherjee SK (2014) Unfolding of activity of anti- ribozyme in presence of a RNA silencing suppressor. FEBS Lett 588(10):1967–1972
Mittal D, Borah BK, Dasgupta I (2008) Agroinfection of cloned Sri Lankan cassava mosaic virus DNA to *Arabidopsis thaliana*, *Nicotiana tabacum* and cassava. Arch Virol 153:2149–2155
Mnari-Hattab M, Zammouri S, Belkadhi MS, Doña BD, Nahia BE, Hajlaoui MR (2015) First report of Tomato leaf curl New Delhi virus infecting cucurbits in Tunisia. New Dis Rep 31
Monga D, Kumar M, Chander S, Singh NP, Meena RA (2008) Identification of cotton leaf curl virus disease (CLCuVD) resistant lines. J Cotton Res Dev 22(2):234–237
Morinaga T, Ikegami M, Miura K (1990) Physical mapping and molecular cloning of mungbean yellow mosaic virus. Intervirology 31:50–56
Mote UN (1978) Effect of few insecticides alone and in combination with agricultural spray oil on the control of whitefly (*Bemisia tabaci* Gennodius) population and incidence of leaf curl virus. Indian J Plant Prot 6:19–22
Mubin M, Briddon RW, Mansoor S (2009) Diverse and recombinant DNA betasatellites are associated with a begomovirus disease complex of Digera arvensis, a weed host. Virus Res 142:208–212
Mukhopadhyay S, Nath PS, Das S, Basu D, Verma AK (1994) Competitive epidemiology and cultural control of some whitefly transmitted viruses in West Bengal. In: Rishi N, Ahuja KL, Singh BP (eds) Virology in the tropics. Malhotra Publishing House, New Delhi, pp 285–308
Mullineaux PM, Donson J, Morris-krsinich BA, Boulton MI, Davies JW (1984) The nucleotide sequence of maize streak virus DNA. EMBO J 3:3063–3068
Muniyappa V. (1980). Whiteflies, Vectors of plant pathogens, Harris K. F, Maramorosch K., Academic, New York, 39–85.
Muniyappa V, Reddy HR (1976) Studies on the yellow mosaic disease of horsegram (*Dolichos biflorus* Linn.). I. Virus vector relationships. Mysore J Agric Sci 10:605–610
Muniyappa V, Veeresh GK (1984) Plant virus diseases transmitted by whiteflies in Karnataka. Proc Anim Sci 93:397–406
Muniyappa V, Reddy HR, Shivashankar C (1975) Yellow mosaic disease of *Dolichos biflorus* Linn. Curr Res 4:176
Muniyappa V, Rajeswari R, Bharathan N, Reddy DVR, Nolt BL (1987) Isolation and characterization of a Gemini virus causing yellow mosaic disease of horsegram (*Macrotyloma uniflorum* (Lam.) Verdc.) in India. J Phytopathol 119:81–87
Muniyappa V, Jalikop SH, Saikia AK, Chennarayappa, Shivashankar G, Bhat AI, Ramappa HK (1991a) Reaction of Lycopersicon cultivars and wild accessions to tomato leaf curl virus. Euphytica 56:37–41
Muniyappa V, Swanson MM, Duncan GH, Harrison GD (1991b) Particle purification, properties and epitope variability of Indian tomato leaf curl geminivirus. Ann Appl Biol 118:595–604
Muniyappa V, Maruthi MN, Babitha CR, Colvin J, Briddon RW, Rangaswamy KT (2003) Characterisation of pumpkin yellow vein mosaic virus from India. Ann Appl Biol 142(3):323–331
Muniyappa V, Venkatesh HM, Ramappa HK, Kulkarni RS, Zeidan M, Tarba CY, Ghanim M, Croznek H (2000) Tomato leaf curl virus from Bangalore (ToLCV-Ban4): Sequence comparison with Indian ToLCV isolates, detection in plants and insects and vector relationships. Arch Virol 145:1583–1598
Muraleedhar ASK, Dikshit HK, Singh D, Singh A, Prapti P (2015) Inheritance of resistance to Mungbean Yellow Mosaic India Virus (MYMIV) in mungbean (Vigna radiata). Indian J Agr Sci 85(9):1144–1147
Murugesan S, Chellaiah S (1977) Transmission of greengram yellow mosaic virus by the whitefly, *Bemisia tabaci*. Madras Agric J 64:437–438

Murugesan S, Chelliah S, Murugesan M (1977) Prediction of whitefly vector *Bemisia tabaci* (Genn.) and yellow mosaic disease incidence in greengram. Madras Agric J 64(1):22–28

Nair NG, Nene YL (1973) Studies on yellow mosaic of urd bean (*Phaseolus mungo* L.) caused by mungbean yellow mosaic virus I. transmission studies. Indian J Farm Sci 1:109–110

Nair SG, Unnikrishnan M, Sheela MN, Abraham K, Vimala B, Sreekumari MT, Easwari Amma CS (1998.; Series) Tuber crops varieties released by CTCRI, Technical Bulletin, vol 24. CTCRI, Trivandrum, pp 6–14

Naqvi AR, Haq QMR, Mukherjee SK (2010) MicroRNA profiling of tomato leaf curl new delhi virus (tolcndv) infected tomato leaves indicates that deregulation of mir159/319 and mir172 might be linked with leaf curl disease. Virol J 7:281

Naqvi AR, Choudhury NR, Mukherjee SK, Haq QM (2011a) In silico analysis reveals that several tomato microRNA/microRNA sequences exhibit propensity to bind to tomato leaf curl virus (ToLCV) associated genomes and most of their encoded open reading frames (ORFs). Plant Physiol Biochem 49:13–17

Naqvi AR, Sarwat M, Pradhan B, Choudhury NR, Haq QM, Mukherjee SK (2011b) Differential expression analyses of host genes involved in systemic infection of tomato leaf curl New Delhi virus (ToLCNDV). Virus Res 160:395–399

Nariani TK (1956) Leaf curl of papaya. Indian Phytopathol 9:151–157

Nariani TK (1960) Yellow mosaic of mung (*Phaseolus aureus* L.) Indian Phytopath 13:24–29

Nariani TK (1968) Enation of leaf curl of tomato. Plant Dis Rep 52:595–596

Narula AM, Monga D, Chauhan MS, Raj S (1999) Cotton leaf curl virus disease in India-the challenge ahead. J Cotton Res Dev 13:129–138

Nateshan MM, Muniyappa V, Swanson MM, Harrison BD (1996) Host range, vector relations and serological relationships of cotton leaf curl virus from southern India. Ann Appl Biol 128:233–244

Nath P, Saikia AK (1992) Relative resistance of okra cultivars to yellow vein mosaic virus. New Agric 3:199–202

Nath P, Gupta MK, Bura P (1992) Influence of sowing time on the incidence of yellow vein mosaic and whitefly population on okra. Indian J Virol 8:45–48

Navot N, Pichersky E, Zeidan M, Zamir D, Czosnek H (1991) Tomato yellow leaf curl virus: a whitefly-transmitted geminivirus with a single genomic component. Virology 185:151–161

Nehra C, Gaur RK (2014) Molecular characterization of chilli leaf curl viruses infecting new host plant *Petunia hybrida* in India. Virus Genes 50:58–62

Nene Y L. (1972) A survey of viral diseases of pulse crops in U.P. G.B. Pant University of Agriculture and Technology. 191 pp, Nene, 1972

Nene YL (1973) Viral diseases of some warm weather pulse crops in India. Pl Dis Rep 57:463–467

Nene YL, Naresh JS, Nair NG (1971) Additional hosts of mungbean yellow mosaic virus. Ind Phytopathol 24:415–417

Padidam M, Beachy RN, Fauquet CM (1995) Tomato leaf curl geminivirus from India has bipartite genome and coat protein is not essential for infectivity. J Gen Virol 76:25–35

Padidam M, Sawyer S, Fanquet CM (1999) Possible emergence of new geminiviruses by frequent recombination. Virology 265:218–225

Pal BP, Tandon RK (1937) Types of tobacco leaf curl in northern India. Indian J Agric Sci 7:363–393

Pal SS, Dhaliwal HS, Bains SS (1991) Inheritance of resistance to yellow mosaic virus in some Vigna species. Plant Breed 106:168–171

Pandey PK, Marathe TS (1986) A leaf crinkle disease of papaya in Maharashtra. J Maharashtra Agric Univ 11:105–106

Panno S, Lacono G, Davino M, Marchione S, Zappardo V, Bella P, Tomassoli L, Accotto GP, Davino S (2016) First report of Tomato leaf curl New Delhi virus affecting zucchini squash in an important horticultural area of southern Italy. New Dis Rep 33

Pant V, Gupta D, Choudhury NR, Malathi VG, Varma A, Mukherjee SK (2001) Molecular characterization of the Rep protein ofthe blackgram isolate of Indian mungbean yellow mosaic virus. J Gen Virol 82:2559–2567. Choudhary et al., 2006

Paprotka T, Metzler V, Jeske H (2010) The first DNA 1-like α satellites in association with New World begomoviruses in natural infections. Virology 404:148–157

Patil BL, Rajasubramaniam S, Bagchi C, Dasgupta I (2005) Both Indian cassava mosaic virus and Sri Lankan cassava mosaic virusare found in India and exhibit high variability as assessed by PCR-RFLP. Arch Virol 150:389–397

Paul S, Roy A, Ghosh R, Das S, Chaudhuri S, Ghosh SK (2008) Molecular characterization and sequence variability of betasatellites associated with leaf curl disease of kenaf (*Hibiscus cannabinus* L.) from different geographical locations of India. Acta Virol 52(4):251–256

Pimpale TR, Summanwar AS (1986) Some observation on biology, behaviour and bionomics of whiteflies (Bemisia tabaci Genn.) – the vector of some plant viruses. In: National Seminar on Whitefly Transmitted Plant Virus Diseases. IARI, New Delhi, p 18

Pooggin M, Shivaprasad PV, Veluthambi K, Hohn T (2003) RNAi targeting of DNA virus in plants. Nat Biotechnol 21:131–132

Pradhan B, Naqvi AR, Saraf S, Mukherjee SK, Dey N (2015) Prediction and characterization of tomato leaf curl New Delhi virus (ToLCNDV) responsive novel microRNAs in *Solanum lycopersicum*. Virus Res 195:183–195

Prajapat R, Marwal A, Gaur RK (2013) Begomovirus associated with alternative host weeds: a critical appraisal. Arch Phytopathol Plant Prot 47:157–170

Pramesh D, Mandal B, Phaneendra C, Muniyappa V (2013) Host range and genetic diversity of croton yellow vein mosaic virus, a weed infecting monopartite begomovirus causing leaf curl disease in tomato. Arch Virol 158(3):531–542

Prasanna HC, Rai M (2007) Detection and frequency of recombination in tomato-infecting begomoviruses of South and Southeast Asia. Virol J 4:111

Prasanna HC, Sinha DP, Verma A, Singh M, Singh B, Rai M, Martin DP (2010) The population genomics of begomoviruses: global scale population structure and gene flow. Virol J 7:220

Prasanna HC, Sinha DP, Rai GK, Krishna R, Kashyap SP, Singh NK, Malathi VG (2015) Pyramiding Ty-2 and Ty-3 genes for resistance to monopartite and bipartite tomato leaf curl viruses of India. Plant Pathol 64(2):256–264

Pratap D, Kashikar AR, Mukherjee SK (2011) Molecular characterization and infectivity of a Tomato leaf curl New Delhi virus variant associated with newly emerging yellow mosaic disease of eggplant in India. Virol J 8:305. doi:10.1186/1743-422X-8-305

Praveen S, Ramesh SV, Mishra AK, Koundal V, Paulkaitis P (2010) Silencing potential of viral derived RNAi construct in Tomato yellow leaf curl virus-AC4 gene suppression in tomato. Trangenic Res 19:45–55. doi:10.1007/s11248-009-9291-y

Priyadarshini CP, Ambika MV, Tippeswamy R, Savithri HS (2011) Functional characterization of coat protein and V2 involved in cell to cell movement of Cotton leaf curl Kokhran virus-Dabawali. PLoS One 6(11):e26929

Pruthi HS, Samuel CK (1937) Entomological investigations on the leaf curl diseases of tobacco in North Bihar. I. Transmission experiments with some suspected insect vectors. II An alternate host of the virus and the insect transmitter. Indian J Agric Sci 7:659–670

Pun KB, Doraiswamy S (1999) Effect of age of okra plants on susceptibility to okra yellow vein mosaic virus. Indian J Virol 15:57–58

Qazi J, Ilyas M, Mansoor S, Briddon RW (2007) Legume yellow mosaic viruses genetically isolated begomoviruses. Mol Plant Pathol 8:343–348

Radhakrishnan, G. (2002) Biological and molecular characterization of cotton leaf curl Geminivirus from India. Ph.D. Thesis. Division of Plant Pathology.IARI, New Delhi, India

Radhakrishnan G, Malathi VG, Varma A (2004a) Biological characterization of an isolate of cotton leaf curl Rajasthan virus from northern India and identification of source of resistance. Indian Phytopathol 57(2):174–180

Radhakrishnan G, Malathi VG, Varma A (2004b) Detection of DNA A and DNA β associated with cotton leaf curl and some other plant diseases caused by whitefly transmitted geminiviruses Indian

Raghavan V, Malik PS, Choudhury NR, Mukherjee SK (2004) The DNA-A component of a plant geminivirus (Indian mung bean yellow mosaic virus) replicates in budding yeast cells. J Virol 78(5):2405–2413
Rahman J, Karjee S, Mukherjee SK (2012) MYMIV-AC2, a geminiviral RNAi suppressor protein, has potential to increase the transgene expression. Appl Biochem Biotechnol 167(4):758–775
Raj SK, Khan MS, Snehi SK, Kumar S, Khan AA (2007) Natural occurrence of a Begomovirus on *Dimorphotheca sinuata* in India. Aust Plant Dis Notes 2:25–26
Raj SK, Snehi SK, Khan MS, Singh R, Khan AA (2008) Molecular evidence for association of with leaf curl disease of papaya (L.) in India. Aust Plant Dis Notes 3(1):152
Raj SK, Snehi ASK, Khan MS, Tiwari AK, Rao GP (2010a) First report of Pepper leaf curl Bangladesh virus strain associated with bitter gourd (*Momordica charantia* L.) yellow mosaic disease in India. Aust Plant Dis Notes 5:14–16
Raj SK, Snehi SK, Khan MS, Tiwari AK, Rao GP (2010b) Detection of *Ageratum enation virus* from cat's whiskers (*Cleome gynandra* L.) with leaf curl symptoms in India. J Gen Plant Pathol 76:292–294
Raj SK, Snehi SK, Tiwari AK, Rao GP (2011) Molecular identification of an *Ageratum enation virus* isolate associated with mosaic disease of pointed gourd (*Trichosanthes dioica*) in India. Phytoparasitica 39:497–502
Rajagopalan PA, Naik A, Katturi P, Kurulekar M, Kankanallu RS, Anandalakshmi R (2012) Dominance of resistance-breaking cotton leaf curl Burewala virus (CLCuBuV) in northwestern India. Arch Virol 157(5):855–868
Rajinimala N, Rabindran R (2007) First report of Indian cassava mosaic virus on bittergourd (*Momordica charantia*) in Tamil Nadu. India Australas Plant Dis Notes 2:81–82
Rajinimala N, Rabindran R, Mohan S, Sethuraman K (2011) Survey for the Occurrence of cassava mosaic disease in Tamil Nadu. J Root Crops 37:197–199
Ramachandran V, Karthikeyan AS, Anuradha S, Veluthambi K (1996) Genome homologies among geminiviruses infecting Vigna, cassava, Acalypha, Croton and Vernonia. Curr Sci 10:65–69
Rao RD, Prasad VJ, Ragunathan VB, Joshi NC (1982) Occurrence of yellow mosaic disease on clusterbean, *Cymopsis tetragonoloba*. Indian J Plant Prot 10:100
Rataul HS, Butter NS (1977) Leaf curl virus of tomato. Indian Hort. January to March
Rathi YPS, Nene YL (1974) Sex of Bemisia tabaci Genn. in relation to the transmission of mungbean yellow mosaic virus. Acta Bot Ind 2:74–76
Rathi YPS, Nene YL (1974) The additional hosts of Mungbean yellow mosaic virus. Indian Phytopathology 27:249
Raychaudhuri SP, Nariani TK (1977) Viruses and Mycoplasma diseases of plants in India. Oxford and IBH Publishing company, New Delhi, pp 160–167
Reddy HR, Ravi KS (1991) A leaf curl disease of Cassia uniflora. Curr Res 20:16–17
Reddy KR, Singh DP (1995) Inheritance of resistance to mungbean yellow mosaic virus. Madras Agric J 88:199–201
Reddy KS, Yaraguntaiah RC (1979) Transmission of tomato leaf curl virus through dodder. Indian Phytopathol 32:653
Reddy KS, Yaraguntaiah RC (1981) Virus vector relationship in leaf curl virus disease of tomato in India. Indian Phytopathol 34:310–313
Reddy RVC, Colvin J, Muniyappa V, Seal S (2005) Diversity and distribution of begomoviruses infecting tomato in India. Arch Virol 150:845–867
Resmi TR, Nivedhitha S, Karthikeyan C, Veluthambi K (2014) Sri Lankan cassava mosaic virus replication associated protein (Rep) triggers transposition of IS426 in Agrobacterium. FEMS Microbiol Lett 360:42–50
Resmi TR, Hohn T, Hohn B, Veluthambi K (2015) The agrobacterium tumefaciens Ti plasmid virulence gene virE2 reduces Sri Lankan cassava mosaic virus infection in transgenic *Nicotiana benthamiana* plants. Virus 7:2641–2653
Rishishwar R, Mazumdar B, Dasgupta I (2015) Diverse and recombinant begomoviruses and various satellites are associated with Bhendi yellow vein mosaic disease of okra in India. J Plant Biochem Biotechnol 24:470–475

Rochester DE, DePaulo JJ, Fauquet CM, Beachy RN (1994) Complete nucleotide sequence of the geminivirus tomato yellow leaf curl virus, Thailand isolate. J Gen Virol 75:477–485

Rojas MR, Hagen C, Lucas WJ, Gilbertson RL (2005) Exploiting chinks in the plant's armor: evolution and emergence of geminiviruses. Annu Rev Phytopathol 43:361–394

Romay G, Chirinos D, Geraud-Pouey F, Desbiez C (2010) Association of an atypical alphasatellite with a bipartite New World begomovirus. Arch Virol 155:1843–1847. doi:10.1007/s00705-010-0760-7

Rothenstein D, Briddon RW, Haible D, Stanley J, Frischmuth JH (2005) Biolistic infection of cassava using cloned components of Indian cassava mosaic virus. Arch Virol 150:1669–1675

Rothenstein D, Haible D, Dasgupta I, Dutt N, Patil BL, Jeske H (2006) Biodiversity and recombination of cassava-infecting begomoviruses from southern India. Arch Virol 151:55–69

Rouhibakhsh A, Malathi VG (2005) Severe leaf curl disease of cowpea – a new disease of cowpea in northern India caused by Mungbean yellow mosaic India virus and a satellite DNA β. Plant Pathol 54:259

Rouhibakhsh A, Haq QMI, Malathi VG (2011) Mutagenesis in ORF AV2 affects viral replication in Mungbean yellow mosaic India virus. J Biosci 36:329–340

Rouhibaksh A, Choudhury NR, Mukherjee SK, Malathi VG (2012) Enhanced nicking activity of Rep in presence of pre-coat protein of Mungbean yellow mosaic India virus. Virus Genes 44(2):356–361

Roy A, Sanchalika A, Subha D, Raju G, Sujay P, Ram Kumar S, Subrata Kumar G (2009) Distribution, epidemiology and molecular variability of the begomovirus complexes associated with yellow vein mosaic disease of mesta in India. Virus Res 141:237–246

Saeed ST, Samad A (2017) Emerging threats of begomoviruses to the cultivation of medicinal and aromatic crops and their management strategies. Virus Dis 28:1–17

Saeed ST, Khan A, Kumar B, Ajayakumar PV, Samad A (2014) First report of chilli leaf curl India virus infecting *Mentha spicata* (Neera) in India. Plant Dis 98(1):164

Sahu PP, Rai NK, Chakraborty S, Singh M, Chandrappa PH, Ramesh B, Chattopadhyay D, Prasad M (2010) Tomato cultivar tolerant to Tomato leaf curl New Delhi virus infection induces virus-specific short interfering RNA accumulation and defence-associated host gene expression. Mol Pl Pathol 11:531–544. doi:10.1111/j.1364-3703.2010.00630.x

Sahu AK, Avinash M, Shahid MS, Chitra N, Gaur RK (2014) First report of a begomovirus and associated betasatellite in *Rosa indica* and in India. Aust Plant Dis Notes 9(1)

Sahu AK, Marwal A, Shahid MS, Nehra C, Gaur R (2014a) First report of a begomovirus and associated betasatellite in *Rosa indica* in India. Aust Plant Dis Notes 9:147

Sahu PP, Sharma N, Puranik S, Muthamilarasan M, Prasad M (2014b) Involvement of host regulatory pathways during geminivirus infection: a novel platform for generating durable resistance. Funct Integr Genomics 14:47–58

Sahu PP, Sharma N, Puranik S, Chakraborty S, Prasad M (2016) Tomato 26S Proteasome subunit RPT4a regulates ToLCNDV transcription and activates hypersensitive response in tomato. Sci Rep 6:27078. doi:10.1038/srep27078

Saikia AK, Muniyappa V (1986) Epidemiology and control of tomato leaf curl virus. In: National Seminar on Whitefly Transmitted Plant Virus Diseases. June, 25–27, IARI, New Delhi, pp 30–31

Saikia AK, Muniyappa V (1989) Epidemiology and control of tomato leaf curl virus in southern India. Trop Agric (Trinidad) 66:350–354

Saklani UD, Mathai PJ (1977) Effect of date of planting on leaf curl disease of tomato. Indian J Hortic 34:64–68

Saleem M, Haris WAA, Malik A (1998) Inheritance of yellow mosaic virus in mungbean (*Vigna radiata* L. Wilczek). Pakistan J Phytopathol 10:30–32

Saleem H, Nahid N, Shakir S, Ijaz S, Murtaza G, Khan AA, Mubin M, Nawaz-ul-Rehman MS, Melcher U (2016) Diversity, mutation and recombination analysis of cotton leaf curl geminiviruses. PLoS One 11(3), e0151161

Samad A, Gupta MK, Shasany AK, Ajayakumar PV, Alam M (2008) Begomovirus related to tomato leaf curl Pakistan virus newly reported in mentha crop in India. New Dis Rep 18:11

Sandhu TS, Brar JS, Sandhu SS, Verma MM (1985) Inheritance of resistance to mungbean yellow mosaic virus in greengram. J Res Punjab Agric Univ 22(1):607–611

Sanwal SK, Venkataravanappa V, Singh B (2016) Resistance to bhendi yellow vein mosaic disease: a review. Indian J Agric Sci 86:835–843

Sastry KSM, Singh SJ (1973) Assessment of losses in tomato by tomato leaf curl virus. Indian J Mycol Plant Pathol 3:50–54

Sastry KSM, Singh SJ (1974) Effect of yellow vein mosaic virus infection on growth and yield of okra crop. Indian Phytopathol 27:294–297

Sastry KSM, Singh SJ, Sastry KS (1978) Studies on epidemiology of tomato leaf curl virus. Indian J Hort 35:269–277

Sathya VK, Malathi VG, Velazhahan R, Rabindran R, Jayamani P, Alice D (2013) Characterization of betasatellite associated with the yellow mosaic disease of grain legumes in Southern India. Acta Virol 57:405–414. 2013

Sattar MN, Kvarnheden A, Saeed M, Briddon RW (2013) Cotton leaf curl disease-an emerging threat to cotton production worldwide. J Gen Virol 94:695–710

Satya VK, Malath VG, Velazhahan RO, Rabindran R, Jayamani P, Alice D (2013) Characterization of betasatellite associated with the yellow mosaic disease of grain legumes in Southern India. Acta Virol 57:405–414

Satya VK, Alice D, Malathi VG (2015) Seed-borne nature of a begomovirus, mung bean yellow mosaic virus in black gram. Appl Microbiol Biotechnol. doi: 10.1007/s00253-015-7188-7

Satyavir (1980) Seasonal incidence of insect pests of mothbean and cowpea crops. Annual Progress Report, Centre for Arid Zone Research Institute, Jodhpur, p 53

Saunders K, Stanley J (1999) A nanovirus-like component associated with yellow vein disease of Ageratum yellow vein disease of Ageratum conyzoides: evidence for interfamilial recombination between plant disease plant DNA viruses. Virology 264:142–152

Saunders K, Stanley J (n.d.) A nanovirus-like DNA component associated with yellow vein disease of Ageratum conyzoides: evidence for interfamilial recombination between plant DNA viruses. Virology 1264(1):142–152

Saunders K, Bedford ID, Briddon RW, Markham PG, Wong SM, Stanley J (2000) A novel virus complex causes Ageratum yellow vein disease. Proc Natl Acad Sci U S A 94:7088–7093

Saunders K, Bedford ID, Stanley J (2002a) Adaptation from whitefly to leafhopper transmission of an autonomously replicating nanovirus-like DNA component associated with Ageratum yellow vein disease. J Gen Virol 83:907–913

Saunders K, Salim N, Mal VR, Malathi VG, Briddon SR, Markham PG, Stanley J (2002b) Characterisation of Sri Lankan cassava mosaic virus and Indian cassava mosaic virus: Evidence for acquisition of a DNA B component by a monopartite begomovirus. Virology 293:63–74

Saunders K, Bedford ID, Yahara T, Stanley J (2003) Aetiology: the earliest recorded plant virus disease. Nature 422:831

Saxena S, Hallan V, Singh BP, Sane PV (1998) Nucleotide sequence and inter geminiviral homologies of the DNA A of Papaya leaf curl geminivirus from India. Biochem Mol Biol Int 45:101–113

Sen PK, Ganguli BD, Malik PC (1946) A note on a leaf curl disease of papaya (*Carica papaya* L.) Indian J Hortic 3:38–40

Senanayake DMJB, Mandal B, Lodha S, Varma A (2007) First report of chilli leaf curl virus affecting chilli in India. Plant Pathol 56:343

Senanayake DMJB, Varma A, Mandal B (2012) Virus-vector relationships, host range, detection and sequence comparison of chilli leaf curl virus associated with an epidemic of leaf curl disease of chilli in Jodhpur, India. J Phytopathol 160:146–155

Seth ML, Raychaudhuri SP, Singh DV (1971) A streak disease of Bajra [*Pennisetum typhoides* (Burm) Stapf and Hubb.] in India. Current Science 40

Sequeira JC, Harrison BD (1982) Serological studies on cassava latent virus. Ann Appl Biol 101:33–42

Serfraz S, Amin I, Akhtar KP, Mansoor S (2015) Recombination among begomoviruses on malvaceous plants leads to the evolution of in Pakistan. J Phytopathol 163(9):764–776

Shafiq M, Asad S, Zafar Y, Briddon RW, Mansoor S (2010) Pepper leaf curl Lahore virus requires the DNA B component of Tomato leaf curl New Delhi virus to cause leaf curl symptoms. Virol J 7:367

Shahid MS, Briddon RW, Al-Sadi AM (2017) Identification of mungbean yellow mosaic Indian virus associated with Tomato Leaf Curl Betasatellite infecting *Phaseolus vulgaris* in Oman. J Phytopathol. doi:10.1111/jph.12551

Shanmugapriya G, Das SS, Veluthambi K (2015) Transgenic tobacco plants expressing siRNA targeted against the Mungbean yellow mosaic virus t. Virus Dis 26(1–2):55–61. ranscriptional activator protein gene efficiently block the viral DNA accumulation

Sharma SR, Varma A (1976) Cowpea yellow fleck a whitefly transmitted disease of cowpea. Indian phytopathol 29:421–423

Sheela MN, Abraham K, Moorthy SN (2012) Development of elite inbreds of cassava and their use in heterosis breeding, Annual Report 2011–2012. CTCRI, Thiruvananthapuram, p 25

Sherry AVM (2016) A new variant of cassava mosaic virus causes mulberry mosaic disease in India. Int J Plant Animal Environ Sci 6:83–93

Shih SL, Tsai WS, Green SK, Hanson PM, Valand GB, Kalloo G, Shrestha SK, Joshi S (2003) Molecular characterization of a new tomato begomovirus from India. Plant Dis 87(5):598–598

Shimizu S, Ikegami M, Negishi K, Suyama K, Fujii H (1987) Comparative studies on bean golden mosaic virus and mungbean yellow mosaic virus. J Agri Sci, Tokyo Nogyo Daigaku 31:151–155

Shivaprasad PV, Akbergenov R, Trinks D, Rajeswaran R, Veluthambi K, Hohn T, Pooggin MM (2005) Promoters, transcripts, and regulatory proteins of MungbeanYellow Mosaic Geminivirus. J Virol 79:8149–8163

Shukla DD, Ram S (1977) Natural occurrence of three different viruses on chillies in Rajasthan. Indian J Mycol Plant Pathol 7:122–125

Shukla S, Pandya GP, B P. (1985) Resistance to yellow mosaic in greengram. SABRAO J Genet Breed 17:165–171

Shukla R, Dalal S, Malathi VG (2013a) Suppressors of RNA silencing encoded by tomato leaf curl betasatellites. J Biosci 38:45–51

Shukla R, Dalal S, Malathi VG (2013b) Isolation and identification of promoter elements of Cotton leaf curl Multan betasatellite associated with Tomato leaf curl viruses. Arch Phytopathol Plant Protect 46(9):1015–1029

Shruti G, Paul Khurana SM, Biswas KK (2017) Three variants of cotton leaf curl begomoviruses with their satellite molecules are associated with cotton leaf curl disease aggravation in New Delhi. J Plant Biochem Biotechnol 26(1):97–105

Shweta, Khan JA (2014) In silico prediction of cotton (Gossypium hirsutum) encoded microRNAs targets in the genome of Cotton leaf curl Allahabad virus. Bioinformation 10:251–255. doi:10.6026/97320630010251

Singh A (2006) Studies of occurrence of papaya viruses in eastern Uttar Pradesh and their possible management approaches. PhD Thesis. Gorakhpur: University of Gorakhpur

Singh AK, Singh S, Awasthi AK, Singh S (1999) Studies on tomato leaf curl virus in Chattisgarh, Madhya Pradesh. Indian J Virol 15:115–117

Singh RN (1979) Natural infection of bean (*Phaseolus vulgaris*) by mungbean yellow mosaic virus. Indian J Mycol Plant Pathol 9:124–126

Singh SR, Allen DJ (1979) Cowpea pests and disease. Manual series No. 2, International Institute of Tropical Agriculture, Ibadan, p 113

Singh SJ, Dutta OP (1986) Enation leaf curl of okra – a new virus disease. Indian J Virol 2(1):114–117

Singh DV, Lal SB (1964) Occurrence of viruses of tomato and their probabales strains at Kanpur. Proc Natl Acad Sci India 34:179–187

Singh D, Patel PN (1977) Studies on resistance in crops to bacterial diseases in India. III. Investigations on inheritance of reactions to bacterial leaf spot and yellow mosaic diseases and linkage, if any, with other characters in mungbean. Indian Phytopathol 30:202–206

Singh JP, Kadian OP, Varma JP (1979) Survey of virus of mung bean and urd bean in Haryana. Harayana Agric Univ J Res 9(4):345–347

Singh G, Kapoor S, Singh K (1988) Multiple disease resistancein mungbean with special emphasis on mungbean yellow mosaic virus. In: S Shanmugasundaram (ed) Proceedings of the Second International Symposium on Mungbean Vegetable Research and Development Centre (pp 290–296). Tainan, Taiwan

Singh J, Sohi AS, Mann HS, Kapur SP (1994) Studies on whitefly, *Bemisia tabaci* (Genn.) transmitted cotton leaf curl disease in Punjab. J Insect Sci 7:194–198

Singh SK, Chakraborty S, Singh AK, Pandey PK (2006) Cloning, restriction mapping and phylogenetic relationship of genomic components of MYMIV from Lablab purpureus. Bioresour Technol 97(15):1807–1814

Singh DK, Malik PS, Choudhury NR, Mukherjee SK (2008) MYMIV replication initiator protein (Rep): roles at the initiation and elongation steps of MYMIV DNA replication. Virology 380:75–83

Singh AK, Chattopadhyay B, Chakraborty S (2012) Biology and interactions of two distinct monopartite begomoviruses and betasatellites associated with radish leaf curl disease in India

Singh-Pant P, Pant P, Mukherjee SK, Mazumdar-Leighton S (2012) Spatial and temporal diversity of begomoviral complexes in papayas with leaf curl disease. Arch Virol 157(7):1217–1232

Singh S, Gill JS, Gumber RK, Singh R, Singh S (2013) Yield and fible quality associated with cotton leaf curl disease of Bt-cotton in Punjab. J Environ Sci 34:113–116

Sinha V, Kumar A, Sarin NB, Bhatnagar D (2016) Recombinant croton yellow vein mosaic virus associated with severe leaf curl disease of papaya in India. Int J Adv Res 4:1598–1604

Sivalingam PN, Varma A (2012) Role of betasatellite in the pathogenesis of a bipartite begomovirus affecting tomato in India. Arch Virol 157:1081–1092

Sivalingam PN, Malathi VG, Varma A (2010) Molecular diversity of the DNA betasatellites associated with tomato leaf curl disease in India. Arch Virol 155:757–764

Sivalingam PN, Satheesh V, John P, Chandramohan S, Malathi VG (2011) Complete genome sequence of an isolate of Clerodendron yellow mosaic virus – a new begomovirus from India. Acta Virol 55(4):357–360

Sivanathan P (1977) Virus disease of crops in Sri Lanka. Trop Agric Res Ser 10:143–150

Snehi SK, Prihar SS, Gupta G, Singh V, Raj SK, Prasad V (2016) The current status of new emerging begomovirus diseases on *Jatropha* species from India. J Plant Pathol Microbiol 7:357. doi:10.4172/2157-7471.1000357

Sohrab SS, Mandal B, Varma A (2004) Characterization of the virus associated with Lufa mosaic disease in northern India. In: National Symposium on Crop Surveillance: Disease Forecasting and Management, Feb 19–21, IARI, New Delhi, p 25

Srivastava SK, Varma A (1988) Virus disease of cowpea in India. In: Proceedings of National Symposium on plant virus problems in India, IARI, New Delhi 5–6 April 14–15

Srivastava KM, Hallan V, Raizada RK, Chandra G, Singh BP, Sane PV (1995) Molecular cloning of Indian tomato leaf curl virus genome following a simple method of concentrating the super coiled replicative form of viral DNA. J Virol Methods 51:297–304

Srivastava KM, Singh BP, Dwadash Shreni VC, Srivastava BN (1977) Zinnia yellow net disease transmission, host range, and agent-vector relationship. Plant Dis Rep 61:550–554

Srivastava A, Raj SK, Kumar S, Snehi SK, Kulshreshtha A, Hallan V, Pande SS (2013) Molecular identification of ageratum enation virus, betasatellite and alphasatellite molecules isolated from yellow vein diseased Amaranthus cruentus in India. Virus Genes 47(3):584–590

Srivastava A, Jaidi M, Kumar S, Raj SK, Shukla S (2014a) Association of Papaya leaf curl virus with the leaf curl disease of grain amaranth (*Amaranthus cruentus* L.) in India. Phytoparasitica 83:97–101

Srivastava A, Kumar S, Raj SK (2014b) Association of Pedilanthus leaf curl virus with yellow mottling and leaf curl symptoms in two jasmine species grown in India. J Gen Plant Pathol 80(4):370–373

Srivastava A, Kumar S, Raj SK, Pande SS (2014d) Association of a distinct strain of Hollyhock yellow vein mosaic virus and ludwigia leaf distortion betasatellite with yellow vein mosaic disease of hollyhock (Alcea rosea) in India. Arch Virol 159:2711–2715

Srivastava A, Kumar S, Raj SK (2014c) First report of ageratum enation virus, betasatellite and alphasatellite causing leaf curl and enation disease of Amaranthus hypochondriacus in India. Plant Dis 98(9):1285

Srivastava A, Kumar S, Jaidi M, Raj SK (2016) First report of tomato leaf curl New Delhi virus on opium poppy (*Papaver somniferum*) in India. Plant Dis 100:232

Stanley J (1985) The molecular biology of geminiviruses. Adv Virus Res 30:139–177

Stanley J, Gay MR (1983) Nucleotide sequences of cassava latent virus DNA. Nature 301:260–262

Storey HH (1936) Virus diseases on East African plants. VI. A progress report on the studies of the diseases of cassava. East Afr Agric J 2:34–39

Sudha M, Karthikeyan A, Anusuya P, Ganesh NM, Pandiyan M, Senthil N, Raveendran M, Nagarajan P, Angappan K (2013) Inheritance of resistance to mungbean yellow mosaic virus (MYMV) in inter and intra specific crosses of mungbean (*Vigna radiata*). Am J Plant Sci 4:1924–1927

Sudhakar R, Prasad Rao DDVJ, Reddy PS (1979) A whitefly transmitted yellow mosaic disease on groundnut (Arachis hypogaea L.). Curr Sci 48:160

Summanwar AS, Ram RD (1993) Virus diseases of papaya. In: Chaddha KL, Pareek OP M (eds) Advances in horticulture. Fruit crops: Part 3, vol 3. Publishing House, New Delhi, pp 1439–1446

Sunitha S, Mahajan N, Veluthambi K (2012) The *TrAP/REn* monodirectional promoter of Mungbean yellow mosaic geminivirus (MYMV) displays root-specific expression in transgenic tobacco. Plant Cell Tissue Organ Cult 109:535

Sunitha S, Shanmugapriya G, Balamani V, Veluthambi K (2013) Mungbean yellow mosaic virus (MYMV) *AC4* suppresses post-transcriptional gene silencing and an *AC4* hairpin RNA gene reduces MYMV DNA accumulation in transgenic tobacco. Virus Genes 46:496–504

Surekha SK, Mathur K, Shukla DD (1977) Virus disease on papaya (*Carica papaya*) in Udaipur. Indian J Mycol Plant Pathol 7:115–121

Surendranath B, Usharani KS, Nagma A, Victoria AK, Malathi VG (2005) Absence of interaction of genomic components and complementation between mungbean yellow mosaic India virus isolates in cowpea. Arch Virol 150(9):833–1844

Suteri BD (1974) Occurrence of soybean yellow mosaic in Uttar Pradesh. Curr Sci 43:689–690

Suteri BD, Srivastava SB (1979) Effect of soybean yellow mosaic ongrowth and yield of soybean. Plant Dis Rep 63:151–153

Swanson MM, Brown JK, Poulos BT, Harrison BD (1992) Genome affinities and epitope profiles of whitefly-transmitted geminiviruses from the Americas. Ann App Biol 121(2):285–296

Swanson MM, Varma A, Muniyappa V, Harrison BO (1992) Comparative epitope profiles of the particle proteins of whitefly-transmitted geminiviruses from nine crop legumes in India. Ann Appl Biol 120:425–433

Swapna Geetanjali A, Shilpi S, Mandal B (2013) Natural association of two different betasatellites with Sweet potato leaf curl virus in wild morning glory (*Ipomoea purpurea*) in India. Virus Genes 47(1):184–188

Tahir M, Haider MS (2005) First report of Tomato leaf curl New Delhi virus infecting bitter gourd in Pakistan. New Disease Report. http://www.bspp.org.uk/ndr/

Tahir M, Amin I, Haider MS, Mansoor S, Rob W, Briddon RW (2015) Ageratum enation virus-a begomovirus of weeds with the potential to infect crops. Viruses 7:647–665

Tarr SAJ (1951) Leaf curl disease of cotton, 1st edn. Commonwealth Mycol. Instit., Kew, pp 20–28

Tarr SAJ (1957) The host range of cotton leaf curl virus in sudan. Emp Cotton Growers Rev 34:258–262

Thakur RP, Patel PN, Verma JP (1997) Genetical relationships between reactions to bacterial leaf spot, yellow mosaic and Cercospora leaf spot diseases in mungbean (*Vigna radiata*). Euphytica 26:765–774

Thomas KM, Krishnaswami CS (1939) Leaf crinkle – a transmissible disease of papaya. Curr Sci 8:316

Tripathi S, Varma A (2003) Identification of sources of resistance in *Lycopersicon* species to tomato leaf curl geminivirus (ToLCV) by agroinoculation. Euphytica 129:43–52

Tsai WS, Shih SL, Lee LM, Wang JT, Duangsong U, Kenyon L (2013) First report of BYVMV associated with yellow vein mosaic of Okra in Thailand. Plant Disease 97(2):291–291

Tiwari N, Padmalatha KV, Singh VB, Haq QMI, Malathi VG (2012) Tomato leaf curl Bangalore virus (ToLCBV): infectivity and enhanced pathogenicity with diverse betasatellites. Arch Virol 155:1343. doi:10.1007/s00705-010-0710-4

Tiwari AK, Sharma PK, Khan MS, Snehi SK, Raj SK, Rao GP (2010) Molecular detection and identification of Tomato leaf curl New Delhi virus isolate causing yellow mosaic disease in Bitter gourd (), a medicinally important plant in India. Med Plants - Int J Phytomed Relat Ind 2(2):117

Trinks D, Rajeswaran R, Shivaprasad PV, Akbergenov R, Oakeley EJ, Veluthambi K, Hohn T, Pooggin MM (2005) Suppression of RNA silencing by a geminivirus nuclear protein, AC2, correlates with transactivation of host genes. J Virol 2005:2517–2527

Unnikrishnan M, Mohan C, Santha VP, Ramanathan S, Anantharaman M, Makeshkumar T, Jeeva ML, Sajeev MS, Byju G (2011) Varietal improvement in cassava for mosaic disease resistance, yield and quality. Annual Report 2010–2011. CTCRI, Thiruvananthapuram, pp 26–29

Uppal BN, Varma PM, Capoor SP (1940) Yellow vein mosaic of bhendi. Curr Sci 9:227–228

Usharani KS, Surendranath B, Haq QMR, Malathi VG (2004a) Yellow mosaic virus infecting soybean in northern India is distinct from the species infecting soybean in southern and western. India Curr Sci 86:845–850

Usharani KS, Surendranath B, Paul-Khurana SM, Garg ID, Malathi VG (2004b) Potato leaf curl-a new disease of potato in Northern India caused by a strain of Tomato leaf curl New Delhi virus. Plant Pathol 53:235

Usharani KS, Surendranath B, Haq QMR, Malathi VG (2005) Infectivity analysis of a soybean isolate of mungbean yellow mosaic India virus by agroinoculation. J Gen Plant Pathol 71:230–237

Usharani KS, Periasamy M, Malathi VG (2006) Studies on the activity of a bidirectional promoter of mungbean yellow mosaic India virus by agroinfiltration. Virus Res 119:154–162

Varma PM (1952) Studies on the relationship of the bhendi yellow vein mosaic virus and vector, the whitefly (*Bemisia tabaci*). Indian J Agric Sci 22:75–91

Varma PM (1955) Persistence of yellow vein mosaic virus of *Abelmoschus esculentus* (L.) Moench in its vector, *Bemisia tabaci* (Genn.) Indian J Agric Sci 25:293–302

Varma A, Malathi VG (2003) Emerging geminivirus problems: a serious threat to crop production. Ann Appl Biol 142:145–164

Varma, A., Aiton, M., Malathi, V.G. and Harrison, B.D. (1989) Antigenic diversityin geminivirus in India. In: Proceedings on Indo-French seminar on Organization and expression of plant genomes, Strasbourg, October 5–7, France

Varma A, Dhar A K, Malathi V G. (1991) Cloning and restriction analysis of munbean. mosaic virus. In: Proceedings of International Conference on Virology in Tropics, Lucknow. pp 114–.115

Varma A, Dhar A K, Mandal B (1992) MYMV transmission and its control in India. In Mungbean yellow mosaic disease. Proceedings of an International Workshop, Bangkok 1991. AVRDC, Shanhua, Tainan, Taiwan. Publication No. 92–373, 54–58

Varma JP, Kadian OP, Singh JP (1973) Occurrence and incidence of virus diseases of mung and urid in Haryana. Indian Phytopathol 26:592–594

Varma A, Malathi VG, Handa A, Aiton M, Harrison BD, Varma JP, Singh RP, Singh M, Srivastava M, Singh J (1993) Occurrence of leaf-curl of cotton and okra in Northern India. In: Abstracts of the 6th International Congress of Plant Pathology, Montreal, pp 17.5.14

Vasudeva RS, SamRaj J (1948) A leaf curl disease of tomato. Phytopathology 38:364–369

Venkataravanappa V (2008) Molecular characterization of bhendi yellow vein mosaic virus. Ph D thesis. GKVK, Bengaluru

Venkataravanappa V, Loganathan M, Saha S, Rai AB, Singh B (2011a) Management of yellow vein mosaic disease of okra through seed treatment and foliar spray of insecticides. J Interacad 15(3):349–356

Venkataravanappa V, Reddy CNL, Swaranalatha P, Jalali S, Briddon RW, Krishnareddy M (2011b) Diversity and phylogeography of begomovirus-associated beta satellites of okra in India. Virol J 8:555

Venkataravanappa V, Reddy LCN, Krishnareddy M (2012a) Begomovirus characterization and development of phenotypic and DNA-based diagnostics for screening of okra genotyperesistance against Bhendi yellow vein mosaic virus. Biotech. doi: 10.1007/s13205-012-0107-z

Venkataravanappa V, Reddy CNL, Jalali S, Krishnareddy M (2012b) Molecular characterization of distinct bipartite begomovirus infecting bhendi (*Abelmoschus esculentus* L.) in India. Virus Genes 44:522–535

Venkataravanappa V, Reddy LCN, Devaraju A, Jalali S, Krishnareddy M (2013a) Association of a recombinant Cotton leaf curl Bangalore virus with yellow vein and leaf curl disease of okra in India. Indian J Virol. doi: 10.1007/s13337-013-0141-4

Venkataravanappa V, Reddy CNL, Jalali S, Krishnareddy M (2013b) Molecular characterization of a new species of begomovirus associated with yellow vein mosaic of bhendi (okra) in Bhubhaneswar, India. Eur J Plant Pathol 136:811–822

Venkataravanappa V, Reddy CNL, Devaraju A, Jalali S, Reddy MK (2013c) Association of Cotton leaf curl Bangalore virus with yellow vein mosaic disease of okra is further evidence for natural recombination of begomoviruses in India. Indian J V 24:188–198

Venkataravanappa V, Reddy LCN, Jalali S, Briddon RW, Krishnareddy M (2014a) Molecular identification and biological characterization of a begomovirus associated with okra enation leaf curl disease in India. Eur J Plant Pathol. doi: 10.1007/s10658-014-0463-0

Venkataravanappa V, Prasanna HC, Lakshminarayana Reddy CN, Krishnareddy M (2014b) Evidence for two predominant viral lineages, widespread recombination and subpopulation structure in begomoviruses associated with yellow vein mosaic disease of okra in India. Plant Pathol 64(3):508–518

Venkataravanappa V, Lakshminarayana Reddy CN, Salil J, Krishnareddy M (2015) Association of tomato leaf curl New Delhi virus DNA-B with bhendi yellow vein mosaic virus in okra showing yellow vein mosaic disease symptoms. Acta Virol 59:125–139

Venkataravanappa V, Reddy CNL, Chauhan NS, Singh B, Sanwal SK, Krishnareddy M (2016) Nucleotide sequencing and an improved diagnostic for screening okra (*Abelmoschus sculentus* L.) genotypes for resistance to a newly described begomovirus in India. J Hort Sci Biotechnol. http://dx.doi.org/10.1080/14620316.2015.1123407

Verma AK (1991) CIHNP, Lucknow, India. Annual Report 1990–1991, pp 35–36

Verma RPS, Singh DP (1988) Inheritance of resistance to mungbean yellow mosaic virus in greengram. Ann Agric Res 9:98–100

Varma A, Reddy DRR (1984) Golden and green mosaic-twi new diseases of cowpea in northern India. Indian Phytopathol 37:409

Verma HN, Srivastava KM, Mathur AK (1975) A whitefly-transmitted yellow mosaic virus disease of tomato from India. Plant Dis Rep 59:494–498

Verma AK, Basu D, Nath PS, Das S, Ghatak SS, Mukhopadhyay S (1989) Relationship between population of whitefly, *Bemisia tabaci* Genn. (Homoptera: Aleyrodidae) and the incidence of tomato leaf-curl virus disease. Indian J Mycol Res 27:49–52

Verma AK, Basu D, Nath PS, Das S, Ghatak SS, Mukhopadhyay S (1989) Some ecological considerations of whitefly and whitefly transmitted virus diseases of vegetables in West Bengal. Indian J Virol 5:79–87

Vinoth-Kumar R, Prasanna HC, Singh AK, Ragunathan D, Garg GK, Chakraborty S (2016) Molecular genetic analysis and evolution of begomoviruses and betasatellites causing yellow mosaic disease of bhendi. Virus Genes. doi: 10.1007/s11262-016-1414-y

Vinoth-Kumar R, Prasanna HC, Singh AK, Ragunathan D, Garg GK, Chakraborty S (2017a) Molecular genetic analysis and evolution of begomoviruses and betasatellites causing yellow mosaic disease of bhendi. Virus Genes 53(2):275–285

Vinoth-Kumar R, Singh D, Singh AK, Chakraborty S (2017b) Molecular diversity, recombination and population structure of alphasatellites associated with begomovirus disease complexes. Infect Genet Evol. doi:10.1016/j.meegid.2017.01.001

Vohra K, Beniwal SPS (1979) Effect of mungbean yellow mosaic virus on yield and seed quality of urdbean. Seed Res 7(2):168–174

Wang HL, Cui XY, Wang XW, Liu SS, Zhang ZH, Zhou XP (2015) First report of Sri Lankan cassava mosaic virus infecting cassava in Cambodia. Plant Dis 100:1029

Warburg O (1894) Die Kulturpflanzen Usambaras. Mitt Dtsch Schutzgeb 7:131–200

Williams FJ, Grewal JS, Amin KS (1968) Serious and new diseases of pulse crops in India in 1966. Plant Dis Rep 52:300–304

Yadav RK, Chattopadhyay D (2014) Differential soybean gene expression during early phase of infection with mungbean yellow mosaic India virus. Mol Biol Rep 41:5123

Yadav RK, Shukla RK, Chattopadhyay D (2009) Soybean cultivar resistant to mungbean yellow mosaic India virus infection induces viral RNA degradation earlier than the susceptible cultivar. Virus Res 144:89–95

Yadav CB, Bhareti P, Muthamilarasan M, Mukherjee M, Khan Y, Rathi P, Prasad M (2015) Genome-wide SNP identification and characterization in two soybean cultivars with contrasting mungbean yellow mosaic India virus disease resistance traits. PLoS One 10(4):e0123897

Yadava P, Suyal G, Mukherjee SK (2010) Begomovirus DNA replication and pathogenicity. Curr Sci 98:360–368

Zaffalon V, Mukherjee SK, Reddy VS, Thompson JR, Tepfer M (2012) A survey of geminiviruses and associated satellite DNAs in the cotton-growing areas of northwestern India. Arch Virol 157(3):483–495

Zaidi SS, Martin DP, Amin I, Farooq M, Mansoor S (2016) Tomato leaf curl New Delhi virus: a widespread bipartite begomovirus in the territory of monopartite begomoviruses Mol Plant Pathol

Zaim M, Kumar Y, Vipin H, Zaidi AA (2011) Velvet bean severe mosaic virus: a distinct begomovirus species causing severe mosaic in Mucuna pruriens (L.) DC. Virus Genes 43(1):138–146

Zerbini FM, Briddon RW, Idris A, Martin DP, Moriones E, Nacas-Castillo J, Bustamante R, Roumagnac P, Varsani A et al (2017) ICTV virus taxonomy profile: Geminiviridae. J Gen Virol 98(2):131–133

Characterisation of Carlaviruses Occurring in India

6

T. Jebasingh and T. Makeshkumar

Abstract

Carlaviruses infects field, vegetable and ornamental crops in India. The genus *Carlavirus* (family *Betaflexiviridae*) has as many as 43 recognised virus species and 13 tentative members. Only five carlavirus species, *Cowpea mild mottle virus, Chrysanthemum virus B, Lily symptomless virus, Potato virus S* and *Garlic common latent virus* and one tentative member, football lily mosaic virus are known in India. In this chapter, characterisation of carlavirus occurring in India is presented.

Keywords

Carlaviruses in India • Carnation latent virus • Cowpea mild mottle virus • Chrysanthemum virus B • Lily symptomless virus • Potato virus S • Garlic common latent virus • Football lily mosaic virus

6.1 Introduction

The name 'carlavirus' has been derived from the type species, acquired from *Carnation latent virus* (CLV). Due to its particle morphology and RNA size, carlaviruses is grouped between potexviruses and potyviruses. The International Committee on Taxonomy of Viruses (ICTV) has categorized carlaviruses as a genus

T. Jebasingh
School of Biological Sciences, Madurai Kamaraj University, Madurai 625021, Tamil Nadu, India
e-mail: jebasinghs@gmail.com

T. Makeshkumar (✉)
ICAR-Central Tuber Crops Research Institute, Thiruvananthapuram 695 017, Kerala, India
e-mail: makeshctcri@gmail.com

B. Mandal et al. (eds.), *A Century of Plant Virology in India*,
https://doi.org/10.1007/978-981-10-5672-7_6

of the family *Betaflexiviridae* derives from the subsequent subdivision of *Flexiviridae*. Currently, 52 virus species have been included in this genus.

Carlavirus has narrow host range and causes mild or no symptoms. In most of the cases they were discovered accidentally with other viruses from mixed infection. Caralviruses are easily spread by vegetative propagation and aphids in a semi-persistent mode. Transmission through seed and *Bemisia tabaci* also has been reported for cowpea mild mottle virus (CPMMV).

The virions of carlaviruses are non-enveloped, filamentous morphology with size of 610–700 × 12 nm (Wetter and Milne 1981). The virus particle contains linear, positive-sense ssRNA of 5.8–9 kb in size. *Potato virus S* (PVS) (Foster and Mills 1991b) and CLV (Meehan and Mills 1991) of carlavirus also contains subgenomic RNA. The RNA genome lacks viral protein-genome linked (VPg) at the 5′ of the genomic RNA (Monis and de Zoeten 1990) but has cap structure (Foster and Mills 1991a, 1991c). The cap structure is absent at the 5′ end of subgenomic RNA (Foster and Mills 1990).

The genome of carlaviruses has six open reading frames (ORF). RNA replicase with methyltransferase and helicase activity is coded from the ORF1. ORFs 2, 3 and 4 codes a Triple Gene Block (TGB) is responsible for virus transport and cell membrane modification. The coat protein (CP) is coded by ORF5 (CP) and the protein from ORF6 is a cysteine-rich protein. This protein of ORF6 contains a zinc finger motif binds to single- and double-stranded nucleic acid *in vitro* (Chen et al. 2002). Between ORF1 and ORF2 and ORF4 and ORF5, the genome of carlaviruses contains two small non-coding regions (Chen et al. 2002). There are as many as six carlaviruses known in India. A brief description of the work on carlaviruses carried out in India is presented here.

6.2 Cowpea Mild Mottle Virus (CPMMV)

CPMMV from cowpea plant was first reported in Ghana (Brunt and Kenten 1973). In India, the infection of CPMMV has been observed in cowpea (Jeyanandarajah and Brunt 1993), groundnut (Iizuka et al. 1984) and soybean (Bhargandi et al. 1987; Manoj et al. 2013). Jeyanandarajah and Brunt (1993) reported that CPMMV is serologically related to tomato pale chlorosis virus, Psophocarpus necrotic mosaic virus, Voandzeia mosaic virus and groundnut crinkle virus (Naidu et al. 1998). Based on the physicochemical properties, Brunt (1995) classified CPMMV as the definite member of genus Carlavirus.

Transmission electron microscopic analysis of CPMM from infected soybean has shown that CPMMV as the flexuous filamentous particles measuring 650 × 12 nm in size (Manoj et al. 2013). CPMMV contains ssRNA of 8127 nucleotides codes for six ORFs. In India, during 2011 and 2012, high disease incidence of CPMMV on soybean occurred from 25.1–38.3 % to 45.1–71.0 % respectively (Manoj et al. 2013). The different symptoms like mottling, mosaic, leaf distortion and plant stunting were observed in soybean (Manoj et al. 2013). Non-persistent transmission of CPMMV is occurred in groundnut and soybean by the whitefly (Jeyanandarajah and Brunt 1993; Muniyappa and Reddy 1983; Mali et al. 1987; Badge et al. 1996; Manoj et al. 2013). The reports are available regarding the transmission of CPMMV in cowpea, soybean and French bean through seeds (Iizuka et al. 1984; Siva et al. 1990; Almeida et al. 2005; Tavasoli et al. 2009).

CPMMV showed mild mosaic mottling symptoms in peanuts grown in Mianpuri in Northern India designated as CPMMV-M and in Hyderabad (Southern India), it showed severe disease symptoms, designated as CPMMV-S (severe) (Naidu et al. 1998). The sequencing of 2.5-kb from 3′-terminal of these isolates (GenBank Accession number for CPMMV-S: AF024628 and for CPMMV-M: AF024629) showed 70 % sequence identity. The organization of ORFs and molecular weight of proteins coded from the respective ORFs are similar between these isolates (Naidu et al. 1998). The TGB proteins of CPMMV-M and CPMMV-S not only similar to TGB of other member of *Carlavirus* but also to TGB of *Potexvirus, Hordeivirus* and *Furovirus* genus (Naidu et al. 1998). In both the isolates, P7 from ORF 3 is substantially overlaps the P12 codes by ORF 2. The initiation codon of the ORF 2 and 5 are partially embedded in the termination codon of the ORF 1 and 4 respectively in both the isolates. An intergenic region separates the ORF 4 and 3 with 56 nt in CPMMV-S, and 19 nt in the case of CPMMV-M (Naidu et al. 1998).

Twenty seven cultivars of soybean were checked for CPMMV infection and all are susceptible to CPMMV (Manoj et al. 2013). The symptoms are appeared in different cultivars of soybean within 9–14 days of inoculation (DAI). CPMMV causes venial chlorosis mottling and leaf distortion in mungbean and urdbean plant, chlorotic lesions, necrotic spot with mild mottling and downward curling in cowpea it (Manoj et al. 2013). CPMMV infection on groundnut showed systemic chlorosis, mottling and leaf distortion with stunted growth where as in bean plants it produce systemic mosaic and stunting (Manoj et al. 2013). In French bean it has showed mosaic, veinal chlorosis, mottling, necrotic lesions and leaf curling (Manoj et al. 2013). *Nicotiana benthamiana* and *N. glutinosa* have shown chlorosis and systemic mosaic and stunting, whereas as in *N. tabacum*, no symptoms were found (Manoj et al. 2013). Inoculation of CPMM on fenugreek plant showed systemic chlorosis at the initial but latter it causes necrotic lesion on the leaves. After the infection, the symptoms started to appear within 15–20 DAI in mungbean, urdbean, cowpea, groundnut, French bean, asparagus bean and fenugreek, but in tobacco plants the symptoms started to appear within 8–10 DAI (Manoj et al. 2013).

Manoj et al. (2013) submitted the sequence of 1.3 kb of 3′-terminal region of CPMMV of D1 isolate containing the complete CP gene, nucleic acid binding protein (NABP) gene and 3′-UTR in NCBI GenBank with Accession number: JX524810. The CP gene of CPMMV D1 isolate has 90–94 % aa identity with other CPMMV isolates (Manoj et al. 2013). NABP gene of D1 isolate has 67–70 % nt identity with CPMMV-M and CPMMV-S and 55–59 % nt identity with other isolates. In the 3′-UTR region it has 77–81 % nt identity with CPMMVM and CPMMV-S isolates and 73–81 % nt identity with other isolates. Phylogenetic analysis of 1300 bp of 3′-terminal region of CPMMV isolates, placed CPMMV D1 in separate position in the phylogenetic tree which is also different from other Indian isolates; CPMMV-S and CPMMV-M (Manoj et al. 2013).

First time in India, Baranwal et al. (2015), reported that CPMMV infects urdbean and mungbean and produce leaf crinkle symptom. The presence of CPMM in urdbean and mungbean was confirmed by DAS-ELISA and RT-PCR. But, by the next generation sequencing of small RNA, they identified the presence of groundnut bud necrosis virus and soybean yellow mottle mosaic virus along with CPMMV. The 3′ terminal sequence contains CP, NABP and 3′ UTR of CPMMV from mungbean (Acc. No. KJ534277) and urdbean (Acc. No. KJ534276) are available (Baranwal et al. 2015). The CP gene of CPMMV isolates from mungbean and urdbean is 867 nt in length and has low sequence identity: 77.9 % at the nucleotide and 90.9 % at amino acid level. The nucleotide length of NABP is varied between urdbean and mungbean isolates. In the urdbean isolates, nucleotide length of NABP is 327 nt whereas mungbean isolate it is 306 nt but has very low identity of 56.4% at the amino acid level (Baranwal et al. 2015). The analysis of amino acid sequences of CP and NB, clustered mungbean isolate with other Indian isolates of CpMMV and urdbean isolate with Brazil, USA and Ghana isolates. But based on 3’UTR nucleotide sequences both these isolates are clustered together with Indian isolates (Baranwal et al. 2015).

Naidu et al. (1998) expressed CP of CPMMV-M in *E. coli* showed same size (32 kDa) of CP purified from infected CPMMV-M infected plant. The polyclonal antiserum raised against the purified virus of CPMMV-M is reacted with recombinant CP (Naidu et al. 1998).

6.3 Chrysanthemum B Virus (CVB)

CVB is the major pathogen of chrysanthemum and it is available wherever chrysanthemum grows in India (Singh et al. 2007) and the rest of the world. CVB has filamentous particle, measuring approximately 680 nm long and 12 nm wide (Verma et al. 2003) (Fig. 6.1a, b). Electron microscopic studies of CVB infected leaf tissues showed the covering of CVB around chloroplast showed abnormalities like loss of chloroplast envelope and extensive invagination by cytoplasm (Fig. 6.1c, d). The association of the virions with the chloroplast and disruption of chloroplast envelope suggest that the chloroplast is necessary for the assembly of virions as suggested earlier in case of PVS (Garg and Hegde 2000).

Verma et al. (2003) reported that CBV infects *N. clevelandii*, *N. glutinosa*, *N. rustica*, *Petunia hybrida* and *Vicia faba* and produce symptoms like banding, mottling, clearing and mosaic in vein and during the severe infection it make the flowers get malformed.

Out of 80 cultivars of chrysanthemum tested, Regol Time and Maghi cultivars were infected by CVB, TAV, CMV, poty- and tospoviruses). The percentage of infection in 80 cultivars by CVB, CMV, TAV, tospo- and potyviruses were 48.7 %, 42.5 %, 26.2 %, 12.5 % and 6.25 % respectively. In 12 cultivars, single infection of

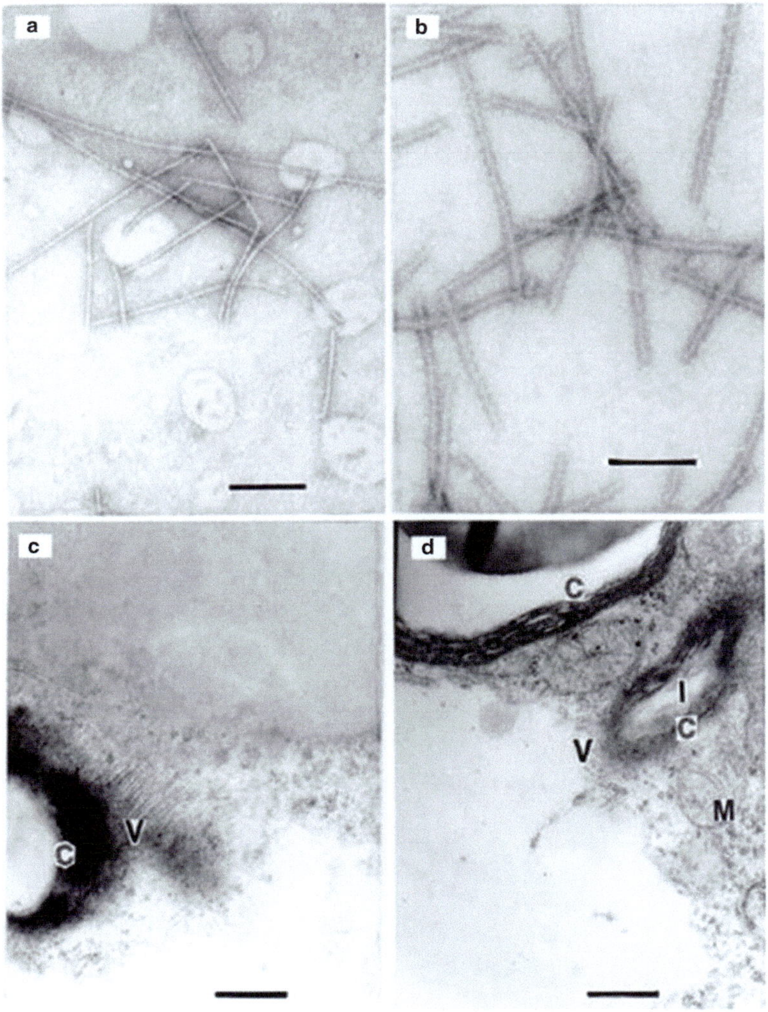

Fig. 6.1 Electron micrograph of CVB from leaf tissue. **a** Clarified virus showing negatively stained virions; **b** Virions coated with CSB carlavirus antiserum; **c** Infected leaf cell with dense chloroplast *c* and large number of virions *v*; **d** Infected leaf cell with normal mitochondria *m* and dense chloroplast *c* has cytoplasmic invagination *I* and virions *v* around it. Bar = 250 nm in (**a**, **b**, Bar = 500 nm in **c**, **d**) (Source: Verma et al. 2003)

CMV was observed with characteristic symptoms and in 9 cultivars as double infection with CVB and other viruses listed earlier. CVB is transmitted primarily by the aphid *Myzus persicae* in a non-persistent manner and secondarily through sap-transmission (Singh et al. 2012).

Singh et al. (2007) have surveyed Chrysanthemums showing symptoms of mosaic and mottling all over the India and identified 29 isolates (Table 6.1). These 29 isolates are grouped into 3 groups based on the phylogenetic analysis of CP gene.

Table 6.1 Source, size of coat protein gene and accession numbers of the various chrysanthemum B virus isolates

Abbreviation (Region)	Collected from state	Gene size (Nucleotides)	Accession number
AP (Hyderabad)	Andhra Pradesh	945	AJ580954
AR (Arunachal Pradesh)	Arunachal Pradesh	939	AJ748852
AS (Guwhati)	Assam	945	AJ812735
BR (Patna)	Bihar	945	AJ580931
CH (Chandigarh)	Chandigarh	945	AJ621814
CHH (Raipur)	Chhattisgarh	945	AM0349442
DL (Delhi)	New Delhi	945	AJ619742
GJ (Gandhinagar)	Gujrat	945	AJ871582
HP1 (Palampur)	Himachal Pradesh	969	AJ564858
HP2 (Chamba)	Himachal Pradesh	945	AJ871365
HP3 (Chail)	Himachal Pradesh	948	AJ876635
HP4 (Sangla)	Himachal Pradesh	939	AJ871583
HR (Haryana)	Haryana	945	AJ629843
HSR (Hissar)	Haryana	945	AM0349440
JK1 (Jammu)	Jammu and Kashmir	945	AJ812569
JK2 (Srinagar)	Jammu and Kashmir	945	AJ871366
JK3 (Leh)	Jammu and Kashmir	945	AJ871367
JH (Ranchi)	Jharkhand	945	AJ580930
KK (Bangalore)	Karnataka	945	AJ585240
KR(Trivandrum)	Kerala	945	AM039441
MH (Nagpur)	Maharashtra	945	AJ581993
MP (Gwalior)	Madhya Pradesh	945	AJ748853
OR (Bhubaneswar)	Orissa	945	AJ879077
PB (Ludhiana)	Punjab	945	AJ580956
RU	Russian isolate of CVB	945	S60150*
RJ (Jaipur)	Rajasthan	945	AJ619743
SK (Sikkim)	Sikkim	945	AJ585514
UA2 (Uttarkashi)	Uttaranchal	939	AJ879078
WB1 (Kolkata)	West Bengal	945	AJ619744
WB2 (Siligudi)	West Bengal	945	AJ621815
Helenium virus S (HelVS; outgroup 1)		897	D10454*
Daphne virus S (DVS; outgroup 2)		954	AJ620300*
Lily symptomless virus (LSV; outgroup 3)		876	AJ585052*

Source: Singh et al. (2007)

Group I contains seven subgroups (named A–G) representing seven lineages diverged from the common ancestral group I. The clustering of isolates was not based on geographical since diversity was observed within the same state as three (HP1, HP3 & HP4; Table 6.1) of the four isolates from Himachal Pradesh are present in group I where as HP2 in group III (Table 6.1).

To assess the role of recombination on CVB diversity in India, the complete genome were obtained for four isolates namely Punjab (PB), Uttarakhand (UK),

Uttar Pradesh (UP) and Tamil Nadu (TN) represents four corners in India (Singh et al. 2012). The complete genome analysis revealed that all these isolates were more similar within themselves than to other isolates from outside the India including the Japanese isolate (CVB-S). N-terminus of replicase is highly conserved among the carlaviruses but surprising replicase of the Indian isolates was 59 % identical to that of the CVB-S whereas the TGB of the Indian isolates showed 81–92 % sequence similarity to CVB-S. NABP has showed the highest degree of similarity of 89–100 % (Singh et al. 2012).

Sixteen unique recombination events were detected in the full length genome of various CVB isolates (Singh et al. 2012). In fact, CVB-S was consider to be a recombinant of UP and TN. The recombination occurred in replicase ORF between 538 and 4260 nucleotide positions (Singh et al. 2012).

Singh et al. (2011a) have observed the homotypic interaction of the CVB CP in the yeast two-hybrid system (Y2H) and co-immunoprecipitation (Co-IP). In the Y2H analysis, the complete coat protein (CP) in the activation domain (AC) interacts strongly with complete coat protein in the binding domain (BD). The complete coat protein interacts with N-terminus of CP having amino acids 1–184 aa and with middle region have 98–184 aa. There was no interaction between complete CP and N-terminus of CP having 1–102 aa. These are clearly indicate that the 86 aa from middle region spanning from 98 to 184 is extremely important for CP-CP interactions (Singh et al. 2011a). This is also confirmed with Co-IP analysis (Fig. 6.2) in which precipitation was observed between the c-Myc tagged 98–184 aa and HA-tagged 1–184 aa with c-Myc and HA antibodies. (Singh et al. 2011a).

Initially CVB was detected with immunological techniques (Raizada et al. 1989; Zaidi et al. 1990). ELISA and other forms of ELISA were extensively used to detect CVB (Verma et al. 2003). Immunosorbant electron microscopy (ISEM) was also used to identify CVB in chrysanthemum (Verma et al. 2003). The polyclonal antibodies were raised against coat protein gene of CVB with Glutathione S-Transferase (GST) tag showed specific reaction to CVB from infected plant which was used to develop ELISA based diagnostic kit for CVB and for indexing of the

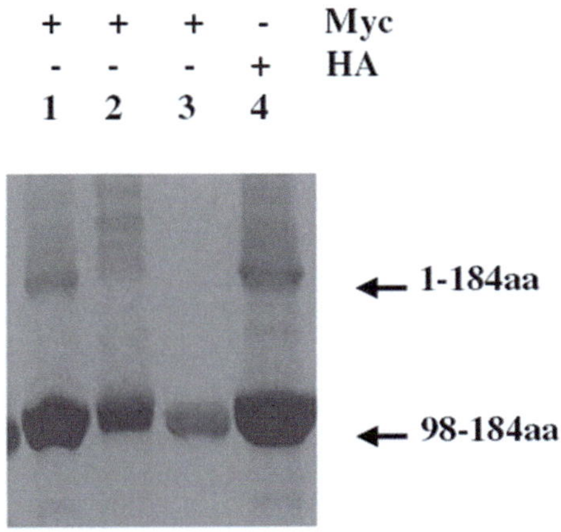

Fig. 6.2 Co-immuno precipitate of 98–184 aa (BD) with 1–184 aa (AD) in Co-IP assay of chrysanthemum B virus. Lane 1: (1–184 AD) + (98–184BD) + c-Myc antibody, Lanes 2, 3: (98–184BD) + c-Myc antibody, Lane 4: (1–184 AD) + (98–184BD) + HA-Tag antibody (Source: Singh et al. (2011a))

chrysanthemum mother stock with virus free (Singh et al. 2011b). This group is currently involved in developing the PCR based detection kit for CVB.

CVB-free chrysanthemum can be produced by chemotherapy, thermotherapy and *in vitro* meristem tip culture (Ram et al. 2005). For chemotherapy, 2-thiouracil at 40 g dm-3 concentration was effective to produce CVB-free plants when compared to other chemicals like Acyclovir, Amantadine, 5-bromouracil, 2-thiouracil and Zidovudine. Incubating the plants at 38 °C for 30 days was also effective to treat the CVB virus (Ram et al. 2005).

6.4 Lily Symptomless Virus (LSV)

LSV, is an important pathogen of Lily, is transmitted by *Myzus persisae, Aulocorthium solani Macrosiphum euphorbiae, Aphis fabae, Aphis gossypii,* or by whitefly (Suhasini et al. 2010). LSV exhibit various symptoms like chlorosis, vein clearing and deformed flowers on various lilies like *Lilium longiflorum*, *Lilium tigrinum*, Asiatic hybrid lily and Oriental hybrid lily (Singh et al. 2005). The isolates obtained from *L. Longiflorum*, *L. tigrinum*, Oriental and Asiatic hybrids and spider lily were designated as LSV-L, –T, –A, –O and -S, respectively. EMBL accession number of these isolates are AJ585052 (LSV-A), AJ748320 (LSV-L), AJ748277 (LSV-O), AJ781318 (LSV-T) and AJ780923 (LSV-S). The comparison of coat protein gene was made between these LSV isolates and isolates from other countries like China (LSV-C), the Netherlands (LSV-N), Japan (LSV-J) and Korea (LSV-K) (Singh et al. 2005). Seventy-eight to 96% homology has been observed between the Indian isolates and 83–98 % homology with isolates from other countries (Singh et al. 2005). The isolates LSV-L and LSV-A has stretches of amino acids in the core region CP protein at position between 129 and 146 in LSV-L and between 129 and 155 in LSV-O (Fig. 6.3) but this is absent in other LSV isolates. A 41 amino acids stretch is found in the C-terminal of LSV-T is unique to this isolate (Fig. 6.3) (Sing et al. 2005).

The isolate LSV-T, which is native to India, has 78–84 % homology with other isolates. Due to its less homology with other isolate, variations at C-terminal (Fig. 6.3) and phylogenetic analysis made LSV-T as a distinct isolate.

Suhashini et al. (2010) analysed the sequence of 79 LSV isolates and observed that the variations in the LSV genome are taking place at a faster rate and these variations can produce new strain which will causes widespread dispersal and damage.

The complete nucleotide sequence for the LSV obtained from the *L. Longiflorum* is available in the GenBank with Accession Number: AM422452 (Singh et al. 2008). The genome size is 8.394 Kb with six open reading frames (ORFs). This isolate is closely similar to Netherland isolate (Accession No: AJ564638) (Singh et al. 2008). This LSV isolate showed 97–98 % nucleotide sequence homology with the other isolates reported by Sing et al. (2005).

Three recombination events have been observed between Indian, Netherlands and Chinese isolates. First recombination event was observed between the Indian

```
LSV-A          IAGLGVPTGARRISNIANGHHVCLREPVQRSLTLKGSIEFENGAVPVDSIAAIMKKHAGL 180
LSV-L          IAGFGVPTGARRISNIANGHSCVLGVSSSAFLDPEGSIEFENGAVPVDSIAAIMKKHAGL 180
LSV-O          IAGLGVPT-EHVASVILQMVIMCACVSSSAFLDPEGSIEFENGAVPVDSIAAIMKKHAGL 179
LSV-S          IAGLGVPT-EHVASVILQMVIMCACVSSSAFLDPEGSIEFENGAVPVDSIAAIMKKHAGL 179
LSV-T          IARLGVPT-EHVASVILQMVIMCACVSSSAFLDPEGSIEFENGAVPVDSIAAIMKKHAGL 179
               ** :****  :  * * :        . .  *  :*************************

LSV-A          RKVCRLYAPIVWNSMLVRNQPPADWQAMGFQYNTRFAAFDTFDYVTNQAAIQPVEGIIRR 240
LSV-L          RKVCRLYAPIVWNSMLVRNQPPADWQAMGFQYNTRFAAFDTFDYVTNQAAIQPVEGIIRR 240
LSV-O          RKVCRLYAPIVWNSMLVRNQPPADWQAMGFQYNTRFAAFDTFDYVTNQAAIQPVEGIIRR 239
LSV-S          RKVCRLYAPIVWNSMLVRNQPPADWQAMGFQYNTRLAAFDTFENVTNQAAIQPVEGIIRR 239
LSV-T          RKVCRLYAPIVWNSMLVRNQP-QLMASYGLPYNTRFAAFDTSLRGLTKRLSNLSRGSSGD 238
               *********************     : *: ****:*****     .:   :  .*

LSV-A          PTSAEVIAHNAHKQLALDRSNRNERLGSLETEYTGGVQGAEIVRNHRYANNG 292
LSV-L          PTSAEVIAHNAHKQLALDRSNRNERLGSLETEYTGGVQGAEIVRNHRYANNG 292
LSV-O          PTSAEVIAHNAHKQXALDRSNRNERLGSLDTKYTGGVQGAEIVRNHRYANNG 291
LSV-S          PTSAEVIAHNAHKQLALDRSNRNERLGSLETEYTGGVQGADFVRNGKYANNG 291
LSV-T          PLQLRSLPTTRTSNLALDRSNRNERLGSLETEYTGGVQGAEIVRNHRYANNG 290
               * . . :. .  .: **************:*:********::*** :*****
```

Fig. 6.3 Coat protein sequence comparison between the various LSV isolates of India. Differences in the C-terminal of LSV-T and middle portion of LSV-A and LSV-L represented by shading. *LSV-A* Asiatic hybrid, *LSV-L Lilium longiflorum*, *LSV-O* Oriental hybrid, *LSV-S* spider lily, *LSV-T* Tiger lily (Source: Singh et al. 2005)

isolate (AM422452) and Chinese isolate (AM263208) which led to a formation of Korean isolate (AJ516059). In the Korean isolate, a region of 5,597–7,796 nt of the Indian Isolate (AM422452) was replaced with C-terminal of RdRp, TGB, almost complete CP region of Chinese isolate (AM263208). Second recombination event was occurred between the Indian isolate (AM422452) and the Netherlands isolate (AJ564638) and produce Korean isolate. In this recombination event, a region of RdRp (2,344–5,189 nt) of the Indian isolate was replaced with the corresponding genome sequence of the Netherlands isolate (AJ564638). Third recombinations observed between Korean isolate (AJ516059) and the Chinese isolate (AM263208) led to Korean isolate. The ending breakpoint could not be identified but detected with a low degree of confidence (Singh et al. 2008).

6.5 Football Lily Mosaic Virus (FLMV)

In India, football lily (*Scandoxus katharine*) is commonly grown in nurseries as ornamental flower. It is naturally occurring plant in Sikkim and Darjeeling hills. Virus-like symptoms were observed in various nurseries at Delhi, Meerut, Sikkim and Kalimpong. Electron microscopic observation revealed that flexuous particles measuring 650 × 13 nm constantly associated with the disease. Samples from naturally affected plants were collected and maintained in glass house. The disease was mechanically transmitted to football lily and *Chenopodium amaranticolor*. A filamentous virus was partially purified from the football lily plants which were grown inside the glass house. Polyclonal antibodies against the virus were developed and used for virus detection in ELISA (Das et al. 2010). Virus associated with football lily mosaic disease has single coat protein of 31 kDa. It was serologically related to

Carnation latent virus. Ultra thin sections of infected tissue revealed large number of virus particles scattered in the cytoplasm but not in nucleus. No inclusion bodies characteristic of potyvirus was observed. Particle morphology, transmission, serological relationship, size of coat protein and absence of characteristic inclusion bodies suggests that the virus may be a member of carlavirus group (Das et al. 2010).

6.6 Potato Virus S (PVS)

Potato in India is known to be infected by two carlaviruses, PVS and *potato virus* M (PVM). PVM was first recorded during late 1970s in India based on serology and transmission of the virus (Khurana and Singh 1980) and since then not much information on PVM has been generated. Recently, the complete genome sequence of one isolate from Delhi, PVM-Del-144 has been generated (GenBank No. KJ194171) and CP gene sequence of 14 isolates from the potato growing region of northern plain has been analysed that showed considerable diversity of the virus (Unpublished results, Dr. Bikash Mandal). Raigond et al. (2013) developed RT-PCR based detection kit to detect both PVY and PVS simultaneously from the potato leaf tissues and tubers. The detection is based on the amplification of coat protein gene. This detection method makes the simultaneous detection of PVY and PVS in a simple and rapid manner and also it reduces the time and cost of the consumables (Raigond et al. 2013).

6.7 Garlic Common Latent Virus (GarCLV)

The CP gene sequence is available for five Indian GarCLV isolates: Northern India (GarCLV-G1; Acc. No. JQ818259), Eastern India (GarCLV-RAU; Acc. No. JQ818256), Southern India (GarCLV-Kolar; Acc. No. JQ818257), Western India (GarCLV-JN; Acc. No. JQ818255 and Anand; Acc. No. JQ818258). Comparative nucleotide sequence analysis of CP gene with other CP sequences available in the database has revealed that there is high sequence diversity among all GarCLV isolates worldwide (11.9 %) and low sequence diversity among the Indian isolates (4.3 %) (Pramesh et al. 2013). The major variability was found at the N-terminus (1–48 aa) whereas the central and the C-terminal regions (49–319 aa) are highly conserved in all the isolates (Pramesh et al. 2013).

The phylogenetic analysis clustered the GarCLV isolates into two major clusters as subgroup I and II. Subgroup I contains isolates from USA and China and subgroup II contain isolates from Australia, Brazil, India, Japan, and South Korea. All the five isolates of India are grouped together with K2 isolate of South Korea and GCLV-BZL isolate of Brazil.

The phylogenetic analysis of CP gene five GarCLV Indian isolates with representative isolates of other 37 Carlavirus species grouped viruses of Carlavirus into two distinct phylogenetic subgroups. The subgroup 1 contains all the Indian isolates in single cluster with shallot latent virus and 10 other Carlavirus species. The subgroup 2 contains remaining 26 Carlavirus species (Pramesh et al. 2013).

During the intra species analysis of CP gene of Indian isolates, Pramesh et al. (2013) observed the potential recombination in the CP gene of the Anand isolate from JN and Kolar isolates and also observed the absence of genetic exchange from Carlavirus species to GarCLV (Pramesh et al. 2013).

6.8 Microsatellites in Carlvirus Genome

The genome wide screening of 32 carlavirus for the presence, abundance, and composition of simple sequence repeat (SSR) has revealed that carlavirus has 18–42 SSRs (Chaudhary et al. 2014) as compared to similar genomes of potyviruses (23–45 SSRs) (Xiangyan et al. 2011) or *Human immunodeficiency virus* isolates (22–48 SSRs) (Chen et al. 2009) but having more SSRs than that of geminivirus (4–19) with a smaller genome. A single nucleotide repeats were observed in all the carlavirus genomes. Among polyrepeats, poly (A/T) repeats being more prevalent than poly (G/C) repeats. According to Karaoglu et al. (2005), the percentage of poly (A/T) repeats (78 %) are significantly higher than poly (G/C) repeats in each complete carlavirus genome. Chaudhary et al. (2014) based on the analysis of 32 carlavirus genome reported that the (A/T) content is slightly higher than G/C content.

Di-nucleotide repeats (GT/TG. AG/GA, AC/CA, AT/TA, CT/TC and CG/GC are found across the carlavirus genomes. GT/TG repeats predominately present in carlaviruses genome whereas GC/CG is rarely available (Chaudhary et al. 2014). Among 64 triplet repeat types, AAG/GAA type repeats codes for lysine/glutamic acid was most abundant followed by GAG coding for glutamic acid (Chaudhary et al. 2014). Di-nucleotide repeats are more prevalent because of higher slippage rate (Katti et al. 2001). Tetra-nucleotide repeats like AAGA, GTAC, TAAA and CAAG are present in many carlavirus genomes and pentanucleotide CCATA repeat is present in one genome. The occurrence of diverse types of repeats in carlavirus genome facilitates genome evolution (Chaudhary et al. 2014).

The search for compound simple sequence repeat (cSSR) in carlavirus genomes yielded a total of 34 compound microsatellites (Chaudhary et al. 2014). In general, cSSR are involved in regulating the gene expression in several species (Kashi and King 2006; Chen et al. 2011) but the function of these cSSR in carlavirus is not clear but possibly of complex regulation at the functional level. In general, the largest compound microsatellite in carlavirus is composed of three cSSRs. Interestingly, cSSRs% in carlavirus genome is varied between 0 % and 11.42 % (Chaudhary et al. 2014). Among 32, 11 of the carlavirus species does not possess even a single compound microsatellite, this may be due to less number of strains present in these species where in non availability of cSSR might restrict their variation and evolution (Chaudhary et al. 2014).

According to Chaudhary et al. (2014), 66.4 % and 67.52 % of SSRs and compound simple sequence repeats (cSSRs) are found in RDRP gene respectively followed by 8.6 % and 8.8 % of SSRs and cSSRs in ORF-6 respectively. The intergenic region of TGB1/TGB2 and TGB2/TGB3 contains very low SSRs and cSSR and RDRP gene contains very high mono-, di- and tri-nucleotide repeats.

6.9 Concluding Remarks

Although, six different carlaviruses has been identified in India, their prevalence in the different crops and regions has not yet been adequately investigated. The reason for this may be due to the fact that these viruses are not known to cause as severe disease that is crop yield limiting. The complete genome information is available from India only for CVB and LSV. The study on the genetic diversity between the isolates of CVB and LSV revealed the occurrence of recombination between the isolates of each species. The polyclonal antisera raised against the coat protein of CPMMV, CVB and FLMV have the potential application in determining their prevalence and also for indexing the plants for the freedom from carlavirus.

References

Almeida AMR, Piuga FF, Marin SR, Kitajima EW, Gaspar JO, Moraes TG (2005) Detection and partial characterization of a Carlavirus causing stem necrosis of soybeans in Brazil. Fitopatol Bras 30:191–194

Badge J, Brunt A, Carson R, Dagless E, Karamagioli M, Phillips S, Seal S, Turner R, Foster GD (1996) A Carlavirus-specific PCR primer and partial nucleotide sequence provides further evidence for the recognition of Cowpea mild mottle virus as a whitefly-transmitted Carlavirus. Eur J Plant Pathol 102:305–310

Brunt AA (1995) Genus *Carlavirus*. In: Murphy FA, Fauquet CM, DHL B, Ghabrial SA, Jarvis AW, Martelli GP, MD MMAS (eds) Virus taxonomy. Classification and nomenclature of viruses. Sixth report of the International Committee on Taxonomy of Viruses. Springer, Wien, pp 474–478. Arch Virol [Suppl] 10

Brunt AA, Kenten RH (1973) Cowpea mild mottle, a newly recognized virus infecting cowpea (*Vigna unguiculata*) in Ghana. Ann Appl Biol 74:67–74

Chaudhary MA, Avadhesh KS, Choudhary S, Safdar A (2014) Genome-wide scan for analysis of simple and imperfect microsatellites in diverse carlaviruses. Infect Genet Evol 21:287–294

Chen J, Zheng H, Chen J, Yang C (2002) Genome organization and variation in the 3′-partial sequence of *garlic latent virus* in China. Sci China 45:441–448

Chen M, Tan Z, Jiang J, Li M, Chen H, Shen G, Yu R (2009) Similar distribution of simple sequence repeats in diverse completed human immunodeficiency virus type 1 genomes. FEBS Lett 583:2959–2963

Chen M, Zeng G, Tan Z, Jiang M, Zhang J, Zhang C, Lu L, Lin Y, Peng J (2011) Compound microsatellites in complete Escherichia coli genomes. FEBS Lett 585:1072–1076

Das DR, Singh L, Pant RP, Ahlawat YS (2010) Characterization of a new Carlavirus isolated from football lily (Scandoxus katharine) in India. Indian Phytopathol 63:433–437

Foster GD, Mills PR (1990) Investigation of the 5' terminal structures of genomic and subgenomic RNAs of potato virus S. Virus Genes 4:359–366

Foster GD, Mills PR (1991a) Cell-free translation of American hop latent virus. Virus Genes 5:327–334

Foster GD, Mills PR (1991b) Evidence for subgenomic RNAs in leaves infected with an Andean strain of potato virus S. Acta Virol 35:260–267

Foster GD, Mills PR (1991c) Translation of Potato virus S RNA in vitro: evidence of protein processing. Virus Genes 6:45–52

Garg ID, Hegde V (2000) Biological characterization, preservation and ultrastructural studies of Andean strain of Potato virus S. Indian Phytopathol 53:256–260

Iizuka N, Rajeshwari R, Reddy DVR, Goto T, Muniyappa V, Bharathan N, Ghanekar AM (1984) Natural occurrence of a strain of Cowpea mild mottle virus on groundnut (*Arachis hypogaea*) in India. J Phytopathol 109:245–253

Jeyanandarajah P, Brunt AA (1993) The natural occurrence, transmission, properties and possible affinities of cowpea mild mottle virus. J Phytopathol 137:148–156

Karaoglu H, Lee CM, Meyer W (2005) Survey of simple sequence repeats in completed fungal genomes. Mol Biol Evol 22:639–649

Kashi Y, King DG (2006) Simple sequence repeats as advantageous mutators in evolution. Trends Genet 22:253–259

Katti MV, Ranjekar PK, Gupta VS (2001) Differential distribution of simple sequence repeats in eukaryotic genome sequences. Mol Biol Evol 18:1161–1167

Khurana SMP, Singh MN (1980) Occurrence of potato virus M in Shimla hills. Indian J Mycol Plant Pathol 10:179–181

Mali VR, Nirmal DD, Nolt BL, Rajeshwari R, Bharathan N, Reddy DVR (1987) Properties of a Cowpea mild mottle virus isolate from groundnut. Indian Phytopathol 40:2–26

Manoj KY, Kajal KB, Sanjay KL, Virendra KB, Rakesh KJ (2013) A distinct strain of cowpea mild mottle virus infecting soybean in India. J Phytopathol 161:739–744

Meehan BM, Mills PR (1991) Cell-free translation of carnation latent virus RNA and analysis of virus specific dsRNA. Virus Genes 5:175–181

Monis J, de Zoeten GA (1990) Characterization and translation of potato virus S RNA. Phytopathology 80:441–445

Muniyappa V, Reddy DVR (1983) Transmission of cowpea mild mottle virus by *Bemisia tabaci* in a non-persistent manner. Plant Dis 67:391–393

Naidu RA, Gowda S, Satyanarayana T, Boyko V, Reddy AS, Dawson WO, Reddy DVR (1998) Evidence that whitefly-transmitted cowpea mild mottle virus belongs to the genus *Carlavirus*. Arch Virol 143:769–780

Pramesh D, Virendra K, Baranwal (2013) Molecular characterization of coat protein gene of Garlic common latent virus isolates from India: an evidence for distinct phylogeny and recombination. Virus Genes 47:189–193

Raigond B, Sharma M, Chauhan Y, Jeevalatha A, Singh BP, Sharma S (2013) Optimization of duplex RT-PCR for simultaneous detection of *Potato virus Y* and *S*. Potato J 40:22–28

Raizada RK, Srivastava KM, Chandra G, Singh BP (1989) Comparative evaluation of serodiagnostic methods for detection of Chrysanthemum virus B in chrysanthemum. Indian J Exp Biol 27:1094–1096

Ram R, Verma N, Singh AK, Singh L, Hallan V, Zaidi AA (2005) Indexing and production of virus-free chrysanthemums. Biol Plant 49:149–152

Singh AK, Hallan V, Ram R, Zaidi AA (2005) Variability in the coat protein of *lily symptomless virus* isolates infecting various lily species. Plant Pathol 54:621–624

Singh L, Hallan V, Jabeen N, Singh AK, Ram R, Martin DP, Zaidi AA (2007) Coat protein gene diversity among *Chrysanthemum virus B* isolates from India. Arch Virol 152:405–413

Singh AK, Mahinghara BK, Hallan V, Ram R, Zaidi AA (2008) Recombination and phylogeographical analysis of Lily symptomless virus. Virus Genes 6:421–427

Singh L, Hallan V, Zaidi AA (2011a) Intermolecular interactions of chrysanthemum virus B coat protein: implications for capsid assembly. Indian J Virol 22:111–116

Singh L, Hallan V, Ram R, Zaidi AA (2011b) Expression of recombinant Chrysanthemum virus B coat protein for raising polyclonal antisera. J Plant Biochem Biotechnol 20:96–101

Singh L, Hallan V, Martin DP, Ram R, Zaidi AA (2012) Genomic sequence analysis of four new chrysanthemum virus B isolates: evidence of RNA recombination. Arch Virol 157:531–537

Siva PV, Sai Gopal DVR, Sreenivasulu P, Nayudu MV (1990) Seed transmission of a carlavirus naturally infecting groundnut (Arachis hypogaea L.) in India. J Plant Dis Protect 97:548–550

Suhasini H, Satya VB, Rameshwar SR, Kamlesh K, Gaurav Z, Anil H (2010) Inference on coat protein evolution of *Lily Symptomless Carlavirus* in India and abroad based on motifs study and phylogenetic analysis. J Proteomics Bioinform 3:204–211

Tavasoli M, Shahraeen N, Ghorbani SH (2009) Serological and RT-PCR detection of cowpea mild mottle carlavirus infecting soybean. J Gen Mol Virol 1:7–11

Verma N, Sharma A, Ram R, Hallan V, Zaidi AA, Garg ID (2003) Detection, identification and incidence of Chrysanthemum B carlavirus in chrysanthemum in India. Crop Prot 22:425–429

Wetter C, Milne RG (1981) Carlaviruses. In: Kurstak E (ed) Plant virus infections: comparative diagnosis. Elsevier/North-Holland, Amsterdam, pp 695–730

Xiangyan Z, Zhongyang T, Haiping F, Ronghua Y, Li M, Jianhui J, Guoli S, Ruqin Y (2011) Microsatellites in different Potyvirus genomes: survey and analysis. Gene 488:52–56

Zaidi AA, Ram R, Zaidi SNH, Mukherjee D (1990) Diagnosis of viruses in some ornamental plants with special reference to serological methods: new developments. Indian Rev Life Sci 13:157–174

7 The Present Status of Carmoviruses Research in India

Nagamani Sandra

Abstract

The genus *Carmovirus* of the family *Tombusviridae* consists of the members having isometric virion particles with positive sense ssRNA genome of 4.0–4.8 kb encoding five proteins. In India, till date four carmoviruses were reported *viz*., blackgram mottle virus (BMoV), carnation mottle virus (CarMV), melon necrotic spot virus (MNSV) and soybean yellow mottle mosaic virus (SYMMV). BMoV and SYMMV are the legume infecting carmoviruses, which differs by serology and symptomatology. The Indian isolate of CarMV is a wide spread and distinct from other isolates. The Indian isoate of SYMMV is distinct from the SYMMV isolates occurring in the other countries being highly sap transmissible to guarbean, French bean, mungbean, soybean and urdbean with the distinct symptoms. Polyclonal antiserum developed against the recombinant coat protein of the Indian isolate of SYMMV can be utilized for successful detection of SYMMV in various plant samples. However, there is a great need to exploit these carmoviruses for further understanding the process of replication, gene expression, and exploiting them as gene expression vector for the expression of heterologous proteins in plant and as virus induced gene silencing vector for studying gene functions in legume crops. This chapter summarises the research work conducted on carmoviruses occurring in India.

Keywords

Carmoviruses • Blackgram mottle virus • Carnation mottle virus • Melon necrotic spot virus • Soybean yellow mottle mosaic virus

N. Sandra (✉)
Division of Seed Science and Technology, Indian Agricultural Research Institute (IARI), New Delhi 110012, India
e-mail: Nagamani.iari@gmail.com

B. Mandal et al. (eds.), *A Century of Plant Virology in India*,
https://doi.org/10.1007/978-981-10-5672-7_7

7.1 Introduction

Carmoviruses belongs to the family *Tombusviridae,* which is one of the largest family having genera including Aureusvirus, Avenavirus, Carmovirus, Dianthovirus, Necrovirus, Panicovirus and Tombusvirus with virions showing icosahedral symmetry of 30 nm; consists of 180 identical coat protein (CP) subunits of about 38–43 kDa. Among these viruses, significant degree of sequence similarity was observed with respect to replicase associated and CP genes. Mild or asymptomatic infections were the most common symptoms observed on relatively restricted natural host ranges. Carmoviruses contain a single-stranded (ss) positive sense RNA genome ranging in size from 4.0 to 4.8 kb. The genomic RNA has neither a 3′ poly (A) tail nor 5′ cap structure which is a most common feature for the members of genus *Carmovirus.* The genome consists of five definitive open reading frames (ORFs) from the 5′ to the 3′ end which encode proteins of about 28, 88, 8, 9 and 38 kDa respectively.

As carmoviruses possesses small genome, they are highly accessible to serve as an interesting model for understanding plant RNA virus genome structure, function and regulation. More than 30 carmoviruses have been reported from all over the world. The first carmovirus identified in India was blackgram mottle virus in 1974. Since then, three more carmovirus species *Carnation mottle virus*, *Melon necrotic spot virus* and *Soybean yellow mottle mosaic virus* were reported from India. Qu and Morris (2008) gave brief description about carmoviruses and Simon (2015) reviewed about the 3′ UTR region of carmoviruses. However, the research work conducted in India has not so far been reviewed. This chapter describes the up to date research work conducted on carmoviruses in India.

7.2 Blackgram Mottle Virus

Blackgram mottle virus (BMoV) responsible for mottling and stunting symptoms in blackgram was identified as a distinct virus based on biological and serological properties (Phatak 1974). An improved method for BMoV purification using magnesium-bentonite was described by Balasubrahmanyam et al. (1997a). BMoV was showed to be a seed borne virus (8% of seed transmission) with ssRNA, containing isometric particles with a diameter of 28 nm belongs to the genus Carmovirus, family *Tombusviridae*. BMoV had a 5×10^{-4}–1×10^{-5} dilution end point, a thermal inactivation point of 90–92 °C and longevity *in vitro* of 40–45 days (Phatak 1974). The host range study of BMoV showed that it was mainly restricted to *Leguminosae* members and induces the symptoms like mosaic, mottling and veinal necrosis in mungbean, blackgram and guarbean (Krishnareddy 1989).

Seed transmission rate of BMoV was also tested in different cultivars of blackgram which showed the highest seed transmission rate in cv. PLU-277 (15.9%), followed by cvs T-9 (11.8%), PLU-213 (7.0%) and UH-81-7 (1.3%). In addition the mean amount of the virus in various tissues was determined and found that highest in embryonic axis (48–1234 ng) followed by cotyledon (15–24 ng) and testa

(12–20 ng). Further small amount of the virus in embryonic axes was observed in the cultivars that resisted seed transmission (Varma et al. 1992).

In vitro translation of BMoV RNA in rabbit reticulocyte lysate resulted in production of five major virus specific polypeptides with molecular weight 90,000 (p90), 82,000 (p82), 42,000 (p42), 39,000 (p39) and 32,000 (p32) respectively. The polypeptide p39 was identified as a CP based on its elctrophoretic mobility and immune-precipitation with BMoV antisera (Balasubrahmanyam et al. 1997a, b). However, natural infection of BMoV in mungbean and other legumes is still not known in India. Further due to the lack of genome sequence information, the molecular characterization of BMoV is pending and therefore, it is considered as a tentative member of the genus *Carmovirus*.

7.3 Carnation Mottle Virus

Carnation mottle virus (CarMV) is the most important and wide spread virus affects carnation cut flower crop and causing severe economic losses to the farmers. Although CarMV infection leads to mild symptoms, it causes severe infection in all types of carnations. Carnation mottle disease caused by CarMV is characterized with chlorotic spots on young apical leaves, followed by mosaic chlorosis, mottling and streaks of yellowish-white colour on infected leaves. The infected plant shows stunted growth and bushy appearance. In addition, CarMV infection results in the poor quality of cut-flower in terms of size, split calyces and reduced vigour, and also results in lesser yields by reducing number of lateral shoots, total number of flowers and fresh weight. CarMV infection not only affects the flower quality and shelf life, but also weakens the plant making it susceptible to infection by other pathogens. It can be easily transmitted by contact and cropping operations (Singh et al. 2005). Singh and Singh (1989) reported that CarMV mainly transmitted through aphid vector *Aphis gossypii* and retained within the vector for up to 8 days, indicating the persistent nature of the virus.

Bansal and Singh (1980) reported that *Chenopodium amaranticolor* and *Chenopodium quinoa* were the local lesion hosts of CarMV. The experimental host range of CarMV includes *Catharanthus roseus*, *C. amaranticolor*, *C. quinoa*, *Cucumis sativus*, *Gomphrena globosa*, *Lycopersicon esculentum*, *Medicago sativa*, *Nicotiana clevelandii* and *Saponaria vaccaria* (Singh et al. 2005). The presence of CarMV in different geographical regions was confirmed through screening of 93 carnation cultivars by DAS-ELISA using polyclonal IgG. Interestingly 90% of the carnation cultivars showed positive result indicating the widespread nature of CarMV in India (Singh et al. 2005).

Sequencing and sequence analysis of CarMV movement and coat protein showed that Indian isolate was distinct and belonged to a new Group PN ($p^{164}N^{331}$) (Singh et al. 2005). The complete genome of CarMV has also been sequenced, which consisted of 4005 bp in length with five ORFs. Multiplex PCR was mainly employed to detect the CarMV from annual and perennial carnations through the amplification of CP and MP. Comparative sequence analysis of CarMV isolates from annual and perennial carnations revealed the 95–99% sequence identity for both CP and MP (Raikhy et al. 2006).

7.4 Melon Necrotic Spot Virus

A new carmovirus infecting *Cucumis melo* was identified as melon necrotic spot virus (MNSV-Hyd) in Hyderabad during 2009. The complete genome of MNSV-Hyd has been sequenced which consisted of 4274 nucleotides. The viral genome encodes five ORFS: ORF1 (p28) and ORF2 (p89) encode putative RNA dependent RNA polymerase; ORF3 (p7A) and ORF4 (p7B) encode movement protein 1 and 2 required for *in planta* movement of the virus, ORF5 (p42) encodes a coat protein (Accession No: JX879088).

7.5 Soybean Yellow Mottle Mosaic Virus

Soybean yellow mottle mosaic virus (SYMMV-Mb) has been recently reported in India from Mungbean exhibiting mosaic, mild mottling and puckering symptoms (Fig. 7.1A). SYMMV-Mb can be easily sap transmitted to French bean cv. Pusa Parvati for maintenance of pure culture and for further studies of this virus. Purified virus preparations revealed the ultraviolet light absorption spectrum with $A_{260/280}$ of about 1.75; 39 kDa protein band in SDS-PAGE analysis; single band of ~4 kb genome size on 1% formaldehyde agarose gel electrophoresis. Host range study revealed that SYMMV was mainly restricted to only *Leguminosae* members but not to *Amaranthaceae*, *Cucurbitaceae* and *Solanaceae* and induced various symptoms *viz*., veinal mottling, mild mottling, chlorotic blotching, local and systemic necrosis in soybean (*Glycine max*), mungbean (*Vigna radiata*), blackgram (*Vigna mungo*), French bean (*Phaseolus vulgaris*) and guar bean (*Cyamopsis tetragonoloba*) respectively (Fig. 7.1B-E). The local symptoms were observed at 7–10 days post inoculation (dpi) where as systemic symptoms were observed at 15–22 dpi. The progeny virions were also stable and mechanical sap inoculation resulted in 100% infection. However, the symptomatology of the Indian isolate of SYMMV-Mb was distinct from South Korean isolate as the later did not induce visible symptoms in any of the legumes other than soybean. Polyclonal antiserum was developed by over expression of coat protein gene as 39 kDa protein in *E. coli* and successfully utilized for detection of SYMMV. Samples tested for understanding the prevalence of SYMMV-Mb in India revealed that SYMMV-Mb was mainly confined to Northern India. Serological studies showed that SYMMV-Mb was serologically related to BMoV but not to CarMV (Sandra et al. 2015).

The CP region of SYMMV-Mb consisting of 1065 bp, showed maximum sequence identity with SYMMV Korean and USA isolates followed by cowpea mottle virus (CPMoV). The phylogenetic analysis based on CP with other carmoviruses showed that SYMMV Indian isolate formed a separate cluster with SYMMV Korean isolate and CPMoV (Fig. 7.2). The complete genome of SYMMV-Mb

Fig. 7.1 Field symptoms of soybean yellow mottle mosaic virus on mungbean (**a**). Symptoms in leguminous plant species followed by mechanical sap inoculation of SYMMV-Mb. Systemic chlorotic blotches on frenchbean cv. Pusa Parvati (**b**). Systemic veinal necrosis on guarbean (**c**). Pinpoint necrotic spots and veinal necrosis on mungbean cv. Pusa Vishal (**d**). Mottling and puckering symptoms on blackgram cv. Bharabanki local (**e**)

isolate was sequenced which consisted of 3974 nucleotides lacking the 3′ poly (A) tail and 5′ cap structure. The viral genome encodes six ORFs: ORF1 (p25) and ORF2 (p83) encode replication associated protein; ORF3 (p8) and ORF4 (p10) encode the double gene block proteins (DGPs) required for virus movement, ORF5 (p39) encodes a coat protein (CP) and ORF6 is of unknown function. An infectious clone of SYMMV-Mb was developed, which successfully produced typical disease symptoms in various plant species similar to that caused by the native virus (Sandra et al. 2017). The infectious clone was utilized for designing a vector for expression of heterologous protein in plant system.

7.6 Concluding Remarks

Identification of new or emerging carmoviruses before they become economically important provides an opportunity to devise control measures in advance. However, the potential economic consequences of these carmoviruses are yet to realise in India, because the research pertaining to carmoviruses was confined to identification, sequencing and host responses upon mechanical inoculation. Seed transmission and infectivity studies are necessary to understand the impact of carmoviruses in India. BMoV, although, was reported as a different carmovirus, it's identity as a distinct

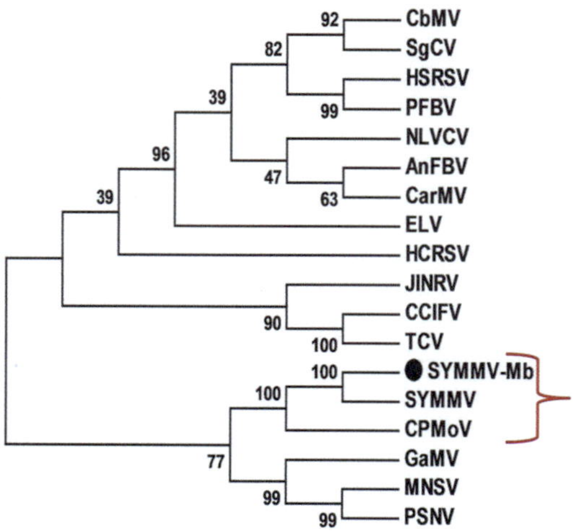

Fig. 7.2 Phylogenetic relationship of SYMMV-Mb isolate with other carmovirus species based on amino acid sequence of CP. The NJ tree was constructed in MEGA 6 programme. AnFBV angelonia flower break virus (NC_007733), CbMV calibrachoa mottle virus isolate California (GQ244431), CCIFV cardamine chlorotic fleck virus (NC_001600), CarMV carnation mottle virus (NC_001265), CPMoV cowpea mottle virus (NC_003535), ELV elderberry latent virus isolate (AY038066), GaMV galinsoga mosaic virus (Y13463), HCRSV hibiscus chlorotic ringspot virus (X86448), HSRSV honeysuckle ringspot virus (NC_014967), JINRV Japanese iris necrotic ring virus (NC_002187), MNSV melon necrotic spot virus (NC_001504), NLVCV nootka lupine vein-clearing virus (NC_009017), PSNV pea stem necrosis virus (NC_004995), PFBV pelargonium flower break carmovirus (AJ514833), SgCV saguaro cactus virus (NC_001780), SYMMV SYMMV (FJ457015), SYMMVMb14 isolate of the present study (KF619242), TCV turnip crinkle virus (M22445)

virus species is not yet clear. The recent studies (Sandra et al. 2015, 2017) showed that SYMMV infected blackgram and shared serological relationships with BMoV. It may be possible that BMoV is a variant of SYMMV, however further studies are necessary to establish the correct identity of BMoV. Although, MNSV has been isolated from melon in southern India, its prevalence is not yet clear. Infectious clone of carmovirus is useful to assess the extent of yield loss due to infection of individual carmovirus as well as mixed infection with the other plant viruses. So far infectious clone has been developed for only one Indian isolate of carmovirus, SYMMV. Further, the members of the genus *Carmovirus* serve as an interesting models to deeply understand RNA plant virus genome structure, function and regulation due to their simplest genome complexity. So, there is a need to exploit the carmoviruses as gene expression and silencing vectors to understand these complex processes.

References

Balasubrahmanyam A, Kapoor HC, Varma A (1997a) Purification of blackgram mottle virus (BMoV) using magnesium-bentonite. Phytochemistry 44:1237–1240

Balasubrahmanyam A, Kapoor HC, Varma A (1997b) Blackgram mottle virus RNA and its in vitro translation. Indian J Exp Biol 35(1):37–41

Bansal PK, Singh A (1980) *Chenopodium quinoa* and *Chenopodium amaranticolor,* two local lesion hosts for carnation mottle virus. Sci Hortic 46:435–436

Krishnareddy M (1989) Studies on yellow mosaic and leaf crinkle diseases of blackgram. PhD Thesis. IARI, New Delhi, 263 pp

Phatak HC (1974) Seed-borne plant viruses-Identification and diagnosis in seed health testing. Seed Sci Technol 2:3–155

Qu F, Morris TJ (2008) Desk encyclopedia of plant and fungal virology. In: Brian W, Regenmortal MHVV (eds). Elsevier, Academic, San Diego

Raikhy G, Hallan V, Kulshrestha S, Ram R, Zaidi AA (2006) Multiplex PCR and genome analysis of Carnation mottle virus Indian isolate. Curr Sci 90(1):74–82

Sandra N, Kumar A, Sharma P, Kapoor R, Jain RK, Mandal B (2015) Diagnosis of a new variant of soybean yellow mottle mosaic virus with extended host-range in India. Virus Dis 26(4):304–314

Sandra N, Abdul Kader Jailani A, Jain RK, Mandal B (2017) Genome characterization, infectivity assays of *in vitro* and *in vivo* infectious transcripts of soybean yellow mottle mosaic virus from India reveals a novel short mild genotype. Virus Res 232:96–105

Simon AE (2015) 3′ UTR of carmoviruses. Virus Res 206:27–36

Singh R, Singh A (1989) *Aphis gossypii* Glover, an insect vector of carnation (*Dianthus caryophyllus* L.) mottle virus- a new record. Curr Sci 58:931

Singh HP, Hallan V, Raikhy G, Kulshreshtha S, Sharma ML, Ram R, Garg ID, Zaidi AA (2005) Characterization of an Indian isolate of Carnation mottle virus infecting carnations. Curr Sci 88(4):594–601

Varma A, Krishnareddy M, Malathi VG (1992) Influence of the amount of blackgram mottle virus in different tissues on transmission through the seeds of *Vigna mungo*. Plant Pathol 41(3):274–281

Closterovirus in India: Distribution, Genomics and Genetic Diversity of Citrus Tristeza Virus

8

K.K. Biswas, Supratik Palchoudhury, and D.K. Ghosh

Abstract

Only one closterovirus species (Family: *Closteroviridae*), *Citrus tristeza virus* (CTV) is known to occur in India. CTV is one of the most important plant viruses in India and extensive studies have been conducted over the last 60 years. The failure of Malta sweet orange on sour orange root stocks provided the evidence of tristeza disease in India. CTV infects nearly all the citrus species and citrus relatives and hybrids showing variables biological symptoms. Most citrus species and cultivars are susceptible to infection but some are tolerant inducing no obvious symptoms. Citrus orchards in Northeast India are severely affected by citrus decline, and several orchards in this region have been wiped out and many Sweet orange orchards in South and Central India are facing problem of decline. *Toxoptera citricida* is an efficient vector for the local natural spread of CTV in India. Stem pitting symptoms caused by CTV are not common in India. The Indian CTV isolates are genetically diverse and seven to ten genetic variants have been recognized in India. The complete genome (19,253 nt) of a mandarin decline inducing CTV strain, Kpg3 from the Darjeeling hills was sequenced. This chapter presents the work conducted on CTV in India.

Keywords

Closterovirus in India • Citrus tristeza virus

K.K. Biswas (✉) • S. Palchoudhury
Advanced Centre for Plant Virology, Division of Plant Pathology, ICAR-Indian Agricultural Research Institute, New Delhi 110012, India
e-mail: drkkbiswas@yahoo.co.in

D.K. Ghosh
ICAR-Central Citrus Research Institute, Amravati road, Nagpur 440010, Maharashtra, India

B. Mandal et al. (eds.), *A Century of Plant Virology in India*,
https://doi.org/10.1007/978-981-10-5672-7_8

8.1 Introduction

The family *Closteroviridae* consists of the three virus genera, *Closterovirus* (Type species: *Beet yellows virus*), *Ampelovirus* (Type species: *Grapevine leafroll-associated virus 3*) and *Crinivirus* (Type species: *Lettuce infectious yellows virus*). The virus particles of the family *Closteroviridae* are helical, flexuous and filamentous having a size ranging from 650 to 2000 nm in length with 11–12 nm in width. The pitch of the primary helix of the virion is 3.4–3.8 nm, containing ~10 protein subunits per turn of the helix with a central hole of 3–4 nm. The virions have two coat proteins (CPs) as a major CP and a minor CP (CPm). The duplication of the CP gene is the only example in the viruses in the family *Closteroviridae* among plant viruses with elongated particles. The CPm encapsidates the 5′-terminal 600–700 nt of the viral RNA that coats 75–100 nm of the virus particle resulting in formation a distinct structure, 'rattlesnake' in this family. The genome of the family *Closteroviridae* contain a single molecule of linear, +ve sense, single stranded RNA of 13–19 kb with 7–13 ORFs (Fig. 8.1). The genomic RNAs are 5–6% of the particle weight and the 5′end of the genomic RNA is likely to be capped. The 3′ end do not have poly (A) tract but contains tRNA-like structure. However, the 3′ end has several hairpin structures, and a putative pseudoknot essential for replication. The structural proteins of most of the closterovirids consist of a major CP (22–46 kDa) and CPm (22–46 kDa). A group of ampeloviruses contain a small genome of 13 kb in length that lacks a true CPm. Most of the virus members of this family require CPm for the assembly of the 5′-extremity of the virions.

The family *Closteroviriae* contains the virus members having the largest genomes among the plant viruses. Because of sequence duplication and acquisition of non-viral coding sequences like protease, HSP70, protein, and genetic recombination, the viruses have largest genome. Recombination is the important cause to make differences in genome organization between the genera and members of the genus. However, the complex ORF-1a to ORF-1b invariably encodes the replication-related proteins, Mtr, Hel, and RdRp. In downstream ORFs at 5–3′ direction, a

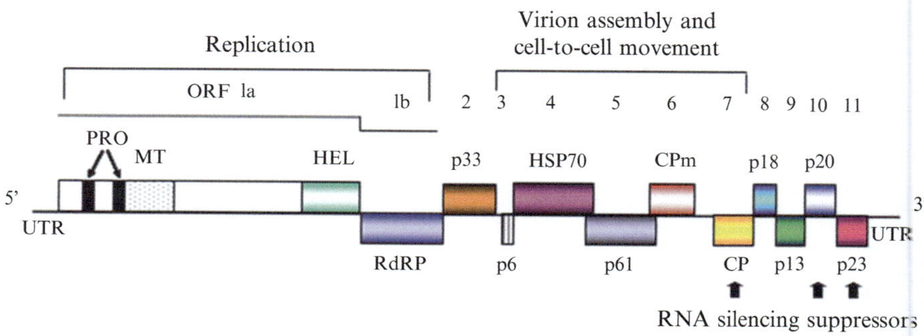

Fig. 8.1 Genome organization of citrus tristeza virus isolate Kpg3

conserved five-gene module is formed with (i) a 6 K small hydrophobic protein, (ii) the HSP70h (for a homolog of the cellular HSP70 heat shock protein), (iii) the 60 kDa protein, (iv) the CP and (v) CPm.

The HSP70h and the 60 kDa proteins are integral virion components present most of the closteroviruses. The HSP70h in the viruses of the family may function for (i) mediation of cell-to-cell movement through plasmodesmata, (ii) involvement in the assembly of multisubunit complexes for genome replication and/or sgRNAs synthesis, and (iii) assembly of virus particles. The 60 kDa proteins are required for incorporation of both HSP70h and CPm to virion heads. The duplication of the capsid protein gene seems to be the only example among viruses with elongated particles. In general, major CP and their homologs CPm show a significant degree of sequence conservation. But ampeloviruses generally do not possess CPm. The genome expression strategy in the closteroviruses of this family is based on (i) proteolytic processing of the polyprotein encoded by ORF1a, (ii) +1 ribosomal frame shift for the expression of the RdRp domain encoded by ORF-1b and this mechanism is not found in other positive sense RNA plant viruses, (iii) expression of the downstream ORFs via the formation of a nested set of 3′ co-terminal sgRNAs. The dsRNA patterns are very complex and variable among species. Different numbers and sizes of the ORFs are present in individual genomes. In some cases in effective RNAs are existed. Replication of the viruses occurs in the cytoplasm, possibly in association with endoplasmic reticulum-derived membranous vesicles and vesiculated mitochondria. The closterovirruses have probably evolved from the smaller filamentous virus. Under the pressure of further modular evolutionary events, the duplication of the CP gene, acquisition of diverse suppressors of RNA silencing and additional genes are acquired in the closteroviruses.

Citrus tristeza virus (CTV) is the only closterovirus known in India. Another closterovirus species *Grapevine leafroll associated virus 2*, which known to infect grapevine has been examined in Himachal Pradesh, India, however, it could not be detected (Kumar et al. 2013). In this chapter the research work on the characterization of closteroviruses carried out in India is summarised.

8.2 CTV in India

CTV, an aphid-transmitted closterovirus, causes devastating 'Tristeza' or decline in mostly all the economically grown citrus species in the world. It causes phenomenal economic damage to citrus production globally destroying about 100 million citrus trees over the last 70 years (Moreno et al. 2008). As citrus is cultivated in diverse ecological conditions, it is exposed to several CTV strains/variants resulting in diverse disease syndromes; decline of citrus species grafted on sour orange, yellowing and growth cessation of many citrus species and stunting and stem pitting with poor yield and quality in citrus species regardless of kind of citrus rootstock used. The serious epidemic causing the decline of large new plantings of sweet orange trees on sour orange rootstock was recorded for the first time in 1930 in Argentina that destroyed about 30 million trees in this country. Similar situations were also reported from Brazil (1937), Venezuela (1980), South Africa (1974) Ghana (1938), Spain (1960), California (1939) and Florida (1951) causing death of millions of citrus trees.

CTV, a phloem-limited, flexuous filamentous virus with particle size of 2000 × 11 nm belongs to the genus Closterovirus under the family Closteroviridae. It is predominantly transmitted by brown citrus aphid (BrCA; *Toxoptera citricida*) in a semipersistent manner (Bar-Joseph et al. 1989). CTV genome is positive sense, ssRNA of 19.3 kb in length and contains 12 ORFs; ORF1a and b and ORFs 2–11 potentially encoding at least 19 putative proteins (Karasev et al. 1995). ORFs 1a and 1b plus the nontranslated termini are all that is required for replication in protoplasts. Ten 3′ ORFs, are expressed by 3′-coterminal subgenomic mRNAs (Karasev et al. 1995) encode proteins, p33, p6, p65, p61, p27 (CPm), p25 (CP), p18, p13, p20 and p23, those are expressed via 3′ co-terminal sub genomic RNAs (sgRNA) (Satyanarayana et al. 2000). These proteins involve in movement and interactions with its insect vectors and hosts. CTV contains two capsid proteins (CP): a major CP of 25 KDa (p25) that encapsidates about 95% and a minor CP (CPm) of 27 KDa (p27) that encapsidates about 5% of total particles (Febres et al. 1996). In addition, the p65 (plant heat shock protein hsp70 homolog) and p61 are required for efficient virion assembly (Satyanarayana et al. 2000). CTV has three RNA silencing suppressors (RSSs), proteins CP (p25), p20 and p23 (Lu et al. 2004). In addition to RSSs activity p23 regulates asymmetrical accumulation of the positive and negative strands RNA during replication (Satyanarayana et al. 2000). CTV requires two proteins p20 and p6 (transmembrane protein) for translocation/systemic infection in citrus host (Tatineni et al. 2011). The p6 homologue in BYV has been shown to be a movement protein and required for systemic invasion of host plants (Peremyslov et al. 2004). Recently it was reported that p33 protein is responsible for superinfection exclusion between genetically related CTV isolates (Folimonova 2012). CTV has three genes (p33, p18, and p13) that are not necessary for infection of most of its hosts, but are needed in different combinations for infection of certain citrus species. These genes apparently were acquired by the virus to extend its host range (Dawsona and Folimonova 2013).

The genome expression strategy in closteroviruses is based on (i) proteolytic processing of the polyprotein by ORF1a, (ii) +1 ribosomal frame shift for the expression of the RdRp domain encoded by ORF1b, (iii) expression of the downstream ORFs via the formation of a nested set of 3′ co-terminal sgRNAs. Replication of the viruses occurs in the cytoplasm, possibly in association with endoplasmic reticulum-derived membranous vesicles and vesiculated mitochondria. From an evolutionary point of view, the closteroviruses might have evolved from a smaller filamentous virus.

Genetic diversity in CTV in different citrus growing regions of the world has been reported earlier. Analysis with several complete genomes of CTV isolates shows extensive sequence variation in CTV genome and determines at least seven CTV genotypes; T36, T3, VT, T30, B165, HA16-5 and RB (resistance breaking) occurring in citrus growing countries in the world (Melzer et al. 2010; Biswas et al.

2012a; Harper 2013). Of them, genotypes VT is biologically severe, T30 as mild and T36 as intermediate strains of CTV (Anonymous 2012).

CTV infect all the citrus species, and citrus relatives and hybrids and a non-rutaceous host *Passiflora* sp. (Moreno et al. 2008). The CTV epidemics are experienced in countries where sour orange is used as rootstock and BrCA vector, *T. citricida* is common and active. This virus is transmitted by aphids in a semi-persistent manner. CTV can be transmitted to other hosts by inoculation of sap, although the sap transmission is very difficult. It can be transmitted through dodder (*Cuscuta sp.*). None of the closteroviruses is transmitted through seeds. Virions are usually found in the phloem tissues. The virus is localized only in phloem tissue and sieve tubes, occasionally in the mesophyll and epidermis of infected trees. The necrosis and degeneration below the bud union result in the decline of trees (Schneider 1959). The virus is transported through sieve elements and occasionally enters an adjacent companion or phloem parenchyma cell where virus replication occurs; in some plants this is followed by cell-to-cell movement into only a small cluster of adjacent cells, while in others there is no cell-to-cell movement.

8.3 History of CTV in India

First indication in occurrence of *Tristeza* or citrus decline disease in Indian subcontinent was given by Brown in 1920, when he observed the failure of Malta sweet orange (*C. sinensis*) on sour orange (*C. aurantium*) root stock in Peshawar (now in Pakistan). After that, decline of citrus orchard was reported from Bombay State (Nagpal 1959; Capoor 1961, 1963; Vasudeva and Capoor 1968) and subsequently from North India (Nariani et al. 1965). Tristeza was shown to be widely spread in almost all the citrus growing regions of India (Raychaudhuri et al. 1977). Transmission of citrus tristeza disease through *Toxoptera citricidus* was demonstrated by Vasudeva and Capoor (1968). It infects all the commercially grown citrus species, cultivars and hybrids of mandarin (*Citrus reticulata*), sweet orange (*C. sinensis*), acid lime (*C. aurantifolia*), sweet lime/lemon (*C. limettoides/limon*) grown in India that has killed more than one million citrus trees (Ahlawat 1997; Biswas 2008). CTV was detected in infected citrus trees for the first time India using ELISA using CTV specific antisera by Chakroborty et al. (1992). Using monoclonal antibodies like MCA 13 and MABs3DF1, the CTV isolate of India has been reported to be different from CTV isolate of USA (Chakroborty et al. 1992). The CTV particle measuring about 2000 × 11 nm in size was observed under electron microscopy for the first time in India by Ahlawat et al. (1992). Cloning and sequencing of CP gene of four South Indian isolates were reported by Manjunath et al. (1993). It was reported that the citrus tristeza disease destroyed about a million of citrus trees in India (Ahlawat 1997).

8.4 Incidence of CTV

Citrus is cultivated in all the four geographical zones of India; Northeast, Northwest, Central and South India and CTV occurs in nearly all the commercial citrus species in India (Ahlawat 1997; Biswas 2008, 2010). CTV is reported to be a century old problem in India but it was un-recognized earlier (Ahlawat 1997). Northeast India is considered to be one of the most important centres of origin of different citrus species (Ghosh 2007) and CTV is an important factor to cause citrus decline in this region (Ahlawat 1997; Bhagabati et al. 1989; Borah et al. 2012). In the Northeast India, CTV is a major problem as occurence of the efficient aphid vector; BrCA is very common. Mandarin orchards in the Darjeeling hills are severely being affected, many of them are been wiped out, due to severe infection of CTV that causes huge economic losses in the citrus industry in this region (Ahlawat 1997; Biswas 2008). Assam, and Meghalaya of Northeast India produce important citrus fruits like Khasi mandarin (*C. reliculata*), Kagzi lime and Assam lemon (*C. lemon*) and CTV was reported to be one of the major factors to cause citrus decline in these Northeast states of India (Bhagabati et al. 1989; Chakroborty et al. 1992).

In the South and Central zones of India, CTV is a chronic problem that occurs in mixed infections with huanglongbing (citrus greening disease) (Ahlawat 1997; Ghosh et al. 2003, 2009a). Tirupati region of South India produces many economically important citrus fruit like, sweet orange, sweet lime (*C. limettoides*) and Kagzilime (*C. aurantifolia*) and association of CTV for causing decline of citrus trees has been reported in this region (Ahlawat 1997; Tarafdar et al. 2013). Recently, based on field survey and detection by ELISA and PCR, overall disease incidence caused by CTV in India has been reported; 26.3% in Central India (Maharastra), 47.1–56.0% in NE India (Assam, Meghalaya, Sikkim and the Darjeeling hills), 36–50% in South India (Andhra Pradesh and Karnataka) and 16–60% in North-Northwest India (Uttarakhand, Delhi, Punjab, Rajasthan) (Biswas et al. 2014a, b).

8.5 Transmission of CTV

Dispersal of CTV is taken place through virus infected planting materials like buds or grafted- or seedling-plants those are responsible for introduction of this virus in new growing areas. Insect vector transmission, subsequently, is important for local spread in many parts of India, as presence of the efficient aphid vector BrCA, is common in most of the citrus growing areas in India, particularly in Northeast India (Ahlawat 1997; Biswas 2008). Transmission studies of citrus decline disease in India began in early sixties of nineteenth century. It was demonstrated that BrCA *(T. citricidus)* is one of the important vectors to transmit the citrus decline disease in citrus trees (Vasudeva et al. 1959; Raychaudhuri et al. 1977). The transmission of this disease through other aphid vectors, *Aphis gossypii* and *Myzus persicae* has also been reported by Verma et al. (1965), and through *A. craccivora* and *Dactynotus jaccae* by Verma et al. (1965). The virus was reported to be non-persistently transmitted from citrus to citrus. Like other citrus growing counties of the world,

T. citiricidus is the most efficient vector of CTV in India (Verma et al. 1965; Capoor and Rao 1967; Biswas 2008). It has also been reported that single BrCA can efficiently transmit multiple genotypes of CTV resulting in changed population dynamics and multiple infection in infected trees (Biswas et al. 2004).

8.6 Symptoms, Host Range, Biological Indexing and Host Resistance

CTV infects nearly all the citrus species and citrus relatives and hybrids in India. Mexican lime/Kagzi lime *(C. aurantifolia)* is commonly used as an indicator host, and trifoliate orange is used to filter tristeza from mixed infection of other citrus viruses (Tanaka et al. 1971). Some citrus relatives are highly resistant or immune to CTV infection (Garnsey et al. 1987). Trifoliate orange (*Poncirus trifoliata*) and its relative are immune to CTV isolates. Many CTV tolerant citrus root stock, *P. trifoliata* and its hybrid, citrumelo (Sweet orange X *P. trifoliata*) and Rangpur lime (*C. limolina*) have been identified (Moreno et al. 2008).

In India, most citrus species and cultivars are susceptible to infection but some are tolerant and do not show obvious symptoms of the disease. Rough lemon (*C. jambhiri*) and trifoliate hybrid (Rangpur lime X *P. trifoliata*) were immune when tested with the CTV isolates in the Darjeeling hills (Biswas 2008). Although, Rangpur lime has reported to be tolerant to CTV, it shows vein clearing and vein corking symptoms after inoculation of a particular CTV isolate of the Darjeeling hills. All the cultivated lime/lemons/Kagzilime, Assam lemon (*C. lemon*), Tahiti lime (*C. latifolia*) and Sweet lime (*C. limettioides*) are reported to be infectde by CTV (Biswas et al. 2012b). Acid limes (*C.aurantifolia*) are most susceptible and show vein clearing, stem pitting and stunting when infected by most isolates of CTV. Earlier it has been reported that mandarin is tolerant or resistant to CTV (Ahlawat 1997). But except Kinnow mandarin, all the cultivated mandarins in India: Darjeeling mandarin, Sikkim mandarin, Khasi mandarin, Nagpur mandarin, Mudkhed mandarin and Coorg mandarin are infected by CTV in field and as well as in greenhouse experimentally (Biswas et al. 2016). In Delhi condition acid lime cv Kagzi Kalan, pumello (*C. paradisi*), Kinnow mandarin were free from CTV. All the Sweet orange orchard in Delhi was highly susceptible to CTV and three CTV variants were reported from these Sweet orange orchard (Sharma et al. 2012).

Four CTV isolates of the Darjeeling hills, Kpg1, 2, 3 and 4 were tested for symptom production on different citrus species (Biswas 2010). All the CTV isolates infected Kagzi lime, Darjeeling mandarin and Mosambi sweet orange with a variety of biological reactions. The major symptoms were vein clearing, vein flecking, vein corking and stunting, depending on the hosts. On Mosambi sweet orange, all of the isolates induced stunting, but Kpg1 additionally induced vein corking. Rangpur lime was not infected by three of the four CTV isolates but Kpg2 induced vein clearing and veincorking in Rangpur lime, indicating that Rangpur lime may not be immune to all the CTV isolates. Rough lemon was infected by Kpg1 and Kpg2, inducing vein clearing and vein corking, but was not infected by Kpg3 and Kpg4;

showing rough lemon is not immune to all the CTV isolates. Although stem pitting caused by Indian isolates CTV-B and CTV-P has been reported previously on Mosambi sweet orange and Kagzi lime (Roy et al. 2005), but none of the Darjeeling CTV isolates studied did not produce stem pitting symptoms on this host (Biswas 2010).

The virus titer in CTV infected citrus hosts in different agroclimatic zones of India has been studied (Tarafdar et al. 2013). The Kagzilime and sweet orange plants contain higher CTV titer compared to mandarin plant, thus Kagzilime and Sweet orange plants are as ideal indicator hosts for CTV in India. The symptom severity in Kagzilime could not be correlated with the high virus titer value in CTV infected tree (Tarafdar et al. 2013). CTV is not evenly distributed in all the infected plant parts of different citrus species. The tender bark, petiole and mid rib of new leaves, and apical bud contain maximum virus titer. The barks obtained from 6 months to 1 year old twigs have higher CTV titer, upto eightfolds in infected Darjeeling mandarin, Sweet orange and Kagzilime trees. Old bark obtained from more than 2 years old twigs contains very less or no virus titre (Tarafdar et al. 2012). The petioles and mid ribs of all kinds of leaves contain higher amount of CTV titer (7 to 20-fold). The apical buds of all the citrus hosts tested contain huge virus titer ranging from 14 to 17-folds. CTV can persists up to 180 days in crude sap in Phosphate buffer (0.05 M; pH 7.0) kept at 4 °C and only up to 2–4 days at 25–32 °C. Accumulation of CTV titer varied in growing seasons; lower titer value, 3.6 to 4.4-folds in the month of February to May and higher titer value, sevenfolds in the month of September to October were estimated (Tarafdar et al. 2012).

8.7 Genetic Diversity, Distribution, Intra-farm Diversity of CTV

Extensive genetic diversity in CTV in citrus growing regions of the world including India has been reported time to time (Rubio et al. 2001; Biswas et al. 2012b; Tarafdar et al. 2013). Complete genome analysis identified seven distinct CTV genotypes internationally and they are VT, T36, T30, T3, B165, HA16-5 and RB (resistance breaking) (Roy and Brlansky 2010; Melzer et al. 2010; Biswas et al. 2012a; Harper 2013). Genetic recombination is a major phenomenon in the evolution of CTV variants (Martin et al. 2009; Biswas et al. 2012a, b; Tarafdar et al. 2013). Genetic diversity and factors responsible for the origin of CTV variants in India have been examined and reported by many different workers time to time (Roy et al. 2005; Ghosh et al. 2009a; Biswas 2010; Sharma et al. 2012; Biswas et al. 2012a; Singh et al. 2013; Tarafdar et al. 2013; Chander et al. 2015). Based on biological indexing, multiple molecular marker (MMM) analysis, heteroduplex mobility assay (HMA) and sequence analysis using CP gene and 5′ORF1a fragment of CTV genome, three CTV variants sharing 89–97% nt identity identified in the Darjeeling hills of the Northeastern Himalayan regions (Biswas 2010).

Near about 114 CTV isolates covering all the citrus growing-geographical zones of India (Table 8.1) have been characterized based on sequencing of complete CP

Table 8.1 Nucleotide sequence diversity and distribution of *Citrus tristeza virus variants* in citrus growing geographical regions of India

	Based on 5′ORF1a(L-Pro domain)			Based on CP gene (ORF7)		
Geographical region	No of isolate studied	% range of nt identity	CTV variant	No of isolate studied	% range of nt identity	CTV variant
Northeast India	63	83–98	Seven (VT, K5, T30, HA16–5, K10, T3 and AR1)	52	88–99	Five (I, III, V, VI and VII))
North India	14	85–98	Three (VT, D13 and T36)	15	88–99	Three (I, V and VII)
Central India	20	91–99	Two (VT and T30)	16	92–99	Five (I, IV, V, VI and VII)
South India	17	86–99	Eight (AR1, VT, BAN1, B165, HA16–5, K5, D13 and T36)	18	88–99	Five (I, III, V, VII and VIII)
Overall India	114	78–99	Ten	101	86–99	Seven
Worldwide	–	77–99	Ten	–	86–99	Eight

Using the data from Biswas et al. (2012a), Tarafdar et al. (2013) and Palchoudhury et al. (2017)

gene and 5′ORF1a (L-Pro domain) fragment gene (Ghosh et al. 2009a; Biswas 2010; Sharma et al. 2012; Biswas et al. 2012b; Singh et al. 2013; Tarafdar et al. 2013; Palchoudhury et al. 2017). Indian CTV isolates are genetically extremely diverse sharing 80–99% identity for 5′ORF1a and 89–99% identity for CP genes analyzed till today (Biswas et al. 2012b; Tarafdar et al. 2013; Palchoudhury et al. 2017). Homologous and non-homologous recombination may be frequent phenomena in the evolution of all of the known CTV genotypes. Identification of several potential recombination events among Indian CTV isolates has been determined (Biswas et al. 2012a; Sharma et al. 2012; Singh et al. 2013). Recombination phenomena among CTV isolates is responsible for evolution of extensive diversity of CTV in India.

Based on sequence analysis of 5’ORF1a gene fragment, occurrence of eight CTV variants in citrus growing areas of India has been reported earlier (Biswas et al. 2012b). Later on, 5′ORF1a of more number of CTV isolates of Northeast (Assam, Manipur, Meghalaya, the Darjeeling hills and Sikkim hills) and South India (Tirupati) were included and analysed and occurrence of overall ten CTV variants; Kpg3/VT, K5 (distinct), AR1(distinct), BAN-1 (distinct), D13(distinct), K38/T3, Kpg2/T30, AG-28/HA16-5, K10/B165 and BAN-2/T36/RB-G90; were determined in India (Fig. 8.2a) (Tarafdar et al. 2013; Palchoudhury et al. 2015, 2017). Based on analysis of CP genes, Biswas et al. (2012b) reported occurrence of six CTV variants in India. Recently, including and analyzing CP genes more numbers of CTV isolates collecting from Northeast and South India , overall seven CTV variants; B165/VT, P14/T36, Kpg2/RB-G90, TP6 (distinct), K10/T3, K5 (distinct) and Kpg3/HA16-5 are determined (Fig. 8.2b) (Tarafdar et al. 2013; Palchoudhury et al.

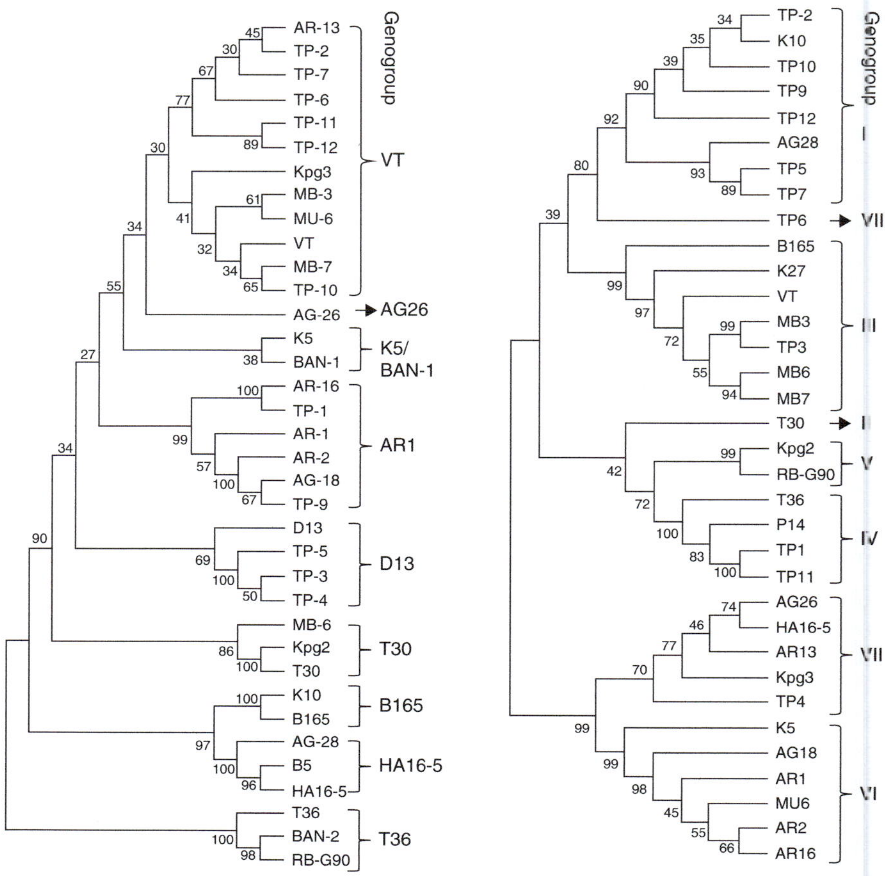

Fig. 8.2 Phylogenetic grouping of Indian CTV isolates based on 5′ORF1a region (**a**) and CP gene (**b**) using maximum likelihood method (Tarafdar et al. 2013). Significance of the nodes was estimated with 1000 bootstrap repetitions

2015, 2017). Therefore, it is concluded that seven to ten CTV variants are prevalent in citrus growing regions of India. Genetically most of the Indian isolates are similar to the Israel severe CTV isolate VT (Biswas et al. 2012a; Tarafdar et al. 2013; Palchoudhury et al. 2015, 2017). The distribution of CTV genotypes in citrus growing different geographical regions of India have also been determined and mentioned in Table 8.1.

Occurrence of several CTV variants in citrus growing areas of Northeastern Himalayan hill region of India has been reported. During the year of 2007 and 2008, analysis of CP genes and 5'ORF1a gene fragment of 26 CTV isolates from this region, four to five CTV variants were reported in this region (Biswas et al. 2012b). Recently, nine more number of CTV isolates of Mirik region of the Darjeeling hills and Rumtek area of the Sikkim hills were analysed and overall five CTV variants;

Kpg3/VT, K10/B165, Kpg2/T30, K5 (distinct) and K38/T3 type in the Darjeeling hills and its surrounding areas of Northeastern Himalayan hill regions of India are reported (Palchoudhury et al. 2017). CTV isolates of Manipur state of Northeast India have been characterized and occurrence of Kpg3/VT and T3 genotypes is common in Manipur state (Palchoudhury et al. 2015).

The p23 gene (ORF11, 630 nt;) of Indian CTV isolates (two from Arunachal Pradesh, two from Nagaland, three from Manipur, six from Assam, three from Delhi, two from Vidarbha, one from the Darjeeling hills and two from South India) was characterized. Extensive sequence diversity ranging from 88–100% nucleotide to 87–100% amino acid identity among the p23 gene were found and made five genetic clusters (Chander et al. 2015).

Occurrence of divergent CTV isolates in individual citrus farms in many citrus growing regions of India is reported. For instance, an sweet orange farm, IARI, New Delhi has shown at least three CTV variants (e.g., VT and D13 types) and another mandarin farm, IARI-Regional Station of the Darjeeling hills has at least three CTV genotypes (e.g., VT and B165). Similarly, Intra-farm diversity of CTV has been observed in many other individual farms in India suggesting that intra farm diversity of CTV might be common in India (Biswas et al. 2012a; Sharma et al. 2012).

8.8 Complete Genome Sequence of Indian Isolate of CTV

Based on biological property and host range study, a isolate Kpg3 was identified as a decline inducing CTV strain in the mandarin growing areas of the Darjeeling hills (Biswas et al. 2012a, b). The complete genome, 19,253 nt in length, of isolate Kpg3 was sequenced, analyzed and submitted in NCBI database as accession number HM 573451 (Biswas et al. 2012a). The Kpg3 genome contains all the 12 putative ORFs similar with the other CTV genomes reported earlier. In phylogenetic relationship, the Kpg3 is closely related to Israel severe CTV isolate VT and it is a recombinant strain (Biswas et al. 2012a). The isolates Kpg3 and VT, both might have originated from distantly related ancestors through a complex evolutionary pathway by multiple recombination events exchanging sequences between diverged CTV variants (Biswas et al. 2012a).

Further, 3′ half of the genome (8398 nt) comprising ten genes (ORFs 2–11) of four other CTV isolates, B5 of Bangalore (HQ912023), D1 of Delhi (HQ912022), G28 of Assam (KJ914661) and Kat1 of Vidarbha (KJ914662) of India were sequenced and compared with other Asian and internationally recognized CTV genotypes (Biswas et al. 2016). All the Asian isolates categorized into six genogroups, whereas the Indian isolates fell into four, and other Asian isolates into three genogroups. Indian isolates B5, D1, Kat1 and Kpg3 grouped together (Kpg3Gr) along with Florida isolate T3. However, the isolate B5 was placed distantly from other members of Kpg3Gr. Thus isolate B5 might be a new isolate. The isolate G28 was found to be distinct lineage.

8.9 Cross Protection

Cross protection is the ability of mild strains or isolates to protect the severe or more virulent strains or isolates of the same virus. The phenomenon of cross protection and use of mild cross protecting strain (MCPS) have been known for a long time for potential management of severe viruses. Cross protection was applied in very few crops and not accepted widely due to lack of pure MCPS. However, this concept of cross protection has been successfully used in management of CTV in many countries like Brazil, Australia, Japan and South Africa (da Graça and van Vuuren 2010; Roistacher et al. 2010). The unique aspect of cross protection in CTV is that the mild strain is easily inoculated to the target plant by grafting scions collecting from mild strain infected mother plants.

Cross protection of CTV was initiated in the years of 1970s and many CTV strains were identified based on vector specificity (Capoor and Rao 1967; Capoor and Chakroborty 1980). The mild and severe CTV strains were identified and effort was made for cross protection of CTV in Tirupathi and Bangalore region (Balaraman and Ramakrishnan 1977). Unfortunately, the experiment was failed and it might be due to appearance of severe strains mixed with the mild strain. Recently, effort was made to identify pure MCPS of CTV with the help of *in silico* molecular-based codon biasness analysis using CP gene of CTV isolates from Northeast India and their biological evaluation (Biswas et al. 2016). Two Indian isolates Mnp1 from Manipur and MB3 from Meghalaya were identified mild CTV type. This finding will be effective for designing a proof concept of MCPS after biological evaluation challenging with severe strain (Biswas et al. 2016).

8.10 Conclusion Remarks

Citrus orchards of many citrus growing areas of India, particularly in Northeast India, are being wiped out due decline or slow death of the citrus trees caused by CTV. The genomics of CTV, genetic diversity, evolution of CTV complex population and geographical distribution of virus variants have clearly been understood in India that has helped in understanding of the disease epidemiology. Nucleotide sequence analysis and phylogenetic relationships of large number of CTV isolates determined occurrence of seven to ten CTV variants in India, some of them are new to India or new to the world. These studies would lead to develop an improved diagnostics targeting diiferent virus strains/variants using specific primers, and also to make molecular-based management strategy targeting conserved sequence of the virus through gene silencing. Recombination events, negative selection and gene flow play major role for evolution of in CTV variants.

In cross protection, super infection exclusion occurs between isolates of the same strain but not between isolates of different CTV strain. As it is known that severe CTV isolates of the virus frequently represent a mixture of different virus strains, for practical applications of cross protection in the field, broad-spectrum mild strains are needed against multiple CTV strains. The technologies for detecting

CTV in planting material by biological, serological and molecular methods have been perfected and are being utilized to detect virus at early stages. These techniques would help to develop a long term disease management. The distribution of the CTV variants in India has been studied. The genes of interest have been cloned and transgene constructs were made for transformation of citrus to develop virus resistant transgenic plants in India in near future.

The old citrus orchards, particularly in Northeast India, are severely infected by CTV. Thus, sanitation and replanting with virus free propagative materials are foremost needed to reduce the economic losses of citrus. The random distribution of CTV infected citrus planting materials as means of virus dissemination is common in India. Thus legislation/notification is needed from the Government sector to develop budwood certification programmes and provide disease-free budwood or seedlings to the growers. As the disease is horizontally spread through planting material and vertically through aphids, establishment of new orchards using disease free planting material and keeping the orchards free from aphids with regular inspection are essential to maintain the citrus industry viable and profitable. In this regard a strategy for production of CTV-free planting material for the Darjeeling hill developed earlier (Biswas et al. 2009) could be followed. Efforts have been made to produce disease free citrus planting materials through seed certification program and shoot tip grafting successfully practiced in new plantations in many citrus growing areas in India.

References

Ahlawat YS (1997) Viruses, greening bacterium and viroids associated with citrus (Citrus species) decline in India. Indian J Agr Sci 67:51–57

Anonymous (2012) In: King AMQ, Adams MJ, Carstens EB, Lefkowitz EJ (eds) Virus taxonomy: classification and nomenclature of viruses. Ninth Report of the International Committee on Taxonomy of Viruses. International Union of Microbiological Societies, Virology Division, Amsterdam, pp 871–879. 11 March

Balaraman K, Ramakrishan K (1977) Studies on strains and strain interaction in *Citrus tristeza virus*, Technical series bulletin. University of Agricultural Sciences, Bangalore, p 62

Bar-Joseph M, Marcus R, Lee RF, (1989) The continuous challenge of citrus tristeza virus control. Annu Rev Phytopathol 27(1):291–316

Bhagabati KN, Ahlawat YS, Chakroborty NK, Borthakur BC (1989) Distribution of greening, tristeza amd mosaic disease of citrus in North eastern staes of India. Indian Phytopath 42:552–555

Biswas KK (2008) Molecular diagnosis of *Citrus tristeza virus* in mandarin (*C. reticulata*) orchards of Darjeeling hills of West Bengal. Indian J Virol 19:26–31

Biswas KK (2010) Molecular characterization of Citrus tristeza virus isolates from the Northeastern Himalayan region of India. Arch Virol 155:959–963

Biswas KK, Manjunath KL, Marais LJ, Lee RF (2004) Single aphids transmit multiple genotypes of citrus tristeza virus, but often with changed population dynamics. Phytopathology 94:S8

Biswas KK, Tarafdar A, Jayakumar BK, Pun KB (2009) Strategy for the production of *Citrus tristeza virus*-free mandarin (*Citrus reticulata*) planting materials in the Darjeeling hills of India. Indian Phytopathol 62:376–380

Biswas KK, Tarafdar A, Diwedi S, Lee RF (2012a) Distribution, genetic diversity and recombination analysis of Citrus tristeza virus of India. Virus Genes 45:139–148

Biswas KK, Tarafdar A, Sharma SK (2012b) Complete genome of mandarin decline *Citrus tristeza virus* of Northeastern Himalayan hill region of India: comparative analyses determine recombinant. Arch Virol 157:579–583

Biswas KK, Godara S, Nayak D (2014a) Distribution of *Citrus tristeza virus* in the Darjeeling hills and their biological symptoms in mandarin orchards. Indan J Hort 71:408–411

Biswas KK, Tarafdar A, Sharma SK, Singh JK, Dwivedi S, Biswas K, Jayakumar BK (2014b) Current status of *Citrus tristeza virus* incidence and its spatial distribution in citrus growing geographical zones of India. Indian J Agric Sci 84:8–13

Biswas KK, Pal Choudhury S, Godara S (2016) Decline of mandarin orange caused by Citrus tristeza virus in Northeast India: conventional and biotechnological management approaches. J Agric Eng Food Technol 3(3):236–241

Borah M, Nath PD, Saikia AK (2012) Serological detection of Citrus tristeza virus affecting citrus tree species in Assam. Indian Phytopath 65:289–293

Brown WR (1920) The Orange: a trial of stock at Peshawar. In: Bulletin 93, Indian Agricultural Research Institute, Pusa, p 7

Capoor SP (1961) Kagzi lime: an indicator plant of citrus 'decline' virus in India. Indian Phytopathol 14:109–112

Capoor SP (1963) Decline of citrus trees in India. Bul Nat Inst Sci India 24:48–64

Capoor SP, Chakraborty NK (1980) Further studies of citrus seedling yellow yellows in Deccan trap country. Indian J Hort 37:97–100

Capoor SP, Rao DG (1967) Tristeza virus infection of Citrus in India. In: Proceedings of the international symposium on subtropical and tropical horticulture. Horticulture Society of India, Bangalore, pp 723–736

Chakroborty NK, Ahlawat YS, Varma A, Chandra KJ, Ramapandu S, Kapur SP (1992) Serological reactivity in *Citrus tristeza virus* strains in India. In: Proceedings of the 12th conference on IOCV, pp 108–112

Chander V, Godara S, Kumar A, Saha B, Biswas KK (2015) Nucleotide sequence of ORF 11-p23 suppressor gene of *Citrus tristeza virus* isolate of different citrus growing states of India and determination of extensive genetic diversity. J Mycopathol Res 53(1):25–29

da Graça JV, van Vuuren SP (2010) Managing *Citrus tristeza virus* losses using cross protection. In: Karasev AV, Hilf ME (eds) Citrus tristeza virus complex and tristeza diseases. APSPress, Eagan, pp 247–260

Dawsona WO, Folimonova SY (2013) Virus-based transient expression vectors for woody crops a new frontier for vector design and use. Annu Rev Phytopathol 51:321–337

Febres VJ, Ashoulin L, Mawasaki M, Frank A, Barjoseph M, Manjunath KL, Lee RF, Niblet CL (1996) The p27 protein is present at one end of citrus tristeza virus particles. Phytopathology 86:1331–1335

Folimonova SY (2012) Superinfection exclusion is an active virus-controlled function that requires a specific viral protein. J Virol 86:5554–5561

Garnsey SM, Civerolo EL, Gumf DJ, Yokomi RK, Lee RF (1987) Toward a standardized evolution of the biological properties of Citrus tristeza virus. Phytophylactica 19:151–157

Ghosh SP (2007) Citrus fruits. ICAR, New Delhi

Ghosh DK, Das AK, Singh S (2003) Individual and mixed infection of CTV and greening Liberobacter and their association with sweet orange decline in the state of Maharashtra. J Mycol Plant Pathol 33(1):69–72

Ghosh DK, Aglave B, Roy A, Ahlawat YS (2009a) Molecular cloning, sequencing and phylogenetic analysis of coat protein gene of a biologically distinct CTV isolate occurring in Central India. J Plant Biochem Biotechnol 18:105–108

Ghosh DK, Das AK, Shivankar VJ (2009b) Implementing citrus budwood certification program. Indian Farming 59(8):41–45

Harper SJ (2013) Citrus tristeza virus: evolution of complex and varied genotypic groups. Front Microbiol 4:1–18

Karasev AV, Boyko VP, Gowda S, Nikolaeva OV, Hilf ME, Koonin EV, Niblet CL, Cline K, Gumpf DJ, Lee RF (1995) Complete sequence of the citrus tristeza virus RNA genome. Virology 208:511–520

Kumar S, Singh L, Ferretti L, Barba M, Zaidi AA, Hallan V (2013) Evidence of Grapevine leafroll associated virus-1–3, Grapevine fleckvirus and Grapevine virus B occurring in Himachal Pradesh. India Indian J Virol 24:66–69

Lu R, Folimonov A, Shintaku M, Li WX, Falk BW, Dawson WO, Ding SW (2004) Three distinct suppressors of RNA silencing encoded by a 20-kb viral RNA genome. Proc Natl Acad Sci U S A 101:15742–15747

Manjunath KL, Pappu HR, Lee RF, Niblett CL, Civerolo E (1993) Studies on coat prtein genes of four Indiam isolates of Citrus tristeza closterovirus: cloning sequencing and expression. In: Proceedings of the 12th conference IOCV, Riverside, CA, pp 20–27

Martin S, Sambade A, Rubio L, Vives MC, Moya P, Guerri J, Elena SF, Moreno P (2009) Contribution of recombination and selection to molecular evolution of Citrus tristeza virus. J Gen Virol 90:527–1538

Melzer MJ, Borth WB, Sether DM, Ferreira S, Gonsalves D, Hu JS (2010) Genetic diversity and evidence for recent modular recombination in Hawaiian Citrus tristeza virus. Virus Genes 40:111–118

Moreno P, Ambros S, Albiach-Marti MR, Guerri J, Pena L (2008) Citrus tristeza virus: a pathogen that changed the course of the citrus industry. Mol Plant Pathol 9:251–268

Nagpal RL (1959) Tristeza found in Bombay state. Calif Citrograph 44:402–405

Nariani TK, Sahambi HS, Chona BL (1965) Occurrence of tristeza virus in citrus in Northern India. Indian Phytopath 18:220

Palchoudhury S, Sharma SK, Biswas MK, Biswas KK (2015) Diversified Citrus tristeza virus causing decline disease in Khasi mandarin in Manipur State of Northeast India. J Mycol Plant Pathol 45:317–323

Palchoudhury S, Ghimiray P, Biswas MK, Biswas KK (2017) Citrus tristeza virus variants and their distribution in mandarin orchards in Northeastern Himalayan hill region of India. Int J Curr Microbiol App Sci 6(5):1680–1690, https://doi.org/10.20546/ijcmas.2017.606.196

Peremyslov VV, Andreev IA, Prokhnevsky AI, Duncan GH, Taliansky ME, Dolja VV (2004) Complex molecular architecture of beet yellows virus particles. Proc Natl Acad Sci U S A 101, 5030l

Raychaudhuri SP, Nariani TK, Ahlawat YS (1977) Dieback of citrus in India. Proc Int Soc Citriculture 3:914–918

Roistacher CN, daGraça JV, Muller GW (2010) Cross protection against Citrustristezavirus-a review. In: Hilf ME, Timmer LW, Milne RG, da Graça JV (eds) Proceedings of conference of the international organization of Citrus Virology, 17th (Riverside, CA: IOCV), pp 1–27

Roy A, Brlansky RH (2010) Genome analysis of an orange stem pitting Citrus tristeza virus isolate reveals a novel recombinant genotype. Virus Res 151:118–130

Roy A, Manjunath KL, Brlansky RH (2005) Assessment of sequence diversity in the 5-terminal region of Citrus tristeza virus from India. Virus Res 113:132–142

Rubio L, Ayllon MA, Kong P, Fernandez A, Polek M, Guerri J, Moreno P, Falk BW (2001) Genetic variation of isolates from California and Spain: evidence for mixed infections and recombination. J Virol 75:8054–8062

Satyanarayana T, Gowda S, Mawassi M, Albiach-Marti MR, Ayllon MA, Robertson C, Garnsey SM, Dawson WO (2000) Closterovirus encoded HSP70 homolog and p61 in addition to both coat proteins function in efficient virion assembly. Virology 278:253–265

Satyanarayana T, Gowda S, Ayllon MA, Albiach-Marti MR, Rabindram R, Dawson WO (2002) The p23 protein of Citrus tristeza virus controls asymmetrical RNA accumulation. J Virol 76:473–483

Schneider H (1959) The anatomy of tristeza-virus-infected citrus. In: Wallace JM (ed) Citrus virus diseases. University of California, Division of Agriculture Sciences, Berkeley, pp 73–84

Sharma SK, Tarafdar A, Khatun D, Sumita K, Biswas KK (2012) Intra-farm diversity and evidence of genetic recombination of *Citrus tristeza virus* isolates in Delhi region of India. J Plant Biochem Biotechnol 21:38–43

Singh JK, Tarafdar A, Sharma SK, Biswas KK (2013) Evidence of recombinant *Citrus tristeza virus* isolate occurring in acid lime cv. pant lemon orchard in Uttarakhand Terai Region of Northern Himalaya in India. Indian J Virol 24:35–41

Tanaka H, Yamada S, Nakanishi (1971) Approach to eliminating tristeza virus from citrus trees by using trifoliate orange seedlings. Bull Hort Res Station Japan Sec 11:157–163

Tarafdar A, Ghosh PD, Biswas KK (2012) *In planta* distribution, accumulation, movement and persistence of *Citrus tristeza virus* in citrus host. Indian Phytopathol 65:184–188

Tarafdar A, Godara S, Dwivedi S, Biswas KK (2013) Characterization of *Citrus tristeza virus* and determination of genetic variability in North-east and South India. Indian Phytopathol 66:302–307

Tatineni S, Robertson C, Garnsey SM, Dawson WO (2011) A plant virus evolved by acquiring multiple nonconserved genes to extend its host range. Proc Natl Acad Sci U S A 108:17366–17371

Vasudeva RS, Capoor SP (1968) Citrus decline in Bombay state. FAO Plant Prot Bul 6:91

Vasudeva RS, Varma PM, Rao DG (1959) Transmission of Citrus decline virus by *Toxoptera citricudus* (Kirk) in India. Curr Sci 28:418–419

Verma PM, Rao DG, Capoor SP (1965) Transmission of tristeza virus by Aphis craccivora Koch and Dactynotus jaceae (L.) India J Entomol 27:67–71

The Progress of Research on Cucumoviruses in India

9

S.K. Raj, S. Kumar, K.K. Gautam, C. Kaur, A. Samad, M. Zaim, V. Hallan, and R. Singh

Abstract

The genus *Cucumovirus* of the family *Bromoviridae* contains only three virus species of which *Cucumber mosaic virus* (CMV) and *Tomato aspermy virus* (TAV) are known to occur in India. The research work on CMV and TAV causing numerous diseases in vegetables, pulses, ornamentals, medicinal and aromatic plants and weeds, which were reported in India during last 65 years (1951–2015) have been summarized in this chapter. Biological, biochemical, serological, and molecular characterization of CMV and TAV, the mode of spread of the diseases in nature through insect vectors, search of alternate hosts/reservoirs and diagnostics methods for sensitive detection of the virus/es at an early stage of infection in plants and in propagating materials are discussed.

Keywords

Cucumber mosaic virus • Important crops of India • Virus transmission • Particle morphology • Serological properties • Biochemical characterization and molecular identification

S.K. Raj (✉) • S. Kumar • K.K. Gautam • C. Kaur
Plant Molecular Virology Laboratory, CSIR-National Botanical Research Institute, Lucknow 226 001, India
e-mail: skraj2@rediffmail.com

A. Samad • M. Zaim
Crop Protection Division, CSIR-Central Institute of Medicinal and Aromatic Plants, Lucknow 226015, India

V. Hallan
CSIR-Institute of Himalayan Bioresource Technology, Post Box No. 6, Palampur 176061, Himachal Pradesh, India

R. Singh
Amity Institute of Biotechnology, Amity University, Lucknow campus, Gomti Nagar extension, Lucknow, Uttar Pradesh, India

B. Mandal et al. (eds.), *A Century of Plant Virology in India*,
https://doi.org/10.1007/978-981-10-5672-7_9

9.1 Introduction

Cucumber mosaic virus (CMV) has been considered as a very important plant virus due to its capability to infect and cause huge losses in a number of plants in India. CMV was first reported by Doolittle (1916) as a causal agent of mosaic disease in cucumber in (country name?). CMV is also reported to causes mosaic, fern leaf, systemic necrosis, blight, ringspot, stunting, infectious chlorosis, heart rot and dwarfing in cucumber and other cucurbits, tomato, melon, spinach, pepper, clover, lupins, Lucerne, soybean, banana and many other species grown all over the world (Palukaitis 1992; Pratap et al. 2008; Raj et al. 2008a). CMV belongs to the genus *Cucumovirus* of the family *Bromoviridae* which contains ~28 nm (in diameter) isometric particles. It is transmissible by aphids in a non-persistent manner, and through the seed in some host plants. It is also transmissible easily by mechanical inoculations.

CMV has a tripartite plus sense RNA genome consisting of four RNA species designated as RNA1, RNA2, RNA3 and a subgenomic RNA4 responsible for the expression of CP. All the four RNAs contain a 7-methyl guanosine cap at 5′ ends. For infectivity, only the three largest RNAs are required. Occasionally a fifth RNA, known as the satellite RNA, about 330–386 nucleotides in length may be present (Palukaitis and Garcia-Arenal 2004). RNA 1 encodes one open reading frame, whereas RNA 2 and 3 each encodes two open reading frames. RNA 1 encodes 1a protein while RNA 2 encodes 2a proteins which are essential for replication. RNA 2 also encodes for 2b protein that affects long distance virus movement and symptomatology of the virus. RNA 3 is dicistronic and encodes two proteins, the 5′ encodes for movement protein (MP) whereas 3′ encodes for coat protein (CP). MP is involved in cell-to-cell movement of the virus while role for CP has been demonstrated in virus encapsidation, systemic movement, host range determination and aphid transmission (Palukaitis and Garcia-Arenal 2004).

Based on serology, peptide mapping of the CP, nucleic acid hybridization, RFLP of RT-PCR amplicon and sequence identity, isolates of CMV are divided into subgroup I and II (Palukaitis 1992; Roossinck 2002). Subgroup I is further splitted into IA and IB based on 5′ non-translated region of RNA 3 and CP gene (Roossinck 2002). Most of the Indian CMV strains have been clustered in IB subgroup (Srivastava and Raj 2004).

The occurrence of CMV was reported first time on chilli in India by Bhargava (1951a, b). The existence of CMV has been reported in India in early years from 1951 to 1965. There were only four reports of natural occurrence of CMV on chilli (Bhargava 1951a), properties of four strains of CMV infecting French bean, sugar beet and *Phytolacca* sp. (Bhargava 1951b) and a mosaic disease of tomato (Das and Raychaudhary 1953; Joshi and Bhargava 1965) are available in literature. These studies were based on symptomatology and transmission of the causal pathogen by sap and by aphids; responses of virus inoculations on various host species and biophysical properties of the virus: dilution end point (DEP), thermal inactivation point (TIP) and longevity in vitro (LIV) of the sap of the infected plant. The similar work has also been reported on CMV from 1966 to 1980 by Indian workers (Anjaneyulu and Apparao 1967; Khatri and Chenulu 1967; Seth and Raychaudhuri 1973; Dhingra et al. 1979; Mali and Rajegore 1980). The detailed characterization of CMV based on host assay, electron microscopy, DEP/TIP/LIV, vector transmission, and serology has also been attempted by Chalam (1982) in India.

In 1983, the comparison of genome sequences of *Brome mosaic virus* (Murthy 1983), and virus particle stability and structural transition of *Tomato aspermy virus* (TAV) with CMV (Savithri et al. 1983) has been done for the first time in India. Further, a CMV isolated from chickpea was identified based on the sedimentation coefficient and a buoyant density of purified virus, and molecular weight of coat protein (25 kDa) by SDS-PAGE. PAGE of CMV nucleic acid revealed the presence of four RNA species (Chalam et al. 1986). The characterization and establishment of relationship of CMV infecting *Petunia* using ELISA (Srivastava et al. 1991), and identification of CMV infecting banana using immuno/nucleic acid probes (Srivastava et al. 1995) has been done for the first time in India. Haq et al. 1996 characterized a CMV-banana isolate at molecular level using CP gene sequence which shared ~90% identity with CMV subgroup I strains. The potential role of satellite RNA in lethal necrosis of tobacco was also depicted for the time in India by Raj et al. (2000).

Srivastava and Raj (2004) investigated the high sequence similarity in three Indian isolates of CMV (Datura, Amaranthus and Henbane) and suggested their common origin based on RT-PCR, RFLP and sequence analysis of CP gene and observed that Indian isolates into subgroup IB. Vishnoi et al. (2013) identified a CMV-banana isolate as a member of CMV subgroup IB by sequence analysis of three RNA genomes. A CMV strain infecting Gladiolus was also identified as a member of subgroup IA based on distinct phylogenetic relationships with Indian isolate of IB (Dubey et al. 2010). The identification of CMV subgroup II isolate causing severe mosaic in cucumber was reported by Kumari et al. (2013) based on its complete genome for the first time in India.

Tomato aspermy virus (TAV; synonym Chrysanthemum aspermy virus), is another species of the genus *Cucumovirus* of the family *Boromoviridae* which causes rolling and twisting of corolla and reduction in size of flowers (Prentice 1952); slight dwarfing of the blooms together with waving and curling of ray florets including severe colour break, dwarfing and distortion of flowers (Hollings 1955; Brierley 1955, 1958; Marani 1969; Chuyan and Krylov 1979; Gupta and Singh 1981) and chlorotic ring mosaic on leaves (Raj et al. 1991) in chrysanthemums (*Chrysanthemum morifolium*). It causes serious losses in chrysanthemum crops and the symptoms include severe "flower-break" or "colour-break" in flowers, distortion of flowers and dwarfing of the chrysanthemum plant. TAV is also transferred from plant to plant by aphids. The yield and flower quality of various cultivars of chrysanthemum has been affected frequently in India by infection of TAV (Gupta and Singh 1981; Raj et al. 1991, 2007a, b, c, 2009; Verma et al. 2007a, b). The natural occurrence of TAV on chrysanthemum has been reported earlier on the basis of biophysical properties (Sastry 1964), aphid transmission, Ouchterlony gel double diffusion tests (Gupta and Singh 1981), double antibody sandwiched-enzyme linked immunosorbent assay (DAS-ELISA; Raj et al. 1991) and reverse transcription-polymerase chain reaction (RT-PCR) of the capsid gene (Raj et al. 2007a, b, c; Verma et al. 2007a, b). Raj et al. (2009) reported biological studies, nucleotide sequence of the complete RNA 3 genome, sequence identities and phylogenetic relationships of three TAV isolates collected from three locations in India.

TAV contains a tripartite genome of messenger plus-sense single-stranded RNAs designated RNA 1, 2 and 3 (in decreasing molecular mass), which are encapsidated in 28 nm isometric particles. RNA 1 and 2 encode 1a and 2a proteins, respectively, responsible for virus replication (Palukaitis and Garicia-Arenal 2004). The overlapping small 2b protein expressed from subgenomic RNA 4A (Ding et al. 1994) influences virulence, and facilitates long distance movement and functions as a suppresser of plant–host mediated gene silencing (Ding et al. 1995, 1996). RNA 3 is dicistronic, and encodes the 3a movement protein (MP), responsible for cell–cell movement and a capsid protein (CP) that is expressed from subgenomic RNA 4 (Palukaitis et al. 1992), the structural protein. CP is multifunctional in addition to having a role in encapsidation; it affects virus movement in plants (Suzuki et al. 1991; Kaplan et al. 1998), aphid transmission, symptom expression and host range. Two novel subgenomic RNAs RNA 3b and RNA 5, putatively the derivatives of RNA 3, have also been recorded (Shi et al. 1997).

Furthermore, the natural occurrence of TAV has also been reported in gladiolus plants the basis of reverse transcription-polymerase chain reaction (RT-PCR) of the capsid gene (Raj et al. 2011). The sequence analysis of cloned PCR product revealed the presence of 657 nucleotide-containing single open reading frames of the coat protein gene of TAV coding for 218 amino acids. The analysis of sequence data showed 98–99% identities and close phylogenetic relationships with several strains of TAV, hence the virus isolates were identified as two isolates of TAV. This was the first molecular detection and identification of TAV naturally occurring on gladiolus in India (Raj et al. 2011).

Although, many of the researchers have worked out the identification and characterization of several strains of CMV and TAV infecting various economically important plants grown in India for the last 65 years (1951–2015), however, the data generated by them has not been compiled at one place either as a review or as a book chapter. Therefore, in this chapter, we described the general features, natural incidence, virus transmission, host range studies, particle morphology, serological relationships, biochemical characterization and molecular identification of CMV and TAV isolates/strains causing diseases and economic losses on various economically important plants cultivated in India and for their possible disease management.

9.2 Cucumber Mosaic Virus

9.2.1 Characterization of CMV Infecting Vegetable Crops

9.2.1.1 Tomato (*Solanum lycopersicumum*)

CMV is one of the important viruses infecting tomato and cause foliar mosaic, shoestring/fern leaf, yellowing and, stunting and malformation of entire plants. A report of CMV tomato strain causing mosaic in tomato was published from India by Das and Raychaudhary (1953) who identified the CMV by symptomatology and virus transmission. The diseased tomato plants appear stunted; the young leaves are small crinkled, deformed pale yellow and brittle. Sometimes small necrotic areas are seen on the leaves, while in a few cases the leaf margins become brown. Occasionally leaves also show 'shoestring' effect due to infection with cucumber mosaic. Transmission of the virus by sap and aphids revealed that CMV infects tomato was sap transmissible and has many important aphid vectors like: *Aphis gossypii*, *Myzus persicae* and *A. craccivora.*

CMV was detected from tomato showing symptoms such as: mosaic mottling of leaves, smaller than normal and have uneven margin, the virus had specific properties such as: TIP between 60 and 65 °C for 10 min and LIV between 3 and 4 days at 20–22 °C. The virus was easily transmitted by *M. persicae* and *A. gossypii* in non-persistent manner and showed positive reaction with antiserum of CMV (Joshi and Bhargava 1965).

There is a report from India on molecular detection of a virus isolate causing shoestring disease in tomato plants, attempted by RT-PCR using CMV-specific primers by Pratap et al. (2008). The ~650 bp products obtained by PCR were cloned and sequenced. The obtained nucleotide sequence data was corrected as 657 bases and translated into 218 amino acid residues. The sequence data was deposited in GenBank database (accession no: DQ141675). BLASTn analysis of sequence data revealed 99% (655/657) identities with coat protein gene of CMV isolate of Amaranth. These results indicated that the virus isolate causing shoe string in tomato possesses the close relationship with Amaranth isolate of CMV and/or tomato may be considered a new host of the CMV Amaranth isolates. Screening of number of tomato plants was performed using CMV-tomato clone as positive probe and the disease incidence was measured. The infection was found more than 60–70% in some cultivars (Pratap et al. 2008).

9.2.1.2 Bell Pepper (*Capsicum annuum* var. grossum)

CMV was also isolated from bell pepper (*Capsicum annuum* var. grossum) plants in two Karnataka districts (Nagaraju and Reddy 1982). George et al. in 1993 did isolation and identification of two major viruses infecting bell pepper in Karnataka. The symptomatology, host assay, particle morphology and serology suggested that the two viruses infecting bell pepper (*Capsicum annuum* L.) were PVY and CMV (George et al. 1993). The reaction of this virus was positive only when tested against CMV antiserum. Taking into account of all these findings, this virus was identified as CMV belonging to cucumovirus group (George et al. 1993).

Sharma et al. 1993 studied the properties of the viruses associated with disease complex of bell pepper. Based on host range, symptomatology, transmission and electron microscopy, at least two different potyviruses and a cucumovirus associated with mosaic disease complex of bell pepper were characterized. Host range of these viruses was restricted. Symptoms induced in bell pepper included mosaic, mottling, vein banding, shoe-stringing, stunting and fruit deformities. All the virus isolates were found to be the sap and aphid transmissible. Virus particles of the potyviruses isolates were flexuous filaments measuring 741–785 × 12 nm while of the particles of cucumovirus were isometric and of 27 nm (in dia). Isolates of potyviruses induced cytoplasmic cylindrical inclusions and that of cucumovirus induced vesicles along the tonoplast. The cucumovirus isolate as serologically related to CMV -K; the isolates of potyviruses were not related to PVY (Sharma et al. 1993).

Kapoor investigated the viruses infecting bell pepper (*Capsicum annuum* L.) in India. The disease incidence in bell pepper growing districts of Himachal Pradesh was observed during 2003–2004 cropping seasons and the yield losses calculated ranged from 1% to 100%. The symptoms observed were leaf deformation, stunting and yellowing in fields. The two representative isolates of CMV (C-I and C-II) were characterized based on host range assay, biophysical properties, ISEM and DAS-ELISA using CMV antisera. Both the isolates were confirmed by RT-PCR by using cucumovirus primers and identified to be the strains of CMV.

Kapoor in 2012 again reported a CMV isolate in bell pepper crop from Himachal Pradesh. The CMV isolate was found to be mechanical as well as aphid transmissible. The identity of the isolate was established based on symptom expression on indicator plant species, biophysical properties, host range, serology, morphological properties of the virus particles by electron microscopy and RT-PCR. In ELISA, the virus isolate gave positive reaction with CMV specific antibodies. Seventeen weed species were tested as an alternate host of CMV and *Bidens pilosa* was added as a new reservoir host. RT-PCR assays with CMV specific primers and total nucleic acid extracted from symptomatic bell pepper leaves yielded the expected ~550 bp amplicon (Kapoor 2012).

9.2.1.3 Brinjal (*Solanum melongena*)

The studies on a new mosaic disease of brinjal (*Solanum melongena* L.) caused by CMV isolate was carried in Andhra Pradesh by Seth and Raychaudhuri (1973). The purified virus particles were observed to be spherical in shape and measured 35–36 mμ in diameter. The virus was identified as a new strain of CMV. It was successfully transmitted by sap inoculation to *Solanum gilo*, *S. indicum*, *S. intergrifoliumr*, *S. khasianum*, *S. sisymbrifolium*, *Coccinia cordifolia*, *Cucumis anguria*, *C. melo*, *Cucurbita maxima*, *Lagenaria siceerarial*, *Spinacia oleracea*, *Calendule officinalis*, *Carthamus tinctorius*, *Zinnia elegans*, *Ocimum sanctum*, *Salvia officinlis* and *Tropaeolum majus*. Virus purification was done by two procedures and the pellet in both cases was obtained after two cycles of high-speed (30,000–40,000 rpm) centrifugation of 2-h duration each. The virus particles were spherical and measured 35–36 mμ in diameter which was similar to reported for CMV–l (Seth and Raychaudhuri 1973).

Kiranmai et al. in 1997 characterized CMV isolates infecting three vegetable crops in Andhra Pradesh. In this study viruses isolated from commercial tomato, chilli and brinjal crops in Chittoor district, Andhra Pradesh were identified as strains of CMV based on host range, transmission, serology and physicochemical properties. *Dolichos lablab* for CMV-tomato isolate, *Cucumis sativus* for CMV-brinjal isolate and *Datura metel* for CMV-chilli isolate were identified as differential diagnostic host plants. The LIV was 4–6 days, TIP was 50–65 °C and DEP 10.3–10″ for the three isolates. The titres of polyclonal antisera produced against three isolates ranged from 1:5,000 to 1:23,000 in DAC-ELISA. CMV-brinjal and CMV-chilli isolates were serologically different but related to CMV-Tomato isolate. The Mr of coat protein of three isolates ranged from 24.5 to 25.7 × 1Q3 d. The Mr of three genomic RNA s and one subgenomic RNA of three isolates was 1.24 to 1.26; 1.05 to 1.12; 0.78 to 0.86 and 0.34 to 0.39 × 106 d. The dsRNA isolated from infected chilli, tomato and brinjal was resolved into three species with Mr ranged from 2.48 to 2.52; 2.10 to 2.26 and 1.56 to 1.72 × 106 d. The tomato, chilli and brinjal isolates of CMV were identified as CMV-To, CMV-Ch and CMV-Br strains, respectively (Kiranmai et al. 1997).

In another study by Bharti et al. (1997), one isolate was found to be the most virulent, sap transmissible and recorded 100% infection with an incubation period of 16 days. Based on the symptomatology in brinjal and reaction of indicator plants like *C. amaranticolor*, *C. sativus, Gomphrena globosa*, *N. glutinosa*, *N. tabacum*, *Physalis floridana*, *Vigna unguiculata* and *Zinnia elegans,* the virus isolate was identified as CMV. Of the 19 weeds, only 3 weeds *viz*., *Amaranthus viridis*, *Digera arvensis* and *Physalis minima* were infected by this CMV isolate. When ten brinjal genotypes were tested against this virus, Pusa Purple Cluster had 10% infection. Selection of virulent isolate of CMV infecting brinjal was done based on the symptoms (mosaic, vein clearing, puckering and leaf distortion), incubation period and percentage of infection. Based on similarity of symptoms on brinjal and reaction on indicator plants, the present virus has been identified as a stain of CMV infecting brinjal (Bharti et al. 1997).

Further, the incidence and progress of CMV were recorded by Kiranmai et al. (1998c) in commercial brinjal, chilli and tomato crops around Tirupati (A.P.) during 1992–1994 Kharif and Rabi seasons which ranged to 9–21% in the young crops (10–15 days after transplanting) and progressed up to 72–86% as the crop aged. Several other crop and weed growing in and around the above crops found infected with CMV could probably act as alternate hosts. The average virus incidence at 10–15 days after transplanting ranged from 13% to 21% in brinjal, 9–16% in chilli and from 13% to 19% in tomato. It increased during 7 weeks period up to 76–85% in brinjal, 74–84% in chilli and 75–86% in tomato in both kharif and rabi seasons. There appears to be no significant difference in the incidence and progress of the virus infection during both seasons in all the 3 years. This indicates that cultural practices followed around Tirupati might have favoured the high level of initial infection as well as subsequent progress (Kiranmai et al. 1998c). The field collected crop plants that were positive for CMV in DAC-ELISA also produced chlorotic and/or necrotic local lesions an assay plants. Crops like *Dolichos lablab, Lagenaria siceraria*, *Luffa acutangula*, *Momordica charantia*. *Sesamum indicum* and *Trichosanthes anguina*, collected in and around the three vegetable crops were found infected with CMV and they probably acted as alternative hosts. Based on the present studies, it is suggested that rouging of alternate virus source plants initially infected crop plants may help in decreasing the primary and secondary spread of Virus (Kiranmai et al. 1998c).

Kumar et al. in 2014 detected the association of CMV with severe mosaic disease of eggplant growing in Lucknow and Kanpur, India by host reaction and serological assay, and confirmed by RT-PCR using CMV specific primers. Furthermore, the complete RNA3 genome was cloned and sequenced which shared 97–99 % identities and close phylogenetic relationships with CMV subgroup IB members therefore identified as isolates of CMV subgroup IB (Kumar et al. 2014).

9.2.1.4 Carrot, French Bean, Sugar Beet and *Phytolacca* sp

Bhargava (1951) described properties of four CMV strains isolated from French bean, sugar beet and *Phytolacca* sp. for the first time in India. The CMV strains differed in their host range, symptoms, and transmissibility by aphids, DEP and TIP (Bhargava 1951).

Afreen et al. (2009) observed severe chlorotic mottle disease of carrot (*Daucus carota*) in fields of in northern Uttar Pradesh, India. The causal pathogen was transmitted through sap inoculations and by aphid to a number of indicator plants. The association of CMV was detected with the disease by RT-PCR an using CMV specific primer which was further identified by sequence analysis of cloned RNA3 genome. Analysis of complete RNA3 sequence revealed 97% identities and close phylogenetic relationship with various CMV strains of subgroup II available in database. The association of CMV of subgroup II with chlorotic mottle disease of *D. carota* was the only report (Afreen et al. 2009).

9.2.2 Characterization of CMV Infecting Pulse Crops

9.2.2.1 Sweet Pea (*Pisum sativum*)

The CMV strain causing mosaic disease of pea (*Pisum sativum*) is reported by Rao et al. (1995). During this study, survey was conducted in the year 1983, around Gorakhpur (India) and widespread occurrence of an apparently unrecorded mosaic disease of pea was observed. The symptoms in infected plants were severe mosaic, mottling, puckering of leaves and stunting. It was systemic in *Arachis hypogea*, *Crotolaria juncea*, *C. sericea*, *Cyamopsis tetragonoloba*, *Dolichos biflorus*, *D. lablab*, *Pisum sativum*, *Vigna mungo*, *V. radiata*, *V. sinensis*, *Vicia faba*, and *Cucumis sativus* and was localized in *Chenopodium amaranticolor* and *C. ambrisoides*. The symptoms in different host varied from mild mosaic, mottling, chlorotic spots, necrotic spots, vein clearing and vein banding to reduction of entire plant. The virus was serologically related to CMV and the particles were isomeric having an average diameter of ~32 nm. *Aphis craccivora*, *Acyrthosiphon pisum* and *Myzus persicae* were able to transmit the virus. Of these *A. pisum* was found to be most efficient vector. The plants grown from the seeds of disease plants, only 5.5% showed mosaic symptoms suggesting a positive seed borne nature of the virus. The nucleic acid content in the virus particles was approximately 19.8%. The virus was identified as strain of CMV and is designated as CMV-pea (Rao et al. 1995).

9.2.2.2 Chickpea (*Cicer arientinum*)

Chickpea (*Cicer arientinum* L.) in India is known to be affected by two viral diseases causing tip necrosis and stunting. Dhingra et al. (1979) observed symptoms of slightly chlorotic, very small, narrow and deeply dentate leaves of chickpea BG-2 and G-113 cultivars. The plants remain stunted giving bushy appearance and phloem of roots and collar region of stem developed necrosis. Diseased plants bore lesser flowers and pods as compared to healthy plants. A sap transmissible virus was isolated and identified as CMV (Dhingra et al. 1979).

Chalam in 1982 identified and characterized the CMV along with a BYMV isolate infecting chickpea in India. CMV is reported to produce twisting of terminal bud initially and with progression of disease cause wilting and death of plants or proliferation and bushiness of branches bearing very small green leaves and very small pods. The virus was transmissible in non-persistent manner by aphids. The shape and size of CMV particles was spherical and measured about 30 nm in diameter, respectively. The CMV WAS revealed the presence of four RNA species in PAGE and was relatively closer to CMV-C 1, followed by CMV-Ix and M-CMV (Chalam 1982; Chalam et al. 1986).

Sporadic occurrence of the disease of chickpea causing stunting of the plants and excessive axillary proliferation having smaller leaflets was noticed by Singh et al. (1996) in several parts of Himachal Pradesh. The virus induced necrotic lesions on *C. amaranticolor* and mosaic on *C. sativus*, *G. max*, and *N. glutinosa* while no symptoms were observed on *G. globosa*, *A. hypogea*, *C. annum*, *C. cajan*, *P. sativum* and *V. faba*. DEP of virus ranged between 1:5,000 and 1:10,000; TIP between 65 and 70 °C and LIV for 2 weeks in desiccated leaves stored at 5–7 °C. The virus isolate in the present study was identified as CMV on the basis of symptoms, reaction on diagnostic hosts and serological tests. Authors suggested that though the frequency of occurrence of CMV on chickpea is less than 1% but due to cropping pattern in India aid in the building up of aphid vector, this virus has the potential of becoming widespread on chickpea in future (Singh et al. 1996).

Singh et al. 2002 studied viral diseases of chickpea in Himachal Pradesh. In this study extensive survey of the chickpea fields in Himachal Pradesh, India was done to determine the prevalence of viral disease which revealed four different types of symptoms. Leaves from symptomatic plants were used to conduct the pathogenicity tests using chickpea cv. HPG-17. Based on transmission tests, host range, physical properties and serological tests, association of four viruses, namely, CMV, BYMV and chickpea stunt caused by bean leaf roll virus and chickpea chlorotic dwarf virus was observed in the infected chickpea plants. The incidence of CMV and BYMV was about 1% in the cv. HPG-17 (bold seeded *desi* type) whereas the incidence of chickpea stunt varied from 3% to 5% (Singh et al. 2002).

9.2.2.3 Cowpea (*Vigna unguiculata*)

To date, there is the only a report of Nagaraju and Murthy (1994) describing the association of CMV with mosaic disease of cowpea (*Vigna unguiculata*). CMV was found to be sap, seed and aphid transmissible. CMV had a DEP of 1:1,000–1:5,000, TIP between 60 and 65 °C and LIV of 24–48 h at room temperature (23–25 °C) and 7 days at 14 °C. It produced chlorotic spot and thickening along veins in *Cucurbita moschata* and *Luffa acutangula*, mosaic in *Cajanus cajan, Cyamopsis tetragonaloba, Phaseolus aureus* and *Phaseolus vulgaris*, chlorotic local lesions on *Chenopdoium amaranticolor* and *C. quinoa* and systemic necrosis on *Dolichos biflorus*.

9.2.3 Characterization of CMV Infecting Fruit Crops

9.2.3.1 Banana (*Musa paradisiaca*)

The production and quality of banana are greatly affected due to several diseases caused by various viral pathogens, and among viruses CMV, banana bunchy top virus (BBTV), banana streak virus and banana bract mosaic virus commonly infect. These diseases of banana are vertically transmitted through planting suckers. Although conventional control measures such as quarantine eradication and certified virus free stocks confirmed by ELISA has been used to reduce crop losses caused by these viruses, but they are not effective for controlling CMV and BBTV. Mali and Rajegore (1980) described a disease of banana caused by CMV in India. Banana heart rot in the Deccan was found to be a syndrome of banana mosaic. The virus isolated, identified as CMV, was transmissible mechanically and by aphid in stylet borne manner. It was inactivated as 65–70 °C and had a dilution end point of 104. It was viable up to 92 h at 27–30 °C. This was the first record of the virus on banana in India (Mali and Rajegore 1980).

Patel and Mali (1986) did comparative studies of three isolates of cucumber mosaic virus from banana. Isolates CMV-1, CMV-2 and CMV-3 were indistinguishable serologically, in particle size and morphologically and transmission characteristics, but differed in reactions of 13 hosts and physical properties, on the basis of which they are regarded as distinct structures of the virus (Patel and Mali 1986).

CMV causing infectious chlorosis in banana plants growing under natural conditions in Andhra Pradesh was observed. The virus culture was established on cow pea and single local lesion was subsequently was maintained on tobacco by sap inoculations and back inoculation in banana by injecting purified virus particles (Kiranmai 1989). Infection of CMV causing mosaic disease of banana was identified by immuno/nucleic acid probe using polyclonal antibodies of CMV-T and slot blot hybridization tests with nucleic acid probe of CMV-P genome by Srivastava et al. (1995).

The comparative CMV detection efficiency of three different tests in banana was done by Kiranmai et al. (1996). Direct ELISA like double antibody sandwich (DAS) ELISA, simplified rapid direct antigen coating (SRDAC)-ELISA and indirect form of DAC-ELISA for detection of CMV in leaf extracts and pseudostem sap exudates and purified virus diluted with antigen buffer or healthy banana leaf sap were compared. The sensitivity levels of three ELISA tests were similar and the virus was detected up to 10–7 dilution with leaf extract, 1 μl/well with sap exudates, 20 ng/well with purified CMV and 100 ng/well with purified virus diluted with healthy banana leaf extract. Of three forms of ELISA compared DAC-ELISA was further evaluated with detection of infectious virus by local lesion bioassay on cowpea (*Vigna unguiculata*) and by double stranded RNA (dsRNA) analysis. The banana leaf and pseudostem sap exudate samples positive in DAC-ELISA were also positive by other two tests. Collection of pseudostem sap exudates by pin pricking from test plant and detection of virus by DAC-ELISA in them appears ideal for mass screening of banana plants (Kiranmai et al. 1996).

RT-PCR/PCR based detection of CMV was attempted by Vishnoi et al. (2006) in severely infecting banana plantations using their specific primers. Cloning and sequencing of CMV-CP gene was done and sequence data were submitted to GenBank (DQ152254, CMV CP-Banana). The work aims to developing a simple and accurate means of rapid detection system in banana and to search of virus-free propagating material of banana to be used for mass propagation or large-scale cultivation (Vishnoi et al. 2006).

Khan et al. (2011) attempted the CP gene based characterization of CMV isolates infecting banana in India. Banana plants exhibiting stripes on leaves in addition to leaf distortion and stunting of plant were collected from Karnataka (KAR), Maharashtra (MH) and Uttar Pradesh (UP) in India. The virus was identified as CMV based on TEM and RT-PCR analysis. CP gene of all isolates was amplified using gene specific primers, cloned and sequenced. Complete CP gene contains 657 nucleotides coding for 219 amino acids. Sequence analysis of CP gene showed 93–98% (at nucleotide) and 94–99% (at amino acid) sequence identity between all three Indian isolates. On comparing CP gene sequences of CMV isolates of KAR, MH and UP with P isolate; 94%, 99% and 96% identity respectively was obtained. High degree identity at nucleotide level between these isolates of banana and *Physalis minima* (a weed) suggest that *Physalis minima* could be an alternate host of CMV banana. Phylogeny of nucleotide along with amino acid sequence of CP gene showed that all our isolates belong to IB subgroup (Khan et al. 2011a). Further, Vishnoi et al. (2013) worked out molecular characterization of complete genome of CMV infecting Banana in three banana farms of Uttar Pradesh, India where it cause severe mosaic accompanied by leaf and fruit deformation with 18–25% disease incidence. The RNA 1a, RNA 2b and RNA 3 genomic fragments were amplified and sequenced for molecular identification of virus. Sequence analyses of these fragments revealed its highest identities and close relationships with Indian strains of CMV of subgroup IB; therefore, identified as an isolate of CMV of subgroup IB (Vishnoi et al. 2013).

Recently, Khan (2015) attempted the development of sensitive diagnostic for the detection of CMV in banana. The 300 diseased samples from 21 banana orchards in 3 bananas producing states: Karnataka, Maharashtra, and Uttar Pradesh were collected, of which 13 were positive for CMV tested by DAS-ELISA. One representative isolate from each state was sequenced and designated as CMV-KAR, CMV-MR, and CMV-UP. Comparison of these isolates with 29 CMV isolates reported from various plants around the world showed that they were most closely related to subgroup-I isolates, sharing up to 95.81% and 96.84% sequence identity at nucleotide and amino acid, respectively and clustered with the CMV subgroup-IB strains. Study suggested that CP gene based RT-PCR assay may be a more sensitive, reliable, and convenient molecular tool for detection of the CMV, and can be used in quarantine, eradication, and tissue culture certification programs (Khan 2015).

9.2.3.2 Papaya (*Carica papaya*)

The infection of CMV in papaya was first reported by Kiranmai et al. (1998a) in India. A virus disease with mosaic, leaf distortion and filiform symptoms on commercially growing papaya (*Carica papaya* L.) around Kovvur, West Godavari district, Andhra Pradesh state was noticed during 1995. The sap transmissible virus from papaya leaves was successfully purified and its yield ranged from 7 to 9 mg/100 g leaf tissue. The purified virus was infectious to papaya. In DAC- ELISA, the virus positively reacted with three CMV antisera. PAGE revealed the molecular weight of CP subunit about 26.2 kDa. The isolated genomic nucleic acid from purified in agarose gel was resolved into three genomic and one sub genomic RNAs with Mr of 1.23, 1.12, 0.77 and 0.3 × 10^6 Daltons, respectively. The double stranded RNA isolated from infected papaya leaves was resolved in agarose into three species with *Mr* of 2.46, 2.24 and 1.54 × 10^6 Daltons. Based on these studies, virus causing leaf distortion in papaya was identified as CMV which was antigenically identical to CMV-banana and CMV-Chilli isolates previously reported from Andhra Pradesh.

9.2.3.3 Cape Goose Berry (*Physalis peruviana*)

A severe mosaic disease of cape goose berry (*Physalis peruviana*) showing severe puckering and blistering of leaf lamina was studied in Kanpur, U.P. The host reactions, physical properties, transmission and serological reactions confirmed it to be caused by a virulent strain (CMVSS) of CMV (Gupta and Singh 1997).

9.2.4 Characterization of CMV Infecting Spice Crops

9.2.4.1 Chillis (*Capsicum annuum*)

Bhargava in 1951 reported the occurrence of CMV on chilli for the first time. Then after, Anjaneyulu and Apparao (1967) reported the occurrence of CMV on chilli. Here, a collection of virus infected chilli plants obtained around Hyderabad and Warangal in Andhara Pradesh revealed that Chilli is affected by a number of viruses. One of these, as judged by the symptoms, appeared to be not described from India and a detailed investigation has been made to characterize the virus. The symptoms in the field consisted of mosaic mottling and various types of leaf distortions mainly exhibited by filiform leaf tip. Infected plants exhibited marked stunting in growth and severe reduction in size of the leaves and fruits. The virus was found to have a DEP of 1:250–1:500, TIP of 55–60 °C and LIV of 12–18 h at room temperature (26 °C) and 36 h at low temperature (10 °C). A comparison of the characters like transmission, host range and physical properties with CMV identified the virus under study as CMV (Anjaneyulu and Apparao 1967).

Prasada Rao (1976) and Bidari and Reddy (1983) identified the occurrence of CMV in chilli based on electron microscopy of purified preparations showing isometric particles with a diameter of 29 nm. Gowda and Reddy (1985) described the distribution of chilli mosaic virus in some parts of Kolar district. A survey in this area of Karnataka revealed five viruses causing mosaic of *Capsicum annuum*: CMV, PVY, pepper vein banding virus, pepper veinal mottle virus and TMV (Gowda and Reddy 1985).

Narayan and Dhawan (1989) characterized and identified viruses causing mosaic disease of *Capsicum. Capsicum* plants in Haryana, India, were found infected by various types of mosaic symptoms. The CMV, PVX, PVY and TMV were identified, associated singly or in combination with other types of mosaic diseases. CMV was found to be the most prevalent virus (Narayan and Dhawan 1989).

Singh et al. (1990) attempted identification of virus(es) associated with chilli mosaic syndrome. During the study, four isolates were obtained from naturally infected *Capsicum* plants showing different symptoms in the field at Ludhiana. Host range, transmission and serological tests established that they were all closely related to cucumber mosaic cucumovirus. Determination of physical properties confirmed the predominance of this virus as a cause of mosaic in *Capsicum,* no other virus being detected (Singh et al. 1990). Singh and Shukla (1990) described the properties of a new strain of cucumber mosaic virus from chilli. The properties of a virus causing necrosis and mosaic in *Capsicum* around Kanpur, Uttar Pradesh, India are described. The virus was transmitted mechanically in seeds and by 2 aphids, *Aphis gossypii* and *Myzus persicae,* and could infect 44 plant species, inducing a range of symptoms. EM of the purified virus showed spherical particles of *c.* 30 nm diameter and serological tests produced a positive reaction to cucumber mosaic cucumovirus antiserum. It is concluded that the virus is a new strain of cucumber mosaic cucumovirus (Singh and Shukla 1990).

Biswas et al. (2013) evidenced the CMV subgroup II infection in *C. annuum* L. in Himachal Pradesh of India associated with virus-like symptoms in the foliage of chilli plants. DAS-ELISA, host assay and RT-PCR suggest the occurrence of CMV in these samples. The CP gene was amplified form several samples, cloned and sequenced. Analysis revealed high (95–99%) homologies at nucleotide level with the sequences of CMV subgroup II and phylogram could assigned the clustering of CMV isolate under study with the other analyzed isolates belonging to subgroup II. These findings helped to understand the role of CMV subgroup II as a major virus-causing disease infecting chilli in western Himalayan region of India (Biswas et al. 2013).

9.2.4.2 Black Pepper (*Piper nigrum*)

The CMV associated with stunt disease of black pepper was recorded in Kerala, South India by Sarma et al. (2001). CMV was found transmissible through grafting. Virus was also sap transmissible from infected black pepper to healthy black pepper and to some assay plants. Virus produced local chlorotic/necrotic lesions on *C. amaranticolor*, *C. quinoa*, *V. unguiculata*, *V. radiata* and *V. mungo* while systemic mosaic on *C. sativus* plants. Sap as well as purified virus preparation reacted positively with polyclonal antisera of CMV-Banana (India), CMV-Brinjal (India),

CMV-Chilli (India), CMV-Tomato (India), CMV-L (USA) and CMV-A (China) in DAC-ELISA. In agreement to other CMV isolates, it also showed non-enveloped isometric virions, 26.1 kDa CP subunits and presence of four species with *M*r 1.21, 1.10, 0.81 and 0.37 × 106 Da. Hence, the virus associated with stunt disease of black pepper was identified as an isolate of CMV (Sarma et al. 2001).

Bhat et al. (2003) reported viral disease – a new threat to black pepper cultivation in India. The symptoms, causal agents [CMV (Cucumovirus) and an uncharacterized virus (Badnavirus)], and transmission and spread of viral disease infecting black pepper in India, particularly in Karnataka, Kerala, and Tamil Nadu are discussed. The different methods (i.e. virus-free vegetative planting material, rouging and eradiation of infected plants and chemical control of insect vectors) for effective viral disease management are also discussed (Bhat et al. 2003).

Bhat et al. (2004a) also characterized a CMV strain infecting black pepper. CMV strain was transmissible to *N. benthamiana* and *N. glutinosa*. The virus was purified by differential and sucrose density gradient centrifugation which showed the presence of isometric cored particles of ~28 nm diameter. Polyclonal antiserum against the virus was raised and DAS-ELISA method was standardized for detection of CMV in diseased black pepper samples collected from different regions of Karnataka, Kerala and Tamil Nadu. The CMV was also detected in other *P. chaba*, *P. colubrinum* and *P. longum* and weeds such as *Ageratum conyzoides*, *Colacasia esculanta*, *Synedrella nodiflora*, *Cynodon dactylon* and *Sonchus oleraceus* found in and around black pepper gardens (Bhat et al. 2004a). Further Bhat et al. in 2005 reported the association of CMV with stunted disease of black pepper based on sequence analysis of CP gene containing 657 nucleotides encoding a protein of 218 amino acids. CP gene showed 92–99% sequence identity with members of CMV subgroup I and close relationship with CMV-H strain isolated from Egyptian henbane in India. The study suggested that virus infecting black pepper is a strain of CMV belonging to subgroup I (Bhat et al. 2005).

Bhat and coworkers in 2013 reported quick and sensitive detection of CMV along with Piper yellow mottle virus (PYMV) in black pepper by loop-mediated isothermal amplification (LAMP) method. The LAMP and RT-LAMP assay for PYMV and CMV, respectively were done using five primer pairs targeting the conserved sequences in the viral genome. The assay detected presence of both the viruses. The detection limit for both LAMP and RT-LAMP assays was up to 100 times as compared to PCR and real-time PCR (Bhat et al. 2013).

9.2.4.3 Betelvine (*Piper betle*) and Long Pepper (*Piper longum*)

Raj and coworkers in 2003 detected the association of CMV and BYMV with mosaic disease of betelvine grown at Mahoba and Banthara (in India) by ELISA and RT-PCR methods using their specific antisera and universal primers. DAC-ELISA could detect only CMV. However, RT-PCR detected both CMV and BYMV infections in several betelvine samples. The association of CMV with betelvine was reported for the first time (Raj et al. 2003).

Bhat et al. (2004b) reported occurrence and identification of a CMV isolate infecting Indian long pepper (*Piper longum*). The virus causing mosaic mottling in long pepper was identified as CMV based on biological, physicochemical and serological properties. The virus isolate was easily sap transmissible to *C. album*, *N. benthamiana*, *N. glutinosa*, *N. tobacum* cv. white burley, *P. floridana* indicator hosts and a few other cultivated plants species including black pepper. The purified virus preparations revealed ~28 nm isometric particles when observed under TEM. The PAGE showed presence of 25.6 kDa capsid protein subunits. The leaf extract from diseased *Piper longum* and virus preparations showed positive reaction with polyclonal antiserum of CMV both in DAC-ELISA and electro-blot immunoassay (EBIA). Based on particle morphology, antigenic relationship and molecular weight of CP subunits suggest that the virus associated with Indian long pepper was an isolate of CMV (Bhat et al. 2004b).

In another study, characterized a CMV strain infecting Indian long pepper (*Piper longum* L.) and betel vine (*Piper betle* L.). The infection of CMV in was detected by RT-PCR using CMV-CP gene specific primers. For virus identification, the resulting amplicons were cloned and sequenced. Analysis showed that both of them consisted of 657 nucleotides and encoding for a protein of 218 amino acids. The sequence comparisons revealed 100% identity with each other suggesting a common origin. Both of them also showed that CMV infecting banana and black pepper were close to CMV isolated from long pepper and betel vine. Based on sequence identities Indian long pepper and betel vine isolates were identified as members of CMV-subgroup I for the first time.

9.2.4.4 Vanilla (*Vanilla planifolia*)

Madhubala et al. (2005) recorded occurrence of CMV causing mosaic, leaf distortion and stunting of vanilla (*Vanilla planifolia* Andrews) for the first from India. The virus was characterized on the basis of biological and coat protein properties. The virus was purified from inoculated *N. benthamiana* plants and purified preparations showed isometric particles of ~28 nm. The viral coat protein subunits were of 25.0 kDa. DAS-ELISA method was also standardized for the detection of CMV infection in vanilla plants. The coat protein gene region of the virus was amplified by RT-PCR, cloned and sequenced. Sequence analysis revealed presence of a single open reading frame of 657 nucleotides coding for 218 amino acids. The isolate of vanilla showed 99% identity and close relationship with black pepper isolate of CMV belongs to subgroup IB (Madhubala et al. 2005).

9.2.5 Characterization of CMV Infecting Ornamental Plants

9.2.5.1 Phlox

Khatri and chenulu (1967) did studies on mosaic disease of phlox. In this study, a virus that caused mosaic disease of phlox had been studied with regard to symptomatology, mode of transmission, host range and biophysical properties. The virus was readily transmitted by juice inoculation and by *Aphis gossypii* Glover. It has been mechanically transmitted to and produce mosaic symptoms on *Phlox drummondii*, *Nicotiana glutinosa*, *N. tabacum* white burley and Harrisons special, *Capsicum annuum*, *Solanum nigrum*, *Datura stramonium* var. tatula, *Lycopersicon esculentum*, *Cucurbita pepo*, *Cucumis melo* and *C. sativus*, *C. amaranticolor* induced local necrotic lesions. The virus was inactivated at 60 °C, at a dilution of 1:50 and after storage for 16 h at room temp. Freezing the virus-infected leaves for 80 h resulted in loss of infection. A comparison of the virus studied with several other viruses reported to cause mosaic symptoms on phlox either naturally or under artificial conditions has shown that the viruses resembles Cumis virus 1, smith orits strain. These studies confirmed an earlier report by Faan and Johnson that phlox serves as an over wintering host of cucumber mosaic virus (Khatri and Chenulu 1967).

9.2.5.2 Amaranths (*Amaranthus* spp.)

Sharma and Chowfla (1986), described a mosaic disease of amaranths (*Amaranthus caudatus* L.) in Himachal Pradesh. They described the serological and biophysical properties of the virus isolates, its host range and transmission indicated that the severe mosaic disease of this green vegetable crop was caused by CMV (Sharma and Chowfla 1986).

A number of *Amaranthus* spp. viz. *A. tricolor*, *A. hypochondriacus*, *A. deflexus*, *A. lividus* and *A. retroflexus* are known as natural hosts of CMV, however, there was no record of CMV infection from India until 1997. The disease was important due to reduction of seed yield in *A. hypochondriacus* and the total loss of ornamental qualities of *A. tricolor* plants. About 10–12% disease incidences were recorded in experimental fields at CSIR-NBRI, Lucknow, in 1992 and 1993, respectively by Raj et al. (1997a, b). The virus was easily transmissible by sap inoculations on a number of test hosts that developed local lesions or systemic symptoms. The virus isolate was purified and observed by TEM which revealed the presence of polyhedral virus particles of ~28.0 nm. SDS-PAGE and Western immunoblot analysis revealed the presence of one band of 26 kDa coat protein in naturally infected *A. tricolor*. The nucleic acid isolated from purified virus particles remained infectious when treated with DNAase. During electrophoresis of the extracted nucleic acid, RNA species separated as three distinct bands which were identified as RNA 1 and RNA 2 altogether, RNA 3 and RNA 4, similar to that of usually observed for CMV. The isolate infecting *A. tricolor* and *A. hypochondriacus* was more closely related to CMV-C than to CMV-D (Raj et al. 1997).

Further, molecular characterization was also done using CP and MP ORFs located in its RNA 3 genome. The complete coat protein gene (657 bp) of CMV-A was cloned and sequenced and phylogenetic status of the Indian strains of CMV vis-à-vis the global CMV was analyzed. The studies based on CP analysis indicated that the Indian CMV strains are more close to the subgroup IB rather than to subgroup IA or II according to the currently accepted phylogeny. However, a high degree of homology depicted between CMV-A and other Indian strains may be reflective of the adaptation of the virus to Indian location or the Indian hosts it infects (Srivastava 2003). The complete movement protein gene (840 bp) of CMV-A was cloned, sequenced and analyzed. The phylogenetic relationship based on MP sequence data also placed the Indian strain into subgroup IB rather than IA. Within subgroup IB, maximum evolutionary relationship was revealed to IA-3a strain, a subgroup IB member from Indonesia. The phylogenetic analysis based on MP had also showed maximum homology of the Indian strain under study with the MP sequence of Indonesian strain except CMV-D, H and Phym, the other Indian strains (Srivastava 2003).

9.2.5.3 Gladiolus (*Gladiolus* spp.)

Several gladiolus cultivars maintained in CSIR-NBRI, Lucknow gardens were found exhibiting the mosaic, stripe on leaves and flower colour-breaking symptoms. The association of CMV was identified as on the basis of aphid transmission, presence of 28–29 nm isometric particles, 26 kDa coat protein subunits, four species of RNA and serological relationship with CMV strains. SDS-PAGE of partially purified preparations from infected gladiolus and tobacco plants revealed one major band of viral coat protein subunits of 26 kDa molecular weight and a minor 52 kDa band (dimer was also observed) as seen in case of CMV strain obtained earlier from *Dianthus barbatus*. Result of Western immunoblot analysis using CMV-T antibodies also confirm the presence of 26 kDa band in infected Gladiolus samples which indicated that the gladiolus isolate of CMV is serologically related to CMV-T (Raj et al. 2002). In order to co-relate these findings at nucleic acid level, total RNA was extracted from infected and healthy gladiolus leaves. These RNA preparations (50–200 μl) corresponding to leaf tissue (25–100 mg) were transferred to a zeta-probe membrane (Bio-Rad) under vacuum using a slot blot apparatus. cDNA clone of CMV-P containing part of CP gene and MP genes was used for preparation of α-^{32}P labeled probe by random primer extension method. Positive signals in all infected gladiolus leaf samples indicated the presence of CMV; however, no signals were obtained with healthy gladiolus leaves. On the basis of slot blot hybridization results the gladiolus strain of CMV seems to carry certain homology with CMV-P strain (Raj et al. 2002).

Dubey et al. in 2010 reported molecular characterization of CMV infecting gladiolus by sequence analysis of cloned 657 bp coat protein region amplified by RT-PCR. The phylogenetic relationships were determined by the comparison of coat protein gene nucleotide (DQ295914) and its amino acid sequences with other CMV isolates reported from India and worldwide. The comparisons of nucleotide and amino acid sequences and phylogenetic tree revealed a close resemblance of the virus isolated from gladiolus with the Fny strain, which was an unusual in the Asian continent (Dubey et al. 2010).

9.2.5.4 Petunia (*Petunia hybrid*)

Petunia plant exhibiting severe mosaic, necrosis and stunting symptoms were observed in various gardens in Lucknow. The association of a strain of CMV was identified on the basis of non-persistent transmission by aphids, molecular weight of protein sub units as 24,500 Da, spherical particles of 29 nm diameter and serological relationship with several CMV isolates as established by enzyme immunoassay. Naturally infected *P. hybrida* plants exhibited necrosis of leaves and petioles and apical growth (Fig. 9.1). The affected plants gave a stunted appearance and did not bear flowers. The virus was easily transmitted by mechanical means from *P.*

Fig. 9.1 Necrotic local lesions and systemic mosaic, blistering, leaf deformations and puckering induced by CMV-Petunia virus isolate (Gautam et al. 2014)

hybrida to *P. hybrida* and by aphids to an extent of 60% (6/10) and 80% (8/10), respectively. Results on DAC-ELISA indicated that the virus isolate from petunia is related to CMV-C, CMV-D, CMV-T, CMV-L and CMV-S. Chrysanthemum aspermy (PVAS-24) and Tomato aspermy virus-N (TAV-N) however, showed a very distant relationship with the isolate (Srivastava et al. 1991).

Praveen et al. (2001) differentiated the biologically distinct CMV isolates of *N. tabacum*, *Commellina* and *Petunia* by double stranded RNA profiling. In this study, four isolates of CMV from *N. tabacum*, *Commellina* and *Petunia*, designated as CMV-T, C, P1 and P2, respectively, differed in their host reaction and temperature sensitivity, at the molecular level. Serologically related isolates of CMV *viz*., CMV-P1 and P2C and T were analyzed on the basis of their symptomatology, temperature tolerance and double stranded RNA (ds-RNA) pattern. The ds-RNA pattern of these CMV isolates representing biological diversity were examined, Two distinct patterns differing by the number and or/position of the ds-RNA bands were obtained for the four isolates. The ds-RNA pattern cannot be necessarily correlated with the symptom severity nor with any specific biological characteristics, however, it can be utilized in distinguishing isolates (Praveen et al. 2001).

Gautam et al. (2012) identified a CMV strain associated with mosaic disease on petunia. The association of CMV was detected based on presence of cored virus particles of ~28 nm and positive amplification of 650 bp during RT-PCR using CMV -CP gene specific primers. Further, the complete RNA 3 genomic fragment of 2.2 Kb of virus isolate was amplified by RT-PCR using RNA 3 specific primers and obtained amplicon was cloned and sequenced. The analysis of sequence data of RNA 3 showed 96% with several CMV strains possessed the closest phylogenetic relationships with banana strain of CMV of subgroup IB reported from India. Therefore, virus isolate of petunia was identified as a new member of CMV subgroup IB which was the first report from India (Gautam et al. 2012).

9.2.5.5 Chrysanthemums (*Chrysanthemum morifolium*)

Naturally infected *Chrysanthemum morifolium* showing diffused chlorosis, chlorotic dots near veins and necrosis of leaves were seen in gardens in CSIR-NBRI, Lucknow. Based on the obvious symptoms observed it may be concluded that the disease was prevalent in about 20% of the chrysanthemum grown in the gardens of NBRI at Lucknow. Naturally infected plant showed conspicuous ring mosaic symptoms of leaves, stunting of the plants and distortion of the florets. The virus was sap transmissible to *N. tabacum*, *C. amaranticolor* and *C. murale*. *C. amaranticolor* and *C. murale* developed necrotic local lesions 7 dpi. *M. persicae* and *A. gossypii* could transmit the virus from chrysanthemum to chrysanthemum and from chrysanthemum to *N. tabacum* var. Samsun NN to an extent of 80% (8/10) and 60% (6/10), respectively, when tested in non persistent manner. In Ouchterlony double diffusion tests, the isolate form precipitin lines with antisera to tomato aspermy virus-N and *Chrysanthemum aspermy virus* (PVAS-24 ATCC). However, the isolate did not react in identical tests carried out using antisera of few strains of CMV viz. CMV-D, CMV-L, CMV-T and CMV-S. The molecular weight of the coat protein sub units was calculated as 24,500 + 500 Da by linear regression based on log of the molecular standard vs. their mobility in SDS-PAGE.

In another study, chrysanthemum plants were observed showing browning of major veins, leading to downward cupping and, finally abscission symptoms. The virus was transmitted by mechanical means to several hosts. The virus in sap of *C. morifolium* remained infectious after it was diluted to 10–3, heated at 55 °C for 10 min, and stored for 2 days at ambient temperature 28–30 °C. The gradient purified virus preparations produced symptoms on *C. morifolium* identical to those observed in nature. Purified virus gave a characteristic peak of nucleoprotein at $A_{max/min}$ at $A_{260/240nm}$ with a ratio of 1.58. Spherical particles averaging 29 nm were observed in preparations stained with uranyl acetate. Molecular weight of viral CP subunits was approximately 24 kDa as determined by SDS-PAGE (Srivastava et al. 1992).

Viral RBA isolated from virus particles was infectious in White Burley tobacco without any further treatment or when treated with DNase. Infectivity was however lost after treatment with RNase and S1 nuclease, which showed ssRNA as the infectious entity. Electrophoresis of nucleic acid preparation revealed five major bands that moved along with two CMV strains isolated from *Physalis minima* L. *and Petunia hybrida* Hort. Vilm-Andr. In ODDT, crude antigen and purified virus preparations were tested which reacted strongly with antisera to CMV-L, CMV-C, CMV-T and CMV-S. No reaction was observed with these antigens and antisera to either TAV-N or CAV (PVAS-24). Chrysanthemum isolates as well as three other CMV strains did react with antisera to CMV-T, but under similar conditions CAV did not react with CMV-T antiserum. Results of ELISA confirm the findings of ODDT. CMV-D antiserum, which was not included in ODDT, showed strong reaction with the chrysanthemum isolates (Srivastava et al. 1992).

Further, Kumar and coworkers in 2005 attempted to develop a molecular tool based sensitive diagnostic method for reliable detection of CMV infection in chrysanthemum cultivars. RT-PCR with CMV-CP gene specific primers resulted in amplification of ~650 bp band in most of the samples. The identities of amplicons were confirmed by Southern hybridization using the α32P radiolabelled probes prepared from the cloned CP gene of CMV-A strains. Positive signal of hybridization with PCR product in CMV and CMV probe proved the identity of PCR amplicons as a fragment derived from the CMV genome in infected chrysanthemum samples (Kumar et al. 2005).

Verma et al. (2007) screened several chrysanthemum cultivars of Northern India for presence of Chrysanthemum virus B (CVB), CMV, TAV, tospo- and potyviruses by ELISA, northern hybridization and RT-PCR methods. During screening of 80 cultivars, 48, 42.5, 26.2, 12.5 and 6.25% cultivars were found infected by CVB, CMV, TAV, tospo- and potyviruses, respectively.

9.2.5.6 Carnation (*Dianthus barbatus*)

The association of CMV with crinkle and stunting disease of carnation (*Dianthus barbatus*) was reported for the first time in India by Raj et al. (1993). The virus isolate was sap transmissible to *D. barbatus* and *Vaccaria pyramidata* which was further confirmed by Western immunoblotting assay. RNA extracted from purified CMV-CR isolate showed four distinct bands of RNA 1, 2, 3, and 4 during electrophoresis. cDNA synthesis of RNA4 of both the strains and their cloning into pBS-KS II (+) vector resulted in several clones carrying insert of about 1 kb when digested with *Pst* I/*BamH* I or *Pvu* II enzymes. Northern hybridization using the *Pst* I/*BamH* I (1 kb), *Sal* I (700 bp) and *BamH* I/*Sal* I (200 bp) fragments as probes confirmed the presence of cDNA of CMV RNA3 and 4 relating to coat protein gene in our clones. Northern hybridization using radiolabelled RNA transcripts produced *in-vitro* in each of the directions from T3 and T7 promoters located on opposite site of the inserts in recombinant clones, gave (+) ve signals with RNA3 and RNA4 by T3 transcripts and (−) ve signals by T7 transcripts. This confirmed that CP gene was located in sense orientation with respect to β-galactosidase gene in recombinant clones (Raj et al. 1993). Clones containing the inserts of CP gene of strains CMV-CR and CMVP were partially sequenced and compared with other CMV strains. Comparison of 249 nucleotide sequences of 5′ coding region of coat protein gene of both clones showed 88–90% similarity with C, O, Y and Fny strains of CMV subgroup I but only 60–64% homology with CMV-Q and CMV-WL strains of CMV subgroup II. Indian isolates CMV-CR and CMV-P also possess 97% similarity with each other in nucleotide sequences of above strains, they showed 91–95% similarity with CMV-C, CMV-O, Fny, CMV-Y and 61–67% homology with CMV-WL and CMV-Q strains respectively. Dendogram based on alignment of coat protein gene of eight CMV strains revealed that both the strains CMV-CR and CMV-P belong to subgroup I to which CMV Fny belong but not to CMV-Q of subgroup II (Raj et al. 1993).

9.2.5.7 Alstroemeria Hybrids

Verma et al. in 2005 reported the CMV strain of subgroup 1 on *Alstroemeria hybrids* for the first time in India based on host range, transmission by aphids, ELISA, electron microscopy, RT-PCR using CMV-CP gene specific primers and dot-blot hybridization. In this study, CMV was detected in nine hybrids and 61% of plants (Verma et al. 2005a, b).

9.2.5.8 Lily

The screening for presence of viruses in lilies (*Lilium longiflorum* and *L. tigrinum*) grown in Himachal Pradesh, India was carried out by Sharma et al. (2005). These viruses were detected by testing bulbs and leaves by ELISA and RT-PCR. During the study, CMV, Lily symptomless virus (LSV) and Lily mottle virus (LMoV) were detected in Asiatic and Oriental hybrids of lily. The viruses were identified by sequence analysis of cloned amplicons. Sequence analysis of CMV infecting lily revealed 96–98% nucleotide and 93–97% amino acid homology with CMV subgroup II sequences confirming presence of CMV of subgroup II in lilies (Sharma et al. 2005).

9.2.5.9 Salvia Splendens

A virus disease of garden sage (*Salvia splendens* Ker-Gawl.) was observed in gardens of Aligarh, UP, India by Ali et al. (2012). Sage plants showed symptoms of severe mosaic, mottling and distortion of leaves and retarded growth. By sap and aphid inoculations, the virus was transmitted to *Salvia* spp. and other hosts. Partially purified virus preparations revealed presence of ~28 nm particles. Crude sap also reacted positively with CMV specific polyclonal anti-serum (PVAS 242a, ATCC, USA) during DAC-ELISA. For molecular detection, RT-PCR was performed using total RNA and CMV-CP gene specific primers which resulted in 650 bp expected size fragments only in the infected samples. Further sequence analysis suggested that virus was a strain of CMV (Ali et al. 2012).

9.2.5.10 Calendula and Marigold

A disease on calendula plants was observed where plants exhibit mosaic disease. From leaf samples virus was purified using 0.05 M phosphate buffer (pH 7.5) and chloroform followed by differential centrifugation. Purified preparations were infectious and gave spectrum typical of nucleoprotein. Virus had a spherical particle about 35 nm diameters, infectious RNA isolated from purified virus preparations. It is concluded that calendula yellow net is a strain of CMV (Naqvi and Samad 1985).

A strain of CMV casing marigold mosaic was partially characterized by Singh et al. (1999). The virus was isolated from diseased marigold (*Tagetes erecta* L.) plant growing at Aligarh. Diseased plants showed mosaic, mottling and stunting Singh et al. (1999).

9.2.5.11 Ornithogalum

Ornithogalum thyrsoides is a cut flower ornamental crop and have long vase. Ornithogalum plants were found exhibiting severe mosaic on the leaves growing in a floriculture field at the Palampur, India. The virus was transmissible by sap and cause local lesions on *C. amaranticolor* and *C. album* whereas mosaic on *N. glutinosa*, *N. clevelandii*, *N. megalosiphon* and *N. tabacum* cv. White Burley. Virus reacted positively with CMV-specific antiserum in ELISA and has particle size of about 29 nm isometric particles (Verma et al. 2005). RT-PCR was performed using total RNA and CPTALL primer pair which resulted in expected size ~950 bp amplicons. Sequencing and analysis of the amplicon (accession no. AJ745092) revealed highest 97% and 93–95% nucleotide and amino acid sequence identity, respectively with the available sequences of CMV subgroup II. These studies confirmed the association of CMV with severe mosaic disease of ornithogalum (Verma et al. 2005).

9.2.5.12 Gerbera

The occurrence of CMV in *Gerbera jamesonii* growing in floriculture fields at the IHBT, Palampur and nearby nurseries was reported by Verma et al. (2004). The infected gerbera plants showed color break symptoms on the petals, asymmetrical ray florets, and deformed flowers. The gerbera virus isolate induced chlorotic local lesions on *C. amaranticolor*, *Chenopodium album* and *C. quinoa* while mosaic on *C. sativus*, *N. clevelandii*, *N. benthamiana*, *N. glutinosa*, and *N. tabacum* cv. Samsun plants. As well the virus could be transmitted in non-persistent manner by aphids. Occurrence of CMV strains was expected as crude sap from infected plants reacted positively with CMV-specific antibodies and size was consistent with the reported CMV strains. Molecular detection by RT-PCR with CMV-specific primers resulted in amplification of predicted size of approximately 540 bp. Sequence alignment of 533 bp amplicon (accession no. AJ634532) revealed 91–99% sequence identity with partial intercistronic region and CP gene of CMV in subgroup I. This was the first report of CMV on gerbera (Verma et al. 2004).

Gautum et al. (2014) also detected the occurrence of CMV in *Gerbera jamesonii* where leaf samples exhibit severe chlorotic mosaic and flower deformation symptoms. The complete RNA 3 genome was amplified by RT-PCR from the three infected gerbera leaf samples, cloned sequenced, and deposited in GenBank under accessions JN692495, JX913531 (from cv. Zingaro) and JX888093 (from cv. Silvester). The RNA3 sequences shared 98–99 % identities to each other and with a strain of CMV-Banana reported from India. Phylogenetic analysis revealed their closest affinity with CMV-Banana strain, and close relationships with several other strains of CMV of subgroup IB (Gautum et al. 2014).

9.2.5.13 Zinnia

CMV isolates of *Zinnia elegans* (CMV-Z), Petunia (CMV-Pet) and Chrysanthemum (CMV-CI) were identified as CMV strains based on aphid transmission, electron microscopy and protein analysis and serological studies. They also showed some biological, serological and satellite RNA based differentiation with other CMV strains isolate earlier from chrysanthemum, petunia and tobacco (Raj et al. 1997). The virus isolates could be transmitted by mechanical inoculation to various plant species. The total viral RNA isolated from the partially purified virus preparations when electrophoressed in agarose gels (1.2%) revealed the four RNA species separated as four distinct bands which were identified as RNA 1, RNA 2, RNA 3 and RNA 4.

In an independent study, the virus isolated from *Z. elegans* and *O. sanctum* were identified as CMV strain based on aphid transmission, presence of isometric cored particle (28 nm), 26 kDa CP subunits and close serological relationships with CMV-C and CMV-D strains. RNA extracted from purified virus particles was ssRNA and infectious, and revealed presence of RNA 1, RNA 2, RNA 3 and RNA 4 but no evidence of Sat-RNA in its genome (Raj et al. 1997).

9.2.6 Characterization of CMV Infecting Medicinal and Aromatic Plants

9.2.6.1 Egyptian Henbane (*Hyoscyamus* sp.)

Zaim and Khan in 1988 reported the occurrence of CMV in *Hyoscyamus niger* and *Rauvolfia serpentine* plants exhibiting a severe green mosaic symptoms. The virus was isometric cored and sized ~28 nm in diameter and was transmitted mechanically by aphid. Based on the mode of transmission, host range, biophysical properties and serological relationship revealed that the causal entity in both the cases was an isolate of cucumber mosaic virus (Zaim and Khan 1988).

Later Samad et al. (2000) also observed the mottle crinkle and severe mosaic symptoms in Egyptian henbane (*H. muticus* L.) in Northern India. The virus associated with the disease was characterized as CMV based on aphid transmission, presence of ~28 nm isometric cored virus particles, 26 kDa CP and single-stranded tripartite RNA genome with a subgenomic RNA (RNA 4) lacking satellite RNA. The CMV henbane isolate showed strong serological relationship with CMV-S and CMV-A strains in ODDT test and therefore identified as an isolate CMV and designated as CMV-H strain (Samad et al. 2000).

A molecular relationship of CMV-A, CMV-D and CMV-H strains isolated from *A. tricolor*, *D. innoxia* and *H. muticus*, respectively was investigated by RT-PCR RFLP and sequence analysis of CP gene by Srivastava and Raj in 2004. The RFLP analysis of CP gene with *Hind*III, *Sal*I, *Alu*I and *Rsa*I restriction enzymes indicated their placement into CMV subgroup I. Sequence analysis and phylogenetic trees of nucleotide and amino acid sequences grouped them with Indian isolates and into subgroup IB. CMV strains revealed high molecular similarity among themselves and appeared as a distinct cluster within subgroup IB, indicating their common origin in relation to other members of the subgroup (Srivastava and Raj 2004).

9.2.6.2 Basil (Ocimum sanctum)

Raj et al. (1997) isolated two CMV isolates from *Ocimum sanctum* and *Zinnia elegans* and identified them on the basis of aphid transmission, 28 nm cored spherical virus particles, molecular weight of 26 kDa of CP subunits and serological relationships with CMV. Besides this, the virus strains showed a few biological, serological and satellite RNA based differentiation with other CMV strains isolated earlier from chrysanthemum, petunia and tobacco (Raj et al. 1997) and suggested that some Indian strains of CMV lacking satellite RNA.

After this, Khan et al. in 2011 also found the occurrence of mosaic disease on *O. sanctum* L.) in Aligarh, India. The associated disease was sap transmissible from infected *O. sanctum* to healthy *O. sanctum* and to some assay species. In RT-PCR using CP gene specific primers a fragment of 650 bp in infected samples was amplified, cloned and sequenced (Accession no. EU600216). CP gene sequence analysis revealed high 97–99% nucleotide and amino acid identities with CMV subgroup II strains and identification as new isolate of CMV subgroup II and designated as CMV-Basil (Khan et al. 2011).

9.2.6.3 Winged Bean (*Psophocarpus tetragonolobus*)

Singh in 1990 recorded a virus causing mosaic symptoms in winged bean (*Psophocarpus tetragonolobus*) for the first time in India. The virus was sap and aphid transmissible and had a TIP of 60–65 °C, DEP of 1:1,000–1:10,000, and LIV for 36 h at room temp. (25–30 °C). During EM studies of leaf-dip preparations of infected leaves showed the presence of ~26 nm isometric cored virus particles. The virus was serologically related to CMV and identified as a strain of CMV (Singh 1990).

9.2.6.4 Datura (*Datura innoxia*)

Severe mosaic, leaf deformation and shoestring symptoms in *D. innoxia* plants were observed in 1999 by Raj et al. (1999). The associated virus with the disease was identified as CMV based on RT-PCR using CMV-CP gene specific primers and authenticity of amplicons was evidenced by Southern hybridization tests with a probe derived from cloned CP cDNA of CMV. This was the first report of CMV in *D. innoxia*.

9.2.6.5 Geranium (*Pelargonium* spp.)

Verma et al. (2006) noticed a viral disease on geraniums plants grown in a greenhouse at IHBT, Palampur, where plants were exhibiting mild mottling and stunting symptoms. The associated virus was transmissible to indicator hosts of CMV by sap and aphid inoculations. The virus reacted positively with CMV-specific antiserum in ELISA and its genome hybridized positively with probe during DNA-RNA hybridization. A CP gene using degenerate primers was amplified by RT-PCR and sequenced which showed 97–98% and 96–99% sequence identity at nucleotide and amino acid residues, respectively and close phylogenetic relationships with the sequences of CMV subgroup II. Based on sequence analysis, the virus infecting geraniums was identified as CMV subgroup II strain (Verma et al. 2006).

9.2.6.6 Lemongrass (*Cymbopogon citratus*)

Raj et al. in 2007 reported the natural occurrence of CMV on lemongrass (*Cymbopogon citratus*). The occurrence of CMV was detected by ELISA using CMV-CP specific antiserum and RT-PCR using CMV-CP gene specific primers. The amplified CP gene sequence was cloned and analyzed. The sequence analysis revealed 93–97% nucleotide sequence identities and close phylogenetic relationship with CMV isolates of subgroup Ib (Raj et al. 2007). This was the first record of CMV on lemongrass.

9.2.6.7 Sarpagandha (*Rauvolfia serpentine*)

Sarpagandha (*Rauvolfia serpentine*), family *Apocynaceae* is cultivated for roots which are used in several herbal formulations. The severe mosaic and stunting symptoms were observed on *R. serpentina* plants growing in cultivated plots of CVSIR-NBRI, Lucknow. The causal agent was transmissible by sap inoculation on to a few tobacco species (*N. tabacum* cv. White Burley, *N. rustica*, and *N. glutinosa*) and indicator plant (*C. sativus*) for CMV which produced mosaic symptoms. The virus reacted positively with antiserum of CMV) (PVAS 242a, ATCC, USA) in gel ODDT tests, suggesting the presence of CMV. RT-PCR with CMV-CP specific primers (AM180922/AM180923) resulted in amplification of ~650 bp band in infected samples which was cloned sequenced (Accession No. DQ914877). BLASTn analysis of the virus sequnece showed highest 99% nucleotide sequence identity with CMV isolates (DQ640743, AF350450, X89652, and AF281864). The virus isolate under study shared close phylogenetic relationships with CMV subgroup IB strains reported from India and identified as CMV-R strain. This was the first record of occurrence of CMV on *R. serpentine* (Raj et al. 2007).

9.2.6.8 Castor Bean (*Ricinus communis*)

A blister and leaf distortion disease of castor bean (*Ricinus communis* L.) was observed by Raj et al. (2010). The association of CMV with castor bean was detected by ODDT tests, WIBA tests using antiserum specific to CMV and confirmed by RT-PCR using CMV-CP gene primers. RT-PCR resulted in expected size band of 650 bp which was sequenced. Sequence analysis of amplicon revealed 96–98% nucleotide sequence identities and close relationships with CMV subgroup IB isolates. Based on these studies, the virus was identified as an isolate of CMV of subgroup IB. This was the first record of the natural occurrence of CMV on *R. communis* (Raj et al. 2010).

9.2.6.9 Periwinkle (*Catharanthus roseus*)

Samad and co-workers in 2008 observed the mosaic, leaf distortion and stunting disease in periwinkle (*Catharanthus roseus*). The virus isolate of periwinkle showed all the characters consistent with the CMV properties as similar host range, biophysical properties, isometric cored particles of ~28 nm, ~26 kDa CP subunit, affection to CMV-S during ODDT and transmission by aphids and sap to indicator assay hosts. The CP gene of ~657 bp was amplified from infected samples by RT-PCR using CMV-CP primers, cloned and sequenced. Sequence analysis of amplicon revealed high 98% nucleotide sequence identity and phylogenetic relationships with CMV-R strains of subgroup IB isolated from *R. serpentine* (Samad et al. 2008).

Afreen et al. in 2011 studied a virus causing mosaic and leaf deformation on *C. roseus* observed in the adjoining areas of Aligarh. The diseased plants showed mosaic, leaf deformation and reduced flower size. The virus was transmitted through sap inoculation and aphids, and showed symptoms on different plant species belonging to families: *Cucurbitaceae*, *Chenopodiaceae*, *Solanaceae* and *Leguminosae*. The size of virus ~28 nm diameter was in agreement with the size of CMV and also showed positive reaction with CMV specific antiserum in DAS-ELISA. Based on these analyses, the presence of CMV on *C. roseus* was confirmed (Afreen et al. 2011).

9.2.6.10 *Acorus calamus* and *Typhonium trilobatum*

Being CMV such a high economically importance virus, the screening for CMV in medicinal plants of *Acorus calamus, Typhonium trilobatum, Hippeastrum hybridum, Dahlia pinnata* and *Hemerocallis fulva*, plants exhibiting severe leaf mosaic symptoms was done by Kumar et al. (2015). The CP gene of ~650 bp was amplified from these samples using CMV-CP gene specific primers, cloned and sequenced and isolates were designated as HH (KP698590), AC (KP698588), DP (JF682239),

HF (KP698589), and TT (JX570732). The sequence data was analyzed for sequence and genetic diversity by BLASTn. The CMV-HH, −AC, −DP, −HF and −TT isolates revealed 82–99% nucleotide sequence identity among them and close relationships with CMV subgroup IB isolates. During alignment analysis of amino acid sequences HH and AC isolates showed fifteen and twelve unique substitutions, suggesting high genetic diversity as compared to HF, DP and TT isolates (Kumar et al. 2015).

9.2.7 Characterization of CMV Infecting Oil Crops

9.2.7.1 Safflower (*Carthamus tinctorius*)

Sangar and Rai (1988) identified a mosaic disease of safflower (*C. tinctorius* L.) from Satpura region of India in 1984. The virus isolate of safflower was identified as CMV based on biological, serological and physical properties similar to as CMV.

9.2.7.2 Jatropha (*Jatropha curcas*)

Raj et al. in 2008b observed the *Jatropha curcas* plants exhibiting severe mosaic disease accompanied by yellow spots in 15% plants in plots of CSIR-NBRI, India in 2006. Sap inoculations from infected to healthy Jatropha seedlings induced mosaic symptoms as were in infected plants. OODT tests suggested the presence of CMV as sap from infected plants reacted only with CMV (PVAS-242a, ATCC, USA). RT-PCR with CMV-CP gene primers resulted in expected size ~650-bp amplicon which was cloned and sequenced (accession no. EF153739). BLASTn analysis showed 98–99% nucleotide sequence identity with several CMV accessions DQ914877, DQ640743, AF350450, AF281864, X89652, AF198622, DQ152254, DQ141675, and DQ028777. Phylogeny revealed its close relationships with Indian isolates of CMV of subgroup Ib. This was the first report of CMV on *J. curcas* (Raj et al. 2008b) besides a report of Jatropha mosaic virus on *J. gossypiifolia* in Puerto Rico (Raj et al. 2008b).

9.2.8 Characterization of CMV Infecting Weed Plants

Trainthema monogyna and *Boerhavia diffusa* were reported as alternate hosts of CMV in 1970 by Khurana at Gorakhpur University and the virus was readily transmitted to eggplant from the weeds (Khurana 1970). In another study by Joshi and Dubey (1975), five weed plants *viz*.,: *A. viridis* L., *Nicotiana plumbginifolia* Vir., *Physalis minima* L., *Salvia plebeian* R. Br. and *Solanum nigrum* L. were found affected by CMV.

The incidence and progress of CMV were recorded by Kiranmai et al. (1998a, b, c) in commercial brinjal, chilli and tomato crops around Tirupati (A.P.) during 1992–1994 Kharif and Rabi seasons. The field collected crop plants that were positive for CMV in DAC-ELISA also produced chlorotic and/or necrotic local lesions an assay plants. The weed plants like *Acalypha indica*, *Amaranthus spinosis*, *A. viridis*,

Commelina bengalensis, *Digeria arvensis*, *Phyllanthus niruri* and *Physalis minima* collected in and around the three vegetable crops were found infected with CMV and they probably acted as alternative hosts. Based on the present studies, it is suggested that rouging of alternate virus source plants initially infected crop plants may help in decreasing the primary and secondary spread of Virus (Kiranmai et al. 1998a, b, c).

CMV was also detected from the naturally infected weed hosts around tarai region in Uttar Pradesh (Gupta et al. 2004). The weeds plants were collected and screened for CMV by using biological indicator host technique. Thirty one out of thirty five tested weeds species of twenty families were found CMV positive such as: *Achyranthes aspera*, *Amaranthus spinosus*, *Amaranthus viridis*, *Boerhavia diffusa*, *Croton bonplandianus*, *Euphorbia hirta* and *Solanum nigrum*. The only six weeds species remained free from CMV infection were: *Ageratum conyzoides*, *Ruellia prostrata*, *Rumex dentatus*, *Salvia officinalis*, *Sonchus oleraceus* and *Xanthium strumarium* (Gupta et al. 2004).

9.3 CMV Strains Reported from India

The reported CMV strains have been clustered into two major subgroups: I and II based on serological properties and nucleotide sequence homology (Palukaitis 1992). Furthermore, the subgroup I has been further been split into subgroup IA and IB by phylogenetic analysis of all reported CMV sequences (Roossinck et al. 1999; Palukaitis and Garcia-Arenal 2004). Most of the Indian strains of CMV have been included in 1B subgroup; however, some members of 1A and II subgroups of CMV have also been reported from India (Table 9.1).

Most of the Indian CMV strains detected from various plants: henbane (Samad et al. 2000), Amaranths (Srivastava et al. 2004), chrysanthemums (Kumar et al. 2005), Jatropha (Raj et al. 2008a, b), caster bean (Raj et al. 2010), tomato (Pratap et al. 2012a, b), banana (Vishnoi et al. 2013), eggplants (Kumar et al. 2014) have been identified as the members of subgroup IB. While other Indian CMV strains reported from *Ornithogalum* (Verma et al. 2005a, b); palargonium (Verma et al. 2006); gladiolus (Dubey et al. 2010); carrot (Afreen et al. 2009); basil (Khan et al. 2011a, b); cucumber (Kumari et al. 2013); and oxalis (Sheikh et al. 2013) were identified as members of subgroup 1A or II.

9.4 Genetic Diversity of CMV and TAV in India

To observe genetic diversity among strains of CMV in India, the phylogenetic analysis of CMV subgroup IA, IB and II strains was done by the aligning the nucleotide sequences of CP gene and phylogenetic tree was constructed by MEGA tool using neighbor joining method (Tamura et al. 2004). The CMV strains: Fny, Tfn and Trk7 strains were selected as reference sequences for subgroup IA, IB and II, respectively and a strain of *Peanut stunt virus* was taken as out-group for rooting the

Table 9.1 CMV strains of subgroup 1 (A and B) and II reported from last decade in India

Sub group	Natural host	Family	References
Subgroup-IB			
CMV-IB	*Hyoscyamus muticus*	*Solanaceae*	Samad et al. (2000)
CMV-IB	*Amaranthus tricolor*	*Amaranthaceae*	Srivastava et al. (2004)
CMV-IB	*Datura innoxia*	*Solanaceae*	Srivastava and Raj (2004)
CMV-IB	*Gerbera jamesonii*	*Asteraceae*	Verma et al. (2004)
CMV-IB	*Piper nigrum*	*Piperaceae*	Bhat et al. (2005)
CMV-IB	*Vanilla planifolia*	*Orchidaceae*	Madhubala et al. (2005)
CMV-IB	*Alstroemeria*	*Alstroemeriaceae*	Verma et al. (2005a, b)
CMV-IB	*Chrysanthemum morifrolium*	*Asteraceae*	Kumar et al. (2005)
CMV-IB	*Piper betle*	*Piperaceae*	Hareesh et al. (2006)
CMV-IB	*Cymbopogon citratus*	*Poaceae*	Raj et al. 2007a, b, c
CMV-IB	*Rauvolfia serpentina*	*Apocynaceae*	Raj et al. (2007a, b, c)
CMV-IB	*Solanum lycopersicon*	*Solanaceae*	Pratap et al. (2008)
CMV-IB	*Jatropha curcus*	*Euphorbiaceae*	Raj et al. (2008a, b)
CMV-IB	*Catharanthus roseus*	*Apocynaceae*	Samad et al. 2008
CMV-IB	*Ricinus communisis*	*Euphorbiaceae*	Raj et al. (2010)
CMV-IB	*Musa paradisiaca*	*Musaceae*	Khan et al. (2011a, b)
CMV-IB	*Petunia hybrida*	*Solanaceae*	Gautam et al. (2012)
CMV-IB	*Solanum lycopersicum*	*Solanacece*	Pratap et al. (2012a, b)
CMV-IB	*Musa paradisiaca*	*Musaceae*	Vishnoi et al. (2013)
CMV-IB	*Oxalis corymbosa*	*Oxalidaceae*	Sheikh et al. (2013)
CMV-IB	*Solanum melongena*	*Solanaceae*	Kumar et al. (2014)
CMV-IB	*Gerbera jamesonii*	*Asteraceae*	Gautam et al. (2014)
CMV-IB	*Musa paradisiaca*	*Musaceae*	Khan (2015)
CMV-IB	*Acorus calamus*	*Acoraceae*	Kumar et al. (2015)
Subgroup-IA			
CMV-IA	*Ornithogalum umbellatum*	*Asparagaceae*	Verma et al. (2005a, b)
CMV-IA	*Pelargonium graveolens*	*Geraniaceae*	Verma et al. (2006)
CMV-IA	*Gladiolus* sp	*Iridaceae*	Dubey et al. (2010)
Subgroup-II			
CMV-II	*Lilium longiflorum*	*Liliaceae*	Sharma et al. (2005)
CMV-II	*Pelargonium* sp	*Geraniaceae*	Verma et al. (2006)
CMV-II	*Daucus carota*	*Apiaceae*	Afreen et al. (2009)
CMV-II	*Ocimum sanctum*	*Lamiaceae*	Khan et al. (2011a, b)
CMV-II	*Solanum lycopersicum*	*Solanacece*	
CMV-II	*Cucumis sativus*	*Cucurbitaceae*	Kumari et al. (2013)
CMV-II	*Capsicum annum*	*Solanaceae*	Biswas et al. (2013)

phylogenetic tree. The generated phylogenetic tree reflected that the CMV strains clustered in subgroups: IA, IB and II (Fig. 9.2). Majority of CMV strains (15 out of 33 taken for the study) from India (DQ141675, AF198622, AF350450, AF281864, EF593026, JQ779842, AM158321, JN642676, HM484375, GU253913, AY545924, AY754359, DQ285569, JF279609 and AY125575) were found to lie in subgroup IB

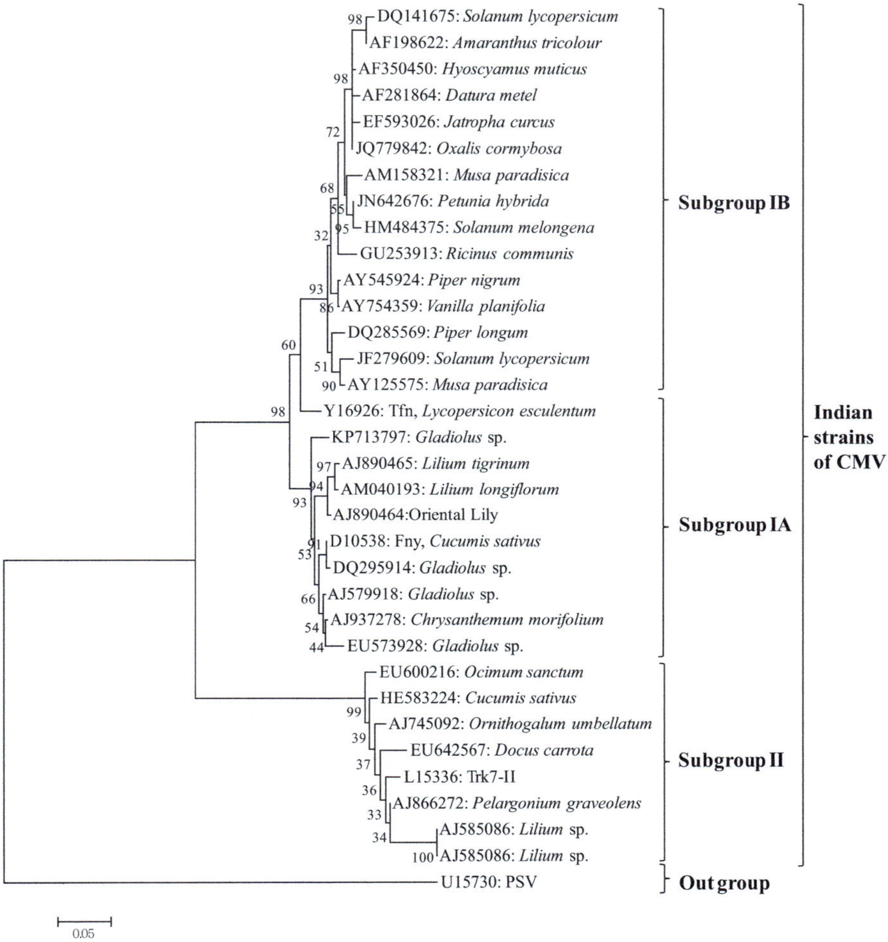

Fig. 9.2 Phylogenetic analysis of different CMV strains of subgroups: *IA*, *IB* and *II*, reported from various parts of India. Fny, Tfn and Trk7 were taken as reference sequences for subgroup *IA*, *IB* and *II*, respectively; whereas, PSV was taken as out-group for rooting

along with Tfn (acc. Y16926) taken as reference. Eight strains of CMV reported from India were clustered in subgroup IA (KP713797, AJ890465, AM040193, AJ890464, DQ295914, AJ937278 and EU573928) along with the reference strain: Fny (acc. D10538). While the six strains of CMV reported from India (acc. EU600216, HE583224, AJ745092, EU642567, AJ866272 and AJ585086) (out of total 33 strains retrieved from NCBI database for study) clustered with the strain of CMV subgroup-II (Trk7, acc. L15336) taken as reference.

These findings suggested that genetic diversity exists among Indian strains of CMV and all CMV strains reported from India belong to subgroups of CMV (IB, IA and II) and the majority of Indian strains belong to subgroup IB. For finding the possibility of recombination, identification of likely parents and localization of possible breakpoints in CMV strains, recombination analysis was performed by a RDP v4 (Martin et al. 2010) using seven methods (RDP, GENECONV, BOOTSCAN, Maximum Chi Square, CHIMEARA, SISCAN and 3Seq). The results eliminated the possibility of recombination in 29 Indian strains of CMV among the 33 sequences retrieved from NCBI database and suggest that the Indian CMV strains may have evolved independently as the subgroups IA, IB and II members.

The genetic diversity of TAV causing severe mosaic, chlorotic ringspots and flower deformation in chrysanthemums (*Chrysanthemum morifolium*) in Lucknow (UP), Dhanbad (MP) and Kolkata (WB), India has also been studied for the first time in India by Raj et al. (2009). TAV was detected by ELISA and by RT-PCR using TAV specific primers. The complete RNA 3 of each TAV isolate was cloned and sequenced and determined to be 2,386 nucleotides (nt) long, and to encode two open reading frames (ORFs): the movement protein (MP) of 741 nt and the coat protein (CP) of 657 nt translating in to 246 and 218 amino acid (aa), respectively. Indian isolates shared 98–99% identities among themselves and with the KC, V, P, B, I and C strains of TAV when aligned with seven other strains of TAV occurring worldwide. During phylogenetic analysis, Lucknow and Kolkata isolates of TAV showed a close relationship with KC-TAV strain of South Korea, while Dhanbad isolate showed closeness with V-TAV strains of Spain and Australia. Recombination events were also observed in the CP region of the Dhanbad isolate, supporting its diverse behavior (Raj et al. 2009). The nucleotide and amino acid sequences of the ORFs of Indian TAV isolates revealed high conservation despite their different geographical origin, and indicated that there are lesser radical radiations of the TAV populations in the natural environment. The nt and aa sequences of the ORFs of Indian TAV isolates revealed high conservation despite their different geographical origin, and indicated that there are lesser radical radiations of the TAV populations in the natural environment. These observations has been strengthened by the finding of Choi et al. (2002) and Moriones et al. (1991) emphasizing lesser or no quasi species of TAV were generated through fitness and recombination mechanisms in host plants, contrary to the other members of Cucumoviruses, as CMV which had a part of an evolutionary lines.

9.5 Concluding Remarks

In view of the literature available on CMV and TAV from India until now, it seems that an ample of research work has been done on biological, biochemical, serological, and molecular characterization of CMV and TAV for its proper identification. For management of the disease caused by CMV and TAV, the significant progress has also been made by control of the vector by conventional method as well as by non-conventional methods. The elimination of virus by culturing of shoot meristem tip, chemotherapy, thermotherapy and electrotherapy of the virus infected explants by many researchers in chrysanthemum (Singh and Gupta 1978; Kumar et al. 2009), gladiolus, carnation, Asiatic lily and gerbera (Gautam 2014). The mixed infection of CMV and TAV was eliminated in *Chrysanthemum morifolium* Ramat. cv. Pooja

(Kumar et al. 2009) by in vitro culturing 0.3 mm long shoot meristem of infected plants on MS medium supplemented with 3.0 mg/l BAP and 0.5 mg/l NAA. The regenerated plants were indexed by DAC-ELISA and confirmed by RT-PCR. A total of 78.1% CMV and TAV-free shootlets were obtained from the regenerated shoot meristem as indexed by DAC-ELISA, of which only 65.6% were found truly virus-free when confirmed by RT-PCR. Virus-free shootlets were rooted on half MS medium and acclamatized under glasshouse. These plants showed better growth and quality of blooms as compared to diseased ones (Kumar et al. 2009).

The significant progress has been achieved in the country for developing virus resistant transgenic plants by expression of antisense RNA, satellite RNA and coat protein that resulted in considerable degree of resistance/tolerance against CMV infection in tobacco (Srivastava and Raj 2008), tomato (Pratap et al. 2012a, b), brinjal (Pratap et al. 2011), chrysanthemum (Kumar et al. 2012) crops.

It is also suggested by many workers that a holistic approach needs to be taken up for designing the disease control/management practices. These include judicious use of integrated management practices such as use of healthy and certified propagating materials, control of vector by use of insecticides, regular surveillance and use of sensitive diagnostic procedures, avoiding alternate hosts and eradication of weeds, inter-cropping with barrier crops, breeding resistant cultivars, cross-protection using mild strains and use of virus resistant transgenic plants using CP mediated resistance.

Acknowledgements Authors are thankful to the Director, CSIR-National Botanical research institute (CSIR-NBRI), Lucknow, India for permission for summarizing the data generated by the researchers of Plant Virus Laboratory and its inclusion as a chapter in this book.

References

Afreen B, Khan AA, Naqvi QA, Kumar S, Pratap D, Snehi SK, Raj SK (2009) Molecular identification of a Cucumber mosaic virus subgroup II isolate from carrot (*Daucus carota*) based on RNA3 genome sequence analyses. J Plant Dis Prot 123:89–94

Afreen B, Baghel G, Khan AA, Naqvi QA (2011) Studies of a virus causing mosaic and leaf deformation on *Catharanthus roseus*. Nat Acad Sci Lett India 80:514–518

Ali S, Akhtar M, Singh KS, Naqvi QA (2012) RT-PCR and CP gene based molecular characterization of a cucumber mosaic cucumovirus from Aligarh, U.P., India. Agric Sci 3:971–978

Anjaneyulu A, Apparao A (1967) Natural occurrence of cucumber mosaic virus on chilli in India. Indian Phytopathol 20:380–381

Bhargava KS (1951a) Some properties of four strains of cucumber mosaic virus. Ann Appl Biol 38:377–388

Bhargava KS (1951b) Natural occurrence of cucumber mosaic virus on chilli in India. Indian Phytopathol 20:380–381

Bharti M, Anandam RJ, Roa AS (1997) Effect of *Cucumber mosaic virus* on indicator plants, non-host crops and brinjal genotypes. Indian J Virol 13:147–151

Bhat AI, Devasahayam S, Anandaraj M (2003) Viral disease – a new threat to black pepper cultivation in india. Indian J Areca Spices Med Plants 5:46–48

Bhat AI, Venugopal MN, Pant RP, Bhai RS (2004a) Occurrence and distribution of viral diseases on vanilla in India. J Spices Aromat Crop 13:143–148

Bhat AI, Faisal TH, Madhubala R, Hareesh PS, Pant RP (2004b) Purification, production of antiserum and development of enzyme linked immunosorbent assay-based diagnosis for Cucumber mosaic virus infecting black pepper (*Piper nigrum* L.) J Spices Aromat Crop 13:16–21

Bhat AI, Hareesh PS, Madhubala R (2005) Sequencing of coat protein gene of an isolate of cucumber mosaic virus infecting black pepper (*Piper nigrum* L) in India. J Plant Biochem Biotechnol 14:37–40

Bhat AI, Siljo A, Deeshma KP (2013) Rapid detection of Piper yellow mottle virus and cucumber mosaic virus infecting black pepper (*Piper nigrum*) by loop-mediated isothermal amplification (LAMP). J Virol Methods 44:901

Bidari VB, Reddy HR (1983) Prevalence of chilli viruses in Dharwad district. Plant Pathol Newslett 1:11–12

Biswas K, Hallan V, Zaidi AA, Pandey PK (2013) Molecular evidence of cucumber mosaic virus subgroup II infecting *Capsicum annuum* L. in the western region of India. Curr Discov 2:97–105

Brierley P (1955) Symptoms induced in chrysanthemums on inoculation with virus of mosaic aspermy and flower distortion. Phytopathology 54:2–7

Brierley P (1958) Tomato aspermy in chrysanthemum from Asia and Europe. Plant Dis Rep 42:61–62

Chalam TV (1982). Identification and characterization of cucumber mosaic and bean yellow mosaic viruses affecting chickpea (*Cicer arietinum* L.) in India. Thesis submitted to the Andhra Pradesh Agricultural University

Chalam TV, Reddy MV, Nene YL, Beniwal SPS, Subbayya J (1986) Some properties of a strain of cucumber mosaic virus isolated from chickpea in India. Plant Dis 70:128–130

Choi SK, Choi JK, Ryn KH, Park WM (2002) Generation of infectious clones of a Korean strain of tomato aspermy virus. Mol Cell 13:52–60

Chuyan AKH, Krylov AV (1979) Properties of tomato aspermy virus from chrysanthemum and its host range in the Priomore region. Byull Glam Bot Kogo Soda 114:84–92

Das CR, Raychaudhary SP (1953) A probable strain of tomato aucuba mosaic virus. Indian Phytopathol 6:116–122

Dhingra KL, Chenulu VV, Verma MDA (1979) A leaf reduction disease of *Cicer arientinum* in India caused by cucumovirus. Curr Sci 48:486–488

Ding SW, Anderson BJ, Haase HR, Symons RH (1994) New overlapping gene encoded by the cucumber mosaic virus genome. Virology 198:593–601

Ding S, Li W, Symons RH (1995) A novel naturally occurring hybrid gene encoded by a plant RNA virus facilitates long distance virus movement. EMBO J 14:5762–5772

Ding S, Shi B, Li W, Symons R (1996) An interspecies hybrid RNA virus is significantly more virulent than either parental virus. Proc Natl Acad Sci U S A 93:7470–7474

Doolittle SP (1916) A new infectious mosaic disease of cucumber. Phytopathology 6:145–147

Dubey VK, Aminuddin, Singh VP (2010) Molecular characterization of cucumber mosaic virus infecting gladiolus, revealing its phylogeny distinct from the Indian isolate and alike the Fny strain of CMV. Virus Genes 41:126–134

Gautam KK (2014). Molecular detection and characterization of strains of cucumber mosaic virus infecting petunia and gerbera and their disease management. PhD thesis, Lucknow University, Lucknow

Gautam KK, Kumar S, Raj SK (2012) Molecular characterization of Cucumber mosaic virus strain causing severe mosaic disease on petunia in India. Phytoparasitica 138:597–640

Gautum KK, Raj R, Kumar S, Raj SK, Roy RK, Katiyar R (2014) Complete sequence of RNA3 of Cucumber mosaic virus isolates infecting *Gerbera jamesonii* suggests its grouping under IB subgroup. Virus Dis 25:398–401

George TE, Anand N, Singh SJ (1993) Isolation and identification of two major viruses infecting bell pepper in Karnataka. Ind Phytopathol 46:24–28

Gowda KTP, Reddy HR (1985) Distribution of chilli mosaic virus in some parts of Kolar district. Curr Res 14:16–17

Gupta RP, Singh BP (1981) Studies on chrysanthemum aspermy virus on chrysanthemum in India. J Mycol Plant Pathol 11:158–160

Gupta SP, Singh BR (1997) Severe mosaic of cape gooseberry due to cucumber mosaic virus. Indian J Virol 12:155–156

Gupta PK, Ansari NA, Gautam A, Pandey A, Tewari JP (2004) Incidence of some cucurbit virus isolates naturally infected weed host around devipatan mandal in tarai region. J Liv World 11:1–6

Haq QMR, Singh BP, Srivastava KM (1996) Biological, serological and molecular characterization of a cucumber mosaic virus isolate from India. Plant Pathol 45:823–828

Hollings M (1955) Investigation of chrysanthemum viruses I. Aspermy flower deformation. Ann Appl Biol 43:83–102

Joshi RD, Bhargava KS (1965) A mosaic disease of tomato (*Cyphomandra betacea* Sendt.) Indian Phytopathol 18:313

Joshi RD, Dubey LN (1975) Some weed reservoirs of cucumber mosaic virus in Gorakhpur and distant area. Indian Phytopathol 28:568

Kaplan IB, Zhang L, Palukaitis P (1998) Characterization of cucumber mosaic virus. V. Cell-to-cell movement requires capsid protein but not virions. Virology 246:221–231

Kapoor P (2012) Detection of cucumber mosaic virus in bell pepper crop from himachal Pradesh. Plant Dis Res 27:71–80

Khan S (2015) Partial characterization and development of sensitive and reliable diagnostic for the detection of cucumber mosaic virus. Turk J Agric For 8:4539–4541

Khan S, Jan AT, Aquil B, Haq QMR (2011a) Coat protein gene based characterization of cucumber mosaic virus isolates infecting banana in India. J Phytol 32:94–101

Khan AA, Sharma R, Afreen B, Naqvi QA, Kumar S, Snehi SK, Raj SK (2011b) Molecular identification of a new isolate of cucumber mosaic virus subgroup II from basil (*Ocimum sanctum*) in India. Phytoparasitica 82:941–945

Khatri HL, Chenulu VV (1967) Studies on mosaic disease of phlox. Plant Dis Rep 51:768–771

Khurana SMP (1970) *Trainthema monogyna* and *Boerhavia diffusa*: unrecorded hosts of brinjal mosaic disease. Plant Dis 54:437–438

Kiranmai GM (1989) M Sc dissertation, S. V. University, Tirupati

Kiranmai G, Sreenivasulu P, Nayudu MV (1996) Comparison of here different tests for detection of *Cucumber mosaic virus* in banana (*Musa paradsiiaca*). Curr Sci 71:764–767

Kiranmai G, Sreenivasulu P, Nayadu MV (1997) Characterization of cucumber mosaic cucumovirus isolates naturally infecting three solanaceous vegetable crops in Andhra Pradesh. Indian Phytopathol 50:416–425

Kiranmai G, Hema M, Sreenivasulu P (1998a) Identification of a virus causing leaf distortion disease of papaya in Andhra Pradesh as cucumber mosaic cucumovirus. Indian Phytopathol 51:130–135

Kiranmai G, Sreenivasulu P, Nayadu MV (1998b) Epidemiology of cucumber mosaic cucumovirus isolates naturally infecting three solanaceous vegetables crops around Tirupati. Indian Phytopathol 51:315–318

Kiranmai G, Sreenivasulu V, Nayudu MV (1998c) Characterisation of cucumber mosaic cucumovirus isolation naturally infecting three solanaceous vegetable crops in Andhra Pradesh. Indian Phytopathol 50:416–425

Kumar S, Srivastava A, Raj SK (2005) Molecular detection of cucumber mosaic virus in chrysanthemum cultivars. Acta Phytopathol Entomol Hung 40:205–211

Kumar S, Khan MS, Raj SK, Sharma AK (2009) Elimination of mixed infection of *Cucumber mosaic* and *Tomato aspermy virus* from *Chrysanthemum morifolium* Ramat. cv. Pooja by shoot meristem culture. Sci Hortic 119:108–112

Kumar S, Raj SK, Sharma AK, Varma HN (2012) Genetic transformation and development of *Cucumber mosaic virus* resistant transgenic plants of *Chrysanthemum morifolium* cv. Kundan. Sci Hortic 134:40–45

Kumar S, Gautam KK, Raj SK (2014) Molecular identification of cucumber mosaic virus isolates of subgroup IB associated with mosaic disease of eggplant in India. Virus Dis 25:129–131

Kumar S, Gautam KK, Raj SK (2015) Sequence analysis and genetic diversity of five new Indian isolates of cucumber mosaic virus. Acta Virol 59:398. (In press)

Kumari R, Bhardwaj P, Singh L, Zaidi AA, Hallan V (2013) Biological and molecular characterization of cucumber mosaic virus subgroup II isolate causing severe mosaic in cucumber. Indian J Virol 60:135–147

Madhubala R, Bhadramurthy V, Bhat AI, Hareesh PS, Retheesh ST, Bhai RS (2005) Occurrence of cucumber mosaic virus on vanilla (*Vanilla planifolia* Andrews) in India. J Biosci 30:339–350

Mali VR, Rajegore SB (1980) A cucumber mosaic virus disease of banana in India. Phytopathol Z 98:127–136

Marani F (1969) Chrysanthemum aspermy virus. Phytopathol Mediterr 8:142–149

Martin DP, Lemey P, Lott M, Moulton V, Posada D, Lefeuvre P (2010) RDP3: a flexible and fast computer program for analyzing recombination. Bioinformatics 26:2462–2463

Moriones E, Roossinck MJ, Gracia-Arenal F (1991) Nucleotide sequence of tomato aspermy virus RNA 2. J Gen Virol 72:779–783

Murthy MRN (1983) Comparison of the nucleotide sequences of cucumber mosaic virus and brome mosaic virus. J Mol Biol 168:469–475

Nagaraju N, Murthy KVK (1994) Studies on a sap transmissible virus associated with mosaic disease of cowpea. Ind Phytopathol 47:23–26

Nagaraju, Reddy HR (1982) Occurrence and distribution of bell pepper viruses around Bangalore. Curr Res 10:155

Naqvi QA, Samad A (1985) Purification and properties of calendula yellow net virus. Indian J Virol 1:143–146

Narayan R, Dhawan P (1989) Characterization and identification of viruses causing mosaic disease of *Capsicum* in nature. Indian Phytopathol 42:221–226

Palukaitis P (1992) *Cucumber mosaic virus*. Adv Virus Res 41:281–348

Palukaitis P, Garicia-Arenal F (2004) Cucumoviruses. Adv Virus Res 62:241–323

Patel KV, Mali VR (1986) Comparative studies of three isolates of cucumber mosaic virus from banana. Indian Phytopathol 36:443–447

Prasada Rao RDV (1976) Characterization and identification of some chilli mosaic viruses. PhD thesis University of Agricultural Sciences, Hebbal, Bangalore, Karnataka, India

Pratap D, Kumar S, Raj SK (2008) First molecular identification of a cucumber mosaic virus isolate associated with shoestring symptom on tomato in India. Aust Plant Dis Notes 3:57–58

Pratap D, Kumar S, Raj SK, Sharma AK (2011) *Agrobacterium*-mediated transformation of eggplant (*Solanum melongena* L.) using cotyledon explants and coat protein gene of *cucumber mosaic virus*. Indian J Biotechnol 10:19–24

Pratap D, Kumar S, Snehi SK, Raj SK (2012a) Biological and molecular characterization of cucumber mosaic virus isolate causing shoestring disease of tomato in India which has closer affinity to European or East Asian isolates of CMV. Indian J Virol 135:193–207

Pratap D, Raj SK, Kumar S, Snehi SK, Gautam KK, Sharma AK (2012b) Coat protein-mediated transgenic resistance in tomato against a IB subgroup *Cucumber mosaic virus* strain. Phytoparasitica 40:375–382

Praveen S, Tripathi S, Giri BK (2001) Differentiation of biologically distinct cucumber mosaic virus isolates by double stranded RNA profile. Indian Phytopathol 54:362–365

Prentice IW (1952) Chrysanthemum flower distortion survey. Plant Pathol 1:77–80

Raj SK, Srivastava KM, Raizada RK, Singh BP (1991) Chlorotic ring mosaic of chrysanthemum caused by a strain of chrysanthemum aspermy virus in India. J Plant Dis Prot 98:640–646

Raj SK, Aminuddin, Srivastava KM, Singh BP (1993) Natural infection of cucumber mosaic virus on *Dianthus barbatus* in India. Plant Pathol 42:811–813

Raj SK, Aminuddin, Singh BP, Pal M (1997a) Characterisation of a cucumber mosaic virus causing leaf crinkle and severe mosaic disease in Amaranthus in India. Can J Plant Pathol 19:97–100

Raj SK, Chandra G, Singh BP (1997b) Some Indian strains of cucumber mosaic virus (CMV) lacking satellite RNA. Indian J Exp Biol 35:1128–1131

Raj SK, Srivastava A, Chandra G, Singh BP (1999) Natural occurrence of cucumber mosaic cucumovirus in *Datura innoxia* in India. OEPP/EPPO Bull 29:455–457

Raj SK, Srivastava A, Chandra G, Singh BP (2000) Role of satellite RNA of and Indian isolate of cucumber mosaic virus in inducing lethal necrosis of tobacco plants. Indian J Exp Biol 38:613–616

Raj SK, Srivastava A, Chandra G, Singh BP (2002) Characterization of cucumber mosaic virus isolate infecting gladiolus cultivars and comparative evaluation of serological and molecular methods for sensitive diagnosis. Curr Sci 83:1132–1137

Raj SK, Srivastava A, Chowdhury MR, Johri JK (2003) Association of cucumber mosaic virus and a potyvirus with betelvine (*Piper betle* L.) as evidenced by ELISA and RT-PCR. Indian J Exp Biol 41:273–276

Raj SK, Kumar S, Choudhari S (2007a) Identification of tomato aspermy virus as the cause of yellow mosaic and flower deformation on chrysanthemums in India. Australas Plant Dis Notes 2:1–2

Raj SK, Kumar S, Pratap D, Vishnoi R, Choudhari S, Chandra S (2007c) Natural occurrence of cucumber mosaic virus on lemongrass (*Cymbopogon citratus*), a new record. Australas Plant Dis Notes 2:95–96

Raj SK, Kumar S, Pratap D, Vishnoi R, Snehi SK (2007b) Natural occurrence of cucumber mosaic virus on *Rauvolfia serpentina*, a new record. Plant Dis 91:322

Raj SK, Khan MS, Kumar S, Pratap D, Vishnoi R (2008a) Cucumber mosaic virus infecting vegetable and pulse crops in India. Characterization, diagnosis & management of plant viruses. In: Rao GP, Lava Kumar P, Holguin RJ, Pena (eds) Vegetable and pulse crops, vol 3. Studium Press LLC, Texas, pp 39–62

Raj SK, Kumar S, Snehi SK (2008b) First report of cucumber mosaic virus on *Jatropha curcas* in India. Plant Dis 92:171

Raj SK, Kumar S, Choudhari S, Verma DK (2009) Biological and molecular characterization of three isolates of tomato aspermy virus infecting chrysanthemums in India. J Phytopathol 157:117–125

Raj SK, Snehi SK, Gautam KK, Khan MS (2010) First report of association of cucumber mosaic virus with blister and leaf distortion of castor bean (*Ricinus communis*) in India. Phytoparasitica 160:106–111

Raj SK, Kumar S, Verma DK, Snehi SK (2011) First report on molecular detection and identification of tomato aspermy virus naturally occurring on gladiolus in India. Phytoparasitica 39:303–307

Rao GP, Singh M, Shukla K, Gupta SN (1995) A new strain of cucumber mosaic virus causing mosaic disease of pea (Pisum sativum L.) in India. Indian J Virol 11:43–48

Roossinck MJ (2002) Evolutionary history of *cucumber mosaic virus* deduced by phylogenetic analyses. J Virol 76:3382–3387

Roossinck MJ, Zhang L, Hellward K (1999) Rearrangements in the 5′ nontranslated region and phylogenetic analyses of *Cucumber mosaic virus* RNA3 indicate radial evolution of three subgroups. J Virol 73:6752–6758

Samad A, Raj SK, Srivastava A, Chandra G, Ajayakumar PV, Zaim M, Singh BP (2000) Characterization of an Indian isolate of cucumber mosaic virus infecting Egyptian henbane (*Hyoscyamus muticus* L.) Acta Virol 476:264–270

Samad A, Ajayakumar PV, Gupta MK, Shukla AK, Darokar MP, Somkuwar B, Alam M (2008) Natural infection of periwinkle (Catharanthus roseus) with cucumber mosaic virus, subgroup IB. Australas Plant Dis Notes 3:30–34

Sangar RBS, Rai MK (1988) A mosaic disease of safflower from Satpura region of India. Farm Sci J 3:185–187

Sarma YR, Kiranmai G, Sreenivasulu P, Anandaraj M, Hema M, Venkatramana M, Murthy AK, Reddy DVR (2001) Partial characterization and identification of a virus associated with stunt disease of black pepper (*Piper nigrum*) in South India. Curr Sci 22:595–605

Sastry KSM (1964) A new strain of CAV from India. Indian J Microbiol 4:37–40

Savithri HS, Devarajan S, Murthy MRN (1983) Stability and structural transition of tomato aspermy virus and cucumber mosaic virus. Virology 134:398–405

Seth ML, Raychaudhuri SP (1973) Further studies on a new mosaic disease of Brinjal. Proc Indian Natl Sci Acad 39:122–128

Sharma PN, Chowfla SC (1986) Studies on a mosaic disease of amaranthus (*Amaranthus caudatus* L.) in Himachal Pradesh. Indian J Mycol Plant Pathol 16:349–350

Sharma PN, Chowfla SC, Garg ID, Paul Khurana SM (1993) Properties of the viruses associated with disease complex of bell pepper. Ind Phytopathol 46:347–353

Sharma A, Mahinghara BK, Singh AK, Kulshrestha S, Raikhy G, Singh L, Verma N, Hallan V, Ram R, Zaidi AA (2005) Identification, detection and frequency of lily viruses in Northern India. Sci Hortic 106:213–227

Sheikh M, Naqvi QA, Mahmood I (2013) Molecular detection of an isolate of cucumber mosaic virus infecting *Oxalis corymbosa* in Aligarh, India. Arch Phytopathol Plant Protect 46:1816–1820

Shi BJ, Ding SW, Symons RH (1997) In vivo expression of an overlapping gene encoded by the cucumoviruses. J Gen Virol 78:237–241

Singh SJ (1990) Studies on virus causing mosaic disease in winged bean. Plant Dis Res 5:1–6

Singh BP, Gupta RP (1978) Control of chrysanthemum aspermy virus by heat therapy. Phytopathology 47:909–910

Singh BR, Shukla P (1990) Properties of a new strain of cucumber mosaic virus from chilli. Ind J Virol 6:58–63

Singh A, Cheema SS, Hundal JS (1990) Identification of virus(es) associated with chilli mosaic syndrome. J Res Punjab Agric Univ 27:75–83

Singh A, Bishnoi SS, Rishi N, Krishi HP (1996) A disease of chickpea caused by a cucumo virus in Himachal Pradesh. Ind J Virol 12:59–61

Singh D, Naqvi QA, Garg ID (1999) A strain of cucumber mosaic cucumovirus causing mosaic in marigold in India. Ind Phytopathol 52:114–117

Singh A, Bishnoi SS, Sirohi A, Rishi N (2002) Viral diseases of chickpea in Himachal Pradesh. Plant Dis Res 17:61–64

Srivastava A (2003) Molecular characterization of cucumber mosaic virus isolate infecting *Amaranthus* spp. and development of transgenic plants utilizing its coat protein gene for resistance against CMV. PhD thesis submitted to University of Lucknow, Lucknow, India

Srivastava A, Raj SK (2004) High molecular similarity among Indian isolates of cucumber mosaic virus suggests a common origin. Curr Sci 87:1126–1131

Srivastava A, Raj SK (2008) Coat protein mediated resistance against an Indian isolate of *Cucumber mosaic virus* of subgroup IB member in *Nicotiana benthamiana*. J Biosci 33:249–257

Srivastava A, Chandra G, Raizada RK, Srivastava KM, Singh BP (1991) Characterization of a virus causing necrosis in *Petunia hybrida* and establishment of its relationship using ELISA. Indian J Exp Biol 29:591–593

Srivastava KM, Raj SK, Singh BP (1992) Properties of a cucumber mosaic virus strain naturally infecting *Chrysanthemum* in India. Plant Dis 76:474–477

Srivastava A, Raj SK, Haq QMR, Srivastava KM, Singh BP, Sane PV (1995) Association of a cucumber mosaic virus strain with mosaic disease of banana, (*Musa paradisiaca*) – an evidence using immuno/nucleic acid probes. Indian J Exp Biol 33:986–988

Srivastava A, Chandra G, Raj SK (2004) Molecular characterization of a strain of cucumber mosaic virus based on coat protein and movement protein genes. Acta Virol 48:229–239

Suzuki M, Kuwata S, Kataoka J, Masuta C, Nitta N, Takanami Y (1991) Functional analysis of deletion mutants of cucumber mosaic virus RNA 3 using an in vitro transcription system. Virology 183:106–113

Tamura K, Nei M, Kumar S (2004) Prospects for inferring very large phylogenies by using the neighbor-joining method. PNAS (USA) 101:11030–11035

Verma N, Singh AK, Singh L, Kulshreshtha S (2004) Occurrence of cucumber mosaic virus in *Gerbera jamesonii* in India. Plant Dis 88:1161

Verma N, Singh L, Singh AK, Kulshrestha S, Raikhy G, Hallan V, Ram R, Zaidi AA (2005a) Ornithogalum: a new host of cucumber mosaic virus (CMV) from India. Plant Pathol 54:256

Verma N, Singh AK, Singh L, Raikhy G, Kulshrestha S, Singh MK, Hallan V, Ram R, Zaidi AA (2005b) Cucumber mosaic virus (CMV) infecting *Alstroemeria* hybrids in India. Austral Plant Pathol 34:119–120

Verma N, Mahinghara BK, Ram R, Zaidi AA (2006) Coat protein sequence shows that cucumber mosaic virus isolate from geraniums (*Pelargonium* spp.) belongs to subgroup II. J Biosci 51:3–10

Verma N, Kumar K, Kulshrestha S, Raikhy G, Hallan V, Ram R, Zaidi AA, Garg ID (2007a) Molecular studies on tomato aspermy isolates infecting chrysanthemum. Arch Phytopathol Plant Protect 40:1–13

Verma N, Mehra A, Singh L, Hallan V, Singh AK, Jabeen N, Singh MK, Ram R, Zaidi AA (2007b) Screening for viruses infecting chrysanthemum cultivars in India. Sci Hortic 86:89–97

Vishnoi R, Kumar S, Khan MS, Raj SK (2006) Molecular diagnosis of banana mosaic virus and banana bunchy top virus affecting banana cultivation India. XVI Annual convention and International symposium on management of vector borne viruses. 7–10 February, ICRISAT, Hyderabad; PP-8/05

Vishnoi R, Kumar S, Raj SK (2013) Molecular characterization of a cucumber mosaic virus isolate associated with mosaic disease of banana in India. Phytoparasitica 71:59–66

Zaim M, Khan MMAA (1988) Green mosaic of Hyoscyamus niger L. and Rauvolfia serpentine Benth. Ex Kurz. Indian J Plant Apathol 6:152–157

Pigeonpea Sterility Mosaic Emaraviruses: A Journey from a Mysterious Virus to Classic Emaraviruses

10

Basavaprabhu L. Patil and P. Lava Kumar

Abstract

Sterility mosaic disease (SMD) of pigeonpea (*Cajanus cajan* (L.) Millsp) first reported in 1931, is an economically most important viral disease, which is endemic to India, Nepal, Bangladesh and Myanmar. SMD was long suspected to be a viral disease, however its causal agent, pigeonpea sterility mosaic virus (PPSMV) was discovered only in 2000. In 2013, the full genome of PPSMV was sequenced and based on the genome organization the virus was assigned to the genus *Emaravirus*. In 2015, association of another distinct emaravirus in SMD affected pigeonpea was discovered and it was named as pigeonpea sterility mosaic virus 2. As per the latest ICTV classification, these two pigeonpea infecting emaraviruses are renamed and recognised as two different virus species, *Pigeonpea sterility mosaic emaravirus 1* and *Pigeonpea sterility mosaic emaravirus 2* under the family *Fimoviridae* of the order *Bunyavirales*. These two emaraviruses involved in SMD are transmitted in a semi-persistent manner by an eriophyid mite, *Aceria cajani* Channabassavanna (Acari: Arthropoda). These viruses and its eriophyid vector are highly specific to pigeonpea and its wild relatives. This chapter presents the review of the studies conducted on SMD of pigeonpea and PPSMVs in India.

Keywords

Pigeonpea • Sterility mosaic disease • Emaravirus • Pigeonpea sterility mosaic virus

B.L. Patil (✉)
ICAR-National Research Center on Plant Biotechnology, IARI,
Pusa Campus, New Delhi 110012, India
e-mail: Basavaprabhu.Patil@icar.gov.in; blpatil2046@gmail.com

P. Lava Kumar
International Institute of Tropical Agriculture, Oyo Road, PMB 5320 Ibadan, Nigeria
e-mail: L.Kumar@cgiar.org

B. Mandal et al. (eds.), *A Century of Plant Virology in India*,
https://doi.org/10.1007/978-981-10-5672-7_10

10.1 Introduction

Sterility mosaic disease (SMD) of pigeonpea (*Cajanus cajan* (L.) Millsp) is caused by two distinct emaraviruses, pigeonpea sterility mosaic virus 1 (PPSMV-1) and pigeonpea sterility mosaic virus 2 (PPSMV-2) (Elbeaino et al. 2014, 2015; Patil and Kumar 2015). However, as per the latest ratification of emaravirus species by the International Committee on Taxonomy of Viruses (ICTV), these two pigeonpea infecting emaravirus species are named as *Pigeonpea sterility mosaic emaravirus 1* and *Pigeonpea sterility mosaic emaravirus 2*. SMD, first reported in 1931 from Pusa in Bihar, is endemic to India, and its neighboring countries such as Bangladesh, Nepal and Myanmar, and is a major constraint for cultivation of pigeonpea resulting in an economic loss of ~300 million US$ in India (Mitra 1931; Nene 1995; Patil and Kumar 2015; Reddy et al. 1998). The nature of symptoms involved in SMD and the extent of yield losses is a manifestation of the pigeonpea genotype, environmental factors and the time of virus infection. Early virus infection (<45 days) can result in more than 95% yield loss, whereas late infections can lead to 26–97% yield losses (Kannaiyan et al. 1984). When the pigeonpea plants are infected by PPSMV during flowering period, the plants show excessive vegetative growth, with no/poor flowering and pod set and under congenial conditions the disease spreads like a plague and hence it is also referred as "Green Plague". The SMD affected pigeonpea plants are also vulnerable to fungal diseases and infestation by spider mites. The characteristic symptoms of SMD are partial to complete cessation of flowering (sterility), enhanced vegetative growth, chlorotic rings or mosaic symptoms on the leaves and a reduction in their size, and stunting (Fig. 10.1, Jones et al. 2004). However the type of pigeonpea cultivar and the time of virus infection largely manifest the symptoms or its severity.

The viruses involved in SMD are transmitted by an eriophyid mite, *Aceria cajani* Channabassavanna (Acari: Arthropoda) but not through seed (Kulkarni et al. 2002; Kumar et al. 2000, 2002b, 2003). Both the virus and its vector are highly host specific and restricted to pigeonpea and some of its wild relatives (Kumar et al. 2008). In the past, multi-location field trials of different pigeonpea genotypes across India had indicated the presence of PPSMV variants (Kumar et al. 2008; Reddy et al. 1993). Some of the studies on PPSMV isolate from Coimbatore (Tamil Nadu state) and Bengaluru (Karnataka state) had shown cytopathological differences and also variation in the size of their nucleocapsid proteins (Kumar et al. 2000, 2003; Patil and Kumar 2015).

After the first report of SMD from India in 1931, it took nearly 85 years to unravel the complete sequence information of the PPSMV genome (Fig. 10.2; Mitra 1931; Elbeaino et al. 2014). Based on the genome organization and sequence information, PPSMV has been included in the genus *Emaravirus* and family *Fimoviridae*, in the order *Bunyavirales* (Mielke and Muehlbach 2007; Mühlbach and Mielke-Ehret 2011). The genus *Emaravirus* is one of the youngest and emerging genera of plant viruses with a negative sense segmented RNA genome, with four to eight RNA segments depending on the emaraviral species (Di Bello et al. 2015; Mielke-Ehret and Muhlbach 2012).

Fig. 10.1 (**a**) A field with sterility mosaic disease (*SMD*)-affected pigeonpea, (**b**) mosaic symptoms on SMD-affected pigeonpea, (**c**) a sprout from the pigeonpea stubble left after harvest showing severe symptoms of SMD

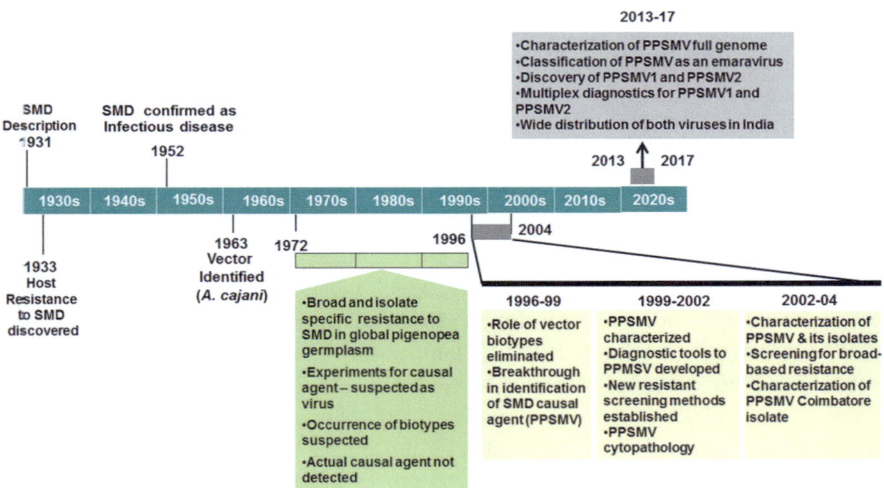

Fig. 10.2 Chronogram of major research findings made in the studies on sterility mosaic disease (*SMD*) of pigeonpea and Pigeonpea sterility mosaic virus-1 (*PPSMV-1*) and Pigeonpea sterility mosaic virus-2 (*PPSMV-2*)

10.2 Vector, Virus Transmission and Host Range

PPSMVs involved in SMD are semi-persistently transmitted by the mite vector *Aceria cajani* (Channabasavanna) (Arthropoda: Acari: Eriophyidae) (Channa basavanna 1981; Kumar et al. 2001; Kulkarni et al. 2002). These eriophyid mite vectors are known to transmit all the emaravirus species and also tritimoviruses of *Potyviridae* family (Kumar et al. 2001). The eriophyid mites are microscopic arthropods which disperse by the wind currents. During the active stages of their life cycle, these eriophyid mites are obligate pests which infest large number of plant species, resulting in severe economic losses (Oldfield and Proeseler 1996). These mites are almost exclusively found on the symptomatic leaves of SMD affected plants, which inhabit on the lower surface of leaves; however mite feeding causes no visible damage. Studies by Kulkarni et al. (2002) demonstrated that the transmission efficiency of a single eriophyid mite (*A. cajani*) can be up to 53% and were able to accomplish 100% transmission when more than five mites were used. To acquire the virus, these eriophyid mites required a minimum acquisition access period of 15 min and minimum inoculation access period of 90 min for transmission of the virus and there was no need of any latent period (Kulkarni et al. 2002). Although PPSMV stays within the mites for about 6–13 h, PPSMV replication within the mite and transovarial transmission has not been found (Kulkarni et al. 2002). There was a significant increase in the proliferation of *A. cajani* in PPSMV infected pigeonpea plants, when compared to the healthy plants suggesting for a beneficial relationship between the virus and its vector (Reddy and Nene 1980). Due to the presence of a smaller sized stylet the eriophyid mites feed on the mesophyll epidermal cells or its adjacent layers on the lower-surface of the leaves and this also affects the transmission efficiency of PPSMV. It is difficult to study the eriophyid mites on the experimental host plants from Chenopodiaceae, Solanaceae and Cucurbitaceae, because of its stringent host specificity and without feeding on pigeonpea the mites can survive only for ~13 h (Kulkarni et al. 2002). The eriophyid mites can proliferate to high densities in a week's time, once they are established on a susceptible pigeonpea genotype. These mites disperse passively, which is facilitated by the wind currents and the relative humidity has a profound impact on the population of mites (Kaushik et al. 2013; Singh and Rathi 1997). A temperature range of about 20–30 °C was found to be favourable for the multiplication of the eriophyid mites, whereas higher temperatures and heavy rains were not favourable. Sequence analysis of ITS region of rDNA did not reveal any significant variation among the populations of *A. cajani*, collected from diverse SMD endemic locations of India, Myanmar and Nepal, and this correlated with conservation of the morphological features of the eriophyid mites (Kumar et al. 2001; Latha and Doraiswamy 2008). These studies probably indicate that there are no biotypes of *A. cajani* and they don't differ much in their ability to transmit PPSMV.

Some success has been obtained in transmission of PPSMV to experimental hosts, such as *Nicotiana benthamiana*, *Nicotiana clevelandii*, *Chrozophora rottleri* and *Phaseolus vulgaris* by sap inoculation, but the efficiency has been low (Kulkarni et al. 2003b; Kumar et al. 2002b, 2003). However, attempts to sap-transmit to the natural host, pigeonpea, have failed and the purified PPSMV preparations were also not infectious (Kumar et al. 2003). However, PPSMV could be experimentally transmitted to pigeonpea by grafting (Ghanekar et al. 1992; Kumar et al. 2002a, b). In natural conditions only a few wild species of *Cajanus* could support the infestation by the eriophyid mite vector (Kumar et al. 2007). Seed or pollen transmission has not been reported for any emaraviruses, including PPSMV (Divya et al. 2005; Mielke-Ehret and Muhlbach 2012), since the virus has been detected in seed coat alone, but not in the cotyledons. Leaf stapling, infector hedge and spreader row inoculation are the three widely used experimental methods for transmission of PPSMV (Nene and Reddy 1976a, b). The leaf stapling method is most commonly used for evaluation in pots and field, whereas for field screening both the infector hedge and spreader row methods are used.

10.3 Epidemiology of SMD

SMD is an endemic disease in most of the pigeonpea cultivating areas of India; however its incidence varies widely from one region to another and from one season to the other (Kumar et al. 2008). SMD is reported from the states of Karnataka, Andhra Pradesh, Telangana, Bihar, Maharashtra, Tamil Nadu, Chhattisgarh, Gujarat, Punjab, Uttar Pradesh and West Bengal (Kannaiyan et al. 1984; Narayana et al. 2000; Singh and Raghuraman 2011; Zote et al. 1991). Since PPSMV is not a seed transmitted virus, SMD is mostly introduced into the new fields through the viruliferous mites which pick up the virus either from the perennial pigeonpea or the volunteer plants. Hence, the incidence of SMD mainly depends on the vicinity of the new fields to the source of inoculum, weather conditions, susceptibility of pigeonpea genotype and the population of eriophyid mites.

Generally the irrigated pigeonpea are more vulnerable to early infection by PPSMV than the rain-fed pigeonpea, as it may provide a favourable environment to the vector mites (Dharmaraj et al. 2004). However the effect of weather parameters on the epidemiology of SMD is not well established and there are varying interpretations. Some of the potential sources of primary virus inoculum are the infected pigeonpea plants remaining in the farms after the harvest, and also the perennial pigeonpea and their wild relatives (Kumar et al. 2008; Narayana et al. 2000). Whereas, in rain-fed agriculture, the pigeonpea stubble left-over in the field after harvest, and the plants thriving around the water bodies, may have sufficient amount of green foliage to harbour the eriophyid mites (Dharmaraj et al. 2004). The left-over pigeonpea stubbles that sprout back to accumulate green foliage, particularly after the summer rains, serves as a primary source of virus inoculum, providing favourable conditions for multiple cycles of virus infection. However because of the vast diversity in the cropping patterns and seasons in India it is hard to recognize the primary sources of SMD inoculum (Patil and Kumar 2015).

10.4 Taxonomy, Genome Organization and Gene Functions

In 2000, a breakthrough was made in diagnosing the causal agent of SMD as a virus with a negative (-ve) strand RNA virus (Kumar et al. 2000). Electron microscopic studies of the virus particles indicated it to be of 8–11 nm in diameter, with undetermined length and showing characteristic features of both the genera *Tenuivirus* (Family: *Phenuiviridae*) and *Orthotospovirus* (Family: *Tospoviridae*) (Kumar et al. 2002a, b, 2003). Both orthotospoviruses and tenuiviruses are negative strand RNA viruses, which are the only plant infecting genera of the order *Bunyavirales*, majority of which infect animals (Kormelink et al. 2011). The viruses of the order *Bunyavirales* are classified into six genera, all of which are transmitted by the arthropod vectors. The virus particles (80–120 nm) are mostly bound by a double membrane, with a tripartite genome. The presence of envelop membrane is unique to orthotospoviruses, emaraviruses, and rhabdoviruses. The crux of the virus particles encompass the ribonucleoproteins which is essentially the viral genomic RNA rigidly associated with the nucleocapsid protein (NP), along with traces of viral RNA-dependent RNA polymerase (RdRp).

Electron micrographs of SMD affected pigeonpea and *N. benthamiana* plants revealed the presence of 100–150 nm DMBs and fibrous inclusions of varying sizes located next to the nucleus (Kumar et al. 2002a, b). The location of these emaravirus DMBs near the endoplasmic reticulum (ER) and Golgi cisterns suggesting that the these organelles are the sites of emaravirus particle morphogenesis, as reported for tospoviruses (Kikkert et al. 1999; Ishikawa et al. 2015). All the emaraviruses have segmented negative sense RNA genomes and are transmitted by the eriophyid mites. The *European mountain ash ringspot-associated emaravirus* (EMARaV) is the type species of this genus *Emaravirus*, consisting of four RNA segments (Mielke and Muehlbach 2007). The other species of the genus *Emaravirus* are *Actinidia chlorotic ringspot-associated emaravirus* (AcCRaV), *Fig mosaic emaravirus* (FMV), *Rose rosette emaravirus* (RRV), *Redbud yellow ringspot associated emaravirus* (RYRSaV), *Raspberry leaf blotch emaravirus* (RLBV), *High plains wheat mosaic emaravirus* (HPWMoV) and *Blackberry leaf mottle-associated emaravirus* (BLMaV) (Table 10.1) (Di Bello et al. 2015, 2016; Hassan et al. 2017; ICTV 2016; Laney et al. 2011; Mielke-Ehret and Muhlbach 2012; Mühlbach and Mielke-Ehret 2011; Patil and Kumar 2015; Tatineni et al. 2014; Zheng et al. 2017). Each RNA segment of the emaraviral genome typically contains only one open reading frame (ORF) encoding for a polymerase, glycoprotein, nucleocapsid, movement protein, and other proteins whose function is yet to be understood.

The first published sequence of pigeonpea sterility mosaic virus; later renamed as pigeonpea sterility mosaic virus 1 (PPSMV-1) was shown to contain five genomic RNA segments of varying sizes (Elbeaino et al. 2014, 2015). The largest of all is referred as RNA1, with 7022 nucleotides length, coding for RNA-dependent RNA polymerase (RdRp, 2295 amino acids); the other four segments are referred as RNA2 (2223 nt) coding for glycoprotein (GP, 649 amino acids); RNA3 (1442 nt) coding for nucleocapsid protein (NP, 309 amino acids); RNA4 (1563 nt) coding for a putative movement protein p4 (MP, 362 amino acids); and RNA5 (1689 nt) coding

Table 10.1 Genome organization of ten definitive members of genus *Emaravirus*, the size of their genomic segments and size of the putative proteins encoded by them

Emaravirus species	RNA-1 RdRp nt (kDa)	RNA-2 Glycoprotein precursor (p2) nt (kDa)	RNA-3 Nucleocapsid Protein (p3 or NP) nt (kDa)	RNA-4 Movement Protein (p4 or MP) nt (kDa)	RNA-5 unknown (p5) nt (kDa)	RNA-6 unknown (p6) nt (kDa)
PPSMV-1	7022	2223	1442	1563	1801	Present[a]
	(268)	74.3	34.6	40.8	55	27
PPSMV-2	7009	2229	1335	1491	1833	1194
	266	74.3	34.9	40.7	55	27
FMV	7093	2252	1490	1472	1752	1212
	264	73	35	40.5	59	22
RRV	7026	2220	1544	1541	–[b]	–
	265	74	36	41	–	–
EMARaV	7040	2335	1559	1348		
	266	75	35	27		–
RLBV	7062	2135	1365	1675	1718	
	269	75	32	42	56	
RYRSaV	7049	2200	1414	1513	1272	–
	267	74	35	42	26	
AcCRaV	7061	2267	1678	1644	1476	–
	226.9	75	34.6	43.6	26.5	
HPWMoV[c]	6981	2211	1439/1441	1682	1715	1752
	266	77	33	42	56	58
BLMaV	7050	2271	1510	1504	1224	–
	268	75	35	41	26	

Modified from Mielke-Ehret and Muhlbach (2012) and Patil and Kumar (2015)
[a]Reported later by Patil et al. 2017
[b]Not detected
[c]HPWMoV genome has eight RNA segments, RNA-7 (1434 nt) and RNA-8 (1339 nt) are predicted to code for 36 kDa and 21 kDa proteins. HPWMoV is also known to have two different sized RNA-3 (1439 and 1441 nt)

for p5 (474 amino acids), a protein with unknown function (Elbeaino et al. 2014). Following the first report of PPSMV-1 genome sequence, another novel emaravirus Pigeonpea sterility mosaic virus 2 (PPSMV-2) was also reported to be associated with the SMD of pigeonpea (Elbeaino et al. 2015). PPSMV-2 has a higher sequence similarity with FMV than the first published sequence of PPSMV-1 (Elbeaino et al. 2014, 2015). Subsequently, the first published PPSMV sequence was renamed as Pigeonpea sterility mosaic virus 1 (PPSMV-1) (Elbeaino et al. 2015), which was further ratified as *Pigeonpea sterility mosaic emaravirus 1* (ICTV 2016). The sequence reports by Elbeaino et al. (2014, 2015) showed that the PPSMV-1 had only five segments, whereas the PPSMV-2 was associated with an additional RNA segment referred as RNA6, with a length of 1194 nt, coding for 27 kDa protein p6, with unknown function. The six RNA segments RNA1 to RNA6 of PPSMV-2 are of the size: 7009, 2229, 1335, 1491, 1833 and 1194 nucleotides, respectively (Patil et al. 2017).

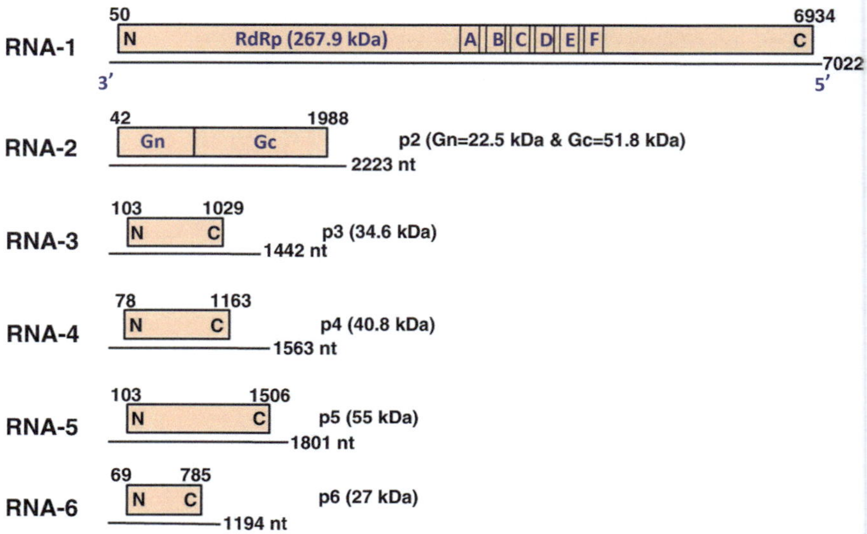

Fig. 10.3 Schematic representation of the genome organization of the six PPSMV-1 and/or PPSMV-2 RNA segments (*linear lines*) and the open reading frames (ORFs) of each RNA segment are indicated (*pink boxes*). Expression products of ORFs of each RNA segment (p1, p2, p3, p4 and p5) and their estimated molecular weights (kDa) are indicated. Letters (**a–f**) represent the conserved motifs on the RdRp protein encoded by RNA-1; Gn and Gc indicate the N- and C-terminal of the glycoprotein of RNA-2; and N and C are N-terminal and C-terminal ends of proteins, respectively, nt, nucleotides (Modified from: Patil and Kumar 2015; Elbeaino et al. 2014, 2015)

All the six RNA segments of PPSMV-1 and PPSMV-2 show maximum sequence identity with the corresponding RNA segments of FMV and RRV (Fig. 10.4). The single ORF of the segment RNA1 of both the PPSMVs encode for RNA dependent RNA polymerase (RdRp), consisting of 2294 amino acids (aa) that accounts for a molecular mass of 267.9 kDa (Fig. 10.3). The RdRp has all the conserved motifs that are common to members of the *Fimoviridae*. The RdRp of PPSMV-1 and PPSMV-2 have amino acid (aa) sequence similarities in the range of 37–54% and 30.9–72.1% respectively, with the RdRp of other emaraviruses (RRV, FMV, RYRV, EMARaV, RLBV). The active site of the RdRp encoded by both the PPSMV-1 and PPSMV-2 has a core polymerase module along with the five conserved motifs A-E (Fig. 10.3). Of the five conserved RdRp motifs the two motifs that are involved in divalent-metal cation binding are: motif A (DASKWS, 1125–1130) and C (SDD, 1183–1185) and these are also parts of the palm domain of the replicase protein (Bruenn 2003). The motif B (QGNLNHLSS, 1210–1218) has the role of RNA binding, whereas the motif D (KK, 1276–1277) with a tertiary structure shows catalytic activity. The cap-snatching activity characteristic of the members of order *Bunyavirales* is present in the motif E (EFLST, 1312–1316) (Duijsings et al. 2001). Cap-snatching is a gene expression strategy of several RNA viruses, wherein the

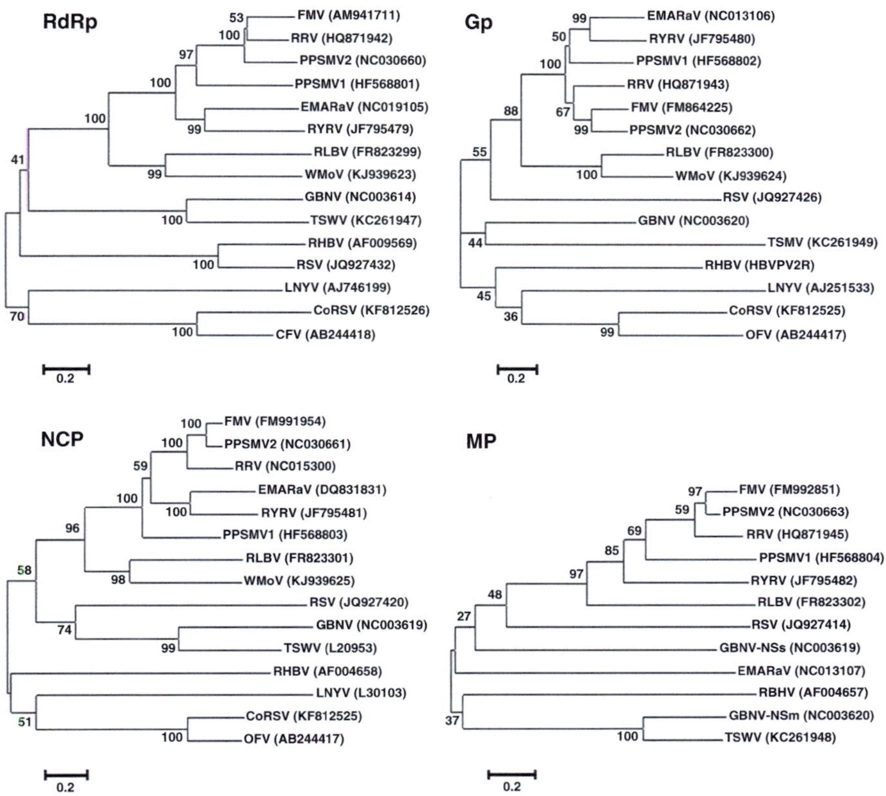

Fig. 10.4 Phylogenetic trees drawn with predicted amino acid sequences of RNA-dependent RNA polymerase (*RdRp*), glycoprotein precursor (*Gp*), nucleocapsid protein (*NP*) and movement protein (*MP*) of PPSMV-1 and PPSMV-2 isolates from Patancheru (Greater Hyderabad) with the orthologues of members of the genus *Emaravirus*, *Tenuivirus*, *Cytorhabdovirus*, *Dichorrhabdovirus* and *Orthotospovirus*. The accession numbers for each emaravirus isolates are given in the parenthesis. The analysis was carried out using the neighbour-joining algorithm and the Poisson-model in MEGA6, using 1000 bootstrap replicates. Bootstrap values are given on each node of each branch and the scale bar represents 0.2 substitutions per amino acid position. Tomato spotted wilt virus (*TSWV*) and Groundnut bud necrosis virus (*GBNV*) are orthotospoviruses; Rice stripe virus (*RSV*) and Rice hojablanca virus (*RHBV*) are tenuiviruses. The two Dichorhabdoviruses (unassigned negative-sense, ssRNA viruses), Coffee ringspot virus (*CoRSV*) and Orchid Fleck Virus (*OFV*); and a Cytorhabdovirus (*Rhabdoviridae* family) Lettuce necrotic yellow virus (*LNYV*) are used as members of out-group (Modified from: Patil and Kumar 2015)

endonuclease property of RdRp helps in the cleavage of the capped-RNA leader sequences of host mRNA, which is eventually used to initiate viral transcription (Kormelink et al. 2011).

The segment RNA2 of both PPSMV-1 and PPSMV-2 encode for glycoprotein precursors (GP) of 648 aa. The PPSMV-1 and PPSMV-2 GP share a sequence identity of 31–45% and 21.3–57.1%, respectively with the glycoprotein precursors of other emaraviruses. The glycoproteins form spike-like projections on the envelop membrane of emaraviruses and show the presence of three putative transmembrane helices and four putative glycosylation sites. Within the golgi complex, the precursor glycoprotein is cleaved into two functional glycoproteins: Gn (22.5 kDa) and Gc (51.8 kDa) at a predicted cleavage site. The PPSMV-1 RNA3 codes for a 34.6 kDa nucleocapsid (NP) protein, with aa identities comparable to NPs of FMV (44%), RRV (43%), EMARaV (35%), RYRSV (37%) and RLBV (27%). Whereas the PPSMV-2 encoded nucleocapsid protein shows highest sequence identity with FMV (Elbeaino et al. 2015). The three conserved amino acid motifs, NVLSFNK (134–140), NRLA (183–186) and GYEF (204–207) present within the nucleocapsid protein is thought to be involved in RNA-binding (Elbeaino et al. 2014, 2015). The RNA4 segment of PPSMV-1 encodes for a 40.8 kDa protein "p4", with sequence homology ranging from 24% to 41% with p4 of other emaraviruses. Whereas the PPSMV-2 p4 shares a higher sequence identity of 75% and 61.2% with FMV and RRV encoded p4, respectively. The amino acid analysis of p4 protein indicates it to be involved in cell-to-cell movement of the virus, similar to the movement proteins (MP) encoded by RNA4 of RRV and RLBV (Yu et al. 2013; McGavin et al. 2012). The movement protein modifies the plasmodesmata interconnecting the plant cells to allow movement of viral genomes from one cell to the other, followed by their systemic transport through the plant vasculature. It may also be possible that P4 functions as a suppressor of gene silencing.

The RNA5 of both PPSMV-1 and PPSMV-2 encodes for a 473 aa long and 55 kDa protein "p5". The PPSMV-1 encoded p5 shares a sequence identity of 33% with p5 of FMV, while the PPSMV-2 encoded p5 had a higher sequence identity with PPSMV-1 (61.1%), rather than FMV (43.1%). When the p5 encoded by RLBV was fused with the reporter gene green fluorescent protein, the fusion protein was localized as aggregated structures within the cytoplasm; however its function is yet to be understood (McGavin et al. 2012). The segment RNA6 of both PPSMV-1 and PPSMV-2 encodes for a 238 aa protein "p6" with a molecular mass of 27 kDa and with an unknown function (Elbeaino et al. 2015; Patil et al. 2017). The RNA6 and the p6 of both PPSMV-1 and PPSMV-2 show more than 96% sequence identity, while they share only 23.9% sequence identity with FMV (Elbeaino et al. 2015; Patil et al. 2017). The 5′ and 3′ termini of all the six RNA segments of both PPSMV-1 and PPSMV-2 possess untranslated regions (UTRs). The first 13 nts at both 5′ and 3′ termini of each emaravirus RNA segment highly conserved and complimentary, except for the two nucleotides at position 8 and 9 (U8-U9), which results in the formation of a panhandle structure, giving a pseudo-circular appearance to the ribonucleoprotein (RNPs). The terminal sequences of the genomic RNA segments are conserved among all the members of the same genus, but they differ among different genera (Elliott and Blakqori 2011).

Sequence variability studies among 23 isolates of PPSMV-1 and PPSMV-2, collected from ten different locations representing six states of India, showed that the isolates of both PPSMV-1 and PPSMV-2 are present across India and also occur as mixed infections (Patil et al. 2017). Detailed sequence analysis indicated the presence of recombination and reassortment among the corresponding RNA segments of both PPSMV-1 and PPSMV-2 isolates (Patil et al. 2017). The segment reassortments mostly involved the swapping of RNA4 segment among PPSMV-1 and PPSMV-2 (Patil et al. 2017). These studies also revealed that the sixth RNA segment (RNA6) previously reported to be associated with PPSMV-2 alone is also associated with the isolates of PPSMV-1 (Patil et al. 2017).

10.5 Diagnostics for PPSMV

Historically SMD diagnosis was based on characteristic mosaic symptoms (Fig. 10.1). However, virus isolation and characterization has paved for the development of serological and nucleic acid-based diagnostic methods. The first diagnostic test was developed using the polyclonal antibodies against the purified ribonucelo-protein particles of PPSMV (Kumar et al. 2003). These polyclonal antibodies were successfully employed for detection of PPSMV by the double antibody sandwich (DAS) and direct antigen coating forms of ELISA and Western immunoblotting (Kumar et al. 2002a, b, 2003). Using the same antibodies direct-immunobinding assay (DIBA) was developed to detect PPSMV in the viruliferous eriophyid mites (Latha and Doraiswamy 2008). Both DAS-ELISA and DIBA were sensitive for detection of PPSMV in the eriophyid mites. However a minimum of 10 mite extracts was required to get a weak signal by DIBA and a strong positive signal could only be obtained from an extract of 180 mites (Kulkarni et al. 2002). First RT-PCR based diagnostic assay was based on amplification about 250 nt RNA3 sequence of PPSMV (Kumar et al. 2003). Based on the current knowledge the primers developed by Kumar et al. (2003) were specific to PPMSV-1. Availability of full-length sequences of several emaraviruses led to the development of emaravirus-specific degenerate PCR primers for amplification of partial RdRp gene sequence (Elbeaino et al. 2013). Recently a diagnostic multiplex-RT-PCR technique for detection and differentiation PPSMV1 and PPSMV2 has also been developed (Patil et al. 2017). The terminal 13 nt sequences of each emaraviral RNA segment are similar and exhibit nucleotide complementarities and this could help in developing primers for amplification of full length RNA segments of emaraviruses by RT-PCR. Next generation sequencing (NGS) methods have also enhanced the speed at which unknown viruses are identified significantly reducing the costs and advancement in its performance. NGS was employed to get the first genome sequences of PPSMV-1 and PPSMV-2 (Elbeaino et al. 2014, 2015).

10.6 Concluding Remarks

SMD is known to occur in South East Asia alone, although pigeonpea is cultivated in almost all the continents. The virus vector, *A. cajani*, has been shown to occur in Southeast Asia, at least as far as Myanmar, but there is no information about its occurrence on pigeonpea in other continents. Based on available evidence that *A. cajani* fail to colonize virus resistant cultivars, it is plausible to conclude that *A.*

cajani may not occur in continents/countries where SMD is not known to occur. Interestingly there are no reports of presence of any other emaraviruses other than PPSMV-1 and PPSMV-2 from the Indian subcontinent. Current knowledge is insufficient to understand the origin and evolution of PPSMVs. This subject gets more complicated considering the high sequence diversity between PPSMV-1 and PPSMV-2 genomes, and close affiliation of PPSMV-2 to *Fig mosaic virus* occurring in Europe, than with PPSMV-1 which appears to be distinct compared to other Emaraviruses reported in Europe and North America. Despite high diversity, both PPSMV-1 and PPSMV-2 can infect same host, can cause SMD, are transmitted by the same insect vector, can also co-infect the same plant and recent evidence suggests genomic segment reassortment between the two viruses (Patil et al. 2017) indicating some co-evolution. It is likely that more emaraviruses in Indian-subcontinent are waiting to be discovered and such discovery studies should focus on host plants of eriophyid mites as these arthropods seems to be the main vector of emaraviruses. Further studies on PPSMV sequence diversity from across India and search for alternative hosts using new diagnostics should help in understanding the evolution of PPSMV-1 and PPSMV-2. Although the PPSMV-1 and PPSMV-2 genome sequences have been unraveled, the proteins encoded by them are yet to be characterized and should help to understand the biology of this novel emaravirus. Studies on virus-host and virus-insect interactions are necessary to find out a durable solution to this menace. Significant advancement in reliable, economical and reproducible diagnostics of SMD is still awaited. Field testing diagnostic methods enables the end users to accelerate their decision making by generating results at the point of sampling of the virus infected plant material. Some of the recently developed diagnostic methods, such as the Lateral Flow Devices (LFDs), Loop-mediated isothermal Amplification (LAMP), and the microarrays offer unique advantages and are promising for diagnostics of emaraviruses (Boonham et al. 2014).

Although broad-based resistance to SMD has been reported in the wild species such as *Cajanus scarabaeoides*, and resistance in some *C. scarabaeoides* accessions was to vector and some accessions were resistant to PPSMV (Kulkarni et al. 2003a; Kumar et al. 2005; Sharma et al. 2012). Studies on inheritance of SMD resistance in different genetic backgrounds of pigeonpea against different PPSMV isolates has led to contrasting interpretations on genetics and inheritance of SMD resistance (Bhairappanavar et al. 2014; Srinivas et al. 1997; Kumar et al. 2005; Gnanesh et al. 2011; Ganapathy et al. 2009, 2012). Hence the previous reports on genetics and inheritance of SMD resistance are not very conclusive. Thus, more efforts are required to map the disease resistance loci by using molecular tools, both in the cultivated and/or wild-type plants. The pigeonpea genome sequence should help in identification and isolation of R genes (NBS-LRR genes) against PPSMV, using different tools and techniques (Singh et al. 2012; Varshney et al. 2012). Significant progress is required in pigeonpea tissue culture and transformation to realize virus resistant transgenic pigeonpea (Krishna et al. 2010). RNA-interference (RNAi) is one of the most successful technologies widely used to accomplish virus resistant transgenics and it has been used to control orthotospoviruses, which have -ve sense RNA genome (Patil et al. 2011; Peng et al. 2014). RNAi-based transgenic resistance could be a promising for control of PPSMV and hitherto this technology has not been employed for transgenic management of emaraviruses (Patil and Kumar 2015).

The discovery of the causal agent of SMD and unraveling the viral genome sequences are important milestones, however, it is a long way to achieve the thorough understanding of the biological significance of single and mixed infections of PPSMV-1 and PPSMV-2 and their impact on the host resistance, virus-vector-host interactions, and development and delivery of broad-based virus resistant pigeonpea genotypes to the farming community of India.

Acknowledgement BLP acknowledges the financial support from ICAR-NRCPB to work on PPSMVs. Authors gratefully acknowledge research contributions of all the students, research scholars and scientists and a number of philanthropic donors for financial assistance to research work on SMD spanning over 85 years in India.

References

Bhairappanavar SB, Fakrudin B, Sambasiva Rao KRS (2014) Inheritance studies of sterility mosaic disease (SMD) resistance cross Gullyal White×BSMR736 in pigeonpea (*Cajanus cajan* (L.) Millsp.) Plant Gene Trait 5:27–32

Boonham N, Kreuze J, Winter S, van der Vlugt R, Bergervoet J, Tomlinson J, Mumford R (2014) Methods in virus diagnostics: from ELISA to next generation sequencing. Virus Res. 186:20–31

Bruenn JA (2003) A structural and primary sequence comparison of the viral RNA-dependent RNA polymerases. Nucleic Acids Res. 31:1821–1829

Channabasavanna GP (1981) Contributions to acarology in India. In: Channabasavanna GP, Mallik B, Ghopardé KD (eds) Proceedings of the 1st All India Symposium in Acarology, Bangalore

Dharmaraj PS, Narayana YD, Kumar PL, Waliyar F, Jones AT (2004) Pigeonpea sterility mosaic disease: an emerging problem in northern Karnataka. Int. Chickpea Pigeonpea Newsl. 11:47–49

Di Bello PL, Ho T, Tzanetakis IE (2015) The evolution of emaraviruses is becoming more complex: seven segments identified in the causal agent of Rose rosette disease. Virus Res. 210:241–244

Di Bello PL, Laney AG, Druciarek T, Ho T, Gergerich RC, Keller KE, Martin RR, Tzanetakis IE (2016) A novel emaravirus is associated with redbud yellow ringspot disease. Virus Res. 222:41–47

Divya P, Kumar PL, Rangaswamy KT, Muniyappa V (2005) Detection of pigeonpea sterility mosaic virus in floral parts and seeds. Indian J. Virol. 16:36

Duijsings D, Kormelink R, Goldbach R (2001) In vivo analysis of the TSWV cap-snatching mechanism: Single base complementarity and primer length requirements. EMBO J. 20:2535–2552

Elbeaino T, Whitfield A, Sharma M, Digiaro M (2013) Emaravirus-specific degenerate PCR primers allowed the identification of partial RNA-dependent RNA polymerase sequences of Maize red stripe virus and Pigeonpea sterility mosaic virus. J. Virol. Methods 188:37–40

Elbeaino T, Digiaro M, Uppala M, Sudini H (2014) Deep sequencing of pigeonpea sterility mosaic virus discloses five RNA segments related to emaraviruses. Virus Res. 188:27–31

Elbeaino T, Digiaro M, Uppala M, Sudini H (2015) Deep sequencing of dsRNAs recovered from mosaic-diseased pigeonpea reveals the presence of a novel emaravirus: pigeonpea sterility mosaic virus 2. Arch. Virol. 160:2019–2029

Elliott RM, Blakqori G (2011) Molecular biology of orthobunyaviruses. In: Plyusnin A, Elliott RM (eds) The Bunyaviridae: Molecular and Cellular Biology. Horizon Scientific Press, Norwich

Ganapathy KN, Byregowda M, Venkatesha SC, Rama Chandra R, Gnanesh BN, Girish G (2009) Identification of AFLP markers linked to sterility mosaic disease in pigeonpea *Cajanus cajan* (L.) Millsp. Int. J. Integr. Biol. 7:145–149

Ganapathy KN, Gowda BM, Ajay BC, Venkatesha SC, Gnanesh BN, Gomashe SS, Prasanth BG, Girish PS, Prasad GN, Veerakumar, Pati JV (2012) Inheritance studies of sterility mosaic disease (SMD) resistance in vegetable type pigeonpea (Cajanus cajan(L.) Millsp.) Aust J Crop Sci. 6:1154–1158

Ghanekar AM, Sheila VK, Beniwal SPS, Reddy MV, Nene YL (1992) Sterility mosaic of pigeonpea. In: Singh US, Mukhopadhyay AN, Kumar J, Chaube HS (eds) Plant Diseases of International Importance, Diseases of Cereals and Pulses, vol 1. Prentice Hall, New Jersey, pp 415–428

Gnanesh BN, Abhishek B, Sarma M, Wesley V, Byre Gowda M, Pande S, Saxena R, Saxena KB, Varshney RK (2011) Genetic mapping and quantitative trait locus analysis of resistance to sterility mosaic disease in pigeonpea [*Cajanus cajan* (L) Millsp.] Field Crop Res. 123:56–61

Hassan M, Di Bello PL, Keller KE, Martin RR, Sabanadzovic S, Tzanetakis IE (2017) A new, widespread emaravirus discovered in blackberry. Virus Res 235:1–5. doi:10.1016/j.virusres.2017.04.006

Ishikawa K, Miura C, Maejima K, Komatsu K, Hashimoto M, Tomomitsu T, Fukuoka M, Yusa A, Yamaji Y, Namba S (2015) Nucleocapsid protein from Fig mosaic virus forms cytoplasmic agglomerates that are hauled by endoplasmic reticulum streaming. J. Virol. 89(1):480–491

Jones AT, Kumar PL, Saxena KB, Kulkarni NK, Muniyappa V, Waliyar F (2004) Sterility mosaic disease-the “green plague” of pigeonpea: advances in understanding the etiology, transmission and control of a major virus disease. Plant Dis. 88:436–445

Kannaiyan J, Nene YL, Reddy MV, Ryan JG, Raju TN (1984) Prevalence of pigeonpea diseases and associated crop losses in Asia, Africa and the Americas. Trop Pest Manag 30:62–71

Kaushik D, Srivastava S, Nath BC, Chauhan VB, Singh R (2013) Correlation between mite population (*Aceria cajani)* and environmental factors causing sterility mosaic disease of Pigeonpea. Int. J. Life Sci. 1:228–232

Kikkert M, van Lent J, Storms M, Bodegom P, Kormelink R, Goldbach R (1999) Tomato spotted wilt virus particle morphogenesis in plant cells. J. Virol. 73:2288–2297

Kormelink R, Garcia ML, Goodin M, Sasaya T, Haenni AL (2011) Negative-strand RNA viruses: the plant-infecting counterparts. Virus Res. 162:184–202

Krishna G, Reddy PS, Ramteke PW, Bhattacharya PS (2010) Progress of tissue culture and genetic transformation research in pigeonpea [*Cajanus cajan* (L.) Millsp.] Plant Cell Rep. 29:1079–1095

Kulkarni NK, Kumar PL, Muniyappa V, Jones AT, Reddy DVR (2002) Transmission of Pigeonpea sterility mosaic virus by the eriophyid mite, *Aceria cajani* (Acari: Arthropoda). Plant Dis. 86:1297–1302

Kulkarni NK, Kumar PL, Muniyappa V, Jones AT, Reddy DVR (2003a) Broad-based resistance to pigeonpea sterility mosaic disease in the accessions of *Cajanus scarabaeoides*. Ind. J. Plant Protect. 31:6–11

Kulkarni NK, Kumar PL, Muniyappa V, Jones AT, Reddy DVR (2003b) Studies on host range of Pigeonpea sterility mosaic virus. J. Mycol. Plant Pathol. 33:141–145

Kumar PL, Jones AT, Sreenivasulu P, Reddy DVR (2000) Breakthrough in the identification of the causal agent of pigeonpea sterility mosaic disease. J Mycol Plant Pathol. 30:249

Kumar PL, Fenton B, Duncan G, Jones AT, Sreenivasulu P, Reddy DVR (2001) Assessment of variation in *Aceria cajani* (Acari: Eriophyidae) using analysis of nuclear rDNA ITS regions and scanning electron microscopy: Implications for the variability observed in host plant resistance to pigeonpea sterility mosaic disease. Ann. Appl. Biol. 139:61–73

Kumar PL, Jones AT, Reddy DVR (2002a) Mechanical transmission of Pigeonpea sterility mosaic virus. J. Mycol. Plant Pathol. 32:88–89

Kumar PL, Duncan GH, Roberts IM, Jones AT, Reddy DVR (2002b) Cytopathology of Pigeonpea sterility mosaic virus in pigeonpea and *Nicotiana benthamiana*: similarities with those of eriophyid mite-borne agents of undefined aetiology. Ann. Appl. Biol. 140:87–96

Kumar PL, Jones AT, Reddy DVR (2003) A novel mite-transmitted virus with a divided RNA genome closely associated with pigeonpea sterility mosaic disease. Phytopathology 93:71–81

Kumar PL, Latha TKS, Kulkarni NK, Raghavendra N, Saxena KB, Waliyar F, Rangaswamy KT, Muniyappa V, Doriswamy S, Jones AT (2005) Broad based resistance to pigeonpea sterility mosaic disease in wild relatives of pigeonpea (*Cajanus*: *Phaseoleae*). Ann. Appl. Biol. 146:371–379

Kumar PL, Jones AT, Waliyar F, Sreenivasulu P, Muniyappa V, Latha TKS, Saxena KB (2007) Pigeonpea sterility mosaic disease. In: Kumar PL, Waliyar F (eds) Diagnosis and detection of viruses infecting ICRISAT mandate crops. ICRISAT's methods manual

Kumar PL, Jones AT, Waliyar F (2008) Virus diseases of pigeonpea. In: Rao GP, Kumar PL, Holguin-Pena RJ (eds) Characterization, Diagnosis and Management of Plant Viruses, Vegetable and Pulse Crops, vol 3. Studium Press, Texas

Laney AG, Keller KE, Martin RR, Tzanetakis IE (2011) A discovery 70 years in the making: characterization of the Rose rosette virus. J. Gen. Virol. 92:1727–1732

Latha TKS, Doraiswamy S (2008) Detection of pigeonpea sterility mosaic virus, the causal agent of sterility mosaic disease of pigeonpea in viruliferous mite vector by DAS-ELISA and DIBA. Arch. Phytopathol. Plant Protect. 41:537–541

McGavin WJ, Mitchell C, Cock PJ, Wright KM, MacFarlane SA (2012) Raspberry leaf blotch virus, a putative new member of the genus Emaravirus, encodes a novel genomic RNA. J. Gen. Virol. 93:430–437

Mielke N, Muehlbach HP (2007) A novel, multipartite, negative-strand RNA virus is associated with the ringspot disease of European mountain ash (*Sorbus aucuparia* L.) J. Gen. Virol. 88:1337–1346

Mielke-Ehret N, Muhlbach HP (2012) Emaravirus: a novel genus of multipartite, negative strand RNA plant viruses. Virus 4:1515–1536

Mitra M (1931) Report of the imperial mycologist. Sci. Rep. Agric. Res. Inst. Pusa 19:58–71

Mühlbach HP, Mielke-Ehret N (2011) Emaravirus. In: AMQ K, Adams MJ, Carstens EB, Lefkovitz EJ (eds) Virus Taxonomy. Classification and Nomenclature of Viruses, Ninth Report of the International Committee on Taxonomy of Viruses. Elsevier, Oxford

Narayana YD, Mahalinga DM, Jayalakshmi SK, Benagi VI (2000) Prevalence of SMD of pigeonpea in Northern Karnataka. Karnataka J. Agric. Sci. 13:470–472

Nene YL (1995) Sterility mosaic of pigeonpea: the challenge continues. Indian J. Mycol. Plant Pathol. 25:1–11

Nene YL, Reddy MV (1976a) Screening for resistance to sterility mosaic of pigeonpea. Plant Dis. Rep. 60:1034–1036

Nene YL, Reddy MV (1976b) A new technique to screen pigeonpea for resistance to sterility mosaic for pigeonpea. Trop. Grain Legume Bull. 5:23–24

Oldfield G, Proeseler G (1996) Eriophiod mites as vectors of plant pathogens. In: Lindquist EE, Sabelis MW, Bruin J (eds) Eriophyoid Mites-Their Biology, Natural Enemies and Control. Elsevier, Amsterdam

Patil BL, Lava Kumar P (2015) Pigeonpea sterility mosaic virus: a legume-infecting Emaravirus from South Asia. Mol. Plant Pathol 16:775–786

Patil BL, Dangwal M, Mishra R (2017) Variability of Emaravirus species associated with sterility mosaic disease of pigeonpea in India provides evidence of segment reassortment. Viruses 9(7):183

Patil BL, Ogwok E, Wagaba H, Mohammed IU, Yadav JS, Bagewadi B, Taylor NJ, Kreuze JF, Maruthi MN, Alicai T, Fauquet CM (2011) RNAi-mediated resistance to diverse isolates belonging to two virus species involved in Cassava brown streak disease. Mol. Plant Pathol. 12:31–41

Peng JC, Chen TC, Raja JA, Yang CF, Chien WC, Lin CH, Liu FL, Wu HW, Yeh SD (2014) Broad-spectrum transgenic resistance against distinct tospovirus species at the genus level. PLoS One 9:e96073

Reddy MV, Nene YL (1980) Influence of sterility mosaic resistant pigeonpeas on multiplication of the mite vector. Ind Phytopathol 33:61–63

Reddy MV, Raju TN, Nene YL, Ghanekar AM, Amin KS, Arjunan G, Astaputre JV, Sinha BK, Reddy SV, Gupta RP, Gangadharan K (1993) Variability in sterility mosaic pathogen in pigeonpea in India. Indian Phytopathol. 46:206–212

Reddy MV, Raju TN, Lenne JM (1998) Diseases of pigeonpea. In: Allen DJ, Lenne JM (eds) The Pathology of Food and Pasture legumes. CAB International, Wallingford, pp 517–558
Sharma M, Rathore A, Mangala UN, Ghosh R, Sharma S, Upadhyay H, Pande S (2012) New sources of resistance to Fusarium wilt and sterility mosaic disease in a mini-core collection of pigeonpea germplasm. Eur. J. Plant Pathol. 133:707–714
Singh J, Raghuraman M (2011) Emerging scenario of important mite pests in north India. Zoosymposia 6:170–179
Singh AK, Rathi YPS (1997) Epidemiology of vector of pigeonpea sterility mosaic virus. Ind. J. Virol. 13:143–145
Singh NK, Gupta DK, Jayaswal PK, Mahato AK, Dutta S, Singh S, Bhutani S, Dogra V, Sing BP, Kumawat G, Pal JK, Pandit A, Singh A, Rawal H, Kumar A, Rama Prashat G, Khare A, Yadav R, Raje RS, Singh MN, Datta S, Fakrudin B, Wanjari KB, Kansal R, Dash PK, Jain PK, Bhattacharya R, Gaikwad K, Mohapatra T, Srinivasan R, Sharma TR (2012) The first draft of the pigeonpea genome sequence. J. Plant Biochem. Biotechnol. 21:98–112
Srinivas T, Reddy MV, Jain KC, Reddy MSS (1997) Studies on inheritance of resistance and allelic relationships for strain 2 of Pigeonpea sterility mosaic pathogen. Ann. Appl. Biol. 130:105–110
Tatineni S, McMechan AJ, Wosula EN, Wegulo SN, Graybosch RA, French R, Hein GL (2014) An eriophyid mite-transmitted plant virus contains eight genomic RNA segments with unusual heterogeneity in the nucleocapsid protein. J. Virol. 88:11834–11845
Varshney RK, Chen W, Li Y, Bharti AK, Saxena RK, Schlueter JA, Donoghue MT, Azam S, Fan G, Whaley AM, Farmer AD, Sheridan J, Iwata A, Tuteja R, Penmetsa RV, Wu W, Upadhyaya HD, Yang SP, Shah T, Saxena KB, Michael T, McCombie WR, Yang B, Zhang G, Yang H, Wang J, Spillane C, Cook DR, May GD, Xu X, Jackson SA (2012) Draft genome sequence of pigeonpea (*Cajanus cajan*), an orphan legume crop of resource-poor farmers. Nat. Biotechnol. 30:83–89
Yu C, Karlin DG, Lu Y, Wright K, Chen J, MacFarlane S (2013) Experimental and bioinformatic evidence that Raspberry leaf blotch emaravirus p4 is a movement protein of the 30K superfamily. J. Gen. Virol. 94:2117–2128
Zheng Y, Navarro B, Wang G, Wang Y, Yang Z, Xu W, Zhu C, Wang L, Serio FD, Hong N (2017) Actinidia chlorotic ringspot-associated virus: a novel emaravirus infecting kiwifruit plants. Mol. Plant Pathol. 18(4):569–581
Zote KK, Mali VR, Mayee CD, Kulkarni SV, Mote TS (1991) Outbreak of sterility mosaic of pigeonpea in Marathwada region Maharashtra. India. Int. Pigeonpea Newsl. 14:19–21

Characterization of Ilarviruses in India 11

M. Krishna Reddy, K. Vemana, and Sunil Kumar

Abstract

Ilarviruses (family *Bromoviridae*) infect a wide range of plants including herbs, shrubs and trees. The type species, *Tobacco streak virus* (TSV) is a problem on many crops worldwide. TSV has a broad host range and is widely distributed. Epidemic occurance of TSV in India has been witnessed on several important crops such as peanut, sunflower, okra, cotton and gherkin. Identification of the extent of ilarvirus spread, host range, thrips-vector species, virus survival and plant-to-plant spread are important to predict the risk factors of ilarvirus epidemics. In India so far three illarviruses have been identified, apple mosaic virus, prunus necrotic ringspot virus and TSV. Of all these ilarviruses, the major research has been focused on TSV. This chapter summarises the research work conducted on illarviruses in India.

Keywords

Ilarviruses in India • Diagnosis • Epidemiology • Management

11.1 Introduction

The members of the family *Bromoviridae* infect a wide range of hosts including herbaceous plants, shrubs and trees. The family is divided into six genera: *Alfamovirus*, *Anulavirus*, *Bromovirus*, *Cucumovirus*, *Ilarvirus* and *Oleavirus*

M. Krishna Reddy (✉) • S. Kumar
Division of Plant Pathology, Indian Institute of Horticultural Research,
Hessaraghatta Lake Po, Bangalore 560089, India
e-mail: mkreddy60@gmail.com

K. Vemana
Agricultural Research Station, Kadiri, Anantapuramu, Andhra Pradesh 515591, India

B. Mandal et al. (eds.), *A Century of Plant Virology in India*,
https://doi.org/10.1007/978-981-10-5672-7_11

(Fauquet et al. 2005). The members of the genus *Ilarvirus* have been known for 100 years. The first report of ilarvirus associated disease was a graft-transmissible variegation in apple, which later was identified as apple mosaic virus. Tobacco streak virus (TSV), the type member of the genus, was first described in 1936 (Johnson 1936). In the 9th report of ICTV, ilarviruses were classified in two 4 subgroups and one unassigned group with the two members that had no relation with the other four subgroups. The subgroup I comprises of TSV and Parietaria mottle virus. The subgroup II comprises of Asparagus virus 2, citrus leaf rugose virus, citrus variegation virus, elm mottle virus, lilac ring mottle virus, spinach latent virus and Tulare apple mosaic virus. Whereas, the subgroup III comprises of apple mosaic virus (ApMV), blueberry shock virus and Prunus necrotic ringspot virus (PNRSV). The subgroup IV comprised of prune dwarf virus and Fragaria chiloensis latent virus. Currently, 19 ilarvirus species have been recognised by ICTV (https://talk.ictvonline.org/files/master-species-lists/m/msl/6776).

In India, so far only three ilarviruses, ApMV, PNRSV and TSV have been identified on different crops. The sunflower necrosis disease (SND) was known to occur for the first time since 1997 on sunflower in Karnataka and Andhra Pradesh. The first experimental evidence of TSV as the cause of sunflower necrosis disease was reported in 2000 (Prasadarao et al. 2000; Ramaiah et al. 2001; Ravi et al. 2001) and subsequently it was reported in groundnut (Reddy et al. 2002). PNRSV known to occur worldwide in temperate regions has been reported from India on almond, apple, cherry, plum, begonia, rose geraniums (*Pelargonium* spp.) and rose (Kulshrestha et al. 2005; Chandel et al. 2008, 2011; Verma et al. 2002; Krishnareddy unpublished). ApMV, which is long been known in India has also been characterized from apple. In this chapter, we present the research workson ilarviruses that have so far been undertaken in India.

11.2 Prevallence and Crop Losses

Among the ilarviruses that infect annual and herbaceous crops, tobacco streak virus is currently distributed throughout the country. Initial disease epidemics of TSV occurred in mostly in southern states like Andhra Pradesh, Karnataka, Tamil Nadu and Maharashtra specially on cucurbits, cotton, Okra, Legumes and Sunflower. Whereas, the PNRSV also occurs through out the country on rose and on temperate fruits in northen India. ApMV is frequently found in stone fruits in Europe, where it is known also as European plum line pattern virus but it is also now found in India.

The occurrence of tobacco streak virus was reported from over 26 countries worldwide. In India, tobacco streak virus was first identified from SND in sunflower and stem necrosis disease-affected groundnut during 1999–2000 from Andhra Pradesh state (Prasadarao et al. 2000; Reddy et al. 2002). Since, then virus was found to be responsible for causing serious damage to groundnut, sunflower and several other annual crops in Andhra Pradesh, Karnataka, Maharashtra and Tamil Nadu states (Krishnareddy et al. 2003a, b; Kumar et al. 2006; Jain et al. 2008). Although, the virus is wide spread around the world, but it is not yet known to be destructive as observed in India (Prasadarao et al. 2000), Australia and USA (Sharman et al. 2008).

TSV appears to be spreading fast in India, as it has also been found in cotton, mungbean, urdbean, sun-hemp, safflower, soybean, pumpkin and bottle gourd (Bhat et al. 2002a, b). At present, TSV has become a menace in states like Andhra Pradesh (Ananthpur, Hyderabad, Kurnool, Cuddapah, Chittor, Warangal, Khammam, Guntur), Karnataka (Bagalkot, Bangalore, Bellary, Bidar, Bijapur, Chitradurga, Dharwad, Gulburga, Haveri, Kolar, Raichur, and Tumkur), Maharashtra (Akola, Aurangabad, Jalna, Osmanabad, Beed, Solapur, and Lathur, Parbhani), and Tamil Nadu (Coimbatore, Dharampuri, Salem, Periyar) on sunflower, peanut, cotton, okra, gherkin and cucurbits. During the 2005 cropping season, TSV affected cotton extensively in the predominant cotton growing regions of Khammam and Warangal regions of Andhra Pradesh. It was reported that an extent of 20,000 ha were affected by TSV. The disease incidence from these regions in cotton due to TSV was varied from 1% to 70%. The epidemics caused substantial losses in sunflower production, forcing many of the farmers to switch over to alternative crops. In early-infected plants, the floral heads are distorted and contain chaffy, ill-filled and poor quality seeds. Maximum disease incidence of up to 90% occurs in *Kharif* (monsoon crop (Ramiah et al. 2001). Similarly a TSV outbreaks in 2001–2002 in gherkin, a major export crop, reduced the cropping area from 257 to 105 ha and also shifted gherkin production into new areas in Andhra Pradesh and Tamil Nadu. The incidence of disease ranges from 31% to 75% both in *Kharif* and summer grown gherkins (Krishnareddy et al. 2003a). TSV incidence is more in *Khraif* grown gherkin in Andhra Pradesh, where as it is high in summer grown gherkin in Karnataka.

PNRSV causes significant crop losses depending on the hosts e.g., up to 15% in sweet cherry and up to 100% in peach. Since 1997, the outbreaks of SND were recognized on sunflower across a large area around Bangalore, Karnataka State (Annual Report, 1998 and 2000). The disease spread rapidly to adjoining regions and caused severe epidemics on sunflower in subsequent years. In 2000, PSND epidemic (Table 11.1) was recognized on groundnut grown in 225,000 ha in Anantapur and Kurnool districts of Andhra Pradesh, resulting in crop losses exceeding Rs 300 crores (Prasadarao et al. 2000; Reddy et al. 2002). The virus has become a menace on sunflower, groundnut, okra, cucurbits and several other annual and horticrops. TSV epidemic in 2002 on sunflower in Karnataka was reported to cause yield loss of Rs 110 crores. An epidemic in 2004 on groundnut in Andhra Pradesh resulted in an estimated yield loss of Rs 93 crores.

Table 11.1 Epidemics of tobacco streak virus and estimated yield loss

Crop	Year	Places occurred	Geographical spread (ha)	Estimated yield loss	Remarks
Groundnut	2002	Anantpuramu, Andhra Pradesh	225,000/700,000	>Rs300 crores (US$65 million)	Severe setback to the marginal farmers who depend solely on this crop
	2004	Anantpuramu	80,000/850,000	Rs 106 crores	Severe setback to the marginal
		Kurnool	69,700/850,000	Rs 93 crores	Farmers who depend solely on this crop
Sunflower	2000	Andhra Pradesh Karnataka	Area reduced to less than 2 m.ha	Yield loss estimated to 40–70%	Regional crop failure and forced farmers to abandon the crop
	2002	Gulburga, Raichur, Koppal, Bijapur, Bagalkot, Haveri, Davenagree, Bellary	135,000/585,000	Rs.110 crores	Regional crop failure and forced farmers
Gherkin	2002	Jalna, Beed, Pune, Solapur, Parbahni	37,000/240,000	Rs.20 crore	Forced farmers to abandon the crop
	2003	Gulburga, Raichur, Bellary, Haveri, Bagalkot, Bijapur	84,000/500,000	Rs 39 crore	Forced farmers to abandon the crop
	2003	Jalna, Latur, Solapur, Beed, Osmanabad	55,000/300,000	Rs.33 crore	Forced farmers to abandon the crop
	2005	Bellary, Dharward, Raichur		50–90% yield loss	More severe on hybrids of sunflower forced farmers to abandon the crop
	2002	Davanagiree, Doddaballpur, Tumkur	84/105	Rs 143.2 crores (US$32.5million)	Farmers who specialized in cultivation of gherkin shifted to cultivation of tomato
	2004	Bellary,Hubili	59/137	Rs 11.7 crores (US$2.65million)	Farmers who specialized in cultivation of gherkin shifted to cultivation of other vegetable crops
	2005	Kupam, V. Kota (A.P.) Bangarupet, Malur (Karnataka)	109/143	Rs 76.5 crores (US$3.75million)	Farmers who specialized in cultivation of gherkin shifted to cultivation of other crops

11.3 ApMV

ApMV is commonly known to infect blackberry, raspberry (*Rubus* sp), apple (*Malus* sp), apricot, cherry, almond, plum and peach (*Prunus* sp) hardnut (*Corylus avellana*), roses (Rosa and hop (*Humulus lupulus*) (Brunt et al. 1996). ApMV was first reported in *Malus domestica* from the USA by Bradford and Joley (1933). In India, apple mosaic were first observed in Uttarakhand during 1957 (Bhargava and Bist 1957). The virus infects most of the apple cultivars and occurs sporadically. No insect vector of ApMV is known, however the virus is transmitted most probably by pollen, vegetative propagation or by mechanical inoculation. ApMV is often found in mixed infections with prunus necrotic ringpost virus and prune dwary virus. Apple tree infected with the ApMV exhibits symptoms of pale to bright cream spots on the leaves. The infection results in to reduction in shoot growth, fruit set, fruit weight, yield per tree and ascorbic acid content of the fruit (Singh et al. 1979) which results in yield reduction of up to 60% (Menzel et al. 2002). Severely infected plants show yellowing of leaves and veins.

The complete genome sequence of one apple isolate of ApMV was determined. The CP was 223 amino acids in length and showed 87–99% identity when compared to 21 ApMV isolates. Whereas, MP (286 amino acids) showed 91–95% identity with other isolates. However, the gene sequences were quite conserved among Indian isolates and grouped together phylogenetically. CP of the Indian isolates showed maximum identity of 95% with Korean isolate (AY 125977) in apple and in other host these showed a maximum identity of 98% to Czech Republic pear isolate. MP showed maximum identity with Chinese isolate i.e., 95% (Lakshmi et al. 2011).

11.4 PNRSV

PNRSV infects most cultivated stone fruit trees, including almond, apple, apricot, begonia, cherry, sour cherry, peach and plum, rose and hops (Mink 1992; Fulton 1970; Verma et al. 2002). The virus causes serious reduction in growth of infected trees (30%) and yield (20–50%) (Uyemoto and Scott 1992). The infection can be latent or expressed through a large range of symptoms such as leaf chlorosis, rugose, mosaic, growth and yield reduction (Mink 1992). Numerous biological variants have been isolated (Aparicio et al. 1999; Aparicio and Pallas 2002; Vaskova et al. 2000; Glasa et al. 2002). Two main pathotypes (Rugose mosaic disease and a mild virus) and three serotypes (CH3, CH9, CH30) were described on sweet cherry tress but no line between the patho and serotypes was found (Mink et al. 1987).

The variability of PNRSV has been evaluated in its prunes hosts with the description of numerous isolates or strains that vary widely in their pathogenic, biophysical, serological properties (Myrta et al. 2001; Mink et al. 1987) and molecular variability (Moury et al. 2001; Vaskova et al. 2000). Further attempts to distinguish pathotypes either on basis of nuclic and hybridization (Crosslin et al. 1992) or evaluation of particle characteristics have also failed. Recently several nucleotide substitutions were identified, that correlated with known sero and pathotypes in Cherry isolates from Washington (Hammond and Crosslin 1998). The CP gene sequence variability was also compared with different PNRSV isolates (Glasa et al. 2002; Hammond 2003) originating from Europe and America (Hammond 2003). PNRSV has been recoded in India on almond, apple, cherry, plum, begonia, Rose geraniums (*Pelargonium* spp.) and Rose (Kulshrestha et al. 2005; Chandel et al. 2008, 2011; Verma et al. 2002).

11.5 TSV

TSV infects >200 plant species belonging to 30 dicotyledonous and monocotyledonous plant families and its widely distributed in Europe, North and South America, South Africa, India, Iran, Japan, Australia, and New Zealand (Fulton 1985; Scott 2001). TSV has caused serious damage to groundnut, sunflower and several other annual crops in India (Kumar et al. 2006). Although, the virus is widespread in the other countries, destructive epidemics have been observed only in India (Reddy et al. 2002). Having wide distribution and severe impact on agricultural as well as horticultural crops, TSV can be considered as one of the economically most important plant viruses. It is unstable in plant extracts but can be transmitted mechanically to a number of hosts. The other modes of transmission are pollen, seed, and thrips.

TSV has many natural and experimental hosts (Tables 11.2 and 11.3). It causes a variety of symptoms ranging from streaks, narrow toothed margins, necrotic and chlorotic spots, mosaic to symptomless infections on horticultural and agricultural crops (Table 11.2). In tobacco, narrow toothed margined leaf is the symptom of

Table 11.2 Natural hosts of tobacco streak virus recorded in India

Crop	Symptoms	References
Pulses		
Vigna radiata (Mungbean)	Necrosis of veins, stem, petioles	Bhat et al. (2002b)
Vigna mungo (Urad bean)	Brown necrotic lesion on young leaves with brown streaks on petioles and stem	Ladhalakshmi et al. (2006)
Cyamopsis tetragonoloba (Guar)	Foliar mosaic and necrotic spotting as well as necrotic streaks on buds and stems	Sivaprasad et al. (2010)
Macrotyloma uniflorum (Horse gram)	Leaves with necrotic spots with wrinkled margins, together with plant stunting and wilting	
Cajanus cajan (Pigonpea)	Symptoms included necrotic spots on young leaves followed by wilting of leaves, petiole and branch/axillary shoot proliferation, with small leaves having mosaic symptoms	Vemana and Jain (2010)

(continued)

Table 11.2 (continued)

Crop	Symptoms	References
Vegetables		
Capsicum annuum (Chili)	Necrosis of leaves and buds	Jain et al. (2005)
Allium cepa (Onion)	Necrotic lesions were observed on the young leaves and flower stalks resulting in flower abortion	Sivaprasad et al. (2010)
Abelmoschus esculentus esculentus (Okra)	Chlorotic streaking, leaf blotches and distortion of fruits	Krishna Reddy et al. (2003b)
Hibiscus cannabinus (Kenaf)	Mosaic, necrotic spotting on leaves and necrosis of bud	Redy et al. (2012)
Cucumis sativa (Cucumber)	Characterized by necrotic lesions on leaves and extending to mid veins, petioles, flower buds and tip. This is followed by die back of vines	Krishna Reddy et al. (2003a)
Cucumis anguria (Gherkin)		
Cucurbita pepo var coimbatore-1 (Pumpkin)	Leaf necrosis, stunting and wilt of plant	Sarover et al. (2010)
Fiber		
Gossypium hirsutum (Cotton)	Necrosis of veins, petioles, stem and other parts	Bhat et al. (2002a, b)
Crotalaria juncea (Sunhemp)		
Oilseed		
Helianthus annuus (Sunflower)	Necrosis of leaf lamina followed by twisting of the leaf and systemic mosaic/necrosis of the lamina, petiole, stem and floral calyx and corolla. Early infected plants remain stunted with malformed heads having chaffy or deformed seeds. Heads were partially twisted and become sterile	Ramaiah et al. (2001)
Arachis hypogaea (Groundnut)	Appears first on young leaves as necrotic lesions and veinal necrosis. The necrosis later spread to the petiole and stem. Necrotic lesions on the stem later spread upwards killing the bud. n the late infection spots appears on pegs and Pods also	Reddy et al. (2002)
Carthamus tinctorious (Safflower)	Veinal and leaf necrosis, necrotic streaks on the stem, necrosis of terminal bud and followed by death of plant	Chander Rao et al. (2003)
Guizotia abyssinica (Nizer)	Leaf and petiole necrosis	Arun kumar et al. (2006)
Ricinus communis (Castor)	Necrotic spots and vein mosaic were observed on the lower side of the leaves	Baskar reddy et al. (2014)
Ornamental		
Jasmine sambac (Jasmine)	Symptoms included severe leaf necrosis followed by wilting, together with necrotic streaking on petioles and branches and axillary shoot proliferation with small leaves	Goud et al. (2013)

Table 11.3 Experimental hosts range of tobacco streak virus recorded from India and other countries

Family	Experimental hosts
Aizoaceae	*Trianthema portulacastrum*
Amaranthaceae	*Amaranthus caudatus, Amaranthus retroflexus, Beta macrocarpa, Beta patellaris, Beta vulgaris, Gomphrena globosa, Spinacia oleracea*
Apocyanaceae	*Catharanthus roseus, Vinca rosea*
Asparagaceae	*Asparagus officinalis*
Caesalpinioideae	*Cassia occidentalis*
Caryophyllaceae	*Stellaria media*
Chenopodiaceae	*Chenopodium album, Chenopodium amaranticolor, Chenopodium capitatum, Chenopodium foetidum, Chenopodium hybridum, Chenopodium murale, Chenopodium quinoa*
Commelinaceae	*Commelina benghalensis*
Compositae	Ageratum *conyzoids, Dahlia rosea, Dahlia pinnata, Helianthus annuus, Lagasca mollis, Parthenium hysterophorus, Tithonia speciosa, Verbacina encelloides*
Cruciferae	*Brassica campestris, Matthiola incana,*
Cucurbitaceae	*Cucumis melo, Cucumis sativus, Cucurbita maxima, Cucurbita moschata, Cucumis pepo* (small sugar), *C. pepo* (Zucchini),, *Momordica balsamina*
Gramineae	*Zea mays*
Labiatae	*Ocimum basilicum, Salvia splendens*
Malvaceae	*Abelmoschus esculentus, Gossypium herbaceum, Gossypium hirsutum*
Papilionoideae,	*Arachis hypogaea, Cajanus cajan, Cicer arietinum, Crotalaria spectabilis, Cyamopsis tetragonoloba, Glycine max, Lens culinaris, Macrotyloma uniflorum, Medicago sativa, Melilotus albus, Phaseolus vulgaris, Pisum sativum, Sesbania exaltata, Trigonella foenum* graecum, *Trifolium incarnatum, Trifolium pretense, Trifolium repens, Vicia faba, Vigna radiata, Vigna mungo, Vigna radiata, Vigna unguiculata, Vigna unguiculata* ssp. *cylindrica, Vigna unguiculata* ssp. *sesquipedalis*
Polemoniaceae	*Phlox drummondii*
Rosaceae	*Fragaria vesca, Rosa setigera*
Scrophulariaeaae	*Antirrhinum majus, Torenia fournieri*
Solanaceae	*Capsicum annuum, Capsicum frutescens, Datura stramonium Hyoscyamus niger, Lactuca sativa, Lycopersicon esculentum, Lycopersicon pimpinellifolium, Nicandra physalodes, Nicotiana benthamiana, Nicotiana bigelovii, Nicotiana clevelandii, Nicotiana debneyi, Nicotiana glutinosa, Nicotiana megalosiphon, Nicotiana tabacum, Nicotiana edwardsonii, Petunia hybrida, Physalis floridana, Physalis peruviana, Physalis ixocarpa, Solanum demissum, Solanum demissum, S. tuberosum, Solanum melongena, Solanum nigrum, Solanum nodiflorum, Solanum rostratum, Solanum tuberosum, Zinnia elegans*

Sources: Fulton (1948), Kaiser et al. (1982), and Cupertino et al. (1984)

TSV. In soybean, systemic necrosis and bud blight were the important symptoms. In asparagus, stunting is the main characteristic symptom. Necrotic shock is the symptom in strawberry (Stace-Smith and Frazier 1971). Mosaic, ring spots, mottling and veinal chlorosis were the chief symptoms exhibited by dahlia (Pappu et al. 2008). The main symptoms in cowpea were mosaic and red node and latent infection in sweet clover. Necrosis of leaf lamina followed by twisting of the leaf and systemic mosaic/necrosis of the lamina, petiole, stem and floral calyx, corolla and stunting with malformed heads having chaffy or deformed seeds were observed in sunflower (Sharman et al. 2008). In groundnut, necrotic lesions appear first on young leaves and the necrosis later spread to the petiole and stem killing the top growing bud. In case of late infection, spots appear on pegs and pods (Reddy et al. 2002). TSV infected *Rubus* plants (black raspberry) (Converse 1972; Jones and Mayo 1975) and *Parthenium hysterophorus* are symptomless (Prasada Rao et al. 2003).

Gomphrena globasa, *Phaseolus vulgaris* and *Vigna unguiculata* have been used as assay hosts for TSV. TSV can also be identified based on the symptoms on indicator hosts such as, *N. tabacum, C. quinoa, V. unguiculata* (cv. C-152), *C. tetragonoloba* (cv. S-51)*, G. globosa* and *P. vulgaris* (cv. Topcrop). Diagnostic hosts such as *V. unguiculata* cv. C-152 and *P. vulgaris* cv. Topcrop help to differentiate TSV from PBNV. TSV produces necrotic local lesions and venial necrosis on inoculated primary leaves of these indicator plants, whereas groundnut bud necrosis virus (GBNV, tospovirus) produces only concentric chlorotic/necrotic local lesions. The symptoms of TSV on several host species can be confused with those caused by cucumber mosaic virus or GBNV. Therefore, confirmatory testing by serological or nucleic acid-based diagnostic assays is necessary.

TSV contains a single-stranded, tripartite RNA genome, virions are quasi-isometric and 27–35 nm in diameter. Virus particle preparations contain at least four ssRNA species of 3.49, 2.96, 2.2 and approx 0.9 kb and a major protein of M_r 28,000.RNA1 and RNA2 encode proteins involved in viral replication, whereas RNA3 encodes a protein required for cell-to-cell movement. The viral coat protein (CP) is expressed by a subgenomic RNA, designated RNA4, collinear with the 3′ end of RNA3 (Scott et al. 1998).

Desired level of resistance to TSV in cultivated species of sunflower and groundnut is not available. One hundred fifty released cultivars of groundnut were found susceptible against TSV in laboratory screening. Besides, 51 wild accessions of groundnut except *Arachis chacoense* (lCG 4983) were found positive to TSV by ELISA (Prasadarao et al. 2003). Large number of sunflower germplasm collection was tested against TSV under field conditions and none were found resistant (Jain et al. 2008). Kalyani and her associates (2007) screened 56 groundnut germplasm accessions from 20 wild *Arachis* spp. and no systemic infection was found in the following accessions 8139, 8195, 8200, 8203, 8205,11550, CG 8144, ICG 13210 even though 0–100% infection was there on inoculated leaves.

Reaction of five spanish (JL 24, TMV 2, Kadiri6, Kadiri 9 and Anantha) and a virginia (Kadiri 7 Bold) groundnut cultivars was studied against TSV by sap inoculation using different age group of plants [7–84 days after sowing (DAS)]. Based on the biological parameters tested including virus titre, the virginia cultivar Kadiri 7

Bold is found to be tolerant to TSV among all the cultivars tested by recording less per cent infection and wilting with high incubation period besides, it supported less virus titre at all stages of infection. Similarly, among spanish cultivars tested, Kadiri 9 is least susceptible to TSV (Vemana and Jain 2013).

The most economical and convenient way to manage TSV is to grow resistant varieties. However, resistance to TSV was not found in cultivated species in India. For instance, 150 groundnut cultivars and advanced breeding lines evaluated for TSV were found to be susceptible (Prasadarao et al. 2003; Kalyani et al. 2005). Three breeding lines, ICGV # 92267, 99029 and 01276 showed delayed symptom expression but they succumbed under high disease pressure (Prasadarao et al. 2003; Kalyani et al. 2005). Large number of sunflower germplasm evaluated for TSV under field conditions at Regional Agricultural Research Station, Nandyal, Andhra Pradesh, were found to be susceptible (Jain et al. 2008). However, difference in symptom expression was observed in various genotypes of groundnut and sunflower indicating that not all the cultivars are equally susceptible to TSV.

Recently, TSV resistance was found in 8 of the 56 wild *Arachis* accessions evaluated (Kalyani et al. 2007). These are, ICG # 8139, 8195, 8200, 8203, 8205 and 11550 belonging to *Arachis duranensis,* ICG 8144 belonging to *A. villosa* and ICG 13210 belonging to *A. stenospenna.* Similarly, resistance to TSV was found in one ancestral soybean line 'Tanner' (Wang et al. 2005). These sources can be used to transfer resistance into cultivated varieties through breeding programs. Resistance against thrips, *F. occidentalis,* was found to offer partial resistance to TSV in chrysanthemum cv. Kan-komichi in Japan (Ohta 2002). Such resistance to thrips may occur in other TSV host species, if found can be exploited to reduce TSV incidence in the fields.

Efforts are being made to develop TSV resistance by transgenic approach. Coat protein mediated approach was undertaken at ICRISAT to engineer resistance to TSV in popular groundnut cultivars JL 24, TMV-2 and ICGV 91114 using a TSV CP gene construct (Saivishnupriya et al. 2006, 2007). So far more than ISO putative transgenics events have been produced and events that resist systemic spread of TSV were identified. This work in progress at ICRISAT and other centers (Bag et al. 2007) may contribute to the development of durable TSV resistant cultivars. Trangenic resistance against PDV has been developed in almond and prunus species using CP gene expression.

The complete genome sequence of four Indian isolates of TSV determined from groundnut, okra, pumpkin and sunflower (Usha Rani 2009). The full-length sequence of TSV from India (TS-pumpkin) consisted of 8639 nt similar to the reference isolate (TSV-WC). The genome organization is in agreement to that of previously published Ilarviruses. The complete nucleotide sequence of RNA 1 showed 63–85% similarity with members of the subgroup I with greatest homology to TSV. Illinois (85%). The aa sequence of N-methyl transferase motif and helicase motif of RNA 1 showed 96.6% and 94.8% homology to TS.Illinois and TS.WC isolate respectively. A conserved octet "GGAGATGC" recognized in the 3′UTR of all three genomic RNAs of pumpkin TSV exists in TSV-WC. ORF 2a possess conserved polymerase signature motif that showed 96.1% homology to both the TSV isolates of USA (TSV-WC and TSV.Illinois). A second overlapping gene, ORF 2b located towards the 3′ proximal of RNA 2 similar to cucumoviruses existed in pumpkin. The deduced amino acid sequence of MP and CP shared highest homology to TSV Indian isolates (TS.India, TSVAPMG and TSVAPGA) than to USA isolates. The analysis of tree topology derived based on the conserved motifs and deduced amino acid sequence of MP and CP revealed separate clustering of TS-Pumpkin, an Indian isolate of TSV from USA isolates depicting sufficient sequence variation existing among TSV isolates of India and USA.

11.6 Concluding Remarks

As many as 19 ilarviruses are known all over the world. So far only ApMV, PNRSV and TSV have been recorded in India. A greater focus on the studies on TSV has been given in India compared to ApMV and PNRSV. This is because TSV has emerged as one of the most prolific plant viruses in India infecting a large number of crops species. The infection of TSV results into a serious impact on the growth and yield of the plant. The disease symptoms of TSV in many crops are similar to tospoviruses specially GBNV and often in the field, it is difficult to distinguish the TSV infection from the GBNV infection on the basis of symptoms alone. In India, many episodes of outbreaks of TSV have been witnessed. It seems that TSV spreads rapidly in the field through pollen and thrips. More studies are necessary to understand the spread and outbreak of TSV. Although, the genome sequence of TSV occurring in India has been generated, so far, this information has not been utilised for developing the infectious clone of TSV to study the infection process.

References

Aparicio F, Palla SV (2002) The molecular variability analysis of the RNA 3 of fifteen isolates of Prunus necrotic ringspot virus sheds light on the minimal requirements for the synthesis of its subgenomic RNA. Virus Genes 25:75–84

Aparicio F, Sanchez-Pina MA, Sanchez-Navarro JA, Pallas V (1999) Location of Prunus necrotic ringspot Ilarvirus within pollen grains of infected nectarine trees: evidence from RT-PCR, dot-blot and in situ hybridization. Eur J Plant Pathol 105:623–627

Arun Kumar N, Lakhmi Narasu M, Zehr UB, Ravi KS (2007) First report of tobacco streak virus infecting *Guizotia abyssinica* from India. Plant Dis 91:330

Arun Kumar N, Lakhmi M, Zehr UB, Ravi KS (2008) Molecular characterization of tobacco streak virus causing soybean necrosis in India. Indian J Biotechnol 7:214–217

Bag S, Singh RS, Jain RK (2007) Agrobacterium-mediated transformation of groundnut with coating protein gene of tobacco streak virus. Indian J Virol 18:65–69

Bhargava KS, Bist NS (1957) Three virus diseases of hill fruits in Kumaon. CurrSci 26:324–325

Bhat AI, Jain RK, Chaudhary V, Krishna Reddy M, Ramaiah M, Chattannavar SN, Varma A (2002a) Sequence conservation in the coat protein gene of tobacco streak virus isolates causing necrosis disease in cotton, mungbean, sunflower and sunn-hemp in India. Indian J Biotechnol 1:350–356

Bhat AI, Jain RK, Kumar A, Ramiah M, Varma A (2002b) Serological and coat protein sequence studies suggest that necrosis disease on sunflower in India is caused by strain of tobacco streak ilarvirus. Arch Virol 147:651–658

Bradford FC, Joley L (1933) Infectious variegation in the apple. J Agric Res 46:901–908

Brunt AA, Crabtree K, Dallwitz MJ, Gibbs AJ, Watson L, Zurcher EJ (1996) Onwards. Plant viruses online: descriptions and lists from the VIDE database. Version: 20th August 1996. http://biology.anu.edu.au/Groups/MES/vide/

Chandel V, Rana T, Hanada A, Thakur PD, Hallan V, Zaidi AA (2008) Incidence of prunus necrotic ring spot virus on *Malus domestica* in India. J Phytopathol 156:382–384

Chandel V, Rana T, Hallan V, Zaidi AA (2011) Detection of Prunus necrotic ring spot virus in plum, cherry and almond by serological and molecular techniques from India. Arch Phytopathol Plant Protect 44:1779–1784

Chander Rao S, Prasada Rao RDVJ, Manoj Kumar V, Raman DS, Raoof MA, Prasad RD (2003) First report of tobacco streak virus infecting safflower (*Carthamus tinctorius*) in Maharashtra. India Plant Dis 87:1396

Converse RH (1972) Tobacco streak virus in black raspberry. Phytopathology 62:1001–1004

Cook G, Miranda RD, Roossinck MJ, Pietersen G (1999) Tobacco streak ilarvirus detected on groundnut in South Africa. Afr Plant Protect 5:13–19

Crosslin JM, Hammond RW, Hammerschlag FA (1992) Detection of Prunus necrotic ringspot virus serotypes in herbaceous and Prunus hosts with a complementary RNA probe. Plant Dis 76:1132–1136

Cupertino FP, Grogan RG, Petersen LJ, Kimble KA (1984) Tobacco streak virus infection of tomato and some natural weed hosts in California. Plant Dis 68:331–333

Fauquet CM, Mayo MA, Manilot TJ, Desselberger U, Ball HA (eds) (2005) Virus taxonomy, VIII Report of the International Committee on Taxonomy of Viruses. Elsevier, New York, p 1259

Fulton RW (1970) Prune dwarf virus, CMI/AAB Descriptions of plant viruses no. 19. Association of Applied Biologists, Wellesbourne

Fulton RW (1985) Tobacco streak virus, CMI/AAB Descriptions of Plant Viruses No. 307. Association of Applied Biologists, Wellesbourne

Glasa M, Betinová E, Kúdela O, Subr Z (2002) Biological and molecular characterization of *Prunus necrotic ringspot virus* isolates and possible approaches to their phylogenetic typing. Ann Appl Biol 140:279–283

Greber RS, Klose MJ, Teakle DS, Milne JR (1991) High incidence of tobacco streak virus in tobacco and its transmission by *Microcephalothrips abdominalis* and pollen from *Ageratum 11OLlstonianum*. Plant Dis 75:450–452

Goud STE, Vemana K, Reddy DL, Mahammed KCS, Padma JG, Shabbir S, Venkateswarlu NC, Naik KSS, Sampath KD, Anthony JAM, Subramanyam K (2013) First report of tobacco streak ilarvirus infecting jasmine and horse gram. New Dis Rep 28:7. https://doi.org/10.5197/j.2044-0588.2013.028.007

Jain RK, Vemana K, Bag S (2008) Tobacco streak virus – an emerging virus in vegetable crops. In: Rao GP, Kumar PL, Holuguin-Pena RJ (eds) Characterization, diagnosis and management of plant viruses volume 3. Vegetable and pulse crops. Studium Press LLC, Texas, pp 203–212

Johnson J (1936) Tobacco streak, a virus disease. Phytopathology 26:285–292

Jones AT, Mayo MA (1975) Further properties of black raspberry latent virus, and evidence for its relationship to tobacco streak virus. Ann Appl Biol 79:297–306

Jones AT, Mc Gavin WJ, Dolan A (2001) Detection of tobacco streak virus isolates in North American Cranberry (*Vaccinium macrocarpon*). Online Plant Health Prog. doi:10.1094/PHP-2001-0717- 01-RS

Kaiser WJ, Wyatt SD, Pesho GR (1982) Natural hosts and vectors of tobacco streak virus in eastern Washington. Phytopathology 72:1508–1512

Kalyani G, Sonali S, Reddy AS, Reddy AGS, Waliyar F, Nigam SN (2005) Sources of resistance to *Tobacco streak virus* in groundnut, *Arachis hypogaea*. J Oil/Seeds Res 22:105–107

Kalyani G, Reddy AS, Kumar PL, Prasada Rao RDVJ, Aruna R, Waliyar F, Nigam SN (2007) Sources of resistance to *Tobacco streak virus* in wild *Arachis* (Fabaceae: Papilionoidae) germplasm. Plant Dis 91:1585–1590

Krishnareddy M, Salil J, Samuel DK (2003a) Fruit distortion mosaic diseases of okra in India. Plant Dis 87:1395

Krishnareddy M, Devaraj LR, Salil J, Samuel DK (2003b) Outbreak of tobacco streak virus causing necrosis of cucumber (Cucumis sativus) and gherkin (Cucumis anguria) in India. Plant Dis 87:1264

Kulshrestha S, Verma N, Hallan V, Raikhy G, Singh MK, Ram R, Zaidi AA (2005) Detection and identification of Prunus necrotic ringspot virus in Pelargonium. Australas Plant Pathol 34:599–601

Kumar A, Lakshmi-Narasu M, Zehr U, Band Ravi KS (2006) Natural occurrence and distribution of *Tobacco streak virus* in South India. Indian J Plant Protect 24:54–58

Ladhalakshmi D, Ramiah M, Ganapathy T, Krishnareddy M, Khabbaz SE, Babu M, Kamalakannan A (2006) First report of the natural occurrence of *Tobacco streak virus* on blackgram *(Vigna mungo)*. Plant Pathol 55:1395

Menzel W, Jelkmann W, Maiss E (2002) Detection of four apple viruses by multiplex RT-PCR assays with coamplification of plant mRNA as internal control. J Virol Methods 99:81–92

Mink GI (1992) Ilarvirus vectors. Adv DisVector Res 9:261–281

Mink GI, Howell WE, Cole A, Regev S (1987) Three serogroups of Prunus necrotic ringspot virus biotypes from rugose mosaic-disease sweet cherry trees in Washington. Plant Dis 68:207–210

Moury B, Cardin L, Onesto JP, Candresse T, Poupet A (2001) Survey of prunus necroitic ringspot virus in rose and its variability in rose and Prunus spp. Phytopathology 91:84–91

Myrta A, Di Terlizzi B, Boscia D, Choueri E, Gatt M, Gavriel I, Caglayan K, Varveri C, Zeramdini H, Aparicio D, Pallas V, Savino V (2001) Serological characterization of Mediterranean *Prunus necrotic ringspot virus* isolates. J Plant Pathol 83:45–49

Pappu HR, Hammett KRW, Draffel KI (2008) Dahlia mosaic virus and tobacco streak virus in dahlia (Dahlia variabilis). Plant Dis 92:1138

Prasada Rao RDVJ, Reddy AS, Chander Rao S, Varaprasad KS, Thirumala-Devi K, Nagaraju MV, Reddy DVR (2000) *Tobacco streak ilarvirus* as causal agent of sunflower necrosis disease in India. J Oilseeds Res 17:400–401

Prasada Rao RDVJ, Reddy AS, Reddy SV, Thirumala-Devi K, Chander-Rao S, Manoj Kumar V, Subramanyam K, Yellamanda Reddy T, Nigam SN, Reddy DVR (2003) The host range of *Tobacco streak virus* in India and transmission by thrips. Ann Appl Biol 142:365–368

Prasadarao RDVJ, Reddy DVR, Nigam SN, Reddy AS, Waliyar F, Yellamanda Reddy T, Subramniam K, John Sudheer M, Naik KSS, Bandopadhyay A, Desai S, Ghewande MP, Basu MS, Somasekhar (2003) Peanut stem necrosis: a new disease of groundnut in India. Information Bulletin, 67. Patancheru 502 324. International Crops Research Institute for the Semi-Arid Tropics, Andhra Pradesh, India, 16 pp

Ramiah M, Bhatt AI, Jain RK, Pant RP, Ahlawat YS, Prabhakar K, Varma A (2001) Partial characterization of an isometric virus causing sunflower necrosis disease. Indian Phytopathol 54:246–250

Ravi KS, Buttgereitt A, Kitkaru AS, Deshmukh S, Lesemann DE, Winter S (2001) Sunflower necrosis diseases from India is caused by an Ilarvirus related to *Tobacco streak virus*. New Dis Rep 3:1–2

Reddy AS, Prasadarao RDVJ, Thirumala-Devi K, Reddy SV, Mayo MA, Roberts I, Satyanarayana T, Subramaniam K, Reddy DVR (2002) Occurrence of *Tobacco streak virus* on peanut *(Arachis hypogaea* L.) in India. Plant Dis 86:173–178

Saivishnupriya K, Arokiasamy S, Kumar PL, Sharma KK (2006) Engineering resistance to combat stem necrosis disease of groundnut: new hope to save groundnut from *Tobacco streak virus* menace in India. Agric Biotech Support Program/I (South Asia) Newsl 1:2–3

Saivishnupriya K, Arokiasamy S, Kumar PL, Reddy AS, Nalla AK, Waliyar F, Nigam SN, Beachy RN, Sharma KK (2007) Development and evaluation of transgenic ground nut for resistance to tobacco streak virus. In: Abstracts of the lOth international plant virus epidemiology symposium, 15–19 October 2007. JCRISAT, Hyderabad, India, 57 pp

Scott SW, Zimmerman MT, Ge X (1998) The sequence of RNA 1 and RNA 2 of tobacco streak virus: additional evidence for the inclusion of alfalfa mosaic virus in the genus Ilarvirus. Arch Virol 143:1187–1198

Sharman M, Thomas JE, Persley DM (2008) First report of *Tobacco streak virus* on sunflower *(Helianthus annuus),* cotton *(Gossypium hirsutum)* and mungbean *(Cicer arietinum).* Aust Plant Dis Notes 3:27–29

Singh BB, Pandey BN, Singh RN, Singh S (1979) Effect of apple mosaic virus onthe growth, yield and quality of apple. Punjab Hortic J 19:80–82

Sivaprasad Y, Bhaskara Reddy BV, Rekha Rani K, Raja Reddy K, Sai Gopal DVR (2010) First report of Tobacco streak ilarvirus infecting onion (*Allium cepa*). New Dis Rep 22:17

Stace-Smith R, Frazier NW (1971) Tobacco streak virus isolated from strawberry infected with necrotic shock. Phytopathology 61:757–758

Usha Rani TR (2009) Molecular characterization and expression of coat protein and movement protein of tobacco streak virus infecting crop plants. PhD thesis, University of Agricultural Sciences, Bangalore, p 174

Uyemoto JK, Scott SW (1992) Important diseases of Prunus caused by viruses and other graft-transmissible pathogens in California and South Carolina. Plant Dis 76:5–11

Vaskova D, Petrzik K, Karesova R (2000) Variability and molecular typing of the woody-tree infecting prunus necrotic ringspot ilarvirus. Arch Virol 145:699–709

Vemana K, Jain RK (2010) New experimental hosts of tobacco streak virus and absence of true seed transmission in leguminous hosts. Indian J Virol 21:117–127

Vemana K, Jain RK (2013) Comparative reaction of popular high yielding spanish and virginia bunch groundnut varieties to *Tobacco streak virus*. Indian J Virol 24:214–219

Verma N, Ram R, Sharma A, Hallan V, Zaidi AA, Garg ID (2002) Detection of prunus necrotic ring spot virus in begonia by RT-PCR. Plant Pathol 51:800

Wang Y, Hobbs HA, Hill CB, Domier LL, Hartman GL, Nelson RL (2005) Evaluation of ancestral lines of U.S. soybean cultivars for resistance to four soybean viruses. Crop Sci 45:639–644

Waqar A, Butt TB, Insan J, Rehman A (2003) Natural occurrence of tobacco streak virus in cotton in Pakistan and screening for its resistance sources. Pak J Bot 35:401–408

The Current Status of Luteovirus and Polerovirus Research in India

12

R. Viswanathan, K. Nithya, B. Parameswari, A. Jeevalatha, and Govind Pratap Rao

Abstract

The family *Luteoviridae* comprises of three genera *Luteovirus*, *Enamovirus* and *Polerovirus,* of which, only two genera *Luteovirus* and *Polerovirus* are known in India. Though luteovirus and polerovirus infects many crops throughout the world, only a few have been documented in India. Luteoviruses were reported in field crops like barley (barley yellow dwarf virus) and chickpea (chickpea stunt virus), while polerovirus was recorded from potato, jute (potato leaf roll virus [PLRV]) sugarcane (sugarcane yellow leaf virus [SCYLV]) and cotton (cotton leaf roll dwarf virus) crops from India. SCYLV and PLRV are the most important poleroviruses as they are of serious constraints in all the sugarcane and potato growing states of India. SCYLV infection causes 39–43% reductions in plant growth and 30–34% loss in yield in sugarcane. The virus is mainly transmitted through infected seed canes. The secondary spread of the virus in the field is by the aphid vectors. PLRV is reported to cause 50–60% yield losses in potato crop and this virus is tuber borne and transmitted mainly by aphid vectors in a circulative non-propagative manner. The complete genome sequences SCYLV and PLRV isolates from sugarcane and potato are available from India. This book chapter deals with an uptodate information available on distribution, biological properties, identification, serological relationships, genetic diversity and

R. Viswanathan (✉) • K. Nithya
ICAR-Sugarcane Breeding Institute, Coimbatore 641007, Tamil Nadu, India
e-mail: rasaviswanathan@yahoo.co.in

B. Parameswari
ICAR-Sugarcane Breeding Institute Regional Centre, Karnal 132001, Haryana, India

A. Jeevalatha
ICAR-Central Potato Research Institute, Shimla 171001, India

G.P. Rao
Advanced Centre for Plant Virology, Division of Plant Pathology, Indian Agricultural Research Institute, New Delhi 110012, India

B. Mandal et al. (eds.), *A Century of Plant Virology in India*,
https://doi.org/10.1007/978-981-10-5672-7_12

transmission of the luteoviruses and poleroviruses reported on different crops from India.

Keywords
Luteovirus • *Polerovirus* • *Sugarcane yellow leaf virus* • *Potato leaf roll virus*

12.1 Introduction

Luteoviridae is a family of plant viruses having common biological properties of persistent transmission by aphid vectors and the induction of leaf rolling, reddening or yellowing symptoms in infected host plants. "Luteo" comes from the Latin luteus, which translates as yellowish. Members of the *Luteoviridae* cause economically important diseases in many food crops, including grains such as wheat and barley, vegetables such as potatoes and lettuce, and other crops such as legumes, sugarcane and sugarbeets. All members of the *Luteoviridae* consist of small (ca. 25–28 nm diameter) icosahedral particles, composed of one major and one minor protein component and a single molecule of messenger sense single-stranded RNA. The family is divided into three genera: *Luteovirus, Polerovirus* and *Enamovirus*. Viruses in *Luteoviridae* are non-enveloped, with icosahedral and spherical geometries of diameter around 25–30 nm. Genomes are linear and non-segmented, around 5.3–5.7 kb in length. Luteoviruses can act as helper viruses for Umbraviruses, providing them with a coat protein (Smith and Barker 1999).

Luteovirus, which resides in phloem tissues of the host plants thereby, inhibits nutrient translocation in the sieve elements. It is a monopartite virus with positive sense single stranded ribonucleic acid (+ssRNA) genome varying in size from 5.6 to 6.0 kb. The geographical distribution of luteoviruses are widespread, with the virus primarily infecting plants via transmission by aphid vectors. The virus only replicates within the host cell and not within the vector .The genus *Enamovirus* (from the type species, *Pea enation mosaic virus 1*) is one of three genera in the family *Luteoviridae* and currently has only one member. Until recently it was considered as an irreversible mixed infection of two autonomously replicating RNAs, but the current taxonomy retains only the old RNA1 (which has some similarity to members of the genus *Polerovirus*) as an enamovirus and places the old RNA2 as a different virus (*Pea enation mosaic virus* 2) in the genus *Umbravirus*. The virions are isometric (polyhedral), not enveloped of two sizes, 25 and 28 nm in diameter. The genome is monopartite, linear, single stranded, positive sense RNA about 5700 nt long. The 3′-terminus has neither a poly(A) tract nor a tRNA-like structure, and the 5′- terminus has a genome-linked protein. The genome (5706 nucleotides) has many similarities to members of the genus *Polerovirus* but lacks the functions for systemic movement and mechanical transmission, which are supplied by pea enation mosaic virus-2. There are currently 17 species in this genus including the type species *Potato leafroll virus* (Smith and Barker 1999), which is one of the most prevalent viral diseases of potato in India (Mukherjee et al. 2003). The virus is tuber borne, not sap transmitted but transmitted efficiently by aphid in a circulative non-propagative manner. Yield loss normally ranges from 50% to 60% in India (Paul Khurana 2000; Mukherjee et al. 2003). PLRV has small, isometric virions measuring about 24–25 nm in diameter. The genome consists of positive sense single stranded RNA with the viral genome linked protein (VPg) at the 5′ end (Mayo and D'Arcy 1999) and an OH group at the 3′ end.

Out of the three genera of the family *Luteoviridae*, *Luteovirus* and *Polerovirus* are emerging as important plant viruses infecting major crop plants all over the world. Though luteovirus and polerovirus occurrence were reported in many crops throughout the world only few have been reported in India based on symptomatology, host range, genome organization and amino acid sequence similarities in field crops like barley (barley yellow dwarf virus), chickpea (chickpea stunt), potato, jute (potato leaf roll virus), sugarcane (sugarcane yellow leaf virus) and cotton (Cotton leaf roll dwarf virus). We have discussed all the existing information about history, distribution, biological properties, identification, genetic diversity and epidemiology on reported luteo- and poleroviruses on different crops in India.

12.2 Luteoviruses and Disease

Luteoviruses are causing leaf yellowing, reddening and/or rolling symptoms in the infected plants, are mostly confined to the phloem tissues, are transmitted by aphids in the persistent (circulative, non-propagative) manner and possess a monopartite linear ssRNA genome, which is packaged in 25 nm diameter isometric particles. Two special feature of luteoviruses seem to set them apart from all other plant viruses. One is possession of coat protein read through domain which interact with the aphid endo-symbiont product, symbionin, in a way that apparently stabilizes the virus particles when they are in the aphid haemocoel. The second feature is the extent of their capacity to interact with other infectious agents, including umbraviruses, satellite RNA and a viroid, in ways that profoundly affect the survival and spread of one or both partners; a key ingredient in such interactions is the ability of luteovirus coat protein(s) to package non-luteoviral DNA (Harisson 1999).

In India luteoviruses have been reported from barley and chickpea which are discussed herein:

12.2.1 Barley yellow dwarf virus (BYDV)

BYDV is the most widely distributed and the most economically important virus disease of wheat and barley. It is caused by a group of luteoviruses called barley yellow dwarf luteoviruses. They are transmitted by aphids in a persistent, circulative but non-propagative manner. Five strains were identified from New York in the

United States based on their transmission phenotypes in an experimental system. The strains and their principal vectors are RPV (*Rhopalosiphum padi*), RMV (*R. maidis*), MAV (*Sitobion avenae*), SGV (*Schizaphis graminum*) and PAV (*R. padi*, *S. avenae* and others). The five strains are also distinguishable serologically (Miller and Rasochová 1997). Barley yellow dwarf luteoviruses symptom includes leaf discoloration from tip to base and from margin to centre, in shades of yellow and sometimes red. Plants are usually stunted, with a decrease in tiller number and biomass and a weak root system. In the field, symptoms usually appear as yellow or red patches of stunted plants. Zhang et al. (1983) reported the strain GPV, DAV and GPDAV from China. It seems that some of the Chinese strains have a serological relation to the US isolates (MAV and PAV) but that they differ slightly in their aphid-transmission patterns. Several strains of BYDV can frequently coexist in the same plant. The resulting symptoms can be more severe when the strains are from different groups, but when from the same group, they may result in the an amelioration of symptoms through the mechanism of cross-protection. Losses in wheat due to BYD can be very serious but differ with the BYDV strains, the growth stage at infection, the wheat varieties and the environmental conditions. Losses of around 11–12% due to natural infection have been reported in Morocco and in Chile. In Australia, yield losses of about 2.2 tonnes/ha in a susceptible wheat and losses of about 1.1 tonnes/ha in tolerant varieties have been reported.

The epidemiology of BYD is influenced by the strains involved, the aphid vectors present in the area, the crop rotation, environmental conditions (temperature and rainfall), and the time of sowing. Barley yellow dwarf alternates from reservoir hosts (grasses, maize, other cereals and volunteer plants) to small grain cereals.

In India, barley and wheat crops are reported to be affected with BYDV. But there are no regular recurrence of this virus disease on wheat in India has been noticed. The first authentic report of the BYD disease occurrence in India was made by Nagaich and Vashistha (1963) from Shimla hills in 1958. The prevalence of BYD on large scale in the hilly regions of Mukteshwar and Bhowali (Uttaranchal) was subsequently recorded by Singh et al. (1979). The disease incidence was also recorded on wheat crop from Karnataka (Kulkarni and Hegde 1980). Later, outbreak of MAV-type BYDV on wheat in the Garhwal Hills in India was reported (Khetarpal et al. 1994). Besides wheat and barley, BYD also infects, oats, triticale and more than 100 other graminaceous hosts. In India, October sown wheat crop is prone to viral infection in hilly regions of Kumaon, whereas December-sown crop remains free from BYD apparently due to lack of vector population (Singh et al. 1979). BYD tolerant wheat genotypes viz., NS 879/4, Arjun, DWR 16, 32, HD 2189, 2278, HW 657 and H-10-5-7 hold great promise in breeding resistant varieties (Singh et al. 1979; Kulkarni and Hegde 1983). Destruction of wheat hosts and volunteer plants may help in reduction of initial inoculums and population of insect vectors.

Not much information on genetic diversity of BYDV isolates from India is available which needs further attention.

12.2.2 Chickpea Stunt Disease

Chickpea (*Cicer arietinum*) is one of the important proteinaceous pulse crop being cultivated in many of the Indian states. Of the several viral diseases, Chickpea stunt is an important viral disease prevalent in all the chickpea growing areas. Internodes shortening, yellowing of leaf (in kabuli types), reddening of leaf (in desi types) and browning in collar region are the general characteristic symptoms of the disease (Nene and Reddy 1987; Nene et al. 1991). The early stage of the infected plants showed the stunted growth and died prematurely, whereas stunting may not be obvious in the infected plants when they are mature. The premature deaths of diseased plants are reducing the chickpea production level in many of the growing areas. It was reported that chickpea plants inoculated with bean leaf roll virus (BLRV) caused cent – percent yield losses (Kaiser and Danesh 1972; Kotasthane and Gupta 1978) with chickpea stunt symptoms. Later, chickpea chlorotic dwarf virus (CCDV) was reported to induce the chickpea stunt symptoms (Horn et al. 1993). Hence, an extensive survey was made to ascertain the exact causal disease in India during 1991 and 1992 in Rajasthan, Madhya Pradesh, Gujarat, Haryana and at Patencheru (ICRISAT Asia Center, Telangana). They were tested positive with polyclonal antisera to BLRV in DAS-ELISA. Simultaneously, luteovirus particles were observed under EM from the symptomatic samples. The polyclonal antiserum was produced from the purified virus particle of chickpea stunt which showed positive reaction with the symptomatic chickpea samples but only few of the samples were reacted with BLRV antiserum. Thus the isolate was referred to as Chickpea luteovirus (CpLV) and the findings of the study confirmed that the existence of two distinct luteoviruses *viz.* BLRV of minor importance and a CpLV as major chickpea viruses in India (Horn et al. 1996). During the same period efforts were initiated on transmissible nature of the viruses.

In order to further confirm the earlier reports of virus association in chickpea stunt disease, chickpea plants showing stunt disease symptoms were collected from the experimental fields at ICRISAT Asia Center (IAC), Hyderabad as well as from different farmers' fields in Junagadh (Gujarat), Khargone (Madhya Pradesh), and Akola (Maharashtra) during 1992–1993 and 1993–1994 growing seasons. Samples were examined Electron microscopically as well with DAS- and TAS-ELISA methods using panel of monoclonal antibodies of potato leaf roll, beet western yellows and barley yellow dwarf (RPV strain) luteoviruses. The results confirmed the presence of virus particles from the ICRISAT field samples. Later, vector transmission work was carried out with different aphid species of which *Myzus persicae* was found to be more efficiently transmitted the virus isolates. Furthermore identification, RT- PCR assay was carried out using universal luteovirus specific primers. The results revealed that one isolate (L) was 94% identical with beet western yellows virus based on their coat protein amino acid sequences whereas the other isolate (IC) was 82% identical to the isolate (L) and 80% or less identical to the coat protein sequences of other luteoviruses (Naidu et al. 1997).

12.3 Polerovirus

So far, polerovirus was recorded from potato, jute (*Potato leaf roll virus*), sugarcane (Sugarcane yellow leaf virus, SCYLV) and cotton (Cotton leaf roll dwarf virus) crops from India and are discussed herein:

12.3.1 Sugarcane yellow leaf virus (SCYLV)

Sugarcane is one of the most important commercially grown field crops cultivated in more than 90 countries. In India it is being cultivated in 5.01 million ha of land area. It is a highly industry centric crop with more than 500 sugar industries and many more cottage industries manufacturing *gur* and *khandsari* are being in operation based on the sole crop in the country. Apart from sugar, sugarcane is one of the most viable alternatives for production of bio-fuels and renewable energy in the world. Because of its huge potential, this crop is being seen as an energy cane/bio fuel crop based on growing demand of fuel and energy in the recent years.

Among the viral diseases of sugarcane, yellow leaf disease (YLD) caused by SCYLV has emerged as a major threat across the states in the country (Viswanathan et al. 2008, 2016). It was referred as yellow leaf syndrome (YLS) previously and typical symptoms of the disease were leaf midrib yellowing and laminar discoloration. This disease was first reported in the year 1989 in Hamakua (Hawaii) on variety H65-0782 (Schenck 1990; Schenck et al. 1997) and subsequently from the United States mainland (Comstock et al. 1994) and many other sugarcane growing countries. In India, it was first time reported during the year 1999 (Viswanathan et al. 1999) and later in 2000 (Rao et al. 2000). Initial reports of its occurrences were from major sugarcane growing states viz. Uttar Pradesh, Tamil Nadu, Haryana, Bihar and Uttarakhand on the basis of visual symptoms, electron microscopic observations of virus particles and serological relationship (Rao et al. 2002).

12.3.1.1 Impact of YLD to Sugarcane Cultivation

YLD occurrence is being widely reported in most of the sugarcane growing places of the country and up to 100% disease incidences were recorded on susceptible varieties in the commercial fields (Viswanathan 2002, 2016; Rao et al. 2001) in India and from some major sugarcane growing countries viz. USA (Comstock et al. 1994, 2001), Reunion Islands (Rassaby et al. 2004) and in Thailand (Lehrer et al. 2008). As this crop occupies 5.01 M ha of land area in India, the losses caused by the disease to this crop is significant and it has worked out to be in several million US dollars. Considering its widespread occurrence and the possibility of epidemic outbreak, an intensive work on YLD had been initiated at ICAR-SBI, Coimbatore right from the reporting year 1999 to investigate in all aspects of the disease. The research findings had shown that SCYLV infection reduces the cane thickness, number of millable canes and leaf photosynthetic rate when compared to the disease free healthy plants, thereby it further reduces yield and sugar potential of the infected crop (Viswanathan 2002; Viswanathan et al. 2006).

Since the virus adversely affects the growth and physiological parameters of the cane, detailed investigation was initiated to access its impacts on physiological efficiency and changes in photosynthates; as both are directly related with source-sink relationship and further by sucrose accumulation in cane. The detailed studies of Viswanathan et al. (2014) had shown that photosynthetic rate, stomatal conductance and SPAD metre values were lower in severely infected YL plants besides it reduced the growth and yield parameters, such as plant height, cane thickness, number of internodes etc. for example the susceptible varieties viz. CoPant 84211, Co 86032, CoC 671 had shown 42.9%, 42.3% and 38.9% reduction in plant growth attributes along with 34.15% reduction in juice yield compared to the disease free healthy plants.

In many susceptible varieties plant growth reduction was noticed when SCYLV was combined with *Leifsonia xyli* sub sp. *xyli* causing ratoon stunting disease and sugarcane leaf yellows associated with phytoplasmas and this suggested that the existence of both these pathogens for a long period in plant crop and subsequent ratoons results varietal deterioration more rapidly, leading to poor performance of the varieties (Viswanathan 2004). Later, planting of infected setts were reported as primary source of the disease in the field and the disease incidence was found more in ratoons and fields with poor maintenance (Viswanathan et al. 2006). Also the latent infection of SCYLV with its asymptomatic stage is considered as an important epidemiological state of the disease and can cause significant yield reduction in susceptible varieties.

12.3.1.2 Biological Properties

YLD symptoms usually appear/visible during 6–8 months stage of the crop in the field. The common YLD symptoms includes intense midrib yellowing on the abaxial surface (Fig. 12.1a), lateral spread of yellow discolouration to the leaf lamina followed by tissue necrosis from the leaf tip spreading downwards along the midrib. In most susceptible varieties, typical yellowing of midribs and laminar region is noticed on upper surface of the leaves. Finally, symptoms of necrosis of discolored laminar region from leaf tip to bottom along the mid rib and subsequent drying of entire leaf is noticed (Fig. 12.1b). The sugarcane varieties showing mild symptoms of midrib yellowing usually records normal cane growth whereas in severely infected clumps cane thickness and stalk height are significantly affected. Severe infection of the disease leads to shortening of internodes on the top. This effect culminates in bunching of leaves at the top. Usually such infection results in drying of leaves in the affected canes (Fig. 12.1b). Usually expression of the symptoms will be more severe in ratoon crops than plant crop (Viswantahan 2002, 2012; Viswanathan et al. 2012).

Variations in pattern of YLD symptom expression was noticed among the susceptible varieties. In order to understand the influence of weather parameters on disease development and symptom expression, a detailed study was undertaken with a set of highly susceptible sugarcane varieties/genotypes viz., Co 419, Co

Fig. 12.1 Yellow leaf disease symptoms on sugarcane in the field (**a**) SCYLV-infected sugarcane exhibits intense yellowing of midribs on the abaxial surface. (**b**) Spread of yellow discolouration to adjoining laminar region parallel to midrib and drying of leaf tissue from leaf tip towards bottom

85019, Co 86010, Co 87269, CoBln 9605, CoPant 84211, CoS 767, CoTl 85441, B 38192 and 57 NG 56 in four planting seasons starting from 2009–2010 to 2012–2013.Of these, progressive increase in symptom expression was noticed up to formative stage of the crop and later, a fluctuation in symptom expressions i.e. intensity of yellowing decreases leaving behind the dried laminar region at the tip was observed among the susceptible varieties. The correlation and regression analyses of symptom expression and different weather parameters had shown that prevailing minimum temperature and RH in the afternoon had a partial relation to symptom expression in the virus-infected plants (Chinnaraja and Viswanathan 2015a). However, further studies are required to assess the influence of weather parameters on disease transmitting vectors, host susceptibility and virus titre of the infected plant.

12.3.1.3 Role of Vectors in Virus Transmission

Assumption of possibility of role of vectors in viral disease development is certainly common. Moreover, the virions of luteovirids are not transmitted mechanically as they are restricted in the vascular regions (phloem) of plants. So, plant to plant transmission of virus was reported through vectors (*Melanaphis sacchari*, sugarcane aphid) in many countries (Singh and Rao 2011). Recently, Chinnaraja and Viswanathan (2015b) conducted studies on vector transmission of SCYLV using virus free meristem derived micro propagated plants of cv. Co 86032 with viruliferous aphids and the transmission was confirmed through RT-PCR and the virus titre was analyzed through RT-qPCR. The results had shown the maximum virus targets of 22.3×10^3, 3.16×10^6 and 4.78×10^6copies in the viruliferous aphid inoculated plants after 7, 180 and 300 days respectively and the results confirmed the *M. sacchari* as an efficient vector to transmit SCYLV from one plant to other in fields.

Subsequently, population dynamics of YLD vector *M. sacchari* was assessed during different growth stages in different resources of germplasm, parents and varieties maintained at SBI, Coimbatore. Some of the genotypes recorded maximum aphid population of up to 621 per plant. It was found that aphid population had shown variations from season to season and genotype to genotype. This study revealed that some of the YLD susceptible varieties harbor more aphids and this needs further confirmation (Viswanathan, personal communication).

12.3.1.4 Identification of the Virus

Plant disease diagnosis is important in order to avoid the spread of diseases and disease causing pathogens to new areas and its importance get doubled in case of vegetative propagated crops like sugarcane, where the planting material i.e. setts should be free from disease. As SCYLV primary mode of transmission takes through infected setts and the asymptomatic plants may also harbor the virus, and YL symptom expression can only be seen after 6–8 months of planting, developing YLD diagnosis was felt most important. Taken in to an account, much emphasis had been given ever since it was suspected in sugarcane using different approaches (Rao et al. 2001; Viswanathan 2002, 2004). The results of DAS -ELISA studies using YLD leaf and plant juice samples as antigens showed juice samples contain more viruses compared to leaf samples and it also revealed that plants raised from YLD plants has high virus titre and it was found to be more in the subsequent ratoons (Viswanathan and Balamuralikrishnan 2004; Gaur et al. 2003).

Later, RT-PCR assay was attempted to detect SCYLV due to some drawbacks in serological methods like cross reactivity and limited antisera availability. RT-PCR based diagnosis was standardized with a set of new primers (Viswanathan et al. 2008) which amplified an amplicon of 513 bp of the virus from infected plants and it was confirmed after sequencing. Subsequently, it was validated with different symptom expression stages of YLD in 44 sugarcane varieties and the results had shown that asymptomatic plants too had detectable level of virus infection (Viswanathan et al. 2009). As the RT-PCR assay has been standardized successfully, it was widely used for virus indexing of tissue culture-derived plantlets and in germplasm (Viswanathan et al. 2009). Although it was successful in large scale diagnosis, the virus titre and its relation with disease severity couldn't be assessable under this method which necessitated real time PCR assays to quantify the virus titre. Accordingly the RT-qPCR assay was standardized to quantify the virus titre in the tissue culture raised plantlets through relative standard curve method. The copies of virus from tissue culture derived plantlets and asymptomatic plants were estimated in the ranges of 20,314.58–4,330.87 and 8.96–0.27 million, respectively. The relative expression level of the virus between in vitro plantlets and asymptomatic plants was in the ratio of 73.7: 243,393.1. The assays led to the conclusion that SCYLV population was significantly reduced in the meristem derived tissue culture plants and the copy numbers of the target virus was efficiently detected through relative standard curve method (Chinnaraja et al. 2014).

Apart from these conventional and real time PCR assays, a multiplex RT-PCR was also developed for simultaneous detection of RNA viruses of sugarcane viz.

Sugarcane mosaic virus (SMV), *Sugarcane streak mosaic virus* (SCSMV), *Sugarcane yellow leaf virus* (SCYLV) using coat protein gene specific primers and the PCR conditions were optimized. Furthermore, it was found efficient as equal as to uniplex- PCR in targeted virus amplification (Viswanathan et al. 2010) in a single reaction through which the targets were specifically detected in suspected varieties and it is considered as viable in large scale applications.

12.3.1.5 Genome Characterization

Molecular research works have been continuing in virology ever since the arrival of PCR and sequencing technologies which make it eases in revealing the genomic structure of an organism; with that more and more information's are being generated by virologists around the world from time to time. In this case, SCYLV has a (+) sense ssRNA genome containing 6- overlapping open reading frames (ORFs) (ORF0 – ORF5) and 3-untranslated regions (UTRs). The genome characterization of SCYLV isolates of different countries has shown the existence of four genotypes (BRA for Brazil, CUB for Cuba, PER for Peru and REU for Reunion Island) based on the geographical locations where they were first detected (Abu Ahmad et al. 2006a, b). Occurrence of fifth genotype of SCYLV i.e. IND from India has been strongly established based on partial sequences of SCYLV encoding for ORF 1 and 2 and complete ORF 3 and 4 sequences other than the existences of CUB and BRA-PER isolates (Viswanathan et al. 2008). Later studies with additional 13 virus isolates from nine states of India also confirmed that the virus population of India has high level of homogeneity and are significantly different from the other reported genotypes (Singh et al. 2011). The coat protein based nucleotide sequences of SCYLV isolates reported across the world were subjected into phylogenetic analysis. The results had shown that all the isolates were separated into three major clusters. The Indian SCYLV isolates from various states viz. Coimbatore, Tamil Nadu, Hyderabad, Andhra Pradesh, Madhya Pradesh, Uttar Pradesh, Pune and Kerala were shown close association with Cuba and Tunisia isolates in cluster one. Majority of the SCYLV isolates from China were grouped in second cluster along with Peru, Hawaii, Reunion, USA, Brazil, Australia, Taiwan, Argentina, Kenya, Guatemala, Colombia and Malaysian isolates. The third cluster consists of South Africa and Mauritius isolates (Fig.12.2).

Recently efforts were made to characterize the full genome of Indian SCYLV isolates for which four SCYLV isolates from Coimbatore were chosen by Chinnaraja et al. (2013) and the complete genome of four virus isolates belonged to the SCYLV-IND genotype (~5875 nt) and has shown close similarity with CHN 1 genotype reported from China. The results of phylogenetic comparison of complete genomic sequences with other isolates and genotypes reported worldwide had shown that IND and CHN 1 originated from Asia grouped together in a cluster and other

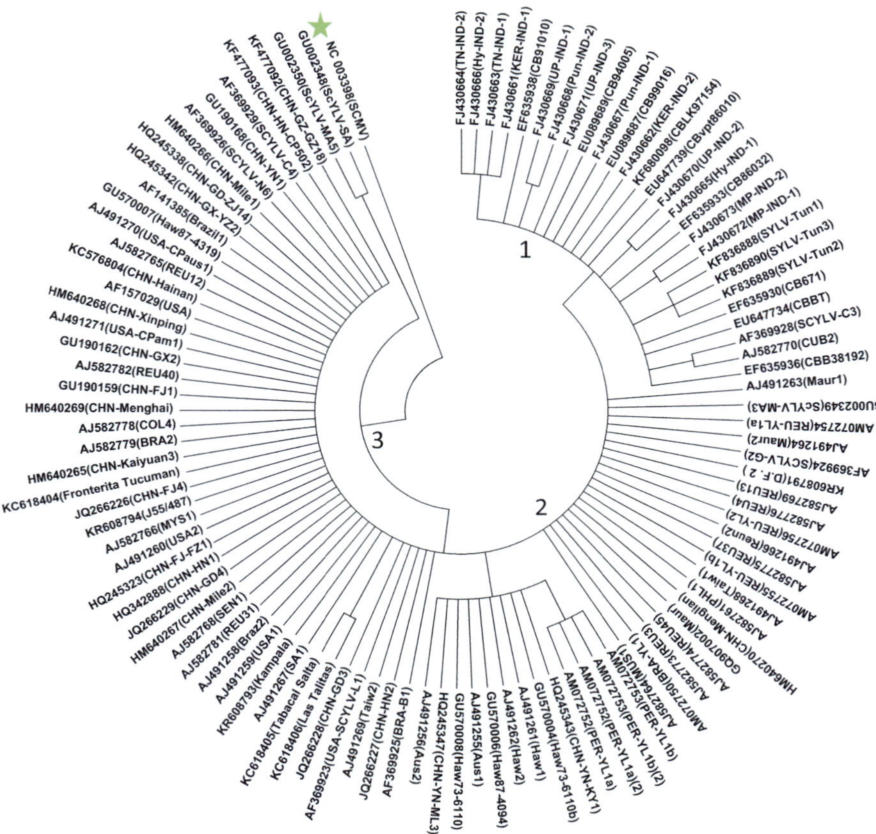

Fig. 12.2 Phylogenetic tree based on coat protein nucleotide sequences of SCYLV isolates (submitted in NCBI) showing evolutionary clustering of Indian isolates with the other globally distributed isolates. Phylograms were generated in MEGA version 4.0.2 using maximum likelihood method Sugarcane mosaic virus (GenBank Accession No. NC 003398) was used as an out group (OG) and is marked with *green color star*

genotypes reported from America and Africa separated in another cluster. These isolates exhibited amino acid sequence differences of 29.2–31.8%, 28.1–34.4% and 30.7–33.4% with REU, HAW-PER and BRA in partial ORF0 sequences, respectively. Further, detailed recombination analyses revealed evidence of recombination in ORF1 to ORF5 with the maximum number of sites occured in ORF2 in one of the four IND isolates (Chinnaraja et al. 2013).

The high incidences of SCYLV recombination suggests that recombination plays a major role in SCYLV evolution and is the driving force in evolution and emergence of new variants of SCYLV. Similarly, the studies conducted from different countries were also suggested that SCYLV genome might have evolved from at least two independent recombination and therefore it is being considered as an emerging virus evolved from recombination between three ancestors genera *Luteovirus*, *Polerovirus* and *Enamovirus* and now under the family Luteoviridae (Moonan et al. 2000; Smith et al. 2000). However, more data needs to be generated from India and other countries to study the genome dynamics of SCYLV. The findings also revealed ORF3 coding for CP in SCYLV is the most conserved; supporting the strategy of SCYLV management through CP gene derived transgenic resistance. Further, works on developments of SCYLV resistant transgenic plants are in progress at ICAR-SBI, Coimbatore, Tamil Nadu.

Sugarcane Breeding Institute (SBI), Coimbatore houses one of the largest collections of sugarcane germplasm and hybrid collections. Recently Viswanathan et al. (2016) conducted detailed surveys on YLD symptom incidence and severity for five seasons in the germplasm resources totaling ~4066 genotypes/varieties maintained by the Institute at Coimbatore and its research centres, Agali, Kannur (Kerala) and Karnal (Haryana). Among the different centres/collections, Agali centre recorded more severity to YLD followed by National Hybridization Garden (NHG), National Active Germplasm (NAG) and 'Co' canes. However, *Saccharum* sp. clones maintained at Kannur recorded low YLD incidence and least severity for the disease symptoms. Overall, the study indicated that most of the parents used for breeding and hybridization were affected by YLD to varying severities. High incidence of vector population and constitution of varietal/parental materials are suspected for the high disease incidence and intensity in the two collections. The study identified 463 resistant sources in the hybrid clones and 773 in *Saccharum* spp. for the first time. The outcome of the study lays foundation for developing YLD resistance in sugarcane progenies in the country. Further, the newly developed disease scale is being used in 21 research centres under All India Coordinated Research Project (AICRP) on sugarcane to identify YLD resistance in new sugarcane varieties. This programme would identify YLD resistance in promising varieties which are promoted for cultivation and will supplement ongoing disease management approaches through virus-free nurseries.

12.4 Potato Leaf Roll Virus (PLRV)

12.4.1 Occurrence and Significance

PLR is one of the most damaging diseases of potatoes throughout the world. It has long been recognized as a major component of potato degeneration. The disease is caused by PLRV, the type species of the genus *Polerovirus* (Mayo and D'Arcy 1999). All Indian potato varieties are susceptible to this virus. Infected plants produce only a few, small to medium tubers (Singh et al. 2015). The virus is tuber borne, not sap transmitted, transmitted efficiently by aphid in a circulative non-propagative manner (Singh et al. 1982; Paul Khurana and Singh 2000). Yield loss normally ranges from 50% to 60% in India (Paul Khurana and Singh 2000; Mukherjee et al. 2003). However, it is lower in autumn season (7–16%) than in spring season (39–60%) (Paul Khurana and Singh 2000). PLRV has small,

Fig. 12.3 Potato leaf roll virus (PLRV) symptoms in the field (**a**) Primary leaf roll symptoms on infected potato plants (**b**) Leaf roll symptoms due to secondary infection

isometric virions (23–25 nm dia) and are primarily confined to the phloem of the infected plants. It occurs in extremely low concentration in infected plants.

The infected plants show two type of symptoms *viz*., primary or secondary, depending upon the age of infection. The primary symptoms are confined to top young leaves, which usually stand upright, roll and turn slightly pale in certain cultivars (Fig. 12.3a). However, reddish/pink colouration occurs in top leaves starting at the margins, sometimes accompanied with slight rolling of the leaflets in most cultivars. Secondary symptoms develop when the plants are grown from infected seed tubers (Fig. 12.3b). Such symptoms are quite prominent in older leaves, i.e. absent or less pronounced on younger top leaves. Infected plants have characteristic pale, dwarfed, and upright appearance with rolling of lower leaves that turn yellow, brittle and are leathery in texture. In some cultivars, a reddish or purple discolouration develops on the margins and underside of the leaves (Paul Khurana and Singh 2000). In storage, the tubers from the infected plants, in certain varieties develop phloem necrosis but most Indian varieties do not develop this necrosis. In India, occurrence of three groups of PLRV (mild, moderate and severe strain) are reported based on their symptom severity on the test plant, *Physalis floridana* plants as well as on potato varieties/hybrids (Singh et al. 1982). However, these strains did not differ antigenetically (Fig. 12.3).

12.4.2 Genome Sequences and Relations

The genome consists of positive sense single stranded RNA with the viral genome linked protein (VPg) at the 5′ end (Mayo et al. 1982) and an OH group at the 3′ end. The genome is divided into two parts by a small non coding RNA and consists of nine open reading frames (ORFs) numbered from 0 to 8 coding for proteins, P0–P7 and Rap1, respectively. Three 5′-proximal ORFs, which are expressed from the genomic RNA, encode the proteins, P0, P1 and P2. Five other ORFs are expressed by translation from two sub genomic RNAs (sgRNAs). Two structural proteins (P3, P5) and P4 are encoded by sgRNA1 and the sgRNA2 encodes two 3′-proximal proteins (P6, P7). ORF 1 harbours a small ORF, ORF 8 which encodes Rap 1 (Jeevalatha et al. 2013).P0 is involved in symptom development and acts as a suppressor of RNA silencing (Pfeffer et al. 2002) and has functional motifs like F-BOX motif to overcome PTGS (Pazhouhandeh et al. 2006). P1 is a proteinase containing polyprotein responsible for the release of virus encoded protein (VPg) (Prufer et al. 1999; Sadowy et al. 2001; van der Wilk et al. 1997). P2 is translated by a rarely occurring ribosomal frame shift from ORF1 and carries the conserved motifs typical of RNA-dependent RNA polymerases (RdRp). P3, P4 and P5 correspond to the major coat protein (CP), the putative movement protein (MP) and the read through domain (RTD), which is translated by suppression of the ORF3 stop codon. The P5 of luteoviruses has been proposed to play a role in interaction between the virus particles and receptors in the aphid vectors (Guilley et al. 1994) especially the five-terminal conserved half may be the site of vector specificity (Brault et al. 2005). The functions of P6 and P7 are not known. But P7 is reported to have nucleic acid binding properties (Ashoub et al. 1998). Rap1 is involved in virus replication (Jaag et al. 2003). Almost all types of modulation mechanism (frame shift, initiation bypass, termination suppression, production of subgenomic (sg) RNA and proteolysis of primary translation products) are used during the expression of the different ORFs (reviewed by Sadowy et al. 2001).

Mukherjee et al. (2003) reported the coat protein gene sequence of a single PLRV isolate from India. The nucleotide and deduced amino acid sequence of the isolate was 97–99% identities to the other reported PLRV isolates. Recently, Jeevalatha et al. (2013) sequenced complete genome of five Indian PLRV isolates each one from North Western plains, Eastern plains, Northern high hills, North Eastern hills and Southern hills representing the different agro-climatic zones of India. The genome of Indian PLRV isolates comprised of 5883 nucleotides and had nine predicted open reading frames (ORF0 to ORF 8) that were similar to the other PLRV isolates. Except the isolate OTNI-2 in which a single nucleotide substitution (A>G) in the stop codon (at position 5742) of ORFs 5 and 7 was observed. The genome was predicted to contain a non coding sequence of 70 nucleotides at the 5′ end, 144 nucleotides at the 3′ end and 197 nucleotides in between the two blocks of coding sequences. About 97.6–98.7% similarities was observed among the Indian isolates and were more close to European, Canadian, African, American and Czech isolates (Group I) with 95.8–98. 6% identities than to an Australian isolate (Group II, 92.9–93.4%). The five Indian PLRV isolates showed maximum similarity to Poland and Egyptian PLRV isolates (Fig. 12.4). The reason for this may be that the potato was introduced from Europe to India and also the germplasms are being imported from the northern hemisphere countries for breeding purpose.

High level of sequence homology has been observed in geographically distinct strains of PLRV worldwide including Indian PLRV isolates (Jeevalatha et al. 2013), except few key changes in different ORFs. At nucleotide level, ORF 3 and ORF 4, corresponding to coat protein and movement proteins are more conserved than other ORFs. The isolates showed more divergence in the ORF 0 region especially the isolate, PBI- 6 was different from other Indian isolates and had only 94.3–95.1% similarity to other four isolates. Recombination analysis using SISCAN method

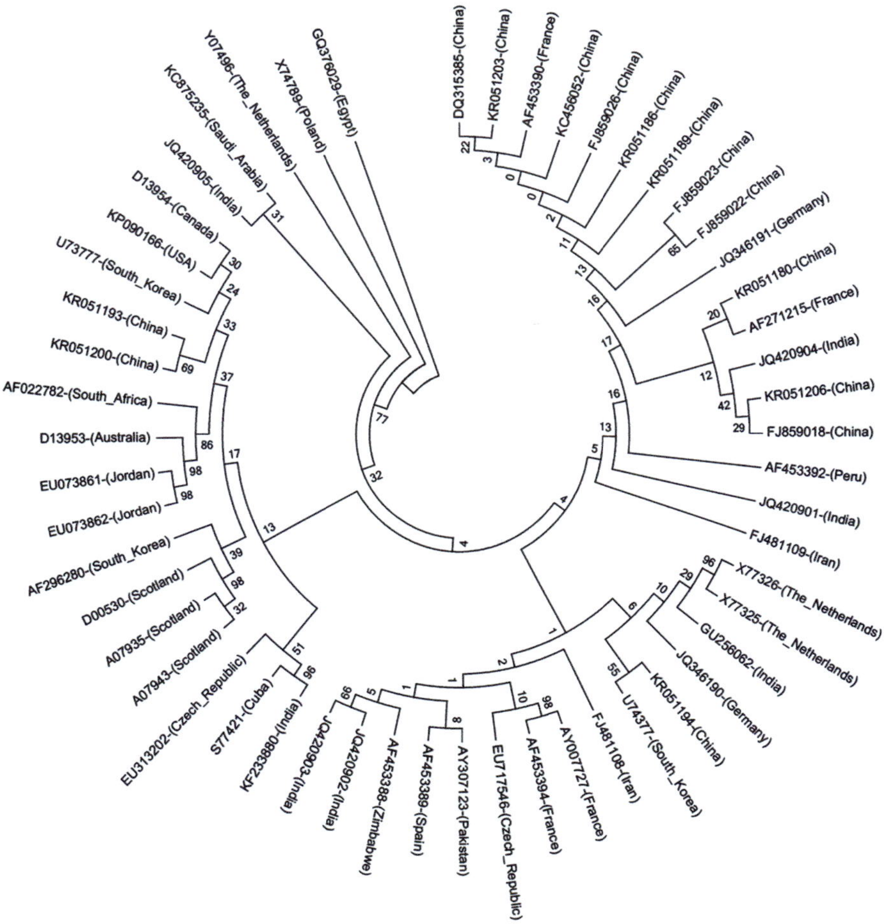

Fig. 12.4 Phylogenetic analysis based on CP gene of seven Indian PLRV isolates with other PLRV isolates reported in NCBI from countries revealed that the Indian isolates were distributed throughout the dendrogram. Two Indian isolates (JQ420904 and JQ420901) were grouped in a single cluster in which most of the Chinese isolate were clustered. Three isolates (GU256062, JQ420902 and JQ420903) formed another cluster along with PLRV isolates from The Netherlands, Germany, South Korea, China, Iran, France, Czech Republic, Pakistan, Spain and Zimbabwe. One isolate (KF233880) grouped with Cuban isolate with 96% bootstrap value and another isolate (JQ420905) was in separate cluster along with Saudi Arabean isolate

showed that the isolate PBI-6 was a recombinant isolate derived from a major parent, EU717546 (Czech isolate) and a minor parent, AF453389 (Spain isolate). The amino acid changes were more in ORF2 region of all five Indian isolates and only few amino acid changes were observed in ORF3 and ORF4 of two isolates. Nucleotide identities of these PLRV isolates with other poleroviruses ranged from 43.7% to 53.1% with a maximum of 52.8–53.1% similarity to CYDV. The identities ranged from 45.3% to 45.6% to CABYV, 45.1–45.2% to MABYV, 46.6–46.9% to

BWYV, 46.4–46.5% to BMYV, 46.8–47.0% to TuYV, 44.2–44.3% to PeVYV, 46.5–46.6% to TVDV, 43.7–44.0% to ScYLV and 44.1–44.3% to CpCSV. The isolates shared 29.1–29.3% similarity to BYDV which belongs to another genus *Luteovirus* of the family *Luteoviridae*.

12.4.3 PLRV Infection in Jute

Jute (*Corchorus olitorius* L.) is a major fiber crop of India grown mainly in west Bengal, Bihar, Assam and Orissa etc. PLRV infection on Jute was first reported from Central Research Institute for Jute and Allied Fibres, Barrackpore (CRIJAF), India, and the disease incidence was less than two percent and diseased plants were stunted in growth and height of the plants were also reduced. Most of the upper leaves showed curling, coiling, puckering and shoe string symptoms on leaf lamina. Stipules and petioles of the infected leaves were exceptionally longer. *Aphis gossypii* vector was often noticed in the field, all the samples were tested by double-antibody sandwich ELISA for common aphid transmitted viruses, e.g., bean common mosaic virus, cucumber mosaic virus, papaya ringspot virus, PLRV, cowpea mosaic virus, potato virus Y and watermelon mosaic virus using commercial ELISA kits available. The symptomatic jute samples showed positive reaction only with PLRV antibody and for further confirmation, reverse transcription PCR was carried out with PLRV CP gene specific primer pair and 627 nucleotide CP gene was sequenced (Accession No. KF233880) that shared 99% sequence identity with the CP gene sequence with PLRV reference strain S77421 (Biswas et al. 2014).

12.5 Cotton leaf roll virus (CLRDV)

Cotton blue disease caused by CLRDV is a serious problem in cotton cultivation in South America causing yield losses up to 80% in susceptible varieties (Silva et al. 2008; Distéfano et al. 2010). This positive-sense, single-stranded RNA virus is transmitted by aphids (*Aphis gossypii*) in a circulative persistent manner. Recently, Mukherjee et al. (2012) reported the occurrence of CLRDV infecting cotton fields at Nagpur, Maharashtra, India with the primers PL4F (5′-GCGACAAATAGTTAATGAATACGGT-3′) and 03R (5′-GTCTACCTATTTBGGRTTNTGGAA-3′). The primers were designed to amplify a region of approximately 600 bp of the capsid protein sequence of CLRDV (Corrêa et al. 2005). Cotton plants affected by this disease show stunting, leaf rolling, intense green foliage, vein yellowing, brittleness of leaves, reduced flower and boll size, sometimes resulting in sterility of plants (Fig. 12.5). PCR from healthy samples did not produce an amplicon. The PCR products were sequenced directly and the resulting sequence was deposited at GenBank. The coat protein sequences derived from the PCR products of Indian

Fig. 12.5 Symptoms of cotton leaf roll virus (**a**) Curling, rolling and vein clearing of leaves; (**b**) leaf vein yellowing and brittleness of leaves (Courtesy: Dr. Arup K. Mukherjee)

isolate of CLRDV (Accession No. JN033875) from symptom-bearing plants showed more than 90% similarity with Cotton leaf roll dwarf virus and chickpea stunt virus (another member of the genus *Polerovirus)* as reported by earlier workers (Corrêa et al. 2005; Silva et al. 2008; Distéfano et al. 2010). This was the first report of the detection of a *Polerovirus* infecting cotton in India. No further report is available for the occurrence and genetic diversity CLRDV in India.

12.6 Concluding Remarks

PLRV and SCYLV are the most established poleroviruses in India but the studies on epidemiology and management of PLRV from potato and sugarcane is limited, which needs a major attention. Similarly, the information on the prevallence, genetic diversity and management of the CLRDV and PLRV is lacking from India. SCYLV is the recently reported emerging polerovirus in India. A lot of information that is available regarding symptomatology, morphological, biological and molecular diversity, genome organization, transgenic resistance is required to be utilized for the management of SCYLV as it is has emerged as a serious threat to sugar industry in causing severe economic losses in yield and sugar recovery. In India, research work should be initiated in the area of whole genome SNP identification and genomic selection to identify the genetic basis of resistance to SCYLV. Further, it will contribute to the knowledge and application of molecular mechanisms governing SCYLV resistance in Indian sugarcane cultivars. The diseases like potato leaf roll and sugarcane yellow leaf are mainly responsible for the elimination of many elite commercial varieties in India. Additionally, these diseases contribute to decline in their performance which is referred as 'varietal degeneration' in vegetative propagated crop. Lack of awareness on tuber/seed cane health and ignoring quarantine regulations resulted in introduction of diseases, their epidemics and varietal degeneration in many parts of the country. To increase potato/sugarcane productivity in

India, supply of healthy tuber/seed canes is to be ensured in the field. Research personnel and development workers should be actively involved in creating awareness on supply of healthy seed. In addition to detect poleroviruses in tubers/seed canes, the recent approaches in the disease diagnosis using serological and molecular approaches have applications in the field of developing virus–free seedlings, germplasm exchange and quarantine, disease surveillance and integrated disease management in India.

References

Abu Ahmad Y, Rassaby L, Royer M, Borg Z, Braithwaite KS, Mirkov TE (2006a) Yellow leaf of sugarcane is caused by at least three different genotypes of *Sugarcane yellow leaf virus*, one of which predominates on the Island of Reunion. Arch Virol 151:1355–1371

Abu Ahmad Y, Royer M, Daugrois JH, Costet L, Lett JM, Victoria JI, Girard JC, Rott P (2006b) Geographical distribution of four *Sugarcane yellow leaf virus* genotypes. Plant Dis 90:1156–1160

Ashoub A, Rohde W, Prufer D (1998) In planta transcription of a second subgenomic RNA increases the complexity of the subgroup 2 luteovirus genome. Nucleic Acids Res 26:420–426

Biswas C, Dey P, Mitra S, Bera A, Satpathy S, Karmakar PG (2014) First Report of Potato leaf roll virus (PLRV) naturally occurring on Jute (*Corchorus olitorius*) in India. Plant Dis 989:1592

Brault V, Perigon S, Reinbold C, Erdinger M, Scheidecker D, Herrbach E, Richards K, Ziegler-Graff V (2005) The Polerovirus minor capsid protein determines vector specificity and intestinal tropism in the aphid. J Virol 79:9685–9693

Chinnaraja C, Viswanathan R (2015a) Variability in yellow leaf symptom expression caused by the *Sugarcane yellow leaf virus* and its seasonal influence in sugarcane. Phytoparasitica 43:339–353

Chinnaraja C, Viswanathan R (2015b) Quantification of sugarcane yellow leaf virus in sugarcane following transmission through aphid vector, *Melanaphis sacchari*. Virus Dis 26:237–242

Chinnaraja C, Viswanathan R, Karuppaiah R, Bagyalakshmi K, Malathi P, Parameswari B (2013) Complete genome characterization of *Sugarcane yellow leaf virus* from India: evidence for RNA recombination. Eur J Plant Pathol 135:335–349

Chinnaraja C, Viswanathan R, Sathyabhama M, Parameswari B, Bagyalakshmi K, Malathi P, Neelamathi D (2014) Quantification of *Sugarcane yellow leaf virus* in in vitro plantlets and asymptomatic plants of sugarcane by RT-qPCR. Curr Sci 106:10

Comstock JC, Irvine JE, Miller JD (1994) Yellow leaf syndrome appears on the United States mainland. Intern Sugar J 56:33–35

Comstock JC, Miller JD, Schnell RJ (2001) Incidence of Sugarcane yellow leaf virus in clones maintained in the world collection of sugarcane and related grasses at the United States National Repository in Miami, Florida. Sugar Tech 3:128–133

Correa RL, Silva TF, Simoes-Araujo JL, Barroso PAV, Vidal MS, Vaslin MFS (2005) Molecular characterization of a virus from the family Luteoviridae associated with cotton blue disease. Arch Virol 150:1357–1367

Distéfano AJ, Kresic IB, Hopp HE (2010) The complete genome sequence of a virus associated with cotton blue disease, cotton leafroll dwarf virus, confirms that it is a new member of the genus Polerovirus. Arch Virol 155:1849–1854

Gaur RK, Singh AK, Singh M, Singh AK, Upadhayaya, Rao GP (2003) Reliability of serological identification of sugarcane mosaic potyvirus and sugarcane yellow leaf luteovirus from cane stalk juice. Sugar Cane Int UK 5:18–21

Guilley H, Wipf-Scheibel C, Richards KE, Lecoq H, Jonard G (1994) Nucleotide sequence of cucurbit aphid-borne yellows luteovirus. Virology 202:1012–1017

Harisson BD (1999) Steps in the development of Luteovirology. In: Smith H, Backer H (eds) The Luteoviridae. CABI Publishing, Wallingford/Oxon, pp 1–14
Horn NM, Reddy SV, Roberts IM, Reddy DVR (1993) *Chickpea chlorotic dwarf virus*, a new leafhopper-transmitted *Geminivirus* of chickpea in India. Ann Appl Biol 122:467–479
Horn NM, Reddy SV, Van den Heuvel JFJM, Reddy DVR (1996) Survey of chickpea (*Cicer arietinum* L.) for chickpea stunt disease and associated viruses in India and Pakistan. Plant Dis 80:286–290
Jaag HM, Kawchuk L, Rohde W, Fischer R, Emans N, Prufer D (2003) An unusual internal ribosomal entry site of inverted symmetry directs expression of a Potato leafroll polerovirus replication associated protein. Proc Natl Acad Sci USA 100:8939–8944
Jeevalatha A, Priyanka K, Shandil RK, Sharma NN, Chakrabarti SK, Singh BP (2013) Complete genome sequence of *Potato leaf roll virus* isolates infecting potato in the different geographical areas of India shows low level genetic diversity. Indian J Virol 24(2):199–204
Kaiser WJ, Danesh D (1972) Biology of four viruses affecting *Cicerm arietinum* in Iran. Phytopathology 61:372–315
Khetarpal RK, Kumar J, Beuve M, Parakh DB (1994) Ram Nath outbreak of MAV-type barley yellow dwarf virus on wheat in the Garhwal Hills in India. Plant Pathol 43(2):415–416
Kotasthane SR, Gupta OM (1978) Yield loss due to Chickpea stunt. Trop Grain Leg Bull 11(12):38–39
Kulkarni S, Hegde RK (1980) Barley yellow dwarf of wheat-a new disease to Karnataka. Curr Res 9:119
Kulkarni S, Hegde RK (1983) Reaction of wheat varieties to barley yellow dwarf disease. Plant Pathol Newsl 1:11–12
Lehrer AT, Kusalwang A, Komar E (2008) High incidence of sugarcane yellow leaf virus (SCYLV) in sugar plantations and germplasm collections in Thailand. Australas Plant Dis Notes 3:89–92
Mayo MA, DArcy CJ (1999) Family Luteoviridae: are-classification of luteoviruses. In: Smith HG, Barker H (eds) The Luteoviridae. Wallingford, CAB International pp, pp 15–22
Mayo M, Barker H, Robinson D, Tamada T, Harrison B (1982) Evidence that Potato leafroll virus RNA is positive stranded, is linked to a small protein and does not contain polyadenylate. J Gen Virol 59:163–167
Miller WA, Rasochová L (1997) Barley yellow dwarf viruses. Annu Rev Phytopathol 35:167–190
Moonan F, Molina J, Mirkov TE (2000) *Sugarcane yellow leaf virus*: an emerging virus that has evolved by recombination between luteoviral and poleroviral ancestors. Virology 269:156–171
Mukherjee AK, Chahande PR, Meshram MK, Kranthi KR (2012) Plant Pathol 25:22
Mukherjee K, Verma Y, Chakrabarti SK, Singh MN, Paul Khurana SM (2003) Cloning and sequencing of coat protein gene of an Indian Potato leafroll virus (PLRV) isolate and its similarity with other members of Luteoviridae. Virus Genes 26:247–253
Nagaich BB, Vashisth KS (1963) Barley yellow dwarf: anew viral disease for India. Indian Phytopathol 16:318–319
Naidu RA, Mayo MA, Reddy SV, Jolly CA, Torrance L (1997) Diversity among the coat proteins of *luteoviruses* associated with chickpea stunt disease in India. Ann Appl Biol 30:37–47
Nene YI, Reddy MV (1987) Diseases of chickpea and their control. In: Saxena MC, Singh KB (eds) The chickpea. CAB International, Wallingford, pp 233–270
Nene YL, Reddy MV, Haware MP, Ghanekar AM, Amin KS (1991) Field diagnosis of chickpea diseases and their control, ICRISAT, Information Bulletin. ICRISAT, Patancheru, p 25
Paul Khurana SM (2000) Potato leafroll virus. In: Paul Khurana SM (ed) Diseases and pests of potato- a manual. Shimla, CPRI p, pp 33–35
Pazhouhandeh M, Dieterle M, Marrocco K, Lechner E, Berry B, Brault V, Hemmer O, Kretsch T, Richards KE, Genschik P, Ziegler-Graff V (2006) F-box-like domain in the Polerovirus protein P0 is required for silencing suppressor function. Proc Natl Acad Sci USA. 103:1994–1999
Pfeffer S, Dunoyer P, Heim F, Richards KE, Jonard G, Ziegler Graff V (2002) P0 of beet western yellows virus is a suppressor of post transcriptional gene silencing. J Virol 76:6815–6824

Prufer D, Kawchuk L, Monecke M, Nowok S, Fischer R, Rohde W (1999) Immunological analysis of potato leafroll luteovirus (PLRV) P1 expression identifies a 25 kDa RNA-binding protein derived via P1 processing. Nucleic Acids Res 27:421–425

Rao GP, Gaur RK, Maneesha S, Srivastava AK, Virk AS, Singh N, Patil AS, Viswanathan R, Jain RK (2000) Existence of sugarcane yellow leaf luteovirus in India. Sugar Tech 2(4):37–38

Rao GP, Gaur RK, Singh M, Viswanathan R, Chandrasena G, Dharamwardhaanhe NMNN (2001) Occurrence of sugarcane yellow leaf virus in India and Srilanka. Proc Intern Soc Sugar Cane Technol 24:469–470

Rao GP, Viswanathan R, Singh SB (2002) Current situation of sugarcane diseases in India. In: Singh SB, Rao GP, Easwaramoorthy S (eds) Sugarcane crop management. SCI Tech Publishing LLC, Houston, pp 1–9. Pp734

Rassaby L, Girard JC, Lemaire O, Costet L, Irey MS, Kodja H, Lockhart BEL, Rott P (2004) Spread of Sugarcane yellow leaf virus in sugarcane plants and fields on the Island of Reunion. Plant Pathol 53:117–125

Sadowy E, Juszczuk M, David C, Gronenborn B, Hulanicka MD (2001) Mutational analysis of the proteinase functions of Potato leafroll virus. J Gen Virol 82:1517–1527

Schenck S (1990) Yellow leaf syndrome – a new disease of sugarcane. Report of HSPA Exp. Sta, p 98

Schenck S, Hu JS, Lockhart BEL (1997) Use of a tissue blot immunoassay to determine the distribution of sugarcane yellow leaf virus in Hawaii. Sugar Cane 4:5–8

Silva TF, Corrêa RL, Castilho Y, Silvie P, Bélot JL, Vaslin MFS (2008) Widespread distribution and a new recombinant species of Brazilian virus associated with cotton blue disease. Virology 5:123

Singh D, Rao GP (2011) Molecular detection of two strains of *Sugarcane yellow leaf virus* in India and their secondary spread in nature through aphids. Acta Phytopathol Entomol Hung 46:17–26

Singh DV, Joshi HC, Singh MN, Nagaidi BB, Jorfa LM (1979) Prevalence of barley yellow dwarf virus on cereals and occurrence of smut in triticales in Kumaon hills. Indian Phytopathol 32:98–99

Singh MN, Paul Khurana SM, Nagaich BB, Agrawal HO (1982) Strains of potato leafroll virus and their aphid transmission. J Indian Potato Assoc 9:121–127

Singh D, Rao GP, Snrehi SK, Rak SK, Karuppaiah R, Viswanathan R (2011) Molecular detection and identification of thirteen isolates of *Sugarcane yellow leaf virus* associated with sugarcane yellow leaf disease in nine sugarcane growing states of India. Australas Plant Pathol 40:522–528

Singh BP, Nagesh M, Sharma S, Sagar V, Jeevalatha A, Sridhar J (2015) A manual on diseases and pests of potato, Technical bulletin No. 101. Central Potato Research Institute, Shimla

Smith HG, Barker H (eds) (1999) The Luteoviridae. CABI Publishing, Wallingford, p 297

Smith GR, Borg Z, Lockhart BEL, Braithwaite KS, Gibbs MJ (2000) *Sugarcane yellow leaf virus*: a novel member of the Luteoviridae that probably arose by inter-species recombination. J Gen Virol 81:1865–1869

Van der Wilk F, Verbeek M, Dullemans AM, van den Heuvel JF (1997) The genome-linked protein of Potato leafroll virus is located downstream of the putative protease domain of the ORF1 product. Virology 234:300–303

Viswanathan R (2002) Sugarcane yellow leaf syndrome in India: Incidence and effect on yield parameters. Sugar Cane Intern 20:17–23

Viswanathan R (2004) Ratoon stunting disease infection favors severity of yellow leaf syndrome caused by sugarcane yellow leaf virus in sugarcane. Sugar Cane Intern 22(2):3–7

Viswanathan R (2012) Need for a paradigm shift in sugarcane disease management. In: Nair NV, Puthira Pratap D, Viswanathan R, Srikanth J, Bhaskaran A, Ram B (eds) Perspectives in sugarcane agriculture. Society for Sugarcane Research and Development, Coimbatore, pp 171–206

Viswanathan R (2016) Varietal degeneration in sugarcane and its management in India. Sugar Tech 18:1–7

Viswanathan R, Balamuralikrishnan M (2004) Detection of sugarcane yellow leaf virus, the causal agent of yellow leaf syndrome in sugarcane by DAS-ELISA. Arch Phytopathol Plant Protect 37:169–176

Viswanathan R, Padmanaban P, Mohanraj D, Ramesh Sundar A, Premchandran MN (1999) Suspected yellow leaf syndrome in sugarcane. Sugarcane Breed Inst Newsl 18:2–3

Viswanathan R, Balamuralikrishnan M, Karuppaiah R (2006) Yellow leaf disease of sugarcane: occurrence and impact of infected setts on disease severity and yield. Proc Sugar Technol Assoc India 67:74–89

Viswanathan R, Balamuralikrishnan M, Karuppaiah R (2008) Identification of three genotypes of sugarcane yellow leaf virus causing yellow leaf disease from India and their molecular characterization. Virus Genes 37:368–379

Viswanathan R, Karuppaiah R, Malathi P, Ganesh Kumar V, Chinnaraja C (2009) Diagnosis of *Sugarcane yellow leaf virus* in asymptomatic sugarcane by RT-PCR. Sugar Tech 11:368–372

Viswanathan R, Karuppaiah R, Balamuralikrishnan M (2010) Detection of three major RNA viruses infecting sugarcane by multiplex reverse transcription-polymerase chain reaction (multiplex-RT-PCR). Australas Plant Pathol 39:79–84

Viswanathan R, Karuppaiah R, Kowsalya V, Chinnaraja C, Malathi P (2012) Yellow leaf disease of sugarcane: symptom, etiology, epidemiology, impact on sugarcane, diagnosis and management. In: Rao GP, Baranwal VK, Mandal B, Rishi N (eds) Recent trends in plant virology. Studium Press LLC, Houston, pp 389–411

Viswanathan R, Chinnaraja C, Malathi P, Gomathi R, Rakkiyappan P, Neelamathi D, Ravichandran V (2014) Impact of *sugarcane yellow leaf virus* (ScYLV) infection on physiological efficiency and growth parameters in sugarcane in India. Acta Physiol Plant 36:1805–1822

Viswanathan R, Chinnaraja C, Parameswari B, Chhabra ML (2016) Status of yellow leaf resistance in sugarcane germplasm and parental clones at Sugarcane Breeding Institute, India. Int Sugar J 115:60–71

Zhang QF, Guan WN, Ren ZY, Zhu XS, Tsai JH (1983) Transmission of barley yellow dwarf virus strains from northwestern China by four aphid species. Plant Dis 67: 895–899

Characterisation of the Macluraviruses Occurring in India

13

S. Vijayanandraj, Bikash Mandal, T. Jebasingh, M.L. Jeeva, T. Makeshkumar, and Yogita Maheshwari

Abstract

The genus *Macluravirus* of the family *Potyviridae* currently contains six recognized and two tentative virus species. In India, so far only two macluraviruses eg., large cardamom chirke virus (LCCV) and cardamom mosaic virus (CdMV) infecting large cardamom (*Amomum subulatum*) and small cardamom (*Elettaria cardamomum*), respectively have been studied well. Recently, a new macluravirus, yam mottling virus has been tentatively identified in mild mosaic disease of yam (*Dioscorea* spp) in southern India. LCCV is distributed in large cardamom cultivated in the North-East sub-Himalayan mountains and CdMV in small cardamom cultivated in southern India. Both these macluraviruses cause chlorotic streak mosaic disease in cardamom. CdMV and LCCV are known in India since long time and considerable infromation has been generated. This chapter summarizes the work on the biological and molecular properties of macluraviruses occurring in India.

Keywords

Macluravirus • India • Large cardamom chirke virus • Cardamom mosaic virus • Cardamom • Chirke • Katte

S. Vijayanandraj • B. Mandal (✉) • Y. Maheshwari
Advanced Centre for Plant Virology, Division of Plant Pathology, Indian Agricultural Research Institute, New Delhi 110012, India
e-mail: leafcurl@rediffmail.com

T. Jebasingh
School of Biological Sciences, Madurai Kamaraj University, Madurai 625021, Tamil Nadu, India

M.L. Jeeva • T. Makeshkumar
Central Tuber Crops Research Institute, Thiruvananthapuram, Kerala 695 017, India

B. Mandal et al. (eds.), *A Century of Plant Virology in India*,
https://doi.org/10.1007/978-981-10-5672-7_13

13.1 Introduction

The name of the genus *Macluravirus* has been derived from the type species *Maclura mosaic virus* (MacMV), which was identified in *Maclura pomifera* in Yugoslavia (Plese and Milicic 1973). The members of the genus *Macluravirus* having many distinguishing features, the genus has been recognised under the family *Potyviridae* by ICTV in 1998 (Pringle and Fauquet 1998). The virions are flexuous filamentous and the size is slightly smaller (650–675 nm) compared to the potyviruses (680–900 nm). Macluraviruses are transmitted by aphids in a non-persistent manner and by mechanical sap inoculation. Generally, they have narrow host range infecting plant species of families eg., *Amaranthaceae, Asteraceae, Iridaceae, Dioscoreaceae, Amaryllidaceae, Moraceae, Aizoaceae, Ranunculaceae* and *Zingiberaceae*. Currently, the genus consists of eight recognised virus species *eg., Alpinia mosaic virus, Artichoke latent virus, Broad-leafed dock virus A, Cardamom mosaic virus* (CdMV)*, Chinese yam necrotic mosaic virus* (ChYNMV)*, Maclura mosaic virus, Narcissus latent virus, Yam chlorotic mosaic virus* and one tentative species large cardamom chirke virus (LCCV).

In India, macluraviruses are known since long time to infect the spices crops, small cardamom (*Elettaria cardamomum*) and large cardamom (*Amomum subulatum)* causing streak mosaic disease known as Katte and chirke, respectively. The virus generally causes light greenish or yellow streaks on the leaf lamina. Infected plants show stunted growth and the yield reduces drastically in the subsequent years and gradually the infected plant becomes unproductive. The katte disease affecting small cardamom was first reported in southern India (Mollison 1900). Thomas (1938) was the first to suggest that the katte disease might be of a virus origin. However, the association of a virus was not proven until the evidence of transmission of the disease by banana aphid (*Pentalonia nigronervosa*) reported by Uppal et al. (1945). Chirke disease was first noticed in 1956 from Algarah, Darjeeling hills of West Bengal, India (Raychaudhuri and Chatterjee 1958). Based on transmission by aphids *Rhopalosiphum maidis* and *Brachycaudus helichrysi* in a non-persistent manner, chirke disease was considered to be a viral disease (Raychaudhuri and Chatterjee 1961, 1965). The virus associated with the katte and chirke disease was identified as two different macluraviruses, CdMV and LCCV (Jacob and Usha 2001; Mandal et al. 2012). In southern India, a new disease of yam (*Dioscorea* spp) was observed recently and preliminary studies showed existence of another macluravirus in India. Although, the macluraviruses in cardamom are known in India since long, the genomic studies have been initiated since 2001 (Table 13.1). In this chapter, the studies on macluraviruses occurring in India are summarised.

13.2 Large Cardamom Chirke Virus

Large cardamom is a perennial plantation crops grown in the natural organic environments of the forests in the North-East sub-Himalayan mountains. The plant grows up to 1.5–2.5 m tall with a large number of tillers forming a robust clump.

Table 13.1 Genome sequence resources of macluraviruses infecting different crops in India

Virus	Accession No.	Region of genome	Crop	Place/isolate	NCBI depositor (Year)
CdMV	AF189125	partial NIb, CP and 3'UTR	*Elettaria cardamomum*	Yeslur	Usha (2001)
	AJ308477	CP and 3'UTR	*E. cardamomum*	Vandiperiyar	Usha (2001)
	AJ308476	Partial CP and 3'UTR	*E. cardamomum*	Kursupara	Usha (2001)
	AJ308475	Partial CP and 3'UTR	*E. cardamomum*	Thalathamane	Usha (2001)
	AJ308474	Partial CP and 3'UTR	*E. cardamomum*	Madikeri	Usha (2001)
	AJ308473	Partial CP and 3'UTR	*E. cardamomum*	Sirsi	Usha (2001)
	AJ308472	Partial CP and 3'UTR	*E. cardamomum*	Appangala	Usha (2001)
	AJ312774	Partial CP and 3'UTR	*E. cardamomum*	Kattappana	Usha (2001)
	AJ345002	Partial NIb	*E. cardamomum*	Yeslur	Usha (2001)
	AJ550379	CI and 6K2	*E. cardamomum*	Yeslur	Usha (2003)
	AJ550378	VPg and NIa prtease	*E. cardamomum*	Yeslur	Usha (2003)
	AY609386	CP and 3'UTR	*E. cardamomum*	Somwarpet	Smitha and Reddy (2004)
	AY609385	Partial NIb, CP and 3'UTR	*E. cardamomum*	Margodu	Smitha and Reddy (2004)
	AY833735	Partial NIb, CP and 3'UTR	*E. cardamomum*	Anemahal	Smitha and Reddy (2004)
	AY823986	Partial NIb, CP and 3'UTR	*E. cardamomum*	SKP-3	Smitha and Reddy (2004)
	AY823985	Partial NIb, CP and 3'UTR	*E. cardamomum*	SKP1	Smitha and Reddy (2004)
	JN544082	CP	*E. cardamomum*	Sakleshpur	Siljo and others (2011)

(continued)

Table 13.1 (continued)

Virus	Accession No.	Region of genome	Crop	Place/isolate	NCBI depositor (Year)
	JN544081	CP	*E. cardamomum*	Appangala	Siljo and others (2011)
	JN544080	CP	*E. cardamomum*	Meppadi	Siljo and others (2011)
	JN544079	CP	*E. cardamomum*	Sirsi	Siljo and others (2011)
	JN544078	CP	*E. cardamomum*	Thadiyankudisai	Siljo and others (2011)
	JN544077	CP	*E. cardamomum*	Pampadumpara	Siljo and others (2011)
	KU513860	CP	*E. cardamomum*	Valparai	Elangovan and Tennyson (2015)
LCCV	JN257715	Partaial NIb, CP ad 3'UTR	*Amomum subulatum*	CHK-1	Mandal and others (2012)
YMoV	KM099684	Partial NIb and CP	*Dioscorea alata*	YMCTCRI-01	Manasa and others (2014)
	KM111541	Partial NIb and CP	*D. alata*	YMCTCRI-02	Manasa and others (2014)
	KM201261	CP	*D. alata*	YMCTCRI-03	Manasa and others (2014)

CdMV cardamom mosaic virus, LCCV large cardamom chirke virus, YMoV yam mottling virus

The aromatic spice capsules are developed at the base of the clump (Fig. 13.1a). India is the largest producer of large cardamom (50%) and the other major producers are Nepal (33%) and Bhutan (13%). In India, Sikkim state (26734 ha) and Darjeeling district of West Bengal state (3305 ha) cover the major large cardamom areas. Large cardamom is affected by several fungal and bacterial diseases, which are of minor importance. Two viral diseases, chirke and foorkey are major constraints in the production of large cardamom in India.

The name 'chirke' was adopted from the Nepali language, which means streak mosaic. The mosaic streak in large cardamom (Fig. 13.1b) initially appears in the emerging tender leaves. In the severely affected plant, the mosaic streaks coalesce and the leaf gradually turns brown and dries up subsequently. The affected clump produces less number of flowers and thus causes a serious loss in yield. Raychaudhuri and Ganguly (1965) experimentally showed that the loss due to the virus was as high as 85.20% and 80.09% in the total number of fruits and seeds, respectively by the end of the third year of the crop. Large cardamom plantations are grown on the slope of the mountains and therefore it is difficult to get access for the regular

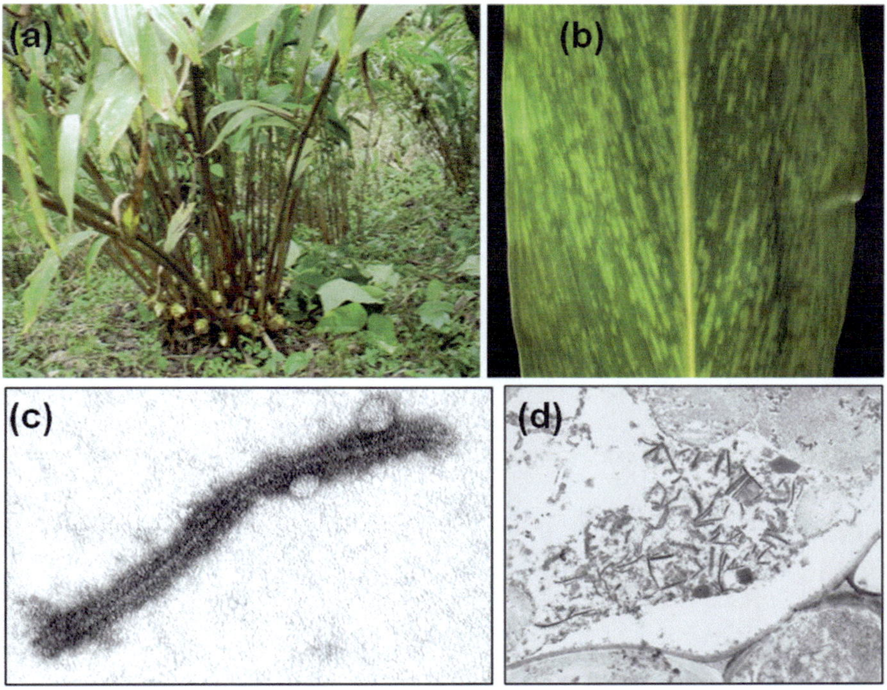

Fig. 13.1 Large cardamom plant and chirke disease. (**a**) Healthy plant bearing flower at the base of plant. (**b**) Streak mosaic symptoms of chirke disease on leaf. (**c**) Decoration of filamentous virion of large cardamom chirke virus with the homologous antiserum. (**d**) Ultrathin sections of chirke affected large cardamom leaf showing inclusions bodies

intercultural operations to manage the diseases. Heat therapy at 30-40^{0}C and chemotherapy with 14 chemicals consisting of growth regulators, antimetabolites and other organic compounds for varying period failed to cure the disease from the planting materials. However, 0.075% hydroquinone treatment of rhizomes and soil drenching with 0.1% of thiouracil showed virus inhibition (Raychaudhuri and Ganguly 1965). Time to time rouging of the diseased plants with the replenishment of certified virus free plants is essential for the management of chirke disease (Raychaudhuri and Ganguly 1965, 1966).

13.2.1 Chirke Distribution

Chirke disease is commonly present in large cardamom plantations. Extensive surveys were conducted in the major large cardamom growing areas in Sikkim and Darjeeling hills during 1998–2003 and the disease distribution was mapped into three categories, the places with no incidence, low (<15%) incidence and high (>15%) incidence (Fig. 13.2) (Mandal et al. 2012). The disease free area was identified in the different places (Tshering Gaon, Mirik Busty and Khaptali Gaon) in

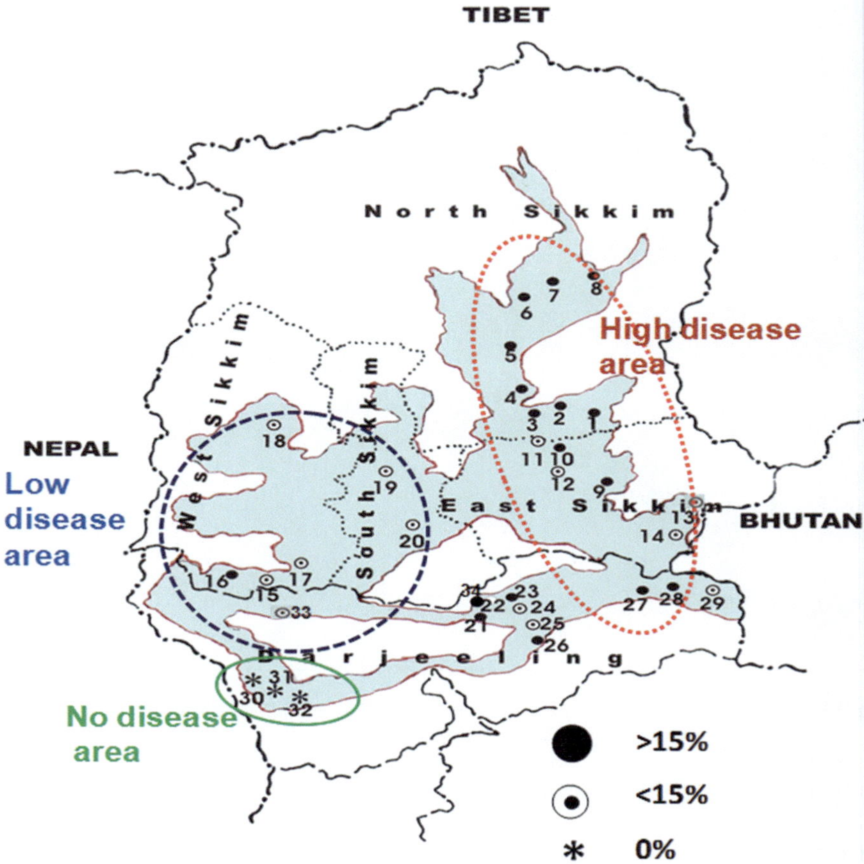

Fig. 13.2 Distribution of chirke disease of large cardamom in Sikkim and Darjeeling, the major large cardamom producing regions of India. Low, high and no disease pressure areas are indicated. *1* Kabi, *2* Chawang, *3* Phodong, *4* Sayem, *5* Namok, *6* Singhik, *7* Manuel, *8* Toong, *9* Assam Lingzey, *10* Lindok, *11* Pangthang, *12* Rumtek, *13*: Rigu, *14* Lingtam, *15* Soreng, *16* Bariakhop, *17* Sribadam, *18* Gezing, *19* Ravangla, *20* Temi, *21* Munsong, *22* Rishop, *23* Geetkolbung, *24* Passabong, *25* Sailungbazar, *26* Kaffer, *27* Suruk, *28* Godak, *29* Thode, *30* Tshering Gaon, *31* Mirik Busty, *32* Khaptali Gaon, *33* Sukhiapokhri, *34* Lava (Adopted from Mandal et al. 2012)

Mirik region. Very low disease incidence (up to 2.5%) was observed in Sukhiapokhri. Low disease pressure was observed in most places in East, West and South Sikkim and some places in Darjeeling district (Rishop, Geetkholbung, Passabong and Thode). The endemic areas were identified in northern Sikkim (Kabi, Chawang, Phodong, Sayem, Namok, Singhik, Manuel and Toong) and eastern parts of Darjeeling hills (Kuwapani, Badey and Sherpatanr), where the high disease incidence (19.2–35%) was recorded.

13.2.2 Host Range

The virus has narrow host range and other than large cardamom no other natural host know so far. Under the experimental conditions, the virus infects *Triticum aestivum* (genotypes NP, HB, HS and K series), *Cucumis angustifolia, Zingiber officinale* and a perennial weed, *Acorus calamus.* The following plant species are non-hosts, *Zea mays, Avena sterilis, Oryza sativa, Hordeum vulgare, Saccharum officinarum*, dwarf wheat, French bean (Alapatri, Gheu and Singtame), and tobacco (*Nicotiana tabacum* cv. *Xanthi*, *N. benthamiana*, *N. rustica* and *N. glutinosa*) (Raychaudhuri and Ganguly 1965, 1966; Ganguly et al. 1970; Mandal et al. 2012). Most of the large cardamom cultivars eg., Golsey, Ramsey, Swaney and Varlangey, and a close relative of large cardamom Churumpa (*Ammomum dealbatum*) are susceptible.

13.2.3 Mode of Dissemination

Under experimental conditions, the chirke was transmitted by several species of aphid, *Rhopalosiphum maidis, Brachycaudus helichrysi* and *Myzus persicae* (Raychaudhuri and Chatterjee 1961, 1964, 1965; Mandal et al. 2012) in a non-persistent manner. A single aphid requires 5 min of acquisition feeding period to transmit the virus. Previous studies although showed that at least three aphid species could transmit chirke but these aphids are generally not found to colonise on large cardamom under natural environments. Recently, studies were undertaken to record the seasonal abundance of aphid species in large cardamom in the different plantations in Sikkim and Darjeeling. The study showed natural occurrence of three aphid species in large cardamom, *Pentalonia nigronervosa, Micromyzus kalimpongenesis* and *Aulacorthum solani* of which only *P. nigronervosa* could transmit the chirke with an efficiency of 3.7% (Ghosh et al. 2016).

The virus is mechanically sap transmitted; however the transmission rate is generally low and perhaps depends on the type of isolates. Previously, chirke was reported to be sap transmissible to the different plant species (Raychaudhuri and Ganguly 1965), but the Kalimpong isolate was sap transmissible to the different cultivars of large cardamom, but not to any other plant species including dwarf wheat (Mandal et al. 2012).

Chirke is not seed transmitted but it spreads through infected sucker, the most commonly used planting material. High incidence (71.1%) of chirke was recorded in plantations that were raised through the suckers from the neighboring plantations. Whereas, low disease incidence (1.6-4.7%) was observed in plantations those were raised through seedlings. Therefore, the infected sucker appears as the major sources of dissemination of chirke.

13.2.4 Virus Structure and Properties

The flexuous filamentous virions were often found associated with the chirke affected leaves of large cardamom. The modal length of the virions in the purified preparation was determined as 625-650 × 12.5 nm. In the ultrathin sections of leaf samples, a few intact pinwheel inclusion bodies and more of incomplete pinwheel inclusion bodies were consistently observed in the electron microscope (Fig. 13.1d). The physical properties of the virus in crude sap was determined, which showed the thermal inactivation point was 50^{0}C for 10 min, dilution end point was 10^{-1} and longevity *in vitro* was 4-8 days at 4.4-33.3 °C (Ganguly et al. 1970). The SDS-PAGE analysis of the purified virus showed two major bands at 40 and 35 kDa. The purified capsid protein appears to be not stable in storage as it showed more degradation at 4°C compared to at -80°C (Mandal et al. 2012).

13.2.5 Purification of the Virus and Serological Relationships

Antiserum to the virus associated with the chirke disease was not available and therefore the initial studies on the serology were undertaken using antisera to potyviruses available at the Advanced Center for Plant Virology, IARI, New Delhi. In the immunosorbent electron microscopy (ISEM) studies, the antisera to potato virus Y (PVY), cowpea aphid borne mosaic virus (CABMV), peanut strive virus (PStV), papaya ringspot virus (PRSV) and soybean mosaic virus (SMV) did not decorate the virus in chirke affected leaf samples, but the antiserum to sugarcane mosaic virus (SCMV) showed partial decoration. Serological relations were further examined by ELISA using nine antisera to different potyviruses, but only SCMV showed moderate positive readings of 0.50-0.65.

Purification of the virions from the large cardamom leaf tissues is cumbersome. The, time of harvest and quantity of tissues are important for obtaining high yield of purified virus. Aggregation, breakage and degradation of virus particles are the other issues while purification. About 100 g of the tender unfurled leaf tissues that were emerging during the beginning of summer season at Kalimpong resulted in 0.94-1.02 mg/ml of virus. The antiserum to the virus purified from the chirke diseased plants decorated the virus particles in crude extract of chirke. The purified virus was used to raise antiserum to chirke, but unfortunately the antiserum had low titer (1:250) and showed considerably high back ground with the healthy large cardamom tissues.

Large cardamom leaves being highly fibrous and readily oxidized to turn brown on storage or after extraction and further, the loss of virus during purification due to aggregation of filamentous particle, it is difficult to prepare adequate quantity of high quality antigen of the virus. These constrains were overcome by expressing highly immunogenic capsid protein (CP) of the virus in *Escherichia coli* (Vijayanandraj et al. 2013). The purification of bacterial expressed recombinant CP with the His-tag based purification using Ni-NTA column resulted in precipitation and thereby loss of protein during the renaturation step. Further, the Ni-NTA column based purification resulted in contamination of additional minor protein bandsin addition to the target CP. A simple and inexpensive method of purification of the recombinant CP of chirke was developed where the sonicated bacterial lysate was resolved on SDS-PAGE and the protein band was extracted by crushing the gel fragments with 5% SDS followed by renaturation in Milli-Q water. The method resulted in no precipitation of the target protein, no contamination of *E. coli* protein and yielded high quantity of the target antigen (1–12.6 mg/liter of induced bacterial culture). Immunization of rabbit with the gel extracted recombinant CP resulted in production of high titer antiserum (1:256,000), which efficiently detected the virus without any significant background reading by ELISA, DIBA and ISEM (Fig. 13.1c). The antiserum also differentiated virus associated with the katte disease of small cardamom. A lateral flow strip was developed using the gold nanoparticles and the polyclonal antibody to the recombinant CP for the rapid detection of the virus in crude leaf sap of large cardamom plant. The assay procedure can be accomplished within 10 min and the results could be evaluated visually (Maheshwari et al. 2014).

13.2.6 Genome Properties

The genomic property of the virus associated with the chirke disease was initially studied based on the 3′ terminal genome containing nuclear inclusion body (NIb), CP and 3′ untranslated region (UTR) sequence (1776 nt), which revealed it as a new member of the genus *Macluravirus,* family *Potyviridae*. The deduced amino acid sequence of CP gene showed the maximum sequence identity of 65.7% with the CdMV. Further, the phylogenetic analysis of CP and 3′UTR showed that the virus related to AlpMV, CdMV and ChYNMV. The differential biological properties and 3′ region polyprotein gene sequence data provided evidence that the virus associated with the chirke disease of large cardamom was different from CdMV of small cardamom and other member of macluraviruses and therefore the virus was identified as a new macluravirus, LCCV (Mandal et al. 2012).

The attempts were made to generate the complete genome sequence information of LCCV and presently 5961 nucleotides sequence was achieved which contained several poly proteins, cytoplasmic inclusion body (CI), 9K, nuclear inclusion body-a and viral protein genome-linked (VPg), NIa-Pro, NIb, CP and the 3′ UTR (Vijayanandraj 2013). The amino acids sequence of these protein showed maximum sequence identity of 41.4–69.2 % with CdMV. The viral polyprotein contains the helicase motif, conserved tyrosine residue, cysteine protease and GDD motif as found in the macluraviruses and potyviruses. Phylogenetic analysis based on the five polyproteins showed that LCCV shared close relations with the subclade containing CdMV, AlpMV and ChYNMV.

Fig. 13.3 Symptoms of the cardamom mosaic virus isolates on small cardamom leaves. (**a**) Yeslur- subgroup I: light green with prominent cholrotic streaks. (**b**) Appangala-subgroup II: continuous dark green stripes along the veins. (**c**) Sirsi-subgroup III: whitish green with discontinuous dark green islands. (**d**) Kursupara-subgroup V: prominent yellow mosaic patches along the veins (Jacob et al. 2003)

13.3 Cardamom Mosaic Virus

The katte disease of small cardamom (*Elettaria cardamomum*) is characterized by light greenish yellowsih streaks on the leaf lamina (Fig. 13.3) and pseudostem. Infected plants show stunted growth and the yield reduces drastically by 26–92 % in the second year and 82–98 % in the third year of infection and subsequently the plant becomes completely unproductive. The virus infected cardamom plant has smaller size capsules with low value. CdMV has been identified to be associated with the katte disease of small cardamom. The disease is prevalent in the Western Ghats of southern India, Sri Lanka and Guatemala (Venugopal 1995).

13.3.1 Host and Transmission

Small cardamom is a natural host for CdMV and it has many experimental hosts from *Zingiberaceae* family-*Amomum connecarpum*, *A. invollucraltum*, *A. microstephanum*, *A. muricatum*, *A. pterocarpum*, *A. subulatum*, *Alpinia neutans*, *Alpinia mutica*, *Curcuma neilgherrensis*, *Hedychium flavescens*, *Zingiber cernuum*; in addition *Maranta arundinacea* of Marantaceae is also an experimental host (Rao and Naidu 1973; Viswanath and Siddaramaiah 1974; Yaraguntaiah 1979).

CdMV is transmitted by aphid, *Pentalonia nigronervosa* in a non-persistent manner (Uppal et al. 1945). About 13 aphid species were reported to transmit CdMV *viz*., *Aphis craccivora*, *A. gossypii*, *A. nerii*, *A. rumicis*, *Brachycaudus helichrysi*,

Greenidia artocarpi, *Macrosiphum pisi*, *M. rosaeformis*, *M. sonchi*, *Schizaphis cyperi*, *S. gramimum*, *Pentalonia nigronervosa* f. *typica* and *P. nigronervosa* f. *caladii* (Varma and Capoor 1958). Mechanical sap transmission of CdMV to small cardamom was not successful but it was successful to *Phaseolus vulgaris*, *Vigna sinensis* and *Crotalaria striata* (Rao and Naidu 1973; Gonsalves et al. 1986).

13.3.2 Virion Structure and Serology

Naidu et al. (1985) observed the flexuous filamentous particles measuring 650 × 10-12 nm and inclusion bodies in the katte affected small cardamom samples and this prompted to suggest the association of a potyvirus with the katte disease. Subsequently, Gonsalves et al. (1986) purified the filamentous virus particles of 700-720 nm and based on the shape and size of the virion, the presence of typical pinwheel inclusion bodies and distinct serological relationships, concluded the association of a new potyvirus, CdMV with the mosaic disease small cardamom in Guatemala. Subsequently, the virus associated with the katte disease in India was also purified and characterized as an isolate of CdMV (Jacob and Usha 2001, 2002).

The virus associated with the katte disease strongly reacted in ELISA with the antiserum to the Guatemalan isolate of CdMV but not with the antisera to the different potyviruses, PStV, peanut mottle virus, pepper vein-banding virus, bean yellow mosaic virus, clover yellow vein virus, Johnsongrass mosaic virus (JGMV), PVY and watermelon mosaic virus. Thus virus causes katte disease shared strong serological relationships with CdMV occurring in Guatemala, but not with the potyviruses (Jacob and Usha 2001).

13.3.3 Genome Characterization

CdMV has a single-stranded linear, positive-sense RNA as its genome (Jacob and Usha 2001). Agarose gel analysis of RNA extracted from the purified virus particles and Northern analysis of the infected leaf samples indicated that CdMV contained the RNA genome of ~8.5 kb in size (Jacob and Usha 2001). The sequence of the 1.8 kb fragment amplified from the 8.5 kb viral RNA consist of partial NIb gene, complete CP gene and the 3' UTR. Further work has resulted in the cloning and sequencing of the regions coding for the partial CI (Cytoplasmic Inclusion), complete 6K2, VPg, NIa protease, and the complete NIb genes (Jebasingh et al. 2008, 2011b, 2013).

The expression of CdMV CP in *E. coli* resulted in filamentous of aggregates having the length of 100–150 nm and breadth of 15–18 nm (Fig. 13.4). The absence of CP mRNA in the filamentous aggregates was also confirmed with immune-captured RT-PCR (Jacob and Usha 2002). Amino acids Arginine and Aspartate present in the central region of CPs are important for proper assembly of virus by making a bond between the helices in the central region (Dolja et al. 1991). Deleting these amino acids caused improper or defective assembly of CP (Jagadish et al. 1991, 1993;

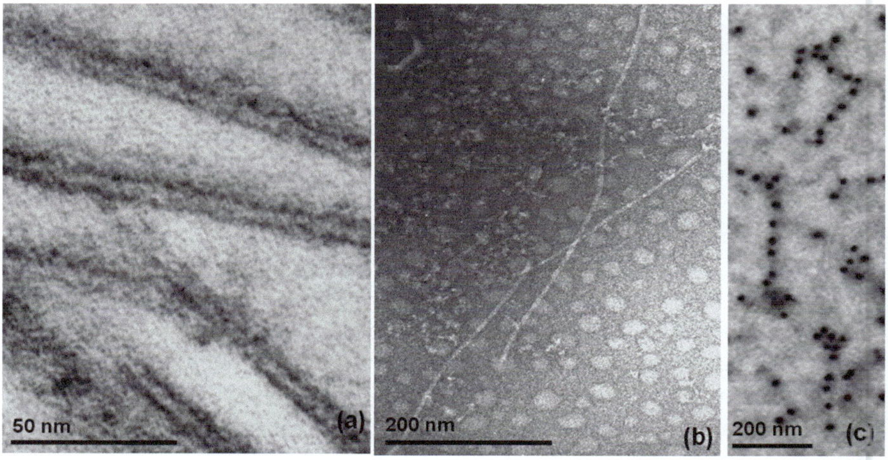

Fig. 13.4 Transmission electron micrographs of the cardamom mosaic virus (CdMV) causing kattte diseases in small cardamom and expression of CdMV coat protein showing filamentous aggregates in *Escherichia coli.* (**a**) Cross-sectioned infected leaf sample (**b**) purified virus (**c**) *E. coli* cell (Jacob and Usha 2002)

Dolja et al. 1994; Jacquet et al. 1998; Varrelmann and Maiss 2000). The deletion of these amino acids in CdMV CP resulted in formation insoluble aggregates in *E. coli* as like the wild type CP (Jacob and Usha 2002).

The CP of potyvirus is a suitable carrier to display the epitopes of the pathogens. The coat protein of several potyviruses like zucchini yellow mosaic virus (ZYMV), plumpox virus (PPV), PVY, JGMV and PRSV have been used to display the epitopes of various antigens (Arazi et al. 2001; Fernández-Fernández et al. 2002; Saini and Vrati 2003; Chatchen et al. 2006; Choudhury et al. 2009; Kalnciema et al. 2011). The coat protein of CdMV was used as a potential system to display the gp41 epitopes of human immuno deficiency virus (HIV). It is the glycoprotein present on the envelope, helps the virus entry into the cell. The cytoplasmic tail of gp41 contains Kennedy peptide (spanning amino acid sequence 735–752), contains three epitopes: 734PDRPEG739, 740IEEE743 and 746ERDRD750. Kennedy peptide and other epitopes of gp41 (2F5 and 4E10) at N and C-terminus of CdMV CP reacted with sera from HIV positive persons. These chimeric proteins cause peripheral blood mononuclear cells to secrete the cytokines in HIV infected persons (Subha et al. 2013). LipL32, the most abundant, immunogenic, and conserved surface lipoprotein present in all pathogenic species of *Leptospira*, is a promising antigen candidate for a recombinant vaccine. In order to develop an effective vaccine for leptospirosis with epitopes of LipL32, they have been displayed at N-, C- and at both termini of CdMV CP. Among these, epitopes at the N-terminus of CdMV CP have shown to be promising vaccine candidates (Kumar et al. 2016).

The conventional breeding program to develop resistance in cardamom against CdMV is a lengthy and difficult process. So far, no naturally occurring resistance gene has been identified in cardamom. Even if the resistance gene is identified, the conventional program to develop resistance will be protracted and expensive. So the only alternative approach is to develop transgenic cardamom. In order to develop transgenic cardamom resistant to CdMV, work is already in progress using CP gene with and without 3' UTR from the Yeslur and Kuruspara isolates (Backiyarani et al. 2007) and NIb gene (Jebasingh 2006). In order to develop a transgenic cardamom for virus resistant, the protocols like induction somatic embryogenesis form rhizome explants, development of shoot and root system from embryogenic calli and gene transformation into embryogenic calli have been developed (Manohari et al. 2008; Jebasingh 2006).

13.3.4 CdMV Encoded Proteins

Nuclear Inclusion b (NIb) The NIb of *Potyviridae* is the RNA dependent RNA polymerase (RdRp). The NIb forms inclusion bodies while it expressed in *E. coli*. The hydrophobic nature, large size and large numbers of cysteines in NIb were thought to be the cause of the aggregation. Therefore, various strategies were attempted to obtain NIb in the soluble form by making truncations, expressing NIb in reduced conditions, expressing with chaperones and trigger factor (TF), making mutations in the amino acid responsible for NIb-NIb interaction, and expressing the C-terminal truncated NIb fused to maltose binding protein (MBP). Out of these, only the MBP-fused truncated NIb could be obtained in a soluble form. Further work showed that the soluble form of full-length NIb could be derived from NIb-transgenic *Nicotiana tabacum* plants (Jebasingh et al. 2008).

The sequences of various DNA and RNA viruses have been found in the various plant genomes. In the case of *Potyviridae*, Tanne et al. (1989) reported the presence of PVY sequences in several grapevine varieties. Jebasingh et al. (2011a) reported the presence of partial 3' end of NIb gene in the genome of cardamom and few other plant species.

Nuclear Inclusion a (NIa) Protease NIa protease is the major protease of *Potyviridae* has typical structural motifs of cellular serine proteases with Cys in the place of Ser at the catalytic site (Gorbalenya et al. 1989; Kim et al. 1996). The CdMV NIa protease with N-terminal His tag (NIa1 protease) has showed an additional band around 20 kDa along with the expected size of the NIa protein (27 kDa) (Jebasingh et al. 2013). The 20 kDa protein could be purified by passing through Ni-NTA column but not the 27 kDa band. The presence of the 20 kDa band was deduced to be due to a C-terminal deletion by the auto catalytic activity of NIa protease (Jebasingh et al. 2013).

NIa protease with C-terminal His tag (NIa2) was found as insoluble aggregates at 37°C, which was solubilized with 8M urea and then refolded using 10% glycerol and 0.5 M arginine hydrochloride (Jebasingh et al. 2013). This purified and refolded NIa protease was used to perform protease and DNA binding assay. The purified NIa protease does not have DNA binding and DNase activity (Jebasingh 2006) but it has *trans* cleavage reaction between NIb and CP and the *cis* protease reactions in VPg-NIa protease and NIa protease-NIb fusions (Jebasingh 2006).

H-D-C is the motif of NIa protease involved in protease catalytic function. The site-specific mutations in the functional motifs of CdMV NIa protease at 42H, 74D and 141C with Ala, Asn and Ala amino acids, respectively in NIa protease-NIb and VPg-NIa protease polyprotein do not affect the *cis* protease function and the release of NIa protease and NIb from NIa protease–NIb fusions and VPg and NIa protease from VPg-NIa protease was observed in the case of CdMV NIa protease (Jebasingh 2006). The structural determination of CdMV NIa predicted the protease activity in the mutants H42A, D74N and C141A could be due to the complementation of amino acids H206, D209 and S162 found the catalytic region (Jebasingh et al. 2013).

Viral Protein Genome-Linked (VPg) The expression of CdMV VPg in *E. coli* resulted in insoluble aggregates. Insoluble aggregates were denatured with high concentration of urea and then refolded with L-arginine hydrochloride (Jebasingh et al. 2011b). The conformational variations between native, denatured and refolded CdMV VPg were noticed with fluorescence spectroscopy and the circular dichroism (CD). The native and refolded VPg showed a CD spectrum with a strong negative peak near 204 nm and 208 nm, respectively. The negative peak obtained for VPg in CD spectrum was typical of intrinsically unordered or disordered proteins, which had little, ordered structures (Jebasingh et al. 2011b).

13.3.5 Genetic Diversity

The cultivation of cardamom for many centuries leads to variations in the mosaic disease. Based on the symptoms, Rao (1977) identified three variants of katte disease. Naidu et al. (1985) conducted an investigation on the strainal variations of katte. Sixty isolates with variation in the mosaic symptoms were studied and six symptomatologically distinct subgroups of CdMV were made based on the inoculation studies on the nine *Zingiberaceae* plant species- *Eletteria cardamomum, Amomum pterocarpum, A. muricatum, A. cannicarpum, Hedychium flavescens, Alpinia mutica, Aframomum melequeta* and *Zingiber cernuum.* The last three plants were immune to all the isolates tested. The subgroups I (Yeslur), II (Appangala, Madikeri and Thalathamane from Coorg district) and III (Sirsi) belong to Karnataka state and subgroup IV (Kattappana and Vandiperiyar from Idukki district), V (Kursupara, Idukki district) and VI (Nelliampathy, Palakkad district) to Kerala state (Jacob et al. 2003). All these isolates of the CdMV showed typical mosaic symptoms but there were clear differences in the mosaic patterns. The subgroup I (Yeslur) showed light green mosaic symptom with the prominent chlorotic streaks (Fig. 13.3a). The subgroup II (Appangala) showed symptoms of continuous dark green stripes along the veins (Fig. 13.3b). The subgroup III (Sirsi) showed whitish green symptoms with discontinuous dark green islands (Fig. 13.3c) and the subgroup IV (Kursupara) showed prominent yellow mosaic patches along the veins (Fig. 13.3d).

For the molecular analysis, ten isolates representing the six subgroups were collected and the virus was purified. The molecular weight of the CP ranged from 38 to 42 kDa. Determination of the CP and 3'UTR sequences of eight isolates: Yeslur

(AF189125), Madikeri (AJ308474), Thalathamane (AJ308475), Appangala (AJ308472), Sirsi (AJ308473), Kursupara (AJ308476), Kattappana (AJ312774) and Vandiperiyar (AJ308477) revealed high variability in the CP (Jacob et al. 2003). The isolates of CdMV showed polymorphism in the length of the CP coding region (from 813 bp to 819 bp) and of the 3' UTR (from 208 bp to 221 bp). The nucleotide sequences of the CP gene of various isolates of CdMV shared identity from 79.1-87.8% (Jacob et al. 2003).

Forty amino acids from the N-terminus of the CP have very low identity between the isolates. The lowest identity (37.5%) was found between Yeslur and Kursupara isolates. The identity between pairs of distinct geographical isolates is also not uniform throughout the sequence of CP and 3'UTR. In the case of the identity between the Vandiperiyar and Thalathamane isolates (pair A) is 42.5% at the N-terminus and that between Kursupara and Kattappana (pair B) is 65%. The pair A and B has 82.6% and 82% identity, respectively in the complete CP. In the case of 3' UTR, pair A has 69.8% and pair B has 84.6%. The non-uniformity found between the isolates could be due to recombination. The isolates of the same state has conserved sequences. However, in order to maintain the structure and function, the variable amino acid residues in the CP of the respective state isolates are chemically similar (Jacob et al. 2003).

The CP and 3'UTR sequences of Sirsi isolate of Karnataka shared the homology with the isolates of Kerala but it differed from other isolates of Karnataka. The phylogenetic tree constructed with the CP and 3'UTR sequences also placed the Sirsi isolate closer to the isolates of Kerala than with isolates of Karnataka. But, Sirsi was geographically far away from Kerala (Jacob et al. 2003). This could have happened while the propagation is done through the CdMV infected rhizomes (Varma 1962). Studying the diversity of CdMV will be helpful to trace the movement of the virus-infected propagules to the different geographical locations.

13.4 Yam Mottling Virus (YMoV)

The occurrence of a mottling and mild mosaic disease in greater yam (*Dioscorea alata*) and lesser yam (*Dioscorea esculenta*) was observed in the experimental farm of Central Tuber Crops Research Institute (CTCRI), Kerala during 2012 (Jeeva et al. 2014). The infected yam plants also showed puckering, cupping, leaf crinkling, mottling, leaf distortion and chlorosis (Fig. 13.5 a, b, c). The association of the virus was tested by DAS-ELISA, DIBA and IC-RT-PCR with the antiserum raised against the homologous virus in the leaf and root samples collected from Thiruvananthapuram, Kollam, Pathanamthitta, Alappuzha and Wayanad districts of Kerala as well as from CTCRI during 2012–2016. Initial screening showed 11% of infection in greater yam and 36% in lesser yam (Manasa 2014; Sudheer 2015). The primers to the partial NIb and CP gene produced amplicons of ~200 and 800 bp, respectively in the infected samples. The virus infection was detected only in the samples collected from Thiruvananthapuram district and CTCRI (Manasa 2014; Sudheer 2015).

Fig. 13.5 Symptoms of yam mottling mosaic virus (YMoV) in *Dioscorea alata* (**a** & **b**) and *D. esculenta* (**c**) observed in Central Tuber Crops Research Institute (CTCRI), Kerala. RT-PCR based detection of YMoV coat protein gene using specific primers in infected samples (**d**) Amplification of partial coat protein gene and (**e**) complete coat protein gene

The partial coat protein and complete coat protein gene were amplified by RT-PCR using specific primers (YMac F: 5'CTGCAACACTCGGAGGCTTA3'/ YMac R: 5' AGGCTCTCTGGACTC CACTT 3') and (YMac F: 5'CTGCAA CACTCGGAGGCTTA3'/YMac R: 5' AGGCTCTCTG GACTCCACTT 3'), respectively (Fig. 13.5d, e). The coat protein gene of the Indian isolates exhibited 65-70% nucleotide sequence identity with the other macluraviruses with the maximum nucleotide sequence identity of 70% with ChYNMV. Phylogenetic analysis based on the complete coat protein sequence showed that the virus associated with yam mild mosaic in India was related to ChYNMV and yam chlorotic mosaic virus (Fig. 13.6). Based on the CP sequence data analysis, it appears that there is a new macluravirus, YMoV associated with yam in India. The limited RFLP analysis with the two restriction enzymes (*Taq*1 and *Mse*1) revealed existence of polymorphism in YMoV (Manasa 2014).

13.5 Concluding Remarks

Limited diversity of macluraviruses is known in India. Two macluraviruses, LCCV and CdMV are economically important and they are restricted only with small and large cardamom, respectively. The economic significance and correct taxonomic

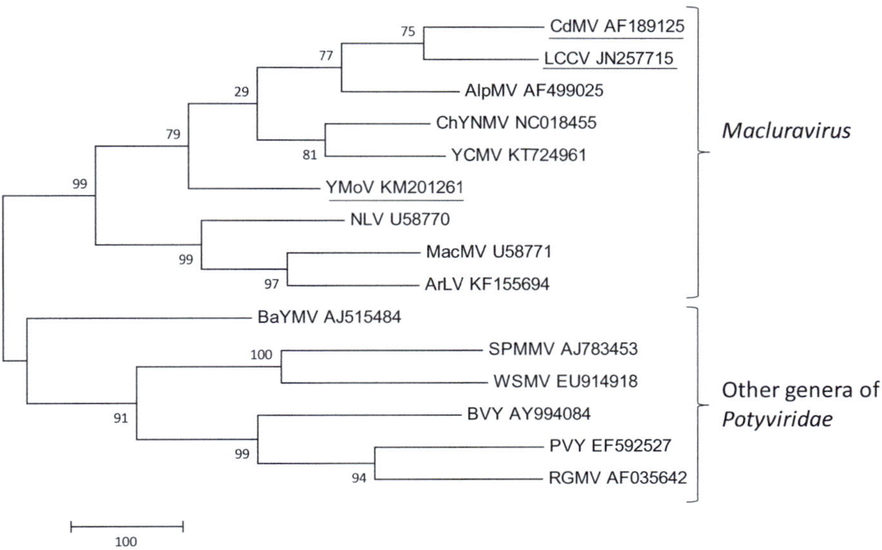

Fig. 13.6 Maximum parsimony phylogenetic analysis of coat protein nucleotide sequence of the macluraviruses occurring in India with the members of the genus *Macluravirus* and other genera of family *Potyviridae*. The macluraviruses occurring in India are underlined. The percentage of replicate trees in which the associated taxa clustered together in the bootstrap test (1000 replicates) is shown next to the branches. The maximum parsimony tree was constructed in MEGA 7 programme

identity of another macluravirus YMoV that has recently been detected in yam in southern India are yet to be established. LCCV and CdMV are close relatives; however, they share distant and diverse ecological niches. LCCV is distributed in the North-East sub-Himalayan mountains, whereas, CdMV in southern India. It seems that field spread of both chirke and katte diseases are slow through the aphid vectors. The disease is widely circulated through the infected planting material raised through either rhizome separation or through tissue culture. Both the macluraviruses have been studied well and diagnostic reagents have been developed that can be utilized for the production of healthy planting materials. The complete genome sequence information for these macluraviruses is yet to be achieved. As breeding for resistance in cardamom is difficult, virus gene sequence based engineering resistance or genome editing will be potentially useful approaches to manage these viruses.

References

Arazi T, Shiboleth YM, Gal-on A (2001) A nonviral peptide can replace the entire N terminus of Zucchini yellow mosaic potyvirus coat protein and permits viral systemic infection. J Virol 75:6329–6336

Backiyarani S, Manohari C, Jebasingh T, Jacob T, Usha R (2007) Cloning of coat protein gene of Kursupara isolate of *Cardamom mosaic virus* for developing transgenic virus-resistant cardamom. Recent Trends Hort Biotech:693–698

Chatchen S, Juricek M, Rueda P, Kertbundit S (2006) Papaya ringspot virus coat protein gene for antigen presentation in Escherichia coli. J Biochem Mol Biol 39:16–21
Choudhury S, Kakkar V, Suman P, Chakrabarti K, Vrati S, Gupta SK (2009) Immunogenicity of zona pellucid glycoprotein-3 and spermatozoa YLP12 peptides presented on Johnson grass mosaic virus-like particles. Vaccine 27:2948–2953
Dolja VV, Boyko VP, Agranovski AA, Koonin EV (1991) Phylogeny of capsid proteins of rod-shaped and filamentous RNA plant viruses: two families with distinct patterns of sequence and probably structure conservation. Virology 184:79–86
Dolja VV, Haldeman R, Robertson NL, Dougherty WG, Carrington JC (1994) Distinct functions of capsid protein in assembly and movement of tobacco etch potyvirus in plants. EMBO J 3:1482–1491
Fernández-Fernández MR, Martínez-Torrecuadrada JL, Roncal F, Domínguez E, García JA (2002) Identification of immunogenic hot spots within Plum pox potyvirus capsid protein for efficient antigen presentation. J Virol 76:12646–12653
Ganguly B, Raychaudhuri SP, Chatterjee SN (1970) Mosaic streak disease of cardamom. In: Proceedings of the first International Symposium on Plant Pathology. Indian Phytopathological Society, IARI, New Delhi, pp 653–661
Ghosh A, Das A, Lepcha R, Mandal B (2016) Identification, distribution and temporal occurrence of aphids infesting large cardamom and their efficiency in transmitting large cardamom viruses in northeastern sub-Himalayan region. Australasian Plant Pathol 45:533–536
Gonsalves D, Trujillo E, Hoch HC (1986) Purification and some properties associated with cardamom mosaic, a new member of potyvirus group. Plant Dis 70:65–69
Gorbalenya AE, Donchenko AP, Blinov VM, Koonin EV (1989) Cysteine proteases of positive strand RNA viruses and chymotrypsin-like serine proteases. A distinct protein superfamily with a common structural fold. FEBS Lett 243:103–114
Jacob T, Usha R (2001) 3'-terminal sequence analysis of the RNA genome of the Indian isolate of cardamom mosaic virus: A new member of genus *Macluravirus* of *Potyviridae*. Virus Genes 23:81–88
Jacob T, Usha R (2002) Expression of cardamom mosaic virus coat protein in *Escherchia coli* and its assembly into filamentous aggregates. Virus Res 86:1333–1141
Jacob T, Jebasingh T, Venugopal MN, Usha R (2003) High genetic diversity in the coat protein and 3' untranslated regions among geographical isolates of cardamom mosaic virus from south India. J Biosci 28:589–595
Jacquet C, Delecolle B, Raccah B, Lecoq H, Dunez J, Ravelonandro M (1998) Use of modified plum pox virus CP genes developed to limit heteroencapsidation-associated risks in transgenic plants. J Gen Virol 79:1509–1517
Jagadish MN, Ward CW, Gough KH, Tulloch PA, Wittaker LA, Shukla DD (1991) Expression of potyvirus CP in Escherichia coli and yeast and its assembly into virus-like particles. J Gen Virol 72:1543–1550
Jagadish MN, Huang D, Ward CW (1993) Site-directed mutagenesis of a potyvirus CP and its assembly in Escherichia coli. J Gen Virol 74:893–896
Jebasingh T (2006) Characterization of the non-structural genes of the Indian cardamom mosaic virus. Ph.D Thesis, Madurai Kamaraj University, India
Jebasingh T, Jacob T, Shah M, Das D, Krishnaswamy S, Usha R (2008) Optimized expression, solubilisation and purification of nuclear inclusion protein b of cardamom mosaic virus. Indian J Biochem Biophys 45:98–105
Jebasingh T, Backiyarani S, Manohari C, Usha R (2011a) Detection of partial genome sequences of cardamom mosaic virus integrated in plant genomes. Indian J Biotechnol 10:369–371
Jebasingh T, Kasin Yadunandam A, Jose M, Backiyarani S, Krishnaswamy S, Usha R (2011b) Molecular modeling and conformational analysis of native and refolded viral genome-linked protein of cardamom mosaic virus. Indian J Biochem Biophys 48:336–340
Jebasingh T, Pandaranayaka EPJ, Mahalakshmi A, Kasin Yadunandam A, Krishnaswamy S, Usha R (2013) Expression, Purification and Molecular modeling of the NIa protease of cardamom mosaic virus. J Biomol Struct Dyn 31:602–611

Jeeva ML, Makeshkumar T, Rajitha M, Manasa VG, Sruthy S (2014) Investigation of greater yam (*Dioscorea alata* L.) viruses in India. In Virocon 2014- IVS- XXIII National conference on Recent trends in Virology Research in the Omics Era, held at TNAU, Coimbatore during 18–20 December 2014. 68.

Kalnciema I, Skrastina D, Ose V, Pumpens P, Zeltins A (2011) Potato virus Y-like particles as a new carrier for the presentation of foreign protein stretches. Mol Biotechnol 52:129–139

Kim DH, Hwang DC, Kang BH, Lew J, Choi KY (1996) Characterization of Nla protease from turnip mosaic potyvirus exhibiting a low-temperature optimum catalytic activity. Virology 221:245–249

Kumar V, Damodharan S, Pandaranayaka EP, Madathiparambil MG, Tennyson J (2016) Molecular modeling and *in-silico* engineering of cardamom mosaic virus capsid protein for the presentation of immunogenic epitopes of Leptospira LipL32. J Biomol Struct Dyn 34:42–56.

Maheshwari Y, Vijayanandraj S, Jain RK, Mandal B (2014) Lateral flow immunoassay for the rapid detection of large cardamom chirke virus in large cardamom plants, In: International conference on Frontier discoveries and emerging opportunities in Life Sciences at Dr. Harisingh Gaur University, Sagar, M.P, p 153

Manasa VG. (2014) Detection and characterization of macluravirus infecting greater yam (*Dioscorea alata* L.). M.Sc thesis, Kerala Agricultural University, p 114

Manasa VG, Rajitha M, Jeeva ML (2014) Nucleic acid based detection of *Maclura virus* infecting greater yam (*Dioscorea alata* L.). In UGC sponsored National Seminar cum Workshop on "Techniques in Biosciences", KKTM College, Kodungallur, Thrissur, February 18–19, 2014.

Mandal B, Vijayanandraj S, Shilpi S, Pun KB, Singh V, Pant RP, Jain RK, Varadarasan S, Varma A (2012) Disease distribution and characterisation of a new macluravirus associated with chirke disease of large cardamom. Ann Appl Biol 160:225–236

Manohari C, Backiyarani S, Jebasingh T, Archana Somanath T, Usha R (2008) Efficient plant regeneration in small cardamom (*Elettaria cardamomum* Maton) through somatic embryogenesis. Indian J Biotechnol 7:407–409

Mollison JW (1900) Cultivation of the betel palm, cardamom and pepper in the Kanara district of Bombay presidency with notes on the manures used in spice gardens of that tract. Department of Land Record and Agriculture, Bombay. Bull No. 20, 24p.

Naidu R, Venugopal MN, Rajan P (1985) Investigation on strainal variation, epidemiology and charatcerization of 'katte' agent of small cardamom. Final Report of Research Project Submited to ICAR, New Delhi, pp 11–14

Plese N, Milicic D (1973) Two viruses isolated from *Maclura pomifera*. Phytopathology 77:178–183

Pringle CR, Fauquet CM (1998) ICTV announcement – ratification of new taxonomic proposals. Arch Virol 143:2504

Rao DG (1977) 'Katte' disease of cardamom and its control. Indian J Horti 34:184–187

Rao DG, Naidu R (1973) Studies on "Katte" or "mosaic" disease of small cardamom. J Plant Crop 1:129–136

Raychaudhuri SP, Chatterjee SN (1958) A preliminary note on the occurrence of a new virus disease of large cardamom *Amomum subulatum* Roxb. in Darjeeling District. Proc. Mycological Research Worker's Conference, I.C.A.R., Shimla, pp 174–176

Raychaudhuri SP, Chatterjee SN (1961) Chirke a new virus threat to cardamom. Indian Farming 9:11–12

Raychaudhuri SP, Chatterjee SN (1964) Aphid transmission of a new mosaic-streak disease of large cardamom. *Phytopathology* 54:904

Raychaudhuri SP, Chatterjee SN (1965) Transmission of chirke disease of large cardamom by aphid species. Indian J Entomol 27:272–276

Raychaudhuri SP, Ganguly B (1965) Further studies on chirke disease of large cardamom (*Amomum subulatum* Roxb.) Indian Phytopathol 18:373–377

Raychaudhuri SP, Ganguly B (1966) Wheat- a host of the virus causing chirke disease of large cardamom. Proceedings Fifth All India Wheat Worker's Conference, I.C.A.R. Jaipur

Saini M, Vrati S (2003) A Japanese encephalitis virus peptide present on Johnson grass mosaic virus-like particles induces virus-neutralizing antibodies and protects mice against lethal challenge. J Virol 77:3487–3494

Subha D, Ravindra G, Sathyamurthy P, Manohar N, Luke EH, Ramanathan DV, Usha R (2013) Expression and immunological characterization of cardamom mosaic virus coat protein displaying HIV gp41 epitopes. Microbiol Immunol 57:374–385

Sudheer KS. (2015). Identification and characterization of viruses infecting lesser yam (*Dioscorea esculenta* (Lour) Burkill). M.Sc thesis, Kerala Agricultural University, p 111

Tanne E, Naveh L, Sela I (1989) Serological and molecular evidence for the complexity of the leafroll disease of grapevine. Plant Pathol 38:183–189

Thomas KM (1938) Detailed Administrative report of the Government Mycologists. Madras:1937–1938

Uppal BN, Varma PM, Capoor SP (1945) Mosaic disease of cardamom. Curr Sci 14:208–209

Varma PM (1962) The banana aphid (*Pentalonia nigronervosa* Coq.). and the transmission of 'katte' disease of cardamom. Indian Phytopathol 15:1–10

Varma PM, Capoor SP (1958) Mosaic disease of cardamom and its transmission by the banana aphid *Pentalonia nigronervosa* Coq. Indian J Agricultural Sci 28:97–108

Varrelmann M, Maiss E (2000) Mutations in the CP gene of *Plum pox virus* suppress particle assembly, heterologous encapsidation and complementation in transgenic plants of *Nicotiana benthamiana*. J Gen Virol 81:567–576

Venugopal MN (1995) Viral diseases of cardamom (*Eletaaria cardamomum* Maton) and their management. J Spices Aromatic Crops 4:32–39

Vijayanandraj S (2013) Molecular diagnosis of virus associated with Chirke and foorkey disease of large cardamom, PhD Thesis, Indian Agricultural Research Institute

Vijayanandraj S, Yogita M, Das A, Ghosh A, Mandal B (2013) Highly efficient immunodiagnosis of large cardamom chirke virus using the polyclonal antiserum against *Escherichia coli* expressed recombinant coat protein. VirusDis 24:227–234

Viswanath S, Siddaramaiah AL (1974) *Alpinia neutans*, a new host of 'Katte' disease of cardamom. Curr Ther Res 3:96

Yaraguntaiah RC (1979) *Curcuma neilgherrensis* WT– a new host of 'Katte' mosaic disease of cardamom. In: Proc. Placrosym II. Central Plantation Crops Research Institute, Kasargod. Kerala, India pp 313–315

Mandarivirus and Allexivirus in India

14

K. Prabha and Virendra Kumar Baranwal

Abstract

The genera *Allexivirus* and *Mandarivrus* belong to the family *Alphaflexiviridae* of the order *Tymovirales*. So far two species of the genus *Mandarivirus*, (*Indian citrus ringspot virus* and *Citrus yellow vein clearing virus*) and four species of the genus *Allexivirus* have been reported from India. Biological characterization and the full genome sequencing of both the mandariviruses have been generated. However, only partial genomic sequences have been reported for the allexiviruses in India. The transmission of mandarivirus occurs mainly through the vegetative propagation, while allexiviruses have been reported to be transmitted by eryophid mites. This chapter describes the symptoms, transmission, diagnostics, genome characterization and management of mandariviruses and allexiviruses reported in India.

Keywords

Alphaflexivirirdae • Allium virus • Citrus virus • Citrus ringspot • Yellow vein clearing

K. Prabha
ICAR-Directorate of Floricultural Research, Shivajinagar, Pune 411005, Maharashtra, India

V.K. Baranwal (✉)
Advanced Centre for Plant Virology, Division of Plant Pathology, ICAR-Indian Agricultural Research Institute, Pusa Campus, New Delhi 110012, India
e-mail: vbaranwal2001@gmail.com

B. Mandal et al. (eds.), *A Century of Plant Virology in India*,
https://doi.org/10.1007/978-981-10-5672-7_14

14.1 Introduction

The genera Allexivirus and Mandarivirus belong to the family *Alphaflexiviridae* of the order *Tymovirales* and comprise of linear positive sense ssRNA(+) genome encased in a flexuous particle. The History of Indian citrus ringspot virus (ICRSV) back to late 1980s (Ahlawat 1989). Once the full genome sequence of ICRSV was available, it was recognised as a type species in the genus *Mandarivirus* in 2000. The genus got expanded with the addition of another species, *Citrus yellow vein clearing virus* in 2014. Natural host of the both mandariviruses is citrus. Allexiviruses are the pathogens of *Allium* species. Since 2007, four allexiviruses have been characterized from alliums in India including the type species, *Shallot virus X*. This chapter describes the progress achieved so far in both the genera in India.

14.1.1 Mandarivirus

The history of the genus *Mandarivirus* in India goes back to 1989, when Psorosis A disease was reported on Mosambi trees (Ahlawat 1989). Initially the symptoms of ICRSV infected trees were confused with psorosis, but the common symptoms with ICRSV were devoid of bark lesions, a characteristic symptom of psorosis disease and consisted of vein flecking and vein clearing in young leaves, unlike in psorosis (Pant and Ahlawat 1998; Ahlawat and Pant 2003). Most characteristic symptoms of the disease are conspicuous yellow ring spots on mature leaves (Fig. 14.1). Pant and Ahlawat (1998), reported 2 types of particles associated with ICRSV measuring 640 × 15 nm and 690 × 9 nm. Based on morphology of particles they suggested that the virus belong to genus *Capillovirus.* After the molecular characterization of Indian isolate of the virus in Italy in 2000, it was named as *Indian citrus ringspot virus*. Later ICRSV was placed in a new plant virus genus *Mandarivirus* of a new plant virus family *Flexiviridae* based on phylogenetic analysis of polymerase and coat protein genome sequences. A new order, Tymovirales was created comprising of *Alphaflexiviridae, Betaflexiviridae, Gammaflexiviridae* and *Tymoviridae* families

Fig. 14.1 Yellow ringspot symptoms on Kinnow leaves caused by ICRSV

in the virus taxonomy list 2009 by International committee on taxonomy of virus (ICTV). According to the new classification, ICRSV belonged to monotypic genus *Mandarivirus* in *Alphaflexiviridae* family of the order *Tymovirales* (http://ictvonline.org/virusTaxonomy.asp?version=2009). Recently, a new member *Citrus yellow vein clearing virus* (CYVCV) has been added to the genus *Mandarivirus*. In India CYVCD was first observed at Abohar in Punjab on a citrus cultivar, Etrog Citron in 1997 (Ahlawat 1997). The viral etiology of citrus yellow vein clearing was established in 2003 (Alshami et al. 2003) and tentatively the virus was named as citrus yellow vein clearing virus, but the genome was not characterized. The complete genome sequencing of CYVCV from Turkey and China, showed only ≈74% similarity with ICRSV. Thus, a new species status for CYVCV in genus *Mandarivirus* was proposed and was accepted in the 2014 release of Virus taxonomy by ICTV. Recently the complete genome of CYVCV has been unravelled by Meena and Baranwal (2016, Unpublished) at Advanced Centre for Plant Virology, IARI, New Delhi.

14.2 Biology, Pathogenicity and Economic Impact

The two viruses, ICRSV and CYVCV are similar with regard to natural host, herbaceous host range, genome organization, particle morphology and symptoms produced in the natural host citrus. In nature, they can be present either singly or together in a single host. Citrus is the only natural host of ICRSV (Pant 1995). The ringspot disease was distributed in sweet orange plantations of Delhi, Maharashtra and Andhra Pradesh (Pant and Ahlawat 1998). The disease is widely distributed in most commercial citrus cultivars such as Malta, Mosambi, Sathgudi sweet orange, Nagpur orange, Kinnow mandarin, Kagzi lime and Kagzikalan in different orchards of Delhi, Haryana, Punjab, Karnataka and Andhra Pradesh. The first report of CYVCV was available on Etrog Citron. Later, similar symptoms were observed on Sour orange, hill lemon (galgal), *C. pectinifera* and Etrog citron in Punjab, Rangpur lime and Lisbon lemon in Gujarat and Sour orange in Maharashtra. The virus was successfully graft transmitted to 16 citrus cultivars mechanically transmitted from infected Etrog Citron to healthy Mosambi seedlings (Alshami et al. 2003).

ICRSV is a limiting factor in Indian citriculture (Ahlawat and Pant 2003). Both ICRSV and CYVCV being graft transmissible virus have serious economic impact, in citrus cultivars, which have vegetative propagation as their only means of commercial propagation. The incidence of ringspot in sweet orange ranged from 5 to 83.8% in Delhi, Haryana, Punjab, Maharashtra, Andhra Pradesh and Karnataka with estimated yield loss of 20.5–98.38% (Byadgi and Ahlawat 1995). Maximum incidence of the ringspot disease was observed in Kinnow mandarin trees. ICRSV has been identified as one of the important potential virus disease of the citrus decline complex (Ahlawat 1997). The health of the affected trees deteriorated year after year and finally the affected trees were collapsed (Ahlawat and Pant 2003). However, it is not clear that the citrus decline was only due to virus or other factors also contributed to citrus decline. In Punjab, efforts were made to study the incidence and behaviour of ICRSV among the commercially grown *Citrus* sp. (Thind et al. 1998). The percentage disease index was highest in Kinnow mandrin (57.2%) followed by Mosambi (43.5%). Average leaf spot size varied from 9.5 to 24.2 mm in Kinnow mandarin and Mosambi and 1–2.1 mm in grape fruit variety Marsh seedless. Large spots in Kinnow mandrin and Mosambi indicated that they are highly susceptible to ICRSV (Thind and Arora 1997). The ICRSV infection brought considerable alteration in the fruit quality of Kinnow mandarin. The average fruit size, net granulation index and juice percentage decreased, where as shape index, peel thickness, peel and rag percent increased in fruits of ICRSV infected kinnow trees (Thind et al. 1998). ICRSV distribution study in citrus plant showed leaves and buds as major pool of the virus. In CYVCV infection, the veins turned yellow called vein clearing (Fig. 14.2) and the affected trees showed very poor growth and bore only few fruits on infected trees showing its economic importance (Alshami et al. 2003).

The infected buds and scions used in grafting are the main mode of virus spread. Except bud transmission, no other mode of natural spread of disease has been established (Byadgi and Ahlawat 1995; Alshami et al. 2003). The ringspot disease was successfully transmitted by wedge grafting to different *Citrus* spp. such as Mosambi Sathgudi and Malta sweet orange, Darjeeling, Nagpur and Kinnow mandarin, Acid limes, lemons, Star Ruby grape fruit, calamondin *Etrog citron*, *Citrus indica*, rough lemon, sour orange and sweet lime (Byadgi et al. 1993; Byadgi and Ahlawat 1995). The graft inoculated cultivars developed variable symptoms, like shock reaction, followed by vein flecking which remained till the leaf become mature and ringspots on mature leaves (Byadgi et al. 1993; Byadgi and Ahlawat 1995).

ICRSV was mechanically transmitted to both citrus and non citrus spp. It showed negative results when mechanically transmitted to mosambi, kinnow mandarin and kagzi lime but it could be successfully transmitted to various herbaceous hosts such as *Phaseolus vulgaris* vars Singtami and Saxa, *Chenopodium quinoa*, cowpea, soybean (Byadgi and Ahlawat 1995; Pant et al. 2000), but not to *Brassica oleracea* var. *botrytis*, var. *capitata*, *Brassica juncea*, *Citrulens vulgaris*, *Vigna mungo*, *Vigna*

Fig. 14.2 Yellow vein clearing symptoms on Etrog citron (Source: Alshami et al. 2003)

sinensis, Vigna unguiculata, Catheranthus roseus, Cucumis melo, C. sativus, Gomphrena globosa, Pisum sativum (Pant 1995). *Chenopodium quinoa* was reported as local lesion host which produced necrotic local lesions in 10 days. The ICRSV also produced local lesions on soybean and cowpea. In case of *P. vulgaris* var. *singtamy* and *saxa*, the inoculated plants showed formation of local lesions followed by systemic symptoms like vein banding, vein clearing, vein necrosis and mosaic mottling (Pant et al. 2000). Among herbaceous hosts, *P. vulgaris* var Singtami had comparatively higher concentration of virus and therefore it was used for multiplication of virus (Pant et al. 2000).

In case of CYVCV, the virus was transmitted by wedge, petiole and bud grafting and also by mechanical inoculations. The comparative efficacy of various grafting methods used for virus transmission was determined and 100%, 80% and 60% transmission was obtained by wedge, petiole and bud grafting, respectively (Alshami et al. 2003). Successful transmission of CYVCV by mechanical inoculation was obtained from Mosambi to french bean (*Phaseolus vulgaris* cv. singtamey, saxa, and gheusemi) and *C. quinoa*.

Attempt on transmission through soil and insects (aphids, nematodes) was negative for ICRSV (Pant and Ahlawat 1998; Byadgi and Ahlawat 1995; Byadgi et al. 1993). Attempts to detect ICRSV in EM from nematode spp., *Xiphenema americanum, Tylenchulus semipenetrans, Tylenchorhynchus spp.* and *Heliotylenchus spp.* were also not successful. ICRSV was successfully transmitted through dodder (*Cuscuta reflexa)* from mosambi to mosambi and mosambi to kinnow mandarin up to 60% (Byadgi et al. 1993). Citrus ringspot virions were trapped in ISEM from the pollen of ringspot affected kinnow trees indicating the possibility of transmission through pollen and seeds (Ahlawat 1997; Pant and Ahlawat 1998). Screening of citrus seeds using DAC-ELISA revealed the presence of virus in seed coats but not in embryo and endosperm (Prabha and Baranwal 2011). This study indicated that ICRSV is not truly seed borne.

14.3 Morphology, Physiology and Genome Properties

Mandarivirus genus contains flexuous viruses with single stranded positive sense RNA genome. ICRSV has flexuous particles of size 640 × 15 nm and 690 × 9 nm and also tubular structures (Byadgi et al. 1993; Byadgi and Ahlawat 1995; Hoa and Ahlawat 2004) (Fig. 14.3). The CYVCV virions ranged from 550 to 820 nm with a modal length of 685 × 14 nm (Alshami et al. 2003).

The genome of *Mandarivirus* is single stranded RNA. ICRSV has a single stranded positive sense RNA consists of 7560 nt excluding 3′ poly A tail. It consists of putative ORF's, encoding putative proteins of 187.3, 25, 12, 6.4, 34 and 23 kDa respectively; a 5′ untranslated region (UTR) of 78 nucleotide and a 3′ UTR of 40 nt followed by a poly A tail (Fig. 14.4). The putative poly peptides encoded by different ORF's are; ORF1 codes for viral helicase, RNA depended RNA polymerase and conserved motifs of putative methyl transferase, ORF 2, ORF3 and ORF 4 are partially overlapping ORF's which form a triple gene block (TGB) a common feature

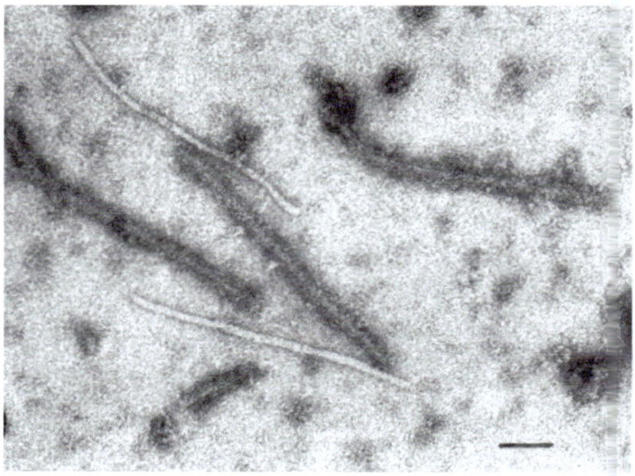

Fig. 14.3 Electron micrograph showing flexuous particles of ICRSV decorated with ICRSV antisera and non decorated flexuous particles of *Citrus yellow vein clearing virus* from mixed infection (Source: Alshami et al. 2003)

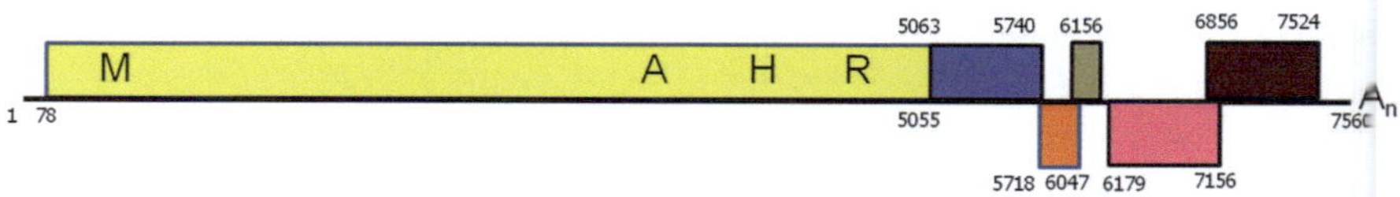

Fig. 14.4 Genome organization of *Mandarivirus*. (ICRSV: 5′UTR – 1 to 78 (78 nt), ORF1 – RNA dependent RNA Polymerase (RdRP): 79–5055 (4,977 nt), ORF2 – triple gene block 1 (TGB1): 5063–5740 (678 nt), ORF3 – TGB2: 5718–6047 (330 nt), ORF 4 – TGB3: 5974–6156 (183 nt), ORF 5- coat protein (CP): 6179–7156 (978 nt), ORF 6 – nucleic acid binding protein (NaBP): 6856–7524 (669 nt), 3′ UTR- 7525–7560 (36 nt))

of several group of plant virus group including *Foveavirus, Carlavirus* and *Allexivirus* (Prabha and Baranwal, 2012).

ICRSV was recorded from different geographical region showing diversity in the coat protein region (Hoa and Ahlawat 2004) in which ICRSV isolates from Pune and Ahmedabad, grouped in different lineage compared to the isolates from north India and later on these isolates were found to share close identity with CYVCV. In a study of diversity at molecular level by analysis of 3′ region of different isolates of ICRSV, partial sequences of NaBP region of all the isolates revealed only 78–80% identity to the reference sequence whereas the isolates were 97–98% identical within themselves in the case of nucleotide sequence identity and a similar trend is observed with sequence identity matrix of amino acids with 96–97% identity among their sequences. The zinc finger domain CX2CX9HX4CXH found highly conserved in the genus *Mandarivirus* where as conservation of CX2CXHX7HXCX5C was throughout the genera of *Alphaflexiviridae*. An arginine rich motif (ARM; KRRRARR) like region was also found conserved in all isolates as well as across the genera. Considerable variability in the 3′UTR secondary structure was observed among these isolates and the reference sequence. Results suggested that diversity do exists in ICRSV genome especially in CP and NaBP genes. Homologous recombination has been observed in CP and NaBP gene sequences of ICRSV and this recombination may be contributing to genetic diversity (Prabha and Baranwal 2011).

Leaf dip electron microscopy and ISEM can be used for the detection of virus particles and to determine the particle size and morphology of *Mandarivirus*. Since both ICRSV and CYVCV are morphologically similar and mixed infection in field can be determined by ISEM decoration. Ahlawat and Chakraborty (1990) purified ICRSV using phosphate buffer 0.05 M (pH 7.6) containing thioglycolic acid for extraction of ICRSV from infected leaves of citrus. Byadgi et al. (1993) developed and standardized the purification method for ICRSV from citrus leaves. Further, Baranwal et al. (2000) and Hoa (2003) standardized the virus purification from propagative host i.e. *P. vulgaris*. ELISA and dot immuno binding assays (DIBA) using polyclonal and monoclonal antibodies are methods for the diagnosis of ICRSV in infected plants (Byadgi et al. 1993; Ahlawat et al. 1995; Pant and Ahlawat 1998; Baranwal et al. 2000; Hoa 2003). The serological reactivity of ICRSV was attempted using indirect PTA-ELISA with goat anti-rabbit IgG and DAS-ELISA (Byadgi et al. 1993; Hoa 2003). For DAS-ELISA optimum concentration of IgG used was 1 μg/ml, antigen and conjugate dilution 1:10 and 1:2000 respectively. For DAC-ELISA IgG concentration of 1 μg/ml antigen dilutions of 1:10 and commercial anti rabbit antibody 1:10,000 was found optimum (Hoa 2003). DAC-ELISA and DAS-ELISA were effectively used to detect ICRSV from field samples of various *citrus* spp. To analyze the serological relationship of ICRSV with other filamentous viruses, a sample of purified virus was tested by EM decoration with antisera of 6 *Allexivirus* and 2 *Crinivirus*, but none reacted with ICRSV. DIBA for detection of ICRSV gave best result with 1:1 dilution of antigen and 1 μg/ml concentration of antiserum. ISEM technique was effectively used for detection of ICRSV by Ahlawat et al. (1995).

14.3.1 Diagnosis and Management

RT-PCR is very sensitive technique, which is used for the detection of viruses with specific primer. Three sets of primers were designed to detect ICRSV (Hoa 2003). One set of primer ICRSV-1096/ICRSV-1420 used to amplify approximately 350 bp and detected four isolates of, ICRSV-Dl, ICRSV-Ab, ICRSV-Ah and ICRSV-Pu. Multiplex RT-PCR is successfully used for simultaneous detection of ICRSV in mixed infection with other viruses such as *Citrus leaf rugose virus* (CLRV), *Citrus psorosis virus* (CPsV), *Citrus tatter leaf virus* (CTLV), *Citrus tristeza virus* (CTV), *Citrus variegation virus* (CVV), *Citrus yellow mosaic virus* (CMBV) (Roy et al. 2005). A degenerate primer pair was designed, targetting the conserved RdRp region of the genus *Mandarivirus*, and both ICRSV and CYVCV could be amplified from total RNA extract from plants infected by two species of *Mandarivirus* genera (Meena and Baranwal 2016). Development of LAMP based detection of the *Mandarivirus* for on field and easy diagnostics is in progress at Central Citrus Research Institute, Nagpur (personal communication).

Various studies have been attempted to remove ICRSV infections in citrus. Heat treatment of citrus was attempted by exposing bud sticks to moist hot air (Cheema et al. 1999). The buds exposed at 50 °C for 30 min or more and at 45 °C for 120 min were most effective in vivo and in vitro for eliminating virus from the infected bud sticks. For the management of the disease protocol for shoot tip culture grafting has been standardized in case of Mosambi and Kinnow and virus free plants were produced and multiplied. Grafts with apical meristem having two leaf primordials were virus free and successful up to 20.83%. Pre-treatment with anti-oxidants and plant growth regulators increased the percentage of successful grafts and among the different methods of grafting, triangular incision gave best result (Hoa et al. 2004). French bean plants were transformed with truncated coat protein gene of ICRSV and proved by PCR and southern hybridization. CP gene expression in transformed plants was also confirmed by DIBA (Hoa 2003). Nucellar embryo culture was used to produce ICRSV-free plants of Kinnow mandarin (*Citrus nobilis* Lourx, *C. Deliciosa* Tenora) from a virus-infected plant (Singh et al. 2005). There are reports of in vitro production of ICRSV free Kinnow plants employing thermotherapy coupled with shoot tip grafting (Sharma et al. 2008). Shoot tips were excised from the nodal sprouts generated from nodal segments from infected mother plant cultured on Murashige and Skoog medium containing 2iP (1 mg/l or 4.9 μM) and malt extract (800 mg/l) and grafted on to rough lemon (*Citrus jambhiri*) under aseptic conditions. Different concentrations of aqueous extracts of *Azadira chtaindica* (Neem) leaves, *Sorghum vulgare* (Jowar) and roots of *Boerhaavia diffusa* (Punarnava) were added to the medium. Fifty percent virus elimination was seen for aqueous leaf extracts of *A. Indica*, while *B. diffusa* root extract and *S. vulgare* leaf extract gave 42.86% and 31.58 % reduction on virus load, respectively and confirmation was done by both indirect ELISA and RT-PCR. In vitro micrografting has been attempted for production of *Mandarivirus* free plants of kinnow mandarin by Singh et al. (2008). Different sized shoot apices (0.2–1.0 mm) excised from the ICRSV-infected kinnow were micrografted onto decapitated rootstock seedlings of rough lemon (*C. jambhiri*) and tested for ICRSV through ELISA and RT-PCR.All the 0.2-mm scion raised plants were negative for ICRSV by both ELISA and RT-PCR while only 20% of the ELISA negative plants raised from 0.3- mm scion were negative for ICRSV with RT-PCR.

14.3.2 Allexivirus

The genus Allexivirus comprises of eight virus species *Garlic Virus A* (GarV-A), *Garlic Virus B* (GarV-B), *Garlic Virus C* (GarV-C), *Garlic virus D* (GarV-D), *Garlic Virus E* (GarV-E), *Garlic Virus X* (GarV-X) and *Garlic mite-borne filamentous virus* (GarMbFV) along with the type species *Shallot virus X* (ShVX). The criteria for demarcation into distinct species is that it should have less than 72% identical nucleotides or 80% identical amino acids between their entire coat protein or replication protein genes.

14.4 Biology, Pathology and Economic Impacts

Mixed infection of *allexiviruses* with potyviruses has a huge impact on garlic production where a reduction of 74% in bulb weight and 37% in bulb perimeter was reported. Yield loss assessment study during multiple infections in garlic showed a yield reduction of 22% in mixed infection of two allexiviruses, *Garlic virus A* and *Garlic virus C*. Reduction in bulb weight (14–32%) and diameter (6–11%) was significantly high with infection of GarV-A compared to GarV-C. It was also observed that garlic virus yield decreased rapidly in plants previously infected with at least one *Allexivirus*. Occurrence of *allexiviruses* along with other viruses like *Onion yellow dwarf virus* (OYDV), *Garlic common latent virus* (GarCLV) and *Shallot latent virus* is common in garlic. Out of the eight species, four species are reported in India. Allexiviruses are reported to be transmitted by eryophid mite, *Acerya tulipae*. Most of the studies on impact of Allexivirus infection have been done in European Countries. In India, only limited studies on biology and proper quantification of economic loss are attempted for *Allexivirus*.

14.5 Morphology, Physiology and Genome Properties

Allexivirus have virions of ~800 nm length with a flexuous morphology (Fig. 14.5a). The genome is a positive sense single stranded RNA of 9.0 kb in size comprises six ORFs and a coat protein (CP) of 26–29 kDa (Fig. 14.5b).

The first *Allexivirus* which was characterized from in India was *Shallot virus X* (ShVX) from shallot plants. The presence of ShVX was confirmed through electron microscopy and sequencing of partial NB gene and un-translated region (UTR) and was identified as ShVX (Majumder et al. 2007). Sequence comparison and phylogenetic analysis of amino acid sequences of CP and Nucleic acid Binding region (NB) of the ShVX associated with Indian shallot indicated it as a distinct isolate of ShVX. The nucleotide sequence of coat protein consisting of 789 nucleotide showed 79–80 % sequence identity to that of ShVX from New Zealand (EU835197) and Russia (NC00379, L76292) and 66–76% with other *Allexiviruses* (Majumder and Baranwal 2011). The detection of *Allexivirus* in cloves has been done by RT-PCR – using degenerate primers ALLEX 1 and ALLEX 2 (Kumar et al. 2010) and ~200 bp

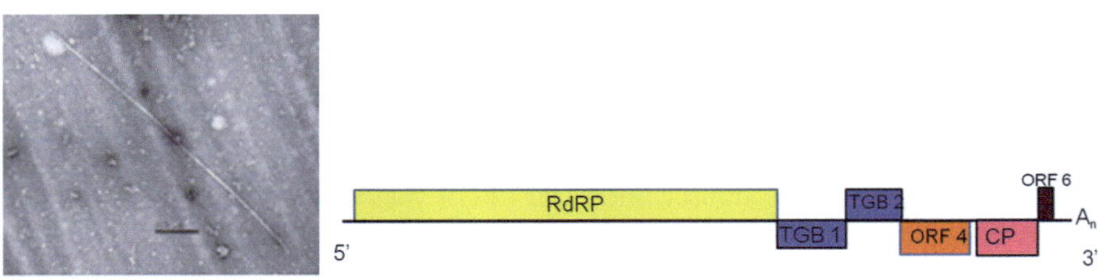

A b

Fig. 14.5 Electron micrograph and genome organization of *Allexivirus* (ShVX)

amplification product was observed in five different cultivars (G-282, IC-375416, Ruag, Yamuna Safed and ACC-40).BLAST analysis of the nucleotide sequences from five garlic cultivars showed identity with different allexiviruses, *Garlic virus A* (GarV-A) (74–83%), *Garlic virus E* (GarV-E) (74–80%), *Garlic virus D* (GarV-D) (76–79%) and *Garlic virus X* (GarV-X) (75–78%).

Garlic virus X is another member of *Allexivirus* genus, reported from both garlic and onion in India (Baranwal et al. 2011). GVX is a single-stranded positive-sense RNA virus consisting of 8.1 kb genome with six open reading frames (ORFs) which encode putative proteins of 174 kDa (ORF1), 26 kDa (ORF2), 12 kDa (ORF3), 32 kDa (ORF4), 26 kDa (ORF5) and 15 kDa (ORF6). A ~ 1.5 kb CP gene of Gar-VX was isolated from cultivar G-282 which originated from the southern part of India. Phylogenetic analysis of CP sequence of allexiviruses, the four isolates of GarV-X from India clustered together and originated from common clade. The other three isolates of GarV-X from Italy, China and Korea grouped together on a separate clade and all these isolates of GarV-X and isolates of Garv-B originated from a common clad and formed a separate group. The second group included GarV-A, GarV-D, GarV-E and GarMbV while third and fourth group constituted of isolates of GarV-C and ShV-X, respectively (Singh et al. 2014). Use of recombinant expression of viral coat protein reduces purification steps and allows generating specific antibody to a particular virus. Antisera were produced by expressing GarV-X full CP (GarV-X-CP^{His6}) ELISA based detection has been standardized. Polyclonal antibodies using truncated CP in which N-terminal was absent (GarV-X CPΔ1-61^{His5}) could not detect GarV-X indicating that the antigenic region was present in N-terminal of coat protein (Singh et al. 2014).

Garlic virus A (GarV-A) of genome with 8660 nucleotides and six ORFs, is reported in garlic from many countries across the world. In India, GarV-A has been reported from garlic cultivars Bhima Omkar, Bhima Purple, G-41, and G-282 showing mild mosaic symptoms (Gawande et al. 2015). Association of eryophid mites were also observed in symptomatic plants. The triple block gene and RNA dependent RNA polymerase genes of GarV-A from India showed more than 80% similarity to Australian GarV-A isolates.

Garlic virus D is one of the most abundant allexivirus across the world. GarV-D has been reported in onion plants showing mild mosaic, or mild chlorotic stripes and some of the symptomless onion from the vegetable research farm of Punjab Agricultural University (Khan et al. 2015). The coat protein and nucleic acid binding regions were showing highest similarity with the Gar V-D from Iran. When infected alone GarV-D showed mild chlorotic stripe, while in mixed infections with OYDV, the common symptoms appeared were mosaic.

Since allium crops are vegetatively propagated, elimination of viruses including *Allexivirus*es was attempted by meristem culture in combination with thermotherapy. A multiplex reverse transcription (RT)-PCR method was developed for the simultaneous detection of allexiviruses along with other viruses infecting garlic accessions in India which is useful in indexing of garlic viruses and production of virus free seed material (Majumder and Baranwal 2014). The garlic cloves with mixed infections of different viruses subjected to meristem tip culture combined with thermotherapy like solar heat treatment and hot air treatment for different time durations. It is to be noted that the longer duration of heat exposure was required to eliminate GarVX from the cloves compared to other viruses (Pramesh and Baranwal 2015). Availability of molecular biology tools has made it possible to detect and characterize allexiviruses in Allium crops and ELISA could be used as detection assay for epidemiological studies and production of virus free garlic materials.

14.6 Concluding Remarks

Only two members are known till now in the genera *Mandarivirus* and existence of more variants and members in the genera needs to be explored. So far citrus is the only reported host of mandarivirus, and mixed infection often occurs providing opportunities for recombination and evolution. The dynamics of mixed infections in the same host and relevance of quasi-species concepts in the variants needs to be elaborately studied. Development of infectious clones from the complete genomes can put light to the functional aspects of the genome of mandariviruses and their introduction to host plant will reveal the mechanism of infection. Mandariviruses are quarantine pathogens as they are restricted to Asia and hence phytosanitary measures are inevitable along with virus certification

The overall picture of the distribution of the different species of the genus *Allexivirus* in India, their biology and economic impact has to be studied systematically. The functional role of ORF4 and ORF 6 is yet to be unravelled in allexivirus genome. Exploration of methods for efficient elimination of allexiviruses from garlic cloves is imperative considering the vegetative nature of propagation.

References

Ahlawat YS (1989) Psorosis-A disease of citrus in India. Indian Phytopathol 42:21–25

Ahlawat YS (1997) Viruses, greening bacterium and viroid associated with citrus decline in India. Indian J Agric Sci 67:51–57

Ahlawat YS, Chakraborty NK (1990) Etiology of Psorosis-A disease of citrus. Indian Phytopathol 43:239

Ahlawat YS, Pant RP (2003) Major viruses and virus like diseases of citrus in India, their diagnosis and management. Ann Rev Plant Pathol 2:447–474

Alshami AAA, Ahlawat YS, Pant RP (2003) A hitherto unreported yellow vein clearing disease of citrus in India and its viral etiology. Indian Phytopathol 56(4):422–427

Baranwal VK, Chakraborty NK, Giri BK, Ahlawat YS (2000) Purification, production of polyclonal antibodies and detection of *Indian citrusringspot* virus (ICRSV). In: Shyam S, Ghosh SP (eds) Hitech citrus management, improvement, biotechnology, production system, pest and disease management, post harvest technology. ISC/ICAR/NRCC, pp 1036–1040

Baranwal VK, Singh P, Jain RK, Joshi S (2011) First report of *Garlic virus X* infecting garlic in India. Plant Dis 95:1197

Byadgi AS, Ahlawat YS (1995) A new viral ringspot disease of citrus (Citrus species) in India. Indian J Agric Sci 65:61–68

Byadgi AS, Ahlawat AS, Chakraborty NK, Verma A, Srivastava M, Mlne RG (1993) Characterization of a filamentous virus associated with citrus ringspot in India. In: Proc. 12th conf. IOCV, Riverside, California. p 155–162.

Cheema SS, Tagjeet SL, Gosal SS (1999) Management of citrus ringspot disease through thermotherapy. Indian Phytopathol 52:354–356

Gawande SJ, Gurav VS, Ingle AA, Gopal J (2015) First report of Garlic virus A in garlic from India. Plant Dis 99:1288

Hoa NV (2003) Molecular characterisation of *Indian citrus ringspot virus* and its management by shoot tip culture. PhD. Thesis, Indian Agricultural Research Institute, New Delhi

Hoa NV, Ahlawat YS (2004) Characterization of four isolates of *Indian citrus ringspot virus*. Indian Phytopathol 57:296–302

Hoa NV, Ahlawat YS, Pant RP (2004) Production of virus-free Kinnow mandarin and Mosambi sweet orange nucleus planting material through shoot tip grafting. Indian Phytopathol 57(4):482–487

Khan I, Sharma A, Dhatt AS, Kang SS, Kaur G (2015) First report of *Garlic Virus D* infecting onion in India. J Plant Pathol. doi:10.4454/JPP.V98I1.007

Kumar S, Baranwal VK, Joshi S, Arya M, Majumder S (2010) Simultaneous detection of mixed infection of *Onion yellow dwarf virus* and an *Allexivirus* in RT-PCR for ensuring virus free onion bulbs. Indian J Virol 21(1):64–68

Majumder S, Baranwal VK (2011) Sequence comparison and phylogeny of nucleotide sequence of coat protein and nucleic acid binding protein of a distinct isolate of shallot virus X from India. Indian J Virol 22(1):63–65

Majumder S, Baranwal VK (2014) Simultaneous detection of four garlic viruses by multiplex reverse transcription PCR and their distribution in Indian garlic accessions. J Virol Methods 15:34–38

Majumder S, Arya M, Pant RP, Baranwal VK (2007) *Shallot virus X* in Indian shallot, a new virus report for India. Plant Pathol 57:2

Meena RP, Baranwal VK (2016) Development of multiplex polymerase chain reaction assay for simultaneous detection of clostero-, badna- and mandari-viruses along with huanglongbing bacterium in citrus trees. J Virol Methods 235:58–64

Pant RP (1995) Studies on ringspot virus and its interaction with a bacilliform virus associated with citrus golden mosaic disease. PhD thesis, Division of Plant Pathology, Indian Agricultural Research Institute, New Delhi

Pant RP, Ahlawat YS (1998) Partial characterization of a filamentous virus associated with ringspot disease of citrus. Indian Phytopathol 51:225–232

Pant RP, Baranwal VK, Singh SP, Ahlawat YS (2000) Mechanical transmission of Indian citrus ringspot virus (ICRSV). In: Hitech citrus management. ISC/ICAR/NRCC, pp 1021–1023

Prabha K, Baranwal VK (2011) *Indian citrus ringspot* virus: localization of virus in seed tissues and evidence for lack of seed transmission. Phytoparasitica 39:491–496

Prabha K, Baranwal VK (2012) The genome sequence of an isolate of *Indian citrus ringspot virus* infecting sweet orange in India. J Virol 86(22):12446–12447

Pramesh D, Baranwal VK (2015) Production of virus-free garlic (*Allium sativum* L.) through meristem tip culture after solar or hot air treatment of cloves. J Hortic Sci Biotechnol 90(2):180–186

Singh P, Prabha K, Jain RK, Baranwal VK (2014) N-terminal in coat protein of *Garlic virus X* is indispensible for its serological detection. Virus Genes 48:128–132

Roy A, Fayad A, Barthe G, Brlansky RH (2005) A multiplex polymerase chain reaction method for reliable, sensitive and simultaneous detection of multiple viruses in citrus trees. J Virol Methods 129:47–55

Sharma S, Singh B, Rani G, Zaidi AA, Vipin K, Hallan K, Nagpal A, Virk GS (2008) In vitro production of *Indian citrus ringspot virus* (ICRSV) free Kinnow plants employing thermotherapy coupled with shoot tip grafting. Plant Cell Tissue Org Cult 92:85–92

Singh B, Sharma S, Rani G, Zaidi AA, Hallen V, Virk GS, Nagpal A (2005) In vitro production of *Indian citrus ringspot virus* free plants of Kinnow mandarin (*Citrus nobilis* Lour ×*C. deliciosa* Tenora) by nucellar embryo culture. Plant Pathol J 5:274–282

Singh B, Sharma S, Rani G, Vipin K, Hallan, Zaidi AA, Virk GS, Nagpal A (2008) In vitro micrografting for production of *Indian citrus ringspot virus* (ICRSV)-free plants of kinnow mandarin. Plant Biotechnol Rep 2:137–143

Thind SK, Arora PK (1997) Incidence and behaviour of ringspot virus in citrus. Indian J Virol 13:159–160

Thind SK, Sharma JN, Arora PK, Cheema SS (1998) Effect of citrus ringspot virus on quality of kinnow mandarin fruits. Indian J Virol 14(1):91–92

Diversity and Pathogenesis of Mastreviruses in India

15

V.G. Malathi and S. Kanakala

Abstract

Mastreviruses (family *Geminiviridae*) are generally known to infect monocotyledon plants and are transmitted by leafhopper. Compared to begomoviruses, number of mastreviruses characterized are far less in India. Only two mastreviruses, chickpea chlorotic dwarf virus infecting dicot plant and wheat dwarf India virus (WDIV) infecting wheat have recently been identified in India. Both the viruses have been characterised at genomic level and infectivity of the cloned DNAs has been demonstrated. Interestingly, WDIV is associated with two alphasatellites and one betasatellite. The satellites contribute to viral pathogenicity and suppress the RNA silencing of the host.

Keywords

Mastreviruses in India • Chickpea chlorotic dwarf virus • Wheat dwarf India virus

V.G. Malathi (✉)
Advanced Centre for Plant Virology, Division of Plant Pathology,
Indian Agricultural Research Institute, New Delhi 110012, India

Tamil Nadu Agricultural University, Coimbatore 641003, Tamil Nadu, India
e-mail: vgmalathi@rediffmail.com

S. Kanakala
Agricultural Research Organisation, The Volcani Centre, P.O.B 15159, Bet Dagan, Israel

Advanced Centre for Plant Virology, Division of Plant Pathology, Indian Agricultural Research Institute, New Delhi 110012, India

B. Mandal et al. (eds.), *A Century of Plant Virology in India*,
https://doi.org/10.1007/978-981-10-5672-7_15

15.1 Introduction

The genus *Mastrevirus* belonging to the family *Geminiviridae* is named after the type species *Maize streak virus* (MSV) that causes maize streak disease in the African continent. The genome of mastreviruses consists of a single component (monopartite) circular ssDNA, 2.5–2.7 kb (Palmer and Rybicki 1998). There are four open reading frames (ORFs), two on the virion sense strand (ORF V1, Coat protein; ORF V2, Movement protein) and two ORF's in the complementary sense strand (ORF C1:C2), from which the replication initiation protein is expressed by transcript splicing (Boulton 2002). The coding sequences are separated by a large intergenic region (LIR) which consists of the highly conserved nonanucleotide sequence (TAATATT↓AC), which represents the origin of replication in all geminiviruses (Orozco et al. 1998). There is a small intergenic region (SIR) at the 3′ end of V1 and C2 ORFs. A short complementary sense DNA primer containing 5′ ribonucleotides is found encapsidated along with the genomic DNA. This primer is complementary to sequences in the SIR region of the genome (Palmer and Rybicki 1998) and is hypothesised to prime the negative strand DNA synthesis.

The genus *Mastrevirus* includes members, which are further subdivided into a major group infecting monocot plants and a minor group infecting dicot plants. Both monocot and dicot infecting mastreviruses share similar features of genome organization. Currently, there are 32 recognized members in the genus *Mastrevirus* (Muhire et al. 2013). More recently, Muhire et al. (2013) reclassified these viruses on the basis of 78% nucleotide identity in the genomic DNA and grouped all the South Asia dicot infecting mastreviruses as chickpea chlorotic dwarf virus including its fourteen strains (Muhire et al. 2013; Kraberger et al. 2015).

15.2 Chickpea Chlorotic Dwarf Virus

15.2.1 Discovery and Distribution

The chickpea chlorotic stunt disease was first described by Horn et al. (1993). The initial survey report showed that the association of three viruses with the stunt affected chickpea samples, leafhopper transmitted geminivirus, bean leaf roll virus (BLRV) like luteovirus and an unidentified luteovirus (Horn et al. 1993, 1994, 1996). Further it was known that BLRV like luteoviruses are of minor importance and the leafhopper transmitted are more widely distributed. The most important observation they made is that the stunt disease is more in *Kabuli* genotypes in Pakistan than the *Desi* types. They designated the virus as chickpea chlorotic dwarf virus (CpCDV) and identified the leafhopper, *Orosius orientalis* (later named as *O. albicinctus*) as the vector of the virus (Horn et al. 1993).

15.2.2 Yield Loss

The stunt disease is widely prevalent in the Indian states of Andhra Pradesh, Gujarat, Haryana, Madhya Pradesh and Punjab; and causes about a 75–95% yield loss (Horn et al. 1995, 1996). More recently, Kanakala et al. (2012) in India characterized CpCDV, which was identified as one of the etiological agents of stunt disease. Kanakala et al. (2013) conducted survey in northern India states and observed high incidence in Rajasthan (67 %) followed by Madhya Pradesh (46%) and New Delhi (45%). In these places equally high disease incidence was recorded in both *Desi* and *Kabuli* types. The yield loss was found to be 75–90% and nearly total loss if the infection occurs at early stage (Horn et al. 1995).

15.2.3 Symptomatology

The symptoms of the chickpea stunt disease caused by CpCDV are stunting, internode shortening, phloem browning in the collar region, leaf reddening in *desi* type plants and leaf yellowing in *Kabuli* type plants (Horn et al. 1993). In the field observations, the symptoms ranged from very mild initiation of reddening in *Desi* and yellowing in *Kabuli* to severe drying of plants. The discoloration appears 45 days after sowing. The typical symptoms that can be easily distinguished in the field are discoloration (red or yellow), smalling of the leaves, proliferation of axillary shoots with small leaves, the bushy phenotype of the plant and stunting (Fig. 15.1). In the final stage, the plant dries, leaves withers and plant appears more like dry-rot–affected plants. Plants that were infected at a later stage showed smalling of leaves and proliferation, but not drying of plants. In certain cases, while smalling of leaves and proliferation of axillary shoots were seen in young plants (60 days after sowing), the new shoots that developed appeared to have recovered from disease (Kanakala et al. 2013).

Fig. 15.1 Field symptoms of chickpea stunt disease. (**a**) Chickpea field showing chickpea stunt disease, (**b**) *desi* type showing reddening and stunting symptoms and (**c**) *kabuli* type showing yellowing and stunting symptoms

15.2.4 Transmission and Host Range

Horn et al. (1993) reported that leafhopper *O. orientalis* successfully transmitted the CpCDV to wide range of hosts belonging to the families Solanaceae, Leguminosae and Chenopodiaceae. The virus was efficiently transmitted with a median acquisition access period (AAP), inoculation access period (IAP) and latent period (LP) of 8 h, 2.3 h and 27.7 h, respectively. In addition to chickpea, more recently, CpCDV was reported from *Capsicum annuum* (Byun HS, unpublished, accession no. KF632712).

15.2.5 Genomic Comparison

The genome of three isolates of the viruses associated with the stunt disease at Delhi were cloned (~2.5 kb length). The genome organization of the virus isolates is similar to that of other mastreviruses, with two ORFs coding for movement and coat protein in the viral sense strand and two on the complementary strand. The Indian isolates revealed 98% sequence identity in the nucleotides with the isolates from Pakistan. The comparisons of the predicted amino acid sequences revealed 98–100% identity between CpCDV isolates from India, and 88 and 93% with chickpea viruses from Sudan and Syria, and bean yellow dwarf virus from both Pakistan and Africa. The percentage identify with chickpea viruses from Australia was only 65% (Kanakala et al. 2012).

15.2.6 Phylogenetic Relationship

Indian CpCDV isolates are closely related to other CpCDV strains noticed at different locations around the Eritrea, Egypt, Iran, Pakistan, Sudan, South Africa, Syria, Turkey and Yemen (Fig. 15.2). However, Indian mastreviruses are more closely related to chickpea infecting mastrevirses in Pakistan. The heat map generated on the basis of pairwise genome wide nucleotide sequence identity gives a clear picture of relationship between monocot and dicot infecting mastreviruses (Fig. 15.3).

15.2.7 Establishment of Koch's Postulates

Kanakala et al. (2012, 2013) performed infectivity analysis. Chickpea plants when agroinoculated with CpCDV complete tandem repeat construct, symptoms were not easily distinguishable from the mock-inoculated plants up to 15 days post-inoculation (dpi), however, it was clear from the third week post-inoculation. The initial symptoms were leaf smalling and then proliferation of axillary shoots, which led to stunting and yellowing of terminal leaves (Fig. 15.4). The agroinoculated plants gradually dried and eventually died before flowering (30–35 dpi). Interestingly,

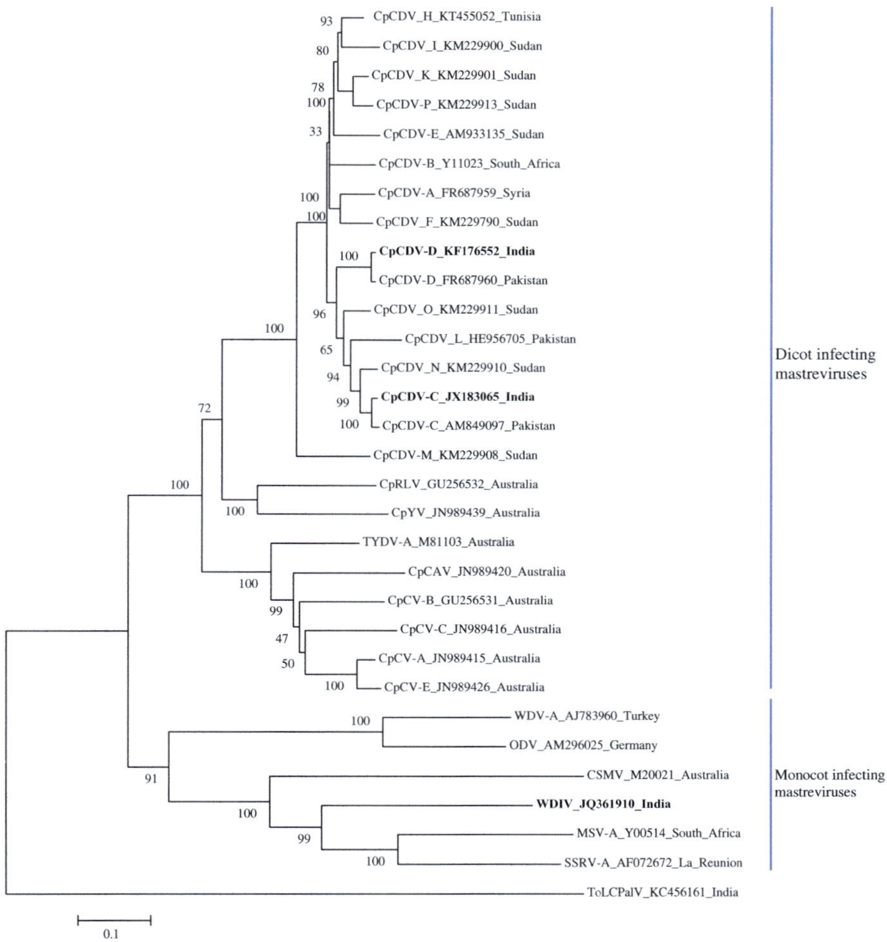

Fig. 15.2 Phylogenetic relationships between mono and dicot infecting mastreviruses. Dendograms were constructed using the neighbor-joining and bootstrap (1000 replications) based on alignment of complete genome sequences of dicot infecting mastreviruses. Alignments were produced with Clustal W (Mega 6, http://www.megasoftware.net/). *Vertical distances* are arbitrary; *horizontal distances* are proportional to genetic distances. The numbers at nodes refer to number of times (as a percentage) in which the branching was supported. Dicot mastreviruses used for this analysis are: *Chickpea chlorotic dwarf virus* (CpCDV-A, B, C, D, E, F, H, I, K, L, M, N, O and P), *Chickpea redleaf virus* (CpRLV), Chickpea yellows virus (CpYV), *Chickpea chlorosis Australia virus* (CpCAV); *Chickpea chlorosis virus* (CpCV A, B, C and E), Tobacco yellow dwarf virus (TYDV-A). Monocot mastreviruses used for this analysis are *Maize streak virus* (MSV), *Chloris striate mosaic virus* (CSMV), *Oat dwarf virus* (ODV), *Sugarcane streak Reunion virus* (SSRV), *Wheat dwarf virus* (WDV-A), *Wheat dwarf India virus* (WDIV)

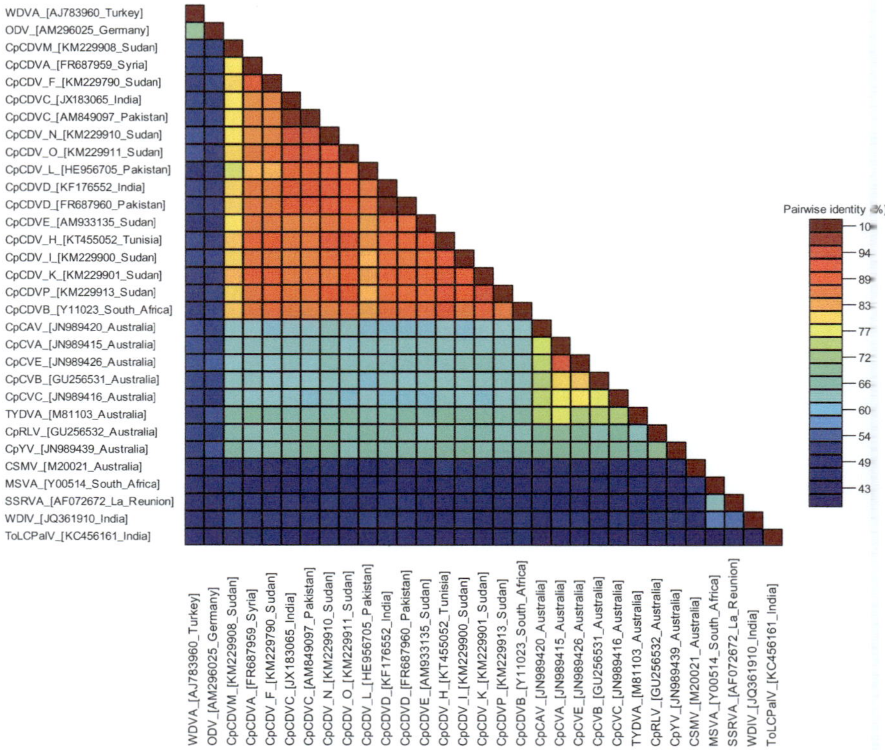

Fig. 15.3 Graphical representation of percentage pairwise genome scores and nucleotide identity plot of full genomes of mastreviruses. The plot was generated using SDTv1.0 (species demarcation tool http://web.cbio.uct.ac.za/SDT)

Fig. 15.4 Infectivity test by agroinoculation with an infectious construct of CpCDV (pBinD320 2.0mer) in chickpea showing leaf smalling, stunting and yellowing of terminal leaves

the reddening symptom seen in *Desi* type is not seen in CpCDV agroinoculated plants (Kanakala et al. 2013).

Cloned DNAs of CpCDV was infectious to *N. benthamiana*, *N. glutinosa* and *N. tabacum*. Symptoms were typically chlorosis in *N. benthamiana* followed by downward marginal folding and a reduction in leaf lamina. In sesame, severe symptoms with thickening of leaves, downward folding, crumpling and reduction of leaf lamina were observed. In mustard, symptoms were typical chlorosis, downward marginal folding and stunting. Interestingly, CpCDV does not infect legumes hosts such as French bean (Sel9), blackgram (Co5), soybean (NRC37) and cowpea (Pusa Komal) (Kanakala et al. 2012). On the other hand, CpCDV is highly infectious to tomato plants. The young unfurling leaves became leathery, were dark green and showed mild backward leaf curling within 7 dpi and the plants remained stunted in growth (Kanakala et al. 2012).

15.3 Wheat Dwarf India Virus

15.3.1 Discovery and Distribution

A survey conducted during 2010 and 2011 for detecting wheat infecting viruses by Kumar et al. (2012) identified a new disease. They described symptoms like dwarfing, yellowing of leaves and sterile spikes in 10% of plants in northern India, and in 15% of wheat cultivated in southern India. They examined the plants initially for the presence of barley yellow dwarf virus and confirmed its absence. They amplified viral genomic DNA and identified to be a Mastrevirus, which was designated as the new species wheat dwarf India virus (WDIV). They also identified the vector to be a leafhopper, *Psammotettix sps*). They further recorded the distribution of the virus in Uttar Pradesh, Bihar, Maharashtra, Madhya Pradesh and Rajasthan, where they recorded 4.5–20 % incidence of the disease (Kumar et al. 2015).

15.3.2 Yield Loss

Kumar et al. (2015) recorded significant reduction in number of tillers at higher severity. Thousand grain weight, at scale 0 was 44 g, but gradually decreased to 35 ± 4 g at higher disease grade. Grain formation was totally lost in the spike at scales 8–9.

15.3.3 Symptomatology and Transmission

WDIV infected wheat plants exhibited symptoms such as dwarfing, yellowing of leaves, sterile spikes and dwarfism. *Psammotettic* sp. (family *Cicadellidae*) was predicted to be the insect vector. Five out of 30 leafhoppers collected from the field tested positive for the virus and resulted in 2.7 kb genome length fragment in rolling circle amplification; however, vector transmission of WDIV was not established. Interestingly, WDIV has also been recorded from the dicot plant, *Solanum sp (Kumar, unpublished, accession no. KJ028209).*

15.3.4 Genome Comparison and Infectivity

The cloned genome from both infected wheat leaves and leafhoppers was ~2.7 kb in size. The whole genome sequence shared only 60 % identify with Panicum streak virus, 59% with MSV and 54% with wheat dwarf virus. There are predictability two ORFs on the virion sense strand (movement protein and coat protein) and two on the complementary sense strand (replication associated protein: Rep and RepA). In a phylogenetic analysis, WDIV though belonged to the main clade of monocot infecting viruses, branched off distinctly indicating its independent origin.

Agroinoculations of wheat seedlings using head to tail repeat construct of WDIV showed dwarf phenotype, the typical streak symptoms that observed under natural conditions in 85 out of 130 plants inoculated.

15.3.5 Association with Begomoviruses and Satellites

Association of two alpha satellites (Cotton leaf curl Multan alphasatellite and a guar leaf curl alphasatellite) and a betasatellite (Ageratum yellow leaf curl betasatellite with WDIV was observed (Kumar et al. 2014a). It was shown that WDIV interacted with satellites, and together induced severe symptoms. In the co inoculated plants, there was increased accumulation of WDIV and reduced accumulation of virus-specific small RNAs. It was shown that the βC1 was a pathogenicity determinant for WDIV and can interact functionally with Ageratum enation virus, a begomovirus (Kumar et al. 2014b). Though, recently, co-infection of mastrevirus and begomovirus has been identified in *Xanthium strumarium* (Mubin et al. 2012) and cotton (Manzoor et al. 2014). Until now either begomovirus or begomo-mastrevirus (co-infection) was not observed in the chickpea stunt disease.

15.4 Concluding Remarks

The occurrence of monocot infecting geminivirus in India has been revealed for the first time in 2012. There may be more mastreviruses infecting monocot and dicot plants in India. Seth et al. (1972) reported a bajra streak disease in northern India, its etiological agent was suspected to be MSV but it is not yet confirmed. Association of satellites with WDIV was an unexpected finding, which again pointer to the possibility of exchange of components between mastre- and begomoviruses.

References

Boulton MI (2002) Functions and interactions of mastrevirus gene products. Physiol Mol Plant Pathol 60:243–255

Horn NM, Reddy SV, Roberts IM, Reddy DVR (1993) Chickpea chlorotic dwarf virus, a new leafhopper transmitted geminivirus of chickpea in India. Ann Appl Biol 122:467–479

Horn NM, Reddy SV, Reddy DVR (1994) Virus-vector relationships of chickpea chlorotic dwarf geminivirus and the leafhopper *Orosius orientalis* (Hemiptera: Cicadellidae). Ann Appl Biol 124:441–450

Horn NM, Reddy SV, Reddy DVR (1995) Assessment of yield losses caused by chickpea chlorotic dwarf geminivirus in chickpea (*Cicer arietinum*) in India. Eur J Plant Pathol 101:221–224

Horn NM, Reddy SV, van den Huevel JFJM, Reddy DVR (1996) Survey of chickpea (*Cicer arietinum* L.) for chickpea stunt disease and associated viruses in India and Pakistan. Plant Dis 80:286–290

Kanakala S, Sakhare A, Verma HN, Malathi VG (2012) Infectivity and the phylogenetic relationship of a mastrevirus causing chickpea stunt disease in India. Eur J Plant Pathol 135:429–438

Kanakala S, Verma HN, Vijay P, Saxena DR, Malathi VG (2013) Response of chickpea genotypes to Agrobacterium-mediated delivery of chickpea chlorotic dwarf virus (CpCDV) genome and identification of resistance source. Appl Microbiol Biotechnol 97:9491–9501. doi:10.1007/s00253-013-5162-9

Kraberger S, Kumari SG, Hamed AA, Gronenborn B, Thomas JE, Sharman M, Harkins GW, Muhire BM, Martin DP, Varsani A (2015) Molecular diversity of chickpea chlorotic dwarf virus in Sudan: high rates of intraspecies recombination a driving force in the emergence of new strains. Infect Genet Evol 29:203–215

Kumar J, Singh SP, Kumar J, Tuli R (2012) A novel mastrevirus infecting wheat in India. Arch Virol 157:2031–2034

Kumar J, Kumar J, Singh SP, Tuli R (2014a) Association of satellites with a mastrevirus in natural infection: complexity of wheat dwarf India virus disease. J Virol 88:7093–7104

Kumar J, Kumar J, Singh SP, Tuli R (2014b) βC1 is a pathogenicity determinant: not only for begomoviruses but also for a mastrevirus. Arch Virol 159:3071–3076

Kumar J, Kumar J, Singh S, Shukla V, Singh SP, Tuli R (2015) Prevalence of Wheat dwarf India virus in wheat in India. Curr Sci 108:260

Manzoor MT, Ilyas M, Shafiq M, Haider MS, Shahid AA, Briddon RW (2014) A distinct strain of chickpea chlorotic dwarf virus (genus Mastrevirus, family Geminiviridae) identified in cotton plants affected by leaf curl disease. Arch Virol 159:1217–1221

Mubin M, Akhtar S, Amin I, Briddon RW, Mansoor S (2012) *Xanthium strumarium*: a weed host of components of begomovirus–betasatellite complexes affecting crops. Virus Genes 44:112–119

Muhire B, Martin DP, Brown JK, Navas-Castillo J, Moriones E, Zerbini M, Rivera-Bustamante RF, Malathi VG, Briddon RW, Varsani A (2013) A genome-wide pairwise-identity-based proposal for the classification of viruses in the genus mastrevirus (family Geminiviridae). Arch Virol. doi:10.1007/s00705-012-1601-7

Orozco BM, Gladfelter HJ, Settlage SB, Eagle PA, Gentry RN, Hanley-Bowdoin L (1998) Multiple cis elements contribute to geminivirus origin function. Virology 242:346–356

Palmer KE, Rybicki EP (1998) The molecular biology of mastreviruses. Adv Virus Res 50:183–234

Seth ML, Raychaudhuri SP, Singh DV (1972) Occurrence of maize streak virus on wheat in India. Curr Sci 41:684

Indian Peanut Clump Virus, a Fungal Transmitted *Pecluvirus* Infecting Both Monocotyledonous and Dicotyledonous Plants in India

16

Vikash Pandey, Bikash Mandal, Rakesh Kumar Jain, and Anirban Roy

Abstract

Indian peanut clump virus (IPCV), a member of the genus *Pecluvirus* under the family *Virgaviridae* has a wide host range, which include monocotyledonous and dicotyledonous plants. Among the diseases caused by IPCV in different crops, clump disease of peanut or groundnut is most important and one of the major limiting factors of groundnut production in sandy and loamy-sand soils in Andhra Pradesh, Tamil Nadu, Gujrat, Rajasthan and Punjab. A comprehensive research on transmission, host-range, serology and molecular characterization of this virus was carried out in India from International Crops Research Institute for the Semi-Arid Tropics (ICRISAT), Hyderabad. The virus is transmitted by sap, seed, and fungus (*Polymyxa graminis*). Besides groundnut, IPCV also infects pigeonpea, cowpea, chili, wheat, barley, sorghum, sugarcane and maize. Virion of IPCV has two rod-shaped particles measuring 249 and 184 nm in length and 24 nm in diameter. ELISA and immunosorbent electron microscopy revealed three serotypes of IPCV, *viz.* IPCV-Hyderabad, IPCV-Durgapura and IPCV-Ludhiana. The genome of IPCV composed of two positive sense ssRNA components, which are encapsidated by a single coat protein. Till now the complete sequence of one RNA-1 and three RNA-2 components of IPCV are available in the sequence database. Sensitive broad spectrum detection of IPCV has been demonstrated using non-radioactive nucleic acid probes. Cultural practices like soil solarization, early planting, and trap cropping with pearl millet are some measures for management of the disease.

V. Pandey • B. Mandal • R.K. Jain • A. Roy (✉)
Advanced Centre for Plant Virology, Division of Plant Pathology,
Indian Agricultural Research Institute, New Delhi 110012, India
e-mail: anirbanroy75@yahoo.com

B. Mandal et al. (eds.), *A Century of Plant Virology in India*,
https://doi.org/10.1007/978-981-10-5672-7_16

Keywords
Indian peanut clump virus • Pecluvirus • Fungal transmission • Peanut • Clump disease

16.1 Introduction

A groundnut disease characterized by severe stunting and clumping of plants was first time described in India by Sundararaman in 1927. However, scientific investigation on the disease and its etiology was initiated during 1977–1978 when Dr. D.V.R. Reddy and his team from ICRISAT, Hyderabad, India noticed severe appearance of the disease in Punjab (Reddy et al. 1979). The disease appeared in patches in the field where crops were mainly grown in sandy and sandy-loam soils. The disease was also observed in groundnut plants where crops were raised in sandy soils in Andhra Pradesh, Gujarat, Tamil Nadu and Rajasthan. A similar disease was reported from Africa (Trochain 1931), where a bipartite, rigid rod-shaped soil borne virus, peanut clump virus (PCV), was reported to be responsible for causing the disease (Thouvenel et al. 1976). However, the serological study of the virus causing similar clump disease in groundnut in India revealed that it was distinct from that known in Africa and named as Indian peanut clump virus (IPCV) (Reddy et al. 1983). Later, based on the similarities on symptoms, particle morphology, mode of transmission and hybridization study indicated IPCV serotypes could be strains of PCV (Reddy et al. 1985). However, when sequence information was available for both PCV and IPCV, it was realized that there were some distinct genomic differences of these viruses from the other fungal transmitted furoviruses. Hence, a separate genus, *Pecluvirus,* was created in the family *Virgaviridae* with type species *Peanut clump virus* (https://talk.ictvonline.org/ictv-reports/ictv_9th_report/). PCV and IPCV are two distinct species under the genus *Pecluvirus* and they differ among themselves in spatial distribution, host-range, symptomatology, serology and nucleotide homology. Information on IPCV generated through research in India has been reviewed and summarized in following sections.

16.2 Virus Morphology

Nolt et al. (1988) observed virus particle in electron microscope as typically bipartite non-enveloped rigid rod with two predominant sizes of 190 × 24 nm and 245 × 24 nm. Both the genomic components are encapsidated by a single type of coat protein. Rao et al (2002) observed presence of ridgid rod shaped paticles measuring 250 × 20 nm from *Chenopodium amaranticocor* showing chlorotic lession after inoculation of virus from sugarcane leaves showing red leaf mottle symptom.

16.3 Host Range

IPCV is known to cause disease in several commercial crops like groundnut, pigeonpea, cowpea, chili, wheat, barley, sorghum, sugarcane and maize. Besides, IPCV has experimental host plants like *Nicotina benthamina*, *N. glutinosa*, *Phaseolus vulgaris* cv. Topcrop, *Vigna unguiculata* cv. Pusa Komal and *Chenopodium amaranticolor*. Monocotyledonous weeds such as *Cynodon dactylon* and *Cyperus rotundus* were reported as source of primary inoculums when main crops are not present in the field (Nolt et al. 1988).

16.4 Symptom Variation in Different Hosts

In case of groundnut, major symptom includes stunting, clumping and smalling of leaves, dark green leaf colour and small size of infected pods (Fig. 16.1). In case of wheat, plants became rossatte like with malformed spikes, which often did not emerge from the flag leaf and they contained few shrivelled seeds. In barley, severe stunting and leaf chlorosis was caused by IPCV. With the aging of the plants, the leaves became necrotic, only few infected plants reached the maturity and produced small spikes. Stunting was the prominent symptom in pearlmillet. In sugarcane, Rao et al. (2002) reported that a PCV isolate was associated with the red leaf mottle disease based on serology and electron microscopy. However, as PCV was reported to be present only in Africa, it is assumed that in India, the red leaf mottle of sugarcane may be associated with IPCV. In sugarcane, along with slight stunting, various degrees of mottling, which later turned into red streaks of varying intensities was observed upon infection of the virus (Rao et al. 2002). Necrotic local lesions were produced in *Chenopodium amranticolor* while chlorotic local lesions were produced in cowpea cv. Pusa Komal. Systemic chlorotic ring spot and leaf

(a) (b)

Fig. 16.1 Groundnut plants showing clump disease in field at Bikaner District, Rajasthan during 2014. (**a**) High incidence of the disease in field, (**b**) comparison of symptomatic (small and bushy) and asymptomatic plants

deformation developed in *Nicotiana benthamiana* infected by IPCV. Systemic mosaic was the main symptom in *N. glutinosa.* Veinal chlorosis and veinal necrosis developed in French bean.

16.5 Impact of the Disease

Peanut clump disease causes significant losses in groundnut crops in West Africa and in the Indian subcontinent (Reddy et al. 1999). Globally, annual losses in peanut due to the disease have been estimated to exceed US$ 38 million (Reddy et al. 1999). Total yield loss was reported when IPCV infection in groundnut occur in an early crop growth stage. Reduction of grain yield occur upto 60% in case of late infection. Besides groundnut, wheat and barley crops were also shown to be naturally infected by IPCV under field conditions. It has been shown that in case of wheat infected with Hyderabad isolate of IPCV, vegetative biomass was reduced up to 33%, which resulted into 36% grain loss (Delfosse et al. 1999). Early infection of Hyderabad isolate of IPCV (IPCV-H) in wheat resulted in severe stunting with dark green appearance of plants and chlorotic streaks on the youngest leaves, which turned necrotic as the plants aged. If wheat plants were infected in later stage of crop growth, their maturity is delayed and such infected plants can readily be identified in the fields because of their dark green appearance. The yield reduction in IPCV-H infected wheat was as high as 58% (equivalent to a yield loss of 1800 kg/ha) (Delfosse et al. 1999).

16.6 Transmission and Perpetuation of the Virus

IPCV was shown to be transmitted by sap, seed, and fungus. Soil-borne fungal pathogen, *Polymyxa graminis*, transmits the virus to dicotyledonous crops, however, in the roots of dicotyledonous plants it does not extensively colonize, which is evident from either absence or presence of less number of sporosori (resting spore clusters or cystosori) in their roots (Ratna et al. 1991; Delfosse et al. 1996; Legrève et al. 1999; Legrève et al. 2000). It has been demonstrated that roots of naturally virus-infected groundnut plants did not contribute to spread of the disease when inoculated into sterile sand, where as roots of sorghum and pearl millet plants infected with IPCV could induce the disease and serve as source of inoculum (Thouvenel et al. 1988; Ratna et al. 1991; Delfosse et al. 1996). This indicates that IPCV can be transmitted to dicotyledonous plants through fungal vector by a matter of chance but roots of such infected plants can not contribute as a source of inoculum for further fungal transmission. Hence, dicotyledonous plants served as 'fortuitous' hosts for IPCV and they do not contribute to building up of inoculum in soil. In contrast, maize, pearl millet, and sorghum are the monocotyledonous hosts which are 'preferred' hosts for *P. graminis* as the fungus can multiply in them and produce more number of sporosori in their roots (Ratna et al. 1991; Delfosse et al. 1996; Legrève 1999). As IPCV also able to multiply in these monocotyledonous hosts, they are supposed to play an important role in the perpetuation and spread of virus inoculum under field conditions (Reddy et al. 1999). However, *P. graminis* was rarely observed in roots of wheat plant and was not detected in barley, thus their role in virus perpetuation has not yet been established.

Besides fungal transmission, IPCV could also be transmitted through seeds from the infected plants. In case of groundnut, under natural field condition, seed transmission varied from 3.5 to 17%, depending on the genotype. However, when seeds from infected groundnut plants were used, the transmission efficiency was increased to 48–55%. In case of monocotyledonous plants, seed transmission was demonstrated with IPCV-H isolate, which was shown to be transmitted by seed in finger millet (*Eleusine coracana*), foxtail millet (*Setaria italica*), and pearl millet (*Pennisetum glaucum*) with 5.2%, 9.7%, and 0.9% efficiency, respectively. In case of wheat, seed transmission efficiency was 0·5–1·3% (Delfosse et al. 1999). No seed transmission was observed in sorghum (*Sorghum bicolor*) (Reddy et al. 1998).

16.7 Epidemiology

High rainfall, allows the free movement of motile zoospore of *P. graminis* and thus increases the transmission of the virus, which results in higher incidences of the disease. A weekly rainfall of 14 mm is sufficient for *P. graminis* to initiate infection. It has been shown that groundnut crops either grown during the post-rainy season or crops grown in the rainy season but sown well beyond the onset of monsoon rains mostly escaped the disease (Reddy et al. 1988). Temperature is another important factor for IPCV infection. Temperature ranged from 23 to 30 °C, which is prevailing during the rainy season were found to be conducive to natural virus transmission (Delfosse et al. 2002). However, under mechanical inoculation condition, IPCV causes infection on wheat at a wider temperatures between 15 and 30 °C (Reddy et al. 1988), indicating the wider temperature adaptation of this virus.

16.8 Serological Properties

Different IPCV isolates were collected and grouped according to the serological properties. The name of IPCV isolates given based on the places from where they were first reported in India. Antiserum was prepared and reactivity was tested through ELISA and ISEM. Based on the serological properties, three distinct serotypes of IPCV were identified, IPCV-D (Durgapura isolate, Rajasthan), IPCV-H (Hyderabad isolate, Andhra Pradesh) and IPCV-L (Ludhiana isolate, Punjab) (Reddy et al. 1983; Nolt et al. 1988).

16.9 Molecular Properties

16.9.1 Genome Size and Sequence

IPCV genome is bipartite positive sense single stranded RNA, which constitutes 4 % of virion mass by weight. Only one RNA-1 genome sequence of Hyderabad isolate of IPCV is available (Miller et al. 1996, accession no. X99149), while three sequences of RNA-2 of IPCV are available in sequence database (Naidu et al. 2000; accession nos. AF239729, AF447396, AF447397). The RNA-1 is 5841 nucleotides and RNA-2 is 4290–4507 nucleotides long.

16.9.2 Genome Organization

RNA-1 sequence of Hyderabad isolate of IPCV have M_7G capping at 5′ end followed by untranslated region (UTR) of 132 nucleotides and code for three proteins (P131, P191, and P15). The protein products of P131 and P191 are involved in replication of the viral genome. P131 is coded by nucleotide sequence from 133 to 3525 and acts as methyl transferase, whereas P191 encoded by translational read-through of UGA stop codon of P131 protein from the nucleotides 133 to 5103, functions as helicase as well as RNA-dependent RNA polymerase. Protein P15 is encoded by a sub-genomic RNA representing nucleotide coordinate 5168 to 5542 of RNA-1 and it appears to be a suppressor of post-transcriptional gene silencing. After P15 ORF there is an untranslated region of 299 nucleotides, which ends with 3′ –OH group.

RNA-2 of IPCV starts with M_7G capping at 5′ end and ends with 3′ –OH group. RNA-2 of IPCV codes for five proteins related with capsid formation, vector transmission and movement of the virus within the host plant. RNA-2 have UTR of variable lengths; 388 nucleotides in Ludhiana isolate (L), 394 nucleotides in Durgapura isolate (D) and 502 nucleotides in Hyderabad isolate (H). The first ORF after the 5′ UTR codes for approximately 24 K Dalton coat protein. The ORF of coat protein is 600 nucleotides in L isolate, 663 nucleotides in D isolate and 627 nucleotides in case of H isolate. The next ORF codes for a vector transmission factor (P40/P39/P38) and translated through the sub-genomic RNA of size 1065 nucleotides (991-1065) in L isolate, 1020 nucleotides (1129-2148) in H isolate and 909 nucleotides (1058-1966) nucleotides in Durgapura isolate. Next three ORFs constitute the triple gene block, which involved in movement of IPCV. After triple gene block, there is a 3′ UTR of 280 nucleotides in L isolate 290 nucleotides in D isolate and 275 nucleotides in H isolate.

16.9.3 Sequence Comparison and Phylogeny

(a) Untranslated Region

5′ and 3′ UTR of RNA-1 are 132 and 299 nucleotides in length, respectively, whereas the RNA-2 has more variable UTR of 388 to 502 nucleotides at 5′ end and approximately 300 nucleotide at 3′ end. Within the 3′ UTR of RNA-2, approximately 100 terminal nucleotides are conserved in all the three isolate, which was used for the development of probe for detection of IPCV (Wesley et al. 1996).

(b) Coat Protein

Coat protein is coded by RNA-2. Nucleotide identity matrix of coat protein sequence retrieved from NCBI of all the isolates of IPCV, PCV and one species of tobacco rattle virus revealed that IPCV isolates are 52–59% similar among themselves, whereas they show 40–61% similarity with the PCV. Whereas, with TRV they show similarity up to 34–35% (Wesley et al. 1994).

(c) Complete RNA-2 Based Phylogeny

In most of the members of the family *Virgaviridae*, coat protein is used as conserved protein for phylogenetic analysis. But in case of IPCV, coat protein sequence is highly diverse and hence complete sequence of RNA-2 is used for the phylogenetic analysis. The Hyderabad and Durgapura isolates of IPCV are more similar than the Ludhiana isolate.

16.10 Management of the Virus

The efforts to develop management practices against IPCV were focused on identifying the resistant sources, and manupulating the cultural practices. Nine thousand *Arachis* germplasm lines were tested to find out resistance against IPCV, but none found resistant (Delfosse 2000). Application of soil biocides and soil solarization to control the disease although found effective, but they were either hazardous or uneconomical (Dhery et al. 1975; Reddy et al. 1999). It has been shown that if sowing of the groundnut crop was done before the onset of monsoon rains and the crop is raised with judicious irrigation, then it can effectively reduce the disease incidence (Delfosse 2000). Use of pearl millet as a trap crop was found successful at different sites in India (Delfosse et al. 1997). As *P. graminis* multiplies intensively in monocots, crop rotation with monocotyledonous plants should be avoided to reduce population of the fungal vector in the soil.

16.11 Concluding Remarks

Pecluviruses were discovered in peanut and so far the genus Pecluvirus contains only two virus species, IPCV and PCV. The geographical distribution of IPCV and PCV is restricted only in India and West Africa, respectively. IPCV is a soil borne and seed-transmitted virus with a broad host range, however, its spread and survival depend on critical interactions of the fungus vector, *P. graminis* with the type of host plant species. It is hypothesized that IPCV possibly coevolved with wild grasses and cereal crops such as millets and sorghum in the tropical and subtropical areas. In monocotyledonous plants, IPCV causes little damage or do not show any symptom but these crops support multiplication of *Polymyxa graminis* and thus may promote virus accumulation. When IPCV is transmitted to the susceptible dicotyledonous crop such as peanut, it causes economically important disease. However, as *P. graminis* does not reproduce in peanut, the IPCV inoculum in the soil is expected to reduce in the peanut field. IPCV and PCV are quarantine viruses to many countries. Rate of seed transmission of IPCV is higher than any of the seed transmitted viruses infecting peanut. IPCV is also transmitted through seeds of millets, wheat, and maize. Therefore, the seeds of these crops specially from the infected regions pose a risk of introduction of IPCV in the new geographical regions. IPCV has unique genome organization that the CP gene is located at the 5' end of RNA2 genome, whereas, in the other members of the family *Virgaviridae*, CP is located at the 3′ end of the genome. A comparison of CP gene among the three RNA2 sequence of IPCV indicated high degree of variability in this ORF, which is contrasting with other members of the family *Virgaviridae*, as CP gene sequence is highly stable within a species. Due to the availability of limited sequence information on IPCV isolates, it is difficult to comprehend the diversity within IPCV, however, it is possible that there could be the existence of new species within the genus *Pecluvirus* in India. The infectivity of the cloned DNAs of IPCV has not yet demonstrated. IPCV having hosts in both mono- and dicotyledonous plants and bipartite genome organization, is attractive for developing virus based vector for inducing gene silencing in plants.

References

Delfosse P (2000) Epidemiology and management of the Indian peanut clump virus. Université catholique de Louvain, Louvain-la-Neuve, PhD thesis

Delfosse P, Devi PS, Reddy AS, Risopoulos J, Doucet D, Legrève A, Maraite H, Reddy DVR (1996) Epidemiology of Indian peanut clump virus transmitted by *Polymyxa* sp. In: Sherwood JL, Rush CM (eds) Proceedings of the third symposium of the international working group on plant viruses with fungal vectors, West Park Conference Centre, Dundee, Scotland, 1996. American Society of Sugar Beet Technologists, Denver, pp 141–144

Delfosse P, Legrève A, Devi PS, Reddy AS, Risopoulos J, Doucet D, Vanpee B, Maraite H, Reddy DVR (1997) Epidemiology of peanut clump virus disease. In: Reddy DVR, Delfosse P, Lenné JM, Subrahmanyam P (eds) Groundnut virus diseases in Africa: summary and recommendations of the sixth meeting of the international working group, 1996. Agricultural Research Council, Plant Protection Research Institute, Pretoria, pp 26–28

Delfosse P, Reddya AS, Legrève A, Devia PS, Thirumala Devia K, Maraiteb H, Reddya DVR (1999) Indian peanut clump virus (IPCV) infection on wheat and barley: symptoms, yield loss and transmission through seed. Plant Pathol 48:273–282

Delfosse P, Reddy AS, Thirumala KD, Legrève A, Risopoulos J, Doucet D, Shoba PD, Maraite H, Reddy DVR (2002) Dynamics of Polymyxa graminis and Indian peanut clump virus (IPCV) infection on various monocotyledonous crops and groundnut during the rainy season. Plant Pathol 51(5):546–560

Dhery M, Germani G, Giard A (1975) Résultats de traitements nématicides contre la chlorose et le rabougrissement en Haute-Volta. Cah ORSTOM Sèrie Biol 3:161–167

Legrève A, Vanpee B, Delfosse P, Maraite H (1999) High temperature during storage favours infection potential of resting spores of *Polymyxa graminis* of Indian origin. Ann Appl Biol 134:163–169

Legrève A, Vanpee B, Delfosse P, Maraite H (2000) Host range of tropical and subtropical isolates of *Polymyxa graminis*. Eur J Plant Pathol 106:379–389

Miller JS, Wesley SV, Naidu RA, Reddy DVR, Mayo MA (1996) The nucleotide sequence of RNA-1 of Indian peanut clump furovius. Arch Virol 141(12):2301–2312

Naidu RA, Miller JS, Mayo MA, Wesley SV, Reddy AS (2000) The nucleotide sequence of Indian peanut clump virus RNA 2: sequence comparisons among pecluviruses. Arch Virol 145(9):1857–1866

Nolt BL, Rajeshwery R, Reddy DVR, Bharathan N, Manoher SK (1988) Indian peanut clump virus isolates: host range, symptomatology, serological relationships, and some physical properties. Phytopathology 78:310–313

Rao GP, Singh M, Gaur RK (2002) Occurrence of red leaf mottle on sugarcane in India. Sugar Cane Int 2:18–21

Ratna AS, Rao AS, Reddy AS, Nolt BL, Reddy DVR, Vijayalakshmi M, McDonald D (1991) Studies on transmission of Indian peanut clump virus disease by *Polymyxa graminis*. Ann Appl Biol 118:71–78

Reddy DVR, Iizuka N, Subrahmanyam P, Rajeshwari R, Mcdonald D (1979) A soil-borne disease of peanuts in India. Proc Am Peanut Res Educ Soc 11:49

Reddy DVR, Rajeshwari R, Iizuka W, Lesemannn DE, Nolt BL, Goto T (1983) The occurrence of Indian peanut clump virus, a soil-borne virus disease of groundnuts (*Arachis hypogaea*) in India. Ann Appl Biol 102:305–310

Reddy DVR, Robinson DJ, Roberts IM, Harrison BD (1985) Genomic properties of Indian peanut clump virus. J Gen Virol 66:2011–2016

Reddy DVR, Nolt BL, Hobbs HA, Reddy AS, Rajeshwari R, Rao AS, Reddy DDR, McDonald D (1988) Clump virus in India: isolates, host range, transmission and management. In: Cooper JL, Asher MJC (eds) Developments in applied biology II. Virus with fungal vectors. Association of Applied Biologists, Wellesbourne, pp 239–246

Reddy AS, Hobbs HA, Delfosse P, Murthy AK, Reddy DVR (1998) Seed transmission of Indian peanut clump virus (IPCV) in peanut and millets. Plant Dis 82:343–346

Reddy DVR, Mayo MA, Delfosse P (1999) Pecluviruses. In: Webster R, Gramoff A (eds) Encyclopedia of virology, 2nd edn. Academic, New York

Sundararaman S (1927) A clump disease of groundnuts. Madras Agricultural Department Year Book, Chennai, pp 13–14

Thouvenel JC, Dollet M, Fauquet C (1976) Some properties of peanut clump, a newly discovered virus. Ann Appl Biol 84:311–320

Thouvenel JC, Fauquet C, Fargette D, Fishpool EDC (1988) Peanut clump virus in West Africa. In: Cooper JI, Asher MJC (eds) Developments in applied biology II. Virus with fungal vectors. Association of Applied Biologists, Wellesbourne, pp 247–254

Trochain J (1931) La "lepre de I'Arachide". Rerue Bot Agric Trop 11:330–334

Wesley SV, Mayo MA, Jolly CA, Naidu RA, Reddy DVR, Jana MK, Parnaik VK (1994) The coat protein of Indian peanut clump virus: relationships with other furoviruses and with barley stripe mosaic virus. Arch Virol 134:271–278

Wesley SV, Miller JS, Devi PS, Delfosse P, Naidu RA, Mayo MA, Reddy DVR, Jana MK (1996) Sensitive broad spectrum detection of Indian peanut clump virus by non-radioactive nucleic acid probes. Phytopathology 86:1234–1237

Potyviruses Infecting Crop Plants in India

17

R. Viswanathan, B. Parameswari, K. Nithya, A. Jeevalatha, and Govind Pratap Rao

Abstract

In India, potyvirus infection is common in Solanaceae, Cucurbitaceae, Poaceae and Leguminosae families. Economically important crops like potato, tomato, papaya, banana, sugarcane and pepper are affected by potyviruses as one of the major production constraints. There are as many as 39 different potyviruses recorded in India of which the virus species *Potato virus Y*, *Papaya ringspot virus, Sugarcane mosaic virus* and *Banana bract mosaic virus* are economically more significant. An up-to-date information on the occurrence, biology, serology and molecular biology of the potyviruses that has been investigated in India is presented in this chapter.

Keywords

Potyviridae • *Potyvirus* • Aphids • *Potato virus Y* • *Papaya ringspot virus* • *Sugarcane mosaic virus*

R. Viswanathan (✉) • K. Nithya
ICAR-Sugarcane Breeding Institute, Coimbatore 641007, India
e-mail: rasaviswanathan@yahoo.co.in

B. Parameswari
ICAR-Sugarcane Breeding Institute Regional Centre, Karnal 132001, India

A. Jeevalatha
ICAR-Central Potato Research Institute, Shimla 171001, India

G.P. Rao
Advanced Centre for Plant Virology, Division of Plant Pathology, Indian Agricultural Research Institute, New Delhi 110012, India

B. Mandal et al. (eds.), *A Century of Plant Virology in India*,
https://doi.org/10.1007/978-981-10-5672-7_17

17.1 Introduction

Plant viral diseases are jeopardising Indian Agriculture in terms of crop production and productivity. The family *Potyviridae* (genus *Potyvirus*) is considered as one of the most important taxonomic groups affecting a wide range of monocotyledons/ dicotyledons species causing mild to severe disease symptoms (Fauquet et al. 2005; Shukla et al. 1994). The members of the genus *Potyvirus* produce a variety of symptoms, including mottling (chilli), mosaic (sugarcane), necrotic or chlorotic lesions (common bean), flower breaking (lily), ring spots, vein clearing (papaya), vein banding (pepper), necrosis (vanilla), stripe stunting (shallot) and wilting (lettuce). The severity of the induced symptoms varies based on the particular host genotype, specific virus strain, associated vectors and environmental conditions (Fauquet et al. 2005). The members of *Potyvirus* have filamentous flexuous particles of 680–900 nm size (helix pitch 3.4–3.5 nm and 11–13 nm diameter), and 10 kb genome encapsidated by the multiple copies of a single coat protein (30–47 kDa size) (Shukla et al. 1994). Potyviruses are transmitted by aphids in a non-circulative, non-persistent and stylet borne manner using a HC-Pro, which facilitates binding of virus particles to the maxillary stylet of aphids (Blanc et al. 1997). The genome of potyviruses is single stranded positive-sense RNA and contains one ORF which is translated as a large polyprotein (340–368 kDa), that is cleaved into ten functional proteins viz., P1, helper component, P3, 6K1, cylindrical inclusion, 6K2, viral genomic protein, nuclear inclusion protein a, nuclear inclusion protein b and capsid protein (Riechmann et al. 1992).

The major potyviral diseases in India are mosaic disease of potato, banana and sugarcane, and ring spot disease of papaya. Potato virus Y may cause losses up to 50% in tuber yield of potato and ring spot disease caused by papaya ringspot virus could affect over 90% of the grown up plants by reducing latex and sugar contents in papaya. Mosaic disease of sugarcane caused by sugarcane mosaic virus causes 10–15% yield loss in susceptible varieties and banana bract mosaic virus causes a loss of Rs. 38.7 crore per annum in banana cultivar Nendran alone. Many reviews are available on potyvirus taxonomy, host range, transmission, infection, genome organization, potyviral proteins and their molecular and cellular mechanism of infection and management strategies (Riechmann et al. 1992; Shukla et al. 1994; Urcuqui-Inchima et al. 2001; Gibbs and Ohshima 2010). Although, a significant amount of information has been generated since long time on the potyviruses occurring in India, a consolidated information is not available so far in Indian context. The present review focuses on the information regarding the biological, serological and molecular properties of important potyviruses reported from India.

17.2 History of Potyviruses Research in India

In India, sugarcane mosaic disease was first reported by Barber in the year 1921, subsequently based on biological and serological characterization, this virus was reported as sugarcane mosaic virus (Bhargava 1975; Kondaiah and Nayudu 1985; Rishi and Rishi 1985). Mosaic disease of papaya was reported by Capoor and Varma in 1958 from Bombay, subsequently confirmed as papaya ringspot virus as causal agent by biological, serological and molecular methods (Khurana 1968; Khurana and Bhargava 1970; Hussain and Varma 1994; Jain et al. 1998a; Roy et al. 1999). Detection of potato virus Y with mosaic and mottling symptoms on potato was first reported from Central Potato Research Institute (CPRI), Shimla in 1960 through biological indexing for PVY on "A6" leaflets (Lal and Khurana 1983). Antiserum sensitized latex was used in indexing against PVY since 1978 onwards. In India, Banana bract mosaic disease was first reported in French plantain cv. Nendran as a Kokkan disease in Kerala (Samraj et al. 1966), which later authentically confirmed as BBrMV (Singh et al. 2000). Radish mosaic caused by turnip mosaic virus was reported from IARI, New Delhi (Ahlawat and Chenulu 1984). Bean yellow mosaic virus was first reported on chickpea from International Crops Research Institute for the Semiarid Tropics (ICRISAT), Hyderabad (Chalam 1982). Sweet potato feathery mottle virus was first reported from sweet potato germplasm material showing chlorotic spot, ring spot and mild mottle symptoms and confirmed by serological methods viz. immunosorbent electron microscopy and direct antigen coated indirect enzyme linked immunosorbent assay at Indian Agricultural Research Institute, New Delhi (Kumar et al. 1991).

17.3 Potyviruses in India

According to potyvirus sequences published in NCBI database, a total of around 39 potyviruses are reported from India. Potyvirus infection is highly recorded in Solanaceae, Cucurbitaceae, Pocaeae and Leguminosae families (www.ncbi.nlm.nih.gov/2016) and the major infected crops are: tomato, potato, brinjal, chickpea, cowpea, pepper, chilli, cardamom, papaya, banana, etc.), ornamentals tuberose, gladiolus, jasmine, sorghum, sugarcane, and some major oilseeds and few forage crops. Papaya ringspot virus, chilli vein mottle virus, banana bract mosaic virus, bean yellow mosaic virus, bean common mosaic virus, sugarcane mosaic virus, potato virus Y, zucchini yellow mosaic virus, dasheen mosaic virus and onion yellow dwarf virus are the important potyviruses reported from different crops followed by potyviruses of minor importance viz., turnip mosaic virus, Ornithogalum mosaic virus, watermelon mosaic virus, sweet potato feathery mottle virus and soybean mosaic virus (Table 17.1).

Table 17.1 The partial gene sequences resources of the different potyviruses reported from India during 1960 to 2017

S. No.	Potyvirus	Host plants	No. of GenBank accession from India
1.	Banana bract mosaic virus (BBrMV)	Banana, Cardamom	71
2.	Bean common mosaic virus (BCMV)	French Bean, Soya bean, cluster bean, Black gram, Cow pea, lablab bean, Tomato	41
3.	Bean yellow mosaic virus (BYMV)	Broad bean, French Bean, Pea, Chilli pepper, Gladiolus, Pepper, Chinese ginger, Soybean, Betel vine, Cape gooseberry, Freesia spp, Iris, Hibiscus rosa-sinensis, Orchid , Canna sp.	59
4.	Carnation vein mottle virus (CVMoV)	Carnation	1
5.	Chilli veinal mottle virus (ChiVMV)	Chilli, Pepper, Brinjal, Chrysanthemum, *Ageratum houstonianum*, Himalayan butterfly bush, *Taraxacum sp*, *Solanum nigrum*, *Physalis floridana*	63
6.	Colombian datura virus (CDV)	Datura	1
7.	Cowpea aphid-borne mosaic virus (CABMV)	Cowpea	2
8.	Cymbidium mosaic virus (CymMV)	Orchid, *Cymbidium elegans*	16
9.	Dasheen mosaic virus (DMV)	Taro, Elephant Foot Yam, Arrowleaf elephant ear	19
10.	Freesia mosaic virus (FreMV)	*Freesia* spp	1
11.	Hippeastrum mosaic virus (HiMV)	*Hippeastrum*	1
12.	Iris mild mosaic virus (IMMV)	Iris	1
13.	Iris severe mosaic virus (ISMV)	Iris	1
14.	Konjac mosaic virus (KoMV)	Colocasia, *Amorphophallus, Caladium*	7
15.	Leek yellow stripe virus (LYSV)	Garlic, *Allium sativum*	6
16.	Lettuce mosaic virus (LMV)	Lettuce, Spinach	7
17.	Lily mottle virus (LMoV)	Lily	4
18.	Narcissus degeneration virus (NDV)	*Narcissus* sp	1

(continued)

Table 17.1 (continued)

S. No.	Potyvirus	Host plants	No. of GenBank accession from India
19.	Narcissus yellow stripe virus (NYSV)	*Narcissus sp*	2
20.	Nerine yellow stripe virus (NeYSV)	*Crinum asiaticum*	6
21.	Onion yellow dwarf virus (OYDV)	Onion, Garlic	9
22.	Ornithogalum mosaic virus (OrMV)	Gladiolus, Vanilla, Iris	10
23.	Papaya ringspot virus (PRSV)	Papaya, Pumpkin, Cucumber, Ash gourd, Watermelon, Sponge gourd, Bottle gourd	136
24.	Pepper mottle virus (PepMoV)		10
25.	Pepper veinal mottle virus (PVMV)	Chilli pepper	1
26.	Plum pox virus (PPV)	Apricot	3
27.	Potato virus A (PVA)	Potato	2
28.	Potato virus Y (PVY)	Potato, Tomato, Chilli, Brinjal, *Physalis floridana*, *Spiraea sorbifolia*	26
29.	Soybean mosaic virus (SMV)	Soybean, Gladiolus	5
30.	Sugarcane mosaic virus (SCMV)	Sugarcane, sorghum, sudan grass	38
31.	Sweet potato feathery mottle virus (SPFMV)	Sweet Potato	8
32.	Tuberose mild mottle virus (TuMMoV)	Tuberose	2
33.	Tulip breaking virus (TBV)	Asiatic hybrid lily	1
34.	Turnip mosaic virus (TMV)	Radish, Turnip, Mustard, Broccoli	14
35.	Vallota mosaic virus (ValMV)	Narcissus	4
36.	Vanilla distortion mosaic virus (VDMV)	Vanilla, Stevia, Cumin, Zinnia	8
37.	Watermelon mosaic virus (WMV)	Watermelon	3
38.	Yam mild mosaic virus (YMMV)	Water Yam	5
39.	Zucchini yellow mosaic virus (ZYMV)	Gherkin, Pumpkin, Cucumber, Snake gourd, Bottle gourd, Bitter gourd, Musk melon, Amaranthus	20

17.4 Description of Major Potyviruses in India

17.4.1 Potato virus Y (PVY)

PVY is one of the most economically important viruses of potato worldwide including India and causes up to 80% yield losses (Khurana and Garg 2000). Five strains of PVY infecting potatoes in Shimla hills were reported by Khurana et al. (1975). The severe strains of *PVY* may reduce yield up to 60–75% (Nagaich et al. 1974). The host range of PVY is very wide, comprising many crop species in the family Solanaceae. It infects economically important crops such as potato, tobacco, tomato, pepper, brinjal and ornamental plants. A necrotic strain of PVY from tomato plants was reported by Behl et al. (1987). A strain of *PVY* causing mosaic mottling in brinjal was identified based on N-terminal serology and coat protein sequence relationship studies (Bhat et al. 1999). The virus is readily transmitted by aphids in a non-persistent manner. There are two main serotypes: the O serotypes, which consist of PVY^{O}, PVY^{C} and PVY^{NW} and the N serotype that includes PVY^{N} and PVY^{NTN}. The strains of PVY have different stability under dilution end point (DEP), thermal inactivation point (TIP) and longevity in vitro (LIV) (Hussain et al. 2016). The potato PVY isolates are classified into different strain groups based on their symptoms (Shukla et al. 1994; Singh et al. 2008). Initially, they were divided into three strains: PVY^{O}, PVY^{N} and PVY^{C} and later two more strain groups (PVY^{Z} and PVY^{E}) were added (Singh et al. 2008). PVY^{O} and PVY^{N} are the most frequent ones in potatoes whereas PVY^{C} is not common in potato fields and has less economic importance (Kerlan et al. 1999). PVY^{Z} and PVY^{E} have also been reported only once and of less significance to potato production (Singh et al. 2008). Bhat et al. (1997) used N-terminal serology and HPLC peptide profiling to differentiate different strains of PVY. In addition to these five strains, recombination led to the evolution of new strains of PVY such as PVY^{NTN}, PVY$^{N\text{-}Wi}$ and PVY$^{N:O}$, PVY$^{NA\text{-}N}$ and PVY-NE11 (Karasev and Gray 2013). These recombinant strains are derived from two primary potato-infecting PVY strains, viz. PVY^{O} and PVY^{N}.

Symptoms vary with the virus strain and the host plant. The PVY^{O} induces mild to severe mosaic and leaf drop streaks symptoms in potato, and systemic mottling in tobacco. The PVY^{N} strain induces mild mottling in most potato cultivars with occasional necrosis (Kerlan et al. 1999), but induces severe systemic leaf vein and petiole necrosis in tobacco. PVY^{C} isolates induce stipple streak symptoms on some potato cultivars. However, PVY^{O} elicits hypersensitive response in potato genotypes carrying *Ny* gene. PVY^{C} in potato genotype carrying *Nc* gene and strains that could overcome both *Ny* and *Nc* genes, did not elicit hypersensitive response towards these two genes were named PVY^{N} strains. PVY^{Z} strains elicit hypersensitive response to *Nz* gene and PVY^{E} strains do not elicit hypersensitive response to *Ny*, *Nc* or *Nz* and do not cause necrosis in tobacco (Kerlan et al. 1999). The recombinant isolates are being reported day by day in many countries indicating the occurrence of recombination, evolution and spread of these new strains. The spread of necrotic PVY strains in wider areas has created serious challenges for seed potato certification programs. PVY^{N}, a tuber necrotic strain originated in South

America and appeared later in Europe in the 1960s (Inoue-Nagata et al. 2001). The tuber necrotic strain, PVY^{N} and PVY^{NTN}, sometimes cause a more mild mosaic symptom on leaves than PVY^{O}, but cause tuber necrosis in some varieties, referred to as potato tuber necrotic ringspot disease. Khurana et al. (1975) reported five isolates based on their varying symptoms on different plant species but none caused systemic veinal necrosis on tobacco var. "Samsun". Later, Sarjeet Singh et al. (1983) reported the presence of PVY^{C} in India based on the reactions on biological indicator. Mukherjee et al. (2004) also confirmed the possible absence of $PVY^{N/NTN}$ strains in India using RT-PCR and reported PVY^{O} strain is the most prevalent one in India. However, recent studies with RT-PCR assay using strain specific primers and ELISA using specific monoclonal antibodies indicates the possible occurrence of PVY^{N} and PVY^{NTN} strains either in single or mixed infection in India in addition to PVY^{O} and PVY^{C} strains. However it needs to be confirmed through biological assays using indicator hosts (unpublished). Several effective protocols have been suggested for purification of *PVY* from various plant sources and production of polyclonal antiserum (Khurana et al. 1987a, b, 1990).

The diagnosis of *PVY* isolates is complex because of existence of several recognised isolates and the incidence of mixed infection. PVY strains cause variety of symptoms and serological diagnosis has limitations of differentiating N from NTN strains and O from N:O strains. Tobacco bioassay remains the best method to identify necrotic isolates (PVY^{N} and PVY^{NTN}), but again limitations in large scale application. For the diagnosis of PVY, ELISA and immunosorbant electron microscopy was used at CPRI, Shimla since 1984 for identification of PVY and nucleus seed production (Khurana 1992; Khurana et al. 1993; Singh and Paul Khurana 1999; Garg and Paul Khurana 1992, 1993, 2001). RT-PCR protocols were used to detect PVY either alone or along with other potato viruses (Ghosh and Bapat 2006; Raigond et al. 2013; Kapoor et al. 2014). The cost of RT-PCR assays is prohibitive for use in seed potato certification and not all the strains are detected by these assays (Nie and Singh 2002). A combination of bioassay, serological tests and RT-PCR assay is required to identify strains and it is difficult diagnose the PVY strains under the seed production system and certification programmes. However, recent reports showed simultaneous detection of all nine PVY strains (PVY^{O}, PVY^{E}, PVY^{N}, PVY^{NTN}, $PVY^{NA\text{-}N}$, PVY^{Z}, $PVY\text{-}^{NE11}$, $PVY^{N:O}$, $PVY^{N\text{-}Wi}$). The coat protein gene of PVY^{o} strain was 801 nucleotides long, coding for 267 amino acids (AY061994) and shared 93.2 and 94.8% nucleic acid and amino acid sequence homology, respectively with an American isolate (Ghosh et al. 2002). The 5′ untranslated region (UTR) and P1 region of the PVY-ordinary Indian strain (PVY^{O}) was determined by Mukherjee et al. (2004) which shared the highest sequence similarity with the PVY^{O} strains of European origin. The use of coat protein gene-mediated resistance in tobacco to an PVY-Indian strain was demonstrated where the resistance level is not correlated to expression level of the PVY CP transgene in tobacco. The same strategy may be applied to impart resistance in potato cultivars (Ghosh et al. 2002).

The complete genome sequence information has been generated only for one isolate PVY-Del-66, which was isolated from a potato plant at IARI experimental farm showing mild mottle symptoms (GenBank accession number JN034046). The Del-66 isolate induces systemic mosaic and veinal necrosis on leaves of *N. benthamiana*, *N. glutinosa* and *N. tabacum* cv. x*anthi*. The isolate reacts strongly with the MAb to PVY^{O} but not with the MAb to PVY^{N} and PVY^{C}. As Del-66 showed necrosis in tobacco and strongly reacted with the MAb to PVY^{O}, it was identified as biologically N-type and serologically O- type PVY (N:O type). The complete genome of PVY-Del-66 9674 nucleotide long and contained a single large open reading frame (ORF) coding a polyprotein of 3061 amino acids whose putative cleavage sites potentially yielded all the characteristic potyviral proteins with estimated size of 275, 465, 365, 52, 634, 52, 188, 244, 519 and 267 aa for P1, HC-Pro, P3, 6K1, CI, 6K2, VPg, NIa-Pro, NIb and CP, respectively. Del-66 shared the closest sequences identity of 97.7–99.9% and a close phylogenetic relatedness with the N:O strain (N-Wi variant) reported from USA and Germany. Cloning of entire genome of *PVY* in *E. coli* often difficult due to the presences of cryptic prokaryotic promoter like element in the 5′ terminal region of the genome (CI or P3 gene) which is assumed to express toxic protein in *E. coli.* As a result functional study of PVY genome is greatly limited by the lack of readily available infectious clone. The overlapping-extension PCR (OE-PCR) was first time successfully employed for the rapid demonstration of infectivity of PVY-Del-66 isolate (Jailani et al. 2017).

17.4.2 Papaya Ringspot Virus (PRSV)

PRSV is a major constraint affecting papaya as well as cucurbits production throughout the world including India (Purcifull et al. 1984a; Varma 1988; Jain et al. 2004). Ring spot disease caused by PRSV could affect over 90% of the grown up plants by reducing latex and sugar contents in papaya. *PRSV* is vectored in a non-persistent mode by numerous aphid species (Khurana and Bhargava 1971; Purcifull et al. 1984a). PRSV isolates are mainly classified into two groups, pathotype-W infecting only members of the families Chenopodiaceae and Cucurbitaceae (Fig. 17.1), and the pathotype-P in addition also infects the members of the family Caricaceae (Purciful et al. 1984a) (Fig. 17.2). The Indian PRSV isolates have been classified on the basis of host range and serological properties. The strain infects papaya was Type- P and non-papaya infecting strain was Type-W (Roy et al. 1999) but they cannot be distinguished by their genomic sequences because of very minor genetic diversity (Mangrauthia et al. 2009). However, they can be distinguished by their host specificity to infect papaya (Parameswari 2009).

The genome organization of *PRSV* was comparable with other potyviruses, except that the first protein (P1) cleaved from N terminus of the polyprotein is 63 kDa, which is 18–34 kDa larger to other reported potyviruses (Yeh et al. 1992; Parameswari et al. 2007). It possessed a single stranded translatable RNA molecule ranging from 10,317 to 10,349 nt (including both 5′ and 3′ UTRs), encapsulated by coat protein sub units of 35 KDa each, coding for a polyprotein of 3341–3347 aa from which all the viral proteins are derived. The entire genome is expressed as a

Fig. 17.1 Characteristic symptoms induced by *Papaya ringspot virus* (PRSV) pathotype W on inoculated squash (**a**–**d**) and pumpkin ((**e**–**h**)) plants: Healthy plant (**a** and **e**); Mottling (**b**); Leaf blistering (**c** and **g**); Vein banding (**d**); Vein clearing (**f**) and Filiformy (**h**)

Fig. 17.2 Characteristic symptoms induced by *Papaya ringspot virus* (PRSV) pathotype P on inoculated papaya (cv. Pusa nanha) seedling: Un-inoculated control (**a**); (**b**–**e**) Inoculated seedling showing mottling (**b**); vein clearing (**c**); leaf blistering (**d**); leaf distortion (**e**) and shoe stringing (**f**)

large polyprotein which is subsequently cleaved co-and/or post translationally by proteases to yield functional proteins. Several PRSV isolates from India have been characterized; the complete genomes of 4 isolates and 136 gene sequences (CP, HC-Pro, NIa etc.) from different parts of the India have been sequenced till date (Tables 17.1 and 17.2), which vary in their size and identities. Among the ten viral proteins derived from polyprotein of PRSV, coat protein (CP) has been studied extensively. CP-based genetic diversity studies have been reported from different parts of the world including China (Wei et al. 2007), India (Jain et al. 2004; Srinivasulu and Sai Gopal 2011), and Brazil (Martinez et al. 2014) with 86.7–100%, 83.2–98.6%, and 90–100%, respectively at amino acid level indicating that the

Table 17.2 The complete genome characterisation of potyviruses occurring in India

Virus	Accession No	Crop	Year	Place/isolate	Reference (NCBI submission)
Banana bract mosaic virus (BBrMV)	HM131454	Banana	2010	Tiruchirapalli	Balasubramanian and Selvarajan (2012)
Bean common mosaic virus (BCMV)	KF114860	Bean	2013	Palampur	Un published
Bean yellow mosaic virus (BYMV)	KF155409	Gladiolus	2013	Lucknow	Un published
	KF155414	Gladiolus	2013	Lucknow	Un published
	KF155420	Gladiolus	2013	Lucknow	Un published
	KF155419	Gladiolus	2013	Lucknow	Un published
	JN692500	*Vicia faba*	2011	Lucknow	Kaur et al. (2013)
Chilli veinal mottle virus (CVMV)/pepper vein banding virus	NC_005778	Chilli pepper	2003	Bangalore	Joseph and Savithri (1999)
	GU170808	Hot pepper	2009	Warangal	Un published
	GU170807	Hot pepper	2009	Jalna	Un published
	AJ 237843		2009	India	Anindya et al. (2004)
Dasheen mosaic virus (DMV)	KT026108	Taro	2015	India	Un published
	KJ786965	Elephant foot yam	2014	India	Un published
Onion yellow dwarf virus (OYDV)	KJ451436	Onion	2014	Sikar	Unpublished
Potato virus Y (PVY)	JN034046	Potato	2011	New Delhi	Jailani et al. (2017)
Papaya ringspot virus (PRSV)	EF017707	Papaya	2006	Delhi	Parameswari et al. (2007)
	EU475877	Ridge gourd	2008	Delhi	Mangrauthia et al. (2008)
	KP743981	Papaya	2015	Hyderabad	Un published
	KJ755852	Papaya	2014	Sikar	
Sugarcane mosaic virus (SCMV)	KT719175	Sugarcane	2016	Coimbatore	Un published
	KT719176	Sugarcane	2016	Coimbatore	Un published
	KT719177	Sugarcane	2016	Coimbatore	Un published
	KT719178	Sugarcane	2016	Coimbatore	Un published
	KT719179	Sugarcane	2016	Coimbatore	Un published
Zucchini yellow mosaic virus (ZYMV)	KT778297	Gherkin	2015	Andhra Pradesh	Un published
Soybean mosaic virus (SMV)	KM979229	Soybean	2014	Sikar	Un published
Leeky yellow stripe virus (LYSV)	KP168261	*Allium sativum*	2014	Delhi	Un published

(continued)

Table 17.2 (continued)

Virus	Accession No	Crop	Year	Place/isolate	Reference (NCBI submission)
Watermelon mosaic virus (WMV)	KM597071	Watermelon	2014	India	Un published
	KM597070	Watermelon	2014	India	Un published
Cowpea aphid-borne mosaic virus (CABMV)	KM597165	Cowpea	2014	India	Mishra et al. (2015)
	KM655833	Cowpea	2014	India	Mishra et al. (2015)
Narcissus yellow stripe virus (NYSV)	KU516386	Narcissus	2016	Lucknow	Un published

PRSV isolates originating from India are most diverse than those from other countries. Besides CP, the HC-Pro and NIa-pro coding regions from 12 PRSV (P and W) isolates originating from different locations were analyzed (Basavaraj 2016) from India. Unlike CP (846–867 bp encoding 282–289 aa), both HC-Pro (1370 bp encoding 456 aa) and NIa-pro (723 bp encoding 241 aa) coding regions were not varied in nucleotide length. However, heterogeneity in aa sequence identity was observed in the coding regions. Among the different isolates, Indian isolates showed only 5% HC-Pro aa sequence variability and 6% with rest of the isolates from different countries. Whereas, the aa sequences of NIa-pro and CP varied by 9% and 12% respectively within Indian PRSV isolates (Fig. 17.3) and rest of the isolates from worldwide. The results indicated that the CP is more diversed than NIa-pro and HC-Pro coding regions. The complete genome of pathotype P from India is 10,317 nt long excluding the 3′ terminal poly (A) tail (Parameswari et al. 2007). The complete genome of pathotype W from India is 10,335 nt with 5′ and 3′ untranslated regions of 85 and 206 nts respectively (Mangrauthia et al. 2009). There is a single large ORF commencing at position 86 in genome of both pathotypes, encoding a polyprotein of 3341 aa in pathotype P or 3347 aa in pathotype W. Comparison of the complete nucleotide sequence of pathotypes P and W indicated that the two genomes shared an overall 89% nt identity and a 94% aa identity. Comparison of individual cistrons revealed that the 5′ UTR was the most variable region of the PRSV genome and P1 was the most variable protein which shared only 67–84% aa sequence identity with other reported PRSV isolates, while other proteins (HC-Pro, P3, 6K1, CI, 6K2, VPg, NIa, NIb and CP) shared 91–100% aa identity. Studies in India confirmed that recombination in Indian P and W pathotypes is a dominant feature in PRSV evolution (Fig. 17.4).

17.4.3 Banana Bract Mosaic Virus (BBrMV)

Banana bract mosaic disease (BBrMD) is becoming serious threat to banana and plantain production. BBrMD caused by BBrMV and was first reported in the Philippines in 1979 on the island of Mindanao (Thomas and Magnaye 1996) and

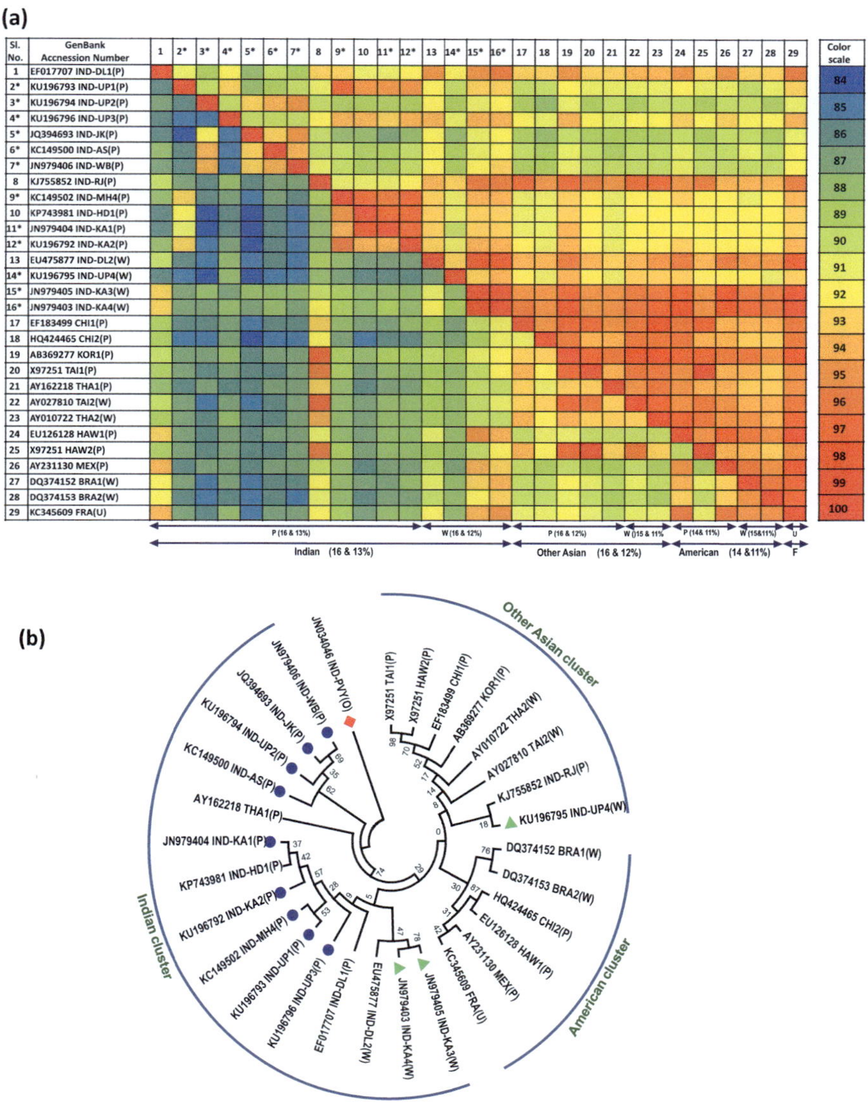

Fig. 17.3 Diversity of *Papaya ring spot virus* (PRSV) isolates originating from India based on coat protein (CP) coding sequences. (**a**) Two-dimensional colour coded graphical representation of pairwise percent [Nucleotide (*above diagonal*) and Amino acid (*below diagonal*)] sequence identities (Basavaraj 2016). (**b**) Phylogenetic tree based on amino acid sequence of CP coding region of PRSV pathotypes P and W showing evolutionary clustering of Indian isolates with globally distributed isolates. Phylogenetic analyses were conducted in MEGA5. Indian Potato virus Y O isoalte IND-PVY(O), GeneBank Accession No. JN034046, as an out group and is marked with the *red solid diamond*. PRSV-P isolates cloned and sequenced in this study marked with *blue solid circle* and PRSV-W with *green solid triangle*

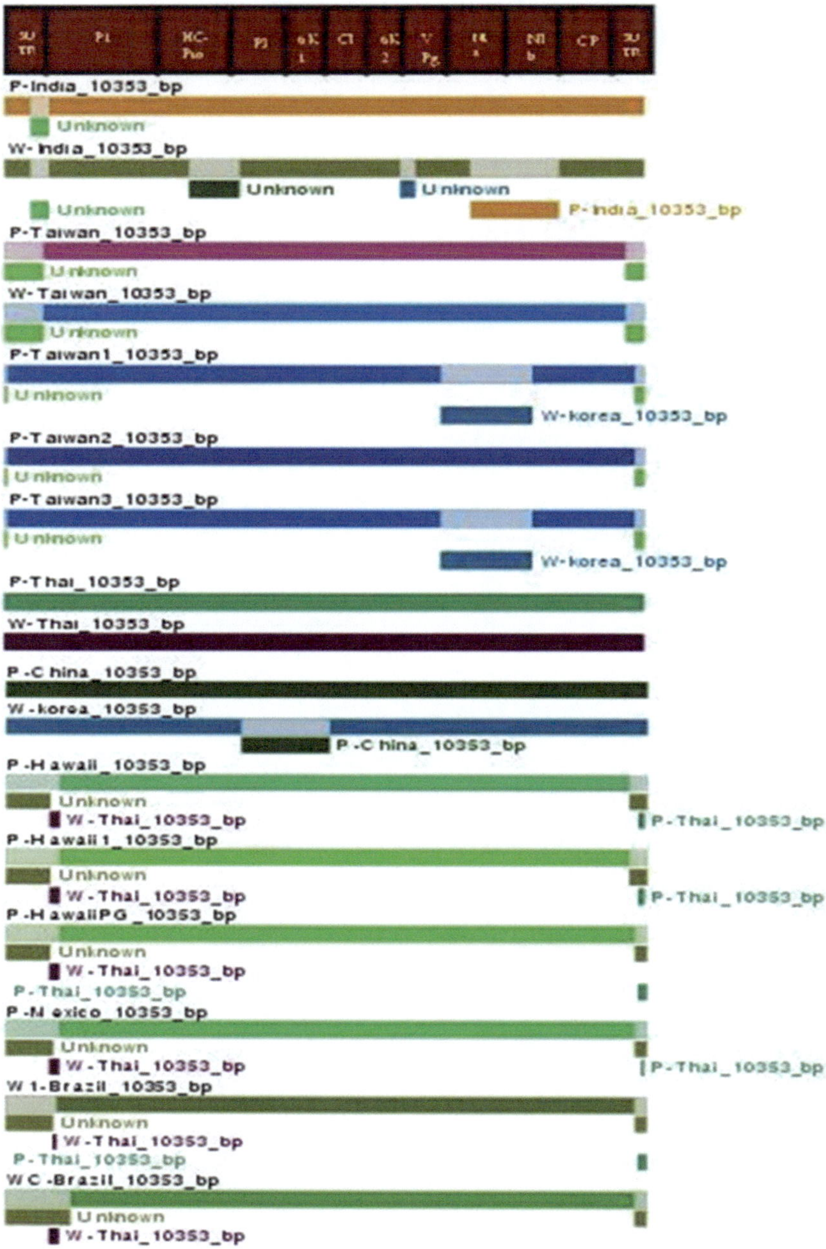

Fig. 17.4 Recombination hot spots detected in 5′ UTR and P1 gene of PRSV genomes sequences bounded by the recombination break points are shade on the graphical representation of PRSV genomes

subsequently shown to widespread throughout the banana growing areas. Now BBrMV is reported to occur in several banana growing states in India viz., Andhra Pradesh, Karnataka, Kerala, Tamil Nadu and Maharastra (Cherian et al. 2002; Kiranmai et al. 2005; Singh et al. 2000; Selvarajan and Jeyabaskaran 2006). In India, banana bract mosaic disease is mostly present in states of Tamil Nadu and Kerala where French plantain cultivar 'Nendran' is particularly susceptible. Around 40% reduction of bunch weight of kokkan diseased plant over healthy was recorded in Kerala (Cherian et al. 2002). The virus infects cvs. Nendran, Poovan, Robusta, Ney Poovan, Rasthali, Red Banana, Karpooravalli and Monthan are also affected but the yield loss is minimal. In Tamil Nadu, 0.5–56.8% incidence of BBrMD in different cultivars was recorded which caused significant reduction in height, girth, leaf area, finger weight, and girth over healthy plants (Selvarajan and Singh 1997). Based on 3 years of survey, per cent infection and yield loss assessment studies, an extrapolation on the loss due to BBrMV was made in cultivar Nendran. An amount of INR387 million per annum loss due to BBrMV (Kokkan) in cultivar Nendran has been estimated by NRCB. The yield loss assessment showed that 68.34, 50.00 and 46.34% reduction in bunch weight in cultivars Nendran, Poovan and Ney Poovan (Thangavelu et al. 2000). The yield loss caused by banana bract mosaic viral disease in Kerala was 52 and 70% in cvs. Nendran and Robusta (Cherian et al. 2002). Selvarajan and Jeyabaskaran (2006) reported that the average yield reduction in cv Nendran due to BBrMV was 30% and the yield loss varied from place to place depending on soil fertility. BBrMV infected plants develop distinct, dark coloured, broad streaks on the bracts of the inflorescence.

The characteristic mosaic symptoms on the flower bracts give the disease its common name. Necrotic streaks on fingers, leaf, pseudostem and mid rib are also recorded (Fig. 17.5). Bunches with unusually very long or with very short peduncle. Sometimes corky growth on peduncle are also observed (Selvarajan and Jeyabaskaran 2006). Though this disease bears the name of bract mosaic and reported that this disease could only be diagnosed when symptom appears on bract, in Thiruvananthapuram district of Kerala, farmers have named the disease as Pola

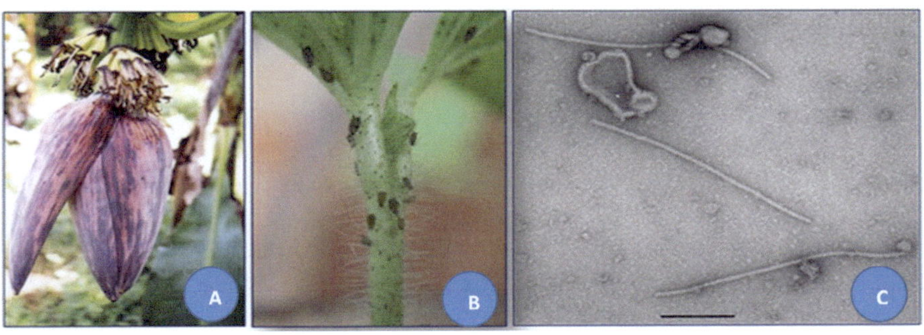

Fig. 17.5 *Banana bract mosaic* infected banana plants. (**a**) Field view of BBrMV infected plants develop distinct, dark coloured, broad streaks on the bracts of the inflorescence. (**b**) The cowpea aphid *Aphis craccivora*- non persistent transmission of BBrMV. (**c**) Electron micrograph of BBrMV

roga, means ‘disease of pseudostem’ in cultivar Nendran (Thangavelu et al. 2000). Singh (2003) reported discontinuous dark red to purple streaks on the bracts and pseudostem of the infected plants. The emerging suckers were deeply pigmented. Foliar symptoms are quite common and appeared as chlorotic streaks parallel to veins and petioles. BBrMV is primarily transmitted through vegetative planting material including suckers and tissue cultured plantlets. The virus is non–persistently transmitted through several aphid species viz., *Pentanlonia nigronervosa, Rhopalosiphum maidis, Aphis gossypii* and in addition the cowpea aphid *Aphis craccivora* has also been reported to transmit the disease (Selvarajan et al. 2006) (Fig. 17.5b). Rodoni et al. (1997) reported BBrMV Indian isolate had 96.6 and 97.2% sequence homology in coat-protein-coding and 3′ untranslated regions with a Philippines isolate of BBrMV. Rodoni et al. (1999) reported variability of 0.3 and 5.6%, and 0.3 and 4.3%, at the nucleotide and amino acid levels, respectively. Sankaralingam et al. (2006) amplified, cloned and sequenced 1062 nucleotides (nt) including 900 nt of the CP coding region and 162 nt from the 3′ UTR from BBrMV infected banana cv. Ney Poovan (AB group) from Coimbatore. Sequence analysis of this complete CP gene showed a variability of 1.0–4.6% at nucleic acid level and 0.7–2.0% at the amino acid level with other Indian isolates and 4.5–5.3% at nucleotide and 1.4–3.0% at amino acid variability level with Southeast Asian isolates, respectively. A total of 71 gene sequences and one complete genome of the BBrMV-TRY isolate have been characterized from India (Tables 17.1 and 17.2). The genome consists of a single large ORF of 9378 nucleotides. When compared with Philippines isolate of BBrMV-PHI, BBrMV-TRY had 94% nucleotide sequence identity and ten mature proteins had amino acid sequence identities ranging from 88 to 98% (Balasubramanian and Selvarajan 2012). Recently, BBrMV causing chlorotic streak disease in small cardamom was identified and characterized based on the coat protein sequence relationships (Siljo et al. 2012).

17.4.4 Chilli Veinal Mottle Virus (ChiVMV)

ChiVMV is the major virus disease infecting chilli crops in eastern Asia. ChiVMV is also known as pepper vein banding virus (PVBV) in India (Anindya et al. 2004; Ravi et al. 1997). In India, ChiVMV was first reported by Prasada Rao and Yaraguntaiah (1979) and is reported to be distributed throughout the chilli growing areas of Karnataka (Bidari and Reddy 1994).

PVBV was the first potyvirus from India whose complete genome sequence was determined and based on the phylogenetic analysis of the potyviral poly protein sequences it was demonstrated to be a distinct species of this genus (Fig. 17.6) (Anindya et al. 2004). However, ICTV has reclassified, the, chilli vein banding mottle virus-Chiengmai isolate (CVbMV-CM1) and PVBV isolates as strains of ChiVMV (Fauquet et al. 2005). Figure 17.7 shows the phylogenetic analysis of all the full length potyviral sequences determined from India (Table 17.2).

The ChiMV virion consists of a linear, positive-sense ssRNA of 9711 nt with a poly (A) tract at its 3′ end and a VPg structure at its 5′ end. The genomic RNA is

Fig. 17.6 Characterization of PVBV: *top panel* shows the symptoms seen on *Capsicum anum* and *Nicotiana benthamiana* upon infection with PVBV. EM of purified PVBV is shown below. The phylogenetic analysis of PVBV polyprotein with other poty viral sequences shows that it is a distinct member of the genus. The sequences used for the alignment obtained from the GeneBank (accession number in parentheses) are as follows: *Bean common mosaic virus* (BCMV, L11890), *Bean yellow mosaic virus* (BYMV, D00490), *Clover yellow vein virus* (ClYVV, D00605), *Dasheen mosaic virus* (DsMV, U00122), *Johnsongrass mosaic virus* (JGMV, Z26920), *Lettuce mosaic virus*(LMV, X65652), *Leek yellow stripe virus* (LYSV, X89711), *Maize dwarf mosaic virus* (MDMV, U07216), *Pepper vein banding virus* (PVBV, AJ237843), *Potato virus A* (PVA, X59985), *Potato virus V* (PVV, NP 734378.1), *Potato virusY* (PVY, U25672), *Plum pox virus* (PPV, D00298), *Pepper mottle virus* (PepMoV, M11598), *Papaya ringspot virus* (PRSV, D00594), *Pea seed-bourne mosaic virus* (PSbMV, D10453), *Soybean mosaic virus* (SMV, D00507), *Sugarcane mosaic virus* (ScMV, D00948), *Tobacco etch virus* (TEV, M11216), *Turnip mosaic virus* (TuMV, D10601), *Tobacco vein mottling* virus (TVMV, NP 056867), *Zucchini yellow mosaic virus* (ZYMV, D00593), *Wild potato mosaic virus* (WPMV, NP 741959) (Anindya et al. 2004)

encapsidated by a single type of coat protein, and a single ORF which encodes a large polyprotein (3088 amino acid residues) that is co-translationally processed into ten functional proteins.

The ChiVMV infected plants usually express dark to light green mottling, vein banding, yellowing and distortion of leaves with reduced fruit bearings (Banerjee et al. 2014b). In India, it is reported on chilli from different parts of the country; from Eastern UP (Prakash et al. 2002), Karnataka and Tamil Nadu (Ravi et al. 1997, Krishnareddy et al. 2001), and Meghalaya (Banerjee et al. 2014b) etc. Dual infection of ChVMV along with a begomovirus, chilli leaf curl virus was found most prevalent in chilli crop in Rajasthan and mixed infection increased disease severity (Sahu et al. 2016).

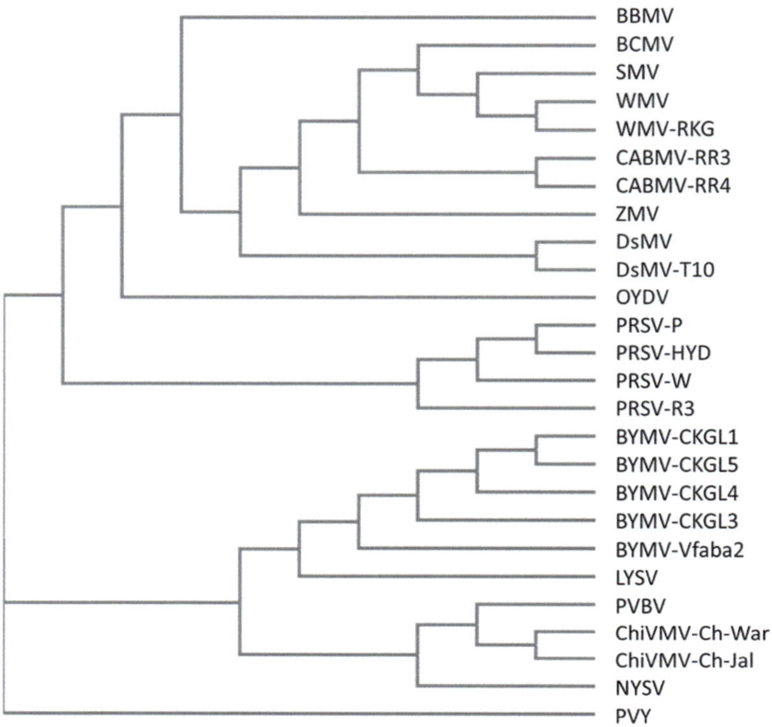

Fig. 17.7 Phylogenetic analysis of the full length potyviral sequences reported from India. The accession numbers for various isolates used for the generation of the cladogram are shown in Table 17.2. Clustal omega was used for generation of the phylogenetic tree. (https://www.ebi.ac.uk/Tools/phylogeny/clustalw2_phylogeny/)

The virus is transmitted by aphids in a non-persistent manner and by mechanical inoculation but not transmitted by the seeds (Sahu et al. 2016). ChiVMV is known to infect different plant species under experimental conditions. The host range of ChiVMV includes *Capsicum annuum*, *C. frutescens*, *Chenopodium* spp., *Datura stramonium*, *Lycopersicon esculentum*, *Solanum melongena* and *Nicotiana* spp. (Green et al. 1999). Apart from *Capsicum* spp., the ChiVMV is known to infect *Solanum nigrum* and *Physalis floridana* in India which could serve as off-season potential reservoirs of ChiVMV. ELISA and RT-PCR methods have been reported for the detection of this virus (Moury et al. 2005). Recently, Banerjee et al. (2015) reported RT-LAMP method for the detection of ChiVMV.

Himalayan butterfly bush (*Buddleja crispa*) and the sequence of part of the coat-protein gene and 3′-UTR of the viral genome showed 95% sequence identity to ChiVMV-Japan strain (Mehra et al. 2006). The cylindrical inclusion (CI) gene sequences of ChiVMV from *S. nigram* (GU294786, GU294787 and GU294788) and *P. floridana* (GU294789 and GU294790) showed 79% and 83–92% nucleotide and amino acid identities, respectively, with corresponding sequences of ChiVMV isolates from India (AJ237843) and South Korea (AJ972878). A distinct strain of

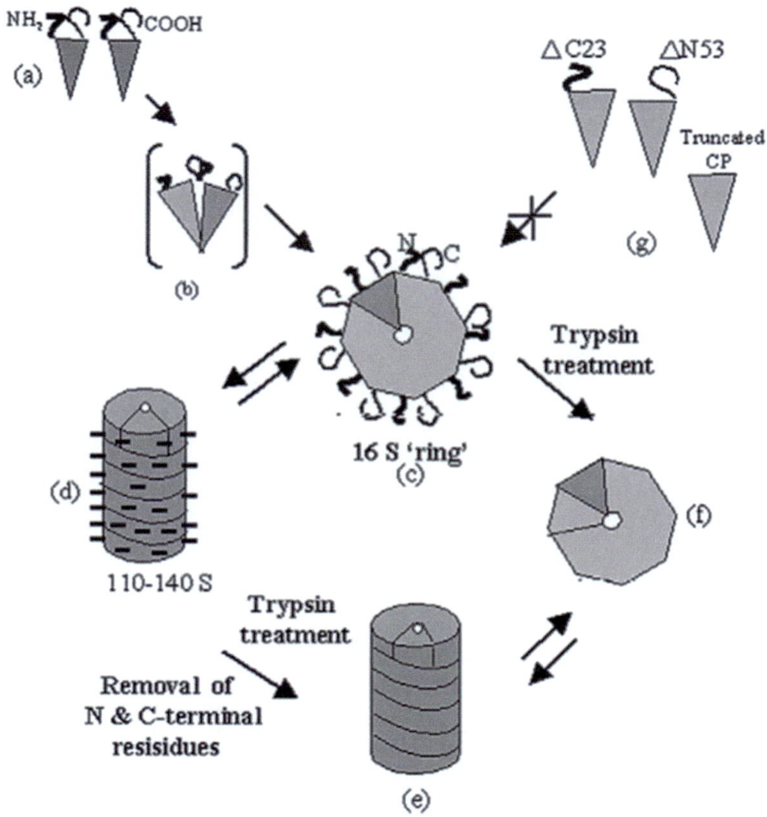

Fig. 17.8 Schematic representation of the assembly and disassembly of ChiMV: the conditions for assembly and disassembly are as indicated

ChiVMV was reported from *Datura inoxia* plants showing severe mottling and distortion disease and the nucleotide sequence of approximately 1.5 kb amplicon comprising of nuclear inclusion b gene, coat protein gene and 3′ un-translated region which had only 83–85% identities with strains of ChiVMV (Kaur et al. 2015a).

The ChiVMV CP gene was overexpressed in *E. coli*, and the assembled particles were shown to encapsidate the CP mRNA (Joseph and Savithri 1999). The mechanism of assembly and disassembly of ChiVMV (Fig. 17.8) was established using purified PVBV and PVLPs. It was shown that the subunits associate in a head to tail manner to form dimers which further assemble to form octomeric rings. The surface exposed amino-terminal and carboxy- terminal regions are essential for the formation of these rings as the N and C-terminal deletion mutants cannot form such intermediates. In the intact virus or the PVLPs, these exposed regions can be removed by trypsin treatment without affecting their assembly.

VPg-Protease, the major protease involved in the polyprotein processing, is composed of two domains, the N-terminal VPg (Viral protein genome-linked), and the

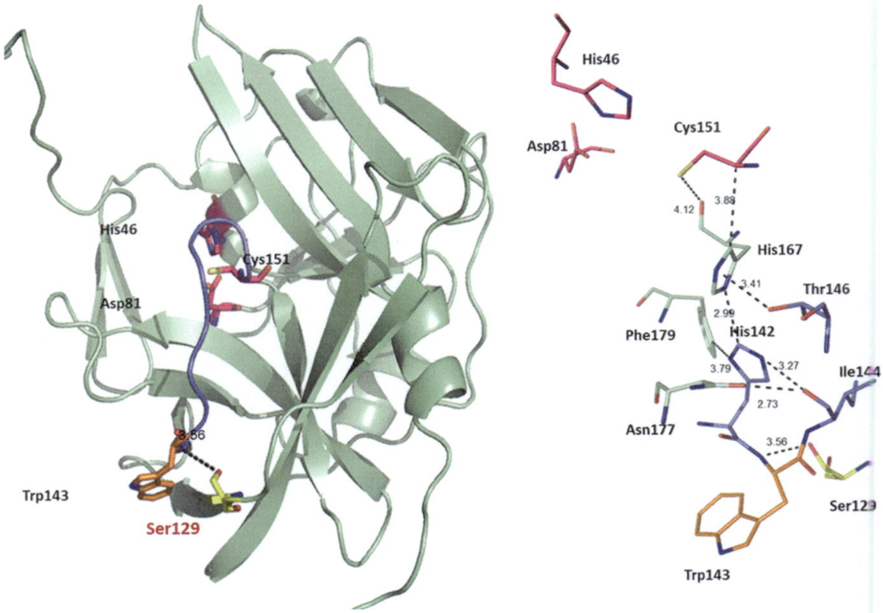

Fig. 17.9 Influence of distal residues on the activity of NIa Pro: modelled structure of ChiVMV NIa pro showing the position of the catalytic triad (His 46, Asp 81 and Cys 154) along with the surface exposed Trp 143 and Ser 129 which are involved in interaction with VPg and phosphorylation respectively. The hydrogen bond network that could relay the conformational changes at W143 or Ser 129 to the active site Cys 154 is also shown

C-terminal proteinase (NIa-Pro) (Riechmann et al. 1992). The ChiVMV NIa-Pro was shown to be related to trypsin-like family of cellular proteases, with Cys151, His 46 and Asp 81 as the active site residues through mutational analysis (Joseph and Savithri 2000). In addition to the protease activity, ChiVMV NIa-Pro was demonstrated to have a novel Mg^{2+} dependent DNase activity. The residue D81 was shown to be crucial for the DNase activity. However, the other protease active site residues, C151 and H46 were found to be dispensable for this activity (Anindya and Savithri 2004a, b). Regulation of NIa-Pro is important for polyprotein processing and successful infection of potyviruses. Interestingly, it was demonstrated that the turnover number of NIa-Pro increases when it interacts with VPg either in *cis* or in *trans*. Further, the protease activity of NIa-Pro could also be modulated by phosphorylation at Ser129. A mutation of this residue either to aspartate or alanine drastically reduced the protease activity. Based on the modeled structure of NIa pro (Fig. 17.9) it was proposed that interaction with VPg as well as phosphorylation of Ser129 could relay a signal through Trp143 and help in modulating protease activity of NIa-Pro (Mathur et al. 2012).

VPg, of ChiVMV is a ~21 kDa protein which gets uridylylated by viral RdRP and serves as the primer for replication. A minimal region, residues 22–100, from the N-terminus was required for in vitro uridylylation and Y66 was shown to be the residue at which the uridylylation occurred (Anindya et al. 2005). VPg was also identified to have a conserved Walker Motif A and Walker Motif B at its N-terminal region, which is a characteristic of NTP binding proteins. The purified VPg did not exhibit ATPase activity. However, VPg-NIa was shown to hydrolyze ATP to release ADP and Pi. Mutation of Lys47 (Walker motif A) and Asp88:Glu89 (Walker motif B) to alanine in the VPg-NIa Pro cleavage site mutant lead to reduced ATPase activity. These results confirmed that this activity was inherent to VPg. It was proposed that potyviral VPg, established as an intrinsically disordered domain, undergoes plausible structural alterations upon interaction with globular NIa-Pro which induces the ATPase activity in VPg (Mathur and Savithri 2012).

17.4.5 Sugarcane Mosaic Virus (SCMV)

Sugarcane mosaic was first reported in India from Pusa in 1921 on sugarcane varieties D 99 and Sathi 131 (Dastur 1923). Mosaic disease, a seed piece transmissible disease with mosaic, chlorotic specks, streaks or stripes on leaf lamina of sugarcane (Fig. 17.10) has been reported widely distributed in almost all sugarcane growing states of India (Bhargava 1975; Agnihotri 1996; Rao et al. 1995, 1998a, 2002a, b; Jain et al. 1998b; Viswanathan and Balamuralikrishnan 2005; Viswanathan and Rao 2011). In India, virus isolates causing mosaic disease in sugarcane continued to be a serious threat to sugarcane industry because of the virus transmission through vegetative propagation. Incidence of mosaic disease is recorded almost on most of the commercial sugarcane varieties in the country and causes significant yield losses (Bhargava 1975; Agnihotri 1996; Viswanathan and Balamuralikrishnan 2005; Viswanathan and Rao 2011).

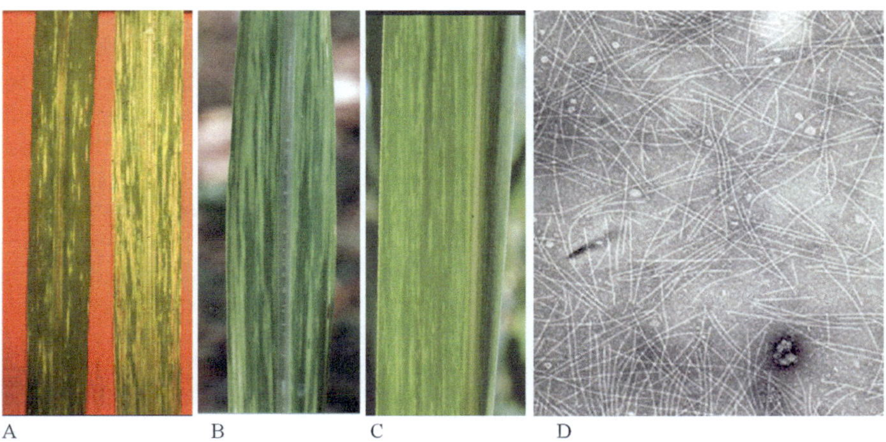

Fig. 17.10 Different types of mosaic symptoms caused by SCMV on sugarcane cultivars. Mosaic mottling (**a**); Chlorotic streaks (**b**); Mosaic (**c**); Long flexuousfilamentous SCMV particles in purified preparation (**d**) (Source: Rao et al. 1996a)

The taxonomic status of SCMV is keeps on changing as many virus strains of this virus are being reported from time to time because of versatility of the virus isolate infecting different hosts viz. sugarcane, maize, sorghum, Johnson grass etc. accordingly, they have been named as SCMV, sorghum mosaic virus, maize dwarf mosaic virus, Johnson grass mosaic virus and also on the basis of biological and serological properties (Shukla et al. 1989a; Rao et al. 1998a, b, 2004; Jain et al. 1998b). During earlier 1990s, 12 SCMV strains (SCMV-A, -B, -C, -D, -E, -F, -H, -I, -J, -K, -M, and -N) were characterized in India based on variations in mosaic symptom pattern on different sugarcane and sorghum differentials (Khurana and Singh 1972; Bhargava 1975; Kondaiah and Nayudu 1985; Rishi and Rishi 1985; Gopal et al. 1991). Since SCMV has many strains, the sensitivity of the detection methods varies depending upon the antibody specificity, type of virus and its strains, virus titre, which led to inconsistent results in quarantine. Hence, to access the efficiency of diagnosis techniques, different immunological assays viz. DAC-ELISA, dot-blot immunoassay, immune sorbent electron microscopy, and electroblot immunoassay were compared with the RT-PCR using SCMV-specific primers from the widely grown *SCMV* infected plants (Rao et al. 2004; Viswanathan 1997). Among them, RT-PCR based diagnostics was found to be very sensitive and was recommended in quarantine and germplasm exchange (Balamuralikrishnan et al. 2004; Viswanathan et al. 2008b).

Hema et al. (1999, 2003) reported sugarcane streak mosaic virus (SCSMV) as the new virus causing mosaic disease in India. But later it was described as a new genus *Susmovirus* under the family *Potyviridae* (Viswanathan et al. 2007, 2008a). This confirms that the virus associated with mosaic disease of sugarcane in India belongs to strains of SCMV subgroup and SCSMV (Rao et al. 1996a; Hema et al. 1999; Viswanathan et al. 2007; Mali 1995, 1996; Mali and Bhagwat 1975; Mali and Garud 1993). Out of these, SCMV and SCSMV have been identified and characterized in India on the basis of serological and molecular studies. Viswanathan et al. (2008a, 2010) developed duplex and multiplex RT-PCR protocols for the simultaneous detection of viruses causing sugarcane mosaics in India. SCSMV is reported as the widely spread and major cause of mosaic disease complex in sugarcane (Rao et al. 2006; Viswanathan et al. 2008b). SCMV and SCSMV cause significant yield losses in promising sugarcane cultivars all over India (Viswanathan and Balamuralikrishnan 2005). Several cases of mixed infection of SCMV and SCSMV were also recorded infecting same sugarcane cultivar in different parts of India (Viswanathan et al. 2007; Viswanathan and Rao 2011).

A shorter and better protocols for purification of SCMV was developed by Rao et al. (1998b, 2001) where the best virus yield was achieved by extracting the virus in borate buffer + thioglycolic acid + diethyl dithiocarbamate followed by treatment with chloroform, butanol (1:1 v/v) + 0.5% Tween-20 and 0.5% Triton X-100. This protocol was reported as less time consuming with one cycle of differential centrifugation. It was also observed that sucrose padding (30%) before density gradient centrifugation helped in removal of host contaminants. The protocol yielded 75–100 mg virus/kg of infected leaf material. The purified preparation yielded long flexuous filamentous particles of 700–760 nm size (Fig. 17.10d) (Rao et al. 2001).

Rao et al. (1996a, b) also developed high titred (1: 2048) polyclonal antibodies against SCMV-Indian strain, which was validated in many labs in India and abroad further for indexing SCMV from leaf, stalk and bud tissues of sugarcane and sorghum at early stage of crop growth in different serological tests.

SCMV was readily transmitted by infected setts, mechanical inoculation and several aphid species viz. *Aphis gossypii, Longiunguis sacchari, Myzus persicae* and *Rhopalosiphum maidis* in a non-persistent manner (Shukla and Rao 1989; Rao et al. 1994, 2001; Singh et al. 2005). The early part of twentieth century was witnessed with mosaic disease epidemics in USA, Brazil and Cuba which led to economic crisis in several sugar Industries. This disease caused 50–100% yield loss and many elite cultivars like Co 740 and CoC 671 in the tropical region and CoLk 8102, CoPant 90223 and CoS 767 from subtropical region were phased out of cultivation due to its severity in India (Agnihotri 1996; Singh et al. 2003a; Viswanathan and Balamuralikrishnan 2005). The impact of SCMV on cane growth and yield was accessed by many Indian workers (Singh et al. 1997; Viswanathan and Balamuralikrishnan 2005) on popularly grown tropical and sub-tropical cultivars. The results had clearly shown that virus infection significantly reduced the number of millable canes, net CO_2 assimilation rate, cane growth parameters viz. cane diameter, cane weight and number of internodes, and other post-harvest juice quality parameters viz. brix, sucrose per cent, purity and commercial cane sugar per cent, sucrose metabolism and accumulation of reducing sugars (Bhargava et al. 1971; Singh et al. 1997; Viswanathan and Balamuralikrishnan 2005) in comparison with mosaic free plants.

To date, five complete genome sequences and 38 gene sequences of SCMV are available from India (Tables 17.1 and 17.2). In order to see the occurrence of any previously reported SCMV strains in India, a molecular based approach was initiated using coat protein (CP) gene sequences of 30 Indian *SCMV* isolates representing seven major states and comparative sequence analysis revealed 88.8–100% amino acid sequence similarities in core region and 51.3–100% in hyper variable region. The results of phylogenetic analysis of SCMV CP gene sequences for both hyper variable and core region had shown that all the isolates taken into comparison were clustered separately according to the geographical origin except the isolate SCMV-IND (clustered along with Australian type strain SCMV-SC), and occurrence of nine new SCMV variants viz. IND-CC1, -CC2, -CC3, -CC4, -CO1, -CO2, -CP, -CS, and –J have been reported in India (Viswanathan et al. 2009). Although occurrence of the type strains SCMV-A, -B, -D, and -E have been previously reported in India based on serology and differential host reaction, the molecular characterization based on hyper variable regions and core region of coat protein sequences had not shown any similarity with these strains. Similarly, the HC-Pro sequence of SCMV-CB isolate originating from India was determined and compared with SCMV isolates reported from different countries (Australia, China, Mexico and Spain). The sequence length of the HC-Pro gene of SCMV-CB was 1380 nucleotides (nt) and encoding a protein of 460 amino acids (aa). The nucleotide sequence SCMV-CB showed 78–94% identity with other SCMV isolates compared in this study (Parameswari et al. 2012). Among the different isolates, those from India and Australia were closely related, with sequence identity of 94%. In the deduced aa sequences, Indian isolate shared 98% identity with Australian isolate and 91–93% with the isolates from other countries. For the first time the complete nucleotide sequence of SCMV-IND infecting popular variety CoC 671 from India was determined was about 9573 nucleotides in length excluding the poly (A) tail. The genome encodes a polyprotein of about 3064 amino acid residues which is processed to form ten mature proteins (P1, HC-Pro, P3, 6K1, CI, 6K2, VPg, NIa-Pro, NIb, and CP). Based on this whole genome study, the Indian isolate showed close identity to the Australian isolate Brisbane of about 98.2% and 94.9% at the protein and nucleotide level, respectively (Bagyalakshmi et al. 2014). The Indian isolate shares the least identity with the American isolate Ohio of about 88.2% and 78.9% at the protein and nucleotide sequences respectively. Within the monocot specific potyvirus, the Indian SCMV isolate shares close identity with sorghum mosaic virus with 69.1% identity followed by maize dwarf mosaic virus with 67.9% identity at the nucleotide level. Further study on recombination event among the SCMV complete genome sequences is in progress which will provide a good insight on SCMV evolutionary biology. In addition, new strains or genome diversity and recombination was reported among SCMV isolates (Viswanathan et al. 2009; Moradi et al. 2016). These indicate that SCMV remains to be a potential threat to sugarcane cultivation and industry in the world.

Virus indexation was carried out by a combination of infectivity test, electron microscopy, serological test and molecular biology assays in India (Viswanathan and Rao 2011). Limited attempts were made on shoot apical meristem culture alone or in combination with heat therapy for the elimination of SCMV. However, in last three decades, in vitro meristem tip culture has been playing significant role in micro-propagation of virus free plants at Vasantsdada Sugar Institute, Pune and ICAR-Sugarcane Breeding institute, Coimbatore which is significantly solving the problems of viral infection in plants (Rao 2002c; Balamuralikrishnan et al. 2002; Viswanathan 1997). Since sugarcane is a vegetative propagated crop and SCMV and SCSMV are seed cane borne transmitted potyviruses, use of virus free material is imperative for raising healthy crops as well as ratoon crops.

17.4.6 Bean Common Mosaic Virus (BCMV)

BCMV is an important potyvirus having wide host range in leguminous crops. *BCMV* is a seed borne virus and it drastically affects the yield and quality of common bean. In India, regular recurrence of the disease was reported in North-western Himalayas with 0.5–77.0% disease incidence (Sharma et al. 1998). High degree of pathogenic variability was reported in the virus all over the world (Hamid et al. 2013). Several BCMV strains viz. NL3, NL5, NL8, NL1, US1, US 2, US3, US4, US5, US 6, US10, NL2, NL4, NL7 belonging to different pathogroup (I–VII) were identified on differential host varieties of bean and have been separated into two serotypes viz. serotype A and B (Shukla et al. 1989b). Later, on the basis of nucleotide sequence, genome size, molecular weight of coat protein, and peptide profiles these strains were reclassified as two separate *Potyvirus* species named as BCMV and bean common mosaic necrosis virus (BCMNV) (Huang and Chang 2005). In India, occurrence of four strains viz. NL-1, NL-1n, NL-7 and NL-7n were reported on the basis of bean host differential reaction and coat protein peptide profiling in Himachal Pradesh (Kapil et al. 2011) of which NL-1n strain belonging to patho-group Ib was found wide prevalent in Himachal Pradesh and Jammu and Kashmir states (Kapil et al. 2011; Sharma et al. 2011; Hamid et al. 2013) and capable of producing bean necrosis at both above and below 30 °C in differential cv. Jubila which was reported as having dominant allele I gene and a recessive bc-1 gene (Anuradha 2014). Thus the temperature insensitive strain, NL-1n was subjected into complete genome sequencing to reveal its genetic characteristics and its relationship with other strains/isolates. Biological and molecular characterization of BCMV associated with vanilla in India was reported by Bhadramurthy and Bhat (2009).

The complete genome of BCMV NL-1n consists of 10,081 nucleotides (nt) (including PolyA tail), long uninterrupted ORF. The ORF yielded a polyprotein of 3222 amino acids. The genome constituted the 5′ UTRs at bases 1–134 and 3′ UTRs at 9801–10,057, P1 protein (protease) at nucleotides position 135–1463, the helper component/protease (HC-Pro) at 1464–2834, a 6-kDa peptide, 6K1 at 3876–4031, the cylindrical inclusion protein (CI) with RNA helicase activity at 4032–5933, a second 6 kDa peptide, 6 K2 at 5934–6092, the viral protein genome-linked NIa–VPg at 6093–6662, the nuclear inclusion 'a' protease NIa–Pro at 6663–7391, the nuclear inclusion 'b' protein (the presumed RNA polymerase) Nib at 7392–8939 and coat protein CP at 8940–9800 bases respectively (Sharma et al. 2016). Further, two potential recombination events were identified in the HC-Pro and 6K2 gene region of BCMV NL-1n isolate. The pairwise comparison and phylogenetic analysis of BCMV NL-1n strain with other reported BCMV strains had shown that it is most closely related to BCMV-NL-1 and differed highly with RU1 and RU1M isolates inducing temperature insensitive whole plant necrosis. The maximum homology with BCMV-NL-1 strain indicates that NL-1n is a variant of NL-1 strain. However, BCMV NL-1n was reported as different from the BCMNV strains because of its inability to cause systemic necrosis in cultivars carrying the dominant allele "I" and induction of mosaic symptoms in cultivars possess recessive "i" alleles. Based on the recombinant analysis it has been noted that BCMV-NL-1 and BCMV-RUI strains are the key players for the BCMV strains evolution including NL-1n in India (Sharma et al. 2016). The study concluded that BCMV NL-1n strain may a recombinant of NL-1 and RU-1 strains and is distinct from the other strains/isolates and might have extensively spread across the bean growing areas.

17.4.7 Dasheen Mosaic Virus (DsMV)

Elephant foot yam (*Amorphophallus paeoniifolius)*, is an important tuber crop used as vegetables and in preparation of indigenous ayurvedic medicines, grown widely in the South – East Asian countries. In India, three major edible aroids were reported to be infected with DsMV namely, *Amorphophallus paeoniifolius*, *Colocasia esculenta,* and *Xanthosomas agittifolium* with characteristic symptoms of mild mosaic, leaf puckering, yellowing of leaves, shoe string and severe leaf deformation. Occurrences of mosaic disease were reported with 24–88% incidence with yield losses of 3.5–38% in Uttar Pradesh (Khan et al. 2006) and 5–10% disease incidence in Kerala, Andhra Pradesh and Orissa (Babu et al. unpublished information). Association of a *Potyvirus* was confirmed with Immuno electron microscopy observations (Ahlawat et al. 2003), and based on the serological tests the causal agent was identified as DsMV (Pandit et al. 2001) and later it was confirmed through characterization of the coat protein (CP) gene (Babu et al. 2011a). The virus is non-persistently transmitted by several aphid vectors viz. *Myzus persicae* and *Aphis gossypii* and also through vegetative propagation. The BLAST analysis of RT-PCR based coat protein gene had shown that DsMV isolate of *A. paeoniifolius* and *C. esculenta* sharing the maximum sequence identity up to 93% and had DVG motif and DTG motif (*X. sagittifolium*) and several varying potential threonine and asparagine rich N-glycosylation motifs in DsMV isolates (Babu and Hegde 2014).

DsMV is a positive sense ssRNA virus with polyadenylation (polyA) at 3′ end. Though the partial genome sequence of DsMV was characterized mainly the NIb, CP and 3′ UTR region (Pappu et al. 1994; Babu et al. 2010, 2011b) the primers designed from the sequences could not amplify the virus from most of the apparently healthy, ELISA positive samples, necessitated to unravel complete genome of the virus. Accordingly, the complete genome of DsMV was sequenced by whole transcriptome method, whereby the whole RNA sequences of the virus along with the host transcriptome sequences were obtained (Kamala et al. 2015). Three DsMV positive samples and in vitro virus free samples were subjected into whole transcriptome sequencing. Around ~7500–25,000 contigs were generated using SOAP-denovo assembly of reads and each sample had varied number of contigs and putative mixed infection of single sample with different DsMV isolates were found during the assembly of contigs to the reference genome and the variations in then ucleotide composition of the contigs were high between the three samples than within the samples. Further analysis revealed that DsMV genome with a size of 9.9 kb (excluding poly-A tail) consisted of one long uninterrupted ORF from 167 to 9754 bases starting with AUG codon at nucleotides 167–169 and terminating with UAA at nucleotides 9752–9754. On BLASTp and BLASTn analysis the genome had shown 89% and 83% identities with DsMV reference genome, China. The ORF yielded a polyprotein of 3195 amino acids and it had all the reported conserved motifs. The phylogenetic analysis of CP sequences had shown ≥90% identity with the selected complete sequences of DsMV infecting subgroups (India) viz. *Soybean mosaic virus* (SMV), *Zucchini yellow mosaic virus* (ZYMV), *Bean common necrosis virus* (BCMNV) and DsMV reported from China. Also, more variations were found in the NIb followed by CP and 3′ UTR based on the sequence identity matrix of 3′ region of the virus. Similarly, no potent recombination event was detected in any of these 3′ regions whereas, five potent recombination events were found in the 5′ UTR in which, first four were with phylogenetically related potyviruses like DsMV (China), *Passion fruit woodiness virus* (*PWV*), *Watermelon mosaic virus*

(*WMV*), *Banana bract mosaic virus* (*BBMV*) as major parents and fifth one was with *Sweet potato feathery mottle virus* (*SPFMV*), an evolutionarily distant one.

17.4.8 Onion Yellow Dwarf Virus (OYDV)

OYDV, is one of the important pathogen, and cause severe economic losses in *Allium* sp. seeds and bulb production. Earlier, epiphytotic incidences of OYDV were reported (Ahlawat and Varma 1997) in India, and after a decade, occurrence of the disease was reported in severe form in Maharashtra, Madhya Pradesh and Gujarat (Gawande et al. 2013). The onion yellow dwarf affected plant shows irregular yellow striping to complete yellowing, downward curling and crinkling of leaves followed by stunting; in case of severe infection it reduces the plant growth and bulb size (Krystyna et al. 2014). Several species of aphids were reported to transmits the virus (Kumar et al. 2011). Earlier ELISA, ISEM assay followed by RT-PCR were used to detect the virus, however, mixed infection of several viruses and variability in N-terminal region of viral CP of different OYDV isolates results, RNA-dependent RNA polymerase gene and 3′- UTR region of viral RNA based detection were standardized for effective detection of OYDV from different geographical locations of India (Arya et al. 2006).

The complete genome of OYDV isolate RR1 from onion (*Allium cepa*) was sequenced to reveal the isolates characteristics further. The genome is 10.5 kb in size (excluding poly-A tail), consisted of one long uninterrupted ORF starting with AUG codon at nucleotides 94 and terminating with stop codon UAA at nucleotide 10,302. The 5′ and 3′ UTR contains 93 and 228 nucleotides, respectively. The RR1 genome has all the characteristic features of potyviruses (Verma et al. 2015). The viral RNA encoded single, large polyprotein which was cleaved into ten smaller functional proteins (P1, HC-Pro, P3, 6K1, CI, 6K2, VPg, NIa-Pro, NIb, and CP) and an additional peptide PIPO, a short ORF found (at nucleotide position 3409–3417) in the P3 protein was reported as plays an important role in cell to cell movement of viruses (Vijayapalani et al. 2012). The RR1 isolate had shown 98% similarity at both nucleotide and amino acid level with Australian garlic OYDV isolate (MS/SW1) and 75–98% similarity at both nucleotide and amino acid sequences of different OYDV isolates from Spain, China, Japan, Argentina and Germany. Amongst the genome, P1 and P3 protein regions were found most variable based on the pairwise comparison of nucleotide and amino acid similarities of the other OYDV polyproteins and the variation was reported as may be due to selection pressure of the isolate coupled with recombination or mutation. In addition, the genome wide microsatellites (SSR) and compound microsatellites (cSSR) analysis of all the *OYDV* isolates had shown that total of 32–47 microsatellites (SSR) and 0–2 compound microsatellites of which, the Indian isolate RR1 had shown 43 SSR and 2 cSSR, wherein high density of SSR was found in P3, CI and HC-Pro regions, and non existence of microsatellites were in the 6K1 and 6K2 regions, amongst the P3 region was found with more SSRs. The Microsatellite distribution plays an important role in evolutionary dynamics of viruses, its genome organization and gene expression (Zhao et al. 2011). Similarly, the entropy analysis of deduced amino acid sequences of RR1 genome had shown P1and P3 regions with greater degree of variations than CI and other genomic regions.

The phylogenetic analysis of the RR1 genome had common ancestry with the garlic isolates of MS/SW1 (Australia) and SG1 (Spain) and the isolates At (Germany) and Se (Argentina) were in separate clade with less similarity with Indian OYDV isolate. Whereas the network structure of the OYDV population using Split decomposition analysis had shown three parallel branches between OYDV isolates of India RR1 (KJ451436), Australia MS/SW1 (HQ258894) and Spain SG1 (JX429964), indicating the possibility of occurrence of recombination among the isolates. Accordingly, six potential recombination breakpoints were detected in Indian isolate RR1 in CP, P3, 3UTR and P1 region, with the major parent related to garlic isolates MS/SW1 (Australia) and SG1 (Spain). In all together, the P1 and P3 regions of RR1 genome of Indian OYDV isolate had high nucleotide and amino acid sequence variability, more number of SSRs distribution, high frequency of recombination break points and high entropy of deduced amino acid sequences. On the basis of phylogenetic analysis and recombination information it was hypothesized as RR1 genome might be a recombinant of garlic isolates MS/SW1 and SG1 of Australia and Spain respectively (Verma et al. 2015).

17.4.9 Bean Yellow Mosaic Virus (BYMV)

In India, French bean (*Phaseolus vulgaris* L.), is an important green vegetable crop being widely cultivated by farmers over an area of about 26.75 thousand hectares. Amongst the plant pathogens causing various diseases, mosaic disease caused by BYMV is widely distributed in leguminaceous crops grown throughout the world. In India, many legumes viz. chickpea (*Cicer arietinum*) (Chalam 1982), green beans (*P. vulgaris*) (Chalam et al. 2005), soybeans (*Glycine max*) (Parakh et al. 2005), faba beans (*Vicia faba*) (Kaur et al. 2013) and non-leguminous plants viz. Cape gooseberry (*Physalis peruviana*) (Kaur et al. 2014), *Fressia* sp. (Kumar et al. 2009), vanilla (*Vanilla planifolia*) (Bhadramurthy et al. 2011) were also reported as to be infected with BYMV. As the symptoms produced by the virus is distinct from the BCMV and it is able to infect mosaic virus resistant varieties like Corbet Refugee, Great 'Northern U1 No.1 and Robust, it was designated as bean virus 2 (BYMV). Though the virus is not seed borne in most of the leguminous crops, sap transmission of the virus was reported in cv. Jawala (Sharma et al. 2015). It induces various types of symptoms viz. yellow mosaic or mosaic spots on infected leaves including stunting, leaf wrinkling and distortion, stunted plant growth with pod deformation and necrosis depending upon the host it infects and virus strain. Association of *BYMV* in the symptomatic plants of French bean (*P. vulgaris*) from Himachal Pradesh (BYMV-HP) isolate was confirmed through DAS-ELISA assay and further by RT-PCR based detection of coat protein of the virus. The aa sequence homology analysis of coat protein with other reported isolates of BYMV had shown that presence of putative proteolytic cleavage site (Q/S) between the NIb protein and CP, and a tripeptide motif asparagine-alanine-glycine (NAG) associated with aphid transmission in potyviruses was also present at a/a position 9 of CP. The isolate had shown 83–99% nucleotide and 89–99% amino acid sequence similarity with other BYMV isolates, amongst the maximum (99%) similarity was with BYMV strain of *Gladiolus* sp. from Japan. The phylogeny analysis of the isolate had also shown the close relationship with BYMV strains of *Gladiolus* sp. (Japan) followed by BYMV isolates of *Vicia faba* from USA and Japan, *P. vulgaris* isolates of Australia and *Freesia* sp. in India (Sharma et al. 2015).

Moreover, a new isolate of BYMV group-IV was found as naturally infecting in ornamental plant gladiolus with leaf mosaic and flower color breaking symptoms. The maximum disease incidence 97.5% and minimum 40% was recorded in Tiger Flame and Snow Princess varieties and reduction of number of florets per spike (7.7–42.1%), cormels (17.6–50.0%) and FW of mother corms (7.2–57.6%) were reported and the reduction % cumulatively increasing in the subsequent years. Among the cultivars, Shagun, Sylvia and Tiger Flame cultivars were reported as most susceptible with ≥90.0% disease incidence, Aldebaron, Regency and Snow Princess were reported as the most tolerant cultivars with ≤48.0% disease incidence. Further keeping in view of the severity of the disease and its effect in exporting the flowers, the complete genome characterization of the virus was taken up after confirming the virus transmission using sap and aphid vectors (*A. craccivora*), presence of flexuous rod-shaped particles of 720 × 11 nm using TEM following the western blot-immunoassay (Kaur et al. 2015b).

The viruses isolate CK-GL2 (*G. dalenii*cv. Sylvia) was subjected into complete genome characterization using primer pairs of overlapping region. The BLASTn analysis had shown 63–85% sequence similarity with BYMV phylogenetic groups I, II, III, V, VI, VII, VIII and IX and 93–96% sequence identity with other BYMV isolates of phylogenetic group IV reported from Japan, USA, India and Taiwan. The isolate had shown 90–94% aa sequence similarity with BYMV strains of phylogenetic group-IV, 89–92% with group-III, 88% (group-I and –II), 86–87% (group-V), 86% (group-VI), 85% (group-VII), 82% and 80% with group-VIII and -IX, respectively. The genome consists of 9532 nucleotide with typical *potyvirus* genome architecture, the 5′ UTR at bases 1–186, P1 (187–1035), HC-Pro (1036–2412), P3 (2413–3456), 6K1 (3457–3615), CI (3616–5520), 6K2(5521–5679), Vpg (5680–6252), Pro (6253–6981), Nib (6982–8538), CP (8539–9357), 3′ UTRs (9358–9532), along with an additional peptide PIPO (at nucleotide position 2589–2699) was found in the P3 region. The pairwise sequence similarity and phylogenetic analysis with other BYMV isolates had shown that the isolate CK-GL2 was in close association with the BYMV isolates of phylogenetic group IV, reported from gladiolus, Japan (MBGP, MB4, GB2, Gla and G1), Faba bean, India (VFaba2), Taiwan and USA; other isolates from Australia, Japan, South Korea and USA were clustered in I–IX phylogenetic groups. Eight recombination events were identified in CK-GL2 isolate of which major were found from HC-Pro to3'UTRregions, Nib followed by P3, CI, 6K2, VPg and NIa-Pro regions. In CP regions, recombination events were identified with groups I, IV, VI and VII. and in the 3′-UTR region groups IV and VI were major and minor parents and no recombination was found in the 5′-UTR, and P1 gene (Kaur et al. 2015b).

17.4.10 Zucchini Yellow Mosaic Virus (ZYMV)

ZYMV is one of the most aggressive and destructive viruses of squash in the world. It was first reported in Italy and France in 1981 (Lecoq et al. 1994; Lisa et al. 1981) and subsequently from other countries (Davis and Yilmaz 1984; Singh et al. 2003b). Within a decade, it spreaded to all the major cucurbit growing areas in the world (Verma et al. 2004). ZYMV causes symptoms resembling those incited by PRSV. The virus reacts with ZYMV antisera but not with PRSV, WMV-2, PVY and CMV (Verma et al. 2004). ZYMV isolates reacts weakly with WMV-2 but not with WMV-1 and PVY (Singh et al. 2003b). In India, it is reported on zucchini (Singh et al. 2003b), cucumber (Verma et al. 2004), gherkin (Anthony Johnson et al. 2013), snake gourd (Nagendran et al. 2015), pumpkin (Nagendran et al. 2016) and other cucurbits. The major symptoms zucchini includes mosaic, vein banding and blotching on leaves with irregularly shaped blisters and filiform leaves. Malformation, discoloration from green to yellow and blistering of fruits occur which make them unmarketable. Early-season infection of cucurbit crops causes up to 100% yield losses and up to 95% losses in marketable fruit.

The DEP of the virus was 10^{-3}, TIP between 50 and 55 °C and LIV was 48 h. The virus particles are long, filamentous, not enveloped usually flexuous virus particles measuring between 700 and 750 × 12 nm (Singh et al. 2003b; Anthony Johnson et al. 2013). Fifteen distinct serotypes were reported using monoclonal antibodies raised against ZYMV isolates from different geographical locations in the world (Desbiez and Lecoq 1997). The host range of the virus was restricted to the members of family cucurbitaceae (Singh et al. 2003b). The virus produces chlorotic local lesions on *Chenopodium amaranticolor*, vein banding and mosaic symptoms on Black Turtle 2 (Singh et al. 2003b), chlorotic spots, veinal chlorosis, mosaic, vein banding and leaf distortion symptoms on *Citrullus lanatus, C. melo, C. sativus, Cucurbita maxima, Cucurbita pepo, Luffa acutangula* and *Trichosanthes anguina* (Verma et al. 2004).

The virus is readily transmitted by mechanical sap inoculation. It is transmitted by seed in some cucurbits (Simmons et al. 2011) and by aphid vectors, *Myzus persicae* and *Aphis gossypii* in a non-persistent manner (Singh et al. 2003b). ELISA and RT-PCR techniques have successfully been utilized for the detection of ZYMV (Simmons et al. 2011; Nagendran et al. 2016). The 3′ terminus of the ZYMV genomes including CP gene sequence have been obtained for many isolates worldwide (Desbiez et al. 2002; Nasr-Eldin et al. 2016). Recently, Coutts et al. (2011) found that based on CP gene sequences, the ZYMV isolates grouped into three major groups (A, B and C), with four subgroups in A (I–IV) and two in B (I–II). In India, ZYMV *isolate* from bottle gourd, 100% identity at the C terminal region of CP and 88.6–99.5% identity at 3′UTR region with other ZYMV isolates (Verma et al. 2004). Molecular analysis of a ZYMV isolate from gherkin showed

98.4–98.9% sequence identity with coat protein gene of the other ZYMV isolates from other countries and India (Anthony Johnson et al. 2013) and pumpkin isolate shared 98–99% identity with ZYMV isolates from India, Iran, Israel, Syria and Germany (Nagendran et al. 2016).

17.5 Description of Minor Potyviruses in India

17.5.1 Turnip Mosaic Virus (TuMV)

TuMV is a member of the family *Potyviridae* that causes diseases in brassicaceae plants. A strain of TuMv is reported to cause mosaic disease in radish in the NE Himalayas (Ahlawat and Chenulu 1984). About 36–63% incidence is reported on different cole crops from Arunachal Pradesh (Singh et al. 2015). TuMv causes mottling, mosaic, irregular chlorotic patches, interveinal chlorosis, and puckering on different crops viz. broad leaved mustard, Indian mustard and broccoli. Severely affected mustard plants showed stunted growth (Singh et al. 2015). A strain of TuMV causing mosaic disease in *Brassica campestris* and *B. juncea* was characterized by Haq et al. (1994) in India. The size of the virus particles was 740 × 12 nm. It induced severe systemic mosaic symptoms in four plant species under *Brassicaceae* family. It also induced chlorotic and necrotic local lesions in *Chenopodium amaranticolor*. The virus was transmitted in a non-persistant manner by *Myzus persicae, Brevicoryne brassicae* and *Aphis gossypii* (Haq et al. 1994). It is also seed transmitted in Radish (Ahlawat and Chenulu 1984). First report on TuMV infection on cole crops (*Brassica* spp) from India was confirmed by symptomatology, electron microscopy and RT-PCR amplification of partial gene of cytoplasmic inclusion protein and CP domains. Phylogenetic analysis of CI domain of three isolates (AR-IndM (Indian mustard), AR-BrLM (broad leaved mustard) and AR-Broc (broccoli)) showed 91–95% nucleotide identity with previously reported TuMV isolates, with a maximum of 95% nucleotide identity with Chinese TuMV isolate. But partial CP gene of these three isolates showed 100% identity with previously reported TuMV isolates worldwide (Singh et al. 2015).

17.5.2 Sweet Potato Feathery Mottle Virus (SPFMV)

Sweet potato is an important vegetatively propagated root crop being consumed widely after potato and cassava in India. It has been reported to be affected by many viruses viz. *Cuccumovirus, Begemovirus, Potyvirus, Enamovirus, Ipomovirus, Tospovirus etc.* Among them, SPFMV belongs to potyvirus is the most common throughout the world and being known with many synonyms viz. sweet potato chlorotic leaf spot virus, sweet potato internal cork virus, sweet potato vein mosaic virus etc. It has restricted host range on *Convolvulaceae*, *Chenopodiaceae* and *Solanaceae* family members and induces various types of symptoms on the infected plants irrespective of strains, such as chlorotic feathering of the leaf mid rib, the expression of chlorotic spots with purple rings, chlorotic specks, yellow netting and mosaic etc. (Thankappan and Nair 1990; Mahesh Kumar et al. 2001; Jeeva et al. 2004). The symptom expression varies depends on cultivar susceptibility and, degree of stress during the crop stage. Occurrence of the disease was reported with 10–60% natural incidence in sweet potato fields of Kerala, Orissa and Andhra Pradesh and the virus was identified initially using antisera supplied by International Potato Centre (Mahesh Kumar et al. 2001), later identified through DAC-ELISA, Dot immune binding assay or NCM-ELISA using antisera produced against the coat protein gene of Indian SPFMV isolate and by RT-PCR based CP gene specific primers (Prasanth and Hegde 2008). On the basis of serological reactions and distinct symptom induction on sweet potato cultivars, four genetic strains viz. C (Common), RC (Russet Crack), O (Ordinary) and EA (East Africa) were reported from different sweet potato growing countries. Based on the complete coat protein gene sequences, occurrence of RC and EA strain groups (Ganga 2009) were reported from Central Tuber crop Research Institute, Thiruvananthapuram, India.

17.5.3 Watermelon Mosaic Virus (WMV)

WMV is an economically important one among the watermelon affecting viruses and drastically reduce the production and productivity of the plant. Occurrence of WMV-1 was first reported in India by Bharagava and Joshi (1960) and on the basis of serological reactions it was reported as synonymous with papaya ring spot virus strain W (Purcifull et al. 1984b). The disease incidence was reported up to 80% in Delhi and Uttar Pradesh states. It induces various types of symptoms includes mosaic mottling, blistering and malformation (Vani 1987), and the bio physical properties of the virus includes it had 10^{-3} to 10^{-4} (dilution end point), 50–55 °C (thermal inactivation point) and longevity in vitro upto 24 h at room temperature (Sharma et al. 2010). Transmission of virus through sap and aphids viz. *Aphis gossypii* and *Myzus persicae* were reported as most efficient vector.

17.5.4 Ornithogalum Mosaic Virus (OrMV)

In India, occurrence of OrMV is being widely reported in ornamental plants. Natural infection of OrMV was found in Iris (*Iris hollandica*) plantation in and around Kangra valley of Himachal Pradesh (Chandel et al. 2006) with mottle, mosaic, chlorotic spots on leaves and deformed flowers symptoms. The RT-PCR based detection of CP gene and 3′ UTR of the OrMV had shown 99% nucleotide sequence identity with Australian isolate of OrMV. Further, the ornamental plant, *Gladiolus* in Uttar Pradesh (Kaur et al. 2011) were reported with OrMV based on symptomatology and by RT-PCR based detection using degenerate oligonucleotide primers designed for the members of the family *Potyviridae*. The comparative sequence analysis showed that the virus shared maximum identity (87–88%) with the amino acid gene sequence of OrMV of *Iris tingitana, Narcissus* spp., *Vanilla planifolia* and

Ornithogalum thyrsoides followed by 78% identity with gladiolus OrMV isolate (GLA-11) from the Netherlands.

17.5.5 Soybean Mosaic Virus (SMV)

SMV is a member of the genus *Potyvirus* and causes mosaic or necrosis in many soybean cultivars. It is easily transmitted by aphids in fields, sap and also transmitted through seeds (Naik and Murthy 1997; Patil and Byadgi 2005a), hence possesses a serious threat to soybean cultivation. It causes up to 25% yield loss in soybean (Patil and Byadgi 2005a). Maximum of 13–16% incidence is reported from Kanpur (Singh et al. 1976). Infection of SMV depletes reducing sugar, non-reducing sugar and starch from the nodule tissues of soybean plants (Gupta et al. 2010). The crop infected at early stage suffered more with severe symptoms like mosaic, dark green mosaic with reduced leaf size and on younger leaves showed clear mottling symptoms along with slight crinkling. Stunting of the infected plants occurs and set only few pods (Patil et al. 2005b).

In India, the occurrence of SMV is reported from New Delhi (Nariani and Pingaley 1960), and Karnataka on soybean (Naik and Murthy 1997; Patil et al. 2005b) and Uttar Pradesh (Singh et al. 1976) on coffee senna based on host range, transmission and physical properties. The first molecular evidence of SMV infection in soybean was reported by Banerjee et al. (2014a). Sequence analysis of partial cytoplasmic inclusion protein (CI) nuclear inclusion protein (NIb) and coat protein (CP) domains of two SMV isolates in soybean from mid-hill of Meghalaya, India, showed 99.0% nucleotide and amino acid identities with previously reported SMV (isolate Ar13) from Iran. These isolates shared 89.0–97.8% nt sequence identity and 91.7–99.0% aa identity with other SMV isolates reported from different countries (Banerjee et al. 2014a). Recently, SMV also reported to infect bottle gourd in West Bengal (d). The infected plants produced mosaic and yellowing symptoms on leaves. The nucleotide sequence of approx. 650 bp amplicon from infected plants showed 94% identical to the SMV coat protein gene.

17.5.6 Maize Dwarf Mosaic Virus (MDMV)

MDMV is one of the most important widely distributed potyvirus associated disease of corn (*Zea mays* L.) in the temperate regions of the world causing yield losses as high as 40%. However, in India only preliminary reports are available on incidence and identification of MDMV. In India, a mosaic disease of maize was reported in UP and Delhi and its incidence in 1970s (Paliwal et al. 1968). Earlier similar mosaic disease was described by Chona and Seth (1960). Seth and Raychaudhari suggested that it may be a strain of sugarcane mosaic virus. Paliwal and Raychaudhari (1965) isolated three distinct strains of virus, namely A, B and C. They reported the first visible symptoms after infection were manifested chlorotic specks on one side of the mid-rib and in rows parallel to it. The diseased plants were rather stunted and

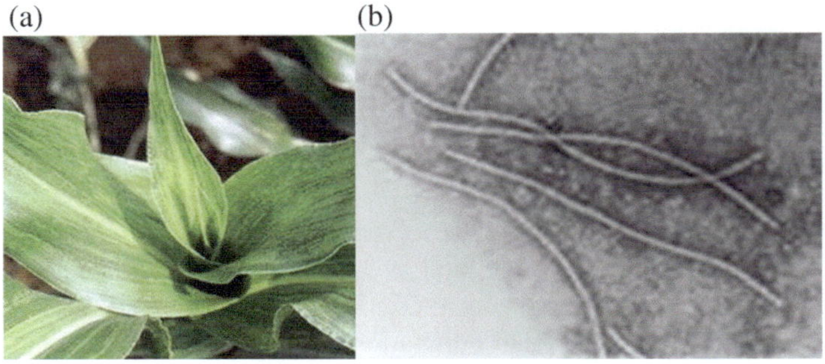

Fig. 17.11 MDMV: symptoms (**a**) and flexuous filamentous virus particles (**b**) (Source: Rao et al. 1996b)

produced poorly filled ears. The virus readily transmitted to maize mechanically but has got low dilutions as well as thermal inactivation points. Five aphid vectors were confirmed as vectors for this maize mosaic disease, viz. *Rhopalosiphum maidis*, *Aphis gossypii,. Macrosiphum granarium, Myzus persicae* and *Schizaphois graminum* (Seth and Raychaudhuri 1967). The disease was reported not seed transmitted.

Biological, serological and molecular confirmation of association of MDMV from maize, sudangrass and sorghum in India was reported by Rao et al. (1996b, 1998b) and Mishra et al. (1998). Symptoms vary widely depending on host genotype and time of infection. Infected plants develop a distinct mosaic—irregularities in the distribution of normal green color, typically appearing at the bases of the youngest leaves. There may also be narrow chlorotic streaks extending parallel to the veins (Fig. 17.11). Later, the youngest leaves show a general chlorosis, and streaks are larger and more abundant. Infected plants are characterized by stunting (of variable severity) and shortening of upper internodes. MDMV can infect other grass and cereal hosts, such as sorghum, sudan grass and Johnsongrass (Rao et al. 1996b, 1998b; Mishra et al. 1998). The MDMV particles are long, flexuous rods about 715 nm in length (Rao et al. 1996b; Mishra et al. 1998). The virus is transmitted mechanically but not by seed. An effective, quick and suitable purification protocol by using Tween-20 for high yield of MDNV has been developed (Rao et al. 1998b). The virus showed positive serological relationship in precipitin test, DAC-ELISA and ISEM test. The MDMV (Indian sorghum isolate) showed different types of symptoms on sorghum genotypes and sudangrass, *Sorghum halepense*, maize but did not show any symptoms on sugarcane, *Oryza sativa, Pennisetum americana, Vigna mungo* and *Chenopodium amaraticolor* (Mishra et al. 1998).

17.6 Concluding Remarks

Potyvirus is one of the largest genera among the 73 genera of plant viruses and also largest genera among the members of the family *Potyviridae*. The potyviruses were investigated in detail in India and there is lot of information available on pathogen identification, characterization, genome organization and vector transmission etc. In recent years, potyvirus diseases as papaya ring spot disease in papaya and cucurbits, mosaics in sugarcane, potato, banana, pepper, cardamom etc. are increasing and causing serious damages to the crop plants. For the effective management of these diseases, potato, banana and sugarcane plants are multiplied through tissue culture. Here molecular diagnostics are employed to confirm only the disease-free seedlings or settlings are produced and distributed to the growers.

In the current scenario, the serology and molecular biology methods are widely used for viral diagnostics. DNA chips or biosensors are also making great impact on the development of rapid and sensitive assays for the detection of viruses. In India, ELISA, RT-PCR using virus specific oligos and NASH for many potyviurses have been standardized. The antibodies, oligos and the viral genes can be exploited to develop highly sensitive biosensors for the identification of virus strains and simultaneous detection of a number of viruses with a single chip.

Using different molecular biological techniques or mutation/deletion studies it was proved that most of the potyviral proteins are multifunctional and the best example is helper component proteinase (HC-Pro). In spite of many experimental studies, functions and the role of potyviral proteins such as 6K1, P3 and 6K2 proteins are unknown. Potyvirus infection requires the interaction of viral proteins with host factors and viral RNA for its replication and systemic spread, for example interaction between viral genome linked protein (VPg) and eIF4E is required for genome translation of potyvirus. Transgenic plants expressing, mutants of eIF4E (host translational initiation factor) that are unable to bind with the VPg protein, could result in broad-spectrum resistance against potyviruses.

References

Agnihotri VP (1996) Current sugarcane disease scenario and management strategies. Indian Phytopathol 49:10–126

Ahlawat YS, Chenulu VV (1984) Radish mosaic: a new disease caused by *turnip mosaic virus* in India. Trop Agric 61:188–192

Ahlawat YS, Varma A (1997) Serological detection of mixed viral infection in onion seed crop and possible measure for its management. Indian Phytopathol 50:137–140

Ahlawat YS, Pant RP, Bhagwati KN (2003) Association of Carla- and potyviruses with mosaic disease of elephant foot yam. Indian Phytopathol 56:300–301

Anindya R, Savithri HS (2004a) Surface-exposed amino- and carboxy-terminal residues are crucial for the initiation of assembly in pepper vein banding virus: a flexuous rod-shaped virus. Virology 316:325–336

Anindya R, Savithri HS (2004b) Potyviral NIa proteinase, a proteinase with novel deoxyribonuclease activity. J Biol Chem 279:32159–32169

Anindya R, Joseph J, Gowri TDS, Savithri HS (2004) Complete genomic sequence of *Pepper vein banding virus* (PVBV): a distinct member of the genus *Potyvirus*. Arch Virol 149:625–632

Anindya R, Chittori S, Savithri HS (2005) Tyrosine 66 of Pepper vein banding virus genom linked protein is uridylylated by RNA-dependent RNA polymerase. Virology 336:154–162

Anthony Johnson AM, Vidya T, Papaiah S, Srinivasulu M, Mandal B, Sai Gopal DV (2013) First report of *Zucchini yellow mosaic virus* infecting Gherkin (*Cucumis anguira*) in India. Indian J Virol 24:289–290

Anuradha NO (2014) Molecular characterization of NL-1n strain of BCMV infecting common bean in Himachal Pradesh and development of diagnostic kits. PhD thesis, Palampur, CSKHPKV, pp 48–49

Arya M, Baranwal VK, Ahlawat YS, Singh L (2006) RT–PCR detection and molecular characterization of *Onion yellow dwarf virus* associated with garlic and onion. Curr Sci 91:10

Babu B, Hegde V (2014) Molecular characterization of *Dasheen mosaic virus* isolates infecting edible aroids in India. Acta Virol 58:34–42

Babu B, Hegde V, Makeshkumar T, Jeeva ML (2010) Rapid detection and identification of potyvirus infecting *Colocasia esculenta* (L.) Schott by reverse transcription-polymerase chain reaction. J Root Crops 36:88–94

Babu B, Hegde V, Makeshkumar T, Jeeva ML (2011a) Characterisation of the coat protein gene of *Dasheen mosaic virus* infecting elephant foot yam. J Plant Pathol 93:199–203

Babu B, Hegde V, Makeshkumar T, Jeeva ML (2011b) Development of non-radioactive probes for specific detection of *Dasheen mosaic virus* infecting *Amorphophallus paeoniifolius*. Curr Sci 100:1220–1225

Bagyalakshmi K, Parameswari B, Chinnaraja C, Brindha S, Viswanathan R (2014) Complete genome sequence of Indian *Sugarcane mosaic virus* and its genetic diversity with other country isolates. In: Proceedings of the symposium on bioenergy for sustainable development- the potential role of sugar crops, 23–25 June. Coimbatore, pp 229–232

Balamuralikrishnan M, Doraisamy S, Ganapathy T, Viswanathan R (2002) Serological specificity and titre of sugarcane mosaic virus polyclonal antisera raised under varying immunization procedures and bleeding time. J Plant Dis Protect 109:646–654

Balamuralikrishnan M, Doraisamy S, Ganapathy T, Viswanathan R (2004) Comparison of antibody and genome based diagnostic techniques for *sugarcane mosaic virus* in sugarcane. Phytoparasitica 32:52–56

Balasubramanian V, Selvarajan R (2012) Complete genome sequence of a *banana bract mosaic virus* isolate infecting the French plantain cv. Nendran in India. Arch Virol 157:397–400

Banerjee A, Chandra S, Swer EKP, Sharma SK, Ngachan SV (2014a) First molecular evidence of *Soybean mosaic virus* (SMV) infection in soybean from India. Australas Plant Dis Notes 9:150

Banerjee A, Dutta R, Roy S, Ngachan SV (2014b) First report of *Chilli veinal mottle virus* (ChiVMV) in Naga chilli *(Capsicum chinense)* in Meghalaya. India Virus Dis 25:142–143

Banerjee A, Roy S, Sharma SK, Dutta SK, Chandra S, Ngachan SV (2016) Reverse transcription loop-mediated isothermal amplification (RT-LAMP) assay for rapid diagnosis of *chilli veinal mottle virus*. Arch Virol 161:1957–1961

Basavaraj (2016) Assessment of genetic diversity within *papaya ringspot virus* population and development of virus resistant transgenic papaya. Ph.D thesis, IARI, New Delhi

Behl MK, Chowfla SC, Paul Khurana SM (1987) Necrotic strain of potato virus Y from tomato plants. Indian Phytopathol 40:14–17

Bhadramurthy V, Bhat AI (2009) Biological and molecular characterization of *Bean common mosaic virus* associated with vanilla in India. Indian J Virol 20:70–77

Bhadramurthy V, Bhat AI, Biju CN (2011) Occurrence and coat protein sequence based characterization of *Bean yellow mosaic virus* (BYMV) associated with vanilla (*Vanilla planifolia* Andrews) in India. Arch Phytopathol Plant Protect 44:1796–1801

Bharagava KS, Joshi RD (1960) Detection of *watermelon mosaic virus* in Uttar Pradesh. Curr Sci 29:443–444

Bhargava KS (1975) Sugarcane mosaic- retrospect and prospect. Indian Phytopathol 18:1–9

Bhargava KS, Joshi RD, Lal KM (1971) Strain D of *sugarcane mosaic virus* in India. Sugarcane Pathol Newsl 8:23

Bhat AI, Varm A, Jain RK, Khurana SMP (1997) Differentiation of potato virus Y strains by N-terminal serology and HPLC peptide profiling. Indian Phytopathol 50:89–96

Bhat AI, Varma A, Pappu HR, Rajamannar M, Jain RK, Praveen S (1999) Characterization of a potyvirus from eggplant (*Solanum melongena*) as a strain of potato virus Y by N-terminal serology and sequence relationships. Plant Pathol 48:648–654

Bidari VB, Reddy HR (1994) A survey on mosaic viruses and their distribution in different chilli cultivars in Karnataka. Mysore J Agric Sci 28:52–59

Blanc S, Lopez-Moya J, Wang R, Gracia-Lampasona S, Thornbury WD, Pirone TP (1997) A specific interaction between coat protein and helper component correlates with aphid transmission of a potyvirus. Virology 231:141–147

Capoor SP, Varma PM (1958) A mosaic disease of papaya in Bombay. Indian J Agric Sci 28:225–235

Chalam V (1982) Identification and characterization of cucumber mosaic and bean yellow mosaic viruses affecting chickpea (*Cicerarietinum* L.) in India. Ph. D Thesis, Andhra Pradesh Agricultural University, Rajendranagar, Hyderabad, p 1–92

Chalam VC, Khetarpal RK, Parakh DB, Maurya AK, Jain A, Singh S (2005) Interception of seed-transmitted viruses in French bean germplasm imported during 2002–2003. Indian J Plant Protect 3:134–138

Chandel V, Kulshrestha S, Hallan V, Zaidi AA (2006) Natural infection of *Ornithogalum mosaic virus* on Iris from India. Plant Pathol 55:284

Cherian AK, Menon R, Suma A, Nair S, Sudheesh MV (2002) Impact of banana bract mosaic diseases on the yield of commercial banana varieties of Kerala. In: Global conference on banana and plantain, Bangalore, 28–31 October. Abstract. p 155

Chona BL, Seth ML (1960) A mosaic disease of maize (Zea mays L.) in India. Indian J Agric Sci 30:25–32

Coutts BA, Kehoe MA, Webster CG, Wylie SJ, Jones AL (2011) *Zucchini yellow mosaic virus*: biological properties, detection procedures and comparison of coat protein gene sequences. Arch Virol 156:2119–2131

Dastur J (1923) The mosaic disease of sugarcane in India. Agric J India 18:505–509

Davis RF, Yilmaz MA (1984) First report of *Zucchini yellow mosaic virus* on watermelon and squash in Turkey. Plant Dis 68:537

Desbiez C, Lecoq H (1997) Zucchini yellow mosaic virus. Plant Pathol 46:809–829

Desbiez C, Wipf-Scheibel C, Lecoq H (2002) Biological and serological variability, evolution and molecular epidemiology of *Zucchini yellow mosaic virus* (ZYMV, *Potyvirus*) with special development of transgenic tobacco for PVY resistance. Curr Sci 82:10

Fauquet CM, Mayo CM, Maniloff J, Desselberger U, Ball LA (2005) Virus taxonomy. Eighth report of the international committee on taxonomy of viruses. Elsevier, Academic, San Diego

Ganga G (2009) Sweet potato viruses: development of molecular based methods for detection. Ph.D thesis, University of Kerala, Thiruvanathapuram, Kerala, India. 224pp

Garg ID, Khurana SMP (1992) Factors governing immune electron microscopy of flexuous potato viruses. Acta Virol 36:435–442

Garg ID, Khurana SMP (1993) Morphological changes in the flexuous potato viruses upon decoration in immunosorbent electron microscopy. Acta Virol 37:407–411

Garg ID, Khurana SMP (2001) Optimization of immuno electron microscopic diagnosis of potato virus A. J Indian Potato Assoc 28(1):94–95

Gawande SJ, Chimote KP, Gurav VS, Gopal J (2013) Distribution and natural incidence of *Onion Yellow Dwarf virus* (OYDV) on garlic and its related *Allium* species in India. Indian J Hortic 70:544–548

Ghosh SB, Bapat VA (2006) Development of RT-PCR based method for detection of *potato virus Y* in tobacco and potato. Indian J Biotechnol 5:232–235

Ghosh SB, Nagi LHS, Ganapathi TR, Khurana SMP, Bapat VA (2002) Cloning and sequencing of potato virus Y coat protein gene from an Indian isolate and development of transgenic tobacco for PVY resistance. Curr Sci 82:855–859

Gibbs A, Ohshima K (2010) Potyviruses and the digital revolution. Annu Rev Phytopathol 48:205–223

Gopal K, Reddy KS, Upadhyaya HD, Reddy DR (1991) Host range of *Sugarcane mosaic virus* (SCMV-B) from Rayalseema. Indian J Virol 7:117–119

Green SK, Hiskias Y, Lesemann DE, Vetten HJ (1999) Characterization of *Chilli veinal mottle virus* as a potyvirus distinct from *Pepper veinal mottle virus*. Petria 9:332

Gupta UP, Srivastava M, Gupta U (2010) Influence of *Soybean mosaic virus* infection on carbohydrate content in nodule of soybean (*Glycine max* L. Merr.) Int J Virol 6:240–245

Hamid A, Ahmad M, Padder BA, Shah MD, Saleem S, Sofi TA, Mir AA (2013) Pathogenic and coat protein characterization confirming the occurrence of Bean common mosaic virus on common bean (*Phaseolus vulgaris*) in Kashmir, India. Phytoparasitica 42:317–322

Haq QMR, Srivastava KM, Raizada RK, Singh BP, Jain RK, Mishra A, Shukla DD (1994) Biological, serological and Ccoat protein properties of a strain of turnip mosaic virus causing a mosaic disease of *Brassica campestris* and *B. juncea* in India. J Phytopathol 140:55–64

Hema M, Venkatramana M, Savithri HS, Sreenivasulu P (1999) Biological, antigenic and genomic relationships among the virus isolates causing mosaic disease of sugarcane in South India. Curr Sci 77:698–702

Hema M, Savithri HS, Sreenivasulu P (2003) Comparison of direct binding polymerase chain reaction with recombinant coat protein antibody based dot-immunobinding assay and immunocapture-reverse transcription-polymerase chain reaction for the detection of sugarcane streak mosaic disease in India. Curr Sci 85:1774–1777

Huang CH, Chang YC (2005) Identification and molecular characterization of *Zantedeschia mild mosaic virus*, a new calla lily-infecting potyvirus. Arch Virol 150:1221–1230

Hussain S, Varma A (1994) Occurrence of *Papaya ringspot virus* from Amritsar (Punjab) India. J Phytopathol Res 7:77–78

Hussain A, Arif M, Abbas A, Hussain B, Ali M, Jaffar S (2016) A review on aphid-borne virus (*Potato Virus Y*). J Entomol Zool Stud 4:189–192

Inoue-Nagata AK, Fonseca EN, Lobo TOTA, De Avila AC, Monte DC (2001) Analysis of the nucleotide sequence of the coat protein and the 3′ untranslated region of two Brazilian *Potato virus Y* isolates. Fitopatol Bras 26:45–52

Jailani AAK, Shilpi S, Mandal B (2017) Rapid demonstration of infectivity of a hybrid strain of potato virus Y. Physiol Mol Plant Pathol 98:62–68

Jain RK, Pappu HR, Pappu SS, Varma A, Ram RD (1998a) Molecular characterization of *Papaya ringspot virus* isolates from India. Ann Appl Biol 132:413–425

Jain RK, Rao GP, Varma A (1998b) Present status of management of *Sugarcane mosaic virus*. In: Hadidi A, Khetrapal RK, Koganezawa H (eds) Plant virus disease control. APS Press, St Paul, pp 495–506

Jain RK, Sharma J, Sivakumar AS, Sharma PK, Byadgi AS, Verma AK, Varma A (2004) Variability in the coat protein gene of *Papaya ringspot virus* isolates from multiple locations in India. Arch Virol 149:2435–2442

Jeeva ML, Balakrishnan S, Edison S, Umamaheswaran K, Makeshkumar T (2004) Characterization, purification and serology of *Sweet potato feathery mottle virus* in India. J Root Crops 30:24–30

Joseph J, Savithri HS (1999) Determination of 3′-terminal nucleotide sequence of pepper vein banding virus RNA and expression of its coat protein in Escherichia coli. Arch Virol 144:1679–1687

Joseph J, Savithri HS (2000) Mutational analysis of the NIa protease from pepper vein banding potyvirus. Arch Virol 145:2493–2502

Kamala S, Makeshkumar T, Sreekumar J, Chakrabarti SK (2015) Whole transcriptome sequencing of diseased elephant foot yam reveals complete genome sequence of *Dasheen mosaic virus*. Virol Rep 5:1–9

Kapil R, Sharma P, Sharma SK, Sharma OP, Sharma OP, Dhar JB, Sharma PN (2011) Pathogenic and molecular variability in *Bean common mosaic virus* infecting common bean in India. Arch Phytopathol Plant Protect 44:1081–1092

Kapoor R, Mandal B, Paul PK, Jain RK (2014) Simultaneous detection of potato viruses Y and X by DAC-ELISA using polyclonal antibodies raised against fused coat proteins expressed in Escherichia coli. J Plant Biochem Biotechnol 23:332–335

Karasev AV, Gray SM (2013) Continuous and emerging challenges of *Potato virus Y* in potato. Annu Rev Phytopathol 51:571–586

Kaur C, Raj SK, Snehi SK, Goel AK, Roy RK (2011) Natural occurrence of *Ornithogalum mosaic virus* newly reported on gladiolus in India. New Dis Rep 24:2

Kaur C, Kumar S, Raj SK (2013) New record of association of *Bean yellow mosaic virus* with mosaic disease of *Vicia faba* in India. Indian J Virol 24:95–96

Kaur C, Raj R, Kumar S, Raj SK (2014) First report of *Bean yellow mosaic virus* on Cape gooseberry in India. New Dis Rep 29:17

Kaur C, Kumar S, Raj SK (2015a) Association of a distinct strain of *Chilli veinal mottle* virus with mottling and distortion disease of *Datura inoxia* in India. Archiv Phytopathol Plant Protect 48:545–554

Kaur C, Kumar S, Raj SK, Chauhan PS, Sharma N (2015b) Characterization of a new isolate of *Bean yellow mosaic* virus group-IV associated with mosaic disease of Gladiolus in India. J Plant Pathol Microb 6:10

Kerlan C, Tribodet M, Glais L, Guillet M (1999) Variability of *Potato virus Y* in potato crops in France. J Phytopathol 147:643–651

Khan NM, Aswathi LP, Singh PK (2006) Survey on symptomatology and assessment of yield losses due to viral diseases in elephant foot yam (*Amorphophallus paeoniifolius* Blume). J Root Crops 32:90–93

Khurana SMP (1968) A study on virus diseases of papaya (Carica papaya L.) in Gorakhpur. PhD thesis, University of Gorakhpur, Gorakhpur. 126 pp

Khurana SMP (1992) Potato viruses and viral diseases, Technical bulletin. No. 35. Central Potato Research Institute, Shimla

Khurana SMP, Bhargava KS (1970) Induced apocarpy and "Double papaya" fruit formation in papaya with distorton ringspot virus infection. Plant Dis Rep 54:181–183

Khurana SMP, Bhargava KS (1971) Three new vectors of papaya viruses. J Hortic Sci 46:209–211

Khurana SMP, Garg ID (2000) Potato mosaics. In: Khurana SMP (ed) Diseases and pests of potato-a manual. CPR, Shimla, pp 36–40

Khurana SMP, Singh S (1972) Sugarcane mosaic strains E & C in India and new sorghum differentials. ISSCT Sugarcane Pathol Newsl 9:6–8

Khurana SMP, Singh V, Nagaich BB (1975) Five strains of *Potato virus Y* affecting potatoes in Shimla hills. J Indian Potato Assoc 57:38–41

Khurana SMP, Singh MN, Kumar S (1987a) An easy schedule for potato virus Y purification and production of its antiserum. Indian J Virol 3:124–131

Khurana SMP, Singh MN, Shiv K (1987b) Purification Y of potato virus and the production of antiserum. Curr Sci 56:420–422

Khurana SMP, Singh MN, Garg ID, Kumar S (1990) Uniform purification scheme for flexuous potato viruses. Asian Potato J 1:46–48

Khurana SMP, Garg ID, Singh MN, Kumar S (1993) Storage and preservation affect immuno electron microscopy of flexuous potato viruses. Asian Potato J 3:33–36

Kiranmai G, Kumar PL, Hema M, Venkatramana M, Kirshna-Prasadji, Mandheva-Rao J, Sreenivasalu P (2005) Partial characterization of a potyvirus causing bract mosaic of banana in Andhra Pradesh. Indian J Virol 16:7–11

Kondaiah E, Nayudu MV (1985) Strain N-A new strains of *sugarcane mosaic virus*. Sugar Cane 4:11–14

Krishnareddy M, Sadashiva AT, Madhavi Reddy K, Chalam C, Deshpande AA, Jagadish Chandra K (2001) Leaf curl and other viruses of tomato and peppers. In: SAVERNET phase II proceedings of final workshop, AVRDC, Bangkok, pp 28–33

Krystyna W, Ewa S, Wojciech S (2014) Prevalence of infections with *Onion yellow dwarf virus*, *Leek yellow stripe virus* and *Garlic common latent virus* in plants from the genus Allium. Acta Sci Polonorum 13:123–133

Kumar CA, Mandal BB, Chandel KPS, Jain RK, Varma A, Srivastava S (1991) Occurrence of *Sweet potato feathery mottle virus* in germplasm of *Ipomoea batatas* (L) in India. Curr Sci 60:321–325

Kumar Y, Hallan V, Zaidi AA (2009) Identification and characterization of *Bean yellow mosaic virus* infecting Freesia. J Plant Biochem Biotechnol 18:253–255

Kumar P, Dhawan P, Mehra R (2011) Characterization, transmission and host range of *Onion yellow dwarf virus*. Pl Dis Res 26:176

Lal SB, Khurana SMP (1983) Detection and identification of viruses and mycoplasmas and evaluation of yield losses. In: Nagaich BB (ed) Potato production, storage and utilization. Central Potato Research Institute, Shimla, pp 334–348

Lecoq H, Wipf-Scheibel C, Desbiez C, Dufour O, Allex D, Hostachy B (1994) Virus de la mosaique jaune de la courgette. Une menace nouvelle pour les cultures de cucurbitacees en Martinique. Phytoma 459:43–45

Lisa V, Boccardo GD, Agostino G, Dellavalle G, d' Aquilio M (1981) Characterization of a potyvirus that causes zucchini yellow mosaic. Phytopathology 77:667–672

Mahesh Kumar T, Sriram S, Nair RR, Edison S, Thankappan M (2001) Symptomatology and characterization of sweet potato virus disease. J Root Crops 27:296–302

Mali VR (1995) Identification and characterization of a potyvirus causing sorghum red stripe in Marathwada. Abstract, tenth Ann Conv IVS, Jan 16–18, 1995, SCTIMST, Dept Microbiology, Trivandrum, Kerala, India. pp 30–31

Mali VR (1996) Identification of sugarcane potyviruses causing mosaic and/or red stripe disease in sorghum. In: Abstract XIth Ann Con IVS, Oct 28–30, 1996, IVRI, Izatnagar-Braeilly, Uttar Pradesh, India, p 61

Mali VR, Bhagwat VY (1975) Virus and virus-like diseases of sorghum in India. Sorghum Newsl 18:41–42

Mali VR, Garud TB (1993) Sorghum red-stripe – a Johnson grass strain of sugarcane mosaic virus in India. FAO Plant Prot Bull Rome Italy., 26 1:28–29

Mangrauthia SK, Parameswari B, Praveen S, Jain RK (2008) Role of genetic recombination in molecular architecture of papaya ringspot virus. Biochem Genet 46:835–846

Mangrauthia SK, Parameswari B, Praveen S, Jain RK (2009) Comparative genomics of papaya ringspot virus pathotypes P and W from India. Arch Virol 154:727–730

Martinez DR, Duarte PSG, Olmedo JG, Figueira AR (2014) Molecular and biological studies of *Papaya ringspot virus* isolates from Brazil and Cuba. Am J Agric For 2:209–218

Mathur C, Savithri HS (2012) Novel ATPase activity of the polyprotein intermediate, viral protein genome-linked-nuclear inclusion-a protease, of pepper vein banding potyvirus. Biochem Biophys Res Commun 427:113–118

Mathur C, Jimsheena VK, Banerjee S, Makinen K, Gowda LR, Savithri HS (2012) Functional regulation of PVBV nuclear inclusion protein-a protease activity upon interaction with viral protein genome-linked and phosphorylation. Virology 422:254–264

Mehra A, Hallan V, Lal B, Zaidi AA (2006) Occurrence of *Chilli veinal mottle virus* in Himalayan butterfly bush (*Buddleja crispa*). Plant Pathol 55:284

Mishra M, Rao GP, Upadhyaya PP (1998) Identification and characterization of *Maize dwarf mosaic virus* in India. Indian J Virol 14(10):75–79

Mishra R, Verma RK, Gaur RK (2015) Characterization of an Indian isolate of cowpea aphid-borne mosaic virus. J Plant Pathol 97:345–347

Moradi Z, Mehrva M, Nazifi E, Zakiaghl M (2016) The complete genome sequences of two naturally occurring recombinant isolates of *Sugarcane mosaic virus* from Iran. Virus Genes 52:270–280

Moury B, Palloix A, Caranta C, Gognalons P (2005) Serological, molecular, and pathotype diversity of *Pepper veinal mottle virus* and *Chilli veinal mottle virus*. Phytopathology 95:227–232

Mukherjee K, Verma Y, Chakrabarti SK, Khurana SMP (2004) Phylogenetic analysis of 5′-UTR and P1 protein of Indian common strain of *Potato virus Y* reveals its possible introduction in India. Virus Genes 29:229–237

Nagaich BB, Shekhawat GS, Paul Khurana SM, Bhattacharyya SK (1974) Pathological problems of the potato cultivation in India. J Indian Potato Assoc 1:32–44

Nagendran K, Balaji CG, Mohankumar S, Manoranjitham SK, Naidu RA, Karthikeyan G (2015) First report of *Zucchini yellow mosaic virus* in snake gourd (*Trichosanthes cucumerina*) in India. Plant Dis 99:558

Nagendran K, Priyanka R, Keerthana U, Mohankumar S, Karthikeyan G (2016) First report of *zucchini yellow mosaic virus* on *Cucurbita moschata* in India. J Plant Pathol 98:173

Naik RG, Murthy KKV (1997) Transmission of *Soybean mosaic virus* through sap, seed and aphids. Karnataka J Agric Sci 10:565–568
Nariani TK, Pingaley KV (1960) A mosaic disease of soyben (*Glycine max* L.) Indian Phytopathol 13:130–136
Nasr-Eldin MA, Abdelkader HS, Abo-Senna AS, Othman BA (2016) Characterization and phylogenetic analysis of *Zucchini yellow mosaic* virus infecting *Cucurbita pepo* in Egypt. J Am Sci 12:93–104
Nie X, Singh RP (2002) A new approach for the simultaneous differentiation of biological and geographical strains of *Potato virus Y* by uniplex and multiplex RT-PCR. J Virol Methods 104:41–54
Paliwal YC, Raychaudhuri SP (1965) Studies on maize mosaic disease proceedings. All India Maize Improvement conference
Paliwal YC, Raychaudhuri SP, Renfero BL (1968) Some properties and behaviour of maize mosaic virus in India. Phytopathology 58:1682–1684
Pandit MK, Mukhopadhyay S, Devonshire BJ, Jones P (2001) First report of *Dasheen mosaic virus* in elephant foot yam in India. Plant Pathol 50:802
Pappu SS, Pappu HR, Rybicki EP, Niblett CL (1994) Unusual amino terminal sequence repeat characterizes the capsid protein of dasheen mosaic potyvirus. J Gen Virol 75:239–242
Parakh DB, Chalam VC, Khetarpal RK, Maurya AK, Jain A, Singh S (2005) Interception of seed-transmitted viruses in soybean germplasm imported during 2003–2004. Indian J Plant Prot 33:119–124
Parameswari B (2009) Genome organization of *Papaya ringspot virus* isolates from India and their biological characterization. PhD thesis, IARI, New Delhi
Parameswari B, Mangrauthia SK, Praveen S, Jain RK (2007) Complete genome sequence of an isolate of *Papaya ringspot virus* from India. Arch Virol 152:843–845
Parameswari B, Chhabra ML, Viswanathan R (2012) Sequence motif comparison and homology modeling of helper component proteinase of sugarcane mosaic virus. In: International symposium on new paradigms in sugarcane research, Coimbatore, 15–18 October, p 244
Patil RG, Byadgi AS (2005) Transmission studies of *Soybean mosaic virus*. Karnataka J Agric Sci 18:52–54
Patil RG, Byadgi AS, Patil PV, Yadahalli KB (2005) Survey and distribution of *Soybean mosaic virus* in few districts of Karnataka. Karnataka J Agric Sci 18:166–168
Prakash S, Singh SJ, Singh RK, Upadhyaya PP (2002) Distribution, incidence and detection of potyvirus on chilli from eastern Uttar Pradesh. Indian Phytopathol 55:284–298
Prasada Rao RDVJ, Yaraguntaiah RC (1979) The occurrence of *Pepper veinal mottle virus* on chilli in India. Mysore J Agric Sci 13:445–448
Prasanth G, Hegde V (2008) Occurrence of *Sweet potato feathery mottle virus* and *Sweet potato leafcurl Georgia virus* on Sweet potato in India. Plant Dis 92:311
Purcifull DE, Edwardson JR, Hiebert E, Gonsalves D (1984a) Papaya ringspot virus. CMI/AAB Descriptions of plant viruses, 292: p 8
Purcifull D, Hiebert E, Edwardson J (1984b) Watermelon mosaic virus 2. CMI/AAB descriptions of plant viruses no. 293. Kew
Raigond B, Sharma M, Chauhan Y, Jeevalatha A, Singh BP, Sharma S (2013) Optimization of duplex RT-PCR for simultaneous detection of *Potato virus Y* and S. Potato J 40:22–28
Rao GP (2002) In: Dube HC (ed) Novel approaches for management of sugarcane diseases, Annual Review Plant Pathology, vol 1. Indian Society of Mycology & Plant Pathology, Udaipur
Rao GP, Singh M, Shukla K, Singh HN (1994) Epidemiological studies on sugarcane mosaic virus in Eastern Uttar Pradesh, India. In: Rishi N, Ahuja KL, Singh BP (eds) Virology in tropics. Malhotra Publishing House, New Delhi, pp. 347–352
Rao GP, Rishi N, Bhargava KS (1995) Status of sugarcane mosaic virus research in India. In: Mishra AS, Prakash O, Agnihotri VP (eds) Disease scenario of plants, vol II. International Books & Periodicals Supply Service, Pitampura, pp 222–234
Rao GP, Jain RK, Lim HS, Ford RE (1996a) Identification of sugarcane mosaic potyvirus infecting sudangrass in India (Abstract no. 514A). Rev Plant Pathol S60

Rao GP, Jain RK, Varma A (1996b) Occurrence of maize dwarf mosaic virus on maize and sudan-grass in India. Plant Dis 80:711

Rao GP, Tosic M, Ford RE (1998a) Tween-20 shortens a protocol for purification of sugarcane mosaic and maize dwarf mosaic potyviruses. Sugar Cane 6:19–22

Rao GP, Jain RK, Varma A (1998b) Identification of sugarcane mosaic and maize dwarf mosaic potyviruses infecting poaceous crop in India. Indian Phytopathol 51:10–16

Rao GP, Ford RE, Tosic M, Jain RK (2001) Advancement in the purification of sugarcane mosaic potyvirus. In: Rao GP, Ford RE, Tosic M, Teakle DS (eds) Sugarcane pathology vol. II: virus and phytoplasma diseases. Science Publishers, Enfield, pp 247–264

Rao GP, Viswanathan R, Singh SB (2002a) Current situation of sugarcane diseases in India. In: Singh SB, Rao GP, Easwaramoorthy S (eds) Sugarcane crop management. SCI Tech Publishing LLC, Houston, pp 1–9

Rao GP, Singh M, Rishi N, Bhargava KS (2002b) Century status of sugarcane virus diseases research in India. In: Singh SB, Rao GP, Easwaramoorthy S (eds) Sugarcane crop management. SCI Tech Publishing LLC, Houston, pp 1–9. pp. 223–254

Rao GP, Singh M, Gaur RK, Jain RK (2004) Antigenic and biological diversity among sugarcane mosaic isolates from different geographical regions in India. Indian J Biotechnol 3:538–541

Rao GP, Chatenet M, Girard JC, Rott P (2006) Distribution of sugarcane mosaic and *Sugarcane streak mosaic virus* in India. Sugar Tech 8:79–81

Ravi K, Joseph J, Nagaraju N, Krishnaprasad S, Reddy H, Savithri HS (1997) Characterization of a pepper vein banding virus from chilli pepper in India. Plant Dis 81:673–677

Riechmann JL, Lain S, Garcia JA (1992) Highlights and prospects of potyvirus molecular biology. Virology 177:710–716

Rishi N, Rishi S (1985) Purification and electron microscopy and serology of strains A and F of *Sugarcane mosaic virus*. Indian J Virol 1:79–86

Rodoni BC, Ahlawat YS, Varma A, Dale JL, Harding RM (1997) Identification and characterization of *Banana bract mosaic virus* in India. Plant Dis 81:669–672

Rodoni BC, Dale JL, Harding RM (1999) Characterization and expression of the coat protein-coding region of banana bract mosaic potyvirus, development of diagnostic assays and detection of the virus in banana plants from five countries in Southeast Asia. Arch Virol 144:1725–1737

Roy G, Jain RK, Bhat AI, Varma A (1999) Comparative host range and serological studies of papaya ringspot potyvirus isolates. Indian Phytopathol 52:14–17

Sahu AK, Nehra C, Mishra R, Verma R, Gaur RK (2016) Molecular evidence of *Chilli veinal mottle virus* and *Chilli leaf curl virus* simultaneously from naturally infected chilli plants (*Capsicum annuum* L). Indian J Biotechnol 15:266–268

Samraj J, Menon MR, Christudas SP, Satyarajan PK (1966) Kokkan, a new disease of banana (*Musa paradisiaca* Linn.) Agric Res J Kerala 4:116

Sankaralingam A, Baranwal VK, Ahlawat YS, Devi R, Ramiah M (2006) RT-PCR detection and molecular characterization of *Banana bract mosaic virus* from the pseudostem and bract of banana. Arch Phytopathol Plant Protect 39:273–281

Selvarajan R, Jeyabaskaran KJ (2006) Effect of *Banana bract mosaic virus* (BBrMV) on growth and yield of cultivar Nendran (Plantain, AAB). Indian Phytopathol 59:496–500

Selvarajan R, Singh HP (1997) Occurrence, geographical distribution and Electron microscopy of BBrMV in India. In: The proceedings of international conference on integrated disease management for sustainable agriculture held at New Delhi on November, pp 11–15

Seth ML, Raychaudhuri SP (1967) Further studies on mosaic disease of maize (*Zea mays* L.) Biol Plant 9:372–376

Sharma OP, Sharma PN, Sharma PK (1998) Comparative occurrence and distribution pattern of French bean mosaic in Himachal Pradesh. Himachal J Agric Res 24:165–167

Sharma NK, Awasthi LP, Singh SK (2010) Biophysical properties of the *Watermelon mosaic virus*-1 causing mosaic in watermelon. J Phytology 2:21–24

Sharma P, Sharma PN, Kapil R, Sharma SK, Sharma OP (2011) Analysis of 3′-terminal region of *Bean common mosaic* virus strains infecting common bean in India. Indian J Virol 22:37–43

Sharma PN, Sharma V, Sharma A, Rajput K, Sharma SK (2015) Identification and molecular characterization of *Bean yellow mosaic virus* infecting French bean in Himachal Pradesh. Virus Dis 26:315–318

Sharma A, Sharma V, Sharma P, Sharma SK, Sharma PN (2016) Molecular characterization and genome organization of temperature insensitive necrosis inducing strain of BCMV infecting common bean in India. Australas Plant Pathol 45:219–227

Shukla K, Rao GP (1989) Spread of Sugarcane mosaic virus- sorghum strain in nature. Indian Sugar 39:409–410

Shukla DD, Tosic M, Jilka J, Ford RE, Taker RW, Langham MAC (1989a) Taxonomy of *Potyvirus* infecting maize, sorghum and sugarcane in Australia and United States as determined by reactivities of polyclonal antibodies directed towards virus specific N-termini of coat proteins. Phytopathology 79:223–229

Shukla DD, Jilka J, Tosic M, Ford RE (1989b) A novel approach to the serology of potyviruses involving affinity-purified polyclonal antibodies directed towards virus specific N termini of coat proteins. J Gen Virol 70:13–23

Shukla DD, Ward CW, Brunt AA (1994) The *Potyviridae*. CAB International, Wallingford

Siljo A, Bhat AI, Biju CN, Venugopal MN (2012) Occurrence of *Banana bract mosaic virus* on cardamom. Phytoparasitica 40:77–85

Simmons HE, Holmes EC, Gildow FE, Bothe-Goralczyk MA, Stephenson AG (2011) Experimental verification of seed transmission of *Zucchini yellow mosaic virus*. Plant Dis 95:751–754

Singh SJ (2003) Viral diseases of banana, 1st edn. Kalyani publishers, Ludhiana

Singh MN, Paul Khurana SM (1999) Evaluation of short reaction times for optimizing das-ELISA schedule for three flexuous potato viruses. J Indian Potato Assoc 26:27–32

Singh BR, Singh DR, Saksena HK (1976) A mosaic disease of soybean at Kanpur. India Sci Cult 42:53–54

Singh S, Paul Khurana SM, Lal SB (1983) An atypical isolate of the *Potato virus Y*^{C}. J Indian Potato Assoc 10:42–49

Singh SP, Rao GP, Singh J, Singh SP (1997) Effect of sugarcane mosaic potyvirus infection on metabolic activity, yield and juice quality. Sugar Cane 5:19–22

Singh SJ, Selvarajan R, Singh HP (2000) Detection of bract mosaic virus (kokkan disease) by electron microscopy and serology. In: Singh HP (ed) Banana-improvement, production and utilization. Proceedings of conference in AIPUB, Tiruchirappali, India, pp 381–383

Singh V, Sinha OK, Kumar R (2003a) Progressive decline in yield and quality of sugarcane due to *Sugarcane mosaic virus*. Indian Phytopathol 56:500–502

Singh SJ, Verma R, Ahlawat YS, Singh RK, Prakash S, Pant RP (2003b) Natural occurrence of a yellow mosaic' disease on zucchini in India caused by a potyvirus. Indian Phytopathol 56:174–179

Singh M, Singh A, Upadhyaya PP, Rao GP (2005) Transmission studies on an Indian isolate of sugarcane mosaic potyvirus. Sugar Tech 7:32–38

Singh RP, Valkonen JPT, Gray SM, Boonham N, Jones RAC, Kerlan C, Schubert J (2008) Discussion paper: the naming of *Potato virus Y* strains infecting potato. Arch Virol 153:1–13

Singh R, Banerjee A, Sharma SK, Bhagawati R, Baruah S, Ngachan SV (2015) First report of *Turnip mosaic virus* occurrence in cole crops (*Brassica spp*) from Arunachal Pradesh. India Virus Dis 26:211–213

Srinivasulu M, Sai Gopal DVR (2011) Coat protein sequence comparison of south Indian isolates of *Papaya ringspot virus* with other Indian subcontinent isolates. Phytopathol Mediterr 50:359–367

Thangavelu R, Selvarajan R, Singh HP (2000) Status of *banana streak virus* and *banana bract mosaic virus* diseases in India. In: Singh HP, Chadha KL (eds) Banana: improvement, production and utilization, Proceedings of the Conference on Challenges for Banana Production and Utilization in 21st century. AIPUB, NRCB, Trichy, pp 364–376

Thankappan M, Nair NG (1990) New disease syndromes in sweet potato. J Root Crops 16:212–215

Thomas JE, Magnaye LV (1996) Banana bract mosaic disease. Musa Dis Fact Sheet; 1

Urcuqui-Inchima S, Haenni A, Bernardi F (2001) Potyvirus proteins: a wealth of functions. Virus Res 74:157–175

Vani S (1987) Studies on viral diseases of muskmelon and watermelon. PhD thesis, I.A.R.I., New Delhi

Varma A (1988) The economic impact of filamentous plant virus – the Indian sub-continent. In: Milne RG (ed) The plant viruses, vol 4. Plenum Press, New York, pp 371–376

Verma R, Prakash S, Tomer SPS (2004) First report of *Zucchini yellow mosaic virus* in Cucumber (*Cucumis sativus*) in India. Plant Dis 88:906

Verma RK, Mishra R, Petrov NM, Stoyanova M, Stoev A, Bakardjieva NV, Gaur RK (2015) Molecular characterization and recombination analysis of an Indian isolate of *Onion yellow dwarf virus*. Eur J Plant Pathol 143:437–445

Vijayapalani P, Maeshima M, Nagasaki-Takekuchi N, Miller WA (2012) Interaction of the trans-frame *Potyvirus* protein P3N-PIPO with host protein PCaP1 facilitates *Potyvirus* movement. PLoS Pathog 8:e1002639

Viswanathan R (1997) Enzyme linked immunosorbent assay (ELISA) for the detection of *Sugarcane mosaic virus*. Madras Agric J 84:377–380

Viswanathan R, Balamuralikrishnan M (2005) Impact of mosaic infection on growth and yield of sugarcane. Sugar Tech 7:61–65

Viswanathan R, Rao GP (2011) Disease scenario and management of major sugarcane diseases in India. Sugar Tech 13:336–353

Viswanathan R, Balamuralikrishnan M, Karuppaiah R (2007) Association of *sugarcane mosaic virus* and *sugarcane streak mosaic virus* with sugarcane mosaic in India. Sugar Cane Intern 25:10–18

Viswanathan R, Balamuralikrishnan M, Karuppaiah R (2008a) Duplex – reverse transcription – polymerase chain reaction (D-RT-PCR) – a technique for the simultaneous detection of viruses causing sugarcane mosaic. Sugar Tech 10:81–86

Viswanathan R, Balamuralikrishnan M, Karuppaiah R (2008b) Characterization and genetic diversity of *sugarcane streak mosaic virus* causing mosaic in sugarcane. Virus Genes 36:553–564

Viswanathan R, Karuppaiah R, Balamuralikrishnan M (2009) Identification of new variants of SCMV causing sugarcane mosaic in India and assessing their genetic diversity in relation to SCMV type strains. Virus Genes 39:375–386

Viswanathan R, Karuppaiah R, Balamuralikrishnan M (2010) Detection of three major RNA viruses infecting sugarcane by multiplex reverse transcription–polymerase chain reaction (multiplex-RT-PCR). Australas Plant Pathol 39:1–6

Wei JY, Liu DB, Li XY, Zhou P (2007) Variation in the coat protein gene of *Papaya ringspot virus* isolates from multiple locations of China. J Integr Plant Biol 49:1062–1069

Yeh SD, Jan FJ, Chiang CH, Doong TJ, Chen MC, Chung PH, Bau HJ (1992) Complete nucleotide sequence and genetic organization of *Papaya ringspot virus*. RNA J Gen Virol 73:2531–2541

Zhao H, Tan Z, Feng H, Yang R, Li M, Jiang J, Shen G, Yu R (2011) Microsatellites in different *Potyvirus* genomes: survey and analysis. Gene 488:52–56

Structural and Functional Characterization of Sesbania Mosaic Virus

18

Arindam Bakshi, G.P. Vishnu Vardhan, M. Hema, M.R.N. Murthy, and H.S. Savithri

Abstract

The genus *Sobemovirus* includes isometric, single-stranded positive sense RNA plant viruses that infect monocots and dicots. Sesbania mosaic virus (SeMV) is the only sobemovirus, which has been studied extensively in India. SeMV, which was identified in southern India, is highly stable and multiplies to high concentration causing mosaic symptoms on the leaves of *Sesbania grandiflora*. SeMV has served as a good model for understanding the molecular mechanism of assembly, replication and movement of sobemoviruses. The structure of the virion and other chimeric VLPs, along with biochemical investigations has resulted in the elucidation of the mechanism of assembly of SeMV. The complete genome sequence, structure and function of non-structural proteins have also been determined. The review presents a summary of all these findings. SeMV being non-toxic and biodegradable and of low immune response, has been shown as a good candidate for biomedical applications.

Keywords

Sesbania mosaic virus • Structure of SeMV • Virus-like particle • Infectious clone

---

A. Bakshi • G.P. Vishnu Vardhan • H.S. Savithri (✉)
Department of Biochemistry, Indian Institute of Science, Bengaluru, India
e-mail: bchss@biochem.iisc.ernet.in

M. Hema
Department of Virology, Sri Venkateswara University, Tirupati, India

M.R.N. Murthy
Molecular Biophysics Unit, Indian Institute of Science, Bengaluru, India

B. Mandal et al. (eds.), *A Century of Plant Virology in India*,
https://doi.org/10.1007/978-981-10-5672-7_18

18.1 Introduction

Sobemoviruses (genus *Sobemovirus*, family unassigned) constitute a group of isometric plant viruses named after the type species *Southern bean mosaic virus* (SBMV). In 1969, Walters proposed classification of all single stranded RNA viruses that are transmitted by beetles as a single group (Walters 1969). Hull (1977) recommended the establishment of sobemovirus group as a separate genus based on the sedimentation coefficient, stability of capsids, distribution of particle morphology and the properties of the capsid protein. International Committee for the Taxonomy of Viruses identified sobemovirus group as an unassigned genus in 1995 and the same taxonomic status has been retained thus far (King et al. 2012). Sobemoviruses have a very narrow host range infecting 15–20 plant families that include both monocots and dicots. Their geographic distribution is wide spread and they are of considerable economic importance, as they cause disease in several commercial crops such as legumes and rice in most parts of the world. Sobemoviruses under natural conditions are transmitted by insects or seeds. Different types of insects, chrysomilidae beetles, aphids, myrids, leafhoppers and leaf miners, have been reported to transmit sobemoviruses, however, the transmission may be due to contamination of insect mouth parts. Under the greenhouse conditions, sobemoviruses can readily be transmitted by mechanical sap inoculation. The infection of sobemoviruses leads to accumulation of large amount of virus in mesophylls and vascular tissues (Tamm and Truve 2000a). The genus *Sobemovirus* contains as many as 19 members of which only sesbania mosaic virus (SeMV) is known to occur in India. SeMV is not a significant viral pathogen in India, although, a considerable body of literature has been generated on virion structure, assembly and gene function.

18.2 Biological Properties

SeMV was identified in *Sesbania grandiflora* showing mosaic disease in the farmer's field around Tirupati, Andhra Pradesh, India (Solunke et al. 1983). SeMV infected sesbania plants show systemic yellowish green irregular patches within 10 days post sap inoculation (Fig. 18.1). At the later stage of infection, the lower leaves appear to be asymptomatic although the younger leaves continue to show symptoms. However, on *Cyamopsis tetragonoloba* plants, chlorotic lesions appear on the primary inoculated leaves but the infection is not systemic in nature. The dilution end point of this virus ranges from 1:300,000 to 1:500,000 and the thermal inactivation temperature is between 80 and 90 °C. The longevity in vitro is up to 21–28 days at 33 °C and 87 days at 8 °C indicating that the virus particles are very stable (Gopinath et al. 1994).

SeMV was purified from the infected sesbania leaves (100 g) harvested 20–30 days post inoculation by sucrose gradient centrifugation. The peak fractions containing the CP of size 29 kDa were pooled, re-centrifuged at 40,000 g for 3 h, resuspended and stored at 4 °C. The virus yield was 70–80 mg/100 g leaf material. Electron microscopy of the purified virus revealed spherical particles of 30 nm diameter (Gopinath et al. 1994).

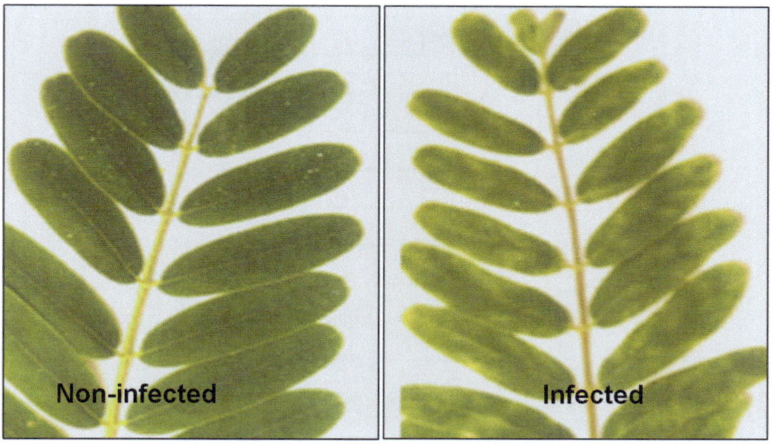

Fig. 18.1 Symptoms of *Sesbania mosaic virus* on *Sesbania grandiflora*

18.3 Genome Sequence Properties

The genomic RNA isolated from the purified virus is of size ~4 Kb. As the genome lacked a poly(A) tail, cDNA synthesis was carried out using an oligo(dT) primer on polyadenylated SeMV genomic RNA as the template. cDNA synthesis was also carried out using various sequence specific primers. The cDNA clones were sequenced either manually or by automated sequencing and the sequences were compiled using the FRAGMENT ASSEMBLY program of Wisconsin GCG package. Figure 18.2 shows the cladogram obtained from the complete genomic sequence of all the 19 species of sobemoviruses annotated thus far. The cladograms obtained using different ORFs are nearly identical suggesting that recombination events are probably not frequent or significant in the sobemovirus group. Also, within the sobemovirus group, viruses that infect dicotyledons cluster as a closely related group (I), while those that infect monocotyledons form another distinct group (II). The analysis clearly indicated that SeMV is a new member of the genus sobemovirus and is not a strain of SBMV (Lokesh et al. 2001).

The SeMV genome is 4148 nts long with a viral protein genome linked (VPg) at the 5′ end and it lacks a poly(A) tail or a tRNA-like structure at the 3′ end (Fig. 18.3). The genome is very compact and encodes for three overlapping open reading frames (ORFs). The 5′ proximal ORF (ORF-1) codes for the movement protein (MP), that helps in the cell to cell movement of the virus. The coat protein (CP), encoded by the 3′ proximal ORF (ORF-3) is involved in the encapsidation of the viral genome. The central ORF (ORF-2) codes for two polyproteins, 2a and 2ab. The ORF-2a codes for polyprotein 2a with the domain arrangement: membrane anchor (TM) – protease – VPg-P10-P8 domain. The ORF-2b codes for RNA-dependent RNA polymerase (RdRp) which is expressed as polyprotein 2ab with the domain arrangement: TM – protease-VPg-RdRp.

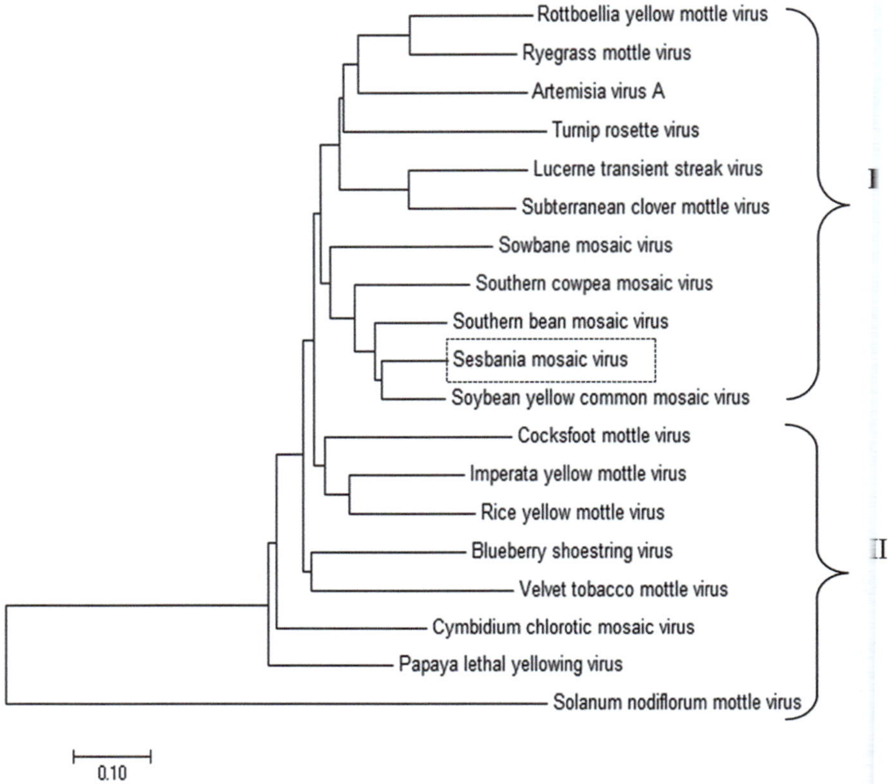

Fig. 18.2 Phylogenetic tree of sobemoviruses based on the complete nucleotide sequence of their genomes obtained using neighbour-joining method. The tree illustrates the relationship between the nineteen species of sobemoviruses

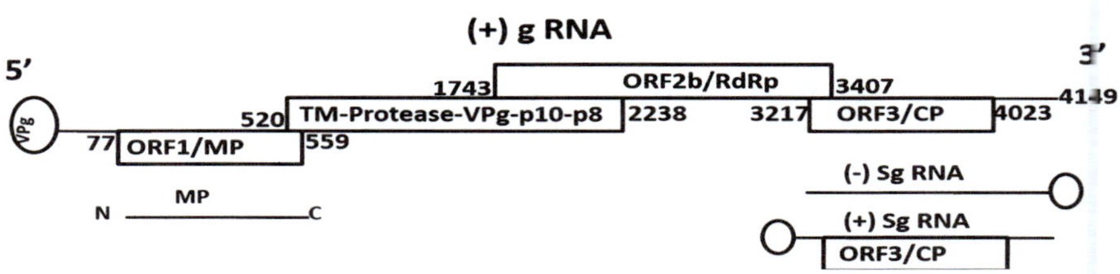

Fig. 18.3 Genome organization of Sesbania mosaic virus

18.4 Translation Mechanism

It has been proposed that only the 5′proximal ORF is translated in a eukaryotic mRNA (Kozak 1991). A comparison of the nucleotide sequence surrounding the ORF1 with the Kozak consensus sequence (A at −3 position and G at +4 position),

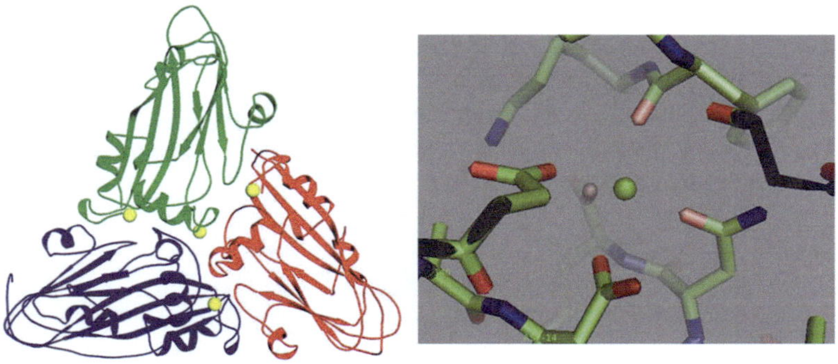

Fig. 18.4 (**a**) Organization of the three subunits A, B and C in the icosahedral asymmetric unit of SeMV capsid. The polypeptide folds are shown as ribbon diagrams. The amino terminal segment ordered only in C subunits is at the left lower edge of the Figure. *Spheres* correspond to ions *bound* to the capsids. The ions bound at the inter-subunit interfaces are calcium ions. (**b**) The geometry of calcium binding. This geometry is common to both T = 3 and T = 1 capsids of SeMV (Savithri and Murthy 2010)

that is suggested to yield a strong initiation site, revealed that the start codon of ORF1 is in a poor context for translation when compared to that of ORF2. ORF1 has a pyrimidine at the −3 position and does not have a G at the +4 position (Lutcke et al. 1987; Cavener and Ray 1991; Lokesh et al. 2001). In contrast, ORF 2 has a purine at the −3 position (Demler and de Zoeten 1991; Miller et al. 1995; Lokesh et al. 2001). Therefore, ribosomes which skip the first AUG codon, scan through the genomic RNA for the next potential start codon for translation initiation (leaky scanning mechanism) (Sivakumaran and Hacker 1998; Tamm and Truve 2000b).

The RdRp (coded by ORF 2b) is translated by a −1 ribosomal frame shift mechanism wherein the tRNAs in the ribosomal A and P sites simultaneously slip back one base on the mRNA before peptide bond formation, and the translation is resumed in the new (−1) reading frame (Fig. 18.4a, b). Pausing of ribosomes required during −1 ribosomal frame shifting is induced by the adjacent pseudoknot or stem-loop structure of genomic RNA. The slippery sequence, UUUAAAC and a stem-loop structure were found to be conserved at the −1 ribosomal frame shift site across all the members of sobemoviruses indicating that this mechanism of frame shifting is common to all sobemoviruses. However, the frequency of the frame shift appears to be very low and it is probably a regulatory mechanism for low expression of the viral encoded RdRp (Makinen et al. 1995; Tamm et al. 2009).

The CP is translated from the 3′ terminal ORF3. The start codon of this ORF is far from the 5′ terminus of the genomic RNA. The initiation of translation is efficient only when the initiation codon is proximal to the 5′ end. Therefore, many viruses follow a specialized mechanism involving synthesis of subgenomic RNAs (sgRNA) for efficient translation of the 3′ proximal ORFs. Sobemoviruses express sgRNA for the translation of CP, which is required in high amounts for encapsidation of the virus genome. Translation of CP via the sg RNA also leads to temporal regulation of the CP during the viral life cycle. Although presence of sgRNA has been reported for sobemoviruses, the mechanism of its expression has not been reported so far. Another strategy used by SeMV and other sobemoviruses is polyprotein processing in which the polyprotein 2a and 2ab are cleaved into functional products by the protease domain autocatalytically. The mechanism by which this processing is regulated is discussed later.

18.5 Structure and Assembly of SeMV

18.5.1 Three-Dimensional Structure of Native Virus

Structural studies on SeMV carried out at IISc, Bangalore using single crystal X-ray diffraction has provided essential details on the size of the virus particles, radial distribution of protein, three dimensional structure of the coat protein, capsid architecture, protein-protein, RNA-protein and metal ion mediated interactions, that account for the stability of the capsids. The three-dimensional structure of SeMV is the first viral structure reported from Asia. SeMV viral capsid comprises of 180 chemically identical subunits arranged in three different conformations/molecular environments termed as A, B and C. The A type subunits form 12 pentamers at the vertices of the icosahedral particle, whereas the B and C subunits form 20 hexamers resulting in icosahedral T = 3 symmetry (Caspar and Klug 1962).

There are two types of dimers, the C/C dimers related by the exact icosahedral twofold symmetry and the A/B dimers related by the quasi- two- fold symmetry. The coat protein consists of two domains viz. the N terminal random (R) domain which is followed by a well ordered shell domain (S domain) consisting of eight antiparallel β-strands connected by loops and a few helices. The amino terminal 71 residues are disordered in the A and B subunits (shown in green and red, respectively), whereas the amino terminal arms of the C subunit shown in blue are ordered from residue 44 (Fig. 18.4a). These additional ordered residues result in the formation of a β-annulus like structure at the quasi -six fold axes. The combination of β-annulus and the dimeric interactions of two- fold related protein subunits result in a scaffold that connects all the C subunits of the capsid. Therefore, it was proposed that the formation of β-annulus might initiate the assembly of T = 3 particles. The asymmetric unit of these icosahedral particles consists of the three CP subunits, A, B and C along with three interfacial calcium ions, resulting in a total of 180 calcium ions per capsid. These calcium ions of the icosahedral asymmetric unit located at A-B, B-C, C-A interfaces are related by the quasi- three- fold symmetry of the T = 3 particles. The calcium ions are coordinated by six ligands (D146 and D149 from one subunit and Y203, N267 and N268 from a neighbouring subunit and a water molecule hydrogen bonded to S116; Fig. 18.4b). The calcium binding motif was found to be conserved across various other sobemoviruses indicating its crucial role in the overall integrity and assembly of the capsids (Subramanya et al. 1993; Bhuvaneshwari et al. 1995).

18.5.2 Biophysical and Structural Studies on the Recombinant Capsids

Over expression of the SeMV CP in *E. coli* resulted in the formation of T = 3 virus-like particles (VLPs) that resembled the native virus particles and were also found to encapsidate CP mRNA as well as *E. coli* 23S rRNA. The thermal stability of the VLPs (Tm ~87 °C) was comparable to that of the wild type particles (Tm ~91 °C) (Lokesh et al. 2002). The three-dimensional structure of these VLPs was found to be very similar to that of the native virus (Sangita et al. 2005). In order to dissect the role of the N-terminal domain of the CP in viral assembly, deletion mutants of CP, CP-NΔ36 and CP-NΔ65 were over expressed *in E. coli* and purified. Electron microscopy of CP-NΔ36 capsids revealed the presence of predominantly T = 1 capsids (with 60 subunits) and a few pseudo T = 2 particles (with 120 subunits). However, deletion of 65 residues from the N terminus of CP (CP-NΔ65) resulted in exclusive formation of T = 1 particles which were less stable compared to the native virus with a Tm of 83 °C (Lokesh et al. 2002). The structure of these T = 1 particles was determined (Sangita et al. 2004) and the recombinant T = 1 particles were found to bind calcium ions in a manner identical to those of native capsids. The organization of the icosahedral dimeric unit in the T = 1 structures is similar to that of the quasi-dimer (AB dimer) of the T = 3 structure. Further, the structure of CP-NΔ31and CP-NΔ36T = 1 particles, were very similar to those of CP-NΔ65 particles (Sangita et al. 2004, 2005). These observations indicated the possible role of the N-terminal disordered domain of CP in the interaction with the RNA, which might be critical for the assembly of T = 3 particles.

There is an arginine rich motif (RRNRRRQR, residues 28–36; N-ARM) within the N-terminal R domain of SeMV CP, which is conserved across sobemoviruses. In order to investigate the role of the N-ARM in viral assembly, substitution mutations were carried out where some (CP-R32-36E) or all of the arginines (CP-R28-36E) were mutated to glutamate residues. These mutant CPs assembled in to empty T = 3 particles and were less stable. These results confirmed the role of the N-ARM in the encapsidation of the RNA (Satheshkumar et al. 2005a). Further, the N-terminal 65 residues of SeMV CP were replaced with polypeptides of similar length, such as the SpA-B-domain (SeMV-B; R-domain replaced by *Staphylococcus aureus* B-domain of Protein A), SeMV-P8 and SeMV-P10, but possessing significantly different structural features. Interestingly, all the chimeric CPs assembled into VLPs, but the particles were heterogeneous. However, TEM analysis of NΔ65-P8 revealed more uniform T = 3 particles. There is no sequence similarity between P8 and the R-domain, however, both these domains are intrinsically disordered and have similar pI and are rich in basic residues. These results suggest that the presence of a positively charged disordered domain promotes the least error-prone particle assembly. The crystal structure of NΔ65-BCP showed that substitution of the R-domain of SeMV CP with the B-domain lead to assembly of T = 1 particles (and not T = 3 particles) and that the B-domain was also disordered. Thus, particle assembly can also induce disorder in the otherwise ordered B-domain (Gulati et al. 2016).

In SeMV, the hydrogen bonding interactions between residues 48–52 from one C-subunit and residues 55–59 of the neighbouring three fold related C-subunit leads to the formation of β-annulus. To examine the importance of the β –annulus in viral assembly, the residues involved were (CP-Δ48–59) deleted. Interestingly, this mutant could assemble into T = 3 particles and the capsid stability was also unaffected. The structure of the CPΔ48–59 capsids showed the absence of the β-annulus in the structure. These results indicated that β-annulus is not essential for capsid assembly. It is plausible that the formation of β-annulus is a consequence of the viral assembly (Pappachan et al. 2008). Further, mutations targeting the interfacial residue (W170) occurring near the five fold axes resulted in complete disruption of T = 3 particle assembly and formation of dimers. Crystal structure of these dimers showed that they are of type A/B related by quasi two fold axes and not of type C/C related by icosahedral two fold symmetry in the T = 3 particles (Pappachan et al. 2009).

The three-dimensional structure of the native virus indicated that D146 and D147 are crucial for calcium binding. Interestingly, CP-NΔ65-D146N-D149N capsids were similar to those of CP-NΔ65 T = 1 capsids. However, similar mutations in the full length CP resulted in heterogeneous particles suggesting that metal ion mediated interactions are more crucial for the stability of the T = 3 particles than for the T = 1 particles (Satheshkumar et al. 2004a).

18.6 Model for SeMV Assembly

Based on extensive mutational analysis of SeMV rCP, a possible mechanism for capsid assembly was proposed. Initially a pentamer of A/B type dimers in which β-annulus is disordered could be formed at one of the icosahedral five-fold axes. Further, assembly could proceed by interaction of ARM with the RNA. This could result in ordering of the amino terminal segments leading to CC dimers added to the 10-mer complex and the formation of β-annulus. Further addition of CP dimers could lead to the formation of swollen T = 3 particles. Calcium binding to the assembling or the assembled particles results in compact T = 3 particles (Savithri and Murthy 2010).

18.7 Polyprotein Processing in SeMV

The ORF 2 of SeMV contains a protease domain which is involved in processing of the polyproteins 2a and 2ab. In order to identify the cleavage sites and also to characterize the products of processing, the gene sequences corresponding to polyprotein 2a and its deletion mutant were cloned from the SeMV full-length cDNA and over expressed in *E. coli.* The cleavage products were identified by western blot analysis using antibodies to P8 and protease. It was concluded that polyprotein 2a undergoes cleavage at four sites E132-S133, E325-T326, E402-T403 and E498-S499 (Satheshkumar et al. 2004b).

Based on the sequences of all four identified cleavage sites, a consensus sequence was derived for the SeMV serine protease cleavage site, i.e., N/Q-E-T/S-X (where X is an aliphatic residue). Interestingly, the mutation of cleavage sites at positions other than E132-S133 did not affect the protease function, whereas the mutant E132A was found to be defective in processing suggesting that the removal of N-terminal membrane anchoring domain from the protease/polyprotein might be essential for the subsequent processing of the polyprotein 2a at the other sites (Nair and Savithri 2010a).

ORF2b is expressed using the −1 reading frame and therefore to translate ORF2a and ORF2b as a single polyprotein 2ab in *E. coli*, an additional nucleotide T was introduced at position 1743 (region of ribosomal frame shifting) by site directed mutagenesis (Fig. 18.4b). Even though the cleavage sites E132-S133, E325-T326 and E402-T403 were retained in the polyprotein 2ab, interestingly, no cleavage was observed between VPg and RdRp (E402-T403). Therefore, the conformation of the domains that determine the context of the cleavage site and not just the sequence could be the determinant of polyprotein processing. The unprocessed VPg-RdRp could be of functional significance in negative (−) strand RNA synthesis. A subsequent cleavage between VPg-RdRp would release RdRp that might perform the positive (+) strand synthesis (Nair and Savithri 2010a).

18.8 Structure and Function of SeMV Protease

As the protease domain of the ORF-2 of SeMV is essential for polyprotein processing, structural studies on the SeMV protease were initiated. Though SeMV protease was identified to be a serine protease of trypsin fold, it showed negligible sequence identity with any other proteases including viral proteases. The crystal structure of SeMV protease was determined at a resolution of 2.4 Å by multiple isomorphous replacement combined with anomalous scattering (MIRAS). The polypeptide fold revealed that it is indeed a serine protease of trypsin fold with a catalytic triad formed by the residues His181, Asp 216 and Ser 284 (Gayathri et al. 2006). Mutation of any of the active site catalytic triad residues completely abolished the polyprotein processing (Satheshkumar et al. 2004b). As depicted in the Fig. 18.5a, the protease structure consists of two β-barrels (domains I and II) that are connected by a long inter-domain loop. The active site and the substrate binding pocket occur in between the two domains and are solvent exposed. Interestingly, four aromatic residues are also solvent exposed. The functional significance of these residues will be discussed later. The crystal structure of SeMV protease is the first report of the structure of a non-structural protein from a plant virus (Gayathri et al. 2006).

As mentioned earlier, the serine protease domain cleaves the polyprotein at four specific E-T/S sites and not only the P1-P1' (E-T/S) residues at these sites, but also the flanking P2 (N/Q) and P2' (A/V/L) residues are conserved or conservatively substituted. In order to identify the structural basis for such stringency in substrate recognition, structure based alignment of SeMV protease was carried out with other Glu/Gln specific proteases. Based on such a comparison, residues T279, A280,

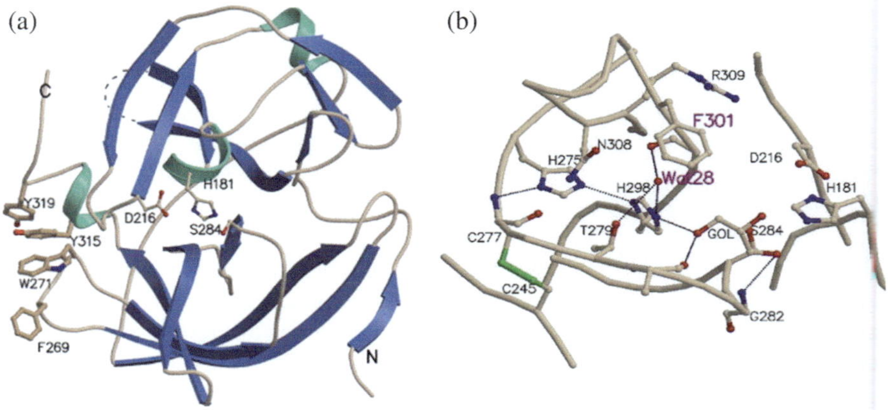

Fig. 18.5 Polypeptide fold of SeMV protease. (**a**) Ribbon drawing of the SeMV serine protease. The active site residues H181, D216 and S284 and the surface exposed aromatic residues are shown in ball-and-stick representation. The stretch of residues 171–173, that could not be traced, is shown as a *broken line*. (**b**) Stick diagram of SeMV serine S1-binding pocket. Residues H298, T279, H275 and N308 form the S1-binding pocket and residue R309 forms the S2-binding pocket (Reproduced with permission)

H298, F301 and N308 were proposed to form the S1-binding pocket that could accommodate the Glu of P1 position in the substrate (Fig. 18.5b). It was also found that these residues were highly conserved across known sobemoviral proteases. The S2-binding pocket of SeMV protease is formed by residue R309 that can recognize Asn/Gln at the P2 position (Gayathri et al. 2006)

It was observed that the protease was inactive, although the catalytic triad residues could be overlaid very precisely with those of trypsin. Interestingly, the protease-VPg fusion protein was found to be active. Therefore, mutational studies were carried out using the protease-VPg cleavage site mutant. These studies clearly indicated that H298, T279 and N308 of the S1 binding pocket were crucial for the protease activity. Also, it was found that the P2 (Asn/Gln) residue recognized by R309 plays an important role in determining the substrate specificity. Further, the crystal structure of the SeMV protease shows the presence of a disulphide bond between C248 and C277, of which C277 lies on the strand forming the S1 specificity site. Disulphides are generally implicated in the stability of the serine proteases. However, the mutation of C277 did not affect the protease activity indicating that the disulphide bond is not important for the protease function or structural stability unlike in the case of trypsin and chymotrypsin. These studies clearly indicate the differences in the structure of the SeMV protease that dictate the substrate specificity (Gayathri et al. 2006) (Fig. 18.5).

18.9 Viral Encoded Intrinsically Disordered Domain and Their Interaction with Other Globular Domains: Functional Implications

Intrinsically disordered or natively unfolded proteins lack defined secondary or tertiary structure. They are rich in charged residues and contain relatively few hydrophobic residues. These proteins attain folded conformations upon interaction with their specific partners or ligands. Their higher structural flexibility permits interaction with a large number of targets and thereby allows such proteins to regulate a number of cellular processes. Viral encoded intrinsically disordered domains/proteins can interact with other viral proteins/host factors and modulate their function. Initially, a bioinformatics analysis of all the proteins encoded by SeMV was carried out to identify the intrinsically disordered domains.

Figure 18.6 shows the fold index analysis of the ORF-1 encoded movement protein (MP), ORF-2 encoded polyproteins 2a and 2ab, and the ORF-3 encoded CP. The N- and C-terminal segments of MP which have negative fold index are predicted to be unfolded/disordered. Similarly, VPg and P8 in polyprotein 2a and VPg and the C terminal domain of RdRp in polyprotein 2ab are predicted to be disordered (Nair et al. 2012). The N-terminal R domain of CP which is structurally disordered as described earlier is indeed predicted to be disordered by the fold index analysis. In this section, the structure, interactions and functions of two of these domains, namely VPg and P8 are discussed.

18.10 Intrinsically Disordered VPg: Implications on Protease Activity

SeMV encoded VPg is a 9 kDa protein and consists of 77 amino acids. Analysis of the sequence using the software PONDR (Predictors of Natural Disordered Regions, available at http://www.pondr.com/) showed that it also falls in the category of disordered domains. (Satheshkumar et al. 2005b). The CD spectral studies with purified recombinant VPg revealed maximum negative ellipticity at 200 nm and absence of the same at 222 nm, typical of random coil structures, consistent with the structural predictions. Furthermore, the fluorescence spectrum of VPg showed a maximum emission at 357 nm even in the native state indicating lack of significant tertiary structure in VPg. In addition, VPg eluted abnormally on size exclusion chromatography as a polypeptide of higher molecular size, confirming the disordered nature of the protein. Interestingly, the CD spectral profile of ΔN70PV-E325A (protease-VPg cleavage site mutant) was characterized by the presence of a positive peak at 230 nm that was absent in the protease as well as the VPg CD spectra suggesting a conformational change in the fusion protein. As mentioned earlier, the protease was active only when expressed as protease-VPg fusion protein. Thus, the disordered VPg, upon interaction, could activate the protease. The positive CD at 230 nm was used as a probe to decipher the nature of interaction between the two domains.

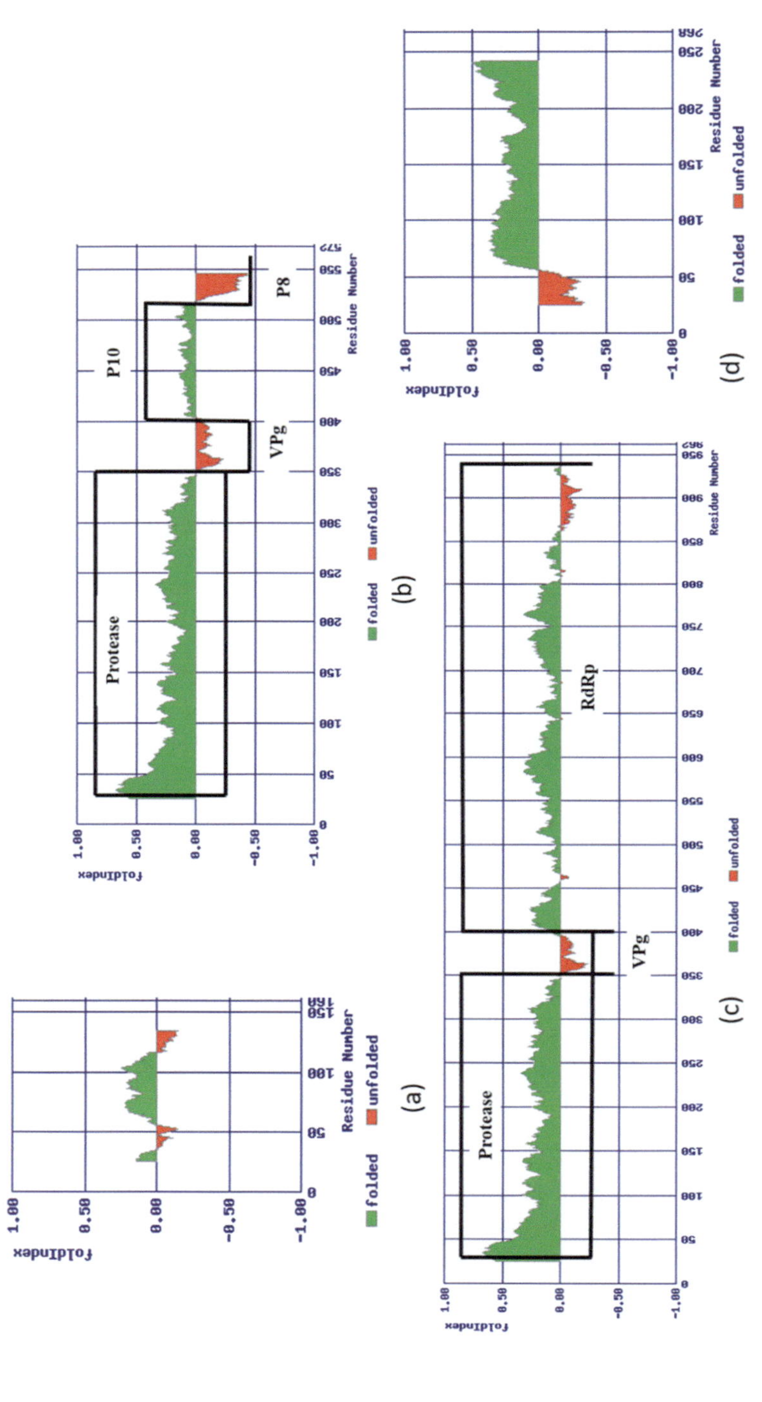

Fig. 18.6 Prediction of disordered domains in SeMV encoded proteins by fold index analysis **(a)** MP, **(b)** polyprotein 2a, **(c)** polyprotein 2ab and **(d)** CP

The presence of positive CD band at 230 nm could be either due to the presence of polyproline II (PPII) like helix (Parrot et al. 2002) or aromatic amino acids (Grishina and Woody 1994; Woody 1994). However, deletion of the C-terminal 28 amino acid residues encompassing a proline-rich sequence did not abrogate the 230 nm peak. VPg has two tyrosine and three tryptophan residues in its amino acid sequence. CD spectral studies of site-directed mutants of all these aromatic residues in the fusion protein (ΔN70PV-E325A) revealed that only the mutation of W43 to F (ΔN70PV-E325A-W43F) abolished the positive CD peak. Interestingly, the protease activity was significantly affected in this mutant (ΔN70PV-E325A-W43F) compared to that of ΔN70PV-E325A. It is possible that W43 of VPg is involved in stacking interaction with an aromatic residue from the protease domain, which leads to the positive CD band at 230 nm and modulates the structure as well as function of the fusion protein (Satheshkumar et al. 2005b). This is the first report of interaction of a natively unfolded domain with the globular protease domain leading to its activation.

As shown in Fig. 18.5a F269, W271, Y315 and Y319 are solvent exposed and form a stack near the C-terminus of the protein (Fig. 18.5a). Site-directed mutants of these residues were generated in the ΔN70PV-E325A (PVEA) mutant to identify their role in mediating the stacking interactions with the W43 of VPg and hence modulate the activity of the protease. H275, was also mutated as it lies close to the W271 in sequence and is conserved in the protease domain of all the known sobemoviruses. *Cis*- and *trans*-cleavage assays suggested that residues W271 and H275 but not Y315 or Y319 are crucial for protease activity. There was a concomitant loss of 230 nm peak in the PVEA-W271A and PV-H275A mutants. Thus, aromatic stacking interactions between W43 of VPg and W271 as well as H275 of the protease results in the positive CD peak at 230 nm in protease-VPg and activation of the protease (Nair et al. 2008).

18.11 Intrinsically Disordered P8 Domain: Activator of ATPase Activity of P10

P8 is another domain in the polyprotein 2a which is predicted to be intrinsically disordered. The far-UV spectra of hexa-histidine tagged P8 showed negative ellipticity at 200 nm, a characteristic of random coil, confirming that it is also natively unfolded. In the domain arrangement of polyprotein 2a, P10 precedes P8 and the presence of P10-P8 (P18) fusion protein was detected in the processing of this polyprotein. It was demonstrated that the ATPase activity of P18 was 2.5-fold higher than that of purified recombinant P10 protein (Nair and Savithri 2010b) further confirming the role of intrinsically disordered domains in the regulation of protein function.

18.12 SeMV Encoded RNA-Dependent RNA Polymerase (RdRp)

All positive stranded RNA viral [(+) RNA] genomes encode a RNA-dependent RNA polymerase (RdRp) that is essential for the replication of their genomes. The sequence analysis of RdRp domain revealed that all the conserved motifs of viral RNA-dependent RNA polymerases are present in SeMV RdRp.

The SeMV gRNA is covalently linked to VPg at the 5′ end (Lokesh et al. 2001). Nucleotidylylation of the VPg by RdRp is the first step of primer dependent replication in positive sense RNA viruses. Although the purified recombinant RdRp failed to nucleotidylylate VPg in vitro, it was able to carry out the polymerization reaction in the absence of the VPg primer when either genomic RNA or subgenomic RNA was used as the template. Further, the active site GDD/GAA mutant was not able to synthesize RNA suggesting that the motif is crucial for the polymerase activity and that the activity was intrinsic to the purified RdRp. The denatured product moved faster compared to the non denatured product and its position corresponded to the single-stranded template showing that the product of the reaction was double stranded. Thus, SeMV RdRp initiates de novo RNA synthesis on (+) gRNA or (+) sgRNA templates. Further, the RNA synthesis with (+) sgRNA as template was significantly reduced when limiting concentrations of ATP and GTP were used, whereas, limiting concentrations pyrimidines had no significant effect suggesting that SeMV RdRp prefers purines at the initiation step (Govind and Savithri 2010).

The secondary structure of 3′ UTR of (+) SeMV RNA was predicted using the M-fold program (Mathews et al. 1999). The UTR folds into stem-loop structures with a ΔG of −29.2 kcal/mol. The *cis* acting elements in the 3′ UTR were identified by systematic deletion analysis. Interestingly, the stem-loop structure 29–57 nt from the 3′ end with 8 bp stem and 12 nt loop seems to play an important role in determining the preferential initiation on the (+) RNA. However, under in vivo conditions, viral or host factors might modulate the polymerase conformation and thereby allow primer (VPg) dependent initiation of replication (Govind and Savithri 2010).

18.13 SeMV Infectious Clone

The in vivo functions of viral encoded genes can be deciphered using an infectious cDNA (icDNA) clone. The delivery of modified Ti-plasmid containing the icDNA via *Agrobacterium* -mediated transformation into the plant cell nucleus allows the transcription of the viral genome (Annamalai and Rao 2005; Citovsky et al. 2007; Lee and Gelvin 2008). The full-length transcripts thus generated enter the cytosol and express the viral encoded proteins leading to the replication of the viral RNA followed by encapsidation and movement (Annamalai and Rao 2005). Therefore, the full-length icDNA clone for SeMV was constructed and agroinfiltrated to *Sesbania grandiflora* plants. However, the efficiency of infection was very low. Therefore, *Cyamopsis tetragonoloba*, the local lesion host, was tested and was found to be more suitable. Time course analysis indicated CP accumulation from

6 days post infection, which increased further after 9 days. Northern analysis of the total RNA revealed complete replication of in vivo generated viral RNA. The SeMV icDNA clone had 4 additional nucleotides at the 5′end and 21 nucleotides at the 3′ end when compared to the wild type SeMV sequence. However, sequencing of the viral RNA isolated from virions obtained from SeMV icDNA infected leaves indicated lack of any extra nucleotides at both 5′ and 3′ ends suggesting that viral progeny RNA was repaired in vivo (Govind et al. 2012).

In order to verify the role of polyprotein processing in planta, mutants of all four cleavage sites (E132A, E325A, E402A and E498A) were generated using the SeMV icDNA as the template and the mutant icDNAs were tested for their ability to infect *Cyamopsis tetragonoloba* plants. There was a drastic reduction in viral infection for all the cleavage site mutants, which was evident from the absence of CP accumulation in the infiltrated cotyledons at 9 days post infection. However, co-infiltration of ORF-2 construct (pEAQ 2ab) *in trans* with all the respective cleavage site mutants could rescue the viral infection and result in CP accumulation similar to that of the wild type icDNA. This further confirmed the role of correct polyprotein processing in viral replication (Govind et al. 2012).

18.14 Viral Movement

Plant viruses have to cross the cell wall barrier while moving from one cell to another. They use resident communication networks in plants that aid in transporting macromolecules through the plasmodesmata and redirect them in such a way that the viral genome is transported and viral infection spreads. ORF-1 of SeMV that codes for the MP has only a limited sequence similarity with other well characterized MPs. One of the characteristics of viral MPs is that they bind to nucleic acids. Purified GST-MP could indeed bind to genomic RNA in a concentration dependent manner. Interestingly, no interaction was observed with other types of nucleic acids such as M13 ssDNA, dsDNA and a nonspecific viral RNA. These results are in contrast to the binding studies reported with other viral MPs, which show that the interaction is non-specific in vitro. More importantly, the interaction with the genomic RNA was lost upon treatment with pronase suggesting that SeMV MP recognizes the genomic RNA via the interaction with VPg. CP is one of the proteins shown to interact with MPs of several plant viruses. It was observed that GST MP could indeed interact strongly with the native virus and that the N-terminal 49 residues of MP were crucial for this interaction (Chowdhury and Savithri 2011a). This was further confirmed by Yeast two hybrid interaction assays as discussed in the next section.

18.15 Role of Protein-Protein Interactions in SeMV Replication and Spread

18.15.1 Interaction of MP with Viral Encoded Proteins

As mentioned earlier, SeMV MP could interact with CP in vitro, it was further validated with Y2H assay using standard Matchmaker system. Yeast cells co-transformed with pGBK T7 MP and pGAD T7 CP could grow on all nutritional selection media similar to the positive control comprising p53 and T-Ag suggesting that MP and CP also interact with each other under the ex vivo conditions of Y2H system.

Similar experiments carried out using systematic deletion constructs of MP clearly indicated that the interaction of MP and CP was significantly reduced on deletion of 49 amino acids from the N-terminus of MP corroborating the results obtained by in vitro studies. However, the C-terminal deletion of MP had no effect on the interaction.

In addition to CP, other viral encoded proteins would also be required for efficient cell-to-cell spread of SeMV genome. Therefore, the interaction of MP with other non- structural proteins/domains of SeMV were also monitored using the Y2H assay. A significant growth of AH109 cells was observed when p53-T Ag (positive control), CP-MP, P10-MP, VPg-MP were co-expressed but not when P8-MP, RdRp-MP and Pro-MP were expressed together. Thus, MP, viral RNA covalently linked to VPg, P10 and CP could all be part of the movement complex (Chowdhury and Savithri 2011b).

18.15.2 Interaction of RdRp and Other Viral Encoded Ancillary Proteins

Complete replication in +sense RNA viruses requires the assembly of the replicase complex, which consists of viral encoded proteins as well as host factors (Kushner et al. 2003; Lin et al. 2009; Vidalain and Tangy 2010). Therefore, interaction of viral encoded ancillary proteins and RdRp is an integral part in the formation of the replication complex. In many viruses it has been reported that one or more of the viral encoded proteins directly interact with RdRp domain and modulate the polymerase activity (Shirota et al. 2002; Shen et al. 2008).

Hence, all the gene segments of SeMV corresponding to individual domains of polyprotein 2a (Pro, VPg, p10, p8, p18 and p27) and RdRp were cloned into the Matchmaker system. The recombinant clones in pGBK/pGAD vectors were co-transformed in pairs into *Saccharomyces cerevisiae* strain AH109 and analysed. In addition to pGBKT7-p53 and pGADT7T antigen (positive control), a significant growth was observed only when RdRp and p10 were co-expressed. Similar analysis of C-terminal deletion mutants of RdRp showed that the C-terminal disordered domain was essential for interaction with P10. It is possible that interaction of this domain with P10 might modulate the function of RdRp. To test this possibility, RdRp and P10 were co-expressed and the RdRp-P10 complex was purified. The complex showed eight to ten fold increase in activity compared to RdRp alone, indicating that P10 may activate the RdRp by masking the C-terminal disordered domain. This was further confirmed by the observation that RdRpCΔ43 was highly active and the activity was comparable to that of the RdRp-P10 complex and it did not increase any further when co-purified with P10 as RdRpCΔ43-P10 complex. The viral replication complex (VRC) is membrane associated and in the case of SeMV, the VRC could contain RdRp, P10 as well as the MP along with the genomic RNA and other yet to be identified host factors (Govind et al. 2014).

18.16 Biodistribution and Toxicity of SeMV in Mice

It has been demonstrated that nanoparticles (NPs) have many favourable properties such as controlled and targeted drug release, improved solubility, reduced dose-related toxicity and better absorption that make them attractive candidates for detection, imaging and treatment of diseases. In order to use these NPs in biomedicine, a detailed biodistribution and toxicity analysis needs to be carried out (Powers et al. 2006). Such studies will also allow modification and optimization of their biodistribution and clearance (Bruckman et al. 2014). Recently, virus nanoparticles (VNPs) are being investigated for various biomedical applications. Plant viruses have a distinct advantage over other NPs as they are generally non-toxic and biodegradable. With a view to use SeMV as a possible nanocarrier, the toxicity and biodistribution of SeMV was examined. Different doses of SeMV were administered intravenously or orally to mice and they were examined for some important clinical signs, body weights, and for haemagglutination. Overall, the animals appeared essentially normal (Vishnu Vardhan et al. 2016). SeMV was localized in liver and spleen up to 72 h post administration in only intravenously administered mice. In the orally administered animals, most of the virus was cleared within 16 h and it was not detected in any of the organs (Vishnu Vardhan et al. 2016). Further, no pathological or biochemical changes were observed in the control as well as treated animals demonstrating that SeMV is non-toxic and does not cause any changes in the vital organs. Further, it was observed that the administration of SeMV did not result in significant immune response indicating that the SeMV can serve as a safe bio-nanoparticle for biomedical applications (Vishnu Vardhan et al. 2016).

18.17 SeMV Virus-Like Particles (VLPs) as Nanocarriers for Antibody Delivery

Antibody based clinical therapy that can disrupt protein-protein interactions and inhibit signalling pathways are considered to be invaluable in cancer therapeutics (Chames et al. 2009; Scott et al. 2012). However, since most of the antibodies cannot cross the cell membrane barrier and internalize in cells, majority of the FDA approved antibodies target the surface exposed receptors. There have been several attempts to internalize antibodies by the use of cell penetrating peptides, liposome conjugation or the use of synthetic gold nanoparticles. The detailed biochemical and structural knowledge available on SeMV capsids prompted the development of SeMV based nanocarriers. A close examination of the structure of SeMV CP revealed a surface exposed eight residue HI loop (238–245 aa) near the five- fold and six -fold axes that might be suitable for genetic manipulations without disturbing particle assembly. The midpoint of the HI loop (Serine 242) was selected for insertion of the antibody binding B domain of *Staphylococcus aureus* protein A. The chimeric CP with the B domain in the loop region (SLB) could assemble into VLPs, although, they were heterogeneous in nature. In addition, antibodies could bind via their Fc region to the exposed B domain of SLB.

18.18 Demonstration of Entry of VLPs into Mammalian Cells

Fluorescent labelling is one of the major techniques for imaging of cellular proteins at various time intervals. SeMV CP and SLB VLPs were fluorescently labelled with Alexafluor 488. Interestingly, CP 488 and SLB 488 were able to enter into the cytoplasm of HeLa cells and the fluorescence reached a maximum in about 4–8 h (Fig. 18.7a). After 8 h, the fluorescence of the particles decreased indicating possible degradation of VLPs. Further, the presence of BSA or sheep serum had no effect on the entry of VLPs (Fig. 18.7b). Similar studies with other mammalian cell lines

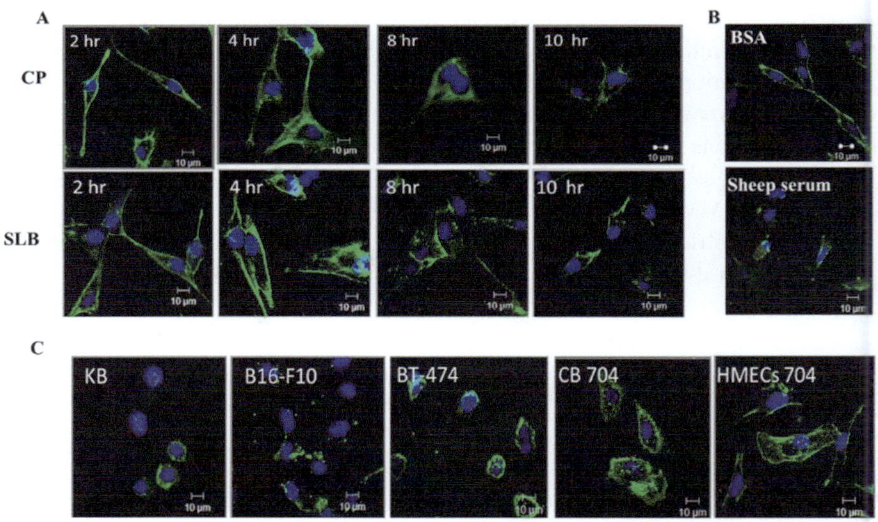

Fig. 18.7 Demonstration of virus like particle entry in to mammalian cells using confocal microscopy. (**a**) Confocal images of HeLa cells incubated with CP 488 or SLB 488 for 2, 4, 8 and 10 h at 37 °C. (**b**) Confocal images showing the entry of CP 488 in HeLa cells for 2 h in the presence of BSA/sheep serum. (**c**) Confocal images showing the entry of 1.58 nM SLB 488 in KB, B16-F10, BT-474, CB 704 and HMECs 704 cells

viz. KB, BT 474 (mammary duct cancer cells), CB 704 etc. showed that SeMV VLPs were highly versatile but non-specific in cellular entry (Fig. 18.7c).

18.19 Intracellular Delivery of Antibodies by SLB

Since SLB has the unique ability of binding the IgGs, it was interesting to check if it could deliver antibodies inside cells. D6F10 (anti-abrin), anti-α tubulin and Herclon (anti-HER2 receptor) were used as cargo for this purpose. As expected D6F10 633 by itself did not enter the cells (Fig. 18.8a). However, D6F10 was delivered inside the cells when SLB 488-D6F10 633 was incubated with HeLa cells (Fig.18.8b). In contrast, similar incubation with CP did not result in the intracellular delivery of antibodies (Fig. 18.8c) further confirming the ability of SLB, but not CP, to deliver the antibodies inside the cell. In addition, the delivered antibodies were shown to be functional as D6F10 delivered through SLB could rescue protein synthesis inhibition as well as apoptosis caused by abrin.

Tubulin is an essential cytoskeletal element. Hence disruption of tubulin network by anti-tubulin antibodies, particularly in cancer cells, could be of use in cancer therapy. It was shown that SLB could deliver anti tubulin antibodies into the cytoplasm that disrupted the tubular network. Herclon (Herceptin/Trastuzumab) is being effectively used in the treatment of HER2 positive breast cancer patients. These antibodies bind to the HER receptor and thereby inhibit downstream signalling pathways that lead to cell proliferation. Surprisingly, Herclon delivered via the SLB

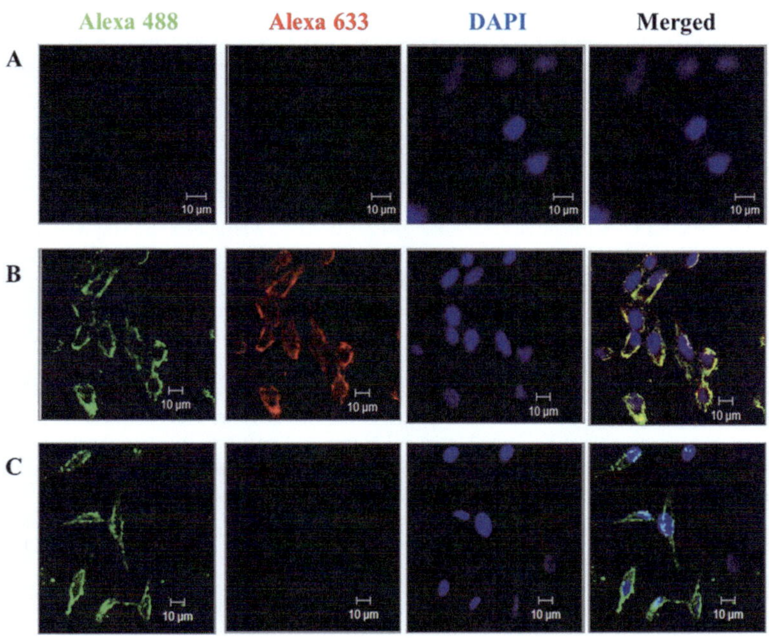

Fig. 18.8 SLB mediated D6F10 delivery in HeLa cells. Confocal images of HeLa cells treated with (**a**) D6F10 633, (**b**) SLB 488-D6F10 633 and (**c**) CP 488-D6F10 633 for 4 h

showed 83% cytotoxicity, which was to 3.33 times higher than that observed with Herclon alone. The increased levels of cytotoxicity were not contributed by CP/SLB as both had minimal effects on cell viability in vitro. These results demonstrate the potential applications of SeMV as a nanocarrier in antibody based therapeutics (Abraham et al. 2016).

18.20 Concluding Remarks

Extensive studies carried out on SeMV has resulted in a much deeper understanding of the molecular mechanism of assembly, polyprotein processing, replication, regulation of protein function by the intrinsically disordered domains encoded by the genome, cell to cell movement, and possible application of plant viruses as nanocarriers for intracellular delivery of antibodies and drugs. These studies could be further augmented by the use of SeMV icDNA which has already been used to demonstrate infectivity and importance of polyprotein processing in replication. The identification of host factors in the viral replication complexes as well as the movement complexes is another area of research that could provide deeper insights into mechanism of plant viral infection.

Acknowledgements We thank late Prof. M.V. Nayudu for initiation of the work summarized in this review on SeMV and Prof. N. Appaji Rao for his encouragement. We acknowledge the contributions made by the following students, research associates and project assistants from HSS and MRN laboratories- Subramanya, Gopinath, Bhuvaneshwari, Lokesh, Satheshkumar, Sangita, Saravanan, Sundareshan, Vijay, Prasad, Gowri, Anju, Subashchandrabose, Smita, Gayathri, Soumya, Govind, Ambily, Usha, Amritha, Ashuthosh, Abhinandan, Kalyani, Sushmitha and Kavitha. The financial support from IISc, DST and DBT is gratefully acknowledged. A.B. and G.P.V.V. acknowledge UGC-CSIR and DST-INSPIRE respectively for their fellowships.

References

Abraham A, Nataraj U, Karande AA, Gulati A, Murthy MRN, Murugesan S, Mukunda P, Savithri HS (2016) Intracellular delivery of antibodies by chimeric Sesbania mosaic virus (SeMV) virus like particles. Sci Rep 6:21803

Annamalai P, Rao AL (2005) Replication-independent expression of genome components and capsid protein of brome mosaic virus in planta: a functional role for viral replicase in RNA packaging. Virology 338(1):96–111

Bhuvaneshwari M, Subramanya HS, Gopinath K, Savithri HS, Nayudu MV, Murthy MRN (1995) Structure of Sesbania mosaic virus at 3Å resolution. Structure 3(10):1021–1030

Bruckman MA, Randolph LN, Van Meter A, Hern S, Shoffstall AJ, Steinmetz NF (2014) Biodistribution, pharmacokinetics and blood compatibility of native and PEGylated tobacco mosaic virus nano-rods and -spheres in mice. Virology 449:163–173

Caspar DLD, Klug A (1962) Physical principles in the construction of regular viruses. Cold Spring Harb Symp Quant Biol 27:1–24

Cavener DR, Ray SC (1991) Eukaryotic start and stop translation sites. Nucleic Acids Res 19(12):3185–3192

Chames P, Van Regenmortel M, Weiss E, Baty D (2009) Therapeutic antibodies: successes, limitations and hopes for the future. Br J Pharmacol 157:220–233

Chowdhury SR, Savithri HS (2011a) Interaction of Sesbania mosaic virus movement protein with the coat protein–implications for viral spread. FEBS J 278(2):257–272

Chowdhury SR, Savithri HS (2011b) Interaction of Sesbania Mosaic virus movement protein with VPg and P10: implication to specificity of genome recognition. PLoS One 6(1):e15609. doi:10.1371/journal.pone.0015609

Citovsky V, Kozlovsky SV, Lacroix B, Zaltsman A, Dafny-Yelin M, Vyas S, Tovkach A, Tzfira T (2007) Biological systems of the host cell involved in Agrobacterium infection. Cell Microbiol 9(1):9–20

Demler SA, de Zoeten GA (1991) The nucleotide sequence and luteovirus-like nature of RNA1 of an aphid non-transmissible strain of pea enation mosaic virus. J Virol 72:1819–1834

Gayathri P, Satheshkumar PS, Prasad K, Nair S, Savithri HS, Murthy MRN (2006) Crystal structure of the serine protease domain of Sesbania mosaic virus polyprotein and mutational analysis of residues forming the S1-binding pocket. Virology 346(2):440–451

Gopinath K, Sundareshan S, Bhuvaneswari M, Karande A, Murthy MRN, Nayudu MV, Savithri HS (1994) Primary structure of Sesbania mosaic virus coat protein: its implications to the assembly and architecture of the virus. Indian J Biochem Biophys 31(4):322–328

Govind K, Savithri HS (2010) Primer-independent initiation of RNA synthesis by SeMV recombinant RNA-dependent RNA polymerase. Virology 401(2):280–292

Govind K, Makinen K, Savithri HS (2012) Sesbania mosaic virus (SeMV) infectious clone: possible mechanism of 3′ and 5′ end repair and role of polyprotein processing in viral replication. PLoS One 7(2), e31190

Govind K, Bakshi A, Savithri HS (2014) Interaction of Sesbania mosaic virus (SeMV) RNA-dependent RNA polymerase (RdRp) with the p10 domain of polyprotein 2a and its implications in SeMV replication. FEBS Open Bio 4:362–369. doi:10.1016/j.fob.2014.03.009

Grishina IB, Woody RW (1994) Contributions of tryptophan side chains to the circular dichroism of globular proteins: exciton couplets and coupled oscillators. Faraday Discuss 99:245–262

Gulati A, Murthy A, Abraham A, Mohan K, Natraj U, Savithri HS, Murthy MRN (2016) Structural studies on chimeric Sesbania mosaic virus coat protein: revisiting SeMV assembly. Virology 489:34–43

Hull R (1977) The grouping of small spherical plant viruses with single RNA components. J Gen Virol 36:289–295

King AMQ, Adams MJ, Carstens EB, Lefkowitz EJ (2012) Virus taxonomy: ninth report of the international committee on taxonomy of viruses. Elsevier Academic, San Diego. ISBN: 978-0-12-384684-6

Kozak M (1991) Structural features in eukaryotic mRNAs that modulate initiation of translation. J Biol Chem 266:19867–19870

Kushner DB, Lindenbach BD, Grdzelishvili VZ, Noueiry AO, Paul SM, Ahlquist P (2003) Systematic, genome-wide identification of host genes affecting replication of a positive-strand RNA virus. Proc Natl Acad Sci U S A 100(26):15764–15769

Lee LY, Gelvin SB (2008) T-DNA binary vectors and systems. Plant Physiol 146(2):325–332

Lin JY, Chen TC, Weng KF, Chang SC, Chen LL, Shih SR (2009) Viral and host proteins involved in picornavirus life cycle. J Biomed Sci 16:103

Lokesh GL, Gopinath K, Satheshkumar PS, Savithri HS (2001) Complete nucleotide sequence of Sesbania mosaic virus: a new virus species of the genus Sobemovirus. Arch Virol 146(2):209–223

Lokesh GL, Gowri TD, Satheshkumar PS, Murthy MRN, Savithri HS (2002) A molecular switch in the capsids protein controls the particle polymorphism in an icosahedral virus. Virology 292:211–223

Lutcke HA, Chow KC, Mickel FS, Moss KA, Kern HF, Scheele GA (1987) Selection of AUG initiation codon differs in plants and animals. EMBO J 6:43–48

Makinen K, Naess V, Tamm T, Truve E, Aaspollu A, Saarma M (1995) The putative replicase of the cocksfoot mottle sobemovirus is translated as a part of the polyprotein by −1 ribosomal frameshift. Virology 207:566–571

Mathews DH, Sabina J, Zuker M, Turner DH (1999) Expanded sequence dependence of thermodynamic parameters improves prediction of RNA secondary structure. J Mol Biol 288(5):911–940

Miller WA, Dinesh-Kumar SP, Paul CP (1995) Luteovirus gene expression. CRC Rev Plant Sci 4:179–211

Nair S, Savithri HS (2010a) Processing of SeMV polyproteins revisited. Virology 396(1):106–117

Nair S, Savithri HS (2010b) Natively unfolded nucleic acid binding P8 domain of SeMV polyprotein 2a affects the novel ATPase activity of the preceding P10 domain. FEBS Lett 584(3):571–576

Nair S, Gayathri P, Murthy MRN, Savithri HS (2008) Stacking interactions of W271 and H275 of SeMV serine protease with W43 of natively unfolded VPg confer catalytic activity to protease. Virology 382(1):83–90

Nair S, Murthy MRN, Savithri HS (2012) Intrinsically disordered domains of Sesbania mosaic virus encoded proteins. In: Vladimir U, Sonia L (eds) Flexible viruses: structural disorder in viral proteins. Wiley, Hoboken., ISBN: 978–0–470-61831-8, pp 257–276

Pappachan A, Subashchandrabose C, Satheshkumar PS, Savithri HS, Murthy MRN (2008) Structure of recombinant capsids formed by the beta-annulus deletion mutant – rCP (Delta48-59) of Sesbania mosaic virus. Virology 375(1):190–196

Pappachan A, Chinnathambi S, Satheshkumar PS, Savithri HS, Murthy MRN (2009) A single point mutation disrupts the capsid assembly in Sesbania Mosaic Virus resulting in a stable isolated dimer. Virology 392(2):215–221

Parrot I, Huang PC, Khosla C (2002) Circular dichroism and nuclear magnetic resonance spectroscopic analysis of immunogenic gluten peptides and their analogs. J Biol Chem 277:45572–45578

Powers KW, Brown SC, Krishna VB, Wasdo SC, Moudgila BM, Roberts SM (2006) Characterization of nanoscale particles for toxicological evaluation. Toxicol Sci 90:296–303

Sangita V, Lokesh GL, Satheshkumar PS, Vijay CS, Saravanan V, Savithri HS, Murthy MRN (2004) T=1 capsid structures of Sesbania mosaic virus coat protein mutants: determinants of T=3 and T=1 capsid assembly. J Mol Biol 342(3):987–999

Sangita V, Lokesh GL, Satheshkumar PS, Saravanan V, Vijay CS, Savithri HS, Murthy MRN (2005) Structural studies on recombinant T = 3 capsids of Sesbania mosaic virus coat protein mutants. Acta Crystallogr D Biol Crystallogr 61:1402–1405

Satheshkumar PS, Lokesh GL, Sangita V, Saravanan V, Vijay CS, Murthy MRN, Savithri HS (2004a) Role of metal ion-mediated interactions in the assembly and stability of Sesbania mosaic virus T=3 and T=1 capsids. J Mol Biol 342(3):1001–1014

Satheshkumar PS, Lokesh GL, Savithri HS (2004b) Polyprotein processing: *cis* and *trans* proteolytic activities of Sesbania mosaic virus serine protease. Virology 318(1):429–438

Satheshkumar PS, Gayathri P, Prasad K, Savithri HS (2005a) "Natively unfolded" VPg is essential for Sesbania mosaic virus serine protease activity. J Biol Chem 280(34):30291–30300

Satheshkumar PS, Lokesh GL, Murthy MRN, Savithri HS (2005b) The role of arginine-rich motif and beta-annulus in the assembly and stability of Sesbania mosaic virus capsids. J Mol Biol 353:447–458

Savithri HS, Murthy MRN (2010) Structure and assembly of Sesbania mosaic virus. Curr Sci 98(3):346–351

Scott AM, Wolchok JD, Old LJ (2012) Antibody therapy of cancer. Nat Rev Cancer 12:278–287

Shen M, Reitman ZJ, Zhao Y, Moustafa I, Wang Q, Arnold JJ, Pathak HB, Cameron CE (2008) Picornavirus genome replication. Identification of the surface of the poliovirus (PV) 3C dimer that interacts with PV 3Dpol during VPg uridylylation and construction of a structural model for the PV 3C2-3Dpol complex. J Biol Chem 283:875–888

Shirota Y, Luo H, Qin W, Kaneko S, Yamashita T, Kobayashi K, Murakami S (2002) Hepatitis C virus (HCV) NS5A binds RNA-dependent RNA polymerase (RdRp) NS5B and modulates RNA-dependent RNA polymerase activity. J Biol Chem 277(13):11149–11155

Sivakumaran K, Hacker DL (1998) The 105-kDa polyprotein of southern bean mosaic virus is translated by scanning ribosomes. Virology 246:34–44

Solunke BS, Eranna S, Nayudu MV (1983) Sesbania mosaic-a new virus disease. Indian Phytopathol 36:568–570

Subramanya HS, Gopinath K, Nayudu MV, Savithri HS, Murthy MRN (1993) Structure of *Sesbania* Mosaic Virus at 4·7 Å resolution and partial sequence of the coat protein. J Mol Biol 229(1):20–25

Tamm T, Truve E (2000a) Sobemoviruses. J Virol 74(14):6231–6241

Tamm T, Truve E (2000b) RNA-binding activities of cocksfoot mottle sobemovirus proteins. Virus Res 66:197–207

Tamm T, Suurvali J, Lucchesi J, Olspert A, Truve E (2009) Stem-loop structure of cocksfoot mottle virus RNA is indispensable for programmed-1 ribosomal frameshifting. Virus Res 146(1–2):73–80

Vidalain PO, Tangy F (2010) Virus-host protein interactions in RNA viruses. Microbes Infect 12(14–15):1134–1143

Vishnu Vardhan GP, Savithri HS, Murthy MRN, Hema M (2016) Biodistribution and toxicity evaluation of sesbania mosaic virus nanoparticles in mice. Arch Virol 161:2673–2681

Walters HJ (1969) Beetle transmission of plant viruses. Adv Virus Res 15:339–363

Woody RW (1994) Contributions of tryptophan side chains to the far-ultraviolet circular dichroism of proteins. Eur Biophys J 23(4):253–262

The Occurrence, Biology and Genomic Properties of Tobamoviruses Infecting Crop Plants in India

19

Alok Kumar, A. Abdul Kader Jailani, Anirban Roy, and Bikash Mandal

Abstract

Tobamoviruses are rod shaped RNA plant viruses of the genus *Tobamovirus* family *Virgaviridae*. Tobamovirus associated plant disease was first recorded in India during 1947. Of the 35 tobamovirus species known in the different parts of the world, only eight have been documented in India: *Cucumber green mottle mosaic virus* (CGMMV), *Frangipani mosaic virus* (FrMV), *Odontoglossum ringspot virus* (ORSV), *Pepper mild mottle virus* (PMMoV), Plumeria mosaic virus (PluMV), *Sunn-hemp mosaic virus* (SHMV), *Tobacco mosaic virus* (TMV) and *Tomato mosaic virus* (ToMV). CGMMV is the most commonly occurring tobamovirus infecting several cucurbits such as bottle gourd, cucumber, watermelon, muskmelon, gherkin, ash gourd, snake gourd and sponge gourd. SHMV infects leguminous plants, TMV and ToMV are important in tomato and pepper and PMMoV is known to infect only pepper, whereas, the other three viruses are known to infect ornamental plants. PluMV, the recently identified in temple tree or frangipani (*P. rubra* f. *acutifolia* and *P. obtusa*) is the only tobamovirus species originally recorded in India. The present review summarises the research work carried out on the characterisation of tobamoviruses occurring in India.

Keywords

Tobamovirus • India • Cucumber green mottle mosaic virus • Frangipani mosaic virus • Odontoglossum ringspot virus • Pepper mild mottle virus • Plumeria mosaic virus • Sunn-hemp mosaic virus • Tobacco mosaic virus • Tomato mosaic virus

---

A. Kumar • A. Abdul Kader Jailani • A. Roy • B. Mandal (✉)
Advanced Centre for Plant Virology, Division of Plant Pathology, Indian Agricultural Research Institute, New Delhi 110012, India
e-mail: leafcurl@rediffmail.com

B. Mandal et al. (eds.), *A Century of Plant Virology in India*,
https://doi.org/10.1007/978-981-10-5672-7_19

19.1 Introduction

The plant virus genus *Tobamovirus* (family *Virgaviridae*) has been derived from tobacco mosaic virus (TMV), the first virus discovered during 1898. Tobamoviruses have typical rigid-rod shapes virion of 300–310 × 18 nm, which is formed by assembling capsid protein (CP) subunits of 17–18 kDa in helical symmetry with pitch of 2.3 nm around a positive sense linear ssRNA genome of 6.3–6.6 nucleotides. The genome of tobamoviruses typically encode four proteins, large replication associated protein (L-Rep), small Rep (S-Rep), movement protein and CP. Tobamoviruses usually cause mosaic, mottle, vein-clearing and ringspot diseases in vegetables, ornamentals and fiber crops in both open field as well as protected cultivations. Tobamoviruses are highly stable and contagious and as a result spread readily in the crop field.

Initially, tobamoviruses were isolated from the solanaceous plants, later they were discovered in plant species belonging to Cucurbitaceae, Leguminosae and Cruciferae. The genome sequence of several tobamoviruses was generated during 1980s–1990s. The sequence information prompted to classify the tobamoviruses based on the origin of virion assembly, sequence of CP and Rep, and organization of the MP and CP open reading frames (ORF) (Fukudu et al. 1981; Gibbs 1986; Lartey et al. 1996). Initially, these viruses were classified into two groups. The subgroup-I included most of the solanaceous infecting tobamoviruses with the origin of assembly located within the MP region and the MP and CP ORFs are separated by two to five nucleotides. The subgroup-II included cucurbit and legume infecting tobamoviruses with the origin of assembly located within the CP sequence and the MP and CP ORFs are arranged in slightly overlapping manner. The discovery of crucifer infecting tobamoviruses, ribgrass mosaic virus and turnip vein-clearing virus suggested the existence of subgroup III, where MP and CP ORFs overlapped to a greater extent (Lartey et al. 1996). Increasing number of tobamoviruses isolated from different hosts prompted the researchers to reassess the basis of tobamovirus classification. Song et al. (2006) proposed an idea to classify the tobamoviruses based on the host families from where the viruses were originally isolated and based on this, the tobamoviruses were classified into six subgroups; namely Solanaceae-, Brassicaceae-, Cactaceae-, Malvaceae-, Cucurbitaceae- and Passifloraceae-infecting tobamovirus.

Limited numbers of tobamoviruses are known to occur in India and they are sporadic and endemic in nature. Of the several genera of plant viruses known in India, the genus *Tobamovirus* is relatively less significant. In India, Sun-hemp mosaic virus is the first tobamovirus recorded as early as in 1947 and then the occurrence of cucumber green mottle mosaic virus was identified in 1948. Subsequently, during1978–2014, six more tobamoviruses were identified in India (Table 19.1). During the past 70 years, a significant body of literature that has been generated on the tobamoviruses occurring in India is summarised in this chapter.

Table 19.1 The first records of tobamoviruses in India

Virus	Crop	Place	Year
Sunn-hemp mosaic virus	Sunn-hemp	Delhi	1947
Cucumber green mottle mosaic virus	Bottle gourd	Pune	1948
Frangipani mosaic virus	Frangipani	Delhi	1978
Tobacco mosaic virus	Tomato/pepper	Kolar	1999
Tomato mosaic virus	Tomato/pepper	Mysore	1999
Odontoglossum ringspot virus	*Cymbidium* sp.	Palampur	2005
Pepper mild mottle virus	Bell-pepper	Palampur	2011
Plumeria mosaic virus	Frangipani	Delhi	2014

19.2 Cucumber Green Mottle Mosaic Virus (CGMMV)

19.2.1 Occurrence and Biology

CGMMV is an economically important viral pathogen in cucurbits. It was first isolated from cucumber (*Cucumis sativus*) in Great Britain (Ainsworth 1935) and subsequently reported from various parts of the world. In India, CGMMV was first observed in bottle gourd (*Lagenaria vulgaris*) in Bombay in 1943 and subsequently, in other cucurbitaceous crops such as muskmelon (*Cucumis melo*), gherkin (*Cucumis anguria*), watermelon (*Citrullus lanatus*), ash gourd (*Benincasa hispida*), snake gourd (*Trichosanthes cucumerina*) and sponge gourd (*Luffa acutangula*) (Capoor and Varma 1948a; Vasudeva et al. 1949; Raychaudhuri and Varma 1978; Varma and Giri 1998; Rashmi et al. 2005; Mandal et al. 2008; Nagendran et al. 2015). The virus has narrow experimental host range restricting mainly to Cucurbitaceae and Chenopodiaceae. In cucurbits, CGMMV generally induces mild greenish mosaic symptoms (Fig. 19.1) The virus is transmitted through contacts of contaminated plant debris, soil, irrigation and river water (Vani and Varma 1993). In Japan, CGMMV was reported to be seed borne, however, the Indian isolate of CGMMV was not transmitted through seeds of bottle gourd or vegetable marrow (*C. pepo*) (Rao and Varma 1984). No insect vector is known for CGMMV transmission except, the cucumber leaf beetle (*Raphidopalpa faeveicollis*), which was reported to transmit the virus at low level (10%) (Rao and Varma 1984). CGMMV is highly prevalent in northern India in bottle gourd, muskmelon and watermelon (Raychaudhuri and Varma 1978; Rao and Varma 1984). Oflate, it has been identified in southern India too (Rashmi et al. 2005; Nagendran et al. 2015).

The other tobamovirus species known to infect cucurbits are *Cucumber fruit mottle mosaic virus* (CFMMV) (Antignus et al. 2001), *Kyuri green mottle mosaic virus* (KGMMV) (Francki et al. 1986; Yoon et al. 2001), *Zucchini green mottle mosaic virus* (ZGMMV) (Ryu et al. 2000; Yoon et al. 2002) and *Cucumber mottle virus* (Orita et al. 2007). Of all these tobamoviruses, CGMMV has been recognized as the most potential threat to cucurbit production. So far, except CGMMV, no other cucurbit infecting tobamoviruses are known in India (Table 19.2).

Fig. 19.1 Symptoms caused by tobamoviruses in the different plant species in India. Cucumber green mottle mosaic virus (*CGMMV*), tobaccomosaic virus (*TMV*), Plumeria mosaic virus (*PluMV*) and frangipani mosaic virus (*FrMV*)

Biological diversity with reference to the plant response has been studied and so far, six variants or strains of CGMMV have been reported from India: the CV2B and CV2C isolate from bottle gourd (Cooper and Varma 1948; Vasudeva et al. 1949; Vasudeva and Nariani 1952), M from muskmelon (Raychaudhuri and Varma 1978), Gh from gherkin (Rashmi et al. 2005), SG from snake gourd (Nagendran et al. 2015) and IW from irrigation water (Vani and Varma 1993). CV2B and CV2C induce similar systemic mosaic symptoms on bottle gourd but differ in host reaction on watermelon and pumpkin (*C. moschata*). In watermelon, CV2B induces mottle

Table 19.2 The genomic resources of tobamoviruses occurring in India

Virus	Accession no.	Region of genome	Crop	Place/isolate	Year of submission to NCBI
CGMMV	DQ767631	Full genome	Bottle gourd	Delhi	2006
	AJ748352	CP	Bottle gourd	Aligarh	2004
	AJ748353	MP	Bottle gourd	Aligarh	2004
	JQ712998	CP	Ash gourd	Delhi	2012
	KF702319	CP	Ridge gourd	Sikar	2013
	KJ729037	CP	Pumpkin	Trichy	2014
	KJ729038	CP	Bottle gourd	Trichy	2014
	KJ729040	CP	Cucumber fruit	Annur	2014
	KJ729039	CP	Snake gourd	Coimbatore	2014
	DQ767636	CP	Bottle gourd	Rajasthan	2006
	DQ364977	5′ UTR, Rep	Bottle gourd	Delhi	2006
	DQ364978	Rep	Bottle gourd	Delhi	2006
	AY866427	Rep	Bottle gourd	New Delhi	2004
	EU366912	CP	Bottle gourd	Bangalore	2007
	AH015694	5′ UTR, Rep	Bottle gourd	Delhi	2006
FrMV	JN555602	Full genome	Frangipani	Delhi	2011
PMMoV	JX173808	CP	Bell-pepper	Hamirpur	2012
	JX173809	CP	Bell-pepper	Bilaspur	2012
	JX185071	CP	Bell-pepper	Bilaspur	2012
	KJ631123	Full genome	Bell-pepper	Hamirpur	2014
PluMV	KJ395757	Full genome	Frangipani	Delhi	2014
TMV	AF126505	MP, CP	Tomato	Bangalore	1999
	GU723495	Rep	Bell-pepper	Mysore	2010
	GU213294	MP	Bell-pepper	Mysore	2009
	EU152113	CP	Tomato	Aligarh	2008
	GU723496	Rep	Tomato	Mysore	2010
	JQ895560	MP, CP	Soybean	Delhi	2012
ToMV	GU723496	Rep	Tomato	Mysore	2010
	AF378152	CP	Tomato	Pune	2001
	GU213293	MP	Tomato	Mysore	2009
	GU723495	Rep	Bell-pepper	Mysore	2010
ORSV	AJ564563	CP	*Cymbidium sp.*	Palampur	2003
	FJ372909	CP	*C. hookerianum*	Sikkim	2009

CP capsid protein, *MP* Movement protein, *Rep* Replicase *UTR* Untranslated region

mosaic and CV2C causes symptomless infection. In pumpkin, CV2C induces yellow spots along the veins, whereas CV2B causes no infection (Vasudeva et al. 1949). The M isolate induces well defined mosaic and green vein banding symptoms on muskmelon leaves, and no symptoms on *C. amaranticolor*. It differs from CV2B and CV2C by not infecting watermelon (Raychaudhuri and Varma 1978).

The IW isolate induces systemic mottle mosaic symptoms in bottle gourd, cucumber, muskmelon and watermelon and induces pinpointed local lesions on *C. amaranticolor*. The IW differs from the other three Indian isolates by inducing mottle mosaic symptoms on *Luffa acutangula* as other isolates causes either symptomless infection or no infection in *L. acutangula*. The Gh isolate, which is reported from southern India, induces yellowing, blistering, mosaic mottling, chlorotic spots, necrotic lesions and leaf cupping on gherkin (Rashmi et al. 2005). The SG isolate, which is also reported from southern India, induces mosaic and mottling symptoms in snake gourd and *Nicotiana glutinosa* (Nagendran et al. 2015).

19.2.2 Genomic Properties

The genomic properties of CGMMV occurring in India have been studied mostly based on the partial genome sequences of the isolate from different plant species (AJ748352, AJ748353, AF015694, AY866427, EU366912, JQ712998, KF7023019, KJ729037–40, DQ364977–78 and DQ767636) (Table 19.2). The complete genome sequence of several isolates of CGMMV is available from different parts of the world; however, the same is available only for one isolate infecting bottle gourd in Delhi (GenBank DQ767631; Mandal 2007). The complete genome of the Indian isolate is 6424 nucleotide long encoding four proteins; 128 kDa S-Rep, 186 kDa L-Rep, 29 kDa MP and 17.3 kDa CP. The MP overlaps with L-Rep by 13 nucleotides. The MP and CP ORF overlap by 25 nucleotides, which is different from all other cucurbit-infecting tobamoviruses (Ugaki et al. 1991; Antignus et al. 2001; Yoon et al. 2001). The 5′ and 3′ non-translated regions are 60 and 176 nucleotides long, respectively. The comparison of the complete genome sequence of the Indian isolate of CGMMV showed more than 97% identity among CGMMV isolates, 59.1–63.6% identity with the other cucurbit-infecting tobamoviruses and less than 50% identity with rest of the tobamoviruses (Mandal 2007).

An infectious clone of the CGMMV-Del has been developed, which could induce systemic mild mottle mosaic symptoms in *N. benthamiana* and cucumber when delivered through *Agrobacterium*. As CGMMV is readily transmitted through sap inoculation, accumulates at high concentration in the infected plants and induces mild symptoms, the genome of CGMMV-Del has been utilized for developing a plant based platform for the production of edible vaccine, antibodies and other useful protein. Two versions of CGMMV genome based vectors have been developed, the one with the full-genome, where at the stop codon of CP, a foreign gene (GFP) was inserted, which resulted in expression of systemic symptom, virion formation and expression of foreign gene in *N. benthamiana* (Fig. 19.2). The other version of vector was designed by deleting most of the CP gene sequence and retaining only first 105 nucleotides. The foreign gene, GFP was inserted in this vector in a way to maintain the original CP ORF. The shortened vector resulted in rapid and higher expression of GFP in *N. benthamiana*. The shortened vector is referred as a replicon vector of CGMMV as it neither caused any disease nor produced virion, however,

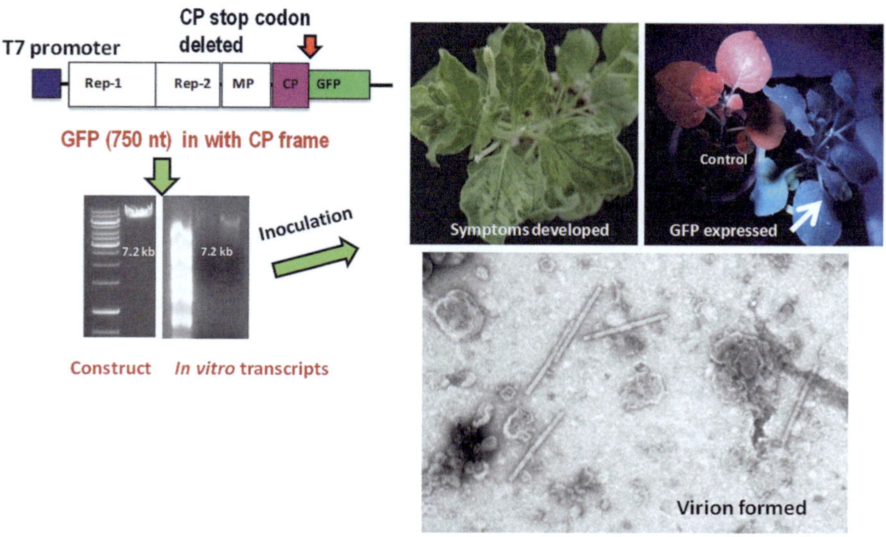

Fig. 19.2 Application of CGMMV in expression of foreign protein in plant. The CGMMV full-genome vector showing expression of GFP in *Nicotiana benthamiana* as well as expression of symptoms and formation of virions

replicated and directed the expression of the foreign protein in *N. benthamiana*. The CGMMV based gene vector has the potential for generation of edible vaccine in cucurbits (Jailani et al. 2017).

19.3 Frangipani Mosaic Virus (FrMV)

19.3.1 Occurrence and Biology

Frangipani or temple tree (*Plumeria* sp.) is grown worldwide as an ornamental tree plant. FrMV infecting frangipani was originally reported from Australia (Francki et al. 1971) and later from India (Varma and Gibbs 1978). The virion of FrMV is 300 × 17 nm in size, although shorter particles similar to CGMMV are also present (Francki et al. 1971). The virus was highly stable, easily transmitted through sap inoculation and vegetative propagation, but not through seeds or any insect vector. The expression of symptoms of FrMV is influenced by temperature. High temperature (30–35 °C) favours rapid expression of symptoms, whereas lower temperature delays symptoms expression. FrMV causes mosaic, vein banding and bronzing in *P. acutifolia* (Fig. 19.1). Three distinct strains of FrMV were described based on symptoms: Adelaide strain (Adel) (Francki et al. 1971), Allahabad strain (Ald) and Delhi strains (Del) (Varma and Gibbs 1978). Recently, a new strain, FrMV-Ind-1 was identified from Delhi, India, which is distinct from all these strains by not causing the systemic infection on *N. tabacum* (Kumar et al. 2015). FrMV-Ind-1 caused local lesions on *Capsicum annuum* (chilli) and *Solanum melongena*, whereas, it caused systemic mosaic symptoms in *N. benthamiana* (Kumar et al. 2015).

19.3.2 Genomic and Serological Properties

FrMV genome was initially characterized based on partial sequence of 2433 nucleotides containing movement and coat proteins (Deng et al. 2000). The complete genome sequence subsequently was reported for FrMV-P isolate (Lim et al. 2010) and FrMV-Ind-1 isolate (Kumar et al. 2015). The FrMV-Ind-1 genome was 6643 (JN555602) nucleotides long with the typical genome organization of tobamoviruses with the four ORFs, encoding four proteins; 130 kDa S-Rep, 187 kDa L-Rep, 29 kDa MP and 19 kDa CP. The replicase and MP genes overlapped by 13 nucleotides whereas, MP and CP gene were 58 nucleotides apart from each other. The untranslated regions at 5′ and 3′ were 77 and 274 nucleotide long, respectively. The comparison of genome sequences of FrMV-Ind-1 with FrMV-P and 25 other tobamovirus species showed that FrMV-Ind-1 is closely related with FrMV-P (98.3%) and highly different from the other tobamovirus species (>50%) (Kumar et al. 2015). Although, biological and genomic property of FrMV was studied previously, the information about the serological relationships of FrMV was not available until recently the Indian isolate, FrMV-Ind-1 was purified from the infected frangipani leaves and used to produce polyclonal antiserum (Kumar et al 2015). The ELISA based diagnosis of FrMV was developed, which detected the virus efficiently in the infected leaf sap of frangipani. FrMV is a serologically distinct tobamovirus as it showed weak serological reactivity with only CGMMV, TMV, bell pepper mottle virus and KGMMV, whereas ORSV, CFMMV, cactus mild mottle virus and ZGMMV did not react at all with the PAb to FrMV-Ind-1(Kumar et al. 2015).

19.4 Plumeria Mosaic Virus (PluMV)

19.4.1 Occurrence and Biology

PluMV was isolated from a leaf sample of frangipani (*P. rubra* f. *acutifolia*) mixed infected with FrMV and showing mosaic, vein banding, bronzing and necrotic ringspots symptoms at IARI, New Delhi in April, 2010. While, conducting mechanical inoculation to study FrMV, one isolate (Gg-1) from *P. rubra* f. *acutifolia* plant showed distinct variation of symptoms on *N. benthamiana* and *Gomphrena globosa* (Kumar et al. 2013a). Further, analysis of host range and complete genome sequence of Gg-1 provided the evidence of existence of a new tobamovirus that was different from FrMV and named as PluMV (Kumar et al. 2013a).

PluMV can be distinguished from FrMV through sap inoculation in the different plant species. *G. globosa* is as an important indicator host of PluMV as it induces prominent reddish local lesions (Fig. 19.1), whereas, FrMV does not produce any symptoms on *G. globosa*. The other plant species that can be used to differentiate PluMV from FrMV are *Nicotiana* spp. In *N. benthamiana*, PluMV induces whitish rings with a spot in the centre on the inoculated leaves followed by wavy whitish lines, mosaic mottling and blistering symptoms in the new leaves, whereas, FrMV induced chlorotic spots on the local leaves and mosaic, mottling and blistering on systemic leaves (Kumar et al. 2013a).

19.4.2 Genomic Properties

The complete genome of PluMV is 6688 nucleotides (KJ395757) long with a genome organization typical of the other tobamoviruses. The four ORFs encode 130 kDa, 188 kDa, 29 kDa and 19 kDa proteins. The replicase and MP gene of PluMV also overlapped by 13 nucleotides similar to FrMV whereas, MP and CP of this virus was separated by 59 nucleotides. The untranslated regions at 5′ and 3′ end were slightly longer than FrMV and contains 90 and 284 nucleotides, respectively. The genome of PluMV closely related to FrMV but it shared only 71.4–71.6% identity. The ICTV guidelines for tobamovirus species demarcation criterion is based on <90% nucleotide sequence identity of the complete genome. Genome sequence identity and phylogenetic analysis revealed PluMV as a new member of the genus *Tobamovirus* (Kumar et al. 2013a).

19.5 Pepper Mild Mottle Virus (PMMoV)

19.5.1 Occurrence and Biology

PMMoV was first reported from USA (McKinney 1952). Subsequently, it was known in Europe (Wetter et al. 1984) and in several other countries (Lamb et al. 2001). However, in India, the occurrence of PMMoV has recently been documented in bell pepper grown under the protected cultivation in Himachal Pradesh (Sharma and Patiyal 2011). PMMoV being externally seed borne and is expected that the virus has been introduced to India through the contaminated seeds imported from China, Japan, Netherlands, USA or Europe (Sharma and Patiyal 2011). The infected bell pepper plants showed mosaic, mottle, leaf and plant deformation symptoms with stunted growth and distorted fruits under polyhouse conditions.

19.5.2 Genomic Properties

Limited genome sequence information of the isolates of PMMoV is available in India. The complete genome sequence is available only for one Indian isolate, HP-1 (Rialch et al. 2015). In addition, partial genome sequence is available for a few isolates from Himachal Pradesh (Rialch et al. 2015). The genome of HP-1 isolate consist of 6356 nucleotides (KJ631123) having the typical genome organization of other tobamoviruses. The first two ORFs encode the small and large replicase proteins of 126 kDa and 183 kDa, respectively. The ORF3 starts just after the stop codon of ORF2 and encodes a 28 kDa MP, whereas, the ORF4 starts after

2 nucleotides of the stop codon of the ORF3 and encodes a 17.2 kDa CP. The 5′ and 3′ UTR of PMMoV-HP1 comprised of 69 and 198 nucleotides, respectively. The Indian isolate of PMMoV is almost identical to one of the Japanese isolate (PMMoV-J; AB000709) as it shares the maximum identity of 99.28% (Rialch et al. 2015).

So far, five pathotypes of PMMoV (P0, P1, P12, P123, and P1234) have been recognized based on their ability to break L gene-mediated resistance (Sawada et al. 2004). The variation in CP amino acids (aa) sequence plays the most important role in pathotypes determination. The PMMoV-HP1 has been tentatively categorized as pathotype P12, as the CP aa sequences shared 100% identity with that of pathotypes P12 (Alonso et al. 1991).

19.6 Odontoglossum Ringspot Virus (ORSV)

ORSV is one of the most commonly occurring viruses infecting orchids worldwide (Zettler et al. 1990). Orchids are commercially grown in the North-eastern region of India. ORSV is one of the most commonly occurring orchid viruses that affects growing orchid industry in India. The virus was first identified in *Odonotoglossum grande* showing ringspot symptoms in California, USA. In India, it was first reported in *Cymbidium sp.* from Palampur (Sherpa et al. 2004) and subsequently, it was found to infect *Cymbidium hookerianum* in Sikkim (Rani et al. 2010). Infected orchids induce chlorotic or necrotic streaks on leaves. The virus is spread through infected planting materials. Limited studies have been conducted so far on the biology, molecular biology and diversity of ORSV occurring in India. The CP gene sequence of two isolates were generated (AJ564563, FJ372909) and utilized for the expression of recombinant antigen in *E. coli.* The ELISA based diagnosis of ORSV has been developed in India that efficiently detected the virus in the field samples of orchids (Rani et al. 2010).

19.7 Sunn-Hemp Mosaic Virus (SHMV)

19.7.1 Occurrence and Biology

SHMV is the oldest known tobamovirus in India and was first described infecting sunn-hemp from Delhi (Raychaudhuri 1947) and later from Pune (Capoor and Varma 1948b; Capoor 1950). The virus was described by different synonyms depending on the host from where it was isolated, e.g., catjang mosaic virus, dolichos enation mosaic virus (Capoor and Varma 1948b), southern sunn-hemp mosaic virus (Capoor 1950) and crotalaria mucronata mosaic virus (Raychaudhuri and Pathanian 1950). SHMV naturally infects several leguminous plants species in India, *Phaseolus vulgaris*, *Vigna catjang*, *Vigna unguiculata*, *Crotolaria mucronata*, *Dolichos lablab* and *Crotalaria juncea* (Nagaich and Vashisth 1963; Capoor and Varma 1948b; Raychaudhuri and Pathanian 1950).

The virus causes mosaic, puckering, blistering, malformation, occasional enations on leaves and stunting of plant (Capoor 1950). The virus is sap transmitted and has a considerably wide experimental host-range infecting 40 plant species from 7 families with almost similar host reactions. *N. glutinosa* and *Cyamopsis tetragonoloba* were considered as the efficient diagnostic hosts as they produce necrotic lesions on the inoculated leaves (Gibbs and Varma 1977; Capoor 1962), whereas, *Crotolaria* sp., produces vein clearing, severe mosaic, blistering and deformation on leaves (Capoor 1962; Niazi et al. 1973). SHMV was reported to be seed transmitted, however, the sunn-hemp isolate from Pune was not transmitted through the seeds of sunn-hemp (Capoor 1962). SHMV was purified and fine glassy acicular crystals were obtained (Raychaudhuri 1947). The high absorption curve and virion structure of SHMV were studied in the crystalline solution (Das Gupta et al. 1951).

19.7.2 Genomic Properties

Despite long history of this virus in India, so far no molecular property has been studied yet. However, the complete genome sequence of SHMV from other country has been reported (Silver et al. 1996).

19.8 Tobacco Mosaic Virus (TMV)

19.8.1 Occurrence and Biology

TMV, the type member of the genus *Tobamovirus* was first reported from the Netherlands (Mayer 1886). Although, TMV is known to occur all over the world and is the most studied plant virus, limited study has been documented in India. In the initial studies, TMV was identified based on symptomatology on indicator plants. The first TMV isolate that has been characterized based on biology, serology and partial sequence of 3′ region of the genome, was obtained from tomato showing mosaic in Kolar district of Karnataka (Cherian et al. 1999). During this period of time, TMV was also reported to cause white mosaic in Egyptian henbane and scotch spearmint (*Mentha gracillis*) in Lucknow (Samad et al. 1999, 2000). Further, the occurrence of TMV in tomato and pepper and seed transmission have been studied in Karnataka (Chitra et al. 1999, 2002). TMV was more prevalent in bell pepper compared to tomato (Chitra et al. 2002). TMV is extremely stable and transmitted mechanically from seeds to seedlings during transplantation at the rate of 1–10% (Chitra et al. 1999). TMV was detected in the commercially available seeds of tomato and bell pepper (Chitra et al. 2002). The inoculation study showed that irrespective of the growth stage of tomato or pepper, the virus became seed borne and the concentration of the virus in seeds was high when the plants were inoculated at the early stage of growth (Chitra et al. 1999). TMV is not generally known to infect pulse crop, recently a new strain of TMV (TMV-SoyIn) was discovered in India infecting soybean (*Glycine max*) (Kumar et al. 2013b). TMV-SoyIn causes chlorosis and puckering symptoms in soybean and local necrotic lesions and systemic mosaic and necrosis symptoms in *N. tabacum* cv. Xanthi (Fig. 19.1).

Inhibition study of TMV with the radioactive molecules, phosphorous (P^{32}) and sulphur (S^{35}) was undertaken in the tissue culture of *N. tabacum* cv. Xanthi. TMV infected tissues were observed to uptake more radioisotopes compared to the healthy tissues. The uptake of P^{32} (15 microcuries/ml) had no effect on the infectivity of TMV, whereas, interestingly at the same dose, S^{35} significantly (90%) inhibited the infectivity of TMV (Chatrath et al. 1970).

19.8.2 Genomic Properties

The complete genome sequence of TMV is available for a number of isolates from all around the world but, in India, the genome has been only partially characterized, and so far only six partial genome sequence from tomato (AF126505, EU152113, GU723496), pepper (GU213294, GU723495) and soybean (JQ895560) are available (Table 19.2). The bioinformatic study was conducted to identify the ligand that interacts with the catalytic residues (GDD) of RdRp of TMV (Prabahar et al. 2015). This study identified several small molecules (Isoproterenol, Riboflavin, Atropine, Albendazole, Neomycin and Ampicillin) that may potentially inhibit the replicase catalytic domain activity of TMV.

19.9 Tomato Mosaic Virus (ToMV)

ToMV is a very close relative of TMV in their genome organization, sequence identity as well as their host reaction. Host range for both the viruses are also similar as they infect pepper, potato, tomato and several other plant species. The field survey of tomato and bell pepper in Karnataka indicated the occurrence of both TMV and ToMV, while ToMV was more prevalent in tomato (Chitra et al. 2002). The cross-reactivity of polyclonal antisera limits sero-diagnosis to differentiate of ToMV and TMV. Bio-assay based on the differential symptomatology on *N. sylvestris*, *N. tabacum* cv. White Burley, *Phaseolus vulgaris* cv. Black Turtle and *Cucumis sativus* is often used for distinguishing TMV and ToMV. For the rapid and simultaneous diagnosis of TMV and ToMV in the seed samples of pepper and tomato in Karnataka, the one step duplex RT–PCR was developed (Kumar et al. 2011).

The complete genome sequence of ToMV occurring in India so far has not been generated. However, the partial sequence is available for the three tomato isolates (AF378152, GU213293, GU723496) and one bell pepper isolate (GU723495) (Table 19.2).

19.10 Concluding Remarks

Several tobamoviruses are known in the different parts of the world, however, only a limited number of tobamoviruses so far has been identified in India. Four tobamovirus (CGMMV, PMMoV, TMV, ToMV) are known to affect cucubitaceous and solanaceous vegetables, two (SHMV and TMV) affect pulse crops and three (FrMV, ORSV and PluMV) affect ornamental plants. As these viruses are highly stable and contagious, they readily spread though the contaminated seed or planting materials. The tobamoviruses like CGMMV, PMMoV, TMV and ToMV are of special concern in the vegetables cultivated in the protected condition. Although, diseases caused by tobamoviruses are known since long time in India, the associated virus species were characterized at molecular level only during the last two decades. CGMMV is the most studied tobamovirus in India. SHMV- the legume infecting tobamovirus, although, the first tobamovirus identified and crystallised in India, so far molecular study has not been carried out. Identification of tobamoviruses are difficult based on only disease symptoms alone. Molecular tests like ELISA and PCR have developed for several tobamoviruses are highly sensitive and useful for the detection of these viruses in the different crops and weeds. There may be many more species of tobamoviruses existing in India, but their presence may not be marked with the distinct symptoms, or they many co-exist in mixed infection with the other virus species. Tobamovirus genome is an excellent source of molecular tools. TMV has been extensively exploited and proved to be ideal for the expression of useful foreign proteins in plant like vaccines and antibodies. For these purposes, infectious clones of tobamoviruses are necessary. In India, the research has been initiated to exploit plant viruses as useful genetic resources with the projects funded by ICAR-World Bank (NAIP) and ICAR-National Agricultural Science Fund. The genome of CGMMV, the cucurbit infecting tobamovirus has been utilized for the expression of foreign gene in plant (Jailani et al 2017). The development of infectious clones of the other tobamoviruses will open up research in India on the functional genomics for encoding pathogenesis as well as the application of new tools in plant biotechnology.

References

Ainsworth GC (1935) Mosaic disease of cucumber. Ann Appl Biol 22:55–67

Alonso E, Garcia-Luque I, de la Cruz A, Wicke B, Rincon AMJ, Serra MT (1991) Nucleotide sequence of the genomic RNA of pepper mild mottle virus, a resistance-breaking tobamovirus in pepper. J Gen Virol 72:2875–2884

Antignus Y, Wang Y, Pearlsman M, Lachman O, Lavi N, Gal-On A (2001) Biological and molecular characterization of new cucurbit-infecting tobamovirus. Phytopathology 91:565–571

Capoor SP (1950) A mosaic disease of sunn hemp in Bombay. Curr Sci 19:22

Capoor SP (1962) Southern sunn hemp mosaic virus: a strain of tobacco mosaic virus. Phytopathology 52:393–397

Capoor SP, Varma PM (1948a) Mosaic disease of *Lagenaria vulgaris* surveyed in the Bombay Province. Curr Sci 17:274–275

Capoor SP, Varma PM (1948b) Enation mosaic of *Dolichos lablab* Linn, a new virus disease. Curr Sci 17:57–58

Chatrath MS, Raychaudhuri SP, Mishra MD (1970) Effect of radioactive phosphorus and sulphur on infectivity of Tobacco mosaic virus in tissue culture. Proc Indian Natl Sci Acad [formerly Proc Natn Inst Sci India, No Part B] 2:126–129, www.dli.gov.in/rawdataupload/upload/insa/INSA.../20005b6c_126.pdf

Cherian S, Joseph J, Muniyappa V, Savithri HS (1999) Characterization of tobacco mosaic virus isolated from tomato in India. Curr Sci 76:1384–1388

Chitra TR, Prakash HS, Albretchsen SE, Shetty HS, Mathur SB (1999) Seed transmission of mosaic viruses in tomato and bell pepper. Trop Sci 39:80–84

Chitra TR, Prakash HS, Albrechtsen SE, Shetty HS, Mathur SB (2002) Indexing of leaf and seed samples of tomato and bell pepper for tobamoviruses. Indian Phytopathol 55(1):84–86

Das Gupta NN, De ML, Raychaudhuri SP (1951) Structure of Sannhemp (*Crotalaria juncea* Linn.) mosaic virus with the electron microscope. Nature 168:114. doi:10.1038/168114a0

Deng XD, Fei XW, Huang JS, Zheng XQ (2000) Cloning and sequencing of the 3′-end of frangipani mosaic virus (FMV). J Trop Subtrop Bot 8:185–192

Francki RIB, Zaitlin M, Grivell CJ (1971) An unusual strain of tobacco mosaic virus from Plumeria acutifolia. Aust J Biol Sci 24:815–818

Francki RIB, Hu J, Palukaitis P (1986) Taxonomy of cucurbit infecting tobamoviruses as determined by serological and molecular hybridization analyses. Intervirology 26:156–163

Fukudu M, Meshi T, Okada Y, Otsuki Y, Takebe I (1981) Correlation between particle multiplicity and location of virion RNA of the assembly initiation site for viruses of tobacco mosaic virus group. Proc Natl Acad Sci USA 78:4231–4235

Gibbs A (1986) Tobamovirus classification. In: van Regenmortel MHV, Fraenkel-Conrat H (eds) The plant viruses. 2. The rod-shaped plant viruses. Plenum Press, New York, pp 167–180

Gibbs AJ, Varma A (1977) Queensland isolate of sunn-hemp mosaic virus, unpublished findings

Jailani AAK, Solanki V, Roy A, Sivasudha T, Mandal B (2017) A CGMMV genome-replicon vector with partial sequences of coat protein gene efficiently expresses GFP in *Nicotiana benthamiana*. Virus Res 233:77–85

Kumar S, Udaya Shankar AC, Nayaka SC, Lund OS, Prakash HS (2011) Detection of tobacco mosaic virus and tomato mosaic virus in pepper and tomato by multiplex RT–PCR. Lett Appl Microbiol 53(3):359–363

Kumar A, Shilpi S, Yogita M, Lal SK, Mandal B (2013a) Identification of a new strain of tobacco mosaic virus infecting soybean in India based on host reactions and 3′ end genome sequence. Crop Prot 53:20–22

Kumar A, Solanki V, Mandal B (2013b) Abstract: Frangipani mosaic virus and Plumeria mosaic virus: identification and comparison of two tobamoviruses infecting frangipani in India. VIROCON, pp 116

Kumar A, Solanki V, Verma HN, Mandal B (2015) Characterisation and diagnosis of frangipani mosaic virus from India. Virus Genes 51:310–314

Lamb EM, Atkins S, Schuler KD, Roberts PD (2001) Pepper mild mottle virus. University of Florida, IFAS Extension Bulletin, p 808

Lartey RT, Voss TC, Melcher U (1996) Tobamovirus evolution: gene overlaps, recombination, and taxonomic implications. Mol Biol Evol 13:1327–1338

Lim MA, Hong JS, Song YS, Ryu KH (2010) The complete genome sequence and genome structure of frangipani mosaic virus. Arch Virol 155:1543–1546

Mandal S (2007) Molecular characterization of *Cucumber green mottle mosaic virus* occurring in India and development of transgenic plant using coat protein gene. PhD thesis, Jamia Millia Islamia, 120 pp

Mandal S, Mandal B, Haq QMR, Varma A (2008) Properties, diagnosis and management of cucumber green mottle mosaic virus. Plant Viruses 2:25–34. file:///C:/Users/m/Downloads/Seema%20PV%202008.pdf

Mayer A. (1886) Concerning the mosaic disease of tobacco (J. Johnson, Trans.). In: Phytopathological classics number 7. American Phytopathological Society, St. Paul, pp 11–24

Mckinney HH (1952) Two strains of tobacco mosaic virus, one of which is seed-borne in an etch-immune pungent pepper. Plant Dis Rep 36:184–187

Nagaich BB, Vashisth KS (1963) A virus causing a typical mosaic disease of beans. Indian J Microbiol 3:113

Nagendran K, Aravintharaj R, Kumar SM, Manoranjitham SK, Naidu RA, Karthikeyan G (2015) First report of cucumber green mottle mosaic virus in snake gourd (*Trichosanthes cucumerina*) in India. Plant Dis 99:4

Niazi FR, Chandra KJ, Prakash N (1973) A new strain of tobacco mosaic virus infecting sunn-hemp. Indian Phytopathol 14:77

Orita H, Sakai J, Kubota K, Okuda M, Tanaka Y, Hanada K, Imamura Y, Nishiguchi M, Karasev AV, Miyata S, Iwanami T (2007) Molecular and serological characterization of cucumber mottle virus, a new cucurbit-infecting tobamo-like virus. Plant Dis 91:1574–1578

Prabahar A, Swaminathan S, Loganathan A and Jegadeesan R (2015) Identification of novel inhibitors for tobacco mosaic virus infection in solanaceae plants. Adv Bioinforma; Article ID 198214, 9 pages, http://dx.doi.org/10.1155/2015/198214

Rani P, Pant RP, Jain RK (2010) Serological detection of cymbidium mosaic and odontoglossum ringspot viruses in orchids with polyclonal antibodies produced against their recombinant coat proteins. J Phytopathol 158:542–545

Rao ALN, Varma A (1984) Transmission studies with cucumber green mottle mosaic virus. Phytopathol Z 109:325–331

Rashmi CM, Reddy CNL, Praveen HM, Nagaraju X (2005) Natural occurrence of cucumber green mottle mosaic virus on gherkin (*cucumis anguria* L.) Environ Ecol 23:781–784

Raychaudhuri SP (1947) A note on mosaic virus of sunn-hemp (*Crotalaria juncea* Linn.) and its crystallisation. Curr Sci 16:26–28

Raychaudhuri SP, Pathanian PS (1950) A mosaic disease of *Crotalaria mucronata* Desv. (C. striata DC.). Curr Sci 19:213

Raychaudhuri M, Varma A (1978) Mosaic disease of muskmelon, caused by a minor variant of cucumber green mottle mosaic virus. Phytopathol Z 93:120–125

Rialch N, Sharma V, Sharma A, Sharma PN (2015) Characterization and complete nucleotide sequencing of pepper mild mottle virus infecting bell pepper in India. Phytoparasitica 43:327–337

Ryu KH, Min BE, Choi GS, Choi SH, Kown SB, Noh GM, Yoon JY, Choi YM, Jang SH, Lee GP, Cho KH, Park WM (2000) Zucchini green mottle mosaic virus is a new tobamovirus; comparison of its coat protein gene with that of kyuri green mottle mosaic virus. Arch Virol 145:2325–2333

Samad A, Zaim M, Ajaykumar PV (1999) White mosaic: a new virus disease in Egyptian henbane in India. Indian J Virol 15:39–42

Samad A, Zaim M, Ajaykumar PV, Garg ID (2000) Isolation and characterization of a TMV isolate infecting Scotch spearmint (*Mentha gracillis*) in India. J Plant Dis Protect 107:649–567

Sawada H, Takeuchi S, Hamada H, Kiba A, Matsumoto M, Hikichi Y (2004) A new Tobamovirus-resistance gene, L1a of sweet pepper (Capsicum annuum L.) J Jpn Soc Hortic Sci 73:552–557

Sharma PN, Patiyal K (2011) Status of viruses infecting sweet pepper under polyhouse cultivation in Himachal Pradesh. Plant Dis Res 26:185

Sherpa AR, Hallan V, Zaidi AA (2004) Cloning and sequencing of coat protein gene of an Indian Odontoglossum ringspot virus isolate. Acta Virol 48(4):267–269

Silver S, Quan S, Deom CM (1996) Completion of the nucleotide sequence of sunn-hemp mosaic virus: a toabamovirus pathogenic to legumes. Virus Genes 13(1):83–85

Song YS, Min BE, Hong JS, Rhie MJ, Kim MJ, Ryu KH (2006) Molecular evidence supporting the confirmation of Maracuja mosaic virus as a species of the genus tobamovirus and production of an infectious cDNA transcript. Arch Virol 151:2337–2348

Ugaki M, Toomiyama M, Kakutani T, Hidaka S, Kiguchi T, Nagata R, Sato T, Motoyoshi F, Nishiguchi M (1991) The complete nucleotide sequence of cucumber green mottle mosaic virus (SH strain) genomic RNA. J Gen Virol 72:1487–1495

Vani S, Varma A (1993) Properties of cucumber green mottle mosaic virus isolated from water of river Jamuna. Indian Phytopathol 46:118–122

Varma A, Gibbs AJ (1978) CMI/AAB descriptions of plant viruses No. 196

Varma A, Giri BK (1998) Virus diseases. In: Nayar NM, More TA (eds) Cucurbits. Oxford and IBH Publishing House Pvt. Ltd., New Delhi, pp 225–245

Vasudeva RS, Nariani TK (1952) Host range of bottle gourd mosaic virus and its inactivation by plant extracts. Phytopathology 2:180–185

Vasudeva RS, Raychaudhuri SP, Singh J (1949) A new strain of Cucumis virus 2. Indian Phytopathol 2:180–185

Wetter C, Conti M, Altschuh D, Tabillion R, Regenmortel MHV (1984) Pepper mild mottle virus, a tobamovirus infecting pepper cultivar in Sicily. Phytopathology 74:405–410

Yoon JY, Min BE, Choi SH, Ryu KH (2001) Completion of nucleotide sequence and generation of highly infectious transcripts to cucurbits from full length cDNA clone of kyuri green mottle mosaic virus. Arch Virol 146:2085–2096

Yoon JY, Min BE, Choi SH, Ryu KH (2002) Genome structure and production of biologically active in vitro transcript of cucurbit infecting zucchini green mottle mosaic virus. Phytopathology 92:156–163

Zettler FW, Ko NJ, Wisler GC, Elliot MS, Wong SM (1990) Viruses of orchids and their control. Plant Dis 74:621–626

The Occurrence, Biology, Serology and Molecular Biology of Tospoviruses in Indian Agriculture

20

Basavaraj, Bikash Mandal, S.J. Gawande, P. Renukadevi, S.K. Holkar, M. Krishnareddy, K.S. Ravi, and Rakesh Kumar Jain

Abstract

Tospoviruses of the family *Bunyaviridae* are very significant viral pathogens in several crops in India. A total of 11 tospovirus species has been recognised worldwide, of which five tospovirus species, *Groundnut bud necrosis virus* (GBNV), *Groundnut yellow spot virus* (GYSV), *Iris yellow spot virus* (IYSV), *Watermelon bud necrosis virus* (WBNV) and *Tomato spotted wilt virus* (TSWV) are known to occur in India. In addition to these tospoviruses, a tentative member, Capsicum chlorosis virus (CaCV) has also been documented in India. Tospoviruses are major constraints in vegetable and pulse crops in India. GBNV is the most prevalent and destructive tospovirus in India, which has been detected in 30 different crops and weed species belonging to eight families and other tospoviruses are restricted in nature with reference to host and geographical region. The present chapter summarises the research work on tospoviruses carried out in India during the past more than half a century.

Basavaraj • B. Mandal (✉) • R.K. Jain
Advanced Centre for Plant Virology, Division of Plant Pathology, Indian Agricultural Research Institute, New Delhi, India
e-mail: leafcurl@rediffmail.com

S.J. Gawande
Directorate of Onion and Garlic Research, Pune, Maharashtra, India

P. Renukadevi
Department of Plant Pathology, Tamil Nadu Agricultural University, Coimbatore, India

S.K. Holkar
Indian Institute of Sugarcane Research, Lucknow, India

M. Krishnareddy
Indian Institute of Horticultural Research, Bengaluru, India

K.S. Ravi
Mahyco Research Center, Jalna, India

B. Mandal et al. (eds.), *A Century of Plant Virology in India*,
https://doi.org/10.1007/978-981-10-5672-7_20

Keywords
Tospoviruses in India • Groundnut bud necrosis virus • Watermelon bud necrosis virus • Capsicum chlorosis virus • Iris yellow spot virus • Groundnut yellow spot virus • Tomato spotted wilt virus

20.1 Introduction

Tospoviruses are classified under the genus *Tospovirus* and family *Bunyaviridae* which include the viruses infecting plant as well as insect vector. Tospoviruses have enveloped isometric virions of 80–120 nm containing three segments of single stranded RNA [small (S), medium (M) and large (L) RNA] coated with nucleocapsid protein (N). They are transmitted naturally by several species of thrips in a persistent propagative manner (King et al. 2012) and emerged as serious constraints for the production of several crops in many countries including India (Pappu et al. 2009).

The occurrence of tospovirus-like diseases has been recorded in India since 1960s on different crops belonging to Cucurbitaceae, Fabaceae, Leguminosae and Solanaceae families (Nariani and Dhingra 1963; Sharma 1966; Reddy et al. 1968; Narayanasamy et al. 1975; Amin et al. 1979; Ghanekar et al. 1979; Prasada et al. 1980, 1985, 1987; Khurana et al. 1989; Krishnareddy and Singh 1993; Bhat et al. 2002). These diseases were thought to be caused by tomato spotted wilt virus (TSWV), which was the only known member of the genus *Tospovirus* during that time. In early 1990s, the virus species, *Groundnut bud necrosis virus* (GBNV) was identified as the first distinct tospovirus species causing bud necrosis disease in groundnut in India (Reddy et al. 1991a). Thereafter, three more tospovirus species e.g., *Watermelon bud necrosis virus* (WBNV) (Jain et al. 1998; Mandal et al. 2003), *Groundnut yellow spot virus* (GYSV) (Satyanarayana et al. 1996a) and *Iris yellow spot virus* (IYSV) (Ravi et al. 2006; Gawande et al. 2010) and one tentative member, Capsicum chlorosis virus (CaCV) (Krishnareddy et al. 2008; Kunkalikar et al. 2011) have been identified in India. Recently, the tospovirus type species, *Tomato spotted wilt virus* (TSWV) has been identified in chrysanthemum in southern India (Renukadevi et al. 2015a, b). The congenial environmental conditions, intensive cultivation of crops, susceptibility in crop genotypes and abundance of thrips vector resulted in emergence of different species of tospoviruses on several vegetable, pulse and ornamental crops in India. Several excellent reviews are available on tospovirus research worldwide (German et al. 1992; Whitfield et al. 2005; Tsompana and Moyer 2008; Mandal et al. 2012). In this chapter, we describe the research work carried out on biological, serological and molecular characterization of the different tospovirus species known to occur in Indian agriculture.

20.2 Prevalence and Significance

The tospoviruses in India have their distribution in diverse geographic locations (Mandal et al. 2012). GBNV and WBNV are widely distributed and endemic in many states including Andhra Pradesh, Gujarat, Haryana, Himachal Pradesh, Karnataka, Kerala, Madhya Pradesh, Maharashtra, Orissa, Punjab, Rajasthan, Tamil Nadu, Uttar Pradesh and West Bengal. IYSV has been detected in Gujarat, Karnataka, Madhya Pradesh, Maharashtra, Tamil Nadu and Uttar Pradesh. CaCV is known in some places of northern, central and southern India (Krishnareddy and Varma 1990; Kunkalikar et al. 2007). GYSV has so far been reported only from Andhra Pradesh.

Economically, the most significant tospovirus in India is GBNV, followed by WBNV and others. GBNV was known to cause several epidemics of bud necrosis disease in groundnut and estimated to cause 70–90% loss (Reddy et al. 1991a, b; Singh and Srivastava 1995). Serious outbreak of GBNV was observed during 2003–2006 in tomato crops in Maharashtra, Karnataka and Andhra Pradesh (Kunkalikar et al. 2011). Besides groundnut and tomato, losses ranging from 15 to 30% have been recorded in potato due to stem necrosis disease caused by GBNV in central and western parts of India (Khurana et al. 1989; Singh et al. 1997). However, the occurrence of GBNV in potato appears to be sporadic. The cultivation of watermelon was affected seriously due to the outbreak of WBNV in southern India. The cent percent losses in watermelon production in several fields were witnessed in Karnataka (Singh and Krishnareddy 1996). The economic impact of CaCV and IYSV has not yet been assessed well, however serious reduction of yield is expected due to widespread occurrence of these tospoviruses on tomato and onion crops in several parts of Maharashtra (Kunkalikar et al. 2011).

20.3 Groundnut Bud Necrosis Virus

20.3.1 Discovery and Distribution

GBNV was initially recorded on groundnut as TWSV (Reddy et al. 1968) and later renamed as GBNV based on its distinct serological properties (Reddy et al. 1992). Subsequently, the virus was recorded on several plant species from different geographical locations suggesting its widespread prevalance in India (Bhat et al. 2002; Thien et al. 2003; Umamaheswaran et al. 2003; Akram et al. 2004; Jain et al. 2004; Raja and Jain 2006; Reddy et al. 2008; Akram and Naimuddin 2010, 2012).

20.3.2 Biology

The field symptoms of GBNV in groundnut are well described. Initially, mild chlorotic spots appear on young quadrifoliate leaves of groundnut, and subsequently chlorotic rings and necrosis develop. Necrosis of terminal bud is the main characteristic symptom under natural conditions. Secondary symptoms such as stunting,

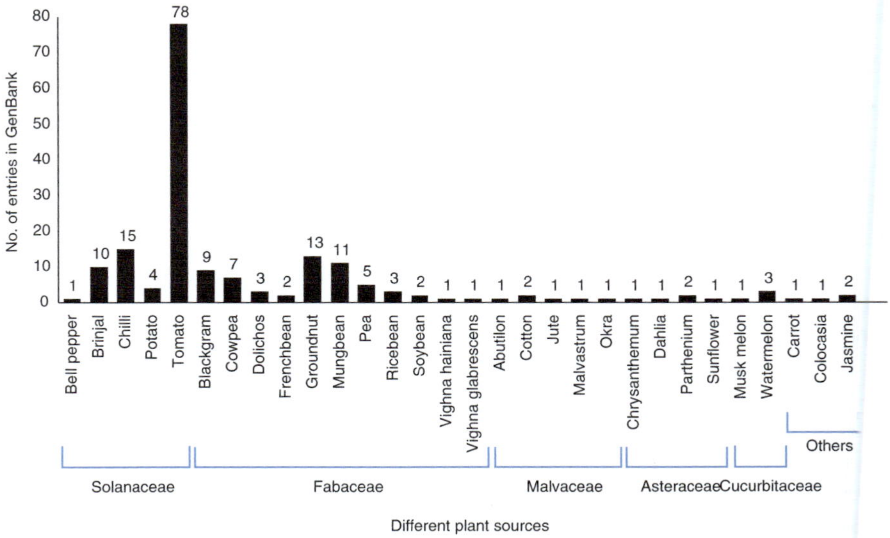

Fig. 20.1 Crop wise sequence data of N gene of groundnut bud necrosis virus occurring in India. The data were obtained from NCBI GenBank Database till 2015

axillary shoot proliferation and leaf malformation are common. Plants infected early are bushy, stunted, and die prematurely. If older plants are infected, the symptoms are restricted to a few branches only (Reddy et al. 1968; Ghanekar and Nene 1976).

Although, GBNV was originally isolated from groundnut, 40% of the isolates characterized so far are from tomato (Fig. 20.1). In addition to groundnut, important plant species like mungbean, urdbean, tomato, cowpea, chilli, brinjal, pea, potato, French bean, onion, jasmine and soybean are also known to be affected by GBNV (Fig. 20.1) (Amin et al. 1979; Prasada et al. 1980, 1985, 1987; Phadtare et al. 1989; Bhat et al. 2002; Jain et al. 2002, 2004, 2007; Thien et al. 2003; Umamaheswaran et al. 2003; Akram et al. 2004; Raja and Jain 2006; Reddy et al. 2008; Akram and Naimuddin 2010, 2012; Sujitha et al. 2012, 2013). Symptoms of GBNV on the other naturally infected crop plants include: bud necrosis or top necrosis on mung and urd beans (Amin et al. 1979); bronzing, severe necrosis of leaves and buds leading to dieback and death of plants in tomato (Fig. 20.2) (Prasada et al. 1980); chlorotic spots on newly produced leaves followed by mosaic mottling and reduction of leaf size with axillary shoot proliferation in chilli (Prasada et al. 1987); chlorotic spots, concentric chlorotic lines followed by necrotic ringspots on leaves leading to stunted plants with distorted leaves and small distorted fruits having pale green blotches or rings in brinjal (Prasada et al. 1987); stem necrosis in potato (Phadtare et al. 1989); bud blight in soybean (Bhat et al. 2002); chlorotic and necrotic local lesions on primary leaves and systemic infection on young growing leaves with necrosis in veins, petiole, stem and growing buds on cowpea (Prasada et al. 1987; Jain et al. 2002); and chlorotic spots with irregular margins, veinal necrosis, slight distortion of leaf lamina stunted growth followed by total collapse on French bean (Akram and Naimuddin 2012).

Fig. 20.2 Symptoms of groundnut bud necrosis virus (GBNV) in tomato plants. (**a**) Necrosis of young leaf and flower buds, (**b**) petiole and stem necrosis, and (**c**) chlorotic concentric circular patches on fruit under field conditions. (**d**) Concentric chlorotic and necrotic rings on leaf at 10–12 days post inoculation (dpi), (**e**) sunken necrotic streaks on stem and leaf petioles at 14–15 dpi, and (**f**) necrosis of growing buds at 20–22 dpi under glasshouse conditions following mechanical sap inoculation by the isolate GBNV-Tom-Del

The experimental host range of GBNV includes several plant species belonging to the families Amaranthaceae, Asteraceae, Chenopodiaceae, Cucurbitaceae, Fabaceae, Malvaceae and Solanaceae (Ghanekar et al. 1979; Prasada et al. 1980; Thien et al. 2003; Raja 2005; Saritha 2007). The cultivars of cowpea (*Vigna unguiculata*, cvs Pusa Komal and C-152) are suitable bio-assay hosts of GBNV as they exhibit localized chlorotic and necrotic lesions as well as occasional systemic symptoms following mechanical sap inoculation (Ghanekar et al. 1979; Prasada et al. 1985; Singh and Krishnareddy 1996; Jain et al. 1998; Raja and Jain 2006; Kunkalikar et al. 2007). The localized symptoms start with initial mild chlorotic spots, which later become necrotic with concentric necrotic rings followed by yellowing, wilting and defoliation of inoculated leaves. Whereas, the systemic symptoms comprised of diffused chlorotic spots, which turn necrotic leading to necrosis of veins, leaf petiole, stem and subsequently the growing bud dries up. *Nicotiana benthamiana* is another important propagative host, which exhibit systemic chlorotic mottling, necrosis of veins, petioles and stem followed by rapid wilting (Fig. 20.3). The symptoms exhibited by the different tospoviruses are very similar and thus the symptom based diagnosis of tospoviruses may mislead their identification at species level.

Though GBNV, under natural conditions, is vectored by thrips, the principal vector of GBNV is not clearly known as there are reports of both *Thrips palmi* (Lakshmi

Fig. 20.3 Symptoms of groundnut bud necrosis virus (GBNV-tomato-Del) isolate in experimental hosts. (**a**) Initial chlorotic local lesions developed at 3–4 days post inoculation (dpi), (**b**) necrotic rings with chlorotic-centre at 6–8 dpi, (**c**) early systemic chlorotic spots and (**d**) systemic veinal necrosis in cowpea cultivar Pusa Komal. (**e**) Initial mild local chlorotic lesions at the site of inoculation and (**f**) systemic puckering and twisting of top leaves in *N. benthamiana*

et al. 1995) as well as *Scirtothrips dorsalis* (Meena et al. 2005) as the vectors of GBNV. The detailed careful investigations are needed to ascertain whether GBNV is predominantly transmitted by *T. palmi* or some other Thrips species.

Limited studies have been conducted on the biological diversity of GBNV isolates. The comparison of disease response of five GBNV isolates (GN, MB, Tom-1, Tom-2 and P) originating from different host plants revealed limited biological diversity. These isolates induced chlorotic or necrotic local lesions followed by systemic necrosis on different hosts. The isolates, irrespective of their origin, induced identical disease response on groundnut and mungbean, however showed some variation on cowpea, soybean, tomato, and cucumber. The host-reactions of Mb isolate originating from mungbean in Delhi was very similar with that of GN isolate originating from groundnut of Hyderabad on most of the hosts except in soybean, where Mb isolate failed to induce systemic necrosis (Thien et al. 2003). Two tomato GBNV isolates, Tom-1 isolated from Kerala and Tom-2 from Hyderabad differed in their host-reactions on cowpea and soybean. Biological diversities of other GBNV isolates originating from potato, chili, brinjal and soybean are yet to know.

20.4 Serological Relations

Polyclonal antibodies (PAbs) were produced against the purified virions of TSWV way back in 1960s and 1970s for the serological detection of the virus (Best 1968; Tas et al. 1977). Subsequently, PAbs against several tospoviruses were produced and initially three serogroups were identified, serogroup I: TSWV (De Haan et al. 1990); serogroup II: TSWV-B, tomato chlorotic spot virus (TCSV), groundnut ring spot virus (GRSV) (de Avila et al. 1993; Pang et al. 1993) and serogroup III: impatiens necrotic spot virus (INSV) (Law et al. 1991). GBNV, however, did not react with the PAbs to either TSWV or INSV (Reddy et al. 1992). Further, PAbs raised against GBNV showed distant serological relationships with the members of serogroup I, II and III, thereby led to the creation of the serogroup IV (Reddy et al. 1992). Subsequently, three other tospoviruses watermelon silver mottle virus (WSMoV), WBNV and CaCV, were described in the literature (Yeh and Chang 1995; Jain et al. 2002, 2005; Thien et al. 2003; Reddy et al. 2008; Gurupad and Patil 2013; Sivaprasad et al. 2015) and these tospoviruses were serologically indistinguishable from GBNV. On the basis of nucleocapsid (N) protein serology, GBNV, WBNV and CaCV were grouped in WSMoV serogroup (Yeh and Chang 1995). As the purification of tospovirus is difficult, recombinant N protein expressed in *Escherichia coli* has been utilized for the production of PAb on renewable basis (Jain et al. 2005), which led to commercialization of an ELISA based diagnostic kit for GBNV (Mandal and Jain 2010). Cocktail of PAbs were also raised in India for multiple tospovirus detection (GBNV, CMV and PRSV) and thereby avoiding individual virus purification, protein expression and protein purification as well as independent immunization in multiple rabbits, which is otherwise a laborious and time consuming process (Kapoor et al. 2014).

Monoclonal antibodies (MAbs) were also raised against N protein of GBNV and a simple dot blot assay was developed for the detection of GBNV in the field samples (Hemalatha et al. 2008). Recently, the immuno capture-reverse transcription polymerase chain reaction (IC-RT-PCR) protocol with the engineered single chain variable fragment (ScFv) was developed and validated for the specific detection of GBNV and differentiation of WBNV (Maheshwari et al. 2015).

20.5 Molecular Biology

20.5.1 Genome Sequence Resource

The extent of genomic characterization of GBNV in India based on NCBI database indicated that, of the three genome segments of GBNV, the S RNA (RNA 3 segment) is extensively studied with 206 out of a total 263 accessions (78.3%) by 2015 (Table 20.1). There are 400 accessions of tospoviruses in the NCBI GenBank database, of which GBNV represents 263 accessions (65.75%). This indicates GBNV is a most studied tospovirus in India (Table 20.1). The first sequence data of GBNV was deposited in the GenBank in the year 2002 (Bhat et al. 2002). Thereafter, the genome sequence information was generated at a slow pace till 2010. Interestingly, there was an upsurge of sequence data between 2010 and 2013. In 2011 alone there were as many as 75 sequence records of N gene in the database (Fig. 20.5). Among the five genes in the three genome segments of GBNV, N gene is the most studied gene with 204 entries in the NCBI GenBank, followed by movement protein (NSm) with 48 entries (Fig. 20.4a). In India, the role of N, non-structural protein (NSs) and NSm encoding genes were studied in detail (Akram and Naimuddin 2009; Bhat and Savithri 2011; Bhushan et al. 2015; Goswami et al. 2012; Lokesh et al. 2010; Singh et al. 2014) as compared to the other genes encoding enveloped glycoprotein (Gn/Gc) and RNA-dependent RNA polymerase (RdRp).

20.5.1.1 Structural Genomics

The complete genome of the groundnut isolate (type isolate) of GBNV has been sequenced. The L, M and S RNA segments are 8.9 kb, 4.8 kb and 3.05 kb long, respectively (Satyanarayana et al. 1996b, c; Gowda et al. 1998). The L RNA is of negative polarity, with a single open reading frame (ORF) located on the viral complementary strand encoding the viral polymerase of 330 kDa L-protein (2877 aa) (Gowda et al. 1998). The M RNA encodes a 34.3 kDa NSm protein in the viral sense and a 127.3 kDa precursor to the two viral membrane glycoproteins, Gn and Gc, in the viral complementary sense (Satyanarayana et al. 1996c). The S RNA encodes a 49.5 kDa NSs protein in the virus sense and a 30.6 kDa N protein in the virus complementary sense (Satyanarayana et al. 1996b). The L-protein had the highest identity in the core polymerase domain with the corresponding regions of the TSWV and INSV (Gowda et al. 1998). The genome sequence of M and S RNA of the mungbean isolate of GBNV was determined (Saritha and Jain 2007) and compared with the type isolate (Satyanarayana et al. 1996b, c). The comparative sequence analysis of GBNV isolates from groundnut, mungbean and tomato revealed that the genome of the M RNA was considerably different in their intergenic regions

Table 20.1 The genomic resources of tospoviruses occurring in India

Virus	L-RNA						M-RNA						S-RNA						Total	
	Complete[a]		Partial		Total		Complete		Partial		Total		Complete		Partial		Total			
	W	In	W	In	W	In	W	In	W	In	W	In	W	In	W	In	W	In	W	In
CaCV	4	1	2	0	6	1	4	1	1	0	5	1	7	1	59	19	66	20	**77**	**22**
GBNV	2	0	1	0	3	0	2	1	0	56	2	57	2	2	4	204	6	206	**11**	**263**
IYSV	2	0	0	0	2	0	3	0	1	0	4	0	2	0	203	44	205	44	**211**	**44**
PYSV	0	0	0	0	0	0	0	0	0	0	0	0	0	1	0	1	0	2	**0**	**2**
TSWV	32	0	409	1	441	1	66	0	314	1	380	1	65	0	1619	1	1684	1	**2505**	**3**
WBNV	0	1	0	1	0	2	0	2	0	2	0	4	0	4	0	56	0	60	**0**	**66**
Total	**40**	**2**	**412**	**2**	**452**	**4**	**75**	**4**	**316**	**59**	**391**	**63**	**76**	**8**	**1885**	**325**	**1961**	**333**	**2804**	**400**

The data were obtained from NCBI GenBank Database till 2015
[a]*W* World, *In* India

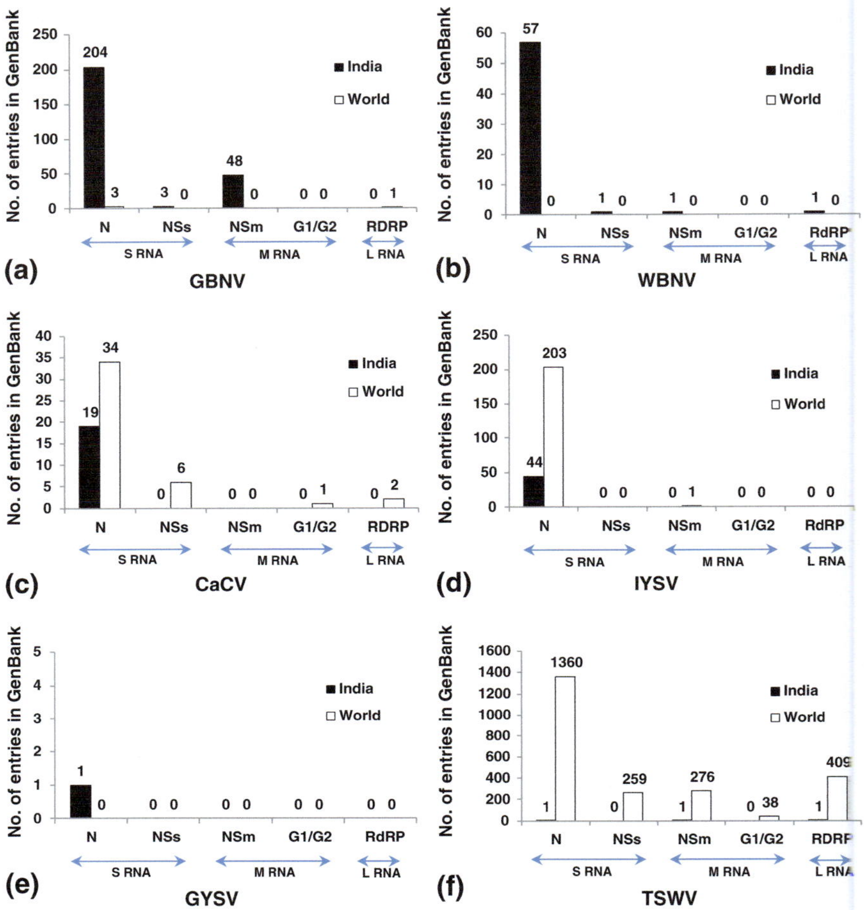

Fig. 20.4 The comparison of sequence data generated on all the five complete genes of each of the six tospoviruses occurring in India and elsewhere in the world: (**a**) GBNV; (**b**) WBNV; (**c**) CaCV; (**d**) IYSV; (**e**) PYSV and (**f**) TSWV. The data were obtained from NCBI GenBank Database till 2015

(56–89% sequence identity) and Gn/Gc protein. The topology of Gn/Gc revealed the presence of both *N*-glycosylation and *O*-glycosylation sites in GBNV-mungbean isolate, whereas only *N*-glycosylation sites were present in GBNV-type isolate. The NSs protein is expressed from a virus sense 1320 nucleotide subgenomic RNA and the N protein from a virus complementary sense 831 nucleotide subgenomic mRNA.

The genetic diversity studies on GBNV so far have been largely based on the N gene, followed by the NSm gene sequences. The amino acid sequence identity (>90%) of N gene is considered as an important criterion for species demarcation of tospoviruses (Goldbach and Kuo 1996). The N gene sequences of more than 200 isolates of GBNV originating from the different hosts and locations are available in

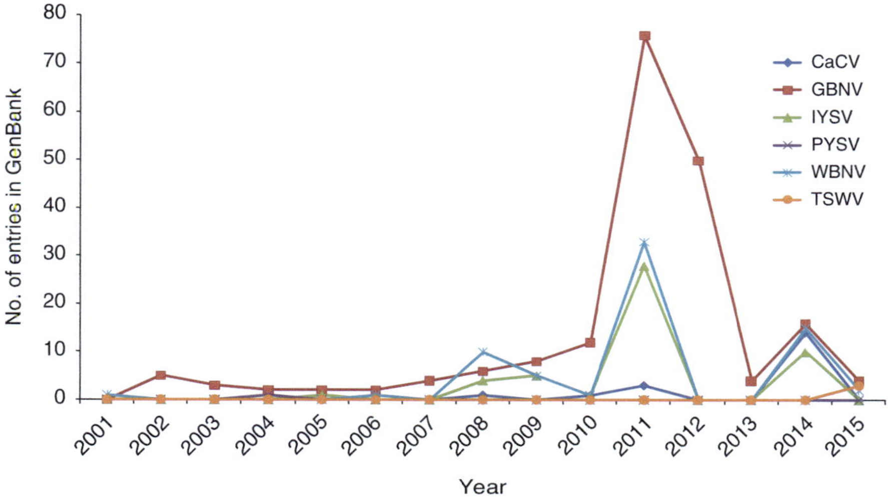

Fig. 20.5 Trend of N gene sequence information generated on the different tospoviruses occurring in India. The data were obtained from NCBI GenBank Database till 2015

the GenBank (Fig. 20.4a). GBNV isolates of mungbean, tomato, cowpea from Kerala, Uttar Pradesh and Delhi were almost identical (99% identity) to the GBNV type isolate (Thien et al. 2003; Jain et al. 2007). GBNV isolates of tomato from Kerala, Tamil Nadu, Maharashtra, Uttar Pradesh showed 96–100% identity with type isolate of GBNV (Raja and Jain 2006; Umamaheswaran et al. 2003). GBNV from potato causing stem necrosis disease in Gujarat, Rajasthan and Madhya Pradesh showed 98–99% identity with the GBNV type isolate (Jain et al. 2004). Comparison of NSm gene also revealed limited diversity (2–3%) in GBNV isolates from cowpea, groundnut, tomato and potato from Kerala, Madhya Pradesh, Rajasthan, Uttar Pradesh and Tamil Nadu (Akram et al. 2004; Raja and Jain 2006; Akram and Naimuddin 2009).

Functional Genomics

The N protein is known to interact with the RNA via a conserved RNA-binding domain present in the C-terminal unfolded region of the protein. Such specific interactions of the N protein by phosphorylation would be necessary for eliciting its multiple functions at the different stages of the viral life cycle, such as replication, assembly and movement (Bhat and Savithri 2011).

The NSs protein of GBNV has been characterized as a bifunctional enzyme containing RNA stimulated ATPase and 5′ phospatase activities, which participates in the suppression of the host defense mechanism (Lokesh et al. 2010). Since the ATP hydrolysis is an essential function of a true helicase, the NSs was tested for the DNA helicase activity and proved that the NSs acts as a non-canonical DNA helicase through the mutant NSs constructs, wherein an alanine mutation in the Walker A (K189A rNSs) and Walker B motifs were induced. Further, both the wild type as well as the mutant NSs could function as efficient suppressors of RNA silencing, indicating that the role of NSs as a RNA silencing suppressor is independent of its helicase or ATPase activity (Bhushan et al. 2015). NSs has also been proved to affect miRNA biogenesis through its RNA silencing suppressor activity and interferes with TCP1 (Teosinte branced/Cycloidea/Pcf) transcriptase-regulated leaf developmental pathways and might also be an important factor affecting pathogenesis of GBNV. The necrotic symptom in the infected plant is attributed to the overproduction of TCP1 (Goswami et al. 2012).

NSm encoded by the M-RNA functions as a movement protein. It was found that the NSm of GBNV associated with the host membranes via the C-terminal coiled coil (CC) domain despite the absence of any putative transmembrane domain. However, the *in planta* expressed NSm failed to interact with membranes when CC domain was deleted. Further, the NSs was found to be associated with host's membrane when the C-terminal CC domain alone was expressed (Singh et al. 2014).

Recently, the levels of biochemical intermediates and stress responsive micro RNAs (miRNAs) with their target transcripts in *V. unguiculata* during GBNV infection has been compared to understand the process of symptom development. The necrosis or cell death in the form of hypersensitive reaction after the infection by GBNV was attributed to the accumulation of reactive oxygen species (ROS). In addition, the increased activities of antioxidant enzymes, such as catalase, peroxidase, superoxide dismutase, ascorbate peroxidase and glutathione reductase that are associated with detoxification of ROS was observed. The down-regulation of miR398a involved in regulating the cytochrome C oxidase subunit 5b (COX-5b) responsible for the oxidative stress resistance was also observed. The other three stress responsive miRNAs; miR319a.2, miR399f, miR482 also down-regulated with respective induction in the expression of transcripts they regulate such as metacaspase-8, PHOSPATE 2 and an NBS-LRR gene, respectively (Permar et al. 2014).

20.6 Watermelon Bud Necrosis Virus

20.6.1 Discovery

WBNV was first recorded during 1991 infecting watermelon in southern India (Krishnareddy and Singh 1993). Since then, a similar disease was observed in other watermelon-growing states of India, including, Karnataka, Andhra Pradesh, Maharashtra and Delhi. This is a serious pathogen on watermelon and other cucurbitaceous hosts causing the incidence ranging from 39 to 100%, with 60–100% of estimated loss in yield (Krishnareddy and Singh 1993; Singh and Krishnareddy 1996; Jain et al. 1998; Mandal et al. 2003, 2012; Pappu et al. 2009). The virus was initially identified as a new tospovirus infecting watermelon on the basis of host range, symptomatology, transmission and serology (Singh and Krishnareddy 1996).

20.6.2 Biology

The bud necrosis disease affected watermelon under field conditions exhibits array of symptoms (Fig. 20.6) including smaller chlorotic leaves, yellowing and necrotic streaks on vines, shortened internodes, stunted growth, severe stem and bud necrosis and dieback of the buds and concentric chlorotic rings on fruits (Jain et al. 1998, 2007; Mandal et al. 2012; Holkar et al. 2016). However, under the greenhouse conditions, WBNV induces chlorotic local lesions followed by veinal chlorosis and necrosis in cowpea, whereas, systemic chlorosis and necrosis in *N. benthamiana* and *N. rustica* (Fig. 20.7) (Singh and Krishnareddy 1996).

WBNV has a narrow host range, predominantly restricted to cucurbitaceous crop species *viz.*, bitter gourd (*Memordica charantia*), cucumber (*Cucumis sativus*), muskmelon (*Cucumis melo*), pumpkin (*Cucumis sativus*), ridge gourd (*Luffa acutangula*), watermelon (*Citrullus lanatus*) (Mandal et al. 2003; Jain et al. 2007; Kunkalikar et al. 2011). Besides these crop species, WBNV infects other hosts belonging to Solanaceae [tomato (*Solanum lycopersicum*) and chilli (*Capsicum annum*)] and Asteraceae [chrysanthemum (*Chrysanthemum indicum*), dahlia (*Dahlia coccinea*)] and weed hosts like *Parthenium hysterophorus*, *Amaranthus* sp., *Euphorbia heterophylla*, *Malvastrum coromandelium* and *Abutilon indicum,* which serve as collateral hosts for perpetuation of the virus (Kunkalikar et al. 2011; Holkar et al. 2016).

Fig. 20.6 Field symptoms of watermelon bud necrosis virus (WBNV-Wm-Del) isolate in watermelon plant. (**a**) Chlorotic and necrotic lesions on leaf, (**b**) necrotic streaks on shoots, (**c**) necrosis of young leaf and buds and (**d**) chlorotic concentric circular patches on fruit

Fig. 20.7 Symptoms of watermelon bud necrosis virus (WBNV-Wm-Del) isolate in experimental hosts. (**a**) Initial chlorotic diffused local lesions at 3–4 days post inoculation (dpi) and subsequent (**b**) chlorotic diffused irregular local lesions, (**c**) early systemic mild veinal chlorosis and subsequent (**d**) systemic severe veinal chlorosis, necrosis and leaf twisting in cowpea cultivar Pusa Komal. (**e**) Initial mild diffused local lesions at the site of inoculation and (**f**) systemic puckering and bending of top leaves in *N. benthamiana*

In India, WBNV was reported to be vectored by *Thrips flavus* (Singh and Krishnareddy 1996). Recently, *Thrips palmi* has been shown as a vector of WBNV based on the species-specific markers for mitochondrial mtDNA and subsequently by deep sequencing (Rebijith et al. 2011, 2016). However, the experimental evidence of transmission of WBNV by *T. palmi* is still awaiting.

20.6.3 Serological Relations

WBNV is serologically indistinguishable from GBNV and CaCV based on the assay with PAbs, suggesting it as a member of the serogroup IV (Jain et al. 1998, 2005; Fauquet et al. 2005). WBNV reacted positively with the PAbs to the N protein of WSMoV isolates from Japan (JPN-W) and Taiwan (TAI-W). However, the moderate reaction with the PAbs to the N protein of GBNV from the Indian groundnut isolate indicated a close serological relationship. Whereas, failure in reaction with the PAbs to N protein of TSWV and INSV indicated the WBNV is distinct from or not related to these tospoviruses (Singh and Krishnareddy 1996).

20.6.4 Molecular Biology

WBNV genome has not been adequately characterized. Molecular characterization of WBNV is predominantly based on the N gene sequence (Jain et al. 1998). Although, WBNV was originally isolated from watermelon, N gene sequences have been generated from the other plant species including tomato (*Solanum lycopersicum*), ridge groud (*Luffa acutangula*), pumpkin (*Cucurbita pepo*), cucumber (*Cucumis sativus*), chilli (*Capsicum annuum*), bitter gourd (*Momordica charantia*) chrysanthemum, parthenium (*Parthenium hysterophorus*), malvastrum(*Malvastrum coromandelianum*), euphorbia (*Euphorbia* sp.), amaranthus (*Amaranthus* sp.) and abutilon (*Abutilon indicum*) (Fig. 20.8). Till 2015, there are 66 sequence accessions of WBNV available in the NCBI database, of these, the S-RNA genome sequences has the major share (Table 20.1). Overall, the N gene sequences constituted the highest share of sequences among all the genes of WBNV genome. There are limited number of comple genome sequences of WBNV available e.g., one for L RNA, two for M RNA and four for S RNA (Table 20.1).

The complete genome sequence of WBNV was generated by Li and his associates (2011). L RNA of WBNV-JT isolate is 8916 nucleotides long, with 247 and 32 nucleotides long 5′- and 3′-UTR, respectively. The L RNA is of negative polarity with coding region of 8637 nucleotides encoding 2878 amino acid long (332 kDa) RdRp protein. M RNA genome is 4794 nucleotides long with two ORFs in an ambisense orientation, and encodes for NSm protein (34.10 kDa; 307 amino acid) on the viral sense strand and Gn/Gc glycoprotein precursors of 127.15 kDa (1121 amino acid) on the viral complementary strand. The two ORFs were separated by an IR of 412 nucleotides (Kumar et al. 2010; Kunkalikar et al. 2011; Li et al. 2011). S RNA has 3405 nucleotides, consisting of NSs and N protein ORFs in an ambisense

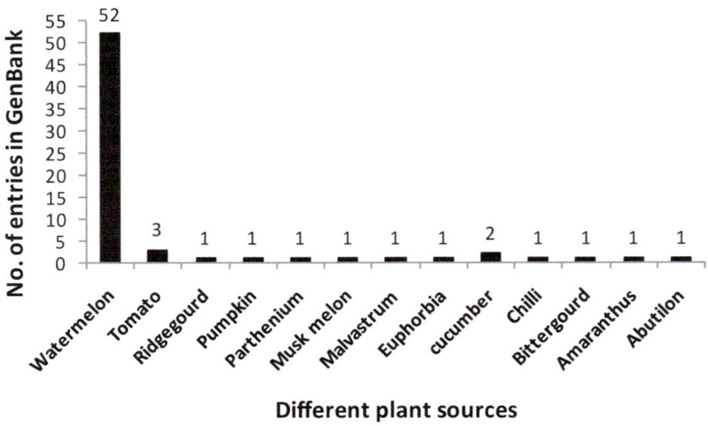

Fig. 20.8 Crop wise sequence data of N gene of watermelon bud necrosis virus occurring in India. The data were obtained from NCBI GenBank Database till 2015

orientation and are separated by 1121 nucleotides A-U rich IR (Li et al. 2011). The N gene of WBNV (accession no. AF045067) shared 81 and 79% identity at nucleotides level and 82 and 84% identity at amino acid level with those of GBNV and WSMoV, respectively. The phylogenetic information of other proteins encoded by WBNV including NSs, NSm, Gn/Gc and RdRp were also consistent with that of N protein (Kumar et al. 2010; Li et al. 2011). In India, during 2011 the nucleotides sequence of the N gene of 28 field isolates of WBNV were determined and compared using pairwise sequence alignment and the nucleotides identities among WBNV N gene sequences ranged from 94.1 to 100% (Kunkalikar et al. 2011).

20.7 Capsicum Chlorosis Virus

20.7.1 Discovery

CaCV was originally described from Australia on tomato and chilli (McMichael et al. 2002). It was identified for the first time in India on tomato from northern India (Kunkalikar et al. 2007) and subsequently, found in northern, central and southern parts of India (Kunkalikar et al. 2007; Krishnareddy et al. 2008). The Australian isolate of CaCV differs from the Indian isolate in host range (Jones and Sharman 2005; Kunkalikar et al. 2010).

20.7.2 Biology

Symptoms induced by CaCV in chilli and tomato are almost similar to that of GBNV. On chilli, CaCV induces initially the yellow spots or patches on younger leaves, which later coalesce to form a mosaic pattern. Later, small concentric rings develop around and within the yellow spots on the mature leaves leading to leaf deformation. The virus infected plant produces small leaves with sparse canopy and shows apical necrosis. Often, necrotic streaks on tender parts of the stem can also be seen, as in case of GBNV infection. On tomato, chlorotic as well as necrotic ringspots develop on leaves and fruits. Later, the systemic infection leads to necrosis in young growing tips followed by bud necrosis. Fruits of CaCV infected tomato plants show discolorations on the surface with circular concentric ringspots (Kunkalikar et al. 2007, 2010; Krishnareddy et al. 2008). On the experimental host like cowpea cv. CO the chlorotic local lesions appeared at 3–4 dpi followed by necrosis (Vijayalakshmi et al. 2016). Whereas, on *N. benthamiana,* chlorotic mottling, chlorotic local lesions appeared at 6–7 dpi leading to systemic necrosis and death of the plants (Kunkalikar et al. 2010).

Like GBNV, CaCV too has a broad host range and they belong to be families *Amaranthaceae, Cucurbitaceae, Apocynaceae, Chenopodeaceae, Fabaceae and Solanaceae* (Kunkalikar et al. 2010; Mandal et al. 2012; Sharma and Kulshrestha 2014; Vijayalakshmi et al. 2016). The isolates derived from the two different naturally infected hosts such as tomato and chilli could induce identical symptoms with no variations on the indicator hosts (Kunkalikar et al. 2010).

Thrips, *Ceratothripoides clartris* was reported as vector of CaCV in Thailand (Premachandra et al. 2005); whereas, *Frankliniella schultzei* and *Thrips palmi* were found to be the vectors of CaCV in Australia (Persley et al. 2006). Investigation is required to identify the thrips vector transmitting CaCV in India.

20.7.3 Serological Relations

CaCV along with GBNV and WBNV belong to the WSMoV serogroup i.e. serogroup IV (Yeh and Chang 1995; Yeh et al. 1996). In India, several isolates of CaCV originating from the different hosts and locations were studied serologically and ascertained that they belong to serogroup IV (Kunkalikar et al. 2007, 2010; Krishnareddy et al. 2008).

20.7.4 Molecular Biology

The first record of CaCV genome sequence in India was made in 2008 with N gene sequence (Krishnareddy et al. 2008). Till 2015, there were only 22 sequence accessions in the NCBI database. So far, the complete sequences for all the three genome components (S, M and L segements) is avaible only for one Indian isolate of CaCV (Kunkalikar et al. 2010; Table 20.1). The sequence information of N gene has been generated from the CaCV isolates of different hosts like chilli, capsicum, tomato and amaranthus (Fig. 20.9).

The S-RNA genome of CaCV was comprised of 3105 nucleotides with an IR of 824 nucleotides separating two non-overlapping ambisense ORFs of N and NSs genes. The IR of the Indian isolate was found to be 372 nucleotides shorter than that of CaCV-Thailand isolate. Similarly, the M-RNA genome is 4821 nucleotides long

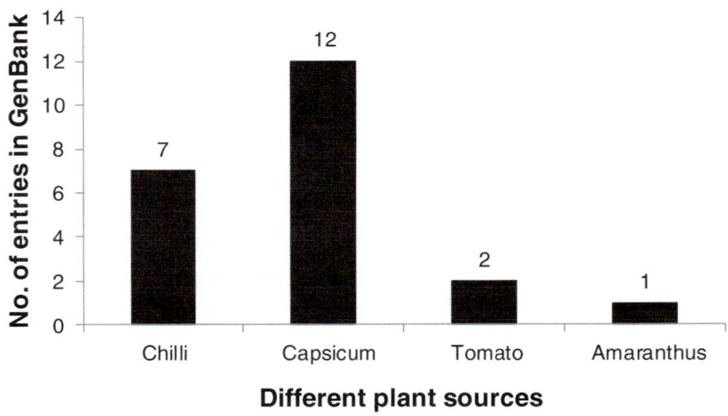

Fig. 20.9 Crop wise sequence data of N gene of capsicum chlorosis virus occurring in India. The data were obtained from NCBI GenBank Database till 2015

having an IR of 425 nucleotides that separates NSm and Gn/Gc genes. The L segment was comprised of 8912 nucleotides, which contained an RdRp gene of 8634 nucleotides. The N gene sequences the Indian isolates of CaCV were analyzed by comparing with those of Australia, Thailand, China, Taiwan and USA. The amino acid sequence identities were 96.7–98.9% among the Indian CaCV isolates and 91.2–96.7% with the isolates from the other parts of the world. The higher sequence identity of the Indian CaCV isolate with Australian isolate (96.75%) was recorded and suggested that the CaCV isolates of these two countries could be classified into one group. Interestingly, based on IR sequence, the Indian CaCV isolate was found closer to GBNV with 76% sequence identity than to CaCV-Thailand isolate (Kunkalikar et al. 2010).

20.8 Iris Yellow Spot Virus

20.8.1 Discovery

IYSV is the major constraint for the production of onion and garlic in some countries (Pappu et al. 2009). It was first identified in the USA in 1989, subsequently the virus was reported in other countries (Hall et al. 1993; Gent et al. 2006). In India, IYSV was first recorded on onions (*Allium cepa*) in 2006, on garlic (*A. sativum*) in 2010 and on chives (*A. tuberosum*) in 2014 (Ravi et al. 2006; Gawande et al. 2010, 2014).

20.8.2 Biology

The symptoms of IYSV appear with the chlorotic spindle or diamond shaped lesions on the leaves and flower bearing stalks (scapes). As the disease progresses, the size of lesion enlarges and coalesces to form large necrotic patches on leaves and scapes.

IYSV infects both seed as well as bulb onion crop, but, the losses are more apparent on seed crop, as necrotic patches on flower/seed causes toppling of stalks prematurely leading to poor seed yield (Fig. 20.10). The symptoms on garlic and garlic chives include straw colored and spindle-shaped spots with less defined ends. As the infection progresses, spots coalesces to form larger patches. These symptoms were more visible on older leaves (Ravi et al. 2006; Gawande et al. 2010, 2014).

In India, the natural hosts of IYSV includes onion, garlic and garlic chives. The experimental host range includes *D. stramonium, C. amaranticolor*, *D. metel, D. alba, N. rustica N. benthamiana, N. tabacum* (varieties 'Xanthi NC', 'White Burley', 'Samsun' and 'GT-4'), *N. clevelandii* and *V. unguiculata.* IYSV induces chlorotic local lesions in *C. amaranticolor* and necrotic lesions in *D. stramonium, D. metel* and *D. alba*, whereas systemic chlorosis, curling and twisting in *N. rustica* (Fig. 20.11) (Ravi et al. 2006).

IYSV is transmitted by thrips in a circulative, propagative manner. Although, *T. tabaci* appears to be the major vector of IYSV in the USA (Diaz-Montano et al. 2011). It has been recently demonstrated that *Frankliniella fusca* is also capable of transmitting IYSV but with low efficiency (Srinivasan et al. 2012). Larval stage L1 (first instar) and L2 (second instar) can only acquire the virus, whereas, all the three stages (L1, L2 and adult) transmit the virus. Variation in transmission efficiency was observed with the differences in *T. tabaci* populations, which has been linked to the presence of mitochondrial haplotypes (S.J. Gawande, unpublished results).

Fig. 20.10 Symptoms of iris yellow spot virus (IYSV) in (**a**) *A. sativum* leaves, (**b**) *A. cepa* flower stalk, (**c**, **d**) seed onion crop [severe symptoms of IYSV causing breaking of flowering stalks] (**d**) sprout of stored onions showing chlorotic patches

Fig. 20.11 Symptoms of iris yellow spot virus in experimental hosts. Systemic chlorosis, curling and twisting of leaves in (**a**) *N. rustica*, and necrotic lesions in (**b**) *D. alba and* (**c**) *D. metel,* and chlorotic local lesions in (**d**) *C. amaranticolor*

20.8.3 Molecular Biology

The virus has been characterized at molecular level in India based on N gene sequence a decade ago (Ravi et al. 2006). The number of sequence entries of the Indian IYSV isolates in GenBank database include 44 sequences of N gene. However, no information is available for the other genomic segments of IYSV (Table 20.1). Although IYSV is known to infect onion, garlic and chives, the N gene sequence was predominantly generated from onion (44 accessions) and only one accession each from garlic and chives (Fig. 20.12).

The complete nucleotide sequence of M- and S-RNA genome of an Indian IYSV isolate was determined very recently (Gawande et al. 2015). These sequences were found to be 95% identical to the IYSV sequences presently available in public database. Phylogenetically, the Indian and Dutch (AF001387) isolates of IYSV were most closely related to tomato yellow ring virus (TYRV), polygonum ringspot virus (PolRSV), hippeastrum chlorotic ringspot virus (HCRV) and spider lily necrotic spot virus (SLNSV) based on sequences of S- and M-RNA genomic segments. Similar pattern of relationship was evident in phylogenetic trees constructed from the individual NSs, NSm and Gn/Gc ORFs. Further, phylogenetic analysis of N gene of IYSV revealed that the Asian and European IYSV isolates formed separate cluster from the US isolates (Gawande et al. 2015).

The possible role of recombinations in the evolution of tospoviruses has been studied with IYSV population and found that the maximum recombination events were in the IR of the S- and M-RNA of IYSV genome (Gawande et al. 2015). These results suggested that the IR of tospovirus may be more predisposed to recombination.

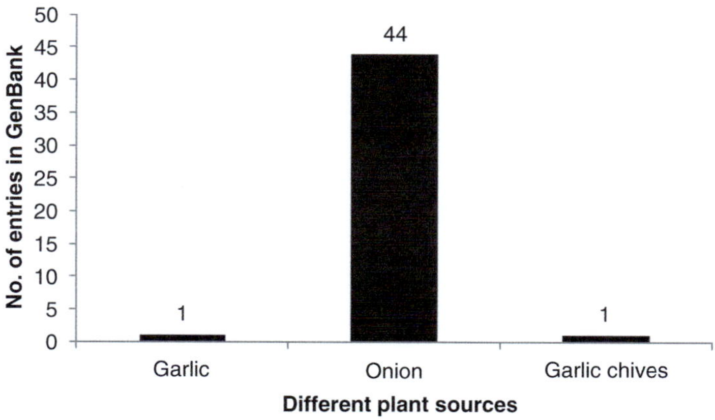

Fig. 20.12 Crop wise sequence data of N gene of iris yellow spot virus occurring in India. The data were obtained from NCBI GenBank Database till 2015

20.9 Groundnut Yellow Spot Virus

20.9.1 Discovery

GYSV was identified as a distinct tospovirus in groundnut in Andhra Pradesh, India based on host-range, particle morphology and serology (Reddy et al. 1991b). The virus was originally described as peanut yellow spot virus. High incidence of GYSV incidence (up to 90%) was recorded from southern India. However, distribution of GYSV in the other places and its impact on yield loss in groundnut are not known yet (Satyanarayana et al. 1996a).

20.9.2 Biology

The disease symptoms of GYSV in groundnut is characterized by initial yellow spots followed by necrosis in leaves (Satyanarayana et al. 1996a). GYSV is transmitted by *S. dorsalis* (Satyanarayana et al. 1996a). Further study shows that the larvae of *S. dorsalis* can cause upto 43.8% transmission of GYSV with the acquisition access period (AAP) of 30 min and the maximum transmission (>75%) resulted following the AAP of 2 days. Single adult thrips could transmit the virus after minimum inoculation access period (IAP) of 30 min. The transmission efficiency increased to 36.1% with an increase in IAP up to 55 h (Gopal et al. 2010).

20.9.3 Serological Relationships

GYSV has been studied for its serological relationships with the antisera to TSWV, INSV and GBNV, which revealed that GYSV is not serologically related to these viruses. Therefore, GYSV was classified as a distinct tospovirus species under the new serogroup-V (Satyanarayana et al. 1996a).

20.9.4 Molecular Biology

GYSV was suggested as a new species based on S RNA sequence characteristics. The N and NSs proteins of GYSV were significantly different from that of the other tospoviruses as these genes shared very low sequence identities of 24–28% and 16–21%, respectively (Satyanarayana et al. 1998). The NCBI database indicates the availability of only two sequence accessions pertaining to the S-RNA in 1998 followed by a N gene sequence in 2004 (Table 20.1, Figs. 20.4e and 20.5). Further, information on genomic properties of GYSV is not available.

20.10 Tomato Spotted Wilt Virus

20.10.1 Discovery

TSWV, the type species of the genus *Tospovirus,* is known to infect a large number of plant species and distributed in the various regions of the world. TSWV is not known to occur in India until 2015. Earlier, tospovirus-like diseases described in the literature from India were assumed to be caused by TSWV, the only known tospovirus at that time. However, the studies on tospoviruses in India during the last two decades suggested the occurrence of a distinct tospovirus, i.e. GBNV. Recently a severe necrosis disease of chrysanthemum (*Dendranthema grandiflora*) was noticed in the Nilgiri hills of Tamil Nadu state during August 2013 and the association of TSWV was demonstrated based on the N gene characteristics (Renukadevi et al. 2015a, b). TSWV was detected in chrysanthemum plants grown in commercial greenhouses in Kothagiri (Nilgris district) and Yercaud (Salem district) during 2013 and later during 2015 in Thandikudi (Dindigul district). It seems, TSWV might have arrived through the stem cuttings of chrysanthemum.

20.10.2 Biology

TSWV symptoms on chrysanthemum includes veinal necrosis and necrotic spots and patches on the foliage and stems, browning of flower petals and malformation of flowers leading to excessive necrosis and complete drying and death of the plants. The mechanical sap inoculation from symptomatic plant to the different plant species resulted in local chlorotic and necrotic concentric rings on the leaves of cowpea (*V. unguiculata* cv. C152) and local as well as systemic chlorosis and necrosis in *D. stramonium, N. glutinosa and S. lycopersici* (Fig. 20.13) (Renukadevi et al. 2015b).

Fig. 20.13 Symptoms of tomato spotted wilt virus (TSWV). Field symptoms in chrysanthemum (**a**) necrotic lesions in leaf, (**b**) veinal necrosis, (**c**) stem necrosis. Greenhouse symptoms in the experimental plants following sap inoculation, (**d**) chlorotic lesions in cowpea C-152 and (**e**) *Chenopodium amaranticolor*, (**f**) necrotic ring-spots in *Solanum lycopersicum* and (**g**) necrotic lesions in *Nicotiana glutinosa* and (**h**) white necrotic rings in *Datura stramonium*. (**i**) The vector of TSWV, *Frankliniella occidentalis* adult in chrysanthemum

The thrips vector of TSWV, *F. occidentalis,* commonly known as Western flower thrips (WFT) was noticed recently in the Nilgiris district of Tamil Nadu and confirmed by morphological keys and mtCOI sequence analysis (Suganthy et al. 2016). The presence of WFT was also recorded on tomato in Bangalore, India (Tyagi and Kumar 2015).

20.10.3 Molecular Biology

Limited information is available on the molecular characterization of TSWV occurring in India. There are only three sequence accessions in the database till 2015, one each representing N, NSm and a partial RdRp gene (Table 20.1, Fig. 20.4f). Besides, three sequences each of N and NSm genes were added in 2016. The nucleotide sequences of N and NSm exhibited the highest nucleotide identity of 98% with the corresponding regions of TSWV isolates from the different hosts and countries (Renukadevi et al. 2015b).

20.11 Concluding Remarks

Tospovirus study in India was intensified since 1991 with the identification of GBNV as a new virus species. Generation of genome sequence information being the most authentic input to virus characterisation led to the identification of half a dozen of tospoviruses in India during past two and a half decades. The generation of genome sequence information of tospoviruses from India began since 2001 and the maximum numbers of sequence accessions were generated during 2010–2012. The analysis of sequence data from India revealed that the maximum numbers of sequence data were generated based on N gene (Fig. 20.4). The complete genome sequence is available only for two tospoviruses, CaCV and WBNV. GBNV being the most studied virus, the L RNA genome is yet to be sequenced. Phylogenetic analysis based on the complete amino acid sequences of N protein of 111 isolates representing all the six tospoviruses known in India showed existence of three major divergent group with GBNV, WBNV and IYSV (Fig. 20.14). The genetic diversities of CaCV, GYSV and TSWV in India have not yet been studied adequately. Transmission and maintenance of isolate of tospovirus on the experimental plants are tedious, which limits identification and comparison of tospoviruses at strain level. More studies are necessary on the easy to use reproducible inoculation methodologies using thrips as well as sap from the infected plant. These methodologies will facilitate not only to study the biological diversity of tospovirus strains but also to screen resistance sources. Limited studies were undertaken on the viral gene function and host interactions. TSWV and western flower thrips (*F. occidentalis*) have recently been documented in India (Tyagi and Kumar 2015; Renukadevi et al. 2015b; Suganthy et al. 2016). TSWV being the most prevalent and destructive tospovirus in many countries, arrival of TSWV and its major vector in India is of a great concern in Indian agriculture.

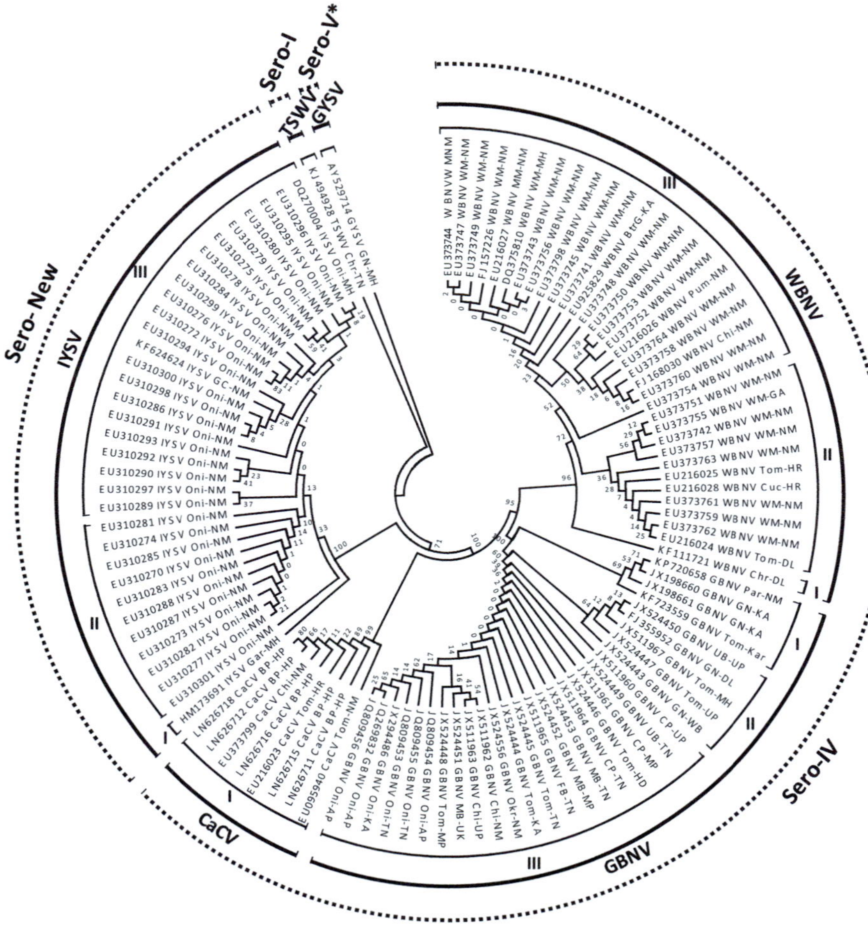

Fig. 20.14 Phylogenetic relations of the six tospovirus isolates occurring in India. The evolutionary history was inferred using the Neighbor-Joining method. The bootstrap consensus tree inferred from 1000 replicates is taken to represent the evolutionary history of the taxa analyzed. Branches corresponding to partitions reproduced in less than 50% bootstrap replicates are collapsed. The evolutionary distances were computed using the Poisson correction method and are in the units of the number of amino acid substitutions per site. The analysis involved complete amino acid sequences of N protein of 111 isolates. All positions containing gaps and missing data were eliminated. There were a total of 204 positions in the final dataset. Evolutionary analyses were conducted in MEGA5. Branch naming is based on accession number followed by virus, host and place of virus origin. Host: *BtG* bitter gourd, *BP* bell pepper, *Chr* Chrysanthemum, *CB* common bean, *CP* Cowpea, *Cuc* cucumber, *FB* French bean, *Gar* garlic, *GC* garlic chives, *GN* groundnut, *MB* mung bean, *Okr* Okra, *Oni* Onion, *Par* Parthenium, *Pum* Pumpkin, *Tom* Tomato, *UB* Urd bean, *WM* watermelon. Place: *AP* Andhra Pradesh, *DL* Delhi, *GA* Goa, *HD* Hyderabad, *HP* Himachal Pradesh, *HR* Haryana, *KA* Karnataka, *MH* Maharashtra, *MP* Madhya Pradesh, *RJ* Rajasthan, *TN* Tamil Nadu, *UK* Uttarakhand, *UP* Uttar Pradesh, *WB* West Bengal, *NM* Location not mentioned in database

References

Akram M, Naimuddin (2009) Movement protein gene sequence based characterization of *Groundnut bud necrosis virus* infecting cowpea. J Food Legumes 22:82–85

Akram M, Naimuddin (2010) First report of *Groundnut bud necrosis virus* infecting pea (*Pisum sativum*) in India. New Dis Rep 21:10

Akram M, Naimuddin (2012) Characterization of *Groundnut bud necrosis virus* infecting French bean. J. Food Legumes 25(1):54–57

Akram M, Jain RK, Chaudhary V, Ahlawat YS, Khurana SMP (2004) Comparison of *Groundnut bud necrosis virus* isolates based on movement protein (NSm) gene sequences. Ann Appl Biol 145:285–289

Amin PW, Ghanekar AM, Reddy DVR (1979) Leaf curl disease of mung and urd bean caused by *Tomato spotted wilt virus*. Indian Phytopath 32:163

Best RJ (1968) Tomato spotted wilt virus. In: Kenneth MS, Lauffer MA (eds) Advances in virus research. Academic, New York, pp 65–146

Bhat AS, Savithri HS (2011) Investigations on the RNA binding and phosphorylation of *Groundnut bud necrosis virus* nucleocapsid protein. Arch Virol 156(12):2163–2172

Bhat AI, Jain RK, Varma A, Lal SK (2002) Nucleocapsid protein gene sequence studies suggest that soybean bud blight is caused by a strain of *Groundnut bud necrosis virus*. Curr Sci 82:1389–1392

Bhushan L, Abraham A, Choudhury NR, Rana VS, Mukherjee SK, Savithri HS (2015) Demonstration of helicase activity in the nonstructural protein, NSs, of the negative-sense RNA virus, *Groundnut bud necrosis virus*. Arch Virol 160(4):959–967

de Avila AC, de Haan P, Kormelink R, de Resende RO, Goldbach RW, Peters D (1993) Classification of tospoviruses based on phylogeny of nucleoprotein gene sequences. J Gen Virol 74:153–159

De Haan P, Wagemakers L, Peters D, Goldbach WR (1990) The S RNA segment of *Tomato spotted wilt virus* has an ambisense character. J Gen Virol 71:1001–1007

Diaz-Montano J, Fuchs M, Nault BA, Fail J, Shelton AM (2011) Onion thrips (*Thysanoptera: Thripidae*): a global pest of increasing concern in onion. J Econ Entomol 104:1–13

Fauquet CM, Mayo MA, Maniloff J, Desselberger U, Ball LA (2005) Virus taxonomy, VIIIth Report of ICTV. Academic, New York, p 1259

Gawande SJ, Khar A, Lawande KE (2010) First report of *Iris yellow spot virus* on garlic in India. Plant Dis 94:1066

Gawande SJ, Gurav VS, Ingle AA, Gopal J (2014) First report of *Iris yellow spot virus* infecting *Allium tuberosum* L. in India. Plant Dis 98:1161

Gawande SJ, Gurav VS, Ingle AA, Martin DP, Asokan R, Gopal J (2015) Sequence analysis of Indian *Iris yellow spot virus* ambisense genome segments: evidence of interspecies RNA recombination. Arch Virol 160:1285–1289

Gent DH, du Toit LJ, Fichtner SF, Mohan SK, Pappu HR, Schwartz HF (2006) *Iris yellow spot virus*: an emerging threat to onion bulb and seed production. Plant Dis 90:1468–1480

German TL, Ullman DE, Moyer JW (1992) Tospoviruses: diagnosis, molecular biology, phylogeny and vector relationships. Annu Rev Phytopathol 30:315–348

Ghanekar AM, Nene YL (1976) A method of assaying *Groundnut bud necrosis virus*. Tropical Grain Legume Bull 5:39

Ghanekar AM, Reddy DVR, Lizuka N, Amin PW, Gibbons RW (1979) Bud necrosis of groundnut (*Arachis hypogaea*) in India caused by *Tomato spotted wilt virus*. Ann Appl Biol 93:173–179

Goldbach R, Kuo G (1996) Introduction. In: Proceedings of international symposium on tospoviruses thrips floral vegetable crops. Taiwan Agricultural Research Institute, Taichung, Taiwan. Acta Hort 431:21–26

Gopal K, Krishnareddy M, Reddy DVR, Muniyappa V (2010) Transmission of *Peanut yellow spot virus* (PYSV) by thrips, *Scirtothrips dorsalis* Hood. in groundnut. Arch Phytopath Plant Prot 43(5):421–429

Goswami S, Sahana N, Pandey V, Doblas P, Jain RK, Palukaitis P, Canto T, Praveen S (2012) Interference in plant defence and development by non-structural protein NSs of *Groundnut bud necrosis virus*. Virus Res 163:368–373

Gowda S, Satyanarayana T, Naidu RA, Mushegian A, Dawson WO, Reddy DVR (1998) Characterization of the large (L) RNA of peanut bud necrosis tospovirus. Arch Virol 143:2381–2390

Gurupad BB, Patil MS (2013) Serological and molecular detection of *Groundnut bud necrosis virus* (GBNV) causing bud blight disease in tomato. Internat J Plant Protec 6(2):320–322

Hall JM, Mohan K, Knott EA (1993) Tospoviruses associated with scape blight of onion (*Allium cepa*) seed crops in Idaho. Plant Dis 77:952

Hemalatha V, Pradnya G, Anjali AK, Krishnareddy M, Savithri HS (2008) Monoclonal antibodies to the recombinant nucleocapsid protein of a *Groundnut bud necrosis virus* infecting tomato in Karnataka and their use in profiling the epitopes of Indian tospovirus isolates. Curr Sci 95:952–957

Holkar SK, Kumar R, Yogita M, Katiyar A, Jain RK, Mandal B (2016) Diagnostic assays for two closely related tospovirus species, *Watermelon bud necrosis virus* and *Groundnut bud necrosis virus* and identification of new natural hosts. J Plant Biochem Biotechnol 26:1–9

Jain RK, Pappu HR, Pappu SS, Krishanareddy M, Vani A (1998) Watermelon bud necrosis tospovirus is a distinct virus species belonging to serogroup IV. Arch Virol 143:1637–1644

Jain RK, Umamaheswaran K, Bhat AI, Thien HX, Ahlawat YS (2002) Necrosis disease on cowpea, mungbean and tomato is caused by *Groundnut bud necrosis virus*. Indian Phytopath 55:354

Jain RK, Khurana SMP, Bhat AI, Chaudhary V (2004) Nucleocapsid protein gene sequence studies confirm that potato stem necrosis is caused by a strain of *Groundnut bud necrosis virus*. Indian Phytopath 57:169–173

Jain RK, Pandey NA, Krishnareddy M, Mandal B (2005) Immunodiagnosis of groundnut and watermelon bud necrosis viruses using polyclonal antiserum to recombinant nucleocapsid protein of *Groundnut bud necrosis virus*. J Virol Methods 130:162–164

Jain RK, Bag S, Umamaheswaran K, Mandal B (2007) Natural infection by tospovirus of cucurbitaceous and fabaceous vegetable crops. J Phytopathol 155:22–25

Jones RAC, Sharman M (2005) *Capsicum chlorosis virus* infecting *Capsicum annuum* in the East Kimberley region of Western Australia. Aust Plant Pathol 34:397–399

Kapoor R, Mandal B, Paul PK, Chigurupati P, Jain RK (2014) Production of cocktail of polyclonal antibodies using bacterial expressed recombinant protein for multiple virus detection. J Virol Methods 196:7–14

Khurana SMP, Phadtare SG, Garg ID, Singh MN, Bharadwaj VP (1989) Potato stem necrosis epidemic due to *Tomato spotted wilt virus* in India. In: Proceedings of IV international plant virus epidemiology workshop, Montpellier, France, 3–8 Sept, pp.301 (Abst.)

King AMQ, Adams MJ, Carstens EB, Lefkowitz EJ (eds) (2012) Virus taxonomy ninth report of the international committee on taxonomy of viruses. Elsevier Academic Press, San Diego. 1327pp

Krishnareddy M, Singh SJ (1993) Immunology and molecular based diagnosis of tospovirus infecting watermelon. In: Golden Jubilee symposium on horticultural research: changing scenario. Indian Institute of Horticultural Research, Bangalore, India, pp 247–248

Krishnareddy M, Varma A (1990) Prevalence of *Tomato spotted wilt virus* in Delhi. Indian Phytopath 43:252

Krishnareddy M, Usha-Rani R, Kumar A, Madhavireddy K, Pappu HR (2008) First report of *Capsicum chlorosis virus* (Genus: *Tospovirus*) infecting chilli pepper (*Capsicum annuum)* in India. Plant Dis 92:1469

Kumar R, Mandal B, Geetanjali AS, Jain RK, Jaiwal PK (2010) Genome organisation and sequence comparison suggest intraspecies incongruence in M RNA of *Watermelon bud necrosis virus*. Arch Virol 155:1361–1365

Kunkalikar SR, Sudarsana P, Rajagopalan P, Zehr UB, Naidu RA, Ravi KS (2007) First report of *Capsicum chlorosis virus* in tomato in India. Plant Health Prog. doi:10.1094/PHP-2007-1204-01-BR

Kunkalikar SR, Sudarsana P, Rajagopalan P, Zehr UB, Ravi KS (2010) Biological and molecular characterization of *Capsicum chlorosis virus* infecting chilli and tomato in India. Arch Virol 155:1047–1057

Kunkalikar SR, Sudarsana P, Arun BM, Rajagopalan P, Chen TC, Yeh SD, Naidu RA, Zehr UB, Ravi KS (2011) Importance and genetic diversity of vegetable-infecting tospoviruses in India. Phytopathology 101:367–376

Lakshmi KV, Wightman JA, Reddy DVR, Ranga-Rao GV, Buiel AAM, Reddy DVR (1995) Transmission of *Peanut bud necrosis virus* by *Thrips palmi* in India. In: Parker BL, Skinner M, Lewis T (eds) Thrips biology and management. Plenum Press, New York, pp 179–184

Law MD, Speck J, Moyer JW (1991) Nucleotide sequence of the 3′ non-coding region and N gene of the S RNA of a serologically distinct tospovirus. J Gen Virol 72:2597–2601

Li TJ, Yeh CY, Yeh DS, Joseph A, Raja J, Rajgopalan AP, Liu YL, Chen CT (2011) Complete genome sequence of *Watermelon bud necrosis virus*. Arch Virol 156:359–362

Lokesh B, Rashmi PR, Amruta BS, Srisathiyanarayanan D, Murthy MRN, Savithri HS (2010) NSs encoded by *Groundnut bud necrosis virus* is a bifunctional enzyme. PLoS One 5(3):e9757. doi:10.1371/journal.pone.0009757

Maheshwari Y, Vijayanandraj S, Jain RK, Mandal B (2015) Engineered single-chain variable fragment antibody for immunodiagnosis of *Groundnut bud necrosis virus* infection. Arch Virol 160:1297–1301

Mandal B, Jain RK (2010) ELISA kit for tospoviruses detection. ICAR News 16:13.

Mandal B, Jain RK, Chaudhary V, Varma A (2003) First report of natural infection of *Luffa acutangula* by *Watermelon bud necrosis virus* in India. Plant Dis 87:598

Mandal B, Jain RK, Krishnareddy M, Krishna Kumar NK, Ravi KS, Pappu HR (2012) Emerging problems of tospoviruses (*Bunyaviridae*) and their management in Indian subcontinent. Plant Dis 96:468–479

McMichael LA, Persley DM, Thomas JE (2002) A new tospovirus serogroup IV species infecting capsicum and tomato in Queensland, Australia. Aust Plant Pathol 31:231–239

Meena RL, Ramasubramanian T, Venkatesan S, Mohankumar S (2005) Molecular characterization of tospovirus transmitting thrips population from India. Am J Biochem Biotech 1:168–173

Narayanasamy P, Kandaswamy TK, Ramiah M (1975) A new disease of groundnut in Tamil Nadu. Madras Agri J 62:371–375

Nariani TK, Dhingra KL (1963) A mosaic disease of groundnut (*Arachis hypogaea* L.) Indian J Agril Sci 33:25–27

Pang SZ, Slightom JL, Gonsalves D (1993) The biological properties of a distinct tospovirus and sequence analysis of its S RNA. Phytopathology 83:728–733

Pappu HR, Jones RAC, Jain RK (2009) Global status of tospovirus epidemics in diverse cropping systems: successes achieved and challenges ahead. Virus Res 141:219–236

Permar V, Singh A, Pandey V, Alatar AA, Faisal M, Jain RK, Praveen S (2014) Tospoviral infection instigates necrosis and premature senescence by micro RNA controlled programmed cell death in *Vigna unguiculata*. Physiol Mol Plant Pathol 88:77–84

Persley DM, Thomas JE, Sharman M (2006) Tospoviruses: an Australian perspective. Aust Plant Pathol 35:161–180

Phadtare SG, Khurana SMP, Garg ID, Bharadwaj VP (1989) Stem-necrosis: a viral disease of potatoes in central India. J Indian Potato Assoc 16:164–165. (Abstr.)

Prasada RRDVJ, Izuka M, Ragunathan V, Joshi NC (1980) Incidence of *Tomato spotted wilt virus* on tomato in Andhra Pradesh. Indian Phytopath 33:436–439

Prasada RRDVJ, Rajeswari R, Rao MVB, Ragunathan V, Joshi NC (1985) Spotted wilt of pea in India. Indian Phytopath 38:90–93

Prasada RRDVJ, Chakraborthy SK, Rajeswari R, Reddy AS (1987) Occurrence of *Tomato spotted wilt virus* on cowpea (*Vigna unguiculata* (*L.*) WALP.), chili (*Capsicum annum L.*) and brinjal (*Solanum melongena L.*) Indian J Plant Prot 15:117–119

Premachandra WTSD, Borgemeister C, Maiss E, Knierim D, Poehling HM (2005) *Ceratothripoides claratris*, a new vector of a *Capsicum chlorosis virus* isolate infecting tomatoes in Thailand. Phytopathology 95:659–663

Raja P (2005) Development of nucleocapsid mediated resistance in tomato against *Groundnut bud necrosis virus*. PhD thesis, Indian Agricultural Research Institute, New Delhi

Raja P, Jain RK (2006) Molecular diagnosis of *Groundnut bud necrosis virus* causing bud blight of tomato. Indian Phytopath 59:359–362

Ravi KS, Kitkaru AS, Winter S (2006) *Iris yellow spot virus* in onions: a new tospovirus record form India. Plant Pathol 55:288

Rebijith K, Asokan R, Kumar NK, Krishna V, Ramamurthy V (2011) Development of species-specific markers and molecular differences in mtDNA of *Thrips palmi* Karny. and *Scirtothrips dorsalis* hood. (*Thripidae:Thysanoptera*), vectors of tospoviruses (*Bunyaviridae*) in India. Entomol News 122(3):201–213

Rebijith KB, Asokan R, Hande HR, Krishna Kumar NK (2016) The first report of miRNAs from a Thysanopteran insect, *Thrips palmi* Karny. Using high-throughput sequencing. PLoS ONE 11(9):e0163635. doi:10.1371/journal.pone.0163635

Reddy M, Reddy DVR, Appa Rao A (1968) A new record of virus disease on peanut. Plant Dis Rep 52:494–495

Reddy DVR, Wightman JA, Beshear RJ, Highland B, Black M, Sreenivasulu M, Dwivedi SL, Demski JW, McDonald D, Smith JW Jr, Smith DH (1991a) Bud necrosis: a disease of groundnut caused by *Tomato spotted wilt virus*, Information Bulletin No. 31. ICRISAT, Patancheru, p 20

Reddy DVR, Sudarshana MR, Ratna AS, Reddy AS, Amin PW, Kumar IK Murthy AK (1991b) The occurrence of yellow spot virus, a member of tomato spotted wilt virus group, on peanut (Arachis hypogaea L) in India. In: Hsu HT, Lawson RH (eds) Virus-thrips-plant interactions of tomato spotted wilt virus. Proceedings of the USDA Agricultural Research Service, ARS-87, pp 77–88.

Reddy DVR, Ratna AS, Sudarshana MR, Kiran Kumar I (1992) Serological relationships and purification of bud necrosis virus, a tospovirus occurring in peanut (*Arachis hypogaea L.*) in India. Ann Appl Biol 120:279–286

Reddy AB, Krishnareddy M, Jalali S, Patil MS, Usha Rani TR (2008) Detection of a tospovirus-infecting tomato (*Solanum lycopersicon* L). Indian J Virol 19:1–5

Renukadevi P, Senthilraja C, Rageshwari S, Nakkeeran S, Suganthy M, Malathi VG (2015a) Emergence of hitherto unreported tomato spotted wilt virus in India. In: XXIV national conference on "Transboundary viral diseases under one health: perspectives & challenges", Virocon-2015, 8–10th October. NEIGRIHMS, Mawdiangdiang, pp 137 (Abstract)

Renukadevi P, Nagendran K, Nakkeeran S, Karthikeyan G, Jawaharlal M, Alice D, Malathi VG, Pappu HR (2015b) First report of *Tomato spotted wilt virus* infection of chrysanthemum in India. Plant Dis 99(8):1190

Saritha RK (2007) Biological characterisation and genome organization of *Groundnut bud necrosis virus* from mungbean (*Vigna radiata*). PhD thesis, Indian Agricultural Research Institute, New Delhi

Saritha RK, Jain RK (2007) Nucleotide sequence of the S and M RNA segments of a *Groundnut bud necrosis virus* isolate from *Vigna radiata* in India. Arch Virol 152:1195–1200

Satyanarayana T, Lakshminaryan Reddy K, Ratna AS, Deom CM, Gowda S, Reddy DVR (1996a) *Peanut yellow spot virus*: a distinct tospovirus species based on serology and nucleic acid hybridization. Ann Appl Biol 129:237–245

Satyanarayana T, Mitchell SE, Reddy DVR, Brown S, Kresovich S, Jarret R, Naidu RA, Damski JW (1996b) Peanut bud necrosis tospovirus S RNA: complete nucleotide sequence, genome organization and homology to other tospoviruses. Arch Virol 141:85–98

Satyanarayana T, Mitchell SE, Reddy DVR, Brown S, Kresovich S, Jarret R, Naidu RA, Damski JW (1996c) The complete nucleotide sequence and genome organization of the M RNA segment of peanut bud necrosis tospovirus and comparison with other tospoviruses. J Gen Virol 77:2347–2352

Satyanarayana T, Gowda S, Lakshminaryan Reddy K, Metals E, Dawson WO, Reddy DVR (1998) *Peanut yellow spot virus* is a member of a new serogroup of Tospovirus genus based on small (S) RNA sequence and organization. Arch Virol 143:353–364

Sharma DC (1966) Studies on bunchy top, chlorosis and ring mottle virus diseases of groundnut (*Arachis hypogaea* L.) Phytopath Z 57:127–137

Sharma A, Kulshrestha S (2014) First report of *Amaranthus* sp. as natural host of *Capsicum chlorosis virus* in India. Virus Dis 25:412–413

Singh SJ, Krishnareddy M (1996) Watermelon bud necrosis: a new tospovirus disease. Acta Hort 431:68–77

Singh AB, Srivastava SK (1995) Status and control strategy of peanut bud necrosis disease in Uttar Pradesh. In: Buiel AAM, Parevliet JE, Lenne JM (eds) Recent studies on peanut bud necrosis disease: proceedings of a meeting, 20 March. ICRISAT, Patancheru, pp 65–68

Singh RB, Srivastava KK, Khurana SMP, Pandey SK (1997) Assessment of yield losses due to potato stem necrosis disease. Indian J Virol 3:135–137

Singh P, Indi SS, Savithri HS (2014) *Groundnut Bud Necrosis Virus* encoded NSm associates with membranes via its C-terminal domain. PLoS ONE 9(6):e99370. doi:10.1371/journal.pone.0099370

Sivaprasad Y, Reddy BVB, Sujitha A, Gopal DVRS (2015) Expression and production of polyclonal antibodies against recombinant coat protein of *Peanut bud necrosis virus*. J Plant Pathol Microbiol 6:1–6

Srinivasan R, Sundaraj S, Pappu HR, Diffie S, Riley DG, Gitaitis RD (2012) Transmission of Iris yellow spot virus by Frankliniella fusca and Thrips tabaci (Thysanoptera: Thripidae). J Econ Entomol 105(1):40–47

Suganthy M, Rageshwari S, Senthilraja C, Nakkeeran S, Malathi VG, Ramaraju K, Renukadevi P (2016) New record of western flower Thrips, Frankliniella occidentalis (Pergande) (Thysanoptera: Thripidae) in South India. Int J Environ Agric Biotechnol 1(4):857–867

Sujitha A, Bhaskara Reddy BV, Sivaprasad Y, Usha R, Sai Gopal DVR (2012) First report of *Groundnut bud necrosis virus* infecting onion (*Allium cepa*). Aust Plant Dis Notes 7:183–187

Sujitha A, Reddy BB, Sivaprasad Y, Usha R, Krishna TG, Gopal DS (2013) First report of *Groundnut bud necrosis virus* in jasmine. J Plant Pathol 95:667

Tas PWL, Boerjan ML, Peters D (1977) Purification and serological analysis of *Tomato spotted wilt virus*. Neth J Plant Pathol 83(2):61–72

Thien HX, Bhat AI, Jain RK (2003) Mungbean necrosis disease is caused by a strain of groundnut bud necrosis tospovirus. Indian Phytopath 56:54–60

Tsompana M, Moyer JW (2008) Tospoviruses. In: Mahy BWJ, Van Regenmortel MHV (eds) Encyclopedia of virology., vol. 5, 3rd edn. Elsevier Ltd., Oxford, pp 157–162

Tyagi K, Kumar V (2015) First report of western flower thrips, *Frankliniella occidentalis* (Pergande) (*Thripidae: Thysanoptera*) from India- a potential havoc to Indian agriculture. Halteres 6:1–3

Umamaheswaran K, Jain RK, Bhat AI, Ahlawat YS (2003) Biological and molecular charecterisation of a tospovirus isolates from tomato and its relationship with other tospoviruses. Indian Phytopath 56:168–173

Vijayalakshmi G, Haokip BD, Karthikeyan G, Alice D, Ganapathy N, Asokan G, Rajendran L, Kennedy JS (2016) First report of field infection of *Capsicum chlorosis virus* on groundnut (*Arachis hypogaea*) in India (Tamil Nadu). Plant Dis 100(11):2339

Whitfield AE, Ullman DE, German TL (2005) Tospovirus-thrips interactions. Annu Rev Phytopathol 43:459–489

Yeh SD, Chang TF (1995) Nucleotide sequence of the N gene of *Watermelon silver mottle virus*, a proposed pew member of the genus *Tospovirus*. Phytopathology 85:58–64

Yeh SD, Sun IJ, Ho HM, Chang TF (1996) Molecular cloning and sequence analysis of the S RNA of *Watermelon silver mottle virus*. Acta Hort 431:244–259

Characteristics of Tungrovirus Occurring in India 21

Shweta Sharma, Gaurav Kumar, Satyendra Mangrauthia, C.N. Neeraja, D. Krishnaveni, and Indranil Dasgupta

Abstract

The research on rice tungro bacilliform virus, the only species in the genus *Tungrovirus* (family *Caulimoviridae*), has a long history in India because of the association of the virus with rice tungro disease, an important rice disease prevalent in the eastern coastal rice growing regions. This research encompasses the transmission characteristics of the vector, characterization of strains, resistance sources, molecular properties of the virus, transgenic resistance and the use of virus-derived gene vector for plant biotechnology. This chapter gives a glimpse of the above research conducted to characterise the RTBV isolates occurring in India.

Keywords

Rice • Tungro • Resistance • Gene silencing

21.1 Introduction

Tungrovirus is a monotypic genus of the family *Caulimoviridae* containing Rice tungro bacilliform virus (RTBV) as the sole member. RTBV was initially considered as a member of the genus *Badnavirus*. In 2004, it was classified under a different genus, *Tungrovirus*. RTBV although resembles to badnaviruses by the similar bacilliform virion morphology and circular dsDNA as genome, it differs by genome

S. Sharma • G. Kumar • I. Dasgupta (✉)
Department of Plant Molecular Biology, University of Delhi South Campus, New Delhi 110021, India
e-mail: indranil58@yahoo.co.in

S. Mangrauthia • C.N. Neeraja • D. Krishnaveni
Indian Institute of Rice Research (Indian Council for Agricultural Research), Rajendranagar, Hyderabad 500030, India

B. Mandal et al. (eds.), *A Century of Plant Virology in India*,
https://doi.org/10.1007/978-981-10-5672-7_21

organisation having an additional (fourth) open reading frame and distinct phylogenetic divergence. Due to these differences, RTBV was classified under a separate genus *Tungrovirus* by The International Committee on Taxonomy of Viruses (ICTV) in 2004. RTBV was isolated from rice plants affected with rice tungro disease (RTD) from various countries of South and Southeast Asia.

The first report of Rice tungro disease from India was published from IARI (Raychaudhuri et al. 1967) in 1967, soon after the first report of this disease was published from the International Rice Research Institute (IRRI), Philippines, half a century ago (Rivera and Ou 1965).

In India, RTD has been recorded in the states of Andhra Pradesh, Odisha, Punjab, Tamil Nadu, Telengana, Uttar Pradesh and West Bengal. Although the loss due to RTD has been estimated to be about 1% at the national level, the local losses can be significant (Muralidharan et al. 2003). RTD is caused by the joint infection of two viruses, RTBV and *Rice tungro spherical virus* (RTSV), having a single-stranded RNA as the genomic material. RTBV is transmitted by the vector green leafhopper (GLH), *Nephotettix virescens* Distant, in which RTSV has a helper role. The symptoms of RTD are stunting and yellow-orange discolouration in the leaves (Fig. 21.1), which is caused mainly by RTBV, as has been demonstrated by inoculating plants with a cloned RTBV DNA (Dasgupta et al. 1991).

This chapter focuses on the research done in India on RTBV, characterizing the genetics of resistance in rice against RTD and the pathological changes in rice plants affected with RTD, which, although as yet unconfirmed, is most likely to be the effect of only RTBV, not RTSV. In addition, a gene silencing vector for rice, developed recently from RTBV is also described. Most of the work described has been carried out at Bidhan Chandra Krishi Viswavidyalaya (BCKV), Kalyani, West Bengal; Central Rice Research Institute (CRRI), Cuttack, Odisha; Directorate of Rice Research (DRR), Hyderabad, recently renamed Indian Institute of Rice Research (IIRR); Indian Agricultural Research Institute (IARI), New Delhi; Tamil Nadu Agricultural University (TNAU), Coimbatore, Tamil Nadu and University of Delhi South Campus (UDSC), New Delhi.

21.2 Leafhopper and Outbreak

Transmission of RTD, and hence, its causative viruses was studied extensively at IARI and at BCKV, which reaffirmed GLH as the principal vector (Mukhopadhyay 1984; Mukhopadhyay and Chowdhury 1970; Raychaudhuri et al. 1967), following the first report from IRRI (Rivera and Ou 1965). In a population of *N. virescens* about 83% are the virulent ones which can transfer Tungro virus (Rivera and Ou 1965). *N. virescens* females were reported to be more efficient than males in transmitting Tungro virus (Shukla 1979). The disease was transmitted by both adults as well as nymphs of all growth stages; but from CRRI, it was reported that the third, fourth and fifth instar nymphs were more efficient than first and second instar (Anjaneyulu 1975). Differential transmission of tungro viruses has been reported by both, different species of *Nephotettix* (Mukhopadhyay 1984) as well as from

Fig. 21.1 Symptoms of Rice tungro disease in the field and green leafhopper, the vector for Rice tungro viruses. Alternate rows of resistant and susceptible plants, the latter showing symptoms of stunting and yellow-orange discolouration of Rice tungro disease

different source plants (Mukhopadhyay and Chowdhury 1973) *N. virescens*, a phloem feeder on rice transmitted both RTBV and RTSV either together or separately. These early studies on transmission laid the foundation of vector control as a strategy for the management of the tungro disease.

An outbreak of rice tungro disease was noticed during kharif 2007 in northern Telangana, India. Over 19,000 acres in Karimnagar and Medak districts were severely affected by the disease. The species composition of vectors present in the tungro damaged area consisted of green leafhopper species *N. virescens* and *N. nigropictus* and zig-zag leafhopper species *Recilia dorsalis* in that order of abundance. *N. virescens* in the population showed distinct colour and morphological variations (Krishnaveni et al. 2011).

21.3 Recognition of the Tungro Disease

Early diagnosis of any disease is based on appearance of its characteristic symptoms under field conditions. The characteristic symptoms of rice tungro disease include distinct stunting of plants, yellow-orange discoloration of leaves and reduction in tiller numbers. The expression of symptoms however, varies depending on the rice variety, growth stage at the time of infection and soil or environmental conditions. Prevalence of GLH is a factor which is used to identify the presence of RTD. In addition, transmission of the viruses from suspected diseased plants to a susceptible variety of rice such as Taichung Native-1 (TN-1), which develops very clear symptoms, can serve as an evidence for diagnosis of rice tungro disease (Anjaneyulu et al. 1994). Confirmatory diagnostic tests based on two approaches; serological and nucleic acid-based were developed. Initially Tests based on biochemical changes occurring in the plant, mainly the Iodine test and the Phloroglucinol test, where used as an indication for the presence of the RTBV; the iodine test being a good indicator for the presence of RTBV even before the appearance of symptoms, as reported by researchers from TNAU (Aiyanathan and Narayanasamy 1989).

Nucleic acid-based diagnostic tests, such as Dot-blot hybridization and Polymerase chain reaction (PCR), has clear advantage over serological tests for diagnosis of rice tungro disease. Dot-blot hybridization has the advantage of being partially quantitative, but suffers from low sensitivity and the requirement of radioactive isotopes or non-radioactive detection kits. Dot-blot using radioactive probes has been used to quantitate RTBV levels in transgenic rice plants (Tyagi et al. 2008) at UDSC. PCR is a very sensitive and convenient method for identification of rice tungro disease. PCR-based methods were designed for the quick and sensitive detection of RTBV from a variety of samples, even those stored at room temperature for several days (Dasgupta et al. 1996). Using a modification of the above, a method for simultaneous detection of both RTBV and RTSV in a single reaction based on multiplex RT-PCR was developed at IARI (Periasamy et al. 2006). Recently, a very sensitive SYBR Green Real Time-PCR based diagnostic method has been reported for the detection and accurate quantification of the RTBV & RTSV titres in infected rice plants at UDSC (Sharma and Dasgupta 2012). Using this method, the copy numbers of RTBV DNA and RTSV RNA were estimated for the first time in infected plants at two time points after inoculation (Sharma and Dasgupta 2012). These diagnostic methods play very important role in timely the detection and containment of this devastating disease in India.

21.4 Biochemical and Physiological Changes in Plant

The orange yellow leaf discoloration in rice and stunting are the major symptoms of RTBV infection. Considerable work was undertaken to study the underlying biochemical and physiological changes, which contribute to the symptom development, at CRRI and IIRR. Tungro infection was seen to reduce the chlorophyll content by approximately 70% (Subbarao et al. 1979) and anthocyanins and flavonols by approximately 40% (Mohanty and Sridhar 1987, 1989) and increase carotenoids considerably (Mohanty and Sridhar 1989). The decrease in chlorophyll content results in the yellow colour of carotenoids to appear more prominent, thus explaining the characteristic orange yellow leaf discoloration symptom of RTD. Since carotenoids are a precursor for abscisic acid (ABA), increased carotenoid content in diseased plants results into more ABA synthesis and accelerates senescence (Mohanty and Sridhar 1989). An increase in chlorophyllase *a* and chlorophyllase *b* upon tungro infection was reported at IARI (Yadav and Mishra 1987) which itself explains the reduction in plant chlorophyll content. At CRRI, a study related the loss of chlorophyll in RTD-affected plants to root infirmity and hormonal (cytokinin and ABA) imbalance (Mohanty and Sridhar 1989). At IARI, a decrease (39.7–69.1%) in photosynthetic rate in susceptible cultivars upon RTD infection was reported (Singh et al. 1990).

It was reported from CRRI, that tungro infection is followed by an increase in ABA and cytokinin (Mohanty and Sridhar 1989; Sridhar et al. 1976) content. Increase in ABA is associated with induction of senescence, enhanced level of free amino acids (Sridhar et al. 1976) decrease in transpiration rate, inhibition of amylase synthesis, hence a decrease in starch hydrolysis and a decrease in the synthesis of growth promoting hormones was observed (Mohanty et al. 1979). Increase in cytokinin however antagonizes the effect of ABA by inhibiting senescence in RTD-affected plants. RTD causes inhibition of biosynthesis of gibberellic acid resulting into low gibberellin content. It is found that a common mevalonate pathway is associated with the synthesis of carotenoids, gibberellins and ABA, hence the increase in carotenoid biosynthesis in RTD-affected plants adversely affects the availability of precursors for the biosynthesis of gibberellins (Srinivasulu and Jeyarajan 1989).

At CRRI, it was found that RTD infection results in a decrease in cell permeability resulting into reduced electrolyte leakage (Mohanty and Sridhar 1986). This is also related to delay or absence of flowering in tungro infected plants by either inhibiting the movement of flowering hormone to floral meristems or by reducing the intake of phosphate ions required for cell differentiation. An increase in the accumulation of phenolic compounds and peroxidase activity in RTD affected plants was also recorded, resulting into stunting and yellowing of leaves, peculiar to tungro symptoms (Yadav and Mishra 1987). Minerals, such as P, K, Ca, Mg, Mn, Fe and Cu were reported to be significantly reduced in infected leaves.

21.5 Host Plant Resistance Against Tungro

Several donor lines for tungro resistance have been identified through national and international screenings. Cultivars Latisail and Kataribhog were identified as resistance sources against tungro and this resulted in the identification of selections combining a high level of resistance to tungro, superior grain type and good plant type. Inheritance studies Resistance to tungro in a cross, IR8/Latisail was shown to be governed by two genes. Field progenies from five crosses involving three resistant (Kataribhog, Kamod 253 and Pankhari 203) and three susceptible cultivars

(TN1, Bala and Cauvery) showed the involvement of three genes for resistance to tungro besides an inhibitor gene. During 1975–1981, a few more donors for resistance were identified. In this study on inheritance and gene action governing resistance to tungro virus in six crosses, a ratio of 15 resistant: 1 susceptible was obtained indicating two pairs of duplicate dominant genes controlling the resistance involving parents ARC5918, ARC11317, ARC 14320, ARC14529 and Churnakati. Similarly, in a cross between IR64 (susceptible) and Nainabachi (resistant), the action of resistance genes was shown to be under dominant duplicate resistance genes (Anjaneyulu et al. 1982; Ghosh and Reddy 1986; Prasad et al. 2004; Shastry et al. 1972).

21.6 Genetics of Resistance

The inheritance of resistance to RTD was studied in seven resistant rice cultivars with field evaluation at hotspot location. The reaction of F_2 progenies from the crosses of resistant cultivars with susceptible TN1 revealed that resistance in Pankhari 203, Ptb8, Utri Merah, Tjempokijik was governed by a single dominant gene. A set of dominant duplicate genes controlled the expression of resistance in TKM6 and ARC 11554. In Utri Rajapan, two dominant and complementary genes were involved in its resistance to tungro virus disease. Allelic studies suggested in most of the donors, the genes governing resistance are different and non-allelic. Only in Ptb8 and Aguiha Anarelo, the genes appeared to be allelic (Prasad et al. 2004).

At IIRR, two sets of quantitative trait loci (QTL) for rice tungro virus resistance were mapped using microsatellites in two resistant genotypes, Utri Rajapan and Vikramarya through molecular mapping (Neeraja et al. 2006). These were located on chromosomes 7 and 2 in 'Utri Rajapan' explaining 40.8% and 21.6% of the phenotypic variance. In variety 'Vikramarya', another two QTLs for RTV resistance were detected on chromosomes 7 and 1 explaining 18.7% and 16.4% of the phenotypic variance (Neeraja et al. 2006).

21.7 Viral DNA Sequence Analysis

RTBV has a double-stranded circular DNA molecule as the genome and carries four Open Reading Frames (ORFs). Nucleotide sequences of two Indian isolates of RTBV were analyzed for the first time at UDSC. The investigators reported that the two Indian isolates analyzed, one from Andhra Pradesh and another from West Bengal (WB isolate), shared high sequence identities with each other but were significantly different from isolates from Southeast Asia, mainly Philippines (Nath et al. 2002). Comparative analysis of these two sequences revealed several substitutions and indels present mostly in the intergenic region as compared to Philippine isolate. Thus, comparison of Indian isolates with Philippine isolate clearly suggested that the two Indian isolates have diverged sufficiently from the Southeast Asian group, and thus belong to a separate South Asian group (Nath et al. 2002).

21.8 Variability in RTBV

Several strains of tungro viruses (RTBV and RTSV) have been reported by various groups in India. On the basis of transmission efficiency by male GLH, it was reported from DRR, that two stains of tungro viruses were present in India namely RTV1 and RTV2 (Shastry et al. 1972). It was observed that RTV2 was more efficiently transmitted as compared to RTV1. In the same year, ten isolates were collected from different regions of Andhra Pradesh, Bihar and West Bengal and the existence of four strains namely RTV1, RTV2A, RTV2B and RTV3 on the basis of differential symptoms produced on rice variety Taichung Native 1 (TN1) were reported (Anjaneyulu and John 1972). RTV1 and RTV3 could be easily differentiated on TN1 as RTV1 produced mild symptoms while RTV3 produced severe symptoms. However RTV2A and RTV2B were indistinguishable on TN1 as both produced very severe symptoms. Inoculation of six different cultivars i.e., Kataribhog, Latisail, Pankhari 203, Ambemohar 102, Ambemohar 159 and Kamod with these four tungro strains produced distinct symptoms, which revealed the striking differences among the strains. At IARI, the occurrence of fifth tungro strain i.e. RTV4 from India was reported, which was characterized as a mild strain (Mishra et al. 1976). Later it was found that all the tungro strains reported from India were similar to the "S" group of strain from Philippines (Basu et al. 1976). Existence of another tungro strain i.e. "T" from West Bengal was reported at BCKV through extensive survey on differential symptoms produced in tungro infected plants (Mukhopadhyay and Bandopadhyay 1984).

To determine the molecular basis of differences among the above tungro strains, investigators at UDSC performed restriction digestion on four cloned RTBV DNA, obtained from different regions of India. Analysis of the digestion pattern revealed distinct restriction profiles, indicating molecular heterogeneity (Joshi and Dasgupta 2001). At UDSC, cloning and sequencing of the complete genome for two Indian RTBV isolates indicated that the two differed from each other by a 30 nucleotide deletion present near the 3′ end of ORFIII. To extend this work further, a PCR-RFLP-based study of six RTBV isolates obtained from different geographical regions of India for the presence or absence of the above deletion, indicated that in addition to the expected pattern, several novel patterns were also present, indicating new viral variants (Joshi et al. 2003). Work performed at IARI and UDSC has revealed additional molecular differences between strains showing differential symptoms (Niazi et al. 2005). Therefore, these findings pointed towards the existence of mixed infections and molecular heterogeneity in the Indian isolates of RTBV.

Recently, sequence analysis of several RTBV isolates collected from different geographical regions of India has been reported from BCKV, DRR, TNAU and UDSC. These include isolates from Chinsurah, Cuttack, Kanyakumari, Punjab and Puducherry (Banerjee et al. 2011; Krishnaveni et al. 2011; Mathur and Dasgupta 2013; Sharma et al. 2011; Mangrauthia et al. 2012). These reports support the earlier observations that Indian RTBV isolate are genetically similar to each other owing to high degrees of nucleotide and amino acid identities among them. Sequence analysis of recent RTBV isolates from India has reinforced the observation that the South Asian and Southeast Asian RTBV isolates are genetically distinct from each other. The Kanyakumari isolate, collected from an area which has seen emergence of RTD in a big way in the last few years, reveals that the RTBV genome is likely to have undergone recombination (Sharma et al. 2011).

21.9 Promoter

Early interest in RTBV promoter developed based on the fact that it belongs to the same family as Cauliflower mosaic virus (CaMV) whose promoter is used routinely to drive expression of heterologus genes in plants. This possibility of it being as useful as CaMV promoter triggered promoter analysis based studies in RTBV. It was found that unlike the CaMV promoter which is constitutive in nature, RTBV promoter was tissue specific expressing mainly in the phloem (Bhattacharyya-Pakrasi et al. 1993).

Structural and functional analysis of the promoter region of the WB isolate of RTBV was carried out at UDSC. Earlier, the promoter region of the Philippine isolate had been analyzed by researchers in USA and several elements had been identified, based on deletion constructs of the promoter region, to play important roles in the promoter activity. Interestingly, except for TATA motif and poly-A signal, none of the elements identified in the Philippine isolate was seen to have an exact counterpart in the West Bengal isolate, indicating the possibility of the existence of novel control elements. Deletion analysis studies of WB isolate promoter revealed that tissue specificity determining regions were present in the downstream region of the promoter (Mathur and Dasgupta 2007). A number of transgenic rice lines were raised containing deletion constructs of the promoter, driving the reporter gene Gus. Analysis of transgenic rice plants obtained using upstream promoter deletion constructs revealed three types of expression patterns i.e. constitutive type, non-expressing type and the full-length type, depending upon the extent of 3′ deletion of the promoter used. The full-length type, representing the native promoter showed a development- and tissue-specific expression pattern, expressing in all tissues at a young stage, but getting restricted mainly to the phloem as the plant matures. Interestingly, these studies revealed the presence of a negative element within the region +58 to +195 and tissue specific elements within the nucleotide region +195 to +460, i.e., downstream to the transcription start site (Fig. 21.2). Thus, compared to the isolate from Philippines, the RTBV isolate from India was observed to contain a different combination of control elements and more interestingly, a negative element in the downstream region of the transcription start site (Mathur and Dasgupta 2007). A recent study at BCKV, involving a detailed comparative analysis of the large intergenic region of 11 complete RTBV genomic sequences available in the database revealed distinct differences in the Southeast Asian and South Asian isolates with regard to the transcription start site and consensus sequence of *cis*-elements. In addition, variations in the number and arrangement of sORFs and length of leader sequence were also observed (Banerjee et al. 2012).

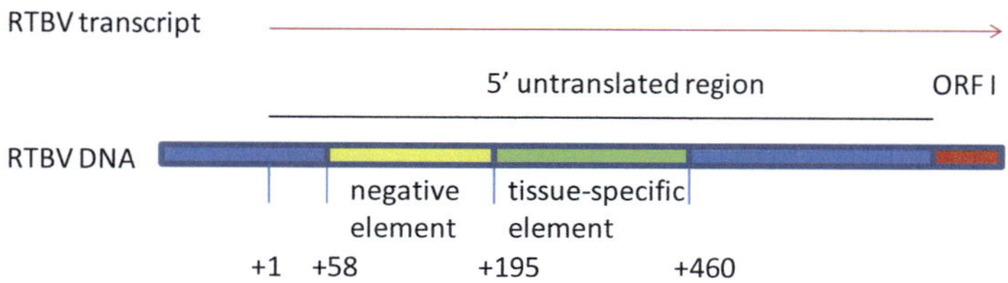

Fig. 21.2 The organization of the downstream control elements in the promoter region of the RTBV-West Bengal. The solid rectangle at the bottom shows the 5′ untranslated region of the RTBV DNA with the transcription start site indicated by +1, followed by the relevant positions of the control elements. The position of the transcript is shown above

The negative element of the WB isolate, described above, was further characterized at UDSC. Transgenic rice plants obtained using promoter deletion constructs revealed the presence of a 37 bp negative element in the un-translated leader region of RTBV-WB promoter present between positions +58 to +94 with respect to the transcription start site (TSS, Mathur and Dasgupta 2007). This 37 bp negative element was found to be functionally active with respect to two heterologus promoters namely CaMV35S and Maize ubiquitin promoter (MUP), which means that the presence of the negative element could completely shut down the expression of reporter genes from the above constitutive promoters. In addition, this negative element was found to be acting in a position- and orientation-independent manner which is suggestive of its functioning at transcriptional level (Purkayastha et al. 2010a). Further delineation of this 37 bp NE has identified the 10 bp region present between position 7476–7485 (i.e. +85 to +94) which was found to be necessary and sufficient for negative regulation of the constitutive RTBV promoter. Similar to the 37 bp NE, this 10 bp delineated NE was also found functional in both monocot as well as dicot systems (Sharma 2011). Interestingly, this 10 bp delineated NE was found to be acting in a position-independent but orientation-dependent manner as its suppressive activity was found compromised in reverse orientation. Site directed mutagenesis of this 10 bp delineated NE revealed that even the first five bases are sufficient to exert the negative regulatory effects (Sharma 2012). These bases possibly represent the binding motif for the host specific interacting protein partner of this negative element. Similar negative elements have not been reported from other Indian isolates of RTBV.

21.10 RTBV-Derived Gene Silencing Vector

Virus-induced gene silencing (VIGS) is a rapid method of gene silencing based on modified viral vectors, which trigger transient silencing of genes if a fragment of the gene is inserted into the VIGS vector (Purkayastha and Dasgupta 2009). At UDSC, realizing the need for a VIGS vector for rice, whose genome has been recently sequenced and a wealth of information is already available, which can be used to functionally characterize several genes using VIGS, the WB isolate of RTBV was modified to obtain a VIGS vector for rice. An agroinfectious clone of RTBV, which was infectious to rice following Agrobacterium-mediated transfer, was first developed, followed by deletions to remove two ORFs, replacement of the tissue-specific promoter by the constitutive maize ubiquitin promoter and insertion of a multicloning site, to develop a VIGS vector for rice, namely pRTBV-MVIGS (Purkayastha et al. 2010b). The vector could silence the gene *phytoene desaturase* (*pds*) almost 80% in emerging leaves within a period of 3 weeks after agroinoculation, resulting in the appearance of white streaks in the leaves due to photo bleaching of chlorophyll because of the silenced *pds* gene. Recently, improved methods for pRTBV-MVIGS based gene silencing in rice have been developed (Kant et al. 2015; Purkayastha et al. 2013). Hence, RTBV-based VIGS system has been shown to be useful in silencing any rice gene, to help decipher its function.

21.11 Concluding Remarks

RTBV, because of its role in the rice tungro disease, has been studied intensively in India. These studies have resulted in a wealth of information on the transmission, strainal variability, vector biology, genetic variability, diagnostics, molecular biology, promoter analysis and its use to derive gene silencing vector. Although resistance sources against rice tungro disease have been identified, their identities and specificity against RTBV are yet to be worked out. The discovery of new gene control elements, both for expression and silencing, have opened the possibility of further investigation in these areas and mining of the RTBV genomes for biotechnology-based applications, especially for cereals.

Acknowledgements Research at IIRR is funded by Indian Council for Agricultural Research and Department of Biotechnology, Government of India (DBT) and at UDSC from DBT and University of Delhi.

References

Aiyanathan KEA, Narayanasamy P (1989) Detection of rice tungro virus disease. Curr Sci 58:262
Anjaneyulu A (1975) *Nephotettix virescens* (Distant) nymphs and their role in the spread of rice tungro virus. Curr Sci 44:357–358
Anjaneyulu A, John VT (1972) Strains of rice tungro virus. Phytopathology 62:1116–1119
Anjaneyulu A, Shukla VD, Rao GM, Singh SK (1982) Experimental host range of rice tungro virus and its vectors. Plant Dis 66:54–56
Anjaneyulu A, Satapathy MK, Shukla VD (1994) Disease detection and diagnosis. Rice Tungro. Oxford and IBH publishing Co. Pvt. Ltd., New Delhi, pp 24–28
Banerjee A, Roy S, Tarafdar J (2011) Phylogenetic analysis of *Rice tungro bacilliform virus* ORFs revealed strong correlation between evolution and geographical distribution. Virus Genes 43(3):398–408
Banerjee S, Roy S, Tarafdar J (2012) The large intergenic region of *Rice tungro bacilliform virus* evolved differentially among geographically distinguished isolates. Virus Genes 44(2):312–318
Basu AN, Mishra MD, Niazi FR, Ghosh A (1976) A proposed key to the strains of rice tungro virus. Rice Pathol Newsletter 2:6–7

Bhattacharyya-Pakrasi M, Peng J, Elmer JS, Laco G, Shen P, Kaniewska MB, Kononowicz H, Wen F, Hodges TK, Beachy RN (1993) Specificity of a promoterfrom the *Rice tungro bacilliform virus* for expression in phloem tissues. Plant J 4:71–79

Dasgupta I, Hull R, Eastop S, Poggi-pollini C, Blakebrough M, Boulton MI, Davies JW (1991) Rice tungro bacilliform virus DNA independently infects rice after *Agrobacterium*-mediated transfer. J Gen Virol 72:1215–1221

Dasgupta I, Das BK, Nath PS, Mukhopadhyay S, Niazi FR, Varma A (1996) Detection of rice tungro bacilliform virus in field and glasshouse samples from India using polymerase chain reaction. J Virol Methods 58:53–58

Ghosh A, Reddy APK (1986) Tungro (RTV) incidence in Andhra Pradesh in 1984. Int Rice Res Newslett 11(3):16

Joshi R, Dasgupta I (2001) Cloning and molecular analysis of DNA of *Rice tungro bacilliform virus* from different rice growing regions of India. Indian Phytopathol 54:469–475

Joshi R, Kumar V, Dasgupta I (2003) Detection of molecular variability in rice tungro bacilliform viruses from India using polymerase chain reaction- restriction fragment length polymorphism. J Virol Methods 109:89–93

Kant R, Sharma S, Dasgupta I (2015) Virus-induced gene silencing (VIGS) for functional genomics in rice using rice tungro bacilliform virus (RTBV) as a vector. Methods Mol Biol 1287:201–217

Krishnaveni D, Laha GS, Shanker C, Katti G, Muralidharan K (2011) Virus-vector composition analysis in rice tungro virus epidemics in Telangana region of Andhra Pradesh, India. J Mycol Plant Pathol 41(3):441–444

Mangrauthia SK, Malathi P, Agarwal S, Sailaja B, Singh J, Ramkumar G, Krishnaveni D, Balachandran SM (2012) The molecular diversity and evolution of rice tungro bacilliform virus from Indian perspective. Virus Genes. doi:10.1007/s11262-012-0751-8

Mathur S, Dasgupta I (2007) Downstream promoter sequence of an Indian isolate of rice tungro bacilliform virus alters tissue-specific expression in host rice and acts differentially in heterologous system. Plant Mol Biol 65:259–275

Mathur S, Dasgupta I (2013) Further support of genetic conservation in Indian isolates of rice tungro bacilliform virusby sequence analysis of an isolate from North–Western India. Virus Genes 46:387–391

Mishra MD, Niazi FR, Basu AN, Ghosh A, Raychoudhuri SP (1976) Detection and characterization of a new strain of rice tungro virus in India. Plant Dis Report 60:23–25

Mohanty SK, Sridhar R (1986) Physiology of rice tungro virus disease: possible cause of carbohydrate and amino acid accumulation due to infection. Acta Phytopatholica Entomolica Hung 21:73–85

Mohanty SK, Sridhar R (1987) Physiology of rice tungro virus disease: changes in leaf pigments due to infection. Intl Rice Res Newslett 8(2–3):11

Mohanty SK, Sridhar R (1989) Physiology of rice tungro virus disease: changes in leaf pigments due to infection. Acta Phytopatholica Entomolica Hung 24:375–385

Mohanty SK, Mohanty SK, Anjaneyulu A, Sridhar R (1979) Physiology of rice tungro virus disease: involvement of abscisic acid-like substances in susceptible host-virus interactions. Physiol Plant 45:132–136

Mukhopadhyay S (1984) Interactions of insect vectors with plants in relation to transmission of plant viruses. Proc Indian Acad Sci (Anim Sci) 93:349–357

Mukhopadhyay S, Bandopadhyay B (1984) Incidence of different strains of rice tungro virus in West Bengal. Int J Trop Plant Dis 2:125–132

Mukhopadhyay S, Chowdhury AK (1970) Incidence of tungro virus of rice in West Bengal. Int Rice Commun Newslett 19(2):9–12

Mukhopadhyay S, Chowdhury AK (1973) Some epidemiological aspects of tungro virus disease of rice in West Bengal. Int Rice Comm Newsl 22:44–57

Muralidharan K, Krishnaveni D, Rajarajeshwari NVL, Prasad ASR (2003) Tungro epidemic and yield losses in paddy fields in India. Curr Sci 85:1143–1147

Nath N, Mathur S, Dasgupta I (2002) Molecular analysis of two complete rice tungro bacilliform virus sequences from India. Arch Virol 147:1173–1187

Neeraja CN, Krishnaveni D, Saivishnupriya K, Prasad GSV, Muralidharan K (2006) Molecular mapping of resistance to tungro virus in rice cultivars Vikramarya and Utrirajapan. Rice Genetics Newsletter 23:80–83

Niazi FR, Dasgupta I, Singh J, Mathur S, Varma A (2005) Characterization of new strains of rice tungro viruses. Indian Phytopathol 58(3):308–313

Periasamy M, Niazi FR, Malathi VG (2006) Multiplex RT-PCR: a novel technique for the simultaneous detection of the DNA and RNA virus causing the rice tungro disease. J Virol Methods 134:230–236

Prasad GSV, Krishnaveni D, Subramanian N, Neeraja CN, Muralidharan K (2004) Genetics of resistance to rice tungro virus disease. J Mycol Plant Pathol 34:846–839

Purkayastha A, Dasgupta I (2009) Virus-induced gene silencing: a versatile tool for discovery of gene functions in plants. Plant Physiol Biochem 47:967–976

Purkayastha A, Sharma S, Dasgupta I (2010a) A negative element in the downstream region of the *Rice tungro bacilliform virus* promoter is orientation- and position-independent and is active with heterologous promoters. Virus Res 153:166–171

Purkayastha A, Mathur S, Verma V, Sharma S, Dasgupta I (2010b) Virus-induced gene silencing in rice using a vector derived from a DNA virus. Planta 232:1531–1540

Purkayastha A, Sharma S, Dasgupta I (2013) Virus-induced gene silencing for rice using agro-inoculation: methods in molecular biology. In: Becker A (ed) Virus-induced gene silencing, methods and protocols, vol 975. Springer Science + Business Media, LLC, Humana Press, New York, pp 33–45. doi:10.1007/978-1-62703-278-0_3

Raychaudhuri SP, Mishra MD, Ghosh A (1967) Preliminary note on transmission of virus, a disease resembling tungro of rice in India and other virus like symptoms. Plant Dis Rep 51:300–301

Rivera CT, Ou SH (1965) Leafhopper transmission of tungro disease of rice. Plant Dis Rep 49:127–131

Sharma S (2011) Engineering tungro resistance, sequence analysis and fine mapping of negative promoter element of *Rice tungro bacilliform virus* Ph.D. thesis, University of Delhi

Sharma A (2012) M.Sc. dissertation, University of Delhi.

Sharma S, Dasgupta I (2012) Development of SYBR Green I based real time PCR assays for quantitative detection of *Rice tungro bacilliform virus* and *Rice tungro spherical virus*. J Virol Methods 181:86–92

Sharma S, Rabindran R, Robin S, Dasgupta I (2011) Analysis of the complete DNA sequence of rice tungro bacilliform virus from southern India indicates it to be a product of recombination. Arch Virol 156:2257–2262

Shastry SVS, Sharma SD, John VT, Seshu DV (1972) Breeding for resistance to rice tungro virus in India, Rice breeding Symposium. IRRI, Manila, pp 239–252

Shukla VD (1979) Studies on the epidemiology of rice tungro virus disease. Ph.D. thesis, Utkal University, Bhubaneswar, p 249

Singh J, Pal M, Mishra MD (1990) Effect of rice tungro virus infection on photosynthetic activity in resistant and susceptible rice cultivars. Indian J Mycol and Pl Patol 20:123–126

Sridhar R, Reddy PR, Anjaneyulu A (1976) Physiology of rice tungro virus disease: changes in chlorophyll, carbohydrates, amino acids and phenol contents. J Phytopathol Z 86:136–143

Srinivasulu B, Jeyarajan R (1989) Biochemical alterations in rice leaf components infected with rice tungro virus. Indian Phytopathol 42:454–455

Subbarao BL, Ghosh A, John VT (1979) Effect of rice tungro virus on chlorophyll and anthocyanin pigments in two rice cultivars. Phytopathol Z 94:367–371

Tyagi H, Rajasubramaniam S, Rajam MV, Dasgupta I (2008) RNA-interferance in rice against rice tungro bacilliform virus results in decreased accumulation in inoculated rice plants. Transgenic Res 17:897–904

Yadav BP, Mishra MD (1987) Metabolic changes induced by rice tungro virus in rice cultivars. Indian Phytopathol 40:139–148

Studies on Viroids Occurring in India

22

Anirban Roy, Ajay Kumar, Yashika Walia, Vipin Hallan, and Padma Ramachandran

Abstract

Viroids are small, single stranded, covalently closed circular, non-encapsidated plant pathogenic RNA molecules which do not code for any protein and thus depend on the host plant for their replication and other functions. After discovery of viroids in 1971, several diseases whose etiology was not known earlier were confirmed to be caused by this new group of plant pathogen. Besides the symptomatic expression, viroids often cause latent infection with no visible symptom in one plant species but when transferred to a new species they may cause severe infection. Indian scenario on researches of viroids-induced diseases started much earlier, even before the discovery of viroids as a plant pathogen. However, due to lack of proper diagnostics and infrastructural support the progress on such diseases were confined to initial level till the end of twentieth century. During this period the research efforts were focused on recording the diseases of unknown etiology, their impact assessment, understanding on the spread of such diseases and screening of resistant sources. With the addition of new infrastructural support and ease in the access of sequencing technologies several laboratories in India and in collaboration with laboratories abroad, viroid research gets its momentum from early twenty-first century. As a result several viroids infecting citrus, tomato, apple, ornamentals, rubber and grapes have been characterized from India and their sequences have been deposited in the sequence database. Association of viroids with yellow corky vein disease of citrus, bunchy top of tomato, tapping panel dryness of rubber, deformation of apple have been estab-

A. Roy • A. Kumar • P. Ramachandran (✉)
Advanced Centre for Plant Virology, Division of Plant Pathology, ICAR-Indian Agricultural Research Institute, New Delhi, India
e-mail: viroidram@yahoo.co.in

Y. Walia • V. Hallan
Plant Virus Lab, CSIR-IHBT, Palampur, Himachal Pradesh, India

B. Mandal et al. (eds.), *A Century of Plant Virology in India*,
https://doi.org/10.1007/978-981-10-5672-7_22

lished. Other viroids which have been detected in ornamentals and grapes do not produce visible symptoms. Till now no major outbreak of viroid disease in the form of epidemic has been reported in India.

Keywords
Viroid disease • Citrus exocortis viroid • Hop stunt viroid • Apple scar skin viroid • Grapevine viroids • Ornamental crop infecting viroids • Tapping panel dryness of rubber • Viroid research historical perspective • India

22.1 Introduction

Among the path breaking researches in the field of plant pathology and molecular biology, the discovery of viroids and prions is considered very significant, as they are regarded as molecules at the threshold of origin of life. Both these unique infectious agents challenge the existing knowledge of biologists on the very concept of pathogenicity. While viroids are composed entirely of ribonucleic acid and infect angiospermic plants, prions are constituted of only protein and cause infections on certain group of animals. The present review would limit to general characteristics of viroids as plant pathogens and progress of viroid research made in our country by illustrating diseases of significance.

22.2 General Characteristics, Yield Loss, Transmission and Detection

Viroids are single stranded, covalently closed, circular RNA molecules existing as highly base-paired rod-like structures and infect only plant tissue. The term was coined by Theodor O. Diener to describe the low molecular weight (LMW) RNA associated with spindle tuber disease of potato (Diener 1971). Since then, viroids are recognized as a novel class of plant pathogens. They are different from viruses in some important respects: (a) the pathogen exists *in vivo* as an unencapsidated RNA, (b) virion-like particles are not detectable in infected tissues, (c) the infectious RNA is LMW consisting of 245–401 nucleotides and (d) the RNA is replicated autonomously in susceptible cells without the need of a helper virus. They totally depend upon the host factors to complete the major steps of their life cycle like replication, trafficking and pathogenicity. They are the etiological agents of diverse diseases affecting food, industrial and ornamental (herbaceous and ligneous) plants. Viroid-induced symptoms range from no symptoms to severe developmental disorders, such as leaf chlorosis, necrosis, stunting, flowering alterations and fruit and seed deformations. They are classified into two families: the *Pospiviroidae*, which replicate in the nucleus, and the *Avsunviroidae*, which replicate (and accumulate) in the chloroplast. To date, there are more than 30 viroid species and about 340 viroid variants known to infect plant species in at least thirteen dicotyledonous plant families worldwide. (More than 1742 viroid sequences are available in "Subviral RNA Database" (http://subviral.med.uottawa.ca/)).

Viroids are as damaging as any other plant pathogens. Diseases caused by viroids were known much before their etiology was established. Nearly a century ago, from 1917, work was being done on a degeneration disease of potato which later came to be known as potato spindle tuber disease in Nebraska in the US (Werner 1924). The disease was known perhaps in other parts of the world too, since losses were reported from different countries. Similar was the case with cadang cadang disease of coconut. Nearly 20–30 million trees of coconut were lost due to the cadang cadang viroid since 1926 in the Phillipine island. In Japan, 17% of the total acreage of hops in 1968 in Fukushima prefecture was infected with hop stunt viroid, with some areas having up to 60% infected plants, the estimated loss being 60–75%. In other crops such as citrus, avocado, grapevine, apple, chrysanthemum, cucumber the viroid induced disease symptoms were known much earlier than the discovery of viroid. However, the exact yield loss data for most viroid diseases is not available.

Viroids are easily transmitted mechanically by contact, infected planting material, agricultural tools, biological seeds and pollen. Owing to the ease of transmission through both vegetative and biological seeds, potato spindle tuber viroid (PSTVd) has been declared a quarantine pathogen and import/export of potato needs to be monitored strictly. Active field transmission by insect vectors had not been established for a long time since their discovery. Two inconclusive reports were made for transmission of PSTVd by aphids (De Bokx and Piron 1981) and tomato planta macho viroid (TPMVd) by *Myzus persicae* (Galindo et al. 1986). Indirect viroid transmission by insects was reported for PSTVd (Syller and Marczewski 2001) and tomato apical stunt viroid (TASVd) (Antignus et al. 2007). The former encapsidates itself into the potato leaf roll virus (PLRV) capsid and is transmitted by aphids, and the latter attaches to bumblebee stylets and is transmitted during pollination. Recently, direct transmission of apple scar skin viroid (ASSVd) by glasshouse whitefly *Trialeurodes vaporariorum* to herbaceous hosts has been demonstrated (Walia et al. 2015) with high transmission efficiency.

Detection of viroids is a challenging task as most viroids remain latent in the hosts they infect, especially in ornamentals like Coleus and Columnea. Visible symptoms are produced in limited hosts. There are not many good indicator hosts. Though based on indicator hosts some bioassay methods *viz.* the citron bioassay for detection of citrus exocortis viroid are available but symptoms appear anywhere from 6 to 24 weeks depending on the temperature in the greenhouse. Viroids generally occur in plant tissues in low concentration, remain localized specifically in cell nucleolus, difficult to separate from host macromolecules and polyphenols and are extremely vulnerable to RNase. Methods used in plant virus detection, such as electron microscopy, ultracentrifugation and protein based immuno assays are of no consequence. Return polyacrylamide gel electrophoresis (R-PAGE), nucleic acid hybridization and polymerase chain reaction (PCR) are the only choice to detect viroids. However, genome sequence is the final step to identify the viroid exactly as minor sequence variations, even one or two nucleotide alterations, can bring about changes in the pathogenicity, a problem which highlights diversity within the viroid population (Roy and Ramachandran 2006).

22.3 A Historical Perspective of Viroid Research

The Indian scenario as regards viroid reports and researches is very interesting since no epidemic level of losses have been reported so far from any crops of economic importance. Active research on viroid detection and characterization has just got initiated since the last two to three decades. A number of viroid like diseases were recorded much before viroids became known as plant pathogen, particularly in citrus and tomato. The first report giving molecular evidence for viroid as causal agent was of tomato bunchy top disease by Saraswati and Mishra (1989), but the disease was reported much earlier by Pandey and Summanwar (1982). Later the causal viroid of this disease was reported as a sequence variant of citrus exocortis viroid (CEVd) and named as tomato variant of CEVd (CEVd-t) (Mishra et al. 1991). This is the first variant of CEVd infecting a herbaceous host. This is a very significant finding as this disease is not known to occur from any other part of the world. In India also, it has remained restricted to its initial region of report i.e. Maharashtra. The work was done in collaboration with Diener's laboratory at Beltsville, Maryland, USA since Indian laboratories at that time were not sufficiently equipped for molecular studies. During the same period a mild variant of PSTVd, earlier reported from Shimla region (Khurana et al. 1989) was also sequenced in the USDA laboratory and its structure reported (Owens et al. 1992).

Since the report of these two viroids, work started to re-examine the viroid like diseases reported earlier. Thus in the last decade of the twentieth century with the improvement and availability of detection techniques like R-PAGE, association of LMW-RNA was reported on a number of hosts such as citrus, chrysanthemum, coleus, grapevine, potato, pigeonpea and tobacco from Advanced Centre for Plant Virology, IARI, NewDelhi (Table 22.1).

The start of twenty-first century saw rapid progress since infrastructure for genome amplification, nucleic acid hybridization and sequencing became available at different institutes. Several viroid sequences have been deposited in the database during this time (Fig. 22.1). This helped in confirming the viroid nature of many diseases by reporting the viroid sequences and their relationship (Table 22.2).

Till now 91 sequences of 13 viroid species occurring in India have been deposited in the sequence data base (Table 22.3). Details of the viroid sequences deposited in the database have been provided in Table 22.4. Most of the viroid research in India was conducted in two laboratories, one at Advanced Centre for Plant Virology, ICAR-Indian Agricultural Research Institute (ICAR-IARI), New Delhi and the other at Plant Virus Laboratory, CSIR-Institute of Himalayan Bioresource Technology (CSIR-IHBT), Palampur, Himachal Pradesh (HP) (Table 22.3). Few isolates of PSTVd were also characterized from ICAR-Central Potato Research Institute (ICAR-CPRI), Shimla, HP. Besides the Indian laboratories, many Indian researchers in collaboration with other laboratories in the World (Hirosaki University, Japan; Potato Research Centre, Canada; USDA, USA) contributed to viroid research in India. Among the sequences deposited in the database maximum number is of apple scar skin viroid (ASSVd) followed by chrysanthemum stunt viroid (CSVd) reported by CSIR-IHBT group from HP. Work on citrus has been

Table 22.1 Viroid diseases reported from India during late twentieth century

Name of the disease	Initial report	Viroid biology estimated by R-PAGE
Citrus exocortis	Patil and Warke (1968), Nariani et al. (1968), and Kapur et al. (1974)	Ramachandran et al. (1993)
Chrysanthemum chlorotic mottle	Singh et al. (1978)	Ramachandran and Mathur (1999)
Citrus yellow vein disease/ Citrus yellow corky vein	Sharma and Pandey (1983)	Ramachandran et al. (1996), Rustem and Ahlawat (1999)
Bunchy top disease of tomato	Pandey and Summanwar (1982)	Saraswati and Mishra (1989), Mishra et al. (1991)
Spindle tuber disease of potato	Khurana et al. (1989)	Khurana et al. (1989), Owens et al. (1992);
Nicotiana glutinosa stunt	Bhattiprolu (1991)	Bhattiprolu (1991)
Coleus viroids	Ramachandran et al. (1991)	Ramachandran et al. (1992)
Viroid disease of pigeon pea	–	Bhattiprolu (1993)
Dapple apple disease of apple	–	Thakur et al. (1995), Handa et al. (1998)
Grapevine latent viroid	–	Ramachandran et al. (1992)

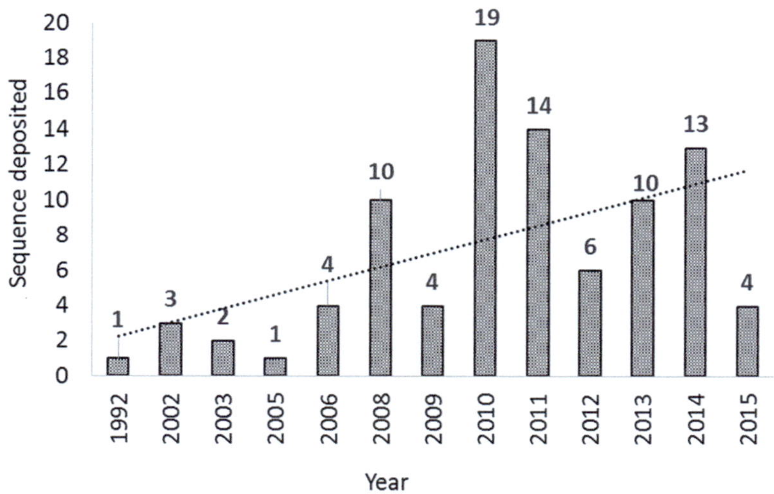

Fig. 22.1 Year-wise progress in submission of viroid sequence (occur in India) in the database. *Dotted line* indicated a linear increasing trend of sequence submission over the years

done extensively at ICAR-IARI, New Delhi. A very important century old disease syndrome in rubber plantations in India called 'Tapping panel dryness' has been investigated and shown to be of viroid etiology for the first time with molecular evidence. (Ramachandran et al. 2000; Kumar et al. 2013, 2015). Interesting details of all these diseases and their associated viorids have been discussed below in four different sections.

Table 22.2 Viroids characterized from India during twenty-first century

Disease	Viroid name	Sequence report
Citrus yellow corky vein	HSVd, CEVd	Roy and Ramachandran (2003, 2006)
Citrus bark scaling and splitting	HSVd	Ramachandran et al. (2005)
Symptomless ornamentals		
Moss verbena (Glandulariapuchella)	CEVd	Singh et al. (2006)
Trailing verbena	CEVd, tomato chlorotic dwarf viroid (TCDVd)	-Do-
Red joyweed (Alternantherasessilis)	Iresine viroid (IrVd)	-Do-
Dapple apple, scar skin and fruit deformation	Apple scar skin viroid (ASSVd)	Walia et al. (2009)
Himalayan wild cherry	ASSVd	Walia et al. (2012)
Coleus irregular chlorotic spots/ patches	Coleus blumei viroid (CbVd)	Adkar-Purushothama et al. (2013)
Color breaking of *Gerbera jamesonii*	ASSVd	Kumar et al. (2014)
Peach latent mosaic viroid	PLMVd	-Do-
Chrysanthemum stunt	CSVd	-Do-
Chrysanthemum chlorotic dwarf	CChMVd	-Do-
Grapevine yellow spot and flecks	Grapevine yellow speckle viroid-1 (GYSVd-1) and Hop stunt viroid (HSVd)	Sahana et al. (2013)
Grapevine pale yellow	Australian grapevine viroid (AGVd)	Adkar-Purushothama et al. (2014)
Tapping panel dryness of Rubber	PSTVd-rubber variant	Kumar et al. (2015)

Table 22.3 The genome sequence resources of viroids occuring in India

	No. of isolates characterized						
	Indian institutes			Foreign collaborators			
Viroid species	IHBT	IARI	CPRI	Hirosaki University, Japan	PRC, Canada	USDA	Total
Apple scar skin viroid	34						34
Australian grapevine viroid				5			5
Chrysanthemum chlorotic mottle viroid	2			3			5
Chrysanthemum stunt viroid	16			4			20
Citrus exocortis viroid		1			2		3
Coleus blumei viroid 1				2			2
Grapevine yellow speckle viroid 1				2			2
Hop stunt viroid		3		2			5
Iresine viroid 1					1		1
Peach latent mosaic viroid	1						1
Potato spindle tuber viroid			5			1	6
Rubber viroid India		6					6
Tomato chlorotic dwarf viroid					1		1
Total	**53**	**10**	**5**	**18**	**4**	**1**	**91**

Table 22.4 Detailed list of viroid sequences reported from India

Viroid sp	Accession No	Size (bp)	Strain/Isolate	Host	Place	Submission year	Institute
Apple scar skin viroid	FM178283	331	Solan (clone Y3)	Malus x domestica	Himachal Pradesh	2008	IHBT, Palampur
Apple scar skin viroid	FM178284	331	Solan (clone Y4)	Malus x domestica	Himachal Pradesh	2008	IHBT, Palampur
Apple scar skin viroid	FM178285	330	Solan (clone Y5)	Malus x domestica	Himachal Pradesh	2008	IHBT, Palampur
Apple scar skin viroid	FM208138	330	Jubbal (clone Y6)	Malus x domestica	Himachal Pradesh	2008	IHBT, Palampur
Apple scar skin viroid	FM208139	331	Jubbal (clone Y7)	Malus x domestica	Himachal Pradesh	2008	IHBT, Palampur
Apple scar skin viroid	FM208140	331	Jubbal (clone Y8)	Malus x domestica	Himachal Pradesh	2008	IHBT, Palampur
Apple scar skin viroid	FM208141	331	Jubbal (clone Y9)	Malus x domestica	Himachal Pradesh	2008	IHBT, Palampur
Apple scar skin viroid	FM208142	331	Jubbal (clone Y10)	Malus x domestica	Himachal Pradesh	2008	IHBT, Palampur
Apple scar skin viroid	AM993159	330	Solan	Malus x domestica	Himachal Pradesh	2009	IHBT, Palampur
Apple scar skin viroid	AM993160	330	Kotkhai	Malus x domestica	Himachal Pradesh	2009	IHBT, Palampur
Apple scar skin viroid	FM877527	331	Rohru	Malus domestica	Himachal Pradesh	2008	IHBT, Palampur
Apple scar skin viroid	FN547406	331	Palampur	Malus domestica cv. Gold Spur	Himachal Pradesh	2009	IHBT, Palampur
Apple scar skin viroid	FN547407	331	Palampur	Malus domestica cv. Red chief	Himachal Pradesh	2009	IHBT, Palampur
Apple scar skin viroid	FN669528	330	Wild Cherry	Prunus sp.	Himachal Pradesh	2010	IHBT, Palampur
Apple scar skin viroid	FN669529	330	Wild Cherry2	Prunus sp.	Himachal Pradesh	2010	IHBT, Palampur
Apple scar skin viroid	FR749995	330	Hatkoti	Malus x domestica	Himachal Pradesh	2010	IHBT, Palampur
Apple scar skin viroid	HE601743	211	Reckong Peo	Malus x domestica	Himachal Pradesh	2011	IHBT, Palampur
Apple scar skin viroid	HE601744	213 (partial)	Kalpa	Malus x domestica	Himachal Pradesh	2011	IHBT, Palampur
Apple scar skin viroid	HE601745	212 (partial)	Rohru	Malus x domestica	Himachal Pradesh	2011	IHBT, Palampur
Apple scar skin viroid	HG764197	329	C1	Cucumis sativus	India	2013	IHBT, Palampur
Apple scar skin viroid	HG764198	329	C2	Cucumis sativus	India	2013	IHBT, Palampur
Apple scar skin viroid	HG764199	329	C3	Cucumis sativus	India	2013	IHBT, Palampur
Apple scar skin viroid	HG764200	331	C4	Cucumis sativus	India	2013	IHBT, Palampur
Apple scar skin viroid	HG764201	329	B1	Phaseolus vulgaris	India	2013	IHBT, Palampur

(continued)

Table 22.4 (continued)

Viroid sp	Accession No	Size (bp)	Strain/Isolate	Host	Place	Submission year	Institute
Apple scar skin viroid	HG764202	329	T1	Solanum lycopersicum	India	2013	IHBT, Palampur
Apple scar skin viroid	HG764203	329	P1	Pisum sativum	India	2013	IHBT, Palampur
Apple scar skin viroid	HG764204	329	A1	Chenopodium amranticolor	India	2013	IHBT, Palampur
Apple scar skin viroid	HG764205	329	N1	Nicotiana benthamiana	India	2013	IHBT, Palampur
Apple scar skin viroid	HG764206	329	Q1	Chenopodium quinoa	India	2013	IHBT, Palampur
Apple scar skin viroid	LK053005	331	DBCH1	Denrocalamus bannaenensis	Himachal Pradesh	2014	IHBT, Palampur
Apple scar skin viroid	LN823954	328	DBCH2	Dendrocalamus sp.	Himachal Pradesh	2015	IHBT, Palampur
Apple scar skin viroid	LN823955	331	DBCH3	Dendrocalamus sp.	Himachal Pradesh	2015	IHBT, Palampur
Apple scar skin viroid	LN823956	331	DHWH2	Dendrocalamus sp.	Himachal Pradesh	2015	IHBT, Palampur
Apple scar skin viroid	LN823957	330	DHWH2	Dendrocalamus sp.	Himachal Pradesh	2015	IHBT, Palampur
Australian grapevine viroid	KJ019300	369	Ind-1	Vitis vinifera	India	2014	Hirosaki University, Japan
Australian grapevine viroid	KJ019301	369	Ind-2	Vitis vinifera	India	2014	Hirosaki University, Japan
Australian grapevine viroid	KJ019302	368	Ind-3	Vitis vinifera	India	2014	Hirosaki University, Japan
Australian grapevine viroid	KJ019303	369	Ind-4	Vitis vinifera	India	2014	Hirosaki University, Japan
Australian grapevine viroid	KJ019304	369	Ind-5	Vitis vinifera	India	2014	Hirosaki University, Japan
Chrysanthemum chlorotic mottle viroid	FN646404	398	Palampur	Chrysanthemum sp.	Himachal Pradesh	2010	IHBT, Palampur
Chrysanthemum chlorotic mottle viroid	FN669541	396	Yellow Star	Chrysanthemum sp.	Himachal Pradesh	2010	IHBT, Palampur
Chrysanthemum chlorotic mottle viroid	KP262531	399	CChMVd-1	Chrysanthemum sp.	India	2014	Hirosaki University, Japan and India

Chrysanthemum chlorotic mottle viroid	KP262532	399	CChMVd-2	Chrysanthemum sp.	India	2014	Hirosaki University, Japan and India
Chrysanthemum chlorotic mottle viroid	KP262533	399	CChMVd-3	Chrysanthemum sp.	India	2014	Hirosaki University, Japan and India
Chrysanthemum stunt viroid	AJ585258	348	Indian	Chrysanthemum sp.	Himachal Pradesh	2003	IHBT, Palampur
Chrysanthemum stunt viroid	AJ969017	348	Maharashtra	Chrysanthemum morifolium	Maharashtra	2005	IHBT, Palampur
Chrysanthemum stunt viroid	FN646407	355	Palampur	Chrysanthemum sp.	Himachal Pradesh	2010	IHBT, Palampur
Chrysanthemum stunt viroid	FN673553	355	Discovery c1	Chrysanthemum sp.	Himachal Pradesh	2010	IHBT, Palampur
Chrysanthemum stunt viroid	FN673554	351	Discovery c2	Chrysanthemum sp.	Himachal Pradesh	2010	IHBT, Palampur
Chrysanthemum stunt viroid	FN669530	354	Poornima	Chrysanthemum sp.	Himachal Pradesh	2011	IHBT, Palampur
Chrysanthemum stunt viroid	FN669531	354	Yellow star	Chrysanthemum sp.	Himachal Pradesh	2011	IHBT, Palampur
Chrysanthemum stunt viroid	FN669532	354	Thicking queen	Chrysanthemum sp.	Himachal Pradesh	2011	IHBT, Palampur
Chrysanthemum stunt viroid	FN669533	354	Yellow spider	Chrysanthemum sp.	Himachal Pradesh	2011	IHBT, Palampur
Chrysanthemum stunt viroid	FN669534	354	Stallion	Chrysanthemum sp.	Himachal Pradesh	2011	IHBT, Palampur
Chrysanthemum stunt viroid	FN669535	354	Shyamal	Chrysanthemum sp.	Himachal Pradesh	2011	IHBT, Palampur
Chrysanthemum stunt viroid	FN669536	354	Majesthic Daisy	Chrysanthemum sp.	Himachal Pradesh	2011	IHBT, Palampur

(continued)

Table 22.4 (continued)

Viroid sp	Accession No	Size (bp)	Strain/Isolate	Host	Place	Submission year	Institute
Chrysanthemum stunt viroid	FN669537	354	Yellow Anemone	Chrysanthemum sp.	Himachal Pradesh	2011	IHBT, Palampur
Chrysanthemum stunt viroid	FN669538	354	Mundiyal	Chrysanthemum sp.	Himachal Pradesh	2011	IHBT, Palampur
Chrysanthemum stunt viroid	FN669539	354	Chandrma	Chrysanthemum sp.	Himachal Pradesh	2011	IHBT, Palampur
Chrysanthemum stunt viroid	FN669540	354	White Daisy	Chrysanthemum sp.	Himachal Pradesh	2011	IHBT, Palampur
Chrysanthemum stunt viroid	KP262534	354	CSVd-1	Chrysanthemum sp.	India	2014	Hirosaki University, Japan and India
Chrysanthemum stunt viroid	KP262535	354	CSVd-2	Chrysanthemum sp.	India	2014	Hirosaki University, Japan and India
Chrysanthemum stunt viroid	KP262536	354	CSVd-3	Chrysanthemum sp.	India	2014	Hirosaki University, Japan and India
Chrysanthemum stunt viroid	KP262537	354	CSVd-4	Chrysanthemum sp.	India	2014	Hirosaki University, Japan and India
Citrus exocortis viroid	AJ490825	370	Yellow corky vein	Citrus aurantifolia	Southern India	2002	IARI, India
Citrus exocortis viroid	DQ846884	197 (partial)		Verbena x hybrida	New Delhi	2006	Potato Research Centre, Canada and IARI, India
Citrus exocortis viroid	DQ846885	201 (partial)		Glandularia puchella	New Delhi	2006	Potato Research Centre, Canada and IARI, India

Coleus blumei viroid 1	AB740017	249	Ind-1	Solenostemon sp.	India	2012	Hirosaki University, Japan and India
Coleus blumei viroid 1	AB740018	249	Ind-2	Solenostemon sp.	India	2012	Hirosaki University, Japan and India
Grapevine yellow speckle viroid 1	AB742222	366	Kar-1	Vitis vinifera	Karnataka	2012	Hirosaki University, Japan
Grapevine yellow speckle viroid 1	AB742223	367	Kar-2	Vitis vinifera	Karnataka	2012	Hirosaki University, Japan and India
Hop stunt viroid	AJ490824	295	Yellow corky vein	Citrus aurantifolia	Southern India	2002	IARI, India
Hop stunt viroid	AY237168	295	HSVd-RL	Nagpur mandarin	Nagpur	2003	IARI, India
Hop stunt viroid	AF517563	295	Yellow corky vein viroid	Citrus aurantifolia	Southern India	2002	IARI, India
Hop stunt viroid	AB742224	298	Ind-1	Vitis vinifera	Karnataka	2012	Hirosaki University, Japan
Hop stunt viroid	AB742225	297	Ind-2	Vitis vinifera	Karnataka	2012	Hirosaki University, Japan
Iresine viroid 1	DQ846886	224 (partial)		Alternanthera sessilis	New Delhi	2006	Potato Research Centre, Canada and IARI, India
Peach latent mosaic viroid	FM955277	337	India	Malus domestica	Himachal Pradesh	2008	IHBT, Palampur
Potato spindle tuber viroid	M88681	359	Mild strain	Solanum sp	Himachal Pradesh	1992	USDA, USA and CPRI, India
Potato spindle tuber viroid	HQ639697	359	PSTVd-1	Solanum sp	Himachal Pradesh	2010	CPRI, Shimla
Potato spindle tuber viroid	HQ639698	359	PSTVd-2	Solanum sp	Himachal Pradesh	2010	CPRI, Shimla

(continued)

Table 22.4 (continued)

Viroid sp	Accession No	Size (bp)	Strain/Isolate	Host	Place	Submission year	Institute
Potato spindle tuber viroid	HQ639699	358	PSTVd-3	Solanum sp	Himachal Pradesh	2010	CPRI, Shimla
Potato spindle tuber viroid	HQ639700	357	PSTVd-4	Solanum sp	Himachal Pradesh	2010	CPRI, Shimla
Potato spindle tuber viroid	HQ639701	359	PSTVd-5	Solanum sp	Himachal Pradesh	2010	CPRI, Shimla
Rubber viroid India	HM107843	360	KER1	Hevea brasiliensis	Kerala, India	2010	IARI, India
Rubber viroid India	HM107844	361	KER2	Hevea brasiliensis	Kerala, India	2010	IARI, India
Rubber viroid India	HM107845	361	KER3	Hevea brasiliensis	Kerala, India	2010	IARI, India
Rubber viroid India	HM107846	361	KER4	Hevea brasiliensis	Kerala, India	2010	IARI, India
Rubber viroid India	HM107847	361	KER5	Hevea brasiliensis	Kerala, India	2010	IARI, India
Rubber viroid India	HM107848	359	KER6	Hevea brasiliensis	Kerala, India	2010	IARI, India
Tomato chlorotic dwarf viroid	DQ846883	195 (partial)		Verbena x hybrida	New Delhi	2006	Potato Research Centre, Canada and IARI, India

22.4 Major Viroids in Different Plants

22.4.1 Citrus

Different viroid induced symptoms were observed in different species of citrus in India even before the discovery of viroid. A list of diseases of citrus supposed to be caused by viroid is mentioned in Table 22.5. Majority of the research efforts were concentrated on two diseases, which are discussed below:

22.4.1.1 Citrus Exocortis Like Diseases

Citrus exocortis disease was recorded in India as early as 1968 much before the viroids became known as plant pathogens. Patil and Warke (1968) first observed this disease during periodical observations on Mosambi rootstock trials in citrus-die-back scheme, Shirampur, Maharashtra. Symptoms like scaling and splitting of bark at bud-union and stunting of entire tree was observed on Rangpur lime stock and

Table 22.5 Viroid diseases of citrus occurring in India

Disease	Distribution	Host		Reference
		Rootstock	Scion	
Citrus exocortis disease "virus"	Pune	Rangpur lime	Mosambi	Patil and Warke (1968)
	New Delhi (IARI)	Karna Khatta	Mandarin grape fruit	Nariani et al. (1968)
	Delhi, Punjab	Rangpur lime	Several scions	Kapur et al. (1974)
	Punjab	Karna Khatta	Malta sweet orange CV "Blood Red"	Cheema et al. (1982)
		Rangpur lime, Troyer and Carrizo citrange	Several scions	Kapur et al. (1984), Kapur et al. (1988)
	Bangalore	Not known	Kinnow mandarins Mosambi	Sharma and Saxena (1989)
Citrus exocortis viroid	Pune	Rangpur lime	Mosambi	Ramachandran et al. (1991, 1993)
	Bangalore	Karna Khatta	Sathgudi	
Citrus yellow corky vein	Andhra Pradesh	Not known	Mosambi Sathgudi	Reddy et al. (1974)
Citrus leaf yellow mid-vein	Maharashtra	Not known	*Kagzi* lime	Sharma and Pandey (1983)
Cachexia like disease	Maharashtra	Several rootstock	Sweet orange	Mali and Chaudhuri (1993)
Citrus yellow vein disease	Andhra Pradesh	Not known	*Kagzi* lime	Rustem et al. (1997)
	Central India	Rangpur lime and rough lemon	Nagpur Mandarin, Mosambi and acid lime	Ghosh et al. (2002)

Fig. 22.2 Symptoms of citrus exocortis viroid. (**a**) Decline of plant, (**b**) Leaf epinasty as seen on grafting infected scion on Etrog citron root stock, indicator host of CEVd, (**c**) bark scaling, (**d**) bunchy top and yellowing symptom on tomato by tomato variant of CEVd (CEVd-t). *H* Healthy, *I* Infected

Mosambi budded plants. Different symptoms of citrus exocortis viroid infected plants are depicted in Fig. 22.2. Nariani et al. (1968) observed similar symptoms on mandarins and grapefruit budded on "Karna Khatta" (*Citrus karna* Raf.) stock plants at ICAR-IARI, New Delhi. Later, in 1973 citrus exocortis disease was reported in some trees from Delhi and Punjab grafted on Rangpur lime (Kapur et al. 1974). However, "Blood red" sweet orange trees budded on rough lemon, citrumelo, Cleopatra mandarin remained free from exocortis infection, whereas trees on "Karna Khatta", Rangpur lime and Troyer and Carrizo citranges showed 100% exocortis incidence (Cheema et al. 1982; Kapur et al. 1984). Nucellar sweet oranges "Blood red" and "Mosambi" budded on rough lemon and Troyer and their seedlings were free from tristeza and exocortis infections (Cheema et al. 1984). Kapur et al.

(1988) observed 2.9% incidence of exocortis viroid in an indexing trial of 135 citrus trees of different species and cultivars using indicator plants. High incidence of exocortis-like symptoms was noticed in Kinnow and Coorg mandarins and Mosambi sweet oranges in rootstock trials at the ICAR-Indian Institute of Horticulture Research Farm, Bangalore (Sharma and Saxena 1989). Saraswati and Mishra (1989) demonstrated the viroid nature of bunchy top disease of tomato. The first molecular evidence of a viroid associated with bunchy top disease of tomato came when the causal viroid of this disease was reported as a sequence variant of citrus exocortis viroid (CEVd) and named as tomato variant of CEVd (CEVd-t) (Mishra et al. 1991). Mali and Chaudhuri (1993) observed cachexia like symptoms in sweet lime and tangelo and stunting in several rootstocks and sweet orange trees at the orchard of All India Coordinated Fruit Improvement Project on Citrus, Shrirampur (Maharashtra). However, the first molecular evidence for the occurrence of CEVd in India was reported by Ramachandran et al. (1991, 1993) by R-PAGE.

22.4.1.2 Yellow Corky Vein Disease of Citrus

During 1970s, yellow corky vein disease of citrus was reported from Andhra Pradesh (AP) state of India in declining sweet orange (*C. sinensis)* cv. Sathgudi (Reddy et al. 1974). Typical symptoms of the disease were yellowing of dorsal veins followed by corking of veins on the ventral side of the leaves (Fig. 22.3). However, the yellowing was restricted to the leaves and did not extend down to petioles and branches. Curling of leaves was observed as the disease progressed. Later the disease was found spreading to acid lime (*C. aurantifolia*) trees in AP (Reddy and Naidu 1989). The disease remained insignificant until it was reported from Assam where it adversely affected lime production (Azad 1993). He reported the comparative yield of two acid lime varieties *Kagzi* and Goal Nemu from 30 trees of each healthy and diseased and found that infected lime trees of var. *Kagzi* yielded 51.3% less fruits with a weight loss of 60.4% while Goal Nimbu produced 60.4% less fruits weighing 89.7% less than from healthy trees.

Fig. 22.3 Yellow corky vein symptom on *Kagzi* lime. Yellow vein on the upper surface and corky vein on the under surface of the leaf

The disease was found to be transmitted by bud grafting to citrus species (Reddy et al. 1974). Reddy and Naidu (1989) studied the transmission of the disease by mechanical means, seed and dodder. It was found that the disease was transmitted mechanically from acid lime to acid lime, where 66% of the inoculated plants developed typical symptoms of yellow corky vein in about 8–9 weeks after inoculation. Detailed study on host range revealed that the disease is transmitted to 19 different hosts of family Rutaceae (Rustem 1998). *Kagzi* lime has been established as the indicator host for yellow corky vein disease. On Etrog citron typical yellowing was developed upon infection. Study also revealed that while one of the important rough lemon rootstock "Jatti Katti" was susceptible to the disease, other two, "Karna Khatta" and "Soh Sorkar" remained symptom less. Initial study on etiology by R-PAGE revealed that the disease is associated with an infectious, heat-stable LMW-RNA similar in characteristics of viroid RNA (Rustem and Ahlawat 1999). The viroid was tentatively given the name as citrus yellow corky vein viroid (CYCVVd). Relationship of this viroid with other viroids was studied by nucleic acid spot hybridization (NASH) test using radiolabelled cDNA probes of PSTVd, CEVd and hop stunt viroid (HSVd) (Rustem et al. 2000). It was observed that sap and nucleic acid extract from infected plants gave faint hybridization signal with PSTVd probe while it gave relatively strong hybridization signal with CEVd and HSVd probes. Finally, two viroids have been characterized from infected samples, one is a variant of citrus exocortis viroid (CEVd-ycv) (Roy and Ramachandran 2006) having high sequence similarity with mild variant but distantly related with CEVd-t and the second one belongs to cachexia isolate of hop stunt viroid (HSVd-ycv) (Roy and Ramachandran 2003) having high sequence similarity with severe variant of HSVd. NASH test using cDNA probe of both viroids efficiently detected respective components from infected sample and can be used as diagnostic tool to detect two major groups of citrus viroid (Ramachandran et al. 2003). A micropropagation protocol for rootstock plant (rough lemon and trifoliate orange) was standardized against viroid infection (Roy and Ramachandran 2008).

Another HSVd variant (HSVd-RL) from Nagpur mandarin (*C. reticulata*) and Mosambi (*C. sinensis*) grafted on rough lemon and Rangpur lime rootstocks showing bark scaling, bark splitting and leaf yellowing symptoms was reported from citrus orchards in certain parts of central India (Ghosh et al. 2002; Ramachandran et al. 2005). This result suggests that HSVd is also a component in producing bark-scaling symptoms in rootstocks which were previously considered to be due to CEVd-infection alone.

22.4.2 Apple, Other Fruits and Ornamentals from Himalayan Region

22.4.2.1 Apple Scar Skin Viroid (ASSVd)

Apple scar skin viroid (type member of genus *Apscaviroid*, family *Pospoviroideae*) has been identified as the major (viroid) pathogen infecting apple in India as well as in the World. Infection causes scarring, dappling and deformation symptoms on

Fig. 22.4 Symptom induced by ASSVd. (**a**) Fruit deformity in apple, (**b**) Gerbera leaves showing mosaic, decreased leaf lamina and curling symptoms while flowers display colour breaking and deformation, (**c**) Chlorotic streak on *Dinochloa bannensis*. *Picture courtesy: Ms. Pooja Bhardwaj, Dr. Tanuja Rana, Ms. Madhvi Ahuja from CSIR-IHBT, HP*

apple fruits. It has genome of 330–331 nt, which is organized as a rod shaped secondary structure, divided into five domains *viz*., the terminal left domain (TLD), pathogenicity domain (PD), central conserved region (CCR), variable domain (VD) and terminal right domain (TRD).

In India, ASSVd was reported in the late nineties mainly on the basis of symptomatology and R-PAGE assays (Behl et al. 1998; Thakur et al. 1995; Handa et al. 1998). However, the molecular evidence was lacking. Later in the year 2009, it was characterized from various apple growing regions of HP and was reported to be associated with the "scar skin" and "dapple apple" diseases of apple fruit (Walia et al. 2009). It was also found associated with fruit deformation and russet symptoms (Fig. 22.4a). Isolates from 13 locations of HP were characterized, of which some were identified as sequence variables. Indian isolates of ASSVd were 329–331 nt in length and they were more variable at the TLD of the ASSVd secondary structure.

Presence of ASSVd was also reported in apple fruit samples showing various deformities and discolorations collected from different areas of HP (Kumar et al. 2014; Bhardwaj 2016). At CSIR-IHBT, Palampur natural occurrence of ASSVd in *Gerbera jamesonii* displaying viroid like symptoms (Fig. 22.4b) was confirmed by slot-blot hybridization with DIG labelled RNA probe, RT-PCR using viroid specific primers and sequencing (Kumar et al. 2014).

Two distinct variants of ASSVd were characterized from bamboo (*Dinochloa bannensis*) showing viroid like symptoms (Fig. 22.4c) by NASH test, RT-PCR and sequencing (unpublished, information provided by Ms. Madhvi Ahuja, CSIR-IHBT). Himalayan wild cherry (*Prunus cerasoides*) was also identified as a host of ASSVd (Walia et al. 2012).

An infectious clone of the Indian isolate of ASSVd was constructed by dimerization of its genome (Walia et al. 2014). Agro-infiltration and mechanical inoculation of dimeric RNA transcripts were identified as best inoculation methods as 100 % infectivity could be achieved (Walia et al. 2014). Agroinfection could produce viroid induced symptom in nine herbaceous plant species, however, different sequence variants were observed in those plants. Different mutational analysis was performed in ASSVd genome to establish the viroid structure function relationship (Walia 2016). ASSVd RNA was found to be directly transmitted by the whitefly (*Trialeurodes vaporariorum*) and no virus was found associated with this transmission experiment (Walia et al. 2015). However, it has been demonstrated that a *Cucumis sativus* phloem protein2 (CsPP2) and ubiquitin ligase mRNA sequence also help in whitefly transmission (Walia et al. 2015).

22.4.2.2 Chrysanthemum Stunt Viroid (CSVd) and Chrysanthemum Chlorotic Mottle Viroid (CCHMVd)

The Chrysanthemum stunt viroid and Chrysanthemum chlorotic mottle viroid were identified and characterized from Chrysanthemum germplasm showing symptoms of chlorotic spots, vein clearing, stunting and flower bleaching at CSIR-IHBT, Palampur.

22.4.2.3 Peach Latent Mosaic Viroid (PLMVd)

Peach latent mosaic viroid (PLMVd) was identified to infect apple plants from Shimla district, HP, India, showing highly dappled fruits. The isolate showed maximum nucleotide sequence identity with Spanish isolate of PLMVd from peach. These sample were also found positive for ASSVd, which is known to cause dappling in apple.

22.4.3 Rubber

Tapping panel dryness (TPD) syndrome (earlier referred to as "brown bast of rubber") is a major limiting factor in rubber (*Hevea brasiliensis*) plantation. The disease was first reported from Malaysia affecting mostly high-yielding clones of rubber (Sharples 1936). It spread to other South Asian countries like China, India and Sri Lanka. In the infected plant the tapping panel gradually dries up and thus cause complete loss in latex yield (Fig. 22.5). TPD syndrome becomes evident, about 6–7 years after transplanting, only when the trees reach maturity for tapping, resulting in complete loss to the grower.

In India, TPD is known to affect rubber plantations since the early years of rubber cultivation. It became important as all the high-yielding clones, including Rubber Research Institute of India (RRII) 105 (developed indigenously by the RRII, Kottayam, Kerala), which cultivated in more than 85% of the area, succumbed to the disease. In Indian rubber plantations, the average incidence of TPD is found to be about 15% which results in a loss of approximately 18–20 billion rupees

Fig. 22.5 Healthy and tapping panel dryness affected rubber tree trunk. A healthy panel showing full latex flow while the infected plant showing dryness syndrome

annually. Many evidences were put forward to show that TPD may be of abiotic nature particularly due to physiological changes. Reports on association of any biotic agents were inconclusive. First evidence for association of a viroid-like RNA with TPD affected trees came from the work done in a collaborative project between ICAR-IARI, New Delhi and RRII, Kerala. During the course of the project extensive survey of rubber plantations in and around Kerala was done and samples from leaf and bark of healthy and TPD affected trees analysed by R-PAGE. Besides latex drying, TPD-affected trees also showed symptoms of bark scaling, cracking, drying, necrotic streaking and browning of internal bark leading to decay of internal tissues. Often, prominent bulges on the lower part of tree trunks was seen where the first panel begins to dry. Investigations also showed absence of association of biotic agents such as fungi, bacteria, virus, mycoplasma-like organism or protozoa. A LMW-RNA was detected from total nucleic acid from TPD-affected rubber tissue by R-PAGE and viroid-like nature of this LMW-RNA was demostrated by its sensitivity to RNase and insensitivity to DNase, phenol and heat (Ramachandran et al. 2000). The LMW-RNA was amplified by RT-PCR using primers specific to PSTVd. During screening of healthy and diseased samples, some of the apparently healthy trees also showed LMW-RNA band. In many trees where LMW-RNA bands were detected in R-PAGE before the trees were tapped, showed TPD when they attained the tapping age (Ramachandran et al. 2006). Kumar et al. (2013) reported an improved R-PAGE procedure and detected LMW-RNA in leaf, bark and root of TPD-affected seedlings and grafted plants of different rubber clones grown in different locations. The infectious nature of the isolated LMW-RNA was shown on healthy tomato seedlings cv. Pusa Ruby by mechanical inoculation. Inoculated tomato showed symptoms of epinasty and inward curling of leaves about 3 weeks post inoculation. The LMW-RNA was reisolated from symptomatic tomato leaves which re-infected healthy tomato seedlings, thus proving the biotic etiology of TPD. The group at ICAR-IARI further reported cloning, sequencing and identification of the LMW-RNA associated with TPD affected rubber trees as well as tomato inoculated with LMW-RNA from rubber that showed symptoms of epinasty and curling. (Kumar et al. 2015). On the basis of nearly 100 % identity with PSTVd isolates M16826 and AY937179, the viroid sequences characterized from rubber samples were considered as variants of PSTVd and the name proposed as PSTVd-Rubber-KER-1(KER- 1 -8), denoting its origin from the rubber tree followed by clone id KER indicating the geographic origin from Kerala. A diagnostic probe for quick detection of TPD has also been developed and validated with field samples of rubber.

22.4.4 Other Viroids

22.4.4.1 Potato Spindle Tuber Viroid (PSTVd)

PSTVd causes deformities in potato tuber (Fig. 22.6a). During routine germplasm screening of potato at ICAR-CPRI, viroid like disease with extensive veinal necrosis was observed in wild species of potato (Khurana et al. 1989). After characterization of the associated viroid at biological and molecular level occurrence of a mild strain of PSTVd and its three sequence variants were detected (Owens et al. 1992). Few more sequences of PSTVd have been submitted from this Institute, however, no detail information are available. Recently a duplex real-time RT-PCR assay for the detection of PSTVd have been developed from ICAR-CPRI which is very sensitive as it could detect the viroid up to 0.025 fg of total RNA from infected tissues. Such method is an alternative to nucleic acid spot hybridization test and could be used in evaluation of potato germplasm during post entry quarantine testing.

22.4.4.2 Ornamental Plants

Ornamental plants are reservoir of viroids, often did not produce any visible symptom in its original host but can cause severe disease when transmitted to other hosts. Among the ornamental plants LMW-RNA was detected from Chrysanthemum

Fig. 22.6 (**a**) Symptom of PSTVd on potato, (**b**) Viroid induced stunting in chrysanthemum, (**c**) Yellow speckle symptom in grapevine leaf. *H* Healthy, *I* Infected

plants showing symptoms of mild chlorosis on young leaves, stunting and delayed flowering (Mathur et al. 2002) (Fig. 22.6b). The infectivity of the LMW-RNA was established but no sequence information is available.

Apparently, many non-symptomatic road side ornamental plants were tested for presence of viroid infection and three different viroids were detected in different combinations in three different ornamental species. Among them moss verbena (*Glandularia puchella*) was infected with a variant of CEVd, trailing verbena (Verbena × hybrid) was co-infected with a variant of CEVd and a variant of Tomato chlorotic dwarf viroid (TCDVd) and red joyweed, (*Alternanthera sessilis*) infected with a variant of Iresine viroid (IrVd) (Singh et al. 2006).

Coleus is another important ornamental host which generally remain asymptomatic, however, LMW-RNA was detected in several such asymptomatic samples (Ramachandran et al. 1991). This LMW-RNA did not produce any symptom in other plants, which are host of columnea latent viroid and Coleus viroid-N, two other viroids reported from asymptomatic Coleus plant. In 2013, a variant of Coleus blumei viroid-1 was characterized from coleus leaf samples with irregular chlorotic spots/patches collected from home gardens of two different districts of Karnataka (Adkar-Purushothama et al. 2013).

22.4.4.3 Pigeonpea

A LMW-RNA was isolated from pigeonpea (*Cajanus cajan*) plants showing mosaic, mottling and reduced size of leaves, stunting and no flowering (Bhattiprolu 1993). The infectivity of this RNA was confirmed in different hosts. This viroid was different from another viroid that infect both *Nicotiana glutinosa* and pigeonpea on the basis of variability in symptom production and electrophoretic mobility (Bhattiprolu 1991). No further characterization was done to reveal the identity of this viroid.

22.4.4.4 Grapevine

Grapevine is reported to be infected by five viroids worldwide, HSVd, Grapevine yellow speckle viroid-1 (GYSVd-1), Grapevine yellow speckle viroid-2 (GYSVd-2), CEVd, and Australian grapevine viroid (AGVd). Among these five viroids, only GYSVd-1 and GYSVd-2 are known to produce symptoms on grapevines. During 2012, symptoms like yellow leaf spots and flecks, shortened internodes, and tiny yellow leaves was observed in vineyards of the Bijapur, Doddaballapur, and Kolar districts of Karnataka State, India (Fig. 22.6c). Characterization of the viroid revealed the presence of GYSVd-1 and HSVd in the symptomatic samples (Sahana et al. 2013). Recently, occurrence of AGVd was detected in from three major grape growing areas of India. Nucleotide sequence analysis showed existence of five major AGVd variants in India along with other 44 minor variants (Adkar-Purushothama et al. 2014).

22.5 Concluding Remark

Viroids have been discovered less than 50 years ego, but they are considered as the living fossil of hypothetical RNA World. So, they were supposed to exist for billions of years. They were adapted to their host in the process of evolution. Some viroids cause economic important diseases to some crops, while some remain latent in some of the hosts. After the discovery of viroids in 1971, several viroid induced diseases have been reported throughout the world, which explained that viroid induced diseases were existed but due to lack of proper diagnostics we could not addressed the issue. In India too there were diseases of unknown etiology, which later were proven to be associated with viroids. Several viroids have been reported from India and a good amount of research have been carried out for their detection and characterization. No major viroid epidemic has happened in India, so, viroid research did not get very much attention except for a few diseases like tapping panel dryness of rubber and scar skin of apple, where economic loss have been recorded. As compared to the progress of viroid research in the world, further research efforts towards understanding the host-viroid interaction needed to be taken up in India. Viroid is a novel molecule to understand basic replication biology and hence some basic research utilizing the viroid genome could be undertaken in future.

References

Adkar-Purushothama CR, Nagaraja H, Sreenivasa MY, Sano T (2013) First report of *Coleus blumei* viroid infecting coleus in India. Plant Dis 97(1):149

Adkar-Purushothama CR, Kanchepalli PR, Yanjarappa SM, Zhang Z, Sano T (2014) Detection, distribution, and genetic diversity of Australian grapevine viroid in grapevines in India. Virus Genes 49:304–311

Antignus Y, Lachman O, Pearlsman M (2007) Spread of tomato apical stunt viroid (TASVd) in greenhouse tomato crops is associated with seed transmission and bumble bee activity. Plant Dis 91:47–50

Azad P (1993) Reduction in yield of acid limes infected with citrus corky vein virus in Assam. In: Proceedings of 12th conference, IOCV. IOCV, Riverside, pp 458

Behl MK, Parakh DB, Khurana SMP (1998) Bumpy fruit and other viroid and viroid-like diseases of apple in HP. Indian Acta Horticult 472:627–630

Bhardwaj P (2016) Molecular characterization and development of diagnostics for *Apple stem grooving virus*. PhD thesis submitted to Academy of scientific and Innovative Research (AcSIR)

Bhattiprolu SL (1991) Studies on a newly recognized disease of *Nicotiana glutinosa* of viroid etiology. Plant Dis 75:1068–1071

Bhattiprolu SL (1993) Occurrence of mosaic mottle, a viroid disease of pigeonpea (*Cajanus cajan*) in India. J Phytopathol 137:55–60

Cheema SS, Kapur SP, Dhillon RS (1982) Reaction of certain "Blood Red" stionic combinations to fungal, bacterial, virus/virus-like diseases. Punjab Hortic J 22:136–139

Cheema SS, Dhillon RS, Kapur SP (1984) Role of stionic combinations in sweet orange decline. II. Susceptibility of old line and nucellar combination to citrus diseases. J Res Punjab Agri Univ Ludhiana 21:199–202

De Bokx JA, Piron PGM (1981) Transmission of potato spindle tuber viroid by aphids. Neth J Plant Pathol 87:31–34

Diener TO (1971) Potato spindle tuber virus: a plant virus with properties of a free nucleic acid II. Subcellular location of PSTVd – RNA and the question of whether virions exist in extracts or in situ. Virology 43:75–89

Galindo J, Lopez M, Aguilar T (1986) Significance of *Myzus persicae* in the spread of tomato planta macho viroid. Fitopatol Bras 11:400–410

Ghosh DK, Mathur S, Gupta KN, Ahlawat YS, Ramachandran P (2002) Viroid infection of citrus in central India. Indian Phytopathol 55(3):290–293

Handa A, Thakur PD, Bhardwaj SV (1998) Status of Dapple apple viroid in India. Acta Hortic 472:631–634

Kapur SR, Weathers LG, Calavan BC (1974) Studies on strains of exocortis virus in citron and gymira aurantiaca. In: Weathers LG, Cohen M (eds) Proceedings of 6th conference on international Organization of Citrus Virologists. University of California, Division of Agricultural Sciences, Richmond, p 105

Kapur SR, Cheema SS, Dhillon RS (1984) Reaction of certain citrus stionic combinations to viral/mycoplasmal diseases. Indian J Hortic 41:142

Kapur SP, Cheema SS, Dhillon RS (1988) Improvement of citrus industry in Punjab: indexing and status of virus/virus-like diseases. J Res Punjab Agric Univ 25:216–220

Khurana SMP, Singh MN, Garg ID, Agarwal HP (1989) PSTV and viroid like infection in solanaceous plants in Shimla hills. In: Proceedings of Indo-US workshop on viroid and disease of uncertain etiology, 15–18 November. IARI, New Delhi, pp 8–9

Kumar A, Pandey DM, Abraham T, Mathew J, Ramachandran P, Malathi VG (2013) Determination of biotic aetiology of tapping panel dryness (TPD) syndrome of rubber tree (Hevea brasiliensis) by return-polyacrylamide gel electrophoresis (R-PAGE) technique. Arch Phytopathol Plant Protect 46(6):710–720

Kumar S, Singh L, Ram R, Zaidi AA, Hallan V (2014) Simultaneous detection of major pome fruit viruses and a viroid. Ind J Microbiol 54(2):203–210

Kumar A, Pandey DM, Abraham T, Mathew J, Jyothsna P, Ramachandran P, Malathi VG (2015) Molecular characterization of viroid associated with tapping panel dryness syndrome of Hevea brasiliensis from India. Curr Sci 108(8):1520–1527

Mali VR, Chaudhuri KG (1993) Occurrence of citrus virus and virus-like disease in Maharashtra, India. In: Proceedings of 12th conference on IOCV IOCV, Riverside, p 466

Mathur S, Garg AP, Ramachandran P (2002) Detection of viroid infecting chrysanthemum in India. Indian Phytopathol 55(4):479–482

Mishra MD, Hammond RW, Owens RA, Smith DR, Diener TO (1991) Indian bunchy top disease of tomato plants is caused by a distinct strain of citrus exocortis viroid. J Gen Virol 72:1–5

Nariani TK, Raychaudhuri SP, Sharma BC (1968) Exocortis in citrus in India. Plant Dis Rep 52:11

Owens RA, Khurana SMP, Smith DR, Singh MN, Garg ID (1992) A new mild strain of potato spindle tuber viroid isolated from wild *Solanum* spp. in India. Plant Dis 76:527–529

Pandey PK, Summanwar AS (1982) Occurrence of tomato bunchy top virus in India. Curr Sci 51:96–97

Patil BP, Warke DC (1968) A note on existence of exocortis virus in India. Curr Sci 37:469–470

Ramachandran P, Mathur S (1999) Viroid diseases of ornamental plants. In: Verma LR, Sharma RC (eds) Diseases of horticultural crops: vegetables, ornamentals and mushrooms. Indus, pp 599–632

Ramachandran P, Smith D, Hammond RW, Diener TO (1991) Studies on coleus viroid from India and USA. In: International conference on Virology in the Tropics, Lucknow, India, 2–6 December, p 230

Ramachandran P, Kumar D, Varma A, Pandey PK, Singh RP (1992) Coleus viroid in India. Curr Sci 62:271–272

Ramachandran P, Pandey PK, Ahlawat YS, Varma A, Kapur SP (1993) Viroid diseases of citrus in India. In: Proceedings of 12th international organization on Citrus Virologists, New Delhi, India, 23–27 November, pp 438–440

Ramachandran P, Ahlawat YS, Varma A (1996) Viroid – The potential plant pathogens. In: 2nd international crop science congress, New Delhi, India

Ramachandran P, Mathur S, Francis L, Varma A (2000) Evidence for association of a viroid with tapping panel dryness syndrome of rubber (*Hevea brasiliensis*). Plant Dis 84(10):1555

Ramachandran P, Roy A, Mathur S, Agarwal J, Ahlawat YS (2003) Diagnostic for *Citrus exocortis* and *Hop stunt* viroids associated with yellow corky vein disease in citrus. Indian Phytopathol 56(4):428–433

Ramachandran P, Agarwal J, Roy A, Ghosh DK, Das DR, Ahlawat YS (2005) First record of a *Hop stunt viroid* variant on Nagpur mandarin and Mosambi sweet orange trees on rough lemon and Rangpur lime rootstocks. Plant Pathol 54(4):571

Ramachandran P, Varma A, Ahlawat YS, Mathew J, Sethuraj MR, Mathew NM, Abraham T, Philip S, Jacob CK (2006) Detection of a low molecular weight viroid-like RNA in Hevea brasiliensis. In: Jacob J, Krishnakumar R, Mathew NM (eds) Tapping panel dryness of rubber trees. Rubber Research Institute of India, Kottayam, pp 183–188

Reddy MRS, Naidu PH (1989) Transmission of citrus yellow corky vein. Indian Phytopathol 42:455–456

Reddy GS, Dakshinamurti V, Reddy VRK (1974) Yellow-corky vein: first report of a new graft – transmissible disorder of 'Sathgudi' in Andhra Pradesh. Indian Phytopathol 27:82–84

Roy A, Ramachandran P (2003) Occurrence of a *Hop stunt viroid* (HSVd) variant in yellow corky vein disease of citrus in India. Curr Sci 85:1608–1612

Roy A, Ramachandran P (2006) Characterization of a *Citrus exocortis viroid* variant in yellow corky vein disease of citrus in India. Curr Sci 91:798–803

Roy A, Ramachandran P (2008) Development of viroid free nucleus material for rootstock of citrus: an approach for management of yellow corky vein disease. Indian Phytopathol 61(2):259–263

Rustem F (1998) Investigations on the citrus yellow corky vein disease to establish its etiology. PhD thesis, Division of Plant Pathology, IARI, New Delhi

Rustem FA, Ahlawat YS (1999) Yellow corky vein – a viroid disease of citrus in India. Ind J Agric Sci 69(11):789–794

Rustem F, Ramachandran P, Ahlawat YS (1997) Association of a viroid with yellow vein disease of citrus in India. In: International conference on integrated plant diseases, Management for Sustainable Agriculture, New Delhi

Rustem FA, Ramachandran P, Ahlawat YS (2000) Association of viroid with yellow vein disease of citrus in India. In: Proceedings of Indian phytopathological society – Golden Jubilee, international conference on integrated plant disease management for sustainable agriculture, pp 778 779

Sahana AB, Adkar-Purushothama CR, Chennappa G, Zhang ZX, Sreenivasa MY, Sano T (2013) First report of grapevine yellow speckle viroid-1 and Hop stunt viroid infecting grapevines (*Vitis vinifera*) in India. Plant Dis 97:1517

Saraswati J, Mishra MD (1989) Bunchy top of tomato – a viroid disease. In: Proceedings of 41st annual meeting Indian phytopathological Society, 28 February to 02 March. IARI, New Delhi (Abst.), p 278

Sharma DC, Pandey PK (1983) Leaf yellow mid vein, a previously undescribed virus disease of citrus. Int J Tropical Pl Dis 1:149–152

Sharma SR, Saxena AK (1989) Effect of some stionic combinations and host nutrition on viral/mycoplasmal diseases in citrus. J Turkish Phytopathol 18:17–19

Sharples S (1936) Brown bast. In: Diseases and pests of the rubber tree. MacMillan, London, pp 229–265

Singh BP, Gupta RP, Abidi SMH, Raizada RK (1978) Chlorotic mottle of chrysanthemum morifoliiim (Ram) Hemsl. A. new record from India. Curr Sci 47:913

Singh RP, Dilworth AD, Baranwal VK, Gupta KN (2006) Detection of citrus exocortis viroid, iresine viroid, and tomato chlorotic dwarf viroid in new ornamental host plants in India. Plant Dis 90(11):1457

Syller J, Marczewski S (2001) Potato leafroll virus-assisted aphid transmission of potato spindle tuber viroid to potato leafroll virus-resistant potato. J Phytopathol 149:195–201

Thakur PD, Ito T, Sharma JN (1995) Natural occurrence of viroid disease of apple in India. Indian J Virol 11(2):73–75

Walia Y, Kumar Y, Rana T, Bhardwaj P, Ram R, Sharma U, Thakur PD, Hallan V, Zaidi AA (2009) Molecular characterization and variability analysis of Apple scar skin viroid in India. J Gen Plant Pathol 75:307–311

Walia Y, Dhir S, Bhadoria S, Hallan V, Zaidi AA (2012) Molecular characterization of apple scar skin viroid from Himalayan wild cherry. For Pathol 42:84–87

Walia Y, Dhir S, Ram R, Zaidi AA, Hallan V (2014) Identification of the herbaceous host range of Apple scar skin viroid and analysis of its progeny variants. Plant Pathol 63:684–690

Walia Y, Dhir S, Zaidi AA, Hallan V (2015) *Apple scar skin viroid* naked RNA is actively transmitted by the whitefly *Trialeurodes vaporariorum.* RNA Biol 12:1–8

Werner HO (1924) Relation of environment to spindle tuber symptoms. Proc Potato Assoc AM Annu 11:102–106

Walia Y (2016) Apple scar skin viroid: infectivity and identification of interacting protein. Ph.D. thesis. CSIR-IHBT, Palampur, p 142

Part II

Virus-Vectors

Aphids as Vectors of the Plant Viruses in India

23

Amalendu Ghosh, Samiran Chakrabarti, Bikash Mandal, and N.K. Krishna Kumar

Abstract

Aphids are important group of vectors of plant viruses, where all three modes of transmission *viz.* non-persistent, semi-persistent and persistent are observed. In India, exploration of aphid species in vector-virus study was started during 1945 and subsequently, several aphid species have been identified as vectors of many important plant viruses. The studies in India was focused in identifying vectors, recording transmission efficiency, determining transmission parameters, recording distribution, and seasonal incidence and developing management options. A summary of these studies conducted in India so far is presented in this chapter.

Keywords

Aphids • Virus • Transmission • Biology • Lifecycle • Virus-vector relationships

A. Ghosh (✉) • B. Mandal
Advanced Centre for Plant Virology, Indian Agricultural Research Institute, New Delhi 110012, India
e-mail: amal4ento@gmail.com

S. Chakrabarti
Department of Zoology, Vidyasagar College, 39 Sankar Ghosh lane, Kolkata 700 006, India

N.K. Krishna Kumar
Bioversity International, National Academy of Agricultural Sciences Complex, New Delhi 110012, India

B. Mandal et al. (eds.), *A Century of Plant Virology in India*,
https://doi.org/10.1007/978-981-10-5672-7_23

23.1 Introduction

Among the 29 orders of insects, seven orders contain vectors of plant viruses. The order Hemiptera includes plant-sucking insects with piercing and sucking type of mouthparts and it contains the highest number of reported insect vectors. In this insect order, aphids, adelgids and phylloxerids are important groups of insects that are known as vectors of many plant viruses. The aphids evolved with a siphuncular pore, although this character has been secondarily lost in some species and other categories. Adelgids and phylloxerids are oviparous and have distinct ovipositors that are lacking in aphids. There exist controversies regarding the higher taxonomic categories of the above extant insect groups. Many of the Aphidologists (Blackman and Eastop 1994; Remaudiere and Remaudiere 1997; Quednau 2010; Nieto Nafria et al. 2011) included these Hemipteran insect groups in one superfamily Aphidoidea and in three different families viz., aphids (family Aphididae), adelgids (family Adelgidae) and phylloxerids (family Phylloxeridae). The family Aphididae is comprised of about 27 subfamilies. Later, Heie and Wegierek (2009) advocated the idea of considering the above insects in three superfamilies *viz*., Aphidoidea, Adelgoidea and Phylloxeroidea and the superfamily Aphidoidea with 16 families including fossil aphids.

Parthenogenetic reproduction, viviparity, polymorphism, host alternation, wide host range in some species, possession of needle-like mouth parts have helped aphids to evolve as the most significant group of insects to exploit higher plants as their food sources and also as one of the most important vectors of plant viruses. The mouthparts of aphids have been modified to form two pairs of stylets containing the food canal and salivary canal held within a groove of modified labium or rostrum. The piercing and sucking type of feeding are accompanied by the modified salivary pump, cibarial pump, and valve. Further, the specialized reproductive behaviour significantly helps them to evolve as a successful vector. Among the 247 viral diseases of plants described by Kennedy et al. (1962), 164 are transmitted by about 200 species of aphids. In India, 31 aphid species have been recorded to transmit around 24 plant viruses of different genera. In this chapter, the information available from the experiments conducted in India on plant virus transmission by aphids have been discussed.

23.2 Host, Biology and Life Cycle of Aphids

Aphids have a peculiar reproductive behaviour of several parthenogenetic generations between each period of sexual reproduction. In fact, there are no known aphid species which have lost the parthenogenetic phase of the life cycle, although it may be shortened to two or three generations. Whereas, loss of sexual generation is a widespread phenomenon and very often it is specific to the species. The loss of sexual reproduction is a dynamic evolutionary process (Tomiuk et al. 1994). The high rate of parthenogenetic reproduction gives them an advantage over sexual reproduction to establish a colony in new habitat. Although, the presence of both sexual and parthenogenetic virginoparae is common in aphids in the temperate region, but the majority of aphids in tropics have only parthenogenetic development.

The bio-geographical conditions of the host have great influence on the life cycle pattern of aphids. Individual species depicts varied life cycle patterns in different bio-geographical conditions (Moran 1992). Some species lead different patterns of life cycles in the tropics and temperate climate (Blackman 1974; Heie 1980). Sexual morphs are produced mainly in the autumn and the oviparous females lay the over-wintering eggs after mating. In the following year, when plants resume growth, the eggs hatch and a series of parthenogenetic generations develop. However, two distinct and different types of life cycles are commonly encountered in aphids, namely, anholocyclic and holocyclic. In anholocyclic mode, the aphids reproduce through parthenogenetic viviparous mode and lay only nymphs. The nymphs also reproduce by parthenogenetic mode subsequently. Whereas in holocyclic mode, the parthenogenetic mode, may be of several generations, is intervened by a sexual phase (production of gametes and laying of fertilized eggs) (Fig. 23.1). It has been observed that some aphid species have an only parthenogenetic mode of reproduction for several years in the laboratory conditions.

Aphids being obligatory plant parasites, association with a particular host or several hosts is an important factor that regulates their biological activities. About 40% of aphid species are monophagous *i.e.* living on a single host plant species or on its congeners. Some are extremely polyphagous which feed on a wide variety of plant species, while some are oligophagous which feed on plants within a genus or species of closely related plant genera. Many plant groups although infested by polyphagous aphids do not have any specific aphid species of their own. Many aphid species are heteroecious demonostrating host alternating behaviour between a woody tree (primary host) and a herbaceous plant (secondary host), while the autoecious feed on the same host plant species throughout the life. This host alternating behaviour mixes with reproductive behaviour of aphids and the life cycles may be autoecious anholocyclic, autoecious holocyclic, heteroecious anholocyclic, and heteroecious holocyclic. Host alternation, although risky, is advantageous to the aphids by avoiding adverse conditions and it helps in optimizing resource utilization. Such an ability to use two hosts also improves the fitness of aphids (Mackenzie and Dixon 1991; Chakrabarti 2007). Host alternation is an eventuality of the historical constraint of aphids (Moran 1992). However, adaptation to a particular host by a specific morph leads to a dead end and a complete switching on to a more favourable secondary host become impossible. Aphids, particularly gall-inducing aphids display seemingly endless complexity and variation in another segment of their life (Wool 2004; Chakrabarti 2007). Among the different genera of plants, Asteridae have more than 800 specific aphids followed by Rosidae, Hamamelidae, Liliatae and Dilenidae. Different Conifer species are recorded as hosts of about 100 aphid species.

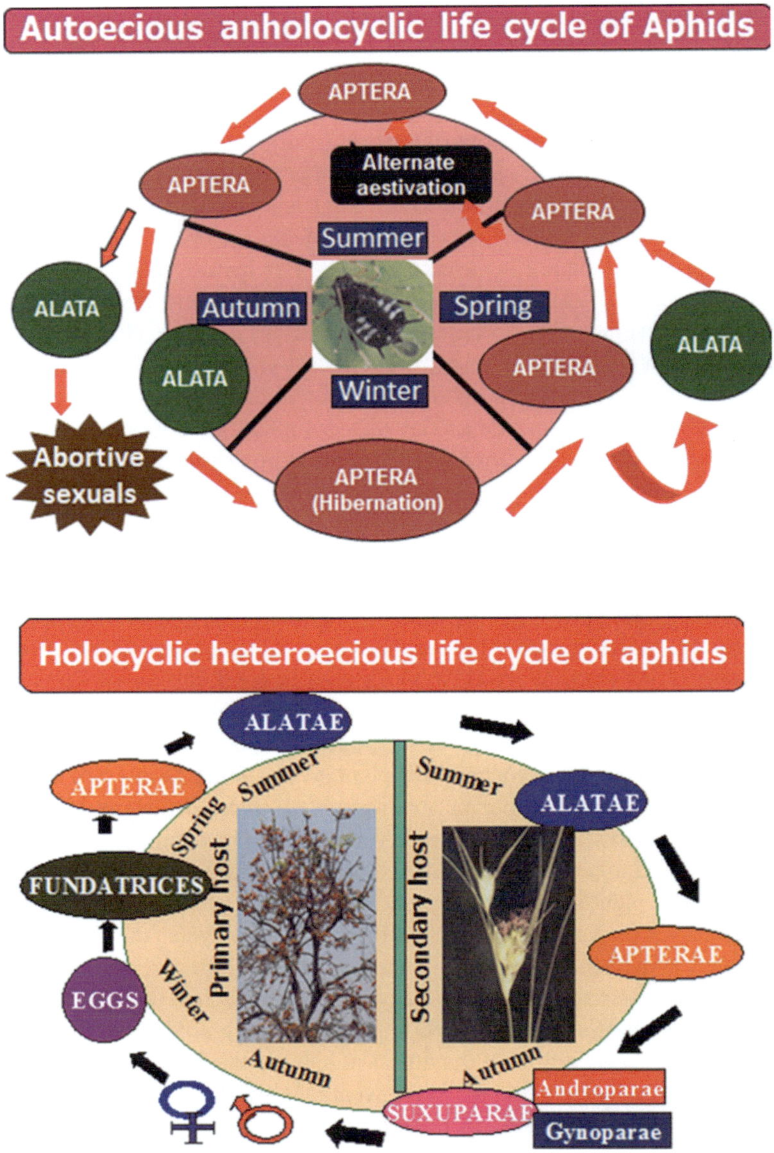

Fig. 23.1 Two types of lifecycles of aphids. In holocyclic life cycle, aphids alternate between two hosts and overwinter as eggs. The holocyclic life cycle is found in *Brachycaudus helichrysi*, *Rhopalosiphum maidis*, *Myzus persicae*, *Rhopalosiphum rufiabdominalis*. In anholocyclic life cycle, aphids overwinter as apterae. *Aphis craccivora*, *Brevicoryne brassicae*, *Pentalonia nigronervosa*, *Lipaphis erysimi* follow anholocyclic life cycle. In Indian conditions, almost all species found in plains have only anholocyclic life cycle.

23.3 Aphids in India

Worldwide little over 5000 aphid species are known (Remaudiere and Remaudiere 1997; Nieto Nafria et al. 2011, http://aphid.speciesfile.org). In the Indian subcontinent, 825 aphid species of 223 genera under 19 subfamilies are known till 2014 (Chakrabarti 2009; Chakrabarti and Das 2014) and out of these 818 species of 216 genera are found in the Himalayan region (Chakrabarti 2015). Only about 43 species are occasionally abundant in different plains in India. Sexual forms of about 120 species (Basu and Raychaudhuri 1980 and subsequent reports) have so far been recorded, mostly from the hilly terrains of the Himalaya. Sexual forms of only a few species namely *Aphis craccivora* Koch, *Aphis gossypii* Glover, *Myzus persicae* (Sulzer) and *Pentalonia nigronervosa* Cocqurell have been reported from the plain in winter season but it is still not confirmed whether these species really lead a holocyclic life in the plains at all.

Aphids exhibit extreme polymorphism and polyphenism and exist in different forms. One species may have different clones or races with or without distinct morphological separations. This variability in life cycle influences their efficiency in transmitting plants viruses. In the context of seasonal incidence of aphids, the pattern in the plains (tropical climate) and in the hills are different. In the plains and at lower altitudes in the hills, aphid species diversity, as well as density, is more during late winter to early spring. However, phenomena change with the elevations. At higher altitudes, aphid diversity and density gradually increase from spring till the advent of winter with a low profile during late summer and rainy season. From late winter to early spring most of the species hibernate either as egg stage or in some other stage (in the case of anholocyclic species). The later pattern is in conformity with the aphids found in the temperate region. Among aphid species recorded from India so far, less than 100 species of aphids have economic importance as crop pests.

23.4 Aphids as vectors of Plant Viruses in India

The earliest record of aphid-transmitted viral disease in India is small cardamom Katte (mosaic) disease (Uppal et al. 1945). In the early era of plant virology in India, the relationships of insect vector and plant virus was not intensely studied. The aphids as vectors of plant viruses was studied only for a few selected aphid species like *Aphis gossypii* and *Myzus persicae* due to their widespread prevalence and convenience in rearing. Hence, only a small portion of possible aphid-virus relationship could be studied. Moreover, the negative results in these type of studies are often not reported. More than eighty reports on aphid transmission of plant viruses are available in India. Maximum numbers of reports are found during 1960s and early 2000s (Fig. 23.2). Eight genera of the plant viruses *viz. Potyvirus, Macluravirus, Babuvirus, Bymovirus, Luteovirus, Cucumovirus, Comovirus*, and *Closterovirus* are transmitted by aphids in India (Table 23.1). Maximum number of plant viruses transmitted by aphids are potyviruses (12) followed by macluraviruses, babuviruses

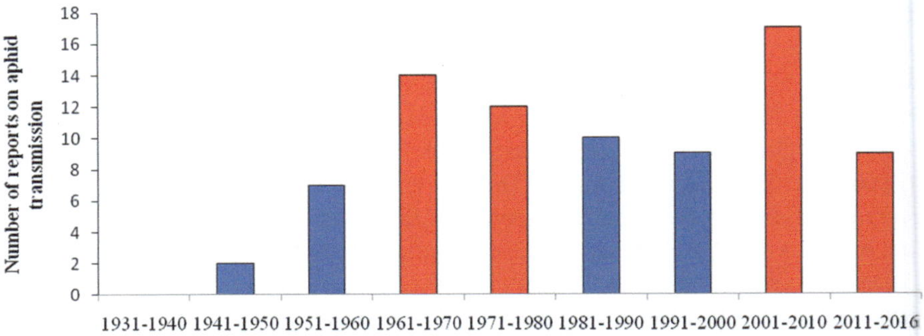

Fig. 23.2 Trend of publications in India on aphid transmission of plant viruses. Maximum numbers of reports are made during 1961–1980 and 2001–2016 which are marked by *red colour bar*

and luteoviruses. *M. persicae* followed by *A. gossypii* and *A. craccivora* are the most widely utilised aphid species for vector-virus study in India. *M. persicae* transmits about 16 plant viruses in different agriculturally important crops. Whereas, records of 15 and 12 plant viruses transmitted by *A. gossypii* and *A. craccivora*, respectively are available. The scientific names of few enlisted aphid species have been revised with time. Such changes in the genus or species level are mentioned within the parenthesis. A phylogenetic relationship among the aphid vector species has been shown in Fig. 23.2. Six hundred fifty-eight nucleotide sequence from cytochrome oxidase subunit I region was considered for this purpose as COI gene sequences are used as a main molecular identification tool for insects (Hebert et al. 2003; Shneyer 2007). The tree has been generated based on the available sequence of the aphid species in NCBI database using MEGA 6.0 software. *Spodoptera mauritia* (order- Lepidoptera) has been taken as an outgroup. COI sequences of three aphid species are not available and hence not considered. Although this tree does not have any connection with virus transmission, the analysis shows the phylogenetic relationship of the aphid vectors (Figs. 23.3).

23.5 Types of Aphid-Plant Virus Relationships

Relationship of aphids with plant viruses has been widely studied and applicable to other arthropod vectors. All the four types of transmission *viz.* non-persistent, semi-persistent, persistent-circulative and persistent-propagative are observed in aphids. The available information on the works conducted in India have been described here based on the type of transmission of the viruses.

23.5.1 Non-persistent Virus Transmission

Non-persistent virus transmission, characterized by very short acquisition and inoculation time, is also known as stylet-borne. In this type of transmission, aphid stylet

Table 23.1 List of aphids transmitting plant viruses in India

Sl no	Name of the virus	Aphid species transmitting the virus	Type of transmission	References
1.	Cardamom mosaic virus	*Pentalonia nigronervosa*	Non-persistent	Uppal et al. (1945), Varma and Capoor (1958), and Rao and Naidu (1974)
		Aphis craccivora		
		Aphis gossypii		
		Aphis nerii		
		Aphis rumicis		
		Brachycaudus helichrysi		
		Greenidia artocarpi		
		Macrosiphum pisi (= *Acyrthosiphon pisum*) *Macrosiphum rosaeformis*		
		Macrosiphum sonchi (= *Uroleucon sonchi*) *Schizaphis cyperi* (= *Toxoptera cyperi*)		
		Schizaphis graminum		
		Pentalonia nigronervosa typica		
		Pentalonia nigronervosa caladii		
2.	Large cardamom chirke virus	*Rhopalosiphum maidis* *Brachycaudus helichrysi*	Non-persistent	Raychaudhuri and Chatterjee (1958, 1961), Raychaudhuri and Ganguly (1965), Ganguly et al. (1970), Capoor (1967), Mandal et al. (2012), and Ghosh et al. (2015a, 2016a, b)
		Rhopalosiphum padi		
		Sitobion avenae		
		Myzus persicae		
		Pentalonia nigronervosa		
3.	Sugarcane mosaic virus	*Rhopalosiphum maidis*	Non-persistent	Seth et al. 1972a), Kondaiah and Nayudu (1984), and Singh et al. (2005a, b)
		Melanaphis sacchari		
		Aphis gossypii		
		Longiunguis sacchari (= *Melanaphis sacchari*)		
		Myzus persicae		
4.	Barley mosaic virus	*Rhopalosiphum maidis*	Non-persistent	Dhanraj and Raychaudhuri (1969)
5.	Potato virus Y	*Myzus persicae*	Non-persistent	Singh et al. (1981)
		Aphis gossypii		
6.	Watermelon mosaic virus	*Myzus persicae*	Non-persistent	Singh (1970) and Raychaudhuri and Varma (1977)
7.	Turnip mosaic virus	*Myzus persicae*	Non-persistent	Ahlawat and Chenulu (1982) and Devi et al. (2004)
		Brevicoryne brassicae		
		Lipaphis erysimi		
8.	Dasheen mosaic virus	*Myzus persicae*	Non-persistent	Babu et al. (2011)
		Aphis gossypii		

(continued)

Table 23.1 (continued)

Sl no	Name of the virus	Aphid species transmitting the virus	Type of transmission	References
9.	Onion yellow dwarf virus	*Aphis gossypii* *Myzus persicae* *Aphis craccivora*	Non-persistent	Hoa et al. (2003)
10.	Pepper vein banding virus	*Aphis gossypii* *Myzus persicae*	Non-persistent	Nagaraju (1981)
11.	Pepper veinal mottle virus	*Aphis gossypii* *Myzus persicae*	Non-persistent	Nagaraju and Reddy (1980)
12.	Bean common mosaic virus was observed	*Aphis craccivora* *Aphis fabae solanella* *Aphis gossypii* *Myzus persicae* *Lipaphis erysimi*	Non-persistent	Muniyappa (1976)
13.	Sunflower mosaic virus	*Aphis gossypii* *Aphis craccivora* *Myzus persicae*	Non-persistent	Nagaraju et al. (1997)
14.	Garlic mosaic virus	*Aphis craccivora* *Myzus persicae*	Non-persistent	Ghosh and Ahlawat (1997)
15.	Cucumber mosaic virus	*Myzus persicae* *Myzus ornatus* *Macrosiphum euphorbiae* *Aphis gossypii* *Myzus ascalonicus* *Acyrthosiphon pisum* *Aphis craccivora,* *Aphis citricola* (= *Aphis spiraecola*) *Macrosiphum euphorbiae*	Non-persistent	Bhargava (1951), Badami (1958), Capoor and Varma (1970), Shukla and Rao (2000), and Singh (2005)
16.	Papaya ringspot virus	*Myzus persicae* *Aphis gossypii* *Aphis craccivora* *Myzus nicotianae* (= *M. persicae*)	Non-persistent	Capoor and Varma (1948, 1958) and Kalleshwaraswamy and Krishna Kumar (2008)
17.	Cowpea mosaic virus	*Aphis craccivora* *Aphis gossypii* *Aphis evonymi* (= *Aphis fabae solanella*) *Myzus persicae*	Non-persistent	Haque and Chenelu (1967)
18.	Banana bract mosaic virus	*Rhopalosipum maidis* *Aphis gossypii* *Pentalonia nigronervosa* *Aphis craccivora*	Non-persistent	Selvarajan et al. (2006)
19.	Urdbean leaf crinckle virus	*Aphis craccivora* *Aphis gossypii*	Non-persistent	Dhingra (1975)

(continued)

Table 23.1 (continued)

Sl no	Name of the virus	Aphid species transmitting the virus	Type of transmission	References
20.	Citrus tirsteza virus	*Toxoptera citricida*	Semi-persistent	Vasudeva et al. (1959), Capoor (1975), Varma et al. (1960, 1965), Nariani and Singh (1971), and Ghosh et al. (2014a, b, 2015b)
		Aphis gossypii		
		Myzus persicae		
		Aphis craccivora		
		Dactynotus jaceae (= *Uroleucon jaceae*)		
		Aphis craccivora		
		Toxoptera aurantii		
		Macrosiphum compositae (= *Uroleucon compositae*)		
21.	Banana bunchytop virus	*Pentalonia nigronervosa*	Persistent-circulative	Capoor (1969)
22.	Cardamom bushy dwarf virus	*Pentalonia nigronervosa*	Persistent-circulative	Varma and Capoor (1964), Basu and Ganguly (1968), and Ghosh et al. (2015a, 2016b)
		Micromyzus kalimpongensis		
23.	Barley yellow dwarf virus	*R. maidis*	Persistent-circulative	Nagaich and Vashisth (1963)
		R. padi		
		Sitobion avenae		
24.	Chickpea stunt disease associated virus	*Aphis craccivora*	Persistent-circulative	Reddy and Lava (2004)
		Myzus persicae		

does not usually penetrate beyond the epidermal cells. As the acquisition period is very brief, the vectors may not be colonizing the host plant. The aphids start losing the infectivity immediately after acquisition. Economically important plant viruses belonging to the genera *Cucumovirus*, *Fabavirus, Macluravirus, Alfamovirus,* and *Potyvirus* are transmitted by aphids in a non-persistent manner.

23.5.1.1 *Macluravirus*

In India, among the available information, the first experimental proof of non-persistent way of aphid transmission was demonstrated with small cardamom mosaic disease by *P. nigronervosa* (Uppal et al. 1945). Intensive work on the relationship of the small cardamom mosaic virus and *P. nigronervosa* was worked out by Varma and Capoor (1958). A negative correlation was recorded between the population of *P. nigronervosa* and spread of the disease. The highest spread of the disease was observed with a low population of the aphids indicating that the spread of the disease was more associated with the aphid's activity than its population (Deshpande et al. 1972). Thirteen aphid species namely, *A. craccivora* Koch, *A. gossypii* Grover, *A. nerii* B.D.F., *A. rumicis Linn.*, *Brachycaudus helichrysi* (Kaltb.), *Greenidia artocarpi* (Westwood), *Macrosiphum pisi* (= *Acyrthosiphon pisum*

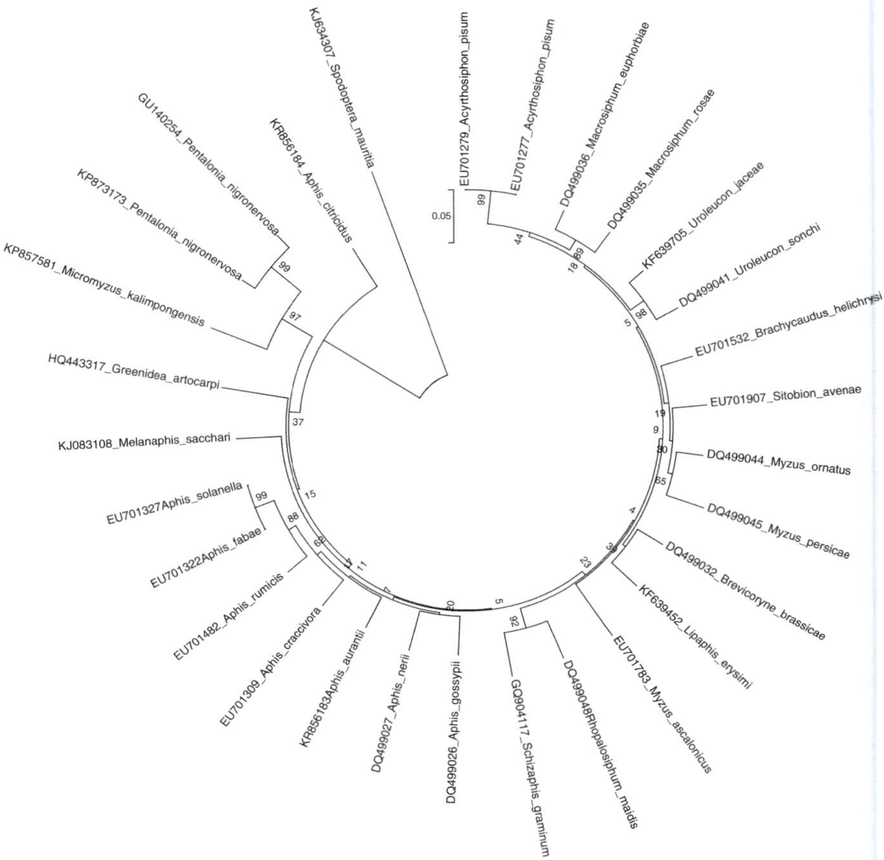

Fig. 23.3 Phylogenetic relationship of the aphid vectors reported in India based on 658 nucleotides of COI gene. *Spodopteramauritia* (order- Lepidoptera) has been taken as outgroup

Harris), *M. rosaeformis* (Das), *M. sonchi* (= *Uroleucon sonchi* Linn.), *Schizaphis cyperi* = *S. rotundiventris* (Signoret), *S. graminum* (Rondani), *P. nigronervosa typica* and *P. nigronervosa caladii* v.d.G. were reported to transmit 'katte' virus (Rao and Naidu 1974). Earlier it was thought that the aphids of cardamom and banana were same. However, it was confirmed that *P. nigronervosa typica* was found to colonize banana (Siddappaji and Reddy 1972; Venugopal 1995). All the four nymphal instars and adult of *P. nigronervosa caladii* were found capable of transmitting this disease. The rate of transmission of the virus showed a positive correlation with increased nymphal age. A range of 26–36 overlapping generations in a year in the conditions of Coorg, Karnataka was reported (Rajan 1981). Mere probing of the aphids was sufficient in the transmission of the virus (Rajan et al. 1989). Relationship of large cardamom chirke virus (LCCV) and aphid species was extensively studied at IARI, Kalimpong station. It was found that virus was non-persistently transmitted by the aphids, *Rhopalosiphum maidis* (Fitch) (Raychaudhuri and Chatterjee 1958, 1961) and *B. helichrysi* (Raychaudhuri and Ganguly 1965). The virus was also transmissible to wheat. Transmission of the virus by the aphid from cardamom to cardamom and wheat to cardamom was extremely poor in comparison to wheat to wheat, where it was far more effective (Ganguly et al. 1970). Capoor (1967) has also mentioned the spread of chirke disease in fields by different aphid species, *R. padi* (Linn.) and *Sitobion avenae* (Fab.) in addition to *R. maidis* and *B. helichrysi*. In recent years, the transmission of LCCV was reported with *R. maidis* and *M. persicae*. A single aphid required 5 min of acquisition feeding period to transmit the virus. Large cardamom cultivar, Varlangey produced typical chirke symptoms in 4 out of 20 seedlings inoculated with *M. persicae* at 173 days post inoculation and 6 of 20 seedlings inoculated with *R. maidis* at 139 days post inoculation (Mandal et al. 2012). In a recent study, all the aphid species infesting large cardamom in Sikkim and Darjeeling hills were collected and identified based on morphological and molecular evidence (Ghosh et al. 2016a). Transmission of LCCV at a rate of 3.7% was observed with *P. nigronervosa*. No natural occurrence of *R. maidis* and *B. helichrysi* in large cardamom was recorded during the study (Ghosh et al. 2015a, 2016a).

23.5.1.2 *Cucumovirus*

Different strains of cucumber mosaic virus (CMV) were observed to differ in transmissibility by aphids. *Myzus persicae* was recorded as more efficient vector than *M. ornatus* Laing followed by *Macrosiphum euphorbiae* Thomas (Bhargava 1951). A change in transmissibility of spinach strain of CMV by aphid species was first noticed in India during 1958 (Badami 1958). The strain was earlier readily transmitted by *M. persicae*, *A. gossypii* and *M. ascalonicus* Doncaster, but afterward became non-transmissible by *M. persicae*. *A. gossypii* was also recorded as a vector of CMV in banana (Capoor and Varma 1970). *M. persicae* was observed to transmit the CMV-tomato, CMV-brinjal and CMV-chilli isolate in a non-persistent manner (Kiranmai et al. 1997). During the early twenty-first century, *A. pisum* was added as a vector of CMV in banana (Shukla and Rao 2000). Two hours of pre-acquisition fasting accompanied by 2 min acquisition threshold was found to increase the transmission efficiency of CMV by *A. pisum* in banana. Another vector transmission study of CMV infecting banana revealed that *Aphis craccivora* and *A. gossypii* could efficiently transmit the virus in a non-persistent manner within 20 min of acquisition followed by 10–15 min of inoculation feeding (Dheepa and Paranjothi 2010). A cowpea strain of CMV was found non-persistently transmissible by *A. craccivora* and *M. persicae* (Nagaraju and Keshava Murthy 1994). Transmission of CMV in salvia plants was observed with *A. gossypii* and *M. persicae* (Ali et al. 2012). Broad specificity of CMV for aphid transmission was experimentally shown by Singh (2005). Several aphid species *viz. A. craccivora, A. gossypii, A. citricola* (=*spiraecola* Patch), *M. euphorbiae* and *M. persicae* were found efficient in transmitting CMV. Globally all the CMV strains were now divided into two subgroups I and II and more than 80 aphid species were recorded to transmit the different strains in a non-persistent manner (Palukaitis et al. 1992).

23.5.1.3 *Potyvirus*

During 1960, aphid transmission of the maize mosaic virus was first observed by Chona and Seth (1960). Later it was identified as sugarcane mosaic virus (SCMV). A mosaic disease of pearl millet caused by SCMV was transmitted by *R. maidis* (Seth et al. 1972). Later on, the differential transmission of SCMV by *R. maidis* and *Melanaphis sacchari* (Zehntner) was described by Kondaiah and Nayudu (1984). *M. sacchari* was found to be an efficient vector to transmit H-strain of SCMV in sorghum. An elaborated work on SCMV- aphid vector relationship was carried out by Singh et al. (2005). They confirmed the varying efficiency of different aphid species (*R. maidis, M. persicae, A. gossypii*, and *Longiunguis sacchari*) in the transmission of SCMV. The longer fasting was found to reduce the transmission efficiency of *R. maidis*. Maximum transmission of SCMV by *R. maidis* was recorded with 2 min of optimum acquisition and 30 min of inculation period. The efficiency of *R. maidis* was also tested for transmission of barley mosaic virus in India (Dhanraj and Raychaudhuri 1969). Transmission of papaya ringspot virus (PRSV) by several aphid species was demonstrated by Capoor and Varma (1948, 1958) and Reddy et al. (2007). Further, the non-persistent transmission of PRSV by aphid species was confirmed by Bhargava and Khurana (1970). PRSV-p infesting gherkin was recorded in Bangalore. The virus transmission studies involving *A. gossypii, A. craccivora, M. persicae*, and *M. nicotianae* Blackman (= *M. persicae*) showed that *M. persicae* was an efficient vector of PRSV-p (Rashmi et al. 2005). A strain of PRSV was identified to infect various cucurbitaceous crops in Bangalore. The strain was transmitted by three aphid species, namely *M. persicae, A. craccivora* and *A. gossypii. M. persicae* was found to be the most efficient vector (Gude et al. 2008). A comprehensive study on the relationship of PRSV and its aphid vectors was undertaken by Kalleshwaraswamy and Krishna Kumar (2008). In single aphid inoculation test, *M. persicae* was observed to transmit PRSV at a rate of 56% followed by *A. gossypii* (53%) and *A. craccivora* (38%). Transmission efficiency of PRSV by all three aphid species increased maximum up to one hundred percent when a batch of five aphids per plant was released. A significant decrease in PRSV transmission rate was recorded with increased sequestration periods in all three species of aphids. Significantly higher transmission occurred when aphids were given inoculation access period immediately after acquisition (Kalleshwaraswamy and Krishna Kumar 2008). In a similar study, five aphid species *viz. A. craccivora, A. gossypii, A. citricola (= spiraecola), M. persicae* and *R. maidis*, were evaluated for transmission of PRSV. *M. persicae* was found as the most efficient vector of PRSV. Transmission efficiency of *M. persicae* was declined after 4 h of pre-acquisition fasting. *M. persicae* could acquire the virus in just 30 s of acquisition feeding with an optimum feeding of 3 min. The transmission efficiency of *M. persicae* showed a decline with a further increase in acquisition feeding. *M. persicae* could readily transmit PRSV after 2 min of inoculation feeding with an optimum transmission after 6 min of inoculation feeding (Singh and Singh 2010). The mutual relation of PRSV and aphid vectors has been taken into account and one model has been established to predict the PRSV infection based on the incidence of aphid vectors and weather factors (Kalleshwaraswamy et al. 2007). Management of PRSV by deterring the aphid vectors from inoculum was evaluated. Timely application of imidacloprid, mineral oil, and deltamethrin was found effective in delaying the PRSV infection by aphids (Kalleshwaraswamy et al. 2009). This might be possible due to the reduced incidence of aphid vectors post insecticide application. Management of chilli mosaic virus (potato virus Y, PVY) transmitted by *A. craccivora* and *M. persicae* by application of neem oil at the rate of 3% and 5% was reported by Mariappan and Samuel (1993). Transmission of PVY in potato by *M. persicae* and *A. gossypii* was reported by Singh et al. (1981). The effect of temperature, relative humidity and light on aphid transmission of PVY and potato leaf roll virus (PLRV) was studied using *M. persicae* and *A. gossypii* as vectors by Singh et al. (1988). They observed that host susceptibility was enhanced by 48 h pre-inoculation exposure at 25 °C and by 48 h post-inoculation exposure at 30 °C coupled with high relative humidity. Continuous fluorescent light at 4000 lux did not demonstrate any alterations in the rate of transmission of the viruses. Increased rate of virus transmission (30–35%) was recorded with high relative humidity (80–90%) in a combination of high temperature (25–30 °C). Study on the relationship of the watermelon mosaic virus (WMV) and its vector, *M. persicae* revealed that the virus was transmitted apparently in a typical 'non-persistent' manner and was 'stylet-borne'. A single aphid was found capable of transmitting the virus. Preliminary fasting of the aphids increased the efficiency of transmission. The infectivity of the vector was lost after 2 h starvation and the percentage of infection decreased with the increase in starvation time. Both alate and apterous forms of the aphids were found to be equally efficient in transmitting WMV (Singh 1970). Transmission factors of marrow mosaic virus, a strain of WMV was studied by Raychaudhuri and Varma (1977). They reported the increase of transmission rate of *M. persicae* with pre-acquisition starving. The minimum time for the acquisition was determined as 30 s to a maximum of 2 min. Single aphid was able to transmit the virus whereas, 100% transmission could achievable with five aphids per plant. Spraying of the marrow plants with 1–2% emulsion of castor, groundnut, and paraffin oil provided good control of the disease. Paraffin at a rate of 2% was found to give the best protection up to 3 days and 90% up to 7 days (Raychaudhuri and Varma 1983). Transmission of pepper vein banding virus on bell pepper by two aphid species, *A. gossypii* and *M. persicae* was observed at a rate of 62% and 75% (Nagaraju 1981). Transmission of pepper veinal mottle virus was also recorded by these two aphid species (Nagaraju and Reddy 1980). A mosaic disease of radish was transmitted by several aphid species and *Lipaphis erysimi* (Kaltenbach) was found to be the most efficient vector (Ahlawat and Chenulu 1982). The virus was later identified as turnip mosaic virus (TuMV). Three aphid species which colonized on the leaf of mustard plants were tested for transmission of TuMV. *M. persicae* was found to be the effective vector for the transmission of the virus (80.66%) followed by *Brevicoryne brassicae* (Linn.) (51.72%) and *Lipaphis erysimi* (29.59%) (Devi et al. 2004). Transmission of bean common mosaic virus by several aphid species like *A. craccivora, A. fabae solanella* Theobald, *A. gossypii, L. erysimi* and *M. persicae* was observed by Muniyappa (1976). Likewise, a mosaic disease was reported in sunflower crop from Karnataka (Nagaraju et al. 1997). They experimentally recorded 20–24%, 14–16% and 20–25% transmission of the virus by *A. gossypii, A.*

craccivora and *M. persicae,* respectively. Further, pre-acquisition starvation of *A. craccivora* was found essential for successful transmission of sunflower mosaic virus. The optimum acquisition and inoculation feeding periods in this relationship were found to be 10–20 and 15 min, respectively. A single viruliferous aphid could transmit the virus and the maximum percent of transmission was obtained with 15 aphids per plant. The inoculation potential of aphid was lost when they were allowed for 2 h of post-acquisition fasting (Singh et al. 2005). Isolates of garlic mosaic virus was found to be transmitted by *A. craccivora* and *M. persicae* in a non-persistent manner to *Chenopodium album*. However, no transmission of this virus was observed with *L. erysimi* (Ghosh and Ahlawat 1997). Transmission of onion yellow dwarf virus by three aphid species *viz. A. gossypii, M. persicae*, and *A. craccivora* were studied. All the three aphid species were recorded to transmit the virus but at a variable percentage ranging from 15 to 20%. Maximum transmission of the virus up to 20% was recorded with *A. gossypii* (Hoa et al. 2003). The spread of dasheen mosaic virus infecting elephant foot yam was reported through a non-persistent manner by several widely distributed aphid species, including *M. persicae* and *A. gossypii* (Babu et al. 2011). Non-persistent transmission of banana bract mosaic virus by *R. maidis*, *A. gossypii, P. nigronervosa* and *A. craccivora* was reported by Selvarajan et al. (2006).

23.5.1.4 Other Viruses

Cowpea mosaic virus (*Comovirus*) was reported to be transmitted by *A. craccivora, A. gossypii, A. evonymi* Fabricius and *M. persicae.* Association of *A. craccivora* with the virus was found more intimate than the other aphid species (Haque and Chenulu 1967). Further study on this relationship revealed that irrespective of the host, *A. craccivora* could transmit the virus with equal efficiency. They also observed that the resistance or susceptibility of host and weather factors influenced the transmission rate of the aphids (Haque and Chenulu 1972). Urdbean leaf crinkle virus was found to be transmitted in a non-persistent manner by *Aphis craccivora* and *A. gossypii* (Dhingra 1975).

23.5.2 Semi-persistent Virus Transmission

Semi-persistent transmission of plant viruses by aphids is well known for *Caulimovirus* and *Closterovirus*. This type of transmission has intermediate characteristics between non-persistent and circulative systems. Generally, an acquisition feeding of few minutes to hours with a retention period of few hours are common phenomena of the semi-persistent type of transmission.

23.5.2.1 *Closterovirus*

The first report of semi-persistently transmitted plant viruses by aphids in India came with *Toxoptera citricida* (Kirkaldy) (Vasudeva et al. 1959) transmitting citrus tristeza virus (CTV). Semi-persistent mode of transmission of CTV by aphids was tested by Capoor (1975). He observed that CTV was not retained by the aphids after moulting (Capoor 1975). *A. gossypii, M. persicae* (Varma et al. 1960), *A. craccivora* and *Dactynotus jaceae* [= *Uroleucon jaceae* (Linnaeus)] (Varma et al. 1965) were studied for transmission of CTV. *T. citricida* was found as the most efficient vector abundant in Southern India, West Bengal, Sikkim, and Assam. *A. gossypii* and *A. craccivora* were not so competent and found abundant in Northern India (Nariani and Singh 1971). In India, six aphid species *viz. T. citricida, T. aurantii, A. gossypii, A. craccivora, M. persicae*, and *Macrosiphum (Dactinotus) compositae* (= *Uroleucon compositae*) have been established as vectors. Very recently all the aphid species infesting citrus in Sikkim and West Bengal were documented and studied for their efficiency in transmitting CTV. Among the nine aphid species recorded (*T. citricida, T. aurantii, A. gossypii, A. craccivora, A. spiraecola, M. persicae, Aulacorthum solani* (Kaltenbach), *B. helichrysi, Macrosiphon euphorbiae), only T. aurantii, T. citricida* and *A. gossypii* were recorded to transmit CTV in Darjeeling hills (Ghosh et al. 2014b, 2015b). Host preference of *T. citricida,* the most efficient vector of CTV was studied. It was observed that *T. citricida* preferred rough lemon for feeding and colonization followed by mandarin orange and kagzi lime (Ghosh et al. 2014a). Incidence and distribution of these aphid species infesting mandarin orange in hills were recorded. *T. citricida* was found predominant in the orchards of lower altitude and responsible for the maximum spread of CTV in those orchards. A higher population of *T. aurantii* was recorded in the citrus orchards located at an altitude more than 500 m (Ghosh et al. 2015b).

23.6 Persistent-Circulative Virus Transmission

In the persistent-circulative type of transmission, the virus has to be ingested by the aphids and reach the salivary gland via haemolymph. There is no demonstration of virus replication in aphids system. Acquisition period generally ranges from hours to days and the aphid remains viruliferous for weeks or lifelong. Among the aphid-transmitted viruses, *Luteovirus* and *Babuvirus* are observed to follow the persistent-circulative pathway.

23.6.1 *Babuvirus*

In India, first persistent type of aphid transmission was recorded with *P. nigronervosa* transmitting banana bunchy top disease (BBTV) (Capoor 1969). Persistent transmission of another babuvirus (cardamom bushy dwarf virus, CBDV) in large cardamom was recorded by Varma and Capoor (1964) and *Micromyzus kalimpongensis* Basu (Basu and Ganguly 1968). Recently, three aphid species viz. *P. nigronervosa, M. kalimpongensis,* and *A. solani* (Kaltb.) were identified in large cardamom based on morphological and molecular evidence (Ghosh et al. 2015a, 2016a, b). *M. kalimpongensis*, a root-feeding aphid was consistently present in large cardamom plantations throughout the year; predominantly on high altitude (>900 m) plantations. *P. nigronervosa* was predominant in plantations lower than 1000 m altitude, and the population was higher during the summer season only. The incidence of *A. solani* was observed during the winter season only. Among these three aphid species, only *M. kalimpongensis* but not the other two species of aphid could transmit CBDV to large cardamom (Ghosh et al. 2015a, 2016a, b). Infection of CBDV in large cardamom was found to alter the behaviour and biology of its vector by regulating its host plant preference and fecundity. The aphids preferred to settle on the infected large cardamom plant or plant extract than healthy plant. The fecundity of a single aphid increased around four times when fed on CBDV- infected host. Infection of the CBDV also altered the nymphal period and adult longevity of *M. kalimpongensis*. A general pattern of spread of the virus by the vector was documented (Mandal and Ghosh 2016). *M. kalimpongensis* migrated to the CBDV-infected plants, colonized there, acquired CBDV, and once the diseased plants withered, migrated to healthy plants, which eventually became diseased (Ghosh et al. 2016b). Enhanced activity of pathogenesis-related proteins was observed when BBTV infected *P. nigronervosa* fed on banana plants (Harish et al. 2009).

23.6.2 *Luteovirus*

Information on the transmission of barley yellow dwarf virus by aphid species in wheat was made available by Nagaich and Vashisth (1963). The virus could be transmitted to oats, barley, and wheat by three aphid species *viz. R. maidis*, *R. padi* and *Sitobion avenae*. Insect vectors of chickpea stunt disease-associated virus were recorded by Reddy and Lava Kumar (Reddy and Lava 2004). The authors tested two commonest aphid species *viz. A. craccivora* and *M. persicae*. Nymphs and winged adults of *A. craccivora* were observed to transmit the virus to healthy chickpea and groundnut plants. *M. persicae* could not acquire the virus from infected chickpea plants but it was able to acquire from infected groundnut plants with only poor efficiency. Persistent-circulative transmission by aphids was observed for PLRV. *M. persicae* was found the most efficient vector of PLRV. Transmission of PLRV was also recorded with *A. gossypii, M. euphorbiae, A. solani* and *A. pisum* (Singh et al. 1981, 1982). The authors recorded the seasonal incidence of the vector population and its relation to the spread of the virus. The incidence of *M. persicae* was maximum during spring season contributing higher spread of the virus in cv. Kufri Sindhuri and Kufri Chandramukhi. The spread of the virus is correlated with the weather factors favourable for aphids' activity and population build-up (Singh et al. 1981, 1982, 1988). Detection of PLRV in *M. persicae* by RT-PCR techniques confirms its vectoring ability (Baswaraj et al. 2014). Recently, the transmission of sugarcane yellow leaf virus by *M. sacchari* was successfully demonstrated. The progress of virus titre after aphid transmission in sugarcane was quantified by RT-qPCR (Chinnaraja and Viswanathan 2015).

23.7 Persistent-Propagative Virus Transmission

In the propagative type of transmission, the virus multiplies in the vector and resembles animal viruses transmitted by mosquitoes. The basic characteristic of this type of transmission is the life long association of the virus with the vector and in some occasion, transovarial passaging of the virus to the offsprings. This type of transmission by aphids is restricted to the family *Rhabdoviridae*. Only eight aphids species transmitting *Cytorhabdovirus* and seven aphid species transmitting *Nucleorhabdoviridae* are reported worldwide (Ng and Perry 2004). Very few works have been undertaken in India on rhabdovirus and aphid relationship because of its limited occurrences in India. Naidu et al. (1989) reported a plant rhabdovirus associated with peanut veinal chlorosis. The transmission of the virus was tried with *A. craccivora*, whitefly and leafhopper. No results on the insect transmission were reported in the study.

23.8 Concluding Remarks

Most of the studies on the aphid vectors in India from 1940 to 2000 were focused on the identity of the vectors, rate, and factors of transmission. Limited efforts are made to understand the mechanisms of transmission in the Indian context. At a global level, the studies on host plant-virus interactions have made impressive advancement, but the studies on aphids-virus interaction yet to achieve significant milestone. As the transmission involves a tritrophic interactions of plant, virus and vector, the elucidation of the process is highly intricate. Study on the cellular interaction of virus within the vector system has not been initiated in our country. The transmission factors such as the aphids' proteins receiving the virus particles, the virus coded helper component proteins so far have not been studied in Indian context. The role of endosymbionts in a whitefly-begomovirus relationships has been studied in India, but the possible role of endosymbionts in Indian aphid population is yet to be addressed. Few studies are there on seasonal incidence of aphid vectors in relation to the spread of the virus. Better understanding is required on the life cycle of aphid vectors with reference to their temporal and seasonal occurrences on different hosts, migration and feeding behaviour. The data on the biology of aphid vectors will aid in understanding the dissemination pattern and developing simulation on the prediction of diseases.

References

Ahlawat YS, Chenulu VV (1982) Studies on the transmission of radish mosaic virus by the aphid *Lipaphis erysimi* (Kalt.) Indian Phytopathol 35:633–638

Ali S, Akhtar M, Singh KS, Naqvi QA (2012) RT-PCR and CP gene based molecular characterization of a cucumber mosaic cucumovirus from Aligarh, U.P., India. Agric Sci 3(8):971–978

Babu B, Hegde V, Makeshkumar T, Jeeva ML (2011) Molecular detection and identification of *Dasheen mosaic virus* infecting *Amorphophallus paeoniifolius*. Arch Hytopathol Plant Protect 44(13):1248–1260

Badami RS (1958) Changes in the transmissibility by aphids of a strain of cucumber mosaic virus. Ann Appl Biol 46:554–562

Basu AN, Ganguly B (1968) A note on the transmission of 'Foorkey' disease of large cardamom by the aphid, *Micromyzus kalimpongensis* Basu. Indian Phytopathol 21:127

Basu RC, Raychaudhuri DN 1980. A study on the sexuals of aphids (Homoptera: Aphididae) in India. Rec Zool Surv India. Occasional paper no. 18: 1–54

Baswaraj R, Venkateswarlu V, Sridhar J, Jeevalatha A, Sharma S, Singh BP (2014) RT-PCR detection of *Potato leaf roll virus* (PLRV) in aphids from Northern and North-Eastern India. Indian J Plant Protect 42(4):430–436

Bhargava KS (1951) Some properties of four strains of cucumber mosaic virus. Ann Appl Biol 38:377–388

Bhargava KS, Khurana SMP (1970) Insect transmission of papaya viruses with special reference to papaya mosaic virus. Zentralblattfür Bakteriologie, Parasitenkunden Infektionskrankeiten und Hygiene 124:688–696

Blackman RL (1974) Invertebrate types-aphids. Ginn & Company, London, p 175

Blackman RL, Eastop VF (1994) Aphids on World's trees. CAB International, Wallingford, p 987

Capoor SP (1967) Viral and virus-like diseases of crops and their control. Indian Council of Agricultural Research, New Delhi

Capoor SP (1969) Banana bunchy top and its control workshop on fruit research, Ludhiana. Indian Council of Agricultural Research, New Delhi

Capoor SP (1975) Role of the Tristezavirus in citrus die-back complex. Indian J Hortic 32(1 and 2):1–6

Capoor SP, Varma PM (1948) A mosaic disease of *Carica papaya* L. in the Bombay province. Curr Sci 17:265–266

Capoor SP, Varma PM (1958) A mosaic disease of papaya in Bombay. Indian J Agric Sci 20:217–235

Capoor SP, Varma PM (1970) Investigation on mosaic disease of babana in Deccan. Indian Phytopathol 28:62–68

Chakrabarti S (2007) Diversity and biosystematics of gall inducing aphids (Hemiptera: Aphididae) and their galls in the Himalaya. Orient Insects 41:35–54

Chakrabarti S (2009) Diversity, distribution and endemism of aphids (Hemiptera) in Indian Subregion of Oriental Realm. Redia 42:78–85

Chakrabarti S (2015) Aphids (Hemiptera: Insecta) in the Himalaya: diversity, distribution and affinity. Glob J Environ Sci Res 2:123–129

Chakrabarti S, Das D (2014) A new species and new records of aphids (Hemiptera) from Bhutan. Orient Insects 48:327–336

Chinnaraja C, Viswanathan R (2015) Quantification of sugarcane yellow leaf virus in sugarcane following transmission through aphid vector, *Melanaphis sacchari*. Virus Dis 26(4):237–242

Chona BL, Seth MD (1960) A mosaic disease of maize in India. Indian J Agric Sci 30:25–32

Deshpande RS, Siddappaji C, Viswanath S (1972) Epidemiological studies on 'katte' disease of cardamom (*Elettaria cardamomum* Maton). Mysore J Agric Sci 6:4–9

Devi PHS, Bhat AI, Devi L, Das ML (2004) Occurrence, transmission and genotype response of a filamentous virus associated with leaf mustard (*Brassica junceavar. rugosa*) in Manipur. Indian Phytopathol 57(4):488–493

Dhanraj KS, Raychaudhuri SP (1969) A note on barley mosaic in India. Plant Dis Rep 53:766–767

Dheepa R, Paranjothi S (2010) Transmission of Cucumber Mosaic Virus (CMV) infecting banana by aphid and mechanical methods. Emirates J Food Agri 22(2):117–129

Dhingra KL (1975) Transmission of urdbean leaf crinkle virus by two aphid species. Indian Phytopathol 28:80–82

Ganguly B, Raychaudhuri, SP, Chatterjee SN (1970) Mosaic streak disease of cardamom. In: Plant disease problems. IPS, Indian Agricultural Research Institute, New Delhi, pp 653–661

Ghosh DK, Ahlawat YS (1997) Filamentous viruses associated with mosaic disease of garlic in India. Indian Phytopathol 50(2):266–276

Ghosh A, Das A, Meena R, Baranwal VK (2014a) Evidence for resistance to Citrus tristeza virus in pomelo (*Citrus maxima* Merr.) grown in Darjeeling and Sikkim hills of India. Phytoparasitica 42:503–508

Ghosh A, Das A, Lepcha R, Meena RK, Baranwal VK (2014b) Occurrence of citrus tristeza virus and its vector in Darjeeling hills and Dooars of India. In: Abstract of International Seminar on Integrating agriculture & Allied Research, 6–9 Nov. CWSS, Kalyani. India. p 196

Ghosh A, Lepcha R, Das A, Mandal B (2015a) Identification, distribution and temporal occurrences of aphids infesting large cardamom and their efficiency in transmitting large cardamom viruses in North Eastern sub-Himalayan region. In: souvenir of National Conference on Virology. 8–10 Oct., Indian Virological Society at Shillong, India, p 137

Ghosh A, Das A, Lepcha R, Majumder K, Baranwal VK (2015b) Identification and distribution of aphid vectors spreading Citrus tristeza virus in Darjeeling hills and Dooars of India. J Asia Pac Entomol 18(3):601–605

Ghosh A, Das A, Vijayanandraj S, Mandal B (2016a) Cardamom bushy dwarf virus infection in large cardamom alters plant selection preference, life stages, and fecundity of aphid vector, Micromyzus kalimpongensis (Hemiptera: Aphididae). Environ Entomol 45(1):178–184

Ghosh A, Das A, Lepcha R, Mandal B (2016b) Identification, distribution and temporal occurrence of aphids infesting large cardamom and their efficiency in transmitting large cardamom viruses in northeastern sub-Himalayan region. Australas Plant Pathol 45(5):533–536

Gude SR, Nagaraju PNS, Ravikumar B, Anjula N (2008) Occurrence of a strain of Papaya ring spot virus on cucurbitaceous crops around Bangalore. Environ Ecol 26(3):1066–1069

Haque SQ, Chenulu VV (1967) Studies on the relationship of Cowpea mosaic virus and its vector *Aphis craccivora* Koch. J Phytopathol 59(3):277–284

Haque SQ, Chenulu VV (1972) Epidemiology of Cowpea Mosaic Virus with special reference to influence of host factors on the transmission efficiency of *Aphis craccivora* Koch. In: Proceedings of the Indian National Science Academy, pp 357–364

Harish S, Kavino M, Kumar N, Samiyappan R (2009) Differential expression of pathogenesis-related proteins and defense enzymes in banana: interaction between endophytic bacteria, Banana bunchy top virus and *Pentalonia nigronervosa*. Biocontrol Sci Tech 19(8):843–857

Hebert PD, Ratnasingham S, deWaard JR (2003) Barcoding animal life: cytochrome c oxidase subunit 1 divergences among closely related species. Proc Biol Sci 270(S1):96–99

Heie O (1980) The Aphidoidea (Hemiptera) of Fennoscandia and Denmark-I. Fauna Entomol Scand 9:1–236

Heie O, Wegierck P (2009) A classification of the Aphidomorpha (Hemiptera, Sternorrhyncha) under consideration of the fossil taxa. Redia 42:69–77

Hoa NV, Ahlawat YS, Pant RP (2003) Partial characterization of onion yellow dwarf virus from onion in India. Indian Phytopathol 56(3):276–282

Kalleshwaraswamy CM, Krishna Kumar NK (2008) Transmission efficiency of papaya ringspot virus by three aphid species. Phytopathology 98(5):541–546

Kalleshwaraswamy CM, Verghese A, Ranganath HR, Krishna Kumar NK, Dinesh MR (2007) Role of transient aphid vectors on the temporal spread of Papaya ringspot virus in South India. Acta Horticult (ISHS) 740:251–258

Kalleshwaraswamy CM, Krishna Kumar NK, Dinesh MR, Chandrashekhar KN, Manjunatha M (2009) Evaluation of insecticides and oils on aphid vectors for the management of papaya ringspot virus (PRSV). Karnataka J Agric Sci 22(3):552–553

Kennedy JS, Day MF, Eastop VF (1962) A conspectus of aphids as vectors of plant viruses. Commonwealth Institute of Entomology, London, p 114

Kiranmai G, Sreenivasulu P, Nayudu MV (1997) Characterization of cucumber mosaic cucumovirus isolates naturally infecting three solanaceous vegetable crops in Andhra Pradesh. Indian Phytopathol 50(3):416–425

Kondaiah E, Nayudu MV (1984) Sugarcane mosaic virus strain H – a new record from India. Curr Sci 53(5):273–275

Mackenzie A, Dixon AFG (1991) An ecological perspective of host alternation in aphids (Homoptera: Aphidoidea: Aphididae). Entomol Gen 16:265–284

Mandal B, Ghosh A (2016) Cardamom bushy dwarf virus (genus Babuvirus) and aphid vector, Micromyzus kalimpongensis: vector behaviour and dissemination of the virus. VirusDisease 27(4):410–411

Mandal B, Vijayanandraj S, Shilpi S, Pun KB, Singh V (2012) Disease distribution and characterisation of a new macluravirus associated with chirke disease of large cardamom. Ann Appl Biol 160(3):225–236

Mariappan V, Samuel DL (1993) Effect of non-edible seed oil in aphid transmitted chilli mosaic vines in Tamilnadu, India. World Neem Conference, 24–28 February, Bangalore, pp 787–791

Mollison JW (1900) Cardamom cultivation in the Bombay Presidency. Agricultural Ledger, p 11

Moran NA (1992) The evolution of aphid life cycle. Annu Rev Entomol 37:321–348

Muniyappa V (1976) Studies on a mosaic disease of French bean (*Phaseolus vulgaris*). Madras Agric J 63:69–70

Nagaich BB, Vashisth KS (1963) Barley yellow dwarf: a new viral disease for India. Indian Phytopathol 16:318–319

Nagaraju RHR (1981) Natural occurrence of pepper vein banding virus on bell pepper. Mysore J Agric Sci 15:50–53

Nagaraju N, Keshava Murthy KV (1994) Studies on a sap transmissible virus associated with mosaic disease of cowpea. Indian Phytopathol 47(1):23–26

Nagaraju N, Reddy HR (1980) Natural occurrence of pepper veinal mottle virus on bell pepper. Indian Phytopathol 33:282–284

Nagaraju MV, Singh SJ, Virupahshappa K (1997) Occurrence of a mosaic virus disease on sunflower in Karnataka. Indian Phytopathol 50(2):277–281

Naidu RA, Monahar SK, Reddy DVR, Reddy AS (1989) A plant rhabdovirus associated with peanut veinal chlorosis disease in India. Plant Pathol 38:623–626

Nariani TK, Singh (1971) Epidemiological studies on the citrus dieback complex in India. Proc Indian Natl Sci Acad 37:365–371

Ng JCK, Perry KL (2004) Transmission of plant viruses by aphid vectors. Mol Plant Pathol 5(5):505–511

Nieto Nafria M, Favret C, Akimoto S, Barbagallo S, Chakrabarti S, Mier, Durante MP, Miller GL, Qiao Ge-xia, Sanom, Hidalgo NP, Stekolshchikov AV, Wegierek P. 2011. Register of genus-group taxa of Aphidoidea. Registers of family-group and genus- group taxa of Aphidoidea. Nieto Nafria M, Favret C, University of Leon, Spain. 21–404

Palukaitis P, Roossinck MJ, Dietzgen RG, Francki RIB (1992) Cucumber mosaic virus. Adv Virus Res 41:281–348

Quednau W (2010) Atlas of the Drepanosiphine aphids of the World Part III. Mem Am Entomol Inst 83:1–361

Rajan P (1981) Biology of *Pentalonia nigronervosa caladii* van der Goot, vector of 'katte' disease of cardamom. J Plant Crop 9(1):34–41

Rajan P, Naidu R, Venugopal MN (1989) Effect of insecticides on transmission and acquisition of 156 'katte' virus of small cardamom and their use in relation to disease spread and vector control. J Plant Crop 16:261–266

Rao DG, Naidu R (1974) Additional vectors of 'Katte' disease of small cardamom. Indian J Hortic 31:380–381

Rashmi CM, Reddy MK, Praveen HM, Nagaraju (2005) Occurrence of papaya ring spot virus (PRSV-p) on gherkin (*Cucumis anguria*). Environ Ecol 23(4):785–788

Raychaudhuri SP, Chatterjee SN (1958) A preliminary note on the occurrence of a new virus disease of large cardamom (*Amomum subulatum* Roxb.) in Darjeeling District. In: Proceedings of mycological research worker's conference, ICAR, held at Shimla from the 20th to 23rd June 1958, pp 74–176

Raychaudhuri SP, Chatterjee SN (1961) Chirke- a new virus threat to cardamom. Indian Farming 9:11–12

Raychaudhuri SP, Ganguly B (1965) Further studies on chirke diseases of large cardamom (*Amomum subulatum* Roxb.) Indian Phytopathol 18:373–377

Raychaudhuri M, Varma A (1977) Virus-vector relationship of marrow mosaic virus with *Myzus persicae* Sulz. Indian J Entomol 37(3):247–250

Raychaudhuri M, Varma A (1983) Effects of oils on transmission of marrow mosaic virus by *Myzus pericae* Sulz. J Entomol Res 7:117–111

Reddy SV, Lava KP (2004) Transmission and properties of a new luteovirus associated with chick-pea stunt disease in India. Curr Sci 86(8):1157–1161

Reddy L, Nagaraju CN, Prasana Kumar MK, Venkataravanappa V (2007) Transmission and host range of the papaya ring spot virus (PRSV). J Plant Dis Sci 2(1):9–13

Remaudiere G, Remaudiere M (1997) "Catalogue des Aphididae du monde of the World's Aphididae (Homoptera, Aphidoidea)". INRA, Paris p 473

Selvarajan R, Balasubramanian V, Sathiamoorthy S (2006) Vector transmission of banana bract mosaic and banana streak viruses in India. In: Abstracts of XVI Annual convention and International symposium on "Management of vector – borne viruses, conducted by IVS, at ICRISAT, 7–10 Feb. p 110

Seth ML, Raychaudhuri SP, Singh DV (1972) A mosaic disease of pearl millet in India. Indian J Agric Sci 42:322–325

Shneyer VS (2007) On the species-specificity of DNA: fifty years later. Biochemistry (Mosc) 72:1377–1384

Shukla K, Rao PG (2000) Studies on Aphid transmission of cucumber mosaic virus- *Pea strain.* Indian Phytopathol 53(2):930–932

Siddappaji C, Reddy DRN (1972) A note on the occurrence of the aphid *Pentalonia nigronervosa* form *caladii* van der Goot (Aphididae- Homoptera) on cardamom (*Elettaria cardamum*). Mysore J Agric Sci 6:194–195

Singh R (1970) Studies on the transmission of Benincasa Mosaic by *Myzus persicae*. Phyton (Austria) 14(1–2):15–21

Singh SJ (2005) Virus and phytoplasma diseases of vegetable crops in India. Kalyani Publishers, p 244

Singh V, Singh D (2010) Studies on natural transmission of papaya ringspot virus disease in Eastern Uttar Pradesh. Ann Plant Protect Sci 18(1):188–192

Singh MN, Khurana SMP, Nagaich BB, Agarwal HO (1981) Epidemiological studies of Potato virus Y and leaf roll in sub-tropical India. In: Proceedings of Epidemiological Conference, Oxford, Jul 27–30, pp 89–90

Singh MN, Khurana SMP, Nagaich BB, Agarwal HO (1982) Efficiency of *Aphis gossypii* and *Acyrthosiphon pisum* in transmitting potato viruses leaf roll and Y. In: Potato in developing countries. Indian Potato Association, Shimla, pp 289–293

Singh MN, Paul Khurana SM, Nagaich BB, Agrawal HO (1988) Environmental factors influencing aphid transmission of potato virus Y and potato leaf roll virus. Potato Res 31:501–509

Singh M, Singh A, Upadhyaya PP, Rao GP (2005a) Transmission studies on an Indian isolate of sugarcane mosaic poty virus. Sugar Tech 7(2):32–38

Singh RK, Singh SJ, Prakash S (2005b) Relationship of Sunflower mosaic potyvirus (SMPV) with its aphid vector *Aphis craccivora* Koch. Indian J Agric Res 39:1–9

Tomiuk J, Loeschke V, Schneiter M (1994) On the origin of polyploid parthenogenetic races in weevil Podruaua mollis (Coleoptera: Curculionidae). J Theor Biol 167:89–92

Uppal BN, Varma PM, Capoor SP (1945) A mosaic disease of cardamom. Curr Sci 14:208–209

Varma PM, Capoor SP (1958) Mosaic disease of cardamom and its transmission by the banana aphid *Pentalonia nigronervosa* Coq. Indian J Agric Res 28:97–107

Varma PM, Capoor SP (1964) 'Foorkey' disease of large cardamom. Indian J Agric Res 34:56–62

Varma PM, Rao DG, Capoor SP (1965) Transmission of tristeza virus by *Aphis craccivora* (Koch.) and *Dactynotus jaceae* L. Indian J Entomol 27:67–71

Varma PM, Rao DG, Vasudeva RS (1960) Additional vectors of tristeza disease of citrus in India. Curr Sci 29:359

Vasudeva RS, Varma PM, Rao DG (1959) Transmission of citrus decline virus by *Toxoptera citricidus* Kirk. in India. Curr Sci 28:418

Venugopal MN (1995) Viral diseases of cardamom (*Elettaria cardamomum* Maton) and their management. J Spices Aromat Crops 4:32–39

Wool D (2004) Galling aphids: Spercialization, biological complexicity and variation. Annu Rev Entomol 49:175–192

Thrips as the Vectors of Tospoviruses in Indian Agriculture

24

Amalendu Ghosh, Debjani Dey, Timmanna, Basavaraj, Bikash Mandal, and Rakesh Kumar Jain

Abstract

Thrips being important sucking pests transmit the deadly tospoviruses in comercial crops in a persistent and propagative manner. Sixteen thrips species are known to transmit 29 tospoviruses worldwide. In India, only six thrips species are so far known to transmit six tospoviruses. The first study in India on thrips is dated in 1856. The research on tospoviruses was initiated in India during 1960s, however, the first study on thrips-tospovirus was published in 1981. Not many studies have been undertaken on the thrips-tospovirus in India. Several reviews have been published on the thrips-tospovirus research, where the Indian works have hardly been reflected. In this chapter, we summarise the studies on thrips-tospovirus undertaken in India.

Keywords

Thrips • Tospovirus • India • Virus-vector • Interaction • Relationship • Transmission

A. Ghosh (✉) • Timmanna • Basavaraj • B. Mandal • R.K. Jain
Advanced Centre for Plant Virology, Division of Plant Pathology, Indian Agricultural Research Institute, New Delhi 110 012, India
e-mail: amal4ento@gmail.com

D. Dey
Division of Entomology, Indian Agricultural Research Institute, New Delhi 110 012, India

B. Mandal et al. (eds.), *A Century of Plant Virology in India*,
https://doi.org/10.1007/978-981-10-5672-7_24

24.1 Introduction

Thrips are minute, slender, and soft-bodied insects, feed on a large variety of plants and animals by puncturing and sucking the cell contents. They belong to order Thysanoptera. The name of the order was derived based on the characteristic fringed wings. In the *Greek*, '*thysanos*' means 'fringe' and '*pteron*' means 'wing'. Thrips are also known as thunder flies, thunder bugs, storm flies, storm bugs, corn flies, corn lice, freckle bugs, and physopods. A majority of the thrips species are considered as agricultural pests as they cause considerable loss to many economically important crops. Some of them are predaceous and feed on other thrips species, mites, aphids, scales and other soft-bodied insects. The beneficial aspect of thrips is that they aid pollination of flowers. The order Thysanoptera is divided into two suborders Terebrantia and Tubulifera. Terebrantia has two Superfamilies, Aeolothripoidea and Thripoidea. The thrips species described in this chapter belong to Superfamily Thripoidea, family Thripidae, and subfamily Thripinae (Richards and Davies 1977). Approximately, 7,700 species of thrips has been recorded to date, of which less than 1% are considered as pests of agricultural crops (Mound 2015). The phytophagous thrips damage leaf tissue by probing into cells and imbibing the sap. Feeding of thrips causes damage to the leaves, flowers or fruits and sometimes induces gall. Besides being important insect pests, thrips act as vectors of deadly tospoviruses of the genus *Orthotospovirus*, family *Tospoviridae* (https://talk.ictvonline.org/). Significant losses by tospoviruses in the yield and quality of vegetables, legumes and ornamentals have been recorded in different countries (Mumford et al. 1996; Pappu 1997; Pearce 2005; Persley et al. 2006). Annual losses due to tospovirus outbreaks are estimated at over $1 billion worldwide (Pappu et al. 2009; Mandal et al.2012). Tospoviruses are not seed-transmitted (except soybean vein necrosis virus, Groves et al. 2016) and thrips play a critical role in the survival and spread of tospoviruses. Sixteen species of thrips as vectors of tospoviruses (Rotenberg et al. 2015) and around 29 tospoviruses are recorded throughout the world (Turina et al. 2016). The associations of thrips with tospoviruses and their ability to transmit specific tospoviruses are distinct (Jones 2005). Thrips transmit the tospoviruses in persistent and propagative manner. Moreover, thrips vectors could be considered as 'reservoirs' for tospoviruses since these viruses replicate inside their thrips vectors (Ullman et al. 1992, 1997; Whitfield et al. 2005a). In India, the studies on different thrips species were initiated during pre-independence period, but the research on tospoviruses gained momentum during 1960s. Since then, several studies have been undertaken to understand the relationship between different thrips species and tospoviruses. This chapter attempts to summarise the thrips-tospovirus studies conducted in India.

24.1.1 General Appearance and Life Cycle

Thrips are mostly yellow, orange, black or whitish-yellow in colour. They crawl leisurely or fast when disturbed. The apex of the abdomen is flexed upward to leap from the plant surface. Adult thrips are usually very small in size, 1–2 mm in length and slender in shape. The head is quadrangular with two prominent compound eyes and three ocelli. The filiform antennae project forward and bear six to nine segments with short, simple or forked sensory organs. The prothorax is broad and freely movable. The mesothorax and metathorax are firmly fused together. The legs are with one to two segmented tarsi and a pair of weak claws. A protrusible vesicle on the tarsus helps them to walk on any types of surfaces. Wings are membranous and fringed with long setae. The base of hind wings bears hooks which engage with an anal portion of forewing to aid in the flight. They are not good fliers and usually make short flights, however long distance migration occurs through the wind, atmospheric convection, and turbulence. The abdomen is 11 segmented, long and tapering. The first and terminal segments are reduced. The genitalia in the adult female is curved downward and saw like. The apex of the abdomen is bluntly rounded in the case of male.

The life cycle of thrips comprises of egg, first and second instar larva, pre-pupa, pupa, and adult. The eggs are mostly reniform, opaque in colour. The mature eggs turn yellowish-orange in colour. The newly hatched larvae resemble adults but lack wing pads, and have lesser number of segments in the antennae. The first instar larvae molt into the second instar within a day's time. The second larval instar is similar to the previous instar but bigger in size. Second instars develop into pre-pupae within 4–5 days (Lowry et al. 1992). Both pupal instars have wing-pads but lack protrusible vesicle on the tarsi of legs. The pre-pupae and pupae do not feed and remain mostly inactive. They move very slowly only if disturbed. The pupation generally occurs in soil or sometimes on plants. The longevity of the adult female is around 4–5 weeks. The adult female makes a slit with their ovipositor in tender plant parts and eggs are embedded in the upper epidermal layer of plant tissues. One female can lay around 50 eggs during its life time (Reitz 2008). The developmental period varies depending on the temperature. The reproduction of thrips is haplodiploid and capable of parthenogenesis. Some species display arrhenotoky and others follow thelytoky. The endosymbiont, *Wolbachia* determines the mode of reproduction in some species (Kumm and Moritz 2008). The sex ratio and rate of reproduction are influenced by the temperature and biotic factors.

24.2 Mouthparts and Feeding Behavior and Virus Transmission

Thrips have very characteristic mouthparts, unlike other insects. Mound (1971), Heming (1978), Milne and Manicom (1978), and Chisholm and Lewis (1984) conclusively demonstrated the piercing-sucking type of mouthparts in thrips. The principal feeding structures are contained within a cone formed by clypeus, labrum, and labium. The mouthpart is asymmetrical with the right mandible of thrips reduced and vestigial or completely absent. The mouthparts include one elongated hypopharynx, two maxillary stylets and one mandibular stylet, supported by cibarial and salivary pumps (Heming 1978). The left mandible is larger and modified to penetrate the cells. The maxillary stylets are held together to form the food canal. There is no separate salivary canal in thrips. The salivary duct directly opens in front of the oesophagus. Some species inject digestive enzymes and suck semi-digested cellular contents. During the feeding process, the head moves towards the feeding site and the tip of the mouthcone is pressed against the plant surface in preparation for feeding. The modified mandible pierces leaf tissues and makes the initial opening for maxillary stylets. The mandible is restricted in movement for its characteristic articulation with the head capsule and musculature. The maxillary stylets can be protracted or retracted singly or together. The labral pad supports the maxillary stylets when they are protracted into the plant tissues (Chisholm and Lewis 1984). The entry to the plant tissue is achieved by the vigorous forward and backward movement of the head (Chisholm and Lewis 1984). The maxillary stylets are interlocked to form a hollow tube for sucking the liquid food from cells (Mound 1971). Due to their feeding, a distinctive silvery or bronze scarring on the surfaces of the stems, leaves or fruits is often visible. The probing of thrips can be either differentiated as shallow probing or penetration (Sakimura 1962). In the case of shallow probing, the thrips make a number of punctures limited to epidermal tissues and few adjacent mesophyll cells (Chisholm and Lewis 1984). On the contrary, penetration deeper in the mesophyll cells tends to be destructive due to extreme plasmolysis or complete disappearance of cells. The emptied space are usually filled with air, giving a silvery, scarred appearance. The brief period of shallow probing favours the transmission of the virus (Chisholm and Lewsi 1984; Hunter and Ullman 1994).

The transmission of tospoviruses by thrips species is resultant of the complex interaction between the two biological organisms. The successful transmission of the virus by thrips involves the acquisition of the virus followed by internalization of the virus in thrips and inoculation of the virus in a susceptible host. Most of the studies undertaken to study the mechanism of the transmission of tospovirus by thrips are based on the western flower thrips–tomato spotted wilt virus (TSWV) biological system owing to its predominace and economic importance. The information on the mechanism of transmission available till date is summarized here.

Thrips transmit tospoviruses in a persistent manner. The tospovirus propagates within the thrips body (Ullman et al. 1992, 1997; Whitfield et al. 2005a). The first instar larvae only acquire the virus (Sakimura 1963). After the multiplication in the thrips body (Ullman et al. 1993a; Wijkamp et al. 1993), the tospovirus can be transmitted by late second instar larvae and adults (Wijkamp and Peters 1993). The adults are the only dispersing agent of tospovirus as they can fly over long distances to infect new host plants. For successful transmission, the tospovirus needs to cross three barriers in the thrips body *i.e.* the midgut barrier, dissemination barrier, and salivary gland barrier. A midgut infection barrier develops during the larval stage of the thrips (Nagata et al. 1999) which limits the virus uptake by the late instar larvae and adults. The virus is allowed to infect the midgut only during early larval stage (Ullman et al. 1992; Ohnishi et al. 1998). A 55 kDa protein has been reported to regulate the barrier (Bandla et al. 1998). The tospovirus glycoproteins are known to interact with the receptors present in midgut apical plasmalemma (Bandla et al.

1998; Ullman 1996; Medeiros et al. 2000; Garry and Garry 2004; Whitfield et al. 2005b). The virus infects the epithelial cells of the first region of the midgut and spreads to the muscle cells attached with the midgut (Nagata et al. 1999). A 94 kDa protein of thrips is reported to assist virus circulation within thrips body (Kikkert et al. 1998). The presence of tospovirus in salivary glands, salivary ducts, and brains of viruliferous larvae and adult thrips has been reported (Ullman et al. 1993a; Wijkamp et al. 1993) which indicates a salivary gland barrier in non-transmitters thrips. The virus particles are injected in the host plant with the saliva during the feeding of viruliferous thrips. As there is only one canal in thrips, food ingestion and saliva egestion are usually accomplished alternately (Chisholm and Lewis 1984; Harrewijn et al. 1996). Multiplication of the virus has been shown in the epithelial cells of midgut, muscle cells surrounding the gut and salivary gland cells (Ullman et al. 1993b; Wijkamp et al. 1993). Once the adult thrips become viruliferous, it remains so throughout its lifespan. A transovarial infection barrier is also reported in thrips which restricts the vertical transmission of the tospovirus (Wijkamp et al. 1996). The next generation of thrips again needs to acquire the virus during their early larval stage. Wijkamp et al. (1996) reported no pathological effect of tospovirus replication on the thrips biological system. Although, lower survival, reduced reproductive potential and slower development of thrips exposed to tospovirus-infected plants have been reported (Robb 1989; De Angelis et al. 1993).

24.3 Historical Perspective of Thrips and Tospovirus Research in India

Research on thrips in India started during the colonial period. From the available records, the first study on thrips was published in 1856 by Newman. However, the first comprehensive account of Indian Thysanoptera was published much later by Karny (1926). Thereafter, the Thysanoptera faunistics in India was enriched by the contributions of Ramakrishna Ayyar, Shumsher Singh, T N Ananthakrishnan and J S Bhatti. In India, altogether 693 species of thrips under 249 genera are known to occur (Alfred et al. 1998). Among them, only six thrips species are involved in transmission of tospoviruses. The most recent addition to the Indian thrips fauna is the *Frankliniella occidentalis*, the well-known vector of TSWV, reported from Karnataka and Tamil Nadu (Tyagi and Kumar 2015; Suganthy et al. 2016). The tospovirus research in India started much later. Since 1960s, the disease symptoms similar to tospovirus has been reported in black gram, brinjal, chili, cowpea, groundnut, mung bean, pea, potato, soybean, and tomato (Amin 1979; Bhat et al. 2001; Ghanekar et al. 1979; Khurana et al. 1989; Krishnareddy and Varma 1990; Narayanasamy et al. 1975; Nariani and Dhingra 1963; Prasada-Rao et al. 1987, 1980, 1985; Reddy et al. 1968; Sharma 1966). In 1992, the first report of groundnut bud necrosis virus (GBNV) was published (Reddy et al. 1992). Subsequently, peanut yellow spot virus (PYSV), (Satyanaryana et al. 1998), watermelon bud necrosis virus (WBNV) (Jain et al. 1998), iris yellow spot virus (IYSV) (Ravi et al. 2006), and Capsicum chlorosis virus (CaCV) (Kulkanikar et al. 2007) were reported from

India. Tomato spotted wilt virus (TSWV) infesting chrysanthemum in Nilgiri district of Tamil Nadu has been reported very recently (Renukadevi et al. 2015).

Not many researches have been undertaken on the thrips- tospovirus study in India. The Indian reports on the thrips-tospovirus study are limited to identification of the vectors and a few transmission parameters. To date, only six thrips vectors of tospoviruses have been reported from India. This chapter attempts to compile the information on general appearance, taxonomy, distribution, host plant diversity, economic impact of six thrips species transmitting tospoviruses in India. Further, all the Indian studies on thrips-tospovirus have been summarised in this chapter. As very less information is available, we have also included some of the important studies undertaken in other parts of the world to provide a comprehensive view.

24.4 Thrips Vector Species

24.4.1 *Thrips palmi*

24.4.1.1 General Appearance and Biology

T. palmi is commonly known as melon thrips. Adults of *T. palmi* are yellow or whitish in colour, measuring about 0.8–1 mm in length. The males appear similar to females but smaller in size. Wings of the adult are fringed and pale in colour. A black line is visible at the dorsal side of the body. Adult longevity is around 10–30 days depending on the temperature. During its life time, a female can lay up to about 200 eggs. The female cuts a slit on the upper layer of leaf tissue and inserts the eggs. The eggs are reniform in shape and opaque to pale in colour. After hatching, larvae start feeding on the leaf tissues. Larvae are pale in colour and resemble adults but smaller in size. The larvae do not bear wings or wing-pads. The mature second instar larvae construct earthen chamber in the soil or leaf litter for pupation. The pre-pupae and pupae resemble adults and larvae but they bear wing-pads.

24.4.1.2 Taxonomic Characters

T. palmi was first described by Karny in 1925 and redescribed by Bhatti (1980). Some of the synonyms reported are as follows:

Thrips clarus Moulton, 1928
Thrips leucadophilus Priesner, 1936
Thrips gossypicola Ramakrishna and Margabandhu, 1939
Chloethrips aureus Ananthakrishnan and Jagadish, 1967
Thrips gracilis Ananthakrishnan and Jagadish, 1968

Few key characters described by Sakimura et al. (1986) for diagnosis of *T. palmi* are summarised here. *T. palmi* has a clear yellow body with no dark areas on the head, thorax or abdomen. The antennae are always seven segmented. The antennal segment I and II are pale in colour while segment III is yellow with apex shaded and segment IV–VII are brown with a pale base. The ocellar setae III stand either just outside of the ocellar triangle or touch the tangent lines connecting the anterior ocellus and each of the posterior ocelli. The postocular setae II and IV are much smaller than the other setae. The metascutum bears medial sculpture which is made of longitudinal lines, often with short connecting lines but never reticulate. The forewings are uniformly and weakly shaded with prominent dark setae. The first vein of forewing has three distal setae. Seta S2 of abdominal terga III and IV is brown in colour and same in size as S3. The sternites III–VII in the male have narrow transverse glandular areas.

24.4.1.3 Distribution

In India, *T. palmi* is recorded from Delhi, Haryana, Punjab, Uttar Pradesh, Rajasthan, Madhya Pradesh, Maharastra, Karnataka, Tamil Nadu, Chhattisgarh, Himachal Pradesh, Jammu and Kashmir, Odisha, and West Bengal (Bhatti 1980, CABI/EPPO 1998, Tyagi and Kumar 2013, 2014).

24.4.1.4 Host Range

T. palmi is a polyphagous pest, attacks more than 50 plant species representing over 20 taxonomic families (Wang and Chu 1986). Walker (1992) listed over 200 records of the plants infested by *T. palmi*. It is a major insect pest in solanaceous and cucurbitaceous vegetables. It occurs in eggplant, potato, pepper, cucumber, pumpkin, squash, watermelon, muskmelon, zucchini, cowpea, soybean, cotton, groundnut, sesame, spinach, amaranthus, chrysanthemum, dahlia, orchid, plumeria and mango etc. (Ananthakrishnan 1955; Palmer et al. 1990; Verghese et al. 1988; CABI 2016a, b). *T. palmi* also attacks different weed species like *Cerastium glomeratum, Vicia sativa*, and *Capsella bursa-pastoris* (CABI 2016a, b).

24.4.1.5 Economic Impact

Limited information is available on the economic loss caused by *T. palmi*. Damage caused by *T. palmi* on cotton, tobacco and wild plants in Java, Sumatra, and India was found insignificant (Bournier 1986). While 50–90% losses on aubergine and cucumber were recorded in Trinidad (Cooper 1991b). *T. palmi* is also responsible for indirect losses by transmitting GBNV, WBNV, CaCV etc. in economically important crop plants. Seventy to 90% losses in groundnut were recorded in India due to infection of GBNV (Singh and Srivastava 1995) while it was around 29% in potato (Singh et al. 1997).

24.4.1.6 Virus Vector Relationship

Palmer et al. (1990) for the first time detected the association of *T. palmi* with tospovirus causing bud necrosis disease in groundnut crop. During the study, the causal organism was considered as TSWV. Later, it was identified as distinct tospovirus named GBNV. Further studies revealed that *T. palmi* could acquire the virus during larval stage and transmit it after becoming adults. The majority of the adult could transmit the GBNV for more than half of their life period. Ten viruliferous thrips per plant were found sufficient for 100% transmission. The male thrips were found to be efficient virus transmitters compared to the females (Vijayalakshmi 1994). The effect of GBNV infection on biology and host-preference of *T. palmi* has been studied (Daimei et al. 2017). *T. palmi* preferred GBNV-infected over healthy cowpea plants. The developmental period of *T. palmi* was decreased after exposure to GBNV (Daimei et al. 2017). In a different study, a group of thrips including *T. palmi* were evaluated for transmission potentiality of PYSV. The laboratory-reared individuals of *T. palmi* could not transmit PYSV at all (Reddy et al. 2010).

24.4.2 *Thrips tabaci*

24.4.2.1 General Appearance and Biology

T. tabaci is also known as Onion thrips. The adult *T. tabaci* is pale yellow to brown measuring about 1.0–1.3 mm in length. Female reproduces parthenogenetically and can lay about 80 eggs. Male is wingless and very rare. The adult female of *T. tabaci* makes an incision in the plant tissue and lay eggs singly under the epidermis of succulent leaf, flower, stem or bulb tissues. Eggs are elliptical in shape and approximately 0.2 mm long. The eggs usually hatch in 5–10 days. The young larvae are white to pale yellow in colour and have an elongated slender body. The larvae resemble adults but lack wings. The two larval stages are completed within 10–14 days depending on the prevalent temperature. The pupation occurs in soil near to the plant base. The pre-pupa and pupa are pale yellow to brown in colour. The antennae are bent to head and bear wing-pads. These pupal stages are inactive and last for 5–10 days. The entire life cycle requires about 3–4 weeks depending on temperature.

24.4.2.2 Taxonomic Characters

T. tabaci was first described by Lindeman (1889). Few synonyms are also reported as follows.

Ramaswamiahiella kallarensis Ananthakrishnan, 1960
Thrips kallarensis Ananthakrishnan, 1963
Thrips kallari Ananthakrishnan, 1963 (error for kallarensis).

T. tabaci is highly variable in colour, but usually with more or less brown or gray markings. All the postocular seate are sub-equal in length. The antennae are seven segmented. The metascutum bears sculpture made of longitudinal lines and partly reticulate. There are no campaniform sensilla in metascutum. The forewing doesnot have any dark cross-bands. The first vein of forewing has more than three distal setae. The abdominal tergite II bears three lateral marginal setae. The abdominal pleurotergites contain numerous ciliate microtrichia arising from lines of sculpture. In males, narrow transverse glandular areas are present only on abdominal sternites III–V.

24.4.2.3 Distribution

T. tabaci is a cosmopolitan species reported from many parts of India, *viz.* Andhra Pradesh, Assam, Bihar, Delhi, Gujarat, Haryana, Himachal Pradesh, Punjab, Jammu and Kashmir, Karnataka, Madhya Pradesh, Maharastra, Rajasthan, and West Bengal (UK, CAB International 1969; Thomas et al. 2008a, b; Patel et al. 1984; Gupta et al. 1992; Hanumantharaya et al. 2008; Aggarwal et al. 2007; Pandey et al. 2008; Ghosh and Senapati 2001).

24.4.2.4 Host Range

Onion thrips are reported from more than 25 plant families including grasses and broad leaves. This is a key pest in onion. The other major hosts are garlic, broccoli, cabbage, cauliflower, leek, carrot, beans, cotton, cucumber, melon, orchids, papaya, pineapple, rose, carnation, eggplant, and tomato etc.

24.4.2.5 Economic Impact

T. tabaci is the most serious pest in onion inflicting 34–43% loss in yield (Krishna Kumar et al. 2001). Kadri and Goud (2005) also recorded around 38% crop loss in onion by *T. tabaci* under unprotected conditions. Yield loss up to 47% was recorded in rabi cultivation of onion crop (Srinivas and Lawande 2004). *T. tabaci* causes an annual yield loss of 10–15% in India (Gupta et al. 1994). The economic impact of IYSV transmitted by *T. tabaci* in India is yet to be estimated.

24.4.2.6 Virus-Vector Relationship

T. tabaci is known to transmit IYSV (Corte^s et al. 1998; Hsu et al. 2010), TSWV (Wijkamp et al. 1995), tomato yellow fruit ring virus (Golnaraghi et al. 2007). However, very less research has been undertaken in India to know the relationship of *T. tabaci* with tospoviruses. *T. tabaci* was taken along with other thrips species for transmission of GBNV in groundnut, tomato, chili, cowpea, sunhemp and cotton. It was not found to be associated with GBNV in any of the host crops (Meena et al. 2005).

24.4.3 *Ceratothripoides claratris*

24.4.3.1 General Appearance and Biology

C. claratris (Shumsher) is also called as oriental tomato thrips. The adult thrips are brown in colour with yellow legs. The adult females have an average width of about 0.2 mm and a length of 1 mm. The males are slightly smaller than females (Rodmui 2002). Like other thrips species, *C. claratris* has two active larval stages and two inactive pupal stages. The adult female inserts the eggs in flowers, young fruits, and leaves. The adult female lives up to 20 days at 25 °C temperature and lays 45–70 eggs during its lifetime (Premachandra et al. 2004). The two larval instars resemble adult but smaller in size. The full grown second instar larvae drop off the plant and pupate in the soil or leaf litters. The developmental period from egg to adult is about 15 days at 25 °C.

24.4.3.2 Taxonomic Characters

The following synonyms of *C. claratris* are reported.

Taeniothrips claratris Shumsher, 1946
Mycterothrips moultoni Seshadri and Ananthakrishnan, 1954
Ceratothrips reticulatus Reyes, 1994

Few key morphological characters of *C. claratris* are discussed here. The head of adult female is slightly longer than width with three pairs of ocellar setae. The ocellar setae III arise close to anterolateral margin and almost twice of the length of the lateral margin of the ocellar triangle. The antennae are eight segmented. The antennal segment III is yellow and IV–V are variably yellow and brown in colour. The forewing is pale but weakly shaded. The first vein of forewing has two setae on distal half while the second vein has 14–16 setae. The pronotum bears two pairs of posteroangular setae. The metanotum is irregularly reticulated. The metanotal campaniform sensilla are absent. The abdominal tergites contain sculptured lines. No ctenidia are there. The abdominal tergite VIII bears long regular marginal comb. No discal setae present on the sternites. The S1 and S2 setae of sternite VII arise well in front of the margin. The males are smaller in size and the sternites III–VII bear about 12 small pore plates in two irregular transverse rows.

24.4.3.3 Distribution

C. claratris is probably originated in Asia and adapted to the hot humid tropical climate of South East Asia. This thrips species also invaded in Africa. In India, *C. claratris* reported from Delhi, Maharashtra, Orissa and Tamil Nadu.

24.4.3.4 Host Range

Tomato is considered as the main host of *C. claratris*. This thrips species is also known to feed on the foliage, stems, and fruits of solanaceous crops like capsicum, eggplants, and tobacco etc. It is also observed on cucurbitaceous, fabaceous and asteraceous crop plants.

24.4.3.5 Economic Impact

C. claratris may cause both direct and indirect damages to host plants. This is one of the most destructive insect pests of tomato causing considerable yield losses in both field and glasshouse conditions. It feeds on the foliage, stems, and fruits. Due to their feeding and oviposition, scarification and malformation occur in tomato. This is the most prevalent thrips species of tomato in Thailand transmitting CaCV (Premachandra et al. 2005). The crop loss due to *C. claratris* has not been estimated so far.

24.4.3.6 Virus-Vector Relationship

C. claratris is enlisted as the thrips species transmitting tospoviruses in India (Mandal et al. 2012). Not much information on the relationship of tospovirus and *C. claratris* is available from India. Premachandra et al. (2005) showed the successful transmission of CaCV in tomato by *C. claratris* in Thailand. The efficiency of transmission by the adult *C. claratris* was found dependent on the larval stage when the virus was acquired. A 69% transmission of CaCV was achieved with adult *C. claratris* after acquiring the virus by freshly emerged larvae. The comparative transmission efficiency between male and female adults did not significantly differ.

24.4.4 *Scirtothrips dorsalis*

24.4.4.1 General Appearance and Biology

S. dorsalis is an important pest of vegetable, ornamental and fruit crops. This thrips species is commonly known as chilli thrips or strawberry thrips. The adult *S. dorsalis* is white on emergence and turn yellowish subsequently. The adult is about 1.2 mm long. The wings are darker and dorsal side of the abdomen has dark spots forming incomplete stripes. The adult females insert the eggs inside plant tissues. The eggs are kidney-shaped, 0.075 mm long, and creamy whitish in colour. The larvae come out of the eggs in 4–6 days. The first instar larva is transparent and smaller in size. The two larval stages are completed within 8–10 days. Unlike other thrips species, the pupation of *S. dorsalis* takes place on leaves or leaf litter. The pre-pua and pupa are yellowish in colour and bear wing-pads. The ocelli are red pigmented and antennae bent over the head. The life span of *S. dorsalis* is influenced by the host plant (Seal et al. 2010).

24.4.4.2 Taxonomic Characters

S. dorsalis was first described as a different species by Hood (1919) from 34 females collected from India infesting castor and chili. The species was subsequently described by Ramakrishna Ayyar (1928) and Shumsher Singh (1944). Following are the synonyms reported for *S. dorsalis.*

Anaphothrips andreae Karny, 1925
Heliothrips minutissimus Bagnall, 1919
Neophysopus fragariae Girault, 1927
Scirtothrips andreae (Karny)
Scirtothrips fragariae (Girault)
Scirtothrips minutissimus (Bagnall)
Scirtothrips padmae Ramakrishna, 1942

Hoddle and Mound (2003) formulated keys to distinguish 21 species of genus *Scirtothrips.* Few key characteristic features for easy identification of *S. dorsalis* are described here. The head and legs of adult *S. dorsalis* are pale in colour. Three pairs of ocellar setae are present on the head. The third pair is situated between posterior ocelli. The median postocular setae are two pairs and equal in length. The antennae are eight segmented. The antennal segments I and II are pale and III and VIII are dark in colour. The pronotum bears four pairs of posteromarginal setae. There is elongated reticulation on the middle of metanotum. The forewing is distally light in color with straight cilia. The first vein of forewing bears three irregular setae distally. The second vein is incomplete and has two setae.

There are numerous microtrichia on the abdominal tergites as well as sternites. The tergites have a median dark patch. The antecostal ridges are also dark. The tergal microtrichial fields of the abdomen have three discal setae. The posteromarginal comb on abdominal segment VIII is complete. In the case of males, no drepanae is present on tergite IX.

24.4.4.3 Distribution

It is believed that *S. dorsalis* originated in the Indian subcontinent and spread worldwide. In India, it has been reported from Andhra Pradesh, Assam, Chhattisgarh, Delhi, Gujarat, Haryana, Himachal Pradesh, Jammu and Kashmir, Karnataka, Kerala, Madhya Pradesh, Maharashtra, Odisha, Rajasthan, Tamil Nadu, Uttar Pradesh, and West Bengal (CABI/EPPO 2010; Ramakrishna Ayyar and Subbiah 1935; Dev 1964; Raizada 1965; Singh et al. 2005; Ananthakrishnan and Sen 1980; Ramakrishna Ayyar and Margabandhu 1931; Ananthakrishnan 1969, 1971).

24.4.4.4 Host Range

S. dorsalis was first reported from chili and castor (Hood 1919). Since then it has been recorded from more than 100 plant species under 40 families. In India, several reports are there on the infestation of *S. dorsalis* in chili (Ramakrishna Ayyar 1932; Ramakrishna Ayyar and Subbiah 1935; Chakraborti 2004). This is a serious problem in groundnut also (Amin 1979, 1980). It is also recorded from cassava and taro (Rajamma et al. 2004). Duraimuragan and Jagadish (2004) recorded *S. dorsalis* on roses. Besides these crops, *S. dorsalis* has also been reported from onion, garlic, okra, cucumber, capsicum, watermelon, soybean, beans, citrus, cashew nut, tea, cotton, sunflower, peach, pomegranate, pears, and grapes etc. *Acacia* species are considered as wild host of *S. dorsalis* (CABI 2016a, b).

24.4.4.5 Economic Impact

S. dorsalis is a key pest of chili and one of the limiting factors in chili production in India. More than 90% yield reduction in chili was experimentally observed due to *S. dorsalis*. Although only 11–32% yield reduction was recorded in sweet pepper, the qualitative yield loss could be up to 92% in sweet pepper (Krishna Kumar 1995). Patel et al. (2009) reported yield loss of 61–74% dedicated to *S. dorsalis*. It also causes indirect yield losses by transmitting different tospoviruses and tobacco streak virus (TSV).

24.4.4.6 Virus-Vector Relationship

At least three tospoviruses are reported to be transmitted by *S. dorsalis* throughout the world *viz*. GBNV (German et al. 1992; Meena et al. 2005), peanut chlorotic fan-spot virus (Chen et al. 1996; Chu et al. 2001), and PYSV (Reddy et al. 2010). In India, *S. dorsalis* was first reported to transmit TSWV causing groundnut bud necrosis (Amin et al. 1981). Subsequent investigations (German et al. 1992) revealed the report of a TSWV-like virus in India to be actually GBNV. The results of another study conducted during 1990 to detect the causal agent of groundnut bud necrosis disease in *S. dorsalis* remained inconclusive (Palmer et al. 1990). The association of *S. dorsalis* with GBNV was reported in groundnut, tomato, and chili. The presence of GBNV in *S. dorsalis* was confirmed by RT-PCR assay (Meena et al. 2005). Transmission of PYSV was successfully demonstrated with *S. dorsalis* (Reddy et al. 2010). The larvae of *S. dorsalis* could acquire the virus within 30 min of AAP resulting 4.71% transmission of PYSV in groundnut. An increase in the AAP up to 1 day significantly increased the transmission rate. A single viruliferous adult required minimum 7 h IAP for transmitting PYSV. The most of the viruliferous adult *S. dorsalis* retained the virus from 76% to 100% of their life period with a transmission efficiency of 40–46.7%.

24.4.5 *Frankliniella schultzei*

24.4.5.1 General Appearance and Biology

F. schultzei is a polyphagous pest feeding different vegetable and ornamental crops. This is also known as the common blossom thrips. *F. schultzei* has two colour morphs- dark and pale (Sakimura 1969). The dark morph is dark brown in colour while the pale morph is yellowish with brown blotches. Both the colour morphs are anatomically similar. A mixed population of both the colour morphs is reported from India (Mound 1968). The distribution of two colour morphs are sympatric and they can interbreed. Both the colour morphs reproduce through arrhenotokous-parthenogenesis (Sakimura 1969). Adult females of *F. schultzei* are 1.1–1.5 mm in length. They make a slit and insert the eggs in flower tissues. The embryonic stage lasts for 4 days. Like other thrips, they undergo two larval instars and inactive pre-pupa and pupal instars in their life cycle. A complete generation may take around 12.6 days at 24.5 °C (Pinent and Carvalho 1998). The longevity of adult is recorded as 13 days.

24.4.5.2 Taxonomic Characters

This thrips species was first described by Trybom (1910) as *Physopus schultzei*. A number of synonyms are reported for *Frankliniella schultzei* as below.

Frankliniella interocellaris Karny
Frankliniella sulphurea Schmutz
Frankliniella delicatula Bagnall
Frankliniella dampfi Priesner (1923)

Frankliniella dampfiinterocellularis Karny (1925)
Frankliniella lycopersici Andrewartha (1937)
Parafrankliniella nigripes Firault (1928)
Frankliniella paucispionosa Moulton (1933)
Frankliniella sulphurea Schmutz (1913)
Physopus schultzei Trybom (1910)
Euthrips gossypii Shiraki (1912)
Frankliniella delicatual Bagnall (1919)
Frankliniella trybomi Karny (1920)
Frankliniella persetosa Karny (1922)
Frankliniella tabacicola Karny (1925)
Frankliniella africana Bagnall (1926)
Frankliniella anglicana Bagnall (1926)
Frankliniella aeschyli Girault (1927)
Frankliniella kellyana Kelly &Mayne (1934)
Frankliniella dampfi nana Priesner (1936)
Frankliniella favoniana Priesner (1938)
Frankliniella pembertoni Moulton (1940)
Frankliniella clitoriae Moulton (1940)
Frankliniella schultzeinigra Moulton (1948)
Frankliniella ipomoeae Moulton (1948)
Frankliniella insularis (Franklin) Morison (1930)

The key diagnostic characters to identify adult *F. schultzei* are described here. The adult bears eight-segmented antennae. Unlike other dark species of the genus *Frankliniella*, the interocellar setae arise along the marginal line connecting the front edges of the two hind ocelli. The postocular setae are shorter than interocellar setae in size and postocular setae IV are pronounced. The anteroangular setae are longer than anteromarginal setae on the anterior prothorax. The abdominal comb is weakly developed on the tergite VIII.

24.4.5.3 Distribution

F. schultzei is distributed worldwide. In India, this is reported to infest different vegetables, legumes, and ornamental plants.

24.4.5.4 Host Range

F. schultzei is polyphagous in nature and prefers to feed on ornamental and vegetable crops. More than 83 plant species representing 35 families are reported to be hosts of *F. schultzei* (Palmer 1990). Although cotton, beans, groundnut, and pigeonpea are reported as the major hosts, *F. schultzei* also infests tomato, chilies, sweet potato, coffee, sorghum, onion and sunflower (Hill 1975). Besides these crops, *F. schultzei* is reported from pineapple, strawberry, bell pepper, watermelon, melon, pumpkin, cowpea, tobacco, rice, carnation, roses, lettuce, spinach, mango, avocado, peach, and orchids etc.

24.4.5.5 Economic Impact

F. schultzei can cause both direct and indirect damage to the crops. Adults and nymphs of *F. schultzei* feed on the pollen and floral tissue of the plants. The pale spots and stripes on flowers are visible due to their sap-sucking nature. Due to their feeding, abortion of the flowers is often noticed. In the case of severe infestation, discoloration and stunted growth of the plant is reported (Amin and Palmer 1985). The indirect damage caused by *F. schultzei* is through transmission of tospoviruses.

24.4.5.6 Virus-Vector Relationship

Due to inaccurate identification of thrips species, Amin et al. (1981) reported *F. schultzei* as a vector of TSWV causing groundnut bud necrosis disease in India. Palmer et al. (1990) tested *F. schultzei* for the presence of a causal agent of groundnut bud necrosis disease. The results remained largely inconclusive. *F. schultzei* was found to be associated with GBNV transmission in cowpea and sunhemp (Meena et al. 2005). Another experiment showed that *F. schultzei* could not transmit PYSV in groundnut plant (Reddy et al. 2010).

24.4.6 *Frankliniella occidentalis*

24.4.6.1 General Appearance and Biology

F. occidentalis is one of the important pests in agriculture, commonly known as western flower thrips. This is native to the Southeastern USA and has successfully invaded to South America, Europe, Africa, Australia and Asia. The adult female of *F. occidentalis* is 1.4 mm in length and slightly longer than the male. The colour of the female varies from yellow to brown to nearly black with median dark brown markings on the abdomen while the males are pale yellow. The reproduction is mostly arrhenotokous parthenogenesis. The adult female inserts the eggs into the epidermis of leaves, flowers, and fruits. The eggs are reniform, measure 0.3 mm in size. Each female may lay up to 100 eggs. The freshly laid eggs are opaque in colour and gradually turn yellowish-orange with maturity. The mature eggs contain two red-eye spots at the apex. The egg incubation period is about 4–5 days at 25 °C temperature. The nymphs are yellowish in colour with red eyes. The larvae are active and tend to hide within leaf folds, flowers or fruit calyx. The full grown larvae fall off the plant and pupate in the ground. The two pupal instars are mostly inactive *i.e.* move only when they are disturbed. They are cream in colour and bear wing-pads. The pre-pupa matures rapidly but the pupa takes more than a week to emerge.

24.4.6.2 Taxonomic Characters

F. occidentalis has different colour morphs depending on the temperature *viz.* black in spring; yellow in summer and intermediate. The colour morphs may differ in the body size and length of setae (Bryan and Smith 1956). Therefore, the colour morphs were mistakenly considered as different species by many of the earlier workers such as the dark morph in California was identified as *F. moultoni* Hood.

More than 18 names are now considered as synonyms of *F. occidentalis* (Nakahara 1997).

Euthrips helianthi Moulton, 1911
Euthrips occidentalis Pergande, 1895
Euthrips tritici var. *californicus* Moulton, 1911
Frankliniella californica Moulton
Frankliniella canadensis Morgan, 1925
Frankliniella chrysanthemi Kurosawa, 1941
Frankliniella claripennis Morgan, 1925
Frankliniella conspicua Moulton, 1936
Frankliniella dahliae Moulton, 1948
Frankliniella dianthi Moulton, 1948
Frankliniella helianthi (Moulton)
Frankliniella moultoni Hood
Frankliniella nubila Treherne, 1924
Frankliniella occidentalis f. *brunnescens* Priesner, 1932
Frankliniella occidentalis f. *dubia* Priesner, 1932
Frankliniella syringae Moulton, 1948
Frankliniella trehernei Morgan
Frankliniella tritici maculata Priesner, 1925
Frankliniella tritici var. *moultoni* Hood, 1914
Frankliniella umbrosa Moulton, 1948
Frankliniella venusta Moulton, 1936

Few key morphological characters of *F. occidentalis* are described here. In adult female, the head is wider and bears three pairs of ocellar setae. The ocellar setae III are longer than the distance between the hind ocelli and do not arise between hind ocelli. The postocular setae I are present. They bear eight-segmented antennae. The antennal segments III–V are yellow and apices are variably brown in colour. The antennal segments III–IV bear forked sensorium. The forewings are pale in colour with two complete rows of dark veinal setae. The pronotum bears four pairs of major setae and four pairs of minor setae on the anterior margin. The metanotum has two pairs of setae at anterior margin. The metanotal campaniform sensilla are usually present. The ctenidium on tergite VIII is anterior to the spiracle. The posteromarginal comb on tergite VIII is not interrupted in the centre. The sternites III–VII do not bear discal setae. The males are smaller in size. The males do not have any posteromarginal comb on tergite VIII. The sternites III–VII bear transverse pore plates in males.

During recent times, the sequences of the D2 domain of 28S and cytochrome oxidase I gene indicated two distinct but sympatric genetic groups of *F. occidentalis* (Rugman-Jones et al. 2010). They are designated as the greenhouse (G) and lupin (L) strains. Therefore, the species, *F. occidentalis* is considered as a complex of two species which may exhibit differing biological characteristics including virus transmission.

24.4.6.3 Distribution

Several reports are available on the widespread occurrence of *F. occidentalis* (Tommasini and Maini 1995; Kirk and Terry 2003; Morse and Hoddle 2006; Reitz 2009). Until 1960, the thrips species *F. occidentalis* was restricted in the western half of North America. It has spread across the USA, Canada, and South America. It entered Europe in 1983 on glasshouse-grown African violet. From Europe and Israel, it has been suspected to spread to the highlands of eastern Africa. Although this was present on lupins tree in New Zealand since 1934 but did not spread to agricultural crops. Further, a pesticide-resistant strain was reported in greenhouses of New Zealand in 1992 (Martin and Workman 1994). This invaded Australia in 1993. This is also present in the Cameron Highlands of Peninsular Malaysia (Fauziah and Saharan 1991).

F. occidentalis recently entered India. It has been reported from Bangalore area of Karnataka (Tyagi and Kumar 2015) and has become a major concern for the quarantine authorities. The occurrence of *F. occidentalis* in Nilgiri hills of Tamil Nadu was confirmed by Suganthy et al. (2016). One earlier report of this species from Andhra Pradesh (Kulkarni 2010) has been found unreliable. The occurrence of this species in other parts of India is yet to be investigated.

24.4.6.4 Host Range

F. occidentalis is a polyphagous pest infesting more than 500 species in 50 plant families. It attacks flowers, fruits and leaves of a wide range of cultivated crops like capsicum, cucumber, beans, strawberry, eggplant, onion, lettuce, watermelon, tomatoes, chrysanthemum, carnation, rose, tuberose and orchids. In India, *F. occidentalis* is collected from tomato.

24.4.6.5 Economic Impact

F. occidentalis can directly reduce the qualitative and quantitative yield of economic crops as well as indirectly by transmitting more than five tospoviruses. This is considered as one of the important pests in glasshouse crops worldwide (Cloyd 2009). Annual losses due to the infestation of *F. occidentalis* were predicted to be US$30 million in the Netherlands. Further, the indirect damage caused by transmission of TSWV was US$19 million (Roosjen et al. 1998). The estimated crop losses due to tospoviruses transmitted by *F. occidentalis* exceeds US$1 billion per year worldwide (Goldbach and Peters 1994).

24.4.6.6 Virus-Vector Relationship

F. occidentalis is one of the most important vectors of tospoviruses. Several tospoviruses like chrysanthemum stem necrosis virus (Nagata and de Aévila 2000; Nagata et al. 2004), groundnut ringspot virus (Wijkamp et al. 1995; Nagata et al. 2004), impatiens necrotic spot virus (De Angelis et al. 1993; Wijkamp et al. 1995; Sakurai et al. 2004), tomato chlorotic spot virus (Nagata et al. 2004; Whitfield et al. 2005a), and TSWV (Medeiros et al. 2004; Nagata et al. 2004, Wijkamp et al. 1995) are known to be transmitted by *F. occidentalis* till date in different parts of the world. Contrastingly, no reports are available from India. *F. occidentalis* was first reported from India during 2015 (Tyagi and Kumar 2015). The occurrence of *F. occidentalis* in Nilgiri hills of Tamil Nadu was confirmed recently by Suganthy et al. 2016. At the same time, a severe necrosis disease noticed in chrysanthemum in the Nilgiris district of Tamil Nadu, which was confirmed to be caused by TSWV (Renukadevi et al. 2015). This was the first report of the appearance of TSWV in India. Although, *F. occidentalis* is known as a vector of TSWV, the experimental evidence relating to the transmission characteristics of chrysanthemum isolate of TSWV occurring in India by *F. occidentalis* is awaited. The further study on the prevallence of *F. occidentalis* and TSWV in different ecosystems in India is yet to be undertaken.

24.5 Concluding Remarks

Only a few reports are available on thrips-tospovirus study in India. Indian studies on thrips-tospovirus are restricted in identifying the thrips vectors and determining the rate of transmission. No studies have been undertaken for understanding the mechanisms involved in the interactions of thrips and tospovirus occurring in India. The mechanisms underlying the interactions of these tospoviruses with thrips allowing replication, protein-protein interactions and intercellular movement are largely unknown.The use of green fluorescent protein tagged viruses will aid visualization of virus uptake, tissue tropism, and intercellular movement in the vector cell. Mutational analysis of infectious clones will also allow identification of gene function and interactions of plant viruses with host proteins in the vector insect cell lines. Successful utilization of immunofluorescence confocal laser scanning microscopy to identify a neurotropic route for maize mosaic virus in its planthopper vector (Ammar and Hogenhout 2008) may also be useful for examining thrips-tospovirus interactions at the cellular level. Increased use of three-dimensional imaging techniques in combination with live cell imaging and correlative microscopy provide powerful tools to visually explore viral entry, movement between tissues, propagation, and egress. While several tospovirus interacting proteins are known, thrips receptor-like molecules are only beginning to be identified.

References

Aggarwal N, Brar DS, Buttar GS (2007) Evaluation of Bt and non-Bt version of two cotton hybrids under different spacings against sucking insect-pests and natural enemies. J Cotton Res Dev 21(1):106–110

Alfred JRB, Das AK, Sanyal AK (1998) Faunal diversity in India. ENVIS Centre, Zoological Survey of India, Calcutta

Amin BW (1979) Leaf fall disease of chilly and pepper in Maharashtra, India. PANS 25:131–134

Amin BW (1980) Techniques for handling thrips as a vectors of tomato spotted wilt and yellow spot virus of groundnut, *Arachis hypogea* L. Occas Pap Groundnut Entomol ICRISAT 80(2):1–20

Amin PW, Palmer JM (1985) Identification of groundnut Thysanoptera. Trop Pest Manag 31:268–291

Amin PW, Reddy DVR, Ghanekar AM (1981) Transmission of tomato spotted wilt virus, the causal agent of bud necrosis of peanut, by *Scirtothrips dorsalis* and *Frankliniella schultzei*. Plant Dis 65(8):663–665

Ammar E-D, Hogenhout SA (2008) A neurotropic route for Maize mosaic virus (Rhabdoviridae) in its planthopper vector Peregrinus maidis. Virus Res 131(1):77–85

Ananthakrishnan TN (1955) Notes on *Thrips palmi* Karny, attacking *Sesamurn indicurn*. J Bombay Nat Hist Soc 52:951–953

Ananthakrishnan TN (1969) Indian Thysanoptera, CSIR Zoology Monograph No.1. CSIR, New Delhi

Ananthakrishnan TN (1971) Thrips in agriculture, horticulture and forestry – diagnosis, bionomics and control. J Sci Ind Res (CSIR), New Delhi 30:130–146

Ananthakrishnan TN, Sen S (1980) Taxonomy of Indian Thysanoptera. Zool Surv. India, Handbook. Ser. 1

Bandla MD, Campbell LR, Ullman DE, Sherwood JL (1998) Inter action of tomato spotted wilt tospovirus (TSWV) glycoproteins with a thrips midgut protein, a potential cellular receptor for TSWV. Phytopathology 88:98–104

Bhat AI, Jain RK, Varma A, Chandra N, Lal SK (2001) Tospovirus(es) infecting grain legumes in Delhi, their identification by serology and nucleic acid hybridization. Indian Phytopathol 54:112–116

Bhatti JS (1980) Species of the genus *Thrips* from India (Thysanoptera). Syst Entomol 5:109–166

Bournier JP (1986) On the geographical distribution of *Thrips palmi* Karny. Coton Fibres Trop 41(1):59–61

Bryan DE, Smith RF (1956) The *Frankliniella occidentalis* complex in California. Univ Calif Publ Entomol 10:359–410

CABI (2016a) Invasive species compendium. http://www.cabi.org/isc/datasheet/49065#20103313776. Accessed on 14 Feb 2017

CABI (2016b) Invasive species compendium. http://www.cabi.org/isc/datasheet/53745. Access: 13 Feb 2017

CABI, EPPO (1998) Thrips palmi. [distribution map]. Distribution maps of plant pests, December (2nd revision), Wallingford, CAB International, Map, p 480

CABI/EPPO (2010) *Scirtothripsdorsalis*. [Distribution map]. Distribution maps of plant pests, no. December. CABI, Map, Wallingford, p 475. 1st revision

Chakraborti S (2004) Sustainable management of apical leaf curling in chilli. J Appl Zool Res 15(1):34–36

Chen CC, Chao CH, Chiu RJ (1996) Studies on host range, transmission and electron microscopy of peanut chlorotic fan-spot virus in Taiwan. Bull Taichung Dist Agric Improv Station Pub 52:59–68

Chisholm IF, Lewis T (1984) A new look at thrips (Thysanoptera) mouthparts, their action and effects of feeding on plant tissue. Bull Entomol Res 74:663–675

Chu FH, Chao CH, Peng YC, Lin SS, Chen CC, Yeh SD (2001) Serological and molecular characterization of peanut chlorotic fanspot virus, a new species of the genus *Tospovirus*. Phytopathology 91:856–863

Cloyd RA (2009) Western flower thrips (*Frankliniellaoccidentalis*) management on ornamental crops grown in greenhouses: have we reached an impasse? Pest Technol 3:1–9

Cooper B (1991) Status of *Thrips palmi* (Karny) in Trinidad. FAO Plant Protect Bull 39(1):45–46

Corte^s I, Livieratos IC, Derks A, Peters D, Kormelink R (1998) Molecular and serological characterization of Iris yellow spot virus, a new and distinct Tospovirus species. Phytopathology 88:1276–1282

Daimei G, Raína HS, Pukhrambam PD, Saurav GK, Renukadevi P, Malathi VG, Senthilraja C, Mandal B, Raman R (2017) Influence of on the life history traits and feeding preference of its vector Karny. Phytopathology

De Angelis JD, Sether DM, Rossignol PA (1993) Survival, development, and reproduction in western flower thrips (Thysanoptera: Thripidae) exposed to impatiens necrotic spot virus. Environ Entomol 22:1308–1312

Dev HN (1964) Preliminary studies on the biology of the Assam thrips, *Scirtothrips dorsalis* Hood on tea. Indian J Entomol 26:184–194

Duraimuragan P, Jagadish A (2004) Population fluctuation of *Scirtothrips dorsalis* Hood on rose in Karnataka. Insect Environ 10(3):112–113

Fauziah I, Saharan HA (1991) Research on thrips in Malaysia. Asian Veg Res Dev Centre Publ 91(342):29–33

Garry CE, Garry RF (2004) Proteomics computational analyses suggest that the carboxyl terminal glycoproteins of Bunyaviruses are class II viral fusion protein (beta-penetrenes). Theor Biol Med Model 1:10

German TL, Ullman DE, Moyer JW (1992) Tospoviruses: diagnosis, molecular biology, phylogeny, and vector relationships. Annu Rev Phytopathol 30:315–348

Ghanekar AM, Reddy DVR, Lizuka N, Amin PW, Gibbons RW (1979) Bud necrosis of groundnut (*Arachis hypogaea*) in India caused by tomato spotted wilt virus. Ann Appl Biol 93:173–179

Ghosh SK, Senapati SK (2001) Evaluation of brinjal varieties commonly grown in terai region of West Bengal against pest complex. Crop Res (Hisar) 21(2):157–163

Goldbach R, Peters D (1994) Possible causes of the emergence of tospovirus diseases. Semin Virol 5(2):113–120

Golnaraghi AR, Pourrahim R, Farzadfar S, Ohshima K, Shahraeen N, Ahoonmanesh A (2007) Incidence and distribution of tomato yellow fruit ring virus on soybean in Iran. J Plant Pathol 6:14–21

Groves C, German T, Dasgupta R, Mueller D, Smith DL, Pappu HR (2016) Seed transmission of soybean vein necrosis virus: the first tospovirus implicated in seed transmission. PLoS One 11, e0147342

Gupta RP, Bhardwaj BS, Srivastava PK (1992) Effect of combinations of different systemic and non-systemic insecticides on the control of onion thrips in Khavit onion. Newsl Assoc Agric Dev Found 12:7–8

Gupta RP, Srivastava KJ, Pandey UB, Midmore DJ (1994) Diseases and insect pests of onion in India. Acta Hortic 358:265–372

Hanumantharaya L, Goud KB, Naik LK (2008) Management of insect pests of sunflower using green lacewing, *Chrysoperla carnea* (Stephens) and neem seed kernel extract. Karnataka J Agric Sci 21(1):130–133

Harrewijn P, Tjallingii WF, Mollema C (1996) Electrical recording of plant penetration by western flower thrips. Entomol Exp Appli 79:345–353

Heming BS (1978) Structure and function of the mouthparts in larvae of *Haplothrips verbasci* (Osborn) (Thysanoptera, Tubulifera, Phlaeothripidae). J Morphol 156:1–38

Hill DS (1975) Agricultural insect pest of the tropics and their control. Cambridge University Press, London

Hoddle MS, Mound LA (2003) The genus *Scirtothrips* in Australia (Insecta, Thysanoptera, Thripidae). Zootaxa 268:1–40

Hood JD (1919) On some new Thysanoptera from southern India. Insecutor Inscitiae Menstruus 7(4–6):90–103

Hsu CL, Hoepting CA, Fuchs M, Shelton AM, Nault BA (2010) Temporal dynamics of Iris yellow spot virus and its vector, *Thrips tabaci* (Thysanoptera: Thripidae), in seeded and transplanted onion fields. Environ Entomol 39:266–277

Hunter WB, Ullman DE (1994) Precibarial and cibarial chemosensila in western flower thrips Frankliniella occidentalis (Thysanoptera: Thripidae). Int J Insect Morphol Embryol 23:69–83

Jain RK, Pappu HR, Pappu SS, Krishanareddy M, Vani A (1998) Watermelon bud necrosis tospovirus is a distinct virus species belonging to serogroup IV. Arch Virol 143:1637–1644

Jones DR (2005) Plant viruses transmitted by thrips. Eur J Plant Pathol 113:119–157

Kadri S, Goud KB (2005) Estimation of yield loss in onion due to onion thrips, *Thrips tabaci* Lindeman Karnataka. J Agric Sci 18(2):513–514

Karny H (1926) Studies on Indian Thysanoptera. Mem Dep Agric India Entomol Ser 9:187–239

Khurana SMP, Phadtare SG, Garg ID, Singh MN, Bharadwaj VP (1989) Potato stem necrosis epidemic due to tomato spotted wilt virus in India. In: Proc. IV Int. Plant Virus Epidemiol. Workshop, Montpellier, France, p 301

Kikkert M, Meurs C, van de Wetering F, Dorfmüller S, Peters D, Kormelink R, Goldbach R (1998) Binding of tomato spotted wilt virus to a 94-kDa thrips protein. Phytopathology 88:63–69

Kirk WDJ, Terry LI (2003) The spread of the western flower thrips Frankliniella occidentalis (Pergande). Agric For Entomol 5:301–310

Krishna Kumar NK (1995) Yield loss in chilli and sweet pepper due to *Scirtoithrips dorsalis* Hood (Thysanoptera: Thripidae). Pest Manag Hortic Ecosyst 1(2):61–69

Krishna Kumar NK, Veeregowda R, Rao SE, Krishna Moorthy PN (2001) Response of onion genotypes for thrips, *T. tabaci* Lindeman resistance. In: Proceedings of the second national symposium on IPM in horticultural crops new molecule pesticides, biopesticides, Bangalore, India, pp 17–19

Krishnareddy M, Varma A (1990) Prevalence of tomato spotted wilt virus in Delhi. Indian Phytopathl 43:252

Kulkarni HD (2010) Little leaf disease of eucalyptus by thrips, *Frankliniella occidentalis* (Pergande). Karnataka J Agric Sci 23(1):203–206

Kumm S, Moritz G (2008) First detection of *Wolbachia* in arrhenotokous populations of thrips species (Thysanoptera: Thripidae and Phlaeothripidae) and its role in reproduction. Environ Entomol 37(6):1422–1428

Kunkanikar SR, Sudarsana P, Rajagopalan P, Zehr UB, Naidu RA, Ravi KS (2007) First report of capsicum chlorosis virus in tomato in India. Plant Health Prog. doi:10.1094/PHP-2007-1204-01-BR

Lindeman K (1889) Die schSdlichsten Insekten des Tabak in Bessarabien. Bull Soc Imp Naturalistes Moscou 1(15):61–75

Lowry VK, Smith JW, Mitchell FL (1992) Life-fertility tables for *Frankliniella fusca* (Hinds) and *F. occidentalis* (Pergande) (Thysanoptera: Thripidae). Ann Entomol Soc Am 85:744–754

Mandal B, Jain RK, Krishnareddy M, Krishna Kumar NK, Ravi KS, Pappu HR (2012) Emerging problems of tospoviruses (Bunyaviridae) and their management in the Indian subcontinent. Plant Dis 96:468–479

Martin NA, Workman PJ (1994) Confirmation of a pesticide-resistant strain of western flower thrips in New Zealand. In: Popay AJ (ed) Proceedings of the forty seventh New Zealand plant protection conference, Waitangi Hotel, New Zealand, 9–11 August, 1994. New Zealand Plant Protection Society, Rotorua, pp 144–148

Medeiros RB, Ullman DE, Sherwood JL, German TL (2000) Immunoprecipitation of a 50-kDa protein: a candidate receptor component for tomato spotted wilt tospovirus (Bunyaviridae) in its main vector, Frankliniella occidentalis. Virus Res 67(2):109–118

Medeiros RB, Resende de OR, De Avila AC (2004) The plant virus tomato spotted wilt Tospovirus activates the immune system of its main insect vector, *Frankliniella occidentalis*. J Virol 78:4976–4982

Meena RL, Ramasubramanian T, Venkatesan S, Mohankumar S (2005) Molecular characterization of tospovirus transmitting thrips population from India. Am J Biochem Biotechnol 1:168–173

Milne DL, Manicom BQ (1978) Feeding apparatus of the South African citrus thrips *Scirtothripsaurantii* Faure. Citrus Subtrop Fruit J 535:6–11

Morse JG, Hoddle MS (2006) Invasion biology of thrips. Annu Rev Entomol 51:67–89

Mound LA (1968) A review of R.S. Bagnall's Thysanoptera collections. Bull Br Mus (Nat Hist) Entomol 11(Supplement):1–181

Mound LA (1971) Gall-forming thrips and allied species (Thysanoptera: Phlaeothripinae) from Acacia trees in Australia. Bull Br Mus (Nat Hist) Entomol 25:387–466

Mound LA (2015) Thysanoptera (Thrips) of the world a checklist. http://www.ento.csiro.au/thysanoptera/worldthrips.php.

Mumford RA, Baker I, Wood KR (1996) The biology of the Tospoviruses. Annal Appl Biol 128(1):1744–7348

Nagata T, de Aevila AC (2000) Transmission of Chrysanthemum stem necrosis virus, a recently discovered Tospovirus, by two thrips species. J Phytopathol 148:123–125
Nagata T, Inoue-Nagata AK, Smid HM, Goldbach R, Peters D (1999) Tissue tropism related to vector competence of *Frankliniella occidentalis* for tomato spotted wilt tospovirus. J Gen Virol 80:507–515
Nagata T, Almeida ACL, Resende RO, de Aevila AC (2004) The competence of four thrips species to transmit and replicate four tospoviruses. Plant Pathol 53:136–140
Nakahara S (1997) Annotated list of the *Frankliniella* species of the world (Thysanoptera: Thripidae). Contrib Entomol Int 2:353–389
Narayanasamy P, Kandaswamy TK, Ramiah M (1975) A new disease of groundnut in Tamil Nadu. Madras Agric J 62:371–375
Nariani TK, Dhingra KL (1963) A mosaic disease of groundnut (*Arachis hypogaea* L.) Indian J Agril Sci 33:25–27
Ohnishi J, Hosokawa D, Fujisawa I, Tsuda S (1998) Tomato spotted wilt tospovirus movement into salivary glands during pupation of thrips vector, Thrips setosus, is associated with its transmissibility. In: Peters D, Goldbach R (eds) Recent progress in Tospoviruses and Thrips research. Wageningen, pp 51–53
Palmer JM (1990) Identification of the common thrips of tropical Africa (Thysanoptera: Insecta). Tropic Pest Manag 36:27–49
Palmer JM, Reddy DVR, Wightman JA, Rao GVR (1990) New information on the thrips vectors of tomato spotted wilt virus in groundnut crops in India. Int Arachis Newsl 7:24–25
Pandey AK, Dwivedi SK, Ahmed SB, Ahmed Z (2008) Spatial distribution of thrips (*Thrips tabaci*) in onion (*Allium cepa*) under cold arid region of Jammu and Kashmir. Indian J Agric Sci 78(1):65–69
Pappu HR (1997) Managing tospoviruses through biotechnology progress and prospects. Biotechnol Dev Monitor 32:14–17
Pappu HR, Jones RAC, Jain RK (2009) Global status of tospovirus epidemics in diverse cropping systems: successes achieved and challenges ahead. Virus Res 141:219–236
Patel BK, Rote NB, Mehta NW (1984) Comparative efficacy of some insecticides against sucking pests of hybrid-4 cotton. Indian J Plant Protect 12(2):139–141
Patel BH, Koshiya DJ, Korat DM (2009) Population dynamics of chilli thrips, *Scirtothrips dorsalis* Hood in relation to weather parameters. Karnataka J Agric Sci 22(1):108–110
Pearce M (2005) 2004 Georgia plant disease loss estimates. University Georgia Cooperative Extension Service, p 24
Persley DM, Thomas JE, Sharman M (2006) Tospoviruses-an Australian perspective. Australas Plant Pathol 35:161–180
Pinent SMJ, Carvalho GS (1998) Biology of *Frankliniella schultzei* (Trybom) (Thysanoptera: Thripidae) in tomatoes. An Soc Entomol Bras Londrina 27(4):519–524
Prasada-Rao RDVJ, Izuka M, Ragunathan V, Joshi NC (1980) Incidence of tomato spotted wilt virus on tomato in Andhra Pradesh. Indian Phytopathol 33:436–439
Prasada-Rao RDVJ, Rajeswari R, Rao MVB, Ragunathan V, Joshi NC (1985) Spotted wilt of pea in India. Indian Phytopathol 38:90–93
Prasada-Rao RDVJ, Chakraborthy SK, Rajeswari R, Reddy AS (1987) Occurrence of tomato spotted wilt virus on cow pea (*Vigna unguiculata* (L.) WALP.), chili (*Capsicum annum* L.) and brinjal (*Solanum melongena* L.) Indian J Plant Prot 15:117–119
Premachandra WTSD, Borgemeister C, Chabi-Olaye A, Poehling HM (2004) Influence of temperature on the development, reproduction and longevity of *Ceratothripoides claratris* (Thysanoptera: Thripidae) on tomatoes. Bull Entomol Res 94:377–384
Premachandra WT, Borgemeister C, Maiss E, Knierim D, Poehling HM (2005) *Ceratothripoidesclaratris*, a new vector of a capsicum chlorosis virus isolate infecting tomato in Thailand. Phytopathology 95(6):659–663
Raizada U (1965) *Scirtothrips dorsalis* Hood. Detailed external morphology of its immature stages. Bull Entomol Loyola College 6:32–49

Rajamma P, Palaniswami MS, Jayaprakas CA (2004) Chilli thrips (*Scirtothrips dorsalis* Hood) (Thripidae: Thysanoptera) as a pest on cassava and taro. J Root Crops 30(1):74–75
Ramakrishna Ayyar TV (1928) Contribution to our knowledge of Thysanoptera of India. Mem Dep Agric India Entomol Ser 10(7):-251
Ramakrishna Ayyar TV (1932) Bionomics of some thrips injurious to cultivated plants in South India. Agric Livest of India 2:391–403
Ramakrishna Ayyar TV, Margabandhu V (1931) Notes on Indian Thysanoptera with brief introduction of new species. J Bombay Nat Hist Soc 34:1029–1040
Ramakrishna Ayyar TV, Subbiah MS (1935) The leafcurl disease of chillies caused by thrips in the Guntur and Madura tracks. Madras Agric J 23:403–410
Ravi KS, Kitkaru AS, Winter S (2006) Iris yellow spot virus in onions: a new tospovirus record form India. Plant Pathol 55:288
Reddy M, Reddy DVR, Appa RA (1968) A new record of virus disease on peanut. Plant Dis Rep 52:494–495
Reddy DVR, Ratna AS, Sudarshana MR, Kiran KI (1992) Serological relationships and purification of bud necrosis virus, a tospovirus occurring in peanut (*Arachishypogaea* L.) in India. Ann Appl Biol 120:279–286
Reddy GK, Reddy MK, Reddy DVR, Muniyappa V (2010) Transmission of peanut yellow spot virus (PYSV) by Thrips, *Scirtothripsdorsalis* Hood in groundnut. Arch Phytopathol Pl Protection 43(4/6):421–429
Reitz SR (2008) Comparative bionomics of *Frankliniella occidentalis* and *Frankliniella tritici*. Fla Entomol 91:474–476
Reitz SR (2009) Biology and ecology of the western flower thrips (Thysanoptera: Thripidae): the making of a pest. Fla Entomol [Western Flower Thrips Symposium Papers, Florida, USA, 15 July 2008] 92(1):7–13
Renukadevi P, Nagendran K, Nakkeeran S, Karthikeyan G, Jawaharlal M, Alice D, Malathi VG, Pappu HR (2015) First report of tomato spotted wilt virus infection of chrysanthemum in India. Plant Dis 99(8):1190
Richards OW, Davies RG (1977) Classification and biology. In: Imms' general textbook of entomology, vol 2, 10th edn. Chapman and Hall Ltd, London, pp 782–792
Robb KL (1989) Analysis of Frankliniella occidentalis (Pergande) as a pest of horticultural crops in California greenhouses. Ph.D Dissertation, University of California Riverside, p 149
Rodmui P (2002) Population dynamics and biological control of thrips, *Ceratothripoides claratris* (Shumsher) (Thysanoptera: Thripidae), on tomato under protected cultivation in Thailand. MSc thesis, Kasetsart University, Bangkok, p 50
Roosjen M, Buurma J, Barwegen J (1998) Verbetering schade-inschattingsmodel quarantaine-organismen glastuinbouw. Verslagen en Mededelingen, Plantenziektenkundige Dienst, Wageningen, 197:1–24
Rotenberg D, Jacobson AL, Schneweis DJ, Whitfield AE (2015) Thrips transmission of tospoviruses. Curr Opin Virol 15:80–8
Rugman-Jones PF, Hoddle MS, Stouthamer R (2010) Nuclear-mitochondrial barcoding exposes the global pest western flower thrips (Thysanoptera: Thripidae) as two sympatric cryptic species in its native California. J Econ Entomol 103(3):877–886
Sakimura K (1962) The present status of thripsborne viruses. In: Maramorosch K (ed) Biological transmission of disease agents. Academic, New York, pp 33–40
Sakimura K (1963) *Frankliniella fusca*, an additional vector for the tomato spotted wilt virus, with notes on *Thrips tabaci*, a thrips vector. Phytopathology 53:412–415
Sakimura K (1969) A comment on the color forms of *Frankliniella schultzei* (Thysanoptera: Thripidae) in relation to transmission of the tomato-spotted wilt virus. Pacafic Insects 11:761–762
Sakimura K, Nakahara LM, Denmark HA (1986) A thrips, Thrips palmi Karny (Thysanoptera: Thripidae), Entomology Circular No. 280. Division of Plant Industry, Dept. of Agriculture and Consumer Services, Florida, p 4

Sakurai T, Inoue T, Tsuda S (2004) Distinct efficiencies of impatiens necrotic spot virus transmission by five thrips vector species (Thysanoptera: Thripidae) of tospoviruses in Japan. Appl Entomol Zool 39:71–78

Satyanaryana T, Gowda S, Lakshminaryan Reddy K, Metals E, Dawson WO, Reddy DVR (1998) Peanut yellow spot virus is a member of a new serogroup of Tospovirus genus based on small (S) RNA sequence and organization. Arch Virol 143:353–364

Seal DR, Klassen W, Kumar V (2010) Biological parameters of *Scirtothrips dorsalis* (Thysanoptera: Thripidae) on selected hosts. Environ Entomol 39(5):1389–1398

Sharma DC (1966) Studies on bunchy top, chlorosis and ring mottle virus diseases of groundnut (*Arachis hypogaea* L.) Phytopathol Z 57:127–137

Singh S (1944) Studies of some Indian Thysanoptera. In: Proceedings of Royal Entomological Society, London (P), 13:139–144

Singh AB, Srivatava SK (1995) Status and control strategy of peanut bud necrosis disease in Uttar Pradesh. In: Buiel AAM, Parevliet JE, Lenne JM (eds) Recent studies on peanut bud necrosis disease: Proceedings of a meeting, 20 March. International Crop Research Institute for Semi-Arid Tropics, Patancheru, pp 65–68

Singh RB, Srivastava KK, Khurana SMP, Pandey SK (1997) Assessment of yield losses due to potato stem necrosis disease. Indian J Virol 3:135–137

Singh V, Thakur BS, Chandraker MK (2005) Bio-efficacy of insecticides against insect pests of chilli. Environ Ecol 23S(Special 3):600–604

Srinivas PS, Lawande KE (2004) Impact of planting dates on *Thrips tabaci* Linderman infestation and yield loss in onion (*Allium cepa* L.) Pest Manag Hortic Ecosyst:10, 11–18

Suganthy M, Rageshwari S, Senthilraja C, Nakkeeran S, Malathi VG, Ramaraju K, Renukadevi P (2016) New record of western flower thrips, Frankliniella occidentalis (Pergande) (Thysanoptera: Thripidae) in South India. Int J Env, Agric Biotechnol 1(4):857–867

Thomas A, Chaudhary B, Singh J, Ramamurthy VV (2008a) Preliminary observations on the non-target insect communities in Bt cotton. Indian J Entomol 70(4):400–402

Thomas A, Chaudhary B, Singh J, Ramamurthy VV (2008b) Preliminary observations on the non-target insect communities in Bt cotton. Indian J Entomol 70(4):400–402

Tommasini MG, Maini S (1995) Frankliniella occidentalis and other thrips harmful to vegetable and ornamental crops in Europe. Wageningen Agricultural University Papers, Wageningen; Landbouwuniversiteit Wageningen (Wageningen Agricultural University), 95(1): iii +1–iii 42

Turina M, Kormelink R, Resende RO (2016) Resistance to tospoviruses in vegetable crops: epidemiological and molecular aspects. Annu Rev Phytopathol 54:347–371

Tyagi K, Kumar V (2013) Thysanoptera fauna of Chhattisgarh India. Rec Zool Surv India 113(4):133–138

Tyagi K, Kumar V (2014) New records of thrips (Thysanoptera, Terebrantia, Thripidae) from Himachal Pradesh, India. Rec Zool Surv India 114(4):591–598

Tyagi K, Kumar V (2015) First report of western flower thrips, *Frankliniella occidentalis* (Pergande) (Thripidae: Thysanoptera) from India -a potential havoc to Indian agriculture. Halteres 06:1–3

UK CAB International (1969) *Thrips tabaci.* [Distribution map]. Distribution maps of plant pests, June. CAB International, Map, Wallingford, p 20. Revised

Ullman DE (1996) Thrips and tospoviruses: advances and future directions. Acta Hort 431:310–324

Ullman DE, German TL, Sherwood JL, Westcot DM (1992) In: Parker BL, Skinner M, Lewis T (eds) Thrips biology and management. Plenum, New York, pp 135–152

Ullman DE, German TL, Sherwood JL, Westcot DM, Cantone FA (1993a) Tospovirus replication in insect vector cells: immunocytochemical evidence that the nonstructural protein encoded by the S RNA of tomato spotted wilt tospovirus is present in thrips vector cells. Phytopathology 83:456–463

Ullman DE, Sherwood JL, German TL, Wescot DM, Chenault KD, Cantone FA (1993b) Location and composition of cytoplasmic inclusions in thrips cells infected with tomato spotted wilt tospovirus (TSWV). Phytopathology 83:1374

Ullman DE, Sherwood J, German TL (1997) Thrips as vectors of plant pathogens. In: Lewis T (ed) Thrips as crop pests. CAB International, New York, pp 539–565

Ullman DE, Whitfield AE, German TL (2005) Thrips and tospoviruses come of age: mapping determinants of insect transmission. PNAS 102(14):4931–4932

Varghese A, Tandon PL, Rao GSP (1988) Ecological studies relevant to the management of *Thripspalmi* Karny on mango in India. Tropic Pest Manag 34(1):55–58

Vijayalakshmi K (1994) Transmission and ecology of *Thrips Palmi* Karny, the vector of peanut bud necrosis virus, Ph.D. (Agril. Entomology) thesis Andhra Pradesh Agril. Univ. Rajendranagar, Hydarabad (India), p 120

Walker AK (1992) Pest status. In: Girling DJ (ed) *Thrips palmi:* a literature survey. International Institute of Biological Control, Ascot, pp 3–6

Wang CL, Chu YI (1986) Review of the southern Thrips, *Thrips palmi* Karny. Chinese J Entomol 6:133–143. (in Chinese)

Whitfield AE, Ullman DE, German TL (2005a) Tospovirus-thrips interactions. Annu Rev Phytopathol 43:459–489

Whitfield AE, Ullman DE, German TL (2005b) Tomato spotted wilt virus glycoprotein GC is cleaved at acidic pH. Virus Res 110:183–186

Wijkamp I, Peters D (1993) Determination of the median latent period of two tospoviruses in *Frankliniella occidentalis*, using a novel leaf disk assay. Phytopathology 83:986–991

Wijkamp I, van Lent J, Kormelink R, Goldbach R, Peters D (1993) Multiplication of tomato spotted wilt virus in its insect vector, *Frankliniella occidentalis*. J Gen Virol 74:341–349

Wijkamp I, Almarza N, Goldbach R, Peters D (1995) Distinct levels of specificity in thrips transmission of tospoviruses. Phytopathology 85:1069–1074

Wijkamp IF, Wetering Van De, Goldbach R, Peters D (1996) Transmission of tomato spotted wilt virus by *Frankliniella occidentalis*: median acquisition and inoculation access period. Ann Appl Biol 129:303–313

Indian Contribution to Whitefly (*Bemisia tabaci*) Research 25

K. Chandrashekar and P.R. Shashank

Abstract

Whitefly, *Bemisia tabaci* (Gennadius) is a destructive insect pest and vector of Begomoviruses (family *Geminiviridae*) which transmits virus diseases in several important crop plants leading to severe losses in yield and quality. Whiteflies also attack and cause direct feeding damage to large number of plant species. The pest has been extensively studied world over by the several research groups. Similarly Indian workers also contributed significantly for the understanding of the pest and diseases transmitted by them. In this chapter we intend to provide overview of contribution of Indian workers to *B. tabaci* research. We cover several aspects of *B. tabaci* including basic studies on pest incidence and its management; identification of biotypes and their host range, season incidence, association with symbiotes, and also advanced studies related to genomics; whitefly management using novel proteins and RNAi in transgenic plants.

Keywords

Whitefly • Bemisia tabaci • Begomoviruses

K. Chandrashekar (✉)
ICAR-IARI, Regional Station, Pune 411067, Maharashtra, India
e-mail: kc_shekar2001@yahoo.com

P.R. Shashank
Division of Entomology, ICAR-IARI, Pusa Campus, New Delhi 110012, India
e-mail: shashank@iari.res.in

B. Mandal et al. (eds.), *A Century of Plant Virology in India*,
https://doi.org/10.1007/978-981-10-5672-7_25

25.1 Introduction

Whitefly, *Bemisia tabaci* (Gennadius) is a complex cryptic species and destructive insect pest which is reported to attack and damage about 600 plant species (Nombela and Muniz 2010). Whitefly cause damage directly by feeding on plants phloem sap and also transmits plant viruses (Berlinger 1986; Singh et al. 1994). Honeydew secreted by them promotes the growth of fungi like sooty mold on fruits, flowers and other economically important plant parts which reduces their market value. Sooty mold on leaves inhibits photosynthesis and reduces the crop yield (Berlinger 1986). Whitefly feeding results in symptoms like chlorosis, withering and premature dropping of leaves and sometimes plant's death. Whiteflies transmitted plant viruses are responsible for over 40 diseases of important crops plants worldwide resulting in yield loss ranging from 10 to 100% which depends upon the factors like type of the crop, growing season, and abundance of the whiteflies (Singh et al. 1994; Singh and Mahant 1989).

B. tabaci is a vector of begomoviruses (family *Geminiviridae*). *B. tabaci* is reported to transmit 111 viruses (Tiwari et al. 2013). About 31 indistinguishable species are reported in *B. tabaci* complex out of which about 9 groups are reported in India, which show differences with respect to their genetic makeup, mating behavior, reproductive characteristics, range of host plants and ability to transmit virus to host plants (Ellango et al. 2015; Crowder et al. 2010). They have been identified mainly based on the techniques like RAPD-PCR (Debarro and Driver 1997; Li et al. 2003), ribosomal internally transcribed spacer 1 sequence known as ITS-1 (Debarro et al. 2000; Li et al. 2005) and sequencing of mitochondrial cytochrome oxidase (mtCO1) (Rekha et al. 2005; Jun et al. 2012). Earlier, *B. tabaci* biotypes were classified as B and Q types based on partial sequence alignment of ITS-1 in which B biotypes shows distinct sequence deletion (Li et al. 2005). B and Q types are also distinguished on the basis of microsatellite site Bem 23 analysis where in both species show distinct band pattern (Bel-Kadhi et al. 2008).

Overall progress of whitefly research in India can be divided in three major categories (Fig. 25.1) viz., virus transmission during pre 1980 and IPM during 1980-1990; biotype determination on the basis of host specificities during 1990-2000 and by using molecular methods, since 2000 onwards and studies related to RNAi, symbionts biodiversity, transcriptomics and whitefly resistant transgenics since 2010 onwards. In this chapter an overview of whitefly research in India is presented.

25.2 *B. tabaci* Species Complex and Biotypes

First report of the occurrence of *B. tabaci* "B" biotype in South India was from the Kolar district in the State of Karnataka, in the year 1999 (Banks et al. 2001). Increase of 1000-fold in population of B-biotype *B. tabaci* was recorded on several vegetables crops which are occasionally infested by whitefly local strains, this resulted in epidemic of tomato leaf curl virus disease (Banks et al. 2001).

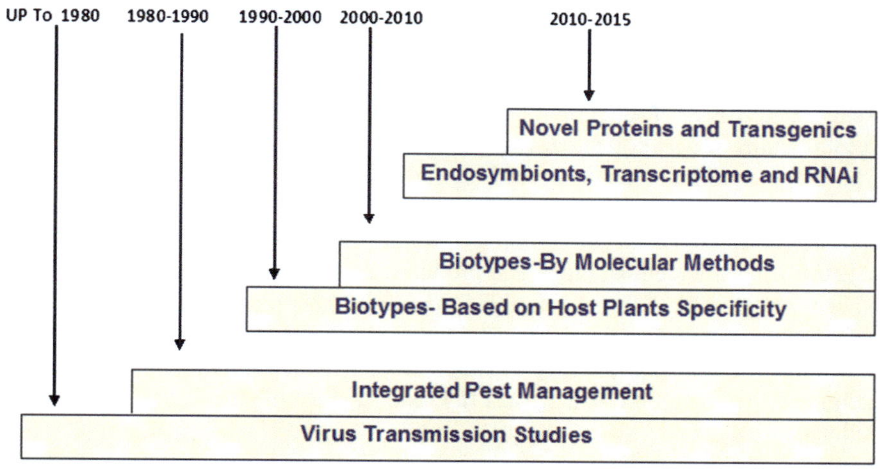

Fig. 25.1 The progress of whitefly research in India

Palaniswami et al. (1996) have reported that extent of spread of *Indian cassava mossaic virus* (ICMV) by the whitefly *B. tabaci* varies in field. They also observed that *B. tabaci* biotype from cassava were able to transmit ICMV but not other biotypes (Palaniswami et al. 2004). Similarly, Lisha et al. (2003) also reported occurrence of biotypes in *B. tabaci* on the basis of preference for host plants, pattern of esterase protein bands, and ability to transmit virus. They identified two strains known as cassava-strain and sweet potato-strain. The cassava-strain of whiteflies could successfully transmit cassava mosaic virus from-infected plants to healthy cassava seedlings but not the sweet potato-strain of whiteflies. Rekha et al. (2005) studied genetic diversity in 108 population of the *B. tabaci* from south India mostly from state of Karnataka using RAPD and partial sequencing of mtCO1 gene. Based on phylogenetic analysis of mtCO1 gene they reported three distinct genotypes indigenous to India but can also be found in China, Malaysia, Nepal, Pakistan and Thailand. Similarly, Sharma et al. (2008) grouped *B. tabaci* collected from cotton, brinjal, potato, tomato, soybean and weed (*Sida* sp) from Punjab in to three distinct groups viz., Group 1 (Cotton, Soybean and Sida), Group 2 (Potato and Brinjal) and Group 3 (Tomato) based on genetic variability study by RAPD- PCR analysis. Perumal et al. (2009) compared (using RAPD markers) 13 whitefly samples collected from several localities of Tamil Nadu and from several plant species. The populations were clustered in to three clusters viz., population collected from okra and cotton, eggplant and cauliflower and eggplant. Jayasekera et al. (2010) establish that the characters of puparia can be used for distinguishing populations of *B. tabaci* on the basis of morphometrics analysis of puparia and adults of *B. tabaci* collected on several host plants in the Delhi region.

Chowda-reddy et al. (2012) collected 266 samples of *B. tabaci* adults from 31 regions across India, grouped them in to host-plant and region of collection and compared using partial mtCOI genes. Majority of these samples were grouped to Asia I, Asia II-5, Asia II-7, and Asia II-8. Biotype-B was clustered into the MEAM-1group and *B. tabaci* collected from pumpkin was close to Asia I group. The biotype-B populations were observed to have more efficient virus transmission characteristics viz., the acquisition access periods, inoculation access periods, latent periods for *Tomato leaf curl Bangalore virus* (ToLCBV). Further they observed MEAM-1 has a wide host range and produces higher quantities of honeydew than other types. Tomato was not found be a preferred host plant for Asia-I because it was associated with low fecundity. Thomas et al. (2014) studied effect of variation in host plant characters specially trichome density influence on the *B. tabaci* population and they recorded Asia 1, Asia II 1, and Asia II 7 biotypes on several cotton cultivars.

Prevalence of *B. tabaci* genetic groups in Karnataka was also determined by Roopa et al. (2015) by analysing mtCOI sequences from 71 samples of *B. tabaci* collected on various host plants. Study revealed the prevalence of new group Middle East Asia Minor-K in addition previously known groups Asia-I, Asia-II-7, Asia-II-8, and MEAM-1. Thomas and Ramamurthy (2014) analyzed *B. tabaci* populations collected from cotton using sequences of MtCOI, and ITS1 and suggest that mtCOI-II region or 3′ end of the gene should be preferred over ITS1 region for comparing species complex of *B. tabaci*. Ellango et al. (2015) sequenced 850 bp of the mtCOI gene of *B. tabaci* collected across India and confirmed the existence of genetic groups Asia I, Asia I-India, Asia II-1, Asia II-5, Asia II-7, Asia II- 8, and Asia II-11 and China-3 group which was only in Birbhum district of West Bengal, India (Fig. 25.2). Similarly, Prasanna et al. (2015) studied the whitefly species complex of soybean and reported existence of Asia I, Asia II 1, and Asia II 7a using *mtCOI* based phylogenetics.

Thomas et al. (2014) studied variations of *B. tabaci* on different cotton genotypes by comparing morphology of the puparia and *mtCOI* sequence analyses and confirmed existence of species Asia1, AsiaII1, and AsiaII7. Chaubey et al. (2015) compared and reported variation ranging between 23.65 and 25.75 days in developmental periods of cryptic species Asia I, Asia II-1 and Asia II-7. Asia I showed longer developmental period compared to Asia II. Principal component analysis of the life history parameters clustered Asia I, Asia II-1 and Asia II-7 into three overlapping clusters. Chaubey and Andrew (2015) recommended that distinct variations existing in the length of operculum, antennae, lingula and pupariae, breadth of vasiform orifice and operculum, and distance of transverse moulting suture, should be used as supplementing evidence in addition to molecular variation for describing different species of *B. tabaci*.

The distribution of *B. tabaci* biotypes differs among different states India. The state wise distribution pattern of different biotypes is represented in Fig. 25.2.

Fig. 25.2 Prevalence of *B. tabaci* genetic groups in India (Modified and adopted from Ellango et al. 2015)

25.3 Virus Transmission Studies

First report of *B. tabaci* as a serious insect pest of crop plants in India was in the 1929 by Misra and Lamba (1929). Then onwards, many Indian workers have reported different aspects of whitefly and virus transmission. Several aspects of virus transmission such as whiteflies numbers required for successfully transmitting virus to healthy plants; effect of pre-acquisition starvation/post-acquisition starvation on transmission efficiency, relative efficacy of male/female whiteflies in virus transmission; host range of whiteflies and viruses transmitted by them have been reported by Indian researchers.

Capoor (1949) reported that whitefly require to probe their stylets for 15–30 min to into the phloem tissues to acquire virus. Varma (1952) shown that a single viruliferous whitefly can successfully transmit viruses like yellow vein mosaic of okra. Capoor and Ahmad (1975) observed 77.3% transmission of pumpkin yellow mosaic virus by 20 whiteflies. Seetharama Reddy and Yaraguntaiah (1981) reported that whitefly need to feed for at least 30 min for successfully transmitting the virus to healthy plants and percentage of successful infection increases with the increase in inoculation feeding period. According to Ragupathi (1989) hundred per cent transmission of TLCV is can be achieved using three whiteflies whereas in case of *Yellow mosaic virus diseases* of urdbean and soybean, but about 15 to 20 whiteflies were required for effective transmission. Jayashee et al. (1999) reported that 21.67% transmission of *Pumpkin yellow vein mosaic virus* (PYVMV) to healthy pumpkin plants can be obtained using single whitefly *B. tabaci* but for getting 100% infection 15 whiteflies needed per plant. It was observed that the number of whiteflies and PYVMV transmission is positively correlated when 20 whiteflies were used for transmission in urdbean and soybean. Pre-acquisition starvation increases per cent transmission, while post-acquisition starvation reduces the transmission efficiency. Number of days required for the symptom expression in host plant also reduces with increasing the number of whiteflies up to 20 whiteflies per plant. Muniyappa et al. (2001) detected ToLCV Ban4 in host tomato and its vector *B. tabaci*. Further a single *B. tabaci* was able to acquire ToLCV-Ban4 from infected plant to and transmit test tomato plant, whereas 100% transmission was achieved when five insects were used. When the insects were exposed to 24 h to the infected plants they could infect tomato test plants for period of up to 12 days. Female whiteflies were 95% efficient and males were only 25% efficient in transmitting the virus.

Sharma and Rishi (2003) performed transmission test of *Cotton leaf curl virus* (CLCuV) from cotton to different plant species by *B. tabaci* and showed that the CLCuV- Haryana could infect plant species belonging to four families viz., Asteraceae (*Ageratum conyzoides*), Fabaceae (*Phaseolus vulgaris cv.* Top crop), Malvaceae (*Abelmoschus esculentus,* cotton) and Solanaceae (*Lycopersicon esculentum, Nicotiana tabacum* var. White Burley, *N. benthamiana*). Further studies on effect of the number of whiteflies on transmission of CLCuV indicate that a single viruliferous whitefly could transmit 15% virus successfully, but 20 whiteflies required for 80% transmission. Female adults found to be twice as efficient as males in the transmission of CLCuV.

Tripathi and Varma (2002a) evaluated *Lycopersicon* genotypes for resistance to the ToLCV by whitefly (*B. tabaci*) transmission and agro-inoculation technique. Infection rate was different between the two methods with agro inoculation technique showing higher rate of infection. The difference between the two methods has been attributed to the different level of resistance in host plants to the vector whiteflies. The authors suggest that that agroinoculation should be more efficient method of delivering ToLCV into plants than the traditional whitefly inoculation method. Verma et al. (2011) highlighted the emergence and spread and threat of

Whitefly-transmitted Gemini viruses (WTGs) in several important crops like cassava, cotton, grain legumes, and vegetables of solanaceous, cucurbitaceous and, malvaceous family. They concluded that the factors such as evolution of variants, of WTGs, aggressive biotypes of *B. tabaci*, adaptation of new cropping systems and crops hosts susceptible genes and of virus infected planting materials as the responsible for outbreak of WTGs.

Transmission studies performed by Senanayake et al. (2012) with chilli leaf curl isolates, collected from Jodhpur area of Rajasthan, India, revealed that for 100% transmission requires eight whiteflies per plant. The whiteflies retains virus for 5 days post acquisition. Minimum acquisition access period (AAP) and inoculation access period (IAP) were of 180 and 60 min, respectively. Host range studies performed using 25 species of plants, but virus infected only capsicum, papaya, tomato and tobacco species *(Nicotiana tabacum, N. benthamiana)*. They showed sequence identity of 96.1% with ChiL- CV from potato. Anokhe et al. (2015) on the basis of transmission abilities reported that Asia-I *B. tabaci* biotype has significantly higher transmission efficiencies for *Tomato leaf curl New Delhi virus* (ToLCNDV) compared to Asia II-1.

25.4 Biotechnological Advances in *B. tabaci* Research in India

25.4.1 Whitefly Bioassays

A simplified bioassay method for testing of oral toxins by mixing with artificial diet has been developed by Upadhyay et al. (2011). This method involves collection of adult whiteflies directly into the bioassay vials from the leaves, without involving steps like aspiration or anesthetization (Fig. 25.3). Artificial diet mixed with test protein was placed in caps of bioassay vials. The caps can be easily replaced when ever required. Using this method, whiteflies can be maintained on artificial diet for about 6 days. Using this method one can also calculate the average amount diet/test protein consumed per insect by measuring initial and finial weight of diet provided in the bioassays vial and dividing it by number of whiteflies in each bioassay vial.

Dixit et al. (2013) modified above method for testing transgenic plants, where they used leaf discs of plant in place of liquid media. According to modified method, agar (1%) was poured up to two-third level in the caps of above bioassay tubes and leaf disc of plants to be tested were placed on the solidified agar. Caps having leaf discs were fitted as above to the bioassay tubes containing insects. Periodic count of dead whiteflies at the bottom of the bioassay tube was used as mortality data.

25.4.2 Contribution to Sequence Database of *B. tabaci* and Its Endosymbiotes

The maximum contribution of Indian worker to sequence data base is for the *B. tabaci* COI sequences (782 sequences) followed by the sequences of *B. tabaci* symbiontes (520 sequences). There is one comprehensive transcriptase data of Indian

Fig. 25.3 Method describing collection and bioassay for whiteflies. (**a**) Cotton leaf showing whiteflies nymph; (**b**) adult whiteflies; (**c**, **d** and **e**) collection of adult whiteflies in bioassay vial from cotton leaves; (**f**) collected whiteflies in bioassay vial; (**g**, **h** and **i**) preparation of diet pouch using sterilized stretched parafilm on the cap of bioassay vial; (**j**) replacement of vial cap with cap containing diet; (**k**) whiteflies feeding on the diet (Adopted from Upadhyay et al. 2011)

species *B. tabci*. There are one sequence each for vitellogenin; vitellogenin receptor, R2D2, argonaute2 and sid1 of *B. tabaci*. The overall contribution of Indian workers for sequence database is depicted in (Fig. 25.4).

25.4.3 Transciptome Related Studies

Upadhyay et al. (2015) performed comprehensive transcriptome sequencing of whitefly (*B. tabaci*). From the transcriptome sequencing 28,716, 25,012 and 12,193 distinct unitigs sequences of H biotype were mapped to 44,425 TSA sequences of MEAM1, 37,365 of AsiaII3, and 21,309 of MED, respectively. On the basis of gene expression analysis they have identified genes encoding for important proteins responsible for growth and development such as vitellogenin, ribosomes, initiation and elongation factors, tubulin, NADH dehydrogenase and malate dehydrogenase as potential targets for insect control through RNAi. Further, on the basis of mapping of unitigs and expression analysis they predicted the integration of genes of

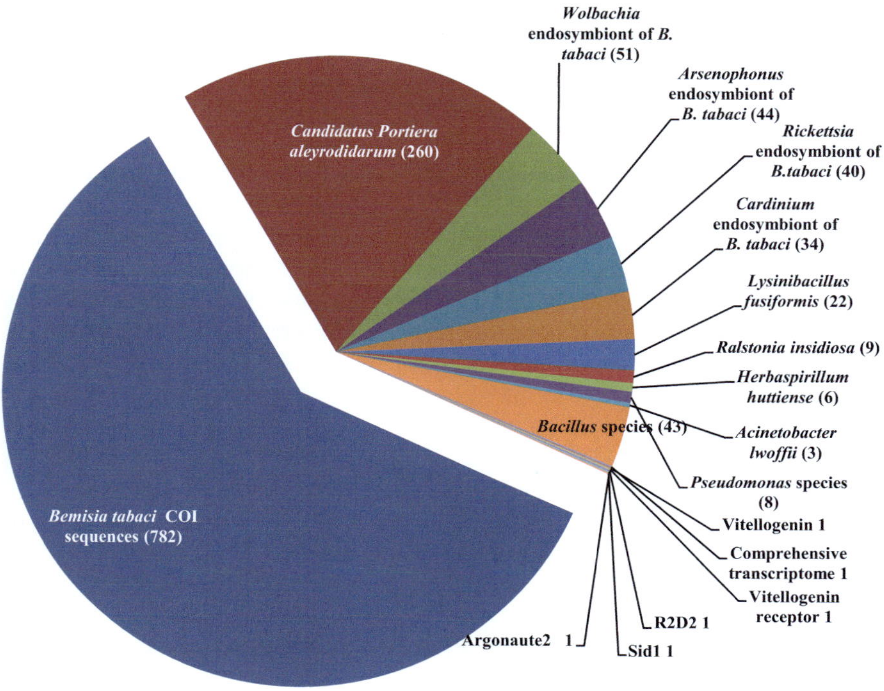

Fig. 25.4 Sequences of *B. tabaci* submitted NCBI database by Indians

insect host and its symbiont in the biosynthetic pathways of amino acids. The expression of genes responsible for biosynthesis of non-essential amino acids was in whitefly but absent in symbionts. Whereas high expression of genes involved in essential amino acids were identified in symbionts but not in host insect. By mapping expression of genes involved in the amino acid biosynthetic pathways, hypothetical amino acid biosynthetic pathway involving genes of the symbiotic and host insect was predicted.

25.4.4 RNAi Related Studies

Whitefly control using RNAi mediated gene silencing of *B. tabaci* through oral rote was first reported by Upadhyay et al. (2011). Synthesized dsRNAs and siRNAs of actin ortholog, ADP/ATP translocase, ±-tubulin, ribosomal protein L9 (RPL9) and V-ATPase A were evaluated using artificial diet. Knocking down the expression of RPL9 (LC_{50} 11.21 μg/ml) and V-ATPase A (LC_{50} 3.08 μg/m) caused higher mortality as compared to other genes. Asokan et al. (2014) showed that continuous application of dsGST of *B. tabaci* delivered through diet to *B. tabaci* resulted in significant mortality. Upadhyay et al. (2015) identified the core siRNA machinery in *B. tabaci* and concluded that universal expression of core siRNA machinery and in good quantity as the reason for susceptibility of *B. tabaci* for siRNA. RNAi machinery of whitefly was found very close to that of aphids.

25.4.5 Whitefly: Endosymbionts and Virus Transmission Ability

Rana et al. (2012) analyzed *B. tabaci* population collected from Sriganganagar district of Rajasthan, India, for the presence of endosymbionts and found presence *Portiera* (primary endosymbiont) *Arsenophonus* (secondary endosymbiont). They further confirmed using pull down and Co-Immuno Precipitation Assays that of GroEL proteins produced by symbionts interact with the CLCuV coat protein. They also localised *Arsenophonus* in the salivary glands and the midgut of *B. tabaci*. Raina et al. (2015) studied role of "*Arsenophonus*" the facultative endosymbiont of *B. tabaci* in *B. tabaci*, elimination of symbiont using tetracycline resulted in increased fecundity and juvenile developmental time of *B. tabaci*. Upadhyay et al. (2015) predicted hypothetical amino acid biosynthetic pathway involving genes of the symbiotic and host insect by mapping their expression.

25.4.6 Novel Whitefly Resistant Proteins and Transgenics

First ever whitefly resistant transgenic plants (tobacco) was developed by Thakur et al. (2014) by expressing long dsRNA as precursor for siRNA for knocking down the V-ATPaseA mRNA in whitefly. The whiteflies feeding on the transgenic plants leaves showed up to 62% reduction in the transcript level of V-ATPaseA. Control plants are heavily infested by whiteflies resulting in significant loss of sugar content and drooping of leaves. There was no drooping seen in the transgenic plants due to fewer no of whiteflies.

The work carried out at CSIR-National Botanical Research Institute, Lucknow, explored large numbers of fern species for insecticidal proteins. They screened species of ferns for whitefly toxic proteins and identified several potential whitefly toxic proteins. They have followed method for isolation of insecticidal which involved extraction of total soluble protein in suitable buffer – precipitate with ammonium sulphate and performing whitefly bioassay. If the protein extract found toxic to whiteflies it was further subjected to heat inactivation and digestion with Protease K and retested for whitefly toxicity. Loss of whitefly toxicity upon heat denaturation and Protease K digestion confirms that the toxicity was due to protein (Upadhyay 2013; Shukla et al. 2016) (Table 25.1).

A protein isolated from *Tectaria macrodonta* (Tma12, 43 kDa) was highly toxic in native form and after heat treatment. Whereas losses activity following proteinase K digestion. The results indicated that protein is thermo-stable. Purified Tma12 was specific to whiteflies and did not cause toxicity to *Aphis gossypii, Spodoptera litura* and *Helicoverpa armigera*. The encoding gene for the protein has been used for

Table 25.1 Toxicity of total soluble protein isolated from different ferns to whiteflies (Adopted from Upadhyay 2013)

S. No	Plant	Plant tissue	% mortality in 72 h		
			Native	Heat denatured	Protenase K digested
1	*Cythea spinulosa*	Leaf	0	0	0
	Polypodium	Rhizome	8	10	10
		Leaf	20	18	18
2	*Polysticum ambile*	Rhizome	28	28	34
		Leaf	30	32	32
3	*Colysis pothifolia*	Leaf	0	0	0
4	*Drynaria quercifolia*	Rhizome	100	89	88
		Leaf	87	18	30
5	*Tectaria macrodonta*	Rhizome	69	60	50
		Leaf	100	0	0
6	*Cyrtomium falcatum*	Rhizome	39	28	30
		Leaf	20	37	36
7	*Diplazium esculentum*	Rhizome	52	49	50
		Leaf	34	42	40
8	*Equisetum*	Leaf	0	0	0
9	*Pteris vittata*	Rhizome	0	0	0
		Leaf	0	0	0
10	*Nephrolepis tuberosa*	Rhizome	0	0	0
		Leaf	60	56	49
11	*Ampelopteris prolifera*	Rhizome	0	0	0
		Leaf	0	0	0
12	*Adiantum capillusveveris*	Rhizome	0	0	0
13	*Psilotum nudum*	Leaf	0	0	0
14	*Paralepto chilus*	Rhizome	0	0	0
		Leaf	0	0	0

development of whitefly resistant transgenic cotton. Transgenic plants were not only resistant to whiteflies but also against CLCuV (Shukla et al. 2016).

Roy et al. (2014) studied 25 kDa tuber agglutinin of *Colocasia esculenta* (CEA) for toxic potentiality against *B. tabaci*. Binding of lectin to insect midgut was observed by confocal microscopic analyses. Binding of CEA to vacuolar ATP synthase and sarcoplasmic endoplasmic reticulum type Ca^{2+} ATPase was observed by ligand blot analysis. Interaction kinetics between ATP synthase of *B. tabaci* with CEA was analyzed by surface plasmon resonance. It was predicted that CEA interaction with insect receptors of target insects results in disruption of cellular processes leading growth retardation and loss of fecundity.

Bhagat et al. (2014) tested *E. coli* expressed *Remusatia vivipara* Lectin (RVL1) and *Sclerotium rolfsii* Lectin (SRL1) against nymphs of *B. tabaci* using the artificial diet. They observed 29.73 and 33.60% of whitefly nymphs in diet supplemented with RVL1 and SRL1, respectively. They suggested these lectins can possibly be deployed in transgenic plants for resistance against whitefly.

25.5 *B. tabaci* Management Strategies in India

Whitefly *B. tabaci* vectored Leaf curl disease (LCD) of vegetables is a serious threat to tomato production worldwide (Varma A 1993; Varma and Malathi 2003). Hence, management of whitefly is crucial for the management of LCD. Tomato cultivation in India is adversely affected due to the high incidence of the LCD disease and losses often exceed 90% (Saikia and Muniyappa 1989). The progress of the disease coincides with the build-up of whitefly and it has been well established that LCD incidence and population of *B. tabaci* are positive correlated (Varma 1986; Borah and Bordoloi 1998; Govindappa 2002). In the Northern Indian minimum temp and minimum relative humidity has been shown to influence the whitefly population (Krishnareddy 1989), whereas in South Indian conditions the maximum temp and rainfall have been found to be more important (Murugesan et al. 1977). Ramappa et al. (1998) observed that weed host-plant species growing near the experimental sites had averages of between 1.5 and 10.0 *B. tabaci* nymphs per plant compared to 0.3 nymphs per plant on the tomato plants. Thomas et al. (2011) reported *leucaena* as a new host for *B. tabaci* and compared *B. tabaci* population from *leucaena* with cotton population with respect to biology, morphometrics and esterases.

Tripathi and Varma (2002b) investigated eco-friendly methods were for the management and epidemiological studies of LCD in tomato. They observed high degree LCD incidence (83%) in the winter crop planted in October as compared to the summer crop planted in February (14%). Least incidence of the LCD was found in polythene mulch treatment and in perforated polythene cover treatment as compared to the other treatments. On the basis of the results they recommended the use of perforated polythene bags to cover plants at the initial stages of the crop and/or polythene mulching along with raising nursery under protective conditions for the eco-friendly management of LCD in tomato.

Krishna Kumar et al. (2004) evaluated different dates of planting of tomato in Akhnoor sector of Jammu for management of leaf curl incidence. Among different dates of transplanting evaluated in between 10th August and 25th of October, maximum yield was observed when crops transplanted on 1st September. Field survey revealed *Lantana camara*, *Malvastrum spp.* and *Hibiscus esculentus*,*Euphorbia hirta*, *Ageratum conyzoides* and *Carica papaya* as alternate hosts for whiteflies (Mandal and Muniyappa 1991; Muniyappa et al. 2003). Chatterjee et al. (2013) reported that when FYM or vermin-compost provided as sources of nutrition it has reducing effect on the population whitefly when compared with inorganic fertilizers. In case of tomato crop, maize acts very good barrier against *B. tabaci* when raised as guard crop, while brinjal acted as a good trap for *B. tabaci*. Both crops were very effective in reducing incidence ToLCV in tomato (Rajasri et al. 2009). Pillai et al. (2014) studied the native parasitoid, *Encarsia transvena* efficiency of the against *B. tabaci* infesting tomato, eggplant and tobacco and reported that parasitisation of *B. tabaci* by *Enc. transvena* was higher on tomato than tobacco or eggplant.

Dahiya et al. (2008) recommended for the management of whitefly on cotton spray of insecticides dimethoate (250–350 ml) or oxydemeton-methyl (300–400 ml) or formothion or imidacloprid (40 ml) or thiomethoxam 25 WG (40 g) in 120–150 l of water. Panduranga et al. (2011) observed that foliar spray of thiomethoxam 25 WS @ 0.005% followed by spirotetramat 150 OD @ 90 g a.i./ha and acetamprid 20% SP @ 0.002% were very effective for management of *B. tabaci and Mung Bean Yellow Mosaic* disease in Mung Bean. Govindappa et al. (2013) evaluated newer chemical against toxicity to whiteflies and incidence of leaf curl and found that cyantraniliprole was effective in reducing both whitefly and disease incidence. Maha Lakshmi et al. (2015) evaluated different pesticides against whiteflies in urdbean and found spiromesifen 240 SC @ 0.4 ml/l followed by buprofezin 10 EC @ 1.0 ml/l and neonicotinoid molecules, acetamiprid 20 SP @ 0.2 g/l as promising against whiteflies. Triazophos 40 EC @ 1.25 was found no-par with the neonicotiniods.

Kranti et al. (2002) recorded levels of resistance to cypermethrin and endosulfan/chlorpyriphos as moderately high and, negligible respectively in the *B. tabaci* field strains. Naveen et al. (2012) suggested the log rank test method involving integration of both dose and time in terms of hazard rate as an improved approach over the current Insecticide Resistance Action Committee (IRAC) susceptibility test methods series no. 15, for conducting bioassays and evaluation of toxicity of insecticide chemicals against *B. tabaci*. Jambhukar et al. (2013) recommended seed treatment and two sprays with imidacloprid as effective measure for control of YVM disease and vector whitefly population in okra. Imidacloprid was also found effective for management of whiteflies and leaf curl disease in chilli (Panday et al. 2010).

Rai et al. (2014) recommended general management practices for whiteflies in vegetables which included early monitoring of pest by installation of yellow sticky traps; cultural practices like removal of weed hosts of whiteflies and associated viral diseases; conservation of natural enemies like parasitoids (*Encarsia brevivena, Eretmocerus corni, Eretmocerus mundus*) and predators (*Chrysoperla zastrowi sillemi, Mallada boninensis, Coccinella septempunctata*); seed treatment with imidacloprid 70WS @ 2.5 g per kg of seed. Protection of seedlings in nursery by nylon net (200 mesh) covering for 25-30 days followed by need based application of systemic insecticides like imidacloprid, thiomethoxam and dimethoate.

Recent outbreak of *B. tabaci* populations on cotton in Punjab during 2015 suggests that insect has developed resistance to commonly used insecticides and insecticides sprays have destroyed of natural enemies. There are several important natural enemies of whiteflies which can provide good control of whiteflies including parasitoids (*Encarsia* sp., and *Eretmocerus* sp) predators like coccinellids and lacewing larvae and fungal pathogens like *Verticillium lecanii*, *Paecilomyces fumosoroseus* and *Beauveria bassiana*. Hence, the whitefly IPM must include cultural practices like use of polythene mulches, destroying and removing the crop residues, alternate weed hosts of whitefly. Use of safer pesticides like neem oil, neem seed kernel extract and fish oil resin soap should be encouraged.

25.6 Concluding Remarks

India has diverse types of agro climatic zones, which categorized into 15 types considering the physical attributes and socio-economic conditions prevailing in the regions. Hence, the occurrence of whitefly biotypes, virus strains, their seasonal incidence and severity are also expected to vary widely in these agro climatic zones. However, so far except for few states (example Karnataka) different biotypes of *B. tabaci* and the geminiviruses transmitted by them is either lacking or insufficient for the most of the states of the country. Similarly there is no information available on their zone wise and seasonal incidence. In several regions of the country commercial cultivation of vegetables is severely affected by incidence of diseases caused by geminiviruses. Hence, there is urgent need to extensively survey and map different *B. tabaci* and geminiviruses transmitted by them in all the agro-climatic zones of the country. Preparation of state wise/zone wise/season wise distribution and severity of occurrence whitefly biotypes and virus disease will be highly useful for policy decision making. Such information can be used for providing advisories to farmers agro climatic zones wise regarding "when and what" to be cultivated and which crop to be avoided from the diseases and whitefly incidence point of view. Finally release of indigenously developed whitefly resistant transgenics varieties of vegetables and cotton will have a huge implications with respect to whitefly and LCD management and expected to considerably improve overall production of cotton and vegetable in the country.

References

Anokhe A, Mandal B, Subramanian S (2015) Characterization of ToLCNDV transmission in Asia-I and Asia-II-1 genotypes of *Bemisia tabaci*. Virocon-15, Shillong, Mehgalaya, 8–10 October 2015

Asokan R, Rebijith KB, Roopa HK, Krishna Kumar NK (2014) Non-invasive delivery of dsGST is lethal to the sweet potato whitefly, *Bemisia tabaci* (G.) (Hemiptera: Aleyrodidae). Appl Biochem Biotechnol. doi:10.1007/s12010-014-1437-6

Banks GK, Colvin J, Chowda Reddy RV, Maruthi MN, Muniyappa V, Venkatesh HM, Kiran Kumar M, Padmaja AS, Beitia FJ, Seal SE (2001) First report of the *Bemisia tabaci* B biotype in India and an associated tomato leaf curl virus disease epidemic. Plant Dis 85:231

Bel-Kadhi MS, Onillon JC, Cenis JL (2008) Molecular charecterization of *Bemisia tabaci* biotypes in South Tunisia. Tunis J Plant Prot 3:79–85

Berlinger MJ (1986) Host plant resistance to *Bemisia tabaci*. Agric Ecosyst Environ 17:69–82

Bhagat YS, Bhat RS, Udikeri SS (2014) Toxicity of E. coli Expressed Remusatia vivipara Lectin and Sclerotium rolfsii lectin delivered via artificial diet against *Bemisia tabaci* (Genn.) and Amrasca biguttula biguttula (Ishida). Trends Biosci 7(20):3178–3187

Borah RK, Bordoloi OK (1998) Influence of planting time on the incidence of leaf curl virus disease and whitefly population on tomato. Indian J Virol 14:71–73

Capoor SP (1949) Feeding method of the whitefly. Curr Sci 18:82–83

Capoor SP, Ahmad RU (1975) Yellow vein mosaic disease of field pumpkin and its relationship with the vector, *Bemisia tabaci*. Indian Phytopath 28:241–246

Chatterjee R, Choudhuri P, Laskar N (2013) Influence of nutrient management practices for minimizing whitefly (*Bemisia tabaci* Genn.) population in tomato (*Lycopersicon esculentum* Mill.) Int J Sci Environ Technol 2:956–962

Chaubey R, Andrew RJ (2015) Geometric morphometrics of puparia of *Bemisia tabaci* (gennadius) (Hemiptera: Aleyrodidae) from western plateau and hill regions agroclimatic zone of India. Indian J Entomol 77:283–289

Chaubey R, Andrew RJ, Naveen NC, Rajagopal R, Ramamurthy VV (2015) Life history traits of three cryptic species Asia I, Asia II-1 and Asia II-7 of *Bemisia tabaci* (Hemiptera: Aleyrodidae) reconfirm their genetic identities. Florida Entomol 98:254–259

Chowda-Reddy RV, Kiran Kumar M, Seal SE, Muniyappa V, Valand GB, Govindappa MR, Colvin J (2012) *Bemisia tabaci* phylogenetic groups in India and the relative transmission efficacy of tomato leaf curl Bangalore virus by an indigenous and an exotic population. J Integr Agric 11:235–248

Crowder DW, Horowitz AR, De Barro PJ (2010) Mating behavior, life history and adaptation to insecticides determine species exclusion between whiteflies. J Anim Ecol 79:563–570

Dahiya KK, Rana RS, Beniwal J, Kumar A (2008) Eco-friendly management of insects and diseases in cotton, Technical Bulletin No.33, Directorate of Extension Education. CCS Haryana Agricultural University, Hisar, p 36

De Barro PJ, Driver F (1997) Use of RAPD PCR to distinguish the B biotype from other biotypes of *Bemisia tabaci* (Gennadius) (Hemiptera: Aleyrodidae). Austr J Entomol 36:149–152

De Barro PJ, Driver F, Trueman JWH, Curran J (2000) Phylo- genetic relationship of world populations of *Bemisia tabaci* (Gennadius) using ribosomal ITS1. Mol Phylogenet Evol 16:29–36

Dixit S, Upadhyay SK, Singh H, Verma PC, Chandrashekar K (2013) Enhanced methanol production in plants provides broad spectrum insect resistance. PLoS One 8(11):e79664

Ellango R, Singh ST, Rana VS, Priya GN, Raina H, Chaubey R, Naveen NC, Mahmood R, Ramamurthy VV, Asokan R, Rajagopal R (2015) Distribution of *Bemisia tabaci* genetic groups in India. Environ Entomol:1–7. doi:10.1093/ee/nvv062

Govindappa MR (2002) Detection and transmission of tomato leaf curl virus and interactions studies with host plants and whitefly *Bemisia tabaci* Genn. In relation to epidemics. Ph.D. thesis, University of Agricultural Sciences, Bangalore, India

Govindappa MR, Bhemanna M, Hosmani A, Ghante VN (2013) Bio-efficacy of newer insecticides against tomato leaf curl virus disease and its vector whitefly (Bemisia tabaci) in tomato. Int J Appl Biol Pharm Technol 4:226–231

Jambhulkar PP, Virendra S, Ramesh BS, Yadav RK (2013) Insecticides and bioproducts against whitefly population and incidence of *Yellow vein mosaic*. Indian J Plant Prot 41:253–256

Jayasekera S, Thomas A, Kar A, Ramamurthy VV (2010) Host correlated morphometric variations in the populations of *Bemisia tabaci* (Gennadius) (Hemiptera: Aleyrodidae). Orient Insects 44:193–204

Jayashree K, Pun KB, Doraiswamy S (1999) Virus – vector relationships of yellow vein mosaic virus and whitefly *(Bemisia tabacis*) in pumpkin. Indian Phytopath 52:10–13

Jun XG, Rao Q, Zhang F, LUO C, Zhang HY, Gao XW (2012) Diversity and genetic differentiation of the whitefly *Bemisia tabaci* species complex in China based on mtCOI and cDNA-AFLP analysis. J Integr Agric 11:206–214

Kranthi KR, Jadhav DR, Kranthi S, Wanjari RR, Ali SS, Russell DA (2002) Insecticide resistance in five major insect pests of cotton in India. Crop Prot 21:449–460

Krishnareddy M (1989) Studies on yellow mosaic and leaf crinkle diseases of blackgram. Ph.D. thesis, IARI, New Delhi, p 263

Kumar K, Mahajan R, Chib HS, Kalha CS (2004) An epidemic of tomato leaf curl disease in Jammu. Indian Phytopath 57(2):181–184

Li QB, Xiang RS, Yang WS, Mandour NS (2003) Biotype Identification of the Populations of *Bemisia tabaci* (Homotera: Aleyrodidae) in China Using RAPD-PCR. Acta Entomol Sin 46:605–608

Li Z, Hu D, Song Y, Shen Z (2005) Molecular differentiation of the B biotype from other biotypes of *Bemisia tabaci* (Homotera: Aleyrodidae), based on internally transcribed spacer 1 sequence. Eur J Entomol 102:293–297

Lisha VS, Antony B, Palaniswami MS, Henneberry TJ (2003) *Bemisia tabaci* (Homoptera: Aleyrodidae) biotypes in India. J Econ Entomol 96(2):322–327

Mahalakshmi MS, Sreekanth M, Adinarayana M, Koteswara Rao Y (2015) Efficacy of some novel insecticide molecules against incidence of whiteflies (*Bemisia tabaci* Genn.) and occurrence of Yellow Mosaic Virus (YMV) disease in urdbean. Int J Pure App Biosci 3:101–106

Mandal B, Muniyappa V (1991) Transmission of croton yellow vein mosaic virus by *Bemisia tabaci*: Virus vector relationship. Fitopatol Bras 16(4):260–262

Misra CS, Lamba SK (1929) The cotton whitefly (*Bemisia gossypiperda* n. sp.) Bull Agric Res Inst Pusa 196:1–7

Muniyappa V, Venkatesh HM, Ramappa HK, Kulkarni RS, Zeidan M, Tarba CY, Ghanim M, Czosnek H (2001) Tomato leaf curl virus from Bangalore (ToLCV-Ban4): sequence comparison with Indian ToLCV isolates, detection in plants and insects, and vector relationships. Arch Virol 145:1583–1598

Muniyappa V, Maruthi MN, Babitha CR, Colvin J, Briddon RW, Rangaswamy KT (2003) Characterization of pumpkin yellow vein mosaic virus from India. Ann Appl Biol 142:323–331

Murugesan S, Cheliiah S, Murugesan M (1977) Prediction of whitefly vector *Bemisia tabaci* (Genn.) and yellow mosaic incidence in green gram. Madras Agric J 64:22–28

Naveen NC, Kumar D, Alam W, Chaubey R, Subramanian S, Raman R (2012) A model study integrating time dependent mortality in evaluating insecticides against *Bemisia tabaci* (Hemiptera: Aleyrodidae). Indian J Entomol 74:384–387

Nombela G, Muñiz M (2010) Host plant resistance for the management of Bemisia tabaci: a multi-crop survey with emphasis on tomato. In: Stansly PA, Naranjo SE (eds) Bemisia: bionomics and management of a global pest. Springer Science Business Media B.V., pp 357–383. doi:10.1007/978-90-481-2460-2

Palaniswami MS, Nair RR, Pillai KS, Thankappan M (1996) Whiteflies on cassava and its role as vector of cassava mosaic in India. J Root Crops 22:1–8

Palaniswami MS, Binu A, Lisha Vijayan S, Henneberry TJ (2004) Whitefly Bemisia tabaci: biotypes, Indian cassava mosaic virus and its biocontrol agents. In: Sixth international scientific meeting of the Cassava Biotechnology Network, 8–14 March 2004, CIAT, Colombia. Abstracts p 87

Panday SK, Mathur AC, Srivastava M (2010) Management of leaf curl disease of chilli (*Capsicum annum* L). Int J Virol 6:246–250

Panduranga GS, Vijayalakshmi K, Loka RK (2011) Evaluation of insecticides for management of *Bemisia Tabaci* and Mymv disease in mung bean [*Vigna Radiata* (L.) Wilczek]. Ann Plant Protec Sci 19:295–298

Perumal Y, Marimuthu M, Salim AP, Ponnusam B (2009) Host plant mediated population variations of cotton whitefly *Bemisia tabaci* Gennadius (Aleyrodidae: Homoptera) characterized with random DNA markers. American J Biochem Biotech 5(1):40–46

Pillai GK, Ganga Visalakshy PN, Krishnamoorthy A, Mani M (2014) Evaluation of the indigenous parasitoid *Encarsia transvena* (Hymenoptera: Aphelinidae) for biological control of the whitefly *Bemisia tabaci* (Hemiptera: Aleyrodidae) in greenhouses in India. Biocon Sci Technol 24:325–335

Prasanna HC, Kanakala S, Archana K, Jyothsna P, Varma RK, Malathi VG (2015) Cryptic species composition and genetic diversity within *Bemisia tabaci* complex in soybean in India revealed by *mtCOI* DNA sequence. J Integ Agricul 14:1786–1795

Ragupathy N (1989) Studies on yellow mosaic virus disease of soybean. M.Sc. thesis, TNAU, Coimbatore, 91 p

Rai AB, Halder J, Kodandaram MH (2014) Emerging insect pest problems in vegetable crops and their management in India: an appraisal. Pest Manag Horti Ecosyst 20:113–122

Raina HS, Rawal V, Singh S, Daimei G, Shakarad M, Rajagopal R (2015) Elimination of Arsenophonus and decrease in the bacterial symbionts diversity by antibiotic treatment leads to increase in fitness of whitefly, *Bemisia tabaci*. Infect Genet Evol 32:224–230

Rajasri M, Vijaya LK, Prasada Rao RDVJ, Loka RK (2009) Management of whitefly transmitted *Tomato leaf curl virus* using guard crops in tomato. Indian J Plant Protect 37:101–103

Ramappa HK, Muniyappa V, Colvin J (1998) The contribution of tomato and alternative host plants to tomato leaf curl virus inoculum pressure in different areas of South India. Ann Appl Biol 133:187–198

Rana VS, Singh ST, Gayatri Priya N, Kumar J, Rajagopal R (2012) Arsenophonus GroEL interacts with CLCuV and is localized in midgut and salivary gland of whitefly *B. tabaci*. PLoS ONE 7(8):e42168

Rekha AR, Maruthi MN, Muniyappa V, Colvin J (2005) Occurrence of three genotypic clusters of *Bemisia tabaci* and the rapid spread of the B biotype in south India. Entomol Exp et Appl 117:221–233

Roopa HK, Asokan R, Rebijith KB, Ranjitha H, Hande RM, Krishna Kumar NK (2015) Prevalence of a new genetic group, MEAM-K, of the whitefly *Bemisia tabaci* (Hemiptera: Aleyrodidae) in Karnataka, India, as evident from *mtCOI* sequences. Florida Entomol 98:1062–1071

Roy A, Gupta S, Hess D, Das KP, Das S (2014) Binding of insecticidal lectin Colocasia esculenta tuber agglutinin (CEA) to midgut receptors of *Bemisia tabaci* and Lipaphis erysimi provides cues to its insecticidal potential. Proteomics 14:1646–1659

Saikia AK, Muniyappa V (1989) Epidemiology and control of tomato leaf curl virus in South India. Trop Agric 66:350–354

Seetharama Reddy K, Yaraguntaiah RC (1981) Virus–vector relationship in leaf curl disease of tomato. Indian Phytopathol 34(3):310–313

Senanayake DMJB, Varma A, Mandal B (2012) Virus–vector relationships, host range, detection and sequence comparison of Chilli leaf curl virus associated with an epidemic of leaf curl disease of Chilli in Jodhpur, India. J Phytopathol 160:146–155

Sharma P, Rishi N (2003) Host range and vector relationships of *Cotton leaf curl virus* from northern India. Indian Phytopath 56:496–499

Sharma RK, Gupta VK, Jindal J, Dilwari VK (2008) Host associated genetic variations in whitefly *Bemisia tabaci* (Genn*)*. Indian J Biotechnol 7:366–370

Shukla AK, Upadhyay SK, Mishra M, Saurabh S, Singh R et al (2016) Expression of an insecticidal fern protein in cotton protects against whitefly. Nature Biotechnol 34:1046–1051

Singh BRS, Mahant SM (1989) Control of yellow vein mosaic of okra checking its vector whitefly through adjusting dates of sowing, insecticidal application and crop barrier. Indian J Virol 5:1–2

Singh J, Sohi AS, Mann HS, Kapur SP (1994) Studies on whitefly *Bemisia tabaci* (Genn.) transmitted cotton leaf curl disease in Punjab. J Insect Sci 7:194–198

Thakur N, Upadhyay SK, Verma PC, Chandrashekar K, Tuli R, Singh PK (2014) Enhanced whitefly resistance in transgenic tobacco plants expressing double stranded RNA of v-ATPase A gene. PloS ONE 9(3):e87235

Thomas A, Ramamurthy VV (2014) Multiple gene markers to understand genetic diversity in the *Bemisia tabaci* (Hemiptera: Aleyrodidae) species complex. Florida Entomol 97:1451–1457

Thomas A, Chaubey R, Naveen NC, Kar A, Ramamurthy VV (2011) *Bemisia tabaci* (Hemiptera: Aleyrodidae) on *Leucaena leucocephala* (Fabaceae): a new host record from India and a comparative study with a population from cotton. Int J Trop Insect Sci 31(4):235–241

Thomas A, Kar A, Rebijith KB, Asokan R, Ramamurthy VV (2014) *Bemisia tabaci* (Hemiptera: Aleyrodidae) species complex from cotton cultivars: a comparative study of population density, morphology, and molecular variations. Ann Entomol Soc Am 107(2):389–398

Tiwari SP, Nema S, Khare MN (2013) Whitefly- a strong transmitter of plant viruses. J Plant Pathol 02:102–120

Tripathi S, Varma A (2002a) Identification of sources of resistance in *Lycopersicon species* to Tomato leaf curl geminivirus (ToLCV) by agroinoculation. Euphytica 129:43–52

Tripathi S, Varma A (2002b) Eco-friendly management of leaf curl disease of tomato. Indian Phytopathol 55(4):473–478

Upadhyay SK (2013) Isolation, development and characterization of novel insecticidal macromolecules. Ph.D. Thesis. Gautam Buddh Technical University, Lucknow

Upadhyay SK, Chandrashekar K, Thakur N, Verma PC, Borgio JF, Singh PK, Tuli R (2011) RNA interference for the control of whiteflies (*Bemisia tabaci*) by oral route. J Biosci 36:153–161

Upadhyay SK, Dixit S, Sharma S, Singh H, Kumar J, Verma PC, Chandrashekar K (2013) siRNA machinery in whitefly (*Bemisia tabaci*). PLoS One 8(12):e83692

Upadhyay SK, Sharma S, Singh H, Dixit S, Verma PC, Chandrashekar K (2015) Whitefly genome expression reveals host-symbiont interaction in amino acid biosynthesis. PLoS One 10(5):e0126751

Varma PM (1952) Studies on the relationship of the bhendi yellow vein mosaic virus and its vector, the whitefly, *Bemisia tabaci* Genn. Indian J Agric Sci 22:75–91

Varma A (1986) Ecology of whitefly transmitted viruses. In: National seminar on whitefly transmitted plant virus diseases, 25–27 June, IARI, New Delhi 41 pp

Varma A (1993) Integrated management of plant viral diseases. In: Crop protection and sustainable agriculture. Wiley, Chichester. CIBA Found Symp 177:140–157

Varma A, Malathi VG (2003) Emerging geminivirus problem: a serious threat to crop production. Ann Appl Biol 142(2):145–164

Varma A, Mandal B, Singh MK (2011) Global emergence and spread of whitefly (Bemisia tabaci) transmitted geminiviruses, page 205–292. In: Winston MO Thompson (ed) The whitefly Bemisia tabaci. (Homoptera: Aleyrodidae) Intraction with geminivirous infected host plants: Bemisia tabaci. Geminivirous and host plants. Springer Publications

Part III

Virus Diagnosis

Application of Immuno-Diagnosis for Plant Viruses Occurring in India

26

Alangar Ishwara Bhat and Yogita Maheshwari

Abstract

Diseases caused by viruses cause heavy crop losses worldwide including India. Correct diagnosis of viruses is the primary requirement for disease management. Of the different methods available for diagnosis, serological methods are useful in the detection of viruses in seeds, plants and insect vectors. The production of virus specific antiserum is the basis for a successful serological method. The earliest serological methods such as precipitin tests are low in sensitivity. With the advent of enzyme linked immunosorbent assay (ELISA) and its subsequent application to plant virus detection during 1980s, some laboratories in India also started working on the same. During 1990s, ELISA and its variants became more popular in many laboratories of India working on plant viruses. During the same period Indian researchers also used other serological methods such as dot immunobinding assay (DIBA), immunosorbent electron microscopy (ISEM), electro-blot immunoassay (EBIA) and florescent antibody technique in the detection of plant viruses. Immunocapture PCR (IC-PCR) which combines both serology and PCR has also been used by many Indian researchers in the identification and characterization of viruses infecting different crops. Lateral flow immunoassay technology is the recent technique that offers several advantages over traditional immunoassays, such as procedural simplicity, limited requirement for special skills or expensive equipment, and rapid results is also gaining importance in the country. During the last decade, recombinant antibody (rAb) engineering has emerged as one of the promising approaches in plant virus diagnosis. rAb fragments in all various formats (Fab, Fv and scFv) can be expressed in different

A.I. Bhat (✉)
ICAR-Indian Institute of Spices Research, Kozhikode 673012, Kerala, India
e-mail: aib65@yahoo.co.in

Y. Maheshwari
Advanced Centre for Plant Virology, Division of Plant Pathology, ICAR-Indian Agricultural Research Institute, New Delhi 110012, India

B. Mandal et al. (eds.), *A Century of Plant Virology in India*,
https://doi.org/10.1007/978-981-10-5672-7_26

systems including bacteria, insect, yeast, plant and mammalian cells, which can be used efficiently in various format of ELISA for diagnosis of plant viruses. India also started a modest beginning in this area now.

Keywords
Agar gel diffusion • Antiserum • Detection • Diagnosis • Dot immunobinding assay • Enzyme linked immunosorbent assay • Electroblot immunoassay • Immunodiagnosis • Immunoflorescence • Immunocapture polymerase chain reaction • Precipitin test • Serology • Tissue blot assay • Virus • Lateral flow immunoassay • Recombinant antibody • Single chain variable fragment

26.1 Introduction

Viral diseases are emerging as serious constraints in improving the productivity of crops in India. The present century faces severe epidemics by newly emerging viruses affecting a variety of crops. The most important viruses, which have emerged during the last two decades, belong to genera *Allexivirus, Badnavirus, Begomovirus, Closterovirus, Cucumovirus, Ilarvirus, Potyvirus* and *Tospovirus*. Major contributory factors for the emergence and spread of viruses include evolution of virus variants, changes in vector biology, changes in the cropping systems, introduction of new crops or varieties, the movement of infected planting materials, introduction of new crop species and new genotypes and introduction of host susceptibility genes through the exchange of germplasm (Rishi 2006).

At least 200 different viruses have been identified that infect and cause disease in different crops of India. Many plant virus diseases that cause significant losses in the production and quality of crops are very difficult to control, and new diseases occur as different crops are introduced or grown in new areas. Many crops are susceptible to multiple viruses, each of which may cause serious economic losses and infected plant material may not be acceptable for export purpose. Some of the viruses in crops do not express symptoms and the virus is often carried as a latent infection leading to substantial yield reduction when exposed to different environment.

Plant viruses residing in the living tissues of plants produce symptoms ranging from virtually undetectable to quite severe. As many viruses remain latent in infections or remain in some special physiological states in propagative plant material, preventive measures demand pathogen detection methods of high sensitivity, specificity and reliability. Unlike bacterial and fungal diseases, detection by means of a sensitive diagnostic method seems to be a promising strategy to control viral diseases. For certification of plant viruses, plants need to be tested for the viruses, in order to prevent the spread of infection and damage of plants from such viruses. Different approaches may be used to analyze the plant viruses. Initially, viral disease could be diagnosed quickly by visual examination of symptoms that they cause in different host species, by studying their vector transmission properties along with the study of their interaction with the hosts. Variants that have broad host range but are often symptomless or produce symptoms showing resemblance with other viruses require molecular tests for diagnosis as these techniques are more sensitive and reproducible and allow quick identification of the viruses. For the development of efficient and stable diagnostic methods, the following factors should be considered: (1) sensitivity, i.e. lower limits a virus may be detected, (2) accuracy and reproducibility, (3) number of samples that can be processed in a given time, (4) cost and sophistication of the apparatus and materials needed, (5) level of training required for operators and (6) adaptability to field conditions.

Earliest reported plant viral disease in India was mosaic disease of cardamom from Karnataka during 1890 (Sastry and Sai-Gopal 2010). This was followed by other reports describing symptoms, yield loss etc. During initial years, studies mainly focused on biological characterization of viruses. During 1970s onwards virus detection and diagnosis slowly gained importance. Virus detection methods of high sensitivity and specificity and reliability are essential to devise effective plant virus disease management practices. Bioassays, electron microscopy, serological tests, dot blot hybridization, PCR, real-time PCR, loop mediated isothermal amplification and microarray based assays are in use for detection, differentiation and diagnosis of plant viruses (Boonham et al. 2014). Polyclonal antisera production for a large number of viruses has been reported from India. The earlier experiments were carried out by using immune diffusion tests which are low in sensitivity. With the advent of ELISA test during 1977 by Clark and Adams (1977), many laboratories in India started working with ELISA and its variants. Recently, there are many reports of recombinant coat protein based polyclonal antibody production on an array of plant viruses. Development of IC-PCR, a combination of serology and PCR has really enhanced sensitivity and specificity of detection (Nayudu 2008).

The present chapter summarizes the development and application of immuno diagnosis for the detection of plant viruses and a brief account of serological methods reported from India on detection and diagnosis of viruses of different genera (Table 26.1).

26.2 Production of Serological Reagents of Plant Viruses Occurring in India

Polyclonal antibodies bind to multiple epitopes of a given antigen and are produced by (1) immunizing a mammal often a goat or rabbit with an antigen, (2) bleeding the animal periodically, and (3) extracting the antibodies directly from the serum. The process is relatively inexpensive, and large quantities of an antibody can be isolated from a single extraction. The general immunization scheme is to give four to five injections at weekly intervals intramuscularly/intravenously (Diano et al. 1987; Noordam 1975; Dijkstra and de Jager 1998).

Monoclonal antibodies are derived from a single clone of antibody-producing B cells. After immunizing a mammal, spleen (location of B cell production) is extracted, and the B cells are fused to immortal myeloma cells; the fused B cells and

Table 26.1 List of viruses for which antibodies and serological tests are reported from India

Genus	Virus	Antibody used	Serological test
Allexivrus	*Garlic virus X*	Polyclonal using in vitro expressed CP	ELISA
Ampelovirus	*Grapevine leafroll-associated virus 1, 2,* and *3*	Polyclonal	DAS-ELISA, IC-RT-PCR
Babuvirus	*Banana bunchy top virus*	Polyclonal	DAS-ELISA
Badnavirus	*Banana streak Mysore virus*	Polyclonal using in vitro expressed CP	DAS-ELISA, IC-PCR
	Citrus yellow mosaic virus	Polyclonal	DAC-ELISA
	Piper yellow mottle virus	Polyclonal	DAS-ELISA
	Sugarcane bacilliform virus	Polyclonal	DAC-ELISA, ISEM
Begomovirus	*Cotton leaf curl virus*	ACMV and ICMV monoclonal antibodies	ELISA, ISEM
	Croton yellow vein mosaic virus	ACMV and ICMV monoclonal antibodies	ELISA
	Dolichos yellow mosaic virus	ACMV and ICMV monoclonal antibodies	ELISA, ISEM
	Horsegram yellow mosaic virus	ACMV and ICMV monoclonal antibodies	ELISA, ISEM
	Indian cassava mosaic virus	ICMV	ISEM, TAS-ELISA
	Pepper leaf curl virus	Polyclonal using in vitro expressed CP	ELISA
	Pumpkin yellow vein mosaic virus	Polyclonal using in vitro expressed CP	ELISA
	Tobacco leaf curl virus	ACMV and ICMV monoclonal antibodies	ELISA
	Tomato leaf curl virus	ACMV and ICMV monoclonal antibodies	TAS-ELISA
	Tomato leaf curl Bangalore virus	Polyclonal	DAC-ELISA
	Tomato leaf curl New Delhi virus	scFV	EBIA, ELISA

(continued)

Table 26.1 (continued)

Genus	Virus	Antibody used	Serological test
Carlavirus	*Cowpea mild mottle virus*	Polyclonal	ELISA, gel diffusion
	Chrysanthemum B virus	Polyclonal	DAS-ELISA, ISEM
	Garlic common latent virus	Polyclonal	DAS-ELISA, ISEM
Carmovirus	*Blackgram mottle virus*	Polyclonal	ELISA
	Carnation mottle virus	Polyclonal	ELISA
	Soybean yellow mottle mosaic virus	Polyclonal using in vitro expressed CP	ELISA
Capillovirus	*Apple stem grooving virus*	Polyclonal	DAS-ELISA
Caulimovirus	*Peanut chlorotic streak virus*	Polyclonal using in vitro expressed CP	Agar gel diffusion, ELISA, EBIA
Closterovirus	*Citrus tristeza virus*	Polyclonal	DAS-ELISA, DAC-ELISA
Cucumovirus	*Cucumber mosaic virus*	Polyclonal using purified virus and in vitro expressed CP	DAS-ELISA, DAC-ELISA, TIBA, EBIA, ISEM
Dianthovirus	*Carnation ringspot virus*	Polyclonal	DAS-ELISA
Emaravirus	*Pigeonpea sterility mosaic* virus	Polyclonal	DAS-ELISA, DIBA
Foveavirus	*Apple stem pitting virus*	Polyclonal	DAS-ELISA
Ilarvirus	*Apple mosaic virus*	Polyclonal	DAS-ELISA
	Prunus necrotic ringspot virus	In vitro expressed CP	DAS-ELISA
	Tobacco streak virus	Polyclonal using purified virus	DAS-ELISA, DAC-ELISA, DIBA, EBIA, ISEM
Luteovirus	*Barley yellow dwarf virus*	Polyclonal	ELISA
	Potato leaf roll virus	Polyclonal	ELISA
	Sugarcane yellow leaf virus	Polyclonal	DAS-ELISA, DAC-ELISA, DIBA, TIBA

(continued)

Table 26.1 (continued)

Genus	Virus	Antibody used	Serological test
Macluravirus	*Cardamom mosaic virus*	Polyclonal using purified virus and *Polyclonal using* in vitro expressed CP	DAS- ELISA
	Large cardamom chirke virus	*Polyclonal using* in vitro expressed CP	ELISA, DIBA, lateral flow immunostrip
Mandarivirus	*Indian citrus ringspot virus*	Polyclonal using purified virus and *Polyclonal using* in vitro expressed CP	DAS- ELISA
Nucleorhabdovirus	*Maize mosaic virus*	Polyclonal	Gel diffusion, ELISA
Pecluvirus	*Indian peanut clump virus*	Monoclonal	TAS-ELISA, DAS-ELISA
Poacevirus	*Sugarcane streak mosaic virus*	Polyclonal	DAC-ELISA, IC-PCR
Potexvirus	*Potato virus X*	Polyclonal using purified virus and *Polyclonal using* in vitro expressed CP	ELISA, Lateral immunostrips
	Cymbidium mosaic virus	Polyclonal	ELISA, ISEM
Potyvirus	*Banana bract mosaic virus*	Polyclonal	DAC-ELISA, EBIA, DIBA, ISEM
	Bean yellow mosaic virus	Polyclonal	DIBA, DAS-ELISA, TIBA, ISEM
	Carnation vein mottle virus	Polyclonal	ELISA, IC-PCR
	Chilli veinal mottle virus	Polyclonal	ELISA
	Cowpea aphid borne mosaic virus	Polyclonal	ELISA, EBIA
	Dasheen mosaic virus	Polyclonal	ISEM
	Henbane mosaic virus	Polyclonal	ELISA, EBIA
	Iris mild mosaic virus	Polyclonal	ELISA, ISEM, IC-RT-PCR
	Konjac mosaic virus	Polyclonal using in vitro expressed CP	ELISA, IC-PCR

(continued)

Table 26.1 (continued)

Genus	Virus	Antibody used	Serological test
	Onion yellow dwarf virus	Polyclonal	DAS-ELISA
	Papaya ringspot virus	Polyclonal using purified virus, rCP against expressed VH and VL fragments	ELISA, EBIA, DIBA
	Pea seed borne mosaic virus	Polyclonal	ELISA
	Peanut green mosaic virus	Polyclonal	ELISA
	Peanut mottle virus	Polyclonal using purified virus and *Polyclonal using* in vitro expressed CP	Precipitin ring interface test, ELISA, ISEM
	Peanut stripe virus	Polyclonal	ELISA
	Pepper veinal mottle virus	Polyclonal	DAS-ELISA
	Plum pox virus	Polyclonal	ELISA, ISEM
	Potato virus Y	Polyclonal	ISEM, ELISA, DIBA, gel diffusion, lateral immunostrips
	Sugarcane mosaic virus	Polyclonal	DAC-ELISA, EBIA, DIBA, ISEM, IC-RT-PCR
	Sweet potato feathery mottle virus	Polyclonal	ISEM, ELISA, DIBA, Gel diffusion
	Tobacco etch virus	Polyclonal	DAS-ELISA
	Turnip mosaic virus	Polyclonal	ELISA, EBIA
	Watermelon mosaic virus	Polyclonal	DAC-ELISA, DIBA
	Zuchini yellow mosaic virus	Polyclonal	ELISA
Tenuivirus	*Maize stripe virus*	Polyclonal	ELISA, EBIA
Tobamovirus	*Cucumber green mottle mosaic virus*	Polyclonal	DAC-ELISA
	Odontoglossum ringspot virus	*Polyclonal using* in vitro expressed CP	DAC-ELISA
	Tobacco mild green mosaic virus	Polyclonal	DAC-ELISA
Tobravirus	*Tobacco rattle virus*	Polyclonal	DAS-ELISA

(continued)

Table 26.1 (continued)

Genus	Virus	Antibody used	Serological test
Tospovirus	*Capsicum chlorosis virus*	Polyclonal	ELISA
	Groundnut bud necrosis virus	Polyclonal, monoclonal, ScFV	ELISA, EBIA, IC-RT-PCR, DIBA
	Iris yellow spot virus	Polyclonal	ELISA
	Watermelon bud necrosis virus	Polyclonal	ELISA, EBIA, IC-RT-PCR
Trichovirus	*Apple chlorotic leaf spot virus*	*Polyclonal using* in vitro expressed CP	DAS-ELISA, TBIA and IC-RT-PCR

myeloma cells or hybridomas are screened via ELISA, and the selected hybridoma line is injected into a second mammal. The myeloma component of the hybridoma induces localized tumor growth in the animal, and an antibody-rich fluid called ascites can be extracted from the tumor, the antibodies are then isolated themselves through column chromatography (Kohler and Milstein 1975).

McCafferty et al. (1990) have developed antibody phage display approach for the preparation of recombinant antibodies (rAbs) by utilizing the phage recombinants that display the antibody in the fused form of coat protein of virus with the biopanning affinity selection technique. rAbs are more popular in recent years and has significant advantages compared with the conventional antibody as no animals are needed and require relatively short time for production. The scFv is a fusion protein of the variable regions of the heavy and light chains of antibody and connected with a short linker peptide of 10 to about 25 amino acids. The linker is usually rich in glycine for flexibility, as well as serine or threonine for solubility. The most frequently used linker today for scFv antibodies is $(Gly_4Ser)_3$, a single 15 aa peptide (e.g. 12 glycines and 3 serines). Nanobodies (or VHHs) are single-domain antigen-binding fragments derived from Camelid heavy chain. Camelid nanobodies, expressed in plants, may offer a solution as they are an attractive tool to bind efficiently to viral epitopes, cryptic or not accessible to conventional antibodies. Ghannam et al. (2015) described a novel, generic approach that might lead to virus resistance based on the expression of camelid specific nanobodies against *Broad bean mottle virus* (BBMV).

26.3 Immunological Methods

Antibodies produced against a virus preparation can be used for the detection of viruses in plants. The basis of entire serological tests is on the specificity of the antibodies to its homologous antigen (Van Regenmortel 1982). A variety of methods have been developed for demonstrating and estimating combination between antibodies and antigens. The traditional methods for using antisera with plant viruses involved direct observation of specific precipitates of virus and antibody, either in liquid media or in agar gels. Over about the past 25 years these methods have been progressively superseded by the use of enzyme-linked immunosorbent assay (ELISA), immunosorbent electron microscopy and dot immunobinding assay. Some of the important methods used in the serological detection of different viruses are as follows: Precipitin tests in liquid media shows the reaction between virus particles and their antibodies, precipitates are formed as a result of reaction between antibodies and multivalent virus antigens. No precipitation will occur when there is extreme excess of one of the reactants. With elongated viruses, precipitate zones are broad, whereas with isometric viruses, they are narrow. Two commonly used methods namely tube precipitin and micro precipitin tests are described by Hill (1984). *Chloroplast agglutination test* is another simple serological technique. It is done by mixing a drop of crude sap from an infected leaf with a drop of antiserum and observing for chloroplast clumping. It is successfully used in the detection of potato viruses (Hill 1984). *Latex agglutination test* sensitivity is increased by using latex particles sensitized with specific antiserum. When a sample containing specific antigen (virus particles) is mixed with latex particles sensitized with specific antiserum, it causes visible agglutination (Talley et al. 1980; Gella et al. 1991). In Agar double diffusion (Ouchterlony) test, antibody-antigen reaction is carried out in an agar gel instead of liquid. As antigen and antibody diffuse into the medium, a precipitation line is formed at a point where they meet in optimum proportions. The test is performed in agar gels in Petri plates or on glass slides. The antigen can be crude sap, clarified extract or purified virus (Ouchterloney 1968; Hampton et al. 1990; Dijkstra and de Jager 1998).

A most useful serological technique called Enzyme linked immunosorbent assay (ELISA), was first developed by Voller et al. (1976) and numerous variation are now available for detection of large numbers of viruses (Van Regenmortel and Burchard 1980; Clark and Adams 1977; Clark and BarJoseph 1984; Clark et al. 1986; Mowat and Dawson 1987; Hampton et al. 1990). Triple antibody sandwich (TAS-ELISA) procedure requires a specific trapping antibody, which is not recognized by the antibody-enzyme conjugate. The trapping antibody is produced in rabbit and the detecting monoclonal antibody in mouse (or rat) and the conjugate is an anti-mouse antibody labeled with enzyme (Koenig and Paul 1982; Clark et al. 1986).

In Dot Immunobinding Assay (DIBA) or Dot ELISA Assays, antibodies or antigens are bound to nitrocellulose or nylon membranes have been used to detect plant viruses. The method has been routinely used for the diagnosis of several plant viruses (Banttari and Goodwin 1985). Tissue blot immunoassay (TIBA) procedure is less labour-intensive than ELISA, rapid, sensitive, simple (no virus extraction is required), inexpensive (minimal equipment is needed), suitable for surveys of 1000–2000 samples per day, and the samples can be taken in the field and processed some time later (Lin et al. 1990). Serologically specific electron microscopy/Immunosorbent electron microscopy (ISEM) techniques involving the detection and identification of plant viruses by combining electron microscopy and serology are highly sensitive. Two techniques namely trapping and decoration are commonly used (Derrick 1973; Milne 1984).

Immunofluorescense provides sensitive assays for the detection of antigens in frozen or fixed tissue sections or viable cells. By staining with specific antibody conjugated to a fluorescent chromophore and illuminating with ultraviolet light the location of specific antigens in a tissue or cell preparation may be studied (Dijkstra and de Jager 1998). Electro-blot immunoassay (EBIA) uses the protein fractionating power of electrophoresis together with sensitivity and specificity of solid-phase immunoassay to identify and assay viral proteins (O'Donnell et al. 1982). Rapid immune-filter paper assay (RIPA) was developed by Tsuda et al. (1992) for the detection of plant viruses using white and colored latex beads coated with antibodies. The technique is simpler, less time-consuming and inexpensive. Simultaneous detection of various viruses TMV, CMV, PVY and Turnip mosaic virus, using different colour bead was reported by Tsuda et al. (1993). Lateral flow technique is suitable for field use by anyone without the need for any sophisticated equipment or technical skills to perform the tests. Commercial kits for the detection of many plant viruses are currently available worldwide (O'Farrell 2009). Immunocapture (IC) PCR or IC-RT-PCR method combines both serology and PCR to enhance sensitivity of detection of viruses. In this method, prior to PCR, virus particles are captured, immobilized and concentrated on a sterile solid surface previously coated with virus specific antiserum (Jain et al. 1998).

26.4 Application of Sero-diagnosis in India

A brief account of protein (serology) based diagnosis of viruses in different genera reported from India is given below. List of viruses for which antibodies and serological tests are reported from India is provided in Table 26.1.

26.5 Allexivirus

Complete coat protein (27 kDa) and truncated coat protein without N-terminal (20 kDa) region of *Garlic virus X* (GarV-X) were expressed in vitro and used for antiserum production. Specific and strong reaction was obtained with antiserum generated against GarV-X full coat protein and GarV-X was detected from field-grown allium crops *viz.,* onion, garlic, leek, and bunching onion and chives in ELISA. Antiserum generated against GarV-X truncated coat protein did not show reaction for GarV-X detection in immunoassay. Epitope mapping also indicated N-terminal as major antigenic determinant region with highest antigenic signal score. The studies confirmed that antigenic signals or epitopes reside in the N-terminal region of GarV-X which can be synthesized and used for production of monoclonal antibodies for specific detection purposes (Singh et al. 2014).

26.6 Ampelovirus

Survey conducted during 2010–2011 in the vineyards of Nashik and Pune regions of India revealed the association of an *Ampelovirus* antigenically related to *Grapevine leafroll-associated virus 3* (GLRaV-3) with seven cultivars of grapevine (Kumar et al. 2012a). Further, the survey and subsequent analysis of samples showing various kinds of virus like symptoms of grapes collected from Himachal Pradesh through DAS-ELISA and RT-PCR, occurrence of *Grapevine leafroll-associated virus* 1, 2, and 3 (GLRaV 1, 2, and 3) was reported (Kumar et al. 2013). Mixed infection with other viruses such as *Grapevine fleck virus* (GFkV) and *Grapevine virus B* (GVB) were also observed in these samples. In view of the uneven distribution of the GLRaV3 in plants, IC-RT-PCR technique where the virus could be detected without isolating the RNA was reported which can be used in the certification program. The modified extraction and RNA release protocol developed in the study was validated for specific detection of the virus in the vines of five infected grapevine cultivars (Kumar et al. 2014b).

26.7 Babuvirus

Selvarajan et al. (2010, 2011) reported detection of *Banana bunchy top virus* (BBTV) in banana sap diluted up to 1:50 in ELISA while PCR could detect the virus in a dilution up to 1:500 and discussed its utility in the production of BBTV-free plants through tissue culture. Natural occurrence of BBTV in *Ensete superbum* in India was reported based on DAC-ELISA using BBTV antiserum and further confirmed through sequencing (Selvarajan and Balasubramanian 2013).

26.8 Badnavirus

Viswanathan et al. (1996) and Singh et al. (2003) reported serological relationship of *Sugarcane bacilliform virus* (SCBV) with *Banana streak virus* (BSV) using DAC-ELISA and ISEM tests. Manoranjitham et al. (2012) reported purification, production of polyclonal antiserum against a Tamil Nadu isolate of BSV and its serological relationship with other badnaviruses through TAS ELISA and ISEM tests. Immunodiagnosis of episomal *Banana streak Mysore virus* (BSMyV) infecting banana using polyclonal antibodies to *in vitro* expressed coat protein of the virus in ELISA and IC-PCR was reported (Sharma et al. 2014). Involvement of a badnavirus was suspected with yellow mottle disease of black pepper based on serological relationship of the diseased leaves with badnaviruses such as BSV in ELISA tests (Bhat et al. 2003). Later Bhadramurthy et al. (2005) reported purification, production of polyclonal antiserum and development of DAS-ELISA based detection for the badnavirus infecting black pepper and related species. Gopi et al. (2010) reported DAC-ELISA based detection of *Citrus yellow mosaic virus* (CYMV) infecting citrus.

26.9 Begomovirus

Characterization of *Horsegram yellow mosaic virus* (HYMV) causing yellow mosaic in horsegram through ELISA and ISEM was reported (Muniyappa et al. 1987). Particles resembling those of geminiviruses were found by ISEM in extracts of plants infected with bhendi yellow vein mosaic, croton yellow vein mosaic, dolichos yellow mosaic, horsegram yellow mosaic, Indian cassava mosaic and tomato leaf curl viruses. All these viruses reacted with at least one out of ten monoclonal antibodies to *African cassava mosaic virus* (ACMV), and all reacted with a probe for ACMV DNA-1 (Harrison et al. 1991). Swanson et al. (1992) compared epitope profiles of the particle proteins of nine begomoviruses infecting legumes in India using 27 monoclonal antibodies produced against *Indian cassava mosaic virus* (ICMV) or *African cassava mosaic virus* (ACMV). The viruses fell into two groups-group 1 comprised isolates of *Dolichos* yellow mosaic; group 2 comprised isolates of HYMV together with isolates from blackgram, cowpea, frenchbean, pigeonpea, soybean, *Indigofera hirsuta* and mungbean. Isolates within each group differed slightly in epitope profile depending on the species or geographical origin. Serological relationship of *Cotton leaf curl virus* isolates from southern India using panel of monoclonal antibodies specific to ACMV and ICMV was reported (Nateshan et al. 1996). Swanson et al. (1998) reported serological detection and antigenic variation of *Tobacco leaf curl virus* and *Croton yellow vein mosaic virus* using ELISA carried out with monoclonal antibodies. Ramappa et al. (1998) reported detection of *Tomato leaf curl virus* (ToLCV) in its whitefly vector, *Bemisia tabaci* and many weed species using TAS ELISA. Garg et al. (2001) reported association of a begomovirus with potato apical leaf curl affected plant using ISEM which showed excellent clumping with ICMV specific antiserum. Occurrence of pumpkin yellow mosaic virus in pumpkin was reported based on ELISA (Muniyappa et al. 2003). Purification, production of antibodies and detection of *Tomato leaf curl Bangalore virus* in tomato and many reservoir hosts *viz*., *Parthenium hysterophorus*, *Acanthospermum hispidum*, *Ageratum conyzoides* using DAC-ELISA was reported (Devaraja et al. 2005). Association of ICMV on bittergourd yellow mosaic disease in Tamil Nadu by ELISA was reported (Rajinimala and Rabindran 2007). Transmission of ICMV through cassava biotype population of *Bemisia tabaci* was confirmed through ISEM and TAS-ELISA (Antony et al. 2009). Detection of *Bittergourd yellow mosaic virus* (BGYMV) in bitter gourd through TAS-ELISA was reported (Rajinimala et al. 2009). A scFv was developed against coat protein (AV1) of *Tomato leaf curl New Delhi virus* (ToLCNDV), which was characterized by western blot, ELISA and surface plasmon resonance spectroscopy (Zakri et al. 2010). Production of polyclonal antiserum against in vitro expressed coat protein gene of the pepper leaf curl virus and its detection through ELISA was reported (Sinha et al. 2011). Coat protein gene of *Pumpkin yellow vein mosaic virus* (PYVMV) was expressed as a fusion protein with maltose binding protein (MBP) in *E. coli* in soluble state. The polyclonal antibodies produced to the purified fusion protein successfully detected PYVMV and other bipartite and monopartite begomoviruses in the field samples at 1:250 dilutions in ELISA (Phaneendra et al. 2014).

26.10 Carlavirus

Two strains of *Cowpea mild mottle virus* (CMMV) infecting groundnut in Andhra Pradesh was identified based on biological and serological properties in ELISA and gel diffusion test using antiserum to CMMV, Cassia mild mosaic and Potato virus M (PVM) (Sivaprasad and Sreenivasulu 1996). Occurrence and incidence of *Chrysanthemum B virus* (CBV) in chrysanthemum in Himachal Pradesh was reported using DAS-ELISA and ISEM using CBV specific antiserum (Verma et al. 2003). Singh et al. (2011) reported expression of recombinant CBV coat protein and production of polyclonal antiserum for indexing mother stock material. Association of a new carlavirus serologically related to *Carnation latent virus* was reported infecting football lily, an ornamental flower plant in Sikkim and Darjeeling hills based on particle morphology and serology (Das et al. 2010). Pramesh et al. (2012) reported widespread association of *Garlic common latent virus* (GarCLV) in garlic accessions including commercial varieties, farmers' cultivars and breeding lines through DAC-ELISA and ISEM using polyclonal antiserum to GarCLV.

26.11 Carmovirus

Seed transmission of *Blackgram mottle virus* (BMoV) in blackgram studied through ELISA showed that seed transmission was highest in cv. PLU-277 (15.9%), followed by cvs T-9 (11.8%), PLU-213 (7.0%) and UH-81-7 (1.3%) (Varma et al. 1992). Seed transmission was correlated with the amount of virus present in the embryonic axis and later in primary leaves. The presence of virus in the testa alone did not result in its transmission through seeds. Virus concentration in different tissues varied; the mean amount of virus in the three cultivars was found to be 48–1234 ng per embryonic axis, 15–24 ng per cotyledon, and 12–20 ng per testa. The infection of primary leaves through the seed also resulted in systemic infection if the amount of virus in primary leaves exceeded 100 ng/100 mg of tissue. Close agreement was found between the percentage of seedlings with systemic infection and the percentage of seeds and embryonic axes containing more than 100 ng of the virus. The cultivars that resisted seed transmission contained relatively small amounts of the virus in embryonic axes (Varma et al. 1992). BMoV was detected in seedlings of urdbean and the extent of seed transmission varied from 5–10% and identity of the virus was confirmed through ISEM test (Dinesh Chand et al. 2004). Testing virus like symptoms on different cultivars of carnation through ELISA showed presence of *Carnation mottle virus* (CarMV) (Singh et al. 2005). Sandra et al. (2015) over expressed CP gene of the *Soybean yellow mottle mosaic virus* (SYMMV) occurring in India in *E. coli* and an antiserum of 1:16,000 titer against the recombinant CP was produced. Serological cross reactivity analysis revealed that SYMMV was serologically related to blackgram mottle virus but not to cowpea mottle virus, the other legume infecting carmoviruses. The antiserum was used to detect prevalence of SYMMV in legume crops by ELISA. Out of 145 field samples of legumes (mungbean, blackgram, French bean and soybean) collected from different places in India, SYMMV was detected only in 16 samples of mungbean and one sample of blackgram.

26.12 Capillovirus

Rana et al. (2011b) used ELISA for the detection of *Apple stem grooving virus* (ASGV) infecting apple. In a survey conducted at Shimla District of Himachal Pradesh and subsequent analysis of samples collected based on DAS-ELISA and RT-PCR indicated that 12 % of the samples were infected with ASGV (Kumar et al. 2012b). During survey of kiwi plantations (*Actinidia deliciosa*) in Palampur, Kangra District, Himachal Pradesh plants exhibiting severe interveinal mottling, severe leaf distortion, ringspots and chlorosis along the leaf margins was identified as caused by a strain of ASGV by DAS ELISA and nucleic acid spot hybridization (NASH) (Bhardwaja et al. 2014).

26.13 Caulimovirus

Association of a strain of *Peanut chlorotic streak virus* with chlorotic vein banding disease of groundnut based on agar gel diffusion, ELISA and EBIA tests was reported (Satyanarayana et al. 1994). Polyclonal antibodies to the coat protein of carnation etched ring virus was expressed in bacterial system and ELISA based kit was developed. The antibody showed good detectability and specificity with infected glasshouse and field material compared to commercially available kit (Raikhy et al. 2007).

26.14 Closterovirus

In planta accumulation of *Citrus tristeza virus* (CTV) in different plant tissues of infected citrus host determined using ELISA revealed that CTV was not evenly distributed in all the plant parts. The tender bark of 6 months to 1 year old, peticle and mid rib of young leaves, and apical bud showed high amount of CTV. Infected acid (Kagzi) lime (*C. aurantifolia*) and sweet orange (Mosambi) (*C. sinensis*) trees accumulated higher amount of virus than Darjeeling mandarin (*C. reticulata*). The severity of symptom caused by CTV in acid lime could not be correlated to virus titer. CTV was found to move faster in acid lime and sweet orange than in mandarin tree. It was observed that virus persisted up to 180 days in crude sap prepared with 0.05 M phosphate buffer, pH 7.0 stored at 4 °C, but only up to 2–4 days in same crude sap kept at 25–32 °C (Tarafdar et al. 2012). Based on DAC-ELISA and RT-PCR, incidence of CTV infecting different species of citrus was 26.% in Vidarbha region, 47.1–56% in northeast, 36–50% in south India and 16–60% in north-north west region (Biswas et al. 2014). The high titre of the virus was observed in Kagzilime (*Citrus aurantifolia*) and sweet orange (*C. sinensis*) but low in mandarin (*C. reticulata*). Borah et al. (2014) reported ELISA based method for the detection of *Citrus tristeza virus* (CTV) infecting *Citrus* spp. in Assam while Ghosh et al. (2014) used DAS-ELISA to study distribution of CTV in different parts of *Citrus* spp. as well as to identify CTV-free *Citrus* spp. in Darjeeling Hills.

26.15 Cucumovirus

Based on the precipitin test with an antiserum of *Cucumber mosaic virus* (CMV), association of CMV was suspected with a mosaic disease of pea (Rao 1986; Rao et al. 1995). Occurrence of CMV infecting chrysanthemum and *Dianthus* spp. based on serological relationship with CMV antiserum was reported (Srivastava et al. 1992; Raj et al. 1993). Kiranmai et al. (1996) compared relative efficacy of three ELISA methods for routine detection of CMV in banana and found that DAS-ELISA is ideal for detection of the virus in pseudostem sap extracts. A virus disease causing mosaic, leaf distortion, curling and stunting in *Amaranthus tricolor* and *A. hypochondriascus* was identified as a strain of CMV based on biological and serological relationship studies (Raj et al. 1997). Virus causing mosaic and leaf distortion on brinjal, chilli and tomato was identified to be caused by different strains of CMV based on its serological reaction with CMV specific antiserum (Kiranmai et al. 1997). A mosaic disease on Egyptian henbane was identified as caused by a strain of CMV based on its serological affinities with CMV antiserum (Samad et al. 2000). Association of CMV with stunt disease of black pepper was reported based on biological and serological properties (Sarma et al. 2001). The serological and molecular methods were evaluated for the sensitive detection of CMV in gladiolous leaf and corms (Raj et al. 2002). Bhat et al. (2004a) reported purification and development DAS-ELISA based diagnostics for the routine detection of CMV infecting black pepper and related species. Occurrence of CMV on Indian long pepper was reported based on biological and serological studies involving ELISA and EBIA (Bhat et al. 2004b). Association CMV with vanilla and lily causing mosaic disease was identified to be caused by a strain of CMV based on both biological and serological relationship studies involving both ELISA and EBIA (Madhubala et al. 2005; Sharma et al. 2005). Occurrence of CMV on alstroemeria and *Tomato aspermy virus* (TAV) on chrysanthemum based on ELISA and RT-PCR was reported (Verma et al. 2005, 2006). Occurrence of CMV on gladiolus was reported based on DAS-ELISA, TIBA and ISEM (Katoch et al. 2003). Occurrence of CMV on polyhouse grown pepper through DAS-ELISA in Pune was reported (Verma et al. 2004c). Verma et al. (2006) reported detection and characterization of TAV infecting chrysanthemums using DAS-ELISA. Kumar et al. (2009) used DAC-ELISA for indexing plants derived from CMV and TAV infected chrysanthemum through meristem culture for freedom from viruses. Khan et al. (2012) produced polyclonal antiserum against recombinant coat protein of CMV expressed in *E. coli*. The antiserum had a titre of 1:8000 and was used in DAC-ELISA and IC-RT-PCR based assay for the detection of CMV isolates. On comparison, they found IC-RT-PCR was more sensitive than ELISA to detect the presence of virus in tissue culture based banana plant as well as in various host plants. Using DAC-ELISA, Suresh et al. (2013) reported occurrence of CMV on various cucurbits grown in the state of Maharashtra.

26.16 Dianthovirus

A survey of 61 carnation cultivars grown in the state of Himachal Pradesh showing varied symptoms on the leaves and flowers, were screened through bioassay and ELISA which revealed the widespread presence of *Carnation ringspot virus* (CRSV). Out of 61 cultivars tested 47 were found infected with CRSV using ELISA (Raikhy et al. 2006). Routine diagnostic techniques like RT-PCR further confirmed infected samples for presence of this virus.

26.17 Emaravirus

Polyclonal antibodies to *Pigeonpea sterility mosaic* virus (PPSMV) preparations were produced which were effective in the detection of the virus in plant tissues by DAS-ELISA (Kumar et al. 2002). Latha and Doraiswamy (2008) developed DAS-ELISA and DIBA for detection of PPSMV from mites and found that DIBA is more sensitive.

26.18 Foveavirus

Occurrence of *Apple stem pitting virus* (ASPV) causing symptoms of curling, puckering and necrosis on leaves in apple in the northwestern Himalayan region was identified based on DAS-ELISA tests using ASPV polyclonal antiserum (Kumar et al. 2012b).

26.19 Ilarvirus

Occurrence, purification, production of polyclonal antiserum and development of serological based assays such as ELISA and EBIA for the detection of *Tobacco streak virus* (TSV) in sunflower infected with necrosis disease in India was reported (Bhat et al. 2001b, 2002). In the same year, Reddy et al. (2002) reported the occurrence of TSV in groundnut infected with stem necrosis virus. Kalyani et al. (2007) used ELISA to identify sources of resistance to TSV in wild *Arachis* germplasm. Vemana and Jain (2010) reported detection of TSV in groundnut pod shell and seed testa using DAC-ELISA using TSV specific antiserum. Their study showed that TSV was absent in cotyledon and embryo and hence concluded that TSV is not a true seed transmitted virus in legumes. Reddy et al. (2012) reported occurrence of TSV in kenaf through DAC-ELISA tests. Kulshrestha et al. (2009) developed ELISA based diagnostic kit for the detection of *Prunus necrotic ringspot virus*

(PNRSV) using polyclonal antibodies obtained against in vitro expressed coat protein of the virus as antigen. Chandel et al. (2007, 2013) reported occurrence of PNRSV in peach, stone and pome fruits using specific antibodies in ELISA tests. In a survey conducted at Shimla District of Himachal Pradesh and subsequent analysis of samples collected based on DAS-ELISA and RT-PCR indicated that 15% of the samples are infected with *Apple mosaic virus* (ApMV) in apple (Kumar et al. 2012b).

26.20 Luteovirus

Presence of *Potato leaf roll virus* (PLRV) in potato, its purification, production of polyclonal antiserum and detection by ELISA was reported (Garg 1987; Dhawan and Rishi 1990). Occurrence of MAV-type *Barley yellow dwarf virus* (BYDV) on wheat in the Garhwal Hills, central Himalaya was demonstrated by using monoclonal antibodies (Khetarpal et al. 1993). Diversity among coat proteins of luteoviruses associated with chickpea stunt disease was studied using monoclonal antibodies (Naidu et al. 1997). Occurrence of *Sugarcane yellow leaf virus* (SCLYV) in Bihar, Haryana, Maharashtra, Uttar Pradesh, Uttarakhand and Tamil Nadu was reported based symptoms and ELISA tests (Rao et al. 2000; Gaur et al. 2003). DAS-ELISA studies were conducted on the detection SCYLV causing yellow leaf syndrome of sugarcane in leaf and juice antigens. Among these, juice antigen showed high titre of the virus. Further, DAC-ELISA studies revealed that plants raised from disease infected planting materials recorded high titre for SCYLV compared to those from symptom-free seed canes (Viswanathan and Balamuralikrishnan 2004). Bharathi et al. (2006) used DAC-ELISA, DIBA and TIBA for the detection of mixed infection of *Sugarcane streak mosaic virus* (SCSMV) and *Sugarcane yellow leaf virus* (SCYLV) infecting sugarcane in Andhra Pradesh.

26.21 Macluravirus

Development and use of DAS- ELISA system was reported for the detection of *Cardamom mosaic virus* (CdMV) infected cardamom plants (Saigopal et al. 1992). A polyclonal antiserum was raised against the purified CdMV. The coat protein (CP) encoding sequence of the virus was expressed in *Escherichia coli* and the protein purified by affinity chromatography under denaturing conditions (Jacob and Usha 2002). The viral nature of the expressed CP was confirmed by positive reaction with anti CdMV IgG in a Western blot. The expressed CP aggregated irreversibly upon renaturation at concentrations above 0.07 mg/ml. The expression of the CP led to the formation of filamentous aggregates in *E. coli* as observed by immunogold electron microscopy. The filamentous aggregates were of 100–150 nm in length. Immuno-capture RT-PCR confirmed the absence of coat protein mRNA in the filamentous aggregates observed in *E.coli*.

A high titer (1:256,000) polyclonal antibody (PAb) to the recombinant coat protein of *Large cardamom chirke virus* (LCCV) CP was produced, which strongly recognized LCCV in crude leaf extract from chirke affected large cardamom plants in ELISA and DIBA tests (Vijayanandraj et al. 2013). The sensitivities of the ELISA and DIBA were 5 and 0.1 ng of expressed protein, respectively. Both the ELISA and DIBA were validated with 100 % accuracy in detecting LCCV in field samples. The PAb differentiated CdMV, another close relative of LCCV. Rapid detection kit of large cardamom chirke virus using lateral flow immunostrip has been developed at the ICAR-Indian Agricultural Research Institute, New Delhi (Maheshwari et al. 2014). The assay procedure accomplished within 10 min and the results could be evaluated visually.

26.22 Mandarivirus

An isolate of *Indian citrus ringspot virus* (ICRSV) from Kinnow mandarin in northern India was characterized based on biological, serological and sequence properties (Rustici et al. 2000). The virus was purified from infected Saxa bean leaves and an antiserum prepared. There was no serological cross-reaction with representative allexi-, capillo-, potex- and trichoviruses, except a faint one-way reaction with *Potato virus X*. Purified virus yielded a major band, the presumed coat protein (CP), of about 34 kDa. The product of the 34 kDa ORF was confirmed as the CP by expression in *E. coli*. The derived amino acid sequence of the CP contained some short motifs similar to those of potex-, fovea-, carla- and allexiviruses but otherwise there was no strong similarity to any of these. Serological and molecular diagnostic reagents were prepared for the routine detection of the virus in field samples. Production of ICRSV-free plants from an infected plant of kinnow mandarin (*Citrus nobilis* Lour × *C. deliciosa* Tenora) was reported (Sharma et al. 2008; Singh et al. 2008). The shoot apices of different sizes (0.2–1.0 mm) excised from the ICRSV-infected plants were micrografted onto decapitated rootstock seedlings of rough lemon (*C. jambhiri*). Micrograft survival depended on the size of shoot apex and the sucrose concentration of the culture medium. Increase in scion size from 0.2 to 0.7 mm resulted in an increase in micrografting success rate from 30.55 to 51.88%. Further, micrograft survival obtained with 0.2 mm was improved from 30.55 to 38.88% by increasing sucrose concentration in the culture media from 5 to 7.5%. The micrografted plants were tested for ICRSV using ELISA and RT-PCR. All plants raised from 0.2-mm scion were found negative with both ELISA and RT-PCR whereas only 20% of the ELISA negative plants raised from 0.3-mm scion were found negative for ICRSV with RT-PCR.

26.23 Nucleorhabdovirus

Naidu et al. (1989) produced polyclonal antiserum to *Maize mosaic virus* (MMV) infecting sorghum and developed DAC ELISA method for its detection in plants. In immuno double gel diffusion tests the virus showed antigenic relationship with MMV from ReUnion and Hawai.

26.24 Pecluvirus

Manohar et al. (1995) reported serological variability in the *Indian peanut clump virus* (IPCV) isolates from India and Senegal based on the reaction to a panel of monoclonal antibodies in TAS-ELISA. A DAS-ELISA procedure developed to test seed transmission of IPCV showed correlation between results of ELISA and grow out tests in peanut (Reddy et al. 1998). Seed transmission of IPCV in the field infected peanut plants ranged from 3.5–17% depending on the genotype. IPCV was also shown to be seed transmitted in finger millet, foxtail millet and pearl millet at frequencies of 5.2%, 9.7% and 0.9% respectively. Further, Delfosse et al. (1999) reported wheat and barley as additional hosts of IPCV based on ELISA tests.

26.25 Poacevirus

Sugarcane streak mosaic virus (SCSMV) was found more commonly associated with mosaic disease affected sugarcane than the *Sugarcane mosaic virus* (SCMV) (Gaur et al. 2006). In addition, mixed infection of SCMV and SCSMV was also detected in sugarcane mosaic infected leaf samples from Maharashtra and Tamil Nadu (Rao et al. 2006). The virus associated with mosaic disease of sorghum growing around the sugarcane fields in Andhra Pradesh state, was found to be serologically related to the SCSMV and Sorghum mosaic Parbhani virus (SMPV) (Srinivas et al. 2010). Reddy et al. (2011) used DAC-ELISA and IC-PCR for the detection of (SCSMV) in meristem derived sugarcane plants to confirm elimination of the virus. Duplex IC-RT-PCR method was developed for the detection of SCSMV and SCMV infecting sugarcane (Viswanathan et al. 2013).

26.26 Potexvirus

Occurrence of *Potato virus X* (PVX) causing mild mosaic symptoms on wild potato species, *Solanum chacoense* was reported based on serology (Sangar et al. 1984). A virus isolated from turnip in Aligarh, Uttar Pradesh, which caused mild mosaic, mottling and curling of leaves followed by overall stunting of plants, was characterized as PVX on the basis of its host range, biological and physical properties, particle morphology, ultrastructural studies, and serological relationship (Samad et al. 1991). Rani et al. (2010) and Sherpa et al. (2012) demonstrated usefulness of polyclonal antiserum prepared from expressed *Cymbidium mosaic virus* (CymMV) coat protein for the detection of the virus in an array of assays including ELISA and ISEM. Polyclonal antibodies were raised against the bacterial expressed fused coat proteins of *Potato virus Y* (PVY) and *Potato virus X* (PVX) detected natural infection of PVY and PVX in potato leaf samples by DAC-ELISA (Kapoor et al. 2014a).

26.27 Potyvirus

A virus disease of tree tomato (*Cyphomandra betacea*) was reported be to be caused by a strain of *Potato virus Y* (PVY) based on host range, physical properties and serology suggesting that tree tomato may be one of the harboring hosts of PVY found prevalent in the potato crop (Bhargava and Joshi 1959). Serological studies of *Peanut mottle virus* (PMoV) infected peanut sample showed a serological relationship with Soybean mosaic virus (SMV) in the precipitin ring interface test (Rajeshwari et al. 1983). ELISA was used to check PMoV in individual part of peanut seed without disturbing its viability. Virus could be detected in extracts of infected cotyledons and embryo at 1/3600 dilution (Bharathan et al. 1984). Using DAS-ELISA, PMoV was shown to be distantly related to adzuki bean mosaic (ABMV), amaranthus leaf mottle (ALMV), clover yellow vein (CYVV) viruses and SMV. Immunosorbent electron microscopy (ISEM) showed PMoV to be closely related to ABMV, ALMV, CYVV and SMV. In the precipitin ring interface, ELISA and ISEM tests no specific serological reaction was noted between PMoV and antisera to groundnut eye spot, peanut green mosaic, pepper veinal mottle, potato virus Y, sugarcane mosaic and turnip mosaic viruses. Three biologically distinct isolates of *Peanut green mosaic virus* in ELISA reacted strongly with antisera to peanut green mosaic and SMV antisera, and moderately with ABMV and peanut stripe virus antisera. All isolates also reacted positively with antisera to peanut eye spot, blackeye cowpea mosaic, pea seed-borne mosaic, PVY and tobacco etch viruses, and did not react with antisera to peanut mottle, bean yellow mosaic, bean common mosaic, CYVV and sugarcane mosaic viruses (Naidu et al. 1991). Khetarpal et al. (1990, 1994) reported production of polyclonal antiserum and use of ELISA for the detection of *Pea seed borne mosaic virus* (PSbMV) in seed embryos and for resistance screening against PSbMV in pea accessions. Occurrence of *Sweet potato feathery mottle virus* (SPFMV) in germplasm of *Ipomoea batatas* in India using ISEM and ELISA was reported (Kumar et al. 1991). Serological relationship of SPFMV with other potyviruses through DAC-ELISA was reported (Jain et al. 1993). A purified viral preparation of a potyvirus from mosaic infected plant of *Brassica campestris* and *B. juncea* in SDS-PAGE analysis showed two major bands of approximately 37 kDa and 31 kDa, a pattern very similar to that of a reference isolate of turnip mosaic virus (TuMV). In Western-blot immunoassay, an antiserum to TuMV reacted with both the coat protein bands of the virus isolate, but not with four other potyviruses. The high performance liquid chromatographic profile of tryptic peptides from the coat protein of these isolates was found to be very similar to that of the reference TuMV, but differed substantially from those of four other potyviruses (Haq et al. 1994). Presence of *Plum pox virus* (PPV) in peach, plum and almond were reported from Himachal Pradesh based on ELISA and ISEM tests (Bhardwaj et al. 1995). Detection of Maize dwarf mosaic virus on maize and sudangras and SCMV on sudangras was reported by DIBA and DAC-ELISA (Rao et al. 1996a, b, 1998b). Comparative host range and serological studies based on ELISA and EBIA of a PRSV isolate from India was reported (Roy et al. 1999). Differentiation of three potyviruses namely PVY, *Henbane mosaic virus* (HMV) and *Cowpea aphid borne mosaic virus* (CABMV) by antibodies directed towards N- or N- and C- terminal peptide domains of coat protein through DAC-ELISA and EBIA was reported (Bhat et al. 1996). A potyvirus causing mosaic mottling in eggplant was identified as a strain of PVY on the basis of serological and partial nucleotide sequence properties. The virus-specific antibodies directed to the N-terminal region of the virus cross-reacted only with PVY in ELISA and EBIA indicating that it is strain of PVY (Bhat et al. 1999). Detection of *Bean yellow mosaic virus* (BYMV) by DIBA, DAS-ELISA, TIBA and ISEM in gladiolus was reported (Katoch et al. 2003). Based on the ISEM techniques such as trapping and decoration using specific antiserum occurrence of *Dasheen mosaic virus* (DsMV) on elephant foot yam was reported for the first time (Pandit et al. 2001). Occurrence of DsMV on ornamental aroids such as *Aglaonema* sp. (mosaic); *Philodendron* sp. (mosaic, vein clearing and leaf malformation); *Colocasia esculenta* (mosaic and chlorotic feathering); and *Zantedeschia* sp. (mosaic and leaves malformation) grown in the Kangra Valley of Himachal Pradesh was confirmed based on DAS-ELISA and ISEM tests using DsMV specific antiserum (Ram et al. 2003). The SPFMV causing feathery mottle disease was purified from infected sweet potato leaves, the antiserum was produced and tested in Ouchterloney agar double diffusion test. SPFMV was detected from different samples through DAC-ELISA and DIBA (Jeeva et al. 2004). Rao et al. (1998a, 2004) used DAC-ELISA, EBIA, DIBA and ISEM to determine antigenic diversity of *Sugarcane mosaic virus* (SCMV) isolates from different parts of India. Based on DAC-ELISA and DIBA, Srivastava et al. (2012) reported association of *Watermelon mosaic virus* (WMV) with mosaic disease of *Catharnathus roseus*. Occurrence of PVY, *Tobacco etch virus* (TEV) and *Pepper veinal mottle virus* (PVMV) on polyhouse grown capsicum in Pune by DAS-ELISA (Verma et al. 2004c). Gaur et al. (2002) suggested that cane stalk juice could be equally satisfactory as virus infected leaf samples and can be used in commercial screening of SCMV in sugarcane. First report of occurrence of *Peanut stripe virus* (PStV) in pea in India, a seed borne virus of quarantine importance was identified based on host range, transmission, purification, particle morphology and serology (Prasada Rao et al. 2004). Association *of Zuchini yellow mosaic virus* (ZYMV) with bottlegourd (*Lagenaria siceraria*) and cucumber plants showing severe mosaic, interveinal chlorosis, and leaf deformation that resulted in fern-leaf appearance and severe fruit distortion in Pune area of Maharashtra was reported based on the strong serological reaction obtained with ZYMV antiserum (Verma et al. 2004a, b). Presence of *Iris mild mosaic virus* (IMMV) and BYMV in cultivated Iris was detected by ELISA, ISEM and IC-RT-PCR (Kulshrestha et al. 2006a, b). Association of a Potyvirus serologically related to WMV was reported causing mosaic disease in *Catharnthus roseus* based on transmission, electron microscopy, DAC-ELISA and DIBA tests. Kiranmai et al. (2005) reported serological relationship of banana infected with *kokkan* disease with *Datura leaf distortion virus* (DLDV) in ELISA indicating association of a *Potyvirus* with the disease which was later identified as *Banana bract mosaic virus* (BBrMV) reported from Philippines (Thomas et al. 1997). Occurrence of a potyvirus causing mosaic symptoms on Himalayan-butterfly bush (*Buddleja crispa*), an ornamental, perennial shrub was reported which was later identified a strain of *Chilli veinal mottle virus* based on sequencing (Mehra et al. 2005). This plant may act as a potential reservoir of virus infecting chilli crop. Raikhy et al. (2006) reported detection *Carnation vein mottle virus* in carnation cultivars based on ELISA and IC-PCR tests. Selvarajan et al. (2011) developed ELISA based diagnostic for the detection of BBrMV in banana especially the mother plants and tissue culture raised banana plants for identification of virus-free plants. Production of polyclonal antibodies using recombinant coat protein of PRSV and its use in immunodiagnosis was reported (Agarwal et al. 2009). Though it was initially reported that host range of BBrMV was restricted only to *Musa* spp., its occurrence was also reported from small cardamom grown in Karnataka and Kerala based on serology and genome properties (Siljo et al. 2012). Association of WMV and ZYMV with cucurbits was reported from Trans-Gangetic plains of India using DAS-ELISA and PCR (Sharma et al. 2012). Duplex IC-RT-PCR method was developed for the detection of SCSMV and SCMV infecting sugarcane (Viswanathan et al. 2013). Recombinant coat protein antibody based IC–RT-PCR was developed for detection and discrimination of SCSMV isolates from southern India and was found more sensitive than DAC-ELISA and dot-blot Immunobinding assay (Hema et al. 2003). Natural association of *Konjac mosaic virus* (KoMV) with mosaic disease of elephant foot yam (*Amorphophallus paeoniifolius*) in India was identified by ELISA using an antiserum raised against KoMV recombinant coat protein (CP) and IC-RT-PCR using CP gene specific primers (Padmavathi et al. 2013). Garlic leaf samples collected from 12 garlic growing states of India showed presence of Onion yellow dwarf virus (OYDV) based on DAS-ELISA, RT-PCR and qRT-PCR tests (Gawande et al. 2013). Out of total 493 samples analyzed, 308 samples from all states were found positive for OYDV in variable proportions. The highest percentage of OYDV positives were recorded from Maharashtra (96%) followed by Gujarat (75%) and Madhya Pradesh (75%). The lowest percentage of OYDV positives were recorded from Rajasthan (25%) followed by Delhi (40%). DAS-ELISA identified BCMV as the causal virus of the mosaic disease in common bean (Hamid et al. 2014). A high titer polyclonal antibody to the expressed core region of coat protein of *Peanut mottle virus* (PeMoV) was produced. The antiserum was useful in the specific detection of PeMoV as it showed negligible cross reactivity with the other potyviruses such as PStV, PVY, PRSV and OYDV. The antiserum was validated in ELISA using 1169 field and greenhouse samples of peanut which showed 1.85–26.3% incidence of PeMoV in peanut seed multiplication field (Soumya et al. 2014). Characterization and confirmation of PRSV strain infecting *Trichosanthese cucumerina* (snake gourd) in Tamil Nadu based on DAC-ELISA and RT-PCR was reported (Kumar et al. 2014a). Indigenous kits for the detection of viruses infecting potato (PVX, PVA, PVS,

PVM, PVY) have been standardized at the ICAR-Central Potato Research Institute, Shimla (CPRI 2010, 2013). Maheshwari et al. (2015b) reported cloning and expression of genes encoding antibody to the recombinant coat protein (rCP) of PRSV and assessed engineered antibody for the detection of PRSV. A 33-kDa rCP of PRSV, which was produced in *E. coli*, generated PRSV specific antibody in immunized mouse. The heavy and light chain variable domain genes (VH and VL) of 351 and 360 nucleotides, respectively, were cloned from the mRNA isolated from the spleen of the immunized mouse with rCP of PRSV. The VH and VL genes were individually expressed in *E. coli*. Both the antibody fragments recognized PRSV in the crude sap; however, the VL antibody fragment showed higher affinity to PRSV. The mixture of VH and VL detected PRSV as effectively as polyclonal antibody. The recombinant antibody fragments mixture detected PRSV in the field samples with 100% accuracy in DIBA and ELISA. The sensitivity of the detection of PRSV using antibody fragments was 1.0 and 10.0 ng in DIBA and ELISA, respectively.

26.28 Sobemovirus

A possible member of the Sobemovirus named as Ginger chlorotic fleck virus (GCFV), was detected in ginger. Purified preparations had major coat protein species of molecular weight 29 kDa. The virus was serologically unrelated to members of the sobemovirus group, including cocksfoot mottle, lucerne transient streak, *solanum nodiflorum* mottle, southern bean mosaic, sowbane mosaic, turnip rosette and velvet tobacco mottle viruses (Thomas 1986).

26.29 Tenuivirus

A disease characterized by chlorotic stripes and bands, named sorghum stripe disease (SStD), was observed on sorghum in India with an incidence of less than 0.5% to nearly 10%. The affected plants were dwarfed and had poor or no panicle formation. The disease was shown to be caused by a tenuivirus serologically related to maize stripe virus (MStV) (Peterschmitt et al. 1991). Virus particles were filamentous, less than 10 nm in width. The purified virus preparation contained only one polypeptide of 34.5 kDa. In ELISA and in EBIA, the virus showed strong serological reaction to the MStV isolates from Florida, Reunion and Venezuela than to a RStV isolate from Japan.

26.30 Tobamovirus

DAC-ELISA based detection of *Odontoglossum ringspot virus* (ORSV) in cymbidium using polyclonal antibodies produced against its recombinant coat protein was reported (Sherpa et al. 2004). A new disease of *Tabernamontana divaricota*, a plant grown for its ornamental value exhibiting chlorotic ringspot, chlorotid banding, oak leaf pattern and mosaic was shown to be associated with a tobamovirus namely, *Tobacco mild green mosaic virus* (TMGMV) based on ELISA test (Cohen et al. 2000). Polyclonal antibodies against ORSV coat protein produced using bacterial expressed recombinant coat protein detected the virus in DAC-ELISA in orchid samples collected from Sikkim which will be useful for virus indexing in orchid certification programmes (Rani et al. 2010). The occurrence of *Cucumber green mottle mosaic virus* (CGMMV) infecting various cucurbitaceous crops in Trans Gangetic plains of India and Maharashtra was reported based on DAC-ELISA tests (Sharma et al. 2012; Suresh et al. 2013).

26.31 Tobravirus

In a survey for viruses infecting gladiolus (*Gladiolus grandiflorus*) around Palampur in the Kangra Valley, plants that showed notched leaf blade margin symptoms reacted positively in DAS-ELISA with polyclonal antibodies specific to *Tobacco rattle virus* (TRV). Further, the presence of TRV confirmed through sap incoculation of *Nicotiana tabacum* that showed characteristic TRV symptoms of necrotic local lesions, followed by systemic mottling which also tested positive for TRV in DAS-ELISA (Katoch et al. 2004).

26.32 Tospovirus

Purification and production of polyclonal antiserum against the tospovirus which causes bud necrosis disease (BND) of peanut in India was described (Reddy et al. 1992). The virus contained three polypeptides of 78 kDa, 54 kDa and 31 kDa. In two ELISA procedures the virus failed to react with antisera to *Tomato spotted wilt virus* (TSWV) obtained from different sources and with an antiserum to *Impatiens necrotic spot virus* (INSV). Additionally, in reciprocal tests TSWV and INSV antigens failed to react with antiserum to the virus infecting peanut in India. In electroblot immunoassay 54 kDa and 31 kDa polypeptides of the virus reacted with the homologous antiserum. None of the heterologous antisera reacted with any of the three viral polypeptides. On the basis of serological differences the virus that caused BND in India was distinct and therefore named as *Peanut bud necrosis virus* (PBNV) (also known as *Groundnut bud necrosis virus,* GBNV). Association of a tospovirus with necrosis disease of sunflower was suspected based serological studies through DAC-ELSA (Jain et al. 2000). Occurrence of a tospovirus infecting grain legumes such as blackgram, greengram, and soybean in Delhi was reported based on serology through DAC-ELISA, EBIA and also by nucleic acid hybridization test (Bhat et al. 2001a). Purification, production of polyclonal antiserum and detection of GBNV infecting mungbean through ELISA, EBIA and IC-RT-PCR reported (Thien et al. 2002). Presence of GBNV on tomato, cowpea and potato from different regions was reported based on the ELISA test using polyclonal antiserum (Jain et al. 2005). Immunodiagnosis of PBNV and *Watermelon bud necrosis virus* (WBNV) using polyclonal antiserum to recombinant nucleocapsid protein of PBNV was reported (Jain et al. 2005). Onion *(Allium cepa)* plants exhibiting chlorotic spindle or diamond shaped lesions on the leaves and scapes, with twisting or bending flower bearing stalks observed in the Jalna and Nasik regions of Maharashtra was identified to be caused by a strain of *Iris yellow spot virus* (IYSV) based on the strong serological reaction obtained with IYSV antiserum which was further confirmed through RT-PCR and sequencing of the coat protein gene of the virus (Ravi et al. 2006). Occurrence of IYSV on garlic plants showing straw-colored, spindle-shaped spots with poorly defined ends was reported from Pune region of Maharashtra using DAS-ELISA and subsequent confirmation through sequencing (Gawande et al. 2010). In the advanced stages, single spindle-shaped chlorotic lesions coalesced, leading to withering of leaves and flower-bearing stalks. Venkat et al. (2008) reported production of monoclonal antibodies and its use in epitope profiling of the GBNV isolates in South India. A dot blot assay was also developed for detection of GBNV from field samples. Occurrence of GBNV infecting onion and *Colacasia esculenta* using DAC-ELISA was reported by Sujitha et al. (2012) and Sivaprasad et al. (2011). In a survey for PBNV, WBNV, *Capsicum chlorosis virus* (CaCV), and IYSV conducted between 2002 and 2009 in the major vegetable-growing areas in India, PBNV was documented widely in tomato and chili peppers in 14 states representing southern, north-western, north-eastern, and central regions and WBNV was predominantly detected in watermelons and cucurbits in all except north-eastern regions. In addition, the expanded host range of PBNV to watermelons and other cucurbits and WBNV to tomato and chili peppers was observed leading to natural mixed infection of the two viruses. IYSV was found in onion in southern, central, and north-eastern regions and CaCV in tomato and chili peppers in northern and southern regions, respectively (Kunkalikar et al. 2011). Detection of PBNV in tomato through DAC-ELISA using polyclonal antiserum was reported (Govardhana et al. 2013). Development of a single-chain variable fragment (scFv) for specific diagnosis of infection by GBNV was reported (Maheshwari et al. 2015a). Heavy chain (372 nucleotide) and light chain (363 nucleotide) variable region clones obtained from a hybridoma were used to make a scFv construct that expressed a 29-kDa protein in *E. coli*. The scFv specifically detected GBNV in field samples of cowpea, groundnut, mungbean, and tomato, and it did not recognize WBNV.

26.33 Trichovirus

ELISA and PCR were used for the identification and characterization of *Apple chlorotic leaf spot virus* (ACLSV) in almond and apple from Himachal Pradesh (Rana et al. 2008, 2011b). Polyclonal antiserum produced against bacterially expressed coat protein gene of ACLSV consistently detected ACLSV in pome and stone fruits as well as herbaceous host plants by DAS-ELISA, TBIA and IC-RT-PCR (Rana et al. 2011a). In a survey of apple orchards of Himachal Pradesh and subsequent analysis of samples through DAS-ELISA revealed the occurrence of ACLSV to tune of 16.8% (Kumar et al. 2012b).

26.34 Detection of Multiple Viruses

Kapoor et al. (2014b) reported production and used cocktail of polyclonal antibodies that will help in multiple virus detection and overcome the limitation of individual virus purification, protein expression and purification as well as immunization in multiple rabbits. A dual fusion construct was developed using conserved coat protein sequences of CMV and PRSV in an expression vector, pET-28a(+). The fusion protein (~40 kDa) was expressed in *E. coli* and purified. Likewise, a triple fusion construct was developed by fusing conserved CP sequences of CMV and PRSV with conserved nucleocapsid protein (N) sequence of GBNV and expressed as a fusion protein (~50 kDa) in pET-28a(+). Antibodies made separately to each of these three viruses recognized the double and triple fusion proteins in Western blot indicating retention of desired epitopes for binding with target antibodies. The fusion proteins (~40 kDa and ~50 kDa) were used to produce cocktail of polyclonal antibodies by immunizing rabbits, which simultaneously detected natural infection of CMV and PRSV or CMV, PRSV and GBNV in cucurbitaceous, solanaceous and other hosts in DAC-ELISA.

26.35 Concluding Remarks

In India plants infected by a wide range of viruses often cause important agronomic and economic losses. Detection of plant viruses is becoming more challenging as globalisation of trade, and the potential effects of climate change facilitate movement of viruses and their vectors, thereby transforming the diagnostic landscape. In vegetatively propagated crops, presence of viruses is a matter of concern as the viruses could be transmitted to generations. Perennial nature of the crops still aggravates the situation as the virus inocula remain in the field for long, rendering the whole plantation vulnerable to disease spread by insect vectors. Under such circumstances, detection of viruses at premature stages of infection by use of rapid, sensitive and accurate detection methods seems crucial to ensure safe and sustainable agriculture and as such reduces economic losses. In addition, some of the crops such as banana, citrus, potato, sugarcane and ornamentals are being multiplied through tissue culture. There are more than 50 established commercial tissue culture units in the country producing more than 200 million plantlets per year. The need for virus testing of mother and tissue culture raised plants is important since inadvertent micro propagation of virus infected plants will not only result in poor stand, but also in spread of viruses wherever such plants are grown. For proper identification, detection technique needs to be rapid, most accurate and inexpensive, as it forms the key step in developing appropriate practical solutions to manage plant virus diseases that pose continuous threats to crops.

Currently immuno-assay and nucleic acid based techniques occupy the leading positions as methods of diagnostics. Among them, various ELISA formats as well as PCR and its modifications are most popular. They are highly sensitive, specific, rapid, user-friendly, and generally excel the conventional methods in these parameters. Combination of PCR and ELISA (IC-PCR) further increase their thresholds, enhance the specificity, and improve results of the analysis. A number of precision instruments and facilities are developed for PCR and ELISA tests; commercial kits, master mixes, additional reagents. To date, the methods elaborated on the basis of these techniques, alone or in conjunction with conventional methods, allow reliable diagnostics of plant viruses as well as more successful control and proper management of many economically important diseases. Recombinant antibodies from synthetic or human antibody libraries are a superior alternative to animal-based methods of monoclonal antibody production. Thus scFv is the best product for medical, diagnostic and research applications. In India some of the virus diagnostics (such as against banana and potato viruses) are being used in routine large scale indexing. Similarly parameters for production of disease-free planting materials of a few crops such as banana, black pepper, potato, sugarcane, vanilla and many ornamentals have been developed. If used properly, this would lead in the production of disease-free certified planting material. But on the contrary, many of the virus diagnostics developed still remain at the laboratory level. Validation, scaling up and large scale use in the certification process is yet to start. Thus, validation of the diagnostics developed, development of easy to use field level diagnostic kits, development of diagnostic centres/clinics, accreditation of diagnostic laboratories, development of standard operating procedures (SOPs) and training of manpower are need of the hour. There is a need to establish a 'National antisera Bank' with a aim to produce, store and distribute antisera to important viruses infecting different crops on demand. There is also a need for awareness creation on the importance of virus-free planting material and capacity building of all stake holders.

Further progress in plant disease diagnostics should be expected, as novel methods and new applications of detecting modes are introduced. Techniques such as micro array, biosensors and antibody engineering seem to be good candidates for the future in this field. Recombinant antibodies offer potential for biosensor applications and will contribute greatly to genomics and proteomics in the future.

References

Agarwal S, Krishnareddy M, Jain RK (2009) Production of polyclonal antibodies using recombinant coat protein of *Papaya ringspot virus* and their use in immunodiagnosis. J Plant Biochem Biotechnol 18:109–111

Antony B, Lisha VS, Palaniswami MS (2009) Evidences for transmission of Indian cassava mosaic virus through *Bemisia tabaci* – cassava biotype. Arch Phytopathol Plant Protect 42:922–929

Banttari EE, Goodwin PH (1985) Detection of potato viruses S, X and Y by enzyme linked immunosorbent assay on nitrocellulose membrane (Dot –ELISA). Plant Dis 69:202–205

Bhadramurthy V, Retheesh ST, Bhat AI, Madhubala R, Hareesh PS, Pant RP (2005) Development of ELISA based technique for the detection of a putative badnavirus infecting black pepper (*Piper nigrum* L.). Indian Phytopathol 58:314–318

Bharathan N, Reddy DVR, Rajeshwari R, Murthy VK, Rao VR, Lister RM (1984) Screening peanut germ plasm lines by enzyme-linked immunosorbent assay for seed transmission of peanut mottle virus. Plant Dis 68:757–758

Bharathi V, Reddy LK, Nagamalleswari P (2006) Mixed infection of sugarcane streak mosaic and sugarcane yellow leaf virus infecting sugarcane crops in AP, India. Indian Sugar 56:37–42

Bhardwaj SV, Kohosla K, Sharma DR, Thakur PD (1995) Detection of plum pox virus in India. Acta Hortic 386:237–240

Bhardwaja P, Ramb R, Zaidi AA, Hallan V (2014) Characterization of apple stem grooving virus infecting *Actinidia deliciosa* (kiwi) in India. Sci Hortic 176:105–111

Bhargava KS, Joshi RD (1959) A virus disease of tree tomato—*Cyphomandra betacea* SENDT.—due to potato virus Y. Am J Potato Res 36:288–291

Bhat AI, Varma A, Jain RK (1996) Differentiation of four potyviruses occurring in northern India by antibodies directed towards N- or N-&C- terminal peptide domains of coat proteins. Int J Trop Plant Dis 14:157–166

Bhat AI, Varma A, Pappu HR, Rajamannar M, Jain RK, Praveen S (1999) Characterization of a potyvirus from eggplant (*Solanum melongena*) as a strain of potato virus Y by N-terminal serology and sequence relationships. Plant Pathol 48:648–654

Bhat AI, Jain RK, Varma A, Chandra N, Lal SK (2001a) Tospoviruses infecting grain legumes in Delhi- their identification by serology and nucleic acid hybridization. Indian Phytopathol 54:112–116

Bhat AI, Kumar A, Jain RK, Rao SC, Ramiah M (2001b) Development of serological based assays for the diagnosis of sunflower necrosis disease. Ann Plant Protect Sci 9:292–296

Bhat AI, Jain RK, Kumar A, Ramiah M, Varma A (2002) Serological and coat protein sequence studies suggest that necrosis disease on sunflower in India is caused by a strain of *Tobacco streak virus*. Arch Virol 147:651–658

Bhat AI, Devasahayam S, Sarma YR, Pant RP (2003) Association of a badnavirus in black pepper (*Piper nigrum* L.) transmitted by mealy bug (*Ferrsia virgata*) in India. Curr Sci 84:1547–1550

Bhat AI, Sarma YR, Sreenivasulu P, Pant RP (2004a) Occurrence and identification of a *Cucumber mosaic virus* isolate infecting Indian long pepper (*Piper longum*). J Med Aromat Plant Sci 26:279–284

Bhat AI, Faisal TH, Madhubala R, Hareesh PS, Pant RP (2004b) Purification, production of antiserum and development of enzyme linked immunosorbent assay-based diagnosis for cucumber mosaic virus infecting black pepper. J Spices Aromat Crop 13:16–21

Biswas KK, Tarafdar A, Sharma SK, Singh JK, Dwivedi S, Biswas K, Jayakumar BK (2014) Current status of *Citrus tristeza virus* incidence and its spatial distribution in citrus growing geographical zones of India. Indian J Agr Sci 84:184–189

Boonham N, Kreuze J, Winter S, van der Vlugt R, Bergervoet J, Tomlinson J, Mumford R (2014) Methods in virus diagnosis: from ELISA to next generation. Virus Res 186:20–31

Borah M, Nath PD, Saikia AK (2014) Biological and serological techniques for detection of citrus tristeza virus affecting citrus species of Assam. India Afr J Agr Res 9:3804–3810

Central Potato Research Institute (CPRI) (2010) Dipstick assay for the detection of potato virus. CPRI Newsletter 43:3

Central Potato Research Institute (CPRI) (2013) Value chain on potato and potato products. NAIP Report: 27

Chand D, Chalam VC, Pant RP, Khetarpal RK (2004) Detection and symptomatology of seed transmitted *Blackgram mottle virus* in *Vigna mungo*. J Mycol Plant Pathol 34:202–205

Chandel V, Rana T, Hallan V, Zaidi AA (2007) Evidence for the occurrence of Prunus necrotic ringspot virus on peach in India by serological and molecular methods. Can J Plant Pathol 29:311–316

Chandel V, Rana T, Hallan V (2013) Prunus necrotic ringspot virus: incidence on stone and pome fruits and diversity analysis. Arch Phytopathol Plant Protect 46:2376–2386

Clark MF, Adams AN (1977) Characteristics of the microplate method of enzyme-linked immunosorbent assay for the detection of plant viruses. J Gen Virol 34:475–483

Clark MF, Bar-Joseph M (1984) Enzyme immunosorbent assays in plant virology. Methods Virol 7:51–85

Clark MF, Lister RM, Bar-Joseph M (1986) ELISA techniques. Methods Enzymol 118:742–766

Cohen J, Rosner A, Kagan S, Lampel S, Beckelman H, Maslenin L, Zeidan M, Ger A (2000) A new disease in Tabernamontana caused by tobacco mild green mosaic virus. Acta Hortic 568(31):103–109

Das DR, Singh L, Pant RP, Ahlawat YS (2010) Characterization of a new Carlavirus isolated from football lily (*Scandoxus katharine*) in India. Indian Phytopathol 63:433–437

Delfosse R, Legrve D, Maraite TD (1999) Indian peanut clump virus (IPCV) infection on wheat and barley: symptoms, yield loss and transmission through seed. Plant Pathol 48:273–282

Derrick KS (1973) Quantitative assay for plant viruses using serologically specific electron microscopy. Virology 56:652–653

Devaraja NK, Savithri HS, Muniyappa V (2005) Purification of tomato leaf curl Bangalore virus and production of polyclonal antibodies. Curr Sci 89:181–183

Dhawan P, Rishi N (1990) Isolation and purification of potato leafroll virus in India. Indian J Virol 6:38–45

Diano M, LeBivic A, Hirn M (1987) A method for production of highly specific polyclonal antibodies. Anal Biochem 166:224–229

Dijkstra J, de Jager CP (1998) Practical plant virology-protocols and exercises. Springer Verlag, Berlin

Garg ID (1987) Degeneration of potato varieties in western Maharashtra. J Indian Potato Assoc 14:127–128

Garg ID, Khurana SMP, Kumar S, Lakra BS (2001) Association of a geminivirus with potato apical leaf curl in India and its immuno-electron microscopic detection. Potato J 28:227–232

Gaur RK, Singh M, Singh AP, Singh AK, Rao GP (2002) Screening of sugarcane mosaic potyvirus (SCMV) from cane stalk juice. Sugar Tech 4:169–171

Gaur RK, Singh AK, Singh M, Singh AK, Upadhyay PP, Rao GP (2003) Reliability of serological identification of sugarcane mosaic potyvirus and sugarcane yellow leaf luteovirus from cane stalk juice. Sugar Cane International. UK. Sept–Oct 18–21

Gaur RK, Singh M, Singh A, Mishra LC, Rao GP (2006) Identification and molecular characterization of Sugarcane streak mosaic virus in North India. Sugar Cane International. UK. Mar-Apr 13–17

Gawande SJ, Khar A, Lawande KE (2010) First report of Iris yellow spot virus on garlic in India. Plant Dis 94:1066

Gawande SJ, Chimote KP, Gurav VS, Gopal J (2013) Distribution and natural incidence of Onion yellow dwarf virus (OYDV) on garlic and its related *Allium* species in India. Indian J Hort 70:544–548

Gella FJ, Serra J, Gener J (1991) Latex agglutination procedures in immunodiagnosis. Pure Appl Chem 63:1131–1134

Ghannam A, Kumari S, Muyldermans S, Abbady AQ (2015) Camelid nanobodies with high affinity for broad bean mottle virus: a possible promising tool to immunomodulate plant resistance against viruses. Plant Mol Biol 87:355–369

Ghosh A, Das A, Pun KB, Kumar R, Meena R, Baranwal VK (2014) Present status of citrus tristeza virus infecting *Citrus* spp. in Darjeeling Hills and its detection in different plant parts. Phytoparasitica 42:381–386

Gopi V, Saakar TG, Kumar AG, Gopal K (2010) Diagnosis of *Citrus yellow mosaic virus* by PCR and ELISA in sweet orange (*Citrus Sinensis* Osbeck). J Plant Dis Sci 5:158–162

Govardhana M, Gopinath K, Vanitha LS, Madd ER, Manjunatha L (2013) Serological detection and host range studies of peanut bud necrosis virus infecting tomato. Trends Biosci 6:75

Hamid A, Ahmad M, Padder BA, Shah MD, Saleem S, Sofi TA, Mir AA (2014) Pathogenic and coat protein characterization confirming the occurrence of Bean common mosaic virus on common bean (*Phaseolus vulgaris*) in Kashmir, India. Phytoparasitica 42:317–322

Hampton R, Ball E, De Boer S (1990) Serological methods for detection and identification of viral and bacterial plant pathogens – a laboratory manual. APS Press, Minnesota. 389pp

Haq QMR, Srivastava KM, Raizada RK, Singh BP, Jain RK, Mishra A, Shukla DD (1994) Biological, serological and coat protein properties of a strain of turnip mosaic virus causing a mosaic disease of *Brassica campestris* and *B. juncea* in India. J Phytopathol 140:55–64

Harrison BD, Muniyappa V, Swanson MM, Roberts IM, Robinson DJ (1991) Recognition and differentiation of seven whitefly-transmitted geminiviruses from India, and their relationships to African cassava mosaic and Thailand mung bean yellow mosaic viruses. Ann Appl Biol 18:299–308

Hema M, Kirthi N, Sreenivasulu P, Savithri HS (2003) Development of recombinant coat protein antibody based IC–RT-PCR for detection and discrimination of sugarcane streak mosaic virus isolates from southern India. Arch Virol 148:1185–1193

Hill SA (1984) Methods in plant virology. Blackwell Scientific, Oxford, p 167

Indian Agricultural Research Institute (IARI) (2014) Novel strategies for molecular diagnosis of plant viruses. NAIP report:61–68

Jacob T, Usha R (2002) Expression of *Cardamom mosaic virus* coat protein in *Escherichia coli* and its assembly into filamentous aggregates. Virus Res 86:133–141

Jain RK, Kumar CA, Bhat AI, Varma A (1993) Serological relationship of sweet potato feathery mottle virus with other potyviruses. Indian Phytopathol 46:162–164

Jain RK, Pappu SS, Pappu HR, Culbreath AK, Todd JW (1998) Molecular diagnosis of tomato spotted wilt *Tospovirus* infection of peanut and other field and greenhouse crops. Plant Dis 82:900–904

Jain RK, Bhat AI, Byadagi AS, Nagaraju SH, Halkeri AV, Anahosur KH, Varma A (2000) Association of a tospovirus with sunflower necrosis disease. Curr Sci 79:1703–1705

Jain RK, Pandey NA, Krishnareddy M, Mandal B (2005) Immunodiagnosis of groundnut and watermelon bud necrosis viruses using polyclonal antiserum to recombinant nucleocapsid protein of *Groundnut bud necrosis virus*. J Virol Methods 130:162–164

Jeeva ML, Balakrishnan S, Edison K, Umamaheswaran K, Makeshkunrar T (2004) Characterization, purification and serology of sweet potato feathery mottle virus in India. J Root Crop 30:24–30

Kalyani G, Reddy AS, Kumar PL, Prasada Rao RDVJ, Aruna R, Waliyar F, Nigam SN (2007) Sources of resistance to *Tobacco streak virus* in wild *Arachis* (Fabaceae: Papilionoidae) germplasm. Plant Dis 91:1585–1590

Kapoor R, Mandal B, Paul PK, Jain RK (2014a) Simultaneous detection of potato viruses Y and X by DAC-ELISA using polyclonal antibodies raised against fused coat proteins expressed in *Escherichia coli*. J Plant Biochem Biotechnol 23:332–335

Kapoor R, Mandal B, Paul PK, Chigurupati P, Jain RK (2014b) *Production* of *Cocktail* of polyclonal antibodies using bacterial expressed recombinant protein for multiple virus detection. J Virol Methods 106:7–14

Katoch M, Abdin MZ, Ram R, Zaidi AA (2003) An overview of diagnostics for viruses infecting gladiolus. Crop Prot 22:153–156

Katoch M, Abdin MZ, Zaidi AA (2004) First report of *Tobacco rattle virus* occurring in gladiolus in India. Plant Pathol 53:236

Khan S, Jan AT, Mandal B, Haq QMR (2012) Immunodiagnostics of cucumber mosaic virus using antisera developed against recombinant coat protein. Arch Phytopathol Plant Protect 45:561–569

Khetarpal RK, Maury Y, Cousin R, Burghofer A, Varma A (1990) Studies on resistance of pea to pea seed borne mosaic virus and new pathotypes. Ann Appl Biol 116:297–303

Khetarpal RK, Kumar J, Beuve M, Parakh DB, Nath R (1993) Outbreak of MAV-like isolate of barley yellow dwarf virus luteovirus on wheat in Gharwal hills of central Himalaya in India. Plant Pathol 43:152–153

Khetarpal RK, Parak DB, Singh S, Nath R, Jain RK, Varma A (1994) Bean common mosaic virus detected by DAC-indirect ELISA in exotic *Phaseolus vulgaris* L. Indian J Virol 10:13–16

Kiranmai G, Sreenivasulu P, Nayudu MV (1996) Comparison of three different tests for detection of cucumber mosaic cucumovirus in banana (*Musa paradisiaca*). Curr Sci 71:764–767

Kiranmai G, Sreenivasulu P, Nayadu MV (1997) Characterization of cucumber mosaic cucumovirus isolates naturally infecting three solanaceous vegetable crops in Andhra Pradesh. Indian Phytopathol 50:416–425

Kiranmai G, Kumar PL, Hema M, Vekatramana M, Prasadji JK, Rao DM, Sreenivasulu P (2005) Partial characterization of a potyvirus causing bract mosaic of banana in Andhra Pradesh. Indian J Virol 16:7–11

Koenig R, Paul HL (1982) Variants of ELISA in plant virus diagnosis. J Virol Methods 5:113–125

Kohler G, Milstein C (1975) Continuous cultures of fused cells secreting antibody of predefined specificity. Nature 256(5517):495–497

Kulshrestha S, Hallan V, Raikhy G, Ram R, Garg ID, Haq MR, Zaidi AA (2006a) Occurrence of Iris mild mosaic potyvirus in cultivated Iris in India. Indian J Biotechnol 5:94–98

Kulshrestha S, Hallan V, Raikhy G, Ram R, Zaidi AA, Garg IG (2006b) Incidence of *Bean yellow mosaic virus* on Iris. Acta Hortic 722:235

Kulshrestha S, Hallan V, Raikhy G, Zaidi AA (2009) Potential uses of in vitro expressed and purified recombinant *Prunus necrotic ringspot virus* coat protein gene. Arch Phytopathol Plant Protect 42:442–452

Kumar CA, Mandal BB, Chandel KPS, Jain RK, Varma A, Srivastava M (1991) Occurrence of sweet potato feathery mottle virus in germplasm of *Ipomoea batatas* (L) in India. Curr Sci 60:321–325

Kumar PL, Duncan G, Roberts IM, Jones AT, Reddy DVR (2002) Cytopathology of Pigeonpea sterility mosaic virus in pigeonpea and *Nicotiana benthamiana*: similarities with those of eriophyid mite-borne agents of undefined aetiology. Ann Appl Biol 140:87–96

Kumar S, Khan MS, Raj SK, Sharma AK (2009) Elimination of mixed infection of *Cucumber mosaic* and *Tomato aspermy virus* from *Chrysanthemum morifolium* Ramat. cv. Pooja by shoot meristem culture. Sci Hortic 119:108–112

Kumar S, Baranwal VK, Singh P, Jain RK, Sawat SD, Singh SK (2012a) Characterization of a grapevine leafroll-associated virus 3 from India showing incongruence in its phylogeny. Virus Genes 45:195–200

Kumar S, Singh RM, Ram R, Badyal J, Hallan V, Zaidi AA, Varma A (2012b) Determination of major viral and sub viral pathogens incidence in apple orchards in Himachal Pradesh. VirusDis 23:75–79

Kumar S, Singh L, Ferretti L, Barba M, Zaidi AA, Hallan V (2013) Evidence of grapevine leafroll associated virus-1–3, grapevine fleck virus and grapevine virus B occurring in Himachal Pradesh, India. Virus Dis 24:66–69

Kumar S, Sankarlingam A, Rabindran R (2014a) Characterization and confirmation of papaya ringspot virus-W strain infecting *Trichosanthese cucumerina* at Tamil Nadu. India J Plant Pathol Microbiol 5:2

Kumar S, Rai R, Baranwal VK (2014b) Development of an immunocapture–reverse transcription–polymerase chain reaction (IC-RT-PCR) using modified viral RNA release protocol for the detection of grapevine leafroll-associated virus 3 (GLRaV-3). Phytoparasitica 42:1–6

Kunkalikar SR, Poojari S, Arun BM, Rajagopalan PA, Chen TC, Yeh SD, Naidu RA, Zehr UB, Ravi KS (2011) Importance and genetic diversity of vegetable-infecting tospoviruses in India. Phytopathology 101:367–376

Latha TKS, Doraiswamy S (2008) Detection of pigeonpea sterility mosaic virus, the causal agent of sterility mosaic disease of pigeonpea in viruliferous mite vector by DAS-ELISA and DIBA. Arch Phytopathol Plant Protect 41:537–541

Lin NS, Hsu YH, Hsu HT (1990) Immunological detection of \plant viruses and a mycoplasma like organism by direct tissue blotting on nitrocellulose membranes. Phytopathology 80:824–828

Madhubala R, Bhadramurthy V, Bhat AI, Hareesh PS, Retheesh ST, Bhai RS (2005) Occurrence of *Cucumber mosaic virus* on vanilla (*Vanilla planifolia* Andrews) in India. J Biosci 30:339–350

Maheshwari Y, Vijayanandraj S, Jain R K, Mandal B (2014). Lateral flow immunoassay for the rapid detection of large cardamom chirke virus in large cardamom plants. International conference on Frontier discoveries and emerging opportunities in life sciences. Dr. Harisingh Gaur University, Sagar, p 153

Maheshwari Y, Vijayanandraj S, Jain RK, Mandal B (2015a) Engineered single-chain variable fragment antibody for immunodiagnosis of groundnut bud necrosis virus infection. Arch Virol 160:1297–1301

Maheshwari Y, Verma HN, Jain RK, Mandal B (2015b) Engineered antibody fragments for immunodiagnosis of papaya ringspot virus. Mol Biotechnol 57:644–652

Manohar SK, Dollet M, Dubern J, Gargani D (1995) Studies on variability of peanut clump virus: symptomatology and serology. J Phytopathol 143:233–238

Manoranjitham SK, Kavino M, Thiribhuvanamala G, Ganapathy T, Rabindran R, Kumar N (2012) Detection of *Banana streak virus* (BSV) Tamil Nadu isolate (India) and its serological relationship with other badnaviruses. Afr J Biotech 11:14632–14637

McCafferty J, Griffiths AD, Winter G, Chiswell DJ (1990) Phage antibodies: filamentous phage displaying antibody variable domains. Nature 348:552–554

Mehra A, Hallan V, Lal B, Zaidi AA (2005) Occurrence of *Chilli veinal mottle virus* in Himalayan butterfly bush (*Buddleja crispa*). New Dis Rep 11:32

Milne RG (1984) Electron microscopy for the identification of plant viruses in in vitro preparation. Methods Virol 7:87–120

Mowat WP, Dawson S (1987) Detection and identification of plant viruses by ELISA using crude sap extracts and unfractionated antisera. J Virol Methods 15:233–247

Muniyappa V, Rajeshwari R, Bharathan N, Reddy DVR, Nolt BL (1987) Isolation and characterization of a geminivirus causing yellow mosaic disease of horsegram (*Macrotyloma uniflorum* (Lam.) Verdc.) in India. J Phytopathol 119:81–87

Muniyappa V, Maruthi MN, Babitha CR, Colvin J, Briddon RW, Rangaswamy KT (2003) Characterisation of pumpkin yellow vein mosaic virus from India. Ann Appl Biol 142:323–331

Naidu RA, Harikrishnan R, Manohar SK, Reddy DVR, Ratna AS, King SB, Bandyopadhyay R (1989) The occurrence of maize mosaic virus on sorghum in India. Ann Appl Biol 114:301–310

Naidu RA, Sreenivasulu P, Manohar SK, Ramakrishna KN, Reddy DVR, Nayudu MV (1991) Identification of peanut green mosaic virus strains in India. J Phytopathol 133:49–56

Naidu RA, Mayo MA, Reddy SV, Jolly CA, Torrance L (1997) Diversity among the coat proteins of luteoviruses associated with chickpea stunt disease in India. Ann Appl Biol 130:37–47

Nateshan HM, Muniyappa V, Swanson MM, Harrison BD (1996) Host range, vector relations and serological relationships of cotton leaf curl virus from southern India. Ann Appl Biol 128:233–244

Nayudu MV (2008) Plant viruses. Tata McGraw-Hill Publishing Company Ltd., New Delhi. p 1249

Noordam D (1975) Identification of plant viruses: methods and experiments. Cen Agric Pub Doc, Wagenigen

O'Donnell IJ, Shukla DD, Gough KH (1982) Electro-blot radioimmunoassay technique for detecting plant viruses. J Virol Methods 4:19–26

O'Farrell B (2009) Evolution in lateral flow–based immunoassay systems. In: Lateral flow immunoassay. Humana Press, Totowa, p 1–33

Ouchterlony O (1968) Handbook of immunodiffusion and immunoelectrophoresis. Ann Arbor Scientific Publications, Michigan

Padmavathi M, Srinivas KP, Hema M, Sreenivasulu P (2013) First report of Konjac mosaic virus in elephant foot yam (*Amorphophallus paeoniifolius*) from India. Australas Plant Dis Notes 8:27–29

Pandit MK, Nath PS, Mukhopadhyay S, Devonshire BJ, Jones P (2001) First report of *Dasheen mosaic virus* in elephant foot yam in India. Plant Pathol 50:802–802

Peterschmitt M, Ratna AS, Sacks WR, Reddy DVR, Mughogho LK (1991) Occurrence of an isolate of maize stripe virus on sorghum in India. Ann Appl Biol 118:57–70

Phaneendra C, Rao KRSS, Kapoor R, Jain RK, Mandal B (2014) Fusion coat protein of pumpkin yellow vein mosaic virus with maltose binding protein: applications in immunodiagnosis of begomoviruses. VirusDis 25:390–393

Pramesh D, Islam S, Singh J, Baranwal VK (2012) Serological and PCR detection of *Garlic common latent virus* associated with garlic accessions in India. Med Plants 4:17–22

Prasada Rao RDVJ, Reddy AS, Chakrabarthy SK, Anitha K (2004) Interception of seed transmitted viruses in peanut germplasm imported into India during 1986 to 2003. Indian J Plant Protect 32:102–104

Raikhy G, Hallan V, Kulshrestha S, Sharma ML, Verma N, Ram R, Zaidi AA (2006) Detection of carnation ringspot and carnation vein mottle viruses in carnation cultivars in India. Acta Hortic 722:247

Raikhy G, Hallan V, Kulshrestha S, Zaidi AA (2007) Polyclonal antibodies to the coat protein of carnation etched ring virus expressed in bacterial system: production and use in immunodiagnosis. J Phytopathol 155:616–622

Raj SK, Aminuddin, Srivastava KM, Singh BP (1993) Natural infection of cucumber mosaic virus on *Dianthus barbatus* in India. Plant Pathol 42:811–813

Raj SK, Aminuddin, Singh BP, Pal M (1997) Characterizatin of a cucumber mosaic virus isolate causing leaf crinckle and severe mosic of Amaranthus in India. Can J Plant Path 19:97–100

Raj SK, Srivastava A, Chandra G, Singh BP (2002) Characterization of cucumber mosaic virus isolate infecting gladiolus cultivars and comparative evaluation of serological and molecular methods for sensitive diagnosis. Curr Sci 83:1132–1137

Rajeshwari R, Iizuka N, Nolt BL, Reddy DVR (1983) Purification, serology and physico-chemical properties of a peanut mottle virus isolate from India. Plant Pathol 32:97–205

Rajinimala N, Rabindran R (2007) First report of Indian cassava mosaic virus on bittergourd (*Momordica charantia*) in Tamil Nadu. India Australas Plant Dis Notes 2:81–82

Rajinimala N, Rabindran R, Ramaiah M (2009) Detection of bittergourd yellow mosaic virus (BGYMV) in bitter gourd (*Momordica charantia*) through triple antibodys Sandwitch (TAS) ELISA and polymerase chain reaction (PCR). Arch Phytopathol Plant Protect 42:559–565

Ram R, Joshi A, Verma N, Kulshrestha S, Raikhy G, Hallan V, Zaidi AA (2003) First report of *Dasheen mosaic virus* infecting four ornamental aroids in India. Plant Pathol 52:411–411

Ramappa HK, Muniyappa V, Colvin J (1998) The contribution of tomato and alternative host plants to tomato leaf curl virus inoculum pressure in different areas of South India. Ann Appl Biol 133:187–198

Rana T, Chandel V, Hallan V, Zaidi AA (2008) Characterisation of *Apple chlorotic leaf spot virus* infecting almonds in India. Aust Plant Dis Notes 3:65–67

Rana T, Chandel V, Hallan V, Zaidi AA (2011a) Expression of recombinant *Apple chlorotic leaf spot virus* coat protein in heterologous system: production and use in immunodiagnosis. J Plant Biochem Biotechnol 20:138–144

Rana T, Negi A, Dhir S, Thockchom T, Chandel V, Walia Y, Singh RM, Ram R, Hallan V, Zaidi AA (2011b) Molecular diagnosis of apple virus and viroid pathogens from India. Arch Phytopathol Plant Protect 44:505–512

Rani P, Pant RP, Jain RK (2010) Serological detection of cymbidium mosaic and odontoglossum ringspot viruses in orchids with polyclonal antibodies produced against their recombinant coat proteins. J Phytopathol 158:542–545

Rao GP (1986) Studies on a virus disease of pea (*Pisum sativum* L.) with reference to identification, host metabolism and varietal susceptibility. Ph D Thesis. Univ of Gorakhpur

Rao GP, Singh M, Shukla K (1995) A new strain of cucumber mosaic virus causing mosaic disease of pea in India. Indian J Virol 11:43–48

Rao GP, Jain RK, Varma A (1996a) Occurrence of maize dwarf mosaic virus on maize and sudangrass in India. Plant Dis 80:711

Rao GP, Jain RK, Lim HS, Ford RE (1996b) Identification of sugarcane mosaic potyvirus infecting sudangras in India. Phytopathology 86:56

Rao GP, Jain RK, Varma A (1998a) Characterization and purification off an Indian isolate of sugarcane mosaic virus. Sugarcane 1:8–10

Rao GP, Jain RK, Varma A (1998b) Identification of sugarcane mosaic and maize dwarf mosaic potyvirus infecting poaceous crops in India. Indian Phytopathol 51:10–16

Rao GP, Gaur RK, Singh M, Srivastava AK, Virk AS, Singh N, Patil AS, Viswanathan R, Jain RK (2000) Existence of sugarcane yellow leaf luteovirus in India. Sugar Tech 2:37–38

Rao GP, Singh M, Gaur RK, Jain RK (2004) Antigenic and biological diversity among sugarcane mosaic isolates from different geographical regions in India. Indian J Biotechnol 3:538–541

Rao GP, Chatenet M, Girard JG, Rott P (2006) Distribution of sugarcane mosaic and sugarcane streak mosaic virus in India. Sugar Tech 8:79–81

Ravi KS, Kitkaru AS, Winter S (2006) Iris yellow spot virus in onions: a new tospovirus record from India. Plant Pathol 55:288

Reddy DVR, Ratna AS, Sudarshana MR, Poul F, Kumar IK (1992) Serological relationships and purification of bud necrosis virus, a tospovirus occurring in peanut (*Arachis hypogaea* L.) in India. Ann Appl Biol 120:279–286

Reddy AS, Hobbs HA, Delfosse P, Murthy AK, Reddy DVR (1998) Seed transmission of Indian Peanut clump virus (IPCV) in peanut and millets. Plant Dis 82:343–346

Reddy AS, Prasada Rao RDVJ, Thirumala Devi K, Reddy SV, Mayo MA, Roberts I, Satyanarayana T, Subramaniam K, Reddy DVR (2002) Occurrence of *Tobacco streak virus* on peanut (*Arachis hypogaea*) in India. Plant Dis 86:173–178

Reddy S, Ch V, Sreenivasulu P (2011) Generation of *Sugarcane streak mosaic virus*-free sugarcane (*Saccharum* Spp. hybrid) from infected plants by in vitro meristem tip culture. Eur J Plant Pathol 130:597–604

Reddy BVB, Sivaprasad Y, Kumar CVMN, Sujitha A, Reddy R, Sai-Gopal DVR (2012) First report of *Tobacco streak virus* infecting Kenaf (*Hibiscus cannabinus*) in India. VirusDis 23:80–82

Rishi N (2006) Significant achievements and current status: virology. In: Chahal SS, Kheterpal RK, Thind TS (eds) One hundred years of plant pathology in India: an overview. Scientific Publishers (India), Jodhpur. pp 143–205

Roy G, Jain RK, Bhat AI, Varma A (1999) Comparative host range and serological studies of papaya ringspot potyvirus isolates. Indian Phytopathol 52:14–17

Rustici G, Accotto GP, Noris E, Masenga V, Luisoni E, Milne RG (2000) Indian citrus ringspot virus: a proposed new species with some affinities to potex-, carla-, fovea-and allexiviruses. Arch Virol 145:1895–1908

Saigopal DVR, Naidu R, Joseph T (1992) Early detection of 'katte' disease of small cardamom through Enzyme Linked Immunosorbent assay (ELISA). J Plant Crop 20(suppl):73–75

Samad A, Naqvi QA, Garg ID, Twhouvenel JC (1991) Potato virus X—a new report from turnip (*Brassica rapa* L.) in India. J Phytopathol 133:65–75

Samad AAS, Raj SK, Srivastava A, Chandra G, Ajayakumar PV, Zaim M, Singh BP (2000) Characterization of an Indian isolate of cucumber mosaic virus infecting Egyptian henbane (*Hyoscyamus muticus* L.) Acta Virol 44:131–136

Sandra N, Kumar A, Sharma P, Kapoor R, Jain RK, Mandal B (2015) Diagnosis of a new variant of soybean yellow mottle mosaic virus with extended host-range in India. Virus Dis 26:304–314

Sangar RBS, Singh RA, Nagaich BB (1984) Potato virus X on wild potato species *Solanum chacoense* from India. Curr Sci 53:549–551

Sarma YR, Kiranmai G, Sreenivasulu P, Anandaraj M, Hema M, Venkatramana M, Murthy AK, Reddy DVR (2001) Partial characterization and identification of a virus associated with stunt disease of black pepper (*Piper nigrum*) in South India. Curr Sci 80:459–462

Sastry KS, Sai-Gopal DVR (2010) Compendium of plant virus, phytoplasma and viroid diseases research in India (1903–2008). Virology Publications, Tirupati. p 782

Satyanarayana T, Sreenivasulu P, Ratna AS, Reddy DVR, Nayudu MV (1994) Identification of a strain of peanut chlorotic streak virus causing chlorotic vein banding disease of groundnut in India. J Phytopathol 140:326–334

Selvarajan R, Balasubramanian V (2013) Natural occurrence of Banana bunchy top virus in Ensete superbum in India. VirusDis 24:97–98

Selvarajan R, Balasubramanian V, Dayakar S, Sathaimoorthy S, Ahlawat YS (2010) Evaluation of immunological and molecular techniques for the detection of different isolates of *Banana bunchy top virus* in India. Indian Phytopathol 63:333–336

Selvarajan R, Balasubramanian R, Sheeba V, Mohan MM, Mustaffa RR (2011) Virus-indexing technology for production of quality banana planting material: a boon to the tissue-culture industry and banana growers in India. Acta Hortic 897:463–470

Sharma A, Mahinghara BK, Singh AK, Kulshrestha S, Raikhy G, Singh L, Verma N, Hallan V, Ram R, Zaidi AA (2005) Identification, detection and frequency of lily viruses in northern India. Sci Hortic 106:213–227

Sharma S, Singh B, Rani G, Zaidi AA, Hallan VK, Nagpal AK, Virk GS (2008) In vitro production of Indian citrus ringspot virus (ICRSV) free Kinnow plants employing thermotherapy coupled with shoot tip grafting. Plant Cell Tiss Org Cul 92:85–92

Sharma A, Kang SS, Kaur SI, Gaikwad A (2012) Prevalence of cucurbit viruses in trans-gangetic plains of India. J Res 49:30–34

Sharma SK, Kumar PV, Baranwal VK (2014) Immunodiagnosis of episomal banana streak MY virus using polyclonal antibodies to an expressed putative coat protein. J Virol Methods 207:86–94

Sherpa AR, Hallan V, Zaidi AA (2004) Cloning and sequencing of coat protein gene of an Indian *Odontoglossum ringspot virus* isolate. Acta Virol. 2004 48:267–269

Sherpa AR, Hallan V, Zaidi AA (2012) In vitro expression and production of antibody against *Cymbidiumm mosaic virus* coat protein. VirusDis 23:46–49

Siljo A, Bhat AI, Biju CN, Venugopal MN (2012) Occurrence of *Banana bract mosaic virus* on cardamom. Phytoparasitica 40:77–85

Singh M, Gaur RK, Rao G (2003) Distribution of bacilliform badnavirus in India. J Mycol Plant Pathol 33:406–410

Singh HP, Hallan V, Raikhy G, Kulshrestha S, Ram R, Garg IG, Zaidi AA (2005) Characterization of an Indian isolate of *Carnation mottle virus* (CarMV) infecting carnation. Curr Sci 88:594–601

Singh B, Sharma S, Rani G, Hallan V, Zaidi A, Virk GS, Nagpal A (2008) In vitro micrografting for production of *Indian citrus ringspot virus*(ICRSV)-free plants of Kinnow Mandarin (*Citrus Nobilis* Lour × *c. Deliciosa* Tenora). Plant Biotech Rep 2:137–143

Singh L, Hallan V, Ram R, Zaidi AA (2011) Expression of recombinant *Chrysanthemum virus B* coat protein for raising polyclonal antisera. J Plant Biochem Biotechnol 20:96–101

Singh P, Prabha K, Jain RK, Baranwal VK (2014) N-terminal in coat protein of garlic virus X is indispensible for its serological detection. Virus Genes 48:128–132

Sinha DP, Saxena S, Kumar S, Singh M (2011) Detection of pepper leaf curl virus through PCR amplification and expression of its coat protein in *Escherichia coli* for antiserum production. Afr J Biotech 10:3290–3295

Sivaprasad S, Sreenivasulu P (1996) Characterization of two strains of cowpea mild mottle virus naturally infecting groundnut (*Arachis hypogaea* L.) in India. J Phytopathol 144:19–23

Sivaprasad Y, Reddy BVB, Kumar CVMN, Reddy KR, Gopal DVRS (2011) First report of *Groundnut bud necrosis virus* infecting taro (*Colocasia esculenta*). Aust Plant Dis Notes 6:30–32

Soumya K, Yogita M, Prasanthi Y, Anitha K, Kishor PBK, Jain RK, Mandal B (2014) Molecular characterization of Indian isolate of *Peanut mottle virus* and immunodiagnosis using bacterial expressed core capsid protein. VirusDis 25:331–337

Srinivas KP, Reddy CVS, Ramesh B, Kumar PL, Sreenivasulu P (2010) Identification of a virus naturally infecting sorghum in India as *Sugarcane streak mosaic virus*. Eur J Plant Pathol 127:13–19

Srivastava KM, Raj SK, Singh BP (1992) Properties of a cucumber mosaic virus strain naturally infecting chrysanthemum in India. Plant Dis 76:474–476

Srivastava D, Panday N, Tiwari AK, Shukla K (2012) Identification of a Potyvirus associated with mosaic disease of *Catharanthus roseus* and its histopathogical effects. Med Plants 4:23–27

Sujitha A, Reddy BVB, Sivaprasad Y, Usha R, Saigopal DVR (2012) First report of groundnut bud necrosis virus infecting onion (*Allium cepa*). Aust Plant Dis Notes 7:183–187

Suresh LM, Malathi VG, Shivanna MB (2013) Serological diagnosis and host range studies of important viral diseases of a few cucurbitaceous crops in Maharashtra. India Arch Phytopathol Plant Protect 46:1919–1930

Swanson MM, Varma A, Muniyappa V, Harrison BD (1992) Comparative epitope profiles of the particle proteins of whitefly-transmitted geminiviruses from nine crop legumes in India. Ann Appl Biol 120:425–433

Swanson MM, Valand GB, Muniyappa V, Harrison BD (1998) Serological detection and antigenic variation of two whitefly-transmitted geminiviruses: tobacco leaf curl and croton yellow vein mosaic viruses. Ann Appl Biol 132:427–435

Talley J, Warren FHJB, Torrance L, Jones RAC (1980) A simple kit for detection of plant viruses by the latex serological test. Plant Pathol 29:77–79

Tarafdar A, Ghosh PD, Biswas KK (2012) *In planta* distribution, accumulation, movement and persistence of *Citrus tristeza virus* in citrus host. Indian Phytopathol 65:184–188

Thien HX, Bhat AI, Jain RK (2002) Serological studies and polyclonal antiserum production against the nucleocapsid protein of mungbean Tospovirus from New Delhi. Omonrice 10:60–66

Thomas JE (1986) Purification and properties of ginger chlorotic fleck virus. Ann Appl Biol 108:43–50

Thomas JE, Geering ADW, Gamblay GF, Kessing AF, White M (1997) Purification, properties and diagnosis of banana bract mosaic potyvirus and its distinction from abaca mosaic potyvirus. Phytopathology 87:698–705

Tsuda S, Kameya-Iwaki M, Hanada K, Kouda Y, Hikata M, Tomaru K (1992) A novel detection and identification technique for plant viruses: rapid immunofilter paper assay (RIPA). Plant Dis 76:466–469

Tsuda S, Kameya-Iwaki M, Hanada K, Fujisawa I, Tomaru K (1993) Simultaneous diagnosis for plants infected with multiple viruses employing rapid immunofilter paper assay (RIPA) with two-step method; multi-RIPA. Ann Phytopathol Soc Jpn 59:200–203

Van Regenmortel MHV (1982) Serology and immunochemistry of plant viruses. Academic, New York

Van Regenmortel MHV, Burchard J (1980) Detection of a wide spectrum of virus strains by indirect enzyme-linked immunosorbent assays (ELISA). Virology 106:327–334

Varma A, Krishnareddy M, Malathi VG (1992) Influence of the amount of blackgram mottle virus in different tissues on transmission through the seeds of *Vigna mungo*. Plant Pathol 41:274–281

Vemana K, Jain RK (2010) New experimental hosts of *Tobacco streak virus* and absence of true seed transmission in leguminous hosts. VirusDis 21:117–127

Venkat H, Gangatirkar P, Karande AA, Krishnareddy M, Savithri HS (2008) Monoclonal antibodies to the recombinant nucleocapsid protein of a groundnut bud necrosis virus infecting tomato in Karnataka and their use in profiling the epitopes of Indian tospovirus isolates. Curr Sci 95:952–957

Verma N, Sharma A, Ram R, Hallan V, Zaidi AA, Garg ID (2003) Detection, identification and incidence of chrysanthemum B carlavirus in chrysanthemum in India. Crop Prot 22:425–429

Verma R, Ahlawat YS, Tomer SPS, Prakash S, Pant RP (2004a) First report of *Zucchini yellow mosaic virus* in bottlegourd (*Lagenaria siceraria*) in India. Plant Dis 88:426

Verma R, Prakash S, Tomer SPS (2004b) First report of *Zucchini yellow mosaic virus* in cucumber (*Cucumis sativus*) in India. Plant Dis 88:906

Verma R, Singh SJ, Singh RK, Prakash S (2004c) Indexing of viruses of polyhouse grown capsicum (*Capsicum annuum* L.) in Pune – causing severe symptoms and epiphytosis. Indian J Agr Res 38:157–163

Verma N, Singh AK, Singh L, Raikhy G, Kulshrestha S, Singh MK, Hallan V, Ram R, Zaidi AA (2005) Cucumber mosaic virus (CMV) infecting Alstroemeria hybrids in India. Australas Plant Pathol 34:119–120

Verma N, Kumar K, Kulshrestha S, Raikhy G, Hallan V, Ram R, Zaidi AA, Garg ID (2006) Detection and molecular characterization of a *Tomato aspermy virus* isolate infecting chrysanthemums in India. Acta Hortic 722:41–53

Vijayanandraj S, Yogita M, Das A, Ghosh A, Mandal B (2013) Highly efficient immunodiagnosis of large cardamom chirke virus using the polyclonal antiserum against *Escherichia coli* expressed recombinant coat protein. VirusDis 24:227–234

Viswanathan R, Balamuralikrishnan M (2004) Detection of sugarcane yellow leaf virus, the causal agent of yellow leaf syndrome in sugarcane by DAS-ELISA. Arch Phytopathol Plant Protect 37:169–176

Viswanathan R, Alexander KC, Garg IG (1996) Occurrence of sugarcane bacilliform virus in sugarcane germplasm. Acta Virol 40:5–8

Viswanathan R, Kumar VG, Karuppaiah R, Scindiya M, Chinnaraja M (2013) Development of duplex-immunocapture (duplex-IC) RT-PCR for the detection of *Sugarcane streak mosaic virus* and *Sugarcane mosaic virus* in sugarcane. Sugar Tech 15:399–405

Voller A, Bartlett A, Bidwell DE, Clark MF, Adams AN (1976) The detection of viruses by enzyme-linked immunosorbent assays (ELISA). J Gen Virol 33:165–167

Zakri AM, Ziegler A, Torrance L, Fischer R, Commandeur U (2010) Generation and characterization of a scFv against recombinant coat protein of the geminivirus tomato leaf curl New Delhi virus. Arch Virol 155:335–342

Nucleic-Acid Based Techniques for the Fine Diagnosis of Plant Viruses in India

27

Susheel Kumar Sharma, Ram Prasnna Meena, D. Pramesh, Sandeep Kumar, Th. Surjit Singh, and Virendra Kumar Baranwal

Abstract

Field and horticultural crops are infected by large number of viruses and virus-like pathogens in India, which can be managed only by preventive measures and for that sensitive diagnosis is a prerequisite. Bioassay was initially utilized in detection of plant viruses. Following this, serological assay enzyme-linked immunosorbent assay (ELISA) was widely employed. In the last two decades, nucleic acid based detection assays were developed for the correct and fine diagnosis of plant viruses in India. Nucleic acid based detection assays in addition to providing sufficient sensitivity, specificity and accuracy are also robust and easily applicable. Correct identity of a large number of viruses which were misdiagnosed earlier was now diagnosed to their correct taxonomic status using nucleic

S.K. Sharma (✉) • T.S. Singh
ICAR Research Complex for NEH Region, Manipur Centre, Lamphelpat, Imphal 795004, India
e-mail: susheelsharma19@gmail.com

R.P. Meena
ICAR-Directorate of Medicinal and Aromatic Plants Research, Boriavi, Anand, Gujarat 387310, India

D. Pramesh
Department of Plant Pathology, University of Agricultural Sciences, Raichur, Karnataka 584104, India

S. Kumar
AICRP on Medicinal & Aromatic Plants and Betelvine, Orissa University of Agriculture & Technology, Bhubaneswar 751003, India

Advanced Centre for Plant Virology, Division of Plant Pathology, ICAR-Indian Agricultural Research Institute, New Delhi 110012, India

V.K. Baranwal
Advanced Centre for Plant Virology, Division of Plant Pathology, ICAR-Indian Agricultural Research Institute, New Delhi 110012, India

B. Mandal et al. (eds.), *A Century of Plant Virology in India*,
https://doi.org/10.1007/978-981-10-5672-7_27

acid based assays. Most commonly used nucleic acid based detection assays employed in India are nucleic acid spot hybridization (NASH), polymerase chain reaction (PCR) and reverse transcription-polymerase chain reaction (RT-PCR). These methods have been utilized for detection of large number of viruses and virus-like pathogens. Multiplex detection assays (multiplex PCR and multiplex–RT-PCR) were also developed for simultaneous detection of more than one virus in same reaction tube. Isothermal detection assay where amplification reaction can be carried out at constant temperature is developed and employed for simple and robust detection as well as characterization of plant viruses in India. The most commonly employed isothermal detection assays includes the rolling circle amplification (RCA) which has been employed to detect DNA viruses of the genera *Begomovirus, Babuvirus* and *Badnavirus*. Loop-mediated isothermal amplification (LAMP) are latest in the application of isothermal detection methods, which provides simple and easy detection of plant viruses and such detection assays have been developed for some viruses in India. The next generation sequencing (NGS) or deep sequencing, which involves massive parallel sequencing approach followed by bioinformatics analysis, has revolutionized the field of novel virus discovery in India. This chapter aims to highlight the application of various nucleic acid based detection assays in diagnosis and characterization of viruses infecting different crops in India. In this chapter, a crop-wise review of the application of different nucleic acid based detection methods in India has been presented.

Keywords
Polymerase chain reaction • Reverse transcription-PCR • Nucleic acid spot hybridization • Rolling circle amplification • Loop-mediated isothermal amplification • Next generation sequencing

27.1 Introduction

Plant viruses are associated with most devastating and economically important diseases of field and horticultural crops. Plant viruses contribute to around 47% of the infectious plant diseases reported worldwide (Anderson et al. 2004). Plant viruses are spread both by infected propagating materials (seeds, cuttings, bulbs, bulblets, rhizomes, suckers, bud woods, tissue culture derived plantlets etc.) as well as through vectors (insects, mites, fungi and nematodes). Management of viral diseases largely depend on timely accurate detection and preventing their introduction and spread to new areas or locations. Definite and fine diagnosis of associated virus species or strains and taking preventive measures to check its spread helps in management of viral diseases under field conditions.

In case of those plant species, which are perennial in nature, once infected the viral load accumulate over the life period of plants and cause serious losses to production and productivity. In case of annuals (vegetable, ornamentals and cereals) viruses can even enter through infected seeds and these initially infected foci can act as source for further spread through vectors. Infected planting materials also act as source for introducing viral diseases to new agro-ecological areas and can have serious epidemiological concerns.

It was in 1890s when plant viruses were first discovered to be associated with plant diseases. Bioassays were first used for the detection of plant viruses. Symptom based detection, and transmission studies were most commonly used assays initially for detection of associated viruses. However, symptom based detection is very crude and exact identity of virus cannot be established. Symptom of viral diseases is often confused with nutrient deficiency. Assay hosts which are susceptible to viruses upon artificial inoculation (mechanical or vector transmission) are used for bioassay based detection of plant viruses. Most commonly used laboratory assay hosts for plant viruses are *Nicotiana benthamiana*, *N. tabacum*, *N. glutinosa*, *N. sylvestris* and *Chenopodium amaranticolor* which are used for sap inoculation of plant viruses. Upon mechanical inoculation either local lesions or systemic symptoms produced on inoculated plants are used to identify the associated virus. However, all viruses are not mechanically transmitted thus limiting the use of these bioassays for transmission of plant viruses. Graft transmission on indexing host (a plant species or a particular variety) is also used as common bioassay in case of woody plants and horticultural crops. Biological assays are however subjective methods of virus detection, are tedious, require large space (under controlled environmental conditions) and long time for typical symptoms to express (Prabha et al. 2013).

After bioassay based detection of plant viruses, assays based on serology and nucleic acid or combination of different methods have been developed and used for more efficient detection of plant viruses. Number of criteria are considered while choosing a detection method: sensitivity (ability to detect minimum quantity of virus), accuracy and reproducibility, number of samples that can be tested at a time, initial investment for apparatus/instruments and running cost per sample, experience, training and technical knowhow required (Boonham et al. 2014). Since the first description of viruses with plant diseases, the detection methods were largely based on the description of symptoms on the hosts, bio-assays on the diagnostic hosts and transmission electron microscopy in crude leaf sap preparations. These two methods were initially used for decades for identification of associated etiological viral agents with new uncharacterized diseases; however, these methods were more of subjective nature and did not characterize the associated virus at species or strain level. Virus diagnostics were revolutionized by the discovery of enzyme-linked immunosorbent assay (ELISA) (Clark and Adams 1977) which in addition to simplify the detection procedures has the ability to screen large number of samples at a time. The results obtained are of semi-quantitative formats, thus giving an indication of virus titre in the infected plant samples.

Recent era of plant virus diagnosis is on the nucleic acid based techniques. In these methods, viral nucleic acid (either DNA or RNA) is detected. Nucleic acid based detection methods were comparatively more sensitive as compared to serological based detection procedures. The methods like nucleic acid spot hybridization (NASH), polymerase chain reaction (PCR) and reverse transcription-PCR (RT-PCR) were developed for more robust detection of plant viruses which detect viral nucleic acid (DNA or RNA whatever is the case). Latest closed tube based diagnostic assays such as quantitative-PCR (qPCR) and real time-PCR (rt-PCR) are the most sensitive assays which reduce the risk of post-PCR contamination allowing the definite and sensitive detection of plant viruses. In the last decade, various methods like macroarray, microarray, lateral flow assays, loop-mediated isothermal amplification (LAMP) and next generation sequencing (NGS) were developed which have specific advantages of either sensitivity, applicability for on-site detection or discovery of novel viruses. In last two decades a large number of next generation technologies have come up which has revolutionized the diagnostics of plant viruses thus leading to efficient detection based management of plant viruses. Classic examples of development and application of nucleic acid based detection methods for viruses infecting different crop plants in Indian context have been discussed in this chapter.

27.2 Significance

Management strategies for plant viral diseases and prevention of their outbreaks depend on the strategies to prevent their spread and introduction to new areas or new susceptible varieties through timely and accurate detection of the etiological viral agents (Kreuze et al. 2009; Sharma et al. 2014b). The management options and their efficient application for the viral diseases in plants largely rely on the definite diagnosis of incitant virus agents followed by taking preventive measures to prevent their spread in field conditions. The accurate and reliable detection of the associated virus pathogens therefore forms the first line of defense in management of these diseases. The biggest challenge faced by the modern virologists is globalization and international trade of agri-horticultural crops, which helps the viral pathogens to spread across geographical boundaries.

In many of the crop plants, particularly in the ornamentals where the latent infection of the viruses exist, the visual inspection for the symptoms in post-entry quarantine is unreliable. In quarantine stations only a limited number of molecular tests can be carried out to test the most probable or potential viral pathogens. Thus the chance of introduction of non-tested viral pathogens or new uncharacterized viruses through infected propagating materials cannot be overruled. After introduction to new places there are chances of spreading the viral pathogens to the new hosts. These threat exists under Indian context also, where probably different banana streak virus (BSV) species and banana bunchy top virus (BBTV) strains occurring in the Southern and North Eastern India if spread to other areas may cause serious losses in banana production.

The availability of sensitive, reliable and robust detection at an effective cost is a prerequisite for devising an effective management strategy for plant viruses. Since the discovery of nucleic acid based detection procedures, virus diagnostics have seen vast improvements in terms of sensitivity, specificity handling cost, ability to handle large number of samples and ease in application (Baranwal and Sharma 2016). Although many nucleic acid based detection systems were developed during last 20 years, yet these approaches are available only for those viruses for which the sequence information is already available. For the uncharacterized viruses, the diagnosis still remains a major challenge. Use of degenerate generic primers is helpful, but not confirmative as they lead to nonspecific amplification and spurious results many times. With the latest developments in molecular virology the new strains, species as well as recombinants viruses are being characterized from many symptomatic as well asymptomatic plant samples. Therefore, indexing of planting material, bud wood sources as well seeds of vegetable crops must include next generation techniques of virus detection.

In India PCR and RT-PCR based detection has been widely developed and employed for the sensitive detection of DNA and RNA viruses, respectively. For some viruses nucleic acid spot hybridization (NASH), Southern blotting and Northern blotting has also been employed. PCR and RT-PCR for detection of DNA and RNA viruses respectively has been widely employed in detection and characterization of plant viruses in India. Multiplex nucleic acid based detection system has also been developed for DNA and RNA plant viruses in the last decade.

Latest development in nucleic acid based detection is closed tube based detection system such as quantitative-PCR (qPCR) and real time-PCR (rt-PCR), which are quite sensitive and also reduces the chances of post-PCR contamination. Isothermal detection assays particularly loop-mediated isothermal amplification (LAMP) where nucleic acid amplification can be carried out at constant temperature has also been developed for some plant viruses in India. Next generation sequencing (NGS) or deep sequencing based methods which are able to detect or characterize novel uncharacterized viruses has also been recently employed in India. The important nucleic acid based diagnostic tools developed and employed for the detection, characterization and discovery of plant viruses in India are discussed here.

The development of nucleic acid based detection methods in India can be divided in to five major catagories: (i) Nucleic acid hybridization and PCR based detection methods, (ii) Multiplex detection methods, (iii) Real time detection methods, (iv) Isothermal amplification based detection methods and (v) Next generation sequencing methods.

27.3 Nucleic Acid Based Methods

Nucleic acid hybridization assay includes nucleic acid spot hybridization (NASH), Southern hybridization and Northern hybridization. These assay involved hybridization of viral nucleic acids with labeled probes and detection of hybridization signals. PCR based diagnostics can be used for the detection of both DNA and RNA genomes of plant viruses utilizing the PCR and reverse transcription-PCR (RT-PCR) respectively. For PCR reaction primers specific to viral genomic fragments are required, which have a prerequisite of prior sequence information of the viral genomes. The increasing number of partial and complete genome sequences of plant viruses are now available in the databases, and it is possible to design either specific or degenerate primers for amplifying the targeted genomic regions of viruses for detection of the specific virus species or strain. In addition to specific detection of the virus, the amplified products can be sequenced either directly or after cloning into a suitable vector. Sequences so obtained can be used for phylogenetic analysis.

PCR based detection methods are very sensitive, and there could be chances of contamination either or both before PCR and post-PCR leading to false positives. Thus it is always recommended to have positive and negative controls to avoid false negatives or positives. In comparison to ELISA, where production of quality antibodies requires a lot of efforts, it is much easier to design and synthesize new primers. At present PCR based diagnostics are the standard diagnostics for detection of many viruses and virus like pathogens in different field and horticultural crops.

Many of the viruses infecting fruit trees, occur at very low concentration in the hosts, such viruses are difficult to detect through normal PCR. In addition, the plant pararetroviruses (genus *Badnavirus*) have their integrated counterparts in the host genome known as endogenous pararetroviruses (EPRV) which have significant sequence homology to cognate episomal viruses. The presence of integrated BSV like sequences (endogenous BSV: eBSV) in the *Musa* genome makes the PCR based diagnosis unreliable. To resolve both the issues mentioned above, nucleo- and immuno- based detection tools are combined in immunocapture-PCR (IC-PCR) where the PCR amplification is carried out on the immuno-trapped virions. In addition to the ability to amplify the episomal viral sequences, IC-PCR also has the ability to concentrate (aggregate) the virus particles from the crude sap thus making it more sensitive. IC-PCR is currently considered as the gold standard for detection of pararetroviruses having integrated counterparts.

27.4 Multiplex Detection Methods

Multiplexing refers to the assay wherein more than one virus is detected in the same reaction. Multiplex assays provide advantage with regard to cost involved, simplicity and ease in handling. Most commonly used multiplex detection methods are:

27.4.1 Multiplex PCR

This technique is able to detect different DNA or RNA targets simultaneously in a single tube. In those cases where plant viruses exist as mixed infection, the multiplex-PCR techniques are very useful. The technique involves the use of many sets of oligonucleotides targeting the genomic regions of multiple viruses (within same genus, strains of same species or viruses from taxonomically different groups) in single PCR amplification. The prerequisite of multiplexing PCR techniques is to design the oligonucleotide primers in such a way that amplification can be carried out at one annealing temperature and different primers do not show any self-complementarity. The amplicon size from different viral genomes should be different so as to differentiate them through agarose gel electrophoresis. The multiplex- PCR has the advantages that it saves the reagents, time, is convenient and cost effective when more than one viruses are to be detected from the same sample. Multiplex-PCR has been standardized for a number of viruses in different crops.

27.4.2 Real Time PCR

Real time-PCR (rt-PCR) or quantitative-PCR (qPCR) are closed-tubes based PCR assays which are quickly adopted in the resource sufficient laboratories. Real time PCR is an improvement over the standard PCR in which the amplification is detected during the reaction process itself through the generated fluorescence. It involves the use of non-specific fluorescent dyes (e.g. SYBR green, LUX etc.) or specific probes labelled with fluorescent dyes (TaqMan, molecular beacon etc.). The SYBR green binds to double stranded DNA molecules and fluoresce. In the TaqMan based rt-PCR, sequence specific internal DNA probes labelled with dye (reporter and quencher) are used and the fluorescent signals are emitted during the amplification reaction. This in addition to providing good specificity and sensitivity, avoids any chances of post-PCR contamination. The fluorescent signals are detected during the amplification process itself.

Real time PCR has more generic applications, and can be quickly standardized compared to the development of antibodies for ELISA. But the cost involved in testing per sample is much higher than ELISA. High-throughput testing of samples, ability to test the viral load and less time involved are the specific advantages of rt-PCR based detection assays. The ELISA and rt-PCR shares many benefits as the detection tools, viz. industry standard format, robustness, high reproducibility and easy to test even by inexperienced users.

The real-time based multiplex techniques can also be used; however as the multiple reporter and quencher dyes need to be used in the rt-PCR, due to the cross-talk between the fluorescent signals being emitted from the reporters, the number of targets is theoretically limited to four only.

27.5 Isothermal Amplification Based Detection Methods

Both PCR and RT-PCR based detection method requires thermal cycler to provide a gradient of temperature for different amplification steps. In many of the resource poor laboratories it is not possible to carry out the PCR based detection, thus the isothermal amplification based detection methods are developed, which are able to carry out the amplification reaction at constant temperature. The most commonly used isothermal methods are:

27.5.1 Rolling Circle Amplification

Rolling circle amplification (RCA) method is used to amplify the circular genomes using exo-resistant random hexamer primers by employing the Phi 29 DNA polymerase which has the strand displacement activity. RCA amplification involves the sequence independent amplification at isothermal temperature (30 °C). Since random hexamer primers are used, the prior sequence information of the targeted viral genomes is not required in RCA. Thus it has the potential to amplify novel circular viral genomes. Application of RCA in the detection and characterization has revolutionized the genomics of begomoviruses and nanoviruses. RCA has been used even to amplify the bigger viral genomes of badnaviruses infecting banana, which typically amplify the episomal viral genomes. By employing this strategy novel badnaviruses associated with the leaf streak disease of banana have been identified. The RCA product after the restriction digestion needs to be sequenced for confirmation of viral origin of the amplified products. However, the restriction analysis of the RCA product, known as RCA-restriction fragment length polymorphism (RCA-RFLP) where specific restriction profile is obtained, can also be used for screening of a large number of samples for the infection of circular viruses particularly begomoviruses and the badnaviruses.

27.5.2 Loop-Mediated Isothermal Amplification (LAMP)

Loop-mediated isothermal amplification (LAMP) is able to amplify from a few copies of DNA to a magnitude of 10^9 in less than an hour. LAMP uses three pairs of primers (internal, external and loop primers) for amplification, which binds at least at six binding sites. The amplification generates a product which is having single stranded loop regions, where the primers bind without the need for denaturation. The amplification reaction is mediated by the DNA polymerase with high strand-displacement activity and the whole process is completed in approximately 1 h at around 60–65 °C. LAMP based detection systems are easy to perform and the final results can easily be visualized, thus easy to interpret. Due to the LAMP reaction, magnesium pyrophosphate (a by-product) is produced, which leads to increase the turbidity that can be observed by the naked eyes. For detection of the amplification product, the intercalating dyes such as SYBR Green or PicoGreen can be added after amplification, and the fluorescence can be detected. As these dyes are added post-amplification, there are chances of contamination. But there are reagents like calcein and $MnCl_2$ or Hydroxyl Napthol Blue (HNB) which can be added in the reaction mixture itself without any inhibition of the amplification. The colour changes due to calcein and $MnCl_2$ can be visualized under ultraviolet light whereas due to HNB can be visualized directly.

The LAMP assays are almost as sensitive as the real-time PCR. With the recent improvements and by inclusion of a reverse-transcription step LAMP assay can be completed in a short period (<30 min) at a constant temperature of 65 °C. The LAMP assay can be carried out on the DNA isolated through the normal procedures and the DNA polymerase used in the LAMP is tolerant to low level of PCR inhibitors which are unavoidable during DNA extraction. The biggest advantage of isothermal amplification based diagnostic methods is that these methods can easily be adopted by resource poor laboratories for the routine indexing and even at the point of decision making that is in the orchards/fields. These methods provide high sensitivity, can be used by laboratories with limited facilities and do not require trained specialists.

27.6 Next Generation Sequencing (NGS) Methods

Next generation sequencing (NGS) or deep sequencing based methods have been developed for the discovery of novel viruses. In response to the virus infection, plant system produces small RNAs (siRNAs) complementary to the viral genomic sequences, which degrades the viral RNAs, known as silencing. Deep sequencing on the siRNAs isolated from the infected samples allows the recovery of the either full or partial genomic sequences of viruses. These siRNA sequencing based methods have been effectively used to discover/detect virus infections in many crops.

The high throughput sequencing technologies involve preparation of the nucleic acid libraries, amplification of these libraries to get sufficient quantity of DNA followed by parallel sequencing of the billions of DNA fragments. These approaches or their steps are now much advanced in terms of simplicity, cost and running time. Increasing number of crop genomes being sequenced, provides good opportunity for deep sequencing based discovery of novel plant viruses, as it is possible in silico to assign the short sequences to be of host or non-host origin thus reducing the volume of data to be analyzed for virome studies. Further it is also important to mention that availability of the genomic data for many viruses is one of the pre-requisite for the sensitive and reliable diagnosis. NGS, in addition to the application in resolving the etiology of viral diseases, characterization, population genetics, has the potential in high throughput diagnosis of plant viruses in both agricultural and horticultural crops. Of the voluminous data obtained from the deep sequencing project, the most important is to analyze in silico and discard those derived from host or other microflora and analyze only the viral derived sequences. Either the enriched viral nucleic acid from infected samples or the nucleic acid obtained from virus like particles (VLP) or partially and completely purified virus particles or rolling circle amplified (RCA) enriched nucleic acid of the circular viruses can be used for NGS based deep sequencing.

This siRNA (generally 21–24 nt) based NGS parallel sequencing of symptomatic and asymptomatic samples followed by *de novo* assembly to long reads has the potential to identify the novel uncharacterized RNA, ssDNA, reverse transcribing dsDNA viruses and viroids without the prerequisite of prior sequence information. In case of some virus groups, the conserved domains have been identified either across the different genera or family, which enables us to design the generic primers targeting the regions which have the potential to identify different variants as well as new or uncharacterized viruses from that family. Small RNA (siRNA) deep sequencing technology has proved to be an efficient generic tool for virus identification in plants. Recent developments in the sequencing technology, high throughput sample processing and automation has drastically reduced the processing cost per sample and these technologies have started finding place in many of the testing/

indexing laboratories. With these ongoing developments, it is further expected that the NGS reaction cost per sample will reduce further to a greater extent possibly making these procedures available for routine testing of the samples in many laboratories. A very big advantage of these methods is the broad spectrum ability to detect number of viruses or viroids from a single sample as well as the uncharacterized virus or virus like agents, which makes them much more robust, and also reliable for strict quarantine testing, indexing and certification programmes.

27.7 Application of Diagnostics in Various Crops

27.7.1 Temperate Fruits and Sub-tropical Fruit Crops

27.7.1.1 Pome and Stone Fruits

Nucleic Acid Hybridization, PCR and RT-PCR Based Detection Methods
Pome and stone fruits are grown in temperate region of Northern Himalaya and North Eastern Himalayan region. Infection of different viruses poses a serious threat to the production and productivity of these fruit crops. Reliable and robust molecular diagnostics are prerequisite for production and supply of disease free planting materials in these fruit crops.

Apple Stem Pitting Virus: ASPV (Genus: *Foveavirus*, Family: *Betaflexiviridae*) In the surveys on different pome and stone fruits (cherry, apple etc.) in Himachal Pradesh and Jammu and Kashmir during 2007, cherry plants with symptoms of vein yellowing and apple plants with symptoms of fruit deformation, leaf fall during early stages and tree decline during later stages were suspected to be infected with viruses. RT-PCR using primers specific to partial coat protein (CP) sequences amplifying ~370 bp fragment specific to apple stem pitting virus (ASPV) genome was used for its molecular detection (Dhir et al. 2009). Sequences were identical to ASPV isolate from China (GenBank Accession no. FM863704). This was the first report on molecular detection of ASPV from India. Later on Dhir et al. (2011) used primers specific to triple gene block (TGB) and the CP genes of ASPV genome to detect the infection in 90 samples collected from pome fruit trees and apple rootstocks, originating from different parts of North Western Himalayan region. In RT-PCR, 36.7% of apple samples and one each apple rootstock and pear were tested positive for ASPV.

Apple Top Working Disease By employing the primers specific to coat protein (CP) gene of viral genomes in PCR assays, association of apple stem grooving virus (ASGV), apple chlorotic leaf spot virus (ACLSV), and apple stem pitting virus (ASPV) with apple top working disease from Himachal Pradesh has been reported (Brakta et al. 2013).

Apple Stem Grooving Virus: ASGV (Genus: *Capillovirus*, Family: *Betaflexiviridae*) Based on nucleic acid spot hybridization (NASH) and RT-PCR using primers specific to CP gene of ASGV, its infection was detected with severe interveinal mottling, severe leaf distortion, ring spots and chlorosis symptoms of kiwi plantations (*Actinidia deliciosa*) in Himachal Pradesh (Bhardwaj et al. 2014). Six out of 15 samples were tested positive in NASH. RT-PCR using CP specific primers gave amplification of around 750 bp; sequencing and sequence analysis of which confirmed the association of ASGV with kiwi plant in India. This was the first documented report on ASGV infection in Kiwi from India (Bhardwaj et al. 2014).

Cherry Virus A (CVA) (Genus: *Capillovirus*, Family: *Betaflexiviridae*) Among the viruses infecting stone fruits, RT-PCR was successfully employed for the detection of CVA infection in sweet cherry (*Prunus avium*) grown in temperate fruit growing belt of India (Himachal Pradesh and Jammu & Kashmir). Based on RT-PCR indexing, CVA incidence of 28% and 13% in J&K and HP, respectively, was reported (Noorani et al. 2010). RT-PCR was performed using primers targeting a part of polyprotein of ORF1 which gave an amplification of 1089 bp in CVA positive samples. Using single RT-PCR amplification, employing primers CVAU and CVAL full length viral genomic first strand cDNA was synthesized and full genome was amplified and sequenced. Full length genome of Indian CVA isolate was 7379 nucleotides (nt) long excluding the poly (A) tail (Noorani et al. 2010).

Apple Chlorotic Leaf Spot Virus: ACLSV (Genus: *Trichovirus*, Family: *Betaflexiviridae*) Quince (*Cydonia oblonga*) plants grown in the apple orchards which were showing symptoms of leaf distortion and yellow spots were tested positive for ACLSV in RT-PCR using degenerate primers targeting full CP and partial 3′-UTR region of viral genome (Rana et al. 2008). Using the designed degenerate primers an amplicon of approximately 800 bp was amplified and sequenced which shared an identity of 84% at CP nucleotides with quince ACLSV isolate from Greece (Rana et al. 2008). Based on the available full genome sequences of ACLSV, degenerate primer pair for CP based RT-PCR detection was developed and standardized (Rana 2010). A protocol for tissue blot hybridization based detection of ACLSV was also standardized and employed for large scale indexing of pome and stone fruit nurseries (Rana 2010). Degenerate primers designed to amplify full CP and 3′-UTR region were used for RT-PCR based detection of ACLSV (Rana et al. 2010). Using the developed diagnostics, complete CP sequences of 26 ACLSV isolates sampled from different pome fruits (apple, pear and quince) and stone fruits (plum, peach, apricot, almond and wild Himalayan cherry) were characterized.

Infection of ACLSV many times remains latent. Five pair of primers was designed for amplification of CP of ACLSV, two sets of primers 1F1R and 1F2R successfully amplified 432 bp CP fragment of viral genome (Watpade et al. 2012). RT-PCR conditions were standardized for detection of ACLSV using designed primers and primers to NAD5 gene (mitochondrial mRNA of apple genome) was used as internal control. Developed RT-PCR based detection was used for indexing of 42 elite mother plants, out of which 40% were positive for ACLSV (Watpade et al. 2012).

Apple Mosaic Virus: ApMV (Genus: *Ilarvirus*, Family: *Bromoviridae*) Bright chlorotic mosaic symptoms on leave of different apple cultivars in Himachal Pradesh was detected to be caused by ApMV based on the amplification of 700 and 800 bp fragments specific to coat protein (CP) and movement protein (MP) genes respectively of viral genome in RT-PCR (Lakshmi et al. 2011). In addition to molecular detection, sequences of CP and MP were also used for genetic diversity and phylogenetic analysis (Lakshmi et al. 2011).

Prunus Necrotic Ringspot Virus: PNRSV (Genus: *Ilarvirus*, Family: *Bromoviridae*) PNRSV is an important virus infecting rose. Based on the multiple sequence comparison of available gene sequences, primers targeting the CP gene and RNA3 were designed and used for detection of PNRSV in RT-PCR amplifying ~680 and ~1700 bp amplicons respectively (Kulshrestha et al. 2013).

Multiple Viruses and Viroids About 250 samples collected from different apple groves of Himachal Pradesh during year 2010 were subjected to molecular detection tools like NASH and RT-PCR using primers specific to fragments of viral genomes. 46.8% of the samples were tested positive for ACLSV, apple mosaic virus (ApMV), apple stem grooving virus (ASGV), apple stem pitting virus (ASPV) and apple scar skin viroid (ASSVd) (Kumar et al. 2012b). RT-PCR for specific detection of ASSVd (based on partial genome); ApMV, ACLSV, ASPV (based on full CP) and ASGV (replicase gene) was standardized. Predominant infection of ASSVd followed by ASPV, ACLSV, ApMV and ASGV as well as mixed infection of different viruses and viroids was detected using molecular techniques (NASH and RT-PCR) (Kumar et al. 2012b). RT-PCR based detection procedures were successfully employed for molecular detection of ApMV, ASGV, ASPV, ACLSV, PNRSV and ASSVd in apple samples collected from different areas of H.P. and J&K during the year 2006–2008 (Rana et al. 2011).

Multiplex Detection Methods

Recently a multiplex RT-PCR (mRT-PCR) based detection system was developed for detection of four cherry viruses: cherry virus A (CVA), cherry necrotic rusty mottle virus (CNRMV), little cherry virus 1 (LChV-1), and prunus necrotic ringspot virus (PNRSV) with nad5 as host internal control (Noorania Md et al. 2013). Earlier reported primers for PNRSV, LChV-1 and nad5 gene of host plant (internal control) (Menzel et al. 2002; Kulshrestha 2005; Bajet et al. 2008) were used, whereas CVA and CNRMV specific primers were designed by Noorania Md et al. (2013). Multiplex RT-PCR conditions for simultaneous detection of all four viruses were standardized. In developed mRT-PCR, amplicon size for internal control was 181 bp whereas for viruses it varied as LChV-1 (300 bp), CNRMV (553 bp), PNRSV

(675 bp) and CVA (1051 bp). Sensitivity of detection in mRT-PCR using ten-fold dilution series of cDNA synthesized from artificially mixed total RNA (1 μg) was found to be 10^{-4} for CVA, whereas for other viruses (LChV-1, PNRSV and CNRMV) it was 10^{-2} dilution (Noorania Md et al. 2013). Out of 20 sweet cherry plants tested using developed mRT-PCR, 13 were having mixed infection of different viruses.

A two-step RT-PCR detection system was developed for simultaneous detection of four apple viruses: apple mosaic virus (ApMV), apple stem pitting virus (ASPV), apple stem grooving virus (ASGV), apple chlorotic leaf spot virus (ACLSV) and a viroid, apple scar skin viroid (ASSVd) with *nad5* gene as internal control (Kumar et al. 2012b). A simplified RNA extraction procedure from apple tissue by employing guanidinium-based extraction buffer and a modified CTAB based procedure was reported, which can be effectively used for multiplex RT-PCR based detection. Five primer pairs specific to ASGV, ASSVd, ASPV, ApMV and ACLSV were successfully used in multiplex RT-PCR amplifying virus specific amplicons of size 198, 330, 370, 547 and 645 bp, respectively. Developed multiplex RT-PCR detection protocol was validated for indexing of 87 samples collected from different apple orchards and found to be more robust compared to simplex PCR and ELISA.

27.7.1.2 Grapevine

Grapevine is an important horticultural crop of India contributing the highest foreign exchange among the fruit crops (Adsule et al. 2011). Production and productivity of grapevine is reduced by many diseases including diseases caused by viruses. Globally, grapevine is attacked by more than 65 viruses of 25 genera and 15 families (Martelli 2012). In India study of grapevine viruses is in its infancy. Works on a few viruses have been published and works on some other viruses are in progress. Grapevine leafroll disease (GLD), a complex disease caused by eleven viruses, is the viral disease of grapevine which has been studied for the first time in India. The virus and virus like pathogens of grapevine in India on which work has been done or is under progress are presented in Table 27.1.

Nucleic Acid Hybridization, PCR and RT-PCR Based Detection Methods

Ampeloviruses and Maculavirus Preparation of good quality nucleic acid, free of PCR inhibitors is a major challenge in diagnosis of viruses associated with grapevine. Most of the standard RNA isolation procedures do not eliminate the chances of contamination with polysachharides and polyphenols present in the tissues and having inhibitory effects on PCR amplification (Minafra et al. 1992; Mackenzie et al. 1997). Globally, several attempts have been made to refine the RNA isolation from woody tissues in general and grapevine in particular. The first studied viruses of grapevine from India i.e. grapevine leafrollassociated virus 3 (GLRaV-3) and grapevine leafroll-associated virus 1 (GLRaV-1) were detected by a simplified sample preparation using crude extracts (instead of RNA isolation) followed by one-step RT-PCR for amplification of the respective genomic regions (Kumar 2013; Rowhani et al. 2000). One step RT-PCR using primers from CP, HSP70h and p23

Table 27.1 List of virus and virus like pathogens infecting different crops on which Indian work has been done

Virus and viroid	Taxonomy	References
Grapevine		
Grapevine leafroll-associated virus 3 (GLRaV-3)	Genus – *Ampelovirus*	Kumar et al. (2012c)
	Family – *Closteroviridae*	Kumar 2013; Kumar et al. (2012c, 2013c)
Grapevine leafroll-associated virus 1 (GLRaV-1)	Genus – *Ampelovirus*	Kumar et al. (2012d)
	Family – *Closteroviridae*	Kumar et al. (2013c)
Grapevine virus B (GVB)	Genus – *Vitivirus*	Kumar et al. (2013c)
	Family – *Betaflexiviridae*	
Grapevine fleck virus (GFkV)	Genus – *Maculavirus*	Kumar et al. (2013c)
	Family – *Tymoviridae*	
Grapevine yellow speckle viroid 1 (GYSVd-1)	Genus – *Apscaviroid*	Sahana et al. (2013)
	Family – *Pospiviroidae*	
Hop stunt viroid (HpSVd)	Genus – *Hostuviroid*	Sahana et al. (2013)
	Family – *Pospiviroidae*	
Australian grapevine viroid (AGVd)	Genus – *Apscaviroid*	Adkar-Purushothama et al. (2014)
	Family – *Pospiviroidae*	
Allium crops		
Onion yellow dwarf virus (OYDV)	Genus – *Potyvirus*	Arya et al. (2006)
	Family – *Potyviridae*	
Leek yellow Stripe virus (LYSV)	Genus – *Potyvirus*	Gupta et al. (2013)
	Family – *Potyviridae*	
Garlic common latent virus (GarCLV)	Genus – *Carlavirus*	Majumder and Baranwal (2009)
	Family – *Betaflexiviridae*	
Shallot latent virus (SLV)	Genus – *Carlavirus*	Majumder et al. (2008)
	Family – *Betaflexiviridae*	
Iris yellow spot virus (IYSV)	Genus – *Tospovirus*	Ravi et al. (2006)
	Family – *Bunyaviridae*	Gawande et al. (2010)
Shallot virus X (ShV-X)	Genus – *Allexivirus*	Majumder et al. (2007)
	Family – *Alphaflexiviridae*	
Gladiolus		
Bean yellow mosaic virus (BYMV)	Genus – *Potyvirus*	Srivastava et al. (1983)
	Family – *Potyviridae*	
Cucumber mosaic virus (CMV)	Genus – *Cucumovirus* Family – *Bromoviridae*	Raj et al. (1998)
Tobacco ringspot virus (TRSV)	Genus – *Nepovirus*	Katoch et al. (2003)
	Family – *Comoviridae*	
Tobacco rattle virus (TRV)	Genus – *Tobravirus*	Katoch et al. (2004)
	Family – Unassigned	
Tomato aspermy virus (TAV)	Genus – *Cucumovirus* Family – *Bromoviridae*	Raj et al. (2011)
Ornithogalum mosaic virus (OrMV)	Genus – *Potyvirus*	Kaur et al. (2011)
	Family – *Potyviridae*	

(continued)

Table 27.1 (continued)

Virus and viroid	Taxonomy	References
Chrysanthemum		
Chrysanthemum virus B (CVB)	Genus – *Carlavirus* Family – *Potyviridae*	Verma et al. (2003)
Cucumber mosaic virus (CMV)	Genus – *Cucumovirus* Family – *Bromoviridae*	Srivastava et al. (1992)
Tomato aspermy virus (TAV)	Genus – *Cucumovirus* Family – *Bromoviridae*	Sastry (1964)
Tomato spotted wilt virus (TSWV)	Genus – *Tospovirus* Family – *Bunyaviridae*	Renukadevi et al. (2015)
Chrysanthemum stunt viroid (CSVd)	Genus – *Pospiviroid* Family – *Pospiviroidae*	Mathur et al. (2002)
Carnation		
Carnation necrotic fleck virus (CNFV)	Genus – *Closterovirus* Family – *Closteroviridae*	Raikhy et al. (2003a)
Carnation latent virus (CLV)	Genus – *Carlavirus* Family – *Betaflexivirdae*	Mangal et al. (2002)
Carnation mottle virus (CarMV)	Genus – *Carmovirus* Family – *Tombusviridae*	Singh et al. (2005)
Carnation etched ring virus (CERV)	Genus – *Caulimovirus* Family – *Caulimoviridae*	Raikhy et al. (2003b)
Carnation vein mottle virus (CVMoV)	Genus – *Potyvirus* Family – *Potyviridae*	Raikhy et al. (2006a)
Carnation ring spot virus (CRSV)	Genus – *Dianthovirus* Family – *Tombusviridae*	Raikhy et al. (2006a)
Orchids		
Odontoglossum ringspot virus (ORSV)	Genus – *Tobamovirus* Family – *Virgaviridae*	Sharma et al. (2005b)
Cymbidium mosaic virus (CymMV)	Genus – *Potexvirus* Family – *Alphaflexiviridae*	Sherpa et al. (2003)
Cymbidium ringspot virus (CymRSV)	Genus – *Tombusvirus* Family – *Tombusviridae*	Sharma et al. (2005b)
Cauliflower mosaic virus (CaMV)	Genus – *Cucumovirus* Family – *Bromoviridae*	Sharma et al. (2005b)
Bean yellow mosaic virus (BYMV)	Genus – *Potyvirus* Family – *Potyviridae*	Sharma et al. (2005b)
Calanthe mild mosaic virus (CalMMV)	Genus – *Potyvirus* Family – *Potyviridae*	Singh et al. (2007b)

genomic regions have been used for the diagnosis of GLRaV-3 while p9 region has been used for GLRaV-1 (Kumar et al. 2012d; 2013c). Kumar et al. (2013c) detected GLRaV-3, GLRaV-1, grapevine fleck virus (GFkV) and grapevine virus B (GVB) from the infected vines of Himachal Pradesh by two step RT-PCR of total RNA extracted using commercial kit. Complementary DNA was synthesized after carrying out reverse transcription of total RNA using random hexamer primers followed by PCR using primers from partial RNA dependent RNA polymerase (GLRaV-3), partial CP gene (GLRaV-1), putative RNA binding protein gene (GVB) and partial CP gene (GFkV) (Kumar et al. 2012c, 2013c).

Viroids Three viroids have been detected in grapevine samples from India. Presence of grapevine yellow speckle viroid 1 (GYSVd-1) was tested by RT-PCR with primer pair PBCVd100C, PBCVd194H and 341M, 342P amplifying 220 bp and full genome, respectively (Sahana et al. 2013). Hop stunt viroid (HpSVd) was detected in two separate sets of RT-PCR using primer pairs HSV78P, HSV83M and HSV7P, HSV220M with the expected size of 303 and 213 bp, respectively. Northern blot assay using digoxygenin-labeled (DIG-labeled) riboprobes further confirmed the RT-PCR results (Sahana et al. 2013). For the diagnosis of Australian grapevine viroid (AGVd) total nucleic acids were extracted using a modified CTAB method followed by low molecular weight RNA enrichment following procedure of Machida et al. (2008). Complementary DNA was synthesized by reverse transcription using random hexamer. Viroid cDNA was amplified by PCR using *Apscaviroid* genus specific primer pair PBCV100C/PBCV194H giving the amplicon of 225 bp size. The species was further narrowed by northern blotting using DIG-labeled riboprobes for apple fruit crinkle viroid (AFCVd; having 85% sequence similarity with AGVd) and RT-PCR amplification by a newly designed primer set (Adkar-Purushothama et al. 2014).

Northern hybridization of six RT-PCR-positive samples using DIG-labeled riboprobe for AFCVd supported the presence of AFCVd-like viroid in Indian grapevines; i.e., AGVd. Since partial AGVd sequence was detected, the six AGVd-positive samples were used for RT-PCR amplification of the full-length cDNA copy using F3/R4 primer set giving the amplicons of 369 bp (Adkar-Purushothama et al. 2014).

27.7.2 Tropical Fruit Crops

27.7.2.1 Papaya

Nucleic Acid Hybridization, PCR and RT-PCR Based Detection Methods

Papaya Ringspot Virus: PRSV (Genus: *Potyvirus*, Family: *Potyviridae*) PRSV is associated with destructive ringspot disease of papaya in India. Jain et al. (1998a) first time used RT-PCR for PRSV detection based on primers amplifying 3′-terminal 1.7 kb genomic region of viral genome (forward primer HRP52 and reverse primers HRP1, 2, 3) comprising of partial NIb gene, full CP gene and untranslated region (UTR). Using this RT-PCR both PRSV-P and PRSV-W virus isolates from India were detected based on amplification, cloning and sequencing of expected amplicons. These primers however could not amplify some virus isolates, therefore another set of primers (from NIb region and CP gene) were designed and successfully used for RT-PCR based detection of PRSV in papaya samples collected from Raipur (Chhattisgarh), Delhi, Solan (Himachal Pradesh), Ranchi (Jharkhand), Dharwad (Karnataka), Pune (Maharashtra) and Lucknow (Uttar Pradesh) (Sharma et al. 2004). Coat protein based primers were also used in RT-PCR for diagnosis of PRSV in samples originating from different parts of India (Andhra Pradesh, Chhattisgarh, Delhi, Himachal Pradesh, Jharkhand, Karnataka, Maharashtra, Uttar Pradesh and West Bengal) (Jain et al. 2004). PRSV-P and PRSV-W infection was also detected by RT-PCR later from South India (Sharma et al. 2005a; Srinivasulu and Sai Gopal 2011; Kumar et al. 2014c). PRSV-P and PRSV-W isolates from India were detected using RT-PCR and characterized at full genome level (Mangrauthia et al. 2008).

Begomovirus Infection Initially Southern blot hybridization using the probes specific to different tomato leaf curl virus species indicated that leaf curl disease of papaya is caused by a bipartite begomovirus (Saxena et al. 1998). PCR using degenerate primers specific to DNA-A and DNA-B also suggested the association of a begomovirus.

Papaya plants showing symptoms of typical leaf curl in Lucknow were detected positive for tomato leaf curl New Delhi virus in PCR using primers from CP gene of begomovirus genome yielding an amplicon of ~770 bp (Raj et al. 2008). The identity of amplified product was confirmed by sequencing and sequence comparison identified it as tomato leaf curl New Delhi virus. Krishna Reddy et al. (2010) detected papaya leaf curl virus (PaLCuV) in papaya leaf samples exhibiting typical leaf curl symptoms based on PCR amplification and sequencing of compete DNA-A of four isolates.

Multiplex Detection Methods

A duplex PCR for detection of mixed infection of PRSV and PaLCuV was developed in India (Usharani et al. 2013). Primer pairs from partial NIb and full CP for PRSV and CP plus partial AC2 gene in case of PaLCuV was designed. In case of PRSV, cDNA was synthesized and total DNA in case of PaLCuV infected sample was used in duplex PCR which amplified 1.4 kb and 1.2 kb fragment specific to PRSV and PaLCuV respectively.

27.7.2.2 Banana

PCR, RT-PCR and RCA Based Detection Methods

Badnaviruses Initially PCR based assay was used for detection of banana streak viruses (BSV) in India (Cherian et al. 2004). Integrated badnavirus sequence counterparts having significant sequence homology to the episomal BSVs are found in banana genome (Harper et al. 1999; Geering et al. 2005), which makes PCR based diagnostics unreliable as both integrated and episomal sequences are equally amplified. Thus immunocapture-PCR (IC-PCR) was developed for detection of episomal banana streak MY virus in India (Sharma et al. 2014a). Further an efficient duplex-immunocapture-PCR (D-IC-PCR) was standardized for routine indexing of field and tissue cultured plantlets (Sharma et al. 2014b). When different detection methods were compared, direct-PCR and direct binding-PCR (DB-PCR) were not found to yield confirmatory and accurate results for the presence of episomal BSV infec-

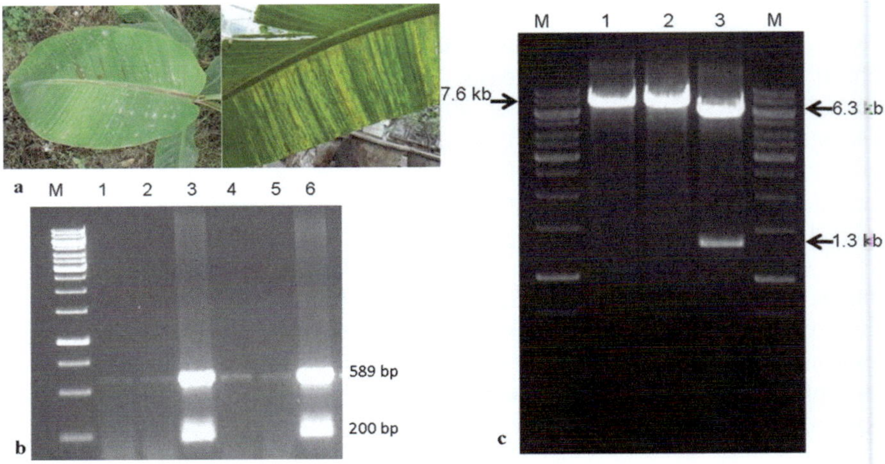

Fig. 27.1 (**a**) Typical streak symptoms on Chini Champa banana cultivars, (**b**) Duplex immuno capture PCR (D-IC-PCR) using MySF1/MySR1 and AGMI 25/26 primers for detection of BSMYV infected banana samples trapped with antiserum prepared against purified virions (Lanes: *1–2*) dilutions of 1:250 and 1:500 respectively, and with anti MBP-p37 antiserum (Lanes *4–5*): at dilutions of 1:500 and 1:1000 respectively, Lanes *3*, *6*: duplex PCR using total DNA of positive samples showing amplification of *Musa* genome specific fragment (~200 bp) in addition to BSMYV specific amplicons, *M*: 1 Kb marker, (**c**) improved random primed RCA of badnaviruses genome. *Kpn*I digested RCA product derived from banana streak MY virus infected banana hybrid plants Lane *1*: Chini champa (AAB), lane *2*: Monthan (ABB), lane *3*: Malbhog (AAB), *M*: 1 Kb maker

tion. However, D-IC-PCR employing *Musa* genome specific internal primers and BSMYV specific primers with an immunocapture time of 3 h was found to be a reliable robust detection method for detection of episomal BSMYV (Fig. 27.1). Improved random primed-RCA was also developed for high amplification of badnaviruses genome (Sharma et al. 2015). DNA isolated from symptomatic banana leaves collected during lower temperature regimes (winter or rainy season) containing high concentration of replicative form of viral DNA (sealed minichromosomal DNA) when used in RCA followed by another addition of phi29 polymerase at the mid of incubation time leads to high badnavirus genome specific amplification (Sharma et al. 2015) (Fig. 27.1). Further sequence independent random primed-RCA coupled with restriction fragment length polymorphism (RCA-RFLP) could also be used for reliable detection of episomal banana streak viruses.

Banana Bunchy Top Virus: BBTV (Genus: *Babuvirus*, Family: *Nanoviridae*) BBTV is a single stranded DNA virus which causes a destructive bunchy top disease of banana in India. PCR amplification of partial and full genome sequences was used for detection of BBTV infection in the symptomatic samples collected from Northern and Southern India. BBTV isolates were classified into PIO group (Karan et al. 1994; Selvarajan et al. 2010; Vishnoi et al. 2009). Primers specific to CP gene and replicase genes (Rep) were designed based on the available sequences in GenBank and used for molecular detection of BBTV infection in the samples collected from different regions of India. Based on the PCR results an incidence of 14–72% in South India on hill banana cultivars was reported (Selvarajan et al. 2010). By using PCR based detection, 20–30% samples from North India were detected positive for BBTV (Vishnoi et al. 2009). BBTV infection was recently detected from North East India. By using abutting primers in PCR targeting all genomic components, full genome sequences were obtained and association of a distinct PIO group BBTV isolate was identified (Banerjee et al. 2014c).

Using primers based on replicase (Rep) and coat protein (CP) gene of BBTV, its infection in *Ensete superbum* plants (showing symptoms of dark green dots, streaks pattern on leaf sheath mid rib, narrowing of leaf lamina) and banana black aphid (*Pentalonia nigronervosa*) was detected (Selvarajan and Balasubramanian 2013). Positive samples gave amplicons of 861 and 513 bp from Rep and CP genes, respectively.

Banana Bract Mosaic Virus: BBrMV (Genus: *Potyvirus*, Family: *Potyviridae*) In India, banana bract mosaic disease (BBrMD) was detected by molecular techniques from different regions including Kerala, Tamil Nadu, Karnataka and Andhra Pradesh using CP based primers (Rodoni et al. 1997, 1999; Sankaralingam et al. 2006). Using genome wide primers from BBrMV genome, its infection in French plantain cv. Nendran (AAB) from Tamil Nadu was detected and full genome sequences were obtained (Balasubramanian and Selvarajan 2012).

Cucumber Mosaic Virus: CMV (Genus: *Cucumovirus*, Family: *Bromoviridae*) CMV causes infectious chlorosis disease in banana which has been reported from many banana growing regions of world including India. Primers specific to CP gene of CMV were used to detect the associated virus in banana leaf samples exhibiting the symptoms of yellow stripes on leaves, distortion and stunting of whole plant in Karnataka, Maharashtra and Uttar Pradesh (Khan et al. 2011). An amplicon of 657 bp was observed in the samples and based on the coat protein sequencing of three CMV isolates from banana, occurrence of CMV subgroup IB was determined from Karnataka, Maharashtra and Uttar Pradesh. Highly similar CP sequences of CMV from *Physalis minima* were amplified which indicated that this weed plant act as weed reservoir of CMV infecting banana (Khan et al. 2011). Recently based on the sequencing of RNA 1a, RNA 2b and RNA 3 genomic fragments of CMV from banana, association of subgroup IB has been confirmed from North India (Vishnoi et al. 2013).

Multiplex Detection Methods

Selvarajan et al. (2011) developed a multiplex-reverse transcription-PCR (mRT-PCR) for the simultaneous detection of BSMYV and BBTV in same sample. The first strand cDNA was synthesized using random primers followed by the mRT-PCR using primers specific the BSMYV (targeting RT/RNase H region), BBTV (partial master rep) and *Musa* genome specific internal sequence tagged microsatellite site (STMS) primers. Developed mRT-PCR was successfully used for indexing of field and tissue culture plants.

27.7.2.3 Citrus

Citrus trees harbor large number of virus and virus like pathogens. Citrus budwood used in vegetative propagation of commercially important citrus cultivars is the major source of the viruses and viroids transmission (Ahlawat 1997; Das et al. 2002; Prabha and Baranwal 2011).

Nucleic Acid Hybridization, PCR and RT-PCR Based Detection Methods

Indian Citrus Ringspot Virus: ICRSV (Genus: *Mandarivirus*, Family: *Alphaflexiviridae*) Citrus ringspot disease caused by Indian ICRSV was first reported in 1989 in sweet orange orchards from Delhi, Maharashtra and Andhra Pradesh states of India (Ahlawat 1989). Later in 1995 the occurrence of ringspot disease in *Citrus sinensis* cultivars was observed from various parts of India (Byadgi and Ahlawat 1995). For molecular detection and diagnosis of ICRSV, RT-PCR assay was developed to amplify 500 bp partial coat protein gene of the virus using gene specific primers (Parveen et al. 1999). Further for the variability study, Hoa and Ahlawat (2004) designed two sets of primer pairs derived from the sequence of ICRSV-D available in the database from the conserved region of viral the genome. The samples showing varying symptoms on different *Citrus* spp. were subjected for RT-PCR amplification. The sequencing results revealed that out of four, two isolates shared ~98% sequence identity while another two ICRSV-Ah and ICRSV-Pu isolates showed 85 and 84.5% identity to ICRSV-D (Hoa and Ahlawat 2004). Further, using RT-PCR genetic diversity of ICRSV isolates from different part of India based on the different genomic regions and full genome was accomplished from India (Prabha and Baranwal 2012).

Citrus Yellow Vein Clearing Virus: CYVCV (Genus: *Mandarivirus*, Family: *Alphaflexiviridae*) Citrus yellow vein clearing virus, causing yellow vein clearing disease earlier though to be a distinct isolate of ICRSV was first observed in lemons and sour orange trees in early 1988 from Pakistan. In India yellow vein clearing disease of citrus (CYVCD) was first observed at experimental research orchard, Abohar in Punjab on a citrus cultivar Etrog citron in 1997 during a survey by Ahlawat (1997) and later by Alshami et al. (2003). The CYVCV was later characterized as distinct member of the genus *Mandarivirus* and shares 72% sequence identity with the ICRSV (Loconsole et al. 2012). Differentiation between both the members of genus *Mandarivirus*; CYVCV and ICRSV, based on the symptoms is a major challenge because both the viruses produce common ringspots and leaf distortion symptoms on infected citrus plants. In the later phases of infection typical veinal necrosis is produced on Etrog citron due to CYVCV infection. For the diagnosis of CYVCV total RNA was isolated from the symptomatic leaf samples, first strand cDNA was synthesized and amplified by PCR assay using the CYVCV genus specific primer pair YVM103/YVM104 (Meena and Baranwal 2016) giving an amplicon size of 333 bp and showing 99% sequence identity with CYVCV-Y1 (JX040635) isolate.

Citrus Tristeza Virus: CTV (Genus: *Closterovirus*, Family: *Closteroviridae*) Tristeza disease caused by citrus tristeza virus (CTV) is one of the most destructive and economically important viral diseases of citrus in India (Biswas 2010). CTV is phloem limited virus and can be transmitted by grafting, budding and insect vectors. Molecular detection techniques have been developed for routine indexing of CTV in India. Biswas (2008) developed a RT-PCR assay using CTV genome specific primers (CP region of viral genome) amplifying an amplicon of 672 bp for detection of CTV in India. Primers from CP region (672 bp), variable partial ORF1a region (404 bp) and other regions of viral genome were used for development of efficient RT-PCR based assays for CTV detection in India (Biswas 2010; Tarafdar et al. 2013; Sharma et al. 2012; Singh et al. 2013). Using overlapping primers covering entire CTV genome, complete genome sequence of decline inducing CTV isolate was reported from India (Biswas et al. 2012a). Citrus samples collected from different citrus growing regions of India were detected using RT-PCR employing primers from CP region and partial ORF1a region and high genetic diversity was reported among the CTV isolates (Biswas et al. 2012b, 2014b).

Ghosh et al. (2009) developed a RT-PCR assay for CTV detection using the virus gene specific primers, derived from the coat protein genomic region. In addition molecular diagnosis of CTV infection in mandarin by using the ORF1a gene specific primers was accomplished and genome characterization from India was also accomplished by various workers.

Citrus Yellow Mosaic Virus: CYMV (Genus: *Badnavirus*, Family: *Caulimoviridae*) PCR based detection has been employed for the detection of CYMV (Baranwal et al. 2003). Baranwal et al. (2003) developed an improved sodium sulphite based DNA extraction procedure which could lead to improved yield, quality and stability of isolated DNA as compared to normal DNA extraction procedure and kit. Even at lower concentration, CYMV was amplified from the DNA isolated through sodium sulphite based procedure (Baranwal et al. 2003). PCR was employed for detection of CYMV infection in Vidarbha region of Maharashtra (Ghosh et al. 2007). Four methods of DNA isolation were further compared by Borah et al. (2008) to determine their suitability for PCR and dot-blot hybridization based detection of CYMV in acid lime and pummelo. Dot-blot hybridization was not influenced by the method of DNA extraction and irrespective of the method followed; similar results were obtained for dot-blot hybridization. However, PCR based detection was much influenced by the method of DNA isolation and sarkosyl method was reported to be the most suitable method when DNA is to be used for PCR detection of CYMV (Borah et al. 2008). Thus method of DNA isolation largely influences the reliability of PCR based detection of CYMV. CYMV specific DNA was amplified from leaf, twig bark, fruit-rind and fruit juice from the infected sweet orange plant, however no amplification was observed in fruit rag (Gopi et al. 2010).

Viroids The major prevailing viroid pathogens in Indian citrus orchards are citrus exocortis viroid (CEVd) and hop stunt viroid (HSVd) variants associated with yellow corky vein disease (Ramachandran et al. 1993). Total RNA isolated from infected leaves of Kagzi lime was used as template for cDNA synthesis with the reverse primers of the two sets. Complementary DNA template used for amplification in RT-PCR with only one set of primer, gave an amplicon (~300 bp) of expected size. The sequencing results revealed around 100% identity with the hop stunt viroid confirming the presence of viroid (Ramachandran et al. 2003). For molecular detection of citrus exocortis viroid with the RT-PCR, a set of the primer RFL CEV(R)/RFL CEV(F) giving an amplification of ~400 bp was used (Baranwal et al. 2003). Sequencing of amplified products confirmed the infection of citrus exocortis viroid in infected samples (Ramachandran et al. 2003; Ghosh et al. 2002a). Nucleic acid spot hybridization (NASH) test is another nucleic acid based method for detection of viroids in India using radiolabelled PCR products of CEVd and HSVd as probes which clearly detected two viroid components in the infected citrus leaves.

Multiplex Detection Methods

A multiplex-PCR based detection was developed to amplify both CYMV (fragment of RT/RNase H region of genome) and citrus greening bacterium, *Candidatus* liberibacter asiaticus; CLas (Baranwal et al. 2005). DNA template from midrib of citrus leaf samples infected with CYMV and greening bacterium could be used as common target for the DNA isolation to be used for simultaneous detection of both CLas and CYMV (Baranwal et al. 2005). Ghosh et al. (2008a) developed a sensitive duplex-PCR technique for the simultaneous detection of CYMV (targeting fragment of ORF3) and CTV (targeting fragment of ORF8) in infected samples. The duplex-PCR could detect the CYMV and CTV nucleic acid up to 10^{-5} and 10^{-6} dilutions.

A multiplex PCR assay was developed for simultaneous detection of clostero (CTV), badna (CYMV) and mandari viruses (CYVCV and ICRSV) along with huanglongbing bacterium in citrus trees from India. The primers used for this assay was designed from the conserved genomic region of the pathogens and successfully employed for multiplex detection (Fig. 27.2) (Meena and Baranwal 2016).

Isothermal Amplification and Real Time Based Detection Methods

Recently for more sensitive, robust and specific detection of CYMV in infected samples, loop-mediated isothermal amplification (LAMP) and SYBR green real-time PCR (SGRTPCR) based detection techniques were developed (Anthony Johnson et al. 2014). In these methods primers targeting putative coat protein domain in ORF3 were used. Both the protocols were successfully used for detection of CYMV in rough lemon (*Citrus jambhiri*), Nagpur Mandarin (*Citrus reticulata*), Pumello (*C. grandis*), sweet orange (*C. sinensis*) and acid lime (*C. aurantifolia*) samples. PCR and LAMP were found equally sensitive in detection of CYMV, capable of detecting in 10 ng of total DNA. However, SGRTPCR assay was ten-fold more sensitive and were able to detect virus in 1 ng of total DNA (Anthony Johnson

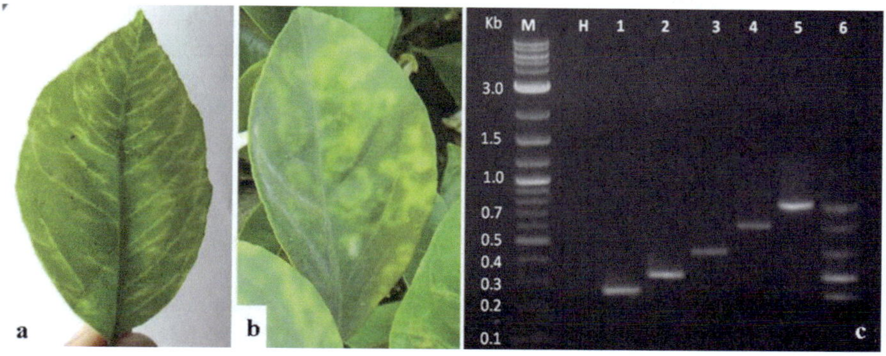

Fig. 27.2 (**a**) Yellow vein clearing symptoms of CYVCV on Etrog citron (**b**) Ringspot symptoms of ICRSV on Sweet orange (**c**) Detection of four viruses and HLB in citrus by simplex polymerase chain reaction using specific primer pairs. The amplified products corresponding to lanes: *M*: 2-log DNA ladder (0.1–10 kb,NEB); H. Healthy control; *1*. 256 bp for citrus tristeza virus (CTV); *2*. 333 bp citrus yellow vein clearing virus (CYVCV); *3*. 451 bp *Candidatus* Liberibacter asiaticus (CLas); *4* 610 bp for citrus yellow mosaic virus (CYMV); *5*. 758 bp Indian citrus ringspot virus (ICRSV); *6*. mPCR. All the five primer pairs did not amplify any product from healthy sweet orange used as negative control (Figure courtesy Meena and Baranwal 2016)

et al. 2014). Using the sensitive SGRTPCR assay, CYMV titres were compared in different citrus species. The highest virus titre was detected in rough lemon and lowest in Nagpur mandarin with 7.98×10^{10} and 5.55×10^{7} of viral load in dsDNA copies/10 ng of total DNA respectively.

27.7.3 Vegetable Crops

27.7.3.1 Bhindi

Nucleic Acid Hybridization, PCR and RT-PCR Based Detection Methods

Bhendi Yellow Vein Mosaic Virus: BYVMV (Genus: *Begomovirus*, Family: *Geminiviridae*) Using primers specific to viral DNA of BYVMV, infection was detected in PCR and association of a β satellite was also confirmed (Jose and Usha 2003). Molecular detection of BYVMV was done by using primers designed earlier to CP gene of viral genome (Jose and Usha 2000). Same primers along with universal beta satellite primers (Briddon et al. 2002) were used for the detection of BYVMV and associated beta satellite in the samples collected from Barrackpore, West Bengal (Ghosh et al. 2008b).

Using primers designed for full length amplification of betasatellite molecules in PCR, sequences of 36 betasatellites sampled from okra plants exhibiting leaf curl and yellow vein symptoms were obtained. This led to identification of okra leaf curl beta satellite (OLCuB) and bhendi yellow vein beta satellite (BYVB) and a novel betasatellite, bhendi yellow vein India beta satellite (BYVIB) with okra plants collected from different locations in India (Venkataravanappa et al. 2011). Bhendi plants exhibiting symptoms of yellow mosaic and vein thickening collected from Palem area of New Delhi were detected positive for bipartite begomovirus (BYVMV), sequences of which were determined using genome specific PCR primers (Venkataravanappa et al. 2012a, b).

Using degenerate geminivirus-specific primer pairs, infection of bhendi yellow vein mosaic virus was detected in PCR which amplified a product of 760 bp (from putative AV1, AC3 and AC2 genomic regions) from *Litsea* species exhibiting curling, yellowing and mosaic symptoms (Roy et al. 2015).

Radish Leaf Curl Virus: RaLCV (Genus: *Begomovirus*, Family: *Geminiviridae*) Association of a monopartite begomovirus with leaf curl disease of okra was detected using specific primers in PCR (Kumar et al. 2012a). Virus was identified to be radish leaf curl virus (RaLCV). Using the specific primers (Kumar et al. 2012a), alphasatellite and betasatellite was also found associated with the disease. Rolling circle amplification was also employed for the detection of radish leaf curl virus.

27.7.3.2 *Allium* species

Alliums (Onion, Garlic, Leek and Shallot) are the important culinary vegetables cultivated in India. Diseases caused by the viruses are the important limiting factors for their lower productivity. In India, as many as six flexuous virus species belonging to three genus (*Potyvirus*, *Carlavirus* and *Allexivirus*) and two spherical virus species belonging to *Tospovirus* genus have been reported to infect allium crops so far (Table 27.1). Different allium crops species are grown in the same geographical region due to similarity in their agroclimatic requirements and also host the similar sucking pest species (aphids, mites and thrips) which leads to the cross infection of many allium virus species between the allium crops. Due to similarity in their symptomatology and lack of differential host, allium viruses are difficult to diagnose symptomatically in field. Nucleic acid based detection has been developed for these viruses in India.

PCR and RT-PCR Based Detection Methods

Onion Yellow Dwarf Virus: OYDV (Genus: *Potyvirus*, Family: *Potyviridae*) OYDV is a flexuous virus infecting multiple allium crops and causes significant yield losses every year. In 2006, Arya et al. has standardized the RT-PCR detection protocol for the diagnosis of OYDV infecting onion and garlic in India. A set of primers, one from the 3′ end of RNA-dependent RNA polymerase gene and another in the 3′- untranslated region (UTR) which amplify 1.1 kb genome were designed. These primers were successfully used to develop the RT-PCR based technique for the detection of OYDV from onion and garlic.

Other Viruses Infecting Allium Crops RT-PCR based methods for the detection of other allium infecting viruses such as garlic common latent virus (GarCLV) (Majumder and Baranwal 2009), shallot latent virus (SLV), (Majumder et al. 2008), shallot virus X (ShV-X) (Majumder et al. 2007), Iris yellow spot virus (IYSV) (Ravi et al. 2006; Gawande et al. 2010, 2014) and leek yellow Stripe virus (LYSV) (Gupta et al. 2013; Gawande et al. 2014) has been developed in India.

Multiplex Detection Methods

Multiplex PCR techniques for detection of multiple virus infection in allium crops were standardized for many virus combinations. Duplex RT-PCR technique was standardized for the simultaneous detection OYDV and SLV in garlic using specific primers which amplify 1.1 kb genome of RNA-dependent RNA polymerase gene and 3′-UTR of OYDV (Arya et al. 2006) and 300 bp CP gene of SLV (Majumder et al. 2008). The developed duplex PCR technique detected both the viruses from the leaves and cloves of garlic. A duplex PCR technique for the simultaneous detection of OYDV and allexivirus in onion was standardized (Kumar et al. 2010a). Quadruplex PCR for the simultaneous detection of four viruses such as OYDV, GarCLV, SLV and allexiviruses was standardized and successfully demonstrated for the virus indexing in tissue culture programme (Pramesh et al. 2012; Majumder and Baranwal 2014; Pramesh and Baranwal 2015).

27.7.3.3 Cucurbits

The important virus groups which infect cucurbit in India are *Tobamovirus, Begomovirus, Cucumovirus* and *Tospovirus.*

Nucleic Acid Hybridization, PCR and RT-PCR Based Detection Methods
Begomoviruses

The begomovirus diseases have been detected time to time in India on cucurbitaceous crops such as bitter gourd, cucumber, muskmelon, winter squash, bottle gourd, sponge gourd, ridge gourd, ivy gourd, pumpkin and watermelon by nucleic acids hybridization tests or by PCR using primers of putative CP gene of tomato leaf curl New Delhi virus (ToLCNDV) (Raj and Singh 1996; Varma and Giri 1998; Singh et al. 2001; Khan et al. 2002; Mandal et al. 2004b; Sohrab et al. 2006; Kumar et al. 2010c). In majority of the cases, the PCR detection of begomoviruses was mainly based on the begomovirus specific degenerate primer pairs which amplify the core segment within the replication initiator protein gene (Rep) (Rojas et al. 1993) and coat protein (described by Deng et al. 1994).

Begomoviruses Infecting Pumpkins Begomoviruses infecting pumpkins: pumpkin yellow vein mosaic virus (PYVMV) (Varma 1955; Muniyappa et al. 2003), squash leaf curl China virus (SLCCNV) (Muniyappa et al. 2003; Maruthi et al. 2007; Singh et al. 2008; Tiwari et al. 2011), tomato leaf curl New Delhi virus (TLCNDV) (Maruthi et al. 2007), pumpkin yellow vein mosaic virus (Sohrab et al. 2006), tomato leaf curl Palampur virus (ToLCPMV) (Tiwari et al. 2010a) have been reported employing molecular detection techniques. PCR detection of PYVMV was described by Muniyappa et al. (2003) using coat protein specific primers of ToLCNDV and movement protein specific primer of ToLCNDV.

Begomoviruses Infecting Gourds Association of begomovirus with the yellow mosaic disease of bitter gourd was shown using PCR with begomovirus specific primers and Southern hybridization with a probe of cloned DNA-A (Raj et al. 2005). The virus was tentatively named as bitter gourd yellow mosaic virus (BGYMV) (Raj et al. 2005). Rajinimala and Rabindran (2007) reported the occurrence of Indian cassava mosaic virus (ICMV) from Tamil Nadu which was found associated with yellow mosaic disease of bittergourd. Based on the degenerate primer pairs and primers specific to ICMV, PCR based diagnostics were developed. In addition, infection of tomato leaf curl New Delhi virus (ToLCNDV) on bottle gourd (Sohrab et al. 2010), sponge gourd (Sohrab et al. 2003) and ridge gourd (Tiwari et al. 2011) has been reported using molecular detection techniques. Cucumber was also found infected with tomato leaf curl Palampur virus (Raj et al. 2011).

Southern blot hybridization has also been used for the detection of begomovirus infecting cucurbits such as squash leaf curl virus (SLCuV) (Chakraborty et al. 2003; Singh 2009a). In recent time, detection and characterization of begomoviruses has been mainly done by rolling circle amplification (RCA) combined with restriction fragment length polymorphism (RFLP) based technique (Kushwaha et al. 2010).

Tospoviruses Watermelon bud necrosis virus (WBNV), groundnut bud necrosis virus (GBNV) and watermelon silver mottle virus (WSMoV) are the three tospoviruses infecting cucurbits in India (Kunkalikar et al. 2010; Mandal et al. 2012c). In India, identification of tospoviruses species is based on N gene sequence derived from first and last 21 bases of coding region of N gene of GBNV (Satyanarayana et al. 1996) and WSMoV (Yeh and Chang 1995). RT-PCR using virus-specific N-gene primers have been standardized and validated for the detection of tospoviruses associated with cucurbits (Jain et al. 1998b, 2005; Mandal et al. 2003, 2012c). Mandal et al. (2003) described the RT-PCR detection of WBNV in cucurbits using primer derived from the conserved sequences of GBNV and WSMoV, targeting part of the nucleocapsid (N) protein gene and the complete noncoding region upstream of the N gene coding sequence.

Cucumber Mosaic Virus: CMV (Genus: *Cucumovirus*, Family: *Bromoviridae*) Infection on Cucurbits CMV is one of the most widespread plant viruses in the world with extensive host range infecting about 1200 species including cereals, fruits, vegetables and ornamentals (Roossinck 1999; Geetanjali et al. 2011). Madhubala et al. (2005) developed the RT-PCR techniques for the detection of CMV infecting vanilla using primer pair specific to 650 bp coat protein gene. RT-PCR techniques for the specific detection of CMV and its subgroups I and II was standardized using genome sequence specific primers (Geetanjali et al. 2011). Shetti et al. (2012) developed a RT-PCR technique for the detection of CMV in cucumber using CMV CP gene primers. Later, for the detection the CMV infection in capsicum, a RT-PCR and RNA-DNA hybridization (Northern hybridization) technique was demonstrated by targeting CP gene of CMV (Biswas et al. 2013). Infection of papaya ringspot virus W in sponge gourd was also reported from India (Verma et al. 2006).

Multiplex Detection Method

Degenerate primers based on the N gene sequence of GBNV, WBNV, capsicum chlorosis virus (CaCV) and WSMoV available in GenBank were designed and used to amplify the N gene sequence for detection of GBNV, WBNV, CaCV, and WSMoV in multiplex PCR (Kunkalikar et al. 2011). Multiplex reverse transcription-polymerase chain reaction (mRT-PCR) detected the CaCV, GBNV and WBNV by this degenerate upstream primer (Mh-F) and species-specific downstream reverse primers (PBNV-R, WBNV-R and CaCV-R). The 494 bp, 647 bp and 764 bp fragments specific to CaCV, GBNV and WBNV respectively, were amplified in single and mixed virus infections. The specificity of virus-specific amplicons was ascertained by cloning and sequencing and comparing with corresponding sequences of GBNV, WBNV and CaCV (Kunkalikar et al. (2011).

27.7.3.4 Tomato

Nucleic Acid Hybridization, PCR, RT-PCR and RCA Based Detection Methods

Begomoviruses Tomato leaf curl disease (ToLCD) is a common disease of tomato all over India. There are many begomoviruses associated with ToLCD in India *viz.*, tomato leaf curl New Delhi virus (ToLCNDV), tomato leaf curl New Delhi virus-Mild (ToLCNDV-Mild), tomato leaf curl Karnataka virus (ToLCKV), tomato leaf curl Bangalore virus (ToLCBV), tomato leaf curl Gujarat virus (ToLCGV), tomato leaf curl Joydebpur virus (ToLCJV), tomato leaf curl Palampur virus (ToLCPMV), tomato leaf curl Bangladesh virus (ToLCBaV), tomato yellow leaf curl China virus (TYLCCNV), tomato leaf curl Kerala virus (ToLCKeV), tomato leaf curl Ranchi virus (ToLCRnV), tomato leaf curl Patna virus (ToLCPaV), tomato leaf curl Rajasthan virus (ToLCRaV), tomato leaf curl Pune virus (ToLCPV), tomato leaf curl Lucknow virus (ToLCLuV) etc. (Padidam et al. 1995; Muniyappa et al. 2000; Kirthi et al. 2002; Chakraborty et al. 2003; Tiwari et al. 2010b, 2012). For species specific detection of four major viruses associated with ToLCD in India *viz.*, ToLCNDV, ToLCKV, ToLCBV and ToLCGV in PCR, Sivalingam (2006) designed a set of common primers and four sets of virus specific primers for sensitive detection. PCR based diagnosis of begomoviruses associated with tomato leaf curl disease was reported (Sivalingam and Varma 2007).

Tospoviruses Groundnut bud necrosis virus (GBNV), watermelon bud necrosis virus and capsicum chlorosis virus (CaCV) are the economically important tospoviruses infecting tomato in India (Kunkalikar et al. 2010, 2011). Degenerate primers from N-gene of tospovirus genome were employed in RT-PCR for detection of CaCV infecting tomato and chilli in India (Kunkalikar et al. 2010).

Cucumovirus Association of CMV with tomato in India based on the previously designed primers flanking the CP gene which amplify approximately 870 bp in RT-PCR was reported (Sudhakar et al. 2006). Similarly a RT-PCR detection system based on the primers from complete RNA-3 of viral genome was developed (Pratap et al. 2008).

Multiplex Detection System

For the simultaneous detection of begomoviruses and tomato big bud phytoplasma infection in tomato, a duplex PCR method has been standardized in India (Swarnalatha and Reddy 2014) using begomovirus specific primer set from the DNA-A component and phytoplasma-specific 16S rDNA universal primers P1/P7 (Deng and Hiruki 1991) and P4/P7 (Smart et al. 1996).

27.7.3.5 Chilli

Nucleic Acid Hybridization, PCR and RT-PCR Based Detection Methods

Infection of begomoviruses particularly chilli leaf curl virus and tomato leaf curl Joydebpur virus were detected in India using molecular techniques (Shih et al. 2006; Senanayake et al. 2006, 2012). Infection of a tospovirus capcicum chlorosis virus was also reported by RT-PCR (Krishnareddy et al. 2008). RT-PCR using primers from different parts of viral genome were used for molecular detection of a potyvirus on chilli in India (Anindya et al. 2004; Joseph and Savithri 1999; Ravi et al. 1997). The characterized virus was later recognized as chilli veinal mottle virus (ChiVMV). Infection of ChiVMV by RT-PCR using potyvirus specific primers and sequencing was reported from Naga chilli in North East India (Banerjee et al. 2014a).

Isothermal Amplification Based Detection Methods

Recently a one-step reverse transcription loop-mediated isothermal amplification (RT-LAMP) assay for sensitive and specific detection of chilli veinal mottle virus (ChiVMV) was developed (Banerjee et al. 2016). LAMP primers based on NIb-CP region (spanning around 1397 bp) were designed and used in RT-LAMP standardization. RT-LAMP incubation at 63 °C for 1 h followed by addition of SYBR Green I successfully detected ChiVMV. Developed RT-LAMP assay was 100-times more sensitive compared to RT-PCR and could detect the virus up to 0.001 ng (in total RNA) (Banerjee et al. 2016).

27.7.3.6 Brinjal

By employing CP specific primers in RT-PCR followed by sequencing, infection of cucumber mosaic virus (CMV) subgroup IB was detected from brinjal plant showing the symptoms of severe mosaic in Lucknow, Uttar Pradesh (Kumar et al. 2014a).

27.7.4 Flowering and Ornamental Crops

27.7.4.1 Gladiolus

Gladiolus is an important cut flower occupying a significant position in export oriented Indian floriculture industry. A wide array of diseases of biological origin is responsible for reducing the quality and quantity of gladiolus flowers drastically. Among various diseases, viruses attain an important status because in addition to causing direct damage they also predispose to secondary invaders and thus affect the floriculture trade in India (Dubey and Aminuddin 2010). The viruses of gladiolus on which Indian researchers have worked are mentioned in the following Table 27.1.

Nucleic Acid Hybridization, PCR and RT-PCR Based Detection Methods
In gladiolus, the majority of viral diseases cause overall stunting, colour breaking, flower distortion, and reduced flower and cormel production (Magie and Poe 1972). Out of these reported viruses, bean yellow mosaic virus (BYMV) and cucumber mosaic virus (CMV) are of prime importance as these are the most frequently occurring viruses (Katoch et al. 2003). Katoch et al. (2002) was the first to diagnose BYMV using nucleic acid method. RT-PCR using primers from CP region of viral genome were employed for detection which gave amplicons of 750 bp in positive samples (Katoch et al. 2002). Dubey and Aminuddin (2009) by employing sodium sulphate based RNA method (Singh et al. 2002) and RT-PCR amplification of CP gene detected infection of BYMV.

CMV was detected in samples from floret, leaf, stem, root, corm and cormule by RT-PCR and Southern hybridization by Raj et al. (1998). Primer specific to CMV-RNA3 were used in RT-PCR reaction which gave an amplicon of 540 bp in CMV infected gladiolus plants (Raj et al. 1998). Further, the same group identified this virus in different sets of gladioli by RT-PCR and southern hybridization (Raj et al. 2002). In recent years Dubey and Aminuddin (2010) detected CMV by RT-PCR amplification of CP gene (657 bp) after isolating the total RNA following the procedure of Singh et al. (2002) and Southern hybridization of RT-PCR gel probed with α-P32-labeled cloned CP gene of CMV.

Recently, Raj et al. (2011) has reported tomato aspermy virus (TAV) from gladiolus by diagnosing it through nucleic acid method. RT-PCR was done using CP gene specific primers of TAV giving the amplification of ~650 bp (Raj et al. 2011). In the same year Kaur et al. (2011) diagnosed ornithogalum mosaic virus (OrMV) in gladiolus. RT-PCR was done using a degenerate pair of primer Pot 1 and Pot 2 for all potyviruses following the protocol mentioned by Gibbs and Mackenzie (1997). Sequence analysis of the resulted amplicon of ~1.5 kb showed 87–88% sequence identity with the polyprotein gene sequence of the OrMV isolates from flowering plants suggesting the identification of an OrMV-related isolate in gladiolus from India (Kaur et al. 2011). TRSV and TRV have been reported on the basis of serology, mechanical transmission and no further works have been done to diagnose them through nucleic acid based methods.

27.7.4.2 Chrysanthemum

Chrysanthemum is one of the most important commercial ornamental crops worldwide and in Indian floriculture industry the annual market value of chrysanthemum is of 299.6 million rupees. Diseases are one of the most important factors responsible for reducing the yield of chrysanthemum. Because of its commercial propagation by vegetative means using suckers and cuttings viruses occupy an important aspect in its cultivation. Viruses reported in chrysanthemum from India are given in Table 27.1.

Nucleic Acid Hybridization, PCR and RT-PCR Based Detection Methods

In RT-PCR and Northern hybridization screening for viruses infecting chrysanthemum in India, Chrysanthemum virus B (CVB) was found to be the most prevalent (48.7%) followed by cucumber mosaic virus (CMV) (42.5%), tomato aspermy virus (TAV) (26.2%), a tospovirus (12.5%) and a potyvirus (6.25%) (Verma et al. 2007). Singh et al. (2007a) studied the diversity of CP gene of CVB after diagnosing it by RT-PCR using the primers flanking the CP gene giving the amplicon size of ~1 kb. Further, in 2012 the same group of authors used RT-PCR and sequencing to diagnose and characterize the four new isolates of CVB. Near-full-length genome sequences of four isolates of CVB from three lineages were amplified using eleven pairs of designed primers (Singh et al. 2012). For the diagnosis of CMV, RT-PCR reaction was done by CP specific primers giving an amplicon of ~650 bp (Kumar et al. 2005).

TAV in chrysanthemum from India was reported long back in 1964 (Sastry 1964) but the work on molecular aspect has not been carried until 2009 when Raj et al. (2009) diagnosed it by RT-PCR and characterized the RNA 3 the virus. Primer pair specific to RNA 3 was designed and used in RT-PCR based detection. In the same year another group of authors detected this virus by RT-PCR (primers specific to CP gene in two step PCR) and slot blot hybridization. Using TAV-specific antibodies immunocapture RT-PCR (IC-RT-PCR) was carried out following the procedure of Jain et al. (1998c). RT-PCR and IC-RT-PCR gave the amplicons of ~660 bp. Three RNAs of TAV i.e. RNA 1, RNA 2 and RNA 3 of could also be amplified together through a single tube multiplex RT-PCR using three primers (P12 and P123 and P3; P123) giving the amplicons of 3.4 kb, 3.0 kb and 2.2 kb, respectively (Verma et al. 2009). In slot blot hybridization the virus was detected in all 11 chrysanthemum cultivars that were positive for TAV (Verma et al. 2009). TAV has also been reported to be the cause of yellow mosaic and flower deformation in the chrysanthemums (Raj et al. 2007).

Very recently tomato spotted wilt virus (TSWV) has been diagnosed from Indian chrysanthemum through RT-PCR using tospovirus degenerate primers specific to the L protein gene. RT-PCR gave an amplicon of ~750 bp which upon sequencing showed 98% identity with the large RNA of TSWV. RT-PCR was again performed with primer pairs specific to N gene and NSm genes of TSWV giving the amplicons of 870 bp and 1000 bp, respectively (Renukadevi et al. 2015).

Mehra et al. (2008) detected a potyvirus from chrysanthemum by RT-PCR and Northern hybridization. A potyvirus group specific degenerate primer pair (p9502 and CPup) was used in RT-PCR. After cloning and sequencing the amplified gene (~850 bp) showed 93% homology with chilli veinal mottle virus (CVMV) and pepper vein banding virus (PVBV) (Mehra et al. 2008).

CSVd was detected by R-PAGE analysis of total nucleic acid extracts from symptomatic leaves. The viroid nature of the RNA was further confirmed by electrophoretic mobility where it showed similarity in mobility with the RNA of potato spindle tuber viroid (PSTVd). Viroid nature was also established by its stability to high temperature treatment as well as sensitivity to RNase and insensitivity to DNase (Mathur et al. 2002). Singh et al. (2010) diagnosed CSVd by RT-PCR and

DNA-RNA hybridization. Using specific primers the isolated RNA was subjected to RT-PCR following the procedure of Nakahara et al. (1999), giving the amplicon of 348 bp. Slot blot hybridization was carried out as per Verma et al. (2007) and the loaded RNA was hybridized with the amplified DNA of viroid after radio labelling it (Singh et al. 2010).

27.7.4.3 Carnation

Carnation is one of the most important cut flowers and it ranks among the top five cut flowers (Singh et al. 2005). It is infected by several viruses causing significant losses (Lovisolo and Lisa 1978). Viruses reported from carnation in India are presented in Table 27.1.

Nucleic Acid Hybridization, PCR and RT-PCR Based Detection Methods
Occurrence of carnation necrotic fleck virus (CNFV) in Indian cultivars was reported by Raikhy et al. (2003a) from Himachal Pradesh. RT-PCR using degenerate primers of Karasev et al. (1994) specific to the closterovirus group giving an amplicon of ~1000 bp size confirmed the presence of this virus. RT-PCR amplifications of coat protein (CP) movement protein (MP) genes and complete genome were used for the diagnosis of an Indian isolate of carnation mottle virus (CarMV) (Singh et al. 2005; Raikhy et al. 2006b). Amplicons of 1050 bp, 200 bp, 250 bp and 4005 bp were observed in the RT-PCR amplification of CP gene, two MP genes (P7 and P9) and complete genome (Singh et al. 2005; Raikhy et al. 2006b). Additionally, IC–RT–PCR was also performed for CP gene amplification of CarMV (Singh et al. 2005).

Carnation etched ring virus (CERV) is the only DNA virus reported from carnation. Infection of this virus was exhibited by the symptoms of necrotic flecks, rings and line patterns "etched" on leaves. Raikhy et al. (2003b) was the first to detect it through nucleic acid based PCR using the primers of Sanchez-Navarro et al. (1999) yielding an amplicon of 1350 bp containing a 3′-terminal part of the coat protein (CP) gene and 5′-part of the enzymatic polyprotein gene (Raikhy et al. 2003b). Further, the same group diagnosed the virus in a multiplex PCR using the various genes-specific primers. Complete genome was also amplified giving an amplicon of ~8 Kb which upon sequencing was found to be of 7924 bp (Raikhy et al. 2006a). In a RT-PCR and IC-RT-PCR detection test the Indian researchers could also detect Carnation vein mottle virus (CVMoV) and CRSV (Raikhy et al. 2006b).

27.7.4.4 Orchids

Orchids are important flowers in international flower trade and are highly priced due to their incredible range of diversity in size, colour, shape, forms, appearance and long lasting qualities of flowers (De et al. 2013). Globally, they are reported to be infected with more than 50 viruses (Chang et al. 2005; Zaidi et al. 2009). In Indian condition, orchids get infected by the different viruses (Table 27.1):

Odontoglossum ringspot virus (ORSV) in orchids was detected by northern slot blot hybridization (Sherpa et al. 2006). Cymbidium mosaic virus (CymMV) was first definitively reported through RT-PCR using RdRp gene specific primer pair giving the amplicon size of ~534 bp (Sherpa et al. 2003; Seoh et al. 1998). Further, in 2007 the complete nucleotide sequence of an Indian isolate of CymMV was determined through RT-PCR using various sets of primers (Sherpa et al. 2007). Detection of cymbidium ringspot virus (CymRSV), CMV and bean yellow mosaic virus (BYMV) in Indian Orchids has been based only on ELISA. Nucleic acid based detection methods have not been used to confirm the presence of these three viruses (Sharma et al. 2005b).

Singh et al. (2007b) employed RT-PCR to detect a potyvirus in orchids using group specific primers described by Van der Vlugt et al. (1999) which gave an amplicon of ~650 bp. The nucleotide sequence of this amplicon showed 80% identity with calanthe mild mosaic virus (CalMMV). The RNA of virus was also detected in northern slot blot hybridization using radiolabelled DNA probe giving the two intensifying screens (Singh et al. 2007b).

27.7.5 Spices and Plantation Crops

Nucleic Acid Hybridization, PCR and RT-PCR Based Detection Methods

Cardamom Bushy Dwarf Virus (CBDV) (Genus: *Babuvirus*, Family: *Nanoviridae*) Symptoms of unusual excessive sprouting and bushy dwarf clump formation at the stem base of large cardamom (*Amomum subulatum*) were observed in Sikkim-Darjeeling area and the disease is known as 'foorkey' disease of large cardamom. Using the primers from Rep gene of viral genome an amplicon of 850 bp was amplified from symptomatic samples suggesting the association of a nanovirus with disease (Mandal et al. 2004a). Viral identity was confirmed by sequencing and sequence analysis. Later on using abutting primers and rolling circle amplification (RCA) based method nine novel DNA components were characterized (both based on complete sequences and sequence analysis) and the associated virus (cardamom bushy dwarf virus: CBDV) was proposed to be a new member of genus *Babuvirus* (Mandal et al. 2013).

Large Cardamom Chirke Virus: LCCV (Genus: *Macluravirus*, Family: *Potyviridae*) Another disease of large cardamom characterized by symptoms of light and dark green streaks on leaf lamina, known as 'chirke' is economically important disease in India endemic to North Sikkim and Eastern region of Darjeeling hills. Mandal et al. (2012b) designed degenerate primer pair from partial NIb region, CP and 3′-UTR amplifying 1.7 kb region which revealed the association of a virus belonging to genus *Macluravirus* with tentative name as large cardamom chirke virus (LCCV). This is the first report on molecular detection and characterization of a new virus with 'chirke' disease of large cardamom (Mandal et al. 2012b).

Cardamom Mosaic Virus: CdMV (Genus: *Macluravirus*, Family: *Potyviridae*) By employing primers from 3′-terminal and CP followed by sequencing, cardamom mosaic virus (CdMV) infecting small cardamom (*Elettaria* *cardamomum*) was detected from India (Jacob and Usha 2001; Jacob et al. 2003). RT-PCR using CP based primers was also used for detection and molecular characterization of CdMV in the cardamom samples collected from Karnataka, Kerala and Tamil Nadu (Siljo et al. 2013). RT-PCR employing primers from conserved CP region (amplifying 800 bp amplicon) have been widely adopted for the indexing of CdMV in India (Biju et al. 2010).

Cucumber Mosaic Virus: CMV (Genus: *Cucumovirus*, Family: *Bromoviridae*) Infection of CMV in black pepper plants showing stunting symptoms were detected (Bhat et al. 2005). CMV infection was detected using primers specific to CP region of viral genome. Based on the sequencing and sequence analysis CMV isolate was identified to be a member of subgroup I (Bhat et al. 2005). Natural infection of CMV in Indian long pepper (*Piper longum* L.) and betel vine (*Piper betle* L.) was detected using RT-PCR by employing primers specific to CP region of viral genome which yielded amplicon of 657 bp (Hareesh et al. 2006). Viral identity was confirmed by sequence analysis and CMV infecting Indian long pepper and betelvine were identified to be belonged to subgroup I.

Banana Bract Mosaic Virus: BBrMV (Genus: *Potyvirus*, Family: *Potyviridae*) Chlorotic streak exhibiting cardamom plants were tested positive for a potyvirus infection in RT-PCR using primers from conserved region of viral genome (Siljo et al. 2012). A potyvirus genome specific amplification of 700 bp was observed, and sequence analysis indicated association of banana bract mosaic virus (BBrMV) with disease. CP gene specific amplification and sequencing of symptomatic leaf samples of cardamom from Kerala, Karnataka and Tamil Nadu indicated widespread occurrence of BBrMV.

Piper Yellow Mottle Virus: PYMoV (Genus: *Badnavirus*, Family: *Caulimoviridae*) Hareesh and Bhat (2008) reported a protocol for DNA isolation from black pepper and PCR detection of PYMoV. RT/RNase H based degenerate primers were used for PCR detection of PYMoV infecting betel vine and Indian long pepper (Hareesh and Bhat 2008). A PCR detection assay was developed for indexing of black pepper planting material for PYMoV (Bhat et al. 2009). Primers for ORF3 genomic fragment were used. This PCR assay was used for indexing of 845 plants of black pepper (14 cultivars) and 82% of which were tested positive for PYMoV. At the time of indexing most of the plants were not exhibiting any external symptoms however, after 1–3 months some positive plants were found to exhibit symptoms of PYMoV, this indicated the suitability of developed PCR assay in routine indexing of black pepper plants even when they are symptomless carrier. A number of PCR based assays for PYMoV and RT-PCR based assays for sensitive detection of CMV has been developed in India (Bhat and Siju 2007; Siju et al. 2007; Bhat et al. 2009).

Multiplex Detection Methods
A multiplex-RT-PCR detection assay was developed for simultaneous detection of PYMoV (primers for partial ORF1) and CMV (primers for CP) in nursery and field samples of black pepper (Bhat and Siju 2007).

Real Time Detection Methods
A SYBR Green based one step reverse transcription-quantitative PCR (RT-qPCR) assay was developed for the sensitive detection of cardamom mosaic virus (CdMV) and banana bract mosaic virus (BBrMV) which infect cardamom (Siljo et al. 2014). Primers derived from CP region of both the viruses were designed and successfully employed in specific detection of CdMV and BBrMV in RT-qPCR. Developed protocol could detect the CdMV up to 16 copies of RNA (in total RNA dilution) and BBrMV upto ten copies of RNA. Developed RT-qPCR protocol was validated in successful detection of both viruses in cardamom samples collected from different regions of India.

Real-time PCR was developed for sensitive detection of PYMoV and CMV in black pepper plants (Bhat and Siljo 2014). In case of PYMoV, primers for real-time assay were designed based on the conserved ORF3 genomic fragment (600 bp) of different PYMoV isolates. In case of CMV, primers from conserved CP region (based on multiple sequence alignment of CMV subgroup 1B isolates) were used. This detection assay was 1000 times more sensitive than the conventional PCR. Real-time PCR could be a method of choice for indexing of mother plants (Bhat and Siljo 2014).

Isothermal Amplification Based Detection Methods
Recently more sensitive loop-mediated isothermal amplification (LAMP) assay was developed for the detection of PYMoV (Bhat et al. 2013). In developed LAMP assay, positive reaction could be detected by turbidity, green fluorescence, formation of pellet in the reaction tube and by gel electrophoresis. This assay was 100 times more sensitive compared to conventional PCR.

27.7.6 Tuber Crops

Nucleic Acid Hybridization, PCR and RT-PCR Based Detection Methods
Begomoviruses A PCR detection system using different forward primers which specifically anneal DNA-A of Indian cassava mosaic virus (ICMV) and Sri Lankan cassava mosaic virus (SLCMV) and common reverse primer which anneal both to ICMV and SLCMV was developed. An amplicon of 900 bp from ICMV and 599 bp from SLCMV was obtained (Patil et al. 2005). A PCR-restriction fragment length polymorphism (PCR-RFLP) detection system was also developed where primers specific to AV1, AC3, BC1 and BV1 were used in PCR amplification followed by restriction digestion using different endonucleases. Based on PCR-RPLP pattern obtained from different ICMV and SLCMV isolates different isolates were characterized (Patil et al. 2005). Duraisamy et al. (2013) designed primers specific to rep-

licase gene of begomoviruses (ICMV and SLCMV) and used for specific molecular detection of these viruses in cassava.

Dasheen Mosaic Virus: DsMV (Genus: *Potyvirus*, Family: *Potyviridae*) Whitish feathery symptoms on leaves of *Colocasia esculenta* were detected positive for potyvirus in RT-PCR using primers specific to core region of CP (amplifying a PCR product of 327 bp) (Babu et al. 2011). Based on sequencing of amplified product and sequencing, association of DsMV was detected which is first report on association of DsMV infection on colocasia in India (Babu et al. 2011).

Konjac Mosaic Virus: KoMV (Genus: *Potyvirus*, Family: *Potyviridae*) Association of KoMV was identified with mosaic symptom exhibiting elephant foot yam (*Amorphophallus paeoniifolius*) from Andhra Pradesh based on IC-RT-PCR detection using primers specific to CP region of viral genome producing an amplicon of 850 bp (Padmavathi et al. 2013). Identity of KoMV was confirmed by sequencing and sequence analysis.

Sweet potato (*Ipomoea batatas*) plants showing symptoms of chlorotic spots, vein clearing, puckering etc. were detected for possible infection of begomovirus and potyvirus in PCR and RT-PCR using group specific primers (Prasanth and Hegde 2008). Nested PCR using primers specific to coat protein gene and sequencing confirmed the association of sweet potato feathery mottle virus (SPFMV) and sweet potato leaf curl Georgia virus (SPLCGV) (Prasanth and Hegde 2008).

Next Generation Sequencing (NGS) Methods

Recently by whole transcriptome sequencing of mosaic infected elephant foot yam from Kerala using Illumina GAII analyzer and based on assembly of sequenced reads, full genome of dasheen mosaic virus (DsMV) was recovered (Kamala et al. 2015). Further based on the primers designed from assembled full genome, sequence of DsMV which shared an identity of 83% with Chinese isolate was identified.

27.7.7 Pulse Crops

Nucleic Acid Hybridization, PCR and RT-PCR Based Detection Methods

Begomoviruses Mungbean yellow mosaic virus (MYMV) is an economically important virus which infects different pulse crops like mungbean, pigeonpea, blackgram, Frenchbean, and soybean. Infection of MYMV was detected in black gram showing symptoms of yellow mosaic using PCR (Mandal et al. 1997). Rani et al. (1996) employed Southern blotting to demonstrate that *Acalypha*, *Croton* and *Vernonia* grown near the fields of *Vigna mungo* and *Vigna radiata* act as reservoir of begomoviruses. MYMV and mungbean yellow mosaic India virus (MYMIV) infection in different pulses was detected by Southern blotting and PCR (Balaji et al. 2004; Usharani et al. 2005).

Karthikeyan et al. (2004) detected the presence of one DNA-A and five DNA-B components of MYMV in the pooled DNA samples of symptomatic *Vigna mungo* plants collected from Tamil Nadu. By restriction digestion of replicative form of virus, association of DNA-A and DNA-B was detected. Southern blotting by using probes specific to DNA-A and DNA-B also confirmed their association with field symptoms on *Vigna mungo*. PCR by employing the primers from conserved region of movement protein gene and nuclear shuttle protein gene also confirmed the co-existence of multiple DNA-B with the yellow mosaic disease under field conditions. Based on the sequencing and sequence analysis of amplified product the DNA-A and DNA-B were identified as MYMV (Karthikeyan et al. 2004).

Girish and Usha (2005) by using overlapping primers from pulse infecting begomoviruses, detected the infection of MYMIV in soybean plants sowing symptoms of yellow mosaic from Central India (Madhya Pradesh) and MYMV in yellow mosaic exhibiting samples of soybean in South India. Multiple DNA-B were found associated with MYMV in South India.

Leaf samples of blackgram plants showing symptoms of yellow mosaic in Tamil Nadu were collected and tested for the infection of *Begomovirus* species (Haq et al. 2011). Total DNA isolated from infected plants was subjected to PCR using abutting primers specific to DNA-A and DNA-B. Amplified DNA-A and DNA-B of associated begomovirus was cloned and sequenced. Sequence comparison indicated that, DNA-A component was a variant of MYMV; however DNA-B of associated begomovirus had more similarity to MYMIV (Haq et al. 2011).

Infection of MYMIV was detected in a wild relative of black gram *Vigna mungo* var. *silvestris* which was showing the symptoms of yellow mosaic, using primers specific to the coat protein (AV1) gene of the begomovirus (Kamaal et al. 2011). Sequencing and sequence analysis confirmed the identity of MYMIV. Based on CP gene amplification in PCR, sequencing and sequence analysis, association of MYMIV was detected with yellow mosaic disease of blackgram in Andhra Pradesh state (Reddy et al. 2015).

Yellow mosaic symptomatic samples of black gram, green gram, cowpea and horse gram collected from different areas of Tamil Nadu were tested for the possible infection of different begomoviruses using CP gene specific primers in PCR (Maheshwari et al. 2014). Symptomatic leaf samples of black gram, green gram and cow pea were found positive for MYMV, whereas, the horse gram samples were detected positive for horsegram yellow mosaic virus (HYMV) (Maheshwari et al. 2014).

Two wild species of *Vigna* (*Vigna hainiana* and *V. trilobata*) exhibiting the symptoms of inter-veinal yellowing and bright yellow spots on leaves were tested in PCR by using primers specific to MYMIV and MYMV (Naimuddin and Pratap 2011). Amplicons specific to DNA-A and DNA-B of MYMIV were amplified in symptomatic samples which confirmed the association of MYMIV with these two wild species of *Vigna* (Naimuddin and Pratap 2011). Restriction pattern of RCA amplified products from these samples also confirmed the association of MYMIV. *Ageratum conyzoides* plants collected from different locations in Uttar Pradesh were tested positive for MYMIV in PCR which indicated that it acts as an alternate host for this virus (Naimuddin et al. 2014).

Cowpea golden mosaic symptoms exhibiting plants in Western India (Gujarat) were tested for the begomovirus infection. Double-stranded replicative form (RF) viral DNA was isolated from infected host plants, cloned and sequenced after restriction digestion (John et al. 2008). Sequence analysis showed that golden mosaic disease of cowpea in Western India is caused by an isolate of MYMIV which is having a distinct DNA-B component which was more similar to other legume infecting begomovirus (mungbean yellow mosaic virus) (John et al. 2008).

Cowpea Aphid-Borne Mosaic Virus: CABMV (Genus: *Potyvirus*, Family: *Potyviridae*) CABMV is another significant constraint in cowpea production in India. Typical symptoms of CABMV in cowpea include green vein banding, leaf deformation and interveinal chlorosis on leaf lamina. Mishra et al. (2015) detected infection of CABMV in field samples of cowpea collected from Rajasthan using genome specific primers in RT-PCR followed by sequencing and sequence analysis. By using the overlapping genome-wide primers, complete genome sequences of two different CABMV isolates was reported (Mishra et al. 2015).

Mastrevirus RCA amplification and sequencing of linear monomer viral genome from chickpea plants showing severe stunt disease (smalling of leaves, axillary shoot proliferation, stunting and overall yellowing) led to identify a mastrevirus (chickpea chlorotic dwarf Pakistan virus) based on 99% similarity at full genome level (Kanakala et al. 2013a). chickpea chlorotic dwarf virus (CpCDV) infection in field samples was also detected using labeled probe in dot-blot hybridisation (Kanakala et al. 2013b).

Pigeonpea Sterility Mosaic Virus: PPSMV (Genus: *Emaravirus*, Family: Unassigned) Pigeonpea (*Cajanus cajan*) is infected by a devastating sterility mosaic disease. Disease is caused by eriophyid mite, (*Aceria cajani*) transmitted PPSMV which comes under the genus *Emaravirus* (Patil and Kumar 2015). Most of the work on detection of causal virus of pigeonpea sterility mosaic disease was carried out at International Crops Research Institute for the Semi-Arid Tropics (ICRISAT), Hyderabad, India in collaboration with other International research institutes.

Initially RNA from purified virus preparations were used for cDNA synthesis using random primers. Based on sequences of random cDNA clone, a sequence fragment of 764 bp from PPSMV was obtained which hybridized with RNA from purified virus preparations (Kumar et al. 2003). Based on these sequences two primers (SM-1 and SM-2) were designed which amplified 321 bp region of RNA-3 of PPSMV genome in RT-PCR (Kumar et al. 2003). Elbeaino et al. (2013) designed degenerate primer pairs from the conserved motifs of RNA-dependent RNA polymerase gene (RdRp) region of emaravirus genomes, which successfully amplified 276 bp and 360 bp genomic fragments of PPSMV.

Next Generation Sequencing (NGS) Methods
Leaf crinkle infected samples of urdbean and mungbean plants from New Delhi and healthy leaf samples were subjected to small RNA (sRNA) based next generation sequencing (Baranwal et al. 2015). Assembled read contigs after subtraction of sequences of host origin were mapped to virus sequence database. Based on contig mapping, sequences of three viruses, cowpea mild mottle virus (CpMMV), groundnut bud necrosis virus (GBNV) and soybean yellow mottle mosaic virus (SoYMMV) were identified. Contigs specific to CPMMV were distributed to entire viral genome including untranslated region. RT-PCR using primers specific to CPPMV (from partial CP region) was able to detect the CPPMV infection in symptomatic field samples (yielding an amplicon of 425 bp). In addition 1.3 kb 3′-terminal part of viral genome (employing primers from CP, NB and 3′-UTR) (Yadav et al. 2013) was employed to detect mungbean and urdbean infecting CPPMV isolates which indicated high genetic variability among these two isolates. Recently Elbeaino et al. (2014) by deep sequencing of sterility mosaic sample obtained from India, identified full genome five RNA segments of PPSMV.

27.7.8 Oil Seed and Cole Crops

The major oilseed crops produced in India includes soybean, groundnut, rapeseed-mustard, sunflower, sesame, safflower and niger which are produced for edible oils and castor and linseed for non-edible oil crops. Several viral diseases become alarming threat in the recent time for the oilseed crops and causes significant losses. Major viruses infecting oilseed crops in India and their molecular detection are discussed below:

Nucleic Acid Hybridization, PCR, RT-PCR and RCA Based Detection Methods
Turnip Yellows Virus (TuYV) (Genus: *Polerovirus*, Family: *Luteoviridae*) Indian mustard (*Brassica juncea*) leaf samples exhibiting symptoms of yellowing and red stripes along with the presence of aphids indicated the possibility of presence of virus in Rajasthan. In RT-PCR using luteovirus group specific primers *Lu1* and *Lu4* (Robertson et al. 1991) an amplicon of ~600 bp was amplified (Sharma et al. 2013). Sequencing of amplified products confirmed the association of the turnip yellows virus which was the first report of molecular characterization and natural infection of TuYV in *B. juncea* from India (Sharma et al. 2013).

Cotton Leaf Curl Kokhran Virus: CLCuKoV (Genus: *Begomovirus*, Family: *Geminiviridae*) Yellow mosaic disease (YMD) of soybean was first reported in from India 1960 (Nariani 1960). Recently the disease incidence of 80–90% was reported on soybean crop from Lucknow (Raj et al. 2006). PCR using CP specific primers from begomovirus genome gave an amplicon of ~800 bp from symptomatic samples, sequences of which shared an identity of 95% with cotton leaf curl Kokhran virus which confirmed its association with disease (Raj et al. 2006).

Begomovirus Infection on Sunflower Sunflower plants exhibiting the symptoms of curling, leaf thickening, enations and severe stunting in Raichur, Karnataka were suspected to be infected with virus (Govindappa et al. 2011). The molecular detection of causal virus was performed by PCR using two sets of begomovirus specific degenerate primers (Deng et al. 1994; Wyatt and Brown 1996). Further the core CP region of 575 bp was amplified from naturally infected as well as white fly inoculated plants and sequenced. The nucleotide sequences of the CP region shares 97.5% similarity with the tomato leaf curl Karnataka virus isolates.

Tobacco Streak Virus: TSV (Genus: *Ilarvirus*; Family: *Bromoviridae*) TSV with its wide host range is an emerging virus in India. TSV epidemic on groundnut (*Arachis hypogaea*) crop was observed during the year 2000–2001 and on sunflower (*Helianthus annuus*) since 1997. Natural occurrence of TSV in India has been reported from sunflower, groundnut, cotton, sunhemp, mungbean, okra, cucumber, gherkin (*C. anguria*), safflower (*Carthamus tinctorious*), chilli, urdbean and soybean (*Glycine max*). First time molecular characterization of TSV on sunflower from Indian subcontinent was done by Bhat et al. (2002) using the RT-PCR employing primers from CP gene which amplified 700 bp genomic fragment in positive samples. In 2008, molecular characterization of TSV infecting soybean from Maharashtra was done by RT-PCR, using primers to amplify the complete CP gene coding region of viral genome (Kumar et al. 2008).

Tobacco Mosaic Virus: TMV (Genus: *Tobamovirus,* Family: *Virgaviridae*) TMV infection on soybean in India was reported recently for the first time from IARI, New Delhi experimental field (Kumar et al. 2013b). The virus was confirmed based on the amplification of different genomic regions viz., movement protein, coat protein and 3′ untranslated region by RT-PCR and sequencing results.

Cucumber Mosaic Virus: CMV (Genus: *Cucumovirus*, Family: *Bromoviridae*) CMV infection in castor was detected by RT-PCR using CP gene specific primer pair which resulted in amplification of 650 bp PCR product (Raj et al. 2010). This study provided molecular evidence for the natural occurrence of CMV on castor and its association with blister and leaf distortion disease. For further identification of a virus isolate at subgroup level, RFLP analysis was also conducted aiding different restriction enzymes (Raj et al. 2010).

Papaya Leaf Crumple Virus: PaLCrV (Genus: *Begomovirus,* Family: *Geminiviridae*) PaLCrV cause natural infection in soybean crop in Uttar Pradesh where soybean found to exhibit the symptoms of yellow mosaic, leaf crumpling and leaf distortion. Total DNA was isolated from the symptomatic leaf samples and molecular characterization was done by PCR using specific begomovirus degenerate primers (Rojas et al. 1993). The expected size of amplicon of ~1.2 kb from samples of leaf crumple diseased plants indicated the presence of begomovirus in the soybean plants identity of which was confirmed by sequencing (Jaidi et al. 2015).

Mungbean Yellow Mosaic India Virus: MYMIV and Mungbean Yellow Mosaic Virus: MYMV (Genus: *Begomovirus,* Family: *Geminiviridae*) The virus associated with yellow mosaic disease of soybean in North India (New Delhi) was identified by Usharani et al. (2004a). Replicative form (RF) of viral DNA was separated on gel and cloned in to pUC vector after restriction digestion. Sequence identity of greater than 89% for DNA-A and DNA-B with the MYMIV indicated that yellow mosaic disease of soybean in North India was caused by MYMIV. Based on the sequence analysis with already reported sequences, it was proposed that yellow mosaic disease of soybean in North and Central India is caused by MYMIV, whereas in Central and South India by MYMV (Varma and Malathi 2003; Usharani et al. 2004a). Based on PCR amplification and sequence analysis of full length bipartite begomovirus genome, infection of Mungbean Yellow Mosaic India Virus was detected in soybean (Yadav et al. 2009).

Cowpea Mild Mottle Virus: CPMMV (Genus: *Carlavirus,* Family: *Betaflexiviridae*) Yadav et al. (2013) detected infection of a distinct strain of cowpea mild mottle virus (CPMMV) in soybean plants showing symptoms of mottling, mosaic and leaf deformation in Delhi. CPMMV was detected by RT-PCR employing the primers Carla-CP and oligo d (T) amplifying 940 bp genomic fragment. Further primers targeting full CP and NABP and 3′-UTR (amplifying a fragment of 1289 bp) were used in RT-PCR. Sequencing of amplified genomic fragments confirmed the association of a strain of CPMMV with soybean in India (Yadav et al. 2013).

Croton Yellow Vein Mosaic Virus: CYVMV (Genus: *Begomovirus,* Family: *Geminiviridae*) CYMV was identified as the causal agent of leaf curl disease in the different germplasms of the rapeseed-mustard in India (Roy et al. 2013). The association of the virus with the disease was confirmed by molecular techniques. To detect the presence of associated virus, PCR was carried out using universal DNA-A (Chowda Reddy et al. 2005) and universal DNA-B specific primers (Rojas et al. 1993). Further complete genome of the begomovirus was amplified following the standard rolling circle amplification (RCA) and sequencing of full genome confirmed the association of *Croton yellow vein mosaic virus* with the leaf curl disease of rapeseed-mustard.

Turnip Mosaic Virus: TuMV (Genus: *Potyvirus*, Family: *Potyviridae*) Cole crops (*Brassica* spp) plants of broad leaved mustard (*B. juncea* var. *rugosa*), Indian mustard (*Brassica juncea*) and broccoli (*B. oleracea* var. *italica*) showing symptoms of mottling, mosaic, chlorosis and puckering were detected positive for turnip mosaic virus (TuMV) infection in RT-PCR (Singh et al. 2015). Partial sequences of cytoplasmic inclusion protein and CP were amplified in RT-PCR and analyzed which confirmed the identity of virus as TuMV.

Soybean Mosaic Virus: SMV (Genus: *Potyvirus*, Family: *Potyviridae*) The molecular evidence and detection of SMV infecting soybean crop was recently provided (Banerjee et al. 2014b). RT-PCR assay was carried out with two sets of potyvirus specific degenerate primers viz., CIFor/CIRev which amplified ~700 bp region of cylindrical inclusion protein (CI) domain (Ha et al. 2008) and POT1/POT2 which amplified a ~1300 bp region encompassing partial nuclear inclusion protein and coat protein (NIb-CP) domain (Colinet et al. 1998).

Groundnut Bud Necrosis Virus: GBNV (Genus: *Tospovirus*, Family: *Bunyaviridae*) GBNV infection has been reported on groundnut and tomato from India. A pair of gene specific primers derived from coding region of N gene of GBNV were employed in RT-PCR amplification. The expected amplication and sequencing results evidenced the molecular detection of the virus (Umamaheswaran et al. 2003). Further the virus was characterized from bud blight symptomatic tomato plants, using the same sets of the primers (Raja and Jain 2006).

Peanut Mottle Virus: PeMoV (Genus: *Potyvirus*, Family: *Potyviridae*) Peanut mottle disease was recorded on imported peanut germplasms as well as its natural occurrence on soybean and peanut in Punjab (Reddy et al. 1978; Prasada Rao et al. 1979). Molecular characterization of the virus was done by RT-PCR using CP gene specific primers (Soumya et al. 2014).

Peanut Clump Virus: PCV (Genus: *Pecluvirus*, Family: *Virgaviridae*) A RT-PCR based detection protocol for efficient detection of PCV was developed by employing the two sets of primers, corresponding to RNA 2 regions. By sequencing the amplified genomic fragment, PCV was characterized (Naidu et al. 2000).

27.7.9 Cereal Crops

Nucleic Acid Hybridization, PCR and RT-PCR Based Detection Methods

Rice Tungro Disease Tungro disease of rice is caused by two virus species; *Rice tungro bacilliform virus* (RTBV), a double stranded DNA virus (Genus: *Tungrovirus*, Family: *Caulimoviriade*) and *Rice tungro spherical virus* (RTSV), a single stranded RNA virus (Genus: *Waikavirus*, Family: *Secoviridae*). Molecular characterization of the RTBV strain was initiated by Dasgupta et al. (1996), using the primers designed from DNA sequences of Indian RTBV isolate which amplified 1.1 kb amplicon of the RTBV isolates. Southern blot analysis and sequencing of restricted fragments was also employed for RTBV detection (Joshi and Dasgupta 2001). Further complete genomic sequences of two geographically distinct isolates of rice RTBV from India were determined (Nath et al. 2002). In an another study, the technique of polymerase chain reaction/restriction fragment length polymorphism (PCR/RFLP) was used to study the presence or absence of the 30 bp deletion in the RTBV genome of tungro-infected leaf samples of rice, representing six geographically diverse isolates from India, as a genetic marker for variability (Joshi et al. 2003). Using overlapping primers in PCR, complete genome sequence of a RTBV isolate from West Bengal was deciphered (Banerjee et al. 2011). CP based primers were used in RT-PCR based detection of RTSV in rice samples collected from Andhra Pradesh, Orissa, Puducherry, West Bengal and Tamil Nadu (Mangrauthia et al. 2012).

Wheat Dwarf India Virus: WDIV (Genus: *Mastrevirus*, Family: *Geminiviridae*) It is a leafhopper (*Psammotettix* sp) transmitted virus that infects wheat in India. Dwarfing or stunting is the typical symptom but many times yellowing of leaf is also associated with the field infection. Its infection was reported first time from India by Kumar et al. (2012e). The molecular detection and characterization of the virus was performed by PCR based on an amplification of ~2.8 kb fragment using WDIV specific primers thus indicating the presence of the virus.

Multiplex Detection Methods
A multiplex RT-PCR based detection protocol for simultaneous detection of RTSV and RTBV was standardized (Periasamy et al. 2006). For cDNA synthesis from RNA either oligo (dT) primer or antisense primers from CP region were used. In case of RTBV, forward primer from 5′-ORF p24 and reverse primer from 3′-ORF p12 were designed. Multiplex PCR was then performed employing cDNA of RTSV and total plant DNA in case of RTBV, where anticipated PCR products of 920 bp (RTBV) and 1032 (RTSV) were observed (Periasamy et al. 2006).

27.7.10 Commercial Crops

27.7.10.1 Sugarcane

PCR and RT-PCR Based Detection Methods
Badnaviruses A groups of badnaviruses known as sugarcane bacilliform viruses (SCBV) are associated with freckling disease of sugarcane and are known to infect a number of species of sugarcane (*Saccharum officinarum*, *S. barberi*, *S. sinense*, *S. robustum* and *Saccharum* interspecific hybrids). PCR based detection was then standardized for screening of sugarcane varieties and germplasm (Singh et al. 2009b; Rao et al. 2014). Generally primers to RT/RNase H region of genome or core RT/RNase H region were used in PCR based detection (Singh et al. 2009b; Karuppaiah et al. 2013; Rao et al. 2014). Since SCBV is not known to occur as integrated form in the host genome, hence PCR is considered as standard indexing procedure. Recently Sharma et al. (2015) reported that RCA concentrated DNA can be used as a template in PCR for sensitive detection SCBV in field samples.

Sugarcane Yellow Leaf Virus: SCYLV (Genus: *Polerovirus*, Family: *Luteoviridae*) SCYLV is another economically important virus infecting sugarcane in India. RT-PCR using different primer pairs from partial ORF1, 2 and complete ORF 3 and 5 of SCYLV genome were used for indexing of 48 sugarcane leaf samples, out of which 36 field samples (both symptomatic and asymptomatic) including 10 tissue cultured samples (derived from meristem) were detected positive (Viswanathan et al. 2008a). The identity of amplified fragments was confirmed by sequencing and sequence comparison. Based on sequence analysis, three SCYLV genotypes (CUB, IND and BRA-PER) were identified to be associated with the yellow leaf disease of sugarcane in India (Viswanathan et al. 2008a).

Viswanathan et al. (2009) standardized a RT-PCR technique for detection of SCYLV in asymptomatic samples of sugarcane based on primer pair which amplify ~613 bp gene fragment (partial ORF3 and 4) of SCYLV genome. This robust RT-PCR based detection protocol could detect the SCYLV infection in sugarcane varieties in initial stages itself when they were asymptomatic. In RT-PCR, 34 varieties out of the 44 were tested positive for SCYLV during pre-symptomatic stage, out of which 33 produced typical symptoms after 2 months. However, after post-symptomatic stage all the varieties (including those which did not exhibit typical symptoms) were tested positive for SCYLV in the standardized RT-PCR detection system (Viswanathan et al. 2009). Same RT-PCR was later used for detection of SCYLV infection in other sugarcane cultivars (Chinnaraja et al. 2013). Another set of specific primers from 5′-UTR, ORF0 and partial ORF1 region of SCYLV genome was also standardized for RT-PCR based detection which yielded a band of ~1025 nt (Chinnaraja et al. 2013). Using different sets of primers full genome sequences of Indian isolate of SCYLV was also obtained.

Sugarcane Streak Mosaic Virus: SCSMV (Genus: *Poacevirus*, Family: *Potyviridae*) A RT-PCR assay for detection of SCSMV based on the primers designed from CP region of viral genome was developed (Viswanathan et al. 2008b). Using this primer pair an amplicon of 690 bp was amplified in RT-PCR (N-terminal region of CP gene). RT-PCR based indexing of 63 sugarcane samples collected from major sugarcane growing belts of India, South Africa, Australia and USA revealed that 92% of the tested samples were positive for SCSMV. For 42 SCSMV isolates partial CP sequences were obtained and were analyzed (Viswanathan et al. 2008b). SCSMV was also detected by RT-PCR amplification and sequencing of two overlapping fragments of 3.1 kb (partial CI, complete 6 K2, VPg-NIa and NIb gene) (Hema et al. 2003).

Molecular detection of SCSMV isolates was also done using primers specific to HC-Pro region of viral genome followed by sequencing and sequence comparison (Bagyalakshmi et al. 2012) and later for full genome characterization based on overlapping primers targeting complete genome (Parameswari et al. 2013). Primers for P1 region of viral genome were recently used for detection and characterization of SCSMV isolates from India (Parameswari et al. 2014).

Sugarcane Mosaic Virus: SCMV (Genus: *Potyvirus*, Family: *Potyviridae*) SCMV was initially detected using immunocapture-RT-PCR (IC-RT-PCR) using primers specific to CP gene of viral genome (Gaur et al. 2003). Specific primers for detection of sugarcane mosaic virus (SCMV), sugarcane streak mosaic virus (SCSMV) and sorghum mosaic virus (SrMV) were used in RT-PCR based detection (Rao et al. 2006). In comparison to SCMV, SCSMV was found to be most commonly associated with the sugarcane mosaic disease in India. Some sugarcane samples collected from Maharashtra and Tamil Nadu were detected positive for both SCMV and SCSMV (Rao et al. 2006). Later on by employing RT-PCR using specific primer pairs, prevalence of SCYLV, SCMV and SCSMV was detected in different sugarcane cultivars grown in 11 states of India (Viswanathan and Karuppaiah 2010).

Multiplex Detection Methods
A sensitive and robust duplex-immunocapture-PCR (D-IC-PCR) for simultaneous detection and discrimination of SCSMV and SCMV employing polyclonal antibodies prepared against purified viruses was developed (Subba Reddy CV et al. 2011). A common reverse primer (5′-d$(T)_{18}$(AGC)-3′ for both viruses and virus genome specific forward primer were employed in RT-PCR which gave 1.4 kb SCSMV specific amplicon and 0.9 kb SCMV specific amplicon (from 3′-termini of viral genomes). Developed D-IC-PCR was validated using large number of field samples where 76% were positive for SCSMV, 37% for SCMV and 16% for both viruses (Subba Reddy CV et al. 2011). Different detection methods were used for indexing of sugarcane planting materials (Viswanathan and Rao 2011).

27.7.10.2 Potato

Potato is an important crop of India and due to its agamic propagation, disease caused by viruses become a major challenge. Potato is known to be infected by a large number of viruses e.g. potato virus A (PVA), potato virus X (PVX), potato virus Y (PVY), potato virus S (PVS), potato virus M (PVM), potato leaf roll virus (PLRV), tomato leaf curl New Delhi virus (ToLCNDV), potato aucuba mosaic virus (PAMV), groundnut bud necrosis virus (GBNV), potato spindle tuber viroid (PSTVd) etc.

Nucleic Acid Hybridization, PCR and RT-PCR Based Detection Methods
Potato Virus Y: PVY (Genus: *Potyvirus*, Family: *Potyviridae*) PVY was detected using primers specific to CP gene of viral genome and sequencing (Ghosh SB. et al.2002b). A RT-PCR based system for efficient detection of PVY (using primers specific to CP gene of PVY genome) in potato and tobacco plants was developed by Ghosh and Bapat (2006). Degenerate primers designed from core region of NIb and CP amplifying a fragment of 323 bp and 353 bp respectively was used for RT-PCR based detection of PVY (Sharma et al. 2013). For efficient detection of PVY in infected samples, immunocapture-PCR (IC-RT-PCR), direct binding RT-PCR (DB-RT-PCR) and print capture RT-PCR (PC-RT-PCR) were developed (Gawande et al. 2011). Primers specific to 5′UTR and P1 region (amplicon size of 969 bp) were used for nucleic acid based detection of PVY. Nucleic acid spot hybridization (NASH) using fluorescein labeled cDNA probes was also used for detection. In IC-RT-PCR, anti-PVY antibodies were used to capture virus particles which were then used for cDNA synthesis after release of viral RNA followed by PCR. In DB-RT-PCR, crude sap was incubated in sterile PCR tubes for 2 h at 37 °C followed by washing and RT-PCR amplification. In PC-RT-PCR, potato leaf samples were used to make tissue prints on nylon membranes and used as template in RT-PCR after putting in sterile water and release of viral RNA. Detection limit of NASH and

DB-RT-PCR was 1 ng, whereas for RT-PCR, IC-RT-PCR and PC-RT-PCR it was found to be 1 pg for PVY virions (Gawande et al. 2011).

Potato Leaf Roll Virus: PLRV (Genus: *Polerovirus*, Family: *Luteoviridae*) Mukherjee et al. (2003) first time used RT-PCR based detection for PLRV isolate from India. Complementary DNA (cDNA) was prepared using oligo (dT)-anchor primer and then CP sequences were amplified by using primers specific to CP region followed by cloning and sequencing. Using two pairs of primer targeting full genome of PLRV with amplicons of 2.8 kb and 3.1 kb respectively, PLRV infection was detected in field samples of potato collected from Punjab, Himachal Pradesh, Bihar, Tamil Nadu and Meghalaya (Jeevalatha et al. 2013a). Using these primers, full genome sequences of the five isolates was reported which indicated the existence of low genetic diversity in Indian PLRV isolates. A RT-PCR based system for detection of PLRV in aphids collected from diseased plants was developed where primers specific to CP gene of PLRV were employed (Raigond et al. 2014). Initially RT-PCR was standardized using primers from mitochondrial cytochrome oxidase I (COI) region of aphids and applicability of standardized protocol was then replicated for detection of PLRV in viruliferous aphids (Raigond et al. 2014).

Potato Virus X: PVX (Genus: *Potexvirus*, Family: *Alphaflexiviridae*) By using RT-PCR amplification employing two primer pairs covering full viral genome, infection of potato virus X (PVX) was detected in symptomatic potato plants which was mechanically transmitted to *Nicotiana tabacum* cv. *Xanthi* and used for amplification and characterization of full genome (Mandal et al. 2012a). Primer pairs based on CP and 3′-UTR of PVX genome were used for RT-PCR detection of virus in asymptomatic and symptomatic potato samples and developed RT-PCR based detection system was validated using field samples (Mandal et al. 2012a). PVX and PLRV have been detected using RT-PCR (Awasthi et al. 2014).

Tomato Leaf Curl New Delhi Virus: ToLCNDV (Genus: *Begomovirus*, Family: *Geminiviridae*) Association of ToLCNDV with the apical leaf curl disease of potato in Northern India was identified by molecular techniques (Usharani et al. 2004b).

Potato Aucuba Mosaic Virus: PAMV (Genus: *Potexvirus*, Family: *Alphaflexiviridae*) For RT-PCR based detection of potato aucuba mosaic virus (PAMV), six set of primers from conserved region of 26 kDa protein, coat protein and RNA polymerase gene were designed (Kumar et al. 2014b). Out of which five primer pairs gave specific amplification with an amplicon size ranging from 191 to 360 bp. Primer pair PAMV1F/1R could lead to high sensitive detection and able to detect upto 10^7 dilutions (0.12 pg of total RNA), whereas other primer pairs successfully detected virus up to a dilution of 10^5 in RT-PCR. A sensitive one step immunocapture RT-PCR was also standardized for robust indexing of PAMV with expected amplicon of 360 bp in positive samples (Kumar et al. 2014b). RT-PCR for detection of PVS^O and PVS^A using primers specific to CP gene yielding an amplicon of 892 bp in positive potato samples was also standardized (Kaushal et al. 2007).

In addition nucleic acid based assay for the detection of different potato viruses like PVX, PVY, PVM, PLRV, ToLCNDV, groundnut bud necrosis virus (GBNV), potato spindle tuber viroid (PSTVd) etc. has been standardized (http://cpri.ernet.in/; http://www.iari.res.in, Anonymous 2013).

Multiplex Detection Methods
For efficient detection of causal virus of potato apical leaf curl disease (tomato leaf curl New Delhi virus-Potato), a duplex PCR detection assay was developed (Jeevalatha et al. 2013b). In the developed PCR assay, primers targeting coat protein (AV1) and replicase (AC1) gene of viral genome were used. Out of the eight sets of primers used in the study, one set from CP region and all sets from replicase region of genome were able to detect the viral DNA from infected samples specifically with amplicon size of 306 bp, 382 bp, 491 bp, 505 bp and 766 bp. Detection limit of simplex PCR assay was 2.4–0.24 pg of total DNA. In addition, a duplex-PCR assay was developed employing primers from CP region of ToLCNDV and internal control (potato urease gene). Developed duplex–PCR assay was validated using large number of field samples as well as micro-plants raised through tissue culture and found suitable for routine indexing of potato samples as well as indexing of mother plants (Jeevalatha et al. 2013b).

27.7.10.3 Cotton

Nucleic Acid Hybridization, PCR and RT-PCR Based Detection Methods
Begomoviruses CP based primers were used in PCR for the detection of begomovirus associated with leaf curl disease of cotton in India (Kang et al. 2004; Monga et al. 2005; Chakrabarty et al. 2005). Using abutting primers designed from the AC1 ORF (replication associated protein ORF) and another sets of primers to amplify overlapping fragments, full length sequences of DNA-A of the begomoviruses associated with leaf curl disease of cotton collected from Haryana and Rajasthan were obtained. Based on compete and partial genomic sequences, associated viruses were found to be *Cotton leaf curl Kokhran virus* (CLCuKV), *Tomato leaf curl Bangalore virus* (ToLCBV) and *Cotton leaf curl Rajasthan virus* (CLCuRaV) (Kirthi et al. 2004).

Kumar et al. (2010b) used CP specific primers from begomoviruses associated with cotton leaf curl for their detection in diseased samples originating from Sriganganagar, Rajasthan. They also employed RCA coupled with restriction digestion and sequencing for the molecular identification of viruses associated with leaf curl disease of cotton in Rajasthan. Leaf samples of cotton showing curling and leaf deformation in Lucknow, Uttar Pradesh were tested for infection of begomovirus using primers specific to DNA-A (Kumar et al. 2011b). The full genome sequence was amplified using two pairs of primers amplifying amplicons of 1.2 and 1.7 kb. Based on assembled sequences overlapping sequences (2.7 kb) of amplified products and sequence comparison, the associated virus was identified to be cotton leaf curl Burewala virus (CLCuBuV) (Kumar et al. 2011b, 2013a). Alphasatellite and betasatellite was also identified using specific primers in PCR.

Leaf curl exhibiting symptomatic and asymptomatic samples of cotton collected from Punjab were tested for the infection of begomovirus by rolling circle amplification (RCA) followed by restriction digestion (Kumar et al. 2015a). The restricted RCA products were cloned and sequenced, and the associated viruses were identified as CLCuBuV, CLCuKoV and cotton leaf curl Multan virus (CLCuMV) (Kumar et al. 2015a). In addition, by using the RCA and sequencing the association of satellites (cotton leaf curl Multan betasatellite: CLCuMB and cotton leaf curl Multan alphasatellite: CLCuMA) was detected. cotton leaf curl Burewala virus (CLCuBuV), cotton leaf curl Alabad virus (CLCuAV), cotton leaf curl Rajasthan virus (CLCuRV), cotton leaf curl Multan virus (CLCuMuV) and cotton leaf curl Kokhran virus (CLCuKV) were detected using nucleic acid based techniques and found to be associated with leaf curl disease in India (Ahuja et al. 2007; Rajagopalan et al. 2012).

A large number of cotton leaf samples collected from North Western India (Punjab, Haryana and Rajasthan) and tested for infection of begomoviruses and associated betasatellites by using RCA enriched DNA in PCR employing DNA-A or satellite specific primers (Rajagopalan et al. 2012). Three hundred samples collected from different parts of North Western India were found positive for cotton leaf curl causing begomoviruses based on positive PCR amplification (904 bp) from 5′-region of CI, full intergenic region and V2 gene. Based on the partial sequences of 258 isolates and full length sequences of 22 isolates, the dominant begomovirus species associated with cotton leaf curl disease were identified as cotton leaf curl Burewala virus (CLCuBuV) and to a lesser extent cotton leaf curl Rajasthan virus (CLCuRV) (Rajagopalan et al. 2012).

27.7.10.4 Jute

Nucleic Acid Hybridization, PCR and RT-PCR Based Detection Methods Corchorus Golden Mosaic Virus: CoGMV (Genus: *Begomovirus*, Family *Geminiviridae*) Jute (*Corchorus capsularis* and *C. olitorius*) is one of the most important commercial fibre crops in India. Among the different biotic constraints, the mosaic disease caused by CoGMV causes severe damage every year. Due to non-availability of the specific diagnostic method, the etiology of the mosaic disease was remained unidentified till 2012. In 2012, Ghosh and associates developed a rapid and sensitive diagnostic method using a non-radiolabelled diagnostic probe against the DNA-A component of the East Indian isolate of CoGMV. Using developed Southern hybridization and nucleic acid spot hybridization method infection of CoGMV was detected in infected plants and viruliferous whiteflies. Presence of the virus was also confirmed when PCR amplification was performed using virus-specific primers designed from the nucleotide co-ordinate 230–270 of DNA-A genome (Ghosh et al. 2012).

Potato Leaf Roll Virus: PLRV (Genus: *Polerovirus*, Family *Luteoviridae*) Jute plants exhibiting symptoms of stunted growth, elongation of upper leaves with curling and coiling of leaf lamina along with puckering and shoestring were tested in RT-PCR using potato leafroll virus (PLRV) coat protein specific primers (Biswas et al. 2014a). An amplicon of 627 bp was observed, sequences of which were similar to PLRV, which was the first report of infection of PLRV on jute from India.

27.7.10.5 Rubber

Nucleic Acid Hybridization, RT-PCR and Real Time Based Detection Methods

Potato Spindle Tuber Viroid (PSTVd) Tapping panel dryness (TPD) of rubber is a serious disorder occurring in all rubber growing countries resulting in severe loss of latex yield. In India, TPD is known to affect rubber plantations since the beginning of commercial cultivation. Association of a viroid RNA (potato spindle tuber viroid: PSTVd) with the TPD syndrome of rubber was first reported in 2000 on the basis of return polyacrylamide gel electrophoresis (R-PAGE), and RT-PCR using primers specific to PSTVd (Ramachandran et al. 2000). Later on RT-PCR technique using PSTVd specific primer pairs was standardized for the detection PSTVd in infected rubber plants (Kumar et al. 2015b). Nucleic acid spot hybridization technique for the specific and quick detection of PSTVd associated with the TPD of rubber in India was also reported (Kumar et al. 2015b). A SYBR Green quantitative RT-PCR (qRT-PCR) detection assay was also standardized for sensitive detection of PSTVd associated with rubber (Kumar et al. 2015b).

27.7.10.6 Tobacco

Tobacco is the big industry in Indian agriculture and India ranks third in terms of production of tobacco and in exports, after Brazil and the US. The science of the plant virology started with mosaic disease of tobacco (tobacco mosaic virus: TMV) in 1886 and TMV is still a model virus for the virological sciences.

PCR, RT-PCR and RCA Based Detection Methods

Tobacco Mosaic Virus: TMV (Genus: *Tobamovirus*, Family: *Virgaviridae*) TMV is one of the most studied and important virus due to its wide host range especially on the solanaceous crops and also infecting soybean in India (Kumar et al. 2013b). In India TMV was first reported on tomato plant (Rao and Reddy 1972). The virus was characterized using a specific primer corresponding to the 3′ end of TMV RNA genome in RT-PCR. Sequencing of the amplified and cloned PCR products confirmed TMV infection (Cherian et al. 1999).

Tobacco Leaf Curl Pusa Virus: TbLCPuV (Genus: *Begomovirus*, Family: *Geminiviridae*) It causes the most destructive disease of tobacco (*Nicotiana tabacum*) and significant yield losses up to 77% in India (Pal and Tandon 1937; Valand

and Muniyappa 1992). Recently, Singh et al. (2011b), detected the association of tobacco leaf curl Pusa virus (monopartite begomovirus) with the leaf curl disease of tobacco in India. PCR by employing the primers specific to alphasatellite were successfully used in detection. Full length DNA-A of the virus was amplified by following the rolling circle amplification (RCA) protocol, for virus characterization (Singh et al. 2011b).

Papaya Leaf Curl Virus: PaLCV (Genus: *Begomovirus*, Family: *Geminiviridae*) Papaya leaf curl disease reported from India in 1939 (Thomas and Krishnaswamy 1939), is caused by a begomovirus (PaLCV) (Saxena et al. 1998). Since 2001, prevalence of this disease in all tobacco growing areas of India has been reported. The molecular detection and characterization of PaLCV was done by Kumar et al. (2009) using begomo- specific primer pair amplifying CP gene. The resulting 850 bp amplicon was subjected to Southern hybridization with DNA probes prepared from the CP of begomovirus associated with cotton leaf curl disease for further confirmation. This was the first report of molecular characterization of papaya leaf curl virus in tobacco (Kumar et al. 2009).

Radish Leaf Curl Virus: RaLCV (Genus: *Begomovirus*, Family: *Geminiviridae*) Association of RaLCV with tobacco yellow leaf curl disease was detected by the amplification of the betasatellite region of genome by PCR following the rolling circle amplification (RCA) of viral genome (Singh et al. 2011a). The sequencing results showed highest nucleotide sequence identity (90%) with radish leaf curl virus evidenced the association of the virus with the yellow leaf curl disease (Singh et al. 2011a).

Multiplex Detection Methods

RT-PCR based detection of tobacco mosaic virus and tomato mosaic virus was done by multiplex-PCR in pepper and tomato. Two sets of the primers were used, one set specific to TMV which amplified the corresponding 880 bp region and another ToMV specific primer pair which generated the 318 bp amplicon in PCR (Kumar et al. 2011a).

27.8 Concluding Remarks

An array of nucleic acid based detection assays have been developed and used for definite detection of different viral agents infecting different agri-horticultural crops in India. Using these PCR, RT-PCR based detection assays a number of viruses have been detected in India and associated viruses have been diagnosed to their exact taxonomic identity. Isothermal detection assays are very recent in the development and use of detection assays for plant viruses in India. Development of on-site detection assays for diagnosis of associated viruses at the point of decision making are required to be taken up in India. Strip based lateral flow assays are likely to become most common for the targeted diagnosis of plant viruses in the planting materials or in the field which need to be developed for most commonly occurring plant viruses. Multiplex diagnostic tools are also gaining importance as routine test in advanced laboratories. The NGS techniques, which are still confined to the centralized sophisticated laboratories, are expected to become cheaper and common in the near future for unknown viruses. It is sure that developments in the nucleic acid based diagnostic tools will lead to correct diagnosis of plant viruses and their detection based management.

Acknowledgements Authors are grateful to various researchers in India for sharing the information. We are grateful to anonymous reviewers for critical screening of manuscript.

References

Adkar-Purushothama CR, Kanchepalli PR, Yanjarappa SM, Zhang Z, Sano T (2014) Detection, distribution, and genetic diversity of Australian grapevine viroid in grapevines in India. Virus Genes 49:304–311

Adsule PG, Upadhyay A, Sawant IS (2011) Vision 2030. National Research Centre for Grapes (Indian Council for Agricultural Research), Pune. http://nrcgrapes.nic.in/zipfiles/NRCG-Vision-2030.pdf

Ahlawat YS (1989) Psorosis- a disease of citrus in India. Indian Phytopathol 42:21–25

Ahlawat YS (1997) Viruses, greening bacterium and viroids associated with citrus (*Citrus* species) decline in India. Indian J Agric Sci 67:51–57

Ahuja SL, Monga D, Dhayal LS (2007) Genetics of resistance to cotton leaf curl disease in *Gossypium hirsutum* L. under field conditions. J Hered 98:79–83

Alshami AAA, Ahlawat YS, Pant RP (2003) A hitherto unreported yellow vein clearing disease of citrus in India and its viral etiology. Indian Phytopathol 56:422–427

Anderson PK, Cunningham AA, Patel NG, Morales FJ, Epstein PR, Daszak P (2004) Emerging infectious diseases of plants: pathogen pollution, climate change and agrotechnology drivers. Trends Ecol Evol 10:535–544

Anindya R, Joseph J, Gowri TDS, Savithri HS (2004) Complete genomicsequence of Pepper vein banding virus (PVBV): a distinct memberof the genus Potyvirus. Arch Virol 149:625–632

Anonymous (2013) Central potato research institute news letter, Shilma-171001, Himachal Pradesh, India; 54

Anthony Johnson AM, Dasgupta I, Sai Gopal DVR (2014) Development of loop-mediated isothermal amplification and SYBR green real-time PCR methods for the detection of Citrus yellow mosaic badnavirus in citrus species. J Virol Methods 203:9–14

Arya M, Baranwal VK, Ahlawat YS, Singh L (2006) RT-PCR detection and molecular characterization of *Onion yellow dwarf virus* associated with garlic and onion. Curr Sci 91:1230–1234

Awasthi LP, Singh SP, Chaubey AN, Abhishek, Kumar R (2014) Molecular characterization of potato viruses through RT-PCR and electron-microscopy. Caribb J Sci Technol 2:405–410

Babu B, Hegde V, Makeshkumar T, Jeeva ML (2011) Detection and identification of Dasheen mosaic virus infecting *Colocasia esculenta* in India. Indian J Virol 22:59–62

Bagyalakshmi K, Parameswari B, Chinnaraja C, Karuppaiah R, Ganesh Kumar V, Viswanathan R (2012) Genetic variability and potential recombination events in the HC-Pro gene of sugarcane streak mosaic virus. Arch Virol 157:1371–1375

Bajet NB, Unruh TR, Druffel KL, Eastwell KC (2008) Occurrence of two little cherry viruses in sweet cherry in Washington State. Plant Dis 92:234–238

Balaji V, Vanitharani R, Karthikeyan AS, Anbalagan S (2004) Veluthambi K Infectivity analysis of two variable DNA B components of *Mungbean yellow mosaic virus*-Vigna in *Vigna mungo* and *Vigna radiata*. J Biosci 29:297–308

Balasubramanian V, Selvarajan R (2012) Complete genome sequence of a Banana bract mosaic virus isolate infecting the French plantain cv. Nendran in India. Arch Virol 157:397–400

Banerjee A, Roy S, Tarafdar J (2011) Phylogenetic analysis of Rice tungro bacilliform virus ORFs revealed strong correlation between evolution and geographical distribution. Virus Genes 43:398–408

Banerjee A, Dutta R, Roy S, Ngachan SV (2014a) First report of Chilli veinal mottle virus in Naga chilli (Capsicum chinense) in Meghalaya, India. Virusdisease 25:142–143

Banerjee A, Chandra S, Swer EKP, Sharma SK, Ngachan SV (2014b) First molecular evidence of *Soybean mosaic virus* (SMV) infection in soybean from India. Australas Plant Dis Notes 9:150

Banerjee A, Roy S, Behere GT, Roy SS, Dutta SK, Ngachan SV (2014c) Identification and characterization of a distinct *Banana bunchy top virus* isolate of Pacific-Indian Oceans group from North-East India. Virus Res 183:41–49

Banerjee A, Roy S, Sharma SK, Dutta SK, Chandra S, Ngachan SV (2016) Reverse transcription loop-mediated isothermal amplification (RT-LAMP) assay for rapid diagnosis of chilli veinal mottle virus. Arch Virol. doi:10.1007/s00705-016-2850-7

Baranwal VK, Sharma SK (2016) Next generation high throughput diagnostics for management of virus and virus-like pathogens in horticultural crops: status and way forward. In: Souvenir of national seminar on integrating agri-horticultural and allied research for food and nutritional security in the era of global climate disruption, pp 168–178, 4–6 March, 2016. Imphal, Manipur

Baranwal VK, Majumdar S, Ahlawat YS, Singh RP (2003) Sodium sulphite yields improved DNA of higher stability for PCR detection of citrus yellow mosaic virus from citrus leaves. J Virol Methods 112:153–156

Baranwal VK, Majumder S, Ahlawat YS, Singh RP (2005) A novel approach for simultaneous detection of *Citrus yellow mosaic virus* and Citrus greening bacterium by multiplex polymerase chain reaction. IJBT 4:228–233

Baranwal VK, Jain P, Saritha RK, Jain RK, Gautam NK (2015) Detection and partial characterization of *Cowpea mild mottle virus* in mungbean and urdbean by deep sequencing and RT-PCR. Crop Prot 75:77–79

Bhardwaj P, Ram R, Zaidi AA, Hallan V (2014) Characterization of Apple stem grooving virus infecting *Actinidia deliciosa* (Kiwi) in India. Sci Hortic 176:105–111

Bhat AI, Siju S (2007) Development of a single tube multiplex RT-PCR for the simultaneous detection of Cucumber mosaic virus and Piper yellow mottle virus associated with stunt disease of black pepper. Curr Sci 93:973–976

Bhat AI, Siljo A (2014) Detection of viruses infecting black pepper by SYBR green-based real-time PCR assay. J Plant Pathol 96:105–109

Bhat AI, Jain RK, Ramiah M (2002) Detection of Tobacco streak virus from sunflower and other crops by reverse transcription polymerase chain reaction. Indian Phytopathol 55:216–218

Bhat AI, Hareesh PS, Madhubala R (2005) Sequencing of coat protein gene of an isolate of Cucumber mosaic virus infecting black pepper (*Piper nigrum* L.) in India. J Plant Biochem Biotechnol 14:37–40

Bhat AI, Siljo A, Jiby MV, Thankamani CK, Mathew PA (2009) Polymerase chain reaction (PCR) based indexing of black pepper (Piper nigrum L.) plants against Piper yellow mottle virus. J Spices Aromat Crops 18:28–32

Bhat AI, Siljo A, Deeshma KP (2013) Rapid detection of *Piper yellow mottle virus* and *Cucumber mosaic virus* infecting black pepper (Piper nigrum) by loop-mediated isothermal amplification (LAMP). J Virol Methods 193:190–196

Biju CN, Siljo A, Bhat AI (2010) Survey and RT-PCR based detection of Cardamom mosaic virus affecting small cardamom in India. Indian J Virol 21:148–150

Biswas KK (2008) Molecular diagnosis of Citrus tristeza virus in mandarin (*Citrus reticulata*) orchards of hills of West Bengal. Indian J Virol 19:26–31

Biswas KK (2010) Molecular characterization of *Citrus tristeza virus* isolates from the Northeastern Himalayan region of India. Arch Virol 155:959–963

Biswas KK, Tarafdar A, Sharma SK (2012a) Complete genome of mandarin decline *Citrus tristeza virus* of Northeastern Himalayan hill region of India: comparative analyses determine recombinant. Arch Virol 157:579–583
Biswas KK, Tarafdar A, Diwedi S, Lee RF (2012b) Distribution, genetic diversity and recombination analysis of Citrus tristeza virus of India. Virus Genes 45:139–148
Biswas K, Hallen V, Zaidi AA, Pandey PK (2013) Molecular evidence of cucumber mosaic virus subgroup II infecting *Capsicum annum* L. in the Western region of India. Curr Discov 2:97–105
Biswas C, Dey P, Mitra S, Bera A, Satpathy S, Karmakar PG (2014a) First report of *Potato leaf roll virus* (PLRV) naturally occurring on jute (*Corchorus olitorius*) in India. Plant Dis 98:1592–1593
Biswas KK, Tarafdar A, Sharma SK, Singh JK, Dwivedi S, Biswas K, Jayakumar BK (2014b) Current status of *Citrus tristeza virus* incidence and its spatial distribution in citrus growing geographical zones of India. Indian J Agric Sci 84:184–189
Boonham N, Kreuzeb J, Winterc S, Vlugtd R, Bergervoetd J, Tomlinsona J, Mumforda R (2014) Methods in virus diagnostics: From ELISA to next generation sequencing. Virus Res 186:20–31
Borah BK, Johnsonb AMA, Sai Gopal DVR, Dasgupta I (2008) A comparison of four DNA extraction methods for the detection of Citrus yellow mosaic badna virus from two species of citrus using PCR and dot-blot hybridization. J Virol Methods 151:321–324
Brakta A, Thakur PD, Handa A (2013) First report of apple top working disease caused by viruses (*Apple stem grooving virus*, *Apple chlorotic leaf spot virus*, and *Apple stem pitting virus*) in apple in India. Plant Dis 97:1001
Briddon RW, Bull SE, Mansoor S, Amin I, Markham PG (2002) Universal primers for the PCR-mediated amplification of DNA α a molecule associated with some monopartite begomoviruses. Mol Biotechnol 20:315–318
Byadgi AS, Ahlawat YS (1995) A new viral ringspot disease of citrus (Citrus species) in India. Indian J Agric Sci 65:61–68
Chakrabarty PK, Sable S, Monga D, Mayee CD (2005) Polymerase chain reaction-based detection of *Xanthomonas axonopodis* pv. *malvacearum* and cotton leaf curl virus. Indian J Agric Sci 75:524–528
Chakraborty S, Pandey PK, Banerjee MK, Kalloo G, Fauquet CM (2003) *Tomato leaf curl Gujarat virus*, a new *begomovirus* species causing a severe leaf curl disease of tomato in Varanasi. India Phytopathol 93:1485–1495
Chakraborty S, Vanitharani R, Chattopadhyay B, Fauquet CM (2008) Super virulent pseudorecombination and asymmetric synergism between genomic components of two distinct species of begomovirus associated with severe tomato leaf curl disease in India. J Gen Virol:818–828
Chang C, Chen CY, Hsu YH, Wu JT, Hu CC, Chang WC, Lin NS (2005) Transgenic resistance to *Cymbidium mosaic* virus in *Dendrobium* expressing the viral capsid protein gene. Transgenic Res 14:41–46
Cherian S, Joseph J, Muniyappa V, Savithri HS (1999) Characterization of tobacco mosaic virus isolated from tomato in India. http://www.iisc.ernet.in/currsci/may25/articles30.htm
Cherian AK, Baranwal VK, Malathi VG, Pant RP, Ahlawat YS (2004) Banana streak virus from India and its detection by polymerase chain reaction. Indian J Biotechnol 3:409–413
Chinnaraja C, Viswanathan R, Karuppaiah R, Bagyalakshmi K, Malathi P, Parameswari B (2013) Complete genome characterization of *Sugarcane yellow leaf virus* from India: Evidence for RNA recombination. Eur J Plant Pathol 135:335–349
Chowda Reddy RV, Colvin J, Muniyappa V, Seal S (2005) Diversity and distribution of begomoviruses infecting tomato in India. Arch Virol 150:845–867
Clark MF, Adams AN (1977) Characteristics of the microplate method of enzyme-linked immunosorbent assay for the detection of plant viruses. J Gen Virol 34:475–483
Colinet D, Nguyen M, Kummert J, Lepoivre P (1998) Differentiation among potyviruses infecting sweet potato based on genus- and virus-specific reverse transcription polymerase chain reaction. Plant Dis 82:223–229

Das AK, Shivankar VJ, Singh S (2002) Presence of citrus (*Citrus* species) greening diseases (Candidatus Liberobacter asiaticum) and its psyllid vector (Diaphorina citri) in maharashtra. Indian J Agric Sci 72:188–191

Dasgupta I, Das BK, Nath PS, Mukhopadhyay S, Niazi FR, Varma A (1996) Detection of rice tungro bacilliform virus in field and glasshouse samples from India using the polymerase chain reaction. J Virol Methods 58:53–58

De LC, Devadas R, Barman D, Chakrabarti S (2013) VISION 2050. National Research Centre for Orchids, Pakyong

Deng S, Hiruki C (1991) Amplification 16S rRNA genes from culturable and non-culturable mollicutes. J Microbiol Methods 14:53–61

Deng D, McGrath PF, Robinson DJ, Harrison BD (1994) Detection and differentiation of whitefly transmitted geminiviruses in plants and vector insects by the polymerase reaction with degenerate primers. Ann Appl Biol 125:327–336

Dhir S, Tomar M, Thakur PD, Ram R, Hallan V, Zaidi AA (2009) Molecular evidence for *Apple stem pitting virus* infection in India. New Dis Rep 19:45

Dhir S, Ram R, Hallan V, Zaidi AA (2011) Molecular characterization of an Indian variant of *Apple stem pitting virus*: evidence of recombination. J Plant Pathol 93:471–478

Dubey VK, Aminuddin SVP (2009) First report of Bean yellow mosaic virus in gladiolus in India. J Plant Pathol 91(S4):110

Dubey VK, Aminuddin SVP (2010) Molecular characterization of Cucumber mosaic virus infecting Gladiolus, revealing its phylogeny distinct from the Indian isolate and alike the Fny strain of CMV. Virus Genes 41:126–134

Duraisamy R, Natesan S, Muthurajan R, Gandhi K, Lakshmanan P, Karuppusamy N, Chokkappan M (2013) Molecular studies on the transmission of *Indian cassava mosaic virus* (ICMV) and *Sri Lankan cassava mosaic virus* (SLCMV) in cassava by *Bemisia tabaci* and cloning of ICMV and SLCMV replicase gene from cassava. Mol Biotechnol 53:150–158

Elbeaino T, Whitfield A, Sharma M, Digiaro M (2013) Emaravirus-specific degenerate PCR primers allowed the identification of partial RNA-dependent RNA polymerase sequences of Maize red stripe virus and Pigeonpea sterility mosaic virus. J Virol Methods 188:37–40

Elbeaino T, Digiaro M, Uppala M, Sudini H (2014) Deep sequencing of pigeonpea sterility mosaic virus discloses five RNA segments related to emaraviruses. Virus Res 188:27–31

Gaur RK, Singh M, Rao GP (2003) Molecular characterization of Sugarcane mosaic virus isolate from North Eastern region of India. Sugar Technol 5:149–153

Gawande SJ, Khar A, Lawande KE (2010) First report of *Iris yellow spot virus* on Garlic in India. Plant Dis 94:1066

Gawande SJ, Shukla A, Chimote VP, Kaushal N, Kaundal P, Garg ID, Chimote KP (2011) Development of PCR-based techniques for the detection of immobilised *Potato virus Y* virions. J Plant Pathol 93:127–132

Gawande SJ, Gurav VS, Ingle AA, Gopal J (2014) First report of *Iris yellow spot virus* infecting *Allium tuberosum* in India. Plant Dis 98:1161

Geering ADW, Olszewski NE, Harper G, Lockhart BEL, Hull R, Thomas JE (2005) Banana contains a diverse array of endogenous badnaviruses. J Gen Virol 86:511–520

Geetanjali AS, Kumar R, Srivastava PS, Mandal B (2011) Biological and molecular characterization of two distinct tomato strains of *Cucumber mosaic virus* based on complete RNA-3 genome and subgroup specific diagnosis. Indian J Virol 22:117–126

Ghosh SB, Bapat VA (2006) Development of RT-PCR based method for detection of *Potato virus Y* in tobacco and potato. Indian J Biotechnol 5:232–235

Ghosh DK, Mathur S, Gupta KN, Ahlawat YS, Ramachandran P (2002a) Viroid infection of citrus in Central India. Indian Phytopathol 55(3):290–293

Ghosh SB, Nagi LHS, Ganapathi TR, Paul Khurana SM, Bapat VA (2002b) Cloning and sequencing of *Potato virus Y* coat protein gene from an Indian isolate and development of transgenic tobacco for PVY resistance. Curr Sci 82:855–859

Ghosh DK, Aglave B, Bhanare K, Baranwal VK (2007) PCR based detection of Citrus yellow mosaic disease from Vidarbha region of Maharastra. Indian Phytopathol 60:520–526

Ghosh DK, Aglave B, Baranwal VK (2008a) Simultaneous detection of one RNA and one DNA virus from naturally infected citrus plants using duplex PCR technique. Curr Sci 94:1314–1318
Ghosh R, Paul S, Palit P, Das S, Das A, Acharyya S, Mir JI, Ghosh SK, Roy A (2008b) Molecular characterization and computational analysis of coat protein gene of an East Indian isolate of *Bhendi yellow vein mosaic virus* and associated DNA-beta satellite. Indian J Virol. 19:173–182
Ghosh DK, Aghave B, Roy A, Ahlawat YS (2009) Molecular cloning, sequencing and phylogenetic analysis of coat protein gene of a biologically distinct CTV isolate occurring in central India. J Plant Biochem Biotechnol 18:105–108
Ghosh R, Palit P, Paul S, Ghosh SK, Roy A (2012) Detection of *Corchorus golden mosaic virus* associated with yellow mosaic disease of jute (*Corchorus capsularis*). Indian J Virol 23:70–74
Gibbs A, Mackenzie A (1997) A primer pair for amplifying part of the genome of all potyvirids by RT-PCR. J Virol Methods 63:9–16
Girish KR, Usha R (2005) Molecular characterization of two soybean-infecting begomoviruses from India and evidence for recombination among legume-infecting begomoviruses from South-East Asia. Virus Res 108:167–176
Gopi V, Kuruba G, Sankar TG, Palanivel (2010) A detection of citrus yellow mosaic virus by PCR and nucleic acid spot hybridisation using non-radioactive probes in commercial citrus species. Arch Phytopathol Pfl 43:892–899
Govindappa MR, Shankergoud I, Shankarappa KS, Wickramaarachchi WART, Reddy BA, Rangaswamy KT (2011) Molecular detection and partial characterization of *Begomovirus* associated with the leaf curl disease of sunflower (*Helianthus annuus*) in Southern India. Plant Pathol J 10:29–35
Gupta N, Prabha K, Islam S, Baranwal VK (2013) First report of Leek yellow stripe virus in garlic from India. J Plant Pathol 95(4):S4.69–S4.77
Ha C, Coombs S, Revill PA, Harding RM, Vu M, Dale JL (2008) Design and application of two novel degenerate primer pairs for the detection and complete genomic characterization of potyviruses. Arch Virol 153:25–36
Haq QMI, Rouhibakhsh A, Ali A, Malathi VG (2011) Infectivity analysis of a blackgram isolate of *Mungbean yellow mosaic virus* and genetic assortment with MYMIV in selective hosts. Virus Genes 42:429–439
Hareesh PS, Bhat AI (2008) Detection and partial nucleotide sequence analysis of Piper yellow mottle virus infecting black pepper (*Piper nigrum* L.) in India. Indian J Virol 19:160–167
Hareesh PS, Madhubala R, Bhatt AI (2006) Characterization of Cucumber mosaic virus infecting Indian long pepper (*Piper longum*:L.) and betel vine (*Piper betle* L.) in India. Indian J Biotechnol 5:89–93
Harper G, Osuji JO, Heslop-Harrison P, Hull R, Harper G, Lockhart BEL (1999) Integration of banana streak badnavirus into the *Musa* genome: molecular evidence. Virology 255:207–213
Hema M, Savithri HS, Sreenivasulu P (2003) Comparison of direct binding polymerase chain reaction with recombinant coat protein antibody based dot-immunobinding assay and immunocapture-reverse transcription-polymerase chain reaction for the detection of sugarcane streak mosaic disease in India. Curr Sci 85:1774–1777
Hoa NV, Ahlawat YS (2004) Characterization of four isolates of Indian citrus ringspot virus. Indian Phytopathol 57:296–302
Jacob T, Usha R (2001) 3′ – terminal sequence analysis of the RNA genome of the Indian isolate of Cardamom mosaic virus, a new member of the genus Macluravirus of Potyviridae. Virus Genes 23:81–88
Jacob T, Jebasingh T, Venugopal MN, Usha R (2003) High genetic diversity in the coat protein and 3′ untranslated regions among geographical isolates of Cardamom mosaic virus from South India. J Biosci 28:589–595
Jaidi M, Srivastava A, Kumar S, Raj SK, Singh R (2015) First report of natural occurence of *Papaya leaf crumple virus* on soyabean in India. New Dis Rep 32:15
Jain RK, Pappu SS, Varma A, Ram RD (1998a) Molecular characterization of papaya ringspot potyvirus isolates from India. Ann Appl Biol 132:413–425

Jain RK, Pappu HR, Pappu SS, Krishanareddy M, Vani A (1998b) Watermelon bud necrosis tospovirus is a distinct virus species belonging to serogroup IV. Arch Virol 143:1637–1644

Jain RK, Pappu SS, Pappu HR, Culbreath AK, Todd JW (1998c) Molecular diagnosis of Tomato spotted wilt tospovirus infection of peanut and other field and greenhouse crops. Plant Dis 82:900–904

Jain RK, Sharma J, Sivakumar AS, Sharma PK, Byadgi AS, Verma AK, Varma A (2004) Variability in the coat protein gene of *Papaya ringspot virus* isolates from multiple locations in India. Arch Virol 149:2435–2442

Jain RK, Pandey NA, Krishnareddy M, Mandal B (2005) Immunodiagnosis of groundnut and watermelon bud necrosis viruses using polyclonal antiserum to recombinant nucleocapsid protein of *Groundnut bud necrosis virus*. J Virol Methods 130:162–164

Jeevalatha A, Kaundal P, Shandil RK, Sharma NN, Chakrabarti SK, Singh BP (2013a) Complete genome sequence of *Potato leafroll virus* isolates infecting potato in the different geographical areas of India shows low level genetic diversity. Indian J Virol 24:199–204

Jeevalatha A, Kaundal P, Venkatasalam EP, Chakrabarti SK, Singh BP (2013b) Uniplex and duplex PCR detection of geminivirus associated with potato apical leaf curl disease in India. J Virol Methods 193:62–67

John P, Sivalingam PN, Haq QMI, Kumar N, Mishra A, Briddon RW, Malathi VG (2008) Cowpea golden mosaic disease in Gujarat is caused by a *Mungbean yellow mosaic India virus* isolate with a DNA B variant. Arch Virol 153:1359–1365

Jose J, Usha R (2000) Extraction of geminiviral DNA from a highly mucilaginous plant (Abelmoschus esculentus). Plant Mol Biol Report 18:349–355

Jose J, Usha R (2003) Bhendi yellow vein mosaic disease in India is caused by association of a DNA β satellite with a *Begomovirus*. Virology 305:310–317

Joseph J, Savithri HS (1999) Determination of 30-terminal nucleotidesequence of pepper vein banding virus RNA and expression of itscoat protein in Escherichia coli. Arch Virol 144:1679–1687

Joshi R, Dasgupta I (2001) Cloning and molecular analysis of DNA of rice tungro bacilliform viruses from different rice growing regions of India. Indian Phytopathol 54:469–475

Joshi R, Kumar V, Dasgupta I (2003) Detection of molecular variability in rice tungro bacilliform viruses from India using polymerase chain reaction–restriction fragment length polymorphism. J Virol Methods 109:89–93

Kamaal N, Mohammad A, Gupta S (2011) Identification of *Mungbean yellow mosaic India virus* infecting *Vigna mungo* var. *silvestris* L. Phytopathol Mediterr 50:94–100

Kamala S, Makeshkumar T, Sreekumar J, Chakrabarti SK (2015) Whole transcriptome sequencing of diseased elephant foot yam reveals complete genome sequence of Dasheen mosaic virus. Virol Rep 5:1–9

Kanakala S, Sakhare A, Verma HN, Malathi VG (2013a) Infectivity and the phylogenetic relationship of a mastrevirus causing chickpea stunt disease in India. Eur J Plant Pathol 135:429–438

Kanakala S, Verma HN, Vijay P, Saxena DR, Malathi VG (2013b) Response of chickpea genotypes to Agrobacterium-mediated delivery of Chickpea chlorotic dwarf virus (CpCDV) genome and identification of resistance source. Appl Microbiol Biotechnol 97:9491–9501

Kang SS, Athar M, Cheema SS, Malathi VG, Radhakrishnan G (2004) Quick detection of cotton leaf curl virus. Indian Phytopathol 57:245–246

Karan M, Harding RM, Dale JL (1994) Evidence for two groups of *Banana bunchy top virus* isolates. J Gen Virol 75:3541–3546

Karasev AV, Nikolaeva OV, Koonin EV, Gumpf DJ, Garnsey SM (1994) Screening of the closterovirus genome by degenerate primer mediated polymerase chain reaction. J Gen Virol 75:1415–1422

Karthikeyan AS, Vanitharani R, Balaji V, Anuradha S, Thillaichidambaram P, Shivaprasad PV, Parameswari C, Balamani V, Saminathan M, Veluthambi K (2004) Analysis of an isolate of *Mungbean yellow mosaic virus* (MYMV) with a highly variable DNA B component. Arch Virol 149:1643–1652

Karuppaiah R, Viswanathan R, Ganesh KV (2013) Genetic diversity of sugarcane bacilliform virus isolates infecting *Saccharum* spp. in India. Virus Genes 46:505–516

Katoch M, Ram R, Zaidi AA, Garg ID (2002) Status of *Bean yellow mosaic virus* on Gladiolus. Crop Prot 21:861–865

Katoch M, Ram R, Zaidi AA (2003) First report of *Tobacco ringspot virus* occurring in gladiolus in India. Plant Pathol 52:789

Katoch M, Abdin MZ, Zaidi AA (2004) First report of *Tobacco rattle virus* occurring in gladiolus in India. Plant Pathol 53:236

Kaur C, Raj SK, Snehi SK, Goel AK, Roy RK (2011) Natural occurrence of *Ornithogalum mosaic virus* newly reported on gladiolus in India. New Dis Rep 24:2. http://dx.doi.org/10.5197/j.2044-0588.2011.024.002

Kaushal N, Gawande SJ, Kaundal P, Garg ID (2007) Reverse transcriptase polymerase chain reaction (RT-PCR) Based detection of PVS and PVS strain by using degenerate primers. Potato J 34:85–86

Khan JA, Siddiqui MK, Singh BP (2002) The association of *Begomovirus* with bitter melon in India. Plant Dis 86:3

Khan S, Jan AT, Aquil B, Haq QMR (2011) Coat protein gene based characterization of *Cucumber mosaic virus* isolates infecting banana in India. J Phytology 3:94–101

Kirthi N, Maiya SP, Murthy MRN, Savithri HS (2002) Evidence for recombination among the Tomato leaf curl virus strains/species from Bangalore. Indian Arch Virol 147:255–272

Kirthi N, Priyadarshini CGP, Sharma P, Maiya SP, Hemalatha V, Sivaraman P, Dhawan P, Rishi N, Savithri HS (2004) Genetic variability of begomoviruses associated with cotton leaf curl disease originating from India. Arch Virol 149:2047–2057

Kreuze JF, Perez A, Untiveros M, Quispe D, Fuentes S (2009) Complete viral genome sequence and discovery of novel viruses by deep sequencing of small RNAs: a generic method for diagnosis, discovery and sequencing of viruses. Virology 388:1–7

Krishna Reddy M, Venkataravanappa V, Madhuvanthi B, Jalali S (2010) Molecular characterization of begomoviruses associated with papaya leaf curl disease in India. Acta Hortic 851:465–472

Krishnareddy M, Usha-Rani R, Kumar A, Madhavireddy K, Pappu HR (2008) First report of Capsicum chlorosis virus (Genus *Tospovirus*) infecting chili pepper (*Capsicum annuum*) in India. Plant Dis 92:1469

Kulshrestha S (2005) Molecular characterization of virus causing rose mosaic dis-ease complex. Ph.D.Thesis, Department of Biosciences, Jamia Millia Islamia, New Delhi

Kulshrestha S, Hallan V, Sharma A, Seth CA, Chauhan A, Zaidi AA (2013) Molecular characterization and intermolecular interaction of coat protein of *Prunus necrotic ringspot virus*: implications for virus assembly. Indian J Virol 24:235–241

Kumar S (2013) Studies on virus(es) associated with grapevine leafroll disease in India. Ph.D. Thesis, Indian Agricultural Research Institute, New Delhi

Kumar PL, Jones AT, Reddy DVR (2003) A novel mite-transmitted virus with a divided RNA genome closely associated with pigeonpea sterility mosaic disease. Phytopathology 93:71–81

Kumar S, Srivastava A, Raj SK (2005) Molecular detection of *Cucumber mosaic virus* in Chrysanthemum cultivars. Acta Phytopathol Entomol Hung 40:205–211

Kumar NA, Narasu ML, Zehr UB, Ravi KS (2008) Molecular characterization of *Tobacco streak virus* causing soybean necrosis in India. IJBT 7:214–217

Kumar J, Kumar A, Khan JA (2009) Aminuddin first report of *Papaya leaf curl virus* naturally infecting tobacco in India. J Plant Pathol 91:112

Kumar S, Baranwal VK, Joshi S, Arya M, Majumder S (2010a) Simultaneous detection of mixed infection of *Onion yellow dwarf virus* and an *Allexivirus* in RT-PCR for ensuring virus free onion bulbs. Indian J Virol 21:64–68

Kumar A, Kumar J, Khan JA (2010b) Sequence characterization of cotton leaf curl virus from Rajasthan: phylogenetic relationship with other members of geminiviruses and detection of recombination. Virus Genes 40:282–289

Kumar J, Kumar A, Roy JK, Tuli R, Khan JA (2010c) Identification and molecular characterization of begomovirus and associated satellite DNA molecules infecting *Cyamopsis tetragonoloba*. Virus Genes 41:118–125

Kumar S, Shankar Udaya AC, Nayaka SC, Lund OS, Prakash HS (2011a) Detection of Tobacco mosaic virus and Tomato mosaic virus in pepper and tomato by multiplex RT–PCR. Appl Microbiol 53:359–363

Kumar A, Snehi SK, Raj SK, Kumar J, Khan JA (2011b) Association of Cotton leaf curl Burewala virus and its satellite molecules with leaf distortion symptoms of cotton in India. New Dis Rep 24:18

Kumar J, Kumar A, Singh SP, Roy JK, Lalit A, Parmar D, Sharma NC, Tuli R (2012a) First report of *Radish leaf curl virus* infecting okra in India. New Dis Rep 25:9

Kumar S, Singh RM, Ram R, Badyal J, Hallan V, Zaidi AA, Varma A (2012b) Determination of major viral and sub viral pathogens incidence in apple orchards in Himachal Pradesh. Indian J Virol 23:75–79

Kumar S, Baranwal VK, Singh P, Jain RK, Sawant SD, Singh SK (2012c) Characterization of a *Grapevine leafroll-associated virus 3* from India showing incongruence in its phylogeny. Virus Genes 45:195–200

Kumar S, Sawant SD, Sawant IS, Prabha K, Jain RK, Baranwal VK (2012d) First report of *Grapevine leafroll-associated virus 1* infecting grapevines in India. Plant Dis 96:1828

Kumar J, Singh SP, Kumar J, Tuli R (2012e) A novel mastrevirus infecting wheat in India. Arch Virol 157:2031–2034

Kumar J, Gunapati S, Singh SP, Kumar A, Lalit A, Sharma NC, Puranik R, Tuli R (2013a) A new betasatellite associated with *Cotton leaf curl Burewala virus* infecting tomato in India: influence on symptoms and viral accumulation. Arch Virol 158:1349–1353

Kumar A, Shilpi S, Yogita M, Lal SK, Mandal B (2013b) Identification of a new strain of Tobacco mosaic virus infecting soybean in India based on host reactions and 3′ end genome sequence. Crop Prot 53:20–22

Kumar S, Singh L, Ferretti L, Barba M, Zaidi AA, Hallan V (2013c) Evidence of Grapevine leafroll associated virus-1–3, Grapevine fleck virus and Grapevine virus B occurring in Himachal Pradesh, India

Kumar S, Gautam KK, Raj SK (2014a) Molecular identification of *Cucumber mosaic virus* isolates of subgroup IB associated with mosaic disease of eggplant in India. Virus Dis 25:129–131

Kumar R, Jeevalatha A, Sharma NN, Sharma S, Chakrabarti SK, Singh BP (2014b) Development of PCR based methods for detection of *Potato aucuba mosaic virus* in India. Potato J 41:166–174

Kumar S, Sankarlingam A, Rabindran R (2014c) Characterization and confirmation of Papaya ringspot virus-W strain infecting *Trichosanthese cucumerina* at Tamil Nadu. India J Plant Pathol Microbiol 5:225

Kumar J, Gunapati S, Alok A, Lalit A, Gadre R, Sharma NC, Roy JK, Singh SP (2015a) Cotton leaf curl Burewala virus with intact or mutant transcriptional activator proteins: complexity of cotton leaf curl disease. Arch Virol 160:1219–1228

Kumar A, Pandey DM, Abraham T, Mathew J, Jyothsna P, Ramachandran P, Malathi VG (2015b) Molecular characterization of viroid associated with tapping panel dryness syndrome of *Hevea brasiliensis* from India. Curr Sci 108:1520–1527

Kunkalikar SR, Sudarsana P, Rajagopalan P, Zehr UB, Ravi KS (2010) Biological and molecular characterization of *Capsicum chlorosis virus* infecting chili and tomato in India. Arch Virol 155:1047–1057

Kunkalikar SR, Sudarsana P, Arun BM, Rajagopalan P, Chen TC, Yeh SD, Naidu RA, Zehr UB, Ravi KS (2011) Importance and genetic diversity of vegetable-infecting tospoviruses in India. Phytopathology 101:367–376

Kushwaha N, Achuit K, Singh AK, Chattopadhyay B, Chakraborty S (2010) Recent advances in *Geminivirus* detection and future perspectives. J Plant Protect Sci 1:1–18

Lakshmi V, Hallan V, Ram R, Ahmed N, Zaidi AA, Varma A (2011) Diversity of *Apple mosaic virus* isolates in India based on coat protein and movement protein genes. Indian J Virol 22:44–49

Loconsole G, Onelge N, Potere O, Giampetruzzi A, Bozan O, Satar S, De Stradis A, Savino VN, Yokomi RK, Saponari M (2012) Identification and characterization of citrus yellow vein clearing virus, a putative new member of the genus Mandarivirus. Phytopathology 102:1168–1175

Lovisolo O, Lisa V (1978) Virus diseases of carnation. Fitopatol Bras 3:219–233

Machida S, Shibuya M, Sano T (2008) Enrichment of viroid small RNAs by hybridization selection using biotinylated RNA transcripts to analyze viroid induced RNA silencing. Gen Plant Pathol 74:203–207

Mackenzie DJ, McLean MA, Mukerji S, Green M (1997) Improved RNA extraction from woody plants for the detection of viral pathogens by reverse transcription-polymerase chain reaction. Plant Dis 81:222–226

Madhubala R, Bhadramurthy V, Bhat AI, Hareesh PS, Retheesh ST, Bhai RS (2005) Occurrence of *Cucumber mosaic virus* on vanilla (*Vanilla planifolia* Andrews) in Indian. J Biosci 30:339–350

Magie RO, Poe SL (1972) In: Koenig N, Crowley W (eds) Disease and pest associates of bulbs and plant, in the world of gladiolus. North American Gladiolus Council, Edgewood Press, Maryland, pp 155–167

Maheshwari R, Panigrahi G, Angappan K (2014) Molecular characterization of distinct YMV (Yellow mosaic virus) isolates affecting pulses in India with the aid of coat protein gene as a marker for identification. Mol Biol Rep 41:2635–2644

Majumder S, Baranwal VK (2009) First report of *Garlic common latent virus* in garlic from India. Plant Dis 93:106

Majumder S, Baranwal VK (2014) Simultaneous detection of four garlic viruses by multiplex reversetranscription PCR and their distribution in Indian garlic accessions. J Virol Methods 202:34–38

Majumder S, Arya M, Pant RP, Baranwal VK (2007) *Shallot virus X* in Indian shallot, a new virus report for India. New Dis Rep 15:52

Majumder S, Baranwal VK, Joshi S (2008) Simultaneous detection of *Onion yellow dwarf virus* and Garlic latent virus in infected leaves and cloves of garlic by duplex RT-PCR. J Plant Pathol 90:369–372

Mandal B, Varma A, Malathi VG (1997) Systemic infection of *Vigna mungo* using the cloned DNAs of the blackgram isolate of mungbean yellow mosaic geminivirus through agroinoculation and transmission of the progeny virus by whiteflies. J Phytopathol 145:505–510

Mandal B, Jain RK, Chaudhary V, Varma A (2003) First report of natural infection of *Luffa acutangula* by *Watermelon bud necrosis virus* in India. Plant Dis 87:598

Mandal B, Mandal S, Pun KB, Varma A (2004a) First report of the association of a *Nanovirus* with foorkey disease of large cardamom in India. Plant Dis 88:428

Mandal B, Mandal S, Sohrab SS, Pun KB, Varma A (2004b) A new yellow mosaic disease of chayote in India. Plant Pathol 53:797

Mandal B, Kumar A, Rani P, Jain RK (2012a) Complete genome sequence, phylogenetic relationships and molecular diagnosis of an Indian isolate of *Potato Virus X*. J Phytopathol 160:1–5

Mandal B, Vijayanandraj S, Shilpi S, Pun KB, Singh V, Pant RP, Jain RK, Varadarasan S, Varma A (2012b) Disease distribution and characterisation of a new *Macluravirus* associated with chirke disease of large cardamom. Ann Appl Biol 160:225–236

Mandal B, Jain RK, Krishnareddy M, Kumar NKK, Ravi KS, Pappu HR (2012c) Emerging problems of tospoviruses (Bunyaviridae) and their management in the Indian subcontinent. Plant Dis 96:468–479

Mandal B, Shilpi S, Barman AR, Mandal S, Varma A (2013) Nine novel DNA components associated with the foorkey disease of large cardamom: evidence of a distinct *Babuvirus* species in *Nanoviridae*. Virus Res 178:297–305

Mangal M, Bhardwaj SV, Sharma DR, Kaur R, Mangal AK (2002) Use of meristem tip culture to eliminate carnation latent virus from carnation plants. Indian J Exp Biol:119–122

Mangrauthia SK, Parameswari B, Jain RK, Praveen S (2008) Role of genetic recombination in the molecular architecture of papaya ringspot virus. Biochem Genet 46:835–846

Mangrauthia SK, Malathi P, Agarwal S, Ramkumar G, KrishnaveniD NCN, Madhav MS, Ladhalakshmi D, Balachandran SM, Viraktamath BC (2012) Genetic variation of coat protein gene among the isolates of Ricetungro spherical virus from tungro-endemic states of the India. Virus Genes 44:482–487

Martelli GP (2012) Grapevine virology highlights 2000–2003. In: Extended abstracts of 17th Meeting of the ICVG. Davis, California, USA. pp 13–31

Maruthi MN, Rekha AR, Muniyappa V (2007) Pumpkin yellow vein mosaic disease is caused by two distinct begomoviruses: complete viral sequences and comparative transmission by an indigenous *Bemisia tabaci* and the introduced B-biotype. EPPO Bull 37:412–419

Mathur S, Garg AP, Ramachandran P (2002) Detection of viroid infecting chrysanthemum in India. Indian Phytopathol 55:479–482

Meena RP, Baranwal VK (2016) Development of multiplex polymerase chain reaction assay for simultaneous detection of clostero-, badna- and mandari-viruses along with huanglongbing bacterium in citrus trees. J Virol Methods 235:58–64

Mehra A, Jabeen N, Singh AK, Hallan V, Zaidi AA (2008) A new chrysanthemum potyvirus: a molecular evidence. Arch Phytopathol Plant Protect 42:436–441

Menzel W, Jelkmann W, Maiss E (2002) Detection of four apple viruses by multiplex RT-PCR assays with co-amplification of plant mRNA as internal control. J Virol Methods 99:81–92

Minafra A, Hadidi A, Martelli GP (1992) Detection of Grapevine closterovirus A in infected grapevine tissue by reverse transcription-polymerase chain reaction. Vitis 31:221–227

Mishra R, Verma RK, Gaur RK (2015) Analysis of genome comparison of two Indian isolates of *Cowpea aphid-borne mosaic virus* from India. Virus Genes 51:306–309

Monga D, Kumar R, Kumar M (2005) Detection of DNA A and satellite (DNA beta) in cotton leaf curl virus (CLCuV) infected weeds and cotton plants using PCR technique. J Cotton Res Dev 19:105–108

Mukherjee K, Verma Y, Chakrabarti SK, Singh MN, Khurana SMP (2003) Cloning and sequencing of coat protein gene of an Indian potato leaf roll virus (PLRV) isolate and its similarity with other members of Luteoviridae. Virus Genes 26:247–253

Muniyappa V, Venkatesh HM, Ramappa HK, Kulkarni RS, Zeidan M, Tarba CY, Ghanim M, Czosnek H (2000) Tomato leaf curl virus from Bangalore (ToLCV-Ban4): sequence comparison with Indian ToLCV isolates, detection in plants and insects, and vector relationships. Arch Virol 145:1583–1598

Muniyappa V, Maruthi MN, Babitha CR, Colvin J, Briddon RW, Rangaswamy KT (2003) Characterization of *Pumpkin yellow vein mosaic virus* from India. Ann Appl Biol 142:323–331

Naidu RA, Miller JS, Mayo MA, Wesley SV, Reddy AS (2000) The nucleotide sequence of *Indian peanut clump virus* RNA 2: sequence comparisons among pecluviruses. Arch Virol 145:1857–1866

Naimuddin AM, Pratap A (2011) First report of natural infection of *Mungbean yellow mosaic India virus* in two wild species of Vigna. New Dis Rep 23:21

Naimuddin AM, Gupta S, Agnihotri AK (2014) *Ageratum conyzoides* harbours *Mungbean yellow mosaic India virus*. Plant Pathol J. doi:10.3923/ppj.2014

Nakahara K, Hataya T, Uyeda I (1999) A simple, rapid method of nucleic acid extraction without tissue homogenization for detecting viroids by hybridization and RT-PCR. J Virol Methods 77:47–58

Nariani TK (1960) Yellow mosaic of mung (Phaseolus aureus L.) Indian Phytopathol 13:24–29

Nath N, Mathur S, Dasgupta I (2002) Molecular analysis of two complete rice tungro bacilliform virus genomic sequences from India. Arch Virol 147:1173–1187

Noorani MS, Awasthi P, Singh RM, Ram R, Sharma MP, Singh SR, Ahmed N, Hallan V, Zaidi AA (2010) Complete nucleotide sequence of cherry virus A (CVA) infecting sweet cherry in India. Arch Virol 155:2079–2082

Noorania Md S, Awasthia P, Sharma MP, Ram R, Zaidi AA, Hallan V (2013) Simultaneous detection and identification of four cherry viruses by two step multiplex RT-PCR with an internal control of plant nad5 mRNA. J Virol Methods 193:103–107

Padidam M, Beachy RN, Fauquet CM (1995) Tomato leaf curl geminivirus from India has a bipartite genome and coat protein is not essential for infectivity. J Gen Virol 76:249–263

Padmavathi M, Srinivas KP, Hema M, Sreenivasulu P (2013) First report of Konjac mosaic virus in elephant foot yam (*Amorphophallus paeoniifolius*) from India. Australas Plant Dis Notes 8:27–29

Pal BF, Tandon RN (1937) Types of tobacco leaf curl in northern India. Indian J Agril Sci 7:363–393

Parameswari B, Bagyalakshmi K, Viswanathan R, Chinnaraja C (2013) Molecular characterization of Indian sugarcane streak mosaic virus isolate. Virus Genes 46:186–189
Parameswari B, Bagyalakshmi K, Chinnaraja C, Viswanathan R (2014) Molecular characterization of Indian Sugarcane streak mosaic virus isolates reveals recombination and negative selection in the P1 gene. Gene 552:199–203
Parveen S, Ahlawat YS, Pant RP, Milne RG (1999) RT-PCR based detection and molecular cloning of coat protein gene of citrus ringspot virus from India. In: International symposium on citriculture. NRCC, Nagpur, p 122
Patil BL, Kumar PL (2015) Pigeonpea sterility mosaic virus: a legume-infecting Emaravirus from South Asia. Mol Plant Pathol 16:775–786
Patil BL, Rajasubramaniam S, Bagchi C, Dasgupta I (2005) Both Indian cassava mosaic virus and Sri Lankan cassava mosaic virus are found in India and exhibit high variability as assessed by PCR-RFLP. Arch Virol 150:389–397
Periasamy M, Niazi FR, Malathi VG (2006) Multiplex RT-PCR, a novel technique for the simultaneous detection of the DNA and RNA viruses causing rice tungro disease. J Virol Methods 134:230–236
Prabha K, Baranwal VK (2011) *Indian citrus ringspot virus*: localization of virus in seed tissues and evidence for lack of seed transmission. Phytoparasitica 39:491–496
Prabha K, Baranwal V (2012) The genome sequence of an isolate of Indian citrus ringspot virus infecting the sweet orange in India. J Virol 86:12446–12447
Prabha K, Baranwal VK, Jain RK (2013) Applications of next generation high throughput sequencing technologies in characterization, discovery and molecular interaction of plant viruses. Indian J Virol 24:157–165
Pramesh D, Baranwal VK (2015) Production of virus-free garlic (*Allium sativum* L.) through meristem tip culture after solar or hot air treatment of cloves. J Hortic Sci Biotechnol 90:180–186
Pramesh D, Islam S, Singh J, Baranwal VK (2012) Serological and PCR detection of *Garlic common latent virus* associated with garlic accessions of India. Med Plant 4:17–22
Prasada Rao RDVJ, Raghunathan V, Joshi NC (1979) Interception of peanut mottle virus. Indian J Pl Prot 7:59–63
Prasanth G, Hegde V (2008) Occurrence of *Sweet potato feathery mottle virus* and *Sweet potato leaf curl Georgia virus*on Sweet Potato in India. Dis Notes 92:311
Pratap D, Kumar S, Raj SK (2008) First molecular identification of a *Cucumber mosaic virus* isolate causing shoestring symptoms on tomato in India. Australas Plant Dis Notes 3:57–58
Raigond B, Venkateswarlu V, Sridhar J, Jeevalatha A, Sharma S, Singh BP (2014) RT-PCR detection of *Potato leaf roll virus* (PLRV) in aphids from Northern and North-Eastern India. Indian J Plant Prot 42:430–436
Raikhy G, Hallan V, Kulshrestha S, Sharma ML, Ram R, Zaidi AA (2003a) First report of *Carnation necrotic fleck virus* (CNFV) infecting carnations in India. Plant Pathol 52:801
Raikhy G, Hallan V, Kulshrestha S, Sharma ML, Ram R, Zaidi AA (2003b) Molecular characterization of a *Carnation etched ring virus* isolate from India. Acta Virol 47:105–111
Raikhy G, Hallan V, Kulshrestha S, Ram R, Zaidi AA (2006a) Detection of carnation ringspot and Carnation vein mottle virus (es) in carnation cultivars in India. Acta Hortic 722:247–258
Raikhy G, Hallan V, Kulshrestha S, Ram R, Zaidi AA (2006b) Multiplex PCR and genome analysis of Carnation mottle virus Indian isolate. Curr Sci 90:74–82
Raj SK, Singh BP (1996) Association of geminivirus infection with yellow green mosaic disease of *Cucumis sativus*: diagnosis by nucleic acid probes. Indian J Exp Biol 34:603–605
Raj SK, Saxena S, Hallan V, Singh BP (1998) Biochem Mol Biol Int 44:89–95
Raj SK, Srivastava A, Chandra G, Singh BP (2002) Characterization of Cucumber *mosaic virus* isolate infecting gladiolus cultivars and comparative evaluation of serological and molecular methods for sensitive diagnosis. Curr Sci 83:1132–1137
Raj SK, Khan MS, Singh R, Nisha K, Prakash D (2005) Occurrence of yellow mosaic geminiviral disease on bitter gourd (*Momordica charantia*) and its impact on phytochemical contents. Int J Food Sci Nutr 56:185–192

Raj SK, Khan MS, Snehi SK, Srivastava S, Singh HB (2006) A yellow mosaic disease of soybean in Northern India is caused by *Cotton leaf curl Kokhran virus*. Plant Dis 90:975

Raj SK, Kumar S, Choudhari S (2007) Identification of *Tomato aspermy virus* as the cause of yellow mosaic and flower deformation of chrysanthemums in India. Australas Plant Dis Notes 157:117–125

Raj SK, Snehi SK, Khan MS, Singh R, Khan AA (2008) Molecular evidence for association of *Tomato leaf curl New Delhi virus* with leaf curl disease of papaya (*Carica papaya* L.) in India. Australas Plant Dis Notes 3:152–155

Raj SK, Kumar S, Choudhary S, Verma DK (2009) Biological and molecular characterization of three isolates of *Tomato aspermy virus* infecting chrysanthemums in India. J Phytopathol 157:117–125

Raj SK, Snehi SK, Gautam KK, Khan MS (2010) First report of association of *Cucumber mosaic virus* with blister and leaf distortion of castor bean (*Ricinus communis*) in India. Phytoparasitica 38:283–289

Raj SK, Kumar S, Verma DK, Snehi SK (2011) First report on molecular detection and identification of *Tomato aspermy virus* naturally occurring on gladiolus in India. Phytoparasitica 39:303–307

Raja P, Jain RK (2006) Molecular diagnosis of Groundnut bud necrosis virus causing bud blight of tomato. Indian Phytopathol 59:359–362

Rajagopalan PA, Naik A, Katturi P, Kurulekar M, Kankanallu RS, Anandalakshmi R (2012) Dominance of resistance-breaking cotton leaf curl Burewala virus (CLCuBuV) in northwestern India. Arch Virol 157:855–868

Rajinimala N, Rabindran R (2007) First report of Indian cassava mosaic virus on bittergourd (Momordica charantia) in Tamil Nadu, India. Australas Plant Dis Notes 2:81–82

Ramachandran P, Pandey PK, Ahlawat YS, Varma A, Kapur SP (1993) Viroid diseases of citrus in India. In: Proceedings of the 12th Cant. IOCV. IOCV. Riverside, pp 438–440

Ramachandran P, Francis L, Mathur S, Varma A, Mathew J, Mathew NM, Sethurajm MR (2000) Evidence for association of a low molecular weight RNA with tapping panel dryness syndrome of rubber. Plant Dis 84:1155

Ramachandran P, Roy A, Mathur S, Agarwal J, Ahlawat YS (2003) Diagnostics for Citrus exocortis and Hop stunt viroids associated with yellow corky vein disease in citrus. Indian Phytopathol 56:428–433

Rana T (2010) Molecular characterization and diagnostic development for *Apple chlorotic leaf spot virus*. Ph.D. Thesis, Institute of Himalayan Bioresource Technology, CSIR, Palampur-176061 (H.P.), India

Rana T, Chandel V, Hallan V, Zaidi AA (2008) *Cydonia oblonga* as reservoir of *Apple chlorotic leaf spot virus* in India. Plant Pathol 57:393

Rana T, Chandel V, Kumar Y, Ram R, Hallan V, Zaidi AA (2010) Molecular variability analyses of *Apple chlorotic leaf spot virus* capsid protein. J Biosci 35:605–615

Rana T, Negi A, Dhir S, Thockchom T, Chandel V, Walia Y, Singh RM, Ram R, Hallan V, Zaidi AA (2011) Molecular diagnosis of apple virus and viroid pathogens from India. Arch Phytopathol Plant Protect 44:505–512

Rani RV, Karthikeyan AS, Anuradha S, Veluthambi K (1996) Genome homologies among geminiviruses infecting Vigna, cassava, *Acalypha Croton and Vernonia*. Curr Sci 70:63–69

Rao MHP, Reddy DVR (1972) A new strain of tobacco mosaic virus on tomato (Lycopersicon esculentum). Indian Phytopathol 24:672–678

Rao RDVJP, Ragunathan V, Joshi NC (1979) Interception of peanut mottle virus. Indian J Plant Prot 7:59–63

Rao GP, Chatenet M, Girard JC, Rott P (2006) Distribution of sugarcane mosaic and sugarcane streak mosaic virus in India. Sugar Technol 8:79–81

Rao GP, Sharma SK, Singh D, Arya M, Singh P, Baranwal VK (2014) Genetically diverse variants of sugarcane bacilliform virus infecting sugarcane in India and evidence of a novel recombinant Badnavirus variant. J Phytopathol 162:779–787

Ravi K, Joseph J, Nagaraju N, Krishnaprasad S, Reddy H, Savithri HS (1997) Characterization of a pepper vein banding virus from chillipepper in India. Plant Dis 81:673–677

Ravi KS, Kitkaru AS, Winter S (2006) *Iris yellow spot virus* in onions: a new *tospovirus* record form India. Plant Pathol 55:288

Reddy DVR, Iizuka N, Ghanekar AM, Murthy VK, Kuhn CW, Gibbons RW, Chohan IS (1978) The occurrence of peanut mottle virus in India. Plant Dis Rep 62:978–982

Reddy BVB, Obaiah S, Prasanthi L, Sivaprasad Y, Sujitha A, Krishna TG (2015) Mungbean yellow mosaic India virus is associated with yellow mosaic disease of blackgram (*Vigna mungo* L.) in Andhra Pradesh, India. Arch Phytopathol Plant Protect 48:345–353

Renukadevi P, Nagendran K, Nakkeeran S, Karthikeyan G, Jawaharlal M, Alice D, Malathi VG (2015) First report of *Tomato spotted wilt virus* infection of Chrysanthemum in India. Plant Dis 99:1190

Robertson NL, French R, Gray SM (1991) Use of group specific primers and the polymerase chain reaction for the detection and identification of luteoviruses. J Gen Virol 72:1473–1477

Rodoni BC, Ahlawat YS, Varma A, Dale JL, Harding RM (1997) The identification and characterisation of *Banana bract mosaic virus* in India. Plant Dis 81:669–672

Rodoni BC, Dale JL, Harding RM (1999) Characterization and expression of the coat protein-coding region of banana bract mosaic potyvirus, development of diagnostic assays and detection of the virus in banana plants from five countries in Southeast Asia. Arch Virol 144:1725–1737

Rojas MR, Gilbertson RL, Russell DR, Maxwell DP (1993) Use of degenerate primers in the polymerase chain reaction to detect whitefly-transmitted geminiviruses. Plant Dis 77:340–347

Roossinck MJ (1999) Cucumoviruses (Bromoviridae)-general features. In: Granoof L, Webster RG (eds) Encyclopedia of virology, 2nd edn. Academic, San Diego, pp 315–320

Rowhani A, Biardi L, Johnson R, Saldarelli P, Zhang YP, Chin J, Green M (2000) Simplified sample preparation method and one-tube RT-PCR for grapevine viruses. In: Extended abstracts of 13th meeting of ICVG. Adelaide, p 148

Roy A, Spoorthi P, Bag MK, Prasad TV, Singh R, Dutta M, Mandal B (2013) A leaf curl disease in germplasm of rapeseed-mustard in India: molecular evidence of a weed-infecting *Begomovirus*-betasatellite complex emerging in a new crop. J Phytopathol 161:522–535

Roy B, Chakraborty B, Mitra A, Sultana S, Sherpa AR (2015) Natural occurrence of Bhendi yellow vein mosaic virus on Litsea spp. in India. New Dis Rep 31:7

Sahana AB, Adkar-Purushothama CR, Chennappa G, Zhang ZX, Sreenivasa MY, Sano T (2013) First report of *Grapevine yellow speckle viroid 1* and *Hop stunt viroid* Infecting Grapevines (*Vitis vinifera*) in India. Plant Dis 97:1517

Sanchez-Navarro JA, Canizares MC, Cano EA, Pallas V (1999) Simultaneous detection of five carnation viruses by nonisotopic molecular hybridization. J Virol Methods 82:167–175

Sankaralingam A, Baranwal VK, Ahlawat YS, Devi R, Ramiah M (2006) RT-PCR detection and molecular characterization of *Banana bract mosaic virus* from the pseudostem and bract of banana. Arch Phytopathol Pfl 39:273–281

Sastry KSM (1964) A new strain of Chrysanthemum aspermy virus from India. Indian J Microbiol. 2007 4:37–40

Satyanarayana T, Mitchell SE, Reddy DV, Kresovich S, Jarret R, Naidu RA, Gowda S, Demski JW (1996) The complete nucleotide sequence and genome organization of the M RNA segment of peanut bud necrosis tospovirus and comparison with other tospoviruses. J Gen Virol 77:2347–2352

Saxena S, Hallan V, Singh BP, Sane PV (1998) Leaf curl disease of Carica papaya from India may be caused by a bipartite Geminivirus. Plant Dis 82:126

Selvarajan R, Balasubramanian V (2013) Natural occurrence of *Banana bunchy top virus* in *Ensete* superbum in India. Indian J Virol 24:97–98

Selvarajan R, Mary Sheeba M, Balasubramanian V, Rajmohan R, Dhevi LN, Sasireka T (2010) Molecular characterization of geographically different *Banana bunchy top virus* isolates in India. Indian J Virol 21:110–116

Selvarajan R, Mary SM, Balasubramanian V (2011) Simultaneous detection of episomal *Banana streak Mysore virus* and *Banana bunchy top virus* using multiplex RT-PCR. Curr Sci 100:31–34

Senanayake DMJB, Mandal B, Lodha S, Varma A (2006) Firstreport of chilli leaf curl virus affecting chilli in India. Plant Pathol 56:343

Senanayake DMJB, Varma A, Mandal B (2012) Virus–vector relationships, host range, detection and sequence comparisonof *Chilli leaf curl virus* associated with an epidemic of leaf curl disease of chilli in Jodhpur. India J Phytopathol. doi:10.1111/j.1439-0434.2011.01876.x

Seoh ML, Wong SM, Zhang L (1998) Simultaneous TD/RT-PCR detection of *Cymbidium mosaic potexvirus* and *Odontoglossum ringspot tobamovirus* with a single pair of primers. J Virol Methods 72:197–204

Sharma J, Jain RK, Varma A (2004) Detection of papaya ringspot virus in naturally infected papaya plants by reverse transcription polymerase chain reaction. Indian Phytopathol 57:237–239

Sharma J, Jain RK, Ramiah M, Varma A (2005a) Natural spread of papaya ringspot virus to new areas: occurrence in Coimbatore, Tamil Nadu. Indian Phytopathol 58:245–249

Sharma A, Zaidi AA, Sood A, Sharma A, Vij SP, Pathak P, Ahuja PS (2005b) Virus analysis of indigenous orchid germplasm collection. J Orchid Soc India 19:41–46

Sharma SK, Tarafdar A, Khatun D, Sumita K, Biswas KK (2012) Intra-farm diversity and evidence of genetic recombination of *Citrus tristeza virus* isolates in Delhi region of India. J Plant Biochem Biotechnol 21:38–43

Sharma P, Verma RK, Mishra RD, Choudhary K, Gaur RK (2013) First report of *Turnip yellow virus* (TuYV) in *Brassica juncea* (Indian mustard) in India. New Dis Rep 27:21

Sharma SK, Vignesh Kumar P, Baranwal VK (2014a) Immunodiagnosis of episomal *Banana streak MY virus* using polyclonal antibodies to an expressed putative coat protein. J Virol Methods 207:86–94

Sharma SK, Vignesh Kumar P, Poswal R, Rai R, Geetanjali AS, Prabha K, Jain RK, Baranwal VK (2014b) Occurrence and distribution of banana streak disease and standardization of a reliable detection procedure for routine indexing of banana streak viruses in India. Sci Hortic 179:277–283

Sharma SK, Vignesh Kumar P, Geetanjali AS, Pun KB, Baranwal VK (2015) Subpopulation level variation of banana streak viruses in India and common evolution of banana and sugarcane badnaviruses. Virus Genes 50:450–465

Sherpa AR, Hallan V, Ram R, Vij SP, Pathak P, Garg ID, Zaidi AA (2003) First report of *Cymbidium* mosaic virus on *Cymbidiums* in India. Plant Pathol 52:788

Sherpa AR, Bag TK, Hallan V, Zaidi AA (2006) Detection and incidence of *Odontoglossum ringspot virus* in orchids. Australas Plant Pathol 35:69–71

Sherpa AR, Hallan V, Pathak P, Zaidi AA (2007) Complete nucleotide sequence analysis of *Cymbidium mosaic virus* Indian isolate: further evidence for natural recombination among potex viruses. J Biosci 34:663–669

Shetti P, Peter A, Jingade P (2012) Serological and molecular detection of an isolate of *Cucumber mosaic virus* (CMV) infecting cucumber (*Cucumis sativus*) and cloning of its coat protein gene. J Biochem Technol 3:198–202

Shih SL, Tsai WS, Green SK, Singh D (2006) First report of tomato leaf curl Joydebpur virus infecting chilli in India. New Dis Rep 14:17

Siju S, Madhubala R, Bhat AI (2007) Sodium sulphite enhances RNA isolation and sensitivity of Cucumber mosaic virus detection by RT-PCR in black pepper. J Virol Methods 141:107–110

Siljo A, Bhat AI, Biju CN, Venugopal MN (2012) Occurrence of *Banana bract mosaic virus* on cardamom. Phytoparasitica 40:77–85

Siljo A, Bhat AI, Biju CN (2013) Analysis of variability in Cardamom mosaic virus isolates occurring in India using symptomatology and coat protein gene sequence. J Spices Aromat Crops 22:70–75

Siljo A, Bhat AI, Biju CN (2014) Detection of *Cardamom mosaic virus* and *Banana bract mosaic virus* in cardamom using SYBR Green based reverse transcription-quantitative PCR. Virus Dis 25:137–141

Singh R, Raj SK, Chanadra G (2001) Association of monopartite *Begomovirus* with yellow mosaic disease of pumpkin in India. Plant Dis 85:1029

Singh RP, Nie X, Singh M, Coffin R, Duplessis P (2002) Sodium sulphate inhibition of potato and cherry polyphenolics in nucleic acid extraction for virus detection by RT-PCR. J Virol Methods 99:123–131

Singh HP, Hallan V, Raikhy G, Kulshrestha S, Sharma ML, Ram R, Garg ID, Zaidi AA (2005) Characterization of an Indian isolate of *Carnation mottle virus* infecting carnations. Curr Sci 88:594–601

Singh L, Hallan V, Jabeen N, Singh AK, Ram R, Martin DP, Zaidi AA (2007a) Coat protein gene diversity among *Chrysanthemum* virus B isolates from India. Arch Virol 152:405–413

Singh MK, Sherpa AR, Hallan V, Zaidi AA (2007b) A potyvirus in *Cymbidium* spp. In Northern India. Australas Plant Dis 2:11–13

Singh R, Raj SK, Prasad V (2008) Molecular characterization of a strain of *Squash leaf curl China virus* from North India. J Phytopathol 156:222–228

Singh AK, Mishra KK, Chattopadhyay B, Chakraborty S (2009a) Biological and molecular characterization of a *Begomovirus* associated with yellow mosaic vein mosaic disease of pumpkin from Northern India. Virus Genes 39:359–370

Singh D, Tewari AK, Rao GP, Karuppaiah R, Viswanathan R, Arya M, Baranwal VK (2009b) RT-PCR/PCR analysis detected mixed infection of DNA and RNA viruses infecting sugarcane crops in different states of India. Sugar Technol 11:373–380

Singh D, Pathania M, Ram R, Zaidi AA, Verma N (2010) Screening of chrysanthemum cultivars for Chrysanthemum stunt viroid in an Indian scenario. Arch Phytopathol Plant Protect 43:1517–1523

Singh MK, Haq QMR, Mandal B, Varma A (2011a) Evidence of the association of *Radish leaf curl virus* with tobacco yellow leaf curl disease in Bihar, India. Virus Genes 23:64–69

Singh MK, Singh K, Haq QMR, Mandal B, Varma A (2011b) Molecular characterization of *Tobacco leaf curl Pusa virus*, a new monopartite *Begomovirus* associated with tobacco leaf curl disease in India. Virus Genes 43:296–306

Singh L, Hallan V, Martin DP, Ram R, Zaidi AA (2012) Genomic sequence analysis of four new chrysanthemum virus B isolates: evidence of RNA recombination. Arch Virol 157:531–537

Singh JK, Tarafdar A, Sharma SK, Biswas KK (2013) Evidence of recombinant *Citrus tristeza virus* isolate occurring in acid lime cv. Pant lemon orchard in Uttarakhand terai region of Northern Himalaya in India. Indian J Virol 24:35–41

Singh R, Banerjee A, Sharma SK, Bhagawati R, Baruah S, Ngachan SV (2015) First report of Turnip mosaic virus occurrence in cole crops (*Brassica* spp) from Arunachal Pradesh, India. Virus Dis 26:211–213

Sivalingam PN (2006) Characterization of begomoviruses affecting tomato in India. Ph.D. Thesis submitted to Indian Agricultural Research Institute, New Delhi-12

Sivalingam PN, Varma A (2007) PCR based diagnosis of begomoviruses associated with tomato leaf curl disease in India. J Biochem Biotechnol 16:17–22

Smart CD, Schneider B, Blomquist CL, Guerra L, Harrison J, Ahrens NA, Lorenz UKH, Seemüller E, Kirkpatrick BC (1996) Phytoplasma-specific PCR primers based on sequences of the 16S-23S rRNA spacer region. Appl Environ Microbiol 62:2988–2993

Sohrab SS, Mandal B, Pant RP, Varma A (2003) First report of association of *Tomato leaf curl New Delhi virus* with the yellow mosaic disease of *Luffa cylindrica*. Plant Dis 87:1148

Sohrab SS, Mandal B, Ali A, Varma A (2006) Molecular diagnosis of emerging begomovirus diseases in cucurbits occurring in northern India. Indian J Virol 17:88–95

Sohrab SS, Mandal B, Ali A, Varma A (2010) Chlorotic curly stunt: a severe *begomovirus* disease of bottle gourd in Northern India. Indian J Virol 21:56–63

Soumya K, Yogita M, Prasanthi Y, Anitha K, Kavi Kishor PB, Jain RK, Mandal B (2014) Molecular characterization of Indian isolate of peanut mottle virus and immuno diagnosis using bacterial expressed core capsid protein. Virus Dis 25:331–337

Srinivasulu M, Sai Gopal DVR (2011) Coat protein sequence comparison of south Indian isolates of Papaya ringspot virus with other Indian subcontinent isolates. Phytopathol Mediterr 50:359–367

Srivastava KM, Raizada RK, Singh BP (1983) Bean yellow mosaic virus on *Gladiolus psittacinus* in India. Indian J Plant Pathol 1:83–88

Srivastava KM, Raj SK, Singh BP (1992) Properties of a Cucumber mosaic virus strain naturally infecting chrysanthemum in India. Plant Dis 76:474–477

Subba Reddy CV, Sreenivasulu P, Sekhar G (2011) Duplex-immunocapture-RT-PCR for detection and discrimination of two distinct potyviruses naturally infecting sugarcane (*Saccharum* spp. hybrids). Indian J Exp Biol 49:68–73

Sudhakar N, Prasad DN, Mohan N, Murugesan K (2006) First report of *Cucumber mosaic virus* subgroup-II infecting *Lycopersicon esculantum* in India. Plant Dis 90:1457

Swarnalatha P, Reddy MK (2014) Duplex PCR for simultaneous detection of Begomovirus and Phytoplasma from naturally infected tomato. Pest Manag Horticult Ecosyst 20:59–68

Tarafdar A, Godara S, Dwivedi S, Jayakumar BK, Biswas KK (2013) Characterization of Citrus tristeza virus and determination of genetic variability in North-east and South India. Indian Phytopathol 66:302–307

Thomas KM, Krishnaswamy CS (1939) Leaf crinkle: a transmissible disease of papaya. Curr Sci 8:316

Tiwari AK, Mall S, Khan MS, Snehi SK, Sharma PK, Rao GP, Raj SK (2010a) Detection and identification of *Tomato leaf curl Palampur virus* infecting *Cucurbita pepo* in India. Guanaxi Agric Sci 41:1291–1295

Tiwari N, Padmalatha KV, Singh VB, Haq QMI, Malathi VG (2010b) *Tomato leaf curl Bangalore virus* (ToLCBV): infectivity and enhanced pathogenicity with diverse betasatellites. Arch Virol 155:1343–1347

Tiwari AK, Snehi SK, Singh R, Raj SK, Rao GP, Sharma PK (2011) Molecular identification and genetic diversity among six *Begomovirus* isolates affecting cultivation of cucurbitaceous crops in Uttar Pradesh, India. Arch Phytopathol Plant Protect 45:1070–1075

Tiwari AK, Snehi SK, Khan MS, Sharma PK, Raj SK, Rao GP (2012) Molecular detection and identification of *Tomato leaf curl New Delhi virus* associated with yellow mosaic and leaf curl disease of *Luffa cylindrica* in India. Indian Phytopathol 65:80–84

Umamaheswaran K, Jain RK, Bhat AI, Ahlawat YS (2003) Biological and molecular characterization of a *Tospovirus* isolate from tomato and its relationship with other *Tospoviruses*. Indian Phytopathol 56:168–173

Usharani KS, Surendranath B, Haq QMR, Malathi VG (2004a) Yellow mosaic virus infecting soybean in Northern India is distinct from the species infecting soybean in southern and western India. Curr Sci 86:845–850

Usharani KS, Surendranath B, Paul Khurana SM, Garg ID, Malathi VG (2004b) Potato leaf curl – a new disease of potato in northern India caused by a strain of *Tomato leaf curl New Delhi virus*. Plant Pathol 53:235

Usharani KS, Surendranath B, Haq QMR, Malathi VG (2005) Infectivity analysis of a soybean isolate of Mungbean yellow mosaic India virus by agroinoculation. J Gen Plant Pathol 71:230–237

Usharani TR, Laxmi V, Jalali S, Krishnareddy M (2013) Duplex PCR to detect both Papaya ringspot virus and Papaya leaf curl virus simultaneously from naturally infected papaya (*Carica papaya* L.) Indian J Biotechnol 12:269–272

Valand GB, Muniyappa V (1992) Epidemiology of Tobacco leaf curl virus in India. Ann Appl Biol 120:257–267

Van der Vlugt RAA, Steffens P, Cuperus C, Barg E, Lesemann DE, Bos L, Vetten HJ (1999) Further evidence that shallot yellow stripe virus (SYSV) is a distinct potyvirus and reidentification of*Welsh onion yellow stripe virus* as a SYSV strain. Phytopathology 89:148–155

Varma PM (1955) Ability of the whitefly to carry more than one virus simultaneously. Curr Sci 24:317–318

Varma A, Giri BK (1998) Virus disease of cucurbits in India. In: Nayer NM, More TA (eds) Cucurbits. Oxford and IBH Publishing House Pvt Ltd, New Delhi. 1998

Varma A, Malathi VG (2003) Emerging geminivirus problems: a serious threat to crop production. Ann Appl Biol 142:145–164

Venkataravanappa V, Reddy CNL, Swaranalatha P, Jalali S, Briddon RW, Krishna Reddy M (2011) Diversity and phylogeography of begomovirus associated beta satellites of okra in India. Virol J 8:555

Venkataravanappa V, Lakshminarayana Reddy CN, Jalali S, Krishna Reddy M (2012a) Molecular characterization of distinct bipartite *Begomovirus* infecting bhendi (*Abelmoschus esculentus* L.) in India. Virus Genes 44:522–535

Venkataravanappa V, Reddy CNL, Jalali S, Reddy MK (2012b) Molecular characterization of distinct bipartite begomovirus infecting bhendi (*Abelmoschus esculentus L.*) in India. Gene 44:522–535

Verma N, Ram R, Sharma A, Hallan V, Zaidi AA, Garg ID (2003) Detection, identification and incidence of Chrysanthemum B carlavirus in chrysanthemum in India. Crop Prot 22:425–429

Verma R, Baranwal VK, Prakash S, Tomer SPS, Pant RP, Ahlawat YS (2006) First report of Papaya ringspot virus W in sponge gourd from India. Plant Dis 90:974

Verma N, Mehra A, Singh L, Hallan V, Singh AK, Jabeen N, Singh MK, Ram R, Zaidi AA (2007) Screening of viruses infecting chrysanthemum cultivars in India. Sci Hortic 111:260–265

Verma N, Kumar K, Kulshrestha S, Raikhy G, Hallan V, Ram R, Zaidi AA, Garg ID (2009) Molecular studies on *Tomato aspermy virus* isolates infecting chrysanthemums. Arch Phytopathol Plant Protect 42:99–111

Vishnoi R, Raj SK, Prasad V (2009) Molecular characterization of an Indian isolate of *Banana bunchy top virus* based on six genomic DNA components. Virus Genes 38:334–344

Vishnoi R, Kumar S, Raj SK (2013) Molecular characterization of a *Cucumber mosaic virus* isolate associated with mosaic disease of banana in India. Phytoparasitica 41:545–555

Viswanathan R, Karuppaiah R (2010) Distribution pattern of RNA viruses causing mosaic symptoms and yellow leaf in Indian sugarcane varieties. Sugar Cane Int 28:202–205

Viswanathan R, Rao GP (2011) Disease scenario and management of major sugarcane diseases in India. Sugar Technol 13:336–353

Viswanathan R, Balamuralikrishnan M, Karuppaiah R (2008a) Identification of three genotypes of sugarcane yellow leaf virus causing yellow leaf disease from India and their molecular characterization. Virus Genes 37:368–379

Viswanathan R, Balamuralikrishnan M, Karuppaiah R (2008b) Characterization and genetic diversity of sugarcane streak mosaic virus causing mosaic in sugarcane. Virus Genes 36:553–564

Viswanathan R, Karuppaiah R, Malathi P, Ganesh Kumar V, Chinnaraja C (2009) Diagnosis of Sugarcane yellow leaf virus in asymptomatic sugarcane by RT-PCR. Sugar Technol 11:368–372

Watpade S, Raigond B, Thakur PD, Handa A, Pramanick KK, Sharma YP, Tomar M (2012) Molecular detection of latent *Apple chlorotic leaf spot virus* in elite mother plants of apple. Indian J Virol 23:359–363

Wyatt SD, Brown JK (1996) Detection of subgroup III geminivirus isolates in leaf extracts by degenerate primers and polymerase chain reaction. Phytopathology 86:1288–1293

Yadav RK, Shukla RK, Chattopadhyay D (2009) Soybean cultivar resistant to *Mungbean Yellow Mosaic India Virus* infection induces viral RNA degradation earlier than the susceptible cultivar. Virus Res 144:89–95

Yadav MK, Biswas KK, Lal SK, Baranwal VK, Jain RK (2013) A distinct strain of *Cowpea mild mottle virus* infecting soybean in India. J Phytopathol 161:739–744

Yeh SD, Chang TF (1995) Nucleotide sequence of the N gene of *Watermelon silver mottle virus*, a proposed new member of the genus *Tospovirus*. Phytopathology 85:58–64

Zaidi AA, Hallan V, Rana T, Raikhy G, Ram R (2009) Viruses in ornamental crops and their management: an Indian scenario. J Ornamental Hortic 12:147–166

Part IV

Virus Management

Phytoproteins and Induced Antiviral Defence in Susceptible Plants: The Indian Context

28

Vivek Prasad and Shalini Srivastava

Abstract

Amongst the many approaches being tried out to contain plant viruses, induced systemic resistance (ISR) is one that finds its basis in the induction of antiviral resistance in susceptible plants against viruses by application of sap from certain non-host plants. Phytoprotein based antiviral researches in India started with the first study published in 1952. Progress on such research in India has been focused on screening plants for potential antiviral activity and/or induction of resistance, purification of the active principles and their characterization, and insights into possible mechanisms of action. However, not much success has been achieved at the field level. Resistance induced by plant proteins has been found to be either local or systemic, and in a majority of cases, is reversed by actinomycin D, suggesting host transcriptional involvement. The major plants harbouring such proteins are *Clerodendrum inerme*, *C. aculeatum*, *Boerhaavia diffusa*, *Bougainvillea spectabilis*, *B. xbuttiana*, and *Celosia cristata*. Most proteins inducing resistance fall in the molecular mass range of 25–35 kDa, are heat tolerant, and basic glycoproteins. A few possess ribosome-inactivating properties and share amino acid sequence homologies with other known ribosome-inactivating proteins. Systemic resistance inducing proteins have been shown to induce the production of a virus inhibitory agent (VIA) in the susceptible plant. The VIAs are also proteins, in the range of 30–65 kDa, and are tolerant to conditions that would degrade normal cellular proteins. One such VIA has exhibited homologies to a lectin. A few studies have suggested that at least some of the resistance inducing proteins suppress virus replication.

V. Prasad (✉) • S. Srivastava
Molecular Plant Virology Lab, Department of Botany, Lucknow University, Lucknow 226007, Uttar Pradesh, India
e-mail: vivprasad@outlook.com

B. Mandal et al. (eds.), *A Century of Plant Virology in India*,
https://doi.org/10.1007/978-981-10-5672-7_28

Keywords
Systemic induced resistance • Virus inhibitory agent • Antiviral activity • Ribosome-inactivating protein • *Clerodendrum inerme* • *Clerodendrum aculeatum* • *Boerhaavia diffusa*

28.1 Introduction

Plant virus control through the use of endogenous virus inhibitory plant proteins holds a lot of promise, as it comes without the involvement of microbial infection or chemical toxicity (Verma et al. 1998; Prasad et al. 2012; Srivastava and Prasad 2014). Plant extracts exhibiting antiviral activity have been known since long, and this topic opened up with the observation that the virus causing a mosaic disease on pokeweed (*Phytolacca* sp.) could be transmitted through sap inoculation to pokeweed but not to tobacco (Allard 1918). Experimental evidence for the same came in 1925 when the transmission of tobacco mosaic virus (TMV) onto healthy tobacco was inhibited following inoculation of TMV mixed with *Phytolacca* sap (Duggar and Armstrong 1925). The studies logically moved towards characterization of these inhibitors, and the virus inhibitor in *Phytolacca esculenta* was identified (Kassanis and Kleczkowski 1948). Studies on the mode of action of these inhibitors were a more challenging area and the importance of the host species in determining the action of the virus inhibitors became apparent (Gendron and Kassanis 1954; Grasso 1977). The virus inhibitors were then classified either as inhibitors of virus infection in vitro/in vivo, or as inhibitors of virus replication (Gianinazzi 1982). Another milestone was achieved with the demonstration of inhibition of polypeptide synthesis *in vitro* by the purified antiviral protein from *P. americana* (PAP) (Obrig et al. 1973; Owens et al. 1973). Thus PAP became the centre of all initial path breaking researches on plant virus inhibitors.

As of now, all well characterized inhibitors of virus infection from higher plants are able to cleave the *N*-glycosidic bond of the adenine residue in the highly conserved sequence (5′AGUACG<u>A</u>GAGGA3′) located on the α-sarcin/ricin loop of both eukaryotic (28S) and prokaryotic (23S) rRNAs, and thereby inhibit protein synthesis on ribosomes. They depurinate eukaryotic ribosomes and are hence classified as ribosome-inactivating proteins (RIPs) with rRNA *N*-glycosidase activity (Endo et al. 1987; Endo and Tsurugi 1988), or more appropriately with polynucleotide:adenosine glycosidase activity since a few RIPs can also deadenylate substrates such as tRNA, mRNA, viral RNA, poly A as well as DNA (Olivieri et al. 1996; Barbieri et al. 1997). RIPs are widespread in the plant kingdom (Barbieri et al. 1993, 2006; Stirpe 2004, 2013; Stirpe and Lappi 2014; Schrot et al. 2015) and were initially discovered as proteins present in different plant tissues that could inhibit viruses (Nielson and Boston 2001; Girbés et al. 2004; Di and Tumer 2015). Their antiviral activity was routinely ascertained by co-inoculating the inhibitor with a plant virus and determining the reduction in lesion number on the test host (Barakat and Stevens 1981; Stevens et al. 1981). This review will focus on inhibitors of plant viruses and inducers of systemic antiviral resistance in susceptible plants, with special emphasis on researches carried out in India in the last few decades.

28.2 Early Reports and Screening for Virus Inhibitors

Given the potential that these inhibitors carried in the realm of plant virus control methods, reports on the presence of antiviral substances in higher plants soon surfaced in India as well. In early screenings, the plant tissue was homogenized in water or buffer and the extract was either co-inoculated with the virus or applied shortly before or after virus challenge. One of the earliest reports was on inactivation of Cucumis virus 2C by plant extracts (Vasudeva and Nariani 1952). A very narrow host range of the virus prompted the investigators to study the effect of leaf extracts of a few solanaceous plants, such as *Nicotiana glutinosa*, *N. tabacum* and *Datura stramonium*, to name a few, on the infectivity of the virus on bottlegourd (*Lagenaria leucantha*). Leaf extracts of *D. stramonium*, *Capsicum annuum* and *Lycopersicon esculentum* markedly inhibited this virus (Vasudeva and Nariani 1952). Initial studies made no attempt towards characterization of these inhibitors but instead, focussed on screening of plants carrying inhibitory principles in their various parts, *viz.*, roots, flowers, fruits and bark, but mostly leaves. In a few cases the physico-chemical attributes of the inhibitors were included, studied through effects on the antiviral activity, of factors like temperature, pH, dilution, chemicals, organic solvents, enzymes, aging in vitro, precipitation by ammonium sulfate and dialysis. The results were a pointer towards the proteinaceous or non-proteinaceous nature of the inhibitors.

Thus, inhibitors reported from several higher plants soon stood partially (Paliwal and Nariani 1965a, b; Verma et al. 1969; Roychoudhury and Basu 1983; Verma and Baranwal 1983; Verma et al. 1984; Baranwal and Verma 1997), or completely characterized. The latter are dealt with separately in this review. Though reported mostly from angiosperms, culture filtrates of fungi, *Trichothecium roseum* and *Aspergillus niger* (Rao and Raychaudhuri 1965; Sharma and Raychaudhuri 1965), and some pteridophytes and gymnosperms, were also found to contain antiviral substances (Pandey and Bhargava 1980, 1983; Rao and Shukla 1985). Plant extracts inhibiting a number of animal viruses are known as well (Vijayan et al. 2004; Bhanuprakash et al. 2007, 2008; Bag et al. 2012; Nutan et al. 2013) along with several instances of antibacterial and antifungal activity (Guleria and Kumar 2006; Sundaram et al. 2011; Yadav et al. 2011).

In the early studies, inhibition of plant viruses by extracts from higher plants, including several medicinal plants such as *Cinchona ledgeriana*, *Emblica officinalis*, *Chrysobalanus icaco*, *Terminalia chebula* and *Ocimum sanctum*, was demonstrated against a number of mechanically transmitted viruses. While an exhaustive list of plants and the viruses inhibited has been variously incorporated in earlier reviews (Verma 1982; Narayanasamy 1990; Verma and Prasad 1992; Verma et al. 1995; Verma and Baranwal 2011), a chronological list of antiviral research in India has been included here as Table 28.1. Plant extracts were reported to inhibit viruses like TMV, potato virus X (PVX), Radish mosaic virus, Chilli mosaic virus, Sunnhemp mosaic virus, Watermelon mosaic virus, Pea top necrosis virus, etc. (Rao and Raychaudhuri 1965; Raychaudhuri and Prasad 1965; Raychaudhuri and Chadha 1965; Paliwal and Nariani 1965a, b; Bhargava and Singh 1965; Singh 1969, 1972; Verma et al. 1969; Dhaliwal and Dhaliwal 1971; Gupta and Raychaudhuri 1971; Lal et al. 1973; Roy et al. 1979).

Virus inhibitory activity was soon reported in seed, root and floral extracts of flowering plants (Khurana and Bhargava 1970; Sharma and Chohan 1973; Verma et al. 1975; Murty and Nagarajan 1980; Rao et al. 1985). Though virus inhibitors were rarely reported in monocotyledons, an inhibitor of TMV infectivity was detected in wheat seed extract that caused a reduction in local lesions on *N. glutinosa* (Verma and Verma 1965). Seed extracts of *Argemone mexicana, Datura fastuosa, Raphanus sativus* and *Rhynchosia asnaris* were inhibitory to PRSV and PapMV on *Carica papaya* (Khurana and Bhargava 1970), while in a related study, the seed extracts of *Syzygium cumini, Callistemon citrinus* and *Mangifera indica* were found to inhibit Cucumis virus I on its local lesion host *Chenopodium amaranticolor* (Sharma and Chohan 1973). Unripe *Lawsonia alba* and germinating seeds of *Phaseolus radiatus* (black gram) showed a strong anti-TMV activity on *N. glutinosa* (Verma et al. 1975; Murty and Nagarajan 1980). Flower extracts of *Azadirachta indica, Euphorbia milii* and *Vinca rosea* were effective at inhibiting PVX lesions on *C. amaranticolor*, whereas only *A. indica* extract prevented systemic PVX infection on *L. esculentum* (Rao et al. 1985). Inhibitors of PVX and watermelon mosaic virus infectivity were also identified in bark extracts of *Ficus elastica*, *Prunus persica* and some other plant species (Singh and Singh 1973, 1975; Tewari 1976). Later, inhibition of turnip mosaic virus (TuMV), peanut green mosaic virus, tobacco ringspot virus, urdbean leaf crinkle virus, bean common mosaic virus, bottle gourd mosaic virus and rice tungro virus by various other plant extracts was also reported (Bose et al. 1983; Chowdhury and Saha 1985; Pandey and Mohan 1986; Saigopal et al. 1986; Srinivasulu and Jeyarajan 1986; Darekar and Sawant 1989). Pumpkin mosaic and Pumpkin yellow vein mosaic could be inhibited by extracts from several medicinal plants as well as their derivatives, such as neem oil and neem seed kernel extract (Louis and Balakrishnan 1996; Jayashree et al. 1999). Upon screening of leaf extracts from several plants, only the extract from *Psidium guajava* was found to inhibit transmission of Brassica isolate of Turnip mosaic potyvirus (TuMV-B) on *Brassica juncea* var. rugosa. Complete inhibition of transmission of TuMV-B and Chilli mosaic virus was recorded when these viruses were mixed with an equal volume of the guava leaf extract (Mandal and Singh 2001). Such screenings for inhibitors continued, and occasionally the focus shifted towards use of inhibitors for plant virus control in glass houses and open fields, where the pathogen stress was

Table 28.1 A chronology of plant antiviral research in India

Year	Source plant[a]	Test host	Virus[b]	Aspect studied[c]	Reference
1952	*Capsicum annuum, Datura stramonium, Lycopersicon esculentum*	*Lagenaria leucantha*	Cucumis virus 2C	AV-A AV-B	Vasudeva and Nariani (1952)
1965	*Cedrus deodara* (cone/fruit)	*Chenopodium amaranticolor*	ChMV	AV-A	Raychaudhuri and Chadha (1965)
1965	*Portulaca grandiflora*	*Cucurbita pepo*	WMV	AV-A AV-C	Bhargava and Singh (1965)
1965	*Triticum aestivum* (seed)	*Nicotiana glutinosa*	TMV	AV-A	Verma and Verma (1965)
1969	*Cinchona ledgeriana* (stem bark extract), *Chrysobalanus icaco* (leaf extract), *Emblica officinalis, Terminalia chebula* (fruit pericarp extract)	–	PVX	AV-A AV-B	Verma et al. (1969)
1970	*Achyranthes aspera, Aloe barbadensis, Argemone mexicana, Capsicum frutescens, Carica papaya, Carum copticum* (leaf), *Argemone mexicana, Datura fastuosa, Raphanus sativus, Rhynchosia asnaris* (seed)	*Carica papaya*	PapMV, PRSV	AV-A AV-C	Khurana and Bhargava (1970)
1971	*Acacia arabica*	*C. amaranticolor*	PVY	AV-B	Gupta and Raychaudhuri (1971)
1971	*Ailanthus altissima, Allium cepa, Allium sativum*	*Phaseolus vulgaris*	TMV	AV-A	Dhaliwal and Dhaliwal (1971)
1972	*Acacia arabica, Callistemon lanceolatus, Syzygium cumini*	*C. amaranticolor*	PVY	AV-M	Gupta and Raychaudhuri (1972)
1972	*Calotropis procera, Carica papaya, Ficus benghalensis, Thevetia neriifolia*	*C. amaranticolor, N. glutinosa*	TMV	AV-A	Khurana and Singh (1972)
1973	*Datura metel, Phaseolus vulgaris, Solanum melongena*	*N. glutinosa, C. amaranticolor, Nicotiana tabacum* cv. NP-31, *N. tabacum* cv. Xanthi	TMV	AV-A	Lal et al. (1973)

(continued)

Table 28.1 (continued)

Year	Source plant[a]	Test host	Virus[b]	Aspect studied[c]	Reference
1973	*Callistemon citrinus*, *Mangifera indica*, *Syzygium cumini* (leaf and seed)	*C. amaranticolor*	Cucumis virus 1	AV-A	Sharma and Chohan (1973)
1974	*Jatropha gossypifolia*, *Jatropha pandurifolia*, *Jatropha podogarica* (latex)	*N. glutinosa*, *C. amaranticolor*, *N. tabacum* cv. NP-31, *N. tabacum* cv. Xanthi	Cucumis virus-2C, TMV	AV-A	Lal and Verma (1974)
1975	*Lawsonia alba* (seed)	*N. glutinosa*	TMV	AV-B AV-M	Verma et al. (1975)
1975	*Madhuca butyraceae*, *Calophyllum tomentosum*, *Aegle marmelos* (flavonoids, coumarins, marmelosin, etc.)	*C. tetragonoloba*	SSMV	AV-A	Chandra et al. (1975)
1975	*Solanum melongena*	*N. glutinosa, N. tabacum*	TMV, TRSV	AV-A	Verma and Mukerjee (1975)
1976	*Dahlia*	*N. glutinosa*	TMV	AV-A	Srivastava et al. (1976)
1976	*Ficus benghalensis*, *Ficus elastica*, *Grewia asiatica*, *Syzygium cumini*, *Tamarindus indica* (bark)	*C. pepo*	WMV	AV-A	Tewari (1976)
1979	*Boerhaavia diffusa* (root)	*C. amaranticolor*, *Datura metel*, *N. glutinosa*, *N. tabacum*, *Crotalaria juncea*, *Vigna sinensis*, *Spinacea oleracea*, *Gomphrena globosa*	TMV, SRV, GMV, TRSV	AV-A AV-B	Verma and Awasthi (1979)
1979	*Boerhaavia diffusa* (root)	*N. glutinosa*	TMV	AV-B	Verma et al. (1979a)
1979	*Boerhaavia diffusa*, *Cuscuta reflexa*, *Datura metel*, *Solanum melongena*	*N. tabacum* cv. NP-31	TMV, TRSV	AV-C	Verma et al. (1979b)

(continued)

Table 28.1 (continued)

Year	Source plant[a]	Test host	Virus[b]	Aspect studied[c]	Reference
1979	*Clerodendrum inerme*, *Gynandropsis pentaphylla*, *Lawsonia alba*, *Nicotiana* spp. *Solanum xanthocarpum*, *Tamarindus indica*	*N. tabacum*	TMV	AV-A	Patel and Patel (1979)
1979	*Datura metel*, *Eucalyptus lanceolatus*, *Euphorbia hirta*, *Solanum melongena*, *Cuscuta reflexa*, *Boerhaavia diffusa*, *Cannabis sativa*	*C. tetragonoloba*, *C. juncea*, *G. globosa*, *N. glutinosa*, *C. amaranticolor*	SRV, GMV, TRSV, TMV	AV-A	Verma et al. (1979c)
1979	*Mirabilis jalapa*	*Solanum tuberosum*	PLRV	AV-C	Verma and Kumar (1979)
1980	*Mirabilis jalapa*	*Cucumis sativus*	CMV, CGMMV	AV-C	Verma and Kumar (1980)
1980	*Boerhaavia diffusa* (root)	–	–	AV-M	Verma and Awasthi (1980)
1980	*Arachis hypogaea*, *Cajanus cajan*, *Cicer arietinum*, *Cyamopsis tetragonoloba*, *Dolichos biflorus*, *Phaseolus aureus*, *Phaseolus radiata*, *Vigna catjang* (germinating pulse and oilseed)	*N. glutinosa*	TMV	AV-A	Murty and Nagarajan (1980)
1981	*Cuscuta reflexa*	*N. glutinosa*	TMV	AV-A AV-B	Awasthi (1981)
1982	*Datura metel*	*C. tetragonoloba*, *C. amaranticolor*, *Vigna sinensis*, *C. juncea*, *G. globosa*	SRV, GMV	AV-A AV-M	Mukerjee et al. (1981)
1983	*Bougainvillea spectabilis*	*Cucumic melo*, *C. juncea*, *Lycopersicon esculentum*	CGMMV, SRV, TMV, TmYMV, PhySMV	AV-C	Verma and Dwivedi (1983)

(continued)

Table 28.1 (continued)

Year	Source plant[a]	Test host	Virus[b]	Aspect studied[c]	Reference
1983	*Chenopodium ambrosoides*	*C. amaranticolor*, *N. glutinosa*, *C. tetragonoloba*, *Datura metel*, *Datura stramonium*, *V. sinensis*	TMV, SRV	AV-A AV-B	Verma and Baranwal (1983)
1984	*Ampelopteris prolifera* (fern leaf)	*N. glutinosa*, *C. amaranticolor*, *N. tabacum* cv. NP-31, *N. tabacum* var. white burley	TMV, CMV	AV-A AV-B	Pandey and Bhargava (1984)
1984	*Boerhaavia diffusa* (root)	*N. tabacum* cv. NP-31, *L. esculentum*, *C. melo*, *C. juncea*, *G. globosa*	TMV, CMV, SRV, CGMMV, GMV	AV-C	Awasthi et al. (1984)
1984	*Bougainvillea spectabilis*	–	–	AV-M	Verma and Dwivedi (1984)
1984	*Clerodendrum aculeatum*, *Clerodendrum fragrans*, *Clerodendrum inerme*, *Clerodendrum indicum*, *Clerodendrum macrosiphon*, *Clerodendrum phlomoides*, *Clerodendrum splendens*, *Clerodendrum viscosum*	*C. amaranticolor*, *N. glutinosa*, *C. tetragonoloba*, *N. tabacum* Ky-58	TMV, SRV, TmYMV, GMV,	AV-A AV-B AV-M	Verma et al. (1984)
1984	*Solanum torvum*	*C. amaranticolor*, *N. glutinosa*, *C. tetragonoloba*	TMV, SRV	AV-A	Roychoudhury (1984)
1985	*Aerva sanguinolenta*	*C. tetragonoloba*, *D. stramonium*	SRV, TMV	AV-A AV-M	Verma and Srivastava (1985)
1985	*Argemone mexicana*, *Azadirachta indica*, *Euphorbia milii*, *Jasminum sambac*, *Lantana indica*, *Nerium indicum*, *Vinca rosea* (Flower)	*C. amaranticolor*, *L. esculentum*	PVX	AV-A AV-C AV-M	Rao et al. (1985)

(continued)

Table 28.1 (continued)

Year	Source plant[a]	Test host	Virus[b]	Aspect studied[c]	Reference
1985	*Clerodendrum aculeatum*, *Aerva sanguinolenta*, *Boerhaavia diffusa* (root)	*Vigna radiata*, *Vigna mungo*	Yellow mosaic disease virus	AV-C	Verma et al. (1985b)
1985	*Capsicum annuum*, *Solanum melongena*, *Lantana camara*, *Datura metel*, *Corchorus capsularis*, *Ipomoea glaberrima*, *Curcuma longa*, *Allium sativum*, *Azadirachta indica*, *Zingiber officinale*, *Callistemon lanceolatus* (leaf, bulb, fruit and rhizome)	*V. mungo*	ULCV	AV-A AV-C	Chowdhury and Saha (1985)
1985	*Pseuderanthemum atropurpureum tricolor*	–	–	AV-B AV-M	Verma and Khan (1985)
1986	*Clerodendrum aculeatum*	*N. tabacum* cv. Samsun NN	TMV	AV-M	Prasad (1986)
1986	*Mirabilis jalapa*, *Cocos nucifera*, *Sorghum vulgare*	*Oryza sativa*	RTV	AV-A AV-C	Srinivasulu and Jeyarajan (1986)
1986	*Peltophorum ferrugenium* (leaf), *Pithecolobium dulce* (twig)	Tobacco	TMV	AV-A AV-C	Murty and Nagarajan (1986)
1986	*Phyllanthus fraternus* (leaf and root)	*C. amaranticolor*, *Phaseolus vulgaris*, *V. sinensis*	TMV, PGMV, TRSV	AV-A	Saigopal et al. (1986)
1986	*Acacia arabica*, *Callistemon lanceolatus*, *Syzygium cumini*	*C. amaranticolor*, *C. album*	TuMV	AV-A	Pandey and Mohan (1986)
1987	*Clerodendrum aculeatum*	*C. tetragonoloba*	SRV	AV-M	Verma and Prasad 1987
1988	*Azadirachta indica*	*C. amaranticolor*	Spinach mosaic virus	AV-A	Zaidi et al. (1988)
1988	*Clerodendrum aculeatum*, *Clerodendrum fragrans*	*C. tetragonoloba*	SRV	AV-M	Prasad (1988)

(continued)

Table 28.1 (continued)

Year	Source plant[a]	Test host	Virus[b]	Aspect studied[c]	Reference
1988	*Boerhaavia diffusa*, *Clerodendrum fragrans*	*C. tetragonoloba*	SRV	AV-M	Verma and Prasad (1988)
1990	*Pseuderanthemum bicolor*	–	–	AV-M	Khan and Verma (1990)
1992	*Celosia cristata*	*D. stramonium*, *D. metel*, *Capsicum pendulum*, *C. tetragonoloba, N. glutinosa, N. rustica, V. sinensis, N. tabacum* cv, Samsun NN,	SRV, TMV, PVX	AV-A	Baranwal and Verma (1992)
1993	*Amaranthus tricolor*, *Aralia balfourii*, *Bougainvillea glabra*, *Celosia cristata*, *Chenopodium ambrosoides*	*C. tetragonoloba*	SRV	AV-A	Baranwal and Verma (1993)
1995	*Clerodendrum inerme*	*N. tabacum* cv, Samsun NN	TMV	AV-B	Prasad et al. (1995)
1995	*Clerodendrum aculeatum*	*C. juncea*	SRV	AC-C	Verma and Varsha (1995a)
1996	*Basella alba*, *Gomphrena globosa* (root) *Glycyrrhiza glabra*, *Phyllanthus fraternus*, *Plumbago rosea* (tuber), *Thespesia populnea*	*C. pepo*	PMV*	AV-A AV-C	Louis and Balakrishnan (1996)
1996	*Clerodendrum inerme*	–	–	AV-B	Olivieri et al. (1996)
1997	*Clerodendrum aculeatum*	*L. esculentum*	TLCV	AV-C	Baranwal and Ahmad (1997)
1997	*Celosia cristata*	*C. tetragonoloba*, *N. glutinosa*	SRV, TMV	AV-B	Baranwal and Verma (1997)
1997	*Clerodendrum aculeatum*	*N. tabacum* cv. Samsun NN	TMV	AV-B AV-M	Verma et al. (1996)
1997	*Clerodendrum aculeatum*	–	–	AV-B	Kumar et al. (1997)
1998	*Bougainvillea spectabilis* (root)	*Vigna unguiculata*	TSWV	AV-A AV-B	Balasaraswathi et al. (1998)

(continued)

Table 28.1 (continued)

Year	Source plant[a]	Test host	Virus[b]	Aspect studied[c]	Reference
1999	*Boerhaavia diffusa, Bougainvillea spectabilis, Croton bonplandianum, Prosopis chilensis, Azadirachta indica*	*C. pepo*	PYVMV	AV-A AV-C	Jayashree et al. (1999)
1999	*Cocos nucifera, Sorghum vulgare, Prosopis chilensis, Croton parsiflorus, Euphorbia thyifolia*	*V. radiata*	TSWV	AV-C	Manickam and Rajappan (1999)
1999	*Bougainvillea spectabilis, Catharanthus roseus, Cocos nucifera, Mirabilis jalapa, Prosopis chilensis, Sorghum vulgare, Vitex negundo*	*Abelmoschus esculentus*	OYVMV	AV-A AV-C	Pun et al. (1999)
1999	*Mirabilis jalapa*	*Solanum melongena*	CMV	AV-C	Bharathi (1999)
1999	*Azadirachta indica, Clerodendrum infortunatum, Ocimum sanctum, Vitex negundo*	*D. stramonium*	Brinjal mosaic virus	AV-C	Surendran et al. (1999)
2000	*Celosia cristata*	*N. tabacum* Samsun NN,	TMV, SRV, ICRSV	AV-A AV-B	Balasubrahmanyam et al. (2000)
2000	*Chenopodium album*	*N. glutinosa, C. tetragonoloba*	TMV, SRV	AV-A AV-B	Dutt et al. (2000)
2001	*Psidium guajava*	*Brassica juncea*	TuMV	AV-C	Mandal and Singh (2001)
2001	*Clerodendrum aculeatum*	*C. tetragonoloba*	SRV	AV-M	Prasad et al. (2001)
2001	*Clerodendrum inerme*	*N. tabacum* cv. White burley, *C. amaranticolor*	CMV, PVY, ToMV	AV-B	Praveen et al. (2001)
2001	*Bougainvillea xbuttiana*	*N. glutinosa, N. tabacum* cv. Samsun nn, *N. tabacum* cv. NP-31	TMV	AV-A AV-C AV-M	Narwal et al. (2001b)
2002	*Celosia cristata*	*In vitro*	BMV, PMV	AV-B	Baranwal et al. (2002)
2003	*Boerhaavia diffusa* (root)	*Cucumis melo*	CGMMV	AV-C	Awasthi and Kumar (2003)

(continued)

Table 28.1 (continued)

Year	Source plant[a]	Test host	Virus[b]	Aspect studied[c]	Reference
2003	*Piper longum*, *Prosopis juliflora*, *Zingiber officinale*	*Vigna mungo*	ULCV	AV-A	Thirumalaisamy et al. (2003)
2004	*Boerhaavia diffusa* (root)	*Vigna radiata*	Yellow vein mosaic virus	AV-C	Singh et al. (2004)
2004	*Celosia cristata*	*N. tabacum* cv. Samsun NN, *C. tetragonoloba*	TMV, SRV	AV-M	Gholizadeh et al. (2004)
2004	*Harpullia cupanioides*, *Mirabilis jalapa*	*L. esculentum*	TSWV	AV-C	Devi et al. (2004)
2005	*Azadirachta indica* (oil)	*C. annuum*, *L. esculentum*, *N. glutinosa*	TMV, ToMV	AV-A AV-C	Madhusudhan et al. (2005)
2005	*Bougainvillea spectabilis*	–	–	AV-B	Rajesh et al. (2005)
2005	*Bougainvillea xbuttiana*	*N. glutinosa*, *C. tetragonoloba*	TMV, SRV	AV-B	Bhatia and Lodha (2005)
2005	*Celosia cristata*	*N. glutinosa*	TMV	AV-B	Gholizadeh et al. (2005)
2006	*Harpullia cupanioides* (seed), *Azadirachta indica* (oil)	Finger millet	Mottle streak virus	AV-C	Saveetha et al. (2006)
2003	*Harpullia cupanioides* (seed)	*Oryza sativa*.....	TSWV, RTV, CABMV	AV-A	Bharathimatha et al. (2003)
2006	*Amaranthus tricolor*	*C. tetragonoloba*	SRV	AV-B	Roy et al. (2006)
2006	*Celosia cristata*	*N. glutinosa*, *C. tetragonoloba*	TMV, SRV	AV-B	Begam et al. (2006)
2007	*Azadirachta indica*, *Boerhaavia diffusa*, *Bougainvillea spectabilis*, *Clerodendrum inerme*, *M. jalapa*, *Psidium guajava*, *Thuja occidentalis*	*Vigna sinensis*	BCMV-BICM	AV-C	Prasad et al. (2007)
2007	*Ocimum sanctum*, *Psidium guajava*, *Thuja occidentalis*, *Tridax procumbens*	*N. glutinosa*	TMV, ToMV	AV-A	Deepthi et al. (2007)
2008	*Bougainvillea xbuttiana*	*C. tetragonoloba*	SRV	AV-B	Choudhary et al. (2008)
2009	*Azadirachta indica*, *Clerodendrum aculeatum*, *Terminalia arjuna*	*Cucumis sativus*	CMV	AV-C	Pardeep and Awasthi (2009)

(continued)

Table 28.1 (continued)

Year	Source plant[a]	Test host	Virus[b]	Aspect studied[c]	Reference
2009	*Boerhaavia diffusa* (root)	*A. esculentus*	OYVMV	AV-C	Singh et al. (2009)
2009	*Boerhaavia diffusa* (root), *Clerodendrum aculeatum*	*Carica papaya*	PRSV	AV-C	Awasthi and Singh (2009)
2009	*Bougainvillea spectabilis*	*Momordica charantia*	BGYMV	AV-C AV-M	Rajinimala et al. (2009)
2009	*Bougainvillea spectabilis*, *Prosopis chilinesis*	*Vigna sinensis*, *Helianthus annuus*	SFNV	AV-C AV-M	Lavanya et al. (2009)
2009	*Bougainvillea spectabilis*, *Mirabilis jalapa*	*Vigna mungo*	ULCV	AV-C	Karthikeyan et al. (2009)
2009	*Clerodendrum aculeatum*	*C. papaya*, *Chenopodium quinoa*	PRSV	AV-C	Srivastava et al. (2009)
2010	*Bacopa monerii*, *Boerhaavia diffusa*, *Catharanthus roseus*, *Calotropis procera*, *Ocimum sanctum*, *Withania somnifera*, *Rauwolfia serpentina*	*Pisum sativum*	PeMV	AV-C	Tiwari et al. (2010)
2010	*Mirabilis jalapa*, *Datura metel*, *Azadirachta indica* (oil)	*V. mungo*	MYMV	AV-C	Venkatesan et al. (2010)
2011	*Azadirachta indica*, *Bougainvillea spectabilis*, *Phyllanthus*, *Pongamia glabra*	*N. glutinosa* *Capsicum anuum*, *L. esculentum*	TMV ToMV	AV-A AV-C	Madhusudhan et al. (2011)
2011	*Azadirachta*, *Clerodendrum*, *Parthenium*	*L. esculentum*	TLCV	AV-C	Srivastav et al. (2011)
2011	*Boerhaavia diffusa* (root), *Clerodendrum aculeatum*	*C. papaya*	PRSV	AV-C	Singh et al. (2011a)
2011	*Boerhaavia diffusa* (root), *Clerodendrum aculeatum*, *Azadirachta indica*	*V. radiata*, *V. mungo*	MYMV	AV-C	Singh et al. (2011b)
2013	*Boerhaavia diffusa* (root)	*C. amaranticolor*, *D. stramonium*, *N. glutinosa*, *C. tetragonoloba*, *V. sinensis*	TMV, SRV, CGMMV, GMV	AV-A AV-M	Awasthi et al. (2013)

(continued)

Table 28.1 (continued)

Year	Source plant[a]	Test host	Virus[b]	Aspect studied[c]	Reference
2013	*Chlorophytum nepalense* (root)	*Solanum tuberosum*	PVX	AV-C	Acharya (2013)
2014	*Clerodendrum inerme*	*C. tetragonoloba*	SRV	AV-M	Prasad et al. (2014)
2015	*Azadirachta*, *Psidium*, *Chrysanthemum*	*A. esculentus*	OYVMV	AV-C	Kumar et al. (2015)
2015	*Boerhaavia diffusa* (root)	*N. tabacum* cv. Xanthi	TMV	AV-B	Srivastava et al. (2015b)
				AV-C	
				AV-M	
2015	*Clerodendrum aculeatum*	*N. tabacum* cv. Samsun NN	TMV	AV-M	Srivastava et al. (2015a)

[a]Antiviral activity associated with leaf tissue, unless specified otherwise in the table
[b]*BCMV-BlCM* Bean common mosaic potyvirus strain blackeye cowpea mosaic, *BGYMV* Bittergourd yellow mosaic virus, *BMV* Brome mosaic virus, *CABMV* Cowpea aphid borne mosaic virus, *CGMMV* Cucumber green mottle mosaic virus, *ChMV* Chilli mosaic virus, *CMV* Cucumber mosaic virus, *GMV* Gomphrena mosaic virus, *ICRSV* Indian citrus ringspot virus, *MYMV* Mungbean yellow mosaic virus, *OYVMV* Okra yellow vein mosaic virus, *PapMV* Papaya mild mosaic virus, *PeMV* Pea mosaic virus, *PGMV* Peanut green mosaic virus, *PhySMV* Physalis shoestring mosaic virus, *PLRV* Potato leaf roll virus, *PMV* Pokeweed mosaic virus, *PMV** Pumpkin mosaic virus, *PRSV* Papaya ringspot virus, *PVX* Potato virus X, *PVY* Potato virus Y, *PYVMV* Pumpkin yellow vein mosaic virus, *RaMV* Radish mosaic virus, *RTV* Rice tungro virus, *SFNV* Sunflower necrosis virus, *SSMV* Southern sunnhemp mosaic virus, *SRV* Sunnhemp rosette virus, *TLCV* Tomato leaf curl virus, *TmYMV* Tomato yellow mosaic/mottle virus, *ToMV* Tomato mosaic virus, *TMV* Tobacco mosaic virus, *TRSV* Tobacco ringspot virus, *TSWV* Tomato spotted wilt virus, *TuMV* Turnip mosaic virus, *ULCV* Urdbean leaf crinkle virus, *WMV* Watermelon mosaic virus
[c]*AV-A* Antiviral-activity, *AV-B* Antiviral-biochemical/molecular characterization, *AV-C* Antiviral-virus control, *AV-M* Antiviral-mode of action

undeniably higher. Extracts from *Psidium guajava*, *Leucas aspera*, *O. sanctum*, *Tridax procumbens*, *Phyllanthus niruri*, *Thuja occidentalis*, *Azadirachta indica*, *Pongamia glabra* and *Bougainvillea spectabilis* were effective in controlling TMV and Tomato mosaic virus (ToMV) on bell pepper and tomato plants (Deepthi et al. 2007; Madhusudhan et al. 2011). Spraying of leaf extracts from *A. indica*, *P. glabra* and *B. spectabilis* on *N. glutinosa* 1 h prior to inoculation, inhibited TMV and ToMV by 53–62%, while reduction in virus concentration, as determined by ELISA, was noted in bell pepper/TMV and tomato/ToMV combinations. Treatment of seeds of bell pepper and tomato with these extracts enhanced seedling vigour and seed germination (Madhusudan et al. 2011). Inhibition of safflower mosaic virus and sunflower mosaic virus by plant extracts from *M. jalapa*, *B. spectabilis* and *Prosopis chilensis* was also noted (Devi et al. 2004; Kulkarni and Byadagi 2004; Lavanya et al. 2009).

28.3 Inducers of Systemic Antiviral Resistance in Plants

It soon became apparent that a few of these inhibitors could also function as inducers of systemic antiviral resistance. Application of pepper extract at some distance from the point of virus inoculation inhibited lesion formation on cowpea and *Chenopodium* was a first in the demonstration of systemic resistance inducing ability associated with virus inhibitory sap (McKeen 1956), a finding that was later substantiated (Apablaza and Bernier 1972). In India, extract from *Solanum melongena* was seen to prevent infection of TMV and mosaic disease of sunnhemp and *Gomphrena* on treated as well as non-treated leaves of the test hosts (Verma and Mukerjee 1975). Thus, a new era of research opened up at the Lucknow University, following the demonstration of induction of resistance in plants against viruses by pre-inoculation treatment with plant extracts from several non-host plants. In all such cases, the extracts or purified proteins were applied onto the lower leaves of a test host, 6–24 h prior to virus challenge on the lower (treated) as well as upper (untreated) leaves of the host plants. Induced resistance was routinely assayed on a test host which gave a hypersensitive response to virus infection. A decrease in the number of local lesions in the untreated leaves was taken as a measure of systemic induced resistance, a host-mediated response that could be reversed by simultaneous application of actinomycin D. In hosts that allowed the virus to spread systemically, a decrease in symptom severity or delay in symptom production post-inoculation was generally noted. The resistance inducing proteins were purified by using a series of column chromatography matrices, which included ion-exchange, hydrophobic-interaction, reverse-phase and molecular sieving, with an occasional use of adsorption chromatography on Hydroxyapatite matrix (Fig. 28.1). Almost all the antiviral resistance inducing proteins were highly stable to thermal denaturation and proteolytic degradation. Some also inhibited protein synthesis, and, hence, at a functional level, were RIPs. Barring very few exceptions, none of the vast majority of RIPs were reported to induce systemic resistance. A few resistance inducers (both systemic and localized inducers) that have been well characterized at the physico-chemical as well as the molecular level are described below and compared in Table 28.2.

28.4 Plant Species as Antiviral Sources

28.4.1 *Clerodendrum inerme*

Leaf extracts from *Clerodendrum inerme* (renamed as *Volkameria inermis,* family Lamiaceae) could inhibit inhibit TMV, SRV and TmYMV on *C. amaranticolor* by nearly 70–75% and TMV infection on *C. amaranticolor*, *D. stramonium*, *D. metel*, *N. glutinosa* and *N. tabacum* var. Ky-58 by almost 100%. A complete inhibition of SRV was noted on *C. tetragonoloba* when the time interval between application of the leaf extract and virus challenge was 24 h. The inhibitor could induce systemic resistance and was reported to be heat stable and non-dialyzable in nature (Verma

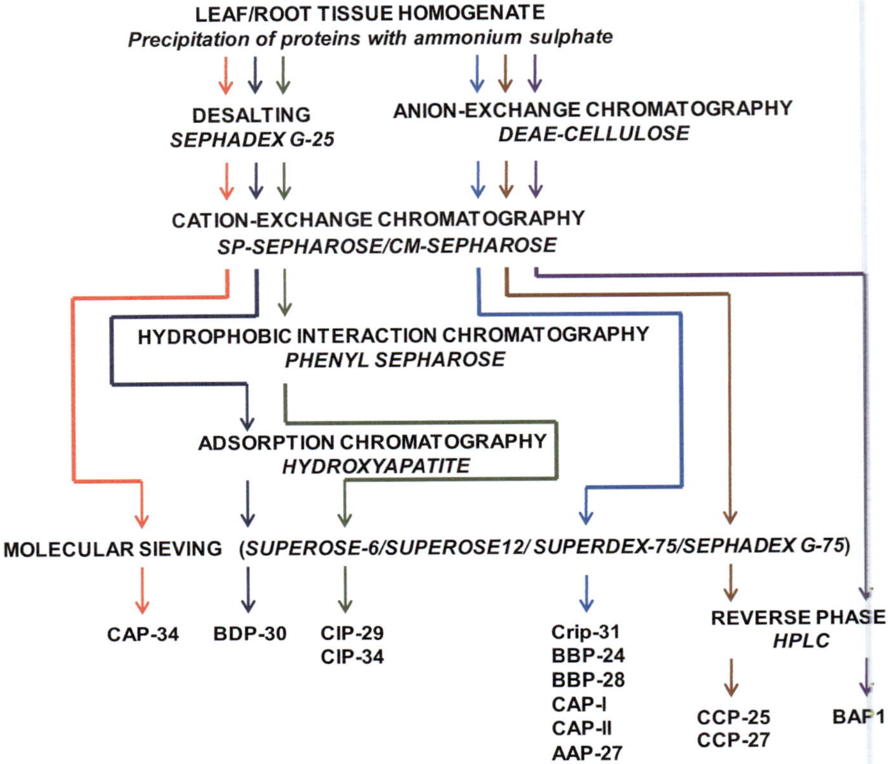

Fig. 28.1 Purification scheme for antiviral proteins from *Clerodendrum aculeatum* (CAP-34), *Boerhaavia diffusa* (BDP-30), *C. inerme* (CIP-29, CIP-34, Crip-31), *Bougainvillea xbuttiana* (BBP-24, BBP-28), *Chenopodium amaranticolor* (CAP-I, CAP-II), *Amaranthus tricolor* (AAP-27), *Celosia cristata* (CCP-25, CCP-27) and *Bougainvillea spectabilis* (BAP1)

et al. 1984). Following these preliminary findings, two antiviral resistance inducing proteins, CIP-29 and CIP-34, were recovered from *C. inerme* leaves (Prasad et al. 1995). These proteins were basic glycoproteins with molecular masses of 29 and 34 kDa, respectively. Stability of both proteins was evident due to their resistance to digestion with proteinase K and exposure to temperatures up to 80 °C. No serological relatedness to RIPs like dianthin, momordin and saporin could be demonstrated (Prasad et al. 1995). Of the two, CIP-29, a monomeric protein, possessed better resistance inducing ability and a concentration as low as 16 μgmL^{-1} effectively induced systemic antiviral resistance (Prasad et al. 1995). Subsequently, CIP-34 was shown to have inhibitory effect at extremely high concentrations and comprised a mixture of proteins, with low levels of RIP activity (Olivieri et al. 1996). CIP-29 inhibited protein synthesis by various cell lines, with BeWo and NB100 being the most sensitive, though at concentrations higher than those required for inhibition of in vitro protein synthesis in a rabbit reticulocyte lysate system. CIP-29 was classified as a polynucleotide:adenosine glycosidase since it released adenine not only

Table 28.2 Comparative characteristics of purified antiviral proteins

Proteins	Source plant/ tissue	M_r (kDa)	pI	Resistance to protease	Glycoprotein	Systemic resistance induction	RIP[a]
CIP-29	*Clerodendrum inerme*/leaf	29	Basic	Yes	Yes	Yes	Yes
CIP-34	*Clerodendrum inerme*/leaf	34	Basic	Yes	Yes	Yes	ND
Crip-31	*Clerodendrum inerme*/leaf	31	Basic	Yes	No	Yes	ND
CAP 34	*Clerodendrum aculeatum*/leaf	34	Basic	Yes	Yes	Yes	Yes
BDP 30	*Boerhaavia diffusa*/root	30	Basic	Yes	Yes	Yes	Yes
BAP1	*Bougainvillea spectabilis*/leaf	28	Basic	ND	ND	Yes	Yes
BBP-24	*Bougainvillea xbuttiana*/leaf	24	Basic	ND	Yes	Yes	Yes
BBP-28	*Bougainvillea xbuttiana*/leaf	28	Basic	ND	Yes	Yes	Yes
CCP-25	*Celosia cristata*/ leaf	25	Basic	Yes	Yes	No	Yes
CCP-27	*Celosia cristata*/ leaf	27	Basic	Yes	Yes	No	Yes
CAP-I	*Chenopodium album*/leaf	24	Basic	ND	No	Yes	Yes
CAP-II	*Chenopodium album*/leaf	24	Basic	ND	No	Yes	Yes
AAP-27	*Amaranthus tricolor*/leaf	27	Basic	ND	Yes	ND	Yes

[a]Ribosome-inactivating protein (RIP) nature confirmed through *N*-glycosidase activity/inhibition of *in vitro* protein synthesis/sequence homology

from rRNA but also from tRNA, poly(A) as well as DNA, with the effects being catalytic (Olivieri et al. 1996). Leaves of *C. inerme* were reported to contain another systemic resistance inducing protein of 31 kDa, named Crip-31. This too was a basic antiviral protein with no effect on its resistance inducing ability following its incubation with proteinase K (Praveen et al. 2001).

28.4.2 *Clerodendrum aculeatum*

Leaf extract from *Clerodendrum aculeatum* was found to offer complete protection against TMV on *D. stramonium*, *D. metel* and *N. glutinosa*, and against SRV on *C. tetragonoloba*. However, its inhibitory effect against viruses like TMV, SRV, TmYMV and GMV on *C. amaranticolor* was less obvious, varying between 43 and 50%. A heat-stable and resistance inducing nature was also indicated through these preliminary studies on the inhibitory principal contained in the extract (Verma et al. 1984). The systemic antiviral resistance inducer present in the *C. aculeatum* leaf extract was later purified as a 34 kDa basic glycoprotein, CA-SRI (CAP-34), possessing a pI of pH 8.65. In early works it was referred to as CA-SRI, however later the purified protein was renamed as CAP-34 on the basis of host from which derived and molecular weight. A concentration of 64 μgml^{-1} of the purified inducer afforded complete protection against TMV infection on *N. tabacum* Samsun NN plants. An overnight incubation of CAP-34 with proteinase K, pronase and trypsin failed to abolish its antiviral resistance inducing activity (Verma et al. 1996). Digestion of CAP-34 with endoproteinase arg-C yielded biologically active fragments of 14, 16, 20, and 28 kDa each. The full length cDNA (1218 bp) with an ORF of 906 bp encoding a 33.9 kDa protein was cloned and sequenced (Kumar et al. 1997). Its N terminus, with highly hydrophobic residues, comprised the secretory signal. The deduced amino acid sequence showed 11–54% homology with other antiviral/ribosome-inactivating proteins, such as PAP, MAP, dianthin, trichosanthin, luffin A chain, abrin A chain, ricin A chain and α-momorcharin, exhibiting a maximum with PAP. However, there was no hybridization seen between CAP-34 gene and *Mirabilis* genomic DNA despite there being 21% homology between MAP and the deduced amino acid sequence of CAP-34. The CAP-34 gene also did not hybridize with the genomic DNA extracted from *Bougainvillea*, indicating absence of significant homology. In vitro protein synthesis was completely inhibited by CAP-34 as well as the recombinant protein in a rabbit reticulocyte lysate system, while it was less efficiently inhibited in a wheat germ lysate system (Kumar et al. 1997). In a separate study, CAP-34 was also produced in consistent amounts in the micropropagated plants of *C. aculeatum* (Srivastava et al. 2004).

28.4.3 *Boerhaavia diffusa*

Boerhaavia diffusa, family Nyctaginaceae, is a perennial herbaceous plant with immense medicinal value (Sreeja and Sreeja 2009). Aqueous root extract of *B. diffusa* could induce systemic resistance and inhibit several viruses on different hosts when applied 24 h prior to virus inoculation (Verma and Awasthi 1979). The extract could check the infectivity of TMV, SRV, GMV and TRSV on *C. amaranticolor* by almost 70–95%. Virus infectivity dipped by almost 85–100% for tubular viruses like TMV and SRV on their several hosts as compared to 55–73% for spherical viruses like GMV and TRSV on their respective hosts (Verma and Awasthi 1979). The inducer was also present in the plantlets established in vitro from the callus cultures of *B. diffusa* (Gupta et al. 2004), and transfer of resistance in the explants regenerated from tobacco plants treated with *B. diffusa* root extract was also demonstrated (Lohani et al. 2007). An attempt at characterization of the inducer was made in an early study which pointed to its heat-stable glycoprotein nature (Verma and Awasthi 1979; Verma et al. 1979a, b, c). Following its purification to homogeneity, the resistance inducing protein, BDP-30, was shown to be a 30 kDa basic glycoprotein, with pI equal to or greater than pH 9.0. BDP-30 showed thermal stability

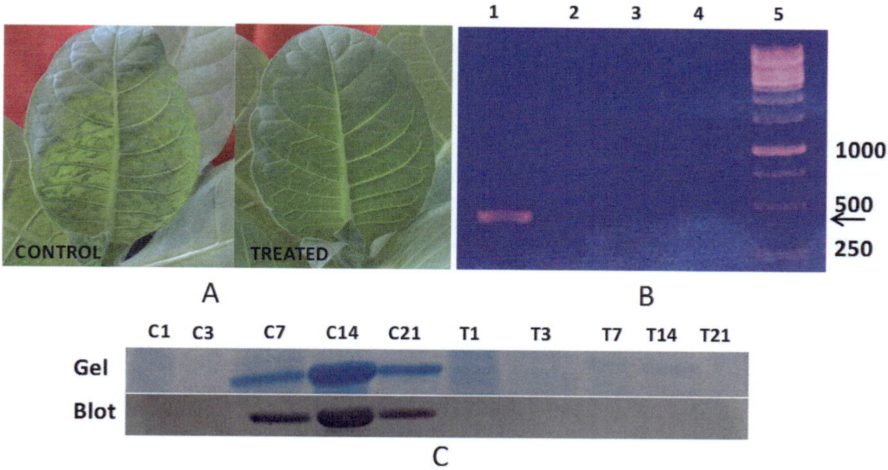

Fig. 28.2 (**a**) Induction of systemic antiviral resistance in *Nicotiana tabacum* cv. Xanthi. Basal leaves of tobacco Xanthi were treated with DW (control) and BDP-30, the purified antiviral protein from *Boerhaavia diffusa* roots (treated). The upper (untreated) leaves of both plants were inoculated with TMV, 18 h post-treatment, and plants observed for development of mosaic after 21 days. (**b**) Detection of TMV RNA by RT-PCR. RT-PCR was carried out with TMV coat protein specific primers using the total RNA extracted from the DW treated set after 7 days of TMV inoculation (lane 1) and BDP-30-treated sets, after 7, 14 and 21 days of TMV inoculation (lanes 2, 3 and 4, respectively). The expected 480 bp amplified product can be seen in the control set (lane 1). DNA ladder (lane 5) (**c**) Detection of TMV coat protein by immunoblot. Leaf sap from control (C) and BDP-30 treated (T) plants was analyzed by SDS-PAGE after Coomassie staining (Gel) and Immunoblotting (Blot). Lanes were loaded with leaf saps from 1, 3, 7, 14, and 21 days after TMV inoculation

upto 80 °C and its antiviral activity remained unaltered even after an overnight incubation with proteinase K, a non-specific protease. Failure to locate TMV coat protein as well as TMV RNA in induced resistant *N. tabacum* cv. Xanthi, strongly suggested inhibition of TMV replication in the treated plants due to BDP-30 application (Srivastava et al. 2015b). The BDP-30-treated plants also remained symptompless (Fig. 28.2). In-gel proteolytic digestion of BDP-30 yielded two peptides, KLYDIPPLR and KVTLPYSGNYER, that shared complete sequence identity with α-Trichosanthin (TCS), an RIP present in the roots of *Trichosanthes kirilowii*, while sharing homologies of 78% and 100%, respectively, with Bryodin, another RIP contained in the roots of *Bryonia dioica* (Srivastava et al. 2015b).

28.4.4 *Bougainvillea spectabilis* and *B. xbuttiana*

Antiviral activity in the leaf extract of *B. spectabilis,* family Nyctaginaceae was first reported in 1983 (Verma and Dwivedi 1983). Multiple sprays of the extract afforded complete protection to *Crotalaria juncea* and *Cucumis melo* against infection by SRV and CGMMV, while TMV, TmYMV and PhySMV were completely inhibited on *L. esculentum* as evident from the local lesion bioassay on *C. amaranticolor* (Verma and Dwivedi 1983). An RIP was purified from the leaves of *B. spectabilis* (Bolognesi et al. 1997), but its ability to induce systemic antiviral resistance was not studied. A 28 kDa basic (pI >8.6) antiviral protein, called BAP I, was purified from the roots of *B. spectabilis* that could inhibit mechanical transmission of TSWV and also interfered with in vitro protein synthesis (Balasaraswathi et al. 1998). A partial BAP-cDNA was synthesized from the leaf mRNA, cloned and sequenced. The probable ORF was translated and showed a poor relatedness to MAP, PAP and CAP-34 (Rajesh et al. 2005).

Antiviral principals and systemic resistance inducers were also detectable in the leaf extract from *B. xbuttiana*. Two highly basic glycoproteins of 24 kDa (pI 10.5) and 28 kDa (pI 10.0), named BBP-24 and BBP-28, respectively, were reported (Narwal et al. 2001a). An antiviral protein (AVP) that could deadenylate rRNA, and hence possibly functioned as a ribosome-inactivating protein, was also detected but its molecular weight was not specified (Narwal et al. 2001b). Subsequently, AVPs from *B. xbuttiana* were reported to exhibit RNase as well as DNase activity against viral RNA and supercoiled plasmid DNA, respectively (Bhatia and Lodha 2005). In this case, though not mentioned, the tests were perhaps conducted using a mixture of the two proteins that had been purified earlier (Narwal et al. 2001a). A full-length cDNA sequence (1364 bp) encoding a 35.49 kDa protein of 319 amino acids was also isolated from the leaves of *B. xbuttiana*. The deduced protein, termed BBAP1, possessed the catalytic RIP domain and was phylogenetically more closely related to RIPs from Nyctaginaceae family, but distantly related to PAP (Choudhary et al. 2008). The purified recombinant protein exhibited rRNA *N*-glycosidase activity (Choudhary et al. 2008) and demonstrated antiviral activity against Groundnut bud necrosis virus, antifungal activity against *Trichoderma harzianum* and *Rhizoctonia solani* as well as insecticidal activity against a voracious insect pest, *Helicoverpa armigera* (Lodha and Choudhary 2011; Lodha et al. 2010).

28.4.5 *Celosia cristata*

Aqueous leaf extract of *Celosia cristata,* family Amaranthaceae, was reported to inhibit plant virus infection, offering localized resistance to the host plant (Baranwal and Verma 1992). The inhibitor was partially characterized as a protease resistant and thermostable glycoprotein of M_r 21–22 kDa (Baranwal and Verma 1997). Subsequently, two glycoproteins, CCP-25 (M_r 25 kDa) and CCP-27 (M_r 27 kDa), were purified from the leaves and shown to inhibit several mechanically transmitted viruses on hosts responding hypersensitively or systemically to virus infection. A growth-stage dependent variation was observed in the concentration of the two proteins and they were able to withstand protease digestion in their native state (Balasubrahmanyam et al. 2000). Depurination studies carried out with CCP-25 yielded a diagnostic fragment from yeast rRNA, indicating its RIP nature. It was also shown to inhibit the in vitro translation of Brome mosaic virus and Pokeweed mosaic virus RNA (Baranwal et al. 2002). Through a modification in the purification protocol, the N-terminally free proteins were obtained which allowed partial N-terminal sequencing of CCP-25 (Gholizadeh and Kapoor 2004). Degenerate primers designed against the conserved RIP domain yielded a small cDNA fragment (150 bp) from *C. cristata* leaves, which when expressed in *E. coli* yielded a fusion protein of 57 kDa. The purified recombinant protein was reported to inhibit plant viruses as well (Gholizadeh et al. 2005). Post-flowering stage of *C. cristata* gave a full-length cDNA clone (1015 bp), encoding an ORF deduced to yield 283 amino acids. The purified recombinant protein (reCCP-27) inhibited in vitro protein synthesis in a rabbit reticulocyte lysate as well as on tobacco ribosomes, in addition to exhibiting antiviral activity towards TMV and SRV (Begam et al. 2006).

28.4.6 *Chenopodium album*

Leaf extract from *Chenopodium* species, family Chenopodiaceae, inhibits plant viruses (Smookler 1971; Alberghina 1976). *C. ambrosoides* aqueous leaf extract inhibited TMV and SRV on different hosts, but not on *C. amaranticolor*. Furthermore, like *C. cristata*, the extract could not induce systemic antiviral resistance (Verma and Baranwal 1983). Antiviral proteins that could impart resistance in hypersensitive hosts against several viruses were purified from *C. album* and also partially characterized (Dutt et al. 2000). This study was followed by a more comprehensive report of two antiviral proteins from the same source, CAP-I and CAP-II, which inhibited virus replication in systemic and hypersensitive host/virus combinations and worked as resistance inducing proteins. Both proteins were basic in nature (pI~10.2), contained no carbohydrate, and inhibited virus infection to different extents, CAP-I was at least 2.5-fold more effective than CAP-II. Despite possessing the same molecular weight of 24 kDa, both proteins differed in their amino acid composition and N-terminal sequence (Dutt et al. 2003). They catalyzed the depurination of rRNA extracted from tobacco ribosomes and hence exhibited *N*-glycosidase activity. Since they could also degrade TMV RNA, they were thought to be associated with RNase activity as well (Dutt et al. 2004).

28.4.7 *Amaranthus tricolor*

Amaranthus tricolor, family Amaranthaceae, antiviral protein (AAP-27) was isolated from dried leaves of the plant and characterized as a 27 kDa basic monomeric glycoprotein (pI 9.8) that could inhibit plant viruses. It also exhibited N-glycosidase and RNase activities. Its full-length cDNA comprised 1058 bp and encoded an ORF of 297 amino acids. The deduced amino acid sequence contained the RIP domain and also exhibited varied levels of homology with other antiviral proteins (Roy et al. 2006).

28.4.8 *Cuscuta reflexa*

Filaments of *Cuscuta reflexa*, family Cuscutaceae, parasitizing *Zizyphus jujube*, contained a proteinaceous inhibitor of plant viruses. The inhibitor was isolated following fractionation of the extract with a series of organic solvents and subsequent precipitation with a saturated solution of ammonium sulphate. Molecular sieving on Sephadex G-200 column yielded a virus inhibitory fraction with characteristics of a protein, and a molecular weight between 14 and 18 kDa. The inhibitor induced both local and systemic antiviral resistance and operated through an AMD-sensitive mechanism (Awasthi 1981, 1982).

28.5 Antiviral Activity of Other Plant Constituents

Apart from proteins, inhibitory activity was also identified in polysaccharides, plant latex, alkaloids, flavonoids, phenolic acids, tannins or essential oils, with effects on viruses, bacteria and fungi. A polysaccharide, T-poly, obtained from culture filtrates of a fungus *Trichothecium roseum*, could inhibit viruses on hypersensitive as well as non-hypersensitive hosts and could induce systemic resistance in these plants (Gupta et al. 1973; Chandra and Gupta 1981). In addition to being proficient in exhibiting antioxidant activity, flavonoids are also known to inhibit animal viruses (Kumar and Pandey 2013). Flavonoids (quercetin) and coumarins inhibited the infectivity of Southern Sunnhemp mosaic virus on its local lesion host *Cyamopsis tetragonoloba* (Chandra et al. 1975). Tannins from *Terminalia chebula* and *Chrysobalanus icacao* also functioned as inhibitors of viruses (Verma and Raychaudhuri 1970, 1972). Plant latex from *Calotropis procera*, *Ficus elastica*, *F. nitida*, *Euphorbia pulcherrima*, inhibited TMV (Khurana and Singh 1972; Lal and Verma 1974; Nagarajan and Murty 1975; Rafiq et al. 1985) and Tobacco necrosis virus (TNV) on bean, Bean yellow mosaic virus (BYMV) on broad bean, and Zucchini yellow mosaic virus (ZYMV) on squash (Mahmoud et al. 2010). The antiviral activity perhaps stemmed from proteases and other defence-related proteins which the latex contain (Kim et al. 2003). Sterols from *Artemisia annua* were identified as virus inhibitory agents (Khan et al. 1991). Effect of neem and custard apple oil was studied on rice tungro virus transmission by *Nephotettix virescens* (Mariappan and Saxena 1983), while neem and *Phyllanthus* oils were also studied for the inhibition of TMV and ToMV on tobacco, bell pepper and tomato (Madhusudhan et al. 2005, 2011). Neem oil could also inhibit PYVMV on pumpkin under glass house conditions, with over 78% reduction in virus transmission (Jayashree et al. 1999). Essential oils present in *Foeniculum vulgare* and *Pimpinella anisum* were also associated with virus inhibitory properties (Shukla et al. 1989) and essential oil extract of *Chenopodium ambrosoides* (EOCA) was found to be effective against *Myzus persicae* (Rajapakse and Janaki 2006).

28.6 Control of Plant Viruses by Inhibitors/Systemic Resistance Inducers

Plasticity of the viral genome poses a major challenge in protecting crops from viral infection. One of the finest approaches for durable resistance is offered by the traditional method of breeding, while development of transgenic crops resistant to plant viruses is a relatively novel approach (Cillo and Palukaitis 2014). Induction of systemic resistance in plants by application of phytoproteins/inhibitors for disease control has also been successfully used for some time now, especially where virus resistant cultivars are not available. Viral mosaic on papaya, cucurbits, urdbean, mungbean, and okra is a very common sight and several efforts are on to prevent infection on these important crops. In general, the resistance inducing proteins or inhibitors are administered as a foliar spray, repeated at specified intervals, often over a period of time till the crop produce is ready for harvesting. Sometimes, the resistance inducers are coupled with primers such as bovine serum albumin, milk proteins, oils or detergents, etc. for enhanced and durable resistance. The treatments work well under both glass house and field conditions, and are effective even under intense pathogen pressure. The effect that these inducers have on the sprayed plants varies, depending upon the host-virus system. In general, the plants show more luxuriant growth, along with a several fold increase in crop yield and improved response to varied stresses. Legumes, in addition, show improved nodulation (Verma et al. 1985b). With the defence responses switched on quickly, the sprayed plants are able to protect themselves, thus lowering the incidence of disease. Treated plants either show absence of virus or a low virus titre in ELISA or immunoblots. Often the viral RNA transcripts remain absent in the RT-PCR reactions set up to detect the viral coat protein genes (Srivastava et al. 2009, 2015a, b).

Aqueous extracts from *B. diffusa*, *Cuscuta reflexa*, *Datura metel* and *Solanum melongena* induced systemic antiviral resistance against TMV and TRSV in *N. tabacum* var. NP-31, with the active virus being assayed on *N. glutinosa*. Such plants were resistant for up to three days (Verma et al. 1979b). This demonstration of antiviral resistance induction in a host showing a non-hypersensitive reaction to virus infection was soon followed by experiments designed to control virus infection on economically important crops. Foliar sprays were given at regular intervals either to prevent natural virus infections or infections following challenge inoculation. Effective control of yellow mosaic disease on mung and urdbeans was reported with aqueous leaf extracts of *C. fragrans* and *Aerva sanguinolenta* and root extract of *B. diffusa*. The extracts were administered as foliar sprays at intervals of 3–4 days spanning a period of 6 weeks (Verma et al. 1985b). The protection exhibited an interesting pattern, not being apparent in the initial stages of the experiment, but checking the virus infection and spread in the treated plants between fourth and fifth week, when the control set of plants displayed a sharp rise in the disease incidence. The protection afforded against viral infection and improvement in growth, yield and nodulation was maximal by *C. fragrans* extract (Verma et al. 1985b). Similarly, a lowering of disease incidence by 90% and the accompanying delay in the onset of yellow mosaic disease together with mitigation of symptom severity was reported on mungbean (*Vigna radiata*) by *B. diffusa* root extract, with plants showing improved nodulation and yield (Singh et al. 2004). *B. diffusa* root extract could also induce resistance against TMV, CMV, SRV, GMV, Cucumber green mottle mosaic virus (CGMMV) on tobacco, tomato, *C. juncea*, *G. globosa* and *Lagenaria spp.*, respectively. The most effective spray regimen consisted of six foliar sprays, spread over three days, followed by virus inoculations 24 h after the last spray. The treated plants showed absolute protection from virus infection even after 45 days of virus inoculation (Awasthi et al. 1984). *B. diffusa* root extract and *C. aculeatum* leaf extract, either used in combination or alone, were effective in inducing resistance against PRSV (Awasthi and Singh 2009; Singh et al. 2011a, b, c). *C. aculeatum* leaf extract was used to control Tomato leaf curl virus (Baranwal and Ahmad 1997), and when primed with proteinaceous additives, it could successfully manage SRV and Tobacco leaf curl virus infection on *C. juncea* and tobacco (Verma and Varsha 1995a, b). CAP-34 (previously called CA-SRI), the purified resistance inducing protein from *C. aculeatum*, could prevent PRSV infection on papaya as well. Only 10% of the CAP-34 treated plants came down with low level symptoms of mild mosaic, with no observable virus, viral protein or viral RNA in the remaining plants of the treated set (Srivastava et al. 2009). Pre-inoculation sprays with extracts from *B. spectabilis* and *Prosopis chilensis* worked against Okra yellow vein mosaic virus (Pun et al. 1999) and Sunflower necrosis virus (Lavanya et al. 2009). The virus titer was slashed in sunflower and cowpea plants treated with these extracts. Reduced disease incidence (33%) was observed in bitter gourd plants treated with *B. spectabilis* and challenge inoculated with bean golden yellow mosaic virus (BGYMV) (Rajinimala et al. 2009). *B. spectabilis* and *M. jalapa* leaf extracts reduced leaf crinkle disease on blackgram caused by the urdbean leaf crinkle virus (Karthikeyan et al. 2009). Leaf extracts of *M. jalapa*, *Datura metel* and *Azadirachta indica* (neem) reduced mungbean yellow mosaic infection on black gram (Venkatesan et al. 2010). Viral diseases on cucumber were prevented by extracts from *A. indica*, *C. auleatum* and *Terminalia arjuna*, and it was also observed that seed treatment with neem extract followed by foliar sprays provided good protection (Kumar and Awasthi 2009). Leaf and seed kernel extract of neem also interfered with aphid transmission of a virus causing mosaic disease on *Brassica juncea*, reducing disease incidence by 46–54%, while extract from *Jatropha curcas* was not as effective (Devi et al. 2008).

28.7 Mode of Action of SRIs

The antiviral resistance induced by plant extracts or purified resistance inducing proteins was sensitive to treatment with the transcription inhibitor, actinomycin D (AMD). Effect of AMD on induced resistance was routinely evaluated by using AMD concomitantly with the inducers or at various time intervals after the treatment of leaves with the inducer (Verma and Awasthi 1979; Verma and Dwivedi 1984; Verma et al. 1984; Prasad et al. 1995). In all cases, a breakdown in the induction of antiviral resistance was seen upon concomitant application of AMD, as evident from an increase in lesion number to the control level. The effect of AMD on reversal of induced resistance diminished with increase in the time interval between the application of inducer and AMD. Furthermore, in vitro incubation of the virus along with the leaf extract from the treated plants, from both site and remote site leaves, could inhibit TMV, SRV, GMV, TRSV, TmYMV and *Physalis shoestring mosaic virus* (PhySMV) on their assay hosts (Verma and Dwivedi 1984). This preliminary observation led to the suggestion that phytoprotein-mediated induction of resistance involved host transcription and that induced resistant tissues possibly carried a virus-inhibitory agent (VIA) that could inactivate viruses in vitro (Verma and Awasthi 1980; Mukerjee et al. 1981; Verma and Dwivedi 1984; Verma et al. 1985a; Prasad et al. 1995). A proteinaceous nature of the VIAs induced by *B. spectabilis* and *B. diffusa* extracts was proposed (Verma and Awasthi 1980; Verma and Dwivedi 1984). *Cyamopsis* plants treated with CAP-34 (CA-SRI) accumulated a 34 kDa basic protein, as compared to the DW-treated fraction, which was also unable to inhibit the virus in vitro (Verma et al. 1996). This 34 kDa protein could have been the elusive VIA that could not be purified to homogeneity. A complete characterization of the VIAs isolated from resistant *C. tetragonoloba* plants pre-treated with CIP-29, the purified inducer from *C. inerme* leaves, has since been reported (Prasad et al. 2014) and the leaf extract from resistant tissues inhibited SRV, TMV and PRSV in vitro. Two VIAs, CT-VIA-32 (M_r 32 kDa) and CT-VIA-62 (M_r 62 kDa) were isolated from such tissues. The CIP-29 inducible CT-VIA-62 displayed better antiviral activity and was characterized as a basic glycoprotein. Its peptides sequenced through LC/MS/MS shared homology with a lectin from *Medicago truncatula* which carried a mannose-binding lectin domain (Prasad et al. 2014). In a related effort, CAP-34 inducible CP-VIA-34, a virus-inhibitory basic protein, was isolated from resistant leaf tissues of *Carica papaya* (Srivastava et al. 2015a). No protease, DNase or RNase activity was found associated with CP-VIA-34. Such proteins, though antiviral but not labelled as VIA, were also detected in *C. tetragonoloba* plants treated with the purified antiviral proteins from *C. album*. Two polypeptides of 17 and 26 kDa accumulated in the un-treated leaves of *C. tetragonoloba* plants whose basal leaves were treated with the inducer (Dutt et al. 2004). It was obvious that VIAs were accumulating in the plants treated with such inducers, but were either absent or present in very low amounts in the un-treated plants (Verma et al. 1996; Prasad et al. 2014). Furthermore, these VIAs differed from the inducers, in that they were unable to induce resistance in plants, though the polyclonal antiserum raised against CIP-29 did recognize CT-VIA-32 (Prasad et al. 2014). Detection of polynucleotide:adenosine *N*-glycosidase activity in CIP-29 and inhibition of in vitro protein synthesis by CAP-34, along with the occurrence of inducible proteins (VIAs), has complicated any thoughts on the probable mode of action of the resistance inducing proteins based on the inactivation of the virus by these VIAs alone.

The role of PAP in plant virus inhibition, and in plant disease resistance, has been dealt with in great detail in a recent review (Di and Tumer 2015). RIPs are associated with depurination linked to ribosome inhibition, cytotoxicity and antiviral activity. Experiments were developed to delink these associations so as to pinpoint the reason for virus inhibition. Antiviral activity of PAP was earlier correlated and attributed to its RNA N-glycosidase enzymic activity on ribosomes. PAP is a secretory RIP and exists in several isoforms. It forms homodimer complexes in the cytosol, which are less active on rRNA as compared to the monomeric form present in the apoplasts (Tourlakis et al. 2010), and this was the presumed way to avoid depurination of pokeweed rRNA, a possible explanation to Allard's findings (Allard 1918). PAP-specific antibodies revealed its extracellular location, with PAP being sequestered in the cell wall matrix (Ready et al. 1986). Inhibition of ribosome-mediated viral protein synthesis by PAP was possible if there was a release of the inhibitor (PAP) from the cell wall into the cytosol following the virus inoculation (Lodge et al. 1993). A positive correlation was also shown between PAP concentration, inhibition of TMV and depurination of ribosomes (Chen et al. 1993).

All this while it was believed that ribosome-inactivation led to virus inhibition. However, this hypothesis was challenged by the finding that the C-terminus of PAP was required for toxicity and depurination of ribosomes but not for antiviral activity. Thus, depurination of ribosomes was not the only mechanism for virus inhibition (Tumer et al. 1997), and soon overwhelming evidence came in support of this possibility (Zoubenko et al. 2000). It was subsequently reported that non-depurinating mutants of PAP could still depurinate capped BMV and PVX viral RNA and inhibit their translation in vitro (Hudak et al. 2000). These mutants had an intact active site but were altered such that they were unable to bind to the ribosomes and cause their depurination. However, given that in a virus infected cell the capped viral RNAs would be present in huge numbers, they would become the preferred substrate for PAP, leading to their depurination. PAP was later shown to directly depurinate BMV viral RNA3, and this depurination was held responsible for decreased efficiency of packaging of the viral genome and consequently generation of fewer infectious particles (Karran and Hudak 2008, 2011). No effect was observed on the quality of the virus. PAP was also able to inhibit uncapped TBSV and satellite panicum mosaic virus (SPMV) in vivo without causing detectable depurination in the viral RNAs (Vivanco and Tumer 2003). Site-directed mutagenesis in the central domain of PAP led to loss of cytotoxicity but not its ability to depurinate indicating that depurination was not the sole reason for cytotoxicity (Hudak et al. 2004). Yeast cells expressing PAP showed apoptotic features and an anti-apoptotic protein reduced cell cytotoxicity of PAP, without affecting ribosomal depurination and translation inhibition (Cakir and Tumer 2015). Finally, RIPs depurinate substrates other than rRNA as well (Olivieri et al. 1996). Thus, RIPs inhibit plant viruses, but not via inactivation of ribosomes, while resistance inducing proteins like CIP-29 and CAP-34, possess RIP function, inhibit viruses and also induce antiviral resistance in plants and VIA. Incubating the virus along with the VIA in vitro was devoid of any detrimental effect on the virus per se, as fully infectious virus particle could be recovered following their separation from the VIA by ultracentrifugation. Hence the VIA participates in vivo in plant defence at some level, its production being triggered by the application of such inducers.

Besides VIA, there are other examples of inducible antiviral proteins in plants. Nagaich and Singh (1970) reported an inhibitory agent in *Capsicum pendulum* inoculated with PVX that could prevent PVX infectivity on *Gomphrena globosa* and

Solanum tuberosum. RIPs called Beetins (BE27 and BE29) were induced in *Beta vulgaris* leaves following infection with Artichoke mottled crinkle virus (AMCV) and application of H_2O_2 and salicylic acid (Iglesias et al. 2005). Its external application was able to control AMCV infection. JIP-60, a jasmonate inducible 60 kDa RIP, was described from barley (Chaudhry et al. 1994; Reinbothe et al. 1994) and shown to alleviate stress in plants and delay the onset of senescence (Rustgi et al. 2014). Induced production of antiviral proteins in tobacco plants displaying SAR is also known, variously termed as the antiviral factor (AVF) (Sela and Applebaum 1962; Mozes et al. 1978) and the inhibitor of replication (IVR) (Loebenstein and Gera 1981; Gera et al. 1990). Tobacco and tomato plants transformed with the IVR gene were resistant to TMV and a variety of fungal pathogens (Loebenstein et al. 2010; Elad et al. 2012). Two forms of AVF, gp22 and gp35, were determined to be closely related to two pathogenesis-related (PR) proteins, PR-5 (chitinases) and PR-3 (1,3-ß-glucanases), although both of these PRs are devoid of any antiviral activity (Edelbaum et al. 1991).

PR-proteins were initially discovered in tobacco plants reacting hypersensitively to TMV infection, and were associated with the development of resistance termed as systemic acquired resistance (SAR) (Gianinazzi et al. 1970; Van Loon and Van Kammen 1970; Stintzi et al. 1993). Induced expression of the PR-proteins in resistant plants, in particular PR 1a, along with an enhanced accumulation of endogenous levels of salicylic acid (SA), was always noticed and hence these came to be viewed as established markers for SAR. SAR engages a number of signalling molecules in a cross-talk and has a huge potential in crop protection due to the broad spectrum nature of resistance conferred on the plants against subsequent invasion by diverse pathogens and even an insect (McIntyre et al. 1981; Kessler and Baldwin 2002; Durrant and Dong 2004; Fu and Dong 2013; Gozzo and Faoro 2013). PR proteins have been assigned to 14 families, some being directly antimicrobial while others with proposed indirect roles in plant defence (Stintzi et al. 1993; Edreva 2005; Van Loon et al. 2006). Enzymic activities attributed to various PRs include β-1,3-glucanase (PR-2), chitinase (PR-3, PR-4, PR-8, PR-11), proteinase (PR-7), peroxidise (PR-9), ribonuclease-like (PR-10), etc. PR-10 family that shows homology to ribonucleases, is of special interest in the context of plant virus control (Park et al. 2004).

Expression of such antimicrobial proteins in plant tissues was expected to curb pathogen invasion and spread. Thus, PR-1, PR-2 and PR-5 genes have been incorporated in crops like cotton, barley, peanut, potato, rice, etc. for resistance against various fungal pathogens (Collinge et al. 2010). Creation of transgenic plants expressing RIPs proved to be relatively difficult because of their cytotoxicity (Lodge et al. 1993). However, a variant of PAP, PAPv, obtained through mutagenesis was less toxic than the wild type PAPw, and was expressed efficiently in transgenic tobacco, although transgenics over-expressing PAP were stunted and exhibited mottled leaves due to apoptosis, while low PAP-expressing plants were normal in appearance (Lodge et al. 1993). Transgenic tobacco and potato plants expressing PAP were resistant to PVX, PVY and CMV, thus PAP provided broad spectrum resistance to plant viruses (Lodge et al. 1993). Similarly, protection against viruses was evident in transgenic plants expressing other RIPs, including type II (Krishna et al. 2002; Chen et al. 2002; Vandenbussche et al. 2004). A few cases of enhanced resistance to important insects species was also reported (Shahidi-Noghabi et al. 2009). In a majority of cases, the transgenic plants expressing RIPs did not produce PR proteins, but PAP was shown to upregulate the expression of PR1 and PR5 genes in transgenic tobacco and such plants displayed enhanced resistance to *Tobacco etch virus* (TEV) and fungal pathogens (Hu and Reddy 1997; Schaad et al. 2000; Zoubenko et al. 2000). However, the accompanying increase in levels of SA was not always noticed and hence RIPs were purported to follow a SA-independent pathway. Induced systemic resistance (ISR) by rhizobacteria may also involve PR proteins (Kim et al. 2015). Volatiles of *Bacillus* sp. strain JS induce the expression of several PR genes, causing an up-regulation of glucanases (PR-2), chitinases (PR-3 and PR-4), peroxidise (PR-9) and PR-14 in tobacco plants exhibiting resistance to fungal pathogens (Kim et al. 2015). Thus the differences in the pathways utilized by the various agents may not always be distinct.

Molecular changes associated with the induction of resistance against viruses by phytoproteins like CIP-29 and CAP-34 have not been studied in detail. In addition to the ubiquity of the inducible VIAs, CAP-34 treated *C. tetragonoloba* plants also yielded two basic isoforms of β-1,3-glucanase (PR-2), possessing an M_r of 34 and 36 kDa (Prasad et al. 2001). Enzymic activities need to be assigned to the induced VIAs as well and probable RIP activity cannot be ruled out. Enhancement in the total protein content and activity profiles of oxidoreductases like catalase, polyphenoloxidase and peroxidises was observed in Samsun NN tobacco treated with *C. aculeatum* leaf extract and *C. tetragonoloba* plants treated with *C. aculeatum*, *C. fragrans* and *B. diffusa* extracts (Verma and Prasad 1988, 1992). However, the RNase and DNase activity profiles in the resistant plants remained unaltered (Verma and Prasad 1992). Increase in the activity of superoxide dismutase and peroxidise and a decrease in catalase activity was noted in tobacco plants treated with AVPs from *B. xbuttiana*, whereas a total reversal in the activity pattern was noted for TMV infected tobacco (Bhatia et al. 2004). Antioxidant activity of *C. cristata* antiviral proteins, CCP-25 and CCP-27, was also reported. Enhancement in the activities of peroxidise, catalase and poly-phenol oxidase were noted in plants treated with these proteins and challenged with TMV (Gholizadeh et al. 2004). Similarly, AVP treated sunflower and black gram plants showed increase in peroxidise, polyphenol oxidase and phenylalanine ammonia lyase activies, along with enhanced phenolic content (Karthikeyan et al. 2009; Lavanya et al. 2009). Thus induction of defence-related enzymes and proteins appears to be generally associated with phytoprotein mediated induced resistance in plants. The observation that virus infection is inhibited in spatially separated tissues that have not received the inducer treatment prompted a study involving calcium-mediated signalling which is thought to be important for viral movement and replication. In a study taken up to determine the effect of a calcium channel blocker, verapamil, on induced resistance, it was noticed that its application increased the calcium efflux from leaf segments, while verapamil itself induced resistance and inhibited TMV infection (Singh et al. 2011c).

28.8 Concluding Remarks

In a multicomponent defence response to virus infection, the inducers/inhibitors and VIAs are envisaged to play a major role in protecting plants from virus infection and enhancing crop yield. It is entirely possible that ISR, SAR and systemic induced resistance by plant proteins, may run as parallel pathways, destined to converge eventually and give identical end results. Tremendous progress has been made in the recent years in the field of induced resistance in plants as a method of plant virus control. However, at a global level, only compounds like benzothiadiazole and chitosan and seed bacterization by PGPR have been occasionally used for effective control in open fields (Faoro and Gozzo 2015). Phytoprotein-mediated induced systemic protection in plants, having proven its worth in biological control, has the potential to become the much sought-after panacea in the management of plant viruses, with a well deserved place in any IPM programme, if not as a stand-alone practice.

References

Acharya S (2013) Control of the potato virus X through application of root extracts of *Chlorophytum nepalense* to potato plants and tubers. Potato Res 56:1–10

Alberghina A (1976) The inhibitory activity of extracts of *Chenopodium amaranticolor* leaves on the infection of tobacco necrosis virus. J Phytopathol 87:17–27

Allard HA (1918) Effect of various salts, acids, germicides, etc., on the infectivity of the virus causing mosaic disease of tobacco. J Agric Res 13:619–637

Apablaza GE, Bernier CC (1972) Inhibition of tobacco mosaic virus infection by plant extracts. Can J Bot 50:1473–1478

Awasthi LP (1981) The purification and nature of an antiviral protein from *Cuscuta reflexa* plants. Arch Virol 70:215–223

Awasthi LP (1982) Characteristics and mode of action of a virus inhibitor from *Cuscuta reflexa* plants. Zentt Mikrobiol 137:509–518

Awasthi LP, Chowdhury B, Verma HN (1984) Prevention of plant virus diseases by *Boerhaavia diffusa* inhibitor. Int J Trop Dis 2:41–44

Awasthi LP, Kumar P (2003) Prevention of infection and multiplication of cucumber green mottle mosaic virus in muskmelon treated with *Boerhaavia diffusa*. Indian Phytopathol 56:362

Awasthi LP, Singh S (2009) Management of ringspot disease of papaya through plant products. Indian Phytopathol 62:369–375

Awasthi LP, Singh SP, Verma HN, Kluge S (2013) Further studies on the antiviral agent isolated from host plants pretreated with *Boerhaavia diffusa* glycoprotein. Virol Mycol 3:124

Bag P, Chattopadhyay D, Mukherjee H, Ojha D, Mandal N, Sarkar MC, Chatterjee T, Das G, Chakraborti S (2012) Anti-herpes virus activities of bioactive fraction and isolated pure constituent of *Mallotus peltatus*: an ethnomedicine from Andaman Islands. Virol J 9:98

Balasaraswathi R, Sadasivam S, Ward M, Walker JM (1998) An antiviral protein from *Bougainvillea spectabilis* roots: purification and characterization. Phytochemistry 47:1561–1565

Balasubrahmanyam A, Baranwal VK, Lodha ML, Varma A, Kapoor HC (2000) Purification and properties of growth stage-dependent antiviral proteins from the leaves of *Celosia cristata*. Plant Sci 154:13–21

Barakat A, Stevens WA (1981) Studies on the mode of action of inhibitors of local lesion production by plant viruses. Microbios Lett 16:7–13

Baranwal VK, Ahmad N (1997) Effect of *Clerodendrum aculeatum* leaf extract on tomato leaf curl virus. Indian Phytopathol 50:297–299

Baranwal VK, Tumer NE, Kapoor HC (2002) Depurination of ribosomal RNA and inhibition of viral RNA translation by an antiviral protein of *Celosia cristata*. Indian J Exp Biol 40:1195–1197

Baranwal VK, Verma HN (1992) Localized resistance against virus infection induced by leaf extract of *Celosia cristata*. Plant Pathol 41:633–638

Baranwal VK, Verma HN (1993) Virus inhibitory activity of leaf extracts from different taxonomic group of higher plants. Indian Phytopathol 46:402–403

Baranwal VK, Verma HN (1997) Characteristics of a virus inhibitor from the leaf extract of *Celosia cristata*. Plant Pathol 46:523–529

Barbieri L, Battelli MG, Stirpe F (1993) Ribosome-inactivating proteins from plants. Biochem Biophys Acta 1154:237–282

Barbieri L, Polito A, Bolognesi A, Ciani M, Pelosi E, Farini V, Jha A, Sharma N, Vivanco JM, Chamberry A, Parente A, Stirpe F (2006) Ribosome-inactivating proteins in edible plants and purification and characterization of a new ribosome-inactivating protein from *Cucurbita moschata*. Biochem Biophys Acta 1760:783–792

Barbieri L, Valbonesi P, Bonora E, Gorini P, Bolognesi A, Stirpe F (1997) Polynucleotide: adenosine glycosidase activity of ribosome-inactivating proteins: effect on DNA, RNA and poly(A). Nucleic Acids Res 25:518–522

Begam M, Kumar S, Roy S, Campanella JJ, Kapoor HC (2006) Molecular cloning and functional identification of a ribosome-inactivating/antiviral protein from leaves of post-flowering stage of *Celosia cristata* and its expression in *E. coli*. Phytochemistry 67:2441–2449

Bhanuprakash V, Hosamani M, Balamurugan V, Gandhale P, Naresh P, Swarup D, Singh RK (2008) In vitro antiviral activity of plant extracts on goatpox virus replication. Indian J Exp Biol 46:120–127

Bhanuprakash V, Hosamani M, Balamurugan V, Singh RK, Swarup D (2007) In vitro antiviral activity of *Eugenia jambolana* plant extract on buffalo poxvirus: conventional and QPCR methods. Int J Trop Med 2:3–9

Bharathi M (1999) Effect of plant extract and chemical inhibitors on cucumber mosaic virus of brinjal. J Mycology Plant Pathol 29:57–60

Bharathimatha C, Doraiswamy S, Rabindran R, Renukadevi P, Velazhahan R (2003) Existence of antiviral principles (AVPs) in seed extracts of *Harpullia cupanioides* (Roxb.) against tomato spotted wilt virus (TSWV), rice tungro virus (TRV) and cowpea aphid borne mosaic virus (CABMV). Acta Phytopathol Entomol Hung 38:109–114

Bhargava KS, Singh R (1965) Inactivation of watermelon mosaic virus by juice of *Portulacca grandiflora*. Curr Sci 11:361

Bhatia S, Kapoor HC, Lodha ML (2004) Modification of antioxidant status of host cell in response to *Bougainvillea* antiviral proteins. J Plant Biochem Biotechnol 13:113–118

Bhatia S, Lodha ML (2005) RNase and DNase activities of antiviral proteins from leaves of *Bougainvillea xbuttiana*. Indian J Biochem Biophys 42:152–155

Bolognesi A, Polito L, Olivieri F, Valbonesi P, Barbieri L, Battelli MG, Carusi MV, Benvenuto E, Del Vecchio BF, Di Maro A, Parente A, Di Loreto M, Stirpe F (1997) New ribosome-inactivating proteins with polynucleotide:adenosine glycosidase and antiviral activities from *Basella rubra* L. and *Bougainvillea spectabilis* Willd. Planta 203:422–429

Bose K, Kulshreshtha K, Joshi RD (1983) Some aspects of the inhibition of bean common mosaic virus on ornamental plants. Agric Sci Dig 3:195–198

Cakir B, Tumer NE (2015) *Arabidopsis* Bax inhibitor-1 inhibits cell death induced by pokeweed antiviral protein in *Saccharomyces cerevisae*. Microbiol Cell 2:43–56

Chandra K, Gupta BM (1981) Acquired local and systemic antiviral (TMV) resistance induced by treatment with T-poly (*Trichothecium* polysaccharide) in non-hypersensitive host plant *Nicotiana tabacum* cv. NP31. Curr Sci 50:69–71

Chandra S, Singh BP, Nigam SK, Srivastava KM (1975) Effect of some naturally occurring plant products on Southern sunnhemp mosaic virus (SSMV). Curr Sci 44:511–512

Chaudhry B, Muller-Uri F, Cameron-Mills V, Gough S, Simpson D, Skriver K, Mundy J (1994) The barley 60 kDa jasmonate-induced protein (JIP-60) is a novel ribosome-inactivating protein. Plant J 6:815–824

Chen ZC, Antoniw JF, White RF (1993) A possible mechanism for the antiviral activity of pokeweed antiviral protein. Physiol Mol Plant Pathol 42:249–258

Chen Y, Peumans WJ, Van Damme EJM (2002) The *Sambucus nigra* type-2 ribosome-inactivating protein SNA-1 exhibits *in planta* antiviral activity in transgenic tobacco. FEBS Lett 516:27–30

Choudhary NL, Yadav OP, Lodha ML (2008) Ribonuclease, deoxyribonuclease and antiviral activity of *Escherichia coli*-expressed *Bougainvillea xbuttiana* antiviral protein 1. Biochem Mosc 73:273–277

Chowdhury AK, Saha NK (1985) Inhibition of urdbean leaf crinkle virus by different plant extracts. Indian Phytopathol 38:566–568

Cillo F, Palukaitis P (2014) Transgenic resistance. Adv Virus Res 90:35–146

Collinge DB, Jorgensen HJL, Lund OS, Lyngkjaer MF (2010) Engineering pathogen resistance in crop plants: current trends and future prospects. Annu Rev Phytopathol 48:269–291

Darekar RN, Sawant DM (1989) Inhibition of bottle gourd mosaic by plant leaf extracts. Indian Phytopathol 42:339

Deepthi N, Madhusudhan KN, Uday Shankar AC, Kumar HB, Prakash HS, Shetty HS (2007) Effect of plant extracts and acetone precipitated proteins from six medicinal plants against tobamovirus infection. Int J Virol 3:80–87

Devi PS, Devi LR, Singh IM (2008) Evaluation of plant products against aphid (*Myzus persicae* Sultz) transmission of mosaic disease of leaf mustard (*Brassica juncea* var. rugosa). Indian Phytopathol 61:514–517

Devi PR, Doraiswamy S, Nakkeeran S, Rabindran R, Ganapathy T, Ramiah M, Mathiyazhagan S (2004) Antiviral action of *Harpulia cupanioides* and *Mirabilis jalapa* against tomato spotted wilt virus (TSWV) infecting tomato. Arch Phytopathol Plant Protect 37:245–259

Dhaliwal AS, Dhaliwal GK (1971) Inhibition of tobacco mosaic virus multiplication by extract from *Allium cepa* and *Allium sativum*. Adv Front Plant Sci 28:305–310

Di R, Tumer NE (2015) Pokeweed antiviral protein: its cytotoxicity mechanisms and applications in plant disease resistance. Toxins 7:755–772

Duggar BM, Armstrong JK (1925) The effect of treating virus of tobacco mosaic with juice of various plants. Ann Mol Bot Gard 12:359–365

Durrant WE, Dong X (2004) Systemic acquired resistance. Annu Rev Phytopathol 42:185–209

Dutt S, Balasubrahmanyam A, Lodha ML (2000) Purification and partial characterization of antiviral proteins from *Chenopodium album* L. leaves. J Plant Physiol 156:808–810

Dutt S, Narwal S, Kapoor HC, Lodha ML (2003) Isolation and characterization of two protein isoforms with antiviral activity from *Chenopodium album* L. leaves. J Plant Biochem Biotechnol 12:117–122

Dutt S, Yadav OP, Kapoor HC, Lodha ML (2004) Possible mechanism of action of antiviral proteins from the leaves of *Chenopodium album* L. Indian J Biochem Biophys 41:29–33

Edelbaum O, Sher N, Rubinstein M, Novick D, Tal N, Moyer M, Ward E, Ryals J, Sela I (1991) Two antiviral proteins, gp35 and gp22 correspond to 1,3-β-glucanase and an isoform of PR-5. Plant Mol Biol 17:171–173

Edreva A (2005) Pathogenesis-related proteins: Research progress in the last 15 years. Gen Appl Plant Physiol 31:105–124

Elad Y, Rav-David D, Leibman D, Vintal H, Vunsh R, Moorthy H, Gal-On A, Loebenstein G (2012) Tomato plants transformed with inhibitor-of-virus-replication gene are partially resistant to several pathogenic fungi. Ann Appl Biol 161:16–23

Endo Y, Mitsui K, Motizuki M, Tsurugi K (1987) The mechanism of action of ricin and related toxic lectins on eukaryotic ribosomes: the site and characteristics of the modification in 28S ribosomal RNA caused by the toxins. J Biol Chem 262:5908–5912

Endo Y, Tsurugi K (1988) The RNA N-glycosidase activity of ricin A-chain. The characteristics of the enzymatic activity of ricin A-chain with ribosomes and with rRNA. J Biol Chem 263:8735–8739

Faoro F, Gozzo F (2015) Is modulating virus virulence by induced systemic resistance realistic? Plant Sci 234:1–13

Fu ZQ, Dong X (2013) Systemic acquired resistance: Turning local infection into global defense. Annu Rev Plant Biol 64:839–863

Gendron Y, Kassanis B (1954) The importance of the host species in determining the action of virus inhibitors. Ann Appl Biol 41:183–188

Gera A, Loebenstein G, Salomon R, Franck A (1990) Inhibitor of virus replication (IVR) from protoplast of a hypersensitive tobacco cultivar infected with tobacco mosaic virus is associated with a 23k protein species. Phytopathology 80:78–81

Gholizadeh A, Kapoor HC (2004) Modification in the purification protocol of *Celosia cristata* antiviral proteins lead to protein that can be N-terminally sequenced. Protein Pept Lett 11:551–561

Gholizadeh A, Kohnehrouz BB, Santha IM, Lodha ML, Kapoor HC (2005) Cloning and expression of small cDNA fragment encoding strong antiviral peptide from *Celosia cristata* in *Escherichia coli*. Biochem Mosc 70:1005–1010

Gholizadeh A, Kumar M, Balasubrahmanyam A, Sharma S, Narwal S, Lodha S, Kapoor HC (2004) Antioxidant activity of antiviral proteins from *Celosia cristata*. J Plant Biochem Biotechnol 13:13–18

Gianinazzi S (1982) Antiviral agents and inducers of virus resistance: analogies with interferon. In: RKS W (ed) Active defence mechanisms in plants. Plenum Publishing Corp, New York, pp 275–296

Gianinazzi S, Martin C, Vallee JC (1970) Hypersensibilite aux virus, temperature et proteins soluble chez le *Nicotiana* Xanthi n.c. Apparition de nouvelles macromolecules lors de la repression de la synthese virale. CR Acad Sci D 270:2382–2386

Girbes T, Ferreras JM, Arias FJ, Stirpe F (2004) Description, distribution, activity and phylogenetic relationship of ribosome-inactivating proteins in plants, fungi and bacteria. Mini-Rev Med Chem 4:461–476

Gozzo F, Faoro F (2013) Systemic acquired resistance (50 years after discovery): moving from the lab to the field. J Agric Food Chem 61:12473–12491

Grasso S (1977) Investigations on the action mechanism of a virus inhibitor from *Phytolacca Americana*. Riv di Patol Veg 13:77–84

Guleria S, Kumar A (2006) *Azadirachta indica* leaf extract induces resistance in sesame against Alternaria leaf spot. J Cell Mol Biol 5:81–86

Gupta BM, Chandra K, Verma HN, Verma GS (1973) Induction of antiviral resistance in *Nicotiana glutinosa* plants by treatment with *Trichothecium* polysaccharide and its reversal by actinomycin D. J Gen Virol 24:211–213

Gupta VK, Raychaudhuri SP (1971) Nature of virus inhibitor in *Acacia arabica*. Ann Phytopath Soc Jpn 37:124–127

Gupta VK, Raychaudhuri SP (1972) Mechanism of inhibition of Potato virus Y by extracts from leaves of some woody plants. J Phytopathol 73:256–262

Gupta RK, Srivastava A, Verma HN (2004) Callus culture and organogenesis in *Boerhaavia diffusa*: a potent antiviral protein containing plant. Physiol Mol Biol Plants 10:263–268

Hu X, Reddy AS (1997) Cloning and expression of a PR5-Like protein from *Arabidopsis*: inhibition of fungal growth by bacterially expressed protein. Plant Mol Biol 34:949–959

Hudak KA, Parikh BA, Di R, Baricevic M, Santana M, Seskar M, Tumer NE (2004) Generation of pokeweed antiviral protein mutations in *Saccharomyces cerevisiae*: evidence that ribosome depurination is not sufficient for cytotoxicity. Nucleic Acids Res 32:4244–4256

Hudak KA, Wang P, Tumer NE (2000) A novel mechanism for inhibition of translation by pokeweed antiviral protein: depurination of the capped RNA template. RNA 6:369–380

Iglesias R, Perez Y, de Torre C, Ferreras J, Antolin P, Jimenez P, Rojo MA, Mendez E, Girbes T (2005) Molecular characterization and systemic induction of single-chain ribosome-inactivating protein (RIPs) in sugar beet (*Beta vulgaris*) leaves. J Exp Bot 56:1675–1684

Jayashree K, Pun KB, Doraiswamy S (1999) Effect of plant extracts and derivatives, butter milk and virus inhibitory chemicals on pumpkin yellow vein mosaic virus transmission. Indian Phytopathol 52:357–361

Karran RA, Hudak KA (2008) Depurination within the intergenic region of Brome mosaic virus RNA3 inhibits viral replication in vitro and in vivo. Nucleic Acids Res 36:7230–7239

Karran RA, Hudak KA (2011) Depurination of brome mosaic virus RNA3 inhibits its packaging into virus particles. Nucleic Acids Res 39:7209–7222

Karthikeyan G, Doraisamy S, Rabindran R, Ganapathy T (2009) Evaluation of antiviral principals for the induction of systemic resistance in black gram (*Vigna mungo*) against urdbean leaf crinkle virus. Arch Phytopathol Plant Protect 42:1172–1186

Kassanis B, Kleczkowski A (1948) The isolation and some properties of a virus-inhibiting protein from *Phytolacca esculenta*. J Gen Microbiol 2:143–153

Kessler A, Baldwin IT (2002) Plant responses to insect herbivory: the emerging molecular analysis. Annu Rev Plant Biol 53:299–328

Khan MMAA, Jain DC, Khakuni RS, Thakur RS (1991) Occurrence of some antiviral sterols in *Artemisia annua*. Plant Sci 75:161–165

Khan MMAA, Verma HN (1990) Partial characterization of an induced virus inhibitory protein associated with systemic resistance in *Cyamopsis tetragonoloba* (L.) Taub. plants. Ann Appl Biol 117:617–623

Khurana SMP, Bhargava KS (1970) Effect of plant extracts on the activity of three papaya viruses. J Gen Appl Microbiol 16:225–230

Khurana SMP, Singh B (1972) Studies on *Calotropis procera* latex as inhibitor of tobacco mosaic virus. Phytopathol Z 73:341–346

Kim JS, Kim YO, Ryu HJ, Kwak YS, Lee JY, Kang H (2003) Isolation of stress-related genes of rubber particles and latex in fig tree (*Ficus carica*) and their expressions by abiotic stress or plant hormone treatments. Plant Cell Physiol 44:412–414

Kim JS, Lee J, Lee C-h, Woo SY, Kang H, Seo S-G, Kim S-H (2015) Activation of pathogenesis-related genes by the rhizobacterium *Bacillus* sp. JS, which induces systemic resistance in tobacco plants. Plant Pathol J 31:195–201

Krishna R, McDonald KA, Dandekar AM, Jackman AP, Falk B (2002) Expression of recombinant trichosanthin, a ribosome-inactivating protein, in transgenic tobacco. J Biotechnol 97:69–88

Kulkarni VR, Byadagi AS (2004) Evaluation of plant extracts against safflower mosaic virus disease. Karnataka J Agric Sci 17:838–840

Kumar P, Awasthi LP (2009) Prevention of infection and spread of viral diseases in cucumber (*Cucumis sativus* L.) through botanicals. J Plant Dis Sci 4:25–32

Kumar S, Pandey AK (2013) Chemistry and biological activities of flavonoids: an overview. Sci World J. doi:10.1155/2013/162750

Kumar H, Singh R, Gupta V, Zutshi SK (2015) Performance of different germplasm, plant extracts and insecticides against yellow vein mosaic of okra (OYVMV) under field conditions. VEGETOS 28:31–37

Kumar D, Verma HN, Tuteja N, Tewari KK (1997) Cloning and characterization of a gene encoding an antiviral protein from *Clerodendrum aculeatum* L. Plant Mol Biol 33:745–751

Lal R, Verma GS (1974) Effect of plant latex on virus infectivity. Zent Bakteriol Parasit Infect Hyg 129:271–277

Lal R, Verma GS, Verma HN (1973) Effect of some plant extracts on infectivity of tobacco mosaic virus. Indian Phytopathol 26:122–128

Lavanya N, Saravanakumar D, Rajendran L, Ramiah M, Raguchander T, Samiyappan R (2009) Management of sunflower necrosis virus through antiviral substances. Arch Phytopathol Plant Protect 42:265–276

Lodge JK, Kaniewski WK, Tumer NE (1993) Broad spectrum virus resistance in transgenic plants expressing pokeweed antiviral protein. Proc Natl Acad Sci U S A 90:7089–7093

Lodha ML, Agarwal S, Biswas K, Vasudev S, Dubey SC (2010) Antimicrobial activity of native and recombinant antiviral proteins from *Bougainvillea xbuttiana* leaves against plant pathogenic fungi and viruses. Indian J Agric Biochem 23:83–90

Lodha ML, Choudhary NL (2011) Purification and evaluation of antiviral proteins from *Bougainvillea xbuttiana* against *Helicoverpa armigera*. Indian J Agri Sci 81:74–78

Loebenstein G, Gera A (1981) Inhibitor of virus replication released from TMV infected protoplasts of a local lesion responding tobacco cultivar. Virology 114:132–139

Loebenstein G, Rav-David D, Leibman D, Gal-On A, Vunsh R, Czosnek H, Elad Y (2010) Tomato plants transformed with the inhibitor-of-virus-replication gene are partially resistant to *Botrytis cinerea*. Phytopathology 100:225–229

Lohani S, Jan A, Verma HN (2007) In vivo and in vitro resistance in tobacco by *Boerhaavia diffusa* systemic resistance inducing protein and transfer of induced resistance in in vitro tobacco plants. Biotechnology 6:389–392

Louis V, Balakrishnan S (1996) Effect of application of selected medicinal plant extracts on the incidence of pumpkin mosaic virus. Indian Phytopathol 49:373–377

Madhusudhan KN, Nalini MS, Prakash HS, Shetty HS (2005) Effect of inducers against tobamovirus infection in tomato and bell pepper. Int J Bot 1:59–61

Madhusudhan KN, Vinayarani G, Deepak SA, Niranjana SR, Prakash HS, Singh GP, Sinha AK, Prasad BC (2011) Antiviral activity of plant extracts and other inducers against tobamoviruses infection in bell pepper and tomato plants. Int J Plant Pathol 2:35–42

Mahmoud SYM, Gad-Rab SMF, Hussein N, Shoreit AAM (2010) Antiviral activity of latex from *Ficus nitida* against plant viruses. Glob J Biotechnol Biochem 5:198–205

Mandal B, Singh B (2001) Inhibition of virus transmission by guava leaf extract. Indian Phytopathol 54:381–382

Manickam K, Rajappan K (1999) Field efficacy of plant extracts and chemicals against green gram leaf curl disease. Indian J Virol 15:35–37

Mariappan V, Saxena RC (1983) Effect of custard apple oil and neem oil on survival of *Nephotettix virescens* (Homoptera Cicadellidae) and on rice tungro virus transmission. J Econ Entomol 76:573–576

McIntyre JL, Dodds JA, Hare JD (1981) Effects of localized infections of *Nicotiana tabacum* by tobacco mosaic virus on systemic resistance against diverse pathogens and an insect. Phytopathology 71:297–301

McKeen CD (1956) The inhibitory activity of extracts of *Capsicum frutescens* on plant virus infections. Can J Bot 34:891–903

Mozes R, Antignus Y, Sela I, Harpaz I (1978) The chemical nature of an antiviral factor (AVF) from virus-infected plants. J Gen Virol 38:241–249

Mukerjee K, Awasthi LP, Verma HN (1981) The inhibitory activity of an interfering agent, extracted from the leaves of host plants treated with Datura leaf extract, on plant virus infections. Z Pflanzenkr Pflanzensch 88:228–234

Murty NS, Nagarajan K (1980) Virus inhibitory effect of extracts from germinating seeds of flowering plants. Indian Phytopathol 33:615–617

Murty NS, Nagarajan K (1986) Role of plant extracts in the control of TMV infection in nursery and field grown tobacco. Indian Phytopathol 39:98–100

Nagaich BB, Singh S (1970) An antiviral principle induced by potato virus X in *Capsicum pendulum*. Virology 40:269–271

Nagarajan K, Murty NS (1975) Effect of certain plant extracts and plant latex on the inhibition of tobacco mosaic virus infection. Tob Res 1:122–132

Narayanasamy P (1990) Antiviral principles for virus disease management. In: Vidhyasekaran P (ed) Basic research for crop disease management, Daya Publishing House, New Delhi, pp 139–150

Narwal S, Balasubrahmanyam A, Lodha ML, Kapoor HC (2001a) Purification and properties of antiviral proteins from the leaves of *Bougainvillea xbuttiana*. Indian J Biochem Biophys 38:342–347

Narwal S, Balasubrahmanyam A, Sadhna P, Kapoor HC, Lodha ML (2001b) A systemic resistance inducing antiviral protein with N-glycosidase activity from *Bougainvillea xbuttiana* leaves. Indian J Exp Biol 39:600–603

Nielsen K, Boston RS (2001) Ribosome-inactivating proteins: a plant perspective. Annu Rev Plant Physiol Plant Mol Biol 52:785–816

Nutan MM, Dezzutti CS, Kulshreshtha S, Rawat AK, Srivastava SK, Malhotra S, Verma A, Ranga U, Gupta SK (2013) Extracts from *Acacia catechu* suppress HIV-1 replication by inhibiting the activities of the viral protease and Tat. Virol J 10:309. doi:10.1186/1743-422X-10-309

Obrig TG, Irvin JD, Hardesty B (1973) The effect of an antiviral peptide on the ribosomal reactions of the peptide elongation enzymes, EF-I and EF-II. Arch Biochem Biophys 155:278–289

Olivieri F, Prasad V, Valbonesi P, Srivastava S, Ghosal-Chowdhury P, Barbieri L, Bolognesi A, Stirpe F (1996) A systemic antiviral resistance inducing protein isolated from *Clerodendrum inerme* Gaertn. is a polynucleotide:adenosine glycosidase (ribosome-inactivating protein). FEBS Lett 396:132–134

Owens RA, Bruening G, Shepherd JR (1973) A possible mechanism for the inhibition of plant viruses by a peptide from *Phytolacca Americana*. Virology 56:390–393

Paliwal YC, Nariani TK (1965a) Effect of plant extracts on the infectivity of sunnhemp (*Crotalaria juncea*) mosaic virus. Acta Virol 9:261–267

Paliwal YC, Nariani TK (1965b) Properties of the inhibitors of sunnhemp (*Crotalaria juncea*) mosaic virus in certain plant extracts. Acta Virol 9:455–458

Pandey AK, Bhargava KS (1980) Antiviral activity of crude extracts of some Pteridophytes. Indian Fern J 3:132–133

Pandey AK, Bhargava KS (1983) Isolation and partial characterization of virus inhibitor from the leaf extract of a fern, *Ampelopteris prolifera* (Retz.) copel. Indian J Plant Pathol 1:193–198

Pandey AK, Bhargava KS (1984) Effect of *Ampelopteris prolifera* leaf extract on the activity of tobacco mosaic and cucumber mosaic viruses. Indian Phytopathol 37:271–277

Pandey BP, Mohan J (1986) Inhibition of turnip mosaic virus by plant extracts. Indian Phytopathol 39:489–491

Pardeep K, Awasthi LP (2009) Prevention of infection and spread of viral diseases in cucumber (*Cucumis sativus* L.) through botanicals. J Plant Dis Sci 4:25–32

Park C-J, Kim K-J, Shin R, Park JM, Shin Y-C, Paek K-H (2004) Pathogenesis-related protein 10 isolated from hot pepper functions as a ribonuclease in an antiviral pathway. Plant J 37:186–198

Patel BN, Patel GJ (1979) Inhibition of tobacco mosaic virus infection by some plant extracts. Tob Res 5:33–36

Prasad V (1986) Alterations in enzyme activity during induced antiviral state by leaf extract. J Indian Bot Soc 65:90–94

Prasad V (1988) Association of polyphenoloxidase with induced antiviral resistance in two host-inducer combinations. J Indian Bot Soc 67:53–55

Prasad V, Ghosal-Chowdhury P, Srivastava S (2001) Purification of two basic 1,3-ß-glucanase isoforms from *Cyamopsis tetragonoloba* (L.) Taub. induced to resist virus infections. Israel J Plant Sci 49:15–19

Prasad V, Mishra SK, Srivastava S, Srivastava A (2014) A virus inhibitory protein isolated from *Cyamopsis tetragonoloba* upon induction of systemic antiviral resistance shares partial amino acid sequence homology with a lectin. Plant Cell Rep 33:1467–1478

Prasad V, Misra SK, Krishna SK (2012) Induced systemic resistance against viruses in susceptible plants: phytoproteins and rhizobacteria provide two fascinating avenues. In: Rao GP, Baranwal VK, Mandal B, Rishi N (eds) Recent trends in plant virology. Studium Press LLC, Houston, pp 467–482

Prasad HP, Shankar UAC, Kumar BH, Shetty SH, Prakash HS (2007) Management of Bean common mosaic virus strain *blackeye cowpea mosaic* (BCMV-BICM) in cowpea using plant extracts. Arch Phytopathol Plant Protect 40:139–147

Prasad V, Srivastava S, Varsha, Verma HN (1995) Two basic proteins isolated from *Clerodendrum inerme* Gaertn. are inducers of systemic antiviral resistance in susceptible plants. Plant Sci 110:73–82

Praveen S, Tripathi S, Varma A (2001) Isolation and characterization of an inducer protein (Crip-31) from *Clerodendrum inerme* leaves responsible for induction of systemic resistance against viruses. Plant Sci 161:453–459

Pun KB, Doraiswamy S, Jeyarajan R (1999) Screening of plant species for the presence of antiviral principles against okra yellow vein mosaic virus. Indian Phytopathol 52:221–223

Rafiq M, Jabri A, Siddiqui KA, Khan TA, Husain SI, Mahmood K (1985) Inhibitory effect of latex on lesion caused by some plant viruses. Indian Bot Rep 4:213–214

Rajapakse RHS, Janaki CR (2006) Insecticidal properties of *Chenopodium*-based botanical against virus vectors, *Myzus persicae* and *Frankliniella schultezi*. Abstract OP 9/14. In: XVI Annual convention of IVS and International symposium on management of vector-borne viruses, ICRISAT, Patencheru, Hyderabad (India)

Rajesh S, Balasaraswathi R, Doraisamy S, Sadasivam S (2005) Synthesis and cloning of cDNA encoding an antiviral protein from the leaves of *Bougainvillea spectabilis* Willd. (Nyctaginaceae). World J Agric Sci 1:101–104

Rajinimala N, Rabindran R, Ramaiah M (2009) Management of bittergourd mosaic virus (BGYMV) by using virus inhibiting chemical, biocontrol agents, antiviral principles (AVP) and insecticide. Arch Phytopathol Plant Protect 42:738–750

Rao DG, Raychaudhuri SP (1965) Further studies on the inhibition of ringspot strain of potato virus x by plant extracts, culture filtrates of *Trichothecium roseum* and chemicals. Indian J Microbiol 5:9–12

Rao GP, Shukla K (1985) Antiviral activity of copra extract and physical properties of the inhibitor. Indian Phytopathol 38:623

Rao GP, Singh RK, Pandey AK, Singh R (1985) Induction of local and systemic resistance of potato virus by flower extracts. Indian J Virol 1:174–182

Raychaudhuri SP, Chadha KC (1965) Deodar fruit extract-an inhibitor of chilli mosaic virus. Indian Phytopathol 18:97–98

Raychaudhuri SP, Prasad HC (1965) Effect of plant extracts and microbial growth products on the infectivity of radish mosaic virus. Indian J Microbiol 5:13–16

Ready MP, Brown DT, Robertus JD (1986) Extracellular localization of pokeweed antiviral protein. Proc Natl Acad Sci U S A 83:5053–5056

Reinbothe S, Reinbothe C, Lehmann J, Becker W, Apel K, Parthier B (1994) JIP60, a methyl-jasmonate induced ribosome-inactivating protein involved in plant stress reactions. Proc Natl Acad Sci U S A 91:7012–7016

Roy S, Sadhana P, Begum M, Kumar S, Lodha ML, Kapoor HC (2006) Purification, characterization and cloning of antiviral/ribosome-inactivating protein from *Amaranthus tricolor* leaves. Phytochemistry 67:1865–1873

Roy AN, Sinha BP, Gupta KC (1979) The inhibitory effect of plant juices on the infectivity of top necrosis virus of pea. Indian J Microbiol 19:198–201

Roychoudhury R (1984) Virus inhibitor from *Solanum torvum*. Indian Phytopathol 37:665–668

Roychoudhury R, Basu PK (1983) Characterization of a plant virus inhibitor from two *Solanum* sp. Indian J Exp Biol 21:213–215

Rustgi S, Pollmann S, Buhr F, Springer A, Reinbothe C, von Wettstein D, Reinbothe S (2014) JIP-60-mediated, jasmonate-and senescence induced molecular switch in translation toward stress and defense protein synthesis. Proc Natl Acad Sci U S A 111:14181–14186

Saigopal DVR, Prasad VS, Sreenivasulu P (1986) Antiviral activity in extracts of *Phyllanthus fraternus* Webst (P. niruri). Curr Sci 55:264–265

Saveetha K, Sankaralingam A, Ramanathan A, Pant R (2006) Symptom, epidemiology and management of finger millet mottle streak disease. Arch Phytopathol Plant Protect 39:409–419

Schaad MC, Anderberg RJ, Carrington JC (2000) Strain-specific interaction of the tobacco etch virus NIa protein with the translation initiation factor eIF4E in the yeast two-hybrid system. Virology 273:300–306

Schrot J, Weng A, Melzig MF (2015) Ribosome-inactivating and related proteins. Toxins 7:1556–1615

Sela I, Applebaum SW (1962) Occurrence of an antiviral factor in virus infected plants. Virology 17:543–548

Shahidi-Noghabi S, Van Damme EJM, Smagghe G (2009) Expression of *Sambucus nigra* agglutinin (SNA-I') from elderberry bark in transgenic tobacco plants results in enhanced resistance to different insect species. Transgenic Res 18:249–259

Sharma YR, Chohan JS (1973) Inhibitors of cucumis virus 1 in extracts of leaf and seeds of different plants. Indian Phytopathol 26:172–173

Sharma DC, Raychaudhuri SP (1965) Antiviral effect of plant extracts and culture filtrate of *Aspergillus niger* on ringspot strain of potato virus x. Indian J Microbiol 5:41–48

Shukla HH, Dubey P, Chaturvedi RV (1989) Antiviral properties of essential oils of *Foeniculum vulgare* and *Pimpinella anisum* L. Agronomie 9:277–279

Singh R (1969) The inhibitory activity of some plant juices on the infectivity of watermelon mosaic virus. Acta Virol 13:244–246

Singh AB (1972) Inhibitory activity of some plant extracts on the infectivity of papaya leaf reduction virus. Acta Phytopathol Acad Sci Hung 7:175–178

Singh S, Awasthi LP, Singh RK (2011a) Induction of systemic resistance through antiviral agents of plant origin against papaya ring spot disease (*Carica papaya* L.) Arch Phytopathol Plant Protect 44:1676–1682

Singh SK, Awasthi LP, Singh S, Sharma NK (2011b) Protection of mungbean and urdbean crops against vector borne mungbean yellow mosaic virus through botanicals. Curr Bot 2:8–11

Singh S, Awasthi LP, Verma HN (2004) Prevention and control of yellow mosaic disease of mungbean by application of aqueous root extract of *Boerhaavia diffusa*. Indian Phytopathol 57:303–307

Singh AK, Najam A, Verma HN, Awasthi LP (2009) Control of natural virus infection on okra (*Abelmoschus esculentus*) by root. Intl J Plant Protect 2:195–198

Singh P, Sharma S, Prasad V (2011c) Verapamil, a calcium channel blocker, induces systemic antiviral resistance in susceptible plants. J Phytopathol 159:127–129

Singh R, Singh R (1973) Properties of an inhibitor of potato virus x from bark of *Ficus elastica*. Indian Phytopathol 26:560–563

Singh R, Singh R (1975) Inhibition of potato virus x by *Rumex hastatus* extract. Sci Cult 43:268–269

Smookler MM (1971) Properties of inhibitors of plant virus infection occurring in the leaves of species of Chenopodiaceae. Ann Appl Biol 69:157–158

Sreeja S, Sreeja S (2009) An in vitro study on antiproliferative and antiestrogenic effects of *Boerhaavia diffusa* L. extract. J. Ethnopharmacol 126:221–225

Srinivasulu B, Jeyarajan R (1986) Effect of leaf extract on the infection of rice tungro virus in rice. Indian J Virol 2:176–180

Srivastava KM, Chandra S, Singh BP, SMH A (1976) Induction of local and systemic resistance in Nicotiana glutinosa against TMV by leaf extract of Dahlia. Ind J Exp Biol 4:377–379

Srivastav SK, Ansari NA, Tewari JP (2011) Eco-friendly management of tomato leaf curl disease. Intl J Plant Protect 4:321–323

Srivastava A, Gupta RK, Verma HN (2004) Micropropagation of *Clerodendrum aculeatum* through adventitious shoot induction and production of consistent amount of virus resistance inducing protein. Indian J Exp Biol 42:1200–1207

Srivastava S, Prasad V (2014) Induction of defence responses for biological control of plant diseases. In: Sharma N (ed) Biological controls for preventng food deterioration. Wiley, UK, pp 321–339

Srivastava A, Srivastava S, Prasad V (2015a) Systemic antiviral resistance induced in papaya by CAP-34, a resistance inducing protein from *Clerodendrum aculeatum*, is associated with a proteinaceous virus inhibitory activity. J Plant Pathol 97:45–54

Srivastava A, Trivedi S, Krishna SK, Verma HN, Prasad V (2009) Suppression of *papaya ringspot virus* infection in *Carica papaya* with CAP-34, a systemic antiviral resistance inducing protein from *Clerodendrum aculeatum*. Eur J Plant Pathol 123:241–246

Srivastava S, Verma HN, Srivastava A, Prasad V (2015b) BDP-30, a systemic resistance inducer from *Boerhaavia diffusa* L., suppresses TMV infection, and displays homology with ribosome-inactivating proteins. J Biosci 40:125–135

Stevens WA, Spurdon C, Onyon LJ, Stirpe F (1981) Effect of inhibitors of protein synthesis from plants on tobacco mosaic virus infection. Experientia 37:257–259

Stintzi A, Heitz T, Prasad V, Wiedemann-Merdinoglu S, Kauffmann S, Geoffroy P, Legrand M, Fritig B (1993) Plant 'pathogenesis-related' proteins and their role in defence against pathogens. Biochimie 75:687–706

Stirpe F (2004) Ribosome-inactivating proteins. Toxicon 44:371–383

Stirpe F (2013) Ribosome-inactivating proteins: From toxins to useful proteins. Toxicon 67:12–16

Stirpe F, Lappi DA (eds) (2014) Ribosome inactivating proteins: ricin and related proteins. Wiley, USA

Sundaram S, Dwivedi P, Purwar S (2011) In vitro evaluation of antibacterial activities of crude extracts of *Withania somnifera* (Ashwagandha) to bacterial pathogens. Asian J Biotechnol 3:194–199

Surendran M, Shanmugam V, Rajagopalan B, Ramanian N (1999) Efficacy of botanicals on brinjal mosaic virus. Plant Dis Res 14:63–66

Tewari JP (1976) Inhibition of three strains of watermelon mosaic virus by bark extract. Curr Sci 45:696–697

Thirumalaisamy PP, Rathi YPS, Tripathi HS (2003) Screening of some plant extracts inhibitory to urdbean leaf crinkle virus. Indian Phytopathol 56:233–235

Tiwari P, Khan H, Ansari NA, Tewari JP (2010) Management of pea mosaic virus by leaf extract of some medicinal plants. Intl J Plant Protect 3:117–119

Tourlakis ME, Karran RA, Desouza L, Siu KW, Hudak KA (2010) Homodimerization of pokeweed antiviral protein as a mechanism to limit depurination of pokeweed ribosomes. Mol Plant Pathol 11:757–767

Tumer NE, Hwang D-J, Bonness M (1997) C-terminal deletion mutant of pokeweed antiviral protein inhibits viral infection but does not depurinate host ribosomes. Proc Natl Acad Sci U S A 94:3866–3871

Van Loon LC, Rep M, Pieterse CMJ (2006) Significance of inducible defence-related proteins. Annu Rev Phytopathol 44:135–162

Van Loon LC, Van Kammen A (1970) Polyacrylamide disc electrophoresis of the soluble leaf proteins from *Nicotiana tabacum* var. "Samsun" and "Samsun NN": changes in protein constitution after infection with tobacco mosaic virus. Virology 40:199–211

Vandenbussche F, Desmyter S, Ciani M, Proost P, Peumans WJ, Van Damme EJ (2004) Analysis of the *in planta* antiviral activity of elderberry ribosome-inactivating proteins. Eur J Biochem 271:1508–1515

Vasudeva RS, Nariani TK (1952) Host range of bottle gourd mosaic virus and its inactivation by plant extracts. Phytopathology 42:149–152

Venkatesan S, Radjacommare R, Nakkeeran S, Chandrasekaran A (2010) Effect of biocontrol agent, plant extracts and safe chemicals in suppression of mungbean yellow mosaic virus (MYMV) in black gram (*Vigna mungo*). Arch Phytopathol Plant Protect 43:59–72

Verma HN (1982) Inhibitor of plant viruses from higher plants. In: Singh BP, Raychoudhary SP (eds) Current trends in plant virology. Today and Tomorrow's printers and publishers, New Delhi, pp 151–159

Verma HN, Awasthi LP (1979) Antiviral activity of *Boerhaavia diffusa* root extract and the physical properties of the virus inhibitor. Can J Bot 57:926–932

Verma HN, Awasthi LP (1980) Occurrence of a highly antiviral agent in plants treated with *Boerhaavia diffusa* inhibitor. Can J Bot 58:2141–2144

Verma HN, Awasthi LP, Mukerjee K (1979b) Induction of systemic resistance by antiviral plant extracts in non-hypersensitive hosts. Z Pflanzenkr Pflansensch 86:735–740

Verma HN, Awasthi LP, Mukerjee K (1979c) Prevention of virus infection and multiplication by extracts from medicinal plants. Phytopathol Z 96:71–76

Verma HN, Awasthi LP, Saxena KC (1979a) Isolation of the virus inhibitor from the root extract of *Boerhaavia diffusa* inducing systemic resistance in plants. Can J Bot 57:1214–1217

Verma HN, Baranwal VK (1983) Antiviral activity and the physical properties of the leaf extract of *Chenopodium ambrosiodes* L. Proc Indian Acad Sci 92:461–465

Verma HN, Baranwal VK (2011) Potency of plant products in control of virus diseases of plants. In: Dubey NK (ed) Natural products in plant pest management. CAB International, India, pp 149–174
Verma HN, Baranwal VK, Srivastava S (1998) Antiviral substances of plant origin. In: Hadidi A, Khetarpal RK, Koganezawa A (eds) Plant virus disease control. American Phytopathological Society (APS) Press, St Paul, pp 154–162
Verma HN, Chowdhery B, Rastogi P (1984) Antiviral activity of different *Clerodendrum* L. species. Z Pflanzenkr Pflansensch 91:34–41
Verma HN, Dwivedi SD (1983) Prevention of plant virus diseases in some economically important plants by *Bougainvillea* leaf extract. Indian J Plant Pathol 1:97–100
Verma HN, Dwivedi SD (1984) Properties of a virus inhibiting agent isolated from plants following treatment with *Bougainvillea spectabilis* leaf extract. Physiol Plant Pathol 25:93–101
Verma HN, Khan MMAA (1985) Occurrence of a strong virus interfering agent in susceptible plants sprayed with *Pseuderanthemum atropurpureum tricolor* leaf extract. Indian J Virol 1:26–34
Verma HN, Khan MMAA, Dwivedi SD (1985a) Biophysical properties of highly antiviral agents present in *Pseuderanthemum atropurpureum* and *Bougainvillea spectabilis* extracts. Indian J Plant Pathol 3:13–20
Verma HN, Kumar V (1979) Prevention of potato from viruses and insect vectors. J Indian Pot Assoc 6:157–161
Verma HN, Kumar V (1980) Prevention of plant virus diseases by Mirabilis jalapa leaf extract. New Bot 7:87–91
Verma HN, Mukerjee K (1975) Brinjal leaf extract induced resistance to virus infection in plants. Indian J Exp Biol 13:416–417
Verma HN, Prasad V (1987) Systemic induced antiviral resistance by plant extract alters physiology of susceptible test host. Indian J Plant Pathol 5:69–72
Verma HN, Prasad V (1988) Metabolic alterations associated with host mediated systemic antiviral resistance. Indian Phytopathol 41:332–335
Verma HN, Prasad V (1992) Virus inhibitors and inducers of resistance: potential avenues for biological control of viral diseases. In: Mukerji KG, Tewari JP, Arora DK, Saxena G (eds) Recent developments in biocontrol of plant diseases. Aditya Book Pvt Ltd., New Delhi, pp 81–110
Verma HN, Rastogi P, Prasad V, Srivastava A (1985b) Possible control of natural virus infection on *Vigna radiata* and *Vigna mungo* by plant extracts. Indian J Plant Pathol 3:21–24
Verma VS, Raychaudhuri SP (1970) Effect of tannins from *Terminalia chebula* on the infectivity of potato virus. Acta Microbiol Pol 19(B):127–132
Verma VS, Raychaudhuri SP (1972) Effect of catechol tannins isolated from medicinal plant *Chrysobalanus icacao* on the infectivity of potato virus x Zent Bakenot Abt II 127:178–179
Verma VS, Raychaudhuri SP, Khan AM (1969) Properties and nature of inhibitors of potato virus X in four medicinal plant extracts. Biologia Plantarum (Praha) 11:384–387
Verma HN, Srivastava A (1985) A potent systemic inhibitor of plant virus infection from *Aerva sanguinolenta* Blume. Curr Sci 54:526–528
Verma HN, Srivastava S, Varsha KD (1996) Induction of systemic resistance in plants against viruses by a basic protein from *Clerodendrum aculeatum* leaves. Phytopathology 86:485–492
Verma HN, Varsha (1995a) Induction of systemic resistance by leaf extract of *Clerodendrum aculeatum* in sunnhemp rosette virus. Indian Phytopathol 48:218–221
Verma HN, Varsha (1995b) Prevention of natural occurrence of tobacco leaf curl disease by primed *Clerodendrum aculeatum* leaf extract. In: Verma JP, Varma A, Kumar D (eds) Detection of plant pathogens and their management. Angkor Publishers (P) Ltd, New Delhi, pp 202–206
Verma HN, Varsha, Baranwal, VK (1995) Agricultural role of endogenous antiviral substances of plant origin. Chessin M, DeBorde D, Zipf A, Antiviral proteins in higher plants, CRC Press, Boca Raton, 23–37
Verma GS, Verma HN (1965) Tobacco mosaic virus inhibitory principle in wheat extract. Indian J Microbiol 5:17–20

Verma GS, Verma HN, Srivastava KM, Mukerjee K (1975) Properties and mode of action of a tobacco mosaic virus inhibitor from seed extract of *Lawsonia alba*. New Botanist 2:109–113
Vijayan P, Raghu C, Ashok G, Dhanaraj SA, Suresh B (2004) Antiviral activity of medicinal plants of Nilgiris. Indian J Med Res 120:24–29
Vivanco JM, Tumer NE (2003) Translation inhibition of capped and uncapped viral RNAs mediated by ribosome-inactivating proteins. Phytopathology 93:588–595
Yadav S, Kumar S, Jain P, Pundir RK, Jadon S, Sharma A, Khetwal KS, Gupta KC (2011) Antimicrobial activity of different extracts of roots of *Rumex nepalensis* Spreng. Indian J Nat Prod Resour 2:65–69
Zaidi ZB, Gupta VP, Samad A, Naqvi QA (1988) Inhibition of spinach mosaic virus by extracts of some medicinal plants. Curr Sci 57:151–152
Zoubenko O, Hudak K, Tumer NE (2000) A non-toxic pokeweed antiviral protein mutant inhibits pathogen infection via a novel salicylic acid-independent pathway. Plant Mol Biol 44:219–229

Management of Viral Diseases Through Conventional Approaches in India

29

Savarni Tripathi and Raj Verma

Abstract

Viral diseases of crop plants cause enormous economic losses as most of the cultivated areas in India are in subtropical and tropical regions which provide congenial environment for multiplication of viruses and their vectors. Intensive agricultural practices for fulfilling the demand of food for the growing population has further contributed to the diseases and pests problems. Under favourable condition, the viral disease incidence can be as high as 100% resulting into serious losses to the farmers and consumers. The control of plant viral diseases has been challenging because of unavailability of effective direct method of control by chemical applications. Hence, indirect methods of managing viral diseases has been utilized such as use of modified cultural practices, use of virus-free planting materials, use of host resistance to viruses and their insect-vectors, cross protection, application of insecticides and oils for the control of virus viruses. Some success in viral disease management has been achieved by using a combination of these approaches in few crops. The range of conventional management approaches that have been studied and applied against plant viral diseases in India are summarized in this chapter.

Keywords

Plant viruses • India • Management

S. Tripathi (✉) • R. Verma
ICAR-Indian Agricultural Research Institute,
Regional Station, 125 ITI Road, Aundh, Pune 411067, Maharashtra, India
e-mail: savarni_pune@iari.res.in; savarni@yahoo.com

B. Mandal et al. (eds.), *A Century of Plant Virology in India*,
https://doi.org/10.1007/978-981-10-5672-7_29

29.1 Introduction

Viral diseases of crop plants cause a great damage in terms of quality and quantity of crop yield, which result in enormous economic losses worldwide (Varma 1976, 1993; Varma et al. 1992; Sastry 2013b; Selvarajan and Balasubramanian 2014). Most of the agriculture production areas of India are in the subtropical and tropical regions where losses caused by viral diseases are much higher because it provides congenial environment for the perpetuation of both viruses and their vectors. The need of ever increasing demand of rapid growing population resulted into intensive agricultural practices, which have further enhance the diseases and pests problems (Varma 1993). Developing effective integrated management practices for plant viral diseases has been the priority and major concern to the plant pathologists, farmers, horticulturists and gardeners.

Plant viral diseases cannot be effectively controlled as no commercial viricides have yet been developed. The viral disease control involves a far greater complexity compare to the fungal and bacterial diseases because viruses are molecular parasites and have a complex disease cycle involving host-virus-vector-environments. However several indirect attempts that are suitable for various regions in India have been attempted with an aim to minimize the impact of virus infection on crop yield. The disease intensity varies with location, crop and season. There are mainly two strategies to manage plant viral diseases: (a) to decrease the sources of infection and (b) to minimize the rate of spread.

The effective viral disease management strategies needs integration of all available strategies which include avoidance of source of infection, vector control by various means, modification in cultivation practices and use of host resistance for virus and vectors. The conventional management approaches for plant viral diseases used in India is discussed in this chapter.

29.2 Avoidance of Source of Infection

Infected planting materials, collateral and weed hosts, adjoining crops, volunteers are the main sources of infection. The use of virus free planting materials and other phytosanitary approaches help in minimizing the source of initial infection which are responsible for further spread of the virus.

29.2.1 Use of Virus Free Planting Materials

The initial goal is to use healthy planting/seed material in the fields for higher yield. About 231 virus diseases are known to be seed transmitted in different plants (Sastry 2013a). Seeds or vegetative propagules carrying the virus provide primary source of infection for establishment of disease in field. This allows: (a) occurrence of infection as soon as possible in young developing seedling and (b) infected seedling act as source of initial virus inoculum for subsequent further spread. One can expect better yield by planting a healthy seed/planting material of a crop. Even after planting the virus-free material/ seed, if regular monitoring and precautionary measures are not taken it will be difficult to achieve expected return. Therefore, it is very important to take regular measures for limiting the spread of disease by other biological, chemical or cultural approaches which will be able to decrease the population of insect vectors responsible for disease spread. Thus, use of virus-free planting materials and controlling the insect vectors of viruses are practical and potent method for viral disease management.

In many regions of the country, certain crops such eggplant, capsicum, chilli, cucumber, rice, tobacco, tomato etc. are grown in nursery or planted in a large scale by the commercial companies or agricultural institutions and these seedlings are transported at the required stage for transplanting in field. Generally, nurseries are examined for the visible symptoms and symptomatic plants are removed. However, suspected seedlings are required to be tested for symptomless carrier of particular virus. The nursery should be tested by easy diagnostic techniques. On the other hand, to produce virus free-seeds, a certification schemes are required for those crops which are grown directly though seeds in the field and may carry seed-borne viruses.

In India, several attempts were made to provide virus free seeds/planting materials to the growers at national and regional levels. In order to ensure the distribution of quality virus free planting materials raised through tissue culture to the farmers under the Seed Act, 1966 of the Ministry of Agriculture, the Department of Biotechnology has established the National Certification System for Tissue Culture Plants in 2008.

Indian Council of Agricultural Research (ICAR) – National Research Center for Citrus at Nagpur has standardized internationally accepted techniques of nursery management for raising disease free citrus planting material. Under this mission mode scheme, plants showing no visual symptoms with high yield were chosen as elite trees for indexing main viral diseases. From the foundation block, bud woods are supplied to nurseries for establishment of nursery under the supervision of designated technician. Thus constant supplies of certified virus-free citrus plants are available to the farmers (http://nrccitrus.nic.in).

Similarly, a methodology for healthy seed potato stocks production in high hills was developed at ICAR-Central Potato Research Institute (CPRI), Shimla. Few pocketed areas above 2000 m in northern temperate hills were found suitable for quality seed potato production. These areas are unfavourable for building up of populations of aphid which are vectors for the potato viral diseases during summer months. However, the production of seed potatoes from the hill is not sufficient for sowing in the plains. Based on the above observation, 'Seed Plot technique' was developed (Pushkarnath 1959; Nagaich et al. 1969). This virus-free nucleus stocks is used by the CPRI seed production centers in hills and at the regional stations in the sub-tropical plains for the production of breeder's seed using 'seed plot technique'. These virus free seeds are used for multiplication by the State Agricultural

Departments and National Seeds Corporation. The healthy seed potatoes are supplied to farmers through seed agencies.

Elimination of the viruses through meristem culture has been demonstrated to eliminate the viruses from the infected planting materials. The technique of in-vitro culturing arises as a powerful technique for elimination of sugarcane mosaic and sugarcane yellow leaf viruses (Mishra and Rao 2011). Meristem culture alone or in combination with heat therapy has also been found useful in generating virus free plants from sugarcane in India (Hendre et al. 1975; Jain et al. 1998; Rao 2002; Mishra et al. 2010; Viswanathan and Rao 2011; Mishra and Rao 2011). Over 90% success was obtained in elimination of SCMV in sugarcane var. Co 740 when shoot apex of 0.05–1.5 mm long as apical meristem tip from SCMV infected plants were used in tissue culture (Hendre et al. 1975). The success of meristem tip culture resides in the ability to dissect the meristematic dome with one or two leaf primordia from the mother plant and its successful regeneration (Parmessur et al. 2002). Meristem tip culture is the most common method of virus elimination in plants, taking advantage of the fact that some viruses are unable to replicate in this region (Faccioli and Marani 1998; Tiwari et al. 2008). The phenomenon of virus elimination through apical meristem is based on the fact that apical meristem of infected plants are generally either free or carrying very low titer of the virus (Wang and Hu 1980; Bhojwani and Razdan 1983; Kartha 1981, 1986). The reasons proposed for absence of vascular virus in meristem are lack of vascular tissue, high metabolic activity of meristematic cells, higher endogenous and exogenous level of auxins in cultured meristems cause inhibition to viral multiplication. Hence, meristem culture could be applied for the successful elimination of viruses and utilized as a potential tool for production of virus-free planting materials.

Numerous chemicals, like phenolic compounds, homeopathic drugs, thiouracil and 2,4-dinitrophenol have been found to inhibit Sugarcane mosaic virus infectivity (Prakash and Joshi 1977; Shah 1972; Shukla 1978; Shukla and Joshi 1979, 1982).

29.2.2 Avoidance of Collateral Hosts and Volunteers

The importance of collateral hosts, volunteers and weeds in crop production is well known, as they play a key role in multiplication and spread of viral diseases. These hosts also harm and weaken the main crops as they exploit the available resources such as moisture, nutrients and solar energy as well as harbor other pests and diseases. Most of the viral diseases of crops have been widely distributed in volunteers/weed host in and around main field. These virus infected collateral hosts act as the primary source of inoculum. Perennial weeds are source of virus infection for long period, hence they are more harmful as compared to annual weeds. In general, the primary introduction of viral diseases into a crop occurs by planting diseased seed/plant material or by insect which acquire virus from other diseased hosts. Wherever healthy planting materials were used, the virus infection comes from the weed hosts or other infected crops. Therefore, control of volunteers and weeds hosts are beneficial in minimizing the losses they caused due to virus infections and through competition as well. In a practical sense, eliminating such sources of virus inoculum is often impossible particularly in tropical regions. The extent for removing these hosts will mainly depend on broadness of virus host range. Controlling alternative hosts is more practical in managing viral diseases where the virus has a narrow host range but in viruses with wide host range this task is usually impossible.

It is generally observed that frequent removal of volunteers and weed hosts during the cropping period indicate promising outcome. For example, the incidence of bhindi yellow vein mosaic virus (BYVMV) was reduced by removal of malvaceous weeds (*Malva sylvestris* and *Sida rhombifolia*) around main plot of bhindi (*Abelmoschus esculentus*) before sowing (Varma 1976). The significance of weed hosts in spread of rice tungro virus (RTV) was also indicated (Mishra et al. 1973; Mukhopadhyay 1980, 1984; Rao and Anjaneyulu 1978; Tarafder and Mukhopadhyay 1979, 1980). Successful elimination of the weeds aided in minimizing the spread of economically important plant viruses. The spread of sugarcane mosaic virus (SMV) in sugarcane was restricted by destroying the volunteer plants like sorghum, maize and other graminaceous hosts (Shukla and Teakle 1989). Use of Stam F herbicides with or without Eptam was useful in restricting the potato virus X (PVX) spread in seed plots of potato (Nagaich et al. 1972).

Crop rotation was also found effective in decreasing the disease spread. For example, a host-free period practice of 2–3 months by not growing the susceptible crop plant can significantly reduce the virus infections in the agro-ecosystem. Field studies by Mukhopadhyay (1980) revealed that nursery beds and infected stubbles play a key role in perpetuating the RTV. Seed beds are normally raised by the farmers close to the main field and leafhoppers carrying the virus from the stubbles may migrate to seedlings in nurseries. Therefore, raising rice nurseries in protected conditions should be prerequisite.

29.2.3 Roguing

Removing of diseased plants is a main strategy widely used to manage viral diseases. This strategy is very effective when the main crop is the primary/sole source of virus infection. For instance, banana bunchy top virus (BBTV) can be managed by frequent rouging the diseased plants from plantations (Sharma 1988). Roguing is not very effective if the sources of infection are occurring from outside the crop or in other word if the virus has wider host range and the spread is relatively fast. Moreover, removal of affected plants based on external symptoms can be difficult for viruses not causing noticeable symptoms.

The effect of roguing was observed in large cardamom (*Amomum subulatum*) infected with aphid (*Pentalonia nigronervosa)* vectored 'Foorkey' disease. The virus infection has been managed by administering liquid 2, 4-D into diseased rhizomes which were removed and replaced with virus free plantlets. Concurrently spraying with 0.04% Folidol E605 at constant gaps of 3–4 weeks was effective in controlling the insect vector (Chattopadhyay and Bhowmik 1965). Same virus management strategies were applied in small cardamom (*Elettaria cardamomum*) for 'Katte' disease spread by the same aphid vector as Foorky disease of large cardamom.

Field management trials conducted under IPM programme for groundnut bud necrosis virus (GBNV) in tomato crop in Tamil Nadu also showed that roguing aids in reducing the incidence of PBNV disease with more yield as compare to control tomato plots.

29.2.4 Selection of Planting Site

Planting site selection is an important strategy for avoidance of disease. Generally, the favourable conditions for disease development or the conditions under which hosts are susceptible to initial infection should be avoided. For example, raising fresh crop of tomato, capsicum or cucurbits should be avoided nearby old fields of same or another crop susceptible to virus infections. Earlier crops should be removed after the harvest as older crops are not appropriate for vector perpetuation. This situation lead the insect vectors of viruses to move in search of more preferred host crop which increases the possibilities of virus spread to adjoining susceptible young host crops.

29.3 Modification in Cultural Practices

Suitable modification in crop cultivation practices can be very helpful in minimizing the impact of virus diseases. This includes any variation in cultivation practices of the crop that reduces the incidence of the viruses either by minimizing or removing the inoculum source or population of spreading vectors. Modification in cultivation practices for viral disease management involves diverse set of practices such as field sanitation, fallow and crop rotation, date of planting, plant density, barrier/border/mix/trap/ cropping, fertilizer use, sowing time etc. The eradication of weeds and residues of crops which act as source of inoculums is very useful in reducing the diseases incidence. Cultural practices such as changing the crop in subsequent planting (crop rotation) and crop fallow can interrupt the normal disease cycle of the virus which greatly helps in reducing its vector population or inoculum source.

29.3.1 Sowing/Planting Dates

Appropriately adjusting the date of sowing or planting significantly reduces the viral infections. Host plants are often more susceptible to insects vectors and virus infections at the early seedling stage as compared to host plant at later stage. Farmer's major concerns have been adopting the suitable sowing or planting time for better crop yield. It is recommended to avoid period for planting the crop when highest number of viruliferous insect vectors are likely to occur. Planting time can be selected based on the vector migration. For example, delay in planting time are recommended if vector migrates early or advance the planting time when vector migrate late. The planting time manipulation mainly depends on the environmental conditions as it has direct influence on plant growth and population of insect vectors. Vector population dynamics studies conducted at each region/place will help to select the date of planting because the climatic factors vary from one region to another.

In India, thrips transmitted GBNV in peanut and tobacco streak virus (TSV) in sunflower are economically important. It has been reported that early sowing in kharif (rainy) season can reduce the incidence of GBNV in peanut (Reddy 1991; Reddy et al. 1983). The incidence of pepper mosaic virus complex in the bell pepper was less (21–28%) when planted between second fortnight of April to first fortnight of May. However, the percentage of disease incidence increases to 88 when the crop was planted in first fortnight of July (Chowfla and Parmar 1995). The incidence GBNV and its thrips vector population were lower in crop sown in first fortnight of May to first fortnight of June (Reddy et al. 1983; Sreenivasulu et al. 2008). However, on the other hand, the GBNV incidence in peanut in Northern India was high when crop was sown in first fortnight of May (Thira et al. 2004). In Central India, thrips activity increased at around 30–35°C along with dry weather during September to October. Therefore, by avoiding crop exposure to thrips vectors, potato planting after October is helpful in minimizing incidence of GBNV (Somani et al. 2007).

Studies conducted by Sreekanth et al. (2002) at ICAR-NBPGR Regional Center, Hyderabad showed that mid May to mid June (early kharif) and late October (rabi) sowing of green gram helps in reducing the thrips infestation and incidence of PBNV. Sowing of sunflower after rainy season (September onwards) was effective in reducing the necrosis disease caused by TSV (Shirshikar 2003). Based on population dynamics of aphid (*Myzus persicae*) vector, the harvesting and planting dates of potato were altered. In the northern hills aphids normally appear in July to August and in the plains in December to January. The summer planted potatoes in the hills and autumn planted potatoes in the plains reach maturity before the onset of the high vector population which helps healthier potato production (Pushkarnath 1959).

In North Indian condition, a high degree of tomato leaf curl incidence (83%) was found in winter crop planted in October as compared to the summer crop planted in February where only 14% of tomato leaf curl incidence was recorded (Tripathi and Varma 2002). Similarly, the seasonal variation in tomato leaf curl incidence was also recorded in South Indian conditions (Sastry et al. 1978; Saikia and Muniyappa 1989). A lower incidence of BYVMV was observed in June sown okra as compared to July sown (Gill et al. 1982).

29.3.2 Barrier Cropping

In the host-recognition period, it has been revealed that aphid vectors immediately resume flight if they have alighted on an unsuitable host. This is helpful for minimizing the virus spread as very soon after acquiring non-persistent viruses, the aphids lose the ability to transmit them. They may do so while searching for suitable host plants. Barrier cropping can be useful to reduce the insect vectored viral diseases. Barrier crops not only reduced the severity of non-persistently transmitted viruses, it has also helped in reduction of diseases caused by persistently transmitted viruses. In order to protect main crops from the virus vectors, the barrier crops should be fast growing and taller than the main crop and it should not be susceptible to the viruses and their vectors of the main crop. Various crop plants such as wheat, maize, sorghum, pearl millet, banana etc. have been tried as barrier crops against several insect transmitted viral diseases.

The incidence of TSV was reduced in groundnut and cowpea by growing maize, pearl millet or sorghum around the main crop as a barrier crop (Prasada Rao et al. 2003a). The disease incidence of BYVMV in okra was reduced using maize and pearl millet as barrier crop (Singh et al. 1979b; Kalita 2003; Pun et al. 2005). The incidence of ToLCV in tomato was also reduced by planting maize as barrier crop (Sastry et al. 1977). Several rows of banana or maize were effective in reducing the *Papaya ringspot virus* (PRSV) incidence in papaya under field conditions (Sharma et al. 2010; Sharma and Tripathi 2014).

29.3.3 Isolation by Distance

Isolation distance can play a key role in reducing the disease incidence specially for non-persistently transmitted viruses. The advancement of non-persistently transmitted viral diseases decreases with distance from the virus inoculum source. However, the reasonable isolation distance depends mainly on the types of virus and its vector involved, direction of wind, insect vector population etc. Isolation strategy for managing the viral disease is effective particularly against non-persistent viruses as they become non-viruliferous during prolonged flights. Whereas this techniques play a minor role for persistently transmitted viruses as the vector remains viruliferous throughout life after acquiring the virus.

29.3.4 Use of physical Barriers

Fine-mesh screens as a physical barrier have been used to protect crops from viral diseases. For example, the initial incidence of leaf curl and mosaic disease was observed to be lower in capsicum, chilli and tomato grown under protected (poly/shed net houses) as compare to open field cultivation in certain area of Maharashtra. Tripathi and Varma (2002) were able to reduce the incidence of ToLCV significantly by covering tomato plants by perforated transparent plastic bags. Krishnakumar and Eswara Reddy (2006) reported ToLCV incidence was 86.72 and 42.50% during September to March and June to December respectively under open field conditions whereas the incidence of the virus was 30.68 and 16.67% in greenhouse.

In order to avoid the virus spread through aerial vectors the general practice is to grow the plants in nursery under insect proof condition. For this purpose, the nurseries are covered with nylon mesh cages. Nilakshi Kakati and Nath (2006) reported no incidence of leaf curl when tomato seedlings were grown under nylon net as against 8% incidence of leaf curl when grown unprotected. The tomato seedlings raised under nylon net with the application 2% Nimbicidine at 20 and 35 days after transplantation showed no disease and a higher yield (318.89 q/ha) as compared to 89.33% disease incidence with low yield of 98.89 q/ha in the control plot grown without net and insecticidal spray. Nylon net covering of nursery beds coupled with two to three applications of monocrotophos/ dimethoate/ cypermethrin after transplanting greatly reduced the spread of disease in tomato field (Muniyappa and Saikia 1983). Venkatesh (2000) showed that the infection was delayed in tomato plot by using deltamethrin treated nylon net painted with yellow colour as it reduced the migration of whitefly vectors into the main plot.

29.3.5 Plant Density

Plant spacing also influence the disease incidence and its role in the spread of viral diseases. In general, higher planting density lessens the number of infections per unit area, for example virus associated wilt in pineapple (Singh and Sastry 1974), sterility mosaic in pigeonpea (Ramakrishnan 1963) and vein banding mosaic in cowpea (Sharma 1975). Increased plant density has also reduced losses in groundnut (Ghanekar 1980), okra (Gill et al. 1982) and bellpepper (Chowfla and Parmar 1995).

29.3.6 Breaking the Disease Cycle

Continuous cultivation of a particular crop in an area usually results into disease epidemics after a period of time. A crop free period before taking the main crop helps in breaking the disease cycle of vectors and viruses having limited host range. PPSMV and its vector both have host range restricted to pigeonpea only but volunteers and crop ratoon are the main source of primary virus inoculum. Therefore, the virus can easily be managed by removing volunteers (Ramakrishnan 1963; Varma 1976).

29.3.7 Use of Plastic Mulches

The use of mulches for the management of viral diseases is gaining more attention. Several reports have shown the successful effect of plastic mulches for the management of insect vectors transmitted viral diseases. Mulches act as repellents which avoid landing of insect vector on crop plants and thus minimize the spread of viral diseases. The ultraviolet light and visible infrared light emitted from different mulches are responsible for repelling insect vectors. Mulches with white surfaces reflect UV or short wave light which is avoided by the alighting aphids as it's unattractive for them (Moericke 1954). Mosaic disease incidence in muskmelon could be reduced by using mulches (Vani et al. 1989). Similarly, use of polythene sheet as mulch was found very effective in decreasing the incidence of tomato leaf curl and promoting the growth and yield of tomato (Tripathi and Varma 2002). Khan and Mukhopadhyay (1985) reported the incidence of BYVMV in bhindi was reduced by 50% by mulching yellow polythene.

29.4 Cross-Protection

Cross-protection is defined as a situation where prior infection with mild or attenuated strain of a virus arrest or hinder the successive infection by severe strain of the same or a closely related virus. This phenomenon has been explored and used as means of protection against severe strains of viruses in some crops to minimize yield losses. As in case of humans and animals, the attenuated virus strains are being used as vaccines for preventing diseases, the cross-protection in plants by a milder virus against a virulent virus strain help in managing the disease. The cross-protection phenomenon was first shown in tobacco plants by McKinney (1929).

This strategy for controlling the viral diseases in certain crops such as apple, banana, citrus, cocao, papaya, passion fruit, peach, tomato were attempted in certain geographic regions worldwide with some success. In India, efforts were made to manage mainly two economically important diseases caused by citrus tristeza virus (CTV) in citrus and PRSV in papaya. The existence of mild and severe strains of CTV has been demonstrated (Capoor and Rao 1967; Balaraman 1981, 1987). Capoor and Rao (1967) were able to identify mild, strong and severe strains based on key lime reactions. Balaraman and Ramakrishnan (1977, 1979a) were able to isolate six strains (s1-very mild, s2-mild, s3-moderate, s4-moderately severe, s5-severe and hd-severe with corky veins). The cross-protection field experiment of 7 years in citrus indicated no behaviour change in the virus strain in immunized Kagzi lime seedlings with mild CTV (Balaraman 1981, 1987; Balaraman and Ramakrishnan 1979a, b; Sharma 1989).

Ram et al. (2006) at IARI Regional Station, Pune have tried to manage PRSV by cross-protection using a naturally occurring mild strain. The results revealed that the pre-immunized seedlings maintained in glasshouse for 30 days and then planted in the field gave higher protection against severe isolates of the PRSV. Pre-immunized papaya plants showed up to 66% of protection against the severe isolate of PRSV under Pune condition. Further, the cross-protection technology was demonstrated on papaya farmer's field and it was reported to be commercially beneficial. However, the cross-protection technique is practiced only in few crops mainly due to the fear that the mild virus may regain potency and become severe as it has been exhibited with virus combinations in fruit trees. Nevertheless, the value of cross-protection using mild strain seems to be beneficial when crops are certain to become infected by the severe strain of the virus at very early plant development stages and there is no effective tolerance or resistance available in commercially acceptable varieties.

29.5 Vector Control

Viral diseases in plants spread very fast through insect vectors such as aphids, leaf-hoppers, thrips, whiteflies etc. due to their lighter body weight and flight advantage. Most of the aerial viral vectors are not active fliers and are carried away by wind to shorter and longer distances. However, the virus transmission by soil borne vectors such as fungi and nematodes are very slow. The insect vectors play key role in spreading the viral diseases very fast which causes heavy yield and economic losses to the main crop. Thus, vector control become necessity for growing the healthy crops for better returns. There are various methods used to control the vector population.

29.5.1 Chemical Control

29.5.1.1 Insect Vectors

Insecticides are commonly used to control the virus spreading insects and hence viruses they transmit. However, insect killing chemicals are generally successful against the dissemination of persistently transmitted viral diseases but not very effective against non-persistently vectored viruses. Non-persistently vectored viruses spread fast even by quick probing of less than a minute by the vector.

Large ranges of insect killing chemicals are available for controlling insect vectors in crops. However, the control of insect vectors to prevent viral infections is difficult because considerable spread of viruses may be caused by few winged vectors. In such cases contact insecticides are of little use unless they are used repeatedly. Systemic pesticides offer more hope for managing the viral diseases. Several non-persistently transmitted viruses are transmitted by winged aphids and they transmit the virus during their first probing prior to getting killed by any insecticide spray. The efficacy of controlling a vector by chemicals depends on the persistence of the pesticide in the plant and the application frequency. Chemicals of systemic nature have been shown to minimize disease incidence of persistently vectored viruses but they are unable to act fast enough to limit infection of non-persistently transmitted ones.

In several cases it has been noted that virus epidemics correlate positively with high vector movement. Efficient vector control and resultant low incidence of disease can be attained if viruliferous vector populations are kept as minimum level as possible. The level of watermelon bud necrosis virus (WBNV) incidence in watermelon was effectively reduced by seed dressing and foliar spray of imidacloprid (Kamanna et al. 2010; Rajasekharam 2010). Krishnakumar et al. (2006a) reported that seed treatment of imidacloprid was effective in controlling WBNV carrying thrips up to 25 days after sowing.

Khurana et al. (2000) showed treatment of imidacloprid was also effective in managing the thrips transmitted stem necrosis disease in potato. Similarly, thrips transmitted TSV in sunflower was significantly managed by treating 1 kg of seed with 5 g imidacloprid followed by three sprays with Confidor 200 SL at the rate of 0.05% after 15, 30 and 45 days of sowing (Shirshikar 2008). Further, the necrosis disease of sunflower was shown to be managed effectively by adopting the integrated disease management approaches which include seed treatment and three sprays (15, 30, 45 days after sowing) of thiamethoxam along with three rows of sorghum as border crop around the field. These treatments showed a reduction in disease incidence up to 3.8% as compared to 21.2% in untreated plot. Moreover, 14% reduction in vector population and up to 41% increment in yield was observed (Bhat Bharati et al. 2012).

Sastry (1989) observed reduction in the incidence of ToLCV by dipping the tomato seedlings in carbofuran liquid along with foliar spray of oil in field. Other examples where insecticidal spray reduced incidence of viral disease and vector populations are: mungbean yellow mosaic virus (MYMV) in mungbean (Agarwal et al. 1979; Borah 1995; Ghosh et al. 2009), PBNV in mungbean (Sreekanth et al. 2003) and in potato (Somani et al. 2007), BYVMV in okra (Sastry and Singh 1973a, b; Dahal et al. 1992), Indian cassava mosaic virus (ICMV) in cassava (Saraswathi et al. 2002), PPSMV in pigeonpea (Rathi 1979) and mesta yellow vein mosaic virus in Mesta (Seetharam et al. 2011).

Incidence of leaf curl in tomato and chilli was effectively reduced by application of carbofuran @ 1.5 kg a.i./ha (Singh et al. 1979a; Sastry et al. 1976). Similar results were observed with tungro disease in rice (Shukla and Anjaneyulu 1980). Palaniswamy et al. (1973) have reported that granular form of carbofuran and aldicarb were effective and resulted in 90% whitefly mortality within 2 days under glasshouse conditions. The soil application of 1.5 kg methyl phosphoro-dithioate (Furatox 10G) per hectare was shown effective in controlling the whitefly population and the incidence of BYVMV in okra (Khan and Mukhopadhyay 1985). The incidence of pigeonpea sterility mosaic was reduced by the application of carbofuran and aldicarb (Rathi 1979; Reddy et al. 1990; Nene 1995). The application of carbofuran and aldicarb (Temik 10G) in soil and the foliar application of dinocap, kelthane, metasystox, monocrotophos and oxythoquinox were effective in managing mite-transmitted sterility mosaic disease in pigeonpea (Ghanekar et al. 1992; Rathi 1979).

Satapathy and Anjaneyulu (1984) have shown good results in controlling the leafhopper vector; *Nephotettix virescens* by application of cypermethrin. Bhaktavatsalam and Anjaneyulu (1984) have indicated that decamethrin was also effective against leafhoppers. Anjaneyulu et al. (1994) have shown that the application of decamethrin in combination with bufrofaziu and tralomethrin were as good as application of cypermethrin and decamethrin in reducing the vector populations. Moreover, it was shown by several researchers (Sinha et al. 1983; Sastry and Zitter 2014) that pyrethroids were effective for reducing thrips vectors along with aphid and leafhoppers.

29.5.1.2 Fungal Vectors

Several plant viral diseases are transmitted through the involvement of soil inhabiting fungi of chytidiomimycota. The primary zoospores released from the resting spores of these fungal vectors penetrate the fresh root hairs and start the infection cycle. These plant virus transmitting fungi can be managed by the application of chemicals and various amendments. For example, reducing the pH up to five of infective soil by the use of sulphur significantly reduced the disease incidence caused by potato mop-top virus (PMTV) and *Spongospora subterranea* in potato. However, the disease was resumed when the pH of treated soil was raised. The application of dibromochloropropdane, carbofuran and aldicarb were found effective in managing Indian peanut clump virus in peanut (Reddy et al. 1988).

29.5.2 Biological Control

The excessive use of agrochemicals often reduce or eliminate the population of beneficial biocontrol agents such as entomo-pathogens, parasitoids and predators which favour the increase of insect-pest populations. Insect vector management using the approach of biological control has been tried by several researches but has met with little success. More attention is required to exploit this non chemical strategy for managing the vector population.

Rizvi and Bhargava (1973) reported parasitization of aphid species colonies by a number of predators. The spread of citrus greening vectored by its psyllid *Diaphirina citri* and *Triosza erytreae* have been successfully restricted in the Reunion Island by releasing eulophid ectoparasite *Tetrastichus dryi* and *T. radiates*. However, *T. radiates* was not able to control the vector (*D. citri*) population in India (Aubert and Quiliei 1984). The whitefly (*B. tabaci*) population was effectively controlled by entomopathogens *Beauveria bassiana* and *Verticillium lecanii* in Indian conditions. The percentage mortality in treated adult whiteflies was 20% after 24 h whereas up to 100% mortality was recorded after 72 h of the treatment (Colvin and Muniyappa 1999). The same entomopathogens were found effective against *Thrips tabaci* in onion. Application of *B. bassiana* at 1×10^9 spores/ml reduced the vector population up to 64.15% in onion crop after weeks of planting (Sudhir Kumar et al. 2012). Natarajan (1990) has proposed to use *Encarsia formosa, E. lutea* and *Eretmocerus mundus* (parasites of whitefly, *B. tabaci*) as a biological control for managing the whitefly populations.

An extensive large field trials studies using different potential biological agents in suitable geographic area need to be tried against insect vectors of plant viruses to have a fair idea about the usefulness of the bio-control agents in minimizing the vector population and incidence of viral diseases in natural field environment.

29.6 Use of Oils for Prevention of Viral Transmission by Insect Vectors

Studies on application of oil spray have shown to lower the disease incidence of non-persistently as well as semi persistently aphid transmitted viruses. The virus transmission can be reduced by lightly coating leaf surface with oil because vectors carrying viruses inhibited transmitting it after probing. The use of oils at recommended doses has many advantages such as: (a) they are environmental friendly, (b) have very good sticking and spreading properties, (c) free from resistance development and (d) cost effective. The novel experiment of Bradley (1956) showed great reduction in subsequent virus transmission when viruliferous aphids were allowed to probe into a low melting paraffin wax membrane. Later on in1962, the reduction in transmission of the virus was shown due to the presence of the oil in the wax and not by wax itself. It was proved that the spread of PVY by *M. persicae* was hindered by coating the plants (source or test) with mineral oil (Bradley 1963). Similar encouraging results in minimizing viral diseases by use of mineral oils have been recorded in several crops (Sharma and Varma 1982; Sastry 1984; Singh 1992).

Aphid transmitted PVY and CMV are very destructive viruses in capsicum and chillies. Khatri and Sekhon (1973) obtained complete inhibition of these viruses by the application of 2% emulsion of paraffin oil. In potato, PVY is the most common virus and its spread could be reduced with the application of various oils (Singh and Nagaich 1976). The transmission of soybean mosaic virus (SMV) by *A. craccivora* was reduced significantly by 1% coconut oil among castor, clove, groundnut, light paraffin, mustard and olive oils tested (Joshi and Gupta 1974). Kalleshwaraswamy et al. (2009) revealed that PRSV incidence in papaya could be reduced by foliar application of 0.1% mineral oil along with imidacloprid and deltamethrin as alternate spray at 15 days interval. A complete reduction of pumpkin mosaic transmission was reported up to 24 h with the application of 2% Sunspray oil (Singh 1981). However, a gradual decrease in effectiveness was observed after 24 h and up to 60% infection was recorded after 168 h in treated plot.

It has been shown that effect of oil application is not virus or vector specific. For example, its application restricts the spread of several viruses in pumpkin transmitted by *Aphis gossypii* (Singh 1981), which indicates a broad spectrum effect. Bhargava and Khurana (1969) showed that the spray of lipids and milk were effective in reducing the incidence of non-persistently aphid transmitted viruses. Out of six vegetable oils tested only coconut oil (0.5–1%) was effective in reducing the transmission of SMV in soybean (Joshi and Gupta 1974). The BYVMV disease incidence in bhindi was reduced effectively by the application of neem oil and neem kernel extracts (Pun et al. 2005). Sastry and Singh (1973b) reported that mineral oil spray (1%) at 10 days interval minimized the BYVMV incidence in okra. Similarly, 100% whitefly mortality within 30 min was recorded with the spray of 2% emulsifiable mineral oil (Nene 1973).

Besides non persistently aphid borne viruses, the oil application was found effective in reducing the transmission of semi persistently whitefly-transmitted viruses such as ToLCV, MYMV, Chilli leaf curl and French bean yellow mosaic (Butter and Rataul 1973; Nene 1973; Singh et al. 1979a).

Neem oil was also reported important for decreasing the spread of RTV and the survival rate of its vector, *Nephotettix virescens* (Mariappan and Saxena 1983). Periodic spray of neem oil emulsion was found effective in reducing the incidence of rice ragged stunt virus spread by *Nilaparvata lugens* in rice (Saxena et al. 1981). Several studies on different oils (castor, clove, groundnut, mustard and neem) have been reported to be effective in reducing the incidence of viral diseases at different levels of success (Bhargava and Khurana 1969; Khatri and Sekhon 1973; Nene 1972; Dubey and Nene 1974; Singh 1981; Singh and Varma 1977).

29.7 Use of Botanicals

The efficacies of certain plant products (botanicals) for reducing the incidence of plant viral diseases have been tested. Sastry and Singh (1982) showed extract of *Dioscorea floribunda* tuber was found to significantly reduce *Tobacco mosaic virus* (TMV) infection in *Chenopodium amaranticolor*. Rao et al. (1984) reported antiviral activity of coralloid root extract of *Cycas revoluta* against tomato viruses. Rao et al. (1985) reported reduction of local and systemic resistance of *Potato virus X* by flower extracts. Two basic proteins involved in systemic resistance induction in susceptible host plants were isolated from *Clerodendrum inerme* (Prasad et al. 1995). Parveen et al. (2001) have isolated and characterized a systemic resistance inducer protein (Crip-31) from *Clerodendrum inerme* leaves for CMV, PVY and ToMV in *Nicotiana tabacum* cv. White Burley. Similarly, CA-SRIP from *Clerodendrum aculeatum* was characterized which induces resistance to PRSV infection in papaya and provide protection (no symptoms) up to 6 months on the CA-SRIP treated papaya plants following challenge inoculation of the virus (Srivastava et al. 2006). Leaf extract of *Azadirachta indica* and root extract of *Boerhaavia diffusa* were also shown to be effective for viral disease management. MYMV incidence in mungbean and urdbean was reduced 66.7 and 63.6% respectively by spraying the leaf extracts of *Clerodendrum* (Singh and Awasthi 2006). A multifunctional antiviral proteins (AVPs)/ribosome-inactivating proteins (RIPs) capable of inducing systemic resistance from *Bougainvillea xbuttiana* have been isolated. The virus inhibitory properties of these antiviral proteins are directly correlated to their enzymatic activities such as DNase, N-glycosidase and RNase (Choudhary et al. 2008).

Volatile constituents (essential oils) of higher plants were also reported to inhibit plant virus activity of PVX, PVY and legume viruses in India (Shukla et al. 1985; Rao et al. 1986, 1987). Essential oils which are the constituents of botanical insecticides can be an option for the harmful chemicals in controlling the viral disease spreading vectors. For example, the application of emusifiable concentrate formulation of essential oil from *Chenopodium ambrosoides* at the rate of 0.5% caused 43.6% mortality in *M. persicae*. An excellent protection of thrips, *Frankliniella schultezi* (95.7%) was provided by foliar spay of *C. ambrosoides* extract at 0.5% concentration as compared to insecticidal soap (84%) and neem oil (18%) (Rajapakse and Janaki 2006). Induction of resistance by 12- Tungsto zincic (II) acid in hypersensitive and non-hypersensitive host against legume viruses was reported (Rao et al. 1989). More detail information on botanicals as plant virus inhibitors is reviewed by Baranwal and Verma (2000).

29.8 Use of Host Resistance

Use of resistant varieties is a practical and very important approach for managing viral diseases as it is comparatively affordable to deploy and has no harmful effect on environment. The use of resistance as control measure of viral diseases in crops may not often be feasible to take on chemical and cultural disease management strategies due to cost, practicality and other considerations. However, in all available approaches, the use of resistant or tolerant varieties has been considered the most important and economical way to manage viral disease in crop plants. Resistant cultivars have been identified and developed for viral diseases and its insect vectors.

Resistance for plant viral diseases may operate through resistance to insect vector, virus multiplication, and symptom development, spread through seed/vector, cell to cell movement of virus, hypersensitivity or immunity to infection. Immunity is the highest form of resistance where no virus multiplication takes place in immune plant. Majority of resistance to plant viruses is generally simple involving single locus which typically follows the Mendelian inheritance of resistance and susceptibility (Khetarpal et al. 1990). However, any general conclusion about resistance and its mechanism is difficult in cases where multiple loci have been involved (Rajamony et al. 1990).

Viral resistance breakdown in cassava, tomato and in number of legumes has been observed because of evolution of new or virulent viral strains. Therefore, continuous resistance breeding for viral disease is required to ensure the practical and effective protection as long as possible. Further reviews on disease resistance against plant viruses are available (Nene 1988; Khetarpal et al. 1998; Varma and Mitter 2001; Varma et al. 2002; Madhavi Reddy et al. 2010).

29.8.1 Resistance for Viral Diseases

29.8.1.1 Resistance in Wild Species

Resistance to viral diseases in wild plants has been explored in the absence of availability of resistance sources in commercial cultivars. In India, diseases caused by *Geminiviruses* and *Tospoviruses* are serious threat to tomato cultivation and several attempts were made to identify the resistance to the virus in different species of tomato including the cultivated one. Tripathi and Varma (2003) have identified a high level of resistance to a severe strain of ToLCV in wild tomato *S. hirsutum*, *S. peruvianum*, *S. pimpinellifolium* and *S. cheesmanii* when ninety genotypes of tomato were screened for resistance through vector whitefly inoculation as well as through agroinoculation of cloned DNAs. The resistance sources to leaf curl have also been identified in wild tomato species and high yielding cultivars have been developed by using these wild sources of resistance in breeding programme (Saikia and Muniyappa 1989; Muniyappa et al. 1991). Venkata Ramana et al. (2006) reported resistant sources to GBNV in wild tomato, *S. pimpinellifolium* and *S. peruvianum*.

Wild papaya (*Vasconcellea cauliflora*) has been identified as immune to local strain of PRSV and the efforts of transferring the resistance into commercial cultivar of cultivated papaya (*Carica papaya*) at IARI Region Station at Pune is being made (Sharma and Tripathi 2014).

Peanut (*Arachis cardenasii)* accession PI11558 was shown to be highly resistant to *Peanut stripe virus* whereas accessions PI4983 (*A. chacoense*), PI11560 (*A. chiquitana*), PI11562 and PI12168 (*A cardenassi*), PI8215 (*A. stenophylla*), and PI8973 (*A. paraguariensii*) were reported resistant (Prasada Rao et al. 1989, 1991). Sources of resistance against GBNV in *Arachis* germplasm accessions have been reported (Gururaj et al. 2002; Reddy et al. 2000)). Eight cross compatible resistant accessions for TSV disease was identified in wild peanut species for resistance breeding in *A. hypogaea* (Kalyani et al. 2007).

The BYVMV resistant varieties of okra namely P7, Parbhani Kranti and Arka Anamika were successfully developed by exploiting the resistance from wild okra (*Abelmoschus manihot)* and released for commercial cultivation (Thakur 1986; Jambhale and Nerkar 1986; Dutta 1978; Arumugam et al. 1975).

29.8.1.2 Resistance in Cultivated Crops

The sources of resistance to RTV in rice and to wheat streak mosaic virus in wheat have been reported by several researchers (Anjaneyulu 1977; Anjaneyulu et al. 1982; Mohanty et al. 1989; Srinivasulu and Jeyarajan 1989; Tarafdar et al. 2001; Prasad et al. 2004; Murugan et al. 2011). Similarly resistance against *Groundnut rosette virus* was identified in the breeding lines of peanut (Van der Merwe and Subrahmanyan 1997; Subrahmanyam et al. 2002). More information on virus resistance breeding in peanut can be obtained from a review on the topic by Nigam et al. 2012. Singh et al. (2006) have identified 17 resistant among 783 entries of cotton evaluated for leaf curl disease. The virus resistance/tolerance has been reported in tuber crop from several tropical regions (Maruthi et al. 2013). A large number of varieties/lines of mungbean and urdbean were screened at different locations for diseases. Singh et al. (2000) reported IPU 94-1 were highly resistant to MYMV. However, the durability of resistance to MYMV in mungbean is limited and most of the varieties succumb within period of 5–6 years (Varma et al. 1992). PRSV tolerant papaya lines viz. Pune Selection-1 and Pune Selection-3 were developed by IARI Regional Station at Pune which were found very effective and productive at farmer's field under high disease pressure (Sharma and Tripathi 2014).

29.8.1.3 Resistance to Seed Transmission

Seed play a key role as the primary virus inoculum source for spreading of disease in case of seed transmitted viruses, hence minimizing the virus spread through seed will have a great effect on development of virus epidemics in field. Attempts to identify the varieties with no or low virus seed transmission have been made. Goodman and Nene (1976) reported some lines of soybean with no seed transmission of SMV. Similarly, EC-76446 (292) and NCAC 17133 (RF) lines of peanut were used in virus resistance breeding as no seed transmission of *Peanut mottle virus* was observed on these lines (Bharatan et al. 1984). A combined resistance to seed transmission of blackeye common mosaic virus, TMV, *Cowpea aphid borne mosaic virus* and CMV in *Vigna unguiculata* genotypes like Pusa-3, N-2-1 and V-16 has been reported (Mali et al. 1987).

29.8.2 Resistance for Vector

The efforts on developing resistance in crop against insect pests that are also vectors of viruses are increasing. The development of resistant variety against insect vectors of economically important viral diseases is promising. For instance, reduction in GBNV incidence was reported due to the host resistance to its vector *Frankliniella schultzei* (Amin 1985). A multiple resistance against virus vectors such as *A. citricola*, *M. persicae* and *Rhopalosiphum maidis* was recorded in soybean genotypes which significantly decrease the spread of SMV in soybean under field conditions (Gunasinghe et al. 1988). Similarly, P-1476 genotype of cowpea was found to be resistant against *A. craccivora* (Mali 1986).

29.9 Concluding Remarks

Plant viral diseases have been studied for a century. However, control of plant viral diseases has been challenging because of the lack of an effective direct method of control by chemical applications. Therefore, indirect methods of managing viral diseases has to be utilized which involves use of modified cultural practices, use of virus-free planting materials, application of insecticides and oils for control of virus spreading insects etc. Some success in viral disease management have been achieved by using a combination of approaches in few crops better than others. The successful strategy for virus management at one location will not necessarily work at another location. Therefore, the studies on epidemiology of viral diseases and its vectors in specific region are needed. Among all methods used, the host plant resistance is being the best solution to the viral disease problem. However this alone will not last longer. Therefore, better crop management along with host resistance should be used for effective viral disease management as well as holding the host resistance longer duration. Moreover, the fast growth of agriculture to ensure food security and the liberalization of trade in agriculture worldwide have its implications and possibilities of imposing new challenges with the risks of introducing either new viral diseases or existing viruses adapting to new hosts or both.

It has been realized that no single method will help in achieving effective viral disease management. Management of viral diseases is effective only through integration of various approaches mentioned in this chapter. Therefore, studies need to be conducted for effective and sustainable disease management by embracing multidisciplinary approach including virus and vector characterization, ecology and epidemiology of virus and vector, chemical methods, genetic methods and implementation of exclusion and eradication techniques along with phytosanitary and quarantine measures.

There are number of new and emerging viral diseases occurring in vegetables, fruits and other commercial crops which require regular surveillance in the ecosystem and need to be managed when they first occur. A better awareness or training about virus disease identification, ecology and management to community/farmer, proactive involvement of national, international, concerned private companies at farm level is required for effective and sustainable plant viral disease management.

References

Agarwal VK, Nene YL, Beniwal SPS, Verma HS (1979) Transmission of bean common mosaic virus through urdbean (*Phaseolus mungo* L.) seeds. Seed Sci Technol 7:103–108

Amin PW (1985) Apparent resistance of groundnut cultivar Robut 33–1 to bud necrosis disease. Plant Dis 69:718–719

Anjaneyulu A (1977) Donors of resistance to rice tungro. IRRN 2:3

Anjaneyulu A, Singh SK, Shenoi MM (1982) Evaluation of rice varieties to tungro resistance by field screening technique. Trop Pest Manag 28:147–155

Anjaneyulu A, Bhaktavatsalam G, Mohanty SK (1994) Prevention and control of rice tungro disease through new synthetic pyrethroids. In: Rishi N, Ahuja KL, Singh BP (eds) Virology in the tropics. Malhotra Publishing House, New Delhi, pp 660–671

Arumugam R, Chelliah S, Muthukrishnan CR (1975) Abelmoschus manihot: a source of resistance to bhindi yellow vein mosaic. Madras Agric J 62:310–312

Aubert B, Quiliei S (1984) Biological control of the African and Asian citrus psyllids (Homoptera-Psyllidea) through eulophid and encyrsid parasites (Hymenoptera: Chalcidaidea) in Reunion Island. In: Proceedings of the 9th conferences of international orgg Citrus virologists, IOCV riverside, CA, pp 100–108

Balaraman K (1981) Pre-immunization for control of Citrus tristeza virus on citron. Z Pfl Krankh Pfl Schtz 88:218–222

Balaraman K (1987) Cross protection performance trials on acid lime against tristeza virus in South India. Mysore J Agric Sci 21:193–195

Balaraman K, Ramakrishnan K (1977) Studies on strains and strain interaction in citrus trizteza virus. Univ Agric Sci Bangalore. Tech Series No 19:1–62

Balaraman K, Ramakrishnan K (1979a) Transmission studies with strains of Citrus tristeza virus on acid lime. Z Pfl Krankh Pfl Schutz 86:653–661

Balaraman K, Ramakrishnan K (1979b) Two new additional hosts of Citrus tristeza virus. Curr Sci 48:453–454

Baranwal VK, Verma HN (2000) Antiviral phytoproteins as biocontrol agents for efficient management of plant virus diseases. In: Upadhyay RK, Mukerji KG, Chamola BP (eds) Biocontrol potential and its exploitation in sustainable agriculture. Springer, Boston, pp 71–79

Bhaktavatsalam G, Anjaneyulu A (1984) Green house evaluation of synthetic pyrethroids for tungro prevention. Int Rice Res Newslett 9:11

Bharatan N, Reddy DVR, Rajeswari R, Murthy VK, Rao VR (1984) Screening of peanut germplasm lines by enzyme linked immunosorbent assay for seed transmission of Peanut mottle virus. Plant Dis 68:757–758

Bhargava KS, Khurana SMP (1969) Papaya mosaic control by oil sprays. Phytapathol Z 64:338–343

Bhat Bharati N, Raja Ram Reddy D, Chander Rao S, Singh TVK (2012) Integrated disease management of sunflower necrosis disease. Indian J Plant Prot 40:99–104

Bhojwani SS, Razdan MK (1983) Plant tissue culture: theory and practice a revised edition. Elsevier, Amsterdam, p 767

Borah RK (1995) Effect of synthetic pyrethroids and organophosphorus insecticides on the incidence of whitefly, Bemisia tabaci (Genn) and yellow mosaic virus in greengram, *Vigna radiata* (L.) Wilczek. Indian J Virol 11:75–76

Bradley RHE (1956) Effects of depth of stylet penetration on aphid transmission of Potato virus Y. Can J Microbiol 2:539–547

Bradley RHE (1963) Some ways in which a paraffin oil impedes aphid transmission of Potato virus Y. Can J Microbiol 9:369–380

Butter NS, Rataul HS (1973) Control of Tomato leaf curl virus (TLCV) in tomatoes by controlling the vector whitefly Bemisia tabaci Gen. by mineral oil sprays. Curr Sci 42:864–865

Capoor SP, Rao DG (1967) Tristeza virus infection of citrus in India. In: Proceedings of the international symposium on sub tropical and horticulture. Bangalore, India, pp 723–736

Chattopadhyay SB, Bhowmik TP (1965) Control of 'foorkey' disease of larger cardamom in west Bengal. Indian J Agric Sci 35:272–275

Choudhary N, Kapoor HC, Lodha ML (2008) Cloning and expression of antiviral/ribosomeinactivating protein from Bougainvillea xbuttiana. J Biosci 33:91–101

Chowfla SC, Parmar GS (1995) Management of bell pepper mosaic complex by cultural practices. In: Gupta VK, Sharma RC (eds) Integrated disease management and plant health. Scientific Publishers, Jodhpur, pp 187–192

Colvin J, Muniyappa V (1999) Sustainable management of the whitefly, Bemisia tabaci, and tomato leaf curl virus (ToLCV) on tomato in India. DFID Crop Protection Programme, Final Technical Report, Project R6627. Natural Resources Institute, University of Greenwich, Chatham

Dahal G, Neupane FP, Baral DR (1992) Effect of planting and insecticides on the incidence and spread of yellow vein mosaic of okra in Nepal. Int J Plant Dis 10:109–124

Dubey GS, Nene YL (1974) Aphid transmission of Cowpea mosaic virus as influenced by oil sprays. Indian Phytopathol 27:325–330

Dutta OP (1978) I.I.H.R. Annual report thermotherapy and chemotherapy. In: Loebenstein G et al (eds) Virus and virus-like disease of potatoes and production of seed-potatoes. Kluwer Academic Publishers, Dordrecht, pp 365–390

Faccioli G, Marani F (1998) Virus elimination by meristem tip culture and tip micrografting. In: Hadidi A, Khetarpal RK, Koganezawa H (eds) Plant virus disease control. American Phytopathological Society, St Paul, pp 346–373

Ghanekar AM (1980) Groundnut virus research at ICRISAT. In: Gibbons RW, Mertin JV (eds) Proceedings of the international workshop on groundnuts. ICRISAT, Patancheru, pp 211–216

Ghanekar AM, Sheila VK, Beniwal SPS, Reddy MV, Nene YL (1992) Sterility mosaic of pigeonpea. In: Singh US, Mukuopadhyay AN, Kumar J, Chaubey HS (eds) Plant diseases of international importance: diseases of cereals and pulses, vol 1. Prentice Hall, Eaglewood Cliffs, pp 415–428

Ghosh D, Latha SK, Biswas NK (2009) Effect of different pesticides on incidence of mungbean yellow mosaic incidence. Int J Plant Prot 2:67–70

Gill SS, Singh H, Sandhu KS (1982) Effect of sowing dates and plant density on the incidence of yellow vein mosaic virus on seed crop of okra (*Abelmoschus esculentus* L. Moench). Indian J Ecol 9:160–162

Goodman RM, Nene YL (1976) Identification of virus-resistant tropically adopted soybean germplasm. Annual report on contract AID/TA-C-1294, 1 April 1977–21 March 1978, pp 73–103

Gunasinghe UB, Irwin ME, Kampmeier GE (1988) Soybean leaf pubescence affects aphid vector transmission and field spread of soybean mosaic virus. Ann Appl Biol 112:259–272
Gururaj S, Kenchanagoudar PV, Naragund VB (2002) Identification of sources for field resistance to peanut bud necrosis disease in groundnut. Karnataka J Agric Sci 15:646–648
Hendre R, Mascarenhas SF, Nadgir AL, Pathak M, Jagannathan V (1975) Growth of mosaic virus free sugarcane plants from apical meristems. Indian Phytopathol 28:175–178
Jain RK, Rao GP, Varma A (1998) Present status of management of sugarcane mosaic virus. In: Hadidi A, Khetrapal RK, Koganezawa H (eds) Plant virus disease control. APS Press, St Paul, pp 495–506
Jambhale ND, Nerkar YS (1986) 'Parbhani Kranti', a yellow vein mosaic resistant okra. Hortic Sci 21:1470–1471
Joshi RD, Gupta UP (1974) Prevention of natural spread of Soybean mosaic virus with some oils. Proc Natl Acad Sci (India) B 44:1–4
Nilakshi Kakati, Nath PD (2006) Integration of physical barrier with bio pesticides against whitefly for management of tomato leafcurl virus. Abstract OP 9/26. In: XVI Annual convention of IVS and international symposium on management of vector-borne viruses, held at ICRISAT, Patancheru, Hyderabad (India)
Kalita MK (2003) Epidemiology and management of yellow vein mosaic and leaf curl begomovirus diseases of okra. Ph.D. thesis, CCS Haryana Agricultural University, Hisar, 129 pp
Kalleshwaraswamy CM, Krishna Kumar NK, Dinesh MR, Chandrashekar KN, Manjunatha M (2009) Evaluation of insecticides and oils on aphid vectors for the management of papaya ringspot virus (PRSV). Karnataka J Agric Sci 22(3-Spl Issue):552–553
Kalyani G, Reddy AS, Kumar PL, Rao RDVJP, Aruna R, Waliyar F, Nigam SN (2007) Sources of resistance to tobacco streak virus in wild Arachis (Fabaceae: Papilionoidae) germplasm. Plant Dis 91:1585–1590
Kamanna BC, Shankarappa TH, Arunkumar GS (2010) Evaluation of fungicides for the management of chrysanthemum leaf blight caused by *Aternaria alternata* (Fr.) Keissler. Plant Arch 10:595–597
Kartha KK (1981) Meristem culture and cryopreservation. Methods and application. In: Thorpe TA (ed) Plant tissue culture. Academic, New York, pp 181–211
Kartha KK (1986) Production and indexing of disease-free plants. In: Withers LA, Alderson PG (eds) Plant tissue culture and its agricultural applications. Butterworths, London, pp 219–238
Khan MA, Mukhopadhyay S (1985) Effect of different pesticide combinations on the incidence of yellow vein mosaic virus disease of okra (*Abelmoschus esculentus*) and its whitefly vector Bemisia tabaci Genn. Indian J Virol 1:147–151
Khatri HL, Sekhon IS (1973) Effect of oil spray on aphid transmission of chilli-mosaic virus. Indian J Agric Sci 43:667–669
Khetarpal RK, Maury Y, Cousin R, Burghoufer A, Varma A (1990) Studies on resistance of pea to pea seed-borne mosaic virus and new pathotypes. Ann Appl Biol 116:297–304
Khetarpal RK, Maisonneuve B, Maury Y, Chaloub B, Dinant S, Lecoq H, Varma A (1998) Breeding for resistance to plant viruses. In: Hadidi A, Khetarpal RK, Koganezawa H (eds) Plant virus disease control. APS Press, St. Paul, pp 14–32
Khurana SMP, Pandey SK, Srivastava KK (2000) Tuber treatment with imidacloprid is effective for control of potato stem necrosis disease. Indian Phytopathol 53:142–145
Krishnakumar NK, Eswara Reddy SG (2006) Incidence and extent of yield loss due to Tomato leafcurl gemivirus (ToLCV) in tomato grown under protected and open field cultivation. Abstract PP 9/16. In: XVI Annual convention of IVS and International symposium on management of vector-borne viruses, held at ICRISAT, Patancheru, Hyderabad (India)
Krishnakumar NK, Venkatesh N, Kalleshwaraswamy CM, Ranganath HR (2006) Seasonal incidence of thrips and bud necrosis virus on watermelon. Pest Manag Hortic Eco-Syst 12:85–92
Madhavi Reddy K, Krishna Reddy M, Kumar R, Rai AB, Mathura R, Singh HB (2010) Breeding for virus resistance. In: Kumar R, Rai AB, Mathura R, Singh AB (eds) Advances in chilli research. Studium Press, New Delhi, pp 119–132

Mali VR (1986) Identification of source of combined resistance to some viruses and seed transmission in cowpea. 2nd annual convention of Indian Virological Society, New Delhi, 3–5 Oct 1986
Mali VR, Kulthe KS, Patil FS, Mundhe GE, Dhond VM, Deshmukh RV (1987) Identification of source of resistance to some viruses and seed transmission in cowpea. Indian J Virol 3:99–109
Mariappan V, Saxena RC (1983) Effect of custard-apple oil and neem oil on survival of Nephotettix virescens(Homoptera: Cicadellidae) and on rice tungro virus transmission. J Econ Entomol 76:573–576
Maruthi MN, Mohammed IU, Hillocks R (2013) Exploiting reversion and tissue culture techniques for eliminating cassava brown streak viruses from infected cassava plants. In: 12th International symposium on plant virus epidemiology, 28th Jan–1st Feb 2013, Arusha, Tanzania (Abstract, pp 051, p 152)
McKinney HH (1929) Mosaic disease in the Canary Islands, West Africa, and Gibraltar. J Agric Res 39:557–578
Mishra S, Rao GP (2011) Regeneration of sugarcane yellow leaf virus free sugarcane plantlets through meristem culture. J Southern Agric 42(1):1–5
Mishra MD, Ghosh A, Niazi FR, Basu AN, Raychudhuri SP (1973) The role of graminaceous weeds in the perpetuation of rice tungro virus. J Indian Bot Soc 52:176–183
Mishra S, Singh D, Tiwari AK, Lal M, Rao GP (2010) Elimination of *Sugarcane* mosaic and maize dwarf mosaic potyviruses. Sugar Cane 6:19–22
Moericke V (1954) New investigations on color in relation to Homoptera. In: Proceedings of the 2nd conference on potato virus diseases, Lisse Wageningen, June 25–29, pp 55–69
Mohanty SK, Bhaktavatsalam G, Anjaneyulu A (1989) Identification of field resistant rice cultivars for tungro disease. Trop Pest Manag 35:48–50
Mukhopadhyay S (1980) Ecology of Nephotettix spp. and its relation with rice tungro virus. Final report Ecology of Indian Council of Agricultural Research, New Delhi
Mukhopadhyay S (1984) Ecology of rice tungro virus and its vectors. In: Mishra A, Polasa H (eds) Virus ecology. South Asian Publishers, New Delhi, pp 139–164
Muniyappa V, Saikia AK (1983) Prevention of the spread of Tomato leaf curl disease. Indian Phytopathol 36:183–186
Muniyappa V, Jalikop SH, Saikia AK, Chennarayappa SG, Bhat AI, Ramappa HK (1991) Reaction of Lycopersicon cultivars and wild accessions to tomato leaf curl virus. Euphytica 56:37–41
Murugan M, Cardona PS, Duraimurugan P, Whitefield AE, Schneweis D, Starkey S, Smith CM (2011) Wheat curl mite resistance: interactions of mite feeding with wheat streak mosaic virus infection. J Econ Entomol 104:1406–1414
Nagaich BB, Pushkarnath Bharadwaj VP, Giri BK, Anand SR, Upreti GC (1969) Production of disease free seed potatoes in the Indo-Gangetic plains. Indian J Agric Sci 39:238–243
Nagaich BB, Chaubey IP, Singh SJ, Upreti GC, Kaley DM (1972) Effect of herbicides on infection, concentration and spread of potato virus X. Indian J Hortic 29:105–108
Natarajan K (1990) Natural enemies of *Bemisia tabaci* (Gennadius) and effect of insecticides on their activity. J Biol Control 4:86–88
Nene YL (1972) A survey of viral diseases of pulse crops in Uttar Pradesh. GB Pant Univ Agric Technol Pantnagar Res Bull 4:1–191
Nene YL (1973) Control of Bemisia tabaci Genn: a vector of several plant viruses. Indian J Agric Sci 43:433–436
Nene YL (1988) Multiple disease resistance in grain legume. Annu Rev Phytopathol 26:203–217
Nene YL (1995) Sterility mosaic of pigeonpea: the challenge continuous. J Mycol Plant Pathol 23:1–22
Nigam SN, Prasada Rao RDVJ, Bhatnagar-Mathur P, Sharma KK (2012) Genetic management of virus diseases in peanut. In: Janick J (ed) Plant breeding reviews, vol 36. Wiley, New York, pp 293–356
Palaniswamy P, Thirumurthi S, Subramaniam TR (1973) Effects of systemic granular insecticides on the incidence of the yellow mosaic disease of okra, *Abelmoschus esculentus* (L). South Indian Hortic 21:104–106

Parmessur Y, Aljanabi S, Saumtally S, Dookun-Saumtally A (2002) *Sugarcane yellow leaf virus* and sugarcane yellows phytoplasma: elimination by tissue culture. Plant Pathol 51:561–566

Parveen S, Tripathi S, Varma A (2001) Isolation and characterization of an inducer protein (Crip-31) from Clerodendrum inerme leaves responsible for induction of systemic resistance against viruses. Plant Sci 161:453–459

Prakash J, Joshi RD (1977) 2,4- dinitrophenol- an inhibitor of sugarcane mosaic virus. Intl Sugar J 79:303

Prasad V, Srivastava S, Verma VHN (1995) Two basic proteins isolated from Clerodendrum inerme Gaertn. are inducers of systemic antiviral resistance in susceptible plants. Plant Sci 110:73–82

Prasad SM, Sharma RB, Kudada N (2004) Effect of density of plant population on incidence of leaf crinkle disease in urd bean. RAU J Res 14:161–162

Prasada Rao RDVJ, Reddy AS, Chakravarthy SK, Sastry KS, Reddy DVR, Ramanath, Moses JP (1989) Peanut stripe virus research in India. II. Peanut stripe coordinator meeting held at ICRISAT, Hyderabad, India

Prasada Rao RDVJ, Reddy AS, Chakravarthy SK, Reddy DVR, Rao VR, Moss JP (1991) Identification of Peanut stripe virus resistance in wild Arachis germplasm. Peanut Sci 18:1–2

Prasada Rao RDVJ, Reddy AS, Reddy SV, Tirumala Devi K, Chander Rao S, Manoj Kumar V, Subramanyam K, Yellamanda Reddy T, Nigam SN, Reddy DVR (2003) The host range of tobacco streak virus in India and transmission by thrips. Ann Appl Biol 142:365–368

Pun KB, Doraiswamy S, Jeyarajan R (2005) Management of okra yellow vein mosaic virus disease and its whitefly vector. Indian J Virol 16:32–35

Pushkarnath VP (1959) Producing healthy seed potatoes in the plains: a new approach. Indian Potato J 1:63–72

Rajamony L, More TA, Seshadri VS (1990) Inheritance of resistance to cucumber green mottle mosaic virus in muskmelon (Cucumis melo L.) Euphytica 47:93–97

Rajapakse RHS, Janaki CR (2006) Insecticidal properties of Chenopodium-based botanical against virus vectors, Myzus persicae and Frankliniella schultezi. Abstract OP 9/14. In: XVI Annual convention of IVS and International symposium on management of vector-borne viruses, held at ICRISAT, Patancheru, Hyderabad (India)

Rajasekharam T (2010) Biological and molecular characterization and management of watermelon bud necrosis virus. PhD thesis, University of Agricultural Sciences, Dharwad, 142 pp

Ram RD, Verma R, Tomer SPS, Prakash S (2006) Management of papaya ring spot virus trough cross-protection strategy. J Maharashtra Agric Univ 31:92–95

Ramakrishnan K (1963) Control of virus disease in plants. Bull Natl Inst Sci India 24:78–91

Rao GM, Anjaneyulu A (1978) Host range of rice tungro virus. Plant Dis Rep 62:955–957

Rao GP (2002) Novel approaches in management of sugarcane diseases. In: Dube HC (ed) Annual review plant pathology, vol 1. Indian Society of Mycology & Plant Pathology, Udaipur, pp 253–274

Rao GP, Baghel AKS, Singh RK, Chatterji KS (1984) Antiviral activity of coralloid root extracts of *Cycas revoluta* against some viruses of tomato plant. Experientia 40:1257–1258

Rao GP, Singh RK, Pandey AK, Singh R (1985) Reduction of local and systemic resistance of Potato virus X by flower extracts. Indian J Virol 1(2):174–182

Rao GP, Pandey AK, Shukla K (1986) Essential oils of some higher plants vis-à-vis some legume viruses. Indian Perfumer 30(4):483–486

Rao GP, Pandey AK, Shukla DN, Shukla K (1987) Inhibition of some plant viruses by essential oil of mango (*Mangifera indica*). Indian Perfumer 31(4):351–353

Rao GP, Gupta VNP, Shukla K (1989) Induction of resistance by 12- Tungsto zincic (II) acid in hypersensitive and non- hypersensitive host against legume viruses. Indian J Virol 4(1–2):64–68

Rathi YPS (1979) Temik treatment of pigeonpea seeds for prevention of sterility mosaic. Acta Bot Ind 7:90–91

Reddy DVR (1991) Groundnut viruses and virus diseases: distribution, identification and control. Rev Plant Pathol 70:665–678

Reddy DVR, Amin PW, Mc Donald D, Ghanekar AM (1983) Epidemiology and control of groundnut bud necrosis and other diseases of legume crops in India caused by tomato spot-

ted wilt virus. In: Plumb RT, Thresh JM (eds) Plant virus epidemiology. Blackwell Scientific Publications, Oxford, pp 93–102
Reddy DVR, Nolt BL, Hobbs HA, Reddy AS, Rajeswari R, Rao AS, Reddy DDR, McDonald D (1988) Clump virus in India: isolates host range, transmission and management. Association of Applied Biologists, Wellesbourne, pp 239–246
Reddy MV, Sharma SB, Nene YL (1990) Pigeonpea: disease management. In: Nene YL, Hall SD, Sheila VK (eds) The pigeonpea. CAB International, Wallingford, pp 303–347
Reddy AS, Reddy LJ, Mallikarjuna N, Abdurahman MD, Reddy YV, Bramel PJ, Reddy DVR (2000) Identification of resistance to peanut bud necrosis virus (PBNV) in wild *Arachis* germplasm. Ann Appl Biol 137:135–139
Rizvi SMA, Bhargava KS (1973) Predators and parasites of vectors of *Sugarcane mosaic virus*. Sugarcane Pathol Newslett 10:18–19
Saikia AK, Muniyappa V (1989) Epidemiology and control of tomato leaf curl virus in Southern India. Trop Agric (Trinidad) 66:350–354
Saraswathi T, Rajamani K, Indra N, Kempuchetty N (2002) Control measures for Indian cassava mosaic disease. J Root Crops 28:76–78
Sastry KS (1984) Management of plant virus diseases by oil sprays. In: Misra A, Polosa H (eds) Virus ecology. South Asian Publishers, New Delhi, pp 31–57
Sastry KS (1989) Tomato leaf curl virus management by carbofuran plus oil combination. J Turkish Phytopathol 18:11–16
Sastry KS (2013a) Seed-transmitted plant virus diseases. Springer, New York. 327 pp
Sastry KS (2013b) Plant virus and viroid diseases in the tropics. Volume-1: introduction of plant viruses and sub-viral agents, classification, assessment of loss, transmission and diagnosis. Springer, New York. 361 pp
Sastry KSM, Singh SJ (1973a) Field evaluation of insecticides for control of whitefly (*Bemisia tabaci*) in relation to the incidence of yellow vein mosaic of okra (*Abelmoschus esculentus*). Indian Phytopathol 26:129–138
Sastry KSM, Singh SJ (1973b) Restriction of yellow vein mosaic virus spread in okra through the control of vector, whitefly Bemisia tabaci. Indian J Mycol Plant Pathol 3:76–80
Sastry KSM, Singh SJ (1982) Effect of plant extracts of some medicinal plants on the infectivity of tobacco mosaic virus. Herba Hungarica 21:101–105
Sastry KS, Zitter TA (2014) Plant virus and viroid diseases in the tropics: epidemiology and management. Springer, New York. 489 pp
Sastry KSM, Singh SJ, Sastry KS (1976) Density of tomato plant population and application of insecticides in relation to incidence of tomato leaf curl virus. Indian J Mycol Plant Pathol 6:8–13
Sastry KSM, Singh SJ, Sastry KS (1977) Effect of border cropping and the use of insecticides in relation to the incidence of tomato leaf curl virus (TLCV). Indian J Hortic 34:319–322
Sastry KSM, Singh SJ, Sastry KS (1978) Studies on epidemiology of Tomato leafcurl virus. Indian J Hortic 35:269–277
Satapathy MK, Anjaneyulu A (1984) Use of cypermethrin, a synthetic pyrethroid in the control of rice tungro virus disease and its vector. Trop Pest Manag 30:170–178
Saxena RC, Waldbauer GP, Liquido NJ, Puma BC (1981) Effect of neem seed oil on rice leaffolder, Cnaphalocrocis medinalis. In: Proceedings of the 2nd international Neem conference, Rottach-Engern, pp 189–204
Seetharam P, Venugopala Rao N, Hari Satya Narayana N, Chenga Reddy V (2011) Evaluation of insecticides to control whitefly, Bemisia tabaci vector of yellow vein mosaic disease on mesta. Indian J Plant Protect 39:1–3
Selvarajan R, Balasubramanian V (2014) Host–virus interactions in banana-infecting viruses. In: Gaur RK, Hohn T, Sharma P (eds) Plant virus–host interaction molecular approaches and viral evolution. Academic, MA, pp 57–78
Shah SS (1972) Sugarcane agriculture, contribution of Sugarcane Breeding Institute. Indian Farming 22:38–43

Sharma SR (1975) Further studies on cowpea viruses. Ph.D. thesis, P.G. School, IARI, New Delhi, 500 pp
Sharma SR (1988) Banana bunchy top virus. Int J Trop Plant Dis 6:19–41
Sharma SR (1989) Performance of a cross protection trial in acid lime (*Citrus aurantifolia*) against severe strain of tristeza virus. Int J Trop Plant Dis 7:203–214
Sharma SK, Tripathi S (2014) Overcoming limitations of resistance breeding in Carrica papaya L. In: Gaur RK, Hohn T, Sharma P (eds) Plant virus-host interaction: molecular approaches and viral evolution. Elsevier Inc, Amsterdam, pp 177–191
Sharma SR, Varma A (1982) Control of virus diseases by oil sprays. Zbl Mycrobiol 137:329–347
Sharma SK, Zote KK, Kadam UM, Tomar SPS, Dhale SPS, Sonawane AU (2010) Integrated management of papaya ringspot virus. Acta Hotric 85:473–480
Shirshikar SP (2003) Influence of different sowing dates on the incidence of sunflower necrosis disease. Helia 26:109–116
Shirshikar SP (2008) Integrated management of sunflower necrosis disease. Helia 31:27–34
Shukla K (1978) Studies on some aspects of a mosaic disease of sorghum from Gorakhpur. PhD Thesis. Univ Gorakhpur, India pp 155–156
Shukla VD, Anjaneyulu A (1980) Evaluation of systemic insecticides for control of rice tungro. Plant Dis 64:79–792
Shukla K, Joshi RD (1979) Influence of some phenol acids on the infectivity of sugarcane mosaic virus. Sugarcane Pathol Newslett 23:19–21
Shukla K, Joshi RD (1982) Inhibition by some homeopathic drugs of sugarcane mosaic virus. Sugarcane Pathol Newslett 29:48–50
Shukla DD, Teakle DS (1989) Johnsongrass mosaic virus. AAB descriptions, No. 340
Shukla HS, Rao GP, Tripathi SC (1985) Reduction in PVX and PVY infectivity by volatile constituents. Anales de Edafol Agrobiol XLIV(11–12):1763–1766
Singh SJ (1981) The effect of different oils on the inhibition of transmission of pumpkin mosaic virus by aphids. Z Pfl Krankh Pfl Schutz 87:86–98
Singh SJ (1992) Management of viral diseases of horticultural crops by use of oils. In: Sokhi SS, Chandel SS, Singh PP (eds) Progress in plant pathological research. Indian Society of Plant Pathologists, Ludhiana, pp 223–240
Singh S, Awasthi LP (2006) Protection of mungbean and urdbean crops against vector-borned Mungbean yellow mosaic virus through botanicals. Abstract, OP 8/07. In: XVI Annual convention of IVS and International symposium on management of vector-borne viruses, held at ICRISAT, Patancheru, Hyderabad (India)
Singh RA, Nagaich BB (1976) Effect of power oil on aphid transmission of potato virus Y. JIPA 3:21–23
Singh SJ, Sastry KSM (1974) Wilt of pineapple—a new virus disease in India. Indian Phytopathol 27:298–303
Singh JP, Varma JP (1977) Effect of some insecticides and oils on incidence of yellow mosaic and grain yield of mung bean cv. Varsha Proc Natl Acad Sci India 47(B):219–225
Singh SJ, Sastry KS, Sastry KSM (1979a) Efficacy of different insecticides and oil in the control of leaf curl virus disease of chillies. Z Pflkrankh Pflschuntz 86:253–256
Singh SJ, Sastry KSM, Sastry KS (1979b) Effect of barrier cropping combined with insecticidal sprays on the control of yellow vein mosaic virus of okra. In: Proceedings of the symposium retrospect and prospects of research in Botany in India, 24–26 Feb, University of Gorakhpur, India
Singh RA, De RK, Gurha SN, Ghosh A (2000) Yellow mosaic diseases of mungbean and urdbean. In: Narain U, Kumar K, Srivstava M (eds) Advances in plant diseases management. Advance Publishing concept, New Delhi, pp 337–348
Singh D, Garg HR, Rathore PK (2006) Elite sources of resistance against cotton leaf curl disease in upland cotton. J Cotton Res Dev 20:304–305
Sinha SN, Chakrabarti AK, Agnihotri NP, Jain HK, Gajbhiye VT (1983) Field evaluation of synthetic pyrethroids against thrip, Thrips tabaci (Lind.) on onion and their residues. Indian J Entomol 45(3):420–423

Somani AK, Kushwah VS, Khurana SMP, Garg ID (2007) Integrated management of stem necrosis disease of potato in India. Potato J 34:231–234
Sreekanth M, Sreeramulu M, Rao RDVJP, Babu BS, Babu TR (2002) Effect of sowing date on Thrips palmi Karny population and peanut bud necrosis virus incidence in greengram (*Vigna radiata* L. Wilczek). Indian J Plant Prot 30:16–21
Sreekanth M, Sreenivasulu M, Rao RDVJP, Babu BS, Babu TR (2003) Relative efficacy and economics of different imidacloprid schedules against Thrips palmi, the vector of peanut bud necrosis virus on mungbean. Indian J Plant Prot 31:43–47
Sreenivasulu P, Subba Reddy CHV, Ramesh B, Kumar PL (2008) Important virus diseases of groundnut. In: Rao GP, Paul Khurana SM, Lenardon SL (eds) Characterization diagnosis and management of plant viruses, vol-1. Industrial crops. Studium Press LLC, Houstan, pp 47–98
Srinivasulu B, Jeyarajan R (1989) Resistance of rice varieties to Rice tungro virus (RTV) and its green leafhopper (GLH) vector in Tamilnadu, India. IRRN 14:14
Srivastava A, Srivastava S, Verma HN (2006) Induction of antiviral resistance in Carica papaya by systemic resistance inducing protein from *Clerodendrum aculeatum*. Abstract, OP 8/05. In: XVI Annual convention of IVS and International symposium on management of vector-borne viruses, held at ICRISAT, Patancheru, Hyderabad (India)
Subrahmanyam P, van der Merwe PJA, Chiyembekeza AJ, Chandra S (2002) Integrated management of groundnut rosette disease. Afr Crop Sci J 10:99–110
Sudhir Kumar S, Ashok Kumar CT, Shivaraju C, Thippaiah M (2012) Efficacy of biochemical agents in the management of onion thrips, thrips tabaci (L.) on onion. J Biol Control 26:173–178
Tarafdar J, Biswas S, Chowdhury AK (2001) Reaction of rice tungro virus and its vector *Nephotettix virescens* (Distant) with some traditional and high yielding varieties of rice. J Int Acad 5:142–149
Tarafder P, Mukhopadhyay S (1979) Potential of weeds to spread rice tungro in West Bengal, India. Int Rice Res Newslett 4:11
Tarafder P, Mukhopadhyay S (1980) Further studies on the potential of weeds to spread tungro in west Bengal, India. Int Rice Res Newslett 5:10
Thakur MR (1986) Breeding for disease resistance in okra. Veg Sci 13:310–315
Thira SK, Cheema SS, Kang SS (2004) Pattern of bud necrosis disease development in groundnut crop in relation to different dates of sowing. Plant Dis Res (Ludhiana) 19:125–129
Tiwari AK, Yadav S, Tripathi S, Rastogi J, Bharti YP, Lal M, Sharma ML (2008) Performance of micropropogated, SCYLV infected and healthy plants of sugarcane. Indian J Sugarcane Technol 23(1&2):41–44
Tripathi S, Varma A (2002) Eco-friendly management of leaf curl disease of tomato. Indian Phytopathol 55:107–112
Tripathi S, Varma A (2003) Identification of sources of resistance in *Lycopersicon* species to *Tomato leaf curl geminivirus* (ToLCV) by agroinoculation. Euphytica 129:43–52
Van der Merwe PJA, Subrahmanyan P (1997) Screening of rosette-resistance early groundnut breeding lines for yield and other characteristics. Int Arachis Newslett 17:23–24
Vani S, Varma A, More TA, Srivastava KP (1989) Use of mulches for the management of mosaic disease in muskmelon. Indian Phytopathol 42:227–235
Varma A (1976) Recent trends in control of plant virus diseases. Proc Natl Acad Sci India 46:193–206
Varma A (1993) Integrated management of plant viral diseases. CIBA Found Symp 177:140–155
Varma A, Mitter N (2001) Durable host-plant resistant, a desirable trait for integrated disease management. In: Peng S, Hardy B (eds) Rice research for food security and poverty alleviation. IRRI, Manila, pp 324–344
Varma A, Dhar AK, Mandai B (1992) MYMV transmission and control in India. In: Green SK, Kim D (eds) Mungbean yellow mosaic disease. Asian Vegetable Research and Development Centre, Taipei, pp 8–27
Varma A, Jain RK, Bhat AI (2002) Virus resistant transgenic plants for environmentally safe management of viral diseases. Indian J Biotechnol 1:73–86

Venkata Ramana C, Prasada Rao RDVJ, Reddy IP, Venkata Rao P, Reddy YN (2006) Screening of tomato germplasm and wild relatives against peanut bud necrosis virus (PBNV) disease. Indian J Plant Prot 34:59–61

Venkatesh HM (2000) Studies on tomato leaf curl geminivirus and Bemisia tabaci (Gennadius): molecular detection farmers perception and sustainable management. PhD thesis, University of Agricultural Sciences, Bangalore

Viswanathan R, Rao GP (2011) Disease scenario and management of major sugarcane diseases in India. Sugar Tech 13(4):336–353

Wang P, Hu C (1980) Regeneration of virus free plant through in vitro culture. In: Fiechter A (ed) Advances in biochemical engineering, 18: plant cell cultures II. Springer, Berlin, pp 61–99

Biosecurity from Plant Viruses in India

30

Vasimalla Celia Chalam, D.B. Parakh, Maurya, and R.K. Khetarpal

Abstract

The liberalized trade of agricultural commodities under World Trade Organisation has necessitated the strengthening of agricultural biosecurity and aligning the national efforts to the International Standards of Phytosanitary Measures. In India an Agricultural Biosecurity Bill has been drafted and is under the consideration of the Government. The strategies for biosecurity for plant viruses include stringent quarantine measures for the imported material, domestic quarantine and use of certified disease-free seed and other planting material within the country. As far as the germplasm is concerned, National Bureau of Plant Genetic Resources has strengthened its quarantine measures for detection and interception of exotic plant viruses. During the last three decades, a number of viruses of serious economic concern have been intercepted including 16 viruses that are not known to exist in India. Besides, interceptions were made for 19 plant viruses that are not known to occur on specific host(s) in India. For facilitating export certification, exclusion of plant viruses from the fields is crucial. Strengthening the quality control of seeds with respect to seed health needs attention though an institutional mechanism for tissue culture-raised plants is in place. Non availability of diagnostic reagents the key challenge in the detection of viruses in quarantine. Also strengthening of infrastructure, capabilities and methodologies for detection of viruses in bulk samples is essential. For virus detection, adoption of a right technique and strategy would help in ensuring biosecurity of Indian

V.C. Chalam (✉) • D.B. Parakh • Maurya
Division of Plant Quarantine, ICAR-National Bureau of Plant Genetic Resources, Pusa Campus, New Delhi 110 012, India
e-mail: celia.chalam@icar.gov.in; mailcelia@gmail.com; dr.parakh@icar.gov.in; ashok.maurya@icar.gov.in

R.K. Khetarpal
CABI – South Asia, 2nd Floor, CG Block, NASC Complex, Pusa, New Delhi 110012, India
e-mail: r.khetarpal@cabi.org

B. Mandal et al. (eds.), *A Century of Plant Virology in India*,
https://doi.org/10.1007/978-981-10-5672-7_30

agriculture from transboundary movement of plant viruses. A need for National Plant Pests Diagnostics and Certification Network is proposed to meet the targets.

Keywords
Biosecurity • Quarantine • Plant viruses • Germplasm • Plant genetic resources • India

30.1 Introduction

Biosecurity encompasses the policy and regulatory frameworks that analyse and manage risks of food safety, animal life and health, and plant life and health, including associated environmental risks (http://www.fao.org/Biosecurity/). The recent years have witnessed a significant growth in trade of agri-horticultural crops due to trade liberalization under the World Trade Organisation (WTO). The global movement of agri-horticultural produce and products has the potential of introducing new pests including viruses which may pose potential risk to the agriculture of the importing country. The unrestricted exchange of seed lots has lead to global distribution of a number of economically important viruses such as *Bean common mosaic virus, Soybean mosaic virus, Pea seed-borne mosaic virus, Wheat streak mosaic virus, Peanut mottle virus,* etc.

Like in other countries, along with imported planting material a number of exotic plant viruses have been introduced into India also causing serious crop losses. These included *Banana bunchy top virus* (BBTV), *Banana streak virus, Peanut stripe virus* etc. BBTV was probably introduced and spread into India from Sri Lanka in 1940 (Magee 1953). The international spread of BBTV is primarily through infected planting material (Wardlaw 1961). In India, an annual loss of Rs.40 crore due to BBTV has been estimated in Kerala alone. These introductions highlight the fact that increased pace of international travel and trade, had exposed countries to the danger of infiltration of exotic viruses harmful to the agriculture.

The National Plant Protection Organizations are entrusted the responsibility to protect the countries from the entry of transboundary pests that include viruses. The strategies for biosecurity from plant viruses include stringent quarantine measures for the imported material and domestic quarantine/use of certified virus-free seed and other planting material within the country. As per norms a workable sample of the bulk samples of seed lots/vegetative propagules/tissue culture-raised plants need to be inspected and tested. The detection of viruses is then carried out by the approved/available techniques, and for that matter availability of specific, sensitive, rapid, reliable and robust methods for detection and identification of viruses become critical.

30.2 Exclusion of Plant Viruses Through Quarantine

30.2.1 International Scenario

For the international trade of agricultural products the establishment of the WTO in 1995 has provided unlimited opportunities. The Sanitary and Phytosanitary (SPS) measures concern the application of food safety and human, animal and plant health regulations. Also, International Standards for Phytosanitary Measures (ISPMs) for regulating plant/planting material are developed by the International Plant Protection Convention (IPPC) of Food and Agriculture Organization (FAO) of the United Nations. These Standards provide guidelines on pest prevention, detection and eradication. To date, 37 ISPMs (https://www.ippc.int/en/core-activities/standards-setting/ispms/) have been developed.

30.2.2 National Scenario

30.2.2.1 Import Quarantine

Plant quarantine is defined as all activities designed to prevent the introduction and/or spread of quarantine pests or to ensure their official control. The Government of India in 1914 enacted Destructive Insects and Pests (DIP) Act, to regulate or prohibit the import of any article into India that are likely to carry any pest, that are potentially harmful to any crop, or can spread to different places. This act has undergone several amendments over the years. In January, 2004 the Plant Quarantine (Regulation of Import into India) Order 2003 [hereafter referred to as PQ Order], came into force and was the first step for complying with the SPS Agreement of WTO (Khetarpal et al. 2006a). A number of amendments of the PQ Order were notified. Of the 12 Schedules in PQ Order, the Schedules IV, V, VI and VIII need special mention. The Schedule IV includes 14 crops and countries from where import is prohibited along with the name of pest(s). Out of 14 crops, 5 crops due to viruses; 3 due to phytoplasmas; 2 due to viruses and phytoplasmas; 1 due to viroids and 1 due to phytoplasma and viroid. The Schedule V deals with 17 crops with restricted import permissible under special conditions with the recommendation of authorized institutions. Whereas, the Schedule VI includes 693 crops permitted to be imported with additional declarations incorporated into phytosanitary certificate. As per the PQ Order, 285 viruses, 18 phytoplasmas and 5 viroids are regulated pests which are of quarantine significance for India. The Schedule IV includes 32 viruses, five phytoplasmas and one viroid affecting 14 crops; Schedule V includes 43 viruses, seven phytoplasmas and two viroids affecting 17 crops; Schedule VI includes 242 viruses, 11 phytoplasmas and five viroids affecting 693 crops. Schedule VIII includes 31 quarantine weed species.

For import commodities such as seed/planting material the PQ Order encompasses "Additional/Special Declarations" free from quarantine pests determined through pest risk analysis as per the international standards (http://plantquarantineindia.nic.in/PQISMain/Default.aspx).

Table 30.1 Plant viruses intercepted in planting material imported into India for commercial purposes

Virus intercepted	Host	Source of import	References
Arabis mosaic virus	Alstromeria	Netherlands	Sushil (2016)
Cymbidium mosaic virus	Dendrobium/ Oncidium/Mokara	Thailand	Sushil (2016)
Pea mosaic virus	Pea	Australia	http://plantquarantineindia.nic.in/PQISPub/html/
Rose mosaic virus	Rose	Australia	http://plantquarantineindia.nic.in/PQISPub/html/
Soybean mosaic virus	Soybean	USA	http://plantquarantineindia.nic.in/PQISPub/html/
Tobamovirus	Vanda	Thailand	Sushil (2016)
Potexvirus	Vanda	Thailand	Sushil (2016)
Potyvirus	Anthurium, chilli, cucumber, oilpalm	Costa Rica, Israel, Netherlands, Singapore, Thailand, USA	Sushil (2016), http://plantquarantineindia.nic.in/PQISPub/html/
Yellow mosaic virus	Chicory	Holland	http://plantquarantineindia.nic.in/PQISPub/html/

The Directorate of Plant Protection, Quarantine and Storage (DPPQS) under the Ministry of Agriculture and Farmers Welfare (MoA&FW) through its 53 Plant Quarantine Stations in the different parts of the country undertakes the quarantine processing of bulk consignments of grain/pulses etc. for consumption and seed/planting material for sowing (https://www.icegate.gov.in/Download/Plant_Quarantine_Stations_and_Lab_Details.pdf). The bulk material for sowing/planting purposes are authorized only through five Plant Quarantine Stations located at New Delhi, Mumbai, Chennai, Kolkata and Amritsar. There are 41 Designated Inspection Authorities spread across the country who inspect the consignment grown in isolation for detection of exotic pests including viruses. In tune with ISPMs the DPPQS has developed 22 national standards on various phytosanitary issues including pest risk analysis. Table 30.1 lists some important viruses detected and intercepted in commercial consignments of imported planting material (Sushil 2016; http://plantquarantineindia.nic.in/PQISMain/Default.aspx).

The ICAR-National Bureau of Plant Genetic Resources (ICAR-NBPGR), which is the nodal institution for exchange of plant genetic resources (PGR) has been empowered under the PQ Order for quarantine processing of all germplasm including transgenic planting material imported for research purposes into the country. Apart from having well-equipped laboratories and post-entry quarantine (PEQ) green house complex it also has a National Containment Facility of level-4 (CL-4) to ensure that no viable biological material/pollen/pathogen enters or leaves the facility during quarantine processing of transgenics. It may be noted that 40 viruses of quarantine significance have been intercepted in exotic germplasm including transgenics in the last three decades (Table 30.2). This has been possible by adopting a workable strategy such as PEQ growing in PEQ greenhouses/containment facility and inspection, PEQ inspection at indenter's site, electron microscopy, enzyme-linked immunosorbent assay (ELISA) and reverse transcription-polymerase chain reaction (RT-PCR). The interceptions include 16 viruses not yet reported from India viz., *Barley stripe mosaic virus* (BSMV), *Bean mild mosaic virus* (BMMV), *Bean pod mottle virus* (BPMV), *Broad bean mottle virus* (BBMV), *Broad bean stain virus* (BBSV), *Broad bean true mosaic virus* (BBTMV), *Cherry leaf roll virus* (CLRV), *Cowpea mottle virus* (CPMoV), *Cowpea severe mosaic virus* (CPSMV), *High plains virus* (HPV), *Maize chlorotic mottle virus* (MCMV), *Pea enation mosaic virus* (PEMV), *Peanut stunt virus* (PSV), *Raspberry ringspot virus* (RpRSV), *Tomato ringspot virus* (ToRSV) and *Wheat streak mosaic virus* (WSMV). Besides, 19 viruses not known to occur on particular host(s) in India have been intercepted and these are also of quarantine significance for India. Twenty viruses have been intercepted in germplasm imported from CGIAR centres (Chalam 2014, 2016; Chalam et al. 2005a, 2007, 2008b, 2009b, 2012a, b, d, 2013b, c, d, 2014a, b, c, 2015a, b, 2016; Chalam and Khetarpal 2008; Khetarpal et al. 1992, 1994, 2001; Kumar et al. 1991; Parakh et al. 1994, 2005, 2006, 2008; Prasada Rao et al. 1990, 2004, 2012; Singh et al. 2003).

Out of 40 viruses detected, 33 viruses either by virus name or as genus viz., *Alfalfa mosaic virus*, *Arabis mosaic virus*, BBSV, BBMV, BBTMV, *Broad bean wilt virus*, BSMV, *Bean yellow mosaic virus*, CLRV, *Cucumber mosaic virus*, GFLV, HPV, MCMV, PEMV, *Peanut stripe virus*, PSV, RpRSV, TAV, TBRV, TNV, TRV, TRSV, ToRSV, TSV, WSMV, SMV as *Potyvirus*, and for cowpea virus genera were mentioned as *Potyvirus* (BCMV, BCMNV and CABMV), *Comovirus* (BPMV, CPMV and CPSMV) and *Carmovirus* (CPMoV) have been mentioned in the PQ Order for legumes and/other crops, in few cases for many crops and require an additional declaration from exporting country that the consignment is free from one or more of these viruses on case to case basis, hence these intercepted viruses are of high quarantine significance for India.

Among the viruses intercepted 16 are not known to occur in India and their potential vectors exist and so also the congenial conditions for them to multiply, disseminate and spread the destructive exotic viruses/strains. The interceptions thus eliminated the risk of introduction of 40 viruses or their strains into India. The inadvertent introductions of which could have caused havoc to our agriculture production. It may be noted that all the virus infected plants were uprooted and incinerated, and the virus-free germplasm was released for further use. Besides, the harvest obtained from virus-free plants also ensured conservation of virus-free exotic germplasm in the National Genebank.

30.2.2.2 Export Quarantine

Export certification is also carried out by DPPQS in accordance with the provision of Article V of IPPC. A Phytosanitary Certificate (as per the model prescribed by IPPC) has to accompany the material meant for export. The export inspection is conducted as per the "*Standard Operating Procedures for Export Inspection and*

Table 30.2 Plant viruses intercepted in germplasm imported into India for research purposes during 1989–2015

Virus intercepted	Host	Source of import	References
Alfalfa mosaic virus (AMV)	*Glycine max*	AVRDC (Taiwan), IITA (Nigeria), Brazil, Myanmar, Sri Lanka, USA	Chalam et al. (2014b), Chalam and Khetarpal (2008), Parakh et al. (2005, 2008)
	Phaseolus vulgaris[b]	CIAT (Colombia), Canada, Kenya, Nepal, Russia	Chalam et al. (2005a, 2014b), Chalam and Khetarpal (2008)
	Pisum sativum[b]	ICARDA (Syria), Russia, Spain, USA	Chalam et al. (2014b)
	Vicia faba	ICARDA (Syria), Spain	Chalam et al. (2009b, 2014b)
	Vigna mungo[b]	Bhutan	Chalam et al. (2014b)
	V. radiata[b]	IITA (Nigeria), Japan	Chalam et al. (2008b, 2014b), Chalam and Khetarpal (2008)
	V. unguiculata[b]	CIAT (Colombia), IITA (Nigeria), USA	Chalam et al. (2008b, 2014b), Chalam and Khetarpal (2008)
Arabis mosaic virus (ArMV)[c]	*G. max*[b]	USA	Chalam et al. (2014b)
Barley stripe mosaic virus (BSMV)[a]	*Hordeum vulgare*	USA	Chalam and Khetarpal (2008), Khetarpal et al. (2006a)
	Triticum aestivum[d]	USA	Chalam et al. (2013b, 2015b, 2016), Chalam (2014, 2016)
	Zea mays[d]	Philippines, USA	Chalam (2014, 2016), Chalam et al. (2016)
Bean common mosaic virus (BCMV)	*G. max*[b]	AVRDC (Taiwan), IITA (Nigeria), Brazil, Thailand, USA	Chalam et al. (2014b), Chalam and Khetarpal (2008); Parakh et al. (2005, 2008)
	P. vulgaris	CIAT (Colombia), CIS, Hungary, Kenya, Nepal, Russia, USA	Chalam et al. (2005a, 2014b), Chalam and Khetarpal (2008), Khetarpal et al. (1994, 2001)
	V. radiata	AVRDC (Taiwan), Japan, USA	Chalam and Khetarpal (2008), Chalam et al. (2014b), Khetarpal et al. (2001)
	V. subterranea[b]	Ghana	Chalam et al. (2014b)
	V. unguiculata	AVRDC (Taiwan), IITA (Nigeria), Guyana, USA	Chalam et al. (2014b)

(continued)

Table 30.2 (continued)

Virus intercepted	Host	Source of import	References
Bean common mosaic necrosis virus (BCMNV)	*P. vulgaris*	Kenya	Chalam et al. (2007, 2014b)
	V. marina[b]	AVRDC (Taiwan)	Chalam et al. (2014b)
Bean mild mosaic virus (BMMV)[a]	*G. max*	Canada, Columbia	Anonymous (2015), Chalam (2016)
	P. vulgaris	USA	Anonymous (2015), Chalam (2016)
Bean pod mottle virus (BPMV)[a]	*G. max*	USA	Chalam et al. (2014b)
Bean yellow mosaic virus (BYMV)	*G. max*	IITA (Nigeria), Myanmar, USA	Chalam et al. (2014b), Chalam and Khetarpal (2008), Parakh et al. (2005)
	P. vulgaris	CIAT (Colombia), Nepal	Chalam et al. (2005a, 2014b) and Chalam and Khetarpal (2008)
	V. faba	ICARDA (Syria), Bulgaria, Spain	Chalam et al. (2007, 2009b, 2014b), Chalam and Khetarpal (2008) and Khetarpal et al. (2001)
Broad bean stain virus (BBSV)[a]	*P. sativum*	Spain	Chalam et al. (2014b)
	V. faba	ICARDA (Syria), Bulgaria	Chalam et al. (2007, 2009b, 2014b), Chalam and Khetarpal (2008), Khetarpal et al. (2001)
Broad bean true mosaic virus (BBTMV)[a]	*P. vulgaris*	USA	Anonymous (2014); Chalam et al. (2015a), Chalam (2016)
Broad bean mottle virus (BBMV)[a]	*P. vulgaris*	AVRDC (Taiwan)	Anonymous (2015), Chalam (2016)
Broad bean wilt virus (BBWV)	*G. max*[b]	USA	Chalam et al. (2014b)
	P. vulgaris[b]	CIAT (Colombia)	Chalam et al. (2014b)
	P. sativum	USA	Chalam et al. (2014b)
	V. faba	ICARDA (Syria)	Chalam et al. (2009b, 2014b)
Cherry leaf roll virus (CLRV)[a]	*G. max*	AVRDC (Taiwan), Sri Lanka, Thailand, USA	Chalam et al. (2007, 2014b), Chalam and Khetarpal (2008)
	P. vulgaris	CIAT (Colombia)	Chalam et al. (2005a, 2014b), Chalam and Khetarpal (2008)

(continued)

Table 30.2 (continued)

Virus intercepted	Host	Source of import	References
Cowpea aphid-borne mosaic virus (CABMV)	*G. max*[b]	AVRDC (Taiwan), IITA (Nigeria), Myanmar, Sri Lanka, Thailand, USA	Chalam et al. (2014b), Chalam and Khetarpal (2008), Parakh et al. (2005, 2008)
	V. radiata[b]	AVRDC (Taiwan)	Chalam et al. (2008b, 2014b), Chalam and Khetarpal (2008)
	V. unguiculata	IITA (Nigeria), Eritrea, Guyana, USA	Chalam et al. (2008b, 2014b), Chalam and Khetarpal (2008) and Khetarpal et al. (2001)
Cowpea mosaic virus (CPMV)	*G. max*[b]	AVRDC (Taiwan)	Chalam et al. (2007, 2014b)
	V. radiata[b]	USA	Chalam et al. (2007, 2014b), Chalam and Khetarpal (2008)
	V. unguiculata	IITA (Nigeria)	Chalam et al. (2007, 2014b), Chalam and Khetarpal (2008)
Cowpea mild mottle virus (CPMMV)	*G. max*	AVRDC (Taiwan), USA	Chalam et al. (2014b)
	P. sativum[b]	Spain	Chalam et al. (2014b)
	V. faba[b]	ICARDA, Syria	Chalam et al. (2014b)
	V. radiata	AVRDC (Taiwan)	Chalam et al. (2014b)
Cowpea mottle virus (CPMoV)[a]	*V. unguiculata*	Philippines	Khetarpal et al. (2001)
	V. subterranea	Ghana	Chalam et al. (2014b)
Cowpea severe mosaic virus (CPSMV)[a]	*G. max*	USA	Chalam et al. (2014b)
	V. radiata	Australia	Chalam et al. (2014b)
	V. unguiculata	Belgium	Chalam et al. (2014b)
Cucumber mosaic virus (CMV)	*Cucumis sativus*	Brazil	Anonymous (2014), Chalam (2016)
	G. max	AVRDC (Taiwan), IITA (Nigeria), Brazil, Myanmar, Sri Lanka, USA	Chalam et al. (2014b), Chalam and Khetarpal (2008), Parakh et al. (2005, 2008)
	P. vulgaris	CIAT (Colombia)	Chalam et al. (2005a, 2014b), Chalam and Khetarpal (2008)
	V. faba	ICARDA (Syria)	Chalam et al. (2014b)
	V. serratifolia	ICARDA (Syria)	Chalam et al. (2014b)
	V. radiata	USA	Chalam et al. (2008b, 2014b), Chalam and Khetarpal (2008)
	V. unguiculata	IITA (Nigeria), Belgium, USA	Chalam et al. (2008b, 2014b), Chalam and Khetarpal (2008)

(continued)

Table 30.2 (continued)

Virus intercepted	Host	Source of import	References
Grapevine fan leaf virus (GFLV)	*G. max*[b]	AVRDC (Taiwan), USA	Chalam et al. (2014b)
	V. mungo[b]	Bhutan	Chalam et al. (2014b)
High plains virus[a]	*Z. mays*[d]	USA	Chalam (2014, 2016), Chalam et al. (2014a, c, 2016)
Maize chlorotic mottle virus[a]	*Z. mays*[d]	Puerto Rico, USA	Chalam et al. (2013b, c, 2016), Chalam (2014, 2016)
Maize dwarf mosaic virus (MDMV)	*Triticum aestivum*[b, d]	USA	Chalam (2014)
	Z. mays	USA	unpublished, V Celia Chalam, DB Parakh and AK Maurya
	Z. mays[d]	Philippines, USA	Chalam (2014, 2016), Chalam et al. (2016)
Pea enation mosaic virus (PEMV)[a]	*G. max*	Costa Rica	Anonymous (2015), Chalam et al. (2015a), Chalam (2016)
	P. vulgaris	Nepal	Chalam et al. (2014b)
	P. sativum	Spain, USA	Chalam et al. (2014b)
	V. faba	ICARDA (Syria)	Chalam et al. (2014b)
Pea seed-borne mosaic virus (PSbMV)	*P. sativum*	AVRDC (Taiwan), Australia, Bulgaria, Columbia, Eritrea, Germany, Nepal, Netherlands, Russia, Spain, Syria, USA	Khetarpal et al. (2001), Parakh et al. (2006), Chalam et al. (2014b), Chalam and Khetarpal (2008)
	V. faba[b]	ICARDA (Syria), Bulgaria, Spain	Kumar et al. (1991), Khetarpal et al. (2001), Chalam et al. (2007), (2009b, 2014b), Chalam and Khetarpal (2008)
Peanut mottle virus (PeMoV)	*A. hypogaea*	Indonesia, Malawi, Philippines, Uganda, USA	Prasada Rao et al. (2004), Khetarpal et al. (2006a), Chalam and Khetarpal (2008)
	G. max[b]	AVRDC (Taiwan), Sri Lanka	Chalam et al. (2014b)
	V. radiata[b]	Sri Lanka	Chalam et al. (2014b)
Peanut stripe virus (PStV)	*Arachis hypogaea*	China, Japan, Myanmar, Philippines, USA	Prasada Rao et al. (1990, 2004, 2012), Khetarpal et al. (2006a), Chalam and Khetarpal (2008)
	G. max[b]	China	Prasada Rao et al. (2012)

(continued)

Table 30.2 (continued)

Virus intercepted	Host	Source of import	References
Peanut stunt virus (PSV)[a]	*P. vulgaris*	USA	Anonymous (2014), Chalam (2016), Chalam et al. (2015a)
	G. max	Costa Rica	Anonymous (2015), Chalam et al. (2015), Chalam (2016)
Raspberry ring spot virus (RRSV)[a]	*G. max*	AVRDC (Taiwan), Costa Rica, Sri Lanka, Thailand, USA	Anonymous (2015), Chalam (2016); Chalam et al. (2007, 2014b), Chalam and Khetarpal (2008)
Red clover vein mosaic virus (RCVMV)[c]	*P. vulgaris*[b]	USA	Anonymous (2014), Chalam et al. (2014a)
Southern bean mosaic virus (SBMV)	*G. max*[b]	AVRDC (Taiwan); IITA (Nigeria), Thailand, USA	Chalam et al. (2014b), Chalam and Khetarpal (2008), Parakh et al. (2005, 2008)
	P. vulgaris[b]	CIAT (Colombia)	Chalam et al. (2005a, 2014b), Chalam and Khetarpal (2008)
	V. unguiculata	Belgium, USA	Chalam et al. (2007, 2014b), Chalam and Khetarpal (2008)
Soybean mosaic virus (SMV)	*G. max*	AVRDC (Taiwan), IITA (Nigeria), Australia, Brazil, Hungary, Nigeria, Sri Lanka, Thailand, USA	Chalam et al. (2014b), Chalam and Khetarpal (2008), Khetarpal et al. (1992, 2001), Parakh et al. (1994, 2005, 2008)
	G. max[d]	USA	Chalam and Khetarpal (2008), Singh et al. (2003)
	P. vulgaris[b]	CIAT (Colombia), Russia	Chalam et al. (2005a, 2014b), Chalam and Khetarpal, (2008)
Tobacco necrosis virus (TNV)	*P. vulgaris*	Nepal	Chalam et al. (2014b)
	P. sativum[b]	Spain, USA	Chalam et al. (2014b)
	V. faba[b]	ICARDA (Syria)	Chalam et al. (2014b)
Tobacco rattle virus (TRV)	*P. vulgaris*[b]	CIAT (Colombia)	Chalam et al. (2014b)
Tobacco ring spot virus (TRSV)	*G. max*	AVRDC (Taiwan); IITA (Nigeria), Brazil, Myanmar	Chalam and Khetarpal (2008), Chalam et al. (2014b); Parakh et al. (2005, 2008)

(continued)

Table 30.2 (continued)

Virus intercepted	Host	Source of import	References
Tobacco streak virus (TSV)	*G. max*	AVRDC (Taiwan), Brazil, Sri Lanka, Thailand, USA	Chalam et al. (2007, 2014b), Chalam and Khetarpal (2008)
	Helianthus annuus	Spain	Prasada Rao et al. (2012)
	P. vulgaris[b]	CIAT (Colombia), Nepal	Chalam et al. (2014b)
	P. sativum[b]	USA	Chalam et al. (2014b)
	V. faba[b]	ICARDA (Syria)	Chalam et al. (2014b)
	V. serratifolia[b]	ICARDA (Syria)	Chalam et al. (2014b)
	V. radiata	AVRDC (Taiwan)	Chalam et al. (2014b)
	V. unguiculata[b]	CIAT (Colombia)	Chalam et al. (2014b)
Tomato aspermy virus (TAV)	*P. vulgaris*	CIAT (Colombia)	Chalam et al. (2014b)
Tomato black ring virus (TBRV)	*G. max*[b]	AVRDC (Taiwan), Brazil, Sri Lanka, USA	Chalam et al. (2014b)
	P. vulgaris[b]	CIAT (Colombia), Brazil, Canada, Kenya, Russia, USA	Chalam et al. (2005a, 2014b); Chalam and Khetarpal (2008)
	V. unguiculata[b]	CIAT (Colombia), IITA (Nigeria), USA	Chalam et al. (2008b, 2014b); Chalam and Khetarpal (2008)
Tomato ring spot virus (ToRSV)[a]	*G. max*	AVRDC (Taiwan), Canada, Costa Rica, Sri Lanka, Thailand, USA	Anonymous (2015), Chalam (2016), Chalam et al. (2007, 2014b), Chalam and Khetarpal (2008)
Wheat streak mosaic virus[a]	*T. aestvum*[d]	USA	Chalam (2013b, 2014, 2016, Chalam et al. (2013d, 2016)
	Z. mays[d]	Philippines, Puerto Rico, South Africa, USA	Chalam (2014, 2016), Chalam et al. (2013b, d, 2016)

AVRDC Asian Vegetable Research Development Center- The World Vegetable Center, *IITA* International Institute of Tropical Agriculture, *ICARDA* International Center for Agricultural Research in the Dry Areas, *CIAT* Centro Internacional de Agricultura Tropical

[a]Virus not reported from India

[b]Viruses present in India but not recorded on the particular host on which intercepted

[c]virus present in India but with restricted distribution, hence it is of quarantine significance

[d]Virus intercepted in transgenics

Phytosanitary Certification" notified by MoA&FW and also in line with the relevant International Standards viz., ISPM 7-Export Certification System; ISPM 12-Guidelines for Phytosanitary Certificates and ISPM 23-Guidelines for Inspection. The exporter or his agent shall submit an application to the Plant Quarantine Station at the designated port.

The export inspections are generally carried out as per the requirement of importing country. The seed for propagation is to be subjected to sampling as per the International Seed Testing Association (ISTA) Rules, 1976 whereas sampling for consumption material (cereals, pulses, oil seeds and others) is as per Bureau of Indian Standards. In certain cases the export inspections are conducted at exporters' premises. Phytosanitary Certificate is not issued if the commodity is found to be affected by quarantine pest could not be disinfested. The MoA&FW, Government of India has notified 161 officers to grant Phytosanitary Certificate for export of plants and plant materials. The ICAR-NBPGR is vested with the authority to issue Phytosanitary Certificate for seed material and plant propagules of germplasm meant for export for research purposes (Chalam and Mandal 2013; Jain and Chalam 2013). In case of seeds/planting materials meant for commercial use the import/export is governed by the Export and Import (EXIM) Policy 2002–2007 and amendments made therein.

30.2.2.3 Domestic Quarantine

Domestic quarantine or internal quarantine is aimed to prevent the spread of introduced exotic species or an indigenous key pest to clean (pest-free) areas within the country and this has its provisions in the DIP Act, 1914 and is enforced by the notification issued by the Central and State Governments. More than 30 pest species seems to have been introduced into India while notifications have been issued against the spread of nine introduced pests only namely fluted scale, San Jose scale, codling moth, coffee berry borer, potato wart disease, potato cyst nematode, apple, BBTV and *Banana mosaic virus* (Khetarpal et al. 2006a). According to notifications issued under the DIP Act, an introduced pest, for example, BBTV, has been declared a pest in states of Assam, Kerala, Orissa, Tamil Nadu (TN), West Bengal (WB) and banana, which come out of these states have to be accompanied by a health certificate from the state Pathologist or other competent authorities that the plants are free from it. However, due to absence of domestic quarantine, BBTV has spread to most banana growing areas in the country. The limitations and constraints of domestic quarantine include lack of basic information on the occurrence and distribution of major key pests in the country, in other words pest distribution maps are lacking for most of the key pests; absence of concerted action and enforcement of internal quarantine regulations by the state governments; lack of interstate border quarantine check-posts at rail and road lines greatly added to the free movement of planting material across the states; lack of close cooperation and effective coordination between state governments and centre for timely notification of introduced pests, organizing pest detection surveys for delineating the affected areas and immediate launching of eradication campaigns in affected areas; lack of public awareness; lack of rapid diagnostic tools/kits for quick detection/identification of exotic pests at the field level; lack of rigorous seed/stock certification or nursery inspection programmes to make available the pest-free seed/planting material for farmers (Bhalla et al. 2014).

There is a dire need to revisit the existing domestic quarantine scenario for strengthening check-posts between the States for preventing spread of viruses. Also, review and update the list of viruses to be regulated under domestic quarantine. For example, BBTV and *Banana mosaic virus* (*Cucumber mosaic virus*) need to be deleted as regulated pests under domestic quarantine as they are widely spread across the country.

The following viruses are known to occur only in certain parts of the country: *Indian citrus ring spot virus*: Known to occur in Haryana, Maharashtra (MH), Punjab and Rajasthan; *Citrus mosaic virus*: Known to occur in Andhra Pradesh (AP), Karnataka and parts of TN; *Tomato spotted wilt virus*: Reported from TN on Chrysanthemum; *Banana bract mosaic virus*: Known to occur in AP, Karnataka, Kerala and TN; *Arabis mosaic virus*: Known to occur in AP, Karnataka and parts of TN; *Red clover vein mosaic virus*: Known to occur on rose in Palampur, Himachal Pradesh (HP).

Thus, there is need to consider the above viruses and others for inclusion as regulated pests for domestic quarantine to prevent their spread to other parts of the country and there is also a need to effectively implement domestic quarantine. India must develop organized services of plant quarantine at state level parallel to Australia and USA.

30.2.2.4 The Agricultural Biosecurity Bill, 2013

In order to meet the challenges of globalization and free trade the Agricultural Biosecurity Bill, 2013 was introduced in the Parliament. The main provisions of the bill is (i) to bring the four sectors of agricultural biosecurity viz., plant health, animal health, living aquatic resources (fisheries, etc.) and agriculturally important micro-organisms under one Umbrella; (ii) to modernize the legal framework to regulate safe movement of plants and animals within the country and in international trade, and (iii) to harmonize the legal requirements of the various sectors of agricultural biosecurity.

The bill repeals DIP Act, 1914 and the Livestock Importation Act, 1898 and will give direct power to the quarantine officers to deport or destroy or confiscate the consignment or lodge complaints under the Indian Penal Code (http://www.indiaenvironmentportal.org.in/files/file/Agricultural%20Biosecurity%20Bill.pdf).

30.3 Exclusion of Plant Viruses Through Certification of Planting Material

30.3.1 Seed Certification

Seed certification for a crop comprises of the legal norms for ensuring the genetic identity, physical purity, germinability and freedom from seed-transmitted pathogens and weeds. ISTA, Association of Official Seed Certifying Agencies (AOSCA) and Central Seed Certification Board, Government of India among others have introduced minimum seed certification standards.

30.3.1.1 Methodology for Quality Control of Seeds

In seed testing stations many seed lots need to be tested and in case of viruses even very low rates of infection have to be detected in large samples. Biological assays require time for standardization, are too laborious, and also time and space consuming for working with bulk samples. The testing of seeds in groups thus becomes imperative. Maury and Khetarpal (1997) discussed in depth the use of ELISA for detecting viruses in single embryo, determination of seed transmission by coupling it with group analysis, mode of eliminating the interference of non-embryonic tissues (which do not play a role in transmission of virus through seeds) in routine assessment of seed transmission rate and its role in seed certification programmes.

30.3.1.2 Group Testing of Seeds for Quality Control of Seed-Transmitted Viruses

A large number of seeds of a bulk seed lot is divided into a number of groups of equal size for group testing. Different groups are tested in ELISA as individual composite samples. The decision on the acceptance or rejection of a seed lot can be taken either on the basis of assessment of seed transmission or by positioning the seed lot in relation to a level of tolerance. These two alternative approaches are (a) Quality control by assessing the seed transmission rate (b) Quality control by positioning the seed lot in relation to a level of tolerance.

30.3.1.3 Seed Health Certification in India

In India, the Seeds Act, 1966 (including the Seeds Bill 2004) does not require a mandatory seed certification against any pathogens including viruses which are the most dangerous pathogens as they cannot be controlled by ordinary physicochemical methods and require sophisticated techniques for proper detection and identification. Besides, seed certification standards are prescribed for 110 crops and seed health standards for seed-borne diseases are available for only 43 crops by seed crop inspection at field stage (59 fungal diseases, 17 bacterial diseases, 14 viral diseases and one phytoplasma disease), for two crops by seed sample analysis at seed stage (two fungal diseases and two bacterial diseases) and by both field inspection and seed analysis for seven crops. Therefore, post-harvest pathology related to seed certification is only in nine crops including potato and sweet potato,. These crops cover 16 fungal diseases, 4 bacterial diseases, one nematode disease and one bacteria + nematode complex only (Khetarpal et al. 2006b). National Seed Research and Training Centre (NSRTC) located at Varanasi, Uttar Pradesh parts seed health testing training to officials working in Seed Certification Agencies (20), Seed Testing Laboratories (101), Seed Law Enforcement Agencies (35 States), Agricultural Universities and other institutes. The NSRTC has full fledged Central Seed Testing Laboratory under clause 4(1) of the Seeds Act, 1966 which acts as a referral lab as and when disputes arise in the court of law with regard to quality of seed. The laboratory performs at par with ISTA w.r.t. seed testing.

About 130 plant viruses are known to be seed-transmitted of which one third have great economic importance, but there are no seed health standards prescribed for viral diseases at seed stage. Also, seed certification for pathogen infection during storage is not mandatory with regard to certified packed seed in store/under storage. Seed analysis is carried out essentially by dry seed examination, though many advanced detection techniques are available.

30.3.1.4 Development of Certification Norms for Seed-Transmitted Viruses of Grain Legumes in India

Certification is an important means of managing seed-transmitted viral diseases which are otherwise not easy to control. Keeping in view the high economic significance of seed-transmitted viruses and the complete absence of seed certification standards for them, initiatives were taken in the year 2000 at ICAR-NBPGR, New Delhi to develop a model system for seed certification for viruses in collaboration with Anand Agricultural University, Anand and University of Mysore, Mysore on important seed-transmitted viruses of grain legumes viz., *Bean common mosaic virus* (BCMV) and urdbean leaf crinkle disease (ULCD) of black gram and green gram, *Black-eye cowpea mosaic virus* (BlCMV, now a strain of BCMV) and *Cowpea aphid-borne mosaic virus* (CABMV) of cowpea, *Soybean mosaic virus* (SMV) of soybean and *Pea seed-borne mosaic virus* (PSbMV) of pea.

Based on testing of 972 seed samples collected from diverse agencies from 21 states and extensive surveys done for 3 years in nine major legume-growing states, a national map on prevalence of the six important seed-transmitted viruses of grain legumes was prepared.

It was inferred that with location and the crop variety the disease incidence varied. The detection and identification of viruses both in leaves and seeds indicated the sites in different states that were found to be free from certain viruses. Besides, a correlation was observed in viral disease incidence with aphid vector population, and yield losses. Based on the intensity of virus spread determined by using a known level of initial seed/seedling infection, the seed standards for certification against viruses of cowpea and soybean were proposed as 0.5% and for pea as 2%. ELISA-based diagnostic kits against BlCMV and SMV were prepared for utilization in quality control of seeds. For testing seed samples in bulk, further studies on group testing of seeds coupled with ELISA is needed. It is expected that the results would contribute in developing a national programme for seed certification of grain legumes (Chalam et al. 2004a, 2008a, 2016; Chalam and Khetarpal 2007).

30.3.2 National Certification System for Tissue Culture-Raised Plants

The Department of Biotechnology (DBT), Government of India has taken a lead in developing a National Certification System for Tissue Culture-raised Plants (NCS-TCP) to facilitate certification of the tissue culture raised plants up to laboratory level. NCS-TCP is the unique quality management system in the world for tissue culture industry. This involves integrated working of a large number of stakeholders; Tissue Culture Certification Agency DBT, Accreditation Unit and Project Management Unit at Biotech Consortium India Limited, Referral Centres namely ICAR-Indian Agricultural Research Institute, New Delhi (for virus indexing) and National Research Centre on Plant Biotechnology, New Delhi (for genetic fidelity testing); Accredited Test Laboratories (ATLs); Recognized Tissue Culture

Production Facilities and State Agriculture/Horticulture Departments to ensure production and distribution of quality tissue culture plants.

Five laboratories viz., Central Potato Research Institute, Shimla; National Research Centre for Banana, Tiruchirapalli; University of Agricultural Sciences, GKVK, Bangalore; Vasantdada Sugarcane Institute, Pune and Indian Institute of Sugarcane Research, Lucknow have been accredited as ATLs by DBT for testing and certification of tissue culture raised plants. Till May 12, 2016, 96 tissue culture production facilities in Assam, AP, Bihar, Chattisgarh, Gujarat, Haryana, HP, Karnataka, MH, Madhya Pradesh, Orissa, Punjab, Rajasthan, TN, Telangana, UP and WB have been recognized based on infrastructure and technical competency. ATLs will test and certify the tissue culture plants produced by recognized tissue culture production facilities (Chalam et al. 2016).

30.4 Challenges in Plant Virus Disease Diagnosis in Quarantine

Testing of seeds and other planting material for viruses is demanding in both cost and labour, compared to, for example, testing for fungi. The issues related to quarantine methodology and the challenges in plant virus disease diagnosis in quarantine were analyzed/reviewed by Khetarpal (2004) and Chalam and Khetarpal (2008).

30.4.1 Challenges Prior to Import

30.4.1.1 Pest Risk Analysis

Under the WTO/SPS regime now Pest Risk Analysis (PRA) is mandatory for import of new commodities. In India the Import Permit will not be issued for the commodities not listed under the Schedule-V, VI and VII of PQ Order. For importing new commodities in bulk for sowing/planting, the importer should apply for conducting PRA to the Plant Protection Adviser. In case of germplasm, Import Permit is issued by the Director, ICAR-NBPGR, after conducting PRA based on international standards.

The IPPC has published 37 ISPMs of which ISPM-2, ISPM-11 and ISPM-21 deals with the Guidelines for PRA, PRA for quarantine pests including analysis of environmental risks and living modified organisms and PRA for regulated non-quarantine pests, respectively (https://www.ippc.int/en/core-activities/standards-setting/ispms/). ICAR-NBPGR has published compilations on pests of quarantine significance for India in cereals (Dev et al. 2005), grain legumes (Chalam et al. 2012c) and edible oilseeds (Gupta et al. 2013). The Crop Protection Compendium of CAB International, UK, is a useful asset to scan for global pest data (http://www.cabi.org/cpc/).

30.4.2 Challenges on Import

30.4.2.1 Applicability of Appropriate Virus Detection Techniques

A combination of techniques including biological tests (growing-on test, infectivity test), electron microscopy, ELISA and PCR-based protocols are being used to detect viruses in plants and planting material. The judicious use and application of a given technique in the context of import are presented below:

Biological Tests The earliest methods used for virus indexing was examination of seedlings/plants raised in isolation/post-entry quarantine greenhouses and infectivity test. These tests are time consuming and labour intensive.

Electron Microscopy The Transmission Electron Microscope is very expensive equipment and is often not available. Moreover, electron microscopy is not suited for routine virus indexing, whereas the highly sensitive immunosorbent electron microscopy (ISEM), developed by Derrick (1973), is occasionally used to detect viruses in seeds or to verify results of other detection methods.

Serological Assays ELISA has been a method of choice for virus detection since the 1970s (Clark and Adams 1977) long before DNA-based techniques were available. There are over 800 different antisera available for plant viruses through the American Type Culture Collection (Schaad et al. 2003). Polyclonal and monoclonal antisera for many viruses are available commercially (Agdia, USA; A.C. Diagnostics Inc., USA; Bio-Reba, Switzerland; Loewe Diagnostics, Germany and Neogen Europe Ltd., UK) and in individual labs. These antibodies are being used in numerous protocols to identify viruses, including ELISA, immunostrips and ISEM. For rapid identification, immunostrips/lateral flow strips are very useful and are user-friendly.

Though specificity of immunoassays can be improved by using monoclonal antibodies, but increased specificity may also imply that some strains may be missed for detection. Also, many viruses exist at sub-liminal levels that are hard to detect using ELISA. Besides, the time needed for good antibody production, the possibilities for encountering false positives/false negatives and the difficulty to differentiate between closely related viruses also minimize the effectiveness of ELISA protocols. However, due to its high throughput capability, simplicity, relative cheapness, suitability for large-scale testing and partial automation ELISA remains the consistent protocol of choice for viruses in diagnostic labs (Khetarpal et al. 2003; Maury and Khetarpal 1989; Chalam and Khetarpal 2008).

Nucleic Acid-based Methods The nucleic acid-based methods, especially PCR and RT-PCR with their high specificity and extremely high sensitivity, are increasingly used in plant virus detection, including seed health testing for viruses. Recently, many took advantage of the speed, sensitivity, and quantitative nature of real-time PCR by which detection of the PCR products takes place during amplification (Siljo et al. 2014; Chalam et al. 2004b). The variant of PCR known as immunocapture RT-PCR has also been used for detection of viruses in seeds (Phan et al. 1998).

The application of multiplex PCR, where several viruses are detected in a single reaction, has been used to detect a number of plant viruses (Bhat and Siju 2007; Chalam et al. 2012a). Array technology has also been used for viral strain detection. In the last decade, various methods like microarray (Boonham et al. 2007), loop-mediated isothermal amplification, LAMP (Arif et al. 2012; Bhat et al. 2013; Siljo and Bhat 2014), Helicase dependent amplification, HDA (Chalam et al. 2012b) and next generation sequencing, NGS (Baranwal et al. 2015; Visser et al. 2016) were developed which have specific advantages of either sensitivity or discovery of novel viruses. Despite the obvious advantages, nucleic acid-based methods still lacks simplicity of use and suitability for large-scale testing in plant quarantine compared with the ELISA. However, intense efforts to improve and simplify molecular detection techniques are underway in many laboratories.

30.4.2.2 Sample Size

The size of consignment in quarantine is very critical for an effective processing. The bulk samples need to be sampled as per prescribed procedures, and thus necessitates the development and adption of protocols for batch testing (Maury et al. 1985; Maury and Khetarpal 1989). For the germplasm consignments which usually represent a small size, extreme precaution is needed to ensure that whatever is the result obtained in the tested part, it should as far as possible not denote a false negative or false positive sample. The importers need to be encouraged to get as much sample as possible to allow effective processing. Attention is thus needed for using non-destructive techniques wherever possible, as in case of groundnut, the whole seed is not used for ELISA testing for viruses and only cotyledonary part of the seed is analysed (Chalam and Khetarpal 2008).

30.4.2.3 Detecting an Unknown/Exotic Virus

It is essential to have a database of viruses present in the country and in other parts of the world to prevent the entry of exotic viruses into the country. It is also important to have the information on different strains/isolates present in the country, so as to prevent the entry of virulent strains. In quarantine the detection of exotic viruses is the prime concern and for specialized detection and identification of exotic viruses, it becomes mandatory to have the corresponding antisera for carrying out serological assays. Similarly for detection by PCR the sequence of the exotic virus or a part of its genome is necessary to know in order to design and synthesize the corresponding primers for their detection. To develop nucleic acid-based protocols, the availability of reference material for exotic viruses is a limiting factor and a very few kits are available commercially. This requires an antisera bank of viruses of quarantine significance and database on sequences of viruses/primers, and so also a repository of seeds of indicator hosts for easy access for quarantine officials as they often have to work with time constraints.

For poorly characterized or unknown viruses, detection methods either are not available or are not cost-effective. Next generation sequencing is a valuable tool for detection of viruses (Baranwal et al. 2015; Visser et al. 2016; Boonham et al. (2007) reviewed the application of microarrays for rapid identification of plant viruses. However, use of NGS and microarrays in quarantine are not cost-effective. Nucleic acid-based methods still lack simplicity of use and suitability for large-scale testing as mentioned earlier. Therefore, detecting an unknown virus in bulk consignments needs proper sampling and development/adaption of protocols for batch testing.

30.4.2.4 Urgency of Clearance of the Sample

Growing of imported germplasm samples in PEQ greenhouses causes an inevitabe delay in the release of seeds. It takes a longer period (the crop season) to release the harvest only from the indexed virus-free plants from PEQ greenhouses/nurseries. This may delay the work of the indenter, and therefore use of non-destructive testing of the seeds could be a good alternative to shorten this time.

30.4.2.5 Maintaining Genebanks Free from Exotic Viruses

Of the germplasm handled by the 15 CGIAR Centres, about 95% is conserved and exchanged in the form of true seeds. Germplasm collections usually contain seeds from regions of the world that are not only the centres of origin of the crop, but may also be centres for genetic diversity of crop-specific pathogens. Seeds from these regions may therefore be contaminated by pathogenic strains or pathotypes exotic to the recipient location (Diekmann 1997; Maury et al. 1998). Similar efforts for establishing virus-free accessions of germplasm seed have been made in the US for cowpea (Gillapsie et al. 1995) and in Brazil for groundnut (Pio-Ribeiro et al. 2000). Unfortunately, plantings with virus-free seeds, in regions with widespread incidence of viruses that are both seed-borne and insect-borne, may also quickly succumb to virus infection and serious crop losses (Albrechsten 2006).

Similarly, the Australian post-entry quarantine found several legume germplasm lines imported from large germplasm seed banks, to be infected (Jones 1987). In this case, among lines introduced in 1978–1981, 29 of 302 (9.6%) of *Vigna* spp. and 54 of 309 lines (17.5%) of *Glycine max* and *Phaseolus* spp. were infected with seed-transmitted viruses.

Testing facilities are inadequate especially in developing and least developed countries as virus testing is cost intensive, and this results in perpetuation of seed-transmitted viruses during germplasm exchange, loss of valuable PGR and unrecognized international distribution of viruses. Therefore, efforts should be made to conserve virus-free germplasm in Genebanks of different countries and CGIAR centres. At ICAR-NBPGR, if seeds would not have been harvested from virus-free plants during post-entry quarantine, the Indian National Genebank could have got exotic germplasm accessions infected by exotic viruses/strain.

30.5 Issues of Biosecurity

Virus detection and diagnosis are crucial for application of mitigation strategies, trade and for exchange of germplasm. ***Accredit diagnostic laboratories*** at central and state level for accurate and quick detection and identification of viruses and there is a need for National Certification Programme for Seed Health in line with NCS-TCP and need to review seed certification standards proposed in Seeds Bill 2004.

Establishment of a ***National Repository of Diagnostics for Viral Diseases*** including antisera Bank, database of primers, seeds of indicator hosts, virus reference collections (lyophilized positive controls), user-friendly diagnostics such as lateral flow strips/dip sticks which can detect multiple viruses, multiplex RT-PCR protocols, LAMP and HDA protocols for detection of viruses in the field and at ports of entry, microarrays, DNA barcoding and ultimately, a cost-effective ***national biosecurity chip*** for diagnosis of all current threats to crop plants would be the backbone for strengthening the programme on biosecurity for plant viruses. Also, ***South Asia Regional Working Group of Detection and Identification of Plant Viruses*** thus need to be formed among SAARC countries for sharing knowledge and facilities.

Initiation of a pre-import inspection mechanism and strengthen the PEQ growing and inspection of imported material.

ICAR-NBPGR has organized Winter School on "*Biosecurity and Biosafety: Policies, Procedures and Issues*" for researchers from State Agricultural Universities and Institutes of ICAR in 2009 (Chalam et al. 2009a). There is a need to organize such programmes from time to time.

Develop a mechanism for off-shore/third country quarantine where the risk is too high.

Regular survey and surveillance programme needs to be undertaken to get status of viral diseases in India and for authentic mapping of endemic viruses. This will help in identification of virus-free areas and to include these viruses under **domestic quarantine**. There is a need for virus eradication strategies for viruses with limited distribution and recently introduced viruses.

This would give a boost to our exports when the importing country is assured by a certification agency that the produce is from a virus-free areas.

Database on all viral diseases, including information on host range, geographical distribution, strains, etc. should be made available for its use as a ready reckoner by the scientists, extension workers and quarantine personnel. Plant viruses of quarantine significance for India in cereals (Chalam et al. 2005b), grain legumes (Chalam et al. 2012d) and edible oilseeds (Chalam et al. 2013a) have been published. The Crop Protection Compendium of CAB International, UK, is an useful asset to scan for global pest data (http://www.cabi.org/cpc/).

Preparation of ***authentic reports of new viruses*** and deposition of voucher specimens as reference cultures in the National Repositories has to be made mandatory. The exports may suffer due to wrong identification of viruses and reporting the same as new reports.

Developing an ***early warning system*** by use of simulation models to predict outbreaks of viral diseases.

Need to develop ***rapid response teams*** at state level to deal with the sudden outbreak of viral diseases.

Develop the ***web-based information portal*** for management of plant viruses and regulations including database on taxonomists and dignosticians.

Need to revisit the national regulatory framework and to develop a national strategy for supply of virus-free seeds/plants/planting material, be it for any purpose viz., seed distribution meant for multilocation testing under All India Coordinated Research Projects of ICAR, inland supply of germplasm by ICAR-NBPGR or seed distribution by the National/State Seed Corporations/private organizations. Also, need to monitor the movement of vegetatively propagated material and tissue culture-raised plants across the states. Requirement for treatment of germplasm/ commercial seed/other material being distributed across the states to be made mandatory. Also, strengthen state certification facilities and procedures to ensure the supply of virus-free nursery material.

Including provisions to check international/interstate movement of invasive pests under the ***Biological Diversity Act 2002*** and to harmonize it with the quarantine and EXIM regulations.

Strengthen Plant Quarantine Stations for dealing with bulk samples in terms of trained manpower, and infrastructure (laboratories, treatment facilities and greenhouses) with focus on advanced techniques for detection of viruses, their strains in bulk samples through regular trainings. ICAR-NBPGR has Organized Training course on "*Diagnostic methods for detection and identification of pests of seed and other planting material and their management*" for plant quarantine officials from DPPQS in 2011 (Gupta et al. 2011). Recently during 2015–2016 ICAR-NBPGR has organized seven Training Workshops on "*Biosafety and transboundary movement of living modified organisms (LMOs)*" including four for customs (113 participants) and three for plant quarantine (43) officials (Bansal et al. 2016). Also, organized an "*Interactive meeting of plant quarantine and customs officials for strengthening the capacities of enforcement agencies for transboundary movement of LMOs*" in 2016.

Need to regularly organize workshops to create ***awareness among stakeholders*** (custom officials, other airport officials, travelers, etc) ***on plant quarantine issues*** with diverse kind of teaching modules.

There is an urgent need to develop a ***National Plant Pests Diagnostic and Certification Network*** (Fig. 30.1) linking the research laboratories with seed/vegetative planting material testing laboratories and quarantine stations, which would be the backbone for strengthening the programme on biosecurity from plant pests including viruses.

The NPPDCN as proposed in Fig. 30.1 can be a store-house of information on pests (fungi, bacteria, viruses, nematodes, insect pests and weeds) diagnostics procedures, national data base on state-wise endemic pests, policies, and related issues. The quality of seed and other planting material will be further enhanced only if seed and other planting material health is integrated into certification procedures. We have the potential for strengthening the system but need to focus in a pragmatic manner.

30.6 Concluding Remarks

Due to the liberalized trade under WTO there is an increasing possibility of introduction of exotic viruses if stringent quarantine measures are not followed. The introduced viruses could get established as virus vectors are present in the country. There is a need to strengthen the domestic quarantine system to prevent the spread

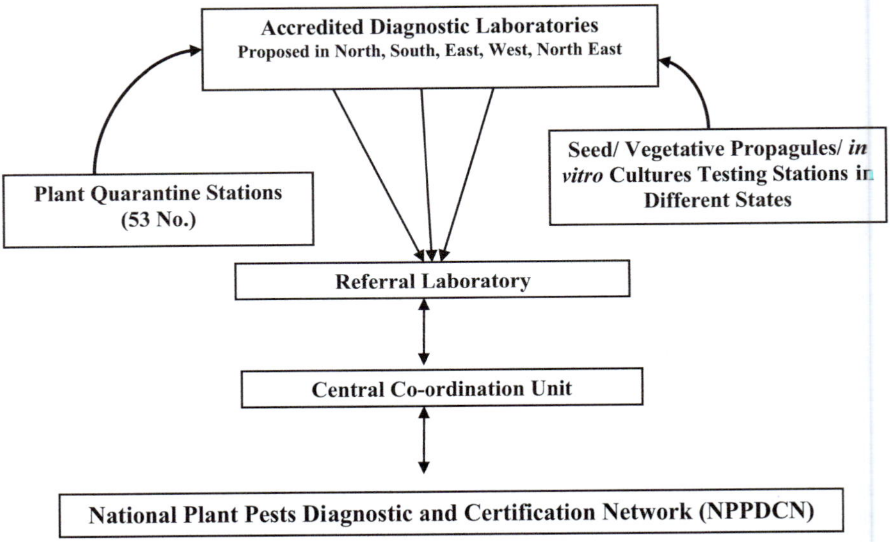

Fig. 30.1 Proposed flow chart of networking by National Plant Pests Diagnostic and Certification Network (Source: Adapted from Chalam 2016)

of viruses with limited distribution within the country. Besides, there are tremendous opportunities to the growers for enhancing export of agricultural commodities if they meet the international quality standards by overcoming phytosanitary constraints, which may involve virus-free areas for production. Various government agencies involved in international and domestic quarantine, certification of seed and other planting material need to work in complete coordination and there is a need to create enough awareness among general public, customs officials and private stakeholders on importance of both international and domestic quarantine.

References

Albrechsten SE (2006) Testing methods for seed-transmitted viruses: principles and protocols. CAB International, Wallingford, 268pp

Anonymous (2014) Annual Report of the National Bureau of Plant Genetic Resources 2013–14, NBPGR, Pusa Campus, New Delhi, 182p

Anonymous (2015) Annual Report of the ICAR-National Bureau of Plant Genetic Resources 2014–15, ICAR-NBPGR, Pusa Campus, New Delhi, 210p

Arif M, Daniels J, Chalam VC, Fletcher J, Ochoa Corona FM (2012) Detection of *High plains virus* with loop-mediated isothermal amplification and qPCR. 2012 American Phytopathological Society meeting, August 4–8, 2012, Rhode Island, USA

Bansal KC, Bhalla S, Chalam VC, Warrier R (2016) Initiatives for strengthening capacities of enforcement agencies on transboundary movement of LMOs under UNEP-GEF supported Phase-II capacity building project on biosafety. PP 13 In: Background document of Regional Conference on Biosafety for Sharing of Experiences, April 7–8 2016, Hyderabad, India 31 p

Baranwal VK, Jain P, Saritha RK, Jain RK, Gautam NK (2015) Detection and partial characterization of *Cowpea mild mottle virus* in mungbean and urdbean by deep sequencing and RT-PCR. Crop Prot 75:77–79

Bhalla S, Chalam VC, Tyagi V, Lal A, Agarwal PC, Bisht IS (2014) Teaching manual on germplasm exchange and plant quarantine. National Bureau of Plant Genetic Resources, New Delhi, 340p

Bhat AI, Siju S (2007) Development of a single tube multiplex RT-PCR for the simultaneous detection of *Cucumber mosaic virus* and *Piper yellow mottle virus* associated with stunt disease of black pepper. Curr Sci 93:973–976

Bhat AI, Siljo A, Deeshma KP (2013) Rapid detection of *Piper yellow mottle virus* and *Cucumber mosaic virus* infecting black pepper (*Piper nigrum*) by loop-mediated isothermal amplification (LAMP). J Virol Methods 193:190–196

Boonham N, Tomlinson J, Mumford R (2007) Microarrays for rapid identification of plant viruses. Annu Rev Phytopathol 45:307–328

Chalam VC (2014) Biotechnological interventions for biosecuring Indian agriculture from transboundary plant viruses. In: Mishra SK (ed) Current trends and future challenges in biotechnology and biomedicine for human welfare and sustainable development Part II. Department of Botany & Biotechnology, Govt. New Science College, Rewa, pp 16–24. 163p

Chalam VC (2016) Challenges for plant virus diagnosis in certification and transboundary movement of planting material, pp 5–17. In: Chalam VC, Akhtar J, Dubey SC (eds) Souvenir and abstracts of zonal annual meeting of Delhi Zone, Indian Phytopathological Society (IPS) & National Symposium on "Biosecurity in Food Value Chain", February 20, 2016, Delhi Zone, IPS and ICAR-National Bureau of Plant Genetic Resources, New Delhi, India, 76 p

Chalam VC, Khetarpal RK (2007) Quality control of seeds for viruses: a case study of legume viruses. In: Seed health tesing and certification: need for marching ahead, November 17, 2007, Indian Phytopathological Society and National Bureau of Plant Genetic Resources, New Delhi, India, pp 23–28

Chalam VC, Khetarpal RK (2008) A critical appraisal of challenges in exclusion of plant viruses during transboundary movement of seeds. Indian J Virol 19:139–149

Chalam VC, Mandal B (2013) Phytosanitary issues for export of ornamental plants. In: Janakiram T, Prasad KV, Singh KP, Swaroop K, Raju DVS, Jain R, Export N (eds) Oriented horticulture. Indian Agricultural Research Institute, New Delhi, 108 pp

Chalam VC, Khetarpal RK, Mishra A, Jain A, Gupta GK (2004a) Seed transmission of soybean mosaic potyvirus in soybean cultivars. J Mycol Plant Pathol 34:86–87

Chalam VC, Pappu HR, Druffel KL, RK Khetarpal (2004b) Detection of potyviruses in legume seeds by RT-PCR and real-time RT-PCR. Proceedings of National Symposium on Molecular Diagnostics for the Management of Viral Diseases, October 14–16, 2004. Indian Agricultural Research Institute and Indian Virological Society, New Delhi, India

Chalam VC, Khetarpal RK, Parakh DB, Maurya AK, Jain A, Singh S (2005a) Interception of seed-transmitted viruses in French bean germplasm imported during 2002–2003. Indian J Plant Prot 33:134–138

Chalam VC, Parakh DB, Singh S, Khetarpal RK (2005b) Viruses of quarantine significance in cereals. In: Dev U, Khetarpal RK, Agarwal PC, Lal A, Kapur ML, Gupta K, Parakh DB (eds) Potential quarantine pests for India: cereals. National Bureau of Plant Genetic Resources, New Delhi, pp 82, 144p–92

Chalam VC, Khetarpal RK, Prasada Rao RDVJ, Parakh DB, Maurya AK (2007) Transboundary movement of plant viruses: a case study of India. Indian J Virol 19(1):97–98

Chalam VC, Khetarpal RK, Prakash HS, Mishra A (2008a) Quality control of seeds for management of seed-transmitted viral diseases of grain legumes in India. In: Kharakwal MC (ed) Food legumes for nutritional security and sustainable agriculture, vol 2, Proceedings of the Fourth International Food Legumes Research Conference (IFLRC-IV). Indian Society of Genetics and Plant Breeding, New Delhi, pp 468–474

Chalam VC, Parakh DB, Khetarpal RK, Maurya AK, Jain A, Singh S (2008b) Interception of seed-transmitted viruses in cowpea and mungbean germplasm imported during 2003. Indian J Virol 19:12–16
Chalam VC, Gupta K, Bhalla S, Rajan, Khetarpal RK (eds) (2009a) Training manual on biosecurity and biosafety: policies, procedures and issues. National Bureau of Plant Genetic Resources, New Delhi, 392p
Chalam VC, Parakh DB, Khetarpal RK, Maurya AK, Pal D (2009b) Interception of seed-transmitted viruses in broad bean (*Vicia faba* L.) germplasm imported into India during 1996–2006. Indian J Virol 20:83–87
Chalam VC, Arif M, Caasi DR, Fletcher J, Ochoa-Corona FM (2012a) Discrimination among CLRV, GFLV and ToRSV using multiplex RT-PCR. 2012. APS Meeting, August 4–8, 2012, Rhode Island, USA
Chalam VC, Bhalla S, Singh B, Rajan (eds) (2012b) Potential quarantine pests for India in grain legumes. National Bureau of Plant Genetic Resources, New Delhi, 324 pp
Chalam VC, Parakh DB, Kaur H, Kumar A, Maurya AK, Khetarpal RK (2012c) Viruses of quarantine significance in grain legumes. In: Chalam VC, Bhalla S, Singh B, Rajan (eds) Potential quarantine pests for India in grain legumes. National Bureau of Plant Genetic Resources, New Delhi, pp 174–211. 324p
Chalam VC, Parakh DB, Kumar A, Maurya AK (2013a) Viruses, viroids and phytoplasma of quarantine significance in edible oilseeds. In: Gupta K, Singh B, Akhtar J, Singh MC (eds) Potential quarantine pests for India in edible oilseeds. National Bureau of Plant Genetic Resources, New Delhi, pp 175–212. 295p
Chalam VC, Parakh DB, Sharma R, Maurya AK (2013b) Maize chlorotic mottle virus: a quarantine pest for India. Technical Bulletin. National Bureau of Plant Genetic Resources, New Delhi, 4p
Chalam VC, Parakh DB, Maurya AK, Khetarpal RK, Agarwal PC (2013c) Quarantine certification of seed and other planting material for viruses, pp 67–69. In: Proceedings of Asia Pacific Congress of Virology, VIROCON 2013, December 17–20, 2013, Amity University, Noida, India, 221p
Chalam VC, Parakh DB, Sharma R, Maurya AK (2013d) Wheat streak mosaic virus: a quarantine pest for India, Technical Bulletin. National Bureau of Plant Genetic Resources, New Delhi, 4p
Chalam VC, Parakh DB, Maurya AK (2014a) Biosecurity umbrella for Indian agriculture against exotic plant viruses: A case study of quarantine of exotic germplasm. In: Proceedings of XXIII National Conference (VIROCON 2014) on Recent trends in Virology Research in the Omics Era, December 18–20 2014, Tamil Nadu Agricultural University, Coimbatore, Tamil Nadu
Chalam VC, Parakh DB, Maurya AK, Singh S, Khetarpal RK (2014b) Biosecuring India from seed-transmitted viruses: the case of quarantine monitoring of legume germplasm imported during 2001–2010. Indian J Plant Prot 42(3):270–279
Chalam VC, Parakh DB, Sharma R, Maurya AK (2014c) High plains virus: a quarantine pest for India, Technical Bulletin. National Bureau of Plant Genetic Resources, New Delhi, 4p
Chalam VC, Parakh DB, Maurya AK (2015a) Biosecuring Indian agriculture from transboundary plant viruses: biotechnological interventions and phytosanitary issues. In: Proceedings of national symposium on Challenges and Management Approaches for Crop Diseases of National Importance – Status and Prospects", February 12–14, 2015, TNAU, Madurai
Chalam VC, Parakh DB, Sharma R, Maurya AK (2015b) *Barley stripe mosaic virus*: a quarantine pest for India, Technical Bulletin. National Bureau of Plant Genetic Resources, New Delhi, 4p
Chalam VC, Khetarpal RK, Agarwal PC (2016) Addressing plant health management using diagnostics and certification issues. In: Katti G, Kodaru A, Somasekhar N, Laha GS, Sarath Babu B, Varaprasad KS (eds) Plant health management for food security issues and approaches. Astral International Pvt. Ltd, New Delhi, pp 39, 230 p–59
Clark MF, Adams AN (1977) Characteristics of the microplate method of enzyme-linked immunosorbent assay for the detection of plant viruses. J Gen Virol 334:475–483
Derrick KS (1973) Quantitaive assay for plant viruses using serologically specific electron microscopy. Virology 56:652–653

Dev U, Khetarpal RK, Agarwal PC, Lal A, Kapur ML, Gupta K, Parakh DB (eds) (2005) Pests of quarantine significance in cereals. National Bureau of Plant Genetic Resources, New Delhi, 142pp

Diekmann M (1997) Activities at the international agricultural research centres to control pathogens in germplasm. In: McGee DC (ed) Plant pathogens and the worldwide movement of seeds. APS Press, St. Paul

FAO (2016) ISPM 5 glossary of phytosanitary terms. https://www.ippc.int/static/media/files/publication/en/2016/06/ISPM_05_2016_En_2016-06-03_c6w6Iq3.pdf

Gillapse AG, Hopkins MS Jr, Pinnow DL, Hampton RO (1995) Seedborne viruses in preintroduction cowpea seed lots and establishment of virus-free accessions. Plant Dis 79:388–391

Gupta K, Chalam VC, Usha D, Bhalla S, Khan Z, Agarwal PC (eds) (2011) Training manual on diagnostic methods for detection and identification of pests of seed and other planting material and their management. National Bureau of Plant Genetic Resources, New Delhi, 165p

Gupta K, Singh B, Akhtar J, Singh MC (eds) (2013) Potential quarantine pests for India in edible oilseeds. National Bureau of Plant Genetic Resources, New Delhi, 295p

Jain RK, Chalam VC (2013) Pests: threat to seed exports, pp 3–8. In: Souvenir of Indian Seed Congress, February 8–9, 2013, organized by National Seed Association of India at Gurgaon, India, 51p

Jones DR (1987) Seedborne diseases and the international transfer of plant genetic resources: an Australian perspective. Seed Sci Technol 15:765–776

Khetarpal RK (2004) A critical appraisal of seed health certification and transboundary movement of seeds under WTO regime. Indian Phytopathol 57:408–421

Khetarpal RK, Parakh DB, Singh S, Nath R (1992) ELISA detection of *Soybean mosaic virus* in testas, embryos and seedlings from mottled and unmottled seeds of imported soybean germplasm. Indian J Virol 8(2):106–110

Khetarpal RK, Parakh DB, Singh S, Nath R, Jain RK, Varma A (1994) *Bean common mosaic virus* detected by DAC-Indirect ELISA in exotic *Phaseolus vulgaris* L. Indian J Virol 10:13–16

Khetarpal RK, Singh S, Parakh DB, Maurya AK, Chalam VC (2001) Viruses intercepted in exotic germplasm during 1991–2000 in quarantine. Indian J Plant Genet Resour 14:127–129

Khetarpal RK, Parakh DB, Chalam VC (2003) Virus indexing of plant germplasm. In: Mandal BB, Chaudhury R, Engelmann F, Bhag Mal, Tao KL, Dhillon BS (eds) Conservation biotechnology of plant germplasm, pp 99–104. National Bureau of Plant Genetic Resources, New Delhi, India/IPGRI, Rome, Italy/FAO, Rome Italy

Khetarpal RK, Lal A, Varaprasad KS, Agarwal PC, Bhalla S, Chalam VC, Gupta K (2006a) Quarantine for safe exchange of plant genetic resources. In: Singh AK, Srinivasan K, Saxena S, Dhillon BS (eds) Hundred years of plant genetic resources management in India. National Bureau of Plant Genetic Resources, New Delhi, pp 83–108

Khetarpal RK, Sankaran V, Chalam VC, Gupta K (2006b) Seed health testing for certification and SPS/WTO requirements. In: Kalloo G, Jain SK, Vani AK, Srivastava U (eds) Seed: a global perspective. Indian Society of Seed Technology, New Delhi, pp 239, 312pp–258

Kumar CA, Singh S, Parakh DB (1991) Detection of pea seed borne mosaic virus (PSbMV) in pea germplasm during quarantine processing. Indian Phytopathol 44:366–369

Magee CJP (1953) Some aspects of the bunchy top disease of banana and other *Musa* spp. J Proc R Soc NSW 87:3–18

Maury Y, Khetarpal RK (1989) Testing seeds for viruses using ELISA. In: Agnihotri VP, Singh N, Chaube HS, Singh US, Dwivedi TS (eds) Perspectives in phytopathology. Today and Tomorrows Printers and Publishers, New Delhi, pp 31–49

Maury Y, Khetarpal RK (1997) Quality control of seed for viruses: present status and future prospects. In: Hutchins JD, Reeves JE (eds) Seed health testing: progress towards the 21st century. CAB International, UK, pp 243–252

Maury Y, Duby C, Bossenec JM, Boudazin G (1985) Group analysis using ELISA: determination of the level of transmission of *Soybean mosaic virus* in soybean seed. Agronomie 5:405–415

Maury Y, Duby C, Khetarpal RK (1998) Seed certification for viruses. In: Hadidi A, Khetarpal RK, Koganezawa H (eds) Plant virus disease control. American Phytopathological Society Press, St. Paul, pp 237–248

Parakh DB, Khetarpal RK, Singh S, Nath R (1994) Post entry quarantine detection of *Soybean mosaic virus* by ELISA in soybean germplasm. Indian J Virol 10:17–21

Parakh DB, Chalam VC, Khetarpal RK, Maurya AK, Jain A, Singh S (2005) Interception of seed-transmitted viruses in soybean germplasm imported during 2003 and 2004. Indian J Plant Prot 33:119–124

Parakh DB, Chalam VC, Maurya AK, Khetarpal RK, Singh S, Kaur A (2006) Interception of *Pea seed-borne mosaic virus* in pea germplasm imported during 2003. Indian J Virol 17:35–38

Parakh DB, Khetarpal RK, Chalam VC (2008) Risk of seed-transmitted viruses associated with exchange of soybean germplasm and the South Asian scenario. Indian J Virol 19(1):47–49

Phan TTH, Khetarpal RK, Le TAH, Maury Y (1998) Comparison of immunocapture-PCR and ELISA in quality control of pea seed for pea seed borne mosaic potyvirus. In: Hutchins JD, Reeves JE (eds) Seed health testing: progress towards the 21st century. CAB International, Wallingford, pp 193–199

Pio-Riberio G, Andrade GP, Cruz SM, Santos RC, Reddy DVR, Andrade AAMX (2000) Virus eradication from peanut germplasm based on serological indexing of seeds and analyses of seed multiplication fields. Fitopatol Bras 25:42–48

Prasada Rao RDVJ, Chakrabarty SK, Reddy AS (1990) Interception of *Peanut stripe virus* in imported groundnut seed from Myanmar. FAO Plant Prot Bull 38:48–49

Prasada Rao RDVJ, Reddy AS, Chakrabarty SK, Anitha K (2004) Interception of seed transmitted viruses in peanut germplasm imported into India during 1986 to 2003. Indian J Plant Prot 32:102–104

Prasada Rao RDVJ, Anitha K, Chakrabarty SK, Girish AG, Sarath Babu B, Rameash K, Abraham B, Varaprasad KS (2012) Quarantine pathogen interceptions on crop germplasm in India during 1986–2010 and their possible economic impact. Indian J Agric Sci 82(5):436–441

Schaad NW, Frederick RD, Shaw J, Schneider WL, Hickson R, Petrillo DM, Luster DG (2003) Advances in molecular-based diagnostics in meeting crop biosecurity and phytosanitary issues. Annu Rev Phytopathol 41:305–324

Siljo A, Bhat AI (2014) Reverse transcription loop-mediated isothermal amplification assay for rapid and sensitive detection of *Banana bract mosaic virus* in cardamom (*Elettaria cardamomum*). Eur J Plant Pathol 138:209–214

Siljo A, Bhat AI, Biju CN (2014) Detection of *Cardamom mosaic virus* and *Banana bract mosaic virus* in cardamom using SYBR Green based reverse transcription-quantitative PCR. Virus Dis 25:137–141

Singh B, Rajan BS, Chalam VC, Pandey BM, Singh SK, Kumar N, Khetarpal RK (2003) Quarantine processing of imported transgenic planting material. Indian J Agric Sci 73(2):97–100

Sushil SN (2016) Agricultural biosecurity system in India, pp 19–25. In: Chalam VC, Akhtar J, Dubey SC (eds) Souvenir and abstracts of zonal annual meeting of Delhi Zone, Indian Phytopathological Society (IPS) & National Symposium on "Biosecurity in Food Value Chain", February 20, 2016, Delhi Zone, IPS and ICAR-National Bureau of Plant Genetic Resources, New Delhi, India, 76p

Visser M, Bester R, Burger JT, Maree HJ (2016) Next-generation sequencing for virus detection: covering all the bases. Virol J 13:85. doi:10.1186/s12985-016-0539-x

Wardlaw CW (1961) Banana diseases including plantain and abaca. Longmans, Green and Co. Ltd.n, London

Transgenic Approaches to Combat Plant Viruses Occurring in India

31

Shelly Praveen, S.V. Ramesh, and Satendra Kumar Mangrauthia

Abstract

Genetic engineering (GE) approaches have been effectively deployed to incorporate foreign genes of economic and/or agricultural importance in crops. Ever since Powell et al. in 1986 showed virus resistance through GE approach, numerous crop plants have been genetically modified to impart virus resistance. Greater understanding of host-virus interactions in the wake of RNA silencing phenomenon have further opened up small non-coding RNAs based virus management strategies. This chapter discusses research priorities, approaches and accomplishments in the field of virus resistant transgenic plants in India. Various genetic modification strategies namely coat protein mediated resistance through RNA silencing have been successfully deployed to develop virus resistance. Transgenic lines have been licensed to private sector, in crops like tomato, and significant progress has been made in crops like potato, rice etc. However, a major bottleneck in developing successful transgenic crop in legumes, cucurbits and other crops, where viral infection is a serious menace is the lack of suitable regeneration and transformation protocols. Hence, this chapter also deliberates upon potential pitfalls of genetic engineering approaches that require intensive research efforts. Further, as a way forward, it is also proposed to explore recently emerging genome editing tools to combat phytopathogenic viruses.

S. Praveen (✉)
Division of Biochemistry, Indian Agricultural Research Institute, New Delhi 110012, India
e-mail: shellypraveen@hotmail.com

S.V. Ramesh
Division of Physiology, Biochemistry & Post Harvest Technology,
ICAR-Central Plantation Crops Research Institute, Kasaragod, Kerala 671124, India

S.K. Mangrauthia
Biotechnology Section, ICAR- Indian Institute of Rice Research, Hyderabad 500030, India

B. Mandal et al. (eds.), *A Century of Plant Virology in India*,
https://doi.org/10.1007/978-981-10-5672-7_31

Keywords
Biosafety • CPMR • Genetic engineering • RNA silencing • Transgenics • Virus resistance

31.1 Introduction

Genetic modification of crops refers to alteration of genome of plants and development of 'transgenics' with a view to obtain altered expression or repression of specific gene(s). The 'transgenic' as the term suggests utilizes the genetic elements of other plants or organisms for traits improvement. Most importantly, it allows harnessing the potential of wide genetic resources, irrespective of their origin or species. It gives opportunity to broaden the genetic diversity by introducing new alleles into the genome, which is very much desired by plant breeders. Transgenic crops are currently grown in more than 27 countries on approximately 179.7 million ha (2015, http://www.isaaa.org/resources/publications/briefs/51/executive summary/default.asp). At the global level India ranks 4th in terms of acreage under transgenic crops. In India insect resistant cotton occupies major share of the area under transgenics. Transgenic crops have already demonstrated substantial benefits to farmers and society at large. One of the greatest contributions of transgenic research in agriculture has been to develop virus resistant crops, which are important to ensure food security and nutrition. Managing viral pathogens in crops is crucial for high productivity as they can impair plant growth, photosynthesis, metabolism, and resource allocation leading to low yield and poor nutrition of agricultural produce. Viruses as pathogen are difficult to manage due to complicated disease cycles, efficient transmission system, limited availability of resistance and non-availability of viricides. In the absence of curative controls, virus diseases can be managed by cultural practices or by development of transgenic resistance.

Application of transgenic approach for virus resistance has been most diverse in terms of gene sources and approaches of transgene expression. Though, a group of researchers have attempted to utilize plants genes having antiviral property for development of resistance, majority of researchers emphasized on use of viral genetic elements for development of virus resistance. This approach was rational because it was very similar to immunization in animals, which had already shown a great success to combat virus disease in humans and animals. However, given the diversity of plant species and infecting viruses, the approaches and deployed genes were not uniform in case of plants. It varied case to case and various approaches made use of all important virus genes encoding coat protein (CP), movement protein (MP), replicase (Rep), pathogenicity factors, and other coding/non coding genes. Further, transgene expression approaches varied from constitutive expression to partial or complete silencing of viral genes. Although CP mediated resistance (CPMR) was successfully used to protect several crops from different viral infections, RNA mediated resistance was more effective and interpreted as an example of homology dependent gene silencing. The concept of gene silencing had started emerging in late 1990s, which later gave birth to the concept of RNA interference in early years of twenty-first century. RNA interference was a Nobel prize winning concept in Physiology and Medicine 2006, which revolutionized the transgenic approaches for virus resistance.

The concept of pathogen-derived resistance (PDR) states that pathogen genes expressed in transgenic plants may confer resistance to infection by the homologous or related pathogens. Numerous studies have shown that PDR is an effective means of producing virus-resistant plants and can be used for a number of different plant virus groups with various viral genes. PDR is mediated either by the transgene protein (protein-mediated) or by the transgene RNA (RNA-mediated). In India, the first demonstration of virus resistance using viral sequences in *Nicotiana tabacum* against potato virus Y (PVY) in 1990, paved way for development of virus resistance. To establish the concept of PDR in model plants, efforts were initiated by the researchers of Advanced Centre for Plant Virology (ACPV) at Indian Agricultural Research Institute in late 1990s (Makeshkumar et al. 2002). The successful demonstration of transgenic resistance in tobacco for PVY at ACPV facilitated in developing first networking program on development of transgenic initiated by the Department of Biotechnology, Government of India in 1999. Under this program, efforts from the Indian scientific community to develop transgenic resistance against different groups of viruses were consolidated at single platform. During this period, the private sector had also shown keen interest in joining hands with the public sector academia to strengthen transgenic research in the country. From early 1990 till date, as the science is advancing in this area, the strategy for development of transgenic resistance have also changed from protein mediated to RNA mediated resistance.

Development of transgenic tomato for leaf curl virus resistance was the first milestone where coat protein mediated and RNA mediated strategies were demonstrated successfully (Praveen et al. 2005; Raj et al. 2005). The coat portein (CP) mediated resistance was developed at National Botanical Research Institute (NBRI), Lucknow. RNA mediated resistance using antisense approach was demonstrated at Indian Agricultural Research Institute (IARI), New Delhi. The investigators demonstrated variable degrees of resistance against tomato leaf curl New Delhi virus (ToLCNDV) in transgenic tomato plants. Later, the stacking of two viral sequences for developing dual resistance (ToLCNDV and cucumber mosaic virus) in tomato was demonstrated by the group at ACPV (Praveen et al. 2006). Researchers at NBRI also demonstrated CP mediated resistance in model plant *N. benthamiana* against cucumber mosaic virus (CMV) (Pratap et al. 2012; Srivastava and Raj 2008). At this juncture, the private sector came forward to translate this research by carrying field trials of the transgenics developed at ACPV. During the last decade, the transgenic program got momentum and transgenic potato for resistance to PVY (Ghosh et al. 2006), cucumber for resistance to CMV, papaya resistance to papaya ringspot virus (PRSV), rice for resistance to rice tungro virus, cotton for cotton curl virus resistance were developed in different Indian laboratories. The efforts carried out in different Research Institutes and Universities are summarized in Table 31.1.

Table 31.1 Summary of virus resistant transgenic development in the Indian laboratories

Sl. No.	Crop	Virus	Gene	Strategy	Organisation/Institute
Horticultural crops					
1.	Tomato	Tomato leaf curl virus	AV2	Antisense	IISc. Bangalore and Rallis India
		Tomato leaf curl virus	CP (coat protein)	Sense	NBRI, Lucknow
		Tomato leaf curl New Delhi virus (ToLCNDV)	AC2 (replicase)	Antisense	IARI, New Delhi
			AC1 (replicase)	RNAi	IARI, New Delhi
			AC4 (RNAi suppressor)	RNAi	IARI and Bejo Sheetal Pvt. Ltd. Jalna
		Cucumber mosaic virus (CMV)	CP (coat protein)	Sense Antisense RNAi	IARI, New Delhi
		Groundnut bud necrosis virus	Np (Nucleo-capsid Gene)	Sense	IARI, New Delhi and IISc, Bangalore
			NSs (RNAi suppessor)	RNAi	IARI, New Delhi
		ToLCNDV + GBNV	AC1 (truncated) + NSs	RNAi	IARI, New Delhi and Advanta India Ltd. Hyderabad
2.	Banana	Banana bunchy top virus	Replicase	RNAi	NRCB, Trichy
3.	Potato	Potato virus-Y (PVY)	CP	Sense	IARI and CPRI Shimla
		Potato leaf roll virus	CP	Sense	IARI and CPRI Shimla
		ToLCNDV	Replicase	RNAi	CPRI Shimla
4.	Papaya	Papaya ring spot virus	CP	Sense RNAi	IARI and Central Institute for Subtropical Horticulture (CISH), Lucknow
			Hc-Pro	RNAi	IARI and Central Institute for Subtropical Horticulture (CISH), Lucknow
5.	Cucumber	CMV	CP	Sense, RNAi	IARI
6.	Citrus	Citrus tristeza virus	CP	Sense	IARI, New Delhi and Delhi University

(continued)

Table 31.1 (continued)

Sl. No.	Crop	Virus	Gene	Strategy	Organisation/Institute
Field crops					
7.	Cotton	Cotton leafcurl virus (CuCLV)	AV2	Antisense	IISc. Bangalore and Mahyco, Jalna
		CuCLV	AC1 (replicase)	Antisense	IARI, New Delhi and CICR, Nagpur
8.	Rice	Rice tungro bacilliform virus (RTBV)	ORF IV	RNAi	Delhi University, IIRR, Hyderabad
		RTSV	–	Sense antisense	Delhi University, IIRR, Hyderabad
9.	Groundnut	GBNV + Tobacco streak virus (TSV)	Np + CP	Sense	IARI, New Delhi and Directorate of Groundnut Research, Junagarh
10.	Sunflower	TSV	CP	Sense	IARI, New Delhi, J.K. Seeds and IIOR, Hyderabad
11.	Soybean	Mungbean yellow mosaic India virus	AC1 (replicase)	Antisense RNAi	IARI, New Delhi and IISR Indore
Model plant					
12.	Tobacco (for proof-of-concept)	CMV	CP	Sense	NBRI, Lucknow
		Physalis mottle virus	CP and 3′ non coding region	RNAi	IISc, Bangaluru
		Indian cassava mosaic virus (ICMV)	AC1	RNAi	CTCRI, Thiruvananthapuram
		Sri Laukan cassava mosaic virus	Part of genome	RNAi	Delhi University
		Chilli leaf curl virus	AC1 AC2	RNAi	JNU, New Delhi
		Dasheen mosaic virus	Coat protein	RNAi	CTCRI, Thiruvananthapuram

During the last decade, researchers from different laboratories (mainly from IARI, IISc, JNU, Delhi University, ICGEB, MKU, CICR, IIRR, IISR) concentrated their efforts on gene silencing approach for management of different plant viruses. In 2005, a transgenic networking project was launched under Indian Council of Agricultural Research (ICAR) for development of virus resistance in different crops including vegetables, fruits, spices, legumes and rice. IARI took initiative in development of transgenic resistance in vegetables including tomato, potato and cucurbits. Under spices, black pepper was addressed by Indian Institute of Spice Research, Calicut and chilli by JNU, New Delhi. Under fruit crops, virus menace in banana, citrus and papaya were addressed by IARI in collaboration with NRCB, Trichy and CISH, Lucknow. Development of rice transgenic for rice tungro virus was initiated by Delhi University. They used both protein as well as RNA-mediated approaches for development of resistance against RTBV and RTSV separately in transgenic rice plants. However, both these strategies confer only moderate level of protection by delaying virus infection (Ganesan et al. 2009). At IISc Bengaluru and CICR Nagpur, transgenic cotton was developed for cotton leaf curl virus resistance. CICR has generated number of transgenic events expressing dsRNAi constructs of CP and AC2 genes of Cotton leaf curl virus. ICRISAT has developed world's first genetically modified groundnut resistant to peanut clump virus. Efforts were initiated by IARI in collaboration with Directorate of Groundnut Research at Junagarh for development of resistance against emerging viral pathogens groundnut bud necrosis virus (GBNV) and tobacco streak virus (TSV).

Number of Indian scientists are engaged in developing virus resistant transgenic plants through expression of transgenes that are detrimental to invading viruses or that strengthen the ability of plants to defend themselves. Last decade has witnessed consolidated efforts by Indian scientists of different organizations and universities in development of transgenic resistance against viruses. During this journey, different groups in the country developed various transgenic events and they were licensed to seed industry for large scale testing and later being utilizing under breeding program after biosafety assessments. In this chapter, the research and developmental efforts that have been addressed during the last two and half decades in India on the transgenic resistance against several important viral diseases of crops are summarized.

31.2 Tomato Leaf Curl and Bud Necrosis Resistance

Tomato is one of the important vegetable crop in India with an annual prodution of 7.2 million tons from 4.58 lakh hectare area. Tomato appears to be an ideal host for viruses being naturally infected by diverse groups of viruses. The most devastating viruses affecting tomato are those with the generic group name tomato leaf curl viruses (ToLCVs) causing tomato leaf curl disease (ToLCD) (Varma and Malathi 2003), which is known in India since 1948 (Vasudeva and Samraj 1948). Bud necrosis disease in tomato is another most important disease in India caused by a virus species, *Groundnut bud necrosis virus* (GBNV) of the genus *Tospovirus* family *Bunyaviridae*. The virus has one of the largest host ranges of most of the plant viruses that includes over 900 plant species and approximately 10 thrips species. It is fast spreading and emerging as a serious pathogen of tomato in India (Umamaheswaran et al. 2003). CMV is one of the most economically important plant viruses in the world and has one of the largest host ranges of any plant virus, infecting plants in approximately 1000 species. Serious CMV infections have occurred in tomato causing important yield losses by reducing fruit production and quality.

Tomato leaf curl disease is known to be caused by large number of begomoviruses, which share only 70–90% homology in their genomes. In India alone, fifteen distinct begomovirus have been reported to cause leaf curl disease in tomato. The effectiveness of replicase gene for developing resistance was illustrated by the transgenic expression in antisense orientation against different begomoviruses. In view of obtaining stable resistance against the ToLCD, antisense gene strategy was adopted. Two overlapping strategies have been used to generate novel resistance against begomoviruses in transgenic plants. The first is RNA-mediated resistance involving blocking the expression of the replicase (*rep*) gene by an antisense gene construct, and the second involves post-transcriptional gene silencing (PTGS) using a shortsense RNA fragment based on homology dependent gene silencing. Tomato transgenics resistant to ToLCD using replicase (*rep*) gene sequences of the tomato leaf curl New Delhi virus (ToLCNDV) in antisense orientation, *via* Agrobacterium-mediated transformation were developed at IARI, New Delhi. A binary vector carrying the antisense *rep* gene (untranslatable full length sequence 1,086 bp) along with the *npt* II gene was used for transformation. This is demonstration of RNA-mediated silencing, since plants carry the untranslatable antisense *rep* gene, and have no detectable protein expression. Progeny analysis of these plants showed a classical Mendelian pattern of inheritance. It is important to note that two of the transgenic lines with a single transgene insertion have shown more than 80 % resistance compared to the non-transformed control plants. These were selfed to produce progeny for resistance evaluation at the T2 stage, which followed the same pattern of resistance as the T1 stage (Fig. 31.1) (Praveen et al. 2004; http://www.isb.vt.edu/news/2006/news06.jun.htm#jun0603). Toxicity and allergenicity assessment of transgenic tomato was undertaken in Balb/c mice. Ovalbumin was used to develop the allergic mice model and hypersensitive patients' sera were used to analyze tomato extract for IgE binding. Genetically engineered tomato with *rep* gene was found to be safe with respect to toxicity and allergenicity (Singh et al. 2009).

Subsequently, as the full genome sequences have started emerging in GenBank, analysis of the *replicase* gene from different variants of tomato leaf curl viruses was carried out and a conserved core of 318 nucleotides at the 3′ end of the gene encoding for Motif III, playing a vital role in viral replication was identified. Later strategies for cross inhibition of ToLCVs replication by siRNAs, targeted to this conserved region of *rep* gene were adopted. The multiple siRNAs have been used to target the *rep* gene, including a small overlapping AC4 gene essential for pathogenicity and RNAi suppression. These strategies imply that ToLCV *rep*-driven RNAi, targeting AC4 and conserved viral sequences, provides a promising approach to suppress a wide spectrum ToLCV infection in the tomato (Ramesh et al. 2007).

For improving the silencing efficacy, Hairpin (hp) RNA expression cassettes carrying the gene sequences of micro RNA were typically constructed. This system allows simple insertion of 21-nt target gene sequences into microRNA backbone, to facilitate the processing of microRNA hpRNA by the endogenous machinery of host, thereby producing artificial microRNA carrying the sequence of target gene(s). The functionality of artificial miRNA mediated viral gene silencing was demonstrated against Tomato leaf curl New Delhi virus (Koundal et al. 2010).

Fig. 31.1 (**a**) Raising nursery of tomato transgenic lines carrying replicase gene sequences from Tomato leaf curl virus. (**b**) Transgenic lines maintained at phytotron facility (*IARI*). (**c**) Resistance evaluation through viruliferous whiteflies; transgenic lines did not develop leaf curl symptoms

For developing dual resistance, leaf curl and bud necrosis disease of tomato were selected. Since no single strategy is available for the management of these viruses individually as well as in combination with each other. It is important to stack the candidate genes from individual viruses. The conserved core of *rep* gene encoding replicase protein (AC1) of ToLCV along with, 830 bp ORF of NP gene of GBNV was deployed as chimeric construct. The chimeric transgenes carrying the conserved viral sequences were employed to induce post-transcriptional gene silencing in the genetically engineered plants, which will be effective for development of multiple virus resistance having broader coverage. Transgenic developed using chimeric constructs showed 65–85% resistance against ToLCV and GBNV (dual infection) and they were licensed to private sector.

31.3 Cotton Leaf Curl Resistance

Cotton leaf curl disease (CLCuD) caused by Cotton leaf curl virus (CLCuV) is a serious threat to cultivation of cotton (*Gossypium* sp.) and several species in the Malvaceae family. The virus affects cotton production by making up to 80% loss in North India and Pakistan (Varma and Malathi 2003; Mansoor et al. 2003; Sattar et al. 2013). CLCuV is a begomovirus of *Geminiviridae* family. This virus has

DNA-A and β DNA, a satellite molecule which acts as the pathogenicity factor or suppressor of RNAi. Genetic engineering for resistance against CLCuD was attempted using antisense movement protein gene (AV2), through transformation of an Indian variety (F846). The transgene AV2 (350 bp) was cloned in antisense orientation in binary vector along with the *npt*II gene. The transformed plants were verified for transgene presence using PCR and Southern hybridization (Sanjaya et al. 2005). In another study, the development of CLCuV resistant transgenic cotton plants with antisense coat protein gene (AV1) was reported by Amudha et al. (2011). Recently, RNAi-mediated approach was followed to develop resistance against CLCuV. Intron hairpin RNA (ihp) capable of expressing dsRNA was developed from intergenic region (IR) of Cotton leaf curl Rajasthan virus (CLCuRV). The construct was transformed in cotton (*G. hirsutum* cv. Narasimha) which resulted in nine independent lines. PCR and Southern hybridization showed transgene presence while inoculation with viruliferous whiteflies, showed high degree of resistance, which was observed even after 90 days post inoculation. This study suggests that RNAi is an effective strategy to combat the CLCuD infection in cotton (Khatoon et al. 2016).

31.4 Papaya Ringspot Resistance

Efforts to develop transgenic papaya for Papaya ringspot virus resistance in India were initiated by Indian Agricultural Research Institute, New Delhi; Central Institute for Subtropical Horticulture, Lucknow; Indian Institute for Horticultural Research, Bengaluru and Tamil Nadu Agriculture University, Coimbatore. Transgenic plants of papaya cultivar Pusa Delicious were developed using shoot tip as explant for genetic transformation. A pyramided gene construct consisting truncated coat protein gene of PRSV and truncated rep gene of Papaya leaf curl virus was transformed in papaya (Chandra et al. 2010).

31.5 Rice Tungro Resistance

Rice tungro disease (RTD) is a serious constraint to rice production in South and South East Asia. In India, rice tungro disease occurs sporadically but causes significant grain yield loss. RTD is a complex disease caused by two unrelated viruses, viz., Rice tungro bacilliform virus (RTBV) and Rice tungro spherical virus (RTSV). The viral complex is transmitted between plants by vector green leafhoppers, *Nephotettix virescens* and *N. nigropictus*. To develop RTBV resistance through transgenic approach, group at Delhi University used CP gene of RTBV. This gene was expressed in rice cultivar Pusa Basmati-1 (PB-1, indica). The progenies representing three independent transformation events were challenged with Indian isolates of RTBV using insect vector. As compared to control non-transgenic plants, two independent transgenic lines showed significantly low levels of RTBV DNA, especially towards next stages of infection and a concomitant reduction of tungro symptoms (Ganesan et al. 2009). The same group attempted RNAi based approach also to develop resistance against RTBV. In this case, they developed transgenic plants expressing ORF-IV of RTBV, both in sense as well as in antisense orientation. The transgenic lines of PB-1 were evaluated for transgene integration and presence of specific dsRNAs. In the two transgenic lines expressing dsRNAs, different resistance responses were observed against RTBV. One of the transgenic line showed initial rapid grownup of RTBV levels after inoculation, however, it recovered after with approximately 50-fold lower level of RTBV as compared to non-transgenic plants. The other trangenic line showed gradual increase of RTBV after inoculation (Tyagi et al. 2008). Further, this RNAi gene construct was later transferred to two high-yielding tungro-susceptible indica rice cultivars (IET4094 and IET4786) from the transgenic rice line PB-1 using back cross breeding till the BC2F3 stage. On challenge inoculation, the progenies (BC2F1) showed mild symptoms of tungro, in contrast to severe symptoms showed by the recurrent parents (Roy et al. 2011). In TNAU, same transgenic line of PB-1 was backcrossed with ASD 16, a popular variety in Southern India, using transgene marker-assisted selection. Variable levels of resistance were observed in BC3F4 and BC3F5 generation rice plants, which was manifested by an average of twofold amelioration in height reduction compared to parent ASD16. Other improved traits in backcrossed rice lines were lesser reduction in chlorophyll content, and 100- to 10,000-fold reduction in the titers of RTBV, but no reduction of RTSV titers (Valarmathi et al. 2016). Same group made efforts to develop RTSV resistance also using untranslatable regions of coat protein (in sense orientation) and antisense of *rep* gene. These constructs were genetically transformed in PB-1, which showed delayed accumulation of RTSV in few selected transgenic lines (Verma et al. 2012).

The research group at Indian Institute of Rice Research (IIRR), Hyderabad also made recent efforts to develop RTSV resistance using transgenic approach. They have used antisense truncated CP gene of RTSV to boost the host resistance. The transformed Taipei 309 plants showed high level of resistance against virus (Fig. 31.2), which was manifested in terms of virus symptoms development, plants growth, and virus accumulation (Mangrauthia et al. 2015; Malathi et al. 2015).

Other alternate strategies also have been deployed to achieve tungro resistance. A novel lectin- ASAL from garlic leaves was expressed under the control of phloem-specific promoters. *In-planta* bioassay studies revealed that transgenic ASAL adversely affect survival, growth and population of green leaf hopper (GLH).GLH resistant TN (1) plants were further evaluated for the incidence of tungro disease. The study showed first report of such GLH mediated resistance to infection by RTBV/RTSV in ASAL expressing transgenic rice plant (Saha et al. 2006).

Overall, although many efforts to develop tungro virus resistance have been made, consistent results have not been obtained. Effective and sustainable tungro management requires a combination of resistant varieties and adoption of synchronous planting. The cultivars already available are unlikely to provide a complete or lasting solution and there is a need for continuing research to identify and exploit new sources of resistance and to deploy the resistant varieties.

Fig. 31.2 (**a**) Development of transgenic rice plants containing antisense truncated CP gene of RTSV. (**b**) Evaluation of transgenic rice lines (25 days post inoculation) for tungro virus resistance. *1* TN1, *2* Taipei-309 (non-transformed control), *3* Vikramarya (GLH resistant) *4*, *5* Transgenic lines. (**c**) Comparison of rice tungro virus inoculated TN1 (*left*) and homozygous RNAi line (*right*) after 8 weeks of inoculation

31.6 Groundnut Bud and Stem Necrosis Resistance

Groundnut bud and stem necrosis is caused by two different viruses, GBNV and tobacco streak virus (TSV, genus *Ilarvirus*; family *Bromoviridae*), respectively. GBNV causes bud necrosis disease in groundnut (GBND) for which no durable resistance source is available in the existing germplasm. Recently, Rao et al. (2013) transformed N gene encoding for the nucleocapsid protein in peanut var. JL24. They developed over 200 transgenic lines, which were evaluated by PCR, Southern hybridization, RT-PCR and western blot analysis. The transgenic plants were phenotyped for GBNV resistance, which showed varying levels of disease incidence and intensity. In another such effort to develop transgenic groundnut for virus resistance, Mehta et al. 2013 transformed antisense coat protein gene for tobacco streak virus (TSV-CP). The transgenic plants showed high level of resistance, which was evident with traces, or no systemic accumulation of the TSV. Research group led by Dr. R.K Jain at IARI, New Delhi had also transformed CP gene of TSV in sense and antisense orientation in cultivar JL24. These transgenic lines were evaluated for presence of transgene and virus resistance (Bag et al. 2007).

31.7 Soybean Yellow Mosaic Resistance

Yellow Mosaic Disease (YMD) is a major constraint in improving the productivity of the legumes in general and soybean in particular. The economic loss caused by the infection of yellow mosaic virus in legume crops including soybean accounts for 300 million US $ (Varma and Malathi 2003). Virus species causing the disease in soybean and other grain legumes are diverse despite no apparent differences in the visible symptoms. In Northern and Western parts of India, mungbean yellow mosaic India virus (MYMIV) and mungbean yellow mosaic virus (MYMV) cause the disease in soybean, respectively (Usharani et al. 2004; Ramesh et al. 2016). In central India, where soybean is a major *Kharif* crop, incidence of MYMIV was observed (Ramesh et al. 2013). In soybean, concerted efforts to map the YMD resistant loci through molecular markers are lacking and non-availability of linked SSR markers severely hampers development of resistant cultivars through breeding approaches (Talukdar et al. 2013). Though, exploitation of wild gene pool such as *Glycine soja* in conferring YMV resistance in cultivated species is an alternative approach but it is a challenging task due to linkage drag issues (Ram et al. 1984). Hence, genetic engineering approaches to incorporate virus resistance trait in soybean is considered as a viable option to counter YMD.

Transient gene silencing assay targeting the promoter region of the MYMV-*vig* (Pooggin et al. 2003) and demonstration of recovery from MYMV infection using RNA silencing based gene construct strategies exemplifies the effectiveness of the technique in engineering resistance against legume infecting begomovirus. Targeted engineering of yellow mosaic virus immunity through antisense RNA has demonstrated the utility of homology based RNA silencing in conferring resistance against the Begomovirus infecting legumes (Haq et al. 2010). Similarly antisense approach has been deployed to target replication initiator protein gene of MYMIV in soybean (Singh et al. 2013). In light of role of viral genome derived suppressors of RNA silencing (VSR), any durable engineered resistance in soybean is feasible with concomitant silencing of VSRs. Ongoing research efforts at ICAR-Indian Institute of Soybean Research, Indore revealed that suppression of MYMIV derived AC2 gene expression was an effective strategy in controlling MYMIV virus accumulation in transient gene silencing assay in soybean (Chouhan et al. unpublished). Further, generation of siRNAs targeting AC2 ORF of MYMIV was correlated with reduction in viral titre quantified by real-time PCR based assay compared to siRNAs targeting movement protein (MP) of MYMIV (Ramesh et al. unpublished). Thus it has been proven that selection of VSR as a target region for engineering virus resistance in soybean is a practical approach.

31.8 Citrus Tristeza Resistance

Citrus tristeza virus (CTV), a closterovirus is one of the important causal agents causing decline disease in citrus worldwide. Over the last ten decades, CTV has killed more than one million citrus trees in India. Citrus are cultivated in diverse ecological conditions in India and affected by several CTV variants/strains causing diverse disease syndromes. Extensive genetic diversity in CTV in citrus growing regions of the world including India has been reported time to time. Genetic engineering approach to modify citrus plant using varieties of CTV derived genes has been used to develop transgenic citrus. Citrus transformation has been successfully performed in many citrus species and hybrids using CP, RdRp, replicase, p23 (suppressor) genes of CTV. Various kinds of explants, epicotyl segment, embryogenic cells, nodal, internodal stem segments, callus and protoplast were used in citrus transformation. The most favored responsive explant is invariably the epicotyl of in vitro grown citrus seedling. However, procedure for development of transgenic citrus through plant transformation techniques is slow.

In India, transformation of citrus to develop plant resistant to CTV is offing using CTV derived genes. In this respect, efficient protocols for regeneration of Indian Kagzilime (*Citrus aurantifolia*) have been developed. The epicotyl segments, 0.75–1.0 cm in length, of in vitro grown 21–28 days old Kagzilime seedlings were found to ideal explants for regeneration (Biswas et al. 2007; Tarafdar et al. 2009).

The conserved sequence of CP and suppressor gene was identified analyzing several CTV isolates from different citrus growing regions of India. Several antisense (RNAi) and sense CP gene constructs from CTV belonging different genogroups occurring in India were developed in pBinAR (Tarafdar et al. 2007; Biswas et al. 2013). Antisense CP gene constructs of mild (K5 and K13) and severe CTV isolates (K9 and K10) of the Darjeeling hills were developed (Tarafdar et al. 2007). Further, one antisense gene construct targeting conserved region of suppressor gene in pBinAR (Saha et al. 2014) and one hairpin (ihpRNAi) gene construct targeting CP in pCAMBIA were developed (Biswas et al. 2016). The constructs were mobilized to *Agrobacterium* cell EHA105. Agrobacterium-mediated transformation protocol was developed using epicotyl explants of in vitro grown 21–28 days old Kagzilime seedlings using one antisense and one sense CP gene constructs (Tarafdar et al. 2009, Biswas et al. 2013, 2016, 2017). For selecting putative transformants, a multiplex PCR-based screening method was developed using primers targeting *tetA* (to detect T-DNA) of pBinAR and CTV-CP gene. Amplification of both *tetA* and CP gene confirmed *Agrobacterium* contamination, whereas, amplification of CP gene only confirmed putative transformants (Tarafdar et al. 2009). The regenerated plants showed a high degree of false transformation. In one experiment of agrobacterium-mediated transformation in Kagzilime, 482 explant were co-cultivated wit antisense construct (K9); of which 36 plants were regenerated in Kanamycin medium. They were tested with multiplex-PCR and majorities of them were found to be false transformants, and only three plants were putative transformants (Tarafdar et al. 2009). In another experiment, several explants were transformed with antisense suppressor gene construct using epicotyl explants, but the regeneration efficiency of transformed plant was as low as 1.38% at 2.0 mg/l BAP in MS medium, whereas regeneration efficiency up to 84% was found in control plant. Maximum number of five micro shoots/explant was regenerated at 2.0 mg/l BAP (Biswas and Ghosh 2015, 2016). However, in the rooting medium, the performance was very poor leading to death of the plant without root formation. To overcome this problem, a new method using explants containing crown region segment of 1.5 cm (junction of root and shoot in 1:1 ratio) were used. In this method the regeneration efficiency is also low (1.06%), the growth of regenerated plant in rooting medium is promising with root formation (Biswas and Ghosh 2015) but to get better success, standardization of this method is needed.

31.9 Potato Virus Resistance

Viral infections in clonally propagated crops such as potato are a serious threat as vertical transmission of viruses across the generations occurs effortlessly (Vasudeva and Lal 1945). Potato virus Y (PVY) and its strains, potato virus X (PVX) and potato leaf roll virus (PLRV) are the major potato viruses that cause serious impediment to the potato seed industry and trade in India. Central Potato Research Institute (CPRI), Shimla in collaboration with Bhabha Atomic Research Centre, Mumbai successfully demonstrated pathogen derived resistance against PVY in tobacco by expressing coat protein gene derived from Indian isolate of PVY (Ghosh et al. 2002). This study paved way for the development of virus resistant transgenic potato. Genetic incorporation of CP gene derived from PVY has also been accomplished in tobacco and potato (Fig. 31.3) at IARI, New Delhi. The collaborative

Fig. 31.3 Tubers of transgenic lines of cv. Kufri Giriraj transformed with the coat protein gene of PVY. Line number and Southern analysis (S) with the transgene copy number are indicated on the picture (Courtesy, Bikash Mandal, IARI)

efforts between IARI and CPCRI has resulted in development of virus resistant transgenics against PVY and PLRV in potato cultivars Kufri Giriraj and Kufri Pukhraj, respectively. Among the six transgenic lines of Kufri Giriraj developed using PVY-CP gene construct, GR-PVY-3.4 was found to be highly resistant to PVY under controlled conditions. Similarly, four stable transgenic lines of Kufri Pukhraj has been developed using PLRV-CP gene construct (Mandal and Garg 2008). Later, CP gene of PVY was cloned in pBIN19 plasmid and vector was used for transformation of popular potato cultivar Kufri Pukhraj. Post transformation, putative transgenic potato lines were screened for transgene integration by performing Southern hybridization. Successful transgene integration was confirmed in few select lines of potato (Chakrabarti et al. 2016).

ToLCNDV (synonym, potato apical leaf curl virus) is another important emerging pathogen infecting potato. An RNAi construct (pBI121-PALCV-rep) to silence replicase gene encoded by ToLCNDV has been developed and transformation of potato has led to transgenic event selection in polyhouse (Pattanayak et al. 2010). Transgenic resistance was deployed by expressing AC1 gene of ToLCNDV-Potato in potato cultivars Kufri Pukhraj and Kufri Badshah. Transformed potato lines were analyzed for transgenic event characterization and two transgenic events, KPLC2–53 and GTLC2–127, were identified using affinity-based genome walking method. Locus-Finding (LF) PCR technique determined that the genomic sequence flanking right border of event KPLC2–53 shared high homology with *Solanum lycopersicum* chromosome 7 and that of event GTLC2–127 was found in chromosome no. 11. Furthermore, computational analysis showed that the site of transgene integration has no adverse implication on function of potato genome (ICAR-CPRI Annual Report 2016- Eds. Chakrabarti et al. 2016).

31.10 Cassava Mosaic Resistance

Cassava mosaic disease (CMD) is a serious threat to Cassava cultivation (Edison et al. 2006) and the disease is attributed to Indian cassava mosaic virus and Sri Lankan cassava mosaic virus in India (Patil et al. 2005). In the absence of genotypes, or donors conferring resistance to Indian cassava mosaic virus (ICMV), development of transgenic resistance has been initiated under ICAR-Network project on transgenics (Makesh and Winter 2010). The concept of PDR, was utilized by developing gene constructs in plant transformation compatible binary vectors using ICMV derived *replicase* (AC1) gene. Besides full length AC1 gene of ICMV (pBin19, pBinAR and pBI121), viral derived partial genes have also been used in developing hairpin RNA gene constructs to target AC1 gene of ICMV, and transcriptional activation protein (TrAP) of SLCMV (pIng IV2). A major impediment in cassava transgenic development is non-existence of regeneration and transformation protocol for Indian cultivars. Hence, protocol has been standardized for production of friable embryogenic callus (FEC) in cultivars such as H226, H165 among others (Beena et al. 2016). FEC initials have been obtained in cultivars H165 and H226 in 3–5 months and it was found that complete plantlets have been regenerated from these initials when grown in regeneration medium. Genetic transformation of FEC has been standardized albeit using reporter/marker gene constructs [Aglc/pOYE 153 construct (GUS gene)], and also hpRNA gene construct (SLCMV-IC-Syn). However, putative transformants obtained were positive for *nptII* gene but negative for virus derived gene. Hence, stringent selection and more efficient selectable marker system are imperative for development of transgenic cassava. Transgenic cassava plants resistant to ICMV have been developed using *rep* gene construct and their resistance status has been evaluated using ICMV infectious clones. In the meanwhile, development of putative cassava transgenic lines resistant to SriLankan cassava mosaic virus (SLCMV) using hpRNA gene constructs has been accomplished (Anuradha et al. 2016). The report states that conserved sequences of SLCMV derived rep gene was used as an sense and antisense arms of hpRNA to achieve transgenic resistance (Anuradha et al. 2016).

31.11 Banana Bunchy Top Resistance

Virus resistant transgenic banana conferring Banana bunchy top virus (BBTV) resistance has been developed in Australia, Hawaii and India following pathogen derived resistance approach. The choice of virus gene for developing gene constructs is restricted to BBTV derived coat protein and *replicase* genes. BBTV derived *rep* gene sequence and its *cis*-regulatory elements were genetically manipulated to express mutated rep protein, antisense rep transcript, inverted repeats, partial *rep* gene to develop BBTV resistance (Borth et al. 2011). However, no more than 13 % of the transgenic plants developed using these gene constructs displayed resistance reaction following banana bunchy top disease (BBTD) bioassays (Borth et al. 2011). Prior to this transgenic ‘Cavendish’ has been developed with potential resistance to Banana bract mosaic virus (BBrMV) at Queensland University of Technology. The transgenic development approach followed expression of coat protein gene of BBrMV under the control of maize polyubiquitin promoter using microprojectile bombardment (Dale and Harding 2003). Virus resistant transgenic banana and hill banana genotypes working on the principle of hairpin RNA mediated suppression of replication initiator protein (rep) gene of BBTV have been developed by Shekhawat et al. (2012) and Elayabalan et al. (2013) respectively. Moreover, ICAR-National Research Centre for Banana (ICAR-NRCB) has been involved in development of putative hill banana transgenics resistant to BBTV employing CP gene mediated and RNAi based replicase gene silencing approaches (Selvarajan et al. Unpublished results).

31.12 Chilli Leaf Curl Resistance

Chilli (*Capsicum annuum* L) is an economically important crop for its versatile use as vegetable, spices and as a source of colouring agent. It is also susceptible to many pathogens including begomoviruses. In order to develop transgenic resistance against begomovirus, *Agrobacterium* mediated transformation of chilli cultivar Pusa Jwala with βC1 gene of betasatellite associated with chilli leaf curl Joydebpur virus was demonstrated (Kumar et al. 2012). The study also reports auxin free, improved regeneration protocol for Indian red pepper cultivars. Eventhough, the resistance status of transgenics was not reported, the study is a major step towards developing virus resistant chilli cultivars. Furthermore, the transgenic genotypes have a greater potential in elucidating host-pathogen interactions especially during the expression of βC1 ORF.

31.13 Cucurbit Virus Resistance

IARI and IIHR have been involved in development of transgenic resistance for viruses infecting cucumber especially, CMV and watermelon bud necrosis virus (WBNV). The coat protein mediated resistance against an Indian isolate of CMV subgroup IB was demonstrated in *N. benthamiana* (Srivastava and Raj 2008). At IARI, transgenic cucumber plant was developed using CP of CMV, which showed high level resistance to the challenge inoculated virus under experimental condition (Fig. 31.4). WBNV is a serious threat to the cultivation of watermelon in India and elsewhere. Hence, in order to develop transgenic resistance to WBNV, full-length

Fig. 31.4 Transgenic cucumber with the coat protein gene of cucumber mosaic virus showing resistance following challenge inoculation of the virus (Courtesy, Bikash Mandal, IARI)

and core nucleocapsid gene of Indian isolate of WBNV was cloned in binary vector pBI121. However, transformation efficiency of watermelon has been found to be very low hence, various combination of tissue culture media constituents have been standardized to increase transformation efficiency (Kumar and Mandal 2012).

31.14 Proof-of-Concept Studies in Experimental Plant

As a proof-of-concept *Nicotiana benthamiana* and *Nicotiana tabacum* have been employed to study the feasibility of suppressing viral gene expression exploiting RNAi mediated gene silencing mechanism in numerous instances (Makeshkumar et al. 2002; Gogoi et al. 2010; Sharma et al. 2015; Makeshkumar et al. unpublished). Transgenic tobacco lines were developed to confer resistance against PVY using coat protein gene. Despite its worldwide economic importance and severity of diseases in potato, tobacco, tomato, chill and brinjal in India, a little has been accomplished to tackle the virus employing recombinant DNA technology. Plant binary vectors harbouring CP gene of PVY in sense and antisense orientation were introduced in to *N. tabacum* cv. White Burley and Xanthi and *N. benthamiana* through *Agrobacterium*. Among putative transformants, 37 plants were found to be positive as identified by nucleic acid spot hybridization. Among the 17 transgenic lines analyzed for PVY resistance, it was found that 5 lines were completely resistant to the disease (Makeshkumar et al. 2002). Resistance against physalis mottle virus in *N. tabaccum* has been accomplished by a research group at IISc, Bengaluru (Kumar et al. 1999).

Chilli-infecting begomoviruses pose serious threat to chilli cultivation in Indian sub-continent. RNAi based broad-spectrum resistance against multiple and diverse chilli-infecting begomoviruses *viz*., Chilli leaf curl virus- Pakistan isolate Varanasi, Tomato leaf curl New Delhi virus-chilliisolate, Chilli leaf curl Vellanad virus was successfully demonstrated in *N. benthamiana*. The target sequences for generation of hpRNA gene constructs were conserved AC1/AC2 genes and betasatellite encoded ORF βC1. Transgenic *N. benthamiana* lines have been shown to accumulate siRNAs specific to target viral genes and concomitant reduction in virus titre (up to 90% reduction compared to control lines) was also proven indicating RNAi based antiviral resistance. In the context of evolving and emerging begomovirus complex causing severe damage to chilli cultivation in almost all the regions of India the study underscores the importance and applicability of RNAi mediated broad spectrum resistance in resisting chilli leaf curl (Sharma et al. 2015).

In order to evaluate and compare hpRNA based gene silencing against coding and non-coding regions of SLCMV transgenic *N. benthamiana* plants were generated targeting DNA-A intergenic (IG) region and AC2 ORF (Gogoi et al. 2010). Transgenic *N.benthamiana* have been developed using hpRNA gene constructs targeting SLCMV rep gene, transcriptional activator protein and using a SLC/IC_ Syn gene constructs. Molecular analysis of transgene (hp constructs) in *N. benthamiana* through PCR, RT-PCR and GUS assay were done (Makeshkumar et al. unpublished results).

Elephant foot yam (*Amorphophallus paeoniifolius*) is tropical tuber crop of economic and nutritional importance (Ravi et al. 2009). Among the various diseases afflicting the crop, mosaic disease caused due to *Dasheen mosaic virus* (DsMV) of *Potyviridae* is of economic importance (Nehalkhan et al. 2006). In order to develop DsMV resistant *A. paeoniifolius*, hairpin RNA gene constructs targeting conserved CP gene of DsMV have been developed. The efficiency of gene constructs in conferring resistance to DsMV has been ascertained in model host *N. benthamiana*. Further, development of transgenic resistance in *A. paeoniifolius* is under progress and is hampered due to the absence of regeneration and transformation protocol (Kamala and Makeshkumar 2013).

31.15 Concluding Remarks

In Indian context, transgenic development for conferring virus resistance has gained momentum during last decade and many research institutes are actively involved in this arena of research. Despite the resounding success in crops like tomato, rice and cotton, development and commercialization of virus resistant transgenic plants in other crops has been lagging behind. The main reason is absence of regeneration and transformation protocol due to their recalcitrant nature in many species of crops like legumes, and tuber crops. Hence, it is imperative to devise suitable and efficient regeneration and transformation techniques to reap the benefits of transgenic virus resistance. Genomic hypervariability of viruses is another major factor that hinders successful transgenic resistance in some crops like cotton, pepper and banana. Hence, identification of conserved viral genomic regions employing Next Generation Sequencing (NGS) technologies and utilizing them for RNA silencing would counter emergence and resurgence of resistance breaking strains of viruses. Finally it is anticipated that, next generation gene modification techniques like genome editing tools would play greater role in developing virus resistance in plants. However, in order to utilize such tools it is also important to explore virus-host interactions and identify host susceptibility factors that are genetic elements suitable for clustered regularly interspaced short palindromic repeats (CRISPR) and CRISPR associated (Cas) mediated genome editing. Exploration of virus resistance conferring genes from closely related species but are not amenable for hybridization and breeding programme is also an impending area of research so as to introgress those R-genes in to cultivated species employing genetic engineering approaches.

References

Amudha J, Balasubramani G, Malathi VG, Monga D, Kranthi KR (2011) Cotton leaf curl virus resistance transgenic with the antisense coat protein gene (AV1). Curr Sci 101:300–307

Anuradha A, Palanisami P, Kumar KK, Balasubramanian P (2016) Development of putative transgenic lines of cassava variety H-226. Afr J Biotechnol 15(13):497–504

Bag S, Singh RS, Jain RK (2007) Agrobacterium mediated transformation of groundnut with coat protein gene of Tobacco streak virus. Indian J Virol 18:65–69

Beena MR, Winter S, Makeshkumar T (2016) Influence of age of explants and genotype on somatic embryogenesis in African and Indian cassava cultivars. J Root Crops 40(2):21–27

Biswas KK, Ghosh DK (2015) Current status of Citrus tristeza *virus* in India: incidence, geographical distribution, genomic diversity and its management through biotechnological tools. National Seminar on Brain Storming Dialogue on Viral and Greening Disease of Citrus: challenges and way forward, ICAR-CCRI, Nagpur, 28 July 2015, pp 50–69

Biswas KK, Sumita Kumari, Avijit Tarafdar (2007) Spatial distribution and characterization of mild and severe strains from diversified Citrus tristeza virus complex for development of transgenic resistant strategy in citrus. In: Proceeding of national symposium on microbial diversity and plant health, Indian Mycological Society and Department of Plant Pathology, Mohanpur, pp 6–7

Biswas KK, Tarafdar A, Sumita Divedi, Sharma SK, Singh JK, Khatun D, Kumar A, Agarwal K, Chandra V, Godara S, Bashir N, Gargi Sen, Anupam Varma (2013) Citrus tristeza virus: diagnostics, distribution, genomics, genetic diversity and biotechnological management in Northeastern region of India. In: National symposium on "Recent Trends in Plant and Microbial Research" in the University of North Bengal, Siliguri, West Bengal during 22–23 March 2013, pp 4–5

Biswas KK, Palchoudhuri S, Tarafdar A, Sharma SK, Singh JK, Chandra V, Godara S, Dwivedi S (2016) Current status of Citrus tristeza virus infecting citrus in India: incidence, spatial distribution and diversity, genomics diversity and biotechnological approach of management. In: ISMPP-37th Annual conference and national Symposia on "Food security through Plant Health Protection", RAU, Pusa Samastipur, April 4–6, 2016, pp 7–8

Biswas KK, Pal Choudhuri S, Shruti Godara (2017) Mandarin decline in Northeast India caused by Citrus tristeza virus: management by conventional and biotechnological approaches. In: National symposium on "Diagnosis and management of plant diseases: Integrated approaches and recent trends organized by IPS, New Delhi as ICAR-Research complex-NEH region, 9–11 January 2017, pp 34–35

Borth W, Perez E, Cheah K, Chen Y, Xie WS, Gaskill D, Khalil S, Sether D, Melzer M, Wang M, Manshardt R, Gonsalves D, Hu JS (2011) Transgenic banana plants resistant tobanana bunchy top virus infection. Acta Hortic 897:449–457

Chakrabarti SK, Govindakrishnan PM, Pandey NK, Singh B, Tiwari J, Singh DK (eds) (2016) ICAR-CPRI annual report

Chandra R, Mishra M, Pati R, Agarwal S, Jain RK (2010) Shoot tip transformation in papaya (*Carica papaya L.*) Acta Hortic 851:219–226

Dale J, Harding R (2003) Strategies for the generation of virus resistant bananas. In: Atkinson H, Dale J, Harding R (eds) Genetic transformation strategies to address the major constraints to banana and plantain production in Africa. INIBAP, Montpellier, p 130

Edison S, Anantharaman M, Srinivas T (2006) Status of cassava in India an overall view. CTCRI Tech Bull Ser 46:11–17

Elayabalan S, Kalaiponmani K, Subramanian S, Selvarajan R, Radhs P, Ramalatha M, Kumar KK, Balasubramanian P (2013) Development of agrobacterium-mediated transformation of highly valued hill banana cultivar Virupakshi (AAB) for resistance to BBTV disease. World J Microbiol Biotechnol. 29:589-596

Ganesan U, Suri SS, Rajasubramaniam S, Rajam MV, Dasgupta I (2009) Transgenic expression of coat protein gene of Rice tungro bacilliform virus in rice reduces the accumulation of viral DNA in inoculated plants. Virus Genes 39:113–119

Ghosh SB, Nagi LHS, Ganapathi TR, Paul Khurana SM, Bapat VA (2002) Cloning and sequencing of Potato virus Y coat protein gene from an Indian isolate and development of transgenic tobacco for PVY resistance. Curr Sci 82:855–859

Ghosh SB, Nagi LHS, Ganapathi TR, Paul Khurana SM, Bapat VA (2006) Development of coat protein gene mediated resistance against potato viruses Y(PVY) in Potato cultivar Kufri Jyoti. Physiol Mol Biol Plants 12:133–138

Gogoi N, Srivastava D, Dasgupta I (2010) Transgenic resistance against Sri Lankan Cassava Mosaic Virus in N. Benthamiana. In: Proceedings of "Conference on Whitefly and Thrips

Transmitted Viruses” at University of Delhi-South Campus, New Delhi. Indian J Virol 21(Suppl 1):A59–A81

Haq QMI, Arif A, Malathi VG (2010) Engineering resistance against Mung bean yellow mosaic India virus using antisense RNA. Indian J Virol 21(1):82–85

Kamala S, Makeshkumar T (2013) Transgenic elephant foot yam (amorphophalluspaeoniifolius (Dennst.) Nicolson) expressing β-glucuronidase reporter gene. J Root Crops 39(2):148–153

Khatoon S, Kumar A, Sarin NB, Khan JA (2016) RNAi-mediated resistance against Cotton leaf curl disease in elite Indian cotton (Gossypiumhirsutum) cultivar Narasimha. Virus Genes 52:530–537

Koundal V, Vinutha T, Haq QMR, Praveen S (2010) Modulation of plant development and MYB down regulation: both during in planta expression of miR159a and in natural ToLCV infection. J Plant Biochem Biotechnol 19(2):171–175

Kumar R, Mandal B (2012) Transgenic constructs and transformation of watermelon cv. Sugar baby for developing transgenic resistance against Watermelon bud necrosis virus. VIRICON, India. Indian J Virol 24(1):99–149. In: Mangrauthia SK, Malathi P (eds) Abstracts of papers presented during XXI National Conference of Indian Virological Society (IVS)

Kumar CTR, Manoharan M, Prasad SK, Cherian S, Umashankar M, Lakshmi Sita G, Savithri HS (1999) Engineering resistance against physalis mottle tymovirus(PhMV) by expression of the coat. protein and 3′non coding region. Curr Sci 77:1542–1548

Kumar RV, Sharma VK, Chattopadhyay B, Chakraborty S (2012) An improved plant regeneration and Agrobacterium – mediated transformation of red pepper (*Capsicum annuum* L.) Physiol Mol Biol Plants 18(4):357–364

Makesh KT, Winter S (2010) Whitefly transmitted viral diseases of tropical tuber crops and their management option through transgenic approach. In: Proceedings of “Conference on Whitefly and Thrips Transmitted Viruses” at University of Delhi-South Campus, New Delhi. Indian J Virol 21(Suppl 1):A59–A81

Malathi P, Muzammil SA, Sailaja B, Krishnaveni D, Balachandran SM, Mangrauthia SK (2015) Development of tungro virus resistant transgenic rice by RNAi driven silencing of coat protein gene of rice tungro spherical virus. Paper presented in “International Rice Symposium on “Rice Science for Global Food and Nutritional Security”, Hyderabad, November 18–20

Mandal B, Garg ID (2008). Engineering viral gene derived resistance in Indian potato cultivars against viruses (Consolidated final progress report-ICAR cess fund)

Mangrauthia SK, Malathi P, Krishnaveni D, Balachandran SM (2015) Development of RNAi-Mediated resistance to Rice Tungro Virus in genetically engineered Oryza sativa. Paper presented in “18th ADNAT convention symposium on genetic engineering of agricultural crops and livestock: Current Status and Social, Ethical and Regulatory Issues” Hyderabad, February 23–25

Mansoor S, Briddon RW, Bull SE, Bedford ID, Bashir A, Hussain M, Saeed M, Zafar Y, Malik KA, Fauquet C, Markham PG (2003) Cotton leaf curl disease is associated with multiple monopartite begomoviruses supported by single DNA β. Arch Virol 148(10):1969–1986

Mehta R, Radhakrishnan T, Kumar A, Yadav R, Dobaria JR, Thirumalaisamy PP, Jain RK, Chigurupati P (2013) Coat protein-mediated transgenic esistance of peanut (*Arachis hypogaea* L.) to peanut stem necrosis disease through Agrobacterium-mediated genetic transformation. Indian J Virol 24:205–213

Nehalkhan M, Aswathi LP, Singh PK (2006) Survey on symptomatology and assessment of yield losses due to viral diseases in elephant foot yam (AmorphophalluspaeoniifoliusBlume). J Root Crops 32:90–93

Patil BL, Rajasubramaniam S, Bagchi C, Dasgupta I (2005) Both Indian cassava mosaic virus and Sri Lankan cassava mosaic virus are found in India and exhibit high variability as assessed by PCR-RFLP. Arch Virol 150(2):389–397

Pattanayak D, Patil VU, Tiwari J, Chakrabarti SKI, Singh BP (2010) Role of biotechnology in potato improvement, December, 2010, CPRI, News Letter

Pooggin M, Shivaprasad PV, Veluthambi K, Hohn T (2003) RNAi targeting of a DNA virus in plants. Nat Biotechnol 21:131–132

Powell Abel P, Nelson RS, De B, Hoffmann N, Roger SG, Fraley RT, Beachy RN (1986) Delay of disease development in transgenic plants that express the tobacco mosaic virus coat protein gene. Science 232:738–743

Pratap D, Raj SK, Kumar S, Kumar S, Snehi SK, Gautam KK, Sharma AK (2012) Coat protein-mediated transgenic resistance in tomato against a IB subgroup Cucumber mosaic virus strain. Phytoparasitica 40:375–382

Praveen S, Dasgupta A, Varma A (2004) Phylogenetic analysis and homologies of the replicase of tomato leaf curl geminiviruses: implications for obtaining pathogen derived resistance. Virus Genes 28(1):197–201

Praveen S, Mishra AK, Dasgupta A (2005) Antisense suppression of replicase gene expression recovers tomato plants from leaf curl infection. Plant Sci 168:1011–1014

Praveen S, Mishra AK, Antony G (2006) Viral gene suppression in transgenic plants expressing chimeric transgene from *tomato leaf curl virus* and *cucumber mosaic virus*. Plant Cell Tissue Organ Cult 84:49–55

Raj SK, Singh R, Pandey SK, Singh BP (2005) Agrobacterium-mediated tomato transformation and regeneration of transgenic lines expressing Tomato leaf curl virus coat protein gene for resistance against TLCV infection. Curr Sci 88:1674–1679

Ram HH, Pushpendra SK, Verma VD (1984) New breeding lines of soybean having a gene for resistance to yellow mosaic virus from Glycine soja. Linn Sieb Zucc Indian J Agric Sci 54:1027–1029

Ramesh SV, Mishra AK, Praveen S (2007) Hairpin RNA-mediated strategies for silencing of tomato leaf curl virus AC1 and AC4 genes for effective resistance in plants. Oligonucleotides 17:251–257

Ramesh SV, Bhaskale R, Admane N, Gupta GK, Husain SM (2013) Multiply primed rolling circle amplification (MPRCA) of yellow mosaic virus genome from infected soybean in central Indian region divulges it as Mungbean yellow mosaic Indian virus-[sb] and its implications for RNAi mediated virus resistance. In: Proceedings of world soybean research conference-IX (WSRC-IX), Durban South Africa

Ramesh SV, Chouhan BS, Gupta GK, Ramteke RK, Chand S, Husain SM (2016) Molecular diversity analysis of coat protein gene encoded by legume begomoviruses and PCR assay to detect yellow mosaic viruses infecting soybean in India. Br Biotechnol J 3:1–10

Rao SC, Bhatnagar-Mathur P, Kumar PL, Reddy AS, Sharma KK (2013) Pathogen-derived resistance using a viral nucleocapsid gene confers only partial non-durable protection in peanut against peanut bud necrosis virus. Arch Virol 158:133–143

Ravi V, Ravindran CS, Suja G (2009) Growth and productivity of elephant foot yam (Amorphophalluspaeoniifolius (Dennst.) Nicolson): an overview. J Root Crops 35(2):131–142

Roy S, Banerjee A, Tarafdar J, Senapati BK, Dasgupta I (2011) Transfer of transgenes for resistance to rice tungro disease into high-yielding rice cultivars through gene-based marker-assisted selection. J Agril Sci 150:610–618

Saha P, Dasgupta I, Das S (2006) A novel approach for developing resistance in rice against phloem limited viruses by antagonizing the phloem feeding hemipteran vectors. Plant Mol Biol 62:735–752

Saha B, Palchoudhury S, Godara S, Biswas KK (2014) Development of gene construct targeting coat protein and supresor gene of Citrus tristeza virus gene and transformation of citrus plant. In: National seminar on "Advances In Plant And Microbial Research" 12–13 December 2014, Organized by Dept of Botany, University of North Bengal, Siliguri-734 013, West Bengal, p 38

Sanjaya VVS, Prasad V, Kirthi N, Maiya SP, Savithri HS, Lakshmi Sita GL (2005) Development of cotton transgenics with antisense AV2 gene for resistance against cotton leaf curl virus (CLCuD) via Agrobacterium tumefaciens. Plant Cell Tissue Organ Cult 81:55–63

Sattar MN, Kvarnheden A, Saeed M, Briddon RW (2013) Cotton leaf curl disease–an emerging threat to cotton production worldwide. J Genl Virol 94(4):695–710

Sharma VK, Basu S, Chakraborty S (2015) RNAi mediated broad-spectrum transgenic resistance in Nicotianabenthamiana to chilli-infecting begomoviruses. Plant Cell Rep 34(8):1389:1399

Shekhawat UKS, Ganapathi TR, Hadapad AB (2012) Transgenic banana plants expressing siRNAs targeted against viral replication initiation gene display high-level resistance to Banana bunchy top virus infection. J Gen Virol 93:1804–1813
Singh AK, Praveen S, Singh BP, Varma A (2009) Arora N Safety assessment of leaf curl virus resistant tomato developed using viral derived sequences. Transgenic Res 18:877–887
Singh VB, Haq QMR, Malathi VG (2013) Antisense RNA approach targeting Rep gene of mung bean yellow mosaic India virus to develop resistance in soybean. Arch Phytopathol Plant Protect 46:2191–2207
Srivastava A, Raj SK (2008) Coat protein-mediated resistance against an Indian isolate of the Cucumber mosaic virus subgroup IB in Nicotianabenthamiana. J Biosci 33:249–257
Talukdar A, Harish GD, Shivakumar M, Kumar B, Verma K, Lal SK, Sapra RL, Singh KP (2013) Genetics of Yellow Mosaic Virus (YMV) resistance in cultivated soybean [Glycine max (L) Merr.] Legum Res 36(3):263–267
Tarafdar A, Sumita Kumari, Biswas KK (2007) Development of management strategy for production of transgenic citrus resistant to Citrus tristeza closterovirus. In: Proceeding of national symposium on plant protection-technology interface organized by Association for Advancement in Plant Protection at BCKV, Mohanpur, 28–29 December 2007, pp 123–124
Tarafdar A, Sumita Kumari, Biswas KK (2009) Agrobacterium mediated transformation of *Citrus aurantifolia* Kagzi lime using coat protein gene of *Citrus tristeza virus*. In: Proceeding of 5th international symposium on plant pathology in the globalized era Indian Phytopathological society, Indian Agricultural Research Institute, New Delhi, p 390
Tyagi H, Rajasubramaniam S, Rajam MV, Dasgupta I (2008) RNA-interference in rice against Rice tungro bacilliform virus results in its decreased accumulation in inoculated rice plants. Transgenic Res 17:897–904
Umamaheswaran K, Jain RK, Bhat AI, Ahlawat YS (2003) Biological and molecular characterization of a tospovirus isolate from tomato and its relationship with other tospoviruses. Indian Phytopathol 56:168–173
Usharani KS, Surendranath B, Haq QMR, Malathi VG (2004) Yellow mosaic virus infecting soybean in northern India is distinct from the species infecting soybean in southern and western India. Curr Sci 86:845–850
Valarmathi P, Kumar G, Robin S (2016) Virus Genes. Evaluation of virus resistance and agronomic performance of rice cultivar ASD 16 after transfer of transgene against Rice tungro bacilliform virus by backcross breeding. Virus Genes 52:521–529
Varma A, Malathi VG (2003) Emerging Gemini virus problem – a serious threat to sustainable crop production. Ann Appl Biol 142:145–164
Vasudeva RS, Lal TB (1945) Studies on the virus diseases of potato in India II. Solanum virus 2 (Orton). Indian J Agric Sci 15:240–242
Vasudeva RS, Samraj J (1948) A leaf curl disease of tomato. Phytopathology 38:364–369
Verma V, Sharma S, Devi SV, Rajasubramaniam S, Dasgupta I (2012) Delay in virus accumulation and low virus transmission from transgenic rice plants expressing Rice tungro spherical virus RNA. Virus Genes 45:350–359

MIX
Papier aus verantwortungsvollen Quellen
Paper from responsible sources
FSC® C105338

FSC
www.fsc.org

If you have any concerns about our products,
you can contact us on
ProductSafety@springernature.com

In case Publisher is established outside the EU,
the EU authorized representative is:
Springer Nature Customer Service Center GmbH
Europaplatz 3, 69115 Heidelberg, Germany

Printed by Libri Plureos GmbH
in Hamburg, Germany